MW01079695

# Fiber-Optic Communications Technology

**DJAFAR K. MYNBAEV**
New York City Technical College of the City University of New York

**LOWELL L. SCHEINER**
Polytechnic University

Upper Saddle River, New Jersey
Columbus, Ohio

**Library of Congress Cataloging-in-Publication Data**
Mynbaev, Djafar K.
Fiber-optic communications technology / Djafar K. Mynbaev, Lowell L. Scheiner.
p. cm.
Includes bibliographical references and index.
ISBN 0-13-962069-9
1. Optical communications. 2. Fiber optics
I. Scheiner, Lowell L. II. Title.

TK5103.59. M96 2001
621.382′75—dc21

00-044092

**Vice President and Publisher:** Dave Garza
**Editor in Chief:** Stephen Helba
**Assistant Vice President and Publisher:** Charles E. Stewart, Jr.
**Production Editor:** Alexandrina Benedicto Wolf
**Production Coordination:** Lisa Garboski, bookworks
**Design Coordinator:** Robin G. Chukes
**Cover Designer:** Dean Barnett
**Cover Image:** FPG International
**Production Manager:** Matthew Ottenweller
**Marketing Manager:** Barbara Rose

This book was set in Times Roman by The Clarinda Company. It was printed and bound by Courier Kendallville, Inc. The cover was printed by Phoenix Color Corp.

10 9 8 7 6 5 4 3 2
ISBN 0-13-962069-9

*To Bronia*

# Preface

## TO THE READER

Fiber-optic communications has been growing at a phenomenal pace over the past twenty years, so rapidly, in fact, that its impact is increasingly felt in nearly all aspects of communications technology. Fiber optics has, in just a couple of decades, metamorphosed from a somewhat exotic research curiosity into a strong commercial reality—to the point where even the general public has some idea of its ever expanding role in communications. But this doesn't mean that this technology, now that it has found its logical (and for many highly profitable) place in telecommunications, will cease developing. On the contrary. As far as we can see, the demand for transmission over the global telecommunications network will continue to grow at an exponential rate and only fiber optics will be able to meet this challenge. (Wireless communications take us in another direction and is a story in itself. For more on this, see Section 1.3 and Chapter 15.)

There are hundreds of books on fiber-optic communications ranging from primers to highly theoretical monographs. There are a number of textbooks aimed at different readership levels as well. And so, you rightly ask, why this book?

### Purpose

This text addresses a broad audience. It is, first and foremost, a textbook for technology and engineering students taking a beginning course in fiber-optic communications. It is also written for those already in the field who may need to pass an exam, for end-users who need to gain an understanding of fiber-optic technology to work with contractors at the professional level, and for students taking advanced courses in fiber-optic technology who need to bone up on some of the basic principles. These readers, too, will find this a helpful guide.

In addition, engineers and technicians who work with fiber optics can use this text as a source of useful everyday information. Finally, people technically trained in other disciplines who simply want to know what this technology is all about—and this number is surprisingly high—will find this book a useful source of comprehensive up-to-date information.

Primarily written for today's students, this textbook, we believe, will provide a marked benefit to you. Traditionally, technical colleges are divided into engineering and technical schools, with the former providing a greater emphasis on the theoretical and the latter stressing the more practical, job-related aspects of the field. Accepting this reality, we have included two levels of discussion in this text. As a result, the technology schools may restrict themselves to the chapters marked "Basics," while engineering colleges may concentrate on the chapters marked "A Deeper Look." We appreciate, of course, that every class—whether in an engineering or a technology school—comprises individuals and so the instructor will certainly base his or her syllabus on the specific needs of the students.

This text is unique in that it responds to the needs and interests of most students seeking to enter the profession and enables them to gain important knowledge whether it be in the theoretical or practical realm.

So how can this textbook help you acquire the essentials you need to succeed in such a vibrant field? First, it provides you with a strong foundation through clear, logical explication of the basic concepts, augmented by lots of examples, graphical presentations, and solutions to problems similar to ones you'll encounter in the workplace. Secondly, it also includes the newest technological innovations in components, systems, and networks. By concentrating our efforts on the truly new and most promising technological achievements, ones we believe you'll be encountering in your career, we've succeeded on both counts.

With so much commercial activity underway and a wealth of exciting research developments in the works, we concluded that the best way to serve your needs in a single volume would be to include only innovations and applications that are commercially available or that, in our judgment, will be within the next few years. Progress in this field has been fast and furious so it is not surprising to see yesterday's R&D project become today's commercial product. For this reason, the book includes discussions of trends in technology and networks to make you aware of today's R&D projects that will likely be commercial realities when you start your career.

## Student Research

A key feature of this text is that it introduces engineering and technology students to the principles of research. Increasingly, undergraduate institutions—both senior and junior colleges—are encouraging students to use research as a powerful teaching tool to help them master a technical subject. The National Science Foundation furthers this practice by spending hundreds of dollars a year to support student research starting at the junior (community) college level.

How can this book help you develop your research skills?

First, it provides a list of topics that are ripe for research activity. For example, consider dispersion in optical fibers. A research project on this topic would require you to determine the real model of dispersion in a singlemode fiber, how nonlinear effects impact dispersion in a singlemode fiber, and what the theoretical limit is to dispersion in singlemode and multimode fibers. These and many other ideas for research will emerge as you read this book.

Secondly, as the scope of this text ranges from the introductory to the intermediate level, a broad area of inquiry is open for students to pursue. As questions and problems arise, the more ambitious students will wish to pursue answers and solutions on their own or perhaps as part of a research team. And since both theoretical and experimental approaches are discussed here, the text can serve as the fulcrum for student research, even to the point of allowing students to adapt the research topic and level to their particular backgrounds, interests, and even future work.

## How Much to Learn

If you hope to become a professional in the components field and concentrate in the area of optical fiber, transmitters, receivers, amplifiers, and active and passive components, you will need a strong understanding of the technology. This knowledge can be acquired only if you have first established a solid background in physics, mathematics, and electronics. There are no shortcuts to this knowledge base.

On the other hand, if you are intent on becoming a fiber-optic communications systems specialist, a knowledge of electronic communications and telecommunications is needed. Those of you planning to work at the fiber-optic-network level must master the intricacies of two disciplines: fiber-optic technology and telecommunications networks. Finally, students interested in becoming telecommunications managers must combine technical knowledge with insight into complex legal regulations and a keen understanding of how a business operates.

Fiber-optic communications technology is a special field, one that integrates knowledge from diverse areas to devise new concepts. As a result, the broader the background and experience you bring to your job, the more valuable you are as a professional.

### Career Opportunities

You have chosen to enter a truly exciting profession, a profession where companies are aggressively pioneering in their efforts to stay in the forefront of leading-edge technology, where yesterday's latest innovation becomes passé today, where professionals with far-reaching ideas still have unlimited opportunities to grow along with a young, fast-paced, rapidly expanding industry and, yes, where talent and success are rewarded with big money. No doubt about that.

Whether you intend to enter the field as a technologist, engineer, technician, or product manager, the future holds unlimited potential for a dynamic technology that was barely a gleam in a far-thinking scientist's eye just twenty-five years ago, let alone the explosive technological business it has become today. The field, in fact, has developed so quickly that in the span of just a few short years it has become the linchpin of the giant telecommunications industry, an industry that itself accounts for one-sixth of the U.S. economy and is still growing—with few seeing an end in sight.

Indeed, your decision to become a professional in some area of telecommunications—which means to become a professional in fiber-optic communications technology—bodes well for your career. Our experience shows that even for recent college graduates, the problem is not finding a job, but choosing the best job from among a plethora of excellent offers. Today corporations serving this industry are competing for engineers, technologists, and product managers even far more aggressively than the job seekers are hunting for positions. As a result, fiber-optic technology professionals are among the highest paid in American industry.

## TO THE INSTRUCTOR

It would be inappropriate to tell you how to use this or any textbook. Thus, we would like to simply share with you the plan of this book to help you prepare your course syllabus.

### General Structure

Most topics are presented on two levels: "Basics" and "A Deeper Look." At the basic level, we introduce the main ideas and the principles behind the devices covered. The basic sections are necessary to give students with little background in fiber-optic communications technology a fundamental understanding of the topic. These sections are presented in a simple manner for the beginning student to grasp the subject matter easily and quickly.

At the deeper-look level, we include a more theoretical, highly detailed discussion of the same material and add new topics. This gives you considerable flexibility based on the technical level of your class and the length of the course. The deeper-look sections have two goals: First, they cover the material in greater depth, thereby involving more theory, assuming, of course, a stronger background in physics and mathematics. Second, they prepare students for further course work that will involve complex theory and the kinds of professional responsibilities they will encounter in the workplace. These sections bring the students to the next level and provides more insight into fiber-optic communications technology. They also provide many topics for student research should this be an integral part of your course.

It should be emphasized that the basic and deeper-look sections are closely related and, if the advanced students should feel the need for a "refresher course," we strongly recommend that they reread the introductory section before proceeding to the more complex material.

## About the Examples

We use examples not only to illustrate how to apply formulas for computing numbers but also to move the discussion in a logical, comprehensive way and to provide additional helpful information.

## Reading the Data Sheets

For practicing engineers, the ability to read data sheets is critical. This ability enables them to assimilate the knowledge they have acquired. In a sense, it is an excellent measure of the level of one's professionalism. For this reason, every topic concludes with a discussion of specific data sheets. In addition, specifications sheets are used as sources of data for making various calculations in the examples given, as an aid in explaining the material, and for solving the problems strewn throughout the text and at the end of each chapter.

## Your Course Syllabus

We cannot imagine a single all-purpose syllabus that could meet the needs of every class using this textbook. Obviously, the volume of material well exceeds the needs of even a two-semester course, let alone a one-semester class. This book gives you wide latitude in building your syllabus.

At New York City Technical College, we always leave time during the term to take up topics suggested or requested by our students. Many part-time students, for example, bring their job experiences and needs to the classroom, where these topics are discussed in an open forum. The design of an installation for local area networks has been one of the most frequently requested topics in recent years, reflecting a significant trend in the communications world. Another important topic raised by students has been fiber-optic networks and their components, particularly erbium-doped fiber-optic amplifiers. Allowing some freedom to revise course content during the semester, without feeling constrained by a rigid course syllabus, brings considerable creativity to the classroom.

A critical aspect of the learning experience is the class format. At New York City Technical College, classes meet once a week for five hours, with each session combining theory and lab experiments. Other schools run classes that meet two or three times a week, separating the lab work from the lectures. As you can see, it is virtually impossible to provide every instructor with a ready-to-use syllabus. Our syllabus for a one-semester course in an engineering-technology program is found in the *Instructor's Manual* accompanying this text. The *Instructor's Manual* also provides you with additional possibilities for preparing course outlines that will meet your course objectives.

## About the Laboratory Exercises

The descriptions of the test and measurement procedures can be used as a guide for building a laboratory course. This has already been done at New York City Technical College. Also the experiments presented are simple to perform yet have a lot of information to convey, they can be carried out as a series of small, independent projects rather than as step-by-step exercises. For example, one of these projects requires the student to measure certain characteristics, such as a multimode fiber's attenuation, but it does not delineate a specific sequence by which to achieve that goal. This approach is recommended because it gives students a feel for the real working environment. The *Instructor's Manual* accompanying this book contains more detailed suggestions on how to set up a program of laboratory exercises.

To sum up, we believe this textbook provides you with all the material you need to devise a course that will be suitable for the technical level of your students.

# Acknowledgments

I am deeply obliged to the many people who helped in the preparation of this book.

I wish to express my thanks, first of all, to my coauthor, Professor Lowell Scheiner, who shared this long, difficult, but ultimately rewarding experience with me every step of the way.

The administration of New York City Technical College cooperated with me fully by allotting me sufficient time to undertake this endeavor, and I am especially indebted to Dr. Emilie Cozzi, the president of the college at the time this project was undertaken.

My colleagues in the Department of Electrical Engineering and Telecommunications Technologies supported and encouraged me throughout this venture.

I am deeply appreciative of the advice given by my close friend and mentor, Dr. Alex Gelman, who introduced me to Bellcore, where I gained invaluable professional experience, the background that played a large role in my plans for this text. Dr. Paul Shumate, also of Bellcore, provided constructive critiques of the text during the writing stage and was always available for consultation when problems arose. David Waring of Bellcore also offered invaluable comments and suggestions.

Dr. Mikhail Levit was kind enough to work out solutions to some problems I discussed with him.

Dr. Karim Mynbaev reviewed Chapters 9, 10, and 11. His contribution helped significantly to improve the material covered there.

Andrei Basov did yeoman work turning rough graphics into polished artwork.

My students at New York City Technical College stimulated me in developing this course. Their constructive feedback introduced new topics and led, I trust, to the overall improvement of the course and the book.

I appreciate the valuable feedback from the following reviewers: Eugene Bartlett, ITT Tech–Florida; John Nawn, Ocean County College; Thomas Shay, New Mexico State University; and Chris Wernicki, New York Institute of Technology.

Finally, Professor Scheiner and I are greatly appreciative of the assistance provided by the various companies that sent us documentation, particularly graphic material, to help illustrate the concepts discussed.

D. K. M.

# Contents

## Chapter 6
## Singlemode Fibers—A Deeper Look 168



## Chapter 7
## Fabrication, Cabling, and Installation 210



## Chapter 8
## Fiber Cable Connectorization and Testing 248

## Chapter 14
## An Introduction to Fiber-Optic Networks 651



## Chapter 15
## Conclusion 712

# 1

# Introduction to Telecommunications and Fiber Optics

This chapter introduces the basic ideas of telecommunications, describes a typical fiber-optic communications system, and reveals some fascinating facts about the early days of this technology. The chapter concludes with a review of the current state of the art and a glimpse into the future.

## 1.1 TELECOMMUNICATIONS

**What It Is**

The word *telecommunications* consists of two parts: the Greek word *tele,* which means "over a distance," and *communications,* which means the "exchange of information." We are all, of course, familiar with *tele* through such terms as *telephone,* which means "speech over a distance," and *television,* which means "vision over a distance." Therefore, *telecommunications* means the "exchange of information over a distance." Since we cannot communicate directly, that is, face to face, over a distance, we have to use some sort of device, such as a telephone, a radio, or a television system. To sum up, then, we can say that *telecommunications is the exchange of information over a certain distance using some type of equipment.*

There are three basic types of information to be exchanged: voice, video, and data. In the not-too-distant-past there were separate branches of telecommunications: The telephone and the radio provided voice delivery, TV delivered the images, and computers processed the data. Today, these systems have coalesced. Telephony transmits voice signals in digital form and uses computers at any level of a telephone network (in access, transmission, switching, and signaling systems). On the other hand, cable TV companies play an increasing role in providing both telephone and computer services in their quest to become serious Internet providers. Increasingly, too, the computer industry is delivering both voice and video. For example, one can make telephone calls over the Internet—in particular, international ones—that are much cheaper than regular telephone calls. What's more, today's computers provide multimedia services (the combination of music, pictures, text, and voice). Thus, today's telecommunications business is a complex industry that provides delivery of any type of information throughout the world.

## Telecommunications: Point-to-Point Systems and Networks

Figure 1.1 shows the general block diagram of a telecommunications system. Information in its original form—that is, voice, video, or data—enters a transmitter. The transmitter converts it into a form suitable for transmission (such as an electrical signal), prepares the converted signal for transmission (in other words, it modulates and multiplexes it), and transmits the signal. The signal travels through a communication link, also called a transmission medium. This link can be a copper wire, a coaxial cable, air, or an optical fiber. A receiver recognizes the signal, prepares it for conversion (it now demodulates and demultiplexes it), and converts the signal into its required form of information; in other words, it changes it from an electrical form back into its original form (sound, image, or a set of characters).

As an example of a telecommunications system, consider a typical telephone system. A microphone converts our voice into an analog electrical signal. This signal is transmitted through copper wires to the nearest switching center (a central office), where it is digitized, multiplexed, and then delivered—through optical fiber—to the destination switching center. From that central office, the signal travels directly, or through a remote terminal, to the calling party's telephone, where it is converted back into sound by a speaker. A variety of telephone network architectures exists today, but this oversimplified example is used to focus on the basic functions of a telecommunications system.

It should be noted that the terms "transmitter" and "receiver" are used here in a more general sense than usual. In Figure 1.1, transmitters and receivers include several components, while in a specific telecommunications system they are the devices responsible only for transmission and receiving operations.

Figure 1.1 depicts a typical *point-to-point* telecommunications system. The first telephone line ever, which ran between the second floor and the basement of Alexander Graham Bell's house, was an example of what's today called a point-to-point link. That dedicated wire gave birth to modern telecommunications. In fact, in any discussion of telecommunications, we bear in mind point-to-point communication as the typical arrangement. But as soon as three or more parties (nodes) become involved in the communication, we need to have a network to connect them. *A network is a combination of nodes connected by links.* Examples range from a simple passive electrical network consisting of resistors, capacitors, and inductors in parallel-series circuits to the complexity of the Internet. Figure 1.2 shows different configurations, or topologies, for telecommunications networks.

The Internet, *the* network of networks, connects computers throughout the world. Internet users are able to exchange voice, video, and data messages with people at any point on the globe. To make these marvels possible, the network must include not only physical connections but also some intelligence. Just as a computer is made functional by the appropriate combination of hardware and software, today's networks are made functional by properly combining physical circuitry and logical layers. The logic elements provide the network with the ability to route your

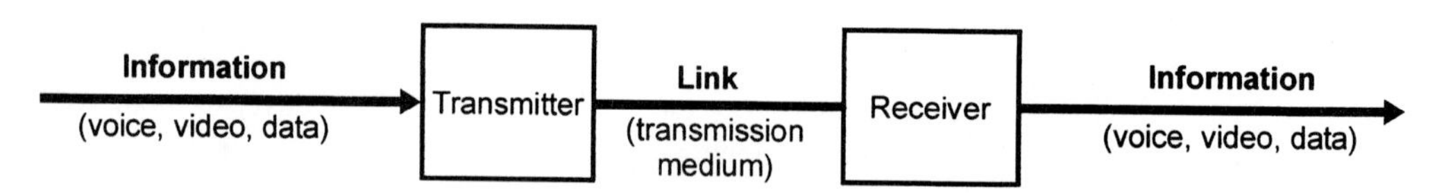


**Figure 1.1** A point-to-point telecommunications system.

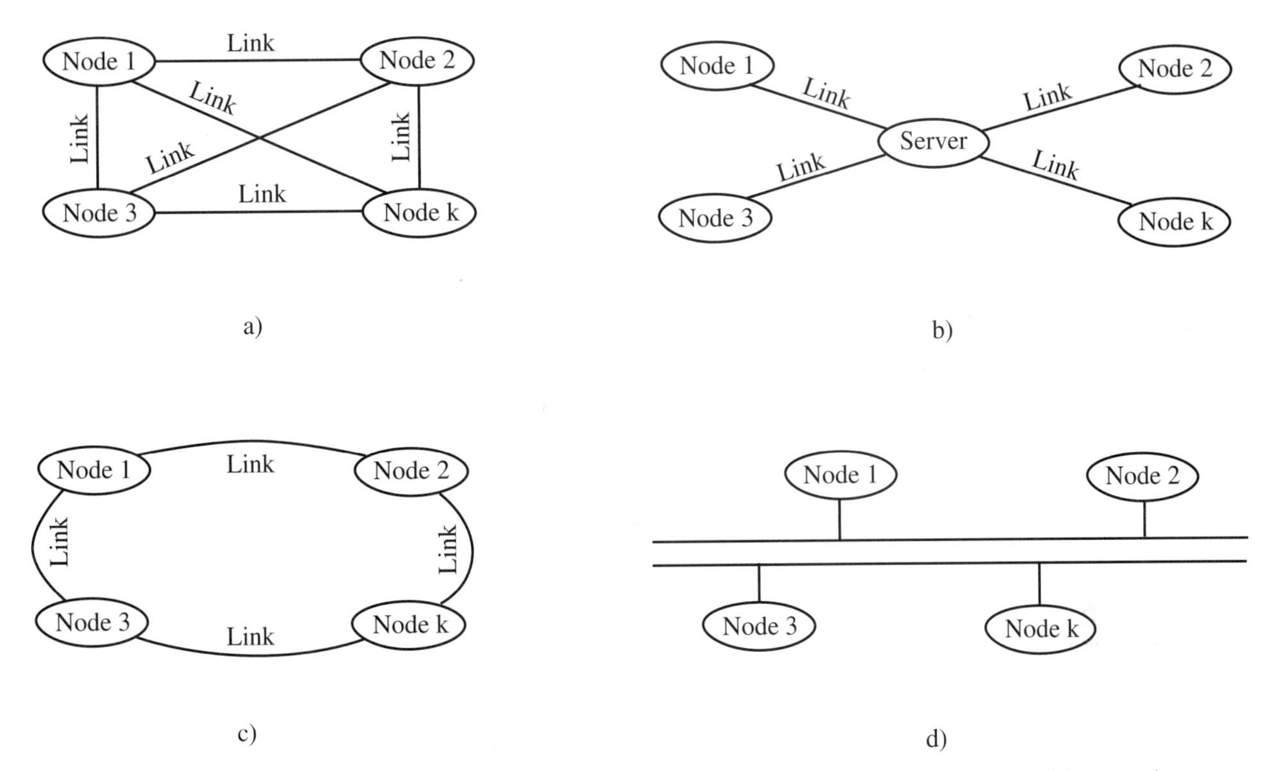


**Figure 1.2** Telecommunications networks: (a) Mesh topology; (b) star topology; (c) ring topology; (d) bus topology.

call—your demand for services, really—to your calling party through the system's extremely sophisticated physical circuitry. The Internet is the perfect example of how sophisticated modern telecommunications networks are.

## Information-Carrying Capacity

What does everyone want from modern telecommunications? Capacity, capacity, and more capacity! We're talking about *information-carrying capacity, which is the ability of a communications link to transmit a certain amount of information per unit of time.* Why do we demand more and ever more capacity? Because, first, we produce information without letup in our society and, second, we transmit most of it in digital form. Let's consider these points in detail.

***Why produce more information?*** It is well known that the developed countries have moved from the postindustrial era to the information era. Incredible as this would have seemed just 30 years ago, these countries now produce more information than they do tangible products, relegating manufacturing to a secondary role in their economies. The more information we produce, the greater the need for its delivery because, obviously, information works only when it is delivered to the right place, at the right time, in the right form. And it is the business of the telecommunications industry to do just that. In short, telecommunications systems must deliver messages instantly and free of distortion.

The importance of telecommunications to modern society cannot be overestimated. Just imagine for a moment that no telephone network existed anywhere in the world. Could we survive without it? Maybe, perhaps, in the days before Alexander Graham Bell, but certainly not today. Like it or not, all aspects of life today depend on telecommunications. Just ponder your

### The Telecommunications Act of 1996

The vital role of telecommunications in our lives is why it is subject to the scrutiny of both government agencies and consumer activists. Take, for example, the far-reaching impact of the heatedly debated Telecommunications Act of 1996, which Congress finally passed and the President signed into law. Now we are watching its implementation. As conceived and finally passed, this act is intended to promote competition in the telecommunications industry to improve the quality of service and keep prices in check. It removes nearly all barriers among the different branches of telecommunications, allowing, for instance, local telephone companies to provide long-distance service and the long-distance companies to penetrate the local markets. Further, it clears the way for cable TV providers to offer telephone service and for telephone companies to provide TV service. All the telecommunications players are allowed to provide Internet service.

Internationally, telecommunications is headed in the same direction as it is in the United States: toward deregulation and privatization. In 1990, only 10 percent of the worldwide telecommunications industry was competitive, but very early into the twenty-first century it is expected that 90 percent of the industry will be competitive, especially in Europe.

own everyday activities. You shop with a credit card, which a cashier swipes through a small machine. In seconds, your payment is approved and your credit account is charged automatically, no matter what bank issued your card, no matter where the bank is headquartered, no matter where the transaction takes place. Today, you can use major credit cards throughout most of the civilized world. Credit-card machines are found in the smallest of retail shops even in the most rural areas of most countries. They're connected to the card-issuing bank through powerful, fast, secure telecommunications networks. And this retailing issue is just one familiar example.

The reader can easily discover many more uses of telecommunications in his or her private and business activities. And, again, the mass production of information—increasing, as it is, by leaps and bounds—requires an even greater ability to deliver it, thus accounting for the importance of the telecommunications industry to modern society.

***Why go digital?*** As you are well aware, there are two basic forms of a signal: analog and digital. The *analog signal* carries information by means of the values of its amplitude, frequency, or phase. Noise and other distortions change these values, resulting very often in an incorrect understanding—a misinterpretation, if you will, of the delivered information. This is why analog technology is so error-prone and requires Herculean efforts to keep it working reliably. The *digital signal* carries information by bits, which can be either logic 1 or logic 0. These logic meanings are represented by electrical pulses. For example, 3.3 volts of electrical pulse represent logic 1 and a 0-volt signal represents logic 0. Any distortion results in a signal-amplitude change but doesn't alter the logical meaning of the signal. For example, a pulse with amplitudes between 2.0 volts and 3.3 volts means logic 1, while a pulse with amplitudes between 0 and 0.8 volt is still logic 0. This is why digital technology is much more reliable than analog and why modern technology is mostly digital. To prove the point, simply refer to today's computers, which have become so powerful because they are based on digital technology.

There is a price to pay, however, for using digital signals: Digital transmission requires more channel capacity than analog transmission. Since we live in what is primarily an analog world, it is necessary to convert analog signals into digital form—another price we have to pay for using digital technology. In telecommunications, a very popular method of analog-to-digital conversion is based on pulse-code modulation (PCM). To understand why digital transmission requires more capacity than analog, see Figure 1.3. Note that every sample of an analog signal requires several bits—four bits in the case shown. The better the transmission quality required, the greater the number of bits required.

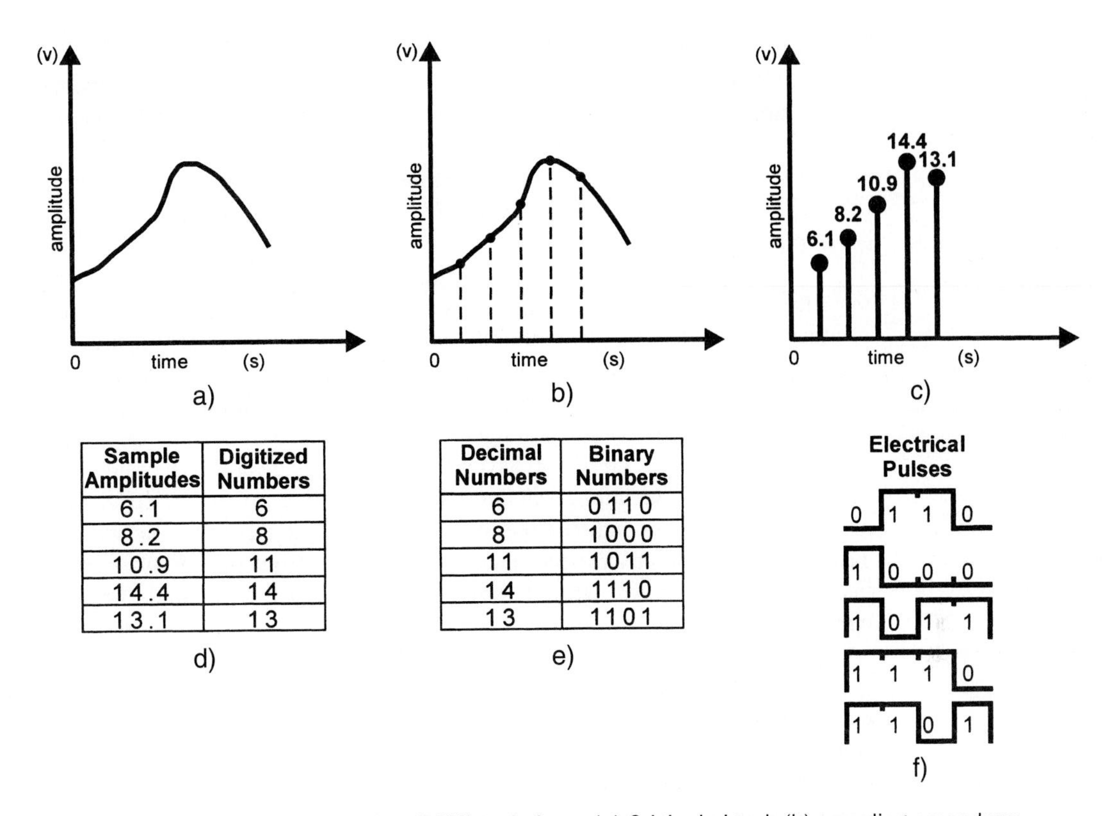


| Sample Amplitudes | Digitized Numbers |
|---|---|
| 6.1 | 6 |
| 8.2 | 8 |
| 10.9 | 11 |
| 14.4 | 14 |
| 13.1 | 13 |

| Decimal Numbers | Binary Numbers |
|---|---|
| 6 | 0110 |
| 8 | 1000 |
| 11 | 1011 |
| 14 | 1110 |
| 13 | 1101 |

**Figure 1.3** Pulse-code modulation (PCM) technique: (a) Original signal; (b) sampling procedure; (c) samples of the sign; (d) quantization; (e) binary coding; (f) electrical signal transmitted in digital form.

To summarize: *The major demand placed on telecommunications systems is for more information-carrying capacity because the volume of information produced increases rapidly. In addition, we have to use digital technology for the high reliability and high quality it provides in the signal transmission. However, this technology carries a price: the need for higher information-carrying capacity.*

## The Need for Fiber-Optic Communications Systems

The major characteristic of a telecommunications system is unquestionably its information-carrying capacity, but there are many other important characteristics. For instance, for a bank network, security is probably more important than capacity. For a brokerage house, speed of transmission is the most crucial feature of a network. In general, though, capacity is priority one for most system users. And there's the rub. We cannot increase link capacity as much as we would like. The major limit is shown by the Shannon-Hartley theorem,

$$C = BW \times \log_2 (1 + SNR), \tag{1.1}$$

where $C$ is the information-carrying capacity (bits/sec), $BW$ is the link bandwidth (Hz = cycles/sec), and $SNR$ is the signal-to-noise power ratio.

Formula 1.1 reveals a limit to capacity C; thus, it is often referred to as the "Shannon limit." The formula, which comes from information theory, is true regardless of specific technology. It was first promulgated in 1948 by Claude Shannon, a scientist who worked at Bell Laboratories. R. V. L. Hartley, who also worked at Bell Laboratories, published a fundamental paper 20 years earlier, a paper that laid important groundwork in information theory, which is why his name is associated with Shannon's formula.

*The Shannon-Hartley theorem states that information-carrying capacity is proportional to channel bandwidth, the range of frequencies within which the signals can be transmitted without substantial attenuation.*

What limits channel bandwidth? The frequency of the signal carrier. The higher the carrier's frequency, the greater the channel bandwidth and the higher the information-carrying capacity of the system. The rule of thumb for estimating possible order of values is this: Bandwidth is approximately 10 percent of the carrier-signal frequency. Hence, if a microwave channel uses a 10-GHz carrier signal, then its bandwidth is about 100 MHz.

A copper wire can carry a signal up to 1 MHz over a short distance. A coaxial cable can propagate a signal up to 100 MHz. Radio frequencies are in the range of 500 KHz to 100 MHz. Microwaves, including satellite channels, operate up to 100 GHz. Fiber-optic communications systems use light as the signal carrier; light frequency is between 100 and 1000 THz; therefore, one can expect much more capacity from optical systems. Using the rule of thumb mentioned above, we can estimate the bandwidth of a single fiber-optic communications link as 50 THz.

To illustrate this point, consider these transmission media in terms of their capacity to carry, simultaneously, a specific number of one-way voice channels. Keep in mind that the following numbers, as well as the frequencies given above, represent only the order of magnitude, *not* the precise value. A single coaxial cable can carry up to 13,000 channels, a microwave terrestrial link up to 20,000 channels, and a satellite link up to 100,000 channels. However, one fiber-optic communications link, such as the transatlantic cable TAT-13, can carry 300,000 two-way voice channels simultaneously. That's impressive and explains why fiber-optic communications systems form the backbone of modern telecommunications and will most certainly shape its future.

To summarize: *The information-carrying capacity of a telecommunications system is proportional to its bandwidth, which in turn is proportional to the frequency of the carrier. Fiber-optic communications systems use light—a carrier with the highest frequency among all the practical signals. This is why fiber-optic communications systems have the highest information-carrying capacity and this is what makes these systems the linchpin of modern telecommunications.*

To put into perspective just how important a role fiber-optic communications will be playing in information delivery in the years ahead, consider the following statement from a leading telecommunications provider: "The explosive growth of Internet traffic, deregulation and the increasing demand of users are putting pressure on our customers to increase the capacity of their network. Only optical networks can deliver the required capacity, and bandwidth-on-demand is now synonymous with wavelength-on-demand." This statement is true not only for a specific telecommunications company. With a word change here and there perhaps, but with the same exact meaning, you will find telecommunications companies throughout the world voicing the same refrain.

## 1.2 A FIBER-OPTIC COMMUNICATIONS SYSTEM: THE BASIC BLOCKS

A modern fiber-optic communications system consists of many components whose functions and technological implementations vary. This is the overall topic of this book. In this section we introduce the main idea underlying a fiber-optic communications system.

## Basic Block Diagram

A fiber-optic communications system is a particular type of telecommunications system (Figure 1.1). The features of a fiber-optic communications system can be seen in Figure 1.4, which displays its basic block diagram.

Information to be conveyed enters an electronic transmitter, where it is prepared for transmission very much in the conventional manner—that is, it is converted into electrical form, modulated, and multiplexed. The signal then moves to the optical transmitter, where it is converted into optical form and the resulting light signal is transmitted over optical fiber. At the receiver end, an optical detector converts the light back into an electrical signal, which is processed by the electronic receiver to extract the information and present it in a usable form (audio, video, or data output).

Let's take a simple example that involves Figures 1.1, 1.3, and 1.4. Suppose we need to transmit a voice signal. The acoustic signal (the information) is converted into electrical form by a microphone and the analog signal is converted into binary form by the PCM circuitry. This electrical digital signal modulates a light source and the latter transmits the signal as a series of light pulses over optical fiber. If we were able to look into an optical fiber, we would see light vary between off and on in accordance with the binary number to be transmitted. The optical detector converts the optical signal it receives into a set of electrical pulses that are processed by an electronic receiver. Finally, a speaker converts the analog electrical signal into acoustic waves and we can hear sound-delivered information.

Figure 1.4 shows that this telecommunications system includes electronic components and optical devices. The electronic components deal with information in its original and electrical forms. The optical devices prepare and transmit the light signal. *The optical devices constitute a fiber-optic communications system.*

***Transmitter*** The heart of the transmitter is a light source. *The major function of a light source is to convert an information signal from its electrical form into light.* Today's fiber-optic communications systems use, as a light source, either light-emitting diodes (LEDs) or laser diodes (LDs). Both are miniature semiconductor devices that effectively convert electrical signals into light. They need power-supply connections and modulation circuitry. All these components are usually fabricated in one integrated package. In Figure 1.4, this package is denoted as an optical transmitter. Figure 1.5 displays the physical make-up of an LED, an LD, and integrated packages.

***Optical fiber*** The transmission medium in fiber-optic communications systems is an optical fiber. *The optical fiber is the transparent flexible filament that guides light from a transmitter to a receiver.* An optical information signal entered at the transmitter end of a fiber-optic communications system is delivered to the receiver end by the optical fiber. So, as with any communications link, the optical fiber provides the connection between a transmitter and a receiver and,

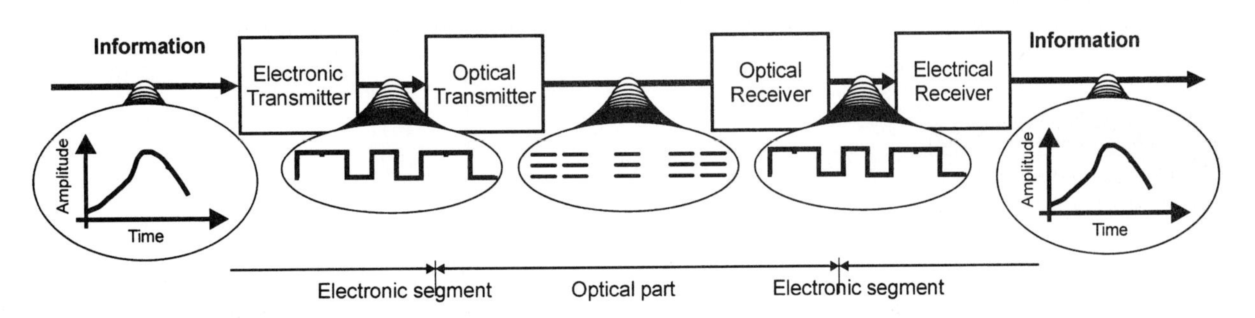


**Figure 1.4** Basic block diagram of a fiber-optic communications link.

## MF430 Datacom, Telecom, General Purpose LED

**MF430 ST**

ST Assembly

| Ordering Information | |
|---|---|
| PART # | RECEPTACLE |
| MF430 ST | ST |
| -40°C to +85°C | |

**Applications**

- Ethernet 10 or 100Mbps
- Token Ring
- Fibre Channel 266Mbps
- Short Wavelength FDDI
- Short Wavelength ATM-SDH/SONET 155Mbps
- Intra-Office Telecom
- General Purpose

**Features**

- 865nm Surface-Emitting LED
- 250MHz Bandwidth
- Designed for 62.5/125µm Fiber
- Aligned in ST® Receptacle
- MTTF >1,000,000 hours

**Description**

This high performance LED has been designed for Datacom, Telecom or General Purpose Applications. The short wavelength LED allows cost-effective links over short distances. This very high speed device has an actively aligned receptacle for optimized coupling of power to 62.5/125µm fiber. A Silicon Photodiode is recommended as Receiver for this LED.

**MF430 Functional Diagram**



**MF430** LED

ST® is a registered trademark of Lucent

13323 11 1997-04-01

a)

**Figure 1.5** Fiber-optic transmitter: (a) Functional diagram and LED assembly *(reprinted with permission from Mitel Semiconductor's* Optoelectronic Solutions Product Catalog *© 1997, Mitel);* (b) DIP transmitter module; (c) butterfly transmitter module. *(Figures b and c courtesy of Nortel Networks).*

SOURCES
LCV 75 SERIES
1300nm Mini-DIL
Transmitter


## FEATURES

- Suitable for 155/622 Mbits/s Long/Intermediate reach SONET AND Long/Short Haul SDH single mode applications
- Wavelength 1300nm
- One piece fibre alignment
- High performance design for high volume feeder and distribution networks
- Nortel's GaInAsP FP laser chip designed for extended temperature operation
- Isolated internal fast InGaAs monitor photodiode
- Proven reliability and Bellcore TA-NWT 000983 capability
- Hermetically sealed package
- 8/125μm cladding mode stripped fibre pigtail
- Maximum data rate in excess of 622 Mb/s
- Footprint compatible with 14 pin DIL series

| RATED OUTPUT POWER (Mean) |
|---|
| 0.1 - 2.0Mw |

b)

**Figure 1.5**
*(continued)*

SOURCES
LC155GC - 20A
Cooled 1550nm DFB
2.5Gbit/s transmitters

## FEATURES

- Operates up to 2.5Gbit/s over long distance single mode fibre
- GaInAsP Distributed Feedback (DFB) single frequency laser chip
- Narrow spectral line-width
- InGaAs monitor photodiode
- Single-mode, cladding mode stripped fibre pigtail
- Internal thermo-electric cooler with precision NTC thermistor for temperature control
- Hermetically sealed 14 pin Butterfly package with optical isolator

| RATED OUTPUT POWER |
|---|
| 2.0mW |

c)

**Figure 1.5**
*(continued)*

very much the way copper wire and coaxial cable conduct an electrical signal, optical fiber "conducts" light.

The optical fiber is generally made from a type of glass called silica or, less commonly nowadays, from plastic. It is about a human hair in thickness. To protect very fragile optical fiber from hostile environments and mechanical damage, it is usually enclosed in a specific structure. Bare optical fiber, shielded by its protective coating, is encapsulated in several other layers that, all together, make up fiber-optic cable—the structure finding widespread use in a host of applications, many of which will be covered in subsequent chapters (Figure 1.6).

***Receiver*** The key component of an optical receiver is its photodetector. *The major function of a photodetector is to convert an optical information signal back into an electrical signal (photocurrent).* The photodetector in today's fiber-optic communications systems is a semiconductor photodiode (PD). This miniature device is usually fabricated together with its electrical circuitry to form an integrated package that provides power-supply connections and signal amplification.

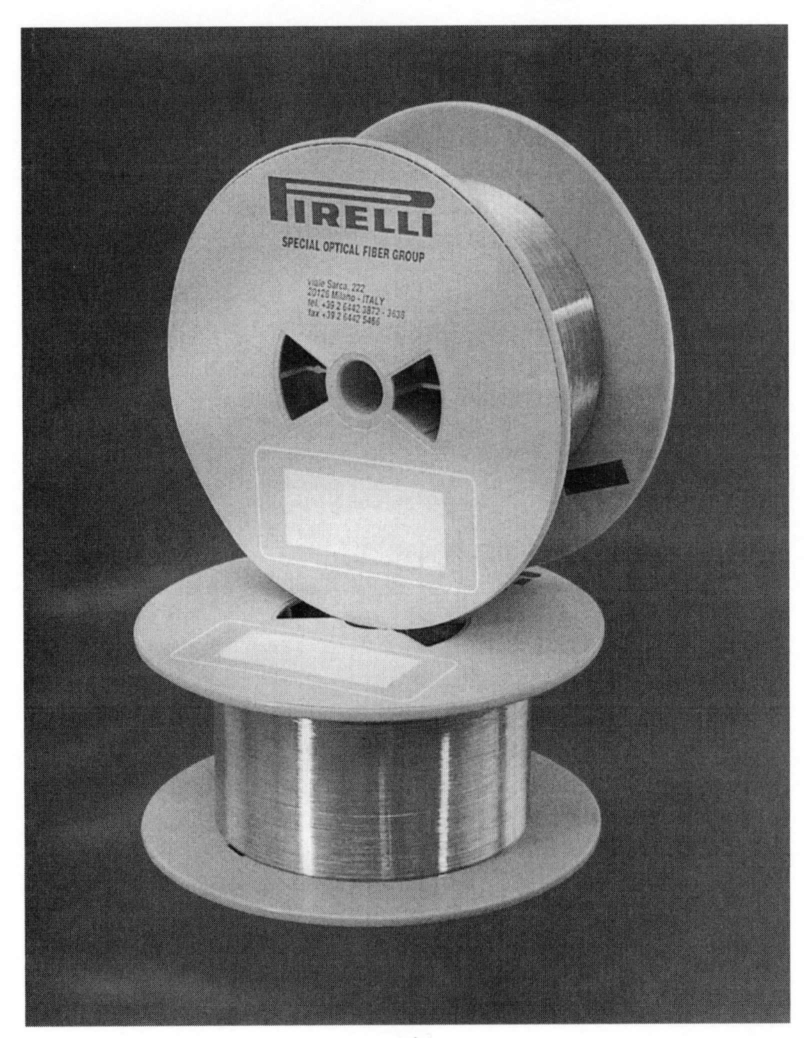


a)

**Figure 1.6** An optical fiber and fiber-optic cable: (a) A reel of a bare optical fiber *(courtesy of Pirelli)*; (b) typical fiber-optic cable *(courtesy of Siecor, Hickory, N.C.)*.

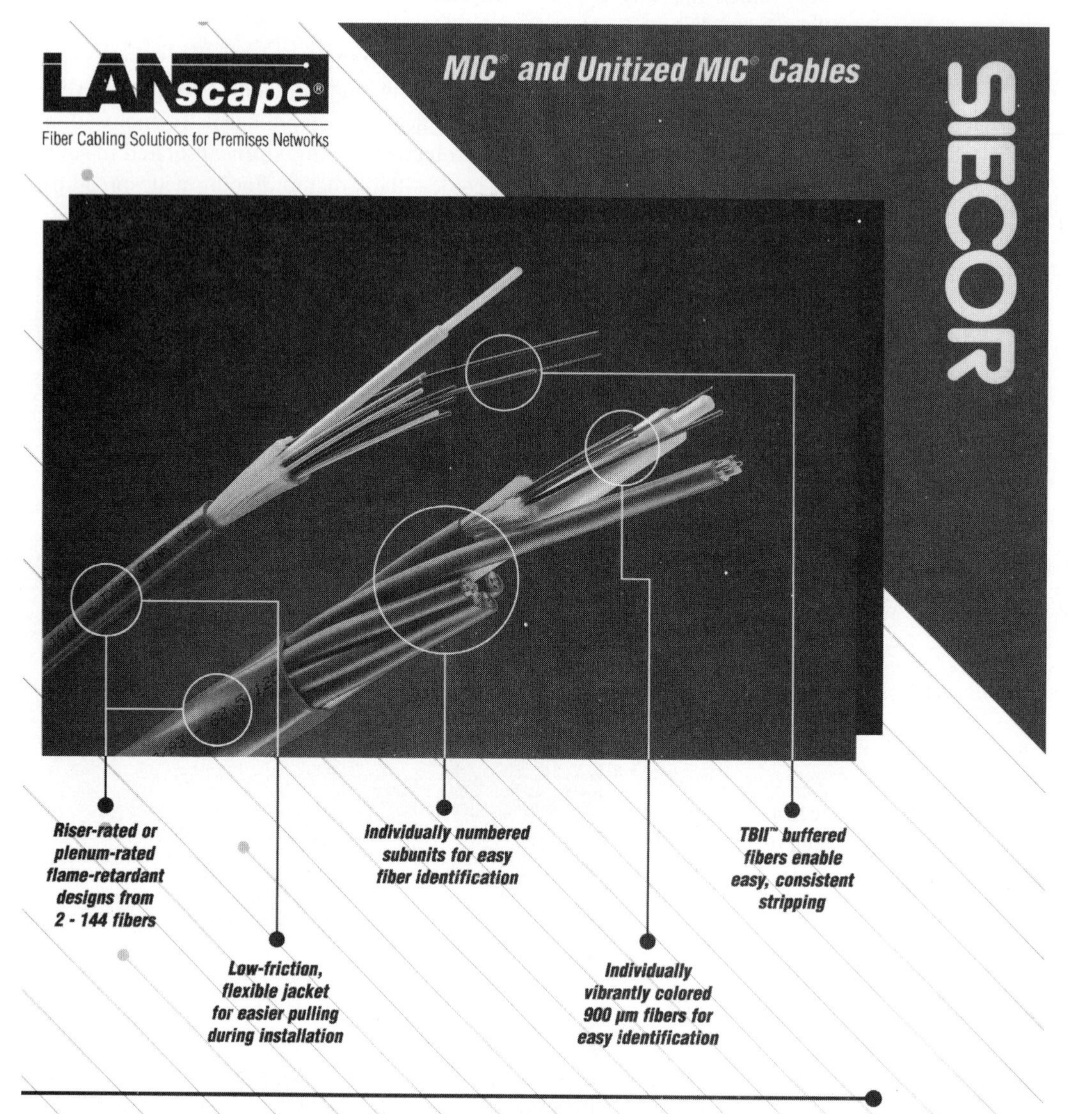


**Fiber counts available:** 2 - 24, up to 144 in unitized design; multimode, single-mode, and hybrid

**Application:** General use indoor plant cable – building backbone, horizontal; riser, plenum

Siecor Operations, LLC • PO Box 489 • Hickory, NC 28603-0489 USA • 1-800-SIECOR5 • 1-800-743-2675 FAX: 828-327-5973 • International: 828-327-5000 • http://www.siecor.com • Siecor reserves the right to improve, enhance, and modify the features and specifications of Siecor products without prior notification. Siecor, LANscape, and MIC are registered trademarks of Siecor Brands, Inc. TBII is a trademark of Siecor Brands, Inc. All other trademarks are the properties of their respective owners. © 1997, 1998 Siecor Operations, LLC. All Rights Reserved. Printed in USA. • CPC-174A / March 1998

**Figure 1.6**
*(continued)*

b)

Such an integrated package is shown in Figure 1.4 as an optical receiver. Figure 1.7 shows samples of a photodiode and an integrated package.

The basic diagram shown in Figure 1.4 gives us the first idea of what a fiber-optic communications system is and how it works. All the components of this point-to-point system are discussed in detail in this book. Particular attention is given to the study of networks based on fiber-optic communications systems.

### The Role of Fiber-Optic Communications Technology

Fiber-optic communications technology has not only already changed the landscape of telecommunications but it is still doing so and at a mind-boggling pace. In fact, because of the telecommunications industry's insatiable appetite for capacity, in recent years the bandwidth of commercial systems has increased more than a hundredfold. The potential information-carrying capacity of a single fiber-optic channel is estimated at 50 terabits a second (Tbit/s) but, from a practical standpoint, commercial links have transmitted far fewer than 100 Gbps, an astounding amount of data in itself that cannot be achieved with any other transmission medium. Researchers and engineers are working feverishly to develop new techniques that approach the potential capacity limit.

Two recent major technological advances—wavelength-division multiplexing (WDM) and erbium-doped optical-fiber amplifiers (EDFA)—have boosted the capacity of existing systems and have brought about dramatic improvements in the capacity of systems now in development. In fact, WDM is fast becoming the technology of choice in achieving smooth, manageable capacity expansion.

The point to bear in mind is this: *Telecommunications is growing at a furious pace, and fiber-optic communications is one of its most dynamically moving sectors.* While this book reflects the current situation in fiber-optic communications technology, to keep yourself updated, you have to follow the latest news in this field by reading the industry's trade journals, attending technical conferences and expositions, and finding the time to evaluate the reams of literature that cross your desk every day from companies in the field.

## 1.3 A LOOK BACK AND A GLANCE AHEAD

### Historical Notes

***Light as the great communicator*** It's not just our sophisticated, technology-driven modern era that is using light to communicate. Since earliest times, man has depended on light—mostly in the form of fire—to send messages. In ancient Israel, for example, people used fire to indicate the beginning of a month. In the early Middle Ages, Russian soldiers relied on fire as part of their defense strategy. They would build a chain of wooden towers at strategic locations some distance apart from one another. If enemy troops approached the sovereign boundary of their state, soldiers at that location set a fire at the top of the tower, signaling the guards at the tower nearest to them in the chain of the impending threat of invasion. The guards at the second station in the chain also set a fire to inform the next station along the link of the danger; this chain reaction of fires was repeated all along the link, giving the town time to make defensive preparations. This was an example—and in its time an ingenious, effective example—of military communication.

Every schoolchild, of course, is familiar with the story of Paul Revere, who signaled by candle lamp from a church belfry the arrival route of British troops ("One, if by land, and two, if by sea," the poem tells us). Well, that was the best he could do with the "technology" existing in his day to communicate his message to a band of scrappy, ragtag American patriots.

Datacom, Telecom, General Purpose PIN Photodiodes **MF432**

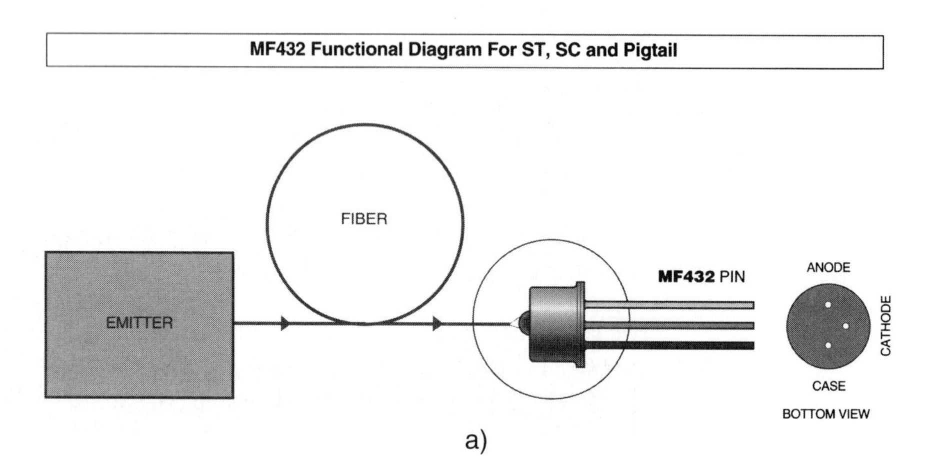


ST

SC

Pigtail

b)

**Figure 1.7** Fiber-optic receiver: (a) Functional diagram; (b) photodiode assembly *(figures a and b reprinted with permission from Mitel Semiconductor's* Optoelectronic Solutions Product Catalog *© 1997, Mitel);* (c) DIP receiver module; (d) butterfly receiver module. *(Figures c and d courtesy of Nortel Networks).*

## FEATURES

- A range of receivers suitable for 155/622 Mbit/s long and Intermediate reach SDH/SONET/ATM single mode applications
- Wavelength range 1200-1600nm
- One piece fibre alignment with no distortion material system
- InGaAs PIN
- Low noise GaAS amplifier
- Proven reliability and Bellcore TA-NWT-000983 capability
- Hermetically sealed package
- High Speed InGaAs PIN photodiode
- The PTAV series is compatible with the LCV mini DIL transmitter

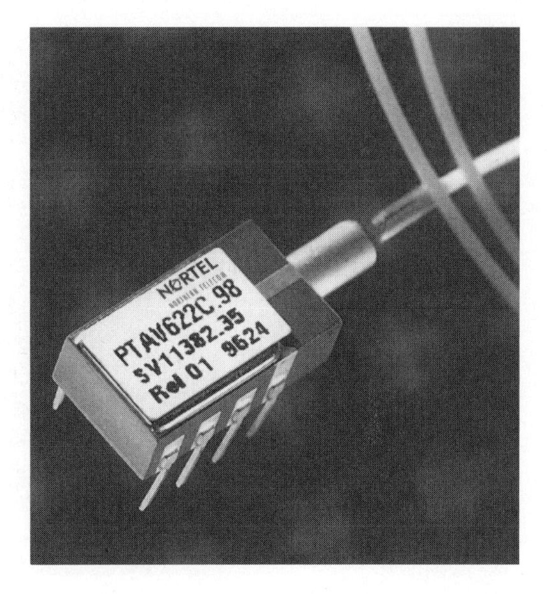


| SENSITIVITY | BANDWIDTH |
|---|---|
| -39dBm min | 100MHz typ |
| -33dBm min | 400MHz typ |

c)

**Figure 1.7**
*(continued)*

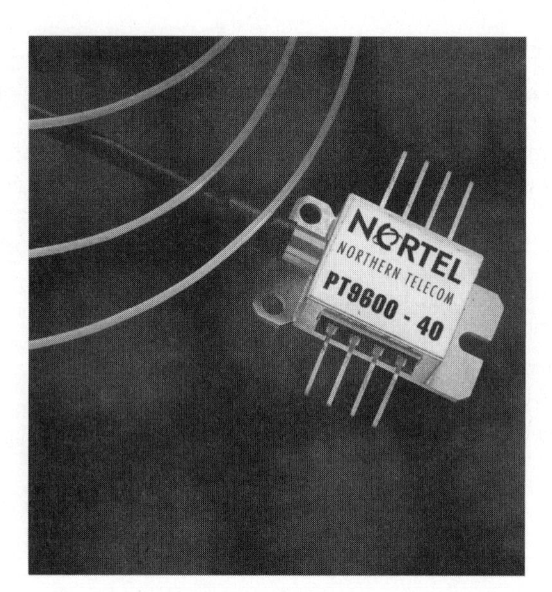


## FEATURES

- Designed for use in cost sensitive applications at transmission rates up to 10Gb/s
- Optimised for high performance at low cost
- Impedance matched microstrip feedthrough for direct PCP attach
- Wide temperature range
- Compact design
- Hermetically sealed butterfly style package
- High performance GaAs HBT preamplifier
- Single mode fibre tail

| RESPONSIVITY | TRANSIMPEDANCE (typ) | BANDWIDTH (min) |
|---|---|---|
| 0.83A/W | 500Ω | 8GHz |

d)

**Figure 1.7**
*(continued)*

Even Arthur Conan Doyle employed fire—here, too, a candle—as a communication medium in one of his famous Sherlock Holmes stories, "The Hound of the Baskervilles." But he took it one step further. The innkeeper's wife opened and closed off the light from the candle lamp in signaling a coded message to her outlaw brother.

***New attempts*** The point we can infer from these anecdotes is that people have, from time immemorial, tried to use light, even in its most primitive form, to deliver some type of information between remote locations. In its visibility and ease of use, light makes such communication extremely practical. But reliability is another matter. The hitch, the catch, the fly in the ointment—call it what you will—is that dependence on atmospheric conditions makes direct optical communication—through the air, that is—so unreliable that it could never be accepted in everyday applications. The advent of the laser in the early 1960s, however, changed all that. The interest of scientists and engineers in optical communications perked up once again and new attempts to build direct, reliable systems for everyday activities were made, but still without commercial success. It became clear that optical communications could be made practical only by using a light conduit. At that time, however, no viable solution could be offered.

In the time since then, scientists and engineers have approached optical communications from another direction. As pointed out previously, communications technology has always tried to use high-frequency signal carriers. The shift from radio frequency to microwaves (a thousandfold leap) allowed engineers to increase a given system's information-carrying capacity tenfold.

This success inspired researchers to seek a solution to the problem of reliable communications by continuing to increase microwave frequency. But at a frequency of more than 100 GHz, where microwaves overlap the infrared zone, microwave attenuation in air reaches such a high level that transmission distance becomes unacceptably short.

The solution appeared to be clear: Use a waveguide structure to transmit ultrahigh frequency (UHF) electromagnetic waves. These waveguides, in the form of steel tubes rectangular in cross section with openings at each end, have now been in use for years in radar and other UHF systems for delivering and distributing microwaves over very short distances (several feet).

In the late 1960s and early 1970s, scientists and engineers at Bell Laboratories achieved significant progress in designing waveguides for long-distance systems. These waveguides boasted a very impressive characteristic: 238,000 voice channels per unit [1]. But the waveguides were still the same old open-end steel tubes about several centimeters in inner diameter, although the newer ones were now circular in cross section. They remained extremely inefficient from cost, installation, maintenance, and other practical standpoints. The results from all these efforts were proving fruitless and any ideas about dotting the country, let alone the world, with a web of thin steel pipes for the purpose of communication were quickly discarded.

Where, then, could these scientists and engineers turn next? What new course should they take to increase information-carrying capacity through reliable, cost-effective systems? The answer was found in reverting to what had been thought about previously but never pursued in depth: increasing the carrier frequency even higher than the highest microwave frequency. In other words, delve into the range of frequencies that cover light. (See Figure 2.3 for a comparison of microwave and light frequencies.) That, in a nutshell, is how the course of development of communication links arrived at light transmission as a serious technology.

***The guided light*** Once this decision was reached, the problem of how to guide the light arose. Could we use tubes? Yes, but that would be as impractical as were the old microwave guided lines. In reality, guiding light over a significant distance is a multifaceted problem. First, there's the matter of developing a practical guide that would carry light in a manner similar to the way copper wire conducts electrical current. That meant developing a guide that would be flexible and easy to install and maintain—not an easy task by any means since it had never before been done. Underscore the word "flexible" in particular, because, for obvious reasons, it would be the key property of the potential guide. (Just imagine a world covered with rigid steel tubes.)

So researchers turned to optical fiber—a transparent, flexible filament made from glass or plastic. And herein lay the next aspect of the problem: How could they preserve light inside a bending fiber if light is a ray propagating along a straight line? The answer was provided by a Dutch mathematics professor long dead: Willebrod van Roijen Snellius, who in 1621 discovered Snell's law, an equation with which every college freshman physics student is quite familiar (see Formula 2.3). Thanks to Snell's law, proponents of optical fibers resorted to the use of total internal reflection.

More than 200 years later, in 1870, British physicist John Tyndall demonstrated that light could be conducted in a curved stream of water. This was the first experiment that proved the existence of the phenomenon of *total internal reflection.* Today, to underscore the milestone role of John Tyndall's achievement for modern fiber-optic communications technology, the Optical Society of America awards an annual prize named after Tyndall.

But it was not until 1951 that the first fiberscope was designed, and that was for delivering an image of human inner organs. In 1953 Narinder Kapany, working at the Imperial College of Science and Technology in London, developed a fiber with cladding, thus giving birth to the structure used in today's optical fiber. It was, in fact, Kapany who coined the term "fiber optics." So at last it looked as if this multifaceted problem had been solved. Or had it? Yes, optical fiber would become the light guide of choice, but no sooner had scientists stopped complimenting one another on resolving one major drawback to optical communications when another arose—light *attenuation.* To transmit light over a significant distance, attenuation—decreasing power—must be low enough so that the detectable signal reaches the receiver end.

Now we come to the turning point in the story of fiber-optic communications—the year 1966. It was then that a Chinese-born British scientist, Charles Kao, delivered a landmark paper, "Dielectric Fibre Surface Waveguides for Optical Frequencies," that was considered the key to unlocking the door to fiber-optic technology as we know it today. Note the terminology in the article's title. You can readily see here the influence of the research actually under way at that time on UHF microwave waveguiding.

Kao predicted the major features of the first practical fiber-optic communications systems. He indicated that the major problem at the time was to reduce light attenuation. His own experiments resulted in an attenuation of more than 1000 dB/km, an unacceptably high loss under any circumstances, a loss level that, if not significantly reduced, would make this technology totally unusable. In fact, he said that the practical use of fiber optics for communications might start when attenuation would be less than 20 dB/km, which meant that light power would have to decrease a hundred times for every kilometer traveled.

***A low-loss fiber at last*** Kao's work was the real breakthrough in this field because he clearly laid out the major technical problems to be solved. At once it became clear that the main task ahead was to find out how to fabricate optical fibers with acceptable attenuation. This difficulty was overcome by Robert Maurer, Donald Keck, and Peter Schultz of the Corning Glass Corporation when, in 1970, they reported attaining the first optical fiber with attenuation less than 20 dB/km. The Corning researchers continued their extensive work to reduce fiber loss and, at the same time, developed an optical fiber manufacturing process that has become the industry's most widely used method [2]. After more than 15 years of extensive work, manufacturers came out with an incredibly low-loss optical fiber with an attenuation less than 0.3 dB/km, which means that light becomes weaker by about 0.933 for every kilometer it travels. (Today, you can buy an optical fiber with attenuation even less than 0.16 dB/km. Just compare this with the 1000 dB/km attenuation that Kao reported only about 30 years earlier.) Manufacturers have developed their manufacturing processes to the point where the price of optical fiber has now become competitive with that of copper. At last, the optical-fiber communications link became available for the mass market.

Refer again to Figure 1.4, the basic block diagram of a fiber-optic communications system. Note that there are two other major components of the system: a light source and a photodetector. Fortunately, by the time optical fiber reached its maturity in terms of acceptable attenuation and cost-efficient mass production, the light sources—light-emitting diodes and laser diodes—also became commercial. The same was true of photodiodes. Obviously, a lot of research and development went into this effort, but the results paid off. All the components of the link were now available, and in 1975 the first field trial of a fiber-optic point-to-point communications system was conducted at Bell Laboratories. Early in 1983 the first intercity fiber-optic link—between New York City and Washington, D.C.—was installed. In 1988 the first transatlantic fiber-optic cable was laid. By the beginning of 1990 about 2 million fiber miles for local telephone service and about 4 million fiber miles for long-distance service had been installed. The R&D story was over and the big business venture into fiber-optic communications began. In fact, by the end of 1999, almost 75 million fiber miles had been installed around the world.

## The Industry Today and Future Trends

***Dynamics of deployment*** What is the current state of the industry? Where is it going? What pitfalls lie ahead? These are some of the critical questions we'll attempt to explore, and we can begin with your own everyday experiences.

For instance, each time you pick up your phone, turn on your TV, transmit documents over a fax, give a cashier your credit card, use a bank ATM, or surf the World Wide Web—in other words, every time you communicate over a significant distance through electronic equipment—you are using fiber-optic communications technology. Future trends will see this technology continue to penetrate all levels of telecommunications networks. The power of fiber optics not only meets the growing demand for information-carrying capacity but also opens whole new vistas of telecommunications possibilities, promising features that we did not even dream about yesterday.

If you'll look once again at the basic block diagram of a fiber-optic communications system (Figure 1.4), you might get the impression that this is simply the powerful link between transmitter and receiver. You're right. It is the most powerful tool we have in terms of information-carrying capacity. But consider again. It is much more than that. Fiber optics not only brought a plethora of benefits to the telecommunications industry, but it also triggered some revolutionary changes. Today, this technology defines access, transmission, signaling, switching, and networking technologies; in other words, it plays the key role in every aspect of modern telecommunications. *Without fiber optics, there is no modern telecommunications.* It's as simple as that. Just attend any general conference or visit any exhibition on telecommunications. You'll find that fiber-optic technology takes up the largest portion by far of exhibit space and that the bulk of the conference papers are devoted to that topic.

It's difficult to mark critical turning points in the development of this technology with specific dates because it has always been in a state of constant flux. And so, as a dynamically growing technology, what was the new kid on the block yesterday has already become commonplace, perhaps even obsolete, today. However, it's generally safe to say that by the early 1990s the first wave of massive deployment of fiber-optic networks—yes, not simply point-to-point connections but actual networks—had been set up and all long-distance and local telephone providers were using optical fiber as a major transmission medium. Telephone networks are still the heart and soul of telecommunications. (In fact, were you aware that the global public switched telecommunications network links about one billion phones?)

Figure 1.8 depicts the deployment of fiber-optic communications networks by local and long-distance operating companies. You can see the exponential growth in fiber miles installed

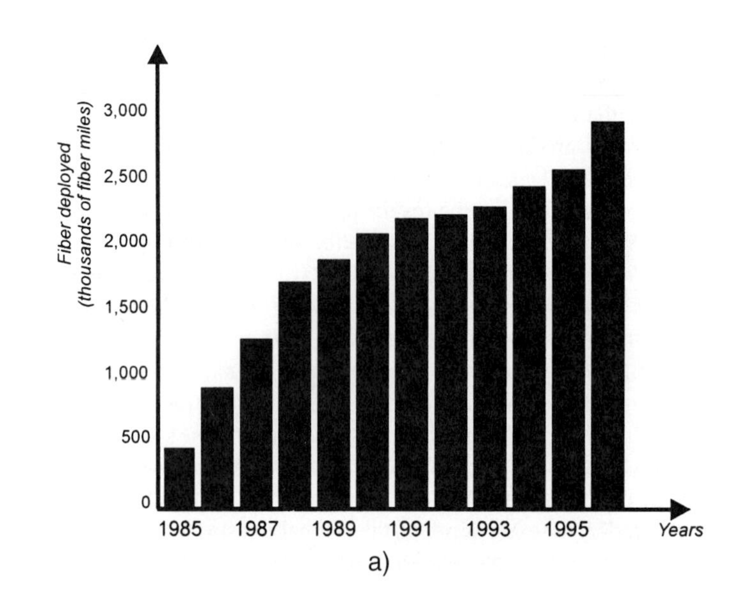


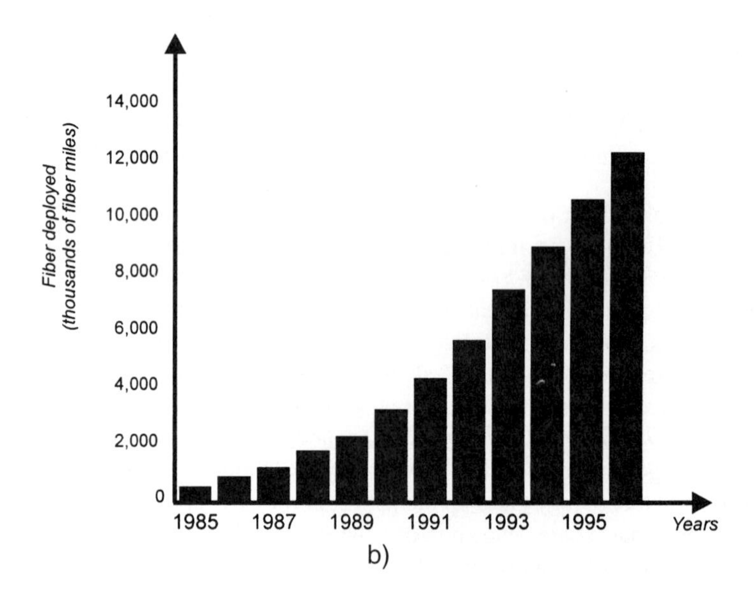


**Figure 1.8** Dynamics of deployment of telephone fiber-optic networks in the U.S.A.: (a) Local telephone networks; (b) long-distance telephone networks. *(Source:* FCC report: *Fiber Deployment Update, End of Year 1996.)*

by the telephone companies (or telcos, as they're known) over the last 10 years. The term "fiber mile" means the length of a single installed fiber. Keep in mind that one fiber-optic cable can carry many single fibers. For example, the first fiber-optic cables in operation before 1980 contained 24 fibers each, which resulted in 45 megabits per second of information-carrying capacity per cable. Today, the average number of fibers per cable is around 48, although a modern cable can carry more than 400 fibers.

To avoid confusion, the industry uses two numbers to describe the amount of fiber installed: *fiber miles,* which refers to the total length of all single fibers, and *sheath miles,* which means the total length of all fiber cables regardless of the number of fibers that each cable contains. You may also encounter the term "route miles." This is the length of an installed fiber-optic communications link regardless of the number of cables in the link. Thus, if you have installed 10 route miles with 12 fiber-optic cables and 50 optical fibers per cable, you simply multiply $50 \times 12 \times 10$ and obtain 6000 installed fiber miles.

A word about the units used in this business: Fiber-optic communications technology uses the International System of Units (SI), that is, meter, second, kilogram, ampere, and their derivatives. But official documents of United States government agencies, such as those issued by the Federal Communications Commission (FCC), for instance, still use traditional American units. This is why Figure 1.8 displays the length of installed fiber networks in miles.

As we've noted, today you find fiber-optic communications technology everywhere, but for discussion purposes we've got to start somewhere, so let's begin with worldwide networks.

***Worldwide submarine networks*** The first international undersea fiber-optic link, which linked England with Belgium, was laid on the floor of the North Sea in 1986. By the end of 1988, the first transatlantic fiber-optic cable, connecting the United States with Europe, was laid. That project, called TAT-8, was the mutual venture of AT&T, British Telecom, and France Telecom, among others [3]. The link runs 5600 km (see the note about units usage) from Tuckerton, New Jersey, to the European Shelf, where it divides into two branches. One branch travels more than 500 km to England and the other continues for more than 300 km to France. The link can carry 80,000 voice channels. But this number just characterizes its information-carrying capacity, since the line obviously carries not only voice but video and data as well. The link uses a singlemode fiber and operates at a wavelength of 1300 nm. Laser diodes with a mean lifetime of $10^6$ hours provide the light sources here. To combat signal attenuation, special devices called repeaters were installed every 50 km along the cable. A repeater, as the word implies, is a device whose function is to keep the signal alive and strong. Functionally, a repeater, also called a regenerator, works with a digital signal. It analyzes the incoming signal, makes a decision as to whether the pulse represents logic 1 or 0, generates a new pulse based on that decision, and sends all of the pulses farther along the link. So the device continuously repeats the logical meaning of the incoming signal and generates a new electrical signal to carry the logic. To do its job, a repeater has to convert an optical signal into an electrical one, process the electrical signal, and convert it back into optical form for continued transmission over the optical fiber.

And herein lies a drawback to fiber optics, although by no means an insurmountable one. Optical fiber is a dielectric and, consequently, it cannot directly carry an electric signal, including power. Thus, a separate conductor must be included in submarine cables to deliver electric power. The conductor in the TAT-8 project carries 1.6 amperes to feed electric power to the system's repeaters.

Recent projects (1996) connecting the United States and Europe, TAT-12 and TAT-13, use the newest wavelength-division multiplexing technique to increase even further the information-carrying capacity of the links. These links carry more than 300,000 voice channels, nearly four times more than the 80,000 voice channels processed by the TAT-8 transatlantic link.

Today, the entire world is connected through this extensive optical network [4]. Figure 1.9 shows the current standing of the submarine optical network developed by Tyco Submarine Systems Company.

A very good example of the developing worldwide network is the international project called the Pan American Cable System [5]. The network links seven countries in South America with North America, Europe, and Asia via the international gateway. The Caribbean undersea portion connects the initial point at St. Thomas in the U.S. Virgin Islands with Venezuela, Aruba, Colombia, and Panama. It's a 2700-km undersea cable with 23 repeaters installed every 120 km. The Pacific submarine portion connects Panama with Ecuador, Peru, and Chile. This link lasts 4400 km with 40 repeaters every 115 km. The total length of the link is 7300 km, with 200 km of cable being land-based, buried in Panamanian soil.

There are several other global submarine projects under development: Project Oxygen aims to build a global undersea optical-fiber cable network. Global Crossing Ltd. (GCL) is building an undersea and terrestrial global fiber-optic communications network. In addition to the Pan American Cable System described above, the Pan American Crossing telecommunications cable system (PAC) is in the process of connecting California, Mexico, Panama, and St. Croix in the U.S. Virgin Islands, where this segment will be joined to the GCL Atlantic submarine network.

***Terrestrial networks*** It is clearly evident that a worldwide fiber-optic network includes not only a submarine segment but a terrestrial portion as well. Apart from the fact that the entire territory of the United States is covered with an extensive web of optical fibers, all other countries are pursuing the same course. One other typical example of an international optical network is the pan-European optical network [6]. This project aims to cover all of Europe, including Eastern Europe and Scandinavia, with fiber-optic networks using wavelength-division multiplexing technology. There have already been several successful field trials in the transmission of $16 \times 10$ Gbit/s signals over more than 500 km. (The number "16" stands for the number of wavelengths used and "10 Gbit/s" indicates the capacity of a single wavelength channel.)

***The United States takes leading role*** The United States, the largest producer and consumer of fiber-optic components going into communications systems, is playing a major role in developing and using fiber-optic technology. American industry manufactures about half of the world's optical fiber, with Europe and Japan producing most of the rest. American companies have invested heavily in both money and expertise to support the worldwide projects described above. Extensive research at American universities and in corporate laboratories has opened new horizons in the development and installation of fiber optics.

***Satellite systems versus fiber-optic networks*** The obvious question at this point is, why do we need a worldwide fiber-optic network if, as everyone knows, we can reach practically any point on the globe using satellite communications systems? The answer lies in the two major advantages fiber-optic communications offers over satellite communications, namely, its higher information-carrying capacity and its speed in transmitting the signal.

As for information-carrying capacity, you'll recall that a single optical fiber has the potential to carry up to 50 Tbit/s. In 1996, the first successful experiments in the transmission of more than 1 Tbit/s over a significant distance—more than 100 km—were made. There is no indication today that satellite-communications technology will ever achieve such an information-carrrying feat. Another disadvantage of satellite communications is the signal delay caused by the long distance that a signal has to travel from earth station to satellite and back again. Fiber optics provides much straighter and shorter connections between transmitters and receivers, virtually eliminating this problem. The last major drawback to satellite communications is its susceptibility to adverse atmospheric conditions.

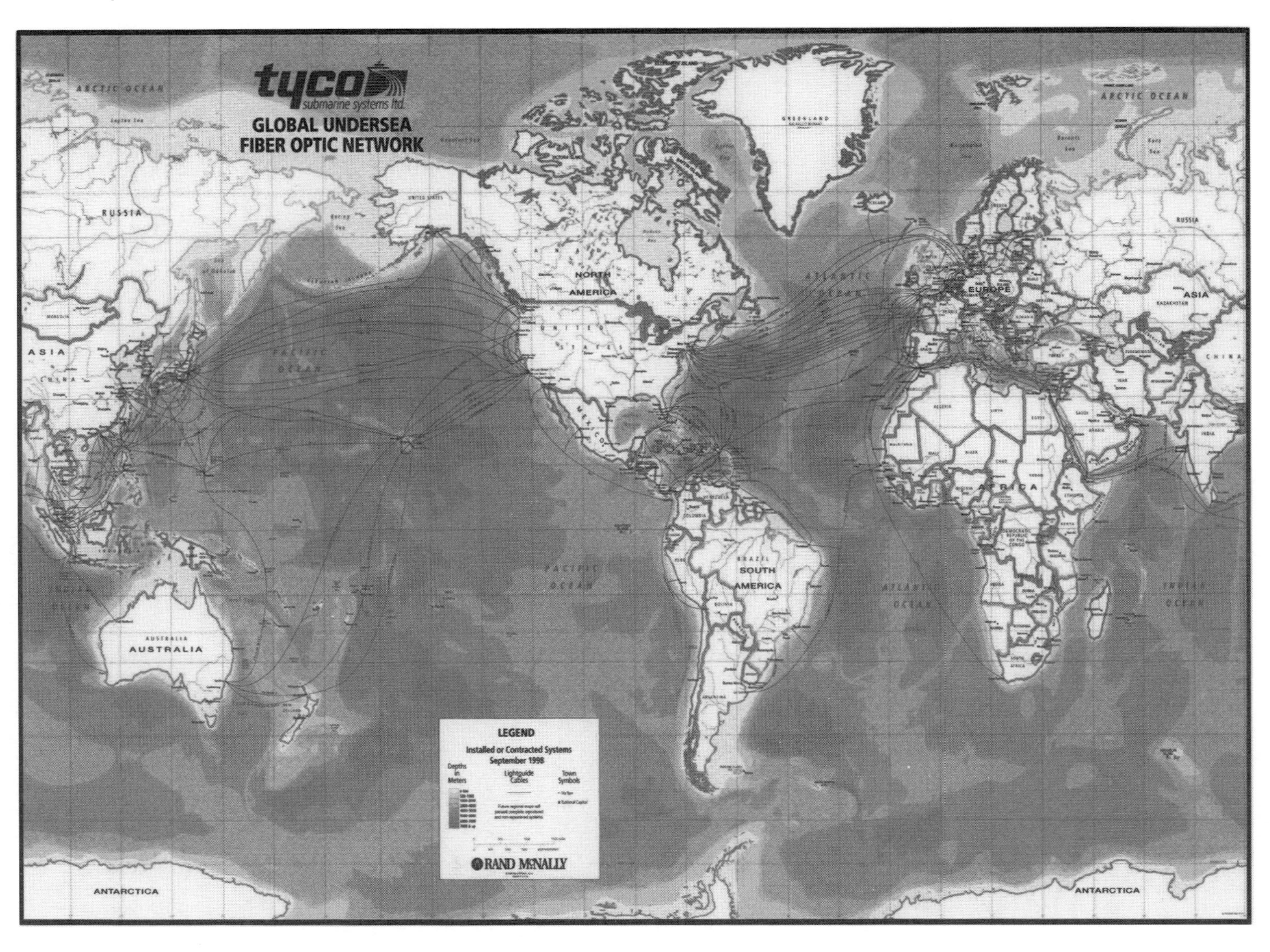


Figure 1.9 Tyco's undersea fiber-optic network. *(Courtesy of Tyco Submarine Systems Laboratories).*

On the other hand, satellite communications has the ability to reach nearly any point on the globe without wiring. The Iridium Project, which is based on a number of low-orbiting satellites, spans the entire globe, giving everyone a global wireless connection.

It is highly unlikely that either of these technologies will eliminate the other. A more likely scenario is that satellite communications and fiber-optic technology will prove to be complementary rather than competitive players in the telecommunications industry.

Today, fiber-optic networks carry 10 times more global traffic than satellite communication systems. Looking at the pace of deployment of both types of communications, we can conclude that this ratio will increase in favor of fiber optics.

***Fiber-to-the-home and fiber-to-the-desk*** For local telephone networks, this means that all connections between central offices and all connections from remote terminals to central offices are optical-fiber cables. The only connections today using copper wire as a transmission medium are the links between the customer's premises and the nearest central office or the nearest remote terminal. This "last mile," if you will, is the bottleneck in the modern telecommunications network only because there is no way to replace, overnight, about a billion twisted pair connections

**The Changing Business Scene**

From the beginning of the last century until 1984, telephone communications—the major form of telecommunications in the United States—was provided by AT&T operating a de facto monopoly, if not on the basis of actual law, since its founding in 1885 as the American Telephone and Telegraph Company. Actually, AT&T dates its history all the way back to the dawn of the telephone, for which Alexander Graham Bell was first awarded a patent in 1876. The original Bell Telephone and a company called New England Telephone were consolidated in 1879 into the National Bell Telephone Company, which in turn became, in 1881, American Bell Telephone, the parent of the Bell System. The American Telephone and Telegraph Company was formed in 1885 as a subsidiary of American Bell to build and operate the first long-distance network. It, in turn, became the Bell System's parent in 1899.

Divestiture of AT&T in 1984 left the company as a long-distance provider and the seven Baby Bells (the Regional Bell Operating Companies—RBOCs) as providers of local service in their respective geographical locations. The 1984 decision by U.S. District Court Judge Harold Greene also allowed for competition in the long-distance market, and two major contenders, MCI and Sprint (previously virtual unknowns outside their limited areas), suddenly burst on the scene to give giant AT&T a run for its money in this extremely lucrative market.

MCI, originally a tiny company known primarily for establising the first independent microwave link between Chicago and St. Louis, and Sprint, another small-time contender, became important players in the long-distance market, but only after they deployed their own massive fiber-optic networks.

The Telecommunications Act of 1996, discussed previously, was the next big step in opening up the former Bell System's exclusive territory by fostering competition at the local level. The act also allows competition among different service providers; for example, cable TV companies now can offer telephone and computer-connection services, while the telcos can offer TV and Internet services. The telecommunications landscape has been changing at an explosive rate since 1996 and will continue doing so as far ahead as anyone can foresee. Many new independent companies have already appeared on the scene, fighting for a piece of the $200-billion telephone market. What's more, the major players are becoming industrial behemoths as they jockey to strengthen their market positions. MCI, for instance, merged with Worldcom and MCIWorldcom made a failed bid to acquire Sprint in its move to challenge AT&T's dominance of the industry. And AT&T, for its part, hasn't been idle, having purchased TCI, the largest American cable company, a step that enables AT&T to gain direct access to your home telephone without having to go through the good auspices of the local phone company. This strong—nay, severe—competition inevitably drives technology innovations. As a result, what was reported yesterday as pure research may very likely become a commercially available product today. It is hard to predict specific technological advances but it is easy to predict their proliferation in this booming, yet still emerging, field.

of copper cable with optical fiber all around the world. To give you some idea of the immensity of this subscriber-line problem, suffice to say that in the United States alone there are more than 250 million lines that would have to be replaced. But this will happen. The industry is constantly abuzz with talk of fiber-to-the-curb (FTTC), fiber-to-the-home (FTTH) and even fiber-to-the-desk (FTTD) systems, largely because the technology exists (after all, just look outside at the cable TV connections), but the cost effectiveness still holds them in abeyance. Will they ever become economical enough to install? Probably, but *exactly* when is still uncertain.

***Fiber optics in LANs and beyond*** Let's look now at the local area network (LAN) market and the role of fiber-optic communications. As you well know, the power of the personal computer increases significantly with every new generation of entries, and these days that seems like an annual event. The real speed of a PC operation is now 100 million instructions per second (MIPS). And remember, the more powerful the computer, the more information it produces and the more information that has to be transmitted per unit of time.

Another move under way is to computer networking, which allows computers to share their resources and information. Local area networks (LANs), metropolitan area networks (MANs), and wide (world) area networks (WANs) have changed computer technology to an even greater extent than have increases in the power of the individual computer.

To make a LAN work effectively, a huge amount of information has to be transmitted over the network at high speed. The minimum transmitting capacity a LAN needs is estimated at 1 Mbit/s per 1 Mips, the so-called Amdahl's law. Today, however, that won't do. We need about 100 Mbit/s, and LANs generally operate at a much lower rate. What's holding back their further development is the networking infrastructure, because LANs are still mostly copper-wired. The solution is obvious: a switch to fiber optics. Until now, extensive use of optical fiber in LANs has been stymied by a higher installation cost than that of copper. But now the cost of all fiber-optic components—optical fiber, suitable plastic fiber, connectors, opto-electronics and interface devices—is declining steadily. This, plus the fact that the electronics for high-speed fiber-optic transmission is easier to design and manufacture (and, consequently, cheaper) because it doesn't require sophisticated encoding and compression, is giving fiber optics the edge in its battle with copper for the LAN business.

## Developments to Watch

Two trends rapidly developing in the United States relate to optical-fiber networking. One push is to use wavelength-division multiplexing (WDM)—putting several signals that differ in wavelength into one fiber—for more efficient use of the existing optical-fiber networks. For example, if you could combine four signals and transmit them simultaneously over the same optical fiber, you would use the link four times more efficiently—it's as simple as that. WDM has changed fiber-optic communications technology dramatically in recent years and continues to do so. (The subject is covered in detail in Chapters 12 through 15.)

The other key trend is the proliferation of long-distance and local networks. Today, besides the many telephone companies, a number of other telecommunications providers are on the scene. Among them are cable TV companies, which use their fiber-optic networks to enter the business of telephone and computer-connection services. Other major contenders are telecommunications companies that specialize in business-oriented communications. They build their own fiber-optic networks with tremendous information-carrying capacity, some with links having a total capacity of tens of Tbit/s. (Don't let these numbers confuse you. This capacity—tens of Tbit/s—means the capacity of the entire link and that includes many fiber-optic cables, not simply single strands of fiber.)

The upshot: From intercontinental transmission lines to intradevice connections, you now see fiber optics everywhere. For example, you find fiber-optic links inside a modern signaling machine (SS7); the utilities and gas-pipeline industries are customers of fiber optics; and the automotive and avionics industries are expanding their use of fiber optics for communications. You can no doubt cite many examples from your own experience.

Talk with scientists, engineers, and business executives in just about any high-tech (and even some low-tech) industries; inevitably, the conversation will turn to fiber optics. In the classic movie *The Graduate,* you may recall, a staid, middle-aged businessman whispered some sage career advice to a young Dustin Hoffman: "Plastics." If that film were being scripted today, surely the writer would reword it: "Fiber optics"—the hot technology of the new century that will most certainly define telecommunications for the foreseeable future. At least that's what the experts tell us, and they all can't be wrong.

To conclude, every day professional journals, popular magazines, newspapers, and TV bring news about innovations in fiber-optic technology. Developments in the new fiber-optic communications projects range in scale from global to intraoffice. Become a part of this exciting world and you'll be on the road to a very successful future.

## PROBLEMS

**1.1.** Define the term "telecommunications."

**1.2.** Why is telecommunications so important in today's society?

**1.3.** What is the projected rate of growth of the telecommunications industry?

**1.4.** Define and give examples of a telecommunications network.

**1.5.** Give examples of telecommunications networks based on your everyday experiences.

**1.6.** The PCM technique involves three major steps: sampling, digitizing, and coding. Identify these steps in Figure 1.3.

**1.7.** A Microsoft vice president and chief information officer said, "Communications is our lifeblood. The only restriction on our insatiable appetite for bandwidth is our budget for bandwidth." How do you interpret that statement, taking into account the Shannon-Hartley theorem?

**1.8.** Draw a block diagram of a fiber-optic communications system and describe the function of each component.

**1.9.** Why are fiber-optic communications systems the backbone of modern telecommunications?

**1.10.** Draw the block diagrams of an optical transmitter and an optical receiver.

**1.11.** What were two major problems encountered in the early development of optical fibers for practical use in communications systems?

**1.12.** The text says that modern communications could not exist without fiber optics. Write a paragraph or two explaining this statement.

**1.13.** What is the growth trend in "installed" fiber optics in the United States? Worldwide?

**1.14.** Which continents are connected by fiber-optic links?

**1.15.** What are the advantages and the drawbacks of fiber-optic communications and satellite communications?

**1.16.** What is the current status of fiber optics in long-distance telephone networks? In local telephone networks?

**1.17.** What are the major trends in the development of fiber-optic communications networks?

**1.18.** Why does fiber optics promise to be the technology of choice for local area networks?

**1.19.** List the advantages of fiber-optic communications over other types of communications technologies.

**1.20.** Psychologists say that a good way to memorize new knowledge is to sketch a memory map—a graphical device that consists of circles or squares in which you list a specific topic (e.g., optical fibers, transmitters, receivers, deployment, Shannon's theorem, etc.). List all the topics, or items, you recall after reading a given chapter. Then connect all the topics according to their relationship. Such a map will help you to review all the material, not simply by names, but—more importantly—by their functions and connections with other entities.

This could be a messy undertaking, so use a large sheet of paper and don't hesitate to sketch the map a couple of times because it is very unlikely you will construct a good map on your first attempt. In the end, however, you'll find it was well worth the effort.

## REFERENCES[1]


1. John Bray, *The Communications Miracle (The Telecommunications Pioneers from Morse to the Information Superhighway),* New York: Plenum Press, 1995.

2. "Recollections of a Leading Fiber-Optics Pioneer," an interview with Dr. Donald B. Keck conducted by Dr. Milton Chang, *Photonics Spectra,* April 1998, pp. 30–31.

3. Ahmed Khan, *The Telecommunications Fact Book and Illustrated Dictionary,* Albany, N.Y.: Delmar Publishers, 1992.

4. Neal Bergano, "WDM Long-Haul Transmission Systems," Paper TuF, *Proceedings of the Optical Fiber Communication Conference (OFC'98),* San Jose, Calif., February 22–27, 1998.

5. Patrick Trischitta, Antonio Medina, and Roberto Remedi, "The Pan American Cable System," *IEEE Communications Magazine,* December 1997, pp. 134–140.

6. Matthias Berger et al., "Pan-European Optical Networking Using Wavelength Division Multiplexing," *IEEE Communications Magazine,* April 1997, pp. 82–87.


[1]See Appendix C: A Selected Bibliography.

# 2

# Physics of Light: A Brief Overview

Fiber-optic communications technology uses light as a signal carrier. Light is electromagnetic radiation. To understand the basics of the technology, we need to examine various aspects of this type of radiation. Therefore, we will consider this phenomenon from three vantage points: as *electromagnetic (EM) waves* (the wave view), as a *ray,* or *beam* (the geometric optics view), and as a *stream of photons* (the quantum view). This section briefly reviews these basic concepts of light theory.

## 2.1 ELECTROMAGNETIC WAVES

Our concern in this section is light as electromagnetic waves. Figure 2.1 is a stop-action picture of an electromagnetic wave.

Let us simplify our discussion by considering light propagation in free space, a vacuum. Clean air is a very good practical approximation for this example. In this environment, light is transverse electromagnetic (TEM) waves. The term *transverse* means that both vectors—electric **E** and magnetic **H**—are perpendicular to the direction of propagation, the $z$ axis in Figure 2.1. Only the electric component of the electromagnetic wave is shown in Figure 2.1, but what follows is also true for the magnetic component of the wave. Observe in particular that the electromagnetic wave is developing with respect to both time and space. To underscore this point, Figure 2.2 shows how the wave changes over time and with respect to space.

Figure 2.2 shows the EM wave at three time moments: $t = 0$, $t = T/4$, and $t = T/2$. It is important to note that the same point of the wave (point M) moves with respect to space while time is changing. So if you were able to sit at point M of the wave, you would see the different instances of the $z$ axis very much the way you see different sights as you gaze out the window of your moving car. But if you were able to stay at the same point on the $z$ axis (point N), you would see the different instances of the EM wave very much the way you see different vehicles passing by when your car is stopped on the shoulder of the road. Now imagine that both events are developing at the same time so that the EM wave changes simultaneously with respect to space and time.

Figure 2.1 is a stop-action picture of an electromagnetic wave in two panels: (a) shows its time dependence and (b) its space dependence.

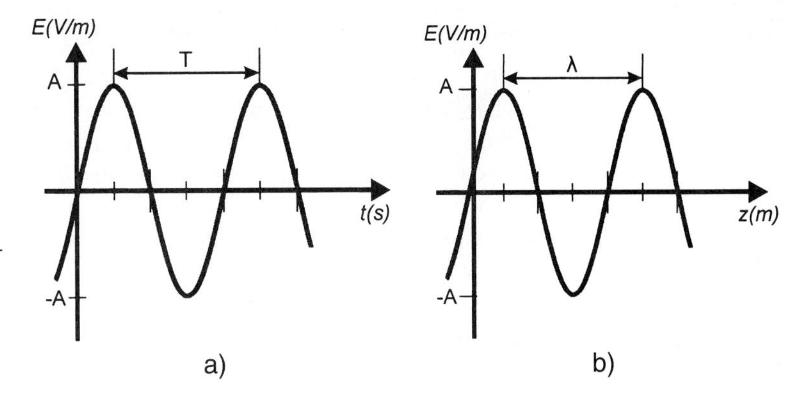


**Figure 2.1** Electromagnetic wave: stop-action picture: (a) Time dependence: $T$ – period (s), $f = 1/T$ – frequency (cycle/s = Hz); (b) space dependence: $\lambda$ – wavelength (m).

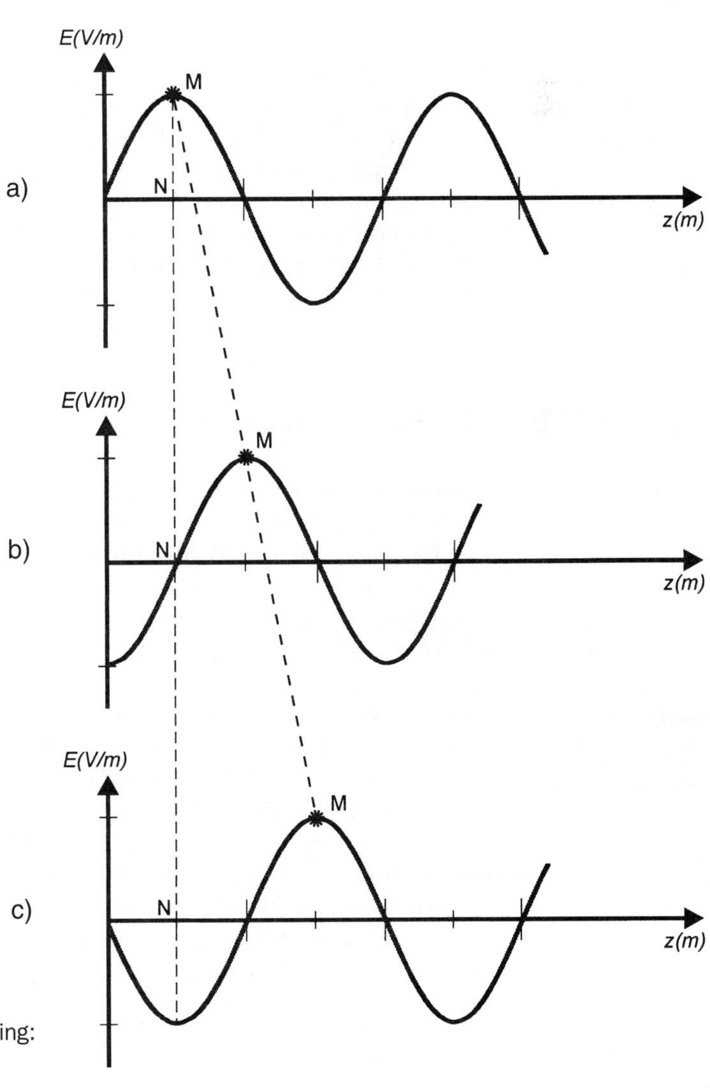


**Figure 2.2** Electromagnetic wave developing: (a) $t = 0$; (b) $t = T/4$; (c) $t = T/2$.

This brings us to two important concepts: wavelength and period. Imagine that you are watching waves rolling over the ocean. Imagine, too, that all movement suddenly freezes, enabling you to measure *the distance between two adjacent crests.* This distance is the *wavelength.* Now suppose that you are in a steady state at an enclosed construction site on the sea floor and that waves are passing you overhead. *The time it takes for two adjacent crests to pass you is the period.*

To put these distinctions into formal terms: *Wavelength is the distance between two identical points (the points having the same phase) of two successive cycles of a wave. Period is the time it takes a wave's two identical points (the points having the same phase) to pass, in sequence, the same space location.* The wavelength and the period of the wave are related through wave velocity. The period, T, *is the time it takes a wave to travel a distance equal to one wavelength,* $\lambda$, *at velocity c*—On the other hand, *a wavelength,* $\lambda$, *is the distance traveled by a wave per one period,* T, *at velocity c.* Therefore, *light velocity is equal to wavelength divided by period,* $c = \lambda/T$. Since the frequency, $f$, is equal to $1/T$, we thus arrive at the very well-known and important formula:

$$\lambda \cdot f = c \qquad (2.1)$$

where $c$ = speed of light (m/s), $\lambda$ = wavelength (m), and $f$ = frequency (Hz).

The speed of light in free space is $c = 3 \times 10^8$ m/s and the central frequency of visible light is about $6 \times 10^{14}$ Hz; hence, the center of visible wavelengths has the order of value $\lambda = 0.5 \times 10^{-6}$ meters, or 0.5 micrometers (μm). Fiber-optic communications technology usually measures wavelength in nanometers (nm); a nanometer is $10^{-9}$ m; thus, this wavelength is 500 nm.

Figure 2.3 shows the part of the electromagnetic spectrum that we are interested in. Refer to the discussion on pages 3 to 6, "Information-carrying capacity," to visualize the frequencies allocation described there. By now you can see that *light is electromagnetic waves that occupy a specific range of the electromagnetic spectrum.* To the layman, the word *light* means visible rays, or radiation, but scientifically this term also includes both ultraviolet and infrared radiation despite the fact that they are invisible to the human eye.

As for the visible spectrum, it is customary to consider this as the range between 400 nm and 700 nm, although there are no hard-and-fast boundary limits to this region. The different wavelengths represent different colors. For example, Figure 2.3(b) shows that green light has a wavelength of about 500 nm and red light a wavelength of about 700 nm. However, if you look at the beam radiated by the helium-neon gas laser—one of the most widely used lasers in physics laboratories—you will see a perfect, intense, saturated red beam despite the fact that its wavelength is 633 nm. As you can also see, however, there is no abrupt boundary between the different colors, and the transition from one to another is very gradual.

## 2.2 BEAMS (RAYS)

### Refractive Index

We know from our everyday experience that light can be treated as a beam (ray). Just look at the light that emanates from a car's headlights or, better yet, at the red beam that emerges from the simple light pointer available at any stationery store. The clearest example of this phenomenon is, of course, light radiated by a laser, which emerges in pinpoint fashion as a ray, or beam.

Light rays propagate within different media at different velocities. It seems, too, as though different media resist light propagation with different strengths. The characteristic that describes this property of a medium is called the *refractive index,* or *index of refraction.* So, if $v$ is the light velocity within the medium, and $c$ is the speed of light in free space, then the refractive index, $n$, can be determined by the following formula:

$$n = c/v \qquad (2.2)$$

The refractive indexes for some media are given in Table 2.1.

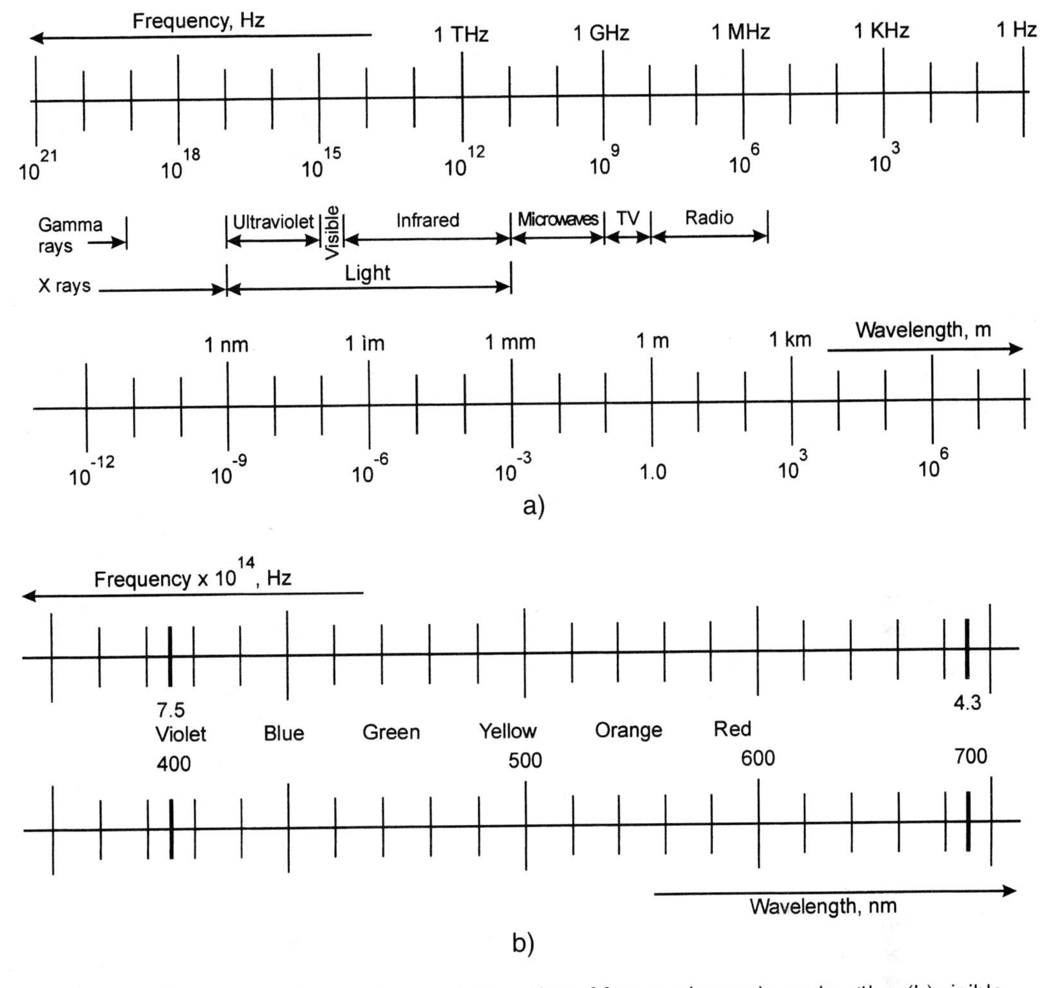


**Figure 2.3** Electromagnetic spectrum: (a) Allocating of frequencies and wavelengths; (b) visible part of electromagnetic spectrum.

**Table 2.1** Refractive indexes of various media

| Material | Air | Water | Glass | Diamond |
|---|---|---|---|---|
| Refractive Index | 1.003 | 1.33 | 1.52–1.89 | 2.42 |

## Example 2.2.1

***Problem:***

What is the light velocity within glass?

***Solution:***

When a light ray from the air strikes *and* penetrates glass, its rate of movement slows. Taking the refractive index of glass, $n = 1.5$, the light velocity within glass will be found as:

$$v = c/n = 3 \times 10^8 \text{ (m/s)}/1.5 = 2.0 \times 10^8 \text{ (m/s)}$$

As you can see, the calculations ignore the slight difference in the refractive indexes of vacuum and air. Obviously, the higher the refractive index, the denser the material from an optical standpoint.

This simple example demonstrates one of the basic ideas of optics: *All characteristics of light in free space are changed inside the material with the refractive index n.* Velocity becomes *c/n,* wavelength becomes *λ/n,* and so forth.

---

When a light ray from the air strikes the surface of, say, glass, it not only slows but also changes its direction within the medium. This is true for any two media. Thus, the refractive index can be a measure of how much light is bent when it penetrates through one medium into another. In general, when a light beam strikes the boundary of two media, the incident beam splits into two beams: reflected and refracted (Figure 2.4).

The question now arises: What are the directions of the reflected and the refracted beams? To find the answer, we need to observe their directions as determined by the appropriate angles: $\Theta_1$ *is the angle of incidence,* $\Theta_3$ *is the angle of reflection, and* $\Theta_2$ *is the angle of refraction.* (*Important note:* These are the angles that light beams make with the line perpendicular to the boundary.) The relationship between these angles depends on the medium the light ray strikes. (Compare Figures 2.4[a] and 2.4[b].) Snell's law[1] gives the rules defining the directions of these beams:

$$\begin{aligned} \Theta_1 &= \Theta_3 \\ n_1 \sin\Theta_1 &= n_2 \sin\Theta_2 \end{aligned} \tag{2.3}$$

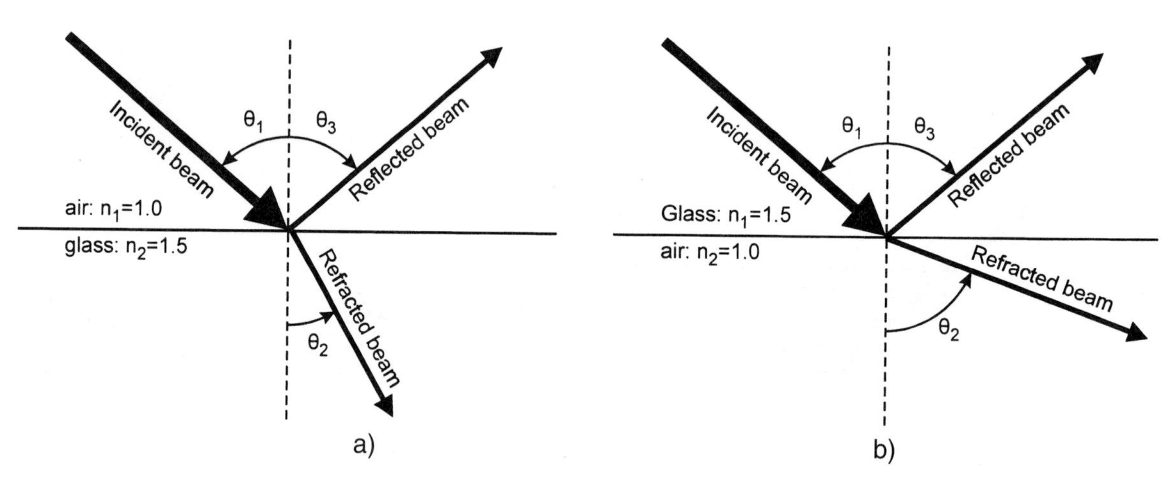


**Figure 2.4** Incident, reflected and refracted beams: (a) Light travels from air to glass; (b) light travels from glass to air.

[1]Willebrod van Roijen Snell (1580–1626), Dutch-born scientist, discovered the law of refraction, which was named after him. Interestingly, he was a professor of mathematics in Leyden and was known in his day for developing the use of triangulation in surveying.

### Example 2.2.2

***Problem:***

Let $n_1 = 1.0$, $\Theta_1 = 30°$, and $n_2 = 1.5$. (See Figure 2.4 [a].) What are $\Theta_3$ and $\Theta_2$?

***Solution:***

$\Theta_3 = 30°$ and $\Theta_2$ is calculated as follows:

$$\sin \Theta_1 = \sin 30° = 0.5; \text{ hence, } \sin \Theta_2 = 0.5/1.5 = 0.333, \text{ and } \Theta_2 = \sin^{-1}(0.333) = 19.5°$$

The same type of calculation made for Figure 2.4(b) results in the following:

$$\Theta_3 = 30° \text{ and } \Theta_2 = \sin^{-1}(0.75) = 48.6°$$

These simple examples bring us to a very important point in our discussion: What happens if we increase the angle of incidence from glass to air? Figure 2.5 demonstrates a sequence of the

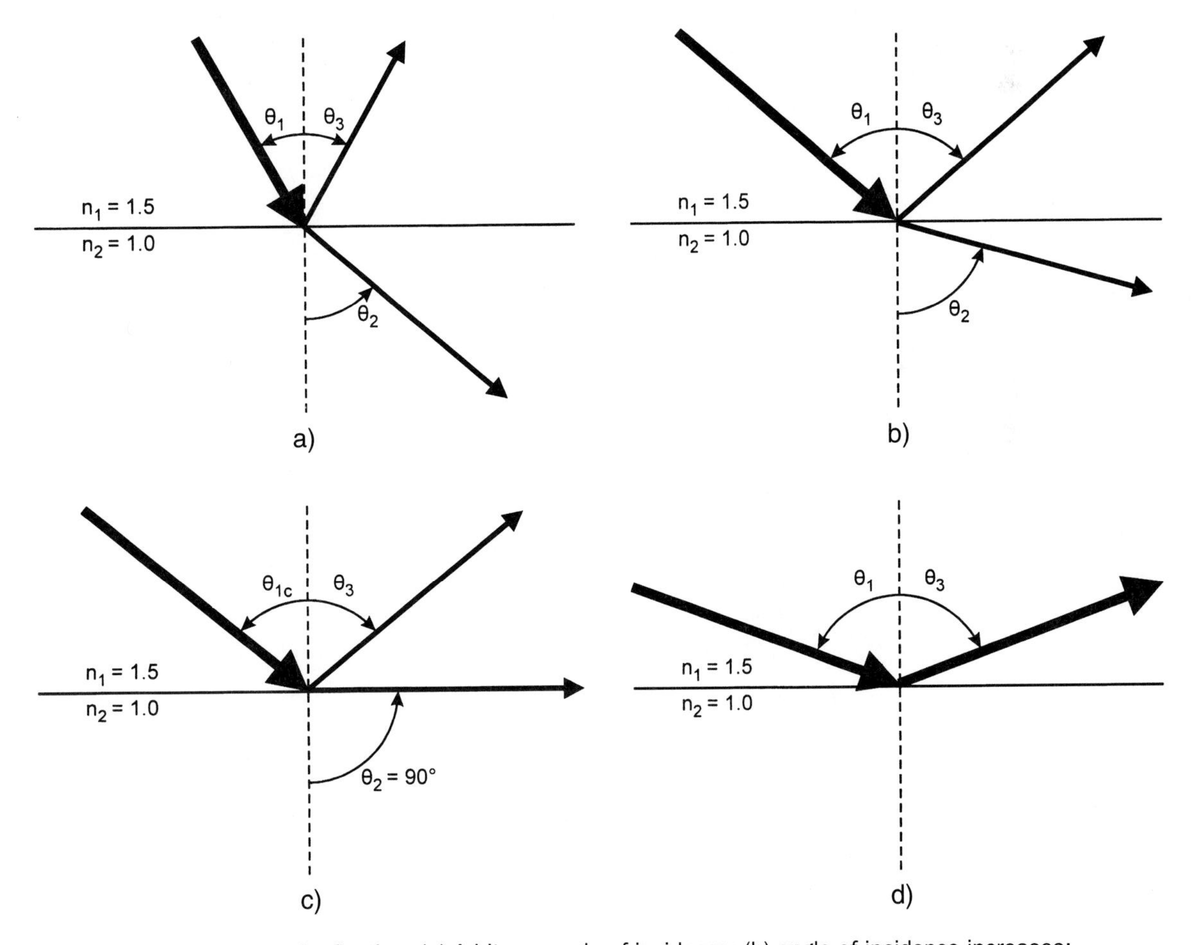


**Figure 2.5** Total internal reflection: (a) Arbitrary angle of incidence; (b) angle of incidence increases; (c) critical angle of incidence—total internal reflection occurs; (d) angle of incidence is more than critical—all light is reflected back.

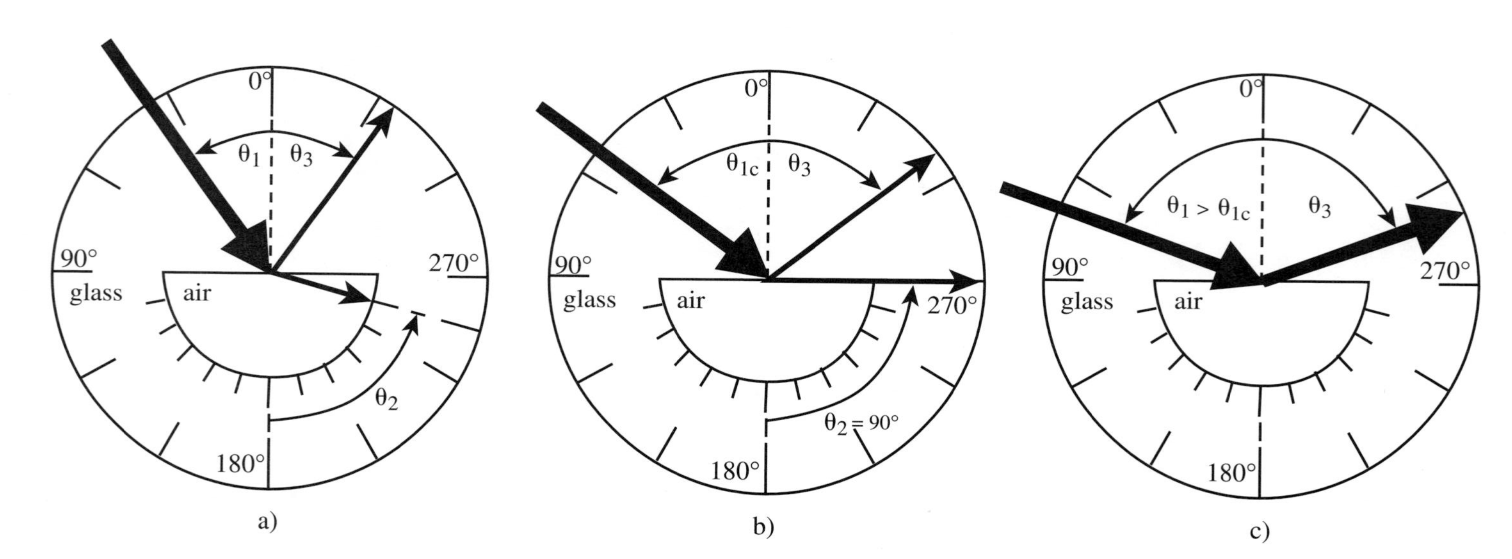


**Figure 2.6** Demonstration of total internal reflection using Snell's wheel: (a) Angle of incidence, $\theta_1$, is less than critical; (b) angle of incidence, $\theta_1$, is critical; angle of refraction, $\theta_2$, is 90°; (c) angle of incidence is more than critical; all light is reflected back.

positions of the ray of light. The most important position is shown in Figure 2.5(c) where the angle of incidence, $\Theta_1$, reaches the critical value of $\Theta_{1C}$—critical because no light penetrates the second medium (in this example, air). The incident angle at which the angle of refraction equals 90° is called the *critical incident angle,* $\Theta_{1C}$. If we continue to increase the angle of incidence so that $\Theta_1 > \Theta_{1C}$, all light will be reflected back into the incident medium. This situation is shown in Figure 2.5(d). This phenomenon is called *total internal reflection* because all light is reflected back to the medium of incidence.

*Total internal reflection is what keeps light inside an optical fiber.* Without this effect, we could not use optical fiber as a light guide over a long distance, as we do now. We can summarize this important concept as it applies to fiber-optic communications technology in this way: *When light travels from a medium with a higher refractive index to a medium with a lower refractive index and it strikes the boundary at more than the critical incident angle, all light will be reflected back to the incident medium,* meaning it will not penetrate the second medium. This phenomenon is called *total internal reflection.*

A very simple device, Snell's wheel, allows us to demonstrate total internal reflection. This wheel and the possible situations one can observe using it are shown in Figure 2.6.

### Example 2.2.3

***Problem:***

Assume you have a glass rod surrounded by air, as shown in Figure 2.7. Find the critical incident angle.

***Solution:***

If you direct a beam of light into the rod so that you reach the critical incident angle or higher, all light will be saved inside the rod. If the refractive index, $n_1$, of the glass rod, equals 1.6, then we can find the critical incident angle, $\Theta_{1C}$, as follows:

$$n_1 \sin\Theta_1 = n_2 \sin\Theta_2$$

Since $n_1 = 1.6$, $n_2 = 1.0$, while $\Theta_2 = 90°$, we have

$$\Theta_{1c} = \sin^{-1}\frac{1}{1.6} = 38.68°$$

That's basically how optical fiber "conducts" light.

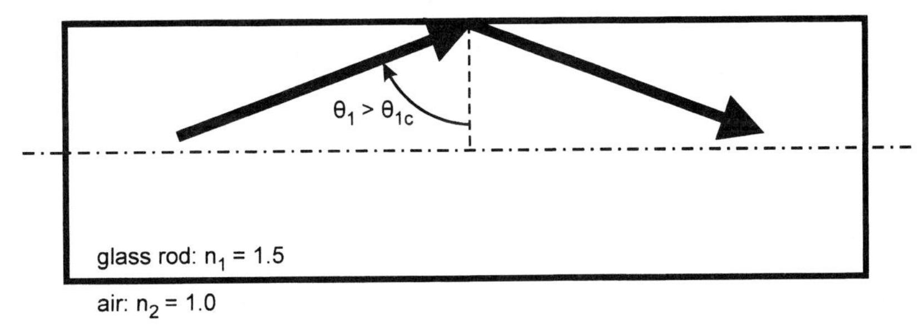


**Figure 2.7** Saving light inside a glass rod.

Total internal reflection is a necessary condition to make optical fiber work as a communications link. Keep that point clearly in mind.

## 2.3 A STREAM OF PHOTONS

### An Energy-Level Diagram

In this section we'll take a different view of light—a quantum view.

All matter ultimately consists of atoms. Each atom, in turn, consists of a nucleus surrounded by electrons. For the sake of simplicity, you can use a solar model of an atom, where electrons rotate on different orbits around the nucleus. The fact that the nucleus is approximately $10^5$ times smaller than the whole atom might help you visualize the model. Bear in mind that this model, developed by Niels Bohr[2] at the dawn of the quantum physics era (1913), does not describe atomic properties as we understand them today, but it does facilitate a presentation of the basic ideas.

Bohr's model assumes that electrons rotate on stationary orbits and therefore possess a stationary value of energy. Bohr's breakthrough was the assumption that rotating electrons do not radiate; that is, they do not change their energy value during rotation, as classic electromagnetic theory suggests. Any change in energy occurs only discretely, such as when electrons jump from one orbit to another. This implies that an entire atom possesses discrete values of energy; in other words, an *atoms's energy is quantized.*

Figure 2.8 introduces an energy-level diagram. Observe that there is only a vertical axis in the diagram and this axis shows an energy value. While it is normal for the position of the horizontal axis to represent an energy level, its length means nothing.

Possible discrete energy values (that is, those that are allowed by the law of quantum physics) are always separated by energy gaps. The lowest energy level is called the *ground state.* An atom can be at any of these levels, or states, and it can change its energy only by jumping from one level to another; in other words, it can change its energy level only discretely. It should be emphasized that there can be no smooth transition between these states. An atom can have as its energy value, say, $E_2$ or $E_3$, but nothing *between* $E_2$ and $E_3$.

### A Photon

What happens if an atom jumps from an upper level to a lower level, say, from level $E_3$ to level $E_2$? There is an energy gap between these two levels, $\Delta E = E_3 - E_2$, and this difference will be released as a *quantum of energy,* which is called a *photon.* You can think of a photon as a particle—not a mechanical particle like a speck of dust, for instance, but as an elementary particle that car-

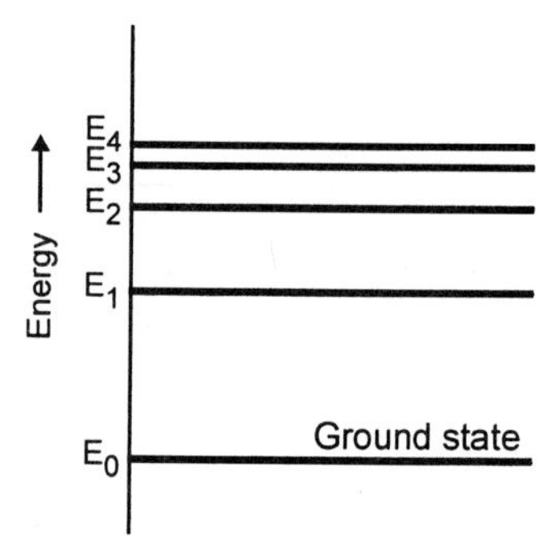


**Figure 2.8** Energy-level diagram.

[2]Niels Henrik David Bohr (1885–1962) was the Danish scientist who explained the spectrum of hydrogen by means of an atomic model and the quantum theory. He was awarded the Nobel Prize for Physics in 1922. During World War II, the American government sent a special submarine to rescue Bohr from Nazi-occupied Denmark. He joined the Manhattan Project, which developed the atomic bomb. After the war he returned to Denmark.

[3]Max Karl Ernst Ludwig Planck (1858–1947), the German physicist after whom the constant was named, introduced the idea of quanta in 1900. He was awarded the Nobel Prize in 1918.

ries a quantum of energy, $E_p$, and that travels with the speed of light, $c$. A photon's energy, $E_p$, is defined as follows:

$$E_p = hf, \quad (2.4)$$

where $h$ is Planck's constant ($h = 6.626 \times 10^{-34}$ J · s) and $f$ is the photon's frequency.[3] Formula 2.4 introduces one of the fundamental concepts of modern physics: *The energy of a photon, which is an elementary particle, depends on its frequency, which we always associate with waves.* Remember, too, that the higher the photon's frequency, the more energy it carries. That's why $x$-rays can penetrate our body but light cannot. (See Figure 2.3.)

Two questions concerning photons must be considered at this point: First, what is the nature of a photon? Simply put, it is electromagnetic radiation. Secondly, if a photon's frequency, $f$, is about $10^{14}$ Hz, what sort of electromagnetic radiation do we mean? Simply put again, it is light. (Again, see Figure 2.3.) Hence, light is a stream of photons.

### Example 2.3.1

***Problem:***

Suppose a laser diode (LD) radiates red light with $\lambda = 650$ nm. What is the energy of a single photon?

***Solution:***

The energy of a single photon can be determined in this way: $E_p = hf = hc/\lambda = \{[(6.6 \times 10^{-34}$ J · s$] \times [3 \times 10^8$ m/s$]\}/650 \times 10^{-9}$ m $= 3.04 \times 10^{-19}$ J.

So a single photon carries an extremely small amount of energy but light radiated by a source consists of a tremendous number of photons.

Typical light-emitting diodes (LEDs) and laser diodes (LDs) radiate at the level of several milliwatts. Let's take an LD with $\lambda = 650$ nm and 1 mW of light power, $P$, and calculate how many photons this source radiates per second.

The total energy $E = P \times 1$ s $= 1 \times 10^{-3}$ W $\times 1$ s $= 1 \times 10^{-3}$ J.

This energy is equal to $E = E_p \times N$, where $N$ is the number of photons.

Thus, $N = E/E_p = [1 \times 10^{-3}$ J$]/[3.04 \times 10^{-19}$ J$] = 3.3 \times 10^{15}$, which is 3.3 thousands of trillions.

Now you have an idea of how many photons it takes to produce visible light. By the way, the human eye is extremely sensitive to light and can visualize even several photons of radiation.

---

## Radiation and Absorption

The next problem to consider is the relationship between a photon's energy, $E_p = hf$, and the energy difference, $\Delta E = E_3 - E_2$, of the energy levels $E_3$ and $E_2$. You'll recall that a photon was created when an atom jumped from $E_3$ to $E_2$ and released energy ($E_3 - E_2$). Therefore,

$$E_p = \Delta E = E_3 - E_2 \quad (2.5)$$

But $E_p = hf$; hence, $hf = E_3 - E_2$ and $f = (E_3 - E_2)/h$. On the other hand, $\lambda = c/f$. Therefore,

$$\lambda = ch/(E_3 - E_2) \quad (2.6)$$

But the product $c \times h$ is the constant, and the only variable in Formula 2.5 is the energy gap $\Delta E = E_3 - E_2$. Since a stream of photons makes light, we arrive at this very important conclusion: *The wavelength (the color) of radiated light is determined by the energy levels of the radiating material.*

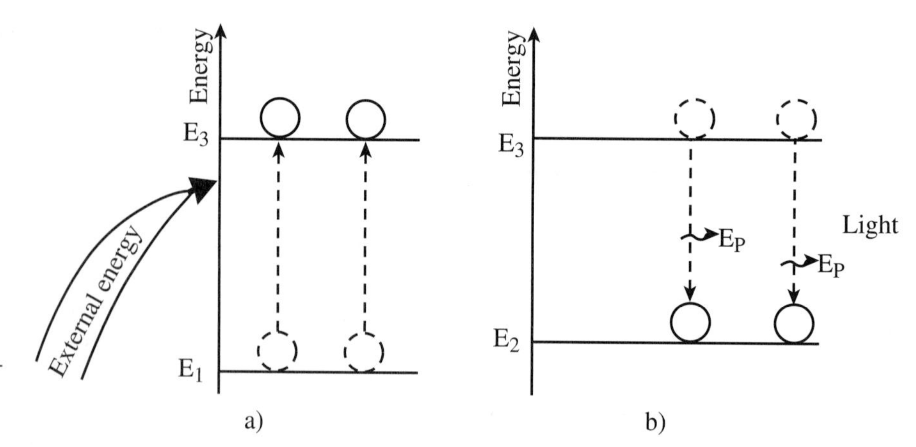


**Figure 2.9** Pumping and radiation processes: (a) Pumping; (b) radiation.

You may wonder whether we can change the energy levels of a given material to obtain a desired color of radiated light. The answer is no. These energy levels are given by nature and we cannot control them. But we can choose another material to achieve different colors of radiated light.

Atoms want to exist at the lowest possible energy levels; that's the law of nature. To raise them to higher levels, which is necessary for atoms to be able to jump down to produce light radiation, we must energize them from an external source. When atoms absorb external energy, they jump to the higher energy levels and then drop to the lower levels, radiating photons—that is, light. The process of making atoms jump to the higher levels by feeding them external energy is called *pumping*.

Figure 2.9 demonstrates these processes. Keep in mind that this illustration is no more than a convenient model and is only a representation of the real pumping and radiation processes.

To visualize these processes in everyday life, look at a lamp in your room. You know, of course, that no light radiates when the switch is off. But did you ever wonder why? When the switch is on, electrical energy is delivered to the light-source material (gas in luminescent lamps and a filament in light bulbs), atoms absorb this energy, jump to the upper energy levels, and then drop down to the lower levels, radiating light.

Why atoms jump to the higher energy levels when they absorb energy from an external source is the next obvious question. Suppose an atom is at level $E_1$ [Figure 2.9(a)]. That means the atom possesses an energy value of $E_1$. Now it absorbs external energy in the amount of $\Delta E = E_3 - E_1$. Its new energy becomes equal to $E_1 + \Delta E = E_1 + (E_3 - E_1) = E_3$. In reality, after absorbing external energy, atoms have a new energy value, which we can demonstrate by placing them at different energy levels in our energy-level diagram. In other words, an energy-level diagram is a convenient model that helps us understand the pumping and radiation processes by visualizing them.

### Example 2.3.2

***Problem:***

Suppose you use an LED whose energy gap equals 2.5 eV. What color will it radiate?

***Solution:***

When we calculate radiating wavelength, we need to know only energy gap $\Delta E$, that is, the energy difference between the upper and the lower energy levels involved in the process. Atomic energy and energy gaps are measured in electron-volts (eV). 1 eV = $1.602 \times 10^{-19}$ J. (Note how small 1 eV is.) Note also that $\Delta E$ (eV) = $\Delta E$ (joules/*e*), where *e*, the electron (unit) charge, equals $1.602 \times 10^{-19}$ coulomb.

Typical energy gaps of semiconductor materials used in fiber-optic LEDs, LDs, and PDs are in units of eV, which means on the order of $10^{-19}$ J. This is why eV is a common energy unit when dealing with atomic processes.

Now, back to the problem. Since $\Delta E = 2.5$ eV and $E_p = \Delta E$, then $E_p = 2.5$ eV. On the other hand, $E_p = hf = hc/\lambda$; therefore,

$$\lambda = h/E_p, \tag{2.7}$$

where $E_p$ is measured in joules.

$$ch = 3 \times 10^8 \text{ m/s} \times 6.6261 \times 10^{-84} \text{ J}\cdot\text{s} = {\sim}20 \times 10^{-26} \text{ m}\cdot\text{J}$$

Converting $E_p$ from eV to J:

$$E_p = 2.5 \text{ eV} \times 1.602 \times 10^{-19} = {\sim}4 \times 10^{-19} \text{ J}$$

Substituting numbers in the above formula, we get

$$\lambda = hc/E_p = 20 \times 10^{-26} \text{ m}\cdot\text{J}/4 \times 10^{-19} \text{ J} = 5 \times 10^{-7} \text{ m} = 500 \text{ nm}$$

Looking at Figure 2.3, we can now say that the color is green.

This type of calculation is made even easier if we simply use three constants: $h$, $c$, and the conversion coefficient from eV to J. Let's redo this calculation but now more quickly:

$$\lambda(m) = hc/1.602 \times 10^{-19}\, E_p(\text{eV}) = 20 \times 10^{-26} \text{ m}\cdot\text{J}/1.602 \times 10^{-19}\, E_p(\text{eV}) = 12.48 \times 10^{-7}/E_p\,(\text{eV}).$$

Hence, you can use the formula

$$\lambda(\text{nm}) = 1248/E_p(\text{eV}), \tag{2.8}$$

where $E_p$ is in eV and $\lambda$ will be in nm.

---

What happens if an external photon (light) strikes a medium? If its energy, $E_p = hf$, is equal to the energy gap, $\Delta E$, the photon will be absorbed by an atom and the atom will jump to the appropriate higher level. If $E_p$ is not equal to $\Delta E$, the photon will pass by the material without interaction. Figure 2.10 demonstrates both absorption and noninteraction processes.

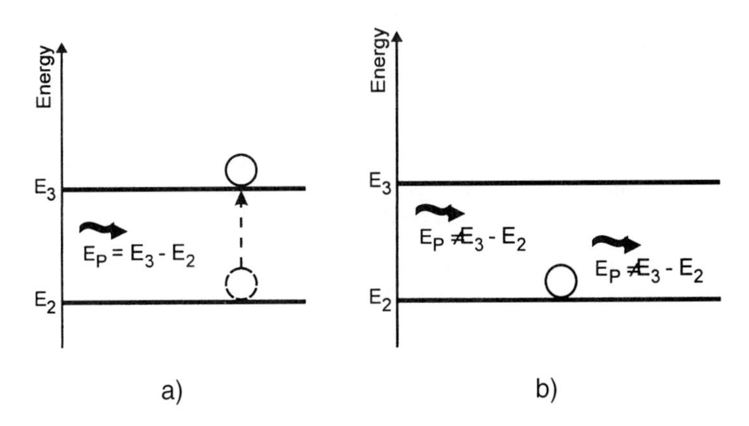


**Figure 2.10** Absorption and noninteraction processes: (a) Absorption; (b) noninteraction.

You know, of course, that sunlight is not as bright inside a room as it is outside. But why? It's because the light is partially absorbed by the window and completely absorbed by the wall. But what happens to the energy that the sun's photons transmit to the window and the wall? This light energy is absorbed by these objects, which become warmer as a result. Analyze your own everyday experience in terms of the absorption and noninteraction processes shown in Figure 2.10.

Optical fiber used as a communications link is made from a highly transparent material. That means a large majority of photons injected into the fiber by an LED or LD will travel through it without interacting with the fiber material. But some impurities have energy gaps close to the energy level of the photons. And that poses a problem. The result of the interaction between the photons and the impurities is discussed in Chapter 3.

Another important point to note about absorption and noninteraction processes is conveyed in Figure 2.10(a), which shows that we can use photons (light, remember) to pump atoms to upper energy levels.

## SUMMARY

- Light is an electromagnetic wave that develops in both time and space. Its development in space is described by a wavelength, which is the distance between two identical points of two successive cycles of a wave. Its development in time is quantified by a period, which is the time it takes a wave's two identical points to pass, in sequence, the same location.
- The wavelength and the period of the light wave are related through wave velocity, which is equal to the wavelength divided by the period. This definition results in an important formula:

$$\lambda \cdot f = c \tag{2.1}$$

- Different wavelengths represent different colors. Visible light is in the range between 400 nm and 700 m. Fiber-optic communications technology works with far-infrared light at wavelengths of 850 nm, 1300 nm, and 1550 nm.
- Light can be treated as a beam (ray). Light rays propagate within different media at different velocities, which is quantified by a refractive index, $n$:

$$n = c/v \tag{2.2}$$

  When a light beam passes from one medium to another, it changes its direction. The refractive indexes of these media determine how much the beam is bent.
- Snell's law gives the rules defining the directions of the incident, reflected, and refracted (transmitted) beams:

$$\Theta_1 = \Theta_3$$
$$n_1 \sin \Theta_1 = n_2 \sin \Theta_2 \tag{2.3}$$

- When a light beam falls from a medium with a greater refractive index to a medium with a smaller refractive index, we can reflect all light back to the medium having the greater refractive index by increasing the angle of incidence. This phenomenon, called total internal reflection, is what keeps light inside an optical fiber.
- Atoms can possess only a discrete amount of energy. An energy-level diagram is a convenient model to show this.
- Atoms aspire to exist at the lowest possible energy level.
- To induce atoms to jump to the upper energy levels, we feed them energy from an external source, a process called "pumping."
- When atoms leap to the upper energy levels, they absorb an exact amount of energy from an external source. This amount is equal to the energy difference between the upper and lower levels between which the jump occurred.
- When atoms drop from an upper energy level to a lower level, they radiate quanta of electromagnetic energy called photons. (This is true for radiative transitions.)
- A photon is an elementary particle that travels at the speed of light, $c$, and carries a quantum of energy, $E_p = hf$, where $h$ is Planck's constant ($6.626 \times 10^{-34}$ J · s) and $f$ is the photon's frequency.
- Light is a stream of photons. Its color is determined by the photon's frequency, $f$, that is, the photon's wavelength, $\lambda$, since $\lambda f = c$, where $c$ is the speed of light in a vacuum.
- A photon's energy, $E_p$, is equal to the energy gap between the radiating upper and lower energy levels. This implies that a photon's frequency (wavelength) is determined by the energy levels—that is, the material—used.

- Energy levels exist naturally; therefore, we can get different colors of light either by using different energy levels of the same material or by using different materials.
- Photons are absorbed by the material whose energy-level gaps are equal to the photon's energy. To make the material transparent, we have to choose either another photon—that is, another color of light—or another material.
- See references [1] through [6] to delve more deeply into the topics that were briefly discussed in this chapter.

## PROBLEMS

**2.1.** a. Define the terms "wavelength" and "frequency."
b. The most popular wavelengths used in fiber-optic communications systems today are 1300 nm and 1550 nm. What are their corresponding frequencies?

**2.2.** a. Define the term "refractive index."
b. What is the refractive index of a vacuum?

**2.3.** A light beam is directed into a glass fiber whose refractive index is 1.5. How long will it take the light to travel through the fiber if its length is 1000 m and light travels in a perfectly straight line along the fiber center?

**2.4.** a. Sketch the diagram of light incidence on a glass–air boundary and give the relevant formulas.
b. When a light beam strikes a glass–air boundary, what is the critical incident angle if $n_{glass} = 1.5$?

**2.5.** Assume that light is traveling from one layer of silica whose refractive index is 1.47 to another layer of silica whose refractive index is 1.45.
a. Find the range of angles at which total internal reflection takes place.
b. Draw a diagram showing what happens to a light ray at the condition of total internal reflection.
c. Write a paragraph explaining the phenomenon of total internal reflection.

**2.6.** Calculate the energy of a single photon at 1300 nm and at 1550 nm.

**2.7.** a. How many photons per second emanate from a laser diode (LD) radiating at 1300 nm if its power is 1 mW?
b. How many photons per second emanate at 1550 nm?

**2.8.** What is the energy gap of silicon used for a light-emitting diode (LED) radiating at 1300 nm?

**2.9.** Suppose an LD material's energy gap equals 0.8052 eV. At what wavelength does this LD radiate?

**2.10.** An optical fiber is made of a very transparent material.
a. How do you interpret this statement in terms of Figure 2.10(b)?
b. Some impurities in an optical fiber have energy gaps close to the energy level of photons. What will be the result of the interaction of photons with these impurities?

**2.11.** Project: Sketch a memory map. (See Problem 1.20.)

## REFERENCES[4]


1. Paul Fishbane, Stephen Gasiorowicz, and Stephen Thornton, *Physics for Scientists and Engineers,* Englewood Cliffs, N.J.: Prentice Hall, 1993.
2. David Halliday and Robert Resnick, *Fundamentals of Physics,* 2d ed., New York: John Wiley & Sons, 1986.
3. Max Born and Emil Wolf, *Principles of Optics,* 6th ed. (with corrections), Cambridge, U.K.: Cambridge University Press, 1980.
4. Christopher Davis, *Lasers and Electro-Optics—Fundamentals and Engineering,* New York: Cambridge University Press, 1996.
5. Shun Lien Chuang, *Physics of Optoelectronics Devices,* New York: John Wiley & Sons, 1995.
6. Joseph Verdeyen, *Laser Electronics,* 3d ed., Englewood Cliffs, N.J.: Prentice Hall, 1995.


[4]See Appendix C: A Selected Bibliography.

# 3

# Optical Fibers—Basics

Optical fiber is the key to fiber-optic communications systems. The real deployment of these systems started only after commercially acceptable optical fibers appeared on the market. This is why we must pay close attention to all aspects of this component, the subject of this and the following chapter. This chapter is devoted to general issues in optical fibers, focusing on multimode fibers.

## 3.1 HOW OPTICAL FIBERS CONDUCT LIGHT

Imagine yourself a researcher working about thirty years ago. Your project: Find a way to transmit a light signal for communications. The concept of optical fibers—thin, transparent, flexible strands—is already known but any attempts to use them for communications have failed because the signal completely disappears after several feet of transmission. So herein lies our problem: We have to determine the conditions needed to transmit light through an optical fiber and resolve how these conditions can be effected in a practical manner.

### Step-Index Fiber: The Basic Structure

***Total internal reflection: refractive indexes of a core and cladding*** *An optical fiber is a thin, transparent, flexible strand that consists of a core surrounded by cladding.* Figure 3.1 shows this structure and the typical dimensions of optical-fiber components.

The core and the cladding of an optical fiber are made from the same material—a type of glass called silica—and they differ only in their refractive indexes. You recall that the refractive index is the number showing the optical property of a material, that is, how strongly the material resists the transmission of light. (See Table 2.1.) The definition of a refractive index, $n$, is given by Formula 2.2, rewritten here:

$$v = c/n, \tag{3.1}$$

where $v$ is the velocity of light inside a material having a refractive index of $n$, and $c$ is the speed of light in a vacuum. The core has the refractive index $n_1$, and the cladding has a different refractive index, $n_2$; thus, different optical properties make up the core and cladding of an optical fiber. If you look at the graph depicting how abruptly the refractive index changes across the fiber (Figure 3.1[a]), you will immediately understand why this structure is called a *step-index fiber.*

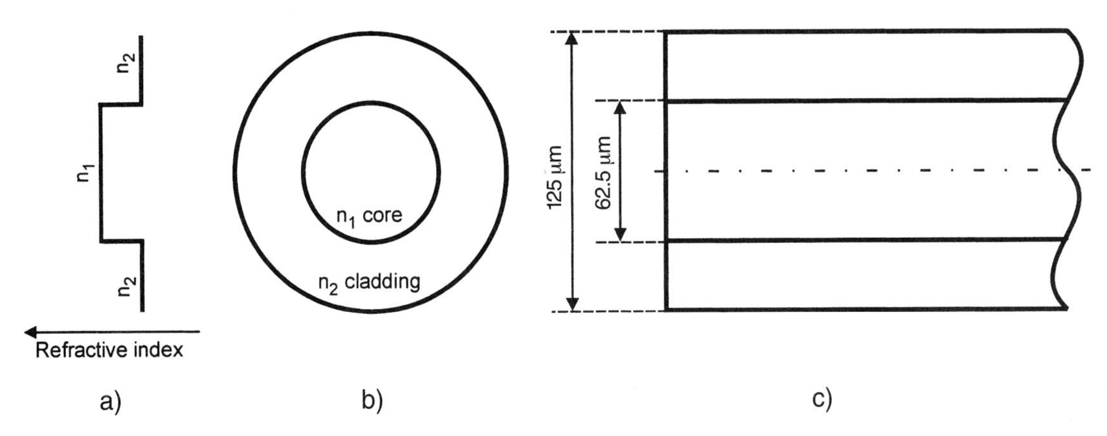


**Figure 3.1** Basic structure of a step-index optical fiber: (a) Refractive-index profile; (b) cross section of an optical fiber—front view; (c) cross section of an optical fiber—right-side view.

The structure is made by applying a layer of cladding over the core. The difference in refractive indexes can be achieved by doping silica with different dopants. Because of the way a refractive index is changed, we can show the strict boundary—an optical boundary—between the core and cladding.

To complete this discussion of the basic makeup of optical fiber, it is necessary to stress that a third layer—a coating—is applied over the cladding to protect the entire structure. The coating is made of a different material from that of the core or cladding. The coating serves, then, as the first line of defense for a very fragile core-cladding structure. Without it, installers and users couldn't work with optical fibers.

To sum up, then, an optical fiber is always manufactured in three layers: core, cladding, and coating. This combination forms a bare fiber.

The question you want to have answered at this point is which layer—core or cladding—has the greater refractive index. The answer can be found in the basic understanding of what an optical fiber is designed for: to be a light conduit, that is, a flexible, transparent strand that transmits light with—*ideally*—no attenuation. Hence, as we saw in Chapter 2, we must make use of the concept of total internal reflection (see Figures 2.5, 2.6, and 2.7) to save light inside the core of an optical fiber. Therefore, *to achieve total internal reflection at the core-cladding boundary, the core's refractive index, $n_1$, must be greater than the cladding's index, $n_2$.* Under this condition, light can travel inside the core not only along its central pathway but also at various angles to this centerline, without leaving the core. Now we have created a light conduit. This conduit—an optical fiber—will save light inside the core even if it is bent. Figure 3.2 shows both situations (By the way, it is commonplace to hear those in the field say that "light bounces inside the core.")

### Example 3.1.1

***Problem:***

a. The refractive index of a core is $n_1 = 1.48$ and the refractive index of a cladding is $n_2 = 1.46$. Under what condition will light be trapped inside the core?
b. Find this condition for a plastic optical fiber where $n_1 = 1.495$ and $n_2 = 1.402$.

***Solution:***

a. This condition is total internal reflection. To attain total internal reflection, we have to direct a light ray to the core-cladding boundary at the critical incident angle (see Figure 2.5). What

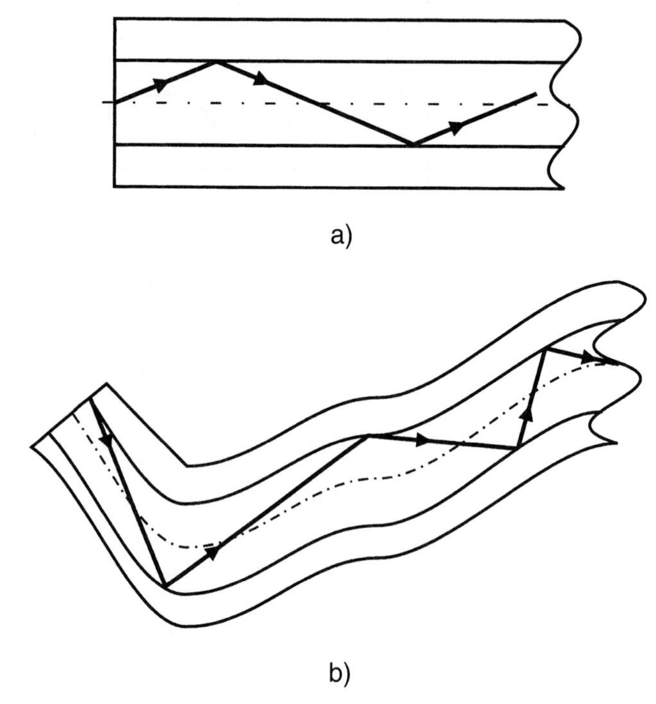


**Figure 3.2** Light propagation inside an optical fiber: (a) Straight fiber; (b) bent fiber.

is this angle? We find it by using Snell's law (Formula 2.3): $n_1 \sin \Theta_1 = n_2 \sin \Theta_2$. The critical angle is reached when $\Theta_2 = 90°$ (again, see Figure 2.5); hence, $n_1 \sin \Theta_{1C} = n_2$ and

$$\sin \Theta_{1C} = n_2/n_1$$

Therefore, $\Theta_{1C} = \sin^{-1}(n_2/n_1) = \sin^{-1}(0.9865) = 80.57°$.

b. Besides silica (glass) optical fiber—the most popular type in today's deployed systems—an optical fiber made from plastic also exists. Let us repeat the same type of calculation for plastic optical fiber where $n_1 = 1.495$ and $n_2 = 1.402$:

$$\Theta_{1C} = \sin^{-1}(1.402/1.495) = \sin^{-1}(0.9378) = 69.68°$$

Observe the difference between the two critical incident angles: Where does this difference come from?

---

The above example also helps us remember this important fact: $n_{core}$ ($n_1$) is always greater than $n_{cladding}$ ($n_2$). Indeed, we found that $\sin \Theta_{1C} = n_2/n_1$; thus, $n_1$ cannot be less than $n_2$ as the property of sine function dictates.

***Total internal reflection: critical incident angle and critical propagation angle*** It is important to point out at this juncture that two key terms often confuse newcomers to this field: *critical incident angle* and *critical propagation angle.* You must distinguish between them. *The critical propagation angle, $\alpha_C$, is the angle the beam makes with the centerline of the optical fiber. (It is very often referred to, in fiber optics parlance, as the "critical angle.") The critical incident angle, $\Theta_{1C}$, is the angle the beam makes with the line perpendicular to the optical boundary between the core and the cladding* (again, see Figure 2.5). Both angles are shown in Figure 3.3.

It is clear from right triangle A-B-C, Figure 3.3, that $\alpha_C = 90° - \Theta_{1C}$. In Example 3.1.1a we found that $\Theta_{1C} = 80.57°$; hence, $\alpha_C = 9.43°$.

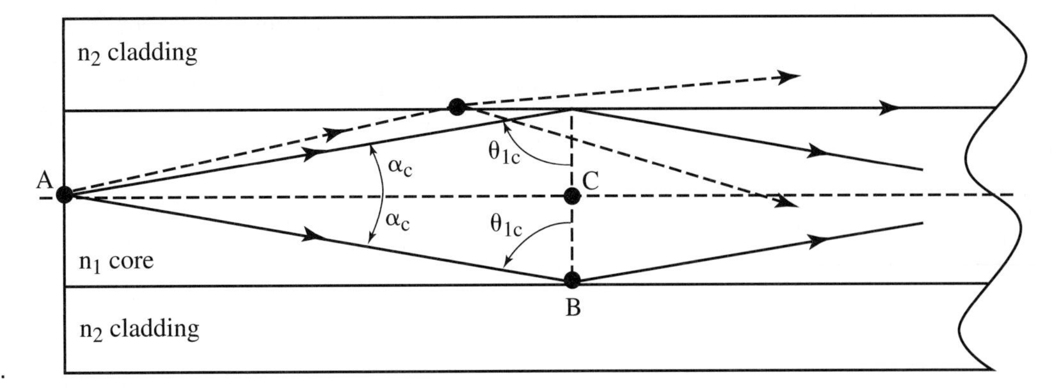


**Figure 3.3** Critical incident angle, $\Theta_{1C}$, and critical propagation angle, $\alpha_C$.

Why is the critical propagation angle, $\alpha_C$, so important? Suppose a beam travels within this optical fiber at $\alpha = 10° > \alpha_C$. Hence, $\Theta_1 = 80° < \Theta_{1C}$, which means that the condition of total internal reflection has been violated. Therefore, the incident beam will divide in two: a reflected beam, which will be saved, and a refracted beam, which will be lost. This beam, which is at $\alpha > \alpha_C$ with the center axis, is shown in Figure 3.3 as a dotted line. (Refer again to Figure 2.4.) Keep in mind that *a beam strikes the core-cladding interface millions and millions of times while traveling through the fiber; therefore, if even a microscopic portion of the beam is lost every time it hits this boundary because of refraction, the beam will be completely lost after traveling only a short distance.* This is what is meant when we speak of unacceptably high attenuation. Thus, *total internal reflection is the condition necessary for using optical fiber for the purpose of communication.* The critical propagation angle, $\alpha_C$, represents the requirement to achieve this condition. In conclusion, then, *to save light inside an optical fiber, it is necessary to direct rays at this critical propagation angle—or even at a lesser angle.*

From here on, we can forget about the critical incident angle, $\Theta_{1C}$, since it does not apply to fiber-optic technology. We are only interested in knowing the critical propagation angle, $\alpha_C$, since this angle dictates how we must direct the light inside the optical fiber. We must never lose sight of the crucial role this angle plays. It is a supplement to the critical incident angle, $\Theta_{1C}$, and therefore represents the condition necessary for achieving total internal reflection.

## Example 3.1.2

***Problem:***

a. The refractive indexes of the core and the cladding of a silica fiber are 1.48 and 1.46, respectively. What is the critical propagation angle?
b. Find this angle for a plastic optical fiber ($n_1 = 1.495$ and $n_2 = 1.402$).

***Solution:***

a. First, let's derive the formula. In Example 3.1.1a we found that $\sin \Theta_{1C} = n_2/n_1$. Since $\alpha_C = 90° - \Theta_{1C}$, $\sin \Theta_{1C} = \cos \alpha_C$; hence, $\cos \alpha_C = n_2/n_1$. Thus, one can derive: $\sin \alpha_C = \sqrt{(1 - \cos^2 \alpha_C)} = \sqrt{(1 - (n_2/n_1)^2)}$. Hence,

$$\alpha_C = \sin^{-1} \sqrt{(1 - (n_2/n_1)^2)}$$

Now let's plug in the numbers:

$$\alpha_C = \sin^{-1} \sqrt{(1 - (1.46/1.48)^2)} = 9.43°$$

This result is clear from Example 3.1.1a, since $\alpha_C = 90° - \Theta_{1C}$.

b. Now calculate the critical propagation angle for a plastic optical fiber:

$$\alpha_C = \sin^{-1}\sqrt{\left(1-(1.402/1.495)^2\right)} = 20.32°$$

---

At this point, it is imperative to bring into our discussion a very important formula: *The critical angle of propagation, $\alpha_C$, is determined by only two refractive indexes, $n_1$ ($n_{core}$) and $n_2$ ($n_{cladding}$):*

$$\alpha_C = \sin^{-1}\sqrt{1-\left(\frac{n_2}{n_1}\right)^2} \tag{3.2}$$

It is important to underscore the logic that led us to this formula: *To save light inside a strand of fiber, we need to have it strike the core-cladding boundary at the critical incident angle, $\Theta_{IC}$, or above it, in order to provide total reflection of this light; to make light fall at or above that angle, we have to direct it so that it is at or below the critical propagation angle, $\alpha_C$, with respect to the centerline of the fiber, as we've already seen.*

## Launching the Light: Understanding Numerical Aperture

***Acceptance angle*** The next question that arises is, how can we direct this beam so that it does indeed fall at or below the critical propagation angle? The light, of course, must come from some source, such as an LED or an LD. This source is outside the fiber; therefore, we have to direct it into the fiber. Figure 3.4 shows how light radiated by a light source is coupled to an optical fiber.

At the gap-fiber interface, the beam at angle $\Theta_a$ is the incident beam and the beam at angle $\alpha_C$ is the launched one, which is the refracted beam with respect to gap-core interface (the reflected beam is not shown here). It will help you to understand this explanation if you look at Figure 3.4. The formal relationship between $\Theta_a$ and $\alpha_C$ can be derived using Snell's law. From Figure 3.4 one can find:

$$n_a \sin\Theta_a = n_1 \sin\alpha_C \tag{3.3}$$

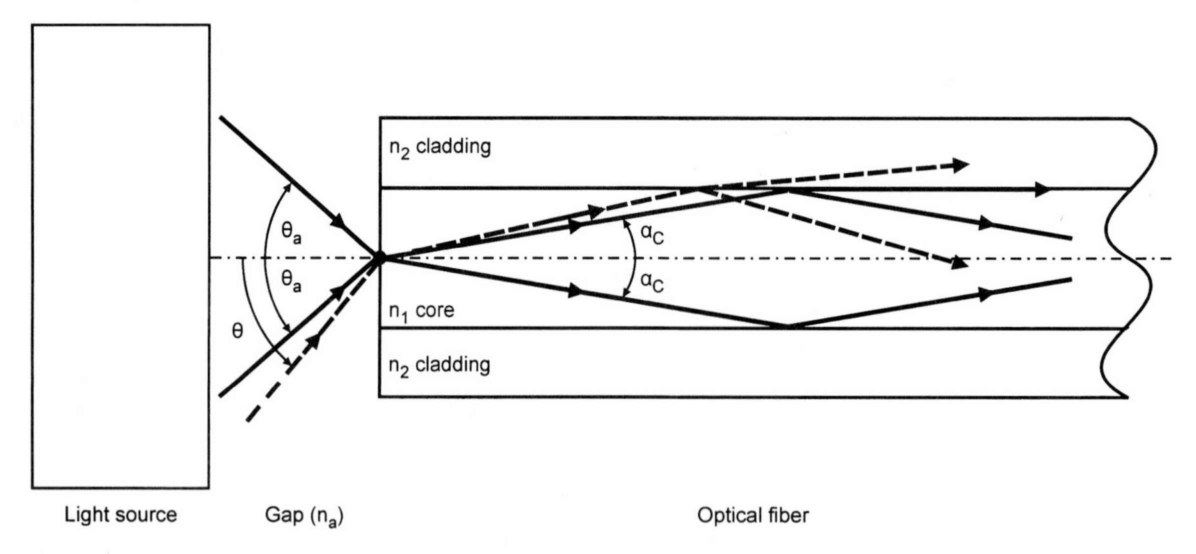


**Figure 3.4** Launching light into an optical fiber.

If the gap between a light source and a fiber is air, then $n_a$ is very close to 1 ($n_a = 1.0003$). Therefore,

$$\sin \Theta_a = n_1 \sin \alpha_C \tag{3.3a}$$

Formula 3.3, in a sense, states the following principle: *To save light inside a fiber (to provide total internal reflection, that is) all rays must propagate at critical angle $\alpha_C$ or less. In order for us to maintain the light inside the fiber at this angle, we have to direct it from outside the fiber (from the light source, remember) at angle $\Theta_a$ or less.*

It's clear from Figure 3.4 that angle $\Theta_a$ is a spatial angle. Light will be saved inside the fiber if it comes from a light source bounded by the cone $2\Theta_a$. This is why we call *angle* $2\Theta_a$ an *acceptance angle.* (Sometimes you might meet an acceptance angle defined simply as $\Theta_a$, without the coefficient 2.)

The dotted line in Figure 3.4 indicates a ray that comes in at an angle exceeding the acceptance angle, $\Theta_a$, outside the fiber. It is obvious the ray will travel inside the fiber at an angle exceeding the critical propagation angle, $\alpha_C$. This will result in the partial refraction of the ray. In other words, if a ray is not within the acceptance cone defined by $2\Theta_a$, it will be lost while traveling inside the fiber. Simply put, *exceeding acceptance angle $2\Theta_a$ is just beyond the requirement for having total internal reflection inside a fiber.*

### Example 3.1.3

***Problem:***

a. What is the acceptance angle for the fiber when $n_1 = 1.48$ and $n_2 = 1.46$?
b. What is the acceptance angle for the plastic optical fiber?

***Solution:***

a. From Snell's law, $n_a \sin \Theta_a = n_1 \sin \alpha_C$. For air, $n_a = 1.00$. From Example 3.1.2, the critical propagation angle $\alpha_C = 9.43°$; hence, $\sin \Theta_a = 1.48 \sin 9.43° = 0.2425$. One half of acceptance angle $\Theta_a = \sin^{-1} 0.2425 = 14.033°$. Therefore, the acceptance angle is $2\Theta_a = 28.07°$.
b. $\Theta_a = \sin^{-1} (1.495 \sin 20.32°) = \sin -1(0.5192) = 31.27°$
Thus, the acceptance angle is $2\Theta_a = 62.54°$.

Observe the difference in values of the acceptance angles for these two fibers.

---

All these considerations serve only to better explain how we can save light inside an optical fiber. Physically, we have two components of a system that have to be connected: an optical fiber and a light source (LED or LD). We don't see any angles—either a critical propagation angle or an acceptance angle—and the only thing that we can do is direct light from the source into the fiber. This is why fiber-optic communications technology does not operate with any angles but, instead, integrates all these factors into one characteristic: numerical aperture (*NA*).

***Numerical aperture*** Numerical aperture, *NA,* is:

$$NA = \sin \Theta_a \tag{3.4}$$

This definition underscores the meaning of the numerical aperture. To compute the numbers, however, it's better to use another form of this expression, which can be derived as follows:

$$NA = \sin \Theta_a$$

$$\sin \Theta_a = n_1 \sin \alpha_C \text{ and } \sin \alpha_C = \sqrt{(1 - (n_2/n_1)^2)} \text{ (see Formula 3.2); hence,}$$

$$NA = n_1 \sin \alpha_C = n_1 \sqrt{(1 - (n_2/n_1)^2)} = \sqrt{(n_1)^2 - (n_2)^2}$$

This is the formula most often used:

$$NA = \sqrt{(n_1)^2 - (n_2)^2} \tag{3.5}$$

## Example 3.1.4

***Problem:***

a. What is the numerical aperture of silica fiber with $n_1 = 1.48$ and $n_2 = 1.46$?
b. What is the numerical aperture of plastic fiber where $n_1 = 1.495$ and $n_2 = 1.402$?

***Solution:***

a. If we plug the numbers into Formula 3.5, we get: $NA = \sqrt{(1.48)^2 - (1.46)^2} = 0.2425$. We can verify our result by using Formula 3.4: $NA = \sin \Theta_a$. We have found in Example 3.1.3 that $\sin \Theta_a = n_1 \sin \alpha_C = 0.2425$; hence, $NA = 0.2425$.
b. $NA = \sqrt{(1.495)^2 - (1.402)^2} = 0.5192$.
Verify the answer: $NA = \sin 31.27° = 0.5192$.

Observe the difference in the values of *NA* for these two fibers.

---

Have you noticed that all the formulas we've used in this chapter depend on only two variables—$n_1$ and $n_2$? This is so because the formulas are mathematical forms of the same basic idea: total internal reflection for light traveling inside an optical fiber.

We can best summarize our discussion by this simple flow diagram: $\Theta_{1C} \rightarrow \alpha_C \rightarrow \Theta_a \rightarrow NA$. What it shows is that *fiber-optic communications technology makes use of numerical aperture,* NA, *which describes the ability of an optical fiber to gather light from a source and then the ability to preserve, or save, this light inside the fiber because of total internal reflection.*

The formula expressing this statement thus becomes:

$$NA = \sin \Theta_a = \sqrt{(n_1)^2 - (n_2)^2} \qquad \text{(3.4) and (3.5)}$$

From here on, all we need to know is the numerical aperture, *NA, which is the only number that you will find in the optical-fiber data sheets.* It is essential that you remember the meaning of this number. It represents the condition of total internal reflection inside the optical fiber, a condition that is absolutely necessary if we want to use optical fiber for communications.

Would we ever want to change the *NA*? Remember, *NA* characterizes the fiber's ability to gather light from a source. Thus, the answer is yes, because for different applications it might be necessary to use fibers with different *NAs*. It would seem that if we wanted to change the *NA* (Formula 3.5), we would have to change either $n_1$ (the core refractive index) or $n_2$ (the cladding refractive index). But let's take a closer look at Formula 3.5.

Fiber-optic communications technology operates not with the refractive indexes of the core and the cladding themselves but with their difference, $\Delta n$. The above discussion made it clear, we trust, why the difference, $\Delta n$, not the values of $n_1$ and $n_2$, is important. We define the difference, $\Delta n$, as:

$$\Delta n = n_1 - n_2 \qquad (3.6)$$

Note that this value is always positive. It is very common to use the *relative difference of the refractive indexes,* $\Delta$, often called the *relative index,* which is defined as follows:

$$\Delta = (n_1 - n_2)/n, \qquad (3.7)$$

where $n$, the average refractive index, equals $(n_1 + n_2)/2$. You can find a formula similar to (3.7) with $n_1$ or $n_2$ in the denominator. The numbers you will calculate with these variations change very slightly because, in reality, $n_1$ is very close to $n_2$.

Using this quantity, we can introduce another formula for numerical aperture, $NA$. This is the simple derivation: $NA = \sqrt{(n_1)^2 - (n_2)^2} = \sqrt{((n_1 - n_2)(n_1 + n_2))} = \sqrt{(\Delta n)(2n)} = \sqrt{((\Delta n/n)(2n)^2)}$. Thus, we arrive at this formula:

$$NA = n\sqrt{(2\Delta)} \qquad (3.8)$$

This formula underscores the following: $n_1$ and $n_2$ are not important in themselves but only in their average and relative difference. Thus, to change $NA$, we need to vary $n$ and $\Delta n$; this is what manufacturers really do. By varying these two parameters, manufacturers are able to change $NA$ over a relatively wide range (from 0.1 to 0.3 for a silica fiber).

Can we measure $NA$? Not directly. We have to first measure the power of light immediately after it is radiated by an LED. Then we make a second measurement by placing a short piece of fiber between an LED and a power meter. The first measurement gives us the power $P_o$, the second measurement, the power $P_{in}$. Now the numerical aperture can be estimated by the simple formula $NA = \sqrt{P_{in}/P_0}$. (See Section 9.1.)

## 3.2 ATTENUATION

Assume that you measure light power before it is directed into an optical fiber and then measure it again as it emerges from the fiber. Would you expect to get the same numbers? Of course not. This is because we understand intuitively that the power coming out of the fiber should be less than the power entering it. But apart from an "intuitive" understanding, we want to have a scientific explanation for this phenomenon. And it is simply this: Every transmission line introduces some loss of signal power. This is the phenomenon of "attenuation." In fiber-optic communications technology, *attenuation is the decrease in light power during light propagation along an optical fiber.*

From this definition, light loss caused by violation of the condition of total internal reflection when launching light into a fiber (see Figures 3.3 and 3.4) is supposed to be included in the total attenuation within the fiber. But, practically speaking, fiber-optic communications technology never considers this loss as a component of total attenuation because, without total internal reflection, optical fiber simply doesn't work as a communications conduit. As emphasized several times in Section 3.1, total internal reflection is an absolutely necessary condition for using optical fiber in communications systems. As an analogy, consider the situation where people ask you about your health. Nobody ever asks whether or not you can breathe, do they? Yet breathing is a necessary condition for living. Therefore, the point to keep clearly in mind as you read what follows is this: *When light is coupled to an optical fiber for the purpose of communication, attenuation in the optical fiber means a power loss for reasons other than failure to achieve total internal reflection initially.* The following discussion explores these other reasons.

## Bending Losses

***Macrobending loss*** One of the most important advantages of today's optical fiber is its flexibility. Just imagine for a moment that you are holding a glass rod that conducts light perfectly but is rigid. Can you use that rod for communications? Certainly not because you'd have to install it in different environments; therefore, you would have to be able to bend it. That is why real optical communications was born with the advent of optical fibers which allow installers to bend cable as necessary. How much this flexible strand can be bent is our next consideration.

Figure 3.5 shows two conflicting situations: (1) The beam forms a critical propagation angle with the fiber's central axis at the straightened, or flat, part of the fiber. (2) But the same beam forms a propagation angle that is more than critical when it strikes the boundary of the bent fiber. The result is failure to achieve total internal reflection in the bent fiber, which means that some portion of the beam is escaping from the core of the fiber. Hence, the power of the light arriving at its destination will be less than the power of the light emitted into the fiber from a light source. In other words, *bending an optical fiber introduces a loss in light power, or attenuation.* This is one of the major causes of the total attenuation that light experiences while propagating through an optical fiber.

At this point you're no doubt wondering how the problem can be overcome. Manufacturers of optical fiber have learned how to reduce a fiber's bending sensitivity by designing refractive-index profiles. Unfortunately, improvement in bending sensitivity can be achieved only at the expense of the degradation of a fiber's other parameters. This is why manufacturers inform users what bending loss can be induced at a certain bending radius. For example, one turn at a 32-mm diameter mandrel causes a 0.5-dB (approximately 11%) bending loss for one popular type of fiber. Sometimes manufacturers include the minimal bending radius in their data sheets.

Thus, we can say that there is no straightforward method to eliminate this cause of attenuation. The only thing we can do about it is to be cautious when bending an optical fiber.

Bending can change not only the optical properties but also the mechanical characteristics of optical fibers. To prevent this, installers and users have to take precautions in bending fibers. The rules of thumb regarding minimum bending radius are these: A bending radius should be more than 150 times the cladding diameter of the fiber for long-term applications and more than

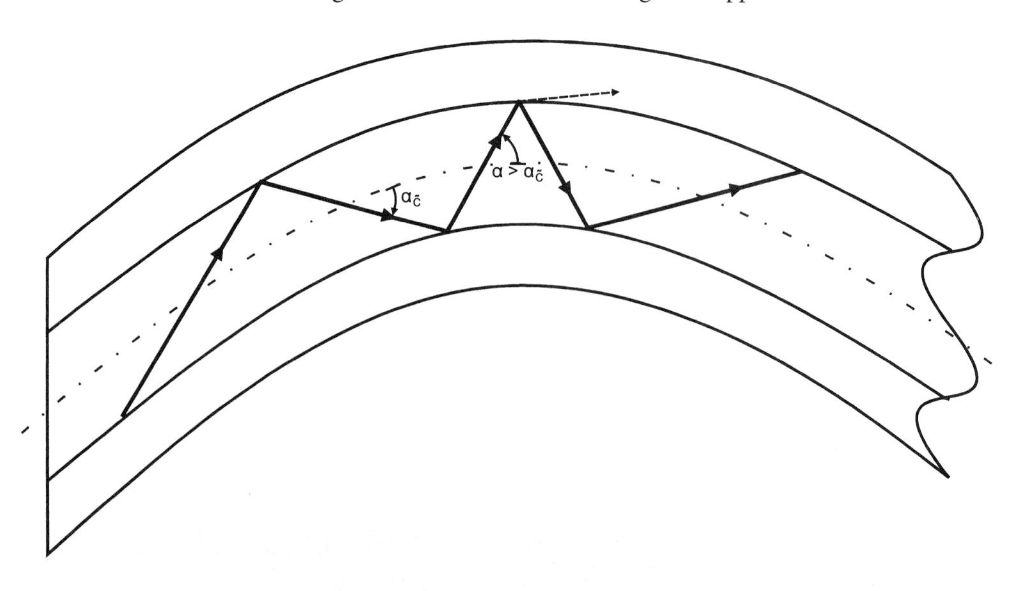


**Figure 3.5** Bending loss.

### How to Make Bending Loss Work for Us

Paradoxically, bending loss has a positive side. Sometimes we need to introduce well-controlled attenuation in a fiber-optic communications link. Specific passive components, called attenuators, actually do this job for us. One type of attenuator is based on the phenomenon of bending loss. Its advantage is that you need only make several turns of the fiber that you are using for transmission, so you don't need to introduce external components to the fiber component. (As we will see, introducing external components always presents problems.) With this type of attenuator, you can easily control attenuation quantitatively by controlling the number of turns the fiber makes around a given bending radius.

Another positive application is to use bent fiber as a mode filter—a device that reduces the number of modes in a fiber. (Modes are taken up in the following section.)

A fusion splicer (see Chapter 8) uses bending losses to control splicing quality. An example of one more positive use of the bending effect is through a device called the *fiber identifier.* This measuring instrument bends an optical fiber slightly and uses escaping light to control the data traffic within the fiber. (See Chapter 8.)

100 times the cladding diameter for short-term applications. Since the cladding diameter for silica fiber is usually 125 μm, we get the numbers 19 mm and 13 mm, respectively. But remember, bending fiber under these radii will damage it.

***Microbending loss*** The type of loss we discussed above is called *macrobending* loss, since it is caused by bending the entire optical fiber. Another type of loss—*microbending* loss—is also caused by failure to achieve the condition of total internal reflection Figure 3.6 shows what this type of loss looks like in an optical fiber.

Some imperfections in the geometry of the core-cladding interface might result in microconvexity, or microdent, in that area. Although light travels along the straight segment of a fiber, the beam meets these imperfections and changes its direction. The beam, which initially travels at the critical propagation angle, after being reflected at these imperfection points, will change the angle of propagation. The result is that the condition of total internal reflection is not attained and portions of the beam will be refracted; that is, they will *leak* out of the core. This is the mechanism of microbending loss.

Now we can give formal definitions to these types of loss: *Macrobending is loss caused by the curvature of the entire fiber axis. Microbending is loss caused by microdeformations of the fiber axis.* To find the connection between the given definitions and the above explanations, we need to realize that the fiber's centerline, or axis, is the imaginary line. *In reality, this line is determined by the core-cladding geometry.* This is why microdeformations of the fiber axis are microdeformations of the core-cladding boundary, as Figure 3.6 shows.

Fiber-optic users can do nothing to overcome microbending loss except ask manufacturers to improve the quality of their optical fibers. Fortunately, the fiber-manufacturing process is so

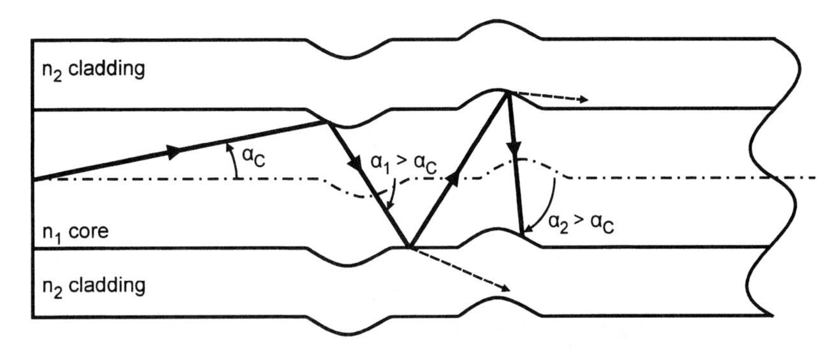


**Figure 3.6** Microbending loss.

### Microbending: Its Origin and Sensitivity

Apart from the microbending loss stemming from the manufacturing process, there is, unfortunately, another cause of this problem: mechanical stress applied directly on a fiber that results in microconvexities, or microdents. This stress might occur during the cabling process—that is, when wrapping a bare fiber into protective layers, thus making a fiber cable. Thermal stress can also result in fiber microbending. And, of course, a user should be careful during installation and maintenance.

To determine the fiber's sensitivity to microbending loss from external sources, microbending tests can be performed. For example, one company winds the fiber, covered with sandpaper, over a drum and applies a calibrated force to the sandpaper. The test enables the user to compare quantitatively different types of optical fibers.

The critical component of an optical fiber that determines its microbending sensitivity is its coating. The technology exists today to produce excellent coatings, which yield substantial improvement in this fiber characteristic. (More about coatings is found in Section 7.1.)

well developed today that microbending loss is not a major problem. Optical-fiber data sheets usually do not even specify microbending loss and so you can assume it is included in the total attenuation specified.

What users can do to reduce macro- and microbending loss is to be sure to handle optical fibers with care, particularly the less-sheathed ribbon fibers, and always remember that the fiber is a very fragile medium. Mechanical and environmental stresses might change the optical properties of a fiber, resulting in deterioration of the transmitting signal.

## Scattering

Suppose there is an imperfection in a core material, as shown in Figure 3.7. A beam propagating at the critical angle or less will change direction after it meets the obstacle. In other words, light will be scattered. This scattering effect prevents attainment of total internal reflection at the core-cladding boundary, resulting in a power loss since some light will pass out of the core. This is the basic mechanism underlying scattering loss.

You might wonder what core imperfections we're referring to and whether some mechanical particles might be found inside the core. A fiber core's diameter can be as small as units of a micrometer, so, knowing this, you can imagine how fine and clean the fiber-optic manufacturing process must be. This is truly one of the prominent achievements of modern technology. Therefore, you can rest assured that absolutely no foreign particles will be found inside the perfectly transparent core of an optical fiber. What might be found there, however, are slight variations in the refractive index.

*Even very small changes in the value of the core's refractive index will be seen by a traveling beam as an optical obstacle and this obstacle will change the direction of the original beam.* This effect will inhibit attainment of the condition of total internal reflection at the core-cladding boundary, as shown in Figure 3.7. The upshot, as noted above, will be scattering loss—light leaving the core.

Can we overcome the problem? Only by making better optical fibers. In fact, manufacturers today fabricate fiber of such a high quality that scattering loss is not a problem users

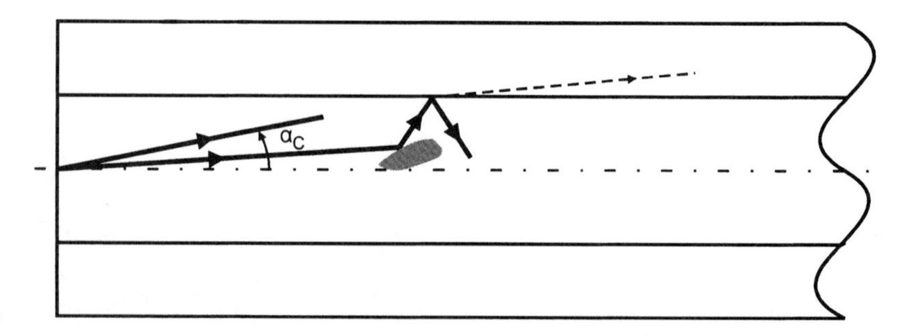


**Figure 3.7** Scattering loss.

need worry about. As is the case with microbending loss, manufacturers' optical-fiber data sheets do not include any specifications on scattering loss. This type of loss is simply included in the total attenuation reported. Incidentally, this type of scattering is called Rayleigh scattering.

*Note:* As you have by now discerned, bending and scattering losses are caused by violation of the condition of total internal reflection. An important point to emphasize one more time is this: Light that *initially* meets the total-internal-reflection requirement might violate this condition when the fiber is bent or its core's refractive index varies.

## Absorption

***Basic mechanism*** You will recall from Chapter 2, that *if an incoming photon has such a frequency* (f) *that its energy ($Ep = hf$) is equal to the energy gap ($\Delta E$) of the material, this photon will be absorbed by the material.* $\Delta E$ is the energy difference between two energy levels. Refer to Figure 2.10. Remember, too, that we learned that we cannot change the energy levels of the material, since they have been predetermined by nature. What we can do, though, to reduce or eliminate absorption is change either the light frequency, $f$, or work with another material. Remember that changing the light frequency, $f$, means also changing the light wavelength, $\lambda$, since $\lambda f = c$, where $c$ is the speed of light in a vacuum.

Now imagine that light (which, you'll recall, is a stream of photons) travels down an optical fiber and encounters a material whose energy level gap is exactly equal to the energy of these photons. Obviously, this impact will lead to light absorption, resulting in a loss of light power. This is the basic mechanism of the third major reason for attenuation in optical fibers.

Does this type of attenuation depend on light wavelength? It follows directly from the above explanations that it does. In other words, there is a spectral dependence of absorption, as shown in Figure 3.8.

We now need to ascertain whether a bulk core material, like silica, absorbs light. Optical fiber, as we've seen, is a transparent strand, that is, a "nonabsorptive" material. Manufacturers make every effort to make their bulk core material as transparent to light as possible. Absorption properties that still remain are caused not by silica atoms but by some molecules of the hydroxide anion $OH^-$, often called high water. These molecules are incorporated in silica during the

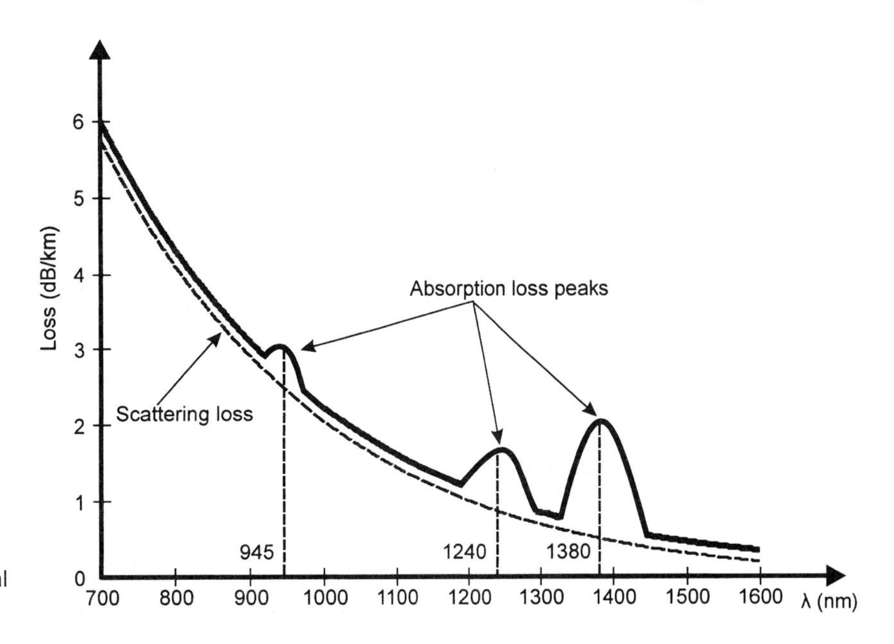


**Figure 3.8** Typical spectral attenuation.

fabrication process and it is very hard to eliminate them. $OH^-$ molecules have major peaks of absorption at 945, 1240, and 1380 nm. (See Figure 3.8.)

***Transparent windows*** While this problem cannot be solved by eliminating the $OH^-$ molecules, we can change the operating wavelength. Again look at Figure 3.8. There are three major regions, called transparent windows, where absorption is low. The first is located near 850 nm, the second near 1300 nm, and the third near 1550 nm. Typically, we can expect attenuation of about 4 dB/km near 850 nm, about 0.5 dB/km near 1300 nm, and about 0.3 dB/km near 1550 nm. The latter is the most widely used wavelength today in long-distance communications.

Observe that the main course of this graph is determined by scattering loss and its dependence on operating wavelength.

Detailed data sheets available from manufacturers might provide a graph of spectral attenuation similar to the one shown in Figure 3.8.

## Calculations of Total Attenuation

Fiber loss is the ratio of power at the output end of a fiber, $P_{out}$, to power launched into the fiber, $P_{in}$.

$$Loss = P_{out}/P_{in}, \tag{3.9}$$

where power is measured in watts.

In communications technology, we measure loss (attenuation) in decibels (dB):

$$Loss\,(\text{dB}) = -10\log_{10}(P_{out}/P_{in}), \tag{3.10}$$

where $P_{out}$ and $P_{in}$ are measured in watts. Since $P_{out}$ is always less than $P_{in}$ (this is why we use the term "attenuation," but not "amplification"), $\log_{10}(P_{out}/P_{in})$ *is always negative. To make the result of the calculations the positive number, the negative sign is used as Formula 3.10 shows.* This is accepted practice in fiber-optic communications technology.

Formulas 3.9 and 3.10 can be used to compute the total attenuation of an optical fiber. It is quite obvious that loss is proportional to fiber length, *L;* therefore, total attenuation characterizes not only the fiber losses themselves but also the fiber length, a fact that makes this characteristic very ambiguous. Indeed, if you know that for one specific fiber $Loss_1$ = 20 dB and for another fiber $Loss_2$ = 30 dB, could you possibly predict which fiber will have the lower loss characteristic? Of course not, because the first fiber could be 100 meters in length and the second 100 km long. This is why fiber-optic communications technology uses another characteristic: attenuation per unit of fiber length, *A*.

$$A\,(\text{dB/km}) = loss\,(\text{dB})/\,fiber\ length\ (\text{km}) \tag{3.11}$$

This quantity, A *(dB/km),* is called *attenuation* and it is one of the most important characteristics of an optical fiber. Attenuation is the number you will see on optical-fiber data sheets. This feature is sometimes called the cable-loss factor, *CLF,* or the attenuation coefficient, but most optical-fiber manufacturers use the term "attenuation."

### Example 3.2.1

***Problem:***

A communications system uses an optical fiber whose attenuation, *A,* is 0.5 dB/km. Find the output light power if the input power is 1 mW and the link length is 15 km.

***Solution:***

We can attack the problem using Formulas 3.10 and 3.11:

$$\begin{aligned} -A\ (\text{dB/km}) &= (10\log_{10} P_{out}/P_{in})\ (\text{dB})/L(\text{km}) \\ \log_{10}(P_{out}/P_{in})\ (\text{dB}) &= [-A(\text{dB/km}) \times L(\text{km})]/10 \\ P_{out}/P_{in} &= 10^{-AL/10} \\ P_{out} &= P_{in} \times 10^{-AL/10}, \end{aligned} \tag{3.12}$$

where $P_{out}$ and $P_{in}$ are given in watts.

For our example, $P_{out} = 1\ (\text{mW}) \times 10^{(-0.5 \times 15)/10} = 1\ (\text{mW}) \times 10^{-0.75} = 1\ (\text{mW}) \times 0.178 = 0.178$ mW.

---

Three important points can be drawn from Formula 3.12:

First, it is a key to understanding the connection between absolute attenuation and attenuation in dB. Indeed, suppose $P_{in}$ is 1 mW and $AL = -3$ dB. Then $P_{out} = P_{in} \times 10^{-0.3} = 0.5$ mW, which means that absolute attenuation equals 0.5. If $AL = -10$ dB, then $P_{out} = P_{in}/10$, and so forth. On the other hand, if you know $P_{in}$ and $P_{out}$, you can find the loss in dB. For example, if $P_{in} = 1$ mW and $P_{out} = 0.001$ mW, then $AL = -30$ dB, and so on.

Second, the negative sign in front of $AL/10$ is still further confirmation that attenuation means decreasing power, that is, that $P_{out}$ is always less than $P_{in}$. The rule: *Loss = 10log $P_{out}/P_{in}$ is always negative but attenuation in dB/km is always positive because of the negative sign in front of the logarithm.* For example, manufacturers display attenuation on their fiber data sheets as $A \leq 0.7$ dB/km at $\lambda = 1300$ nm.

Third, Formula 3.12 allows us to calculate the fiber-link length if given $P_{in}$, $P_{out}$, and $A$. The following formula can be easily derived from Formula 3.12:

$$L = (10/A)\log_{10}(P_{in}/P_{out}) \tag{3.13}$$

Formula 3.13 allows us to calculate the maximum transmission distance imposed by attenuation, bearing in mind that the minimum value of $P_{out}$ is determined by the sensitivity of the receiver.

## Example 3.2.2

---

***Problem:***

Calculate the maximum transmission distance for a fiber link with an attenuation of 0.5 dB/km if the power launched in is 1 mW and the receiver sensitivity is 50 μW.

***Solution:***

Just plug the numbers into Formula 3.13:

$$L_{max}\ (\text{km}) = (10/A)\log_{10}(P_{in}/P_{out}) = (10/0.5)\log_{10}(20) = 26\ \text{km}$$

At first glance, this is not a very impressive distance, but fiber with such a level of attenuation is designed for short- and intermediate-distance applications.

---

In conclusion, remember that *total attenuation encompasses bending, scattering, and absorption losses* and bending losses are usually shown separately on the optical-fiber specification sheets.

## Measuring Attenuation

There is a device called a power meter that allows us to measure the power of light. (For a detailed discussion of power meters, see Section 8.5.) The result is displayed in dBm, which is a specific unit of power in decibels when the reference power is 1 mW:

$$1 \text{ dBm} = -10\log(P_{out}/1 \text{ mW}) \tag{3.13a}$$

(Some power meters enable us to display the readings both in dBm and mW; others display readings in negative dBm.)

A diagram of an experimental arrangement for an attenuation measurement is shown in Figure 3.9. The procedure looks—and is—very simple: Connect a test fiber to the source and to the power meter and record the reading.

The key point here is this: Since *fiber connections to the source and to the power meter inevitably introduce additional losses,* your reading will reflect a sum of these losses and fiber attenuation. To exclude connection losses from your results, you need to take these measurements twice—with the fiber under the test and with a short piece of the same fiber serving as the reference point. You then subtract the second reading from the first. In this case, you can eliminate connection losses. The precision of this method is mainly determined by two factors: how accurately you can reproduce connection losses and how negligible is the attenuation introduced by a short piece of fiber.

This method is known as a cut method because it can be done by simply cutting a fiber under the test to get its short piece; this enables you to exclude the loss introduced by a light-source connector.

To measure macrobending loss, simply wrap the fiber undergoing the test several times with a certain bend radius and observe the difference.

When measuring attenuation in a multimode fiber, special care should be taken to use a light beam filling the entire cross-sectional area of the core (called *overfilled launching*) to make sure that all possible modes are excited. (See the next section and Section 4.4.)

It is evident that with the arrangement shown in Figure 3.9, we can measure fiber loss. To calculate loss in dB when obtaining readings in dBm, use the following obvious formula:

$$\text{Loss (dB)} = P_{in}(\text{dBm}) - P_{out}(\text{dBm})$$

Be careful about signs; always remember that you want to present the fiber loss as a positive number. For example, if your readings are $P_{in} = -1.0$ dBm and $P_{out} = -1.5$ dBm, the fiber loss is 0.5 dB. To calculate attenuation based on your measurement, measure the fiber length and use Formula 3.11.

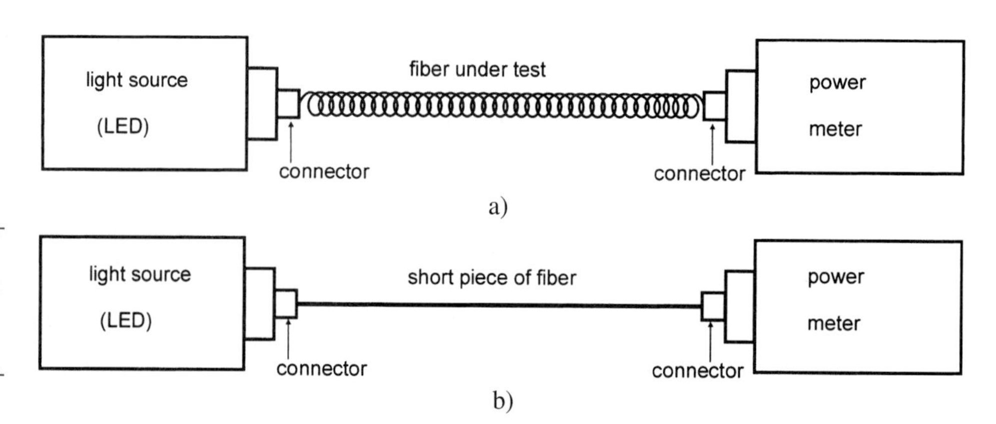


**Figure 3.9** Experimental arrangement for measuring attenuation: (a) Measuring fiber attenuation with connection losses; (b) measuring connection losses.

## 3.3 INTERMODAL AND CHROMATIC DISPERSION

From its inception, fiber-optic communications technology promised the highest possible information-carrying capacity of any medium simply because its signal carrier—light—has the highest frequency among all the practical carriers. But as soon as the first fiber-optic communications systems appeared, it immediately became clear that the capacity of these systems was very far from theoretical expectations. The reason for this disappointment was modal, or intermodal, dispersion, which is the main subject of this section. We'll also consider another important phenomenon—chromatic dispersion.

### Modes

***What they are*** Numerical aperture, we have seen, is the number that characterizes the ability of a specific optical fiber to gather light. The larger the numerical aperture, the easier it is to direct light into an optical fiber (not simply to direct it, but to direct it in such a way as to save light inside the fiber). In other words, the greater the numerical aperture, the larger the amount of light that can be directed into and saved inside an optical fiber. It would seem, therefore, that we would want to have a numerical aperture as large as possible. From this point of view, plastic fiber, with an *NA* of 0.5192, looks better than silica fiber, whose *NA* is 0.2425, as calculated in Example 3.1.4. But this is not always true. There is a stumbling block that prevents us from making the numerical aperture larger. To understand this obstacle, we have to consider the modes in an optical fiber.

The fact is that light can propagate inside an optical fiber only as a set of separate beams, or rays. In other words, if we were able to look inside an optical fiber, we would see a set of beams traveling at distinct propagating angles, $\alpha$, ranging from zero to the critical value, $\alpha_C$. This picture is shown in Figure 3.10(a).

These different beams are called modes. We distinguish modes by their propagating angles and we use the word *order* to designate the specific mode. The rule is this: *The smaller the mode's propagating angle, the lower the order of the mode.* Thus, the mode traveling precisely along the fiber's central axis is the zero-order mode and the mode traveling at the critical propagation angle is the highest order mode *possible* for this fiber. (The zero-order mode is also called the fundamental mode.)

Many modes can exist within a fiber, and so a fiber having many modes is called a multimode fiber.

***The number of modes*** How many modes an optical fiber can carry depends on the optical and geometric characteristics of a fiber. It's reasonable to expect that the larger the core diameter, the more light the core can accommodate and so there will be a greater number of modes. It is also reasonable to think that the shorter the wavelength of light, the more modes a fiber can accommodate. As for numerical aperture, the greater it is, the more light a fiber can gather and the more modes we would expect to see inside the fiber. We can therefore conclude that the number of modes inside a specific fiber should be proportional to the fiber diameter, *d,* and the numerical aperture, *NA,* and inversely proportional to the wavelength of the light used, $\lambda$.

The number of modes in an optical fiber is determined by the normalized frequency parameter, *V,* which is often called, simply, the *V* parameter. In fact, in your career you'll run across many terms for this parameter, such as *normalized cut-off frequency, characteristic waveguide parameter,* and others. We will just call it the "V number."

This number is equal to:

$$V = \frac{\pi d}{\lambda}\sqrt{(n_1)^2 - (n_2)^2}, \tag{3.14}$$

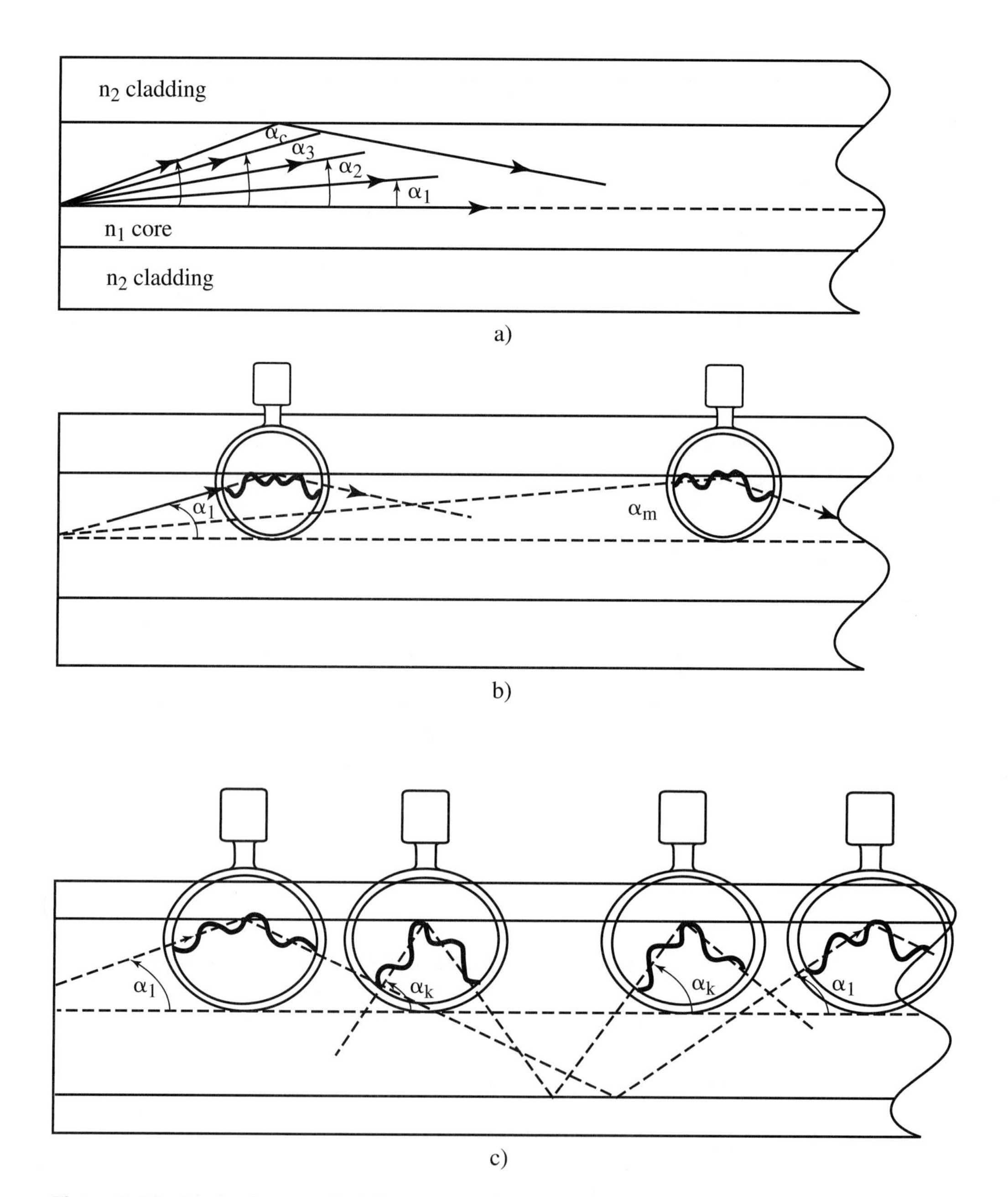


**Figure 3.10** Modes in an optical fiber: (a) Modes as different beams; (b) different beams experience different phase shifts; (c) optical fiber supports only those modes ($\alpha_1$) that complete the full zigzag at the same phase.

where $d$ is the core diameter, $\lambda$ is the operating wavelength, and $n_1$ and $n_2$ are refractive indexes of the core and cladding, respectively.

You may come across Formula 3.14 in different forms as, for example,

$$V = \frac{\pi d}{\lambda} NA, \tag{3.14a}$$

where $NA$ is the numerical aperture. Another popular form of this same equation is:

$$V = \frac{\pi d n}{\lambda} \sqrt{2\Delta}, \tag{3.14b}$$

where $n = (n_1 + n_2)/2$ is the average refractive index and $\Delta = (n_1 - n_2)/n$ is the relative refractive index. All these forms follow from Formulas 3.5 and 3.8.

How can we calculate the number of modes? For a large $V$ number (>20), the following formula for a step-index fiber can be applied:

$$N = V^2/2 \tag{3.15}$$

For a graded-index fiber (which is discussed on page 63) the formula is:

$$N = V^2/4 \tag{3.15a}$$

Formulas 3.14 and 3.15 confirm our discussion: The number of modes is directly dependent on the core diameter and the numerical aperture and inversely dependent on the wavelength.

### Example 3.3.1

***Problem:***

Calculate the number of modes for a graded-index optical fiber if its core diameter $d = 62.5$, its numerical aperture $NA = 0.275$, and its operating wavelength $(\lambda) = 1300$ nm.

***Solution:***

Applying Formula 3.14a to calculate the $V$ number, we get:

$$V = (\pi\, d NA)/\lambda = (3.14 \times 62.5 \times 10^{-6} \times 0.275)\text{ m}/1300 \times 10^{-9}\text{ m} = 41.5.$$

Applying Formula 3.15a to calculate the number of modes, we get:

$$N = (V)^2/4 = 431.$$

The actual number, $N$, we get from these calculations is 430.5625 but, obviously, the number of modes can only be an integer. That is why the calculation is given as 431.

***The physics and importance of modes*** Why do we need to know about modes in an optical fiber? Because *the light beam emerging from a light source into the fiber breaks down into a set of modes inside the fiber. Within the fiber, total light power is carried by individual modes so that, at the fiber output, these small portions combine, producing an output beam with its power.*

One may wonder: Why does continuous light outside a fiber convert into discrete modes inside the fiber? The answer can be found in Figures 3.10(b) and 3.10(c), where “magnifiers” show the points of interest on a larger scale. There are three points we need to bring out here:

First of all, you will recall that light is made up of electromagnetic waves. The phases at which specific waves meet the core-cladding interface are different and depend on the distance the waves travel. But the distance inside a fiber is determined by the propagation angle. Thus, different waves traveling within the fiber at different propagation angles will strike the core-cladding interface at different phase angles, as Figure 3.10(b) shows for two waves.

A second critical point to understand is this: A wave experiences a phase shift when it is reflected; this shift depends on the propagation angle. This is shown in Figure 3.10(b), where the waves traveling at propagation angles $\alpha_1$ and $\alpha_k$ have different phase shifts.

The third and most crucial point is that, after completing a full zigzag, wave $\alpha_1$ strikes the core-cladding interface having the same phase as it had on the previous strike while wave $\alpha_k$ has a new phase. In other words, *wave $\alpha_l$ reproduces itself after the whole cycle of propagation but wave $\alpha_k$ does not.* All these phase shifts depend on the propagation angles of specific waves traveling inside the fiber. Therefore, optical fiber supports only certain waves and the criterion for their selection is the propagation angle. For a detailed discussion of this intriguing topic see Section 4.4.

The concept of modes is a principal concern of ours because you will run across phenomena associated with it many times in the course of our discussions in this book. For now, however, this concept is of interest because it explains intermodal, often called modal, dispersion.

## Modal (Intermodal) Dispersion

***How input pulse is delivered within a fiber*** Let's consider a beam propagating inside a fiber, taking into account the mode concept. Don't forget that we are discussing fiber-optic *communications* technology; therefore, we are looking to use light to carry a communications signal. For the most popular form of digital transmission, *a light pulse represents logic 1, and no light pulse (darkness) represents logic 0. Such light pulses, radiated by a light source, enter a fiber, where each pulse breaks down into a set of small pulses carried by an individual mode. At the fiber output, individual pulses recombine and, since they are overlapping, the receiver sees one long light pulse whose rising edge is from the fundamental mode and whose falling edge is from the critical mode.* This explanation is depicted in Figure 3.11, where four modes are shown as an example.

*Pulse widening caused by the mode structure of a light beam inside the fiber is called modal (intermodal) dispersion.* This text uses the terms *intermodal* and *modal* interchangeably.

***Calculations of pulse spread*** To ascertain why these individual light pulses arrive at the receiver end at different times, let's do some simple calculations. A zero-order mode traveling along the central axis needs time,

$$t_0 = L/v,$$

to reach the receiver end. Here, $L$ is the link length and $v = c/n_1$ is the light velocity within the core having refractive index $n_1$, while $c$ is the speed of light in a vacuum. The highest-order mode propagating at the critical angle—the critical mode—needs time,

$$t_C = L/(v \cos \alpha_C),$$

to complete its path. Reminding ourselves that $\cos \alpha_C = n_2/n_1$ (see Example 3.1.2), we can derive the formula for pulse widening stemming from intermodal dispersion:

$$\Delta t_{SI} = t_C - t_0 = \frac{Ln_1}{c}\left(\frac{n_1 - n_2}{n_2}\right), \tag{3.16}$$

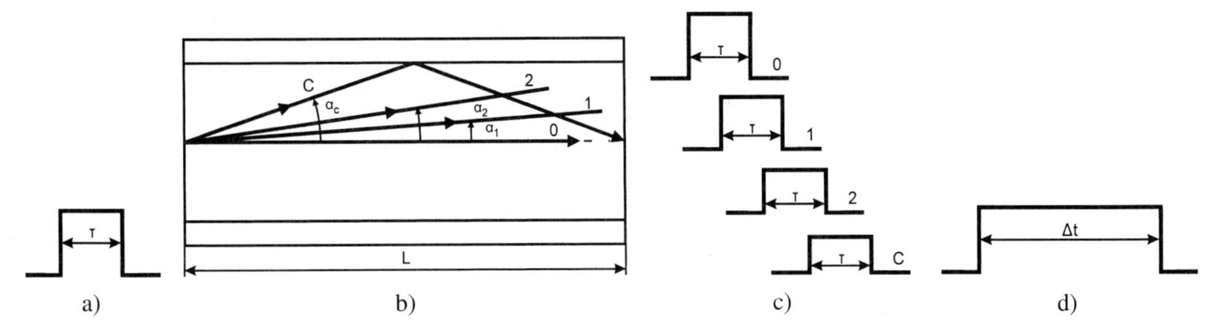


**Figure 3.11** Intermodal (modal) dispersion: (a) Original pulse; (b) modes in an optical fiber; (c) pulses delivered by an individual mode; (d) resulting pulse.

where SI stands for step-index fiber. Using the relative refractive index $\Delta = (n_1 - n_2)/n$, Formula 3.16 can be rewritten as:

$$\Delta t_{SI} = t_C - t_0 = (Ln_1/c)\Delta, \tag{3.16a}$$

where approximation $n_2 \approx n$ has been used.

If the precision of your calculations allows you to neglect the difference between $n_1$ and $n_2$, you can derive this expression in still another form:

$$\Delta t_{SI} = t_c - t_0 = \frac{L}{2cn_2}(NA)^2, \tag{3.16b}$$

where $NA$ is the numerical aperture. This version of the pulse-spreading formula is important because manufacturers provide you with a numerical aperture number, not with numbers $n_1$ and $n_2$.

### Example 3.3.2

***Problem:***

How much will a light pulse spread after traveling along 5 km of a step-index fiber whose $NA = 0.275$ and $n_1 = 1.487$?

***Solution:***

From Formula 3.16b, replacing $n_2$ with $n_1$, we get

$$\Delta t_{SI} = (L \times NA^2)/(2\ cn_1) = (5 \times (0.275)^2)/(2 \times 3 \times 10^5 \times 1.487) = 423.8\,ns$$

Three things to make note of here:

(1) The fiber length is expressed in km and the speed of light in km/s.
(2) The unit used to measure pulse spreading is nanoseconds, ns (1 ns is equal to $10^{-9}$ s).
(3) We can assume $n_1 \sim n_2$ because their difference is about 0.02, which is much less compared with 1.5—their order of value.

Since pulse spreading is proportional to fiber length, it is sometimes useful to operate in terms of pulse spreading per unit length. If we do so using the above example, we get:

$$\Delta t_{SI}/L = 84.76\ \text{ns/km}$$

***How intermodal dispersion restricts bit rate*** The importance of intermodal dispersion in pulse spreading cannot be overestimated. Let's see why. Suppose you need to transmit information at 10 Mbit/s (megabits per second). This means you want to transmit $10 \times 10^6$ pulses every second; in other words, the duration of each cycle is 100 ns. For simplicity's sake, assume that the duration of the input pulses is negligibly short. Nevertheless, these pulses will spread due to intermodal dispersion. For illustrative purposes, let's refer to the numbers discussed in Example 3.3.2, where we found that each pulse will spread up to 84.76 ns every kilometer. Therefore, the duration of each pulse will be 84.76 ns after the first kilometer transmission and 169.52 ns after the second. Figure 3.12 shows this situation.

As you can see, after the second kilometer pulses become so wide that they overlap and the light no longer carries any information. The same is true when you try to increase bit rate even for short-distance transmission. Consider the problem in Example 3.3.3.

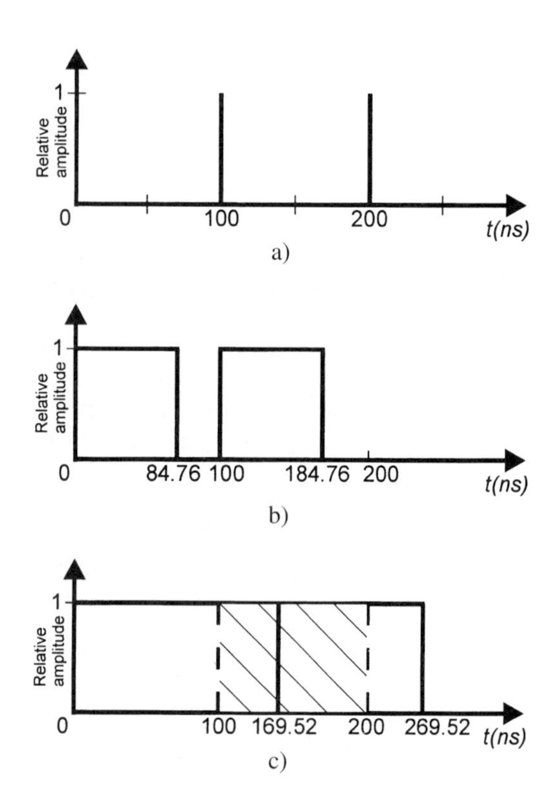


**Figure 3.12** Pulse spreading after transmission: (a) Input pulses; (b) pulses after 1 km transmission in Example 3.3.2; (c) pulses after 2 km transmission in Example 3.3.2. Bit rate: 10 Mbit/s.

### Example 3.3.3

***Problem:***

Find the maximum bit rate for the fiber discussed in Example 3.3.2 if the transmission length is 1 km.

***Solution:***

The solution is based on Formula 3.16a and Example 3.3.2. There are two key points to keep in mind as we work out this problem: First, we are able to distinguish pulses until they overlap. Secondly, let $\Delta t$ be the width of an individual pulse. Then a 1-second interval can accommodate a certain number of these pulses before they will overlap. This number is equal to 1 second divided by

$\Delta t$. If you divide this number by 1 second, you'll find the maximum bit rate. In other words *the maximum bit rate is equal to 1/*$\Delta$*t(s)*. We've computed in Example 3.3.2: $\Delta t_{SI} = 84.76$ ns; thus, the maximum bit rate is $1/(84.76 \times 10^{-9}$ ns) = 11.8 Mbit/s.

Obviously, we want to have a time gap between adjacent pulses to ensure their separation. If we take 25% of the cycle gap, we come up with a lower number for the maximum bit rate. In our example, a 25% gap results in the following: The maximum pulse width is $84.76 \times 1.25 = 105.95$ ns, which in turn gives 1/105.95 ns = 9.44 Mbit/s (the maximum bit rate).

Draw a picture similar to the one in Figure 3.12. It will help you visualize the phenomenon of pulse spreading after transmission, which is the concept of intermodal dispersion.

---

Intermodal dispersion severely limits the bit rate of a fiber-optic link. Indeed, our examples show that the maximum bit rate might not be more than 12 Mbit/s. This is not a very impressive number. We certainly don't need fiber optics to transmit information at this bit rate; a coaxial cable can do that quite easily. In fact, this was the problem that telecommunications companies faced when fiber optics first became a serious contender as an information-carrying medium.

## The First Solution to the Modal-Dispersion Problem—Graded-Index Fiber

***The basic idea and the structure of a graded-index fiber*** Can the problem be overcome? To answer this question, we first have to recall the physical reason for the problem. Within a core, the zero-order mode travels along the central axis and the higher-order modes travel at, or less than, the critical propagation angle. Thus, the beams travel at the same velocity but over different distances and they arrive at the receiver end at different times. If we could arrange it so that they would arrive simultaneously, we would solve the problem. But is that possible? In a word, yes. Recall that the velocity of light, $v$, within a material is defined by its refractive index, $n$: $v = c/n$, where $c$ is the speed of light in a vacuum. Thus, we have the solution: *We can design the core with different refractive indexes so that the beam traveling the farthest distance does so at the highest velocity and the beam traveling the shortest distance propagates at the slowest velocity. Such fibers are called graded-index (GI) multimode fibers.* The principle of this action is clear from Figure 3.13.

Refer to the refractive-index profile in Figure 3.13(a). Observe how the refractive-index value varies gradually from $n_1$ at the core center to $n_2$ at the core-cladding boundary. This is why the fiber is called *graded index*. The higher-order modes move from the higher to the lower refractive indexes at each point along their path. This results in a change of direction in their propagation, shown as curve paths in Figure 3.13(b).

The core of a graded-index fiber can also be seen as a set of thin layers whose refractive indexes change slightly from one to another so that the layer at the central axis has refractive index $n_1$ and the layer at the cladding boundary has index $n_2$. This is how manufacturers physically make the fiber. The fabrication process consists of the deposition of molecular-thin layer after layer with a given refractive index, thus assembling the core and the cladding. A change in the refractive index is achieved by doping a certain number of atoms with material other than silica. (For more details on the fabrication process, see Chapter 7.)

One can understand the principle of light propagation in graded-index fiber by considering the behavior of light at the boundary of the two layers. Each individual interaction results in a small change of direction of the propagation. (The definition of the term *refractive index* also implies that it is a measure of how much a ray of light is bent when propagating from one medium into another.) This is illustrated in Figure 3.13(c). By making these layers smaller and smaller, we arrive at the gradually changed refractive index shown in Figure 3.13(a).

***How well does a graded-index fiber reduce modal dispersion?*** You will recall that an input pulse is delivered within a fiber core in fractions and each of these fractions is carried by a different mode (Figure 3.11). The mode propagating along the centerline of a graded-index fiber—the

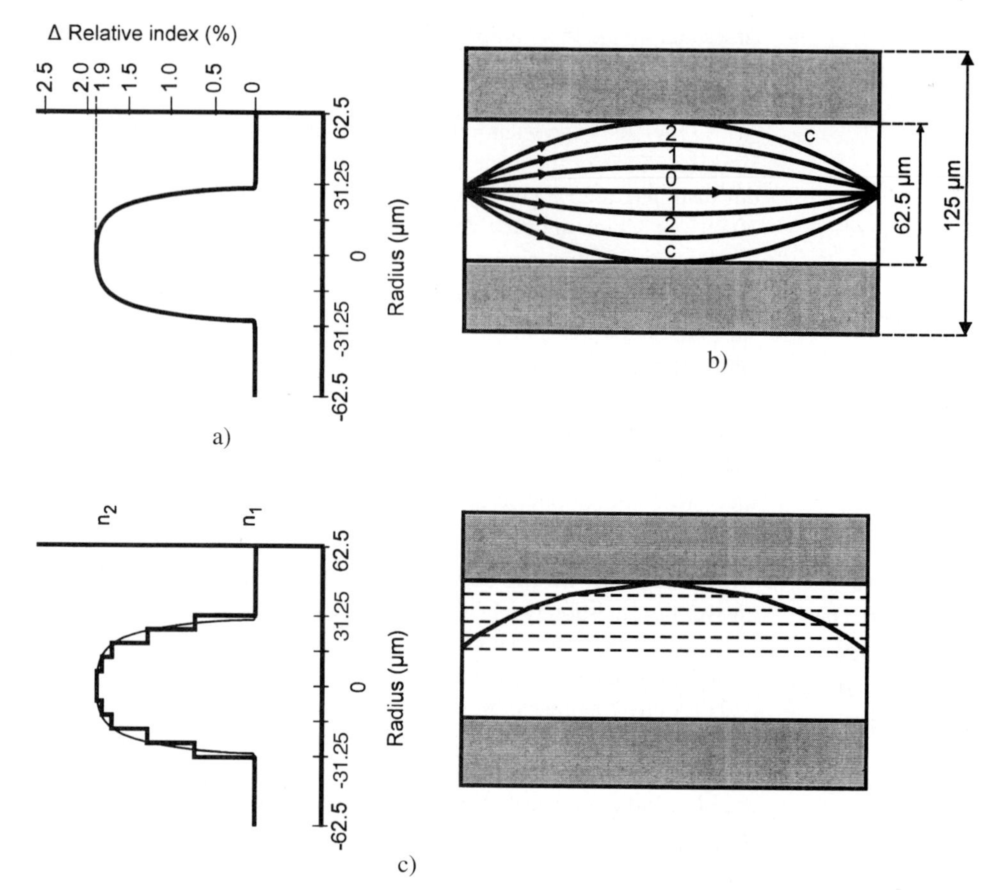


**Figure 3.13** Graded-index (GI) multimode fiber: (a) Refractive-index profile; (b) mode propagation; (c) principle of action of graded-index multimode fiber.

shortest distance—travels at the lowest speed because it meets the highest refractive index, as Figure 3.13(a) demonstrates. The mode traveling closer to the fiber cladding—the longer distance—propagates at the higher speed because it meets a lower refractive index. Hence, *the fractions of an input pulse delivered by the different modes arrive at the receiver end more or less simultaneously.* Therefore, intermodal dispersion will be reduced and the bit rate will be increased.

The formula for calculating pulse spreading ($\Delta t$) for graded-index fiber is given by [1]:

$$\Delta t_{\mathrm{GI}} = \left(LN_1\Delta^2\right)/(8c), \tag{3.17}$$

where *GI* stands for the graded-index fiber, $\Delta$ is the relative index, $c$ is the speed of light in a vacuum, and $N_1$ is the core group index of refraction. It follows from our definition of a graded-index fiber that its core refractive index is variable (see Figure 3.13). However, we can summarize the optical properties of the core as they are seen by the light propagating through a graded-index fiber by introducing one generalized number—$N_1$.

### Example 3.3.4

***Problem:***

A graded-index fiber has $N_1 = 1.487$ and $\Delta = 1.71\%$. For a link 5 km in length, compute pulse spreading due to modal dispersion and determine the maximum bit rate.

***Solution:***

Formula 3.17 yields:

$$\Delta t_{GI} = (LN_1\Delta^2)/(8c) = (5 \text{ km} \times 1.487 \times (0.0171)^2)/(8 \times 3 \times 10^5 \text{ km/s}) = 0.9 \text{ ns}.$$

Again, operating with pulse spreading per km length, one can compute:

$$\Delta t_{GI}/L = 0.18 \text{ ns/km}$$

Compare this answer with the numbers obtained in Example 3.3.2 for step-index fiber with similar parameters to see the much better dispersion characteristics of a graded-index fiber. Maximum bit rate is $1/\Delta t$. For the graded-index fiber of 1 km, one can get 5.5 Gbit/s, which is much better than the 11.8 Mbit/s obtained for a step-index fiber in Example 3.3.3. For a 5-km link, the maximum bit rate equals 1.1 G bit/s. Now you can see how the concept of reducing dispersion in a graded-index fiber works.

Two notes:

(1) Using Formula 3.8, it is easy to derive:

$$\Delta t_{GI} = (L\ NA^4)/(32\ c\ N_1^3), \tag{3.17a}$$

where approximation $n_1 \approx N_1$ was used.

(2) Compare $\Delta t_{SI}$ (Formula 3.16a) and $\Delta t_{GI}$ (Formula 3.17). You can see that $\Delta t_{GI} = \Delta t_{SI}(\Delta/8)$, again, assuming $n_1 \approx N_1$. Thus, a graded-index fiber has a modal dispersion $\Delta/8$ times less than that of a step-index fiber. You can verify this result using the numbers in Examples 3.3.2 and 3.3.4.

---

Graded-index fiber was the first solution to the modal-dispersion problem, but at a price: cost. This was because manufacturers had to expend more effort to control the complex index profile during the mass-production process. However, today this is no longer a problem and graded-index fiber is a popular transmission medium for short- and intermediate-distance networks.

**A Better Solution to the Modal-Dispersion Problem—Singlemode Fiber**

***The structure of a singlemode fiber*** There is another, even better solution to the modal-dispersion problem. The underlying reason for the problem is the existence of many modes that deliver the same light pulse. So researchers asked themselves, "Why not limit the light beam inside the core to only one mode?" Doing so, they reasoned, would eliminate the problem completely. The result: advent of the singlemode fiber.

But just how was this accomplished? Refer again to Formulas 3.14, 3.15, and 3.16. They show that the number of modes is directly dependent on the core diameter, *d,* and the difference between refractive indexes $n_1$ and $n_2$. Hence, the simplest way to restrict the number of modes propagating inside the core to just one is to reduce the core diameter and relative refractive index. This approach is illustrated in Figure 3.14.

Pay attention to the core diameter, *d,* and the relative index, $\Delta$, of a singlemode fiber; typically, $d$ and $\Delta$ are as small as 8.3 μm and 0.37%, respectively. Compare these numbers with 62.5 μm and almost 2% of a graded-index fiber and you will see how one can make a fiber carry only one mode.

A word of caution: Don't try to insert $N = 1$, where $N$ is the number of modes, into Formula 3.15 to obtain the critical $V$ number. Remember, Formulas 3.14 and 3.15 work only for a large ($V > 20$) number of modes. A real singlemode condition is:

$$V \leq 2.405 \tag{3.18}$$

This condition was obtained from considerations that will be discussed in Chapter 6.

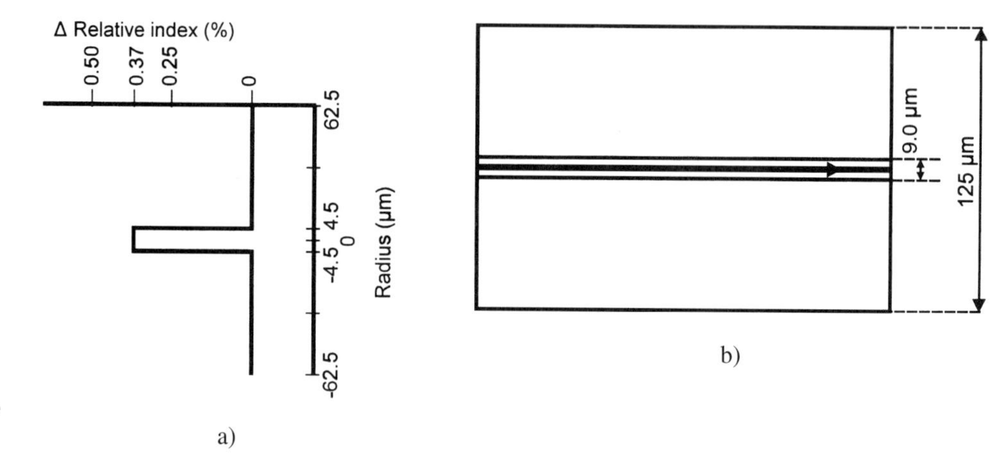


**Figure 3.14** Step-index (SI) singlemode fiber: (a) Refractive-index profile; (b) mode propagation.

***Review of the modal-dispersion problem*** To review the concept of dispersion for all three types of optical fiber, see Figure 3.15.

It is therefore clear that the singlemode fiber affords the best solution to the modal-dispersion problem. The drawback is that it is the most expensive fiber to manufacture and the most difficult to maintain, largely because of the difficulty in maintaining an accurate core size. Indeed, the core size of a singlemode fiber may vary from 4 to 11 μm. You can imagine how difficult it is to maintain this size with accuracy, yet avoid microbending and scattering problems during the mass-production process. What's more, a singlemode fiber is more prone to macro- and microbending losses and many other problems during installation and operation. However, the singlemode fiber is now the most popular type of link, particularly for long-distance communications, and it will surely penetrate other sectors of telecommunications. We will discuss this subject in more detail in Chapters 5 and 6.

## Chromatic Dispersion

***What it is*** Modal dispersion is not the only impediment to fiber bandwidth, or bit rate. Another type of dispersion, chromatic dispersion, also contributes to this drawback. The word *chromatic* is associated with colors, of course. You'll recall that the basic mechanism of dispersion involves different light beams carrying light pulses. The beams arrive at the receiver end at different times, causing the output light pulses to spread. In the case of modal dispersion, these different beams are different modes. But even within a single mode we might have the same problem if this mode were composed of light comprising different colors. Obviously, color is no more than an image and, in reality, we have to talk about wavelength.

Let's consider the zero-order mode, which travels precisely along the fiber's central axis. This beam is composed of light having several wavelengths simply because there is no source in nature that can radiate a single wavelength. And the key point to note here is that refractive index depends on wavelength; thus, $n = n(\lambda)$. In other words, for each specific wavelength, the refractive index is a specific—and different—number. You'll recall that the velocity of light, $v$, within a material is $v = c/n$, where $c$ is the speed of light in a vacuum; therefore, *light of different wavelengths travels along the fiber at different velocities. Even if all of these beams propagate along the same path, they will arrive at the receiver end at different times. This results in the spreading of the output light pulse—chromatic dispersion.*

Chromatic dispersion plays the major role in limiting the bandwidth of a singlemode fiber, since modal dispersion is not a consideration here. (We will consider this in detail in Chapters 5 and 6.) This type of dispersion is important, too, for multimode fibers even though modal dispersion is the major factor limiting multimode-fiber bandwidth.

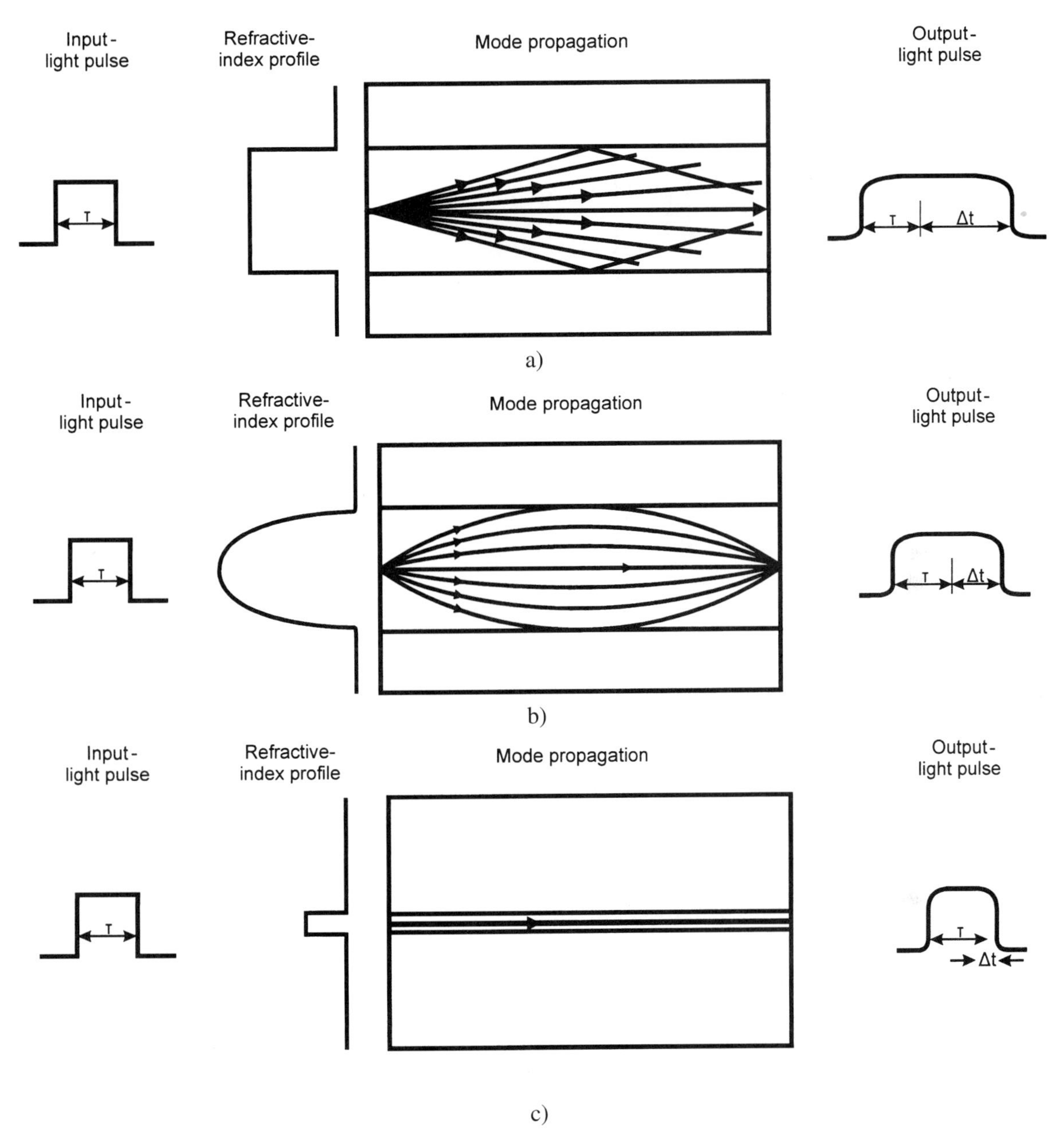


**Figure 3.15** Dispersion in three types of optical fiber: (a) Step-index multimode fiber; (b) graded-index multimode fiber; (c) step-index singlemode fiber.

***Calculating pulse spreading caused by chromatic dispersion*** Pulse spreading caused by chromatic dispersion can be calculated as follows:

$$\Delta t_{\text{chrom}} = D(\lambda) L \, \Delta\lambda, \tag{3.19}$$

where $D(\lambda)$ is the chromatic-dispersion parameter measured in picoseconds (ps) per nanometer (nm) and kilometer (km); thus, we have ps/nm · km; $L$ is the fiber length in km and $\Delta\lambda$ is the spectral width of a light source in nm, the characteristic of how many wavelengths this source radiates.

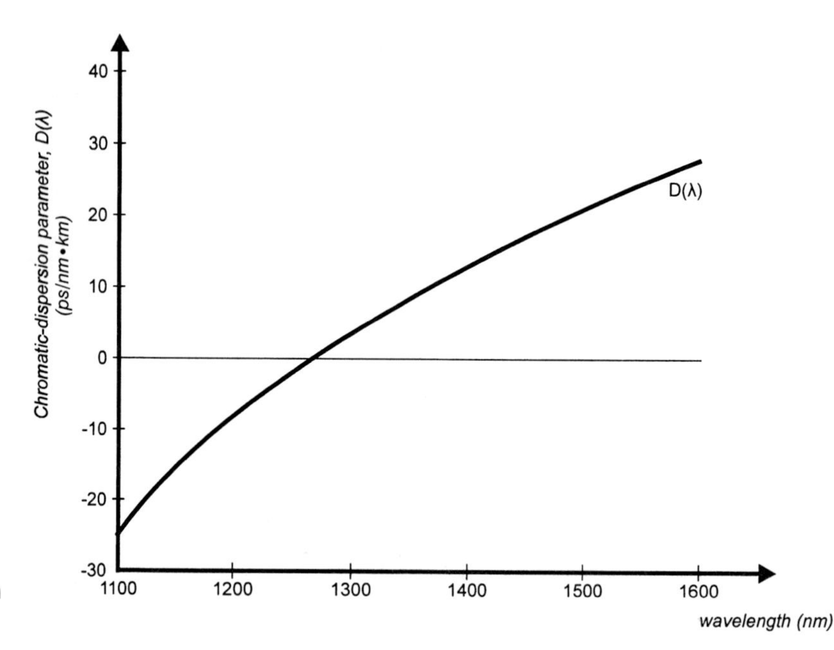


**Figure 3.16** Chromatic-dispersion parameter.

The chromatic-dispersion parameter, $D(\lambda)$, is zero at the specific wavelength called the zero-dispersion wavelength. The graph of $D(\lambda)$ as a function of $\lambda$ is shown in Figure 3.16.

Manufacturers specify the chromatic-dispersion parameter for multimode fibers either by giving its value or by giving the formula:

$$D(\lambda) = \frac{S_0}{4}\left[\lambda - \frac{\lambda_0^4}{\lambda^3}\right], \tag{3.20}$$

where $S_0$ is the zero-dispersion slope in ps/(nm$^2$ · km), $\lambda_0$ is the zero-dispersion wavelength, and $\lambda$ is the operating wavelength.

## Example 3.3.5

***Problem:***

What is the chromatic dispersion for a graded-index fiber if $S_0 = 0.097$ ps/(nm$^2$ · km), $\lambda_0 = 1343$ nm, and $\lambda = 1300$ nm?

***Solution:***

Inserting the numbers into Formula 3.20, one gets:

$$D(\lambda) = -4.38\,\text{ps}/(\text{nm} \cdot \text{km})$$

The minus sign comes from the formula and indicates that pulse spreading decreases as wavelength increases. For practical calculations, we can neglect this negative sign. This result tells us that the pulse spreading of this specific fiber is 4.38 ps (pico means $10^{-12}$) per nm of wavelength radiated by a light source and per km of fiber length.

If we use an LED whose $\Delta\lambda = 50$ nm, we can calculate:

$$\Delta t_{\text{chrom}}/L = 219\,\text{ps/km} = 0.22\,\text{ns/km}$$

This number is the same order of value as the number 0.18 ns/km that we calculated for the modal dispersion of the same fiber. (See Example 3.3.4).

---

***Total pulse spreading caused by modal and chromatic dispersion*** Total pulse spreading from both types of dispersion is calculated using the following formula:

$$\Delta t_{total} = \sqrt{(\Delta t^2_{modal} + \Delta t^2_{chrom})} \tag{3.21}$$

where $\Delta t_{modal}$ is the pulse spreading caused by modal dispersion and $\Delta t_{chrom}$ is the pulse spreading that results from chromatic dispersion.

## 3.4 BIT RATE AND BANDWIDTH

Modal dispersion in optical fibers causes a significant restriction of bit rate. In response, two different types of optical fibers—graded-index multimode and singlemode—were developed to facilitate the problem. This is also true for chromatic dispersion. The result of these efforts is the variety of types and specifications of optical fibers that are available today designed specifically to satisfy a range of customer requirements. The major factor in these requirements is the information-carrying capacity of a link or network that can be characterized by bit rate or bandwidth. What these characteristics are and how we can calculate and use them are the topic of this section.

### Bit Rate and Bandwidth Defined

*Bit rate (some say "data rate") is the number of bits that can be transmitted per second over a channel. It is measured in bit/s. It is the direct measure of information-carrying capacity of a communications link or network for* digital *transmission.* This is why it is also called the "information-transmission rate." *Bandwidth is the frequency range within which a signal can be transmitted without significant deterioration. It is measured in Hertz. It is the information-carrying capacity characteristic of a communications channel used for* analog *transmission.* These two characteristics, then, are obviously quite different. Bit rate—for digital transmission—and bandwidth—for analog transmission—are shown in Figure 3.17.

There is a difference between electrical bandwidth and optical bandwidth, which is discussed in Section 4.6.

What is the relationship between bit rate, *BR,* and bandwidth, *BW*? The simplest approach is to assume that the number of bits per second, bit/s, and the number of cycles per second, Hz, are the same; hence, $BW = BR$.

You often will find another relationship between bandwidth, *BW*, and bit rate, *BR*:

$$BW = BR/2 \tag{3.22}$$

This stems from the following consideration: Let's take the worst-case scenario, that is, when digital transmission is the sequence 1-0-1-0-1. . . . If we represent pulse waveform by sine waveform, we find that one period of sine covers two bits. This is shown in Figure 3.17(c). It is quite obvious that the bit rate is twice as high as the frequency, which results in Formula 3.22.

Which relationship—$BW = BR$ or $BW = BR/2$—we must use depends on the line codes. For instance, Figure 3.17c shows the non-return-to-zero (NRZ) format, which is the simplest line code. Here we use Formula 3.22. There are many other line codes for which the relationship between *BW* and *BR* is different. In general, *one can transmit several bits per second per hertz of a channel bandwidth (bit/s/Hz) by using various forms of modulations.* We will treat this topic in later chapters, but for the remainder of this discussion we will concern ourselves solely with $BW = BR$. Since fiber-optic communications technology uses the terms *bandwidth* and *bit rate* interchangeably, we will follow this pattern.

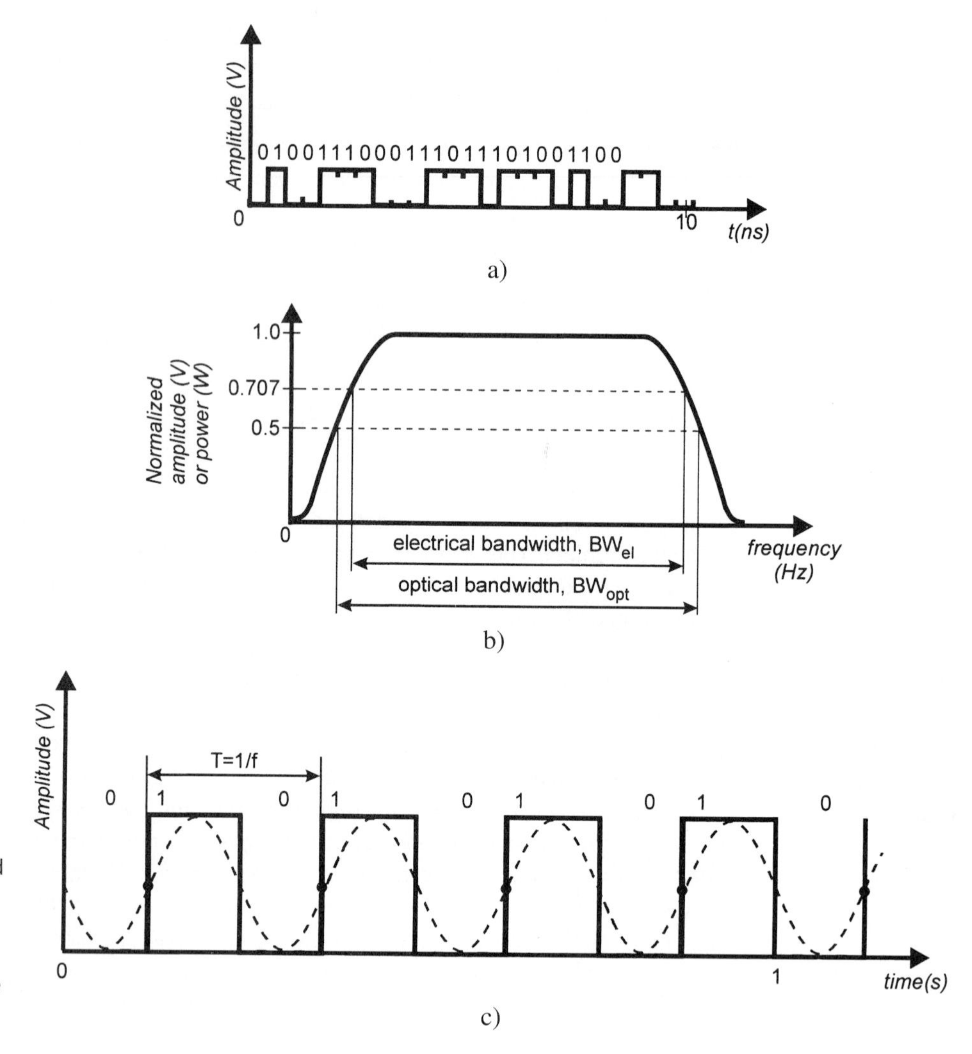


**Figure 3.17** Bit rate and bandwidth: (a) Bit rate of a digital signal; (b) bandwidth of an analog signal; (c) digital pulses and sine signal.

## Dispersion and Bit Rate

***Calculating bit rate*** We considered the importance of dispersion for data rate on pages 62 and 65 and basic definitions were given in Example 3.3.3. Let's now make some calculations based on that discussion. Again, it is quite obvious that the data rate has to comply with the pulse width. For example, if the pulse width is 1 ns, the data rate cannot exceed $10^9$ bit/s; otherwise, the pulses will overlap. But, from a practical standpoint, we always want to have an interval between pulses to ensure reliable transmission.

We assume, for discussion purposes, that the input data rate of a fiber-optic communications system has been designed properly and we need to calculate what signal distortion will be introduced by the system. If the pulse spreading due to dispersion is designated as $\Delta t$, then the bit rate of a system is:

$$BR < 1/(4\Delta t) \tag{3.23}$$

The coefficient 1/4 is generally accepted in the industry.

What $\Delta t$ do we have to plug into Formula 3.23? It depends on our goal. To calculate bit rate caused by modal dispersion in step-index multimode fibers, plug in $\Delta t_{SI}$ from Formulas 3.16, 3.16(a), or 3.16(b):

$$BR_{SI} = 1/(4\Delta t_{SI}) = cn_2/(4(Ln_1)(n_1 - n_2)) = c/(4Ln_1\Delta) = cn_2/(2L\ NA^2) \tag{3.24}$$

For a graded-index fiber, we can derive the bit-rate estimate by substituting Formula 3.18 into Formula 3.23:

$$BR_{GI} = 2c/(N_1 L\Delta^2) \tag{3.25}$$

To calculate bit rate restricted by chromatic dispersion, we simply need to plug Formula 3.19 into Formula 3.23:

$$BR_{chrom} = 1/(4D(\lambda)L\,\Delta\lambda) \tag{3.26}$$

Finally, we can estimate total bit rate by substituting Formula 3.21 into Formula 3.23:

$$BR_{total} = 1/\left(4\sqrt{(\Delta t^2_{modal} + \Delta t^2_{chrom})}\right) \tag{3.27}$$

**Example 3.4.1**

***Problem:***

A graded-index fiber's $NA = 0.275$ and $N_1 = 1.487$. What is the bit rate restricted by modal dispersion for a 1-km link?

***Solution:***

Substitute Formula 3.17a into Formula 3.23 and obtain:

$$BR_{GI} = (8cN_1^{\ 3})/(L\ NA^4) = 1.38 \times 10^9 \text{ bit/s} = 1.38 \text{ Gbit/s}$$

As was shown in Example 3.3.5, $\Delta t_{GI} = \Delta t_{SI}\ (\Delta/8)$. Therefore, one can expect $BR_{SI} = (\Delta/8)BR_{GI}$. For $NA = 0.275$, one finds that $\Delta = 0.0171$ (see Formula 2.8). Thus, the $BR_{SI}$ for a step-index fiber with similar parameters should be 2.95 Mbit/s. The reader can verify this result by direct calculations of $BR_{SI}$ using Formula 3.24.

Look at these numbers again: $BR_{GI} = 1.38$ Gbit/s and $BR_{SI} = 0.00295$ Gbit/s. No further comment is required.

In reality, the bit rate (bandwidth) of a multimode fiber is much higher than the calculations of this section show. The reasons for such a discrepancy are discussed in Section 4.6.

***How manufacturers specify bandwidth*** Manufacturers specify fiber bandwidth as bandwidth-length product. They measure the bandwidth of a specific piece of fiber and multiply the result by the fiber length. If the measured bandwidth of a 2-km-long optical fiber is 300 MHz, then the manufacturer specifies the fiber bandwidth-length product as 600 MHz · km. Remember, this number serves to compare the bandwidth of bare fiber (the number will change for cabled fiber) and is measured under certain conditions, which are discussed in Section 4.6.

## 3.5 READING A DATA SHEET

For the practicing engineer, the ability to read data sheets is critical. This ability is obviously gained through acquisition of a thorough knowledge of one's field and, not to be overlooked, an understanding of the meaning of each specification shown on the data sheet. This is why the first part of our discussion of optical fibers concludes with the analysis of a specification sheet given for a multimode optical fiber.

## Where to Begin

It is important to realize that all the data given in the manufacturer's specifications are measured by the manufacturer under certain standard conditions. These conditions and the characteristics to be measured are specified by the Fiber Optic Test Procedure (FOTP) standards, which were developed by the Telecommunications Industry Association (TIA) and the Electronic Industries Alliance (EIA) in the United States. Similar standards have been developed in Europe and Japan.

The data sheet for a specific fiber is given in Figure 3.18. To start with, there are three items you should concentrate on: the title of the sheet, which gives the official name of the product; corporate-contact information, and the date the document was issued. So looking at Figure 3.18, we see that the title is "Graded-Index Multimode Fibre." The numbers 62.5 and 125 are the core and cladding diameters, respectively. Code 457E is the product code. It assists the buyer when ordering the product or when requesting specific information from the manufacturer. The sheet's issue date is very important since it tells you what version of the product is currently available. If you are a customer, the manufacturer will normally keep you updated on all product modifications and upgrades.

## General Section

***Fibre*** This subsection provides you with a wide range of information about the product, such as the type of fiber (in Figure 3.18, a graded-index multimode fiber) and the operating wavelength (850 and 1300 nm in our illustration). In specifying the operating wavelengths, the manufacturer is telling the buyer that the fiber has been designed to display optimum performance characteristics at these particular wavelengths. The producer accomplishes this by choosing the bulk material and dopants for the core and cladding and tailoring them to make a specific profile of a core refractive index. This subsection also delineates the typical applications for the fiber, thereby enabling the user to choose the most suitable fiber for his or her specific needs.

***Coating*** A bare fiber—its typical core-cladding structure is shown in Figures 3.1 and 3.20—has to be protected from a hostile environment. This is done by coating the bare fiber with a specific material. The importance of the coating (buffer) in reducing a fiber's microbending sensitivity was discussed in Section 3.2. Bear in mind that in order to be used in the field, a bare fiber has to be cabled and spliced—procedures to be considered in Chapters 7 and 8. For these operations, a knowledge of coating characteristics is extremely important. We'll discuss coating in more detail in Section 7.1.

***Process*** In this subsection, the manufacturer introduces its fabrication process. This topic will also be considered in Section 7.1.

Pay attention to the topographic profile of a fiber's refractive index given in this section. This is a three-dimensional representation of a profile shown in Figures 3.13(a) and 3.20.

## "Optical Characteristics" Section

***Attenuation*** This section spells out the optical characteristics of the fiber. Note that under the heading "Attenuation coefficient," the general attenuation at two operational wavelengths is specified. The first listing shows attenuation at not more than 2.7 dB/km at 850 nm and not more than 0.5 dB/km at the 1300-nm operating wavelength. The difference in attenuation at different wavelengths is determined by the decrease in scattering loss as operating wavelength increases. Optical fiber with three values of attenuation for each operating wavelength is available from the manufacturer.

***Minimum modal bandwidth*** This characteristic specifies the fiber's capacity, which is restricted by modal dispersion. (Bandwidth is given in MHz · km.) Modal dispersion does not depend directly on operating wavelength (see Formula 3.17). However, the longer the operating wavelength, the better the bandwidth, as you can see from Figure 3.18. We'll discuss this phenomenon in later chapters.

**Product Code: 457E** **Graded - Index Multimode Optical Fibre**

*Issue date: 11/95 Supersedes: 1/94* *Type: 62.5/125 µm*
Double Layer Primary Coating (DLPC7)

## FIBRE

**The 62.5/125 µm multimode optical fibre product code 457E, is a graded-index multimode optical fibre with a 62.5 µm core diameter and a 125 µm cladding diameter. The optical fibre is designed for use at 850 nm and/or 1300 nm.**

**Applications for this optical fibre are local area networks (LAN), video, data and/or voice transmission using laser or light emitting diodes at 850 nm or 1300 nm. Because of the nature of the manufacturing process, the optical fibre offers the highest bandwidth available in the market.**

**The optical fibre complies with or exceeds the IEC 793-2 type A1b Optical Fibre Specification.**

## COATING

The optical fibre is coated with a double layer of UV curable acrylate, type DLPC7. Designed for more stringent tight-buffer cable applications, the optical fibre also performs perfectly in loose buffer constructions and demonstrates a high resistance to microbending.

The coating offers an excellent stable coating strip force over a wide range of environmental conditions. In tight buffer applications the entire coating construction (tight buffer and primary coating) is in general very easy to strip off. Coating stripping leaves no residues on the bare glass fibre.

The DLPC7 coated optical fibres show high and stable values for the dynamic stress corrosion susceptibility parameter ($n_d$), which offers a greatly improved mechanical protection to the optical fibre when used in harsh environments.

## PROCESS

The optical fibres are manufactured using the advanced PCVD process. Because of the inherent high quality of the graded refractive index profile, multimode optical fibres manufactured with the PCVD process show excellent bandwidth performance.

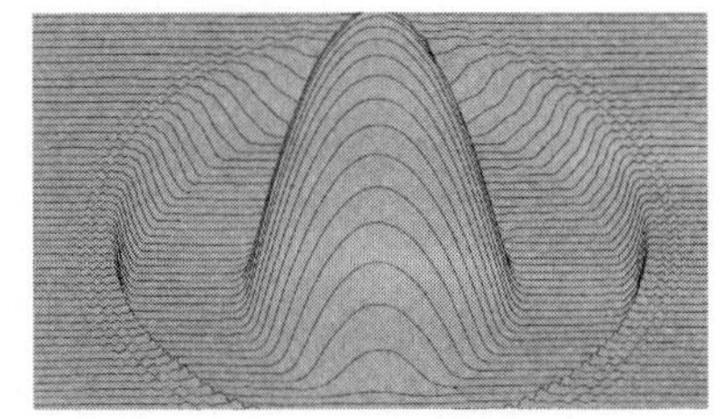

**PLASMA OPTICAL FIBRE B.V.**
Building TAY
P.O. Box 1136
5602 BC
Eindhoven
The Netherlands

**Tel: +31-40-292 3861**
Fax: +31-40-292 3866

**Figure 3.18** Data sheet of a graded-index multimode optical fiber. *(Courtesy of Plasma Optical Fibre.)*

# Graded - Index Multimode Optical Fibre

**Product Code: 457E**

*Type: 62.5/125 μm*
Double Layer Primary Coating (DLPC7)

*Issue date: 11/95 Supersedes: 1/94*

| *Characteristics* | *Conditions* | *Specified Values* | | | *Unit* |
|---|---|---|---|---|---|
| **OPTICAL CHARACTERISTICS** | | | | | |
| Attenuation Coefficient | 850 nm | ≤ 2.7 | ≤ 2.8 | ≤ 3.0 | [dB/km] |
| | 1300 nm | ≤ 0.5 | ≤ 0.6 | ≤ 0.7 | [dB/km] |
| Minimum Modal Bandwidth [1,2] | 850 nm | | | 160 to > 300 | [MHz.km] |
| | 1300 nm | | | 500 to > 1000 | [MHz.km] |
| Numerical Aperture | | | | 0.275 ± 0.015 | |
| Chromatic Dispersion | | | | FDDI spec | |
| Backscatter Characteristics [3] | | | | | |
| Step (mean of bidirectional measurement) | | | | ≤ 0.1 | [dB] |
| Irregularities over fibre length | | | | ≤ 0.1 | [dB] |
| Reflections | | | | Not Allowed | |
| Group Index of Refraction (Typical) [4] | 850 nm | | | 1.496 | |
| | 1300 nm | | | 1.491 | |
| **GEOMETRIC CHARACTERISTICS** | | | | | |
| Core Diameter | | | | 62.5 ± 2.5 | [μm] |
| Core Non-Circularity | | | | ≤ 6.0 | [%] |
| Core/Cladding Concentricity Error | | | | ≤ 1.5 | [μm] |
| Cladding Diameter | | | | 125 ± 2.0 | [μm] |
| Cladding Non-Circularity | | | | ≤ 1.0 | [%] |
| Coating Diameter | | | | 245 ± 10 | [μm] |
| Coating Non-Circularity | | | | ≤ 6 | [%] |
| Coating Concentricity Error | | | | ≤ 12.5 | [μm] |
| Length | | | | 2200 + > 10 | [m] |
| | | | | 4400 + > 10 | [m] |
| | | | | 6600 + > 10 | [m] |
| | | | | 8800 + > 10 | [m] |
| **ENVIRONMENTAL CHARACTERISTICS** | | | | | |
| Temperature Dependence at 850 nm and 1300 nm Induced Attenuation - 60°C to + 80°C | | | | ≤ 0.1 | [dB/km] |
| Watersoak Dependence at 850 nm and 1300 nm Induced Attenuation at 20°C for 30 days | | | | ≤ 0.2 | [dB/km] |
| Damp Heat Dependence at 850 nm and 1300 nm Induced Attenuation at 85°C, 85% R.H., 30 days | | | | ≤ 0.2 | [dB/km] |
| **MECHANICAL CHARACTERISTICS** | | | | | |
| Proof Test | (off line) | | | ≥ 8.8 | [N] |
| | | | | ≥ 1.0 | [%] |
| | | | | ≥ 100 | [KPSI] |
| Bend Induced Attenuation at 850 nm and 1300 nm 100 turns around a mandrel of 75 mm diameter | | | | ≤ 0.5 | [dB] |
| Dynamic Stress Corrosion Susceptibility Parameter (Typical) | | | | ≥ 27 | |
| Coating Strip Force (Typical) | | | | 1.4 | [N] |

*1. The modal bandwidth is referred to a reference length of 1 km using a linear relationship.*
*2. Dual window bandwidth specifications are selectable; possibilities are:*

| 850 nm | 1300 nm | |
|---|---|---|
| 160 | 500 | MHz.km |
| 200 | 600 | MHz.km |
| 250 | 1000 | MHz.km |
| 300 | 800 | MHz.km |

*3. Measurement at 1300 nm with 0.5 μs pulse width.*
*4. In the case that an OTDR is used with c rounded to 3*10^8m/s, add 0.001 to the indicated value.*

plasma optical fibre
A MEMBER OF THE DRAKA HOLDING GROUP

**Figure 3.18** *(continued)*

***Numerical aperture*** Refer back to pages 47 to 49, where this characteristic is discussed. Compare $NA = 0.275$ given here with the numerical aperture calculated in Example 3.1.4. Note that this characteristic not only specifies the value of *NA*, but also defines the preciseness of this specification ($\pm 0.015$).

***Chromatic dispersion*** The manufacturer refers to the FDDI specification. "FDDI" stands for the Fiber Distributed Data Interface standard. This standard, in particular, specifies that the bandwidth-distance product at the 1300-nm operating wavelength is to be a minimum of 500 MHz · km. (We will consider FDDI and other transmission standards in later chapters.)

Some manufacturers, such as Corning Inc. and SpecTran Specialty Optics Co., specify chromatic dispersion by giving the zero-dispersion wavelength, $\lambda_0$, and the zero-dispersion slope, $S_0$. These numbers allow a user to calculate the chromatic-dispersion parameter, $D(\lambda)$, by means of Formula 3.20. Typical numbers are [2]: $1332 \text{ nm} \le \lambda_0 \le 1354 \text{ nm}$ and $S_0 \le 0.097 \text{ ps/(nm}^2 \cdot \text{km)}$. (Also see Figure 3.20.) A manufacturer usually gives the operating-wavelength ($\lambda$) range within which the formula for the dispersion parameter, $D(\lambda)$, can be applied. Typically, Formula 3.20 can be used for $750 \text{ nm} \le \lambda \le 1450 \text{ nm}$ [2].

***Backscatter characteristics*** These characteristics are measured by an optical time-domain reflectometer, OTDR. We'll discuss this device in Section 8.5. Suffice to say here that the principle of an OTDR's operation entails measuring parameters of light scattered backward toward the source. This is the reason for the names of the following characteristics:

***Step*** This means a difference in the same characteristic measured in two directions along the same optical fiber. For example, if you measure losses along the same span of a fiber in two directions, the difference between your two measurements must not be more than 0.1 dB.

***Irregularities over the fiber length*** Here the manufacturer is referring to the difference in the same characteristic measured along different sections of the same fiber. Corning [2] specifies a similar property, calling it a *point discontinuity,* which is a one-time localized loss within the fiber.

***Group index of refraction,* also called *effective group index of refraction* ($N_{eff}$)** The meaning of this characteristic is best explained as follows: Since the refractive index of this fiber is a variable (because it is a graded-index fiber—see Figure 3.13 [a]), the beams propagating along the core center and along the core-cladding boundary will meet different refractive indexes. What's more, all the beams will change their directions while traveling within the fiber; thus, they'll *see* various refractive indexes. (See Figure 3.13 [b].) In addition, beams of different wavelengths will meet different refractive indexes because $n = n(\lambda)$. (See Section 3.3.) But all these phenomena taking place are considered from the standpoint of looking out from *inside* the fiber. For a user, however, there is only one consideration: launching a beam into the fiber and seeing it emerge. From this *outside* standpoint, a fiber is described by one integrated refractive index—the effective group index of refraction, $N_{eff}$. Note that the values of $N_{eff}$ are quite different for the 850-nm and 1300-nm operating wavelengths.

### "Geometric Characteristics" Section

This section details the preciseness of the core, cladding, and coating geometry. Figure 3.19 will help you to visualize all these characteristics.

***Core diameter, cladding diameter, and coating diameter*** These characteristics give you not only the exact values but also the tolerances allowed by a manufacturer, that is, the deviations from the exact values.

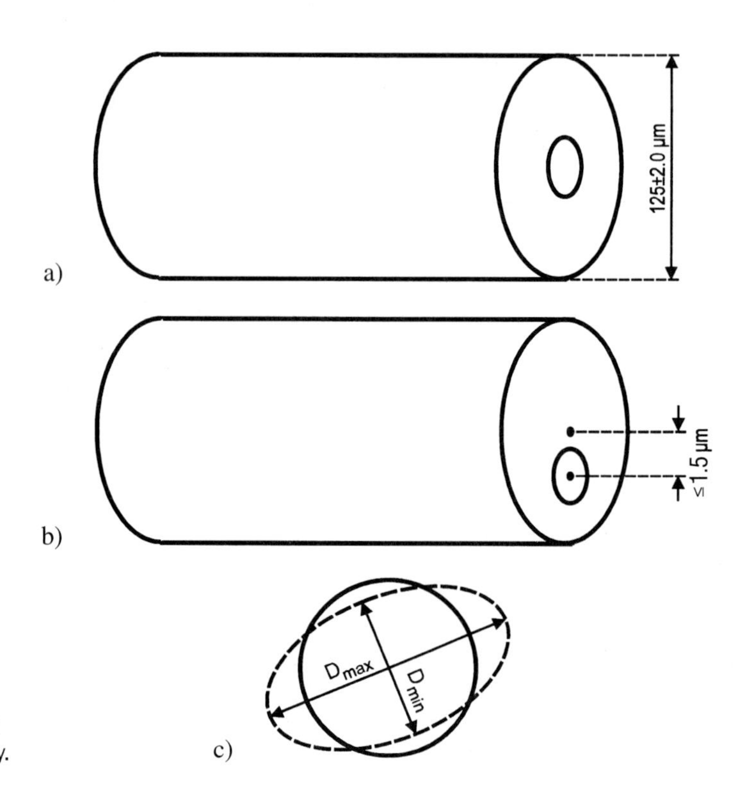


**Figure 3.19** Geometric characteristics of multimode optical fiber: (a) Cladding diameter; (b) core-cladding concentricity; (c) noncircularity.

***Core Noncircularity, Cladding Noncircularity, and Coating Noncircularity*** These describe how much the core's cladding, or coating front cross sections, deviate from a perfect circle. (See Figure 3.19 [c].) The formula for calculating this characteristic is [2]:

$$\text{Noncircularity (\%)} = [1 - (\text{minimum diameter/maximum diameter})] \times 100$$

***Core/Cladding and Coating Concentricity Error*** This is the maximum possible distance between the core and cladding geometric centers. The same is true for the coating-cladding concentricity. (See Figure 3.19 [b].)

All these geometric characteristics are extremely important. Imagine, for instance, that you have bought two reels of fiber and need to connect them. (The permanent connection of fibers is called "splicing," a topic we will consider in Section 8.1.) If geometric parameters—diameters, concentricity, noncircularity—vary from fiber to fiber, the fibers simply will not be compatible. The upshot: huge connection losses and an unacceptably low-quality transmission signal.

***Standard length*** The manufacturer ships its fiber in a certain length per reel. These are the numbers you can see in this subsection of the data sheet.

## "Environmental Specifications" Section

This section shows how much attenuation increases owing to changes in environmental conditions. The test conditions are specified, so you can see how they meet your needs.

***Temperature dependence*** Attenuation of the fiber was measured while the ambient temperature was being changed from – 60° C to + 80° C. The increase did not exceed 0.1 dB/km at either the 850-nm or 1300-nm operating wavelengths. If you consider the attenuation as 0.7 dB/km at 1300 nm, you can readily see that this increase is about 14 %.

***Watersoak dependence*** The same measurement was performed while the fiber was being soaked in water at 20° C for 30 days. The result: The attenuation increase was not more than 0.2 dB/km.

***Damp heat dependence*** Attenuation was measured at 85°C and 85% relative humidity for 30 days. The increase did not exceed 0.2 dB/km.

Note that induced attenuation is given in absolute values, not as a percentage. The attenuation increases under these environmental conditions because such changes affect the values and homogeneity of the refractive indexes and the dimensional characteristics of the core and cladding.

## "Mechanical Specifications" Section

***Proof test*** This subsection tells us that the entire length of the fiber was tested to a tensile stress of more than 100 kilopounds per square inch (kpsi). If you convert this number to International System of units (SI), you will come up with 0.7 giganewtons per square meter (GN/m$^2$). Think about these numbers. Compare the strength of a fiber with the same characteristic of copper wire. The latter has a typical strength of around a hundred pounds (*not* kilopounds) per square inch. In other words, an optical fiber is an extremely strong filament even though it is made from fragile glass. Incidentally, FOTP # 76 covers this test.

***Bend-induced attenuation*** The manufacturer specifies how much fiber attenuation will increase under certain bending conditions. (See Section 3.2, the subsection "Bending Losses.") Some manufacturers give minimal bending radii for short-distance and long-distance applications (see Figure 3.20).

***Coating strip force*** To connectorize and splice the fiber, one has to remove the coating; this is why the coating strip force is given here. (See the subsection "Coating" above and Sections 7.1 and 8.1.)

## Other Characteristics

Certain other characteristics can be found in the optical-fiber data sheets of the various manufacturers. For example, Corning [2] and SpecTran (Figure 3.20) give the refractive-index profile (see Figures 3.13 [a] and 3.20) and the spectral attenuation (see Figures 3.8 and 3.20). Look closely at each specification on the data sheet that crosses your desk and evaluate it in comparison with the numbers supplied by other manufacturers for that specification.

## Conclusion

The best way for you to review the material introduced in this chapter is to study the data sheet (Figure 3.18) carefully. If you feel you understand clearly what each specification means and the significance of the number(s) associated with it, then you have acquired a good grasp of optical fibers, a major first step in your effort to master the larger subject of optical-fiber technology.

An optical-fiber data sheet of still another manufacturer, SpecTran Specialty Optics Co., is given in Figure 3.20. It is strongly recommended that you familiarize yourself with this document as well and then create a memory map of this chapter.

We have summarized in Table 3.1 the typical characteristics of 62.5/125 graded-index multimode fibers. These numbers give you a general idea of what today's multimode fibers look like.

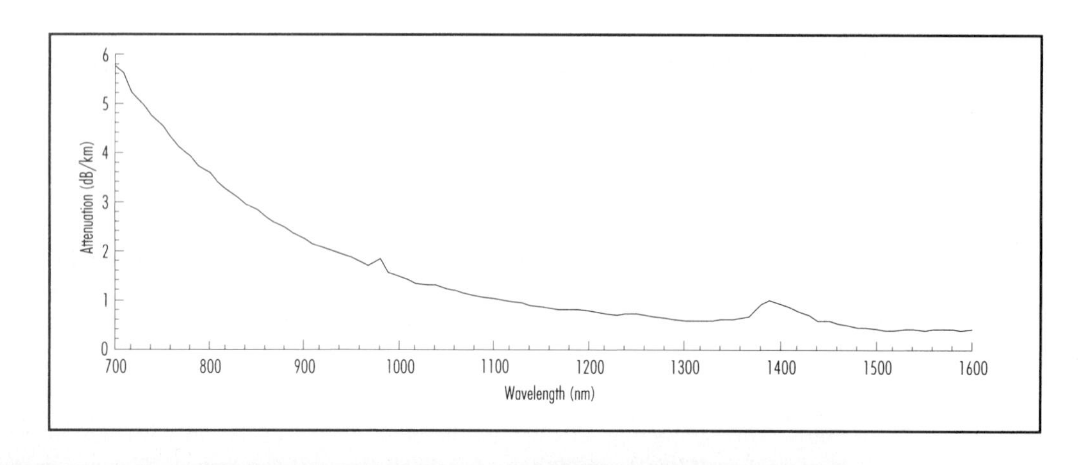


## MULTIMODE GRADED-INDEX FIBER

| **Part Number**<br>**Product Code** | **BF04430-01**<br>**ACU-ME050C** | **BF04430-02**<br>**ACU-ME050D** | **BF04433**<br>**TCU-ME050H** | **BF05179**<br>**TCU-ME050J** | **BF04431-01**<br>**ACU-MD062C** | **BF04431-02**<br>**ACU-MD062D** |
|---|---|---|---|---|---|---|
| Core Diameter (μm) | 50 ± 3 | 50 ± 3 | 50 ± 3 | 50 ± 3 | 62.5 ± 3 | 62.5 ± 3 |
| Cladding Diameter (μm) | 125 ± 2 | 125 ± 2 | 125 ± 2 | 125 ± 2 | 125 ± 2 | 125 ± 2 |
| Coating/Buffer Diameter (μm) | 250 ± 15 | 500 ± 25 | 155 ± 5 | 155 ± 5 | 250 ± 15 | 500 ± 25 |
| Coating Concentricity (%) | ≥ 80 | ≥ 80 | ≥ 80 | ≥ 80 | ≥ 80 | ≥ 80 |
| Numerical Aperture | 0.2 ± 0.015 | 0.2 ± 0.015 | 0.2 ± 0.015 | 0.2 ± 0.015 | 0.275 ± 0.015 | 0.275 ± 0.015 |
| Attenuation @ 850 nm (dB/km) | ≤ 2.7 | ≤ 2.7 | ≤ 3.2 | ≤ 4.0 | ≤ 3.2 | ≤ 3.2 |
| Attenuation @ 1300 nm (dB/km) | ≤ 0.8 | ≤ 0.8 | ≤ 1.2 | ≤ 2.0 | ≤ 0.9 | ≤ 0.9 |
| Proof Test Level (kpsi) | ≥ 100 | ≥ 100 | ≥ 100 | ≥ 100 | ≥ 100 | ≥ 100 |
| Operating Temperature (°C) | -40 to +85 | -40 to +85 | -65 to +300 | -65 to +300 | -40 to +85 | -40 to +85 |
| Coating Type | UV Acrylate | UV Acrylate | PYROCOAT™ | Hermetic/PYROCOAT™ | UV Acrylate | UV Acrylate |
| Bandwidth @850 nm (MHz-km) | ≥ 400 | ≥ 400 | ≥ 400 | ≥ 400 | ≥ 160 | ≥ 160 |
| Bandwidth @1300 nm (MHz-km) | ≥ 400 | ≥ 400 | ≥ 400 | ≥ 400 | ≥ 500 | ≥ 500 |
| Zero Dispersion Wavelength (nm) | 1306.5 ± 9.5 | 1306.5 ± 9.5 | 1306.5 ± 9.5 | 1306.5 ± 9.5 | 1342.5 ± 22.5 | 1342.5 ± 22.5 |
| Zero Dispersion Slope (ps/[nm²-km]) | ≤ 0.101 | ≤ 0.101 | ≤ 0.101 | ≤ 0.101 | ≤ 0.097 | ≤ 0.097 |
| Group Refractive Index @ 850 nm | 1.483 | 1.483 | 1.483 | 1.483 | 1.496 | 1.496 |
| Group Refractive Index @ 1300 nm | 1.479 | 1.479 | 1.479 | 1.479 | 1.491 | 1.491 |
| Short Term Bend Radius (mm) | ≥ 10 | ≥ 10 | ≥ 10 | ≥ 8.5 | ≥ 10 | ≥ 10 |
| Long Term Bend Radius (mm) | ≥ 17 | ≥ 17 | ≥ 17 | ≥ 12 | ≥ 17 | ≥ 17 |
| Core/Clad Offset (μm) | ≤ 3 | ≤ 3 | ≤ 3 | ≤ 3 | ≤ 3 | ≤ 3 |
| Core Noncircularity (%) | ≤ 5 | ≤ 5 | ≤ 5 | ≤ 5 | ≤ 5 | ≤ 5 |
| Clad Noncircularity (%) | ≤ 2 | ≤ 2 | ≤ 2 | ≤ 2 | ≤ 2 | ≤ 2 |

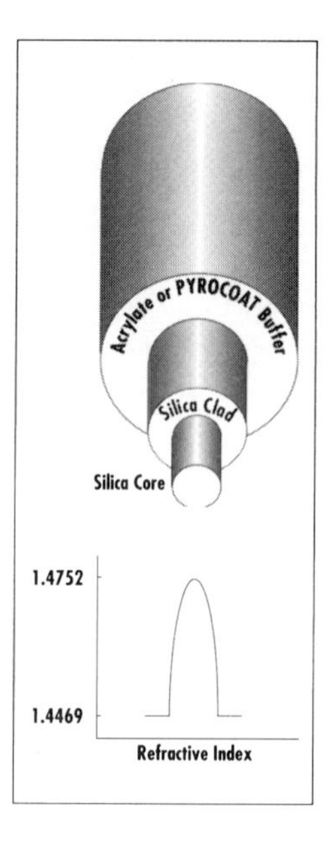


**Figure 3.20** Data sheet of multimode graded-index fiber. *(Courtesy of SpecTran Specialty Optics Co.)*

**Table 3.1** Typical characteristics of multimode fibers

| Fiber type | Maximum attenuation (dB/km) at 850 nm | at 1300 nm | Minimum bandwidth (MHz·km) at 850 nm | at 1300 nm |
|---|---|---|---|---|
| 62.5/125 μm | 2.7–10<br>3.5 typical | 0.6–2<br>1.0 typical | 100–400<br>160 typical | 100–1500<br>500 typical |

Source: *Lightwave 1999 Worldwide Directory of Fiber-Optic Communications Products and Services,* March 31, 1999, pp. 18–22.

## SUMMARY

- An *optical fiber* is a transparent flexible filament designed to conduct a light signal over a significant distance. This is a transmission medium in fiber-optic communications systems.
- Light can be induced to travel within the fiber in a zigzag path and be completely trapped by *total internal reflection.* But two conditions must exist for this phenomenon to occur:
  (1) The fiber must be constructed from a core concentrically surrounded by a cladding. The core's refractive index, $n_1$, must be greater than the cladding's refractive index, $n_2$.
  (2) A light beam must strike the core-cladding interface at an angle no less than a critical incident angle, $\Theta_C$, or—equivalently—the light beam has to form an angle at the fiber centerline that is no greater than the critical propagation angle, $\alpha_C = 90° - \Theta_C$, which is equal to

$$\alpha_C = \sin^{-1}\sqrt{1-\left(\frac{n_2}{n_1}\right)^2} \tag{3.2}$$

- Total internal reflection is the condition necessary for an optical fiber to serve as a transmission link.
- To launch light into a fiber to achieve total internal reflection, it is necessary to direct the outside beam at an angle no greater than the acceptance angle, $\Theta_a$. This condition is usually expressed through numerical aperture (*NA*) so that:

$$NA = \sin\Theta_a = \sqrt{(n_1)^2 - (n_2)^2} \qquad \text{(3.4) and (3.5)}$$

- Optical fibers are made from glass (silica) or—very seldom—plastic. In either case, the material introduces losses caused by light absorption and scattering. These losses are called intrinsic. Another reason for the loss of light—extrinsic losses—is bending of the fiber, which leads to disruption of the condition of total internal reflection. Both macro- and microbending will result in additional loss.
- Loss in light power is described as *attenuation, A,* measured as loss per unit of length in dB/km.

$$A(\text{dB/km}) = \left(-10\log_{10}\left(P_{out}/P_{in}\right)\right)/L, \tag{3.11}$$

where output ($P_{out}$) and input ($P_{in}$) light power are in watts, and fiber length, *L,* is in km.

- Attenuation is a wavelength-dependent entity. The spectral-attenuation curve for silica fibers shows there are wavelength ranges—called transparent windows—at which attenuation reaches minimum values. These windows exist at around 850 nm, 1300 nm, and 1550 nm. There are absorption peaks between these windows. The operating-wavelength range is restricted by these numbers.
- The first solution to the attenuation problem is the use of an appropriate operating wavelength. For multimode fibers, these wavelengths are 850 nm and 1300 nm. The latter is the optimum wavelength for minimizing both attenuation and dispersion problems.
- Light entering an optical fiber breaks down into discrete beams called *modes,* which are stable electromagnetic field patterns existing within the fiber. Each pattern, or mode, can be seen as a separate beam traveling at a certain propagation angle. The mode traveling exactly along the fiber's central axis is called the fundamental, or zero-order, mode. The larger the propagation angle of a mode, the higher its order. The highest-order mode is the mode traveling at the critical propagation angle, $\alpha_C$.
- A fiber that supports only one fundamental mode is called a singlemode fiber. A fiber that can support many modes is called a multimode fiber; multimode fibers can carry hundreds of modes.
- The power of launched light is delivered by separate modes within the fiber. Total output power develops from an accrual of the small quantities of power carried by these separate modes.

- Optical fibers are needed to transmit information signals, mostly in digital form. Bits are represented in the simplest form as light flashes or total darkness (on–off modulation). It appears that light pulses tend to spread while traveling along the fiber. This phenomenon—pulse spreading during transmission along the fiber—is called *dispersion.* If severe, dispersion restricts the optical-fiber bandwidth, since a time interval can accommodate fewer wider pulses.

- Dispersion is caused by several phenomena. In multimode fibers, the dominant mechanism is *intermodal (modal) dispersion.* Since pulse power is delivered by separate modes and these modes travel different distances within the fiber, fractions of the pulse arrive at the receiver end at slightly different times with respect to one another. When these fractions combine, the resulting output pulse will be much wider than the original input pulse. The amount of pulse spreading, $\Delta t_{modal}$, is shown by:

$$\begin{aligned}\Delta t_{modal} &= [(Ln_1)/c][(n_1 - n_2)/n_2] \\ &= [L/(2cn_2)]\,(NA)^2, \end{aligned} \tag{3.16}$$

where $c$ is the speed of light in a vacuum.

- Two solutions to the modal-dispersion problem are possible. The first is to make the core's refractive index variable so that it gradually changes from $n_1$ at the fiber center to $n_2$ at the cladding boundary. This fiber is called graded index and it makes the higher-order modes, which travel the longer distance within the fiber, propagate faster. At the same time, the lower-order modes, which travel closer to the center of the fiber, will propagate more slowly. This approach equalizes the modes' arrival time, reducing pulse spreading.

  The second solution to the modal-dispersion problem is to restrict the number of fiber modes to just one. Such a fiber is called a singlemode fiber and it is inherently free of intermodal dispersion.

- Another important mechanism of pulse spreading, *chromatic dispersion,* is caused by the dependence of the optical properties of fiber materials on wavelength. The term *intramodal,* often associated with this dispersion, underscores the fact that it occurs even within an individual mode. Two phenomena contribute to chromatic dispersion: material dispersion and waveguide dispersion. With regard to the multimode fibers considered in this chapter, waveguide dispersion is negligible. Material dispersion occurs because the fiber's refractive index depends on the light wavelength.

- Pulse spreading caused by chromatic dispersion, $\Delta t_{chrom}$, is described by the following formula:

$$\Delta t_{chrom} = D(\lambda)L\Delta\lambda, \tag{3.19}$$

where $D(\lambda)$ (ps/nm · km) is the chromatic dispersion parameter (a characteristic of the dispersive properties of a fiber material listed in the manufacturers' data sheets), $L$ (km) is the fiber length, and $\Delta\lambda$ (nm) is the spectral width of a light source (LED or laser diode), also found in the data sheets.

- The solution to the chromatic (here, material) dispersion problem is to use a wavelength near which $D(\lambda)$ is, theoretically, zero. Such a wavelength is called the zero-dispersion wavelength. Fortunately, for silica fiber this wavelength is close to 1300 nm, making it possible to minimize the signal distortion caused by both attenuation and chromatic dispersion in multimode fibers.

- The total amount of pulse spreading in multimode fibers, $\Delta t_{total}$, is defined as:

$$\Delta t^2_{total} = \Delta t^2_{modal} + \Delta t^2_{chrom} \tag{3.21}$$

  The bit rate, *BR,* or bandwidth, *BW,* of a multimode fiber is defined as:

$$BW = 1/(4\Delta t_{total}) \tag{3.23}$$

- Attenuation and dispersion put limits on the fiber length that can be used without additional in-line devices like amplifiers or repeaters.

  The material presented in this chapter arms you with the knowledge you'll need on the job to understand the data in the manufacturers' optical-fiber data sheets and to work with optical fibers.

## PROBLEMS

**3.1.** Describe an optical fiber. What is its function in a telecommunications system?

**3.2.** Describe the functions of the core and the cladding in an optical fiber. Why are their refractive indexes different? Which one has to be greater and why?

**3.3.** Example 3.1.1 shows that $n_1$ must be greater than $n_2$ under the critical conditions needed to keep sin $\Theta$ not more than 1. Formally speaking, $n_1$ can be equal to $n_2$ since sin $\Theta$ in this case is not more than 1. Is it possible to realize this equality in an optical fiber?

**3.4.** Why is it necessary to meet the total internal-reflection requirement inside an optical fiber?

**3.5.** What is meant by the term *critical propagation angle?* What fiber parameters does this angle depend on?

**3.6.** What is an acceptance angle? Why do we need to know what this angle is?

**3.7.** The core refractive index is 1.4513 and the cladding index is 1.4468. What is (1) the critical propagation angle? (2) the acceptance angle? (3) the numerical aperture?

**3.8.** The numerical aperture, *NA*, is the characteristic of the optical fiber to gather light from a source. It seems that *NA* should be larger when the core diameter is larger but *NA* doesn't depend on the core diameter at all. (See Formula 3.5.) Why not?

**3.9.** Formula 3.5 gives you another hint about the relationship between $n_1$ and $n_2$. What is this relationship?

**3.10.** Why does neither the acceptance angle nor the propagation angle appear in the data sheets of an optical fiber? Look at a specific data sheet. How do you know that the condition of total internal reflection has been met?

**3.11.** For a specific fiber, $NA = 0.275$ and $n_1 = 1.490$. Find the critical propagation angle.

**3.12.** For a specific fiber, $NA = 0.2375$ and $n_1 = 1.4860$. Find $n_2$ ($n_{cladding}$).

**3.13.** *NA* depends *not* on the core's and cladding's refractive indexes themselves but *on* their difference. In other words, NA might be the same number for different combinations of $n_1$ and $n_2$. (See Example 3.1.4 and Problem 3.7.) Why is this so?

**3.14.** What is attenuation in an optical fiber? What is loss in an optical fiber?

**3.15.** List three major causes of attenuation in an optical fiber and explain their mechanisms.

**3.16.** What is macrobending loss? microbending loss?

**3.17.** A company specifies bending performance for an optical fiber at $\lambda = 1550$ nm as $\leq 0.10$ dB for 100 turns around a 75-mm mandrel and $\leq 0.50$ dB for 1 turn around a 32-mm mandrel. In a few sentences, explain what these specifications mean.

**3.18.** Two optical fibers that differ only in coatings were tested as follows: Fibers were wound on a drum, covered with sandpaper, and a constant winding force was applied. Under these conditions the first fiber showed increased attenuation at 0.8 dB/km and the second at 3.2 dB/km. Why were these numbers different?

**3.19.** Do manufacturers specify scattering loss? Why?

**3.20.** Do manufacturers specify absorption loss? If yes, how?

**3.21.** What does the term *transparent windows* mean? Specify three peak wavelengths for the transparent windows in modern optical fibers.

**3.22.** An optical fiber with attenuation of 0.25 dB/km is used for 20-km transmission. The light power launched into the fiber is 2 mW. What is the output power?

**3.23.** A specification sheet gives fiber attenuation as 0.5 dB/km. Is this a positive or negative number? Explain your answer.

**3.24.** Derive Formula 3.13.

**3.25.** Find the maximum transmission distance for a fiber link with an attenuation of 0.3 dB/km if the power launched in is 3 mW and the receiver sensitivity is 100 μW.

**3.26.** Suppose you need to measure the attenuation of a span of a specific optical fiber. Explain how you are going to do that. Include a drawing of your experimental arrangement and a description of the measuring procedure in your explanation.

**3.27.** What are modes in an optical fiber?

**3.28.** Name the order of modes shown in Figure 3.10.

**3.29.** What is the number of modes for a graded-index fiber if *d* is 50 μm, *NA* is 0.200, and the operating wavelength is 1300 nm?

**3.30.** How many modes can support a step-index optical fiber whose $d = 8.3$ μm, $n_{1core} = 1.4513$, $n_{2clad} = 1.4468$, and $\lambda = 1550$ nm?

**3.31.** The text says that light entering an optical fiber breaks down into a set of separate beams called modes. What physical mechanism causes this effect?

**3.32.** Express the pulse-spreading time caused by modal dispersion in terms of the relative refractive index, Δ.

**3.33.** Consider modal dispersion. For a step-index multimode fiber with $NA = 0.200$ and $n_1 = 1.486$: **a.** Evaluate pulse spreading per 1 km length: **b.** Calculate the maximum number of bits per second that can be transmitted over 1 km.

**3.34** Calculate the pulse spreading due to modal dispersion and the maximum number of bits per second that can be transmitted over 1 km with a graded-index fiber if $NA = 0.200$ and $n_1 = 1.486$.

**3.35.** Calculate the pulse spreading due to modal dispersion and the maximum number of bits per second that can be transmitted over 1 km of a step-index fiber whose parameters are given in Problem 30 ($n_1 = 1.4513$, $n_2 = 1.4468$).

**3.36.** A graded-index multimode fiber has the maximum relative index, $\Delta = 1.0\%$. Calculate the cladding refractive index if the core's maximum index is 1.486. Draw the relative-index profile.

**3.37.** A step-index singlemode fiber has a core refractive index of 1.467. Calculate the cladding refractive index assuming the relative index as $\Delta = 0.3\%$. Depict the relative index profile, keeping the scale.

**3.38.** Suppose you are a manager responsible for choosing optical fibers for your company. You have three types of networks: short distance, intermediate distance, and long distance. What types of optical fiber will you choose for each of your networks and why?

**3.39.** What are the solutions to the modal-dispersion problem?

**3.40.** What is chromatic dispersion?

**3.41.** Why do such different phenomena as intermodal dispersion and chromatic dispersion carry the common name *dispersion?*

**3.42.** Calculate pulse spreading caused by chromatic dispersion for a graded-index fiber if the operating wavelength is 850 nm, the dispersion parameter is that given in Figure 3.20, the distance is 5 km, and the source's spectral width is 70 nm.

**3.43.** Are there any means to decrease or eliminate chromatic dispersion in an optical fiber? If so, discuss them.

**3.44.** Calculate the total dispersion of an optical fiber whose data sheet is given in Figure 3.20.

**3.45.** What are *bandwidth* and *bit rate* in a communications system?

**3.46.** How do manufacturers specify the bandwidth of an optical fiber?

**3.47.** A graded-index fiber has $n_1 = 1.486$ and $NA = 0.200$. What is the bit rate for a 1-km link?

**3.48.** Calculate the product of the bit rate and length for a graded-index fiber if $NA = 0.200$ and $n_1 = 1.486$, taking into account only modal dispersion.

**3.49.** Why do manufacturers specify bending loss in dB but not in dB/km? (See Figures 3.18 and 3.20.)

**3.50.** Project 1. Analyze the data shown in the specifications sheet in Figure 3.20. You may not be familiar with every term and number used; however, analyze this data sheet to the best of your ability. (Be creative. Since this is a project, refer to the index to find sections of this book to help you.)

**3.51.** Project 2. Build a memory map (see Problem 1.20). Leave space to upgrade this map after you read Chapter 4.

## REFERENCES[1]

1. Joseph C. Palais, *Fiber Optic Communications,* 4th ed., Englewood Cliffs, N.J.: Prentice Hall, 1998.
2. *Corning 62.5/125 CPC6 Multimode Optical Fiber* (data sheet), Corning Incorporated, Corning, N.Y., 1996.

[1]See Appendix C: A Selected Bibliography

# 4

# Optical Fibers—A Deeper Look

The basics of two major phenomena in optical fibers—attenuation and dispersion—have been discussed in Chapter 3. But optical fibers are specific types of waveguides, and light propagation within optical fibers is covered by the more general theory of electromagnetic field propagation within guided structures. Thus, developing the light-propagation theory in optical fibers from a general point of view helps one to better understand the specific topic. On the other hand, the goal of teaching any theory is to provide the student with the opportunity to strengthen his or her problem-solving techniques, that is, to perfect approaches that require presenting reasonable explanations and predictions based on sound evidence and conclusions. This requires us to develop very specific theories that will result in formulas you can readily apply in your own everyday work experiences. In this section we'll summarize the theoretical and practical approaches.

## 4.1 MAXWELL'S EQUATIONS[1]

It is impossible to overestimate the importance of Maxwell's equations to modern physics and technology. They describe all the phenomena associated with the electromagnetic (EM) field. In this section, we'll concentrate on one aspect of the many applications of Maxwell's equations: a formal description of the propagation of EM waves. It will provide us with the key to the theoretical explanations of the basic phenomena we studied in Chapter 3 and will facilitate our understanding of the optical-fiber field.

### Set of Maxwell's Equations

In 1873 James Clerk Maxwell, a Scottish-born professor of experimental physics at Cambridge, published *A Treatise on Electricity and Magnetism,* a work that made him the leading theoretical physicist of the century. It brought him not just fame; it linked his name forever with our understanding of nature. His work established a solid foundation for Albert Einstein's theory of relativity—and for the entire era of modern physics, for that matter.

[1]This section can be skipped by the reader who feels comfortable with his or her background in Maxwell's equations.

As with any fundamental work, Maxwell's theory gave theoretical explanations of well-established experimental facts on electromagnetics and provided, on the other hand, a tool that would predict future phenomena. While experiments performed by Michael Faraday and others gave insight into the basics of electromagnetics, the theory developed by Maxwell combined and applied these basics and—what is especially important to us—paved the way to modern communications technology.

Maxwell's theory is based on a set of four equations, known as Maxwell's equations. This set, in differential form (also known as a point form), is:

$$\nabla \cdot \mathbf{D} = \rho_v$$
$$\nabla \cdot \mathbf{B} = 0$$
$$\nabla \times \mathbf{E} = -\frac{\partial \mathbf{B}}{\partial t} \tag{4.1}$$
$$\nabla \times \mathbf{H} = \mathbf{J} + \frac{\partial \mathbf{D}}{\partial t},$$

where the meanings of the terms are as follows.

- Boldface denotes the vectors; regular font denotes the scalars.
- $\nabla$ is the nabla operator, defined as: $\nabla = \mathbf{e}_x\, \partial/\partial x + \mathbf{e}_y\, \partial/\partial y + \mathbf{e}_z\, \partial/\partial z$, with $\mathbf{e}_x$, $\mathbf{e}_y$, and $\mathbf{e}_z$ being the unit vectors of the *x, y,* and *z* axes, respectively, of the Cartesian coordinate system.
- $\partial/\partial t$ is the symbol of a partial derivative with respect to time.
- $\rho_v$ is the volume-charge density ($C/m^3$, where *C* stand for coulombs).
- **E** (V/m) is the electric-field intensity, defined as the force, **F,** produced by the electric field per unit charge, *q,* so that $\mathbf{F} = q\,\mathbf{E}$. On the other hand, moving a unit charge in a static electric field with intensity **E** from point 1 to point 2 causes a change in the electric potential so that $V_2 - V_1 = -\int_1^2 \mathbf{E}\, d\mathbf{l}$ (volts, V,) and here **l** is the unit vector of length. The same relationship given in differential form states that the electric-field intensity, **E,** is the gradient of the potential, *V,* that is, $\mathbf{E} = -\nabla V$. One can find the value of **E** as $E = V/l$ (V/m). **E** is also often called the strength of the electric field.
- **D** is the electric-flux density or electric displacement ($C/m^2$). The electric-flux density, **D,** is related to the electric-field intensity, **E,** through the permitivity, $\varepsilon$, of the medium, where the density vector, **D,** has been developed after the electric field, **E,** has been applied. The formula is $\mathbf{D} = \varepsilon\,\mathbf{E}$. The permitivity of the vacuum, $\varepsilon_0$, one of the universal constants, is equal to $8.854 \times 10^{-12}$ *F/m,* where *F* stands for farad.
- **J** is the current density ($A/m^2$, where *A* stands for ampere); $\mathbf{J} = \sigma\,\mathbf{E}$, where $\sigma$ is the conductivity of the medium with the ampere units divided by volt times meter, A/V · m.
- $\mathbf{J}_d = \partial \mathbf{D}/\partial t$ is the displacement current density ($A/m^2$).
- **H** is the magnetic-field intensity (*A/m*). Probably the best way to introduce **H** is to refer to Ampere's circuital law, $\oint_L \mathbf{H}\, d\mathbf{l} = I(A)$, where *L* is the contour (closed path) bounding the surface, *S,* and *I* is the total free current passing through *S.* One can find the value of **H** from the simple formula $H = I/l$ (*A/m*).
- **B** is the magnetic-flux density (*H/m,* where *H* stands for henry). The magnetic-flux density, **B,** is related to the magnetic-field intensity, **H,** through the permeability of the medium, $\mu$: $\mathbf{B} = \mu\,\mathbf{H}$. This formula is similar to the relationship between the electric-field parameters

**E** and **D** given above. The permeability of the vacuum, $\mu_0$, is another universal constant; its value is $4\pi \times 10^{-7}$ *H/m.* These two universal constants are related to the most fundamental constant, *c,* as: $1/\sqrt{\varepsilon_0 \mu_0} = c$, where *c,* the speed of light in a vacuum, equals $3 \times 10^8$ *m/s.*

It is not our purpose here to engage in a treatise on the vector-analysis problems associated with Maxwell's equations. Those interested in pursuing this course of study should check out the list of references at the end of this chapter. (See references [1] through [6].)

## Interpretation of Maxwell's Equations

Let's discuss the physical meaning of Maxwell's equations to give you some insight into the kinds of operations we are going to perform with them:

- The first equation, $\nabla \cdot \mathbf{D} = \rho_v$, represents Gauss's law. Let's take the volume integral from both sides of the equation:

$$\int_v \nabla \cdot \mathbf{D}\, dv = \int_v \rho_v\, dv$$

The right-hand side is simply the charge, *Q,* bounded by the volume v:

$$\int_v \rho_v\, dv = Q$$

The left-hand side can be transformed by using divergence theorem:

$$\int_v \nabla \cdot \mathbf{D}\, dv = \oint_S \mathbf{D}\, d\mathbf{S}$$

Thus, we get Maxwell's first equation in integral form:

$$\oint_S \mathbf{D}\, d\mathbf{S} = Q \tag{4.2}$$

Both forms—differential and integral—express the same idea: *The flow source of an electric field is a free charge,* Q.

Gauss's law, you'll recall, states that the total electric flux through any closed surface, *S,* is equal to the total charge, *Q,* enclosed by this surface (that is, the charge, *Q,* found in the volume, v, bounded by the surface, *S*). This is exactly what the integral form says, since the integral over the closed surface, *S,* of the electric-flux density, **D** (that is, $\oint_S \mathbf{D}\, d\mathbf{S}$), equals the total electric flux through the surface, *S*. The differential form, $\nabla \cdot \mathbf{D} = \rho_v$, says that the divergence of the electric-flux density, **D,** is equal to the volume-charge density, $\rho_v$.

- Maxwell's second equation, $\nabla \cdot \mathbf{B} = 0$, states that there is no flow source for a magnetic field. If we compare this statement with the above discussion of an electric field, we can say that there is no free magnetic charge in nature or that a magnetic field has no source or sink. One can easily derive the integral form of this equation by following the same technique we used above for an electric field. The integral form of this equation is:

$$\oint_S \mathbf{B}\, d\mathbf{S} = 0 \tag{4.3}$$

- The third equation, $\nabla \times \mathbf{E} = -\partial \mathbf{B}/\partial t$, represents Faraday's law, or the law of electromagnetic induction. Let's take the surface integral from both parts of the equation:

$$\int_s \nabla \times \mathbf{E}\, d\mathbf{S} = \int_s (-\partial \mathbf{B}/\partial t)\, d\mathbf{S}$$

The left-hand side can be rewritten by using Stokes' theorem in this form:

$$\int_S \nabla \times \mathbf{E}\, d\mathbf{S} = \oint_L \mathbf{E}\, d\mathbf{l},$$

where $d\mathbf{l}$ is the vector differential length.
The right-hand side can be rewritten in this form:

$$\int_S (-\partial \mathbf{B}/\partial t)\, d\mathbf{S} = -d/dt \int_S \mathbf{B}\, d\mathbf{S},$$

where the partial derivative is replaced by the derivative, since the integration of magnetic-flux density over the surface eliminates space dependence. By definition, integral $\int_S \mathbf{B}\, d\mathbf{S}$ is the magnetic flux, $\Psi$. Thus, we arrive at the integral form of Maxwell's third equation:

$$\oint_L \mathbf{E}\, d\mathbf{l} = -d\Psi/dt \tag{4.4}$$

As usual, both forms of this equation represent the same idea: *The vortex source of an electric field is a time-varying magnetic field.* The differential form, $\nabla \times \mathbf{E} = -\partial \mathbf{B}/\partial t$, says that the curl **E** is equal to the time derivative of the magnetic-flux density. The integral form says that a time-changing magnetic flux crossing a conductive closed loop induces electromotive force in this loop. Recalling that integral $\oint_L \mathbf{E}\, d\mathbf{l}$ is units in volts, we arrive at an understanding of why this relationship is called the *law of induction.*

- The fourth equation, $\nabla \times \mathbf{H} = \mathbf{J} + \partial \mathbf{D}/\partial t$, shows where a time-varying magnetic field comes from: *The vortex source of a time-varying magnetic field is the time-changing density of conduction and displacement currents.* As you see, there are two sources here: $\mathbf{J}(t)$, which is the conduction-current density $(A/m^2)$, and $\mathbf{J}_d = \partial \mathbf{D}/\partial t$, which is the displacement-current density $(A/m^2)$. Let's derive the integral form for a better understanding of the meaning of this equation. Taking the surface integral from both sides of the equation,

$$\int_S \nabla \times \mathbf{H}\, d\mathbf{S} = \int_S (\mathbf{J} + \partial \mathbf{D}/\partial t)\, d\mathbf{S},$$

we get the left-hand side in this form: $\int_S \nabla \times \mathbf{H}\, d\mathbf{S} = \oint_L \mathbf{H}\, d\mathbf{l}$. The closed path, $L$, bounds the surface, $S$, through which total current, $I$ (see below), passes. The integral $\oint_L \mathbf{H}\, d\mathbf{l}$ describes the circulation of the magnetic-field intensity, **H**, around the closed loop, $L$.

The right-hand side of the equation consists of the sum of two integrals. The first, $\int_S \mathbf{J}\, d\mathbf{S}$, is the integral of the current density over the surface, which by definition is the current, $I$ $(A)$. Thus, the first source of a time-changing magnetic field is the conduction current. This relationship, $\oint_L \mathbf{H}\, d\mathbf{l} = \boldsymbol{I}$, is known as Ampere's circuital law, after A.M. Ampere, the French scientist who made some prominent experiments and calculations establishing the foundation for electromagnetics at the beginning of the nineteenth century.

The second integral, on the right-hand side, $\int_S \partial \mathbf{D}/\partial t\, d\mathbf{S}$, gives the displacement current, $I_d$ $(A)$. This is the quantity responsible for alternating-current (ac) flow through a capacitor that provides a-c continuity over the closed electrical circuit that includes the different components. It was Maxwell who introduced this term. As we see from his fourth equation, this term predicts the existence of an electromagnetic field even without the presence of any conductor. In other words, Maxwell's equations predict that an electromagnetic field can exist in free space. This concept was so revolutionary in Maxwell's time that the scientific community rejected it. It wasn't accepted until Heinrich Hertz, the German physicist, experimentally proved the existence of electromagnetic waves some fifteen years after Maxwell's *Treatise* appeared. What's more, it took another ten years before Guglielmo Marconi, the Italian-born scientist and engineer who worked mostly in England, built the first radio, based, of course, on Maxwell's and Hertz's work. In fact, the first words Marconi transmitted using electromagnetic waves were "Heinrich Hertz." What better proof of EM field existence!

Putting all the integrations together, one can easily obtain Maxwell's fourth equation in integral form:

$$\oint_L \mathbf{H}\, d\mathbf{l}\ (A) = I + I_d \tag{4.5}$$

This equation is interpreted as follows: The circulation of the magnetic-field intensity, **H,** around the closed loop, *L,* is equal to the sum of the conduction current and the displacement current.

There you have it—a review of the brilliant work that laid the foundation for electromagnetics as we know it today. A word of advice: It might be a good idea for you to go over this section again, especially if it's been a long time since you studied basic electricity and magnetism.

## Wave Equations

If you read the above discussion carefully, you might get the impression that Maxwell's equations describe the sources of an electromagnetic field. True enough, but Maxwell's equations contain much more information than that. They enable us to derive wave equations describing the propagation of EM waves.

***Plane TEM waves*** To simplify such derivations, we will consider plane transverse electromagnetic (TEM) waves traveling in free space. Figure 4.1 displays these waves. The term *plane* means the waves are polarized in one plane. In Figure 4.1, electric field **E** is polarized in an $x$-$z$ plane so that vector **E** changes its magnitude but doesn't change its orientation: It never leaves the $x$-$z$ plane. The magnetic component of the EM field, **H,** is always in a $y$-$z$ plane. You can also say that **E** is $x$-polarized and **H** has $y$ polarization.

The term *transverse* means that vectors **E** and **H** are perpendicular to the direction of propagation of the $z$ axis in Figure 4.1. Another assumption here is that these plane TEM waves propagate in free space. Free space, or vacuum, means that the permitivity of the medium is $\varepsilon_0 = 8.854 \times 10^{-12}$ *F/m,* the permeability of the medium is $\mu_0 = 4\pi \times 10^{-7}$ *H/m,* and $\mathbf{J} = 0$ since there is no conduction medium in free space. In other words, $\sigma = 0$. This is why we are talking about the propagation of electromagnetic waves but not about current flow and voltage drop.

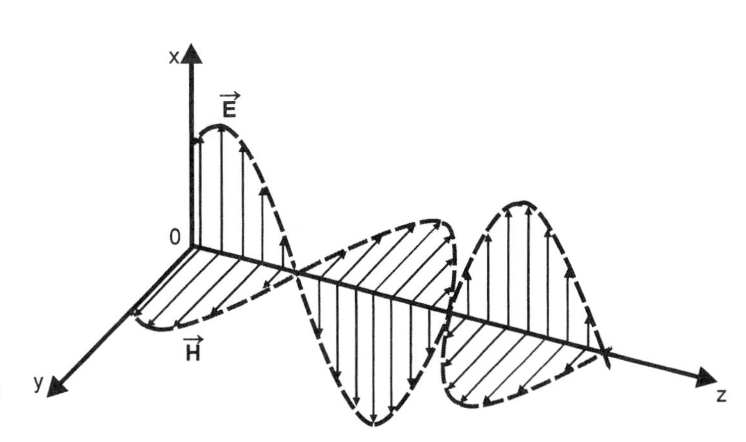


**Figure 4.1** Plane transverse electromagnetic waves.

The above discussion allows us to write formulas for plane TEM waves in the following form:

$$\left.\begin{aligned}\mathbf{E} &= \mathbf{e}_x\, E_x\,(z,t)\\ \mathbf{H} &= \mathbf{e}_y\, H_y\,(z,t)\end{aligned}\right\} \quad (4.6)$$

These formulas show that EM waves change with respect to space (here, axis $z$) and time simultaneously. (See Chapter 2 and Figure 2.2.)

***Wave equations for plane TEM waves*** To make it easier for you to follow the derivation of the wave equation, let's break down the entire derivation into its sequence of steps:

Step 1. Consider Maxwell's fourth equation, known as Ampere's law:

$$\nabla \times \mathbf{H} = \mathbf{J} + \partial\mathbf{D}/\partial t \quad (4.7)$$

In our case it now becomes

$$\nabla \times \mathbf{H} = \partial\mathbf{D}/\partial t$$

since $\mathbf{J} = 0$ for free space. Recalling that $\mathbf{D} = \varepsilon_0\,\mathbf{E}$, we arrive at the following form of Ampere's law:

$$\nabla \times \mathbf{H} = \varepsilon_0 \partial\mathbf{E}/\partial t$$

Calculating cross-product $\nabla \times \mathbf{H}$ for our plane TM wave results in $\nabla \times \mathbf{H} = -\mathbf{e}_x\, \partial H_y/\partial z$. The right-hand side of the equation, after substituting $\mathbf{E} = \mathbf{e}_x\, E_x\,(z,\,t)$, becomes $\mathbf{e}_x\, \varepsilon_0\, \partial E_x\,(z,\,t)/\partial t$. Thus, Ampere's law, in our case, takes the following scalar form:

$$\partial H_y(z,\,t)/\partial y = -\varepsilon_0 \partial E_x(z,\,t)/\partial t \quad (4.7a)$$

Step 2. Consider Maxwell's third equation, known as Faraday's law:

$$\nabla \times \mathbf{E} = -\partial\mathbf{B}/\partial t \quad (4.8)$$

Let's follow a procedure similar to that in Step 1. Calculating cross-product $\nabla \times \mathbf{E}$ for our case results in: $\nabla \times \mathbf{E} = -\mathbf{e}_y\, \partial E_x/\partial z$. After substituting $\mathbf{B} = \mu_0\,\mathbf{H}$ and $\mathbf{H} = \mathbf{e}_y\, H_y\,(z,\,t)$, the right-hand side becomes $\mathbf{e}_y\, \mu_0\, \partial H_y(z,\,t)/\partial t$. Thus, Faraday's law, in our example, takes the following scalar form:

$$\partial E_x(z,\,t)/\partial z = -\mu_0\, \partial H_y(z,\,t)/\partial t \quad (4.8a)$$

Step 3. Let's take the partial derivative $\partial/\partial t$ from Equation 4.7a:

$$\partial/\partial t[\partial H_y\,(z,\,t)/\partial z] = -\varepsilon_0\, \partial^2 E_x(z,\,t)/\partial t^2$$

Since $z$ and $t$ are independent variables, we can change the order of operation in the left-hand-side expression: $\partial/\partial z[\partial H_y(z,\,t)/\partial t]$. Again, refer to Formula 4.8a. It's clear that

$$\partial H_y\,(z,\,t)/\partial t = -\,(1/\mu_0)\,\partial E_x\,(z,\,t)/\partial z$$

Substituting this equation for Formula 4.8a, we arrive at the following wave equation:

$$\partial^2 E_x\,(z,\,t)/\partial t^2 - 1/(\varepsilon_0\mu_0)\, \partial^2 E_x\,(z,\,t)/\partial z^2 = 0 \quad (4.9)$$

This is the wave equation we are looking for. By a similar procedure, the wave equation for a magnetic component can be found in this form:

$$\partial^2 H_y(z, t)/\partial t^2 - 1/(\varepsilon_0 \mu_0)\, \partial^2 H_y(z, t)/\partial z^2 = 0 \qquad (4.10)$$

Now it's time to find the solutions for these equations.

## Solving Wave Equations

Wave Equations 4.9 and 4.10 are called second-order partial differential equations. Their general solutions can be obtained based on boundary conditions. Finding the solutions for these types of differential equations is always a difficult and time-consuming task. We can simplify matters if we consider not the general EM field but the plane harmonic TEM waves shown in Figure 4.1. The general expressions for an EM field are:

$$\mathbf{E} = \mathbf{E}\,(\mathbf{r}, t) \text{ and } \mathbf{H} = \mathbf{H}\,(\mathbf{r}, t),$$

where, again, the boldface type shows vectors **E, H,** and **r** and the italic typeface shows scalar *t;* **r** represents space variation regardless of whether Cartesian or cylindrical coordinate systems are used.

In the case of plane harmonic TEM waves propagating in free space, we can write the expressions for electric and magnetic components in the following form:

$$\begin{aligned} \mathbf{E} &= \mathbf{e}_x\, E_{x0} \cos k\,(z + vt) \\ \mathbf{H} &= \mathbf{e}_y\, H_{y0} \cos k\,(z + vt), \end{aligned} \qquad (4.11)$$

where $E_{x0}$ (*V/m*) and $H_{y0}$ (*A/m*) are amplitudes of electric and magnetic waves, respectively; $k = 2\pi/\lambda$ is the wave number, also called the propagation number (1/*m*); and $v$ is the velocity of the EM waves' propagation (*m/s*).

So let's see whether these expressions provide solutions for Equations 4.9 and 4.10. Let's substitute $E_x = E_{x0} \cos k(z + vt)$ into Equation 4.9 and $H_y = H_{y0} \cos k(z + vt)$ into Equation 4.10. After taking derivatives, we get the following formula for Equation 4.9:

$$-E_{x0} k^2 v^2 \cos\, k\,(z + vt) + l/(\varepsilon_0 \mu_0)\, E_{x0} k^2 \cos\, k\,(z + vt) = 0$$

We obtain a similar expression for Equation 4.10. The result is:

$$-v^2 + 1/(\varepsilon_0\, \mu_0) = 0$$

Therefore, the expressions in Formula 4.11 are the solutions for Equations 4.9 and 4.10 provided that $v^2 = l/(\varepsilon_0\, \mu_0)$. But since $v$ is the velocity of light and since we consider free space a medium of propagation, then $v = c = 3 \times 10^8$ *m/s*—the speed of light in a vacuum. All this brings us at last to one of the fundamental relationships in electromagnetics:

$$c(m/s) = 1/\sqrt{(\varepsilon_0 \mu_0)} \qquad (4.12)$$

Since $\varepsilon_0$ and $\mu_0$ are constants, $c$ is also constant. In other words, EM waves propagate in a vacuum at speed $c$, which is a universal constant. This point—that $c$ is always constant and the highest velocity of any signal—put Albert Einstein on the road to the theory of relativity.

Let's rewrite the expressions in Equation 4.11 in a slightly different form:

$$E_x = E_{x0} \cos(\beta z + \omega t)$$
$$H_y = H_{y0} \cos(\beta z + \omega t), \tag{4.13}$$

where $\beta = 2\pi/\lambda$ is the phase constant, also known as the propagation constant, and $\omega = kv = (2\pi/\lambda)\, c$. You'll recall that since $\lambda f = c$, we get $\omega = 2\pi f$ as the angular frequency (rad/s); therefore, if $c = 1/\sqrt{(\varepsilon_0 \mu_0)}$, which is always true, the expressions in Formula 4.13 are the solutions to Equations 4.9 and 4.10. We can rewrite these solutions in even more explicit form:

$$E_x = E_{x0} \cos((2\pi/\lambda)z + 2\pi ft)$$
$$H_y = H_{y0} \cos((2\pi/\lambda)z + 2\pi ft) \tag{4.14}$$

To summarize: Maxwell's equations describe all the properties of an electromagnetic field. They explicitly show sources of the field and implicitly tell how an electromagnetic field propagates. Even though we have proved the last point only for the simplest case—transverse time-harmonic electromagnetic waves in free space—this statement is true in general. We're now ready to consider EM field propagation in a material.

## 4.2 PROPAGATION OF EM WAVES[2]

We are interested at this juncture in describing, theoretically, how light propagates inside an optical fiber. The key to this description is to apply Maxwell's equations to the conditions that are specific to an optical fiber. As simple as doing this may sound, implementing this idea is quite difficult in practice. We will start with a general consideration of the propagation of EM waves within unbounded lossy dielectric. Then we'll consider light (EM waves, you'll recall) propagation within a dielectric guide, which is a more realistic model of an optical fiber. The ultimate goal is to achieve some formalism to describe the basic phenomena of an optical fiber—attenuation and dispersion, both of which are discussed in Chapter 3.

### Wave Equations for a Time-Harmonic EM Field

Wave Equations 4.9 and 4.10 were derived for the general EM field. We now know that harmonic EM waves are the solutions for these equations. This means that we can make use of the time-harmonic EM field to significantly simplify these equations. The time-harmonic electric and magnetic fields can be written in phasor form as follows:

$$\mathbf{E} = \mathbf{E}(\mathbf{r})\, e^{j\omega t}$$
$$\mathbf{H} = \mathbf{H}(\mathbf{r})\, e^{j\omega t}, \tag{4.15}$$

where $j$ is an imaginary unit and $\omega$ is the angular frequency (rad/s). This form allows us to separate the space dependence and time dependence of electric and magnetic fields. For our plane TEM waves depicted in Figure 4.1, Formula 4.15 takes an even simpler form:

$$\mathbf{E} = \mathbf{e}_x E_x(z)\, e^{j\omega t}$$
$$\mathbf{H} = \mathbf{e}_y H_y(z)\, e^{j\omega t} \tag{4.15a}$$

[2]This section also provides the reader with some general theory, making this book self-sufficient.

If you consider the real part of the expressions in the above formula, you will get a cosine reference for the time dependence of the EM field. (See Formulas 4.11 and 4.13.)

Maxwell's equations will take the simpler form for a time-harmonic field. You can easily derive this form by taking the expressions in Formula 4.15 and inserting them into the set of equations given in Formula 4.1.

By seeking solutions to the appropriate wave equations, our goal here is to learn how EM waves propagate in a medium. To demonstrate this technique, we will continue to work with our simple example—plane TEM waves—by inserting the expressions from Formula 4.15a into Equations 4.9 and 4.10. This results in:

$$-\omega^2 E_x(z) - 1/(\varepsilon_0 \mu_0)\, \partial^2 E_x(z)/\partial z^2 = 0$$

Recalling that $1/(\varepsilon_0 \mu_0) = v^2$, where $v$ is the velocity of EM waves inside the medium of propagation, we get the wave equation for our example:

$$d^2E_x(z)/dz^2 + (\omega/v)^2 E_x(z) = 0 \tag{4.16}$$

Similarly, for the magnetic field, we get:

$$d^2H_y(z)/dz^2 + (\omega/v)^2 H_y(z) = 0 \tag{4.16a}$$

It is evident that partial derivatives are replaced by ordinary derivatives since $E_x$ is now the function of only the $z$ variable. If we now substitute $\omega = 2\pi f$ and $v = \lambda f$, the result is $(\omega/v)^2 = (\beta)^2$, where $\beta = 2\pi/\lambda$, so that Formulas 4.16 and 4.16a take the form:

$$d^2E_x(z)/dz^2 + (\beta)^2 E_x(z) = 0 \tag{4.16b}$$

$$d^2H_y(z)/dz^2 + (\beta)^2 H_y(z) = 0 \tag{4.16c}$$

The solution for the above second-order ordinary differential equation is well known:

$$E_x(z) = E_{x0} \cos(\beta z + \theta),$$

where the amplitude, $E_{x0}$, and the initial phase, $\theta$, are determined by the initial conditions. Obviously, one can easily receive a similar result for the magnetic component of an EM field. Returning to Formulas 4.15 and 4.15a, we can now write the result as:

$$\mathbf{E} = \mathbf{e}_x E_{x0} \cos(\beta z + \theta) e^{j\omega t}$$

$$\mathbf{H} = \mathbf{e}_y H_{y0} \cos(\beta z + \phi) e^{j\omega t} \tag{4.17}$$

Consequently, we have obtained the formal solution for the wave equations that, in turn, had been obtained from Maxwell's equations. These solutions show that an EM field exists in free space in the form of EM waves harmonically dependent on space and time simultaneously. Armed with this information, it would now be a good idea for you to review Section 2.1,"Electromagnetic Waves," in Chapter 2.

### EM Waves: Propagation in a Lossy Medium

Let's apply the technique shown above to a more general case: wave propagation in a lossy medium. We'll repeat the basic steps followed in the previous subsection for time-harmonic EM waves but with other assumptions: $\mathbf{J} = \sigma\,\mathbf{E}$, $\mathbf{D} = \varepsilon\,\mathbf{E}$, and $\mathbf{B} = \mu\,\mathbf{H}$. Here, $\sigma$, $\varepsilon$, and $\mu$ are conductivity, permittivity, and permeability of the medium, respectively. To simplify the mathematical operations, we'll continue to explore the plane TEM waves model shown in Figure 4.1. Maxwell's fourth equation,

$$\nabla \times \mathbf{H} = \mathbf{J} + \partial\mathbf{D}/\partial t,$$

**Solving Wave Equations: Mathematics to the Rescue**

On the one hand, solutions to the wave equations show that an EM field exists in free space in the form of TEM waves; on the other hand, however, our initial assumption stated exactly the same thing. Puzzling? Not really. When we seek to solve an equation, we start with an assumption about the form its possible solution will take. If we are right and our assumed solution is correct, the search for the solution is over since one of the great mathematical theorems has proved the existence and uniqueness of the solution to a differential equation. This is exactly what happened here: We initially assumed that an EM field takes the form of plane TEM waves. It appears we are right; hence, we do not need to look further for a solution since no other solution for this differential equation exists.

after substituting the time-harmonic form given in Formula 4.15a, becomes

$$-dH_y(z)/dz = (\sigma + j\omega\varepsilon)\, E_x(z)$$

Maxwell's third equation,

$$\nabla \times \mathbf{E} = -\mu\, \partial\mathbf{H}/\partial t$$

becomes

$$dE_x(z)/dz = -j\,\mu\omega H_y(z)$$

Taking derivative $d/dz$ of the last equation and incorporating $dH_y(z)/dz = -1/(j\mu\omega)\, d^2E_x(z)/dz^2$ into the previous equation, one obtains this wave equation:

$$d^2E_x(z)/dz^2 - \gamma^2E_x(z) = 0, \tag{4.18}$$

where $\gamma = j\mu\omega\,(\sigma + j\omega\varepsilon)$ is called the *propagation constant.* It is evident that $\gamma$ is the complex quantity; hence, one can write:

$$\gamma = \alpha + j\beta, \tag{4.19}$$

where the real part, $\alpha$, is called the *attenuation constant* and the imaginary part, $\beta$, is called the *phase (propagation) constant* (we'll see why shortly).

Can you suggest the form the solution to Equation 4.18 will take? It is well known that this form is:

$$E_x(z) = E_{x0}e^{-\gamma z} \tag{4.20}$$

One can easily verify that this is the solution for Equation 4.17 by inserting it into that equation.

Recalling Formulas 4.15 and 4.15a, one arrives at

$$E_x(z, t) = E_{x0}e^{-\alpha z}e^{j(\omega t-\beta z)} \tag{4.21}$$

Taking the real part of the time-dependent term of the solution to the above equation, we arrive at its more common form:

$$E_x(z, t) = E_{x0}e^{-\alpha z}\cos(\omega t - \beta z) \tag{4.21a}$$

It is obvious that we can get a similar expression for a magnetic component:

$$H_y(z, t) = H_{y0}e^{-\alpha z}\cos(\omega t - \beta z) \tag{4.21b}$$

This is the result to be analyzed: *An EM field propagating in a medium takes the form of damping waves.* If you compare these forms with an EM field in free space given in Formula 4.13, you will see a major difference: EM waves die out while propagating through a lossy medium. A sketch of these damping waves is shown in Figure 4.2.

How fast do EM waves die out? The answer is determined by the envelope $e^{-\alpha z}$. This is why $\alpha$ is called the *attenuation constant.* The reason why $\beta$ is called the *phase constant* is clear from the expression in Formula 4.21. If you recall that $\beta = \omega/v$, you'll understand why another popular name for this parameter is *propagation constant.*

Let's pause briefly to summarize the discussion so far: *The solutions of wave equations obtained from Maxwell's equations prove that an electromagnetic field propagates in a lossy medium in the form of damping waves.*

True, we have considered only one particular case: plane transverse electromagnetic waves. But our results apply as well to a general situation. That's why we could safely draw the conclusion we did about wave equations.

It is interesting to know what parameters of a medium determine the attenuation and phase constants. The general formulas for these constants might be derived from their definitions. They show that $\alpha$ and $\beta$ depend on $\sigma$, $\varepsilon$, $\mu$, and $\omega$. The relationship among these parameters determines whether a medium is a dielectric or a conductor. What is especially important is the reliance of these parameters on a frequency. Depending on the frequency range of the EM waves, the medium behaves differently with respect to these waves. For example, *glass is a perfect dielectric for a low-frequency alternating current but the same glass perfectly conducts light, which is EM radiation in a very high frequency range.* We don't have to delve too deeply into this matter at this time, but it is necessary to emphasize just one point here: If a medium is lossy—which is always the case in fiber-optic technology—the EM waves will dampen. We previously described this effect as attenuation.

## EM Waves: Propagation in Waveguides

***A rectangular waveguide*** So far, we have considered EM field propagation in unbounded media, but in reality optical fiber confines light mostly within the core, thus guiding the light. To closely examine how optical fiber functions in the real world, we need to consider how waveguides conduct EM waves. To do this, we will briefly study the classic example of waveguiding theory, rectangular waveguides. Although a rectangular waveguide is far from an optical fiber, the main ideas and terminology used are similar. Therefore, this discussion should help you to better understand how optical fibers function in the field.

A rectangular waveguide is a structure whose walls are made from perfect conductors ($\sigma \rightarrow \alpha$). It is filled with lossless dielectric ($\sigma = 0$) containing no sources. A rectangular waveguide with width *a* and depth *b* is shown in Figure 4.3. The thickness of the walls is negligible.

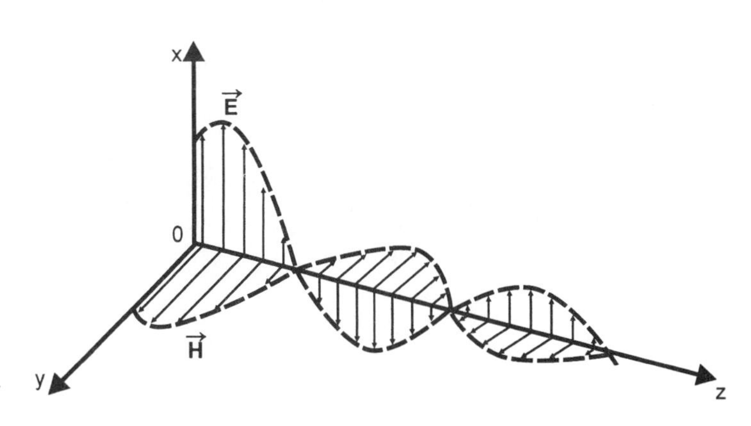


**Figure 4.2** Damping plane transverse electromagnetic waves.

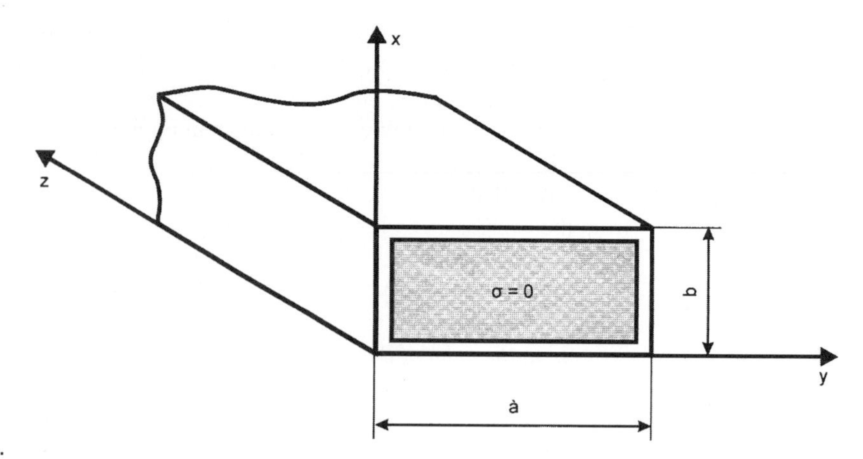


**Figure 4.3** A rectangular waveguide.

The wave equation derived from Maxwell's equations for the rectangular waveguide can be written in the following general vector form:

$$\nabla^2_{xy}\mathbf{E} + h^2\,\mathbf{E} = 0, \tag{4.22}$$

where $\nabla^2_{xy}$ is the three-dimensional Laplacian operator (see Section 4.1) for the cross-sectional Cartesian coordinates and $h^2 = (\gamma^2 + k^2)$. Here $\gamma = \alpha + j\beta$ is the propagation constant in an unbounded medium and a wave (or propagation) number, $k = \omega\sqrt{\epsilon\mu} = \omega/v$.

***Modes*** The solutions to Equation 4.22 exist only for discrete values of $h$ called *eigenvalues.* For the rectangular waveguides, these eigenvalues are:

$$h^2 = \left(\frac{l\pi}{a}\right)^2 + \left(\frac{m\pi}{b}\right)^2, \tag{4.23}$$

where $l$ and $m$ are integers and $a$ and $b$ are the width and depth of the waveguide, respectively. Thus, the solution to Equation 4.22 takes the form:

$$H_z(x, y) = H_0 \cos(l\,\pi x/b)\cos(m\pi y/a)$$

$$E_x(x, y) = (j\omega\mu/h^2)(m\pi/a)\,H_0\cos(l\pi x/b)\sin(m\pi y/a) \tag{4.24}$$

and so forth. Recall that $x$ and $y$ are cross-sectional axes and $z$ is the direction of propagation. (See Figure 4.3.)

Generally, the solutions to waveguide equations take the following forms:

- Transverse electric (TE) waves with no $E_z$ but with $H_z$ components.
- Transverse magnetic (TM) waves with no $H_z$ but with $E_z$ components.
- Hybrid (*HE* and *EH*) waves with both $E_z$ and $H_z$ components, where $z$ is the direction of propagation.

Analyzing Formula 4.24, one can see the meaning of the integers $l$ and $m$. They are the number of half cycles that EM waves make across the waveguide. The situation for $TE_{10}$—a transverse electric wave making a half cycle along the $y$ axis and a zero half cycle along the $x$ axis—is shown in Figure 4.4. Field lines for this $TE_{10}$ wave are shown in Figure 4.5.

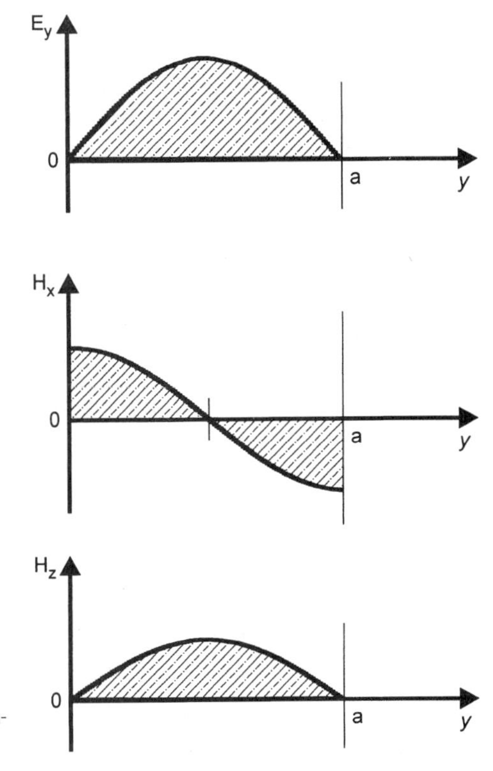


**Figure 4.4** Variations of field components for $TE_{10}$ mode. Wave propagates out of paper. *(Adapted from John D. Krauss,* Electromagnetics, *4th ed., New York: McGraw-Hill, 1992. Reprinted with permission.)*

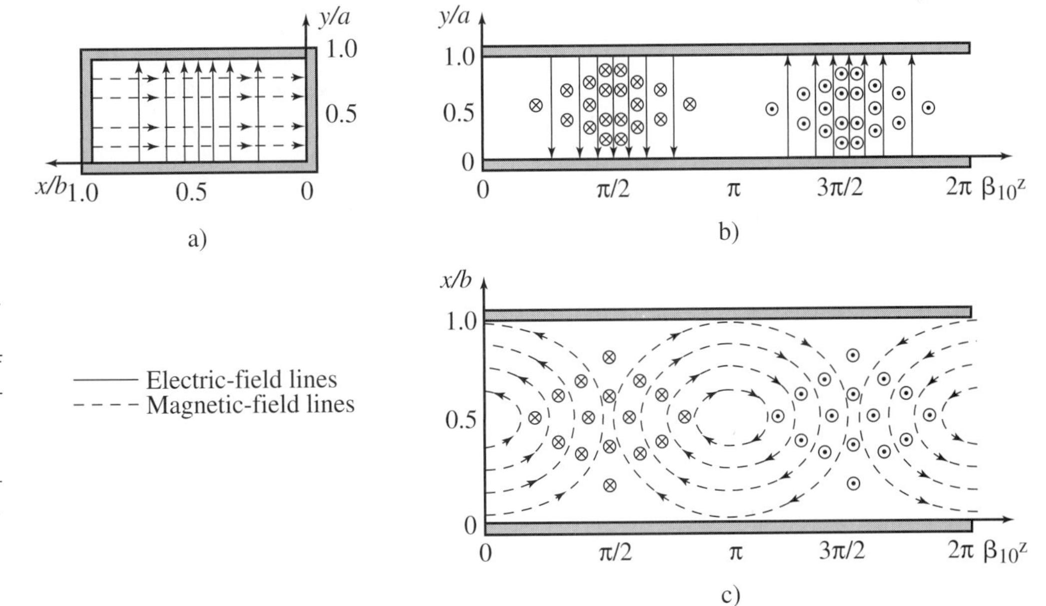


**Figure 4.5** Field lines for $TE_{10}$ mode in rectangular waveguide. *(*FUNDAMENTALS OF ENGINEERING ELECTROMAGNETICS *by Cheng, © 1993. Reprinted by permission of Prentice-Hall, Inc., Upper Saddle River, NJ.)*

It is necessary to study these figures very closely. Two important conclusions can be drawn from analyzing Formula 4.24 and Figures 4.4 and 4.5. First, an EM field will propagate along the waveguide in the form of *stable field patterns.* These field patterns are called *modes.* You'll recall that we met this term when we discussed multimode optical fibers, where we defined a *mode* as a beam making a specific angle with the central axis of the fiber. (See Section 3.3.) Now we come upon another interpretation of this term.

Secondly, it now becomes clear why one can solve the waveguide equation only as a set of discrete waveforms. The point is that *not every harmonic wave can exist in the waveguide.* The criterion for the harmonic wave's existence is this: *The half wavelength of the wave must fit the waveguide's width and depth an integral number of times.* These requirements, called resonant conditions, determine the possible set of waves that the waveguide can support for propagation.

***Cutoff condition*** Finally, one more key term should be explained here. We know, of course, from basic electricity that the rectangular waveguide shown in Figure 4.3 cannot conduct alternating current, since there is only one conductor here. But it can conduct light, as everybody's everyday experience has proved. But what, you may ask, is the difference between alternating current and light? Both are forms of electromagnetic radiation and their difference lies only in their frequencies. Thus, the same waveguide can support high-frequency radiation but not low-frequency radiation and, therefore, there is a frequency below which a waveguide cannot support EM radiation. This frequency is called the *cutoff frequency.*

From Formula 4.23, from the definition of the term $h$, and from the formulas $h^2 = (\gamma^2 + k^2)$ and $k = \omega\sqrt{\epsilon\mu}$, one can find:

$$\gamma = \sqrt{\left(\frac{l\pi}{a}\right)^2 + \left(\frac{m\pi}{b}\right)^2 - \omega^2\epsilon\mu} \tag{4.25}$$

Recall that $\gamma = \alpha + j\beta$, where $\alpha$ is the attenuation constant and $\beta$ is the propagation (phase) constant. Now look at Formula 4.25. When the EM field frequency is low, $\gamma$ is the real number ($\gamma = \alpha$); thus, the EM field will damp, as Formula 4.20 predicts. When the EM field frequency is high, $\gamma$ becomes the imaginary number ($\gamma = j\beta$) and, as a result, the EM field exists in a form of time-harmonic waves propagating without attenuation.

Note that attenuation in this instance has nothing to do with the property inherent in a material filling the waveguide. We have been assuming up to now that a dielectric is a perfect lossless material. *This attenuation originates from the property of the waveguide itself. The lower the EM field frequency, the longer its wavelength. Eventually, the wavelength becomes so long that it cannot meet resonance conditions and so the waveguide can no longer support the propagation of this wave.*

It follows from the above discussion, then, that the cutoff frequency, $f_c$, can be found by inserting $\gamma = 0$ in Formula 4.25. Thus we obtain:

$$f_C(\text{Hz}) = \frac{\omega_C}{2\pi} = \frac{1}{2\pi\sqrt{\epsilon\mu}}\sqrt{\left(\frac{l\pi}{a}\right)^2 + \left(\frac{m\pi}{b}\right)^2} \tag{4.26}$$

To define the cutoff wavelength, we need to distinguish among three possible wavelengths:

- First, *unbounded wavelength* $\lambda = v/f$, where $v = 1/\sqrt{\epsilon\mu}$ is the velocity of light in the unbounded medium with $\epsilon$ and $\mu$, $\lambda = c/f$.
- Second, *guide wavelength* $\lambda_g = 2\pi/\beta$, where $\beta$ is the propagation (phase) constant. If we express $\beta$ in terms of $\lambda$, $f$, and $f_C$, we can obtain:

$$\lambda_g = \lambda/\sqrt{[1 - (f/f_C)^2]}$$

- Third, *cutoff (critical) wavelength,* defined as:

$$\lambda_C = v/f_C \text{ or } \lambda_C = 2\pi/\sqrt{[(l\pi/a)^2 + (m\pi/b)^2]} \tag{4.27}$$

There is a relationship among these three wavelengths given by this formula:

$$l/\lambda^2 = l/\lambda_C^2 + l/\lambda_g^2 \tag{4.28}$$

The cutoff conditions discussed above are based on the results of work carried out at the end of the nineteenth century by Lord Rayleigh, who demonstrated that a waveguide can support—that is, allow to propagate through—only EM radiation whose wavelength is much smaller than the cross-sectional dimensions of the waveguide.

The terms and their definitions covered up to now will be referred to again and again in this and subsequent chapters so it would be advisable for you to return to this material from time to time to refresh your memory.

## 4.3 MORE ABOUT TOTAL INTERNAL REFLECTION

Maxwell's theory provides the formalism necessary to describe the behavior of an EM field—light—in any situation. But the actual implementation of this general statement in everyday practice was a conundrum that baffled many generations of scientists and engineers. Fortunately, one of the first breakthrough steps in this direction is closely related to our topic. Soon after Maxwell's equations finally had been accepted by the scientific community, Lord Rayleigh, in 1897, found some very interesting solutions to the equations. It follows from these solutions that EM waves can propagate within a restricted medium. Rayleigh's results had shown that it was possible to confine EM waves in hollow metallic tubes provided that the EM wavelength was much smaller than the cross-sectional dimension of the tubes. Rayleigh also predicted that EM waves would propagate within the tubes in the form of modes. This research turned out to be the theoretical foundation for radar and other UHF devices.

We can generalize Rayleigh's idea and say that Maxwell's equations are the basis for the description of light propagation within an optical fiber. True, but to make all the derivations and present this statement in formulas would take up the rest of this book and this is certainly not our aim since we're concerned primarily with application, not theory. The reader interested in detailed mathematical exposition of light propagation within an optical fiber is referred to the references at the end of this chapter (See references [7] through [14].)

This and the following sections provide more details and some theoretical explanations of the phenomena related to light propagation through the multimode optical fibers discussed in Chapter 3.

### Boundary Conditions

***Boundary of two lossless dielectrics*** Let's investigate more closely the phenomenon of total internal reflection. To do so, we have to introduce the boundary between two media. Let's take the same plane TEM waves we investigated previously (Figure 4.1) and direct them onto the boundary of two dielectrics. The situation is shown in Figure 4.6 where the normal incidence of TEM waves was taken for the sake of simplicity. Observe the change in polarization of the reflected **H** vector: vectors **E** and **H** must constitute a right-hand coordinate system with respect to the direction of propagation. We apply the same permeabilities to both media, since we know that the optical fiber's core and cladding—the two media we are interested in—are made from the same glass. But we know that permitivities are different because by doping this glass differently,

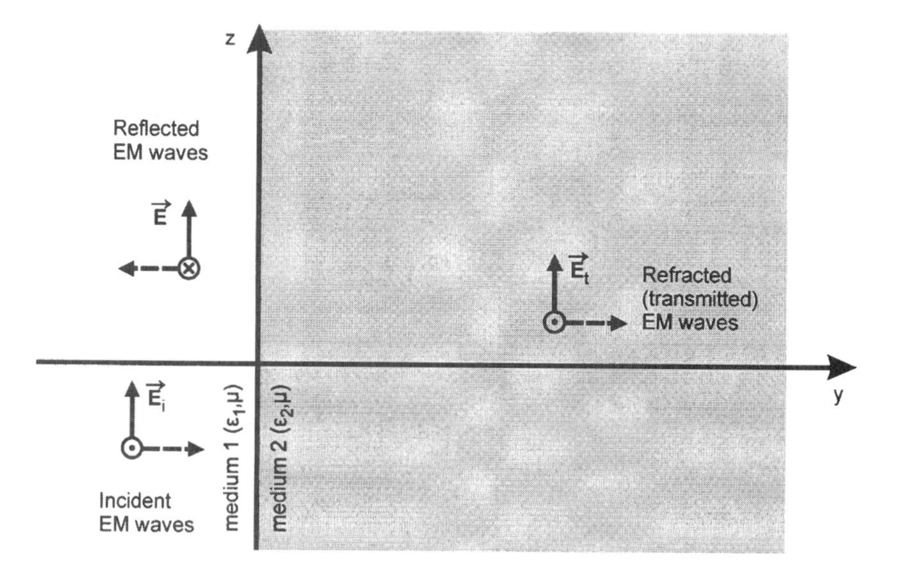


**Figure 4.6** Plane TEM waves at the boundary of two dielectrics: incident, reflected, and refracted (transmitted) waves. ⊙ Vector $\vec{H}$ goes out of the plane. ⊗ Vector $\vec{H}$ goes into the plane. *(*FUNDAMENTALS OF ENGINEERING ELECTROMAGNETICS *by Cheng, © 1993. Reprinted by permission of Prentice-Hall, Inc., Upper Saddle River, NJ.)*

manufacturers make different refractive indexes of the core and cladding. These refractive indexes relate to permitivities of media, as in the formulas

$$n_1 = \sqrt{\epsilon_{r1}} \qquad \text{and} \qquad n_2 = \sqrt{\epsilon_{r2}}\,, \tag{4.29}$$

where $\epsilon_{r1}$ and $\epsilon_{r2}$ are relative permitivities of medium 1 and medium 2, respectively. They are defined through absolute permitivities $\epsilon_1$ and $\epsilon_2$ by the following formulas: $\epsilon_1 = \epsilon_0\, \epsilon_{r1}$ and $\epsilon_2 = \epsilon_0\, \epsilon_{r2}$, where $\epsilon_0$, the permitivity of a vacuum, equals $8.854 \times 10^{-12}$ *F/m.*

As to the question of what happens to electric and magnetic fields at the boundary of two dielectric media, it is quite evident there should be some rules of continuity. These rules, called boundary conditions, state that *tangential components of electric and magnetic field intensities are the same on both sides of the boundary of two lossless dielectrics.* The core and cladding are low-loss media; thus, these conditions are applied to our case with good precision.

Taking the general case of the oblique incidence of plane TEM waves at the boundary of two dielectrics, one can derive Snell's law and the condition of total internal reflection.

***Evanescent wave*** The first consequence of these boundary conditions is the following: Even with total internal reflection, some portion of the incident field will penetrate the second medium. This portion, called the *evanescent wave,* and its amplitude decay exponentially in accordance with the simple formula $E_{t0}\, e^{-\alpha_e y}$, where $E_{t0}$ is the amplitude at the boundary, $\alpha_e$ is the attenuation constant of the evanescent wave, and $y$ is the distance from a boundary into the second medium. The attenuation constant, $\alpha_e$, is defined by the following formula:

$$\alpha_e = \frac{2\pi}{\lambda} \sqrt{n_1^2 \sin^2 \Theta_i - n_2^2} \tag{4.30}$$

The more the incident angle $\Theta_i$ exceeds the critical angle of incidence $\Theta_{1C}$, the less the amplitude of an evanescent wave. Observe from the above formula that $\Theta_i$ cannot be less than $\Theta_{1C} = \sin^{-1}(n_2/n_1)$. In other words, *the evanescent wave exists only under the condition of total internal reflection; otherwise, the refracted wave would propagate through the second medium.* If you find a contradiction here, you're right. On the one hand, there is *total* internal reflection but, on the other hand, some field penetrates the second medium. The explanation lies in the unique property of the evanescent waves. These waves do not transport any power in the radial (core-to-cladding) direction (along the $y$ axis in Figure 4.6). All light power is reflected at the core-cladding interface back into the core and an evanescent wave carries its power toward the direction of propagation (the $z$ axis in Figure 4.6).

Thus, an evanescent wave travels along the $z$ axis as does the core wave. In view of the existence of an evanescent wave, you should be able to determine the additional requirement for the cladding material apart from its having a refractive index less than that of the core: This material should also be transparent to light at the operating wavelength since some light is transmitted through the cladding.

**Reflectances**

***Fresnel formulas, reflectances, and transmittances*** As mentioned above, Snell's law can be derived from wave equations with boundary conditions. We will not spend our time making these derivations, which can be found in any comprehensive book on optics or electromagnetics. Instead, we will consider a general approach to the phenomenon of total internal reflection.

You'll recall that Snell's law, $n_1 \sin \Theta_1 = n_2 \sin \Theta_2$, determines the angle of refraction, $\Theta_2$, if we know the angle of incidence, $\Theta_1$, and the refractive indexes, $n_1$ and $n_2$. From Snell's law, we can find the critical angle of incidence ($\Theta_{1C} = \sin^{-1} n_1/n_2$, when $\Theta_2 = 90°$), which determines the condition of total internal reflection. But this information is not enough. *A light wave is completely defined by its amplitude, its phase, and its state of polarization.* Polarization here means the direction of the vector **E** or **H** with respect to its frame of reference. We have so far considered plane, or linear, polarized waves but, in general, light polarization might be quite arbitrary and changeable over time. If we introduce ρ—a reflection coefficient defined as the ratio of field intensities (field strengths) $E$ and $H$ of incident and reflected waves—then the desired information can be extracted from Fresnel equations [5]:

$$\rho_p = \frac{-n_2^2 \cos\Theta_i + n_1\sqrt{(n_2^2 - n_1^2 \sin^2 \Theta_i)}}{n_2^2 \cos\Theta_i + n_1\sqrt{(n_2^2 - n_1^2 \sin^2 \Theta_i)}} \quad (4.31)$$

$$\rho_s = \frac{n_2^2 \cos\Theta_i + n_1\sqrt{(n_2^2 - n_1^2 \sin^2 \Theta_i)}}{n_2^2 \cos\Theta_i + n_1\sqrt{(n_2^2 - n_1^2 \sin^2 \Theta_i)}}, \quad (4.31a)$$

where $\rho_p$ and $\rho_s$ are reflection coefficients of waves polarized in the plane of incidence and perpendicular to this plane, respectively; $n_1$ and $n_2$ are refractive indexes, $n_1$ being the medium from which light emanates and $n_2$ the medium in which light propagates; $\Theta_i$ is the angle of incidence.

Since intensities of light are proportional to the square of the field strengths, new quantities—reflectances—should be introduced:

$$R_p = |\rho_p|^2 \quad \text{and} \quad R_s = |\rho_s|^2 \quad (4.32)$$

Reflectances are the ratios of the intensities of the incident and reflected beams; subscripts $_p$ and $_s$ show what polarization they describe. Similarly, we can introduce transmittances, the ratios of the intensities of incident and transmitted (refracted) light:

$$T_p = 1 - R_p \quad \text{and} \quad T_s = 1 - R_s \quad (4.32a)$$

The significance of the Fresnel formulas is that one can calculate the intensities of reflected and refracted (transmitted) light striking the boundary of two media. These intensities depend on the refractive indexes of the media, the angle of incidence—which we know about from Snell's law—and the plane of polarization. Since any state of polarization can be described as a combination of two perpendicular polarizations, the Fresnel equations enable us to calculate the state of reflected and refracted (transmitted) beams.

We can draw two important conclusions from an analysis of the Fresnel equations:

First, there is an angle of incidence, $\Theta_B$, called the Brewster angle, at which a plane polarized wave has zero reflectance. (See Formula 4.31.) In other words, a wave polarized in the plane of incidence will be fully transmitted to the second medium. Solving Formula 4.31 for a zero-equal numerator, one obtains:

$$\tan \Theta_B = n_2/n_1 \tag{4.33}$$

Secondly, there is an angle of incidence, $\Theta_{iC}$, called the critical incident angle, at which and above which the reflectance becomes 1. This angle determines the phenomenon of total internal reflection, which we learned from Snell's law is the angle $\Theta_{1C}$. The reflectances versus the angle of incidence at the glass-to-air interface are shown in Figure 4.7.

***Phase shifts of totally reflected waves*** Finally, it is important to know the phase shifts of totally reflected waves. For multimode optical fibers, different modes travel inside the core at different angles; therefore, different modes experience different phase shifts.

The reflection coefficients given in Formulas 4.31 and 4.31a become complex numbers under the condition of total internal reflection. Hence, they can be presented in common magnitude-phase form, that is, $\rho = |\rho|e^{j\delta}$. The phase shifts for parallel and perpendicular polarized waves are defined by the following formulas [5]:

$$\tan\frac{\delta_p}{2} = \frac{\frac{n_1}{n_2}\sqrt{\left(\frac{n_1}{n_2}\right)^2 \sin^2\Theta_i - 1}}{\frac{n_1}{n_2}\cos\Theta_i} \tag{4.34}$$

$$\tan\frac{\delta_s}{2} = \frac{\sqrt{\left(\frac{n_1}{n_2}\right)^2 \sin^2\Theta_i - 1}}{\frac{n_1}{n_2}\cos\Theta_i}, \tag{4.34a}$$

where $\delta_p$ is the phase shift of a wave polarized parallel to the plane of incidence and $\delta_s$ is the phase shift of a wave polarized perpendicular to the plane of incidence. The graphs in Figure 4.8

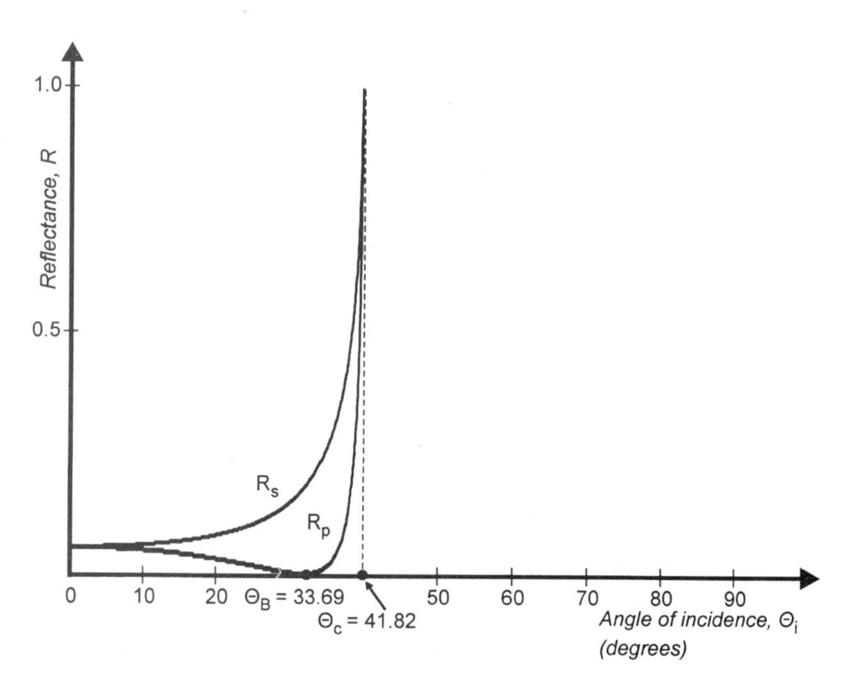


**Figure 4.7** Reflectance versus angle of incidence at the glass ($n_1 = 1.5$)-to-air ($n_2 = 1.0$) interface.

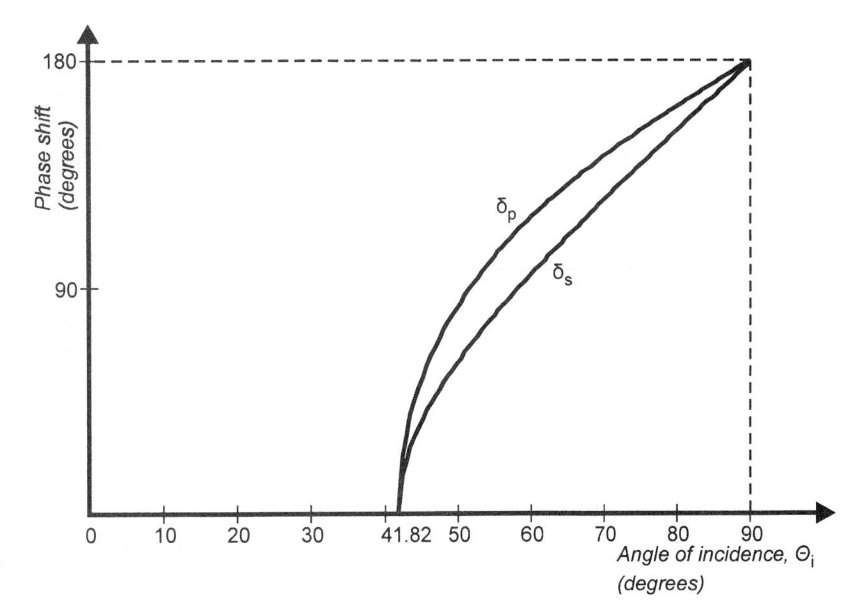


**Figure 4.8** Phase shifts of totally reflected waves polarized in parallel, $\delta_p$, and perpendicular, $\delta_s$, to the plane of incidence at the glass ($n_1$ = 1.5)-to-air ($n_2$ = 1.0) interface.

show phase shifts versus the angle of incidence. Remember, these waves are totally reflected; hence, we can apply these data to modes in multimode optical fiber.

The material covered up to this point provides the essential information you will need for future discussions of the components of fiber-optic communications systems.

## 4.4 MORE ABOUT MODES

For the beginning student who reads Chapter 3, often the first obvious question is: If a single-mode fiber provides the ultimate solution for intermodal dispersion, why do we need to study modes and multimode fibers?

Multimode optical fibers find wide applications in the intermediate- and short-distance networks. This is why the theory underlying these fibers is still an area of great interest to scientists and engineers. For users, however, it is more important to know how to operate with these fibers and what performance characteristics can be expected. For example, bending multimode fibers diminishes the higher-order modes. This change affects the mode's power distribution and dispersion. Such an occurrence might even result in unwanted modification of the bandwidth specified by the manufacturer, a costly consequence that will lead to deterioration of the optical network's performance. For this reason, this section pays close attention to the behavior of modes in multimode optical fibers.

"Well," the student might say, "if we need to study multimode fibers, explain why light exists in the form of a continuous EM field in air, yet breaks down into separate modes within an optical fiber. Also, what are these modes, anyway?" Good point—and now for the answers.

### Some Words About Mode Theory and Important Results

***Natural (true or exact) modes*** Theoretical considerations of light propagation in optical fibers are based on the general theory of the propagation of EM waves in guided structures. (Refer back to Section 4.3 if you feel you need to brush up on this theory before continuing.)

The specific theory of EM field propagation in optical fibers is by now fairly well developed and, obviously, harkens all the way back to Maxwell's equations. It is not our intention to discuss this theory in detail here. (See [7], [9], [10], [11], and [14]. The last reference provides a good review of this theory.) Instead, we'll discuss the most important results of the theory.

Wave equations, together with the boundary conditions (which we discussed in Sections 4.1 and 4.2), are the formalism we need to describe the propagation of EM waves in optical fibers. They allow one to obtain waveguide, or modal, eigenvalue equations whose eigenfunctions are Bessel functions. *The main result of solutions of these modal equations is that the EM field can propagate within a lightguide structure not as a continuum but as a set of discrete field patterns called* natural modes.

These natural modes (you might also meet the terms *true* or *exact*) can be completely transverse (TE and TM) or may have longitudinal (that is, along the direction of propagation) components (hybrid HE and EH). Notations with double subscripts are common. For example, $TE_{lm}$ means transverse electric natural mode, where l is the value of the mode order and m stands for mode rank, or radial mode number. For transverse modes, l = 0; hence, all TE and TM modes have the designations $TE_{0m}$ and $TM_{0m}$. The transverse electric mode, $TE_{0m}$, has a longitudinal magnetic component, while the $TM_{0m}$ mode has a longitudinal electric component. The designations $HE_{lm}$ and $EH_{lm}$—they're hybrid modes—mean that both fields—magnetic and electric—have longitudinal components. Thus, $EH_{0m}$ and $HE_{0m}$ do not exist since $l$ never equals zero for these modes. The field-lines picture of several lower-order modes is shown in Figure 4.9. Compare this figure with Figure 4.5.

## Linear-Polarized (LP) Modes

The most important practical results are obtained from this theory by applying the so-called *weakly guided-mode approximation.* This approximation is based on the assumption $n_1 - n_2 \ll 1$, which is always true for practical optical fibers. (You learned in Section 3.1 that $n_1 - n_2$ is about 0.02 or less. Also, refer again to the data sheets.) *Under this condition, natural modes will combine (degenerate) into linear-polarized (LP) modes that truly exist in optical fibers.*

Figures 4.10 and 4.11 depict this situation. Figure 4.10 shows examples of how natural modes make up linear-polarized modes; Figure 4.11 illustrates the intensity plots and visual patterns of the six LP modes. Observe these pictures very carefully: They provide you with a clear understanding of what the term *mode* means. Bear in mind that linear-polarized mode designations mean the following: l is one-half the number of the maxima (or minima) intensity that occurs while the angular coordinate changes from 0 to $2\pi$ radians; m is the number of maxima intensity that occur while the radial coordinate changes from zero to infinity. Look again at Figure 4.11. Note that for one maximum, l = 0 since l must be an integer and cannot be 1/2. See Figures 4.11a and 4.11d. Observe and analyze other patterns.

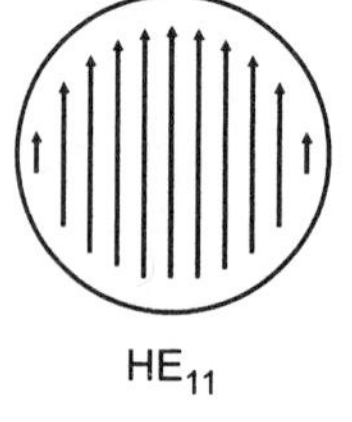


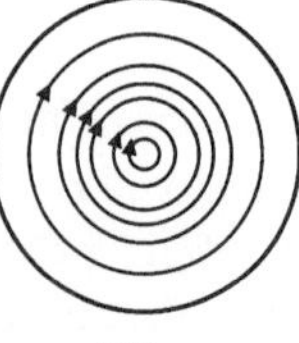


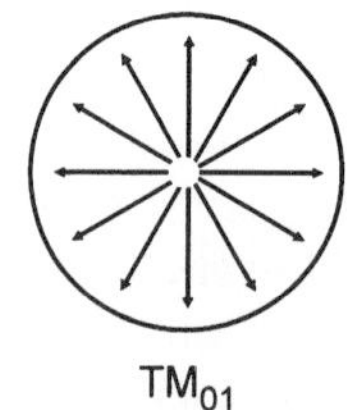


**Figure 4.9** Field lines for four natural lowest-order modes in a step-index fiber. *(Courtesy of Gerd Keiser, Optical Fiber Communications, 2nd ed., New York: McGraw-Hill, 1991. Reprinted with permission.)*

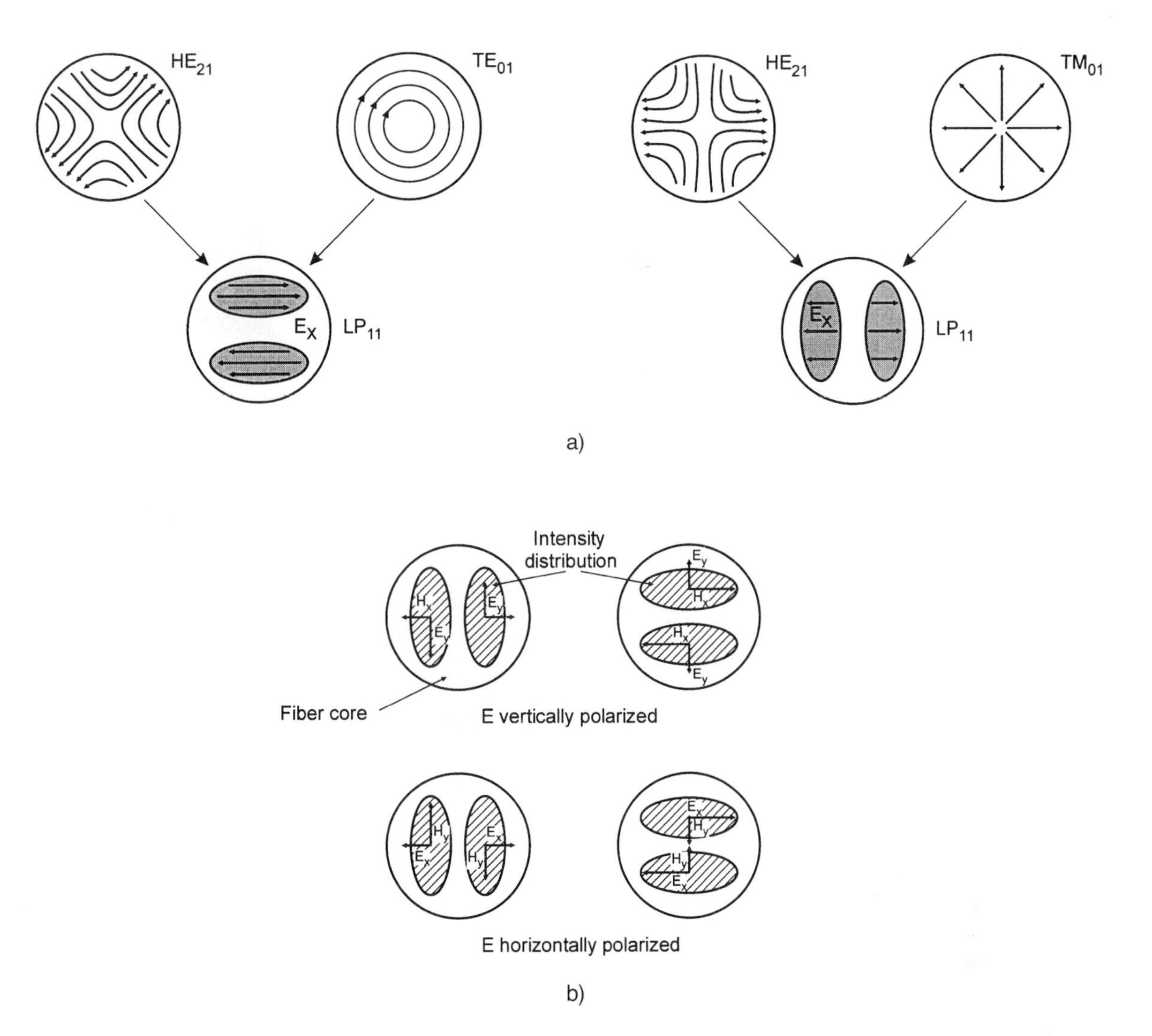


**Figure 4.10** Examples of how modes $HE_{21}$ + $TE_{01}$ and $HE_{21}$ + $TM_{01}$ compose linear-polarized modes $LP_{11}$ (dark spots show intensity distribution; arrows show transverse electric and magnetic fields): (a) Composition of two $LP_{11}$ modes from true modes and their transverse electric field and intensity distributions; (b) the four possible transverse electric-field and magnetic-field directions and the corresponding intensity distributions for the $LP_{11}$ mode. *(Courtesy of Gerd Keiser,* Optical Fiber Communications, *2nd ed., New York: McGraw-Hill, 1991. Reprinted with permission.)*

The formal presentation of LP modes is beyond the scope of this text. It is only worth mentioning that the longitudinal components of LP modes are very small so that LP modes, in most cases, can be treated as transverse modes.

From this discussion and observations of given patterns, the meaning of the term "mode" should now be crystal clear. What should also be crystal clear is that *not natural but linear-polarized (LP) modes really exist within an optical fiber.*

***Why modes, not a continuous EM field, exist within an optical fiber: The physical explanation*** Again, the key point is this: *An optical fiber supports only distinguishable separate field patterns that we designated as linear-polarized (LP) modes.* The physical reason is that an EM

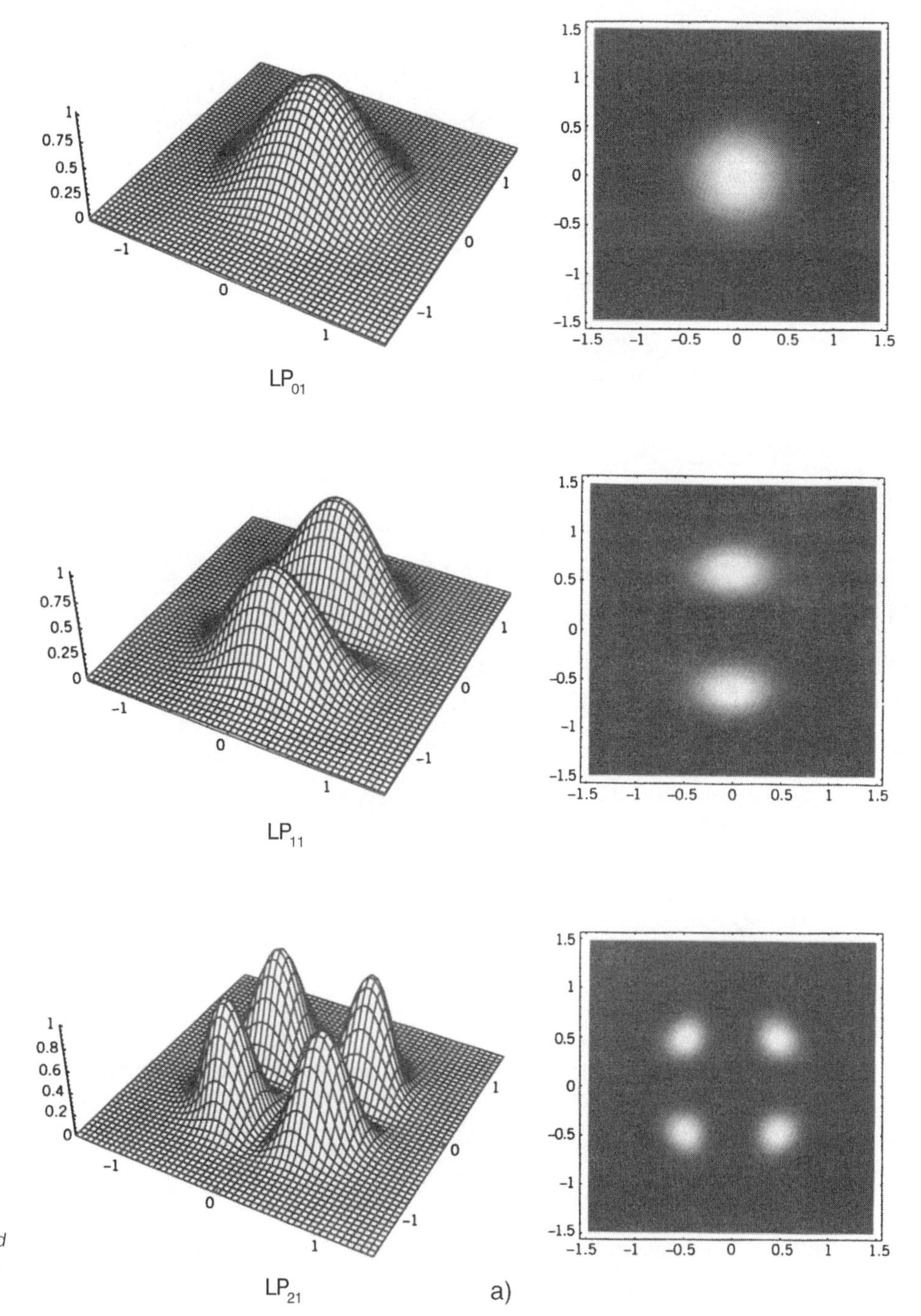


**Figure 4.11** Intensity plots and visual patterns for six LP modes. *(John A. Buck,* Fundamentals of Optical Fibers, *New York: John Wiley & Sons, 1995. Reprinted by permission of John Wiley & Sons, Inc.)*

wave propagating inside an optical fiber has to meet the boundary-condition requirements. Suppose the wave meets these requirements when it strikes the core-cladding interface the first time. To meet these requirements at all other times, the wave must repeat itself when striking the core-cladding boundary again. In other words, its phase $(\omega t - \beta z)$, where $z$ is the direction of propagation, has to be equal to $2\pi \times k$, where $k$ is an integer, at the same distance between any two zigzags. (See Figure 2.2.) EM waves that meet this requirement will exist as a stable pattern, or mode. EM waves that do

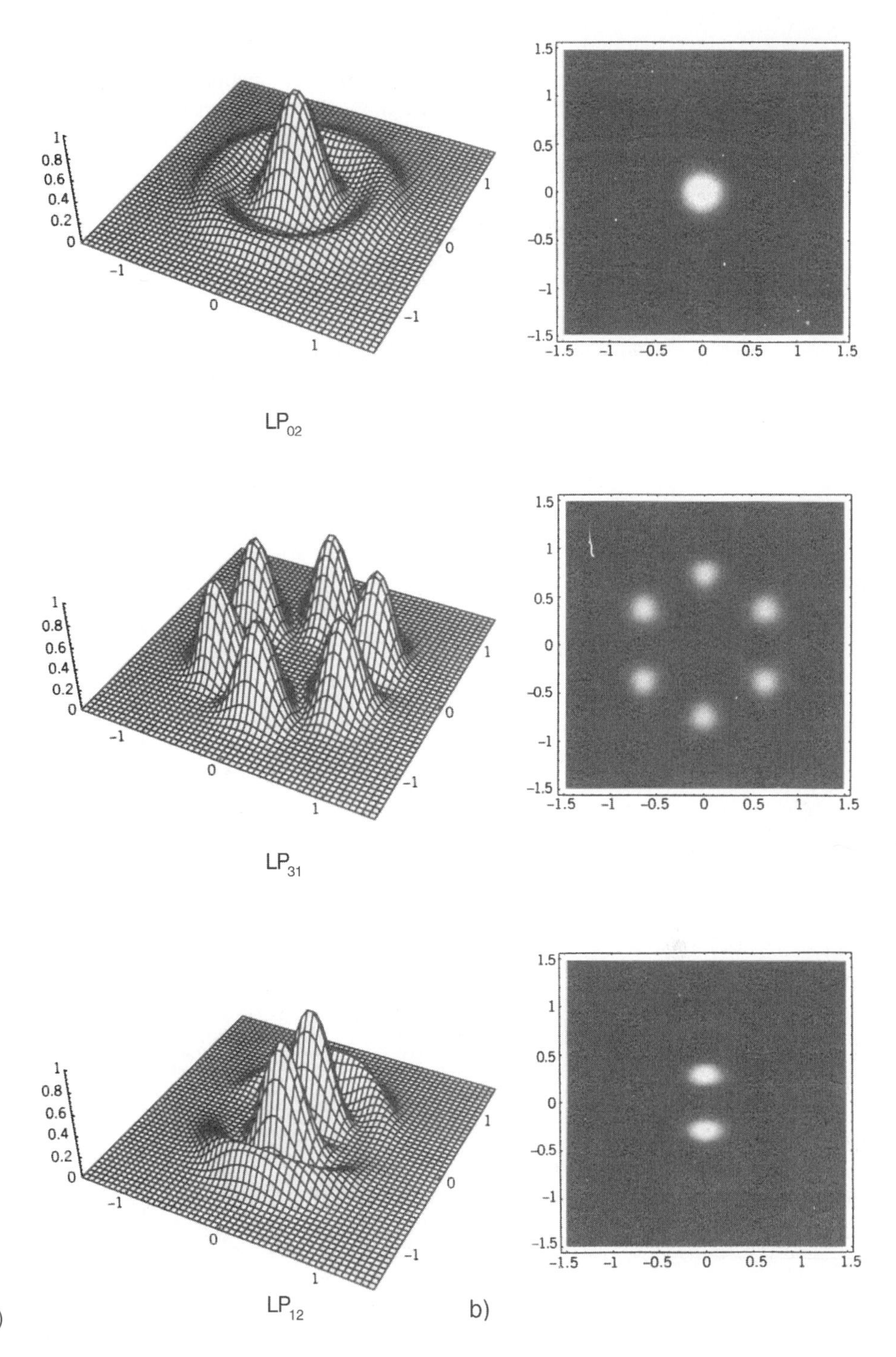


**Figure 4.11** (*Continued*)

not meet this requirement will not appear. This is why optical fiber supports EM waves—modes—that make stable patterns and does not support others. There is a close relationship between this discussion and the resonant conditions obtained for the rectangular waveguide in Section 4.2.

These stable EM field patterns—again, they're modes—are associated with the direction of their propagation within an optical fiber. This means they can be distinguished by the angles that their propagation vectors make with the central axis of the optical fiber. Thus, we arrive at the first definition of modes given in Section 3.3: Modes are beams making different angles with the centerline, or axis, of the fiber. There is a unique correspondence between geometric and wave presentations in terms of how light propagates inside an optical fiber. The key to this correspondence was given in Section 3.3. (See Figure 3.10[b].)

***Meridional and skew beams*** All beams—modes—propagating within an optical fiber are divided into two categories: meridional beams and skew beams. Meridional beams are those that intersect the centerline of the fiber; skew beams propagate without intersecting the fiber's central axis. (See Figure 4.12.) So far, we have discussed only meridional beams. These have only two coordinate components—radial and axial. They can be composed of transverse natural modes $TE_{0m}$ and $TM_{0m}$. Skew rays, on the other hand, are more sophisticated since they have azimuthal coordinate components. Theoretical analysis shows they are composed of modes that must include longitudinal elements. Thus, skew beams correspond to hybrid natural modes $EH_{1m}$ and $HE_{1m}$.

Refer to Figure 4.10 to see how these natural modes are combined in LP modes.

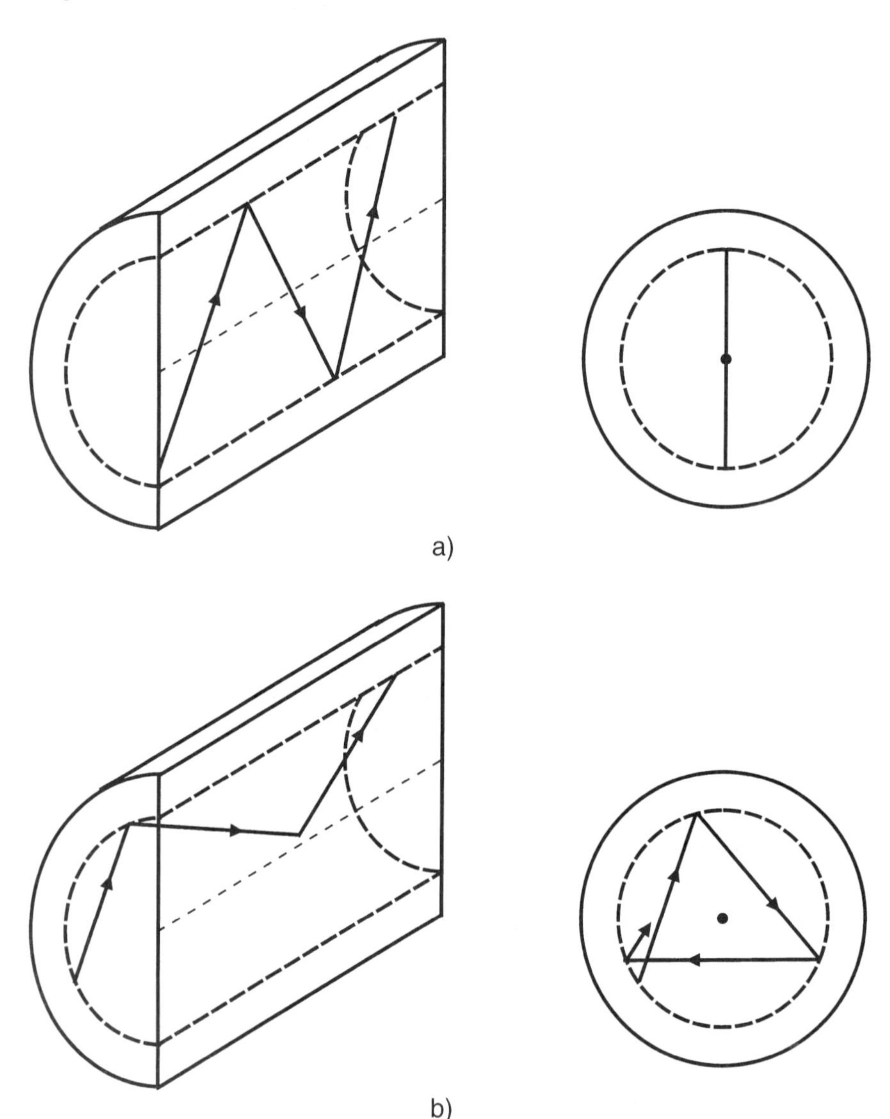


**Figure 4.12** Meridional and skew beams: (a) Meridional beam trajectory in fiber: longitudinal (left) and transverse (right) cross sections; (b) skew-beam trajectory in fiber: longitudinal (left) and transverse (right) cross sections.

## Three Types of Modes: Guided, Radiation, and Leaky

The modes described to this point are called *guided modes.* As the term suggests, those modes are guided by the optical fiber, which means they are experiencing total internal reflection. Theoretically, an optical fiber can support their propagation indefinitely unless we take into account the attenuation introduced by the fiber material. But not all of the light injected into the fiber will experience total internal reflection. What happens to those beams whose incident angles are less than the critical incident angle? You may recall that in Section 3.2 we said that those beams will eventually be lost, since each beam experiences millions and millions of reflections and that even a microscopic loss at each reflection will result in the complete disappearance of the beam after only a short propagating distance. This is true in general but a more detailed analysis is needed for a better understanding of this effect.

Theoretical analysis shows that an optical fiber forms modes regardless of the condition of total internal reflection. In other words, if the EM field inside the fiber forms a stable pattern, an optical fiber will support this type of radiation. One group of these modes will experience total internal reflection and these modes will be trapped by the fiber core; thus, the term *guided modes* is justified.

Another group of modes will not experience total internal reflection and these modes, called *radiation modes,* will propagate outside the fiber core.

Theoretically, solutions to the waveguide equations that describe EM field propagation within a fiber include both guided and radiation modes. In other words, these solutions can be presented as an expansion of the guided and radiation modes. Radiation modes, in contrast to guided modes, do not have the $2\pi \times k$ requirement and they make a continuum. In physical terms, radiation modes originate from light power injected into the fiber at a less-than-critical incident angle. They are partially propagated inside the core and partially transmitted into the cladding. Those modes that are refracted (transmitted) into the cladding will propagate in the cladding. They will meet the cladding-coating interface, be reflected back into the cladding, and perhaps even be transmitted back into the core, where they will couple with the higher-order guided modes. The result: increasing power loss of the core modes.

Another consideration is the optical-property requirement of the coating material, which must be lossy enough to make radiation modes disappear after they strike the cladding-coating boundary.

The third type of mode possible in an optical fiber is called a *leaky mode.* Such modes are not a part of the solutions to Maxwell's equations as applied to a waveguide. These modes are characterized by having the $2\pi \times k$ condition but they are not totally reflected. As a result, their amplitudes change while they propagate along the fiber. This kind of field structure does not form modes as stable patterns; we consider them as modes nonetheless since they are stable with respect to time but, again, unstable with respect to space. Most of them disappear very shortly after being excited; however, a few of them might propagate over a significant distance, carrying an essential portion of the total light power transmitted by the fiber.

In conclusion, it is necessary to emphasize that one property is crucial in a fiber for it to be capable of supporting modes: It must have an $n_1 - n_2$ boundary at the core-cladding interface. This property does not take into account the properties of the core or cladding materials that will, of course, result in attenuation because of absorption and scattering. However, this complex mode structure of the EM field in multimode fiber compels optical fiber manufacturers to take special measures when choosing cladding and even coating material since their optical properties determine the effectiveness of mode propagation in the whole fiber.

## Phase and Group Velocities

***Definitions and explanations*** The x-polarized component of an electric wave takes the general form $E_{xo}\, e^{-\alpha z} \cos(\omega t - \beta z)$, where $E_{xo}$ is the amplitude, $\omega$ the angular frequency (rad/s), $\beta$ the phase constant (rad/$m$)—also called the longitudinal propagation constant—and $z$ the direction of propagation. (See Section 4.2 and Figure 4.1) Let's take point M of the wave and determine at what velocity this point moves with respect to reference frame $x$-$y$-$z$ (See Figure 2.2). Since point M is constant with respect to the wave, its phase ($\omega t - \beta z$) is also constant:

$$(\omega t - \beta z)_M = \text{constant for any } t$$

Taking derivative $d/dt$ of the above expression, one can find, provided that $\omega$ and $\beta$ are time independent,

$$dz/dt = \omega/\beta$$

But $dz/dt$ is the velocity, $v$, at which point M moves with respect to the $z$ axis, that is, with respect to reference frame $x$-$y$-$z$. Hence,

$$v = \omega/\beta, \tag{4.35}$$

where $v$ is called the *phase velocity,* since this expression defines the velocity of a phase point of the EM wave.

Strictly speaking, there is no such thing as a pure sine or cosine wave. We can only approximate this ideal. Therefore, phase velocity is the ideal characteristic of the ideal signal. In reality, a group of signals always combine. Thus, we have to define the velocity of this group (the so-called *group velocity*). The most common example for explaining the difference between phase and group velocities is the amplitude-modulated signal. In this example, group velocity is the velocity of the envelope. Generally, the group velocity, $v_{group}$, is defined as:

$$v_{group} = \partial\omega/\partial\beta \tag{4.36}$$

The concept of phase and group velocities is a general one. It has been customary to use the $\omega - \beta$ plane as a graphical device to explain this concept. It is worth noting that *any information signal and power travel at the group, not at the phase velocity.* For our application, it is important to understand that *group velocity is the speed at which light power propagates along the fiber in a specific mode.*

***Guided, radiation, and leaky modes on $\omega - \beta$ plane*** Usage of the $\omega - \beta$ plane for summarizing this discussion of the three types of modes in optical fibers is given in Figure 4.13.

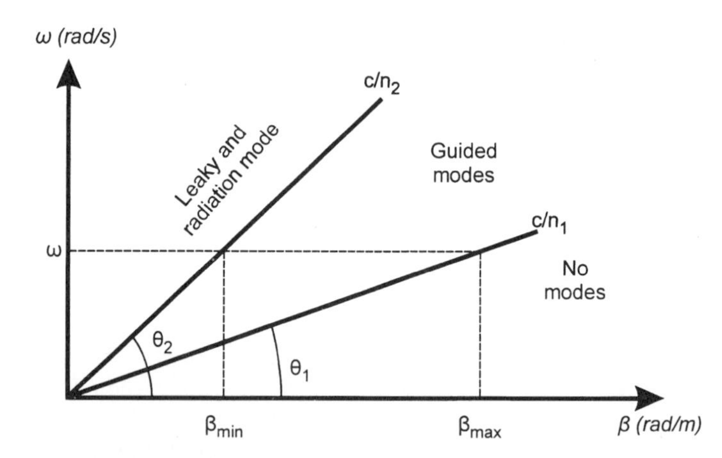


**Figure 4.13** Three types of modes. *(John A. Buck,* Fundamentals of Optical Fibers, *New York: John Wiley & Sons, 1995. Reprinted by permission of John Wiley & Sons, Inc.)*

This figure is true not for regular optical fiber but for a dielectric slab waveguide. The latter structure consists of a thin dielectric layer with refractive index $n_1$ sandwiched between two other identical layers with refractive index $n_2$. The difference between an optical fiber and a slab is that a fiber has a circular structure and a slab has a plane structure without boundaries in the plane. However, a slab waveguide model allows us to introduce the basic ideas of light propagation in a guided structure and leads to a comprehensive model of a real optical fiber.

Observe in Figure 4.13 that the slope of lines determines the phase velocity. Thus, $c/n_1$ is the phase velocity within the slab-guiding layer (which is analogous to the core of a fiber) and $c/n_2$ is the phase velocity within the slab's surrounding layer (analogous to the cladding of a fiber). Hence, the velocities of guided modes are restricted by $c/n_1$ and $c/n_2$. Figure 4.13 shows that phases of guided modes are given by:

$$\begin{aligned} \beta_{min} &= \omega/\tan\theta_2 = \omega n_2/c = 2\pi(f/c)n_2 = (2\pi/\lambda)n_2 \\ \beta_{max} &= \omega/\tan\theta_1 = \omega n_1/c = 2\pi(f/c)n_1 = (2\pi/\lambda)n_1, \end{aligned} \tag{4.37}$$

where $\lambda$ is the wavelength in free space. This idea is commonly expressed in this form:

$$(2\pi/\lambda)\, n_2 \leq \beta \leq (2\pi/\lambda)n_1 \tag{4.38}$$

The concept of phase and group velocities, being important in itself, helps us to better understand mode structure in an optical fiber. It is evident that the main point of our interest is guided modes, but knowing the total mode picture is important for determining the right design and proper maintenance of optical fibers.

## Power Confinement

***Poynting vector*** An electromagnetic field transports energy, as our everyday experience proves. In telecommunications we use low-energy EM signals, whereas a laser weapon can deliver an EM field at a very high energy level. The power per unit of a cross-sectional area that is transported by an EM field is given by Poynting vector **S**:

$$\mathbf{S} = \mathbf{E} \times \mathbf{H}, \tag{4.39}$$

where **S** is measured in W/m$^2$. Since **E** and **H** are instantaneous values, calculating the value of the Poynting vector requires averaging over the period of field oscillation.

***Power transport mechanism and power distribution between core and cladding*** All light power is transported within a fiber by modes that the fiber supports. In other words, *there is no other mechanism to deliver power from a source to a detector except through the individual modes.* This leads to modal dispersion, as discussed in Section 3.3. But now we have to state that *the higher the order of the mode, the less the percentage of total power it carries.* In view of the mode analysis given above, therefore, it is clear that a substantial amount of light power is transported by the cladding, while most of the power is confined in the core. Thus, the total power flow is the sum of power carried by both the core and by the cladding. The formula expressing this statement is [10]:

$$\frac{P_{clad}}{P_{total}} = \frac{2\sqrt{2}}{3V}, \tag{4.40}$$

where $V$ is the $V$-number. Depending on the type and characteristics of a fiber, as much as 20% of the total power can be carried by the cladding.

### Example 4.4.1

***Problem:***

The graded-index fiber has the following characteristics: the core/cladding diameters are 62.5/125μm and the *NA* is 0.275. The fiber operates at a 1300-nm wavelength. What power is carried by the fiber's cladding?

***Solution:***

Recall Example 3.3.1 where the $V$-number for this fiber and this wavelength was calculated as follows: $V = 41.5$. Hence, $P_{clad}/P_{total} = 0.023$. In other words, more than 2% of the total power resides in the cladding.

Formula 4.40 gives us the order of value but not very precise numbers. There are many sources of its approximate nature. For example, Formula 4.40 is based on the assumption that a light source injects an equal amount of power into every mode, which is, obviously, the ideal that is never realized.

How about a singlemode fiber? Since $V \leq 2.405$, $P_{clad}/P_{total} \leq 39.2$. In other words, almost 40% of the total power, by this calculation, might be carried by the cladding. Now you can see how important the optical properties of the cladding are. In reality, a typical singlemode fiber carries about 20% of the power in its cladding.

The result obtained in this example might have surprised you if you had thought that the singlemode fiber confines light more tightly than does the multimode fiber. A closer examination undertaken in Chapters 5 and 6 will explain this result in greater detail.

---

***Number of modes and measurement of attenuation*** The mode structure of light residing within a fiber raises a very interesting question: What power can one measure at the multimode-fiber output? The answer to this question is critical, since it determines the accuracy of the measurement of fiber attenuation. Depending on the light source, the radiation patterns, and the emitter-fiber coupling efficiency, one can obtain substantially different results measuring fiber output power. The source of possible discrepancies is the different number of modes that might be excited at each measurement. This is why manufacturers of optical fiber take steps to excite all available modes when measuring fiber attenuation. This technique is known as *overfilled light launch.*

Another important point is the fiber length. Since the higher-order modes tend to disappear, the longer fiber exhibits less attenuation in dB/km than the shorter one simply because the entire light propagating within the fiber becomes more confined. Any mode-filtering device, like a connector, causes the same effect as increasing fiber length.

We will return to this discussion in Section 4.5.

## Cutoff Wavelength (Frequency)

***Cutoff condition*** An optical fiber, as well as any waveguide for that matter, can support EM radiation—here, light—if, and only if, the radiation's wavelength is much smaller than the fiber's core diameter. That's why an optical fiber conducts light but can't guide a radio-frequency signal. Thus, we arrive at the necessity of defining the cutoff condition for an optical fiber, which determines the cutoff wavelength (frequency) that fiber can support.

The cutoff condition can be obtained as the solution for eigenvalue equations describing mode distribution in an optical fiber. The cutoff condition is presented in many forms but, essentially, all these forms can be reduced to this typical requirement [12]:

$$V \geq ka, \tag{4.41}$$

where $V$ is the normalized cutoff frequency, or $V$-number, $a$ $(m)$ is the fiber radius, and $k$ $(1/m)$ is the wave propagation constant along the transverse direction (Figure 4.14).

***Cutoff condition and total internal reflection*** To examine the cutoff condition in Formula 4.41 more closely, we need to introduce, besides $k$, two more propagation constants: $k_1$ is the propagation constant of a plane wave in the core and $\beta$ is the propagation constant of the guided wave along the $z$ axis so that the field dependence on $z$ is given by $\exp(-j\beta z)$. A plane wave travels within the fiber at angle $\alpha$ to the centerline.

The geometric relationship among these three constants is displayed in Figure 4.14. These constants are also called wave vectors, a fact that justifies the graphical presentation given in this figure.

From Figure 4.14, one can find:

$$k = \sqrt{[k_1^2 - \beta^2]} \tag{4.42}$$

and

$$\beta = k_1 \cos \alpha \tag{4.42a}$$

The constant $k_1 = 2\pi/\lambda_1$, where $\lambda_1$ is the wavelength within the core whose refractive index is $n_1$; hence, $\lambda_1 = \lambda/n_1$, where $\lambda$ is a wavelength in a vacuum. Substituting Formulas 4.42 and 4.42a in Formula 4.41, one can obtain:

$$V \geq (2\pi a/\lambda)n_1 \sqrt{[1-\cos^2 \alpha]} \tag{4.43}$$

Using $V = [(2\pi a)/\lambda]\ NA$, the cutoff condition can be rewritten in this form:

$$NA \geq n_1 \sqrt{[1-\cos^2 \alpha]} \tag{4.44}$$

Substituting $NA = \sqrt{(n_1)^2 - (n_2)^2}$, after simple manipulations one can obtain:

$$\cos \alpha \leq n_2/n_1 \tag{4.45}$$

Since $\alpha = 90° - \Theta$, the above formula can be rewritten using an equality sign for critical angles:

$$\cos \alpha_C = \sin \Theta_C = n_2/n_1 \tag{4.46}$$

This formula represents the condition of total internal reflection. (See Section 3.1.) Thus, the conclusion we draw is this: *The cutoff condition at the upper level is equivalent to the condition of total internal reflection.* This explains why an optical fiber does not support waves that do not meet the cutoff requirement.

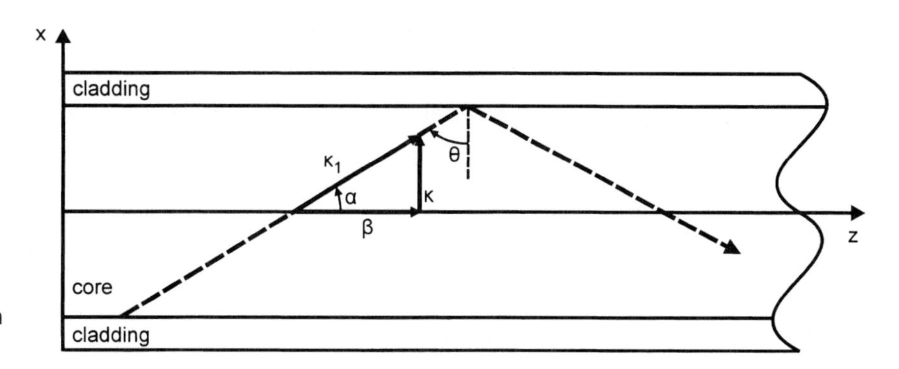


**Figure 4.14** Wave propagation constants within a fiber core.

**Helmholtz Equations and Propagation Constants**

The relationship between propagation constants can be directly obtained from the wave equation: Recalling that $\sigma = 0$ for a dielectric, which is what a fiber core really is, one can find $\gamma = \omega^2\epsilon\mu$. (See Formula 4.18.) The wave equation obtained from Maxwell's equations is similar to that given in Formula 3.18 for the $x$-polarized wave and now becomes:

$$\partial^2 E_x/\partial x^2 + \partial^2 E_x/\partial y^2 + \partial^2 E_x/\partial z^2 = -\omega^2\epsilon\mu E_x \qquad (4.47)$$

This expression is known as the Helmholtz equations. They are usually presented in vector form for both electric and magnetic fields. The Helmholtz equations describe in implicit form any type of waveguide whose material is defined by the constants $\epsilon$ and $\mu$. The solutions for the Helmholtz equations present electric- and magnetic-field distributions within a waveguide; that is, they explicitly describe modes in an optical fiber.

From the definition of $\beta$ given above, we can find:

$$\partial^2 E_x/\partial z^2 = -\beta^2 E_x$$

The slab model implies the field is uniform in the $y$ direction; hence,

$$\partial^2 E_x/\partial y^2 = 0$$

Thus, Equation 4.47 becomes:

$$\partial^2 E_x/\partial x^2 = (\beta^2 - \omega^2\epsilon\mu)E_x \qquad (4.48)$$

On the other hand, propagation constant $k$ can be defined through the transverse $E$-component as $E_x(x) = E_{x0} \cos kx$ or $E_x(x) = E_{x0} \sin kx$. Substituting either case in Formula 4.48 yields:

$$k^2 = \omega^2\epsilon\mu - \beta^2 \qquad (4.49)$$

The last step is to prove that Formula 4.49 is equivalent to Formula 4.42. Indeed, $\epsilon = \epsilon_0\, \epsilon_r$ and $\mu = \mu_0$ for an optical fiber; hence, $\epsilon_0\, \mu_0\, \epsilon_r = (1/c^2)\, n_1^2$. Since $\omega = 2\pi f$, one can obtain $\omega^2\epsilon\mu = 2(\pi/\lambda)\, n_1 = k_1$. Thus, $k^2 = k_1^2 - \beta^2$.

The reason for this derivation is to demonstrate that Formula 4.42 has physical meaning, rather than merely showing a simple geometric relationship.

Formula 4.44 tells us that before the cutoff is reached, light is trapped inside the fiber by the phenomenon of total internal reflection. When we go beyond the cutoff condition, light experiences partial reflection, with the result that some power is transmitted into the cladding. Leaky modes then appear.

***Power confinement and cutoff condition*** Cutoff condition essentially determines the highest mode that a fiber can support. But the term *support* means that the power of this mode will be confined within a core. This point of view is usually presented in graphical form, as Figure 4.15 shows.

In a sense, Figure 4.15 is the graphical representation of Formula 4.40. The figure shows a fraction of the total light power that resides in the cladding as a function of the $V$-number. A close analysis reveals the cutoff conditions for each mode presented here. For example, if $V < 4$, only the $LP_{01}$ and $LP_{11}$ modes can exist. In fiber-optic communications technology, the term "cutoff" usually means the condition necessary to sustain singlemode operation. Looking at Figure 4.15, we can conclude that if $V \le 2.4$, only one fundamental mode, $LP_{01}$, is supported by the optical fiber. This is where the crucial number $V = 2.405$ comes from. The reader can easily identify the cutoff condition for other modes. Thus, Figure 4.15 gives the cutoff condition in graphical form.

***Limiting the number of modes and the role of cladding*** To restrict the number of modes, we have to decrease the $V$-number, as Figure 4.15 shows. The $V$-number determines the number of modes supported by the fiber. (Recall that $N = V^2/2$ or $N = V^2/4$ for step- or graded-index fibers when $V > 20$.) On the other hand, $V = [(2\pi a)/\lambda]\, NA$ and $NA = n\sqrt{2\Delta}$; hence:

$$V = [(2\pi a n)/\lambda]\sqrt{2\Delta} \qquad (4.50)$$

Thus, to reduce the number of modes, we need to decrease the core diameter, $d = 2a$, increase wavelength, decrease $n = (n_1 + n_2)/2$, and decrease $\Delta = (n_1 - n_2)/n$, that is, decrease $(n_1 - n_2)$. The

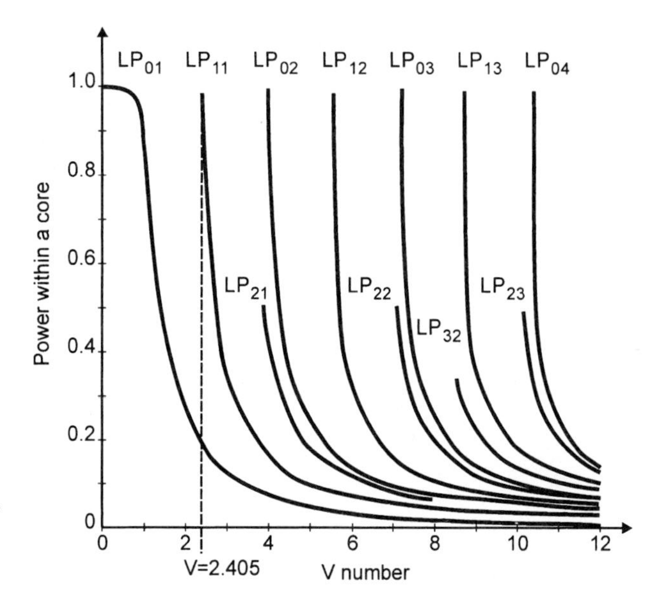


**Figure 4.15** Power-confinement as a function of V-number. *(Adapted from D. Gloge, "Weakly Guiding Fibers,"* Applied Optics, *October 1991, pp. 2252–2258 with permission.)*

straightforward method is to reduce the core diameter, $d = 2a$, but this simple measure makes it difficult to maintain the integrity of the optical fiber, particularly when launching light into it. Formula 4.50 offers a hint at a solution: Make $n_2$ as close to $n_1$ as possible.

Thus, the role of the cladding becomes more apparent. Theoretically, the effect of total internal reflection can be achieved in a fiber by using a glass–air interface, as the examples in Section 2.1 show. However, from a practical standpoint this cannot be done. (Can you explain why?) To have a fiber that is flexible, that is protected from the environment, and that meets the condition of total internal reflection, we must enclose a core with a cladding. To control the number of modes, we need to control $n_1 - n_2$.

Strictly speaking, Formula 4.50 should be rewritten in this form:

$$V = [\pi(2a/\lambda_1)]\sqrt{2\Delta}, \tag{4.50a}$$

where $\lambda_1 = \lambda/n_1$ and $n_1 \approx n$. Thus, there is a tradeoff between $2a/\lambda_1$ and the $\Delta$ or $n_1 - n_2$ fiber parameters that enables us to increase the core diameter by manipulating $\Delta$. This is particularly important for singlemode fibers, which are discussed in Chapters 5 and 6.

***Cutoff wavelength*** Even though the above derivation does not show it explicitly, the cutoff condition is defined by the light frequency, or wavelength. *The lower the frequency (that is, the longer the light wavelength), the greater the propagation angle* $\alpha$ *(that is, the lesser the incident angle* $\Theta$*).* When the angles reach their critical values, the cutoff frequency, or cutoff wavelength, is reached. In the same way we derive cutoff frequency for a rectangular waveguide (see Formulas 4.26 and 4.27), we can derive the cutoff wavelength for each mode supported by the fiber.

On the other hand, we can find the cutoff wavelength, $\lambda_C$, by determining the $V$-number from Figure 4.15 and then calculate $\lambda_C$ from Formula 4.50. Hence,

$$\lambda_C = [(2\pi a n)/V]\sqrt{2\Delta} \tag{4.51}$$

The application of this formula is discussed in Chapters 5 and 6.

***Effective refractive index*** We will conclude our discussion of cutoff wavelength by considering the term *effective refractive index.* As we have already learned, the refractive index is the ratio of free-space velocity to the velocity within a medium, $n = c/v$. The effective refractive index is the ratio of free-space velocity to the velocity of propagation, or guided velocity ($v_{guide}$). The latter is the velocity whose direction is given by the vector β in Figure 4.14. Thus:

$$n_{eff} = c/v_{guide} \tag{4.52}$$

Since, by definition, $v_{guide} = \omega/\beta$ and, by Formula 4.42a, $\beta = k_1 \cos\alpha$ and $k_1 = 2\pi\, n_1/\lambda$, one can obtain for $n_{eff}$:

$$n_{eff} = n_1 \cos\alpha \tag{4.53}$$

This formula is true for step-index multimode fiber. For graded-index fiber, where the refractive-index profile is usually parabolic, the effective refractive index is [10]:

$$n_{eff} = n_1 - (p + q + 1)\,[\lambda(\sqrt{2\Delta})/2\pi a], \tag{4.54}$$

where $p$ and $q$ are the number of times transverse electric- and magnetic-field intensities cross the zero-intensity line (another approach to defining modes), Δ is the relative refractive index, and all other notations are the same as given previously.

It is important to note that the effective refractive index is different for each individual mode. It follows from the definition of $n_{eff}$ given in Formula 4.53, then, that since every mode travels at a different velocity $v_{guide}$, a mode can be understood as a beam traveling at a certain angle α with respect to the fiber's core (Figure 3.9). The propagation angle, α, is unique for each mode. Thus, we arrive, once again, at the relationship between wave and beam interpretation of modes.

It follows from the above discussion that to describe the optical properties of the entire fiber with respect to all the light propagating through it, we have to use the term *group effective index of refraction.* We'll discuss this term in later sections.

**Computer Simulation**

Solving the Helmholtz equations for any possible combination of fiber parameters is a difficult undertaking but, nevertheless, necessary to do in order to design a fiber with the characteristics required. This is why many software companies have developed programs to simulate EM field distribution within an optical fiber. Such programs allow the user to manipulate core and cladding refractive indexes, dimensional characteristics, operating wavelength, and other specifications as may be necessary to suit his or her purposes. This software also gives one the option to obtain results in numerical or graphical form—or both. Graphical presentation allows one to visualize the modes in a fiber in the form of visual field patterns. Adding color makes these pictures very informative and expedites the design process significantly. The chapter references can help the reader find sources of these simulation packages ([15], [16], and [17]).

## 4.5 ATTENUATION IN MULTIMODE FIBERS

As we have seen, attenuation was the major problem in the early years of the development of fiber-optic technology. Since that time, two major breakthrough events occurred. First, research and development work in optical fibers enabled the industry to reduce losses to values close to the theoretical limit. Secondly, the advent of optical amplifiers minimized this problem significantly. However, optical fibers still introduce various losses and we need to understand them to make fiber-optic communications systems meet the modern, high-level performance requirements demanded today. This section provides a more detailed view of the problem.

## General Approach

All losses in optical fibers can be classified in two general categories: intrinsic and extrinsic. *Intrinsic losses* are those associated with a given fiber material and cannot be removed by any improvements in the fabrication and operating processes. *Extrinsic losses* are those associated with fabrication, cabling, and installation processes and, theoretically, can be eliminated under ideal conditions.

Optical networks today use glass fiber almost exclusively. (The only other type—plastic fiber—has not found wide acceptance so far. It will be discussed in detail in Chapter 8.) Glass fibers are made from fused silica, or silicon dioxide ($SiO_2$). The raw material from which it is made is simply sand. To get different refractive indexes, atoms of other materials, called *dopants,* are inserted. Atoms of germanium dioxide, $GeO_2$ increase the refractive-index value, while boron oxide, $B_2O_3$, decreases it. (There are many other dopants but we won't discuss them here.)

The theoretical approach to attenuation is based on the following classic EM theory: The electric flux density, **D,** is related to the electric field intensity, **E,** through permitivity, $\epsilon$, as: $\mathbf{D} = \epsilon\, \mathbf{E}$. When an external electric field is applied to the dielectric, vector **D** is presented as a sum of two parts,

$$\mathbf{D} = \varepsilon_0\, \mathbf{E} + \mathbf{P}, \tag{4.55}$$

where $\varepsilon_0$ is the permitivity of a vacuum, $8.854 \times 10^{-12}$ *F/m,* and **P** is the polarization-density field, or polarization vector ($C/m^2$). The external electric field forces the small displacement of positive and negative bounded charges, thus making electric dipoles. Polarization vector **P** is the electric dipole *moment* per unit volume. In other words, *this vector reflects the response of the dielectric material to the external electric field.* Note that electric flux density **D** will be greater inside the dielectric than in a vacuum since the external electric field induced **P.**

The polarization vector is related to the electric-field strength as

$$\mathbf{P} = \varepsilon_0\, \chi_e\, \mathbf{E}, \tag{4.56}$$

where $\chi_e$ is called the electric susceptibility—the measure of how sensitive to the electric field the given dielectric is. The basic terms you will come across regarding the different media—linear, isotropic, homogeneous—are defined based on their electric susceptibility. *A medium is called* linear *if its electric susceptibility, $\chi_e$, does not depend on the field strength.* (This is a general definition of the term *linear.*) *The term* isotropic *means that the value of the electric susceptibility, $\chi_e$, does not depend on the electric-field orientation. By* homogeneous *we mean that the value of $\chi_e$ is the same regardless of where within the medium we measure the value.*

Strictly speaking, only the isotropic property can be applied with good approximation to fused silica. (Nonlinearity is a topic of interest in today's singlemode fibers and will be discussed in Chapter 5.) In general, electric susceptibility is the complex quantity whose real part is the refractive index and whose imaginary part is the absorption coefficient. Combining Formulas 4.55 and 4.56 and the formula $\mathbf{D} = \epsilon\mathbf{E}$ makes it easy to derive $\epsilon = \epsilon_0(1 + \chi_e)$ or, since $\epsilon = \epsilon_0\, \epsilon_r$, one can obtain:

$$\epsilon_r = (1 + \chi_e) \tag{4.57}$$

This is where $n = \sqrt{\epsilon_r}$ comes from.

The above approach enables us to derive the main attenuation formulas. Calculations based on these formulas allow us to predict attenuation depending on fiber parameters. This theory is developed enough to give quantitative results that coincide fairly closely with experimental measurements. Another possible theoretical approach is based on quantum theory, but since classic theory works well enough, this approach is not very popular.

## Intrinsic Losses

***Material resonances and Rayleigh scattering*** Two factors cause intrinsic losses: (1) material resonances in the ultraviolet (UV) and infrared (IR) regions and (2) Rayleigh scattering. *Material resonances* are associated with the imaginary part of electric susceptibility and can be explained as follows: Molecules, atoms, or even a single electron can be set in oscillation because they experience regenerating forces that originate from neighboring particles. Thus, the model of the classic oscillator—a weight suspended on a spring—can be applied to this case. When an external force is applied to this oscillator, the interaction depends on the relationship between the frequencies of the force and the oscillator. When these frequencies coincide, a resonance condition occurs that results in extensive absorption of energy by the oscillator from the external source. The farther apart these frequencies are, the less the energy that is absorbed from the external source. In our case, the external source is the light propagating through a fiber material and the oscillators are molecules or atoms of this material. Their resonant interaction induces light absorption and, thus, attenuation.

The above explanation is based on the classic model of material and light. If you will recall our discussion in Chapter 2, light is a stream of photons and a material's property is represented by the energy-level diagram (Figure 2.10). Thus, UV and IR intrinsic losses can be explained in terms of a photon's absorption by the proper energy gaps.

Whatever the model one chooses, the fact is simply this: *Silica displays heavy absorption in the UV and IR regions and both absorptions are wavelength dependent.* These absorptions are shown in Figure 4.16. The numbers presented there give an idea of the typical loss range. It's hard to give precise attenuation and wavelength numbers since they vary with different types of fiber and fiber materials.

The other mechanism of intrinsic losses, *Rayleigh scattering,* can also be seen in Figure 4.16. Here's why it occurs: Molecules of silicon dioxide have some freedom when adjacent to one another. Thus, they set up at irregular positions and distances with respect to one another

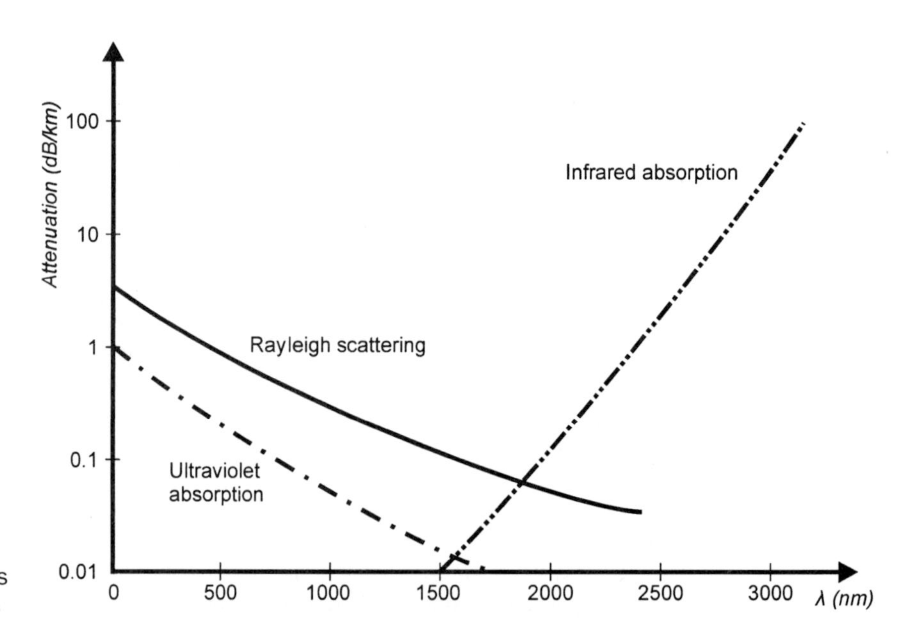


**Figure 4.16** Intrinsic losses in optical fibers.

when the glass is rapidly cooled during the final stage of the fabrication process. Those structural variations are seen by light as variations in the refractive index, thus causing the light to reflect—that is, to scatter—in different directions. (See Section 3.2.) This model does not take into account the size of the scattering centers, which is actually smaller than the operating wavelength. More precisely, the scattering occurs because the EM field excites these irregularities, which work as dipoles. These dipoles absorb the energy of the external field and radiate the new EM waves with the same frequency in any direction. This is the classic model of Rayleigh scattering. Several formulas describe the experimental results. These formulas differ in coefficients and they are different for different types of fiber. What they all have in common is that attenuation is inversely proportional to the fourth power of the wavelength.

***Choice of operating wavelength*** Figure 4.16 helps us to understand the choice of operating wavelength for optical fiber. On the one hand, it would be very convenient to work in the visible spectrum (between 380 and 720 nm), but UV absorption and Rayleigh scattering require the use of a wavelength longer than 2500 nm. On the other hand, using wavelengths longer than 1600 nm causes severe IR absorption. Therefore, *intrinsic losses restrict the range of a practical operating wavelength to between 800 nm and 1700 nm.* If we take into account extrinsic losses, this range becomes even narrower, a matter of great importance for wavelength-division multiplexing (WDM) systems, which we'll discuss in subsequent chapters.

***Other types of fiber*** Ultraviolet and infrared absorptions and Rayleigh scattering sharply limit the degree of control we have to reduce losses in silica-made fiber. Therefore, the only possibility to improve the performance characteristics of a fiber is to make it from other materials. Such a search has been ongoing for years. The most promising is fluoride fiber, whose theoretical intrinsic loss is estimated as low as 0.001 dB/km. The first practical application for this recent development has been in an optical amplifier and broadband light source [18]. Recently, a totally new way of constructing optical silica fiber was developed that has attracted the attention of the industry [19]. This fiber, which has a hole in its center surrounded by an array of microholes is called *photonic bandgap, microstructured* fiber, or even (by some wits) a *holey* fiber [20]. At any rate, with some 75-million miles of the traditional silica-made fiber cable already in place, it is safe to predict that silica fiber will serve as the major transmission medium for the foreseeable future.

## Extrinsic Losses—Absorption

*Extrinsic losses include absorption and bending losses.* Absorption is due to some imperfection introduced during the fabrication process. The major culprit is the hydroxyl group, OH. These molecules enter the silica in the form of water vapor. Their presence results in absorption peaks at 2750 nm. Unfortunately, their oscillations are nonlinear, which means they have subharmonics at 945, 1240, and 1380 nm. These absorption peaks are in the range of working wavelengths, shown in Figure 3.8. This problem raises the necessity of choosing the proper operating wavelengths—so-called transparent windows—from among a wavelength range fraught with loss areas. Fortunately, manufacturers have improved their fabrication process to the extent where they can largely control these losses. Corning says it is able to reduce the absorption peak to 0.92 dB/km while absorption at the 1300-nm operating wavelength is 0.52 dB/km [21]. For singlemode fibers and longer wavelengths, these numbers are even lower. Manufacturers have achieved these results by reducing the hydroxyl-group concentration to less than one part per million by weight.

The above discussion is true for pure silica. However, dopants change the characteristics of this type of extrinsic loss, generally widening the absorption peaks. Germanium-doped silica—

the most popular dopant—is less sensitive to the presence of hydroxyl impurities, while others require an even sharper reduction in the OH level.

## Extrinsic Losses—Bending Losses

Bending losses fall into two categories: macrobending and microbending. (See Section 3.2.) Let's consider them in some detail:

***Macrobending loss*** These losses are caused by curvature of the fiber axis. Theoretically, any fiber bending causes loss but, from a purely practical standpoint, we are interested only in the same loss levels produced by other processing flaws. Theoretical analysis is based on solving the Helmholtz equations for curved fiber. This solution for a slab waveguide shows that when the bending radius is above some critical value, the EM field is oscillatory; the EM field begins to decay exponentially when the bending radius is less than critical. From a practical standpoint, of course, we never reach this critical value. The physical mechanism of macrobending loss is the radiation of light power from the fiber core. Therefore, *the more light power confined within the core, the less sensitive the fiber is with respect to bending.*

It follows, then, that macrobending loss is associated with modes. When the fiber is bent, the higher-order modes disappear first. As Figure 4.17 shows, bending at a radius of 30 mm hardly affects the fundamental mode but introduces severe loss in the second-order mode.

The result shown in Figure 4.17 is intuitively clear from the simple geometric consideration employed in Figure 3.5. The higher the order of the mode, the closer its propagation angle is to the critical angle. Hence, even for a small curvature of fiber axis, the higher-order modes will leak out of the core. This point is clarified in Figure 4.18 where two extremes—fundamental and critical modes—are considered.

Figure 4.18 shows that for higher-order modes some light power radiates into the cladding, where it might be partially transmitted, returned into the core, or escape into the coating.

Surprisingly, there are positive applications of bending loss: (1) filtering and (2) coupling. *Filtering means the elimination of higher-order modes.* When curved, the fiber works as a mode filter. *Coupling means the power from the higher-order mode can be transmitted into the lower-order mode.* (We'll see a possible positive result of this mode "scrambling" when we discuss dispersion in multimode fibers in Section 4.6.)

The key point for you to come away with from this discussion is that, in general, *macrobending loss is inversely proportional to the bending radius.*

***Microbending loss*** Theoretically, microbending loss stems from microdeformations of the fiber axis. Since this axis is the imaginary line that runs along the center of the core, we can say

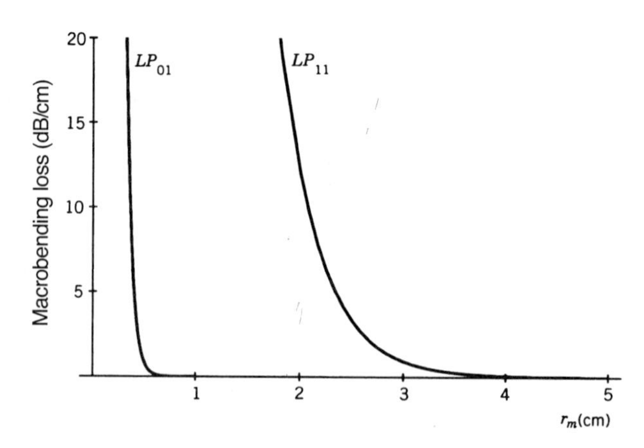


**Figure 4.17** Macrobending loss (dB/cm) versus bending radius. *(John A. Buck,* Fundamentals of Optical Fibers, *New York: John Wiley & Sons, 1995. Reprinted by permission of John Wiley & Sons, Inc.)*

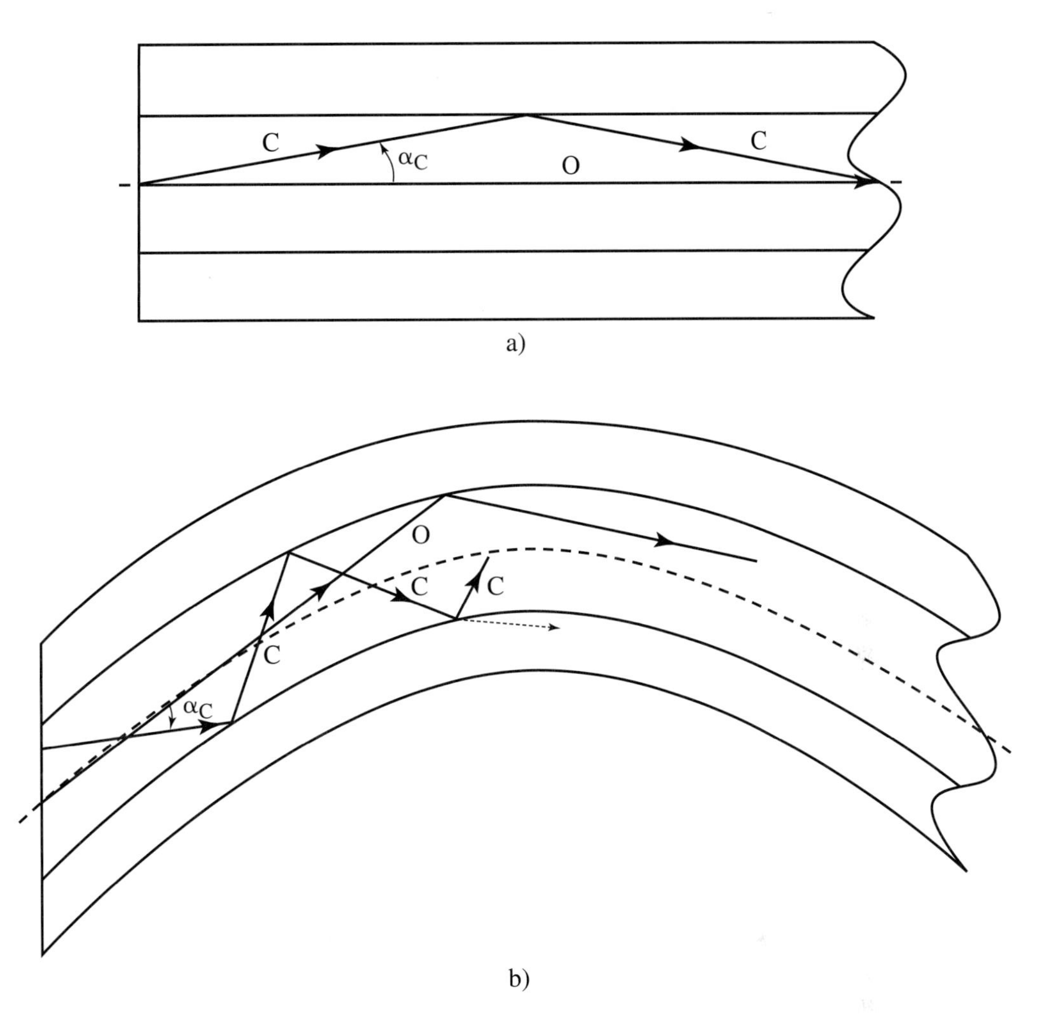


**Figure 4.18** Macrobending losses for fundamental (0) and critical (C) modes: (a) Straight fiber; (b) bent fiber.

that this loss is caused by microdeformations of the core-cladding boundary. A theoretical analysis of this type of loss becomes extremely complex since it has to include the random nature of microscopic imperfections. Nevertheless, formulas for calculating this kind of loss exist, [10], as does simulation software [15].

The physics of microbending loss can be understood by studying Figure 3.6. It is worth adding to the explanation given there about microbending losses that microbending results in transferring power from guided modes to radiation modes.

Surprisingly, a multimode fiber is less sensitive to microbending than is a singlemode fiber. This is because only the higher-order modes are subject to microbending at the core-cladding boundary, and these modes carry only a small portion of the total power. Another important point to note is this: Microbending loss in multimode fiber is practically wavelength independent.

Microbending can be caused by an imperfection in the fabrication process or even induced by mechanical forces during the cabling of bare fiber. Fortunately, however, both technologies are so well developed today that microbending loss is not a major concern in multimode fibers.

## Modes, Attenuation, and Attenuation Constant

***Modes and attenuation*** The fact that many modes exist in a multimode (MM) fiber cannot be overemphasized. This is what causes intermodal dispersion and so severely restricts the bandwidth of an MM fiber. Can we experimentally prove that light inside a fiber is really a set of discrete modes and, therefore, that pulse broadening is really caused by the light's modal structure? Let's perform a simple experiment: Measure the attenuation of an MM fiber, as Figure 3.9 shows. If you measure the attenuation of a 20-m-long cable made from a multimode fiber at 850 nm, for example, you should come up with 4 dB/km. But if you measure the attenuation of the same type of fiber cable that is 60 m long, your result will probably be 3.6 dB/km. If you measure 100-m-long cable, its attenuation should be 3.4 dB/km. If you then measure 200-m-long cable, its attenuation should be 3.3 dB/km. We cannot guarantee these precise numbers, but there is no doubt about the trend: *The longer the fiber, the* less *will be its attenuation (measured in dB/km) until some steady-state value is reached.* The experiment described above can be performed in any college laboratory. In the field, you will measure the attenuation of an MM fiber span in kilometers, not meters; however, you will observe this well-known effect [22].

At this point, let's ask ourselves a simple question: We've introduced attenuation, $A$ (dB/km), in order to have a characteristic independent of the fiber length, as Formula 3.11 states. Now we've discovered that $A$ depends on the length. Why?

The answer is that although Formula 3.11 is correct, it does not take into account the modal structure of light inside a fiber. In reality, *the higher-order modes disappear faster than the lower-order modes, thus causing attenuation to change over the fiber length.* Two major physical mechanisms provoke this disappearance. First, the higher-order modes strike the core-cladding interface much more often than the lower-order modes do; thus, they experience microbending losses much more frequently than the lower-order modes. Secondly, the higher-order modes travel a much longer distance within a fiber; therefore, they have countless more chances to experience scattering and absorption events than do the lower-order modes. Since the higher-order modes carry less power than the lower-order modes, attenuation decreases in a complicated way along the fiber length.

What we want to know is whether attenuation is really independent of fiber length.

Formulas 4.21, 4.21a, and 4.21b show that an electromagnetic wave in a lossy dielectric, such as an optical fiber, dampens exponentially. (See Figure 4.2.) Therefore, we have to expect that light power propagating within a fiber should also decrease as

$$P_{out} = P_{in} \exp(-\alpha L), \tag{4.58}$$

where $P_{out}(W)$ and $P_{in}(W)$ are output and input powers, respectively; $\alpha$ (1/km) is the attenuation coefficient, and $L$ (km) is the fiber length. But looking at Formula 4.58, you might say, "If we keep $P_{in}$ constant and measure $P_{out}$ from fibers of different lengths, we have to have different values. This is why attenuation changes with a change in fiber length, and such changes have nothing to do with fiber modes." True enough, so we seem to have run into a dilemma here.

To resolve it, let's return to the definition of attenuation given in Formulas 3.10 and 3.11:

$$A \text{ (dB/km)} = (1/L)[-10 \log(P_{out}/P_{in})] \tag{4.59}$$

Now substitute $P_{out}$ from Formula 4.58 and obtain:

$$A \text{ (dB/km)} = (1/L)\ [-10 \log(P_{in}e^{-\alpha L}/P_{in})] = 10\,\alpha \log e = 4.34\ \alpha(1/\text{km}) \tag{4.60}$$

Since $\alpha$ is the constant for a given fiber, *the attenuation measured in dB/km is really independent of the fiber length.* Consequently, our conclusion that the decrease in $A$ is caused by the damping of higher-order modes is correct. If you have a better hypothesis about this phenomenon, express it.

***Attenuation and attenuation constant*** Let's take another look at the relationship between attenuation ($A$) and attenuation constant ($\alpha$). The important point here is this: In practice, we measure attenuation ($A$) in dB/km, but for some calculations we use the attenuation constant ($\alpha$) in 1/km. The relationship between them is given by Formula 4.60. Let's study this relationship in more detail.

In Section 2.2 attenuation is defined by Formulas 3.9 and 3.10 (rewritten here):

$$Loss = P_{out}/P_{in}, \tag{4.61}$$

where power ($P_{out}$ and $P_{in}$) is measured in watts. This formula implies that attenuation measured in 1/km is given by

$$A\ (1/\text{km}) = [P_{out}/P_{in}]/L(\text{km}) \tag{4.62}$$

and the input/output power relationship is governed by the following formula:

$$P_{out} = P_{in}\ (AL) \tag{4.63}$$

*This formula shows that output power is proportional to attenuation* A *(1/km).* Usually, as we've seen, attenuation is measured in dB/km. To do so, the following formulas were introduced (Formulas 3.10 and 3.11, rewritten here):

$$Loss\ (\text{dB}) = -10\ \log(P_{out}/P_{in}) \tag{4.64}$$

$$A\ (\text{dB/km}) = [-10\ \log(P_{out}/P_{in})]/L \tag{4.65}$$

This is the most common definition of attenuation. Example 2.5 and the following discussion clarify this definition, which is the basis of fiber-optic test procedure FOTP-78, accepted by fiber manufacturers as the industry standard.

Another quantity describing the loss property of an optical fiber is the attenuation constant introduced in Formulas 4.21, 4.21a, and 4.21b and discussed above. This constant shows that magnitudes of electric and magnetic fields in a lossy dielectric degrade at an exponential rate. Developing this idea, we arrive at the following expression for power attenuation along the fiber [12]:

$$dP/dz = -\alpha\ \text{P}, \tag{4.66}$$

where $\alpha$ is the (power) attenuation constant measured in 1/km.

Let's consider a fiber whose length is $L$ (km) and where $P_{out}$ and $P_{in}$ are the output and input light power, respectively, of this fiber. Then

$$\int_{P_{in}}^{P_{out}} dP/P = -\alpha \int_0^L dz, \tag{4.67}$$

which, after integration, yields

$$P_{out} = P_{in}\ e^{-\alpha L} \tag{4.68}$$

That's where Formula 4.58 comes from. Compare this formula with Formula 4.63. In this definition, *output power depends, exponentially, on the attenuation constant measured in 1/km.*

If we express the attenuation constant, $\alpha$, in dB/km, as

$$\alpha\ (\text{dB/km}) = [-10\ \log(P_{out}/P_{in})]/L, \tag{4.69}$$

we arrive at Formula 4.60:

$$A \text{ (dB/km)} = \alpha \text{ (dB/km)} = 4.34\ \alpha \text{ (1/km)} \quad (4.70)$$

Thus, for $A = \alpha = 0.2$ dB/km we can compute $\alpha = 0.23 \times 0.2 = 0.046$ 1/km. We'll need this conversion in future calculations.

The real question is this: Which quantity—attenuation ($A$) or attenuation constant ($\alpha$)—do we need to use in our considerations? It follows from Formulas 4.63 and 4.70 that

$$\alpha \text{ (1/km)} = -1/\text{L (km)} \ln [A\text{(1/km)}\ L \text{ (km)}] \quad (4.71)$$

or

$$A \text{ (1/km)} = [1/L \text{ (km)}]\ e^{[-\alpha \text{ (1/km) L (km)}]} \quad (4.72)$$

The final criterion is measurement: For $\alpha = A = 0.2$ dB/km, we find $A = 0.95$ 1/km; thus, if $L = 1$ km and $P_{in} = 1$ mW, the power meter shows $P_{out} = 0.95$ mW. On the other hand, under the same conditions, $\alpha = 0.046$ 1/km and $P_{out} = P_{in}\ e^{-\alpha L} = 0.95$ mW.

The conclusion is this: *When using attenuation in dB/km, either* A *or* $\alpha$ *can be used, so Formulas 4.65 and 4.69 are therefore identical. However, when measuring attenuation in 1/km, distinguish between attenuation and attenuation constant and use either Formula 4.62 (for attenuation) or Formula 4.71 (for attenuation constant).*

## 4.6 DISPERSION IN MULTIMODE FIBERS

Dispersion in multimode fibers falls into two categories: intermodal (modal) and intramodal. In this section we will consider both phenomena. The theoretical approach, key results, and practical solutions to existing problems will be given. It is strongly recommended that you reread Section 3.3 to strengthen your understanding of the basics of this topic.

### General Comments

***Definition*** The term *dispersion* describes the dependence of the refractive index, $n$, of a medium on the wavelength, $\lambda$, of light traveling through the medium so that $n = n(\lambda)$. By the definition of the term "refractive index" ($n = v/c$), dispersion implies changing the light velocity inside the medium, depending on its wavelength. Since another definition of refractive index states that $n$ is the measure of the bending of a beam within a medium, dispersion means that different wavelengths are bent at different angles. Thus, the famous Newton experiment involving the separation of sunlight into different colors by a prism—known to every high school student—was the discovery of the dispersion phenomenon.

In optical fibers, dispersion—light traveling at different velocities, depending on the wavelength—causes pulse spread. This is why it is customary in fiber-optic communications technology to gather all the phenomena associated with pulse spreading under one general term: dispersion. This is also why intermodal dispersion is called dispersion, even though we neglected to consider wavelength dependence in our discussion. (See Section 3.3.) You will encounter the term *distortion* throughout our analysis of modal dispersion, the term underscoring the strict sense of the word *dispersion.*

Keep in mind that we distinguish between two types of dispersion: *intermodal,* caused by the presence of many modes within a fiber, and *intramodal,* caused by the effects occurring from the actions of components within a single mode.

We need to understand that dispersion restricts a fiber's bandwidth, or its information-carrying capacity.

***Total dispersion and pulse width*** As mentioned above, dispersion results in the spreading of light pulse (which is an information signal) while it's traveling along the fiber. If we denote the degree of pulse spreading as $\Delta t$, then the sense of Formula 3.21, repeated here, would be clear:

$$\Delta t^2_{\text{total}} = \Delta t^2_{\text{modal}} + \Delta t^2_{\text{chrom}} \tag{4.73}$$

The sum of the squares appears because of the assumption that both components of total dispersion are linear independent.

The obvious question that now arises is this: How can one define pulse width? (Without this definition, pulse spreading, $\Delta t$, remains a qualitative characteristic rather than a number since the real pulses are very far from the ideal rectangular form.) The answer: In fiber-optic communications technology, *pulse width is usually measured as full width at half the maximum of the pulse power (FWHM).* This definition is clarified in Figure 4.19. Of course, pulse width can also be defined in terms of rise and fall time, as is usually the case in digital technology.

***Electrical and optical bandwidth*** Pulse spreading caused by dispersion restricts the information-carrying capacity of a fiber link in the following way: The wider a pulse, the fewer pulses a time-unit interval can accommodate; thus, the bit rate is smaller. (See Example 3.3.3.) The number of pulses per unit of time—the bit rate—is the direct measure of information-carrying capacity, since each pulse carries one bit of information. The Shannon-Hartley theorem (Formula 1.1) states that information-carrying capacity, $C$, is directly proportional to bandwidth, $BW$. This is why it is customary in communications technology to use the terms *bandwidth* and *information-carrying capacity* interchangeably. Manufacturers, however, determine this characteristic of optical fibers as bandwidth (see Figures 3.18, 3.20, and 3.21), the term we will use in our discussion (see Figure 3.17).

Bandwidth, $BW$, by definition, is the frequency range, $\Delta f$, within which either the power or current output–input ratio declines to 3 dB of the maximum value. This is shown in Figure 4.20(a).

To express the result in dB, the following commonly used procedure should be followed:

$$H_{\text{optical}}\text{ (dB)} = 10 \log_{10} P_{\text{out}}/P_{\text{in}}, \tag{4.74}$$

where power is measured in watts. In optics, it's common practice to measure bandwidth at half the power level, that is, $P_{\text{out}}/P_{\text{in}} = \frac{1}{2}$. Formula 4.74 gives: $H_{\text{optical}}\text{ (dB)} = 10 \log_{10} (\frac{1}{2}) = -3$ dB. Thus, the *bandwidth of an optical fiber is the frequency range within which the output power of the information signal drops to half the input value.* (Be careful about the sign. Sometimes you

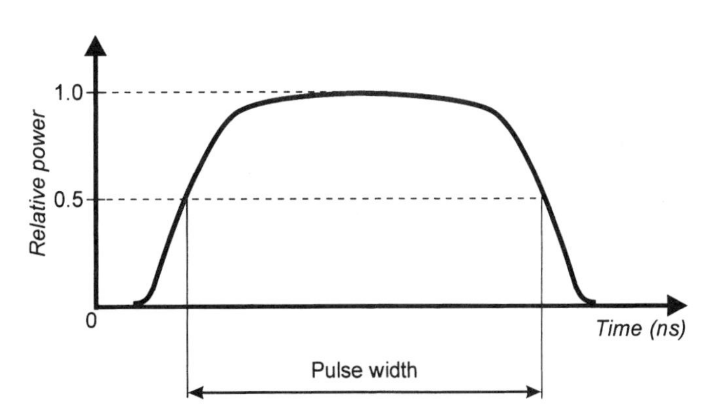


**Figure 4.19** Pulse width defined as full width at half maximum (FWHM) power.

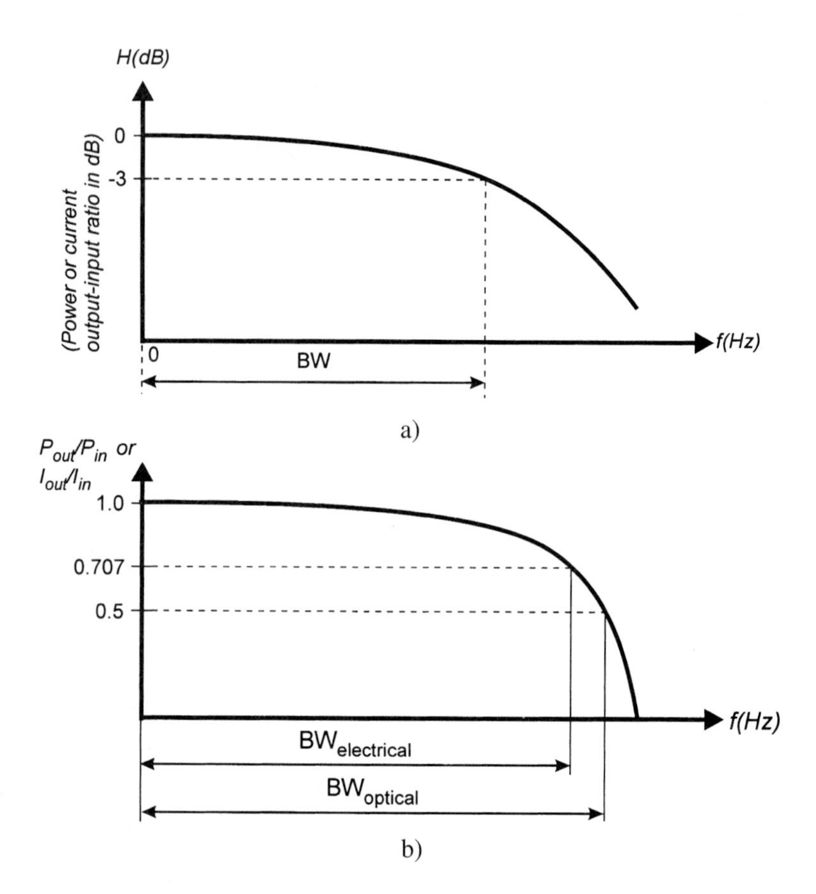


**Figure 4.20** Bandwidth: (a) Definition of bandwidth; (b) electrical and optical bandwidth.

will come across a reference to a +3-dB bandwidth, which means that one is using a minus sign in Formula 4.74.)

Now refer again to Figure 1.4, the basic block diagram of a fiber-optic communications system. The system includes a light source and photodetector. If the bandwidth of the entire system is to be determined, the ratio of electrical output power to electrical input power must be known. But electrical power is proportional to the square of the current; thus, Formula 4.74 becomes:

$$H_{electrical}\,(\text{dB}) = -10\,\log_{10} P_{out}/P_{in} = -10\,\log_{10} I_{out^2}/I_{in^2} = -20\,\log_{10} I_{out}/I_{in} \qquad (4.75)$$

To obtain $H_{electrical}$ = 3 dB, the ratio $I_{out}/I_{in}$ must be 0.707, where both currents are measured in amperes. Figure 4.20(b) shows both situations. Therefore, using the definition of bandwidth determined by Formula 4.74, we come to the conclusion that the optical and electrical bandwidths are different. This is because the light source converts electrical current into optical power and the photodetector does just the reverse. Thus, to measure electrical bandwidth—the system bandwidth—we use the current ratio; to measure optical bandwidth—the bandwidth of the fiber itself—we use the power ratio. *This is why the bandwidth of the optical fiber itself,* $BW_{optical}$, *is greater than the total bandwidth of the entire system.*

The formal approach to this problem is based on defining the optical-fiber transfer function. This attempt has led to rather sophisticated theory without essential results. Today, it is much more common to measure, rather than calculate, the bandwidth of multimode fibers.

## Intermodal (Modal) Dispersion—A Closer Look

***The mechanism of intermodal dispersion*** The primary reason for dispersion in multimode fibers is usually intermodal dispersion. If you'll recall our discussion of this phenomenon in Section 3.3, you know that this type of dispersion is caused by the fact that light inside the fiber propagates in different modes. *The higher-order modes travel a longer distance and arrive at the receiver end later than the lower-order modes. Thus, one modes travels more slowly than another, which means the different modes have different group velocities.*

Since the power is delivered by individual modes, the output-light pulse builds up through a merging of these individual mode pulses. Just because these pulses have a delay time with respect to one another, the output pulse is always much wider than the input pulse. To visualize this explanation, look again at Figure 3.10.

The simple Formula 3.16 was derived from these geometric considerations:

$$\Delta t = (Ln_1/c)\,(n_1 - n_2)/n_2 = (Ln_1/c)\,\Delta \tag{4.76}$$

Since we are always interested in dispersion per unit length, it's worth rewriting Formula 4.76 in its widely accepted form:

$$\Delta t/L(\text{ns/km}) = (n_1\,\Delta)/c \tag{4.77}$$

The calculations using Formula 4.77 shown in Example 3.3.2 for the real fiber give the dispersion at 86.57 ns/km, while the manufacturer measured this number at about 30 ns/km or less. This is a huge discrepancy, so we need to understand what Formula 4.77 does not count and what it does count—but incorrectly. This problem, in turn, leads us to a closer investigation of the assumptions on which Formula 4.77 is based.

***Analysis of Formula 4.77*** The assumptions implicitly made for the derivation of Formula 4.77 are as follows:

- All modes that a given fiber can support are excited.
- Each existing mode carries equal power.
- Modes do not interact with one another.
- All modes travel at the same velocity.
- The refractive index of the fiber core does not depend on the wavelength of the light.

But here's the kicker: *In reality, these assumptions are not true.* To excite all the modes the fiber supports, special measures should be taken. (Remember "overfilled light launch" from Section 4.4?) When the light source is an LED, it's a reasonable assumption that all the modes become excited but, for a laser diode, this assumption is far from reality. Thus, we have here the first reason for the discrepancy. (As a matter of fact, the fewer the modes, the less severe will be the spreading of the output pulse.)

The higher-order modes carry much less power than the lower-order modes. Since the former are prone to diminishing over a long distance, their contribution to pulse widening becomes negligible. This is the second reason for the discrepancy.

Modes interact with one another during propagation so that the higher-order modes exchange energy with lower-order ones and vice versa. (Recall our discussion of "mode coupling" in Section 4.4.) Because of this mode coupling, the power of the initially critical mode does not travel the longest distance, as Figure 3.10 shows, but some intermediate path between the longest and the shortest distances. Formula 4.77 does not take this effect into account; thus, we have the third cause of the discrepancy. In fact, as a result of the mode-coupling effect, pulse spreading is not directly proportional to the fiber length but, starting with some critical value, depends on the square root of the length.

Different modes cannot travel at the same velocity (we mean the velocity determined by the vector $k_1$ in Fig. 4.14) because of the inevitable variations of the core refractive index, $n_1$. Hence, the fourth reason for the discrepancy.

The effective refractive index—the real index that each mode "sees" while traveling along the fiber—depends on wavelength. This is the source of chromatic dispersion. But for modal dispersion, this dependence is the fifth reason for the discrepancy between the theoretical calculations and the experimentally observed dispersion numbers.

Despite the obvious drawbacks of Formula 4.77, it gives us a first-order estimation of modal dispersion. You know that the actual pulse spread cannot be more than that calculated from Formula 4.77.

Much more sophisticated theory does not provide us with a more precise explicit formula for intermodal dispersion. However, more accurate calculations are possible involving graphical solutions of the Helmholtz equations [10]. Computer simulation can help, too ([15], [16], and [17]).

## Chromatic Dispersion—Material Dispersion

***Basic definitions*** As the terms suggest, *intermodal* dispersion is due to the interaction among the modes and *intramodal* dispersion occurs within one mode. Here we reduce intramodal dispersion to *chromatic* dispersion, which is caused by the fact that *an individual mode includes light consisting of different wavelengths, each traveling along the fiber at a different velocity.* This is the classic definition of dispersion; hence, we can use the term *dispersion* here without reservations.

Where do these different wavelengths come from? A light source, even the best one, radiates a bunch of wavelengths—the effect we described using the spectral width.

In optical fibers, chromatic dispersion is composed of two mechanisms: material dispersion and waveguide dispersion. Waveguide dispersion plays an essential role in singlemode fibers, but is negligible in multimode fibers. Material dispersion is the major contributor to chromatic—and consequently, to intramodal—dispersion in multimode fibers.

*Material dispersion is pulse spreading due to the dispersive properties of the material.* The silica refractive index is different for various wavelengths, as shown in Figure 4.21(a). This is the physical reason for material dispersion, which causes pulse spreading only because every light pulse includes several wavelengths radiated by the source. Thus, material dispersion should depend on $n(\lambda)$ and be proportional to the spectral width of the light source.

It is necessary to underscore again that each spectral component (that is, each individual wavelength) travels at a different velocity within a fiber even though all these components propagate along the same path; in other words, they are components of the same mode. Thus, even singlemode fiber exhibits material dispersion.

***Derivation of the formula for material dispersion*** The formula for calculations of material dispersion can be derived as follows: A group of spectral components travels at group velocity: $v_g = \partial w/\partial\beta$. (See Formula 4.36.) Let $\tau_g = 1/v_g$ be a propagation delay per kilometer of the path length—a unit propagation delay (ns/km). Both $v_g$ and $\tau_g$ are wavelength dependent and $\tau_g$, as a function of wavelength, is shown in Figure 4.22.

The graph shows that the pulse at $\lambda = 800$ nm arrives 10 ns later than the pulse at $\lambda = 900$ nm. It also shows that the pulses at wavelengths around 1300 nm have very small variations in arrival time, which means they are almost independent of $\lambda$. There is a wavelength at which propagation delay is zero. This wavelength is called the zero-dispersion wavelength.

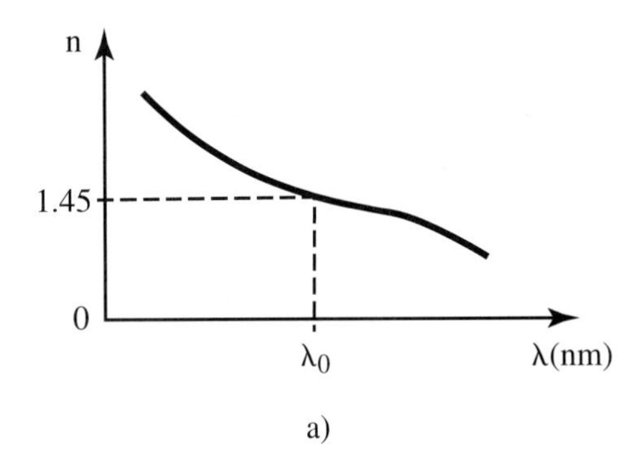


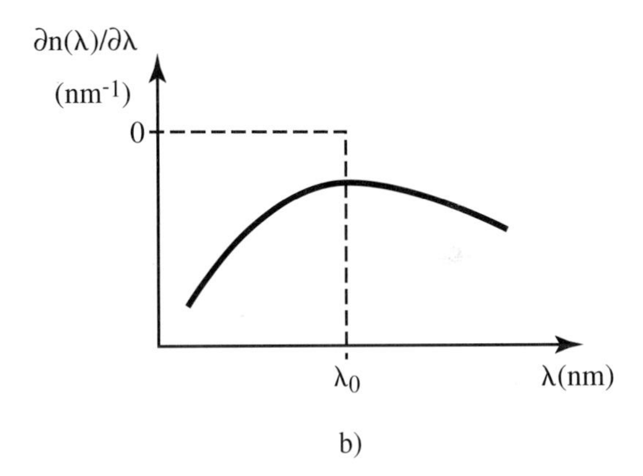


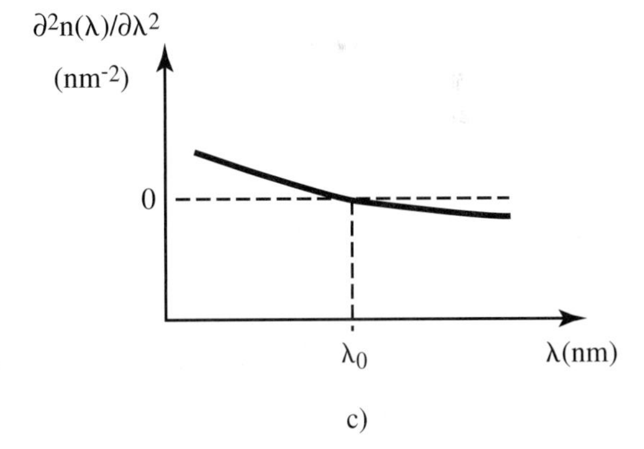


**Figure 4.21** Refractive index of pure silica and its derivatives as a function of wavelength (n = 1.458 at $\lambda$ = 850 nm): (a) $n = n(\lambda)$; (b) $\partial n(\lambda)/\partial\lambda$; (c) $\partial^2 n(\lambda)/\partial\lambda^2$. *(FIBER OPTIC COMMUNICATIONS, 4th ed. by Palais, © 1998. Reprinted by permission of Prentice-Hall, Inc., Upper Saddle River, NJ.)*

To express this physical idea mathematically, let's expand $\tau_g$ into the Taylor series [23]:

$$\tau_g(\lambda) = \tau_g(\lambda_0) + \tau_g(\lambda - \lambda_0)\,\partial\tau_g/\partial\lambda + {}^1\!/_2\,(\lambda - \lambda_0)^2\,\partial^2\tau_g/\partial\lambda^2 + \ldots, \tag{4.78}$$

where $\tau_g(\lambda_0)$ is the unit propagation delay for the chosen wavelength, $\lambda_0$. Let's restrict ourselves to linear approximation and denote:

$$D_{mat}(\lambda) = \partial\tau_g/\partial\lambda, \tag{4.79}$$

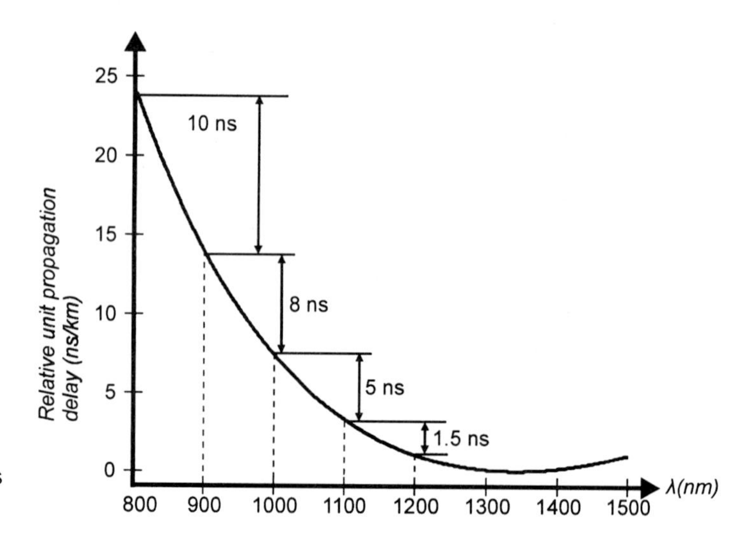


**Figure 4.22** Relative unit propagation delay as a function of wavelength.

where $D_{mat}(\lambda)$ is called the material dispersion parameter, or coefficient. Also, we'll use the notations $\Delta t_{gmat} = \tau_g(\lambda) - \tau_g(\lambda_0)$ and $\Delta\lambda = (\lambda - \lambda_0)$. Now Formula 4.78 is reduced to:

$$\Delta t_{gmat} = D_{mat}(\lambda)\,\Delta\lambda \tag{4.80}$$

Since $\tau_g = 1/v_g$, the parameter $D_{mat}(\lambda)$ can be expressed as follows:

$$D_{mat}(\lambda) = \partial\tau_g/\partial\lambda = \partial[1/v_g]/\partial\lambda \tag{4.81}$$

But $v_g = \partial\omega/\partial\beta$ (see Formula 4.36); thus,

$$D_{mat}(\lambda) = \partial[1/v_g]/\partial\lambda = \partial[\partial\beta/\partial\omega]/\partial\lambda \tag{4.82}$$

Since propagation constant $\beta = \omega/v = \omega n/c$ (see Formula 3.35) and refractive index $n = n(\omega)$, we find

$$\partial\beta/\partial\omega = (1/c)\,\partial[n(\omega)]/\partial\omega = [n + \omega\,\partial n/\partial\omega]/c \tag{4.83}$$

Recalling that $\omega = 2\pi c/\lambda$, we obtain

$$\partial\beta/\partial\omega = [n - \lambda\,\partial n/\partial\lambda]/c, \tag{4.84}$$

which is an intuitively plausible result because Formulas 4.83 and 4.84 say essentially the same: *Group velocity,* $v_g = \partial\omega/\partial\beta$, *is equal to the speed of light in a vacuum,* c, *divided by the group effective index of refraction,* $N_{eff}$, *given by*

$$N_{eff} = n + \omega\,\partial n/\partial\omega, \text{ or } N_{eff} = n - \lambda\,\partial n/\partial\lambda \tag{4.85}$$

Substituting Formula 4.84 into Formula 4.82, we obtain

$$D_{mat}(\lambda) = -(\lambda/c)\,(\partial^2 n/\partial\lambda^2) \tag{4.86}$$

Thus, an explicit expression for material dispersion has been obtained. Graph $D(\lambda)$ versus $\lambda$ is shown in Figure 4.23. Look at Formula 4.86 very closely and compare the graphs in Figures 4.23 and 4.21(c).

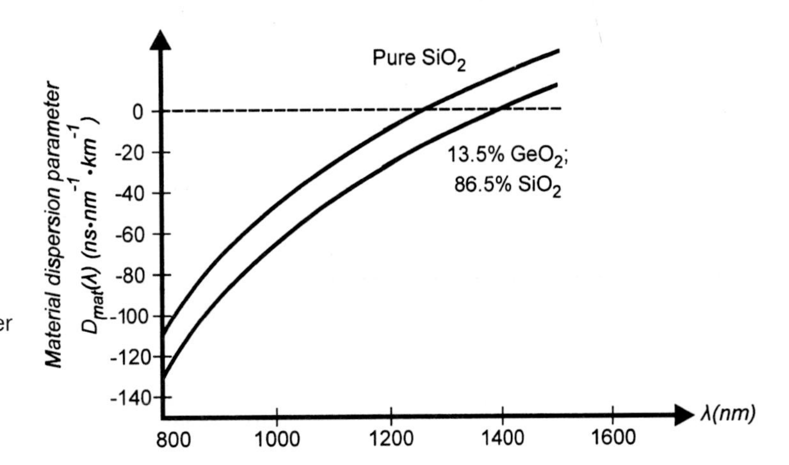


**Figure 4.23** Material dispersion parameter as a function of wavelength (for pure silica, $SiO_2$, and germanium-doped silica, $SiO_2$: $GeO_2$). *(Gerd Keiser,* Optical Fiber Communications, *2nd ed., New York: McGraw-Hill, 1991.)*

The material-dispersion parameter is measured in picoseconds of pulse widening per nanometer of the signal spectral width and per kilometer of the path length.

***Making practical calculations of material dispersion*** Formula 4.86 looks like just another mathematical expression and, for practical calculations, we need to know what $\partial^2 n/\partial\lambda^2$ is. Figures 4.21(b) and 4.21(c) show qualitatively the graphs of the first and the second derivatives of the silica refractive index with respect to wavelength. We skip here all formal derivations. (See "Sellmeier Equations," page 130.)

The result is that the manufacturers of optical fiber use the following formula for calculating the dispersion parameter, $D(\lambda)$ (Formula 4.20, repeated here):

$$D(\lambda)(ps/\text{nm}\cdot\text{km}) = \frac{S_0}{4}\left[\lambda - \frac{\lambda_0^4}{\lambda^3}\right], \tag{4.87}$$

where $\lambda_0$ is the zero-dispersion wavelength (nm), $\lambda$ is the operating wavelength (nm), and $S_0$ is the zero-dispersion slope ($ps$/nm$^2$ · km). The graph $D(\lambda)$ shown in Figure 4.24 (see also Figure 3.16) is essentially the same graph as in Figure 4.23 but calculated for specific parameters. *It is necessary to emphasize that we use a chromatic-dispersion parameter,* D($\lambda$)*, instead of a material-dispersion parameter,* D$_{mat}$($\lambda$)*, because we know that in multimode fibers the material-dispersion parameter is almost equivalent to that of chromatic dispersion.*

It is important to stress again that each mode consists of a band of wavelengths, $\Delta\lambda$, and experiences its own material dispersion. As a result, parameters $\lambda_0$ and $S_0$ depend on the core material and its diameter and on the refractive-index profile. Examples of these parameters for

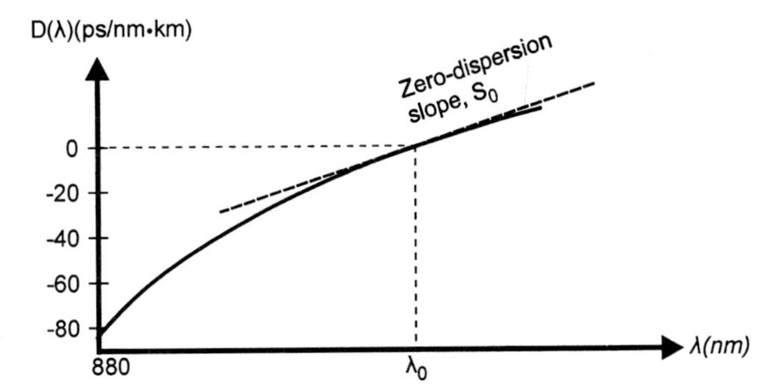


**Figure 4.24** Dispersion parameter, D($\lambda$), as a function of wavelength for $S_0$ = 0.091 ps/nm$^2$ · km$^1$ and $\lambda_0$ = 1341 nm.

### Sellmeier Equations

To step from Formula 4.86 to Formula 4.87, we need to know formula $n = n(\lambda)$ in explicit form. Building this formula theoretically, that is, deriving it based on a theoretical description of fiber material, is not productive. A graph describing this dependence and shown in Figure 4.21 was obtained experimentally. In such a case, the common approach is to find the mathematical expression describing the experimental graph as accurately as possible; in other words, we have to draw the proper curve. Such a formula is known as a Sellmeier equation.

There are several forms of the Sellmeier equation. The industry has accepted as its standard "Recommendation 455–80" of the Electronic Industries Alliance, which defines three-term

$$\tau_g = A + B\lambda^2 + C\lambda^{-2} \tag{4.88}$$

and five-term Sellmeier equations

$$\tau_g = A + B\lambda^4 + C\lambda^2 + D\lambda^{-4} + E\lambda^{-2}, \tag{4.89}$$

where $\tau_g$ is a unit propagation delay in ns/km and coefficients should be determined experimentally. A three-term equation is used for a regular fiber at 1300 nm and a five-term is used at 1550 nm.

Let's derive Formula 4.87 from a three-term Sellmeier equation.

Considering $\partial\tau_g/\partial\lambda = 0$, we can find the zero-dispersion wavelength, $\lambda_0^4 = C/B$. Then $\partial\tau_g/\partial\lambda$ takes the form $\partial\tau_g/\partial\lambda = 2B\,(\lambda - \lambda_0^4/\lambda^3)$. Let's take $\partial D/\partial\lambda$, which is the slope $S_0$ by definition. We get $S_0 = \partial D/\partial\lambda = \partial^2\tau_g/\partial\lambda^2 = 8B$ at $\lambda_0$. Substitute $S_0 = 8B$ and we finally obtain Formula 4.87:

$$D(\lambda) = \partial\tau_g/\partial\lambda = S_0/4[\lambda - (\lambda_0^4/\lambda^3)]$$

Note that we can define $\tau_g$ as

$$\tau_g\ (\text{ns/km}) = \tau\ (\text{ns})/L\ (\text{km}), \tag{4.90}$$

where $\tau$ is a total delay and $L$ is the fiber length. Now Formula 4.81 can be rewritten in this form:

$$D(\lambda) = (\partial\tau/\partial\lambda)/L \tag{4.91}$$

You might often run across this formula in books and manuals.

the popular graded-index fibers are given in Table 4.1. Consult the fiber data sheets given in Figure 3.20 and reference [21].

Now we know what numbers to plug into Formula 4.87.

***Spectral width*** We need to know how much pulse will be spread because of material dispersion. Formula 4.80 allows us to calculate this spread per unit length and Formula 4.90 takes into account the fiber length. Thus, total pulse spreading caused by material dispersion is given by

$$\Delta t_{mat}\ (\text{ns}) = D(\lambda)\, L\Delta\lambda \tag{4.92}$$

Material dispersion is, obviously, an unwanted effect, and Formula 4.92 tells us how we can reduce $\Delta t_{mat}$: Decrease $D(\lambda)$, $L$, and $\Delta\lambda$. It is quite evident that $L$ is given. We'll discuss $D(\lambda)$ in more detail in Chapter 5. We will concentrate here on $\Delta\lambda$.

The number $\Delta\lambda$ is the spectral width of the light signal traveling along the fiber. This width is determined by the spectral width of the light source but we won't distinguish between these two quantities.

**Table 4.1** Parameters for popular graded-index fibers

| Fiber type | $\Delta$ | $\lambda_0$ (nm) | $S_0$ (ps/nm$^2$·km) |
|---|---|---|---|
| 50/125 | 1.0% | 1305 | 0.096 |
| 62.5/125 | 1.9% | 1341 | 0.091 |
| 100/140 | 2.1% | 1349 | 0.090 |

Source: James J. Refi, *Fiber Optic Cable: A LightGuide*, Geneva, Ill.: abc TeleTraining, 1991.

As was mentioned, any real light source radiates a band of wavelengths. These wavelengths concentrate near the peak (central) wavelength, $\lambda_p$. The more a wavelength deviates from the peak wavelength, $\lambda_p$, the less its amplitude. A typical spectral line is shown in Figure 4.25.

For the current discussion, the most important characteristic of this line is its *spectral width, $\Delta\lambda$, which is the width in nanometers at half of maximum power. The greater the spectral width, the more wavelengths emitted by the light source.* The result is broader material dispersion and increased pulse spreading. To reduce pulse spreading, it is necessary to decrease $\Delta\lambda$ as much as possible. We will discuss the physics causing this shape of the spectral line in Chapters 9 and 10. At this point, however, it is important to mention that an LED's typical spectral width is tens of nanometers; a laser diode's is about one nanometer and even less.

## Example 4.6.1

### Problem

Calculate pulse spreading caused by material dispersion for a graded-index multimode fiber working at $\lambda = 850$ nm if the fiber's length is 100 km and the light source is an LED whose $\Delta\lambda = 70$ nm. The given parameters are [21]: $S_0 \leq 0.097$ ps/nm$^2 \cdot$ km and $1332_{nm} \leq \lambda_0 \leq$ 1354 nm.

### Solution

Let's first calculate $\Delta t_{gmat} = D(\lambda)\,\Delta\lambda$. We can compute $D(\lambda)$ using Formula 4.87. Taking $\lambda_0 =$ 1343 nm and $S_0 = 0.097$ ps/nm$^2 \cdot$ km, we obtain:

$$D(\lambda) = S_0\lambda/4[1 - (\lambda_0/\lambda)^4] = -105.484 \text{ ps/nm} \cdot \text{km}$$

The minus sign indicates only that the pulse at, say, $\lambda = 815$ nm travels more slowly than the pulse at $\lambda = 885$ nm. For practical calculations, the sign is meaningless. Plugging the given numbers into Formula 4.80, we obtain:

$$\Delta t_{gmat} = D(\lambda)\,\Delta\lambda = 105.484 \text{ ps/(nm} \cdot \text{km) } 70 \text{ nm} = 7.38 \text{ ns/km}$$

Compared with pulse spread $\Delta t = 0.18$ ns/km caused by intermodal dispersion in a graded-index fiber (see Example 4.4.1), one notices that the material dispersion may be more than modal dispersion in graded-index multimode fibers.

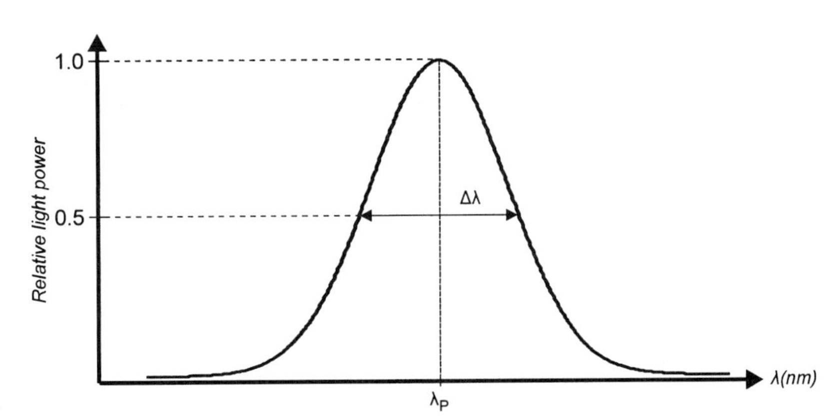


**Figure 4.25** Typical spectral line.

Total pulse spread over the entire fiber is:

$$\Delta t_{mat} = D(\lambda)\, L\, \Delta\lambda = 738 \text{ ns},$$

which is absolutely unacceptable.

We can significantly decrease material dispersion by selecting an operating wavelength, $\lambda$, close to the zero-dispersion wavelength, $\lambda_0$. For an operating wavelength in the second transparent window, $\lambda = 1300$ nm, and the same fiber, one can obtain:

(a) $D(\lambda) = -2.802$ ps/nm · km
(b) $\Delta t_{gmat} = 196.2$ ps/km $\approx 0.20$ ns/km
(c) $\Delta t_{mat} = 20$ ns

These numbers are quite comparable to those from pulse spreading caused by intermodal dispersion. Therefore, to achieve reasonable pulse spreading, one has to choose an operating wavelength as close as possible to the zero-dispersion wavelength (but bear in mind our spectral attenuation graph, Figure 3.8).

Pay attention to the fact that we have used formulas and numbers given in the data sheet for chromatic dispersion in order to calculate material dispersion. Again, this is because in multimode fibers, chromatic dispersion is practically equivalent to material dispersion.

---

## Waveguide Dispersion

Another mechanism contributing to intramodal dispersion is waveguide dispersion, which develops when the propagation constant is dependent on wavelength. In multimode fibers, waveguide dispersion is a small fraction of the total dispersion, so it is quite common to see the terms *chromatic* and *material* dispersion used interchangeably when applied to multimode fibers. For our purposes, we can neglect this effect, but waveguide dispersion is an important dispersion component in singlemode fibers; consequently, we will postpone a detailed discussion until Chapters 5 and 6.

Strictly speaking, material and waveguide dispersions are dependent on each other and so we should consider them together. But since we are dealing here with approximations, we can neglect their mutual dependence and consider them separately, as we have done. This is also true for modal and intramodal dispersions. We saw a direct relationship between them; however, we obtained reasonable results when we dealt with them separately.

## Bandwidth of Multimode Fibers

The basics of bandwidth were covered in Section 3.4, you'll recall, where the difference between bandwidth and bit rate was emphasized. For our current discussion, we'll use bandwidth as a characteristic of the information-carrying capacity of fiber-optic communications systems. Bandwidth is inversely proportional to pulse spreading, $\Delta t_{total}$. The industry dictum is:

$$BW = 1/(4\, \Delta t_{total}) \tag{4.93}$$

***Intermodal (modal) bandwidth*** For step-index fibers, the modal bandwidth is given by Formula 2.23 (repeated here):

$$BW_{modal} = (\text{MHz} \cdot \text{km}) = c\Delta/(4\, L\, n_1) \tag{4.94}$$

For graded-index fibers, we obtained Formula 3.24 (repeated here):

$$BW_{modal}\ (\text{MHz} \cdot \text{km}) = 2c/(N_1\, L\, \Delta^2) \tag{4.95}$$

While for step-index fibers we can use Formula 4.95 to get a rough estimation of modal bandwidth, the above discussion convinced us that this simple approach is not adequate for graded-index fibers. *The key point is that the group effective index of refraction,* $N_1$, *depends on a wavelength.* Therefore, the maximum achievable *bandwidth* depends not only on $\Delta$ but also on a specific wavelength Figure 4.26 shows a theoretical modal bandwidth as a function of wavelength for graded-index multimode optical fibers.

These graphs divulge a very important point: Even for modal dispersion, wavelength dependence is crucial. The graphs in Figure 4.26 were calculated by an empiric formula [24]:

$$BW_{\text{modal}}\ (\text{MHz} \cdot \text{km}) = 14.6/\{\Delta[28(\lambda p - \lambda)^2 + 66\Delta(\lambda p - \lambda)^2 + 150\ \Delta^2\}^{0.5}], \tag{4.96}$$

where $\lambda p$ is the peak wavelength at which the modal bandwidth has been optimized and $\lambda$ is the operating wavelength. The measured real bandwidth is almost ten times smaller than the theoretical limits shown in Figure 4.26; therefore, one should be very careful in applying Formula 4.96.

***Material bandwidth*** Material bandwidth is given by the following self-explanatory formula:

$$BW_{\text{material}} = 1/(4\ \Delta t_{\text{mat}}) = 1/[4\ D(\lambda)L\Delta\lambda] \tag{4.97}$$

Again, we can use Formula 4.97 to get a rough estimation but, taking into account the interaction of light and fiber, we can achieve a more detailed description, as seen in Figure 4.27.

We can well expect to see wavelength dependence here. These graphs were constructed in accordance with the following formula [24]:

$$BW_{\text{material}}\ (\text{MHz} \cdot \text{km}) = (400 \times 10^3)/\Delta\lambda\ \{[\text{D}(\lambda)]^2 + 0.09\ [\Delta\lambda\ D'(\lambda)]^2\}^{0.5}, \tag{4.98}$$

where $D'(\lambda) = S_0/4[1 + 3(\lambda_0/\lambda)^4]$. The operating wavelength, $\lambda$, is considered to be the central emitting wavelength of a light source. Formula 4.98 is also empiric and the measured bandwidth is less than these calculations predict.

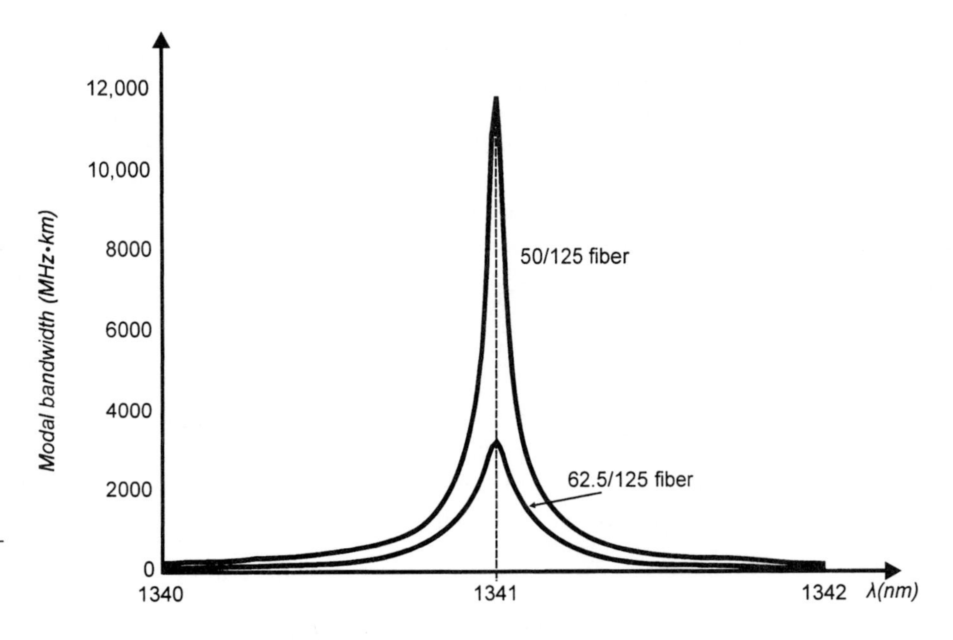


**Figure 4.26** Theoretical modal bandwidth as a function of wavelength for two graded-index fibers.

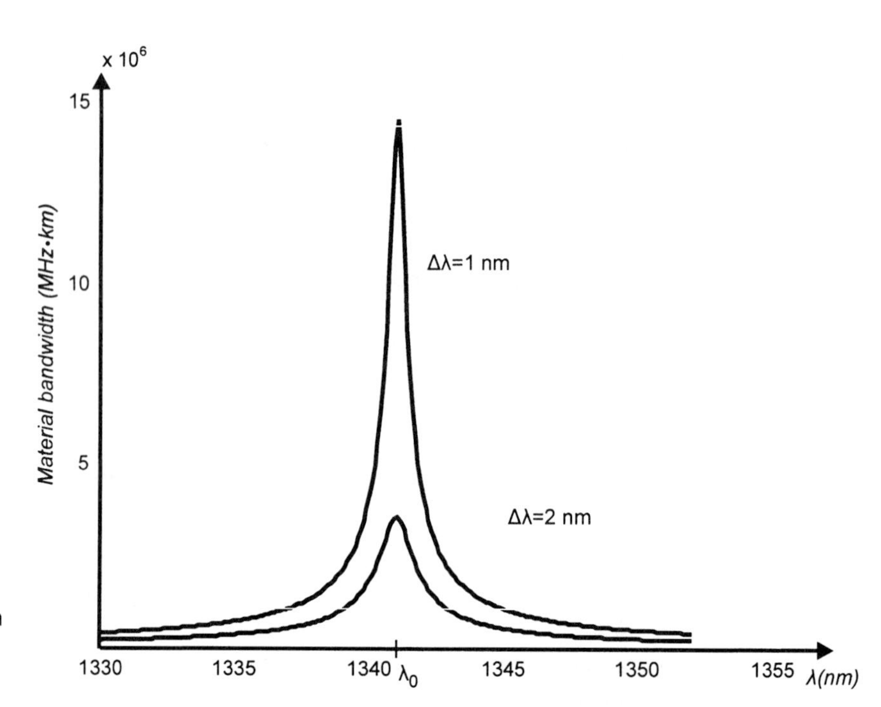


**Figure 4.27** Material bandwidth as a function of light-source wavelength.

***Choice of operating wavelength*** Examining graphs in Figures 4.26 and 4.27, which summarize all dispersion considerations, one observes that the best bandwidth figures can be achieved at a wavelength around 1300 nm. This is the natural result of the dispersion properties of silica displayed in Figure 4.21. Combining this result with our discussion of attenuation phenomena (see Sections 3.2 and 4.5), where the wavelength of 1300 nm was shown as the center of the second transparent window, we can conclude that 1300 nm is the wavelength of choice for operation with multimode fibers.

***Dispersion power penalty*** The dispersion-bandwidth limitation that we considered so far is caused by pulse spreading, which results in intersymbol interference. But for pulse on–off transmission, another source of bandwidth limitation exists in the presence of dispersion.

*In general, because of attenuation, pulses may become so small in amplitude that a detector may misread a pulse-carrying logic 1 "on" for a logic 0 "off" pulse, thus resulting in bit errors. Increasing bit error rate (BER) caused by attenuation can be blocked by increasing the transmitting power. The amount of power required for such compensation is called* power penalty.

Intersymbol interference can be removed at the receiver end by filtering, provided that the amplitudes of adjacent bits are large enough for the success of this operation. However, when dispersion is combined with attenuation—as always happens with real fibers—a specific dispersive power penalty appears. This power penalty, $P_D$, can be calculated by the following formula [10]:

$$P_D(\text{dB}) = -10 \log_{10}\{\exp[-(1/4) (\Delta t_{\text{total}})^2 (\pi BR)^2]\}, \tag{4.99}$$

where $\Delta t_{\text{total}}$ is the total dispersion-caused pulse spreading and $BR$ is the bit rate relating to radian frequency, $\omega$, as $\omega = \pi BR$. By transferring to a natural logarithm by means of $\log_{10}(x) = 0.434 \ln(x)$, introducing *rms pulse spread,* $\sigma_t$, as $\sigma_t = \Delta t/\sqrt{2}$, and having the power penalty, $P_D$, equal to 1 dB, one can arrive at this familiar relationship:

$$BR \leq 1/(4\,\sigma_t), \tag{4.100}$$

which differs from Formula 3.22 by using rms instead of the magnitude value of pulse spreading.

***Bandwidth-length product limit*** Pulse spreading is proportional to the fiber length, causing a dispersion limit in the length of the transmission link. This limit can be estimated by using the bandwidth-length product, which is a very general characteristic of a communications system. The following formula can be obtained from Formula 4.100 and the discussion developed in this section:

$$(BR \times L)_{max} = 1/[4\, \sigma_\lambda\, D(\lambda)], \tag{4.101}$$

where the bandwidth is represented by the bit rate; $\sigma_\lambda$ is the rms of the spectral width; $\sigma_\lambda = \Delta\lambda/\sqrt{2}$; and the chromatic dispersion parameter, $D(\lambda)$, is given by Formula 4.87. Calculations for $|D(\lambda)| = 2.8$ ps/nm · km and $\sigma_\lambda = 49.5$ nm (see Example 4.6.1 above) give

$$(BR \times L)_{max} = 1.8 \text{ Gbps} \cdot \text{km}$$

Formula 4.101 can be rewritten in this form:

$$BR \times L \times D(\lambda) \times \sigma_\lambda < 1/4 \tag{4.101a}$$

Given $D(\lambda)$ and $\sigma_\lambda$, both formulas allow us to calculate the limitation of the bit-rate-length product.

## SUMMARY

- Maxwell's equations are a set of four differential equations describing all the properties of an electromagnetic field. These equations can also be presented in the integral form. They explicitly show sources of the field and implicitly tell how an electromagnetic field propagates. These equations laid the foundation for electromagnetics as we know it today.

- Maxwell's equations enable us to derive wave equations describing the propagation of EM waves. The use of the time-harmonic EM field significantly simplifies these equations. Solutions to the wave equations for free space show that an EM field exists in the form of EM waves harmonically dependent on space and time simultaneously. Solutions to the wave equations for a lossy medium prove that an electromagnetic field propagates in the form of damping waves.

- It's important to study an EM field not only in unbounded media but also within waveguides because an optical fiber confines light largely within the core, thus guiding the EM waves. Maxwell's equations are the basis for the waveguide theory. Our brief study of a rectangular waveguide, the classic example of a waveguide, results in two main conclusions: (1) A waveguide can support only the specific set of EM waves that meet the resonant conditions and (2) there is a cutoff frequency below which a waveguide cannot support EM radiation.

- Wave equations, together with the waveguide-boundary conditions, describe the propagation of EM waves in a waveguide. Solving these equations shows that the EM field can propagate within an optical fiber not as a continuum but as a set of discrete field patterns called modes. Light power propagates along the fiber in a specific mode at a speed called "group velocity." Total light power is delivered from a transmitter to a receiver in small fractions carried by individual modes.

- A close examination of the phenomenon of total internal reflection reveals two important points: First, in an optical fiber, even with total internal reflection, some portion of the incident field will penetrate the second medium (the cladding). This portion is called the *evanescent wave*. Second, totally reflected waves experience phase shifts. For multimode optical fibers, different modes travel inside the core at different angles; therefore, different modes experience different phase shifts.

- A deeper study of attenuation shows that there are intrinsic and extrinsic losses in an optical fiber. The intrinsic losses restrict the range of a practical operating wavelength to between 800 nm and 1700 nm because of ultraviolet and infrared absorption and Rayleigh scattering. Extrinsic losses, consisting of absorption and bending, contribute significantly to the total attenuation of an optical fiber.

- Dispersion in multimode fibers consists of intermodal and chromatic dispersion. An actual pulse spread caused by intermodal dispersion is less than Formula 4.77 predicts. Chromatic dispersion is due primarily to material dispersion, and pulse spread caused by this dispersion is comparable to that of intermodal dispersion. Dispersion restricts the bandwidth of an optical fiber. Another factor restricting a fiber's transmission capacity is the increase in bit-error rate due to attenuation. To compensate for this signal deterioration, an additional amount of power called *power penalty* has to be transmitted.

## PROBLEMS

**4.1.** Derive the integral form of Maxwell's equations from a given set of these equations in differential form (Formula 4.1). Show your work step by step.

**4.2.** Explain the physical meaning of each of Maxwell's equations.

**4.3.** Maxwell's fourth equation introduces two currents—conduction and displacement—but in circuit analysis we operate with only one current. Why?

**4.4.** A plane transverse electric wave is polarized along the $y$ axis. Derive wave equations for this EM field. (See Formulas 4.9 and 4.10.)

**4.5.** Solve the equations you derived in Problem 4.4.

**4.6.** What is a time-harmonic EM field? Why is this an important consideration?

**4.7.** Solve the equations you derived in Problem 4.4 for time-harmonic EM waves.

**4.8.** The EM waves described in Problem 4.4 propagate in a lossy dielectric. What will be a solution for the wave equations in this case?

**4.9.** What is the difference between an EM field propagating in free space and one propagating within a rectangular waveguide?

**4.10.** What are *eigenvalues?* What parameters do they depend on? Are they discrete or continuous?

**4.11.** What are the differences between TE, TM, and HE modes?

**4.12.** Sketch variations of field components for the $TM_{10}$ mode in a rectangular waveguide.

**4.13.** What are modes in a rectangular waveguide? Why are they only discrete?

**4.14.** Analyze the formula for the cutoff condition. What is its physical meaning?

**4.15.** Does a cutoff frequency depend on an eigenvalue? If so, how?

**4.16.** What are unbounded, waveguide, and cutoff wavelengths? Is there any relationship among them?

**4.17.** If a plane TEM wave strikes the boundary of two dielectrics, what happens to this wave?

**4.18.** Describe an evanescent wave.

**4.19.** Light falls at the glass–air interface. What power of this light will be reflected and transmitted if the refractive index of the glass is 1.5? Make these calculations for both polarizations and the following values of the angle of incidence: 0°, 45°, 85°.

**4.20.** What are the phase shifts of light at the glass–air boundary if the angle of incidence is 85° and the refractive index of the glass is 1.5? Make these calculations for both polarizations.

**4.21.** What are natural modes in an optical fiber?

**4.22.** What kind of modes—natural or linear polarized—truly propagate in an optical fiber? Explain your answer.

**4.23.** How can we visually distinguish among the different types of LP modes?

**4.24.** Why do discrete modes, not a continuous EM field, exist within an optical fiber?

**4.25.** The text introduced the meridional and skew beams. Is there any relationship between these beams and LP modes?

**4.26.** What are the differences between guided, radiation, and leaky modes?

**4.27.** What is the difference between phase and group velocities?

**4.28.** Explain the physical meaning of Formula 4.38.

**4.29.** How is light power delivered within a fiber?

**4.30.** A graded-index fiber has the following characteristics: $NA = 0.200$, $d_{core} = 50\ \mu m$ and $\lambda = 1300$ nm. What power is carried by the fiber's cladding?

**4.31.** The number of excited modes in a multimode fiber affects the measured attenuation. Why? What means must be taken to reproduce the result of your measurement?

**4.32.** What is meant by *cutoff condition?* What is the relationship between the cutoff condition and the condition of total internal reflection?

**4.33.** What is the relationship between power confinement and cutoff condition in an optical fiber?

**4.34.** Why do we need cladding?

**4.35.** Describe the role of cladding in terms of power propagation within a fiber.

**4.36.** We defined *cutoff wavelength* twice: in Formula 4.27 and in Formula 4.51. Is there any relationship between these definitions? If so, in what way?

**4.37.** Manufacturers do not specify the refractive index of a fiber's core, but, rather, its effective refractive index. Why?

**4.38.** What are extrinsic and intrinsic losses in an optical fiber?

**4.39.** What are the inherent restrictions on the range of operating wavelengths for a silica optical fiber?

**4.40.** Name and explain all the sources of intrinsic losses in optical fibers.

**4.41.** Name and explain all the sources of extrinsic losses in optical fibers.

**4.42.** Is the value of macrobending loss different for various modes or not? Explain.

**4.43.** What phenomena put an absolute limit on decreasing loss in an optical fiber? Do you have any suggestions to resolve the problem?

**4.44.** Which type of fiber—multimode or singlemode—is more sensitive to microbending and why?

**4.45.** What is the difference between *dispersion* and *distortion?*

**4.46.** What does "FWHM" stand for?

**4.47.** Define "bandwidth." Why are electrical and optical bandwidths different?

**4.48.** What is the relationship between bit rate and bandwidth?

**4.49.** Pulse spreading caused by modal dispersion in a step-index fiber is given by Formula 4.77. How accurate is this formula?

**4.50.** What is chromatic dispersion? What phenomena contribute to it?

**4.51.** Calculate the pulse spreading caused by chromatic dispersion for BF04431-02 ACU-MDO62D multimode graded-index fiber from SpecTran (see Figure 3.20) operating at 1300 nm.

**4.52.** A manufacturer specifies the chromatic-dispersion parameter in multimode fibers by Formula 4.87. Derive this formula and explain the physical meaning of each of your steps.

**4.53.** Why is the spectral width of a light signal important? How do you measure spectral width?

**4.54.** What wavelength is the most suitable for multimode fibers and why?

**4.55.** What is meant by the term *dispersion power penalty*? Calculate the dispersion power penalty for the pulse spread obtained in Problem 4.51. The *BR* is 2.5 Gbit/s.

**4.56.** What is the bit-rate length limitation caused by chromatic dispersion for the SpecTran fiber referred to in Problem 4.51?

**4.57.** Add all the new topics and relationships obtained from Chapter 4 to the memory map you sketched in response to Problem 3.30.

[3]See Appendix C: A Selected Bibliography

## REFERENCES[3]


1. Paul Fishbane, Stephen Gasiorowicz, and Stephen Thornton, *Physics for Scientists and Engineers,* Englewood Cliffs, N.J.: Prentice Hall, 1993.
2. Matthew N. O. Sadiku, *Elements of Electromagnetics,* 2nd ed., New York: Oxford University Press, 1995.
3. David K. Cheng, *Fundamentals of Engineering Electromagnetics,* Reading, Mass.: Addison-Wesley, 1993.
4. John D. Krauss, *Electromagnetics,* 4th ed., New York: McGraw-Hill, 1992.
5. Max Born and Emil Wolf, *Principles of Optics,* 6th ed. (with corrections), Cambridge, U.K.: Cambridge University Press, 1980.
6. Stephen Lipson, Henry Lipson, and David Tannhauser, *Optical Physics,* 3rd ed., Cambridge, U.K.: Cambridge University Press, 1995.
7. Ajoy Ghatak and K. Thyagarajan, *Introduction to Fiber Optics,* New York: Cambridge University Press, 1998.
8. Joseph C. Palais, *Fiber Optic Communications,* 4th ed., Englewood Cliffs, N.J.: Prentice Hall, 1998.
9. Gerd Keiser, *Optical Fiber Communications,* 2nd ed., New York: McGraw-Hill, 1991.
10. John A. Buck, *Fundamentals of Optical Fibers,* New York: John Wiley & Sons, 1995.
11. Govind P. Agrawal, *Fiber-Optic Communication Systems,* 2nd ed., New York: John Wiley & Sons, 1997.

12. Leonid Kazovsky, Sergio Benedetto, and Alan Willner, *Optical Fiber Communication Systems,* Boston: Artech House, 1996.

13. Dietrich Marcuse, "Selected Topics in the Theory of Telecommunications Fibers," in *Optical Fiber Telecommunications -II,* ed. by S.E. Miller and I.P. Kaminow, San Diego: Academic Press, 1988, pp. 55–119.

14. Rajappa Papannareddy, *Introduction to Lightwave Communication Systems,* Boston: Artech House, 1997.

15. Apollo Photonics, Inc., *Optical Waveguide Mode Solver,* version 1.1 (product catalog), Waterloo, Can., 1998.

16. RSoft, Inc., *LinkSim 1.1* (product catalog), Ossining, NY., 1999.

17. Zkom, *Photoss* (*The Photonic Systems Simulator*), Dortmund, Germany, September 1999.

18. Galileo Corporation, *FlouroAmp* and *FluoroLight* (data sheets), Sturbridge, Mass., 1998.

19. Barbara Goss Levi, "A New Way to Guide Light in Optical Fibers," *Physics Today,* December 1999, pp. 21–23.

20. "Microstructured Optical Fiber," in the Advance Program of *Optical Fiber Communication Conference, OFC 2000,* March 5–10, 2000, Baltimore, Md., pp. 54–56.

21. *Corning 62.5/125 CPC6 Multimode Optical Fiber* (data sheet), Corning Incorporated, Corning, NY., 1996.

22. James Hayes, "Testing the Fiberoptic Cable Plant for Gigabit Ethernet," *Fiberoptic Product News,* March 1999, pp. 25–28.

23. Max Ming-Kang Liu, *Principles and Applications of Optical Communications,* Chicago: Irwin, 1996.

24. James J. Refi, *Fiber Optic Cable: A LightGuide,* Geneva, Ill., abc TeleTraining, 1991.

25. D. Gloge, "Weakly Guiding Fibers," *Applied Optics,* October 1971, pp. 2252–2258.

# 5

# Singlemode Fibers—Basics

It is impossible to overestimate the importance of singlemode fibers in today's fiber-optic communications systems. Developed as a possible solution to the intermodal-dispersion problem, singlemode fibers passed through a difficult period of research, a period when it was impractical to use a fiber with a core diameter less than 10 μm. But today singlemode fibers are the dominant optical links found in the field. Not only do long-distance connections use them exclusively but, increasingly, intermediate- and even short-distance communications links are being based on singlemode fibers. In fact, were you to attend a conference today on fiber-optic communications, you would hear presentations in the optical-fibers section that were devoted almost exclusively to singlemode fibers. What's more, if speakers on other topics at the event were to mention the term "optical fiber," they wouldn't even bother to explain that they meant singlemode fiber. They would assume that this fact was quite obvious to every professional in attendance. By some estimates, 88% of all installed fibers are of the singlemode type. This, then, is why we are devoting two chapters to singlemode fibers and their properties.

## 5.1 HOW A SINGLEMODE FIBER WORKS

A singlemode fiber conducts light, as any optical fiber does, but there are many differences between multimode and singlemode fibers. These differences start with the principle of action and run through all the fiber characteristics. In this section, we will discuss how a singlemode fiber conducts light.

### The Principle of Action

***Making a fiber that supports only one mode*** As the term "singlemode" suggests, this fiber supports only one mode, that is, one beam propagating along the fiber's centerline. (In older literature, you find the term "monomode fiber" used; it reflects the same idea.) Why do we need this type of fiber? Refer to Section 3.3, "Intermodal Dispersion," where we found that multimode fiber suffers from high pulse-signal spreading (dispersion). Thus, the singlemode fiber is the solution to the intermodal-dispersion problem.

How can we make the fiber conduct only one mode? You'll recall from Section 3.3 that the number of modes, $N$, is defined by Formulas 3.15 and 3.15a:

$$N = V^2/2 \text{ (for step-index fiber)} \quad \text{or} \quad N = V^2/4 \text{ (for graded-index fiber)}$$
$$\text{with } V = (\pi d/\lambda)\sqrt{[(n_1)^2 - (n_2)^2]}, \tag{5.1}$$

where $d$ is the core diameter, $\lambda$ is the operating wavelength, and $n_1$ and $n_2$ are the core and cladding refractive indexes, respectively. Theory shows and experiments confirm that for singlemode operation we need to have $V = 2.405$ or less. Therefore, to arrange the fiber to conduct only one mode, we can either decrease the core diameter, $d$, increase the operating wavelength, $\lambda$, or make $n_2$ as close to $n_1$ as possible.

Manufacturers use all these means to achieve their goal. The core diameter of a singlemode fiber is around 10 μm and less, the range of operating wavelengths typically starts at 1300 nm, and the relative index, $\Delta = (n_1 - n_2)/n$, is less than 0.4%. The result is a fiber that rejects all higher-order modes and conducts only one fundamental mode—a beam traveling exactly along the centerline of the fiber.

To refresh your memory, the same characteristics of typical multimode fibers are a core diameter of 50, 62.5, or even 1000 μm; an operating wavelength range starting in the visible light region (650 nm); and a relative refractive index, $\Delta$, that is a minimum of 1% and, typically, 2% and higher. One can easily see the contrast between these types of fiber.

***How a singlemode fiber conducts one beam*** When a fiber is straight, conducting one beam along the central axis is not a problem. But when a fiber is bent, the beam has to experience the condition of total internal reflection to be completely trapped. This situation is shown in Figure 5.1.

Let's consider these typical values: refractive index $n_1 = 1.4675$ and refractive index $n_2 =$ 1.4622. Thus, the critical propagation angle—defined by Formula 2.2, $\alpha_C = \sin^{-1}\sqrt{[1 - (n_2/n_1)^2]}$ —is equal to 3.45°. This angle is 2.7 times less than the critical propagation angle, 9.43°, we calculated for the multimode fiber. (See Example 3.1.2.) Consequently, the requirement to trap the beam in a singlemode fiber is much tougher than it is to trap the beam in a multimode fiber. We will refer to this number—the critical propagation angle—when attenuation becomes the point of discussion.

### Example 5.1.1

***Problem:***

Calculate the numerical aperture, $NA$, of a singlemode fiber where $n_1 = 1.4675$ and $n_2 = 1.4622$.

***Solution:***

Recall Formula 3.5: $NA = \sqrt{[(n_1)^2 - (n_2)^2]}$. Hence, $NA = 0.125$. One can see how difficult it is to direct a beam from a light source into a singlemode fiber compared with doing the same thing using a multimode fiber. Since an LED radiates light as a wide beam, this type of source obviously doesn't work well with singlemode fibers. It is therefore not surprising to find that laser diodes are the light sources of choice in singlemode-fiber systems.

## Gaussian Beam

A beam of light does not have strict cross-sectional boundaries. Observing a cross-sectional beam's structure, we readily see that the beam is most intense in the center, with its intensity declining gradually from the center outward. The reader can easily arrange a simple experiment that

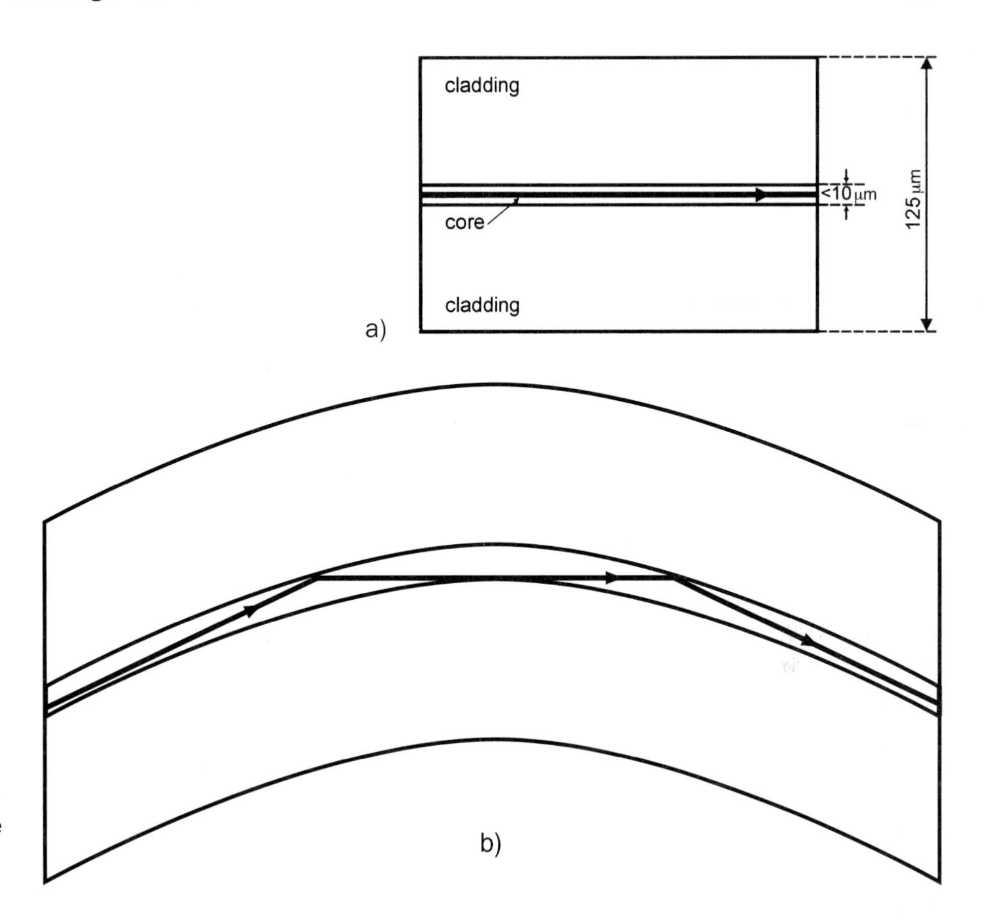


**Figure 5.1** Beam propagation within a singlemode fiber: (a) Straight fiber; (b) bent fiber.

directs an LED's or LD's beam toward a white screen. The result will look like what is shown in Figure 5.2(a). This physical appearance is depicted by the curve in Figure 5.2(b).

This curve describes the change of a beam's intensity as a function of the beam's cross-sectional radius. The most popular model used in singlemode fibers is a Gaussian curve, given by:

$$I(r) = I(0)\exp(-2r^2/w_0^2), \tag{5.2}$$

where $I(r)$ is the current value of the beam's intensity at the radius $r$, $I(0)$ is the maximum beam intensity at $r = 0$, and $w_0$ is the mode-field radius.

How can we compare different beams? One beam has a wider curve, another a narrower curve. But these are qualitative, not quantitative, characteristics. Observe in Figure 5.2 that there is no boundary to a beam's intensity and so, theoretically, a beam's light spreads to infinity. Hence, there is no natural measure of distribution of the beam's intensity. As usual in such a case, we have to introduce some measure by convention. This measure, known as the *mode-field diameter (MFD),* is equal to $2w_0$.

If you plug $r = w_0$ into Formula 5.2, you obtain:

$$I(r) = I(0)/e^2 = 0.135\ I(0) \tag{5.3}$$

Thus, *mode-field diameter, MFD, is the cross-sectional dimension* $2w_0$, *where the beam's intensity drops to* $1/e^2 = 0.135$ *of its peak value.*

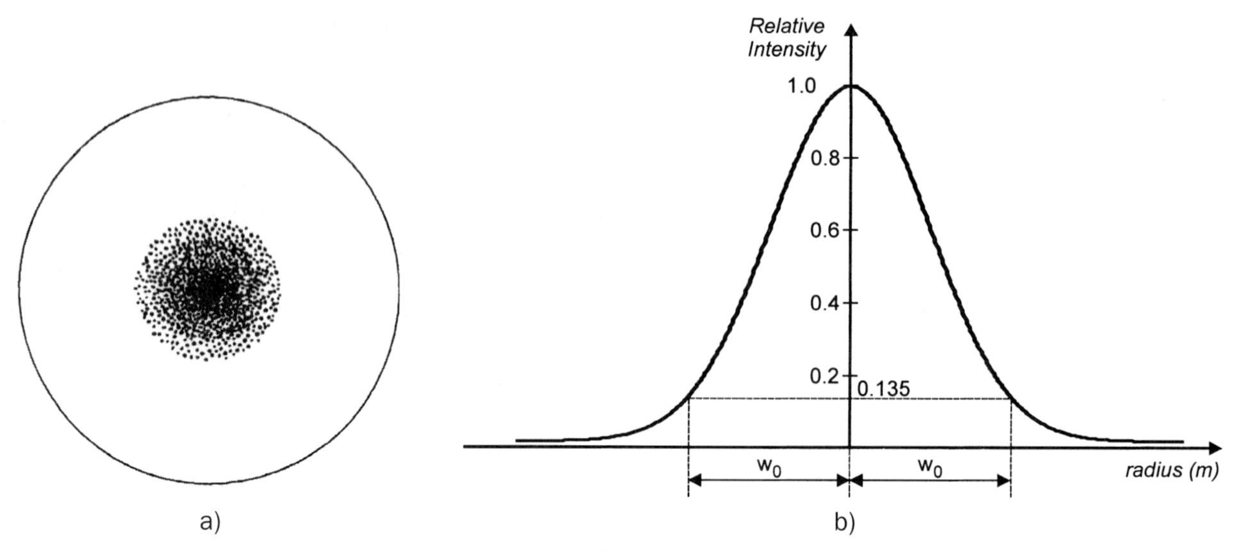


**Figure 5.2** Gaussian beam: (a) Physical appearance of the cross-sectional beam's structure; (b) Gaussian model of the beam's intensity.

The Gaussian model of distribution of a beam's intensity is the most popular because, first, it is close to the measured results, and second, it is easy to use in theoretical calculations.

## Core, Cladding, and Mode-Field Diameter (MFD)

The type of singlemode fiber we are discussing here is the step-index fiber. You'll recall from Section 3.1 that "step index" means a step-like change in the refractive index profile at the core-cladding interface.

Manufacturers use the mode-field diameter, MFD, rather than the core diameter as a parameter that describes singlemode fibers. This is in contrast to multimode fiber, where the geometric sizes of both the core and the cladding are given in the data sheets. The reason for using MFD as the major singlemode-fiber parameter becomes clear from looking at Figure 5.3.

In multimode fibers, we need to know the core diameter because all the light is confined within the core. In singlemode fibers, however, this statement is not true, as Figure 5.3 shows. The essential portion of the light (typically, about 20%) is carried by the cladding, thereby making the core diameter just an auxiliary parameter of a singlemode fiber. Very often, manufacturers do not even specify the core diameter at all because, as we will see in later sections of this chapter, the refractive-index profile might be so complex that it is difficult, if not impossible, to detect the border between the core and the cladding.

Typical dimensions of MFD and the core diameters in step-index singlemode fibers are 9.3 and 8.3 μm, respectively. (Of course, you may come across other dimensions.)

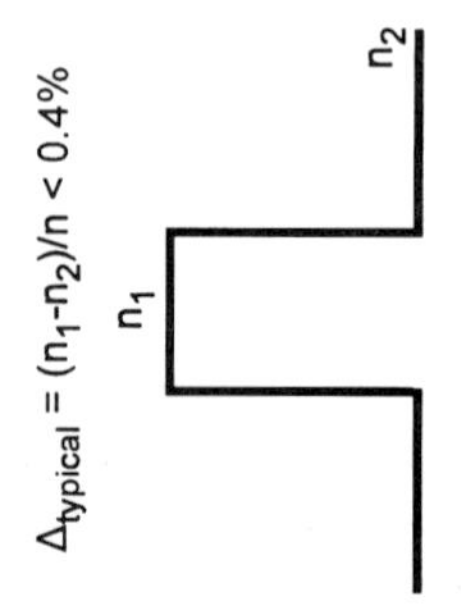


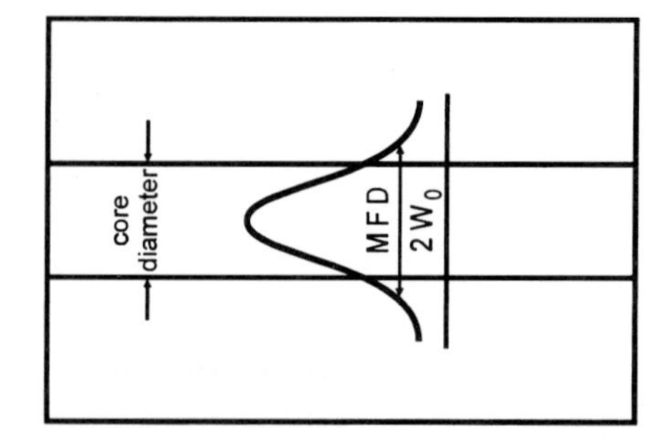


**Figure 5.3** Distribution of the beam's intensity in a singlemode fiber. Core diameter is 8.3 μm and MFD is typically 9.3 μm.

Thus, a light beam propagates in a singlemode fiber mostly within the core and partially within the cladding. This means *the fiber's effective refractive index,* $n_{eff}$, *is the combination of the core and cladding refractive indexes.* This is another significant characteristic of a singlemode fiber.

It is important to know that the mode-field diameter depends on operating wavelength. The shorter the wavelength, the less the MFD. The physical reason for this is that shorter wavelength radiation has more pronounced light properties as a stream of protons; as a consequence, the light beam is more focused, which, in turn, results in more stringent confinement of the light beam at the fiber's central axis. This property is described by the MFD. Typical graphs of MFD as a function of wavelength are shown in Figure 5.4.

It is worth noting that the two graphs represent two fiber-design options, which will be discussed in Section 5.3. The graphs have almost the same slope; hence, the different designs change the specific numbers but not the tendency in wavelength dependence.

MFD is a very important characteristic of a singlemode fiber. When you need to connect two fibers, you must take all measures necessary to eliminate connection losses caused by geometrical misalignments. You do so by relying on cladding dimensions. Suppose, for example, you connect two fibers as perfectly as geometry allows. Still, if these two fibers have different mode-field diameters, you will have extra insertion loss. This insertion loss can be calculated by the following formula:

$$Loss_{\text{coupling MFD}}(\text{dB}) = -10\log\left[4/(\text{MFD}_1/\text{MFD}_2 + \text{MFD}_2/\text{MFD}_1)^2\right] \quad (5.4)$$

This is an example of how important MFD is for a singlemode fiber.

## Cutoff Wavelength

It follows from Formulas 3.14 and 3.15 and from Figure 4.15 that there is a cutoff wavelength, $\lambda_C$, when the *V*-number is exactly 2.405, that is, at the brink of singlemode operation. It is therefore easy to derive from Formula 5.1:

$$\lambda_C \le \left\{\pi d\sqrt{[(n_1)^2 - (n_2)^2]}\right\}/2.405 \quad (5.5)$$

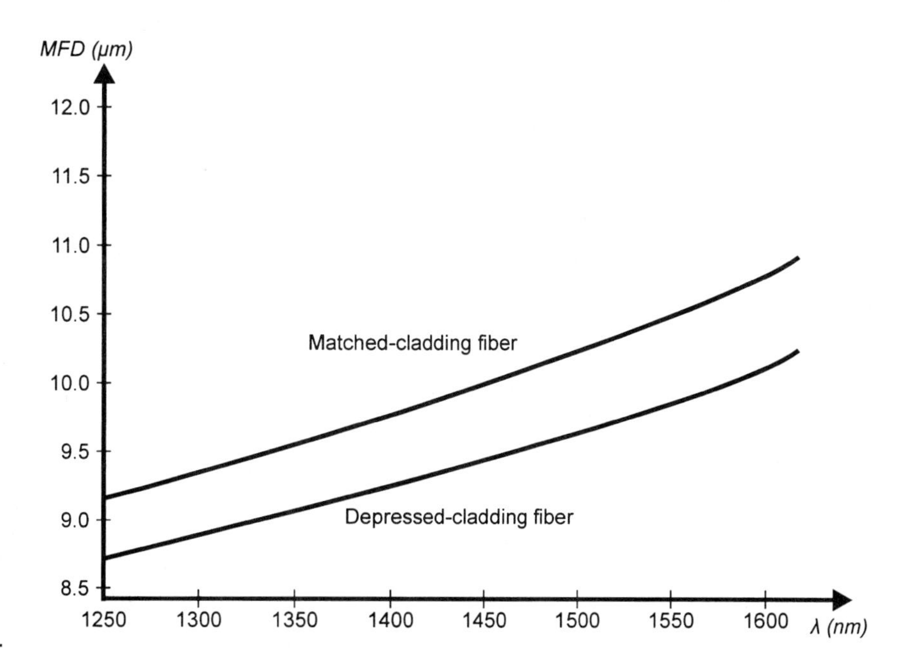


**Figure 5.4** Mode-field diameter, MFD, as a function of wavelength.

If we recall that $\sqrt{[(n_1)^2 - (n_2)^2} = NA$, where $NA$ is numerical aperture, we can rewrite Formula 5.5 in another form:

$$\lambda_C \leq 1.306\ d\ NA \tag{5.5a}$$

Here cutoff wavelength is the new parameter that refers to singlemode fibers only. This is the shortest wavelength at which a fiber can support singlemode operation. If we work with light at a wavelength shorter than $\lambda_C$, two, three, or more modes will propagate along the fiber. In other words, *the same optical fiber can be singlemode or multimode, depending on operating wavelength.* The transition between singlemode and multimode operation occurs gradually. In practice, the fibers are designed specifically for either multimode or singlemode operation; for the latter, the operating wavelength is always longer than $\lambda_C$.

Thus, when we refer to any wavelength dependence of a singlemode fiber, we imply wavelengths above the cutoff.

## 5.2 ATTENUATION

Attenuation is one of the first transmission characteristics a user looks at when choosing a fiber. As we saw in Section 3.2, attenuation in multimode fibers stems from a multiplicity of causes, including bending, scattering, and absorption. Here we will discuss attenuation in singlemode fibers in the same sequence, focusing on new phenomena that we did not meet in our previous considerations.

### Bending Losses

As you will recall from Section 3.2, there are two types of bending losses: macrobending, which is caused by the curving of the fiber's central axis (see Figure 3.5), and microbending, which is caused by microdeformations of the fiber's central axis (see Figure 3.6). A singlemode fiber experiences both types of losses.

***Macrobending loss*** A singlemode fiber is more sensitive to bending than is a multimode fiber. (Refer again to Figure 5.3 and its accompanying discussion). Any fiber bend causes, first, a shift of the center of the Gaussian curve to the outer side of the cladding and, second, redistribution of the mode field so that the curve will no longer be Gaussian. The result is that the longer "tail" appears at the outer side of the cladding, as shown in Figure 5.5.

In order to continue to be a part of the entire mode, the longer tail should keep pace with the beam's propagation. Since it has to cover a much longer distance, the tail must travel at a much higher velocity than the rest of the beam. The farther the tail is from the fiber's centerline, the higher its velocity should be. Eventually, this velocity must exceed the speed of light, which means the tail will be lost. Therefore, a portion of the beam's power will be lost, which means attenuation will increase. This model helps us to understand the mechanism of macrobending loss in a singlemode fiber.

It is evident that macrobending loss is larger when the MFD is larger. This statement is equivalent to saying that the more tightly the mode field is confined within the fiber core, the less sensitive a singlemode fiber is to bending. (There are other restrictions on decreasing mode-field diameter.) It is also quite evident that the less the bending radius is, the greater will be the macrobending loss, which is true for multimode fibers as well.

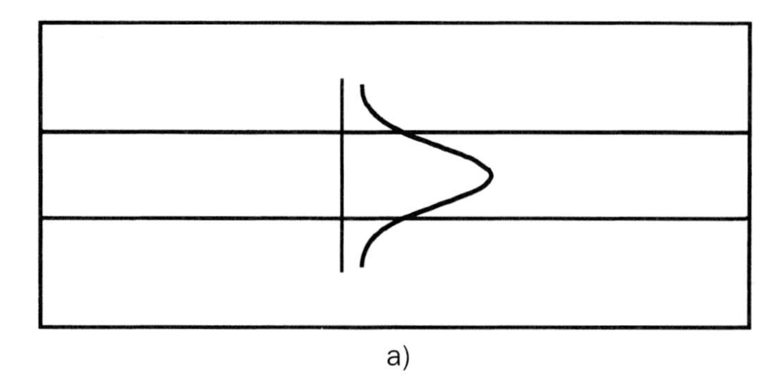


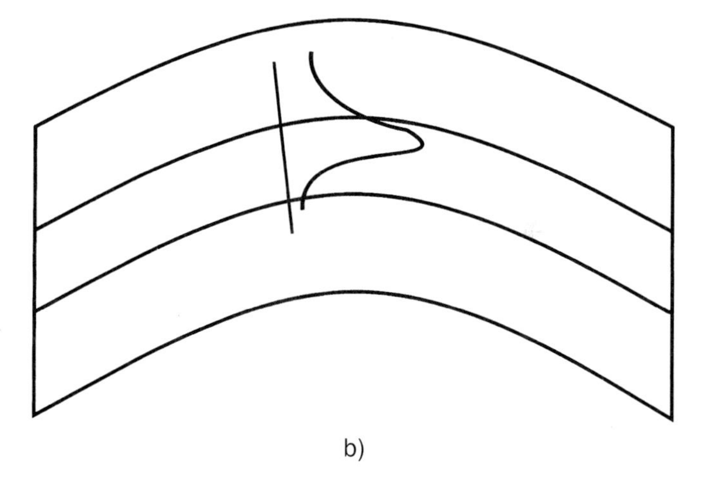


**Figure 5.5** Macrobending loss in a singlemode fiber: (a) Mode-field distribution in a straight fiber; (b) mode-field distribution in a bent fiber.

The MFD depends on wavelength, as we saw in Section 5.1 (see Figure 5.4); therefore, macrobending loss also depends on wavelength. The shorter the wavelength, the smaller the mode-field diameter; hence, the macrobending sensitivity of a singlemode fiber is lessened. But to make the fiber carry only one mode, we need to use an operating wavelength above $\lambda_C$. (See Section 5.1.) On the other hand, the longer an operating wavelength, the less attenuation a fiber experiences. (This is true up to the range of 1600 nm.) As you can see, the need for compromise arises when choosing an operating wavelength.

To sum up, then, *macrobending loss in a singlemode fiber increases as operating wavelength increases and bend radius decreases.* For empirical evaluation of a fiber's bending performance, the MAC number is introduced as MAC = MFD/$\lambda_C$. Typically, this number is in the range of 7 to 9 for a standard step-index singlemode fiber. Where the MAC number equals 4, a singlemode fiber can operate at 1650 nm with the same level of bending loss as at 1550 nm.

***Microbending loss*** Microbending loss in a singlemode fiber occurs for the same reason it does in a multimode fiber: microvariations of the fiber's central axis around its theoretically straight position. (See Section 3.2.) Microbending loss in a singlemode fiber is wavelength dependent.

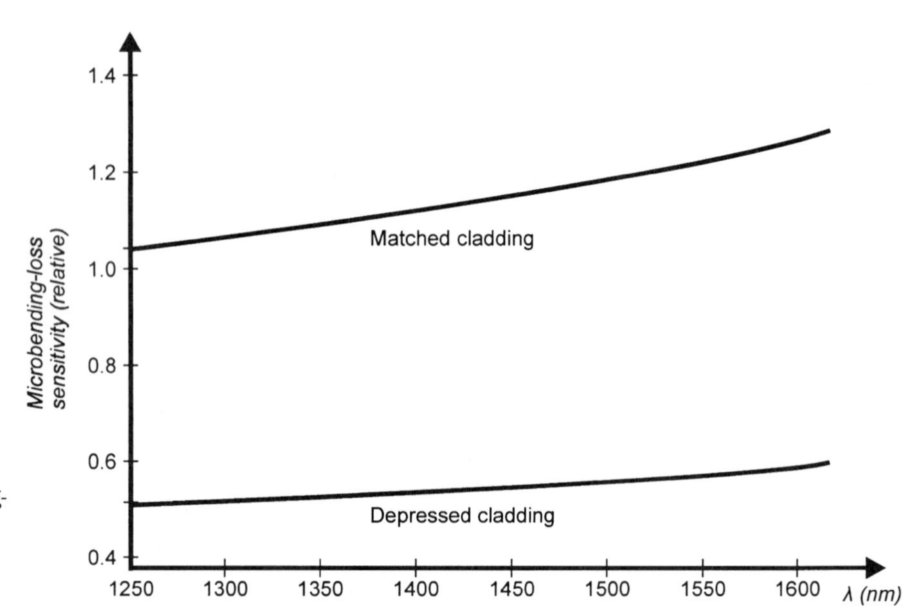


**Figure 5.6** Microbending-loss sensitivity as a function of wavelength.

In other words, the sensitivity of a singlemode fiber to microbending loss increases slightly with increasing wavelength. The physical reason for this wavelength dependence is the MFD's increase with a longer $\lambda$, a phenomenon that makes more power radiate outside the core, as shown in Figure 5.6.

You may wonder where these microvariations of the central axis come from. The answer: As in the case with multimode fiber, they appear mostly because of stress exerted on the fiber during the cabling process and because of thermal stress. The latter implies that microbending loss changes randomly as temperature and thermal stress do.

Formulas for calculations of bending losses in singlemode fibers exist [1] but they can't help you very much in your practical work because they rely on many unknown constants. Manufacturers specify macrobending loss in their data sheets and include microbending loss in their calculation of total attenuation.

## Scattering

The same mechanism that we discussed in Section 3.2 causes scattering loss in a singlemode fiber. (See Figure 3.7.) Scattering loss determines spectral dependence of attenuation, and Rayleigh scattering (see Section 3.2) limits minimum attenuation in general.

## Absorption

Absorption in singlemode fibers is very similar to that in multimode fibers. It differs only in the numbers: Attenuation caused by absorption is less in singlemode fibers. Figure 5.7 shows two spectral attenuation graphs: multimode and singlemode fibers.

These graphs show the similarities and differences between singlemode and multimode fibers. As you can see, the key difference is in the lower level of attenuation introduced by a singlemode fiber. This is because in a singlemode fiber less light goes through less of the silica and, therefore, there is less chance for absorption and scattering events to occur. Small attenuation is another important advantage of a singlemode fiber, as is the absence of intermodal dispersion.

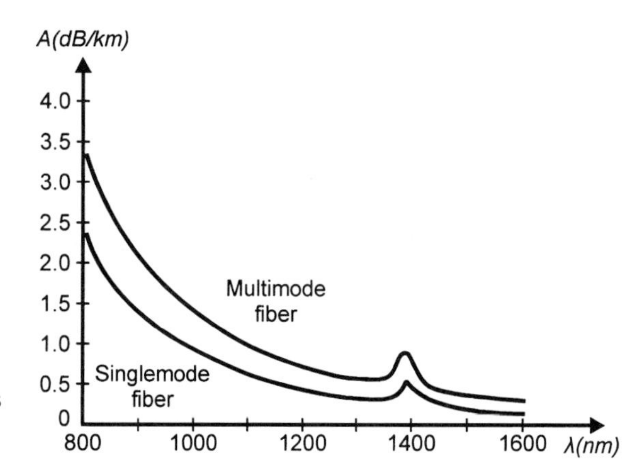


**Figure 5.7** Typical spectral attenuation curves of singlemode and multimode fibers.

In conclusion, let's consider the following example.

**Example 5.2.1**

***Problem:***

Calculate the limitation in transmission length caused by fiber loss if $A = 0.2$ dB/km, $P_{in} = 0.029$ mW, and $P_{out} = 0.001$ mW, where $A$ is fiber attenuation, $P_{in}$ is light power launched into the fiber, and $P_{out}$ is power coupled to a photodiode.

***Solution:***

Refer to Formula 3.13: $L = (10/A) \log_{10}(P_{in}/P_{out})$. Plugging all the numbers into the formula, we obtain:

$$L = (10/A)\log_{10}(P_{in}/P_{out}) = 10/0.2 \times \log(0.029/0.001) = 73.12 \text{ km}$$

The typical span between two adjacent amplifiers in the latest system is about 80 to 90 km; hence, our calculations are realistic.

## 5.3 DISPERSION AND BANDWIDTH

A singlemode fiber carries only one mode and therefore doesn't experience intermodal dispersion—the major contributor to bandwidth limitation in multimode fibers. Thus, one would expect a singlemode fiber to have much lower dispersion or, equivalently, much higher bandwidth than a multimode fiber. This is true. However, a singlemode fiber is still a dispersive transmission medium. Because of the nature of a singlemode fiber, its dispersion is often called "intramodal." The major mechanism that causes dispersion in a singlemode fiber is *chromatic dispersion.* Another important form of dispersion is *polarization-mode dispersion.* This section discusses these phenomena and their impact on the bandwidth of singlemode fibers.

### Chromatic Dispersion

The term "chromatic dispersion" covers all the phenomena associated with wavelength-dependent pulse spreading. We discussed this dispersion in covering multimode fibers. (See Sections 3.3 and 4.6.) Therefore, we will concentrate here only on certain features of singlemode fiber. Two mechanisms play important roles in chromatic dispersion: material dispersion and waveguide dispersion.

***Material dispersion*** Material dispersion is caused by the wavelength dependence of the silica's refractive index. Refractive index defines the velocity of light within a medium, $v = c/n$, and $n$ depends on $\lambda$. An information-carrying light pulse contains different wavelengths because a light source radiates light of a finite spectral width. (See Figure 4.26.) Therefore, the components of the pulse with different wavelengths will travel within a fiber at different velocities and will arrive at the fiber end at different times, thus causing the spread of the pulse. This explanation is illustrated in Figure 5.8.

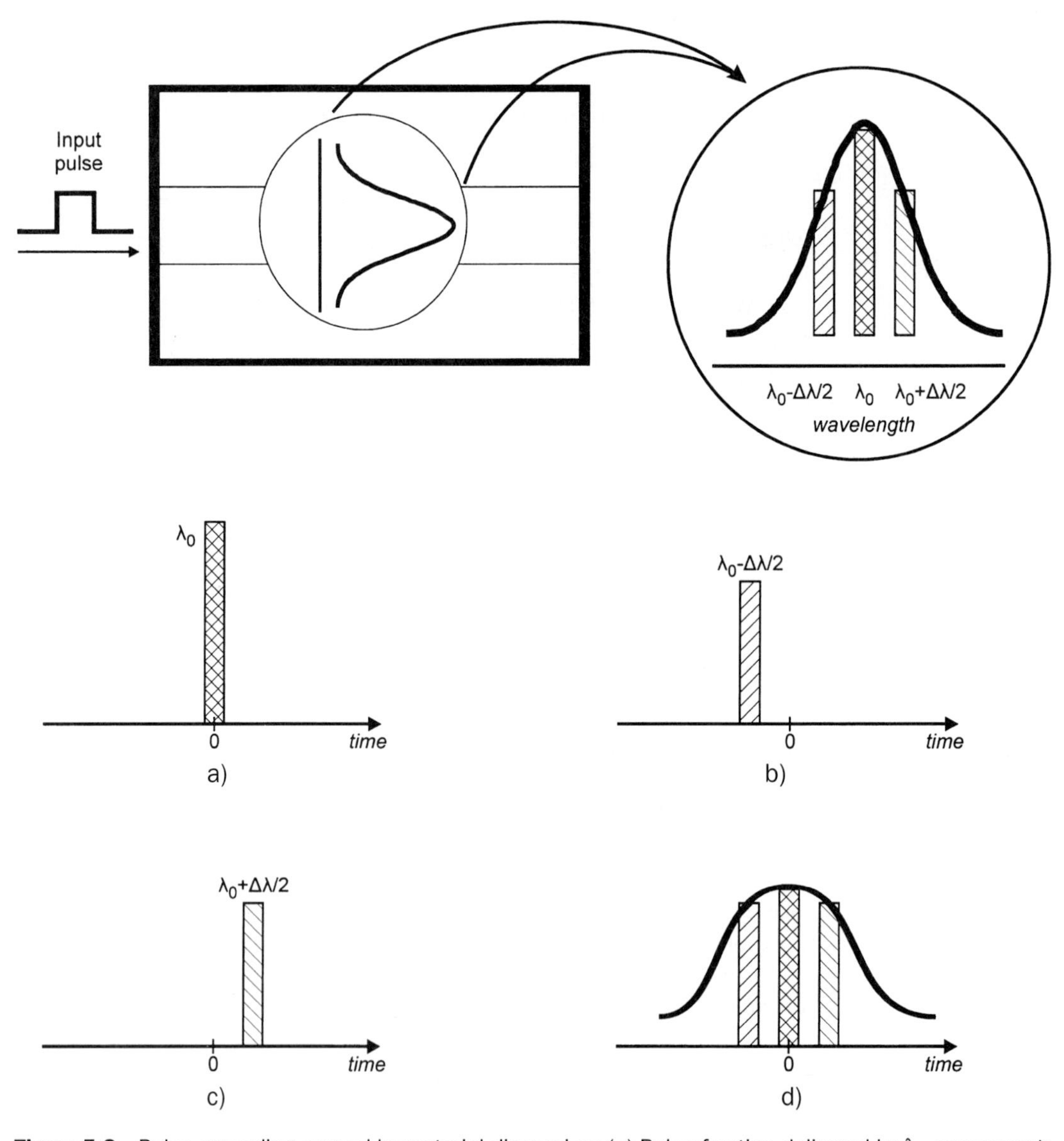


**Figure 5.8** Pulse spreading caused by material dispersion: (a) Pulse fraction delivered by $\lambda_0$ component; (b) pulse fraction delivered by $\lambda_0 - \Delta\lambda/2$ component; (c) pulse fraction delivered by $\lambda_0 + \Delta\lambda/2$ component; (d) output pulse.

In this figure, the arrival time of the pulse fraction carried by the central wavelength, $\lambda_0$, is taken as the zero (reference) point at the time axis. Thus, the pulse fraction carried by the longer wavelength component of the injected light, $\lambda_0 + \Delta\lambda/2$, arrives before the central-wavelength portion. This is because the fiber's refractive index is less for longer wavelengths (see Figure 4.22), and therefore *light at longer wavelengths travels faster within this medium.* The inverse statement is true for shorter wavelengths: The pulse portion carried by the component $\lambda_0 - \Delta\lambda/2$ arrives later. All these accelerations and delays result in a type of pulse spreading called material dispersion.

Singlemode fibers are made from the same material—fused silica—as multimode fibers; therefore, both experience the same material dispersion. The amount of pulse spreading caused by material dispersion, $\Delta t_{mat}$, per length is given by:

$$\Delta t_{mat}/L(\text{ps/km}) = D_{mat}(\lambda)\,\Delta\lambda, \tag{5.6}$$

where $D_{mat}(\lambda)$ is the dispersive characteristic of the material called the material-dispersion parameter, $\Delta\lambda$ (nm) is the spectral width of the light source, and $L$ (km) is the fiber length. The material-dispersion parameter, $D_{mat}(\lambda)$, is measured in picoseconds of pulse spreading per nanometer of the spectral width of the light source and per kilometer of fiber length (ps/nm·km). The graph in Figure 5.9 typifies this parameter.

Keep clearly in mind that material dispersion equals zero at a specific wavelength—around 1300 nm. This phenomenon stems from the dispersive properties of silica. (See Section 4.6.) Another point to stress is that for $\lambda < 1300$ nm, $D_{mat}(\lambda)$ is negative. This means that light at a wavelength, say, of 1000 nm travels more slowly than light at a wavelength of 1100 nm, as we discussed above.

Our discussion of attenuation in singlemode fibers in Section 5.2 concluded that only laser diodes with a spectral width of 1 nm and narrower should be used as a light source. We reach the same conclusion here since pulse spreading, $\Delta t_{mat}$, is proportional to $\Delta\lambda$. Thus, the light source preferred in singlemode-fiber communications systems is a laser diode. This is another reason why pulse spread in singlemode fibers is much less than in multimode fibers. It is clear that $D_{mat}(\lambda)$ is about the same order of value for both fiber types.

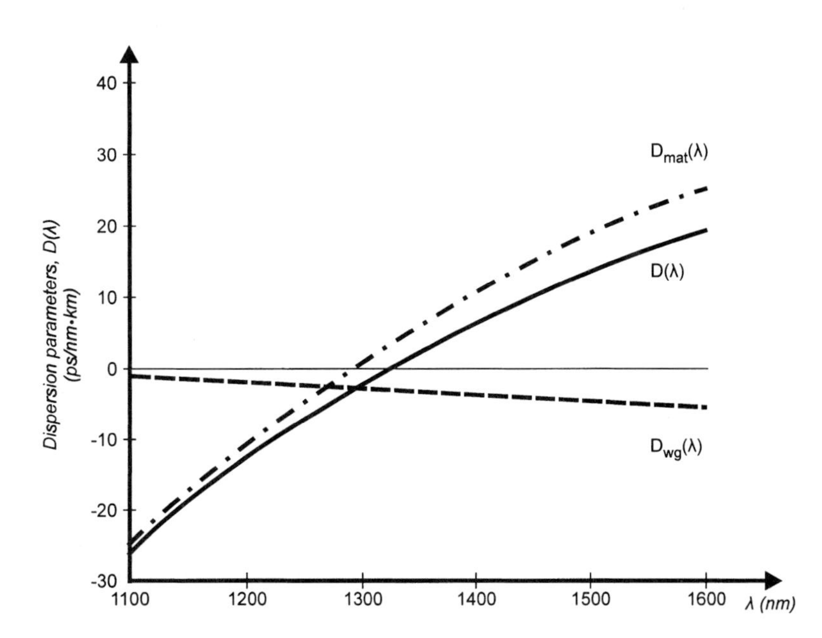


**Figure 5.9** Chromatic-dispersion parameters of a singlemode fiber as a function of wavelength. $D_{mat}$ = material dispersion; $D_{wg}$ = waveguide dispersion; $D$ = total dispersion.

### Example 5.3.1

***Problem:***

Calculate pulse spreading due to material dispersion in a singlemode fiber operating at $\lambda = 1310$ nm if the LD's $\Delta\lambda = 1$ nm and $L = 1$ km.

***Solution:***

From the graph in Figure 5.9, we find that $D_{mat}(\lambda)$ at 1310 nm is approximately 2 ps/nm·km. Substituting the numbers in Formula 5.6, we get:

$$\Delta t_{mat}/1\text{ km} = D_{mat}(\lambda)\Delta\lambda = 2\ (\text{ps/nm·km}) \times 1\text{ nm} = 2\text{ ps/km} = 0.002\text{ ns/km}$$

Let's compare this number with those calculated for multimode fibers (see Example 3.3.5), where we obtained 5.6 ns/km and—at best—0.21 ns/km. A singlemode fiber provides an improvement of at least 120 times over multimode fibers in its material-dispersion characteristic because we use an operating wavelength at the near-zero dispersion zone and an LD with a much narrower spectral width. Don't become too alarmed about the emergence of $\Delta t_{mat}$; it is only an offshoot of chromatic dispersion.

It is important to recall the amount of pulse spreading caused by intermodal dispersion in multimode fibers. For step-index fiber, we computed $\Delta t/L = 86.57$ ns/km (Example 3.3.1), and for graded-index fiber, we obtained $\Delta t/L = 0.18$ ns/km (Example 3.3.4). You can draw your own conclusion from looking at these numbers.

Since pulse spreading in singlemode fibers is small, we measure it in ps/km. It is advisable to present all the above calculations in these units.

---

If light travels within a medium whose refractive index is wavelength dependent, the light will experience material dispersion regardless of whether the medium is enclosed or not. This is why the material-dispersion parameter, $D_{mat}(\lambda)$, in singlemode and multimode fibers is the same. But if light travels within an enclosed medium (in other words, if light is guided), then another phenomenon—waveguide dispersion—comes into play.

***Waveguide dispersion*** Waveguide dispersion is caused by the fact that light is guided by a structure—here, an optical fiber. Putting it another way, this type of dispersion doesn't exist in an open medium. Strictly speaking, waveguide dispersion occurs in multimode fibers, but the resultant pulse spreading it causes is negligible compared with the pulse spreading produced by intermodal and material dispersion. This is not the case with a singlemode fiber, where intermodal dispersion does not exist and material dispersion is very small. Waveguide dispersion, it should be noted, is one of the major components of chromatic dispersion in singlemode fibers.

The mechanism that causes waveguide dispersion can be explained as follows: After entering a singlemode fiber, an information-carrying light pulse is distributed between the core and the cladding. Its major portion travels within the core, the rest within the cladding. (See Figure 5.3.) Both portions propagate at different velocities since the core and the cladding have different refractive indexes. The pulse will spread simply because light is confined within a structure having different refractive indexes—the core-cladding combination of the fiber. It is important to realize that waveguide dispersion will occur even if the fiber material has no dispersive properties. Pure waveguide dispersion occurs merely from restricting light within a certain structure.

Waveguide dispersion in a singlemode fiber is relatively small compared with material dispersion. What makes this phenomenon so important in singlemode fibers is its wavelength dependence. Indeed, waveguide dispersion depends on the mode-field distribution between the core and the cladding, that is, on the MFD, but the MFD depends on wavelength. (See Figure 5.4.) Figure 5.9 shows waveguide dispersion, $D_{wg}(\lambda)$, as a function of wavelength.

In what way does waveguide dispersion depend on wavelength? The longer the wavelength, the larger the MFD and the larger the fraction of total pulse power that will travel in the cladding. The cladding portion of the pulse travels faster than the core portion since the cladding refractive index, $n_2$, is less than the core index, $n_1$. If one takes the arrival time of the core portion of the pulse as a zero reference point at the time axis, then the cladding portion will arrive at a negative time. This consideration is visualized in Figure 5.10.

The longer the wavelength, the greater the pulse spreading in the negative half of the time axis. This interpretation helps us to understand the $D_{wg}(\lambda)$ graph shown in Figure 5.9. The amount of pulse spreading caused by waveguide dispersion, $\Delta t_{wg}$, per unit of length is given by:

$$\Delta t_{wg}/L = D_{wg}(\lambda)\Delta\lambda \quad (5.7)$$

## Example 5.3.2

***Problem:***

Calculate pulse spread caused by material and waveguide dispersions at the 1550-nm operating wavelength if $\Delta\lambda = 1$ nm and $L = 1$ km.

***Solution:***

From Figure 5.9 one can find $D_{mat}(\lambda) = 20$ ps/nm·km and $|D_{wg}(\lambda)| = 5$ ps/nm·km. Thus, one can compute $\Delta t_{mat} = 20$ ps/nm·km × 1 nm × 1 km = 20 ps and $\Delta t_{wg} = 5$ ps/nm·km × 1 nm × 1 km = 5 ps.

Formally speaking, $D_{wg}(\lambda)$ is negative, which means nothing when doing computations of $\Delta t$. To avoid any ambiguity, the sign of absolute value is used. This is how we will deal with the value of D(λ) in subsequent chapters.

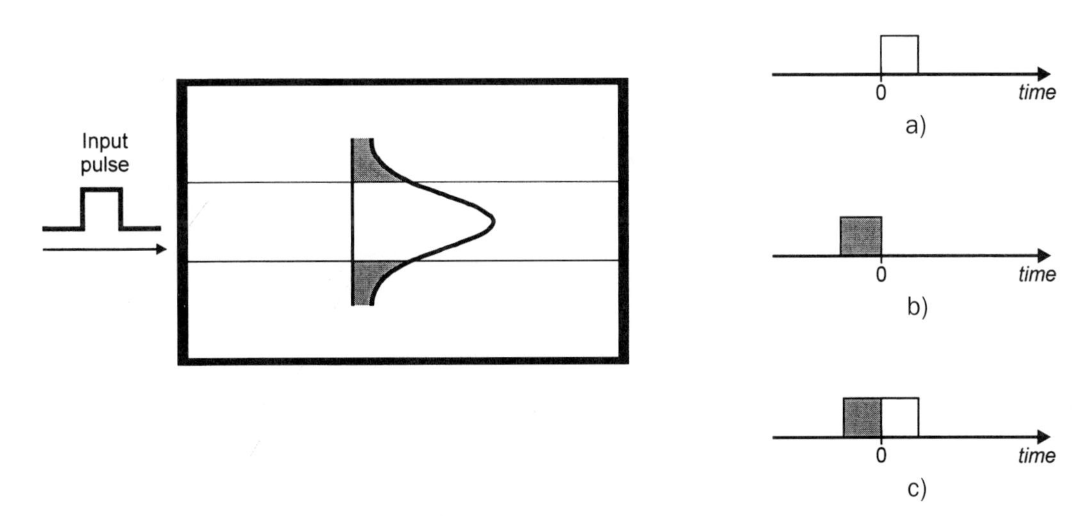


**Figure 5.10** Pulse spreading caused by waveguide dispersion: (a) Core fraction of the pulse; (b) cladding fraction of the pulse; (c) total pulse.

To clarify the picture of dispersion in fibers, look again at Figures 3.10, 3.14, 5.8, 5.9, and 5.10.

***Chromatic dispersion in a singlemode fiber*** The chromatic-dispersion parameter in a singlemode fiber is the sum of the material and waveguide dispersions, so that:

$$D(\lambda) = D_{mat}(\lambda) + D_{wg}(\lambda) \tag{5.8}$$

This relationship is very important since the material-dispersion parameter above 1300 nm becomes positive while the waveguide-dispersion parameter stays negative (Figure 5.9); the fact is that they cancel each other out, resulting in a zero chromatic-dispersion parameter, $D(\lambda) = 0$. This occurs around 1310 nm, the customary operating wavelength for standard singlemode fiber.

Pulse spreading caused by chromatic dispersion, $\Delta t_{chrom}$, is given by

$$\Delta t_{chrom}/L = D(\lambda)\Delta\lambda, \tag{5.9}$$

where $D(\lambda)$ is the chromatic-dispersion parameter of the fiber and $\Delta\lambda$ is the spectral width of the light source, as is customary. Manufacturers give the dispersion parameter, $D(\lambda)$, in their fiber data sheets either as a number—they call it the "dispersion coefficient"—or as a formula. We discussed this parameter and its usage in Section 3.5. Let's recall Formula 3.25:

$$D(\lambda) = \frac{S_0}{4}\left[\lambda - \frac{\lambda_0^4}{\lambda^3}\right]$$

To calculate the dispersion parameter near the zero-dispersion wavelength, $\lambda_0$, one can use a simplified version of the above formula:

$$D(\lambda) = S_0\,(\lambda - \lambda_0), \tag{5.10}$$

where the zero-dispersion slope, $S_0$, can be found in a fiber data sheet.

### Example 5.3.3

***Problem:***

Calculate chromatic dispersion in a singlemode fiber at the 1550-nm operating wavelength with $\Delta\lambda = 1$ nm and $L = 1$ km.

***Solution:***

Figure 5.9 shows that $D(\lambda) = 15$ ps/nm·km. Plugging the numbers into Formula 5.9, we obtain:

$$\Delta t_{chrom} = D(\lambda)\Delta\lambda = 15 \text{ ps/nm·km} \times 1\text{km} = 15 \text{ ps}$$

Compare this number with the 20 ps computed for material dispersion in Example 5.3.2. It looks ridiculous because the ingredient—material dispersion—is more than the sum, that is, the total chromatic dispersion. The clue is in Figure 5.9. At 1550 nm, the material dispersion is positive and its coefficient is about 20 ps/nm·km, while the sum of the material and negative waveguide dispersions is 15 ps/nm·km. Again, by choosing the operating wavelength at 1310 nm, we can almost eliminate the effect of chromatic dispersion.

### Conventional, Dispersion-Shifted, and Dispersion-Flattened Fibers

From the above discussion, we saw that the zero-dispersion region is around 1310 nm. But the spectral attenuation curve shows that minimum attenuation occurs around 1550 nm. Therefore, to obtain a singlemode fiber with the best properties, it is necessary either to move the minimum attenuation point to 1310 nm or to shift the minimum dispersion point to 1550 nm. The attenuation is defined by Rayleigh scattering, that is, by the properties of the fabricated fiber material, and it cannot be changed by manipulating the fiber design. (See Section 4.5.) The dispersion itself is chromatic, that is, it comprises both the material and waveguide dispersions. Material dispersion is determined by the wavelength dependence of the silica's refractive index. As long as we use silica, we cannot change this property significantly, although doping helps a little. Waveguide dispersion is controlled by the mode-field distribution in the core-cladding guiding structure. Mode-field distribution is, in turn, determined by the profile of the core's refractive index. *Here is the key to shifting the minimum chromatic dispersion to the 1550-nm region: Change the waveguide-dispersion value by redesigning the refractive-index profile of the fiber core.*

***Conventional fiber*** The singlemode fiber is called *conventional* if its zero-dispersion wavelength is around 1300 nm. The fiber is so named because this zero-dispersion wavelength stems from the natural properties of the fiber material—silica—and the waveguiding property of a simple step-index fiber, whose refractive-index profile is shown in Figure 5.3.

But there is another possible design of a refractive-index profile for a conventional fiber. This is when the cladding's refractive index is depressed immediately around the core index and then returned to its basic value. See Figure 5.11. The traditional step-index design is called *matched cladding,* while the alternative design is called *depressed cladding.*

Pay particular attention to the relative refractive index, $\Delta = (n_1 - n_2)/n$. The value of $\Delta$ is very small—0.37%—compared with that of a multimode fiber—more than 2%. The relative

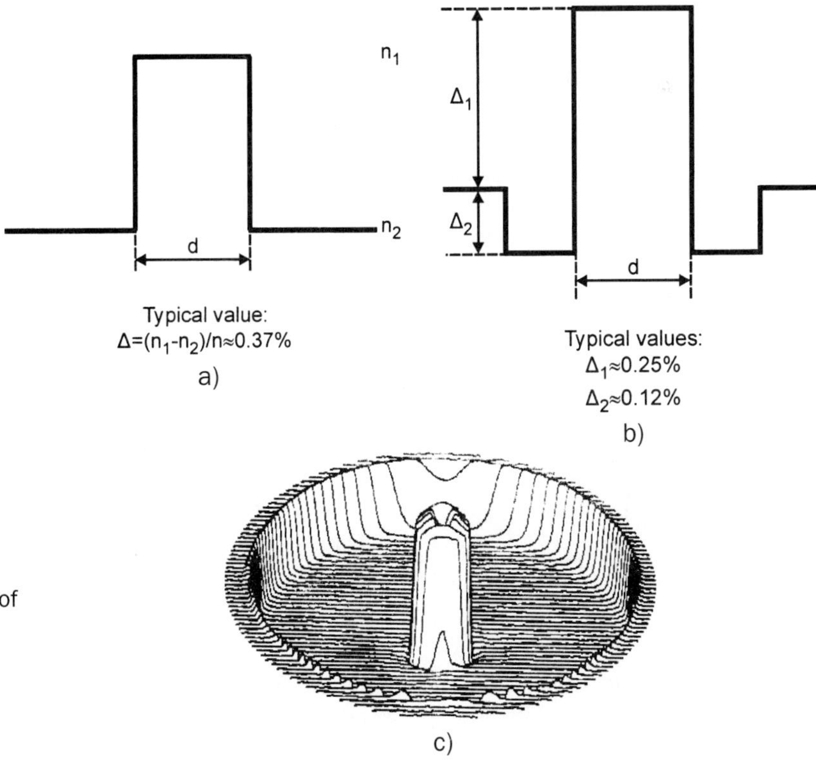


**Figure 5.11** Refractive-index profiles of a conventional fiber: (a) Matched-cladding design; (b) depressed-cladding design; (c) example of topographic profile of depressed-cladding design.

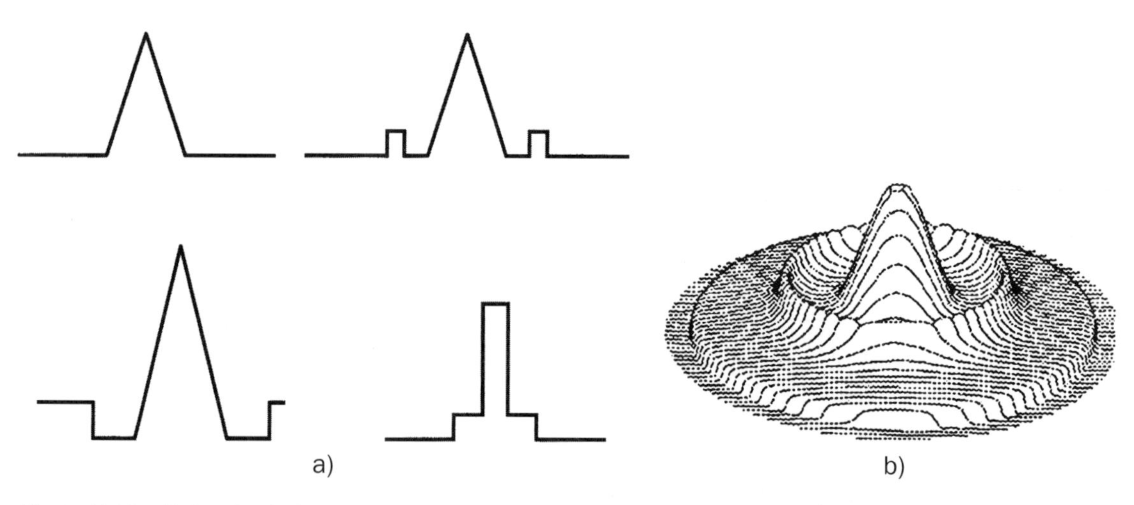


**Figure 5.12** Refractive-index profiles of a dispersion-shifted fiber: (a) Refractive-index profiles; (b) example of topographic profile.

index of a singlemode fiber can, theoretically, be even smaller. By increasing Δ up to 0.37%, manufacturers decrease the macrobending sensitivity of a singlemode fiber. But they can't increase Δ much more because they have to keep their fibers in a singlemode state of operation [2].

How can manufacturers create such a complicated index profile as the depressed-cladding one? Recall that the refractive index is determined by the optical properties of the material and that these can be changed simply by doping. So the manufacturers dope the pure silica with atoms of another material, thereby changing the value of the refractive index. Since the manufacturers fabricate fibers by putting one layer of material on another, they can control the exact index profile of the core and the cladding. We'll discuss fiber fabrication in Section 7.1.

***Dispersion-shifted fiber*** To shift the occurrence of minimum dispersion to the 1550-nm region, manufacturers fabricate fibers with sophisticated index profiles, some of which are shown in Figure 5.12.

Again, remember that the purpose of creating a sophisticated profile of the core refractive index is to increase the occurrence of waveguide dispersion at the longer wavelength so that, at approximately 1550 nm, material and waveguide dispersions cancel each other out. (See Figure 5.9 and draw the waveguide-dispersion graph for the dispersion-shifted fiber.) To achieve this goal, designers want to control mode-field distribution by changing the profiles of the core and

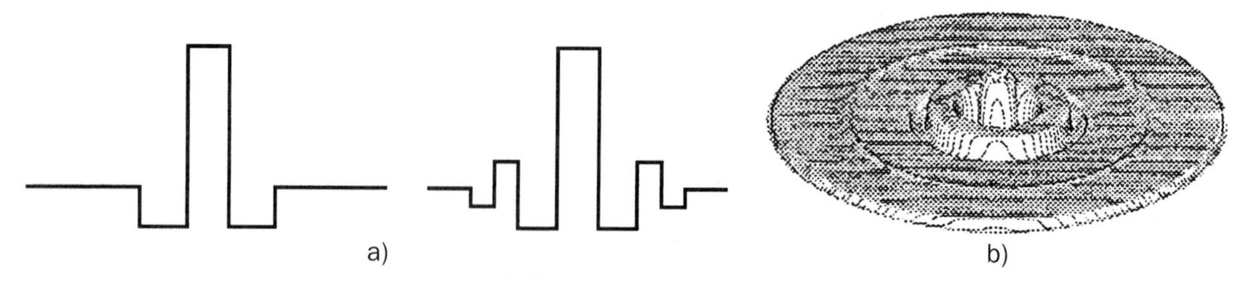


**Figure 5.13** Refractive-index profiles of a dispersion-flattened fiber: (a) Refractive-index profiles; (b) example of topographic profile.

cladding. The variety of index profiles for dispersion-shifted fibers is surprisingly diverse, with the triangular-shaped core refractive index being the most popular.

***Dispersion-flattened fiber*** Wavelength-division multiplexing technology (see Section 1.3 and Chapters 12 through 15) was a new challenge to fiber designers. Since this technology uses many wavelengths simultaneously, it needs fibers with a zero-dispersion property, not at a single wavelength but distributed along a region of wavelengths. Thus, the need for dispersion-flattened fibers arose. In 1988, Phillips Corporation reported the first commercially manufactured dispersion-flattened singlemode fiber. The barrier had been hurdled. Now manufacturers fabricate such fibers in a wide range of profiles, some of which are shown in Figure 5.13.

Once manufacturers learned how to control waveguide dispersion by changing the refractive-index profile, fabricating fibers with specific characteristics became possible. It wasn't easy but at last it could be done. Today, we have a multiplicity of commercially available fibers to satisfy the demands of nearly every customer. In fact, manufacturers often custom-tailor a fiber's design to meet the most challenging of customer requests. The graphs in Figure 5.14 sum up these characteristics of chromatic dispersion for all three types of fiber.

## Polarization-Mode Dispersion (PMD)

Light, as an electromagnetic wave, is determined by three parameters: amplitude, phase—including frequency (wavelength) and initial phase—and the state of polarization. The first two parameters are discussed in Chapter 2 and Section 3.2. Here we introduce the idea of polarization.

Let's start by considering a plane electric wave. (See Figure 4.1.) The word "plane" means the wave always shows one linear arrangement, such as the $x$–$z$ plane in Figure 4.1. Vector **E** changes its magnitude over time and moves along the $z$ axis, but it is always in the $x$–$z$ plane. We call such a wave *linear polarized.*

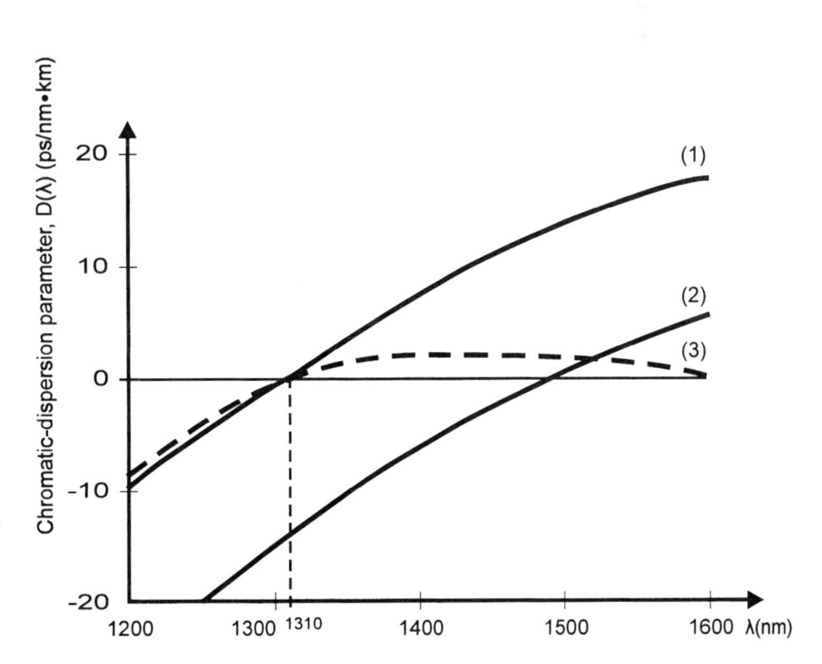


**Figure 5.14** Chromatic-dispersion parameters for (1) conventional, (2) dispersion-shifted, and (3) dispersion-flattened fibers. *(Adapted from Gerd Keiser,* Optical Fiber Communications, 2d ed., *New York: McGraw-Hill, p. 117. Reprinted with permission of McGraw-Hill.)*

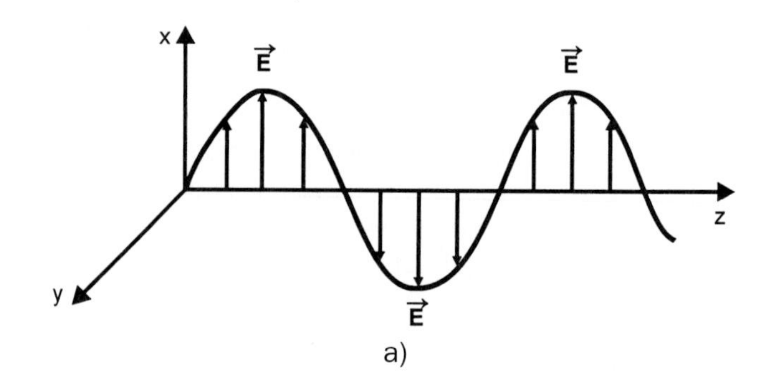


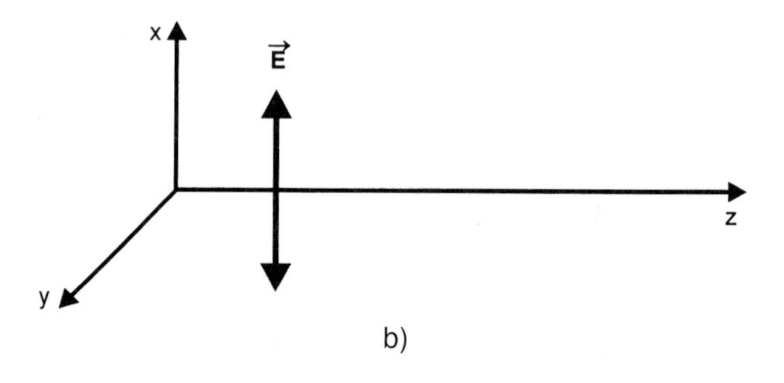


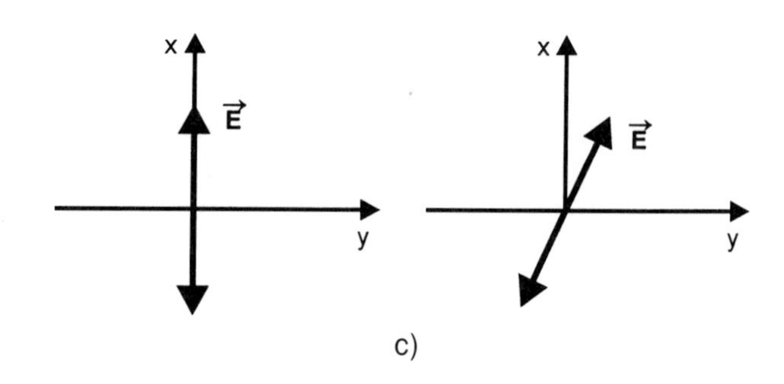


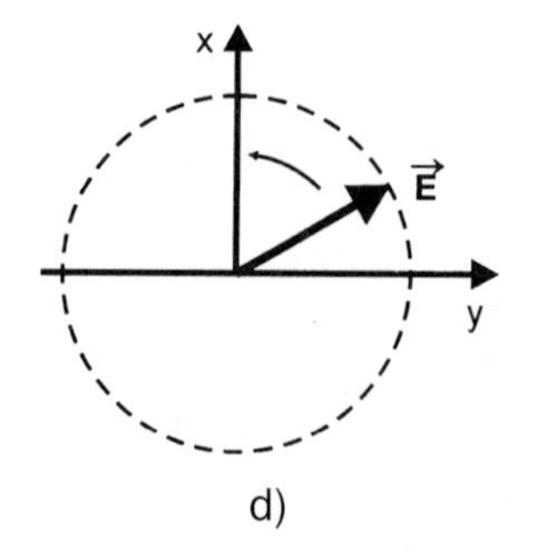


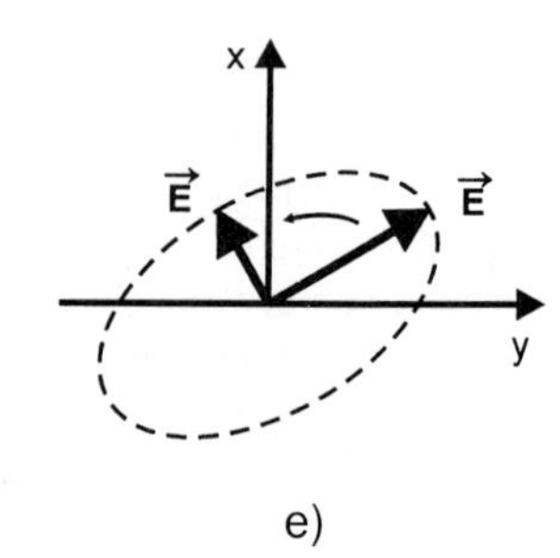


**Figure 5.15** Polarized waves: (a) Linear-polarized electric wave traveling along *z* axis; (b) and (c) other forms of presentation of a linear-polarized wave; (d) circular-polarized wave; (e) elliptically polarized wave.

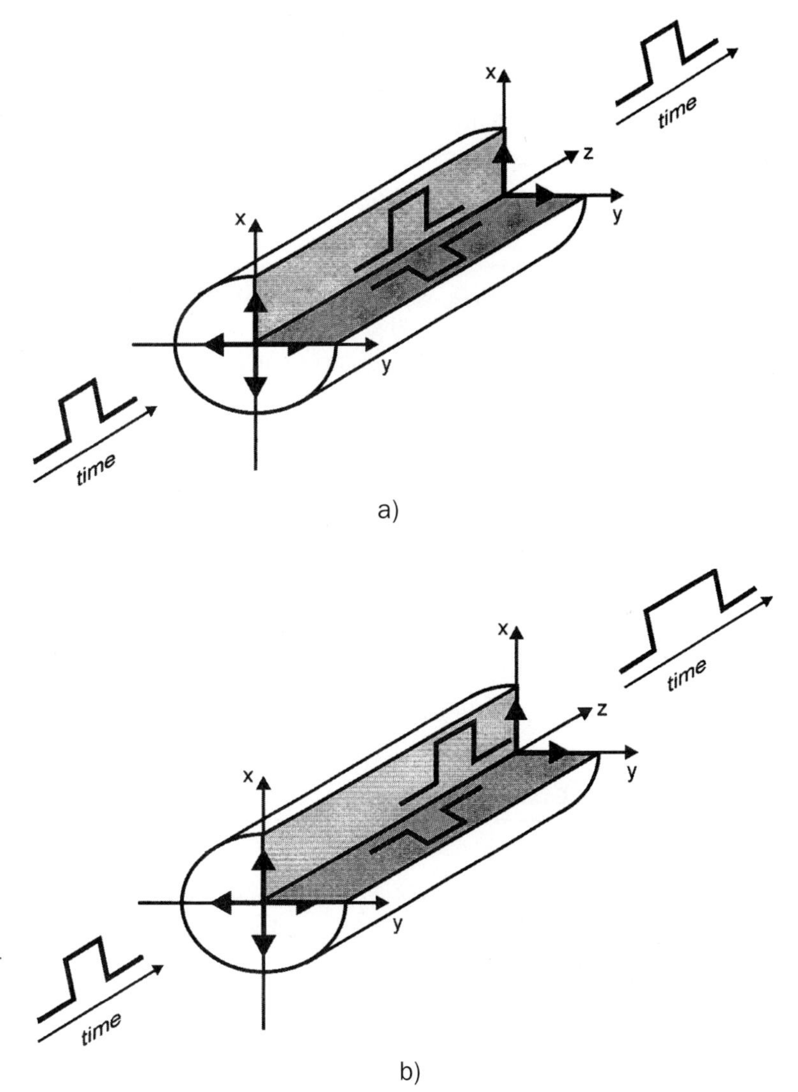


**Figure 5.16** Pulse spreading caused by changing of the state of a mode's polarization: (a) Two linear-polarized modes traveling within an ideal fiber; (b) two linear-polarized modes traveling within a real fiber.

Linear-polarized waves, shown in Figure 5.15(b) and 5.15(c) always propagates in one plane, but this plane can make a number of angles with either the $x$ or the $y$ axis, as Figure 5.15(c) shows.

Linear polarization is the simplest state of polarization of electromagnetic waves. These waves can also be circularly and elliptically polarized. The latter is the most general form of polarization. (See Figures 5.15(d) and 5.15(e).) Light may also be unpolarized when its **E** and **H** vectors randomly change their orientation.

Even though we call the fiber "singlemode," it actually carries two modes under one name. These modes are linear-polarized waves that propagate within a fiber in two orthogonal planes, as shown in Figure 15.16(a). Ideally, each of the modes carries half of the total light power. If the fiber is perfect, both of them propagate at the same velocity and arrive at the fiber end simultaneously. Thus, signal travel along the fiber remains undisturbed and the presence of the polarized modes goes unnoticed. It is common to say, then, that these modes are degenerated, which means we will not see the effect of their presence at the output signal. If, however, the fiber exposes the modes to property changes along the $x$ and $y$ planes of their propagation, the velocity of the modes will be affected. They will then begin to travel at different velocities, resulting in pulse spreading. This situation is shown in Figure 5.16(b).

Pulse spreading caused by a change of fiber polarization properties is called *polarization-mode dispersion* (*PMD*). This pulse spreading, $\Delta t_{PMD}$, can be calculated as follows:

$$\Delta t_{PMD} = D_{PMD}\sqrt{L}, \tag{5.11}$$

where $D_{PMD}$ is the coefficient of polarization-mode dispersion measured in ps/$\sqrt{\text{km}}$, and $L$(km), as usual, is the fiber length. Be aware in particular of two key facts regarding polarization-mode dispersion: (1) $D_{PMD}$ does not depend on wavelength and, (2) it does depend on the square root of the fiber length.

What physical reasons bring about PMD? To keep the state of polarization undisturbed, a fiber must have ideal symmetric cross-sectional properties; in other words, the refractive indexes along the fiber's $x$ and $y$ axes—$n_x$ and $n_y$, respectively—must be equal. Obviously, there is some asymmetry in every fabricated fiber, but the most likely times for serious asymmetry to occur are during the fiber-cabling and -splicing processes. Under the pressure used in these operations—no matter how slight—fiber symmetry inevitably changes to become $n_x \neq n_y$. This inequality causes polarization-mode dispersion.

Polarization-mode dispersion is relatively small compared with chromatic dispersion. But when one operates at zero-dispersion wavelength, chromatic dispersion drops to such small numbers that PMD becomes a significant component of the total dispersion.

### Example 5.3.4

***Problem:***

Calculate the pulse spread caused by polarization-mode dispersion for a singlemode fiber with a PMD coefficient of 0.5 ps/$\sqrt{\text{km}}$ and a fiber length of 100 km.

***Solution:***

Plug the given numbers into Formula 5.11 to obtain:

$$\Delta t_{PMD} = D_{PMD}\sqrt{L} = 0.5 \text{ ps}/\sqrt{\text{km}}\sqrt{100\,(\text{km})} = 5\text{ps}$$

In Example 4.3.2, we found that pulse spreading due to chromatic dispersion was 15 ps/km, which resulted in 1500 ps for 100 km of fiber length.

It might seem as though we do not need to take polarization-mode dispersion into account but this is not true. Figures 5.9 and 5.14 show that there is a possibility of reducing chromatic dispersion almost to zero, but there is no way to eliminate polarization-mode dispersion completely. Thus, this phenomenon—PMD—puts a limit on the minimum degree of dispersion we can achieve with singlemode fibers.

## Bandwidth (Bit Rate) of a Singlemode Fiber

Section 3.4 introduced the idea of bandwidth and bit rate in fiber-optic communications systems. It was said, you'll recall, that for our purposes we could use the terms "bandwidth" and "bit rate" interchangeably, although they are not truly synonymous.

The bit rate, $BR$, of a fiber link was defined in Formula 3.22 as:

$$BR < 1/(4\Delta t),$$

where $\Delta t$ is dispersion-induced pulse spreading.

Let's now calculate $BR$ limited by chromatic dispersion. Formula 5.6 can be rewritten:

$$\Delta t_{\text{chrom}} = D(\lambda)\ \Delta\lambda\ L \tag{5.12}$$

Substituting Formula 5.12 into Formula 3.22, we obtain:

$$BR_{\text{chrom}} = 1/[4\ D(\lambda)\ \Delta\lambda\ L], \tag{5.13}$$

where $BR_{\text{chrom}}$ is the maximum bit rate limited by chromatic dispersion.

### Example 5.3.5

***Problem:***

Calculate $BR_{\text{chrom}}$ for a singlemode fiber with a zero-dispersion wavelength at 1310 nm. A laser diode with a peak wavelength of 1300 nm and a spectral width of 1 nm is used.

***Solution:***

To make this calculation, we need to know $D(\lambda)$. We observe that the given fiber is the conventional one. (Can you explain why?) The appropriate curve from Figure 5.14 gives the number for $D(\lambda)$ as –2 ps/nm·km.

Our discussion of the meaning of the negative sign in $D(\Delta)$ (see Section 4.6 and Example 5.3.2) resulted in the following form of Formula 5.11:

$$BR_{\text{chrom}} = 1/[4\ |D(\lambda)|\ \Delta\lambda\ L], \tag{5.14}$$

Thus, we have

$$BR_{\text{chrom}} \times L = 1/[4 \times 2\ \text{ps/nm} \times \text{km} \times 1\ \text{nm}] = 125\ \text{Gbps} \times \text{km}$$

Actually, we've calculated a $BR$-distance product; this is why the units are given in Gbps·km. To compute bit rate, we need to know the distance. Taking a 100-km link, we obtain:

$$BR_{\text{chrom}} = 1.25\ \text{Gbps}$$

---

A bit-rate-distance product of 125 Gbps·km is very impressive, particularly compared with that of 1.38 Gbps·km we calculated for a graded-index fiber. However, the bit-rate limit of a singlemode fiber is affected by the PMD phenomenon. Let's see how much by referring to Example 5.3.6.

### Example 5.3.6

***Problem:***

Compute how much the bit rate is limited by PMD if a fiber's PMD coefficient is 0.5 ps/$\sqrt{\text{km}}$.

***Solution:***

In Example 5.3.4, we computed $\Delta t_{PMD} = 5$ ps for 100 km of fiber length. Thus, the bit rate is:

$$BR_{PMD} = 1/(4\ \Delta t_{PMD}) = 50 \text{ Gbps}$$

---

Now everyone can see the numbers: $BR_{chrom} = 1.25$ Gbps and $BR_{PMD} = 50$ Gbps. It seems as though chromatic dispersion limits bit rate, but this is not so. By choosing an operating wavelength equal to the zero-dispersion wavelength, chromatic dispersion can be—theoretically—eliminated. But there is no such means to exclude PMD. This is why for long-haul links, PMD becomes the bandwidth-limiting factor. More about coping with dispersion will be given in Section 6.3.

To conclude, in fiber-optic communications technology, the highest bandwidth can be achieved using singlemode fibers. For this reason—together with their attenuation rate, which is the lowest attainable—singlemode fibers are the links of choice in long-distance terrestrial and underwater telecommunications systems.

## 5.4 READING A DATA SHEET

Data sheets of singlemode (SM) fibers contain information similar to what's in the data sheets of multimode (MM) fibers discussed in Section 3.5. Rather than repeat what was covered in that section, we will concentrate here on the particulars of singlemode-fiber data sheets. This section incorporates all the knowledge you've acquired from the previous discussions; hence, it is advisable to refer to the appropriate sections if you encounter any difficulties in understanding this material.

### General Section

The data sheet of a singlemode fiber is shown in Figure 5.17. The general section contains, of course, introductory information about this fiber. Pay close attention to the paragraph where optimal operating wavelength is specified. Is this a conventional or dispersion-shifted fiber? Why is it so significant to emphasize that this fiber has low PMD? The manufacturer refers to the specific standards that this fiber meets. See Appendix D to gain an understanding of the given designations.

The subsections "Coating" and "Process" contain information similar to that discussed in Section 3.5. Pay special attention to the subsection designated "Profile." First, what is meant by "matched cladding type?" Secondly, note that the mode-field diameter—and not a core—is specified here. Can you tell why?

### Specifications Section

As we concentrate on new specifications here, be sure to refer back to Section 3.5 to brush up on those you are not yet thoroughly familiar with. To simplify this discussion, we have gathered all related characteristics under one heading rather than follow the format of the data sheet.

***Attenuation*** It is useful to compare the specified values of attenuation of the fiber types. (See Figure 3.18.). For example, one can achieve attenuation as low as 0.19 dB/km for a singlemode (SM) fiber, while the lowest attenuation figure for a multimode (MM) fiber is 0.5 dB/km. All the

**Product Code: 267E** **Low-PMD Singlemode Optical Fibre**

*Issue date: 11/95 Supersedes: -* *Type: Synthetic Quartz Matched Cladding*
Double Layer Primary Coating (DLPC7)

## FIBRE

**The low-PMD singlemode optical fibre, product code 267E, is a 1310 nm optimized (non-dispersion shifted) optical fibre, also suitable for use at 1550 nm.**

**This optical fibre offers low Polarisation Mode Dispersion and is especially suitable for high bit-rate, long distance digital transmission links and analog CATV networks.**

**The optical fibre complies with or exceeds the ITU Recommendation G.652 or the IEC 793-2 type B.1.1 Optical Fibre Specification.**

## COATING

The optical fibre is coated with a double layer UV curable acrylate, type DLPC7. Designed for more stringent tight-buffer cable applications, the optical fibre also performs perfectly in loose buffer constructions and demonstrates a high resistance to microbending.

The coating offers an excellent stable coating strip force over a wide range of environmental conditions and the coating stripping leaves no residues on the bare glass fibre.

Ribbon tests show excellent performance in 60°C watersoak tests, exceeding 100 days.

The DLPC7 coated optical fibres show high and stable values for the dynamic stress corrosion susceptibility parameter ($n_d$), which offers a greatly improved mechanical protection to the optical fibre when used in harsh environments.

## PROFILE

The optical fibre is of the matched cladding type with a nominal mode field diameter of 9.3 μm. It has a high level of splice compatibility in applications with optical fibres manufactured by other processes (OVD, MCVD, VAD).

## PROCESS

The optical fibres are manufactured using the PCVD process.

**PLASMA OPTICAL FIBRE B.V.**
Building TAY
P.O. Box 1136
5602 BC
Eindhoven
The Netherlands

**Tel: +31-40-292 3861**
Fax: +31-40-292 3866

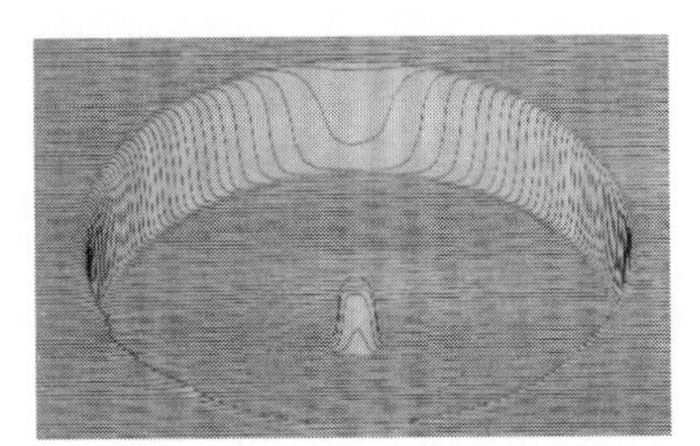

*The Plasma Optical Fibre policy of continuous improvement and updating means that specifications can be altered without prior notice.*

**Figure 5.17** Data sheet of a singlemode fiber. (*Courtesy of Plasma Optical Fibre.*)

## Low-PMD Singlemode Optical Fibre

**Product Code: 267E**

*Type: Synthetic Quartz Matched Cladding*
Double Layer Primary Coating (DLPC7)

*Issue date: 11/95 Supersedes: -*

| *Characteristics* | *Conditions* | | *Specified Values* | | *Unit* |
|---|---|---|---|---|---|
| **OPTICAL CHARACTERISTICS** | | | | | |
| Attenuation Coefficient | 1310 nm | ≤ 0.34 | ≤ 0.36 | ≤ 0.38 | [dB/km] |
| | 1285 - 1330 nm | ≤ 0.36 | ≤ 0.38 | ≤ 0.40 | [dB/km] |
| | 1550 nm | ≤ 0.19 | ≤ 0.21 | ≤ 0.23 | [dB/km] |
| Mode Field Diameter | 1310 nm | | | 9.3 ± 0.5 | [μm] |
| Fibre Cut-Off Wavelength | | | | 1215 ± 65 | [nm] |
| Chromatic Dispersion | | | | | |
| Zero Dispersion Wavelength | | | ≥ 1300 | ≤ 1324 | [nm] |
| Zero Dispersion Slope | | | | ≤ 0.093 | [ps/(nm^2.km)] |
| Dispersion Coefficient | 1285 - 1330 nm | | | ≤ 3 | [ps/(nm.km)] |
| | 1550 nm | | | ≤ 18 | |
| Polarisation Mode Dispersion | 1550 nm | | | ≤ 0.2 | [ps/km^0.5] |
| Backscatter Characteristics [1] | | | | | |
| Step (mean of bidirectional measurement) | | | | ≤ 0.05 | [dB] |
| Irregularities over fibre length | | | | ≤ 0.05 | [dB] |
| Difference Backscatter Coefficient | | | | ≤ 0.03 | [dB/km] |
| Attenuation Uniformity | | | | ≤ 0.05 | [dB/km] |
| Reflections | | | | Not Allowed | |
| Group Index of Refraction (Typical) [2] | 1310 nm | | | 1.467 | |
| | 1550 nm | | | 1.467 | |
| **GEOMETRIC CHARACTERISTICS** | | | | | |
| MFD Non-Circularity | | | | ≤ 6 | [%] |
| MFD / Cladding Concentricity Error | | | | ≤ 1.0 | [μm] |
| Cladding Diameter | | | | 125 ± 2.0 | [μm] |
| Cladding Non-Circularity | | | | < 2.0 | [%] |
| Coating Diameter | | | | 245 ± 10 | [μm] |
| Coating Non-Circularity | | | | ≤ 6 | [%] |
| Coating Concentricity Error | | | | ≤ 12.5 | [μm] |
| Length | | | Standard lengths up to | 25600 | [m] |
| **ENVIRONMENTAL CHARACTERISTICS** | | | | | |
| Temperature Dependence at 1310 nm and 1550 nm<br>Induced Attenuation - 60°C to + 80°C | | | | ≤ 0.05 | [dB/km] |
| Watersoak Dependence at 1310 nm and 1550 nm<br>Induced Attenuation at 20°C for 30 days | | | | ≤ 0.05 | [dB/km] |
| Damp Heat Dependence at 1310 nm and 1550 nm<br>Induced Attenuation at 85°C, 85% R.H., 30 days | | | | ≤ 0.05 | [dB/km] |
| **MECHANICAL CHARACTERISTICS** | | | | | |
| Proof Test | (off line) | | | ≥ 8.8 | [N] |
| | | | | ≥ 1.0 | [%] |
| | | | | ≥ 100 | [KPSI] |
| Bend Induced Attenuation at 1550 nm<br>100 turns around a mandrel of 60 mm diameter | | | | ≤ 0.2 | [dB] |
| Dynamic Stress Corrosion Susceptibility Parameter (Typical) | | | | ≥ 27 | |
| Coating Strip Force (Typical) | | | | 1.4 | [N] |

*1. Measurement at 1310 nm and 1550 nm with 1 μs pulse width.*
*2. In the case that an OTDR is used with c rounded to 3*10^8m/s, add 0.001 to the indicated value.*

plasma optical fibre
A MEMBER OF THE DRAKA HOLDING GROUP

**Figure 5.17** *(continued)*

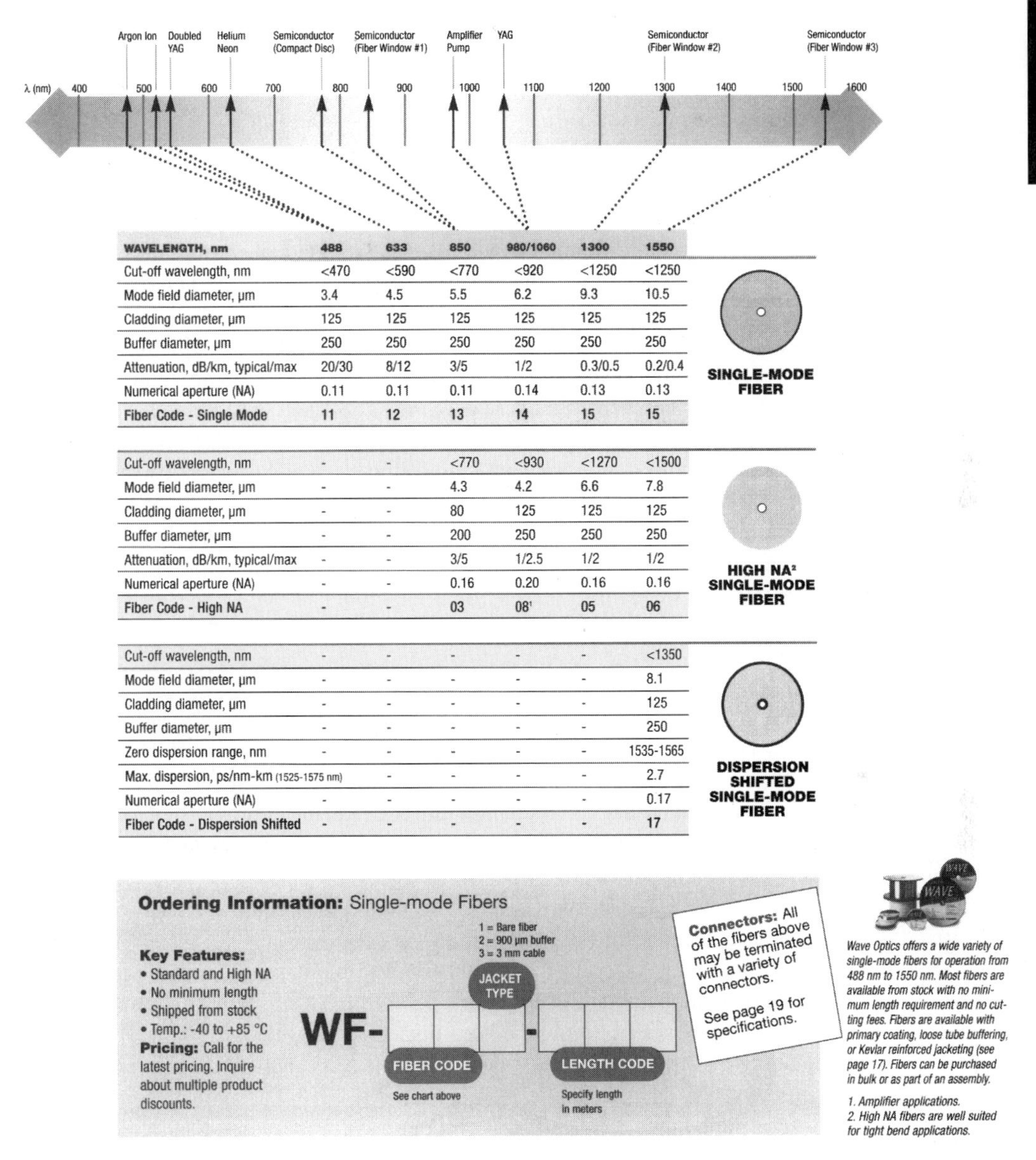


| WAVELENGTH, nm | 488 | 633 | 850 | 980/1060 | 1300 | 1550 |
|---|---|---|---|---|---|---|
| Cut-off wavelength, nm | <470 | <590 | <770 | <920 | <1250 | <1250 |
| Mode field diameter, µm | 3.4 | 4.5 | 5.5 | 6.2 | 9.3 | 10.5 |
| Cladding diameter, µm | 125 | 125 | 125 | 125 | 125 | 125 |
| Buffer diameter, µm | 250 | 250 | 250 | 250 | 250 | 250 |
| Attenuation, dB/km, typical/max | 20/30 | 8/12 | 3/5 | 1/2 | 0.3/0.5 | 0.2/0.4 |
| Numerical aperture (NA) | 0.11 | 0.11 | 0.11 | 0.14 | 0.13 | 0.13 |
| Fiber Code - Single Mode | 11 | 12 | 13 | 14 | 15 | 15 |

SINGLE-MODE FIBER

| | | | | | | |
|---|---|---|---|---|---|---|
| Cut-off wavelength, nm | - | - | <770 | <930 | <1270 | <1500 |
| Mode field diameter, µm | - | - | 4.3 | 4.2 | 6.6 | 7.8 |
| Cladding diameter, µm | - | - | 80 | 125 | 125 | 125 |
| Buffer diameter, µm | - | - | 200 | 250 | 250 | 250 |
| Attenuation, dB/km, typical/max | - | - | 3/5 | 1/2.5 | 1/2 | 1/2 |
| Numerical aperture (NA) | - | - | 0.16 | 0.20 | 0.16 | 0.16 |
| Fiber Code - High NA | - | - | 03 | 08[1] | 05 | 06 |

HIGH NA[2] SINGLE-MODE FIBER

| | | | | | | |
|---|---|---|---|---|---|---|
| Cut-off wavelength, nm | - | - | - | - | - | <1350 |
| Mode field diameter, µm | - | - | - | - | - | 8.1 |
| Cladding diameter, µm | - | - | - | - | - | 125 |
| Buffer diameter, µm | - | - | - | - | - | 250 |
| Zero dispersion range, nm | - | - | - | - | - | 1535-1565 |
| Max. dispersion, ps/nm-km (1525-1575 nm) | - | - | - | - | - | 2.7 |
| Numerical aperture (NA) | - | - | - | - | - | 0.17 |
| Fiber Code - Dispersion Shifted | - | - | - | - | - | 17 |

DISPERSION SHIFTED SINGLE-MODE FIBER

**Ordering Information:** Single-mode Fibers

**Key Features:**
- Standard and High NA
- No minimum length
- Shipped from stock
- Temp.: -40 to +85 °C

**Pricing:** Call for the latest pricing. Inquire about multiple product discounts.



**Connectors:** All of the fibers above may be terminated with a variety of connectors.

See page 19 for specifications.

*Wave Optics offers a wide variety of single-mode fibers for operation from 488 nm to 1550 nm. Most fibers are available from stock with no minimum length requirement and no cutting fees. Fibers are available with primary coating, loose tube buffering, or Kevlar reinforced jacketing (see page 17). Fibers can be purchased in bulk or as part of an assembly.*

*1. Amplifier applications.*
*2. High NA fibers are well suited for tight bend applications.*

**Figure 5.18** Data sheet of a singlemode fiber. (*Courtesy of Wave Optics, Mountain View, CA.*)

numbers given are for uncabled (bare) fiber because cabling changes attenuation, mainly by increasing microbending loss.

Comparing the parameter "Attenuation Uniformity," one can see that uniformity of attenuation through an SM fiber is very low (0.05 dB), while the typical value for an MM fiber is 0.2 dB. Attenuation (loss) induced by macrobending under even more stringent wrapping conditions—100 turns around a 60-mm mandrel—is less for this SM fiber than for its MM counterpart (0.2 dB versus 0.5 dB). However, note that these values were measured at different operating wavelengths. (See the subsection "Mechanical Characteristics.")

***Mode-field diameter and cutoff wavelength*** Instead of a core diameter, the manufacturer specifies the mode-field diameter (MFD) and its precision. Sometimes manufacturers specify MFD values for different operating wavelengths; for example, MFD = 9.30 ± 0.5 μm and 10.50 ± 1.00 μm [3]. (See Section 5.1.)

Another new characteristic is the fiber cutoff wavelength. This fiber will operate in a singlemode regime if the wavelength used is longer than 1215 ± 65 nm, which means that only 1300 nm and 12550 nm can realistically be used with this fiber. (See Section 5.1.)

Cutoff wavelengths are different for a bare fiber and for a cabled fiber because cabling introduces microbending, thereby changing a fiber's characteristics. Customers are interested in knowing the cable cutoff wavelength, the parameter they meet in installed fiber cables. The manufacturer measures cutoff wavelength using a bare fiber. By introducing additional fiber loops, the manufacturer simulates the cabling effect, which is the number for $\lambda_{ccf}$ in a typical data sheet [3].

***Numerical aperture*** Typically NA varies from 0.10 to 0.17 for an SM fiber; these numbers range from 0.2 to 0.3 and even higher for an MM fiber.

***Bandwidth*** Note that no bandwidth is specified in the data sheet. This is because the manufacturer is certain that an SM fiber's bandwidth exceeds the bandwidth of "electronics," that is, the bandwidth of the transmitter and receiver. (More about the higher-level bandwidth of an SM fiber will be found in Chapter 6.)

Let's now consider two bandwidth-related characteristics: chromatic dispersion and polarization-mode dispersion.

Chromatic dispersion is presented here by two means: (1) zero-dispersion wavelength and zero-dispersion slope and (2) dispersion parameter (coefficient). The information provided is redundant because if you know the zero-dispersion wavelength and slope, you will be able to calculate the dispersion parameter using Formula 3.25. This is the same formula manufacturers give in their MM data sheets. (See, for example, [3].) The only difference between the data sheets of Plasma Optical Fibre and Wave Optics is the numbers we have to plug in for the calculations. You should be able to easily identify all the differences between MM and SM fibers. Remember that in an SM fiber, $D(\lambda)$ is the sum of the material and waveguide dispersions, while in an MM fiber it is determined primarily by material dispersion. Bear in mind that $\lambda_0$ depends on temperature while $S_0$ doesn't, even though this fact is not noted in the data sheets.

(As an exercise, can you explain why the dispersion parameter [coefficient] of this fiber at 1300 nm is much less than it is at 1550 nm?)

Polarization-mode dispersion (PMD) is another characteristic we haven't yet met in multimode-fiber specifications. Such a fiber is called *low PMD* because the value of a PMD parameter equals 0.2 ps/$\sqrt{\text{km}}$, while, more typically, this value is about 0.5 ps/$\sqrt{\text{km}}$.

**Table 5.1** Typical characteristics of singlemode fibers

| Maximum attenuation (dB/km) at 1300 nm | at 1550 nm | Maximum mode field diameter (μm) at 1300 nm |
|---|---|---|
| 0.35–0.8 | 0.1–0.7 | 8.75–11 |
| 0.5 typical | 0.25 typical | 9.3 typical |

Source: *Lightwave 1999 Worldwide Directory of Fiber-Optic Communications Products and Services,* March 31, 1999, pp. 10–15.

***Geometric characteristics*** The significance of geometric characteristics is discussed in Section 3.5. Under this category, we come upon a new parameter, fiber curl [3], a term describing curvature along the length of the fiber. Curl appears during the draw phase of the fabrication process. To determine its extent, a short piece of uncoated fiber is stretched along a restraining fixture and its deflection from the horizontal position is measured. This deflection, in microns, is then converted into radius of curvature in meters.

Fiber curl, MFD/cladding concentricity, and the outer-cladding diameter determine the quality and pace of splicing (see Section 8.2) and, therefore, are extremely important. A fiber manufacturer has to think not only about reducing the cost of the fiber itself but also about reducing the installation cost. Since approximately 30% of the installation cost falls under the category of *labor associated with splicing,* good geometry enhances fiber-manufacturing competitiveness by rewarding its practitioners with lower production costs.

***Other characteristics*** In addition to the characteristics listed in Figure 5.17, you will run across some others in data sheets available from other manufacturers. (See reference [3] and Figure 5.18.) These include core diameter, numerical aperture, refractive-index difference (also called the relative refractive index), and the spectral-attenuation graph. Each of these has already been discussed, so there is no need to describe them here.

*To summarize:* The data sheet for a singlemode fiber displays all the information necessary for a user to evaluate the performance of this fiber and make the right buying decision for his or her application.

Table 5.1 gives you a general idea of what today's singlemode fibers look like.

## SUMMARY

- A singlemode (SM) fiber has been a major telecommunications workhorse in long-haul networks since the early 1980s. This has been the main transmission medium in terms of millions of installed miles. Today, its usage pervades all optical networks—intermediate and local—because, in addition to its performance superiority over multimode fiber, the cost of manufacturing and installing this fiber has become competitive. This is why we direct your attention to the singlemode fiber.
- A singlemode (SM) fiber conducts only one mode. To eliminate higher-order modes, manufacturers make a core diameter smaller than 9 μm and a core-cladding relative refractive index, Δ, less than 0.37%.
- Another condition for a fiber's singlemode operation is using an operating wavelength longer than the cutoff wavelength. Cutoff wavelength, $\lambda_C$, is the minimal wavelength at which a fiber supports only one propagating mode. This parameter is specified by the manufacturer as an important characteristic of an SM fiber. Typically, $\lambda_C > 1260$ nm. The $\lambda_C$ and $V$-number, which is less than 2.405 for SM fiber, can be used interchangeably.
- A mode of an SM fiber is characterized by its field distribution over the fiber's cross section. The Gaussian curve is the most accepted model of this distribution, even though there is a discrepancy between Gaussian and measured curves.

- The mode-field diameter (MFD), at which intensity drops to $1/e^2$ of its peak value, is the parameter used to characterize mode-field distribution. It is an important parameter because mode field is distributed between the core and the cladding (typically as 80% and 20%, respectively); hence, the core diameter is not a representative characteristic of an SM fiber. MFD increases with increasing operating wavelength. The typical MFD value is about 10 μm at 1550 nm. MFD is always larger than the core diameter.

- Attenuation in an SM fiber is smaller than in an MM fiber simply because in the SM fiber less light encounters fewer absorption and scattering spots. The spectral-attenuation graphs for SM and MM fibers are very similar and differ only in their values.

- Attenuation in a singlemode fiber is caused by the same mechanisms that produce it in a multimode fiber: bending, absorption, and scattering. Macrobending loss in a singlemode fiber increases as operating wavelength increases and bend radius decreases. The sensitivity of a singlemode fiber to microbending loss increases slightly with increasing wavelength. Scattering and absorption in a singlemode fiber are very similar to the same phenomena found in a multimode fiber. They differ only in that their values are smaller.

- Linear dispersion in an SM fiber is mainly caused by chromatic dispersion, which is a combination of the material and waveguide dispersions. Fortunately, material and waveguide dispersions above 1300 nm have opposite signs and compensate for each other at the zero-dispersion wavelength, $\lambda_0$. Unfortunately, the zero-dispersion wavelength for a *conventional* SM is at 1310 nm, while the minimum attenuation is at 1550 nm. By manipulating the refractive-index profile, the designer is able to shift $\lambda_0$ to 1550 nm. This type of fiber is called a "dispersion-shifted fiber" (DSF).

- To make an SM fiber useful for wavelength-division multiplexing (WDM) systems, the designer will make the fiber's chromatic dispersion low throughout the range of wavelengths. Such a fiber is called *a dispersion-flattened fiber.*

- Pulse spread caused by chromatic dispersion is estimated by the following formula:

$$\Delta t_{chrom}/L(\text{ps/km}) = D(\lambda)\Delta\lambda, \tag{5.9}$$

where $D(\lambda)$ (ps/nm·km) is the dispersion parameter of an SM fiber, $\Delta\lambda$ (nm) is the spectral width of the light source, and $L$ (km) is the transmission length. D($\lambda$) is given by:

$$D(\lambda)\ (\text{ps/nm·km}) = S_0/4[\lambda - \lambda_0^4/\lambda^3],$$

where $S_0$ is the zero-dispersion slope, $\lambda_0$ is the zero-dispersion wavelength, and $\lambda$ is the operating wavelength for SM fiber. Typically, $D(\lambda) \sim 17$ ps/nm·km, $S_0 \sim 0.092$ ps/nm$^2$·km, and $\lambda_0 \sim 1310$ nm for a conventional fiber. All these numbers can be found in any manufacturer's data sheets. *Chromatic dispersion is a time-stable phenomenon,* which means that $\Delta t_{chrom}$ varies slightly with respect to time.

- The *bit rate (BR)* that can be transmitted over the fiber is defined as:

$$BR\ (\text{Gbit/s}) < 1/[4\Delta t(\text{ns})], \tag{5.11}$$

where $\Delta t$ is dispersion-caused pulse spreading.

- An SM fiber experiences *polarization-mode dispersion (PMD).* Its mechanism is as follows: A mode in an SM fiber is combined from two orthogonally polarized waves. Because of the difference in the values of refractive index along the two perpendicular cross-sectional axes (birefringence), those two waves travel inside the fiber at different velocities. This results in pulse spreading, $\Delta t_{PMD}$, which can be calculated as follows:

$$\Delta t_{PMD} = D_{PMD}\sqrt{L}, \tag{5.10}$$

where $D_{PMD}$ is the polarization-mode-dispersion coefficient measured in ps/$\sqrt{\text{km}}$.

- Fortunately, $D_{PMD}$ is relatively small (0.5 ps/$\sqrt{\text{km}}$ versus 17 ps/nm·km for $D[\lambda]$); unfortunately, though, PMD is a random process. This is why there is no real means for its compensation.

## PROBLEMS

**5.1.** A singlemode fiber has the following parameters: core diameter ($d$) = 8.3 μm, core refractive index ($n_1$) = 1.4692, and relative index ($\Delta$) = 0.36%. Calculate the $V$-number at the 1550-nm operating wavelength.

**5.2.** A singlemode fiber has the following parameters: numerical aperture ($NA$) = 0.125 and relative index ($\Delta$) = 0.36%. Calculate $n_1$.

**5.3.** Is MFD always a representative characteristic of a field distribution? Explain.

**5.4.** What is the portion of the maximum intensity Gaussian beam at $r = 0.5\ w_0$? At $r = 0.75\ w_0$?

**5.5.** Two singlemode fibers with MFD $2w_{01} = 10.5$ and $2w_{02} = 9.3$ are spliced. What is the coupling loss?

**5.6.** a. What is the cutoff wavelength for a singlemode fiber with $2a = 8.3$ μm and $NA = 0.125$?
b. If the operating wavelength is 1310 nm, does this fiber carry only one mode?

**5.7.** Given two singlemode fibers with MFDs of 5.0 μm and 10.0 μm, respectively, which one is more sensitive to bending and why?

**5.8.** Analyze the physical mechanisms leading to similarities and differences in the spectral attenuation of multimode and singlemode fibers, as Figure 5.7, shows.

**5.9.** Explain the difference between intermodal and intramodal dispersion.

**5.10.** What is the pulse spread caused by material dispersion if $\Delta\lambda = 0.5$ nm, $L = 1$ km, and $\lambda = 1550$ nm?

**5.11.** Light travels within a cladding faster than it does within a core. Why?

**5.12.** Calculate the pulse spread caused by waveguide dispersion at the 1550-nm operating wavelength if $\Delta\lambda = 0.5$ nm and $L = 1$ km.

**5.13.** Compute the pulse spread caused by chromatic disperion if a fiber has a zero-dispersion wavelength at 1312 nm, a zero-dispersion slope of 0.090 $ps/nm^2 \cdot km$, a length of 100 km, and operates at 1310 nm. The LD's spectral width is 1 nm.

**5.14.** What do manufacturers want to achieve by designing sophisticated refractive-index profiles like those shown in Figures 5.11, 5.12, and 5.13?

**5.15.** What is the difference between conventional, dispersion-shifted (DSF), and dispersion-flattened fibers?

**5.16.** a. What does the term "the state of polarization" mean?
b. Does it describe the fiber's properties? Explain.

**5.17.** Calculate pulse spread caused by polarization-mode dispersion if $D_{PMD} = 0.2$ ps/$\sqrt{km}$ and $L = 120$ km.

**5.18.** A fiber-optic communications link uses 120 km of a conventional fiber at 1550 nm and has an LD with a spectral width of 1 nm. Calculate the maximum bit rate imposed by:
a. Chromatic dispersion
b. PMD

**5.19.** *Special Project:* Figure 5.18 is a data sheet for a singlemode optical fiber manufactured by Wave Optics Co., Mountain View, CA. Explain the meaning of each characteristic listed and analyze the conditions and specified values.

**5.20.** *Project:* Memory grid. (See Problem 1.20.)
Sketch a grid summarizing what you have learned about singlemode fibers.

## REFERENCES[1]


1. James J. Refi, *Fiber Optic Cable—A LightGuide,* Geneva, Ill.: abc TeleTraining, Inc., 1991.
2. Peter Kaiser and Donald B. Keck, "Fiber Types and Their Status," in *Optical Fiber Telecommunications-II,* ed. by S.E. Miller. San Diego, Calif.: Academic Press, 1988, pp. 29–55.
3. *Corning SMF-28 CPC6 Single-Mode Fiber* (data sheet), Corning Inc., Corning, N. Y., 1997.


[1] See Appendix C: A Selected Bibliography.

# 6

# Singlemode Fibers—A Deeper Look

Today's singlemode fiber-optic communications systems demonstrate brilliant performance characteristics. But the high level of achievement in this area causes, in turn, new, even higher requirements of fiber performance. This means taking into account the most minuscule details of each phenomenon in a singlemode fiber. What yesterday might have been thought of as relatively insignificant areas of interest applicable only to the pure scientist today have become practical limitations in the usage of a singlemode fiber as a transmission link. Such "secondary" matters, once relegated to the back burners of the research and development laboratories, are now hotbeds of manufacturing activity because of their recognized influence on the performance characteristics of singlemode fibers.

## 6.1 MODE FIELD

Mode-field distribution is of great importance to singlemode-fiber performance; it is the parameter that describes how the mode really propagates within the fiber. The main phenomena in a singlemode fiber—attenuation and dispersion—depend on mode-field distribution. This section is devoted to a closer investigation of mode-field distribution and its associated effects.

### Gaussian Model and Real Mode-Field Distribution

***Refractive-index profiles*** As you saw in Section 5.4, singlemode fibers can have a variety of refractive-index profiles. (See Figures 5.11, 5.12, and 5.13.) There are formulas describing each of these profiles with a certain level of accuracy. For simple matched-cladding profiles, the following general description can be used:

$$\begin{aligned} n(r) &= n_1\sqrt{\left[1-2\Delta(r/a)^{\alpha}\right]} \qquad r \leq a \\ n(r) &= n_1\sqrt{[1-2\Delta]} = n_2 \qquad r \geq a \end{aligned} \tag{6.1}$$

In this formula, $a$ is the core radius, $\Delta$ is the relative refractive index, and $\alpha$ is the profile parameter ranging from 1 to infinity. The role of this parameter is clear from Figure 6.1.

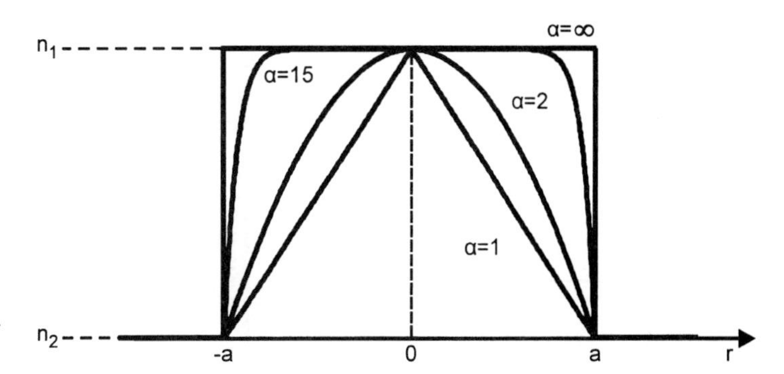


**Figure 6.1** Refractive-index profiles for matched-cladding fibers.

For $\alpha = 1$, the profile is a simple triangle; for $\alpha = 2$, the profile is close to the parabolic, which is used in graded-index fibers; for $\alpha \to \infty$, the profile is a simple step index.

It is quite evident that the mode-field distribution is a function of a refractive-index profile.

***Gaussian model and other alternatives*** The Gaussian model shown in Figure 5.2 is a good approximation of the real mode-field distribution. Given in Formula 5.2 as $I(r) = I(0) \exp(-2r^2/w_0^2)$, this model is characterized by a single parameter: mode-field diameter (MFD), $2\,w_0$. This formula complies with the experimental measurement for an $LP_{01}$ mode field when the fiber has a step-index or simple parabolic-index profile. For these fibers, the Gaussian model shows MFD to be the major parameter governing singlemode-fiber operation. Indeed, MFD is as important to a singlemode fiber as core diameter and *NA* are to a multimode fiber because MFD determines the power confinement within a singlemode fiber. Another view of MFD: The closer the operating wavelength is to the cutoff wavelength, the more the Gaussian model will approximate the real mode-field distribution.

Thus, if you use the Gaussian model, the only parameter you need to know is MFD = $2\,w_0$. How can we find $w_0$? Either read a fiber's data sheet or compute it by the following approximate formulas [1]:

$$w_0/a = 0.65 + 1.619/V^{3/2} + 2.879/V^6 \tag{6.2}$$

$$w_0/a = 1/\sqrt{(\ln V)} \tag{6.3}$$

Ratio $w_0/a$ is known as a normalized MFD, or normalized spot size. These formulas show how MFD (essentially the mode-field radius, $w_0$) changes with respect to the core radius, which is constant for a given fiber, as a function of the *V*-number. Formulas (6.2) and (6.3) were derived using different approaches. To make a decision as to which formula to use, take a look at the graphs for both Formulas 6.2 and 6.3 shown in Figure 6.2.

To use Formulas 6.2 and 6.3, we need a *V*-number. Manufacturers provide us not with a *V* value but with a cutoff wavelength. Fortunately, we know their relationship through Formula 5.5.

There is, however, a major drawback to the Gaussian model: systematic inconsistency when compared with the results of very precise professionally made measurements. This is shown in Figure 6.3.

Pay attention especially to the fact that MFD, at a wavelength of 1550 nm, is larger than it is at 1310 nm. This is another indication of how MFD depends on operating wavelength. (See Figure 5.4.)

The Gaussian model is not the only expression of mode-field distribution. For example, mode field is described by the Bessel function within the core and by the exponential function within the cladding. Bessel functions are the form of solutions of Helmholtz equations describing the behavior of modes within a fiber, and the exponent is a common model for the evanescent wave.

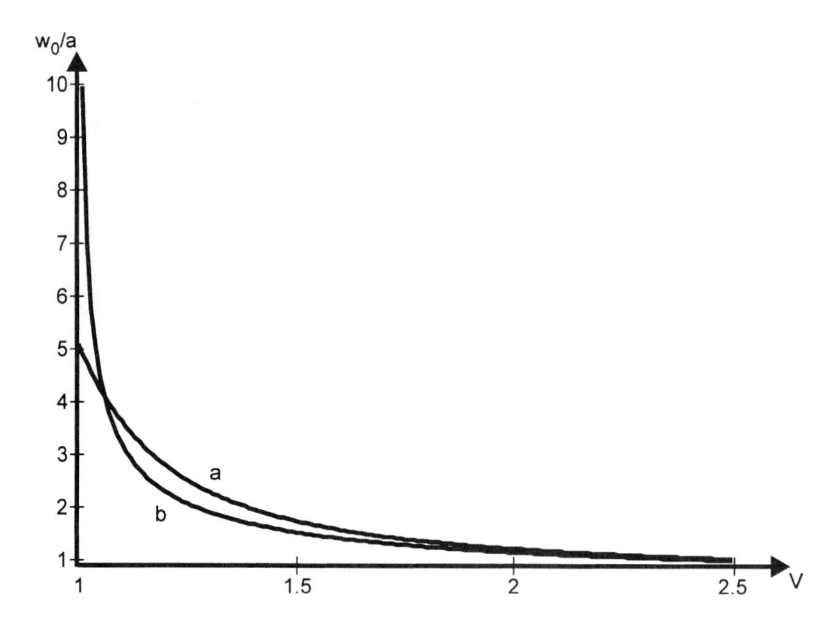


**Figure 6.2** Normalized MFD as a function of V number: (a) $w_0/a = 0.65 + 1.619/v^{-3/2} + 2.871/v^{-6}$; (b) $w_0/a = \ln(v)^{-1/2}$.

The general approach is to find the formulas for describing mode-field distributions in both the core and cladding and then seek the solution based on the coincidence of both formulas at the core-cladding interface.

Mode-field intensity distribution can be measured directly by scanning the image of the fiber-mode field using a photodetector with a small active area.

It is worth pointing out that you can encounter mode-field distribution given in terms of field-strength—electric-field intensity, $E(r)$—distribution rather than in the distribution of mode-field intensity, $I(r)$, so that:

$$E(r) = E(0)\exp\left(-r^2/w_0^2\right) \tag{6.4}$$

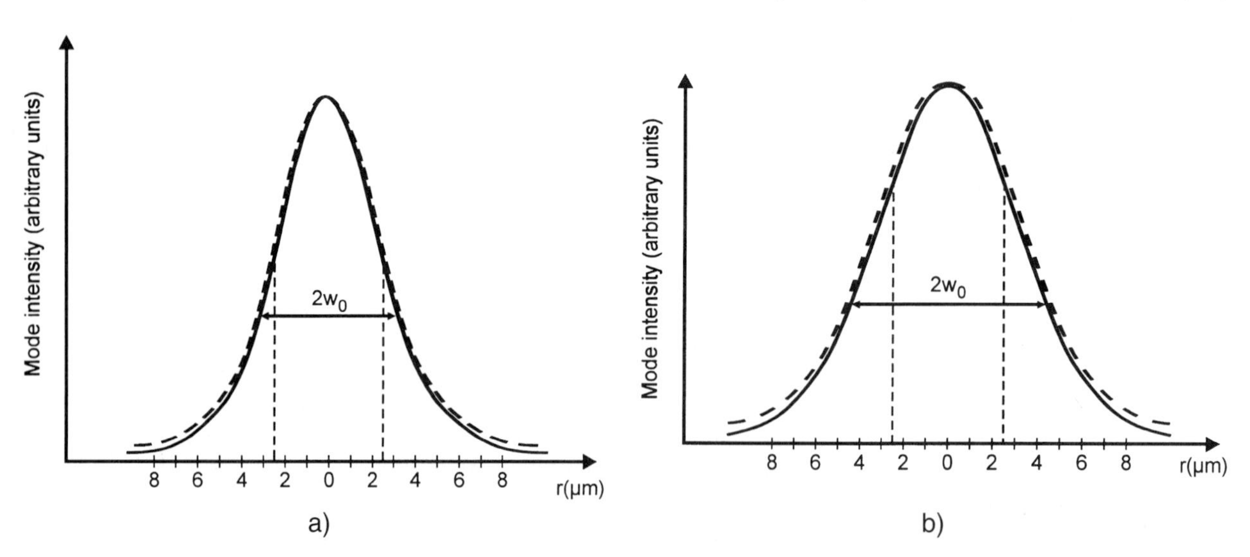


**Figure 6.3** Gaussian (solid curve) and experimental (dotted curve) intensity distributions: (a) $\lambda$ = 1320 nm; (b) $\lambda$ = 1550 nm. *(J-S Jang et al, "Fundamental Mode Size and Bend Sensitivity of Graded and Step-Index Single-Mode Fibers with Zero Dispersion Near 1, 5 μm,"* IEEE Journal of Lightwave Technology, *vol. LT-2, 1984, pp. 312–316.)*

Remember, intensity represents optical power. Since intensity is proportional to the square of the electric-field strength, $I \propto E^2$, $2w_0$ is defined as the diameter at which $E(r)$ becomes $e^{-1}$ of $E(0)$. This is equivalent to the definition of $2w_0$ as the diameter at which $I(r)$ becomes $e^{-2}$ of $I(0)$.

## Cutoff Wavelength and *V*-number

***Their relationship*** As you saw in Section 5.4, manufacturers give the value of the cutoff wavelength, $\lambda_C$, in their fiber data sheets. You will find the value for $\lambda_C$ in any singlemode-fiber data sheet, a fact that tells you that manufacturers consider this parameter important.

In Section 5.1 we learned that $\lambda_C$ is the wavelength below which a fiber operates as a multimode. Let's rewrite Formula 5.5 in this form:

$$\lambda_C = \left[2\pi a n \sqrt{\Delta}\right]/V \tag{6.5}$$

One can compute $\lambda_C$ by substituting the fiber's parameters and assuming $V = 2.405$, the number obtained from the fiber-mode theory.

It seems, therefore, that we can vary the cutoff wavelength simply by changing the fiber core's radius, $a$, the relative refractive index, $\Delta$, or the $V$-number. In reality, the core radius and relative index cannot be changed arbitrarily because their values are determined by the bending-sensitivity and dispersion-compensation requirements. Consequently, the only degree of latitude we have in varying the cutoff wavelength is changing the $V$-number in the range of 0 to 2.405.

The $V$-number determines the power distribution within the fiber, which follows from Figure 4.15. Practically all manufacturers choose a $V$-number between 1.8 and 2.2. As a rule, however, manufacturers prefer not to specify the $V$-number but, rather, the cutoff wavelength. Given the cutoff wavelength, the fiber user can easily compute the $V$-number because the core radius and the relative index are given in the data sheets. (See Section 5.4.) What this means is that the $V$-number and the cutoff wavelength can be used interchangeably.

How does one go about choosing the right cutoff-wavelength range? This rule of thumb should answer the question: The longer the cutoff wavelength, the closer the fiber is to the bimodal state with its attendant excess modal noise. A shorter cutoff wavelength, on the other hand, does not confine the mode securely within the core. If the mode is not confined sufficiently, absorption and bending losses will increase. So, as you can see, the ideal operational cutoff wavelength must lie within some reasonable window. For a $V$-number between 1.8 and 2.2 and for typical singlemode-fiber parameters, this window is between 1000 nm and 1250 nm. Typically, singlemode fibers have $\lambda_C < 1250$ nm.

An interesting situation arises from the availability of very cheap light sources with operating wavelengths in the 780- to 900-nm range. Most of these light sources are laser diodes manufactured for CD players that work at 780 nm. To take advantage of this commercial availability, manufacturers have made extensive efforts to fabricate singlemode fiber with a short cutoff wavelength. The result: Today we have commercially available singlemode fibers with $\lambda_C$ as low as $< 720$ nm, a real breakthrough. These fibers are used for specific communications links and in fiber sensors.

***Effective cutoff wavelength*** The above discussion reaffirms that the cutoff wavelength plays a twofold role. On the one hand, $\lambda_C$ determines the minimum wavelength feasible for singlemode-fiber operation. For example, the Plasma Optical Fibre data sheet (Figure 5.17) specifies $\lambda_C < 1215 \pm 65$ nm. This means that for all wavelengths longer than 1280 nm, this fiber supports only one mode. On the other hand, the value of $\lambda_C$ determines the power confinement within a core, as noted above.

Questions arise at this point from our definition of cutoff wavelength: What happens if $\lambda_C$ will be, for example, 1281 nm or 1279 nm for, say, the Plasma Optical SM fiber? Will we see a dramatic difference in the fiber's performance by gradually varying $\lambda_C$ from 1279 to 1261 nm? Absolutely not. Fiber operation does not change abruptly when the operating wavelength varies around the cutoff value. In fact, the transition from dual mode to singlemode operation (and back) occurs gradually so that the second mode continues to exist but its power becomes less and less while $\lambda$ increases over $\lambda_C$.

At this point we need to introduce another important criterion for singlemode operation: *effective cutoff wavelength.* This is the wavelength at which the second mode, $LP_{11}$, experiences 19.3 dB more attenuation than the fundamental mode, $LP_{01}$. The CCITT Recommendation G.652 and Standard EIA-455-80 define effective cutoff wavelength for a 2-m fiber loop with a single 14-cm radius loop. (The technique for measuring effective cutoff wavelength is described in later chapters. For more on CCITT and EIA, see Appendix B.)

Effective cutoff wavelength depends on certain conditions. For example, *the effective cutoff wavelength of cabled fiber is shorter than that of uncabled fiber. The smaller the bend radius, the shorter the effective cutoff wavelength; the longer the fiber, the shorter the effective cutoff wavelength.*

To sum this all up succinctly, we have introduced three cutoff wavelengths: theoretical $\lambda_C$, defined by Formula 5.4; effective $\lambda_C$, discussed above; and cable $\lambda_C$, considered in Section 5.4. In practice, however, we will always be dealing with the cutoff wavelength of cabled fiber.

Finally, we need to say a few words about so-called *modal noise.* When the fiber's operating wavelength is close to $\lambda_C$, the second-order mode demonstrates a significant sign of its presence. In other words, the power of the $LP_{11}$ mode becomes high enough to exert a detrimental effect on the fiber's operation. This effect is an increase in both attenuation and dispersion because the fiber takes on an almost dual mode—no longer singlemode—link. This additional degradation in fiber-performance characteristics when the fiber operates on the edge of the singlemode state is called modal noise.

Let's pause here to underscore the two major points of our discussion of mode field:

1. Mode-field diameter, MFD, is the representative characteristic of mode-field distribution if, and only if, the Gaussian model is used. But the Gaussian model consistently shows discrepancies in its measured results, thus warning us not to rely on MFD as an exclusive characteristic. The more precise your model of a singlemode fiber, the less justification you have to use the Gaussian model.
2. The cutoff wavelength and the *V*-number are interchangeable fiber parameters determining the minimum operating wavelength and power confinement within a core.

## 6.2 MORE ABOUT ATTENUATION IN A SINGLEMODE FIBER

Attenuation in singlemode fibers is much smaller than in multimode fibers, as we saw in Section 5.2. However, since singlemode fibers are used as major transmission links for long-distance networks, we want to reduce their attenuation to an even smaller value. We know the limits for attenuation in a silica fiber; hence, we realize that attenuation can never be zero. Today's fibers are made with attenuation close to the theoretical limits. What's more, with the advent of optical amplifiers, the problem of attenuation plays a secondary role, which makes modern optical-network designers feel quite comfortable. However, the goal of an optical-fiber designer is to reduce attenuation as much as possible to minimize any extra intrusion in an optical link. This is why this section will focus on a deeper analysis of attenuation problems not covered in Section 5.2.

## Intrinsic and Extrinsic Losses

***Intrinsic losses*** The general consideration of attenuation in optical fibers given in Section 4.5 is true for singlemode fibers too. Intrinsic losses in a singlemode fiber are defined by IR and UV resonances and Rayleigh scattering. These losses put absolute limits on minimizing attenuation of a silica-made singlemode fiber. (Refer to Figure 4.16 for the specific numbers.) These losses are determined by the properties of silica—the fiber material—and, therefore, they are the same for singlemode and multimode fibers.

***Extrinsic losses—absorption*** Extrinsic losses include absorption and bending losses. Absorption, again, is caused by the properties of fabricated silica; therefore, we can't expect any new phenomena in a singlemode fiber. (Refer to Figure 5.7.)

***Extrinsic losses—bending losses for step-index fibers*** The difference between a singlemode fiber and a multimode fiber is bending—macro- and microbending—losses. Both types of losses, as discussed in Section 5.2, are determined by mode-field confinement within the core. The more the mode field is tightened within the core, the less the bending loss. For the Gaussian model of mode-field distribution, this is equivalent to saying that the less the MFD, the less the bending loss. Since we know that the Gaussian model is not exact, we have to use the general mode-field-distribution approach for more accurate results.

The real question is this: How can we achieve the *desired mode-field confinement?* Let's refer to Formula 6.5, now in this form:

$$V = \left[2\pi a n\sqrt{\Delta}\right]/\lambda_C \tag{6.6}$$

We know that the higher the $V$-number, the more confined the mode field is within the core. But our discussion of $V$-number and cutoff wavelength in Section 6.1 shows that the $V$-number should be between 1.8 and 2.2 and that the $\lambda_C$ is also predefined. Therefore, analyzing Formula 6.6, we see that there are only two parameters left to manipulate: the core radius, $a$, and the relative refractive index, $\Delta$.

The first—and obvious—step to take to decrease the mode-field diameter is to decrease the core radius, $a$. However, this action gives rise to two obstacles: First, the core radius, or diameter, of a singlemode fiber is small enough to pose considerable problems when launching light into the fiber and when splicing. Second—and more important—this step entails increasing the relative index, $\Delta$, because we have to keep $V$ and $\lambda_C$ constant.

Theoretically, it is possible to increase $\Delta$ by doping the core to a greater degree than is customary. This measure follows from the definition of the relative index: $\Delta = (n_1 - n_2)/n$. Since the cladding index, $n_2$, is usually the index of pure silica, the only way to increase $\Delta$ is to raise $n_1$ by increasing its dopant, $GeO_2$. But silica, $SiO_2$, heavily doped with germanium oxide, $GeO_2$, exhibits other problems. For one thing, Rayleigh scattering for this material increases, resulting in a rise in attenuation. Secondly, material dispersion changes, as Figure 4.23 shows, resulting in a shift in the zero-dispersion wavelength. We also have to take into account that—as follows from Formula 5.5—$\Delta$ is inverse to the square of $a$, which means we need to dope the core very heavily to compensate for the decrease in $a$.

It seems that we have another way to decrease bending loss: increase numerical aperture, $NA$. Indeed, Formula 6.6 can be rewritten in the form given by Formula 3.14a:

$$V = [2\pi a\, NA]/\lambda_C$$

Therefore, increasing $NA$ will result in increasing $V$, thus leading to the desired mode-field confinement within the core. Figure 6.4 shows bending loss as a function of bending radius for two $NA$s.

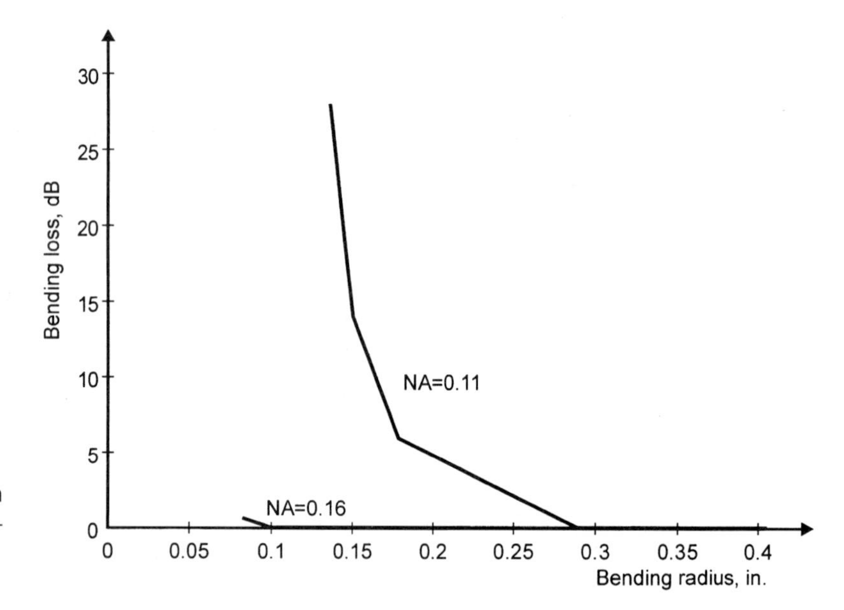


**Figure 6.4** Bending loss as a function of bending radius for two fibers with different NA. *(Courtesy of SpecTran Specialty Optics Co.)*

In a sense, Figure 6.4 gives us the idea of the bending sensitivity of a singlemode fiber as a function of *NA*. Since $NA = n\sqrt{\Delta}$, *NA* could be increased by increasing $\Delta = (n_1 - n_2)/n$ and $n = (n_1 + n_2)/2$. Thus, we arrive at the same problem of increasing the difference and sum of $n_1$ and $n_2$.

***Depressed-cladding-index profile*** All the above considerations are true for fiber with a simple step-index profile, that is, a matched-cladding fiber. One solution to the problem—decreasing the fiber's bending sensitivity—has been found in the depressed-cladding profile. A comparison of the two designs is given in Figure 6.6. (See also Figure 5.11.)

Because, with the depressed-cladding design, the core index, $n_{1DC}$, is smaller than it is for matched cladding, the core material requires less doping. The goal of increasing $\Delta$ is achieved by decreasing the cladding index immediately around the core. With this design, the ring of inner

### Manipulating Fiber Parameters

When fiber manufacturers need to produce a product having special properties, they do so by modifying the fiber's parameters. Let's look at a typical example. Figure 6.5 shows data sheets of two singlemode fibers manufactured by SpecTran Specialty Optics Co.: a bend-insensitive singlemode fiber and a communications fiber. As the name indicates, a bend-insensitive singlemode fiber has very low-induced bending loss: 0.05 dB with 100 turns around a mandrel of 10-mm diameter. In comparison, Corning's singlemode fiber SMF-28 has the same induced attenuation—0.05 dB—with 100 turns, but the diameter of the mandrel is 75 mm. The price for making such bend-insensitive fiber is to increase the general attenuation significantly, which means as high as 0.75 dB/km. The spectral-attenuation graph underscores this point.

In comparing SpecTran's communications and bend-insensitive singlemode fibers, you can see that bend insensitivity has been achieved by the means we discussed above: increasing the relative refractive index, $\Delta$, to 0.6% versus 0.3% for communications fiber. That increase has been done by increasing the difference and sum of $n_1$ and $n_2$. For bend-insensitive fiber, $n_1 = 1.4618$ and $n_2 = 1.4529$, while, for communications fiber, $n_1 = 1.4513$ and $n_2 = 1.4468$. The larger refractive index values for bend-insensitive fiber were obtained by doping both the core and cladding at the expense of attenuation, as we noted previously. Pay particular attention to the small size of the core diameter of this type of fiber. It starts at 3.7 μm, whereas for communications fiber it is typically 8.3 μm.

Bend-insensitive fiber finds application in many areas, including fiber amplifiers and pump fibers used in communications systems.

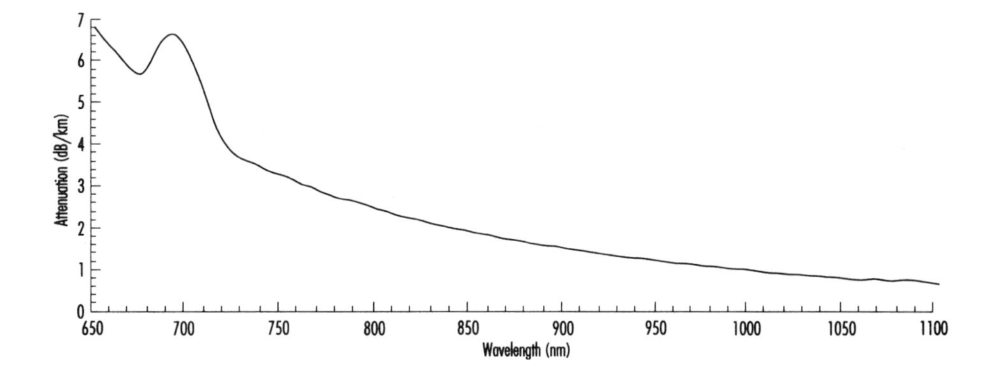


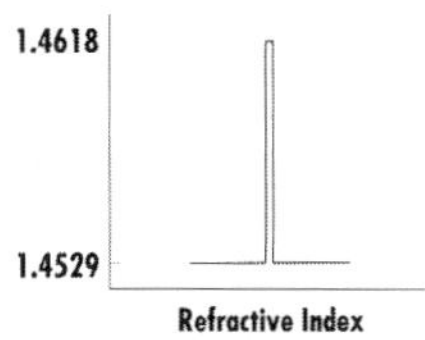


**BEND INSENSITIVE SINGLE-MODE FIBER**

| Part Number<br>Product Code | CF04246-03<br>SMM-D0820A | CF04246-04<br>SMB-D0820B | CF04246-10<br>SMB-D0980B | CF04246-30<br>SMB-D1310B | CF04246-31<br>SMM-D1310A | BF05285-02<br>GyroSil™-02 |
|---|---|---|---|---|---|---|
| Operating Wavelength (nm) | 820 | 820 | 980 | 1310 | 1310 | 1550 |
| Core Diameter (μm) | 3.7 | 3.7 | 4.5 | 6.0 | 6.0 | 6.5 |
| Mode Field Diameter (μm) | 4.1 ± 1.0 | 4.1 ± 1.0 | 5.0 ± 1.0 | 6.7 ± 1.0 | 6.7 ± 1.0 | 7.5 ± 0.75 |
| Cladding Diameter (μm) | 80 ± 2 | 125 ± 2 | 125 ± 2 | 125 ± 2 | 80 ± 2 | 80 ± 2.0 |
| Coating/Buffer Diameter (μm) | 135 ± 5 | 245 ± 15 | 245 ± 15 | 245 ± 15 | 135 ± 5 | 130 ± 4.0 |
| Cut-off Wavelength (nm) | 770 ± 40 | 770 ± 40 | 930 ± 40 | 1250 ± 60 | 1250 ± 60 | 1450 ± 50 |
| Numerical Aperture | 0.16 ± 0.02 | 0.16 ± 0.02 | 0.16 ± 0.02 | 0.16 ± 0.02 | 0.16 ± 0.02 | 0.17 ± 0.02 |
| Attenuation @ Operating Wavelength (dB/km) | ≤ 5 | ≤ 5 | ≤ 3 | ≤ 0.75 | ≤ 0.75 | ≤ 0.75 |
| Attenuation Increase Due to 100 Turns on a 10 mm Radius Mandrel (dB) | ≤ 0.05 | ≤ 0.05 | ≤ 0.10 | ≤ 0.6 | ≤ 0.6 | ≤ 0.4 |
| Refractive Index of Primary Coating | — | 1.54 | 1.54 | 1.54 | — | — |
| Refractive Index of Secondary Coating | 1.53 | 1.53 | 1.53 | 1.53 | 1.53 | 1.53 |
| Proof Test Level (kpsi) | ≥ 100 | ≥ 100 | ≥ 100 | ≥ 100 | ≥ 100 | ≥ 100 |
| Operating Temperature (° C) | -40 to +85 | -40 to +85 | -40 to +85 | -40 to +85 | -40 to +85 | -54 to +92 |
| Coating Type | UV Acrylate | UV Acrylate | UV Acrylate | UV Acrylate | UV Acrylate | UV Acrylate |
| Short Term Bend Radius @ ambient (mm) | ≥ 7 | ≥ 10 | ≥ 10 | ≥ 10 | ≥ 7 | ≥ 7 |
| Long Term Bend Radius @ ambient (mm) | ≥ 11 | ≥ 17 | ≥ 17 | ≥ 17 | ≥ 11 | ≥ 11 |
| Core/Clad Concentricity (μm) | ≤ 1.0 | ≤ 1.0 | ≤ 1.0 | ≤ 1.0 | ≤ 1.0 | ≤ 0.75 |
| Clad Noncircularity (%) | ≤ 2.0 | ≤ 2.0 | ≤ 2.0 | ≤ 2.0 | ≤ 2.0 | ≤ 2.0 |

a)

**Figure 6.5** Data sheets for (a) bend-insensitive singlemode fiber and (b) communications fiber. *(Courtesy of SpecTran Specialty Optics Co.)*

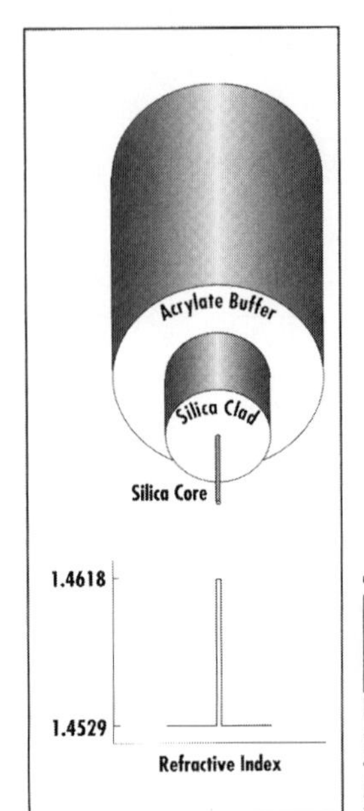


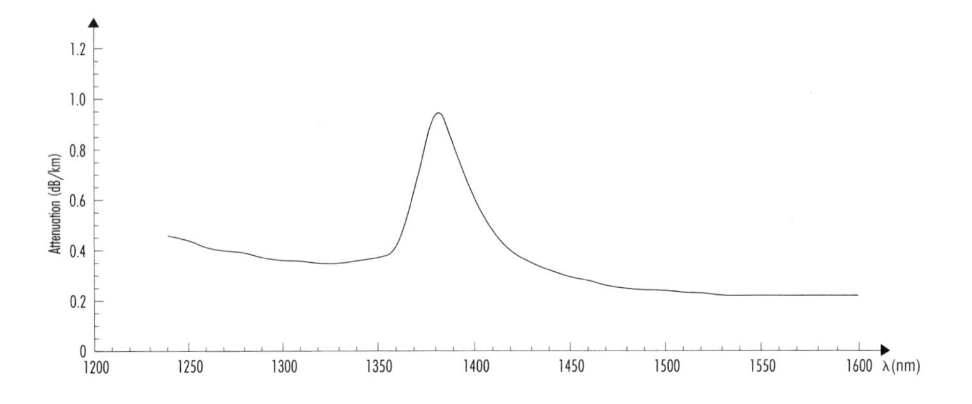


## COMMUNICATIONS FIBER

| Part Number<br>Product Code | BF04445-01<br>SMT-A1310B | BF04446<br>SMT-A1310H | BF04447<br>SMT-A1310J |
|---|---|---|---|
| Mode Field Diameter (μm) | 9.3 ± 0.5 | 9.3 ± 0.5 | 9.3 ± 0.5 |
| Cladding Diameter (μm) | 125 ± 2 | 125 ± 2 | 125 ± 2 |
| Coating/Buffer Diameter (μm) | 250 ± 15 | 155 ± 5 | 155 ± 5 |
| Coating Type | UV Acrylate | PYROCOAT™ | Hermetic/PYROCOAT™ |
| Operating Temperature (°C) | -40 to +85 | -65 to +300 | -65 to +300 |
| Operating Wavelength (nm) | 1310/1550 | 1310/1550 | 1310/1550 |
| Cut-off Wavelength (nm) | 1260 ± 70 | 1260 ± 70 | 1260 ± 70 |
| Numerical Aperture | 0.11 ± 0.02 | 0.11 ± 0.02 | 0.11 ± 0.02 |
| Attenuation @ 1310 nm (dB/km) | ≤ 0.35 | ≤ 0.70 | ≤ 0.70 |
| Attenuation @ 1550 nm (dB/km) | ≤ 0.25 | ≤ 0.60 | ≤ 0.60 |
| Proof Test Level (kpsi) | ≥ 100 | ≥ 100 | ≥ 100 |
| Zero Dispersion Wavelength (nm) | 1311 ± 11 | 1311 ± 11 | 1311 ± 11 |
| Zero Dispersion Slope (ps/[nm²·km]) | ≤ 0.092 | ≤ 0.092 | ≤ 0.092 |
| Core/Clad Concentricity Error (μm) | ≤ 1.0 | ≤ 1.0 | ≤ 1.0 |
| Clad Noncircularity (%) | ≤ 2.0 | ≤ 2.0 | ≤ 2.0 |
| Coating Concentricity (%) | ≥ 70 | ≥ 70 | ≥ 70 |
| Minimum Short-Term Bend Radius | ≥ 10 | ≥ 10 | ≥ 8.5 |
| Minimum Long-Term Bend Radius | ≥ 17 | ≥ 17 | ≥ 12 |

**Figure 6.5** *(Continued)* b)

cladding, with a refractive index of $n_3$, encloses the core. Now $\Delta = (n_1 - n_3)/n$, while the outer cladding index, $n_2$, remains the same. Technically, decreasing the refractive index to the value of $n_3$ is accomplished by doping the silica with appropriate material (e.g., fluorine, F, or boron oxide, $B_2O_3$).

The relative index has been defined as $\Delta = \left(n_{\text{core}}^2 - n_{\text{cladding}}^2\right)/2\,n_{\text{core}}^2$. The depressed-cladding design has two relative indexes: $\Delta_{12} = (n_1^2 - n_2^2)/2\,n_1^2$ and $\Delta_{23} = (n_2^2 - n_3^2)/2\,n_2^2$; as a result, its general relative index is $\Delta = \Delta_{12} + \Delta_{23}$. It is easy to verify that the expression $\Delta = (n_1 - n_3)/n$ is obtained from the above formulas. Typically, $\Delta_{12} = 0.25\%$ and $\Delta_{23} = 0.12\ \%$, so that $\Delta = 0.37\%$.

*The depressed-cladding design decreases bending losses through tighter confinement of the mode field within the core without the need for heavy doping.* So, is it a panacea? Not quite. This design gives rise to a problem of its own. Since the refractive index of the outer part of the cladding, $n_2$, is higher than the refractive index of the inner part, $n_3$, there is now the danger of

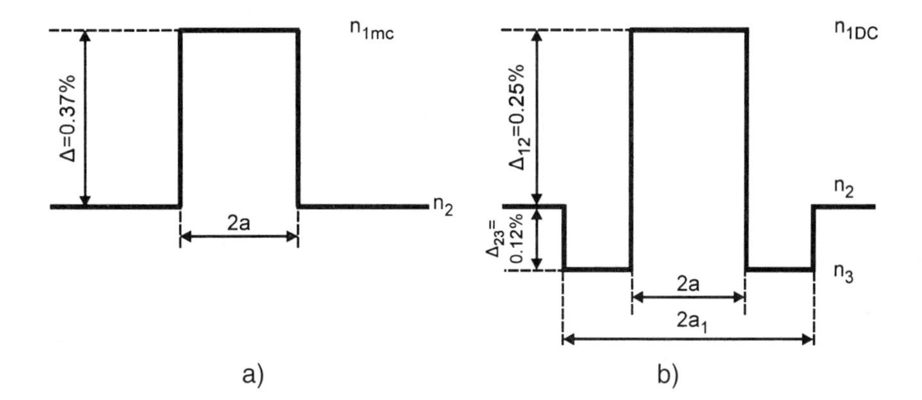


**Figure 6.6** Matched-cladding and depressed-cladding fibers: (a) Matched-cladding design; (b) depressed-cladding design.

power leaking through the inner part into the outer part. (Indeed, the phenomenon of total internal reflection can never be achieved at the boundary of the inner and outer parts of cladding because $n_2 < n_3$.) This difference in the cladding indexes might, in fact, result in an increase in fiber attenuation. To resolve this problem, the radius of the inner cladding ring, $a_1$, must be large enough to allow an evanescent wave to be damped completely.

Bear in mind, too, that since MFD becomes larger with a longer wavelength, increasing the operating wavelength raises the risk of increasing power leakage in depressed-cladding fibers. Thus, in theory at least, the operating wavelength for depressed-cladding fibers should be chosen between the cutoff and some upper value.

### Example 6.2.1

***Problem:***

Calculate the possible power leakage when light penetrates the inner–outer cladding interface. Figure 6.7 shows the mode-field distribution within a depressed-cladding fiber.

***Solution:***

Let's assume core diameter $2a = 8.3$ μm and inner–cladding diameter $2a_1 = 24.9$ μm. Assuming the Gaussian model, the intensity (power) of light is distributed according to Formula 5.2: $I(r) = I(0)\exp(-2r^2/w_0^2)$. Figure 5.4 shows that MFD reaches 10.3 μm at $\lambda = 1600$ nm. Thus, if we substitute $w_0 = 5.15$ μm and $r = a_1 = 12.45$ μm, we get $I(r)/I(0) = \exp[-2(12.45)^2/(5.15)2] = 8.39 \times 10^{-6}$. This is the fraction of light power that reaches the inner–outer cladding interface and that could be lost by transmission into the outer parts of cladding.

In reality, some portion of this light will be reflected and some portion will be refracted (transmitted) at the inner–outer cladding boundary. You can calculate these numbers using Formulas 4.31 and 4.31a (Section 4.3).

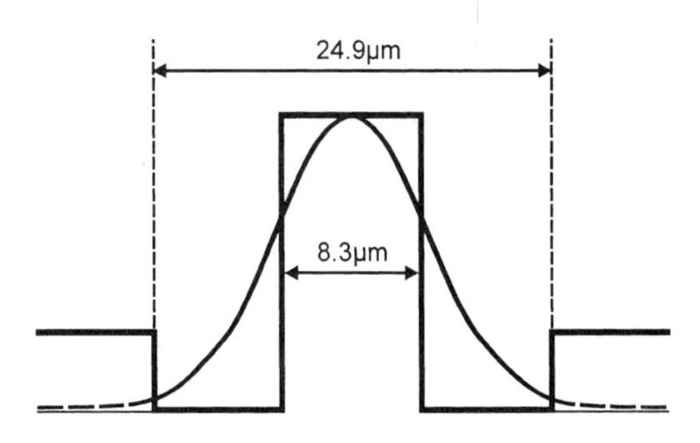


**Figure 6.7** Mode-field distribution within a depressed-cladding fiber.

Assuming all the power that reaches the outer parts of cladding will be lost—a very conservative assumption, to be sure—we determine the loss as $-10 \log[I(0) - I(r)]/I(0) = 36.44 \times 10^{-6}$ dB.

How real is this leakage? By analyzing Figure 5.4, we find that at 1600 nm, the MFD could reach 10.5 μm. This results in $I(r)/I(0) = 13.04 \times 10^{-6}$ and in a $56.63 \times 10^{-6}$ dB loss with the same assumptions. On the other hand, as Figure 4.16 shows, attenuation at 1600 nm becomes 0.1 dB/km. Taking the unit-length fiber for comparison, we see that, in reality, *IR* absorption is actually the factor limiting just how much the operating wavelength can be increased.

---

In summarizing our discussion of extrinsic and intrinsic losses, we must emphasize that intrinsic losses in singlemode fiber occur, as we'd expect, for the same reasons they take place in multimode fiber; these losses differ only in absolute numbers. In terms of extrinsic losses, only the bending characteristic exhibits substantial peculiarities in singlemode fiber. Painstaking consideration of how to reduce bending sensitivity led researchers to a new type of singlemode fiber—depressed cladding. The possibility of varying the depth and radius of the inner-cladding ring gives the fiber manufacturer more freedom in designing fibers with special characteristics.

## 6.3 COPING WITH DISPERSION IN A SINGLEMODE FIBER

A singlemode fiber was created as a solution to the intermodal-dispersion problem found in multimode fibers. But—as usually happens in a technical field—one solution gives rise to many new problems. Dispersion in a singlemode fiber is much less than in a multimode fiber but telecommunications users are demanding less and less dispersion. In fact, users want more bandwidth, but we know that it is dispersion that limits bandwidth in optical fibers. Therefore, reducing dispersion is still one of the major problems confronting the industry, and so in this section we will discuss the matter in considerable detail.

### Chromatic Dispersion

Singlemode fibers experience only intramodal dispersion, that is, dispersion existing within one mode. Two major effects produce this phenomenon: chromatic dispersion and polarization-mode dispersion (PMD). The basic mechanism and simple calculations for these effects were described in Section 5.3. We have restricted our discussion to the linear approximation of these effects. Now we'll consider chromatic dispersion and PMD in greater detail, including the nonlinear approach, focusing on how we can cope with these effects.

***Mechanism of chromatic dispersion*** Chromatic dispersion, the spreading of a light pulse as it propagates along the fiber, is caused by wavelength-dependent phenomena. Such dispersion is the sum of three effects: material dispersion, waveguide dispersion, and profile dispersion. Our discussion of material and waveguide dispersion (Section 5.3) was based on the assumption that the material and waveguide effects are independent, which is not completely true. Material dispersion, caused by the dispersive properties of the core and cladding materials, also depends on the mode-field distribution between the core and cladding, thus bringing the waveguide effect into this mechanism. What's more, waveguide dispersion, caused by the distribution of a mode field between the core and cladding, also depends on the dispersive properties of both materials, which is how material dispersion enters the picture. This mutual dependence is known as the second-order effect.

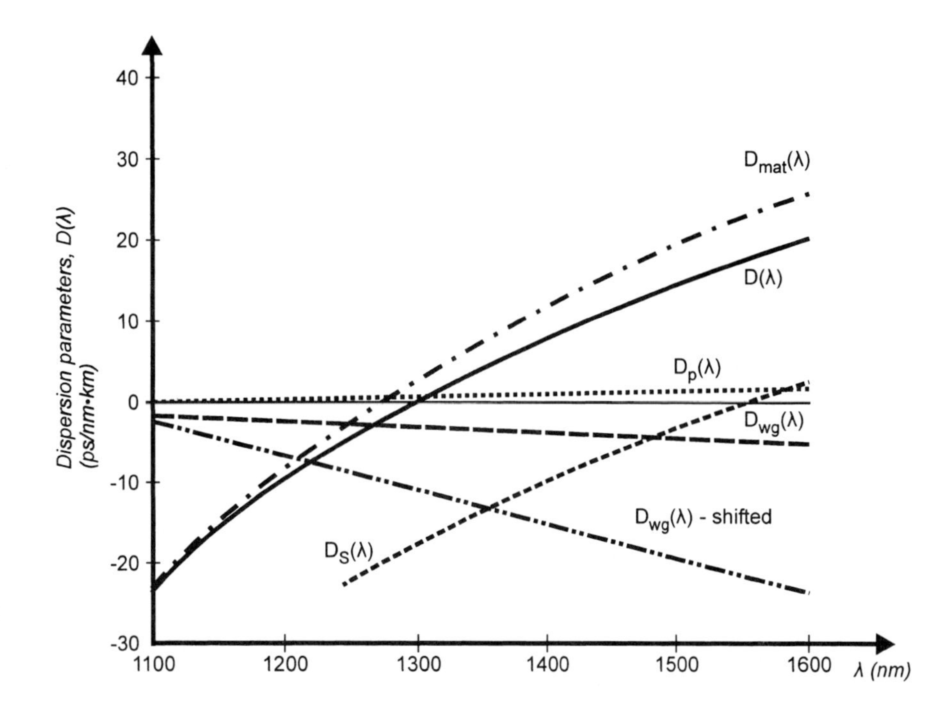


**Figure 6.8** Chromatic-dispersion parameters of a conventional fiber, $D$ (λ), and a dispersion-shifted fiber, $D_S$(λ).

Profile dispersion characterized by parameter $D_p(\lambda)$ is the result of taking into account the derivative of the relative index with respect to the operating wavelength, $d\Delta/d\lambda$. The absolute value of this parameter usually doesn't exceed 2 ps/nm·km and it is very slightly dependent on wavelength. $D_p(\lambda)$ should also be included in the total chromatic-dispersion mix, so that Formula 5.8 should be rewritten in this form:

$$D(\lambda) = D_{mat}(\lambda) + D_{wg}(\lambda) + D_p(\lambda) \tag{6.7}$$

The course of $D(\lambda)$ is shown in Figure 6.8.

***Producing dispersion-shifted and dispersion-flattened fibers*** Chromatic dispersion is zero at the specific zero-dispersion wavelength, where the condition $D_{mat}(\lambda) + D_p(\lambda) = D_{wg}(\lambda)$ is met. To control this condition, we have only one option: waveguide dispersion. Of course, we can control material dispersion by doping the core and cladding materials, but our discussion in Section 6.2 revealed that this approach won't be of very much help. Waveguide dispersion, you'll recall from Section 5.3, is defined by the mode-field distribution between the core and the cladding. The refractive-index profile of the fiber core controls this distribution. Therefore, to make a dispersion-shifted or a dispersion-flattened fiber, one has to have the specific profile of a refractive index.

Sample profiles of such a refractive index are given in Figures 5.12 and 5.13. Let's briefly discuss how these profiles control waveguide dispersion.

Theoretically, the waveguide-dispersion parameter, $D_{wg}$, is proportional to the relative index, $\Delta$, and the $V$-number and is inversely proportional to the operating wavelength [1]. To increase $D_{wg}$, we need to increase $\Delta$ because $V$ and $\lambda$ are given. Raising $\Delta$ necessitates reducing the core radius, $a$, to the order of 2 or 3 μm, as you can calculate from Formula 6.6. With such a small core radius, we will encounter a lot of problems because of increased scattering and connection losses and significant nonlinear effects caused by the high power density. Thus, the triangle and other versions of a graded-index profile with depressed-cladding options appear as the solution to these problems. (See Figure 5.12.)

To make a dispersion-flattened fiber, the W-like profile is usually used (Figure 5.13.) This design results in two zeros on the dispersion curve (Figure 5.14)—the first one for shorter wavelengths, when light concentrates primarily within a core, and the second one for the longer wavelengths, when the role of the outer part of cladding in light propagation increases. Another option appears when outer rings are added to a simple depressed-cladding design (again, see Figure 5.13). The depth and width of these rings allow designers to control the range of dispersion-flattened wavelengths and the bending sensitivity of a singlemode fiber.

The current trend in designing dispersion-shifted fibers is to continue to improve each fiber's characteristics, such as reducing its Rayleigh scattering loss and bending sensitivity in conjunction with increasing the core area.

All this aside, however, the big question confronting fiber producers boils down to this: If $D(\lambda)$ is zero at a specific wavelength, can we ever hope to eliminate pulse broadening caused by chromatic dispersion? The answer is no, as it follows from Formula 5.9 that $\Delta t_{chrom}/L = D(\lambda)\Delta\lambda$. At the zero-dispersion wavelength, $\lambda_0$, dispersion parameter $D(\lambda)$ is really zero but we have to take into account two finer phenomena. First, Formula 5.9 gives us only the first-order approximation of $D$ as a function of $\lambda$. Hence, the derivative $dD(\lambda)/d\lambda = S_0$ should be considered. $S_0$ is familiar to us as the zero-dispersion slope. Secondly, using a real light source, $\Delta\lambda$ never equals zero. Thus, a dispersion-wavelength pulse spread near zero may be calculated by the following formula:

$$\Delta t_{chrom}/L = S_0(\lambda - \lambda_0)\Delta\lambda, \tag{6.8}$$

where $S_0$ (ps/nm$^2$·km) is given in the fiber's data sheet, $(\lambda - \lambda_0)$ is the deviation of operating wavelength from the zero-dispersion wavelength, and $\Delta\lambda$ is the light source's spectral width.

## Coping with Chromatic Dispersion

Chromatic dispersion is the major limiting factor in today's long-haul communications systems, all of which are based on singlemode fibers. This is why so much research and development effort is being expended to cope with this phenomenon.

***The search for solutions*** Today, 75 million miles of conventional singlemode optical fiber blanket not only the United States but the entire world. The problem is that the need to increase the information-carrying capacity of these fibers has begun to conflict with their inherent dispersion-limited bandwidth. As a result, coping with dispersion has become one of the major stumbling blocks in the drive to expand the global application of fiber-optic communications systems.

What's more, the active use of WDM systems reveals a new phenomenon: four-wave mixing (FWM), which will be discussed in the next section. FWM results in the degradation of system performance. It is most troublesome when the range of operating wavelengths, which obviously coincides with the gain band of optical amplifiers, covers the zero-dispersion wavelength. The first means to facilitate the impact of FWM is to shift the zero-dispersion wavelength farther outside of this range, thus introducing a certain well controlled amount of dispersion—*residual dispersion*—into the system. Such a fiber is called "non-zero dispersion-shifted fiber" (NZ-DSF). Its chromatic dispersion parameter is depicted in Figure 6.9.

The data sheet for such a fiber is shown in Figure 5.18. We leave the analysis of this data sheet as an exercise for you.

At the 1550-nm operating wavelength, the regular—that is, the conventional or unshifted—fiber has a negative dispersion coefficient $D(\lambda)$ in the range of 17–18 ps/nm·km; a non-zero dispersion-shifted fiber has a negative dispersion coefficient of about 2–4 ps/nm·km. These are the orders of magnitude of dispersions we have to cope with.

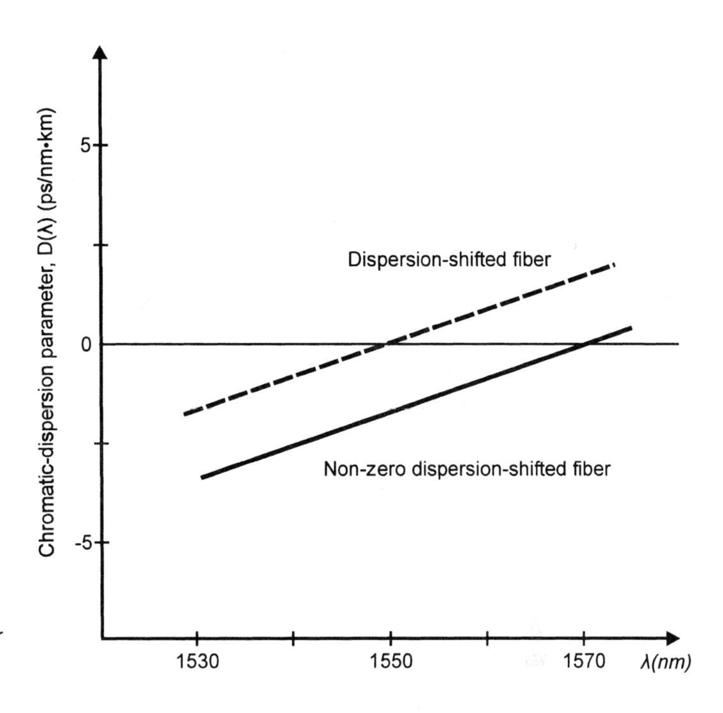


**Figure 6.9** Non-zero dispersion-shifted fiber (NZ-DSF): Chromatic-dispersion parameter.

## Compensation for Chromatic Dispersion with Dispersion-Compensating Fiber

***What it is*** If there is anything good about chromatic dispersion, it's its time stability. For installed cabled fiber, the chromatic-dispersion coefficient is a very stable number. It gives us an opportunity to compensate for chromatic dispersion along the entire span of the fiber-optic link.

There are two basic techniques for dispersion compensation: using dispersion-compensating fiber (DCF) and dispersion-compensating grating (DCG), which is discussed later in this section. DCF is a relatively mature technology in this relatively young industry and large-volume manufacturing of dispersion-compensating fibers is a growing trend [2].

The basic idea is simple: The positive dispersion described above can be compensated for by inserting a piece of a singlemode fiber with a negative-dispersion characteristic so that the total dispersion of the link will be almost zero. Figure 6.10 demonstrates the idea in terms of pulse-spread compensation.

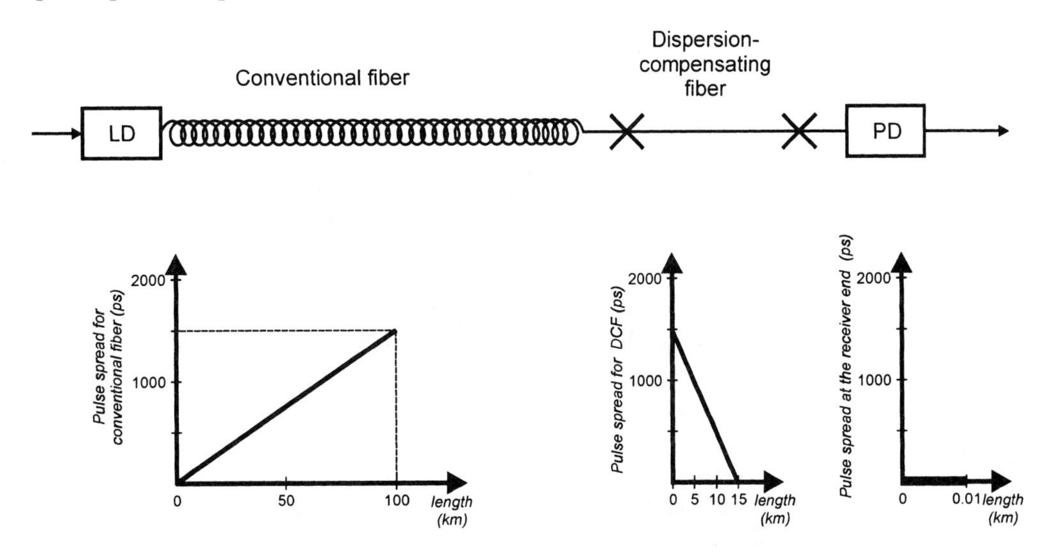


**Figure 6.10** Pulse-spread compensation with a dispersion-compensating fiber (DCF) for $\Delta\lambda = 1$nm.

## Example 6.3.1

***Problem:***

What DCF does one need to use in order to compensate for dispersion in a conventional singlemode fiber link of 100 km?

***Solution:***

**Step 1.** Let's clarify the statement of the problem. First, to compute any number, we need to know the numerical characteristics of a conventional (nonshifted) fiber. These characteristics are available from the manufacturers' data sheets. (See, for example, Figures 5.17 and 5.18 and reference [5.3].) Next, the question of what DCF to use implies that we need to find out the dispersion coefficient of the DCF. Finally, as users, we are interested in the end result, that is, the total pulse spread over the entire fiber link. This is the number we have to calculate. Thus, the approach to finding the solution is as follows: Compute the $D(\lambda)$ of the specific fiber, determine the pulse spread caused by chromatic dispersion, and find out the required characteristics of DCF.

**Step 2.** Let's compute the dispersion coefficient, $D(\lambda)$, of SMF-28 fiber [5.3]. We can find $D(\lambda) = (S_0/4)\lambda[1 - (\lambda_0/\lambda)^4]$ by plugging into this formula $S_0 = 0.092$ ps/nm$^2$·km and $\lambda_0 = 1310$ nm, where 1310 nm is the average of the given zero-dispersion wavelengths. Hence, for a 1550-nm operating wavelength, $\lambda$, we compute $D(\lambda) = 17.46$ ps/nm·km.

The pulse spread caused by chromatic dispersion, $\Delta t_{chrom}$(ps) $= D(\lambda)\ \Delta\lambda\ L$, for $\Delta\lambda = 1$ nm and $L = 100$ km, is equal to $\Delta t_{chrom} = 17.46$ (ps/nm·km) $\times$ 1 (nm) $\times$ 100 (km) = 1746 ps.

**Step 3.** We need to compensate for the 1746-ps pulse spread by inserting DCF with a high negative-dispersion coefficient. Thus, we have to make $\Delta t_{chrom} + \Delta t_{compens} = 0$, where $\Delta t_{compens}$ is the compensated pulse spread. It is evident that $-\Delta t_{compens} = -D_{DCF}(\lambda)\ \Delta\lambda\ L_{DCF}$ (ps), where $L_{DCF}$ is the length of DCF. The typical ratio of $L/L_{DCF}$ is in the range of 6 to 7. Thus, taking $L_{DCF}$ = 15 km, we compute $D_{DCF}(\lambda) = -116.4$ ps/nm·km.

This result obviously follows from these simple relationships:

$$\Delta t_{chrom} = -\Delta t_{compens}$$

and

$$D(\lambda)\,\Delta\lambda\, L = -D_{DCF}(\lambda)\Delta\lambda\, L_{DCF}$$

so that

$$-D_{DCF}(\lambda) = D(\lambda)\, L/L_{DCF} \tag{6.9}$$

Assuming $L/L_{DCF} = 6.66$, Formula 6.9 yields $-D_{DCF}(\lambda) = -6.66\ D(\lambda)$ (ps/nm·km)

Thus, to make this technology work, we need to have a DCF with a negative-dispersion coefficient on the order of 115 to 120 ps/nm·km.

---

To achieve a DCF with a high negative-dispersion coefficient, manufacturers have to manipulate the waveguide dispersion, that is, modify the refractive-index profile and the relative-index value as may be necessary for a specific application. Samples of a typical DCF design with accompanying data are shown in Figure 6.11.

Typically, a designer lowers the refractive index of the inner part of the cladding by doping silica with fluorine. Figure 6.11 shows the relationship between radii of the core and cladding trench: The cladding trench's radius is 2.5 times larger than the core's.

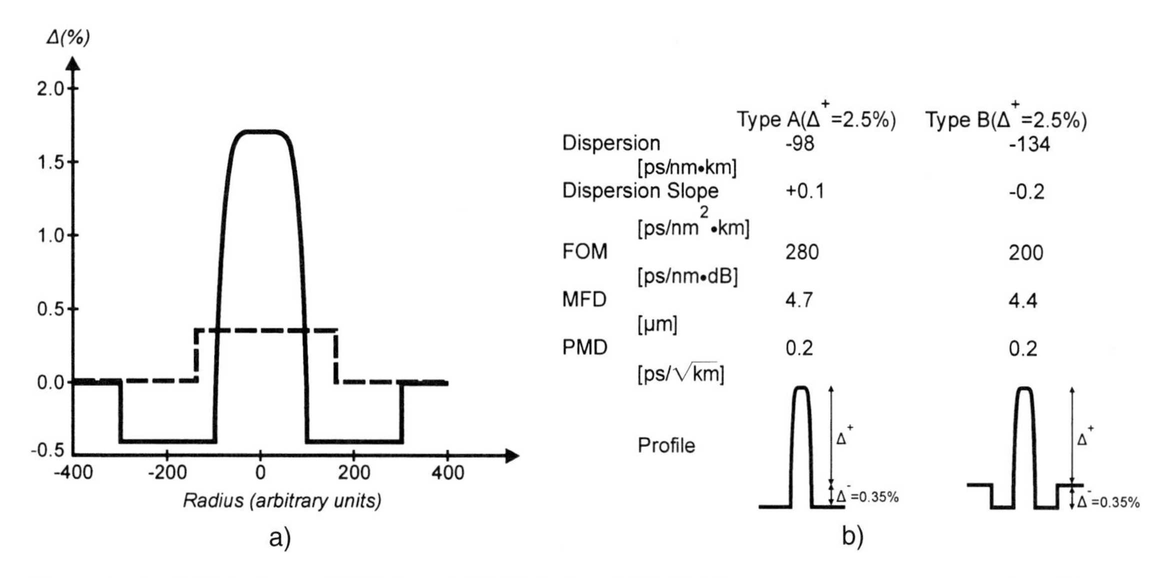


**Figure 6.11** DCF design and DCF data: (a) DCF design; (b) DCF data. *(Ashish M. Vengsarkar, "Tutorial on Dispersion Compensation," Paper TuL,* Proceedings of the Optical Fiber Communication Conference, *San Jose, Calif., February 22–27, 1998. Reprinted with permission.)*

The first problem encountered in using DCF is its high attenuation. Indeed, attenuation around 0.4 dB/km is considered good. Thus, a new characteristic—figure of merit, FOM—is used to describe DCF quality.

*Figure of merit* (ps/nm·dB) = *dispersion* (ps/nm·km)/*attenuation* (dB/km)

If we take $-D_{DCF}(\lambda) = 116$ ps/nm·km and $A = 0.44$ dB/km, we can compute, approximately, that FOM = 265 ps/nm·dB. Typical FOMs are about 200 ps/nm·dB, with the best reaching 400 ps/nm·dB. FOM reveals the existence of a trade-off between the negative-dispersion coefficient and the attenuation of a DCF.

***Designing DCF systems*** One encounters several hurdles in designing DCF systems. First, to make the dispersion highly negative, a designer has to increase the relative index, Δ, to 2.5% compared with 0.37% for a conventional fiber. The depressed-cladding design can be used to facilitate achieving this number but $\Delta^+$ still should be very high (on the order of 2.3%). To obtain such a relative index, the designer has to dope the fiber's core very heavily. In fact, up to 25% of the core should be germanium, added to the silica during the manufacturing process. Doping of this magnitude increases Rayleigh scattering losses.

Manufacturers may also get high negative dispersion by decreasing the core radius. DCF data show that MFD of around 4.7 μm is used, resulting, again, in increasing scattering loss. For this reason—heavy doping and small core size—attenuation in a DCF is much higher than in a regular fiber. Also, the power density in DCF is much higher than in regular fiber because the same power is channeled over a much smaller cross section. The result is a high level of nonlinear effects, causing deterioration in system performance. (Nonlinearities are discussed in the next section.)

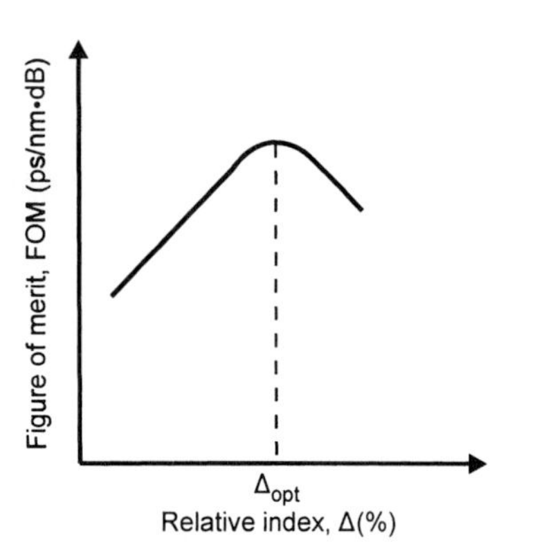


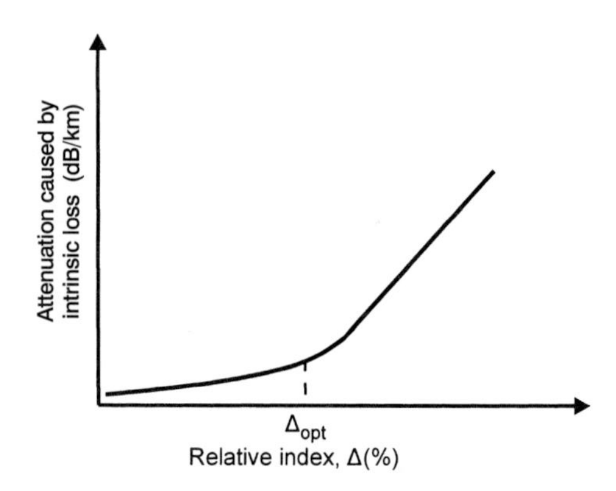


**Figure 6.12** FOM and intrinsic loss of a DCF as a function of relative index. *(Ashish M. Vengsarkar, "Tutorial on Dispersion Compensation," Paper TuL,* Proceedings of the Optical Fiber Communication Conference, *San Jose, Calif., February 22–27, 1998. Reprinted with permission.)*

Still another obstacle is that the FOM reaches its maximum at the specific value of $\Delta = \Delta_{opt}$ [3]. Figure 6.12 shows, qualitatively, FOM and intrinsic loss of a dispersion-compensating fiber as a function of $\Delta$. Thus, the designer is restricted by the specific range of $\Delta$.

Extrinsic loss in DCF is bending loss. Induced bending loss of less than 0.1 dB with 50 bends around a 60-mm diameter mandrel at 1550 nm has been reported. To keep bending loss on the order of 0.1 dB for DCF, the designer has to arrange the MFD/$\lambda_{cutoff}$ ratio (see Section 5.2) to about 15 at 1300 nm, 6 at 1550 nm, and 3.5 at 1650 nm [3].

Since the major problem in using DCF is high attenuation, this dispersion-compensation technique was not widely used until the optical amplifier became the working tool in fiber-optic communications systems. Typically, 17 to 20 km of a DCF is required to compensate for a 120-km span, which results in 8 to 10 dB of additional attenuation. It is quite evident that without extra amplification this technique cannot be used. For a system designer, the combination of an optical amplifier and DCF poses a new challenge. If you place the DCF immediately after the EDFA, the strength of the amplified light will produce nonlinear effects even more severe than usual. Hence, the quandary arises as to the ideal placement of the DCF in the link.

Another challenge for fiber-optic communications system designers arose with the advent of WDM technology. As you can see from the above consideration, all the effects in the dispersion-compensation technique are wavelength dependent. Therefore, the DCF has to be designed to compensate for dispersion—not at a single wavelength but over the entire band. Dispersion parameter of a DCF, as a function of wavelength, is shown in Figure 6.13.

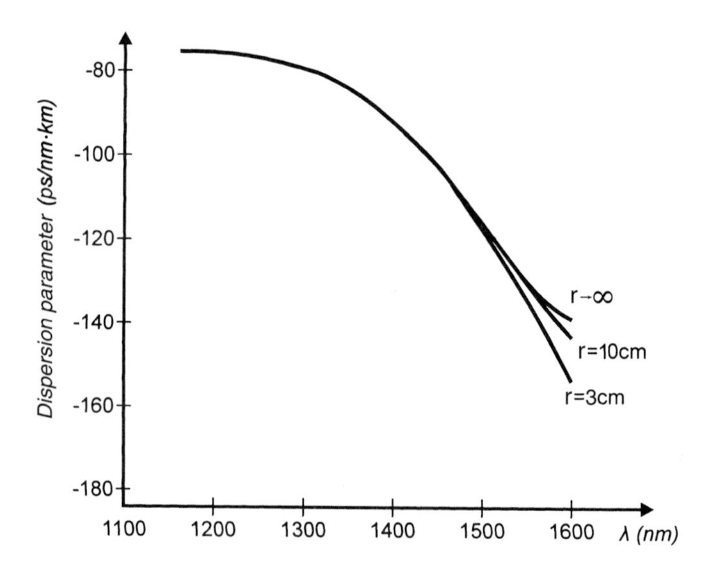


**Figure 6.13** Dispersion parameter of a DCF as a function of wavelength for various bending radii.

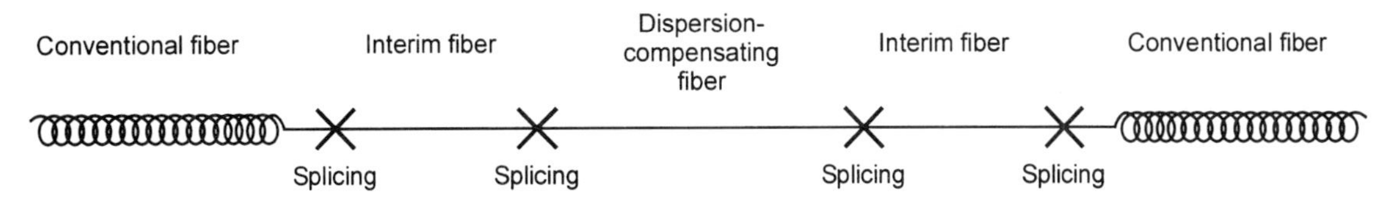


**Figure 6.14** Using an interim fiber to mitigate splicing loss in dispersion-compensation technique.

It is interesting to note that the value of dispersion, the slope, and even the shape of this graph depend on bending [4].

A system designer trying to insert a DCF will meet a splicing problem arising from the fact that MFDs of a conventional fiber and a DCF are quite different. Indeed, at the 1550-nm operating wavelength, the MFD of a regular singlemode fiber is about 10.5 μm (see, for example, Figure 5.17) and the MFD of a DCF is about 4.7 μm. (See Figure 6.11.) The common means of resolving the problem is to insert a piece of interim fiber between the conventional and the DCF fibers, as shown in Figure 6.14.

With direct splicing, one can expect splicing loss on the order of 0.8 dB per splice because of the enormous MFD mismatch. Using interim fiber reduces this loss to about 0.4 dB per splice [3].

*Dispersion management* is the term used to refer to the management of dispersion compensation from the communications-system standpoint. In particular, one of the major tasks of this management program is to choose the right place for the DCF, the sequence for placing the DCF and the regular fiber, and the lengths of these fibers. The trick is to keep the dispersion from reaching zero at any given point along the system span while achieving the zero-dispersion effect at the receiver end. The goal is to maximize the bandwidth of a system, bearing in mind the system's ability to be reconfigured and upgraded.

***Two-mode dispersion-compensating fiber (TM-DCF)*** Singlemode dispersion-compensating fiber suffers from several drawbacks: high attenuation, low negative dispersion, and a high level of nonlinearities. Two-mode dispersion-compensating fiber (TM-DCF) has been developed to overcome these problems [5]. This fiber is designed such that $D(\lambda)$ has a large negative value for the second-order, $LP_{11}$, mode. A negative-dispersion coefficient as high as –770 ps/nm·km has been achieved with this fiber, while attenuation is almost the same as it is for a singlemode DCF. This means a much shorter length of TM-DCF is required to compensate for the same amount of dispersion.

Since TM-DCF exploits the second-order mode, a mode-conversion (MC) device is needed to insert TM-DCF into a singlemode fiber link. Several types of these devices are available; of these, fiber-based MC seems to be the most promising for communications systems.

## Dispersion-Compensating Gratings (DCG)

***Fiber Bragg gratings*** These devices are based on the well-known principle of diffraction gratings. The most developed DCGs—at this time—are *chirped-fiber Bragg gratings (FBGs).*

Let's discuss each word used in this term. *Grating* implies the periodic structure—that is, a periodic change in the value of the refractive index of the core. A small portion of light is reflected at each change of the refractive index. All these reflected portions of light combine into one reflected beam provided that the Bragg condition is met.

W. L. Bragg was the physicist who used the crystal structure as a diffraction grating for x rays. The *Bragg condition* is:

$$2\Lambda n_{\mathrm{eff}} = \lambda_{\mathrm{B}}, \tag{6.10}$$

where $\Lambda$ is the grating period (that is, the distance between two adjacent maximum points of the periodic refractive index), $n_{\mathrm{eff}}$ is the effective core refractive index, and $\lambda_B$ is the Bragg central

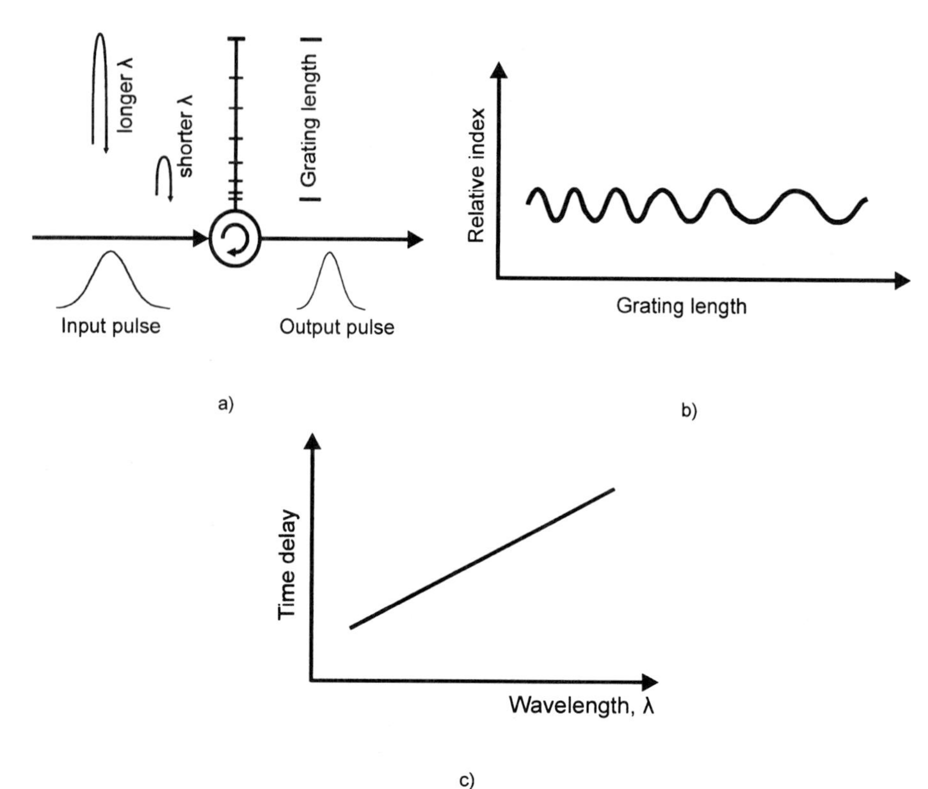


**Figure 6.15** Using chirped-fiber Bragg grating for dispersion compensation: (a) General diagram; (b) variations of a refractive index along the grating length; (c) time delay versus wavelength.

wavelength. This grating works as a mirror, selectively reflecting only one wavelength, $\lambda_B$, and transmitting all the others.

The word *fiber* signifies that the grating is implanted in it.

*Chirped* means that the optical grating period, $\Lambda\ n_{eff}$, changes linearly over the length of the grating. Thus, *chirped FBG reflects not a single wavelength but a set of wavelengths.* Its principle of action is visualized in Figure 6.15. An optical circulator is used to direct pulses into and out of the FBG.

An input pulse, dispersed after propagating along a telecommunications fiber, is directed to the grating, where the shorter wavelengths are reflected almost immediately upon entering while the longer wavelengths penetrate deeper into the grating before they will be reflected. (See Figure 6.15[a].) This effect is achieved by shortening the grating period at the grating entrance and lengthening it at the grating end. That's why it is called "chirped." (See Figure 6.15[b].) Thus, the device ensures less of a delay for shorter wavelengths but creates more delay for longer ones, as Figure 6.15(c) shows. This is exactly the opposite of the delay introduced by a singlemode fiber itself. (See Section 5.3.) Therefore, pulse spread caused by chromatic dispersion in telecommunications fiber is compensated for by a chirped FBG. In other words, FBG compresses the pulse.

These are compact devices, usually about 10 to 20 cm long. Since they're fabricated on a fiber, it is easy to couple FBG with telecommunications fiber.

It follows that chirped FBG works well for the $\lambda_{Bragg}$ and its small variations. To make this type of DCG usable in WDM systems, where the band of wavelengths could be up to 40 nm, manufacturers offer two solutions: (1) commercially available FBG with a wide bandwidth (actually, this is a set of gratings on one fiber) or (2) multichannel FBG with a bandwidth of 0.5 nm per channel, allowing a configuration of up to 16 channels. (The term *channel bandwidth,* in WDM parlance, means the spectral width of light carrying an individual channel—hence, compensated by an FBG.) An FBG's bandwidth is the nonlinear function of its length, so that a 1-cm-long FBG has a bandwidth of 0.2 nm; to get to 40 nm, one has to use a device about 1m in length. (Such FBG devices have indeed been fabricated.)

There is a trade-off between an FBG's bandwidth and its delay, that is, its compensation ability. For example, suppose an FBG introduces a 1400-ps/nm delay. This number means the FBG can introduce, theoretically, a delay of 1400 ps/nm over a 1-nm bandwidth or a delay of 140 ps/nm over a 10-nm bandwidth.

***Fabrication of an FBG*** In 1978, Canadian scientist Kenneth Hill discovered fortuitously that the refractive index of the core of a fiber can be changed under exposure to ultraviolet light. This phenomenon, called *fiber photosensitivity,* is the physical basis for grating fabrication. There are two basic fabrication methods today: The original one—directly exposing a fiber's core to a pair of interfering UV beams—provides radiation of both maximum and minimum intensity. The minimum intensity leaves the refractive index unchanged and the maximum intensity changes the refractive index.

Hill and his colleagues also developed a second fabrication method—the phase-mask technique. Based on essentially the same interference principle, it gives much better results because of the higher grating precision it imposes.

In general, then, FBGs are considered the most promising dispersion-compensation devices. There are already many commercially available types of chirped-fiber Bragg gratings, and these seem to represent the direction in which the industry is going. FBGs are well on their way to giving DCFs stiff competition ([3], [6]). You can get an idea what commercial level the characteristics of chirped-fiber Bragg gratings have reached today by looking at the following data of a modern FBG [7]: bandwidth—from 4 to 10 nm; dispersion—from –700 ps/nm to –1400 ps/nm; insertion loss < 4 dB; PMD < 4 ps; PDL < 0.25 dB, and return loss (see Chapter 8) > 50 dB. The unit can compensate for third-order dispersion. The physical dimension is 212 × 155 × 20 mm.

Dispersion compensation is so important for long-haul systems that scientists continue to search for effective new solutions. Several competitive dispersion-compensating techniques are currently in the research stage.

## Dispersion Compensation: The System Viewpoint

We have to keep in mind that we need to compensate for dispersion only for the purpose of increasing system bandwidth, or bit rate. But evaluation of dispersion-compensating devices should also take into account all other performance characteristics of a fiber-optic communications system. Let's consider dispersion compensation from the system standpoint.

*In modern fiber-optic communications systems, dispersion—not loss—becomes the distance-limiting factor.*

If we rearrange Formula 5.13 in such a way that

$$L_{max} = 1/\left[4BR|D(\lambda)|\Delta\lambda\right], \qquad (6.11)$$

we can calculate the maximum length of a fiber link limited by chromatic dispersion. Let's take these typical parameters: $D(\lambda) = 17$ ps/nm·km at $\lambda = 1550$ nm, $\Delta\lambda = 0.2$ nm, and $BR = 2.5$ Gbps. One finds $L_{max}$ (dispersion) = 29.4 km. Our calculations of loss-distance limitations in Example 5.2.1 showed $L_{max}$ (loss) = 73.12 km. These numbers give you an idea of how dispersion-distance and loss-distance limitations relate to each other.

It seems that with improved dispersion-compensating techniques, this problem could be overcome. For example, for a DSF fiber, $D(\lambda)$ is not more than 2.5 ps/nm·km, which gives $L_{max} = 199.9$ km. Thus loss-distance limitation becomes the major restriction again. On the other hand, if one arranges the transmission of a high-power signal with in-line optical amplifiers, it looks as though the loss-distance limit disappears. In reality, however, both approaches do not allow us to achieve our goal. Pumping too much power into a fiber and working near a zero-dispersion wavelength cause other restrictions associated with nonlinear effects. These are discussed later in this chapter.

In modern fiber-optic communications systems, the real distance-limiting barrier is the combination of nonlinear effects, high-order dispersion, and PMD.

High-order chromatic dispersion at zero-dispersion wavelength is represented by the dispersion slope, $S_0$. This effect is also called third-order, or nonlinear, dispersion because it is proportional to the third derivative of phase constant $\beta$ with respect to angular frequency $\omega$. In the modern high-bit-rate long-haul WDM systems, only *slope compensation*—in addition to dispersion compensation itself—enables the achievement of significant results.

Another promising recent development has been the technology of *dynamic (time-changing) compensation*. Dispersion can change because of changing operating conditions and/or system configuration. Dispersion-compensation techniques must dynamically react to any new situation by providing adequate compensation. Such a dynamic technique is in the process of development in university laboratories [8].

In system design, it's not only to dispersion compensation alone that researchers turn for solutions to the distance-limitation problem but also to the development of techniques and devices allowing the recovery of a widened signal at the receiver end. Such "post-compensation" can help to increase the distance of the signal's range up to three times at the bit rate of around 5 Gbps. Another direction being taken is to develop a technique to prepare the signal at the transmitter end for spreading throughout its transmission. This "precompensating" approach includes special coding and other means to precompress the signal [9].

We sum up our discussion of dispersion compensation by urging you to keep these key points in mind:

- Dispersion compensation and management can be quite effective, allowing system operators to improve system performance significantly.
- The specific implementation of this technique depends, essentially, on the type of communications system used and its architecture; today, for instance, a DCF compensates over a wide range of wavelengths, while an FBG is more suitable for a very narrow band of wavelengths.Today dispersion-compensating fiber is the most widely used tool for dispersion compensation.
- These techniques have produced impressive results. For example, a $17 \times 20$ Gbit/s WDM system successfully works over 300 km [10], while a single-channel dispersion-managed system transmits a signal over 12,000 km [11].
- To achieve even more capacity, not only dispersion compensation but also dispersion-slope compensation must be included. In the latest experiment, a $20 \times 10.66$ Gbit/s WDM system is transmitting over a 9000-km range with dispersion and dispersion-slope compensations [12].
- We need to use dispersion compensation and management only for conventional fiber-based networks installed years ago. The new generation of fibers already has dispersion-coping capabilities, so these new fibers not only satisfy today's bandwidth requirements but they also have upgrading potential to meet future needs.

## Coping with PMD

All the techniques discussed above work only to compensate for chromatic dispersion. But, as discussed in Section 5.3, there is another type of dispersion—polarization-mode dispersion (PMD). From the user's point of view, the major difference between chromatic dispersion and PMD is that chromatic dispersion can be compensated for because of its stability, but, because of its random nature, compensation for PMD is not so straightforward. Obviously, this is true for a regular singlemode fiber.

***Polarization parameters*** To describe PMD quantitatively, we need to introduce a few parameters: PMD is caused by the difference in refractive indexes along the *x* axis and the *y* axis, $n_x$ and $n_y$, respectively. (See Figure 5.16.) This difference is called *birefringence, B:*

$$B = n_x - n_y \tag{6.12}$$

This formula implies that $n_x > n_y$; in other words, *y* is the fast axis, *x* the slow axis. It should be evident that the notations "x" and "y" for these two axes have been chosen quite arbitrarily; usually the terms *fast* and *slow* are used to denote the appropriate axis.

Another way to define birefringence is to introduce a *beat length.* This is the length over which the phase difference between the *x*- and *y*-polarized waves changes by $2\pi$. If we introduce another, but equivalent, formula for birefringence, such as

$$\beta = [2\pi/\lambda](n_x - n_y), \tag{6.13}$$

where $\beta$ is a phase constant, we can define beat length $l_p$ as:

$$l_p(m) = 2\pi/\beta = \lambda/(n_x - n_y) = \lambda/B \tag{6.14}$$

Thus, beat length is inversely proportional to the fiber's birefringence, *B,* and proportional to the operating wavelength, $\lambda$. If $B \approx 10^{-3}$, then $l_p \approx 1.55$ mm at $\lambda = 1550$ nm.

If we excite a mode polarized only along, say, the *y* axis and measure the output power in both polarizations, we will see some power in the mode polarized along the *x* axis. This is because the fiber does not hold light polarization over its length, and power from the initially excited mode will penetrate to the orthogonally polarized mode. A parameter used to evaluate the polarization-holding ability of a fiber is called the *extinction ratio, ER.* It is defined as:

$$ER\ (\text{dB}) = -10 \log(P\perp/P\parallel), \tag{6.15}$$

where $P\perp$ and $P\parallel$ are light power in unexcited and excited modes, respectively (the *x* and *y* axes in the above example). You can also use other formulas to describe the same idea: *h parameter* (for polarization holding) and *polarization cross talk:*

$$h\ (1/m) \cong 1/L(P\perp/P\parallel) \tag{6.16}$$

and

$$\text{polarization cross talk (dB)} = 10 \log(P\perp/P\parallel) \tag{6.17}$$

We can easily see that

$$ER(\text{dB}) = -\text{polarization cross talk(dB)} \tag{6.18}$$

and, more significantly,

$$ER(\text{dB}) = -10 \log(P\perp/P\parallel) = -10 \log(hL) \tag{6.19}$$

For $L = 100\ m$ and $h = 10^{-5}\ 1/m$, we can compute $ER = 30$ dB. It follows that *the higher fiber's birefringence implies its stronger polarization-holding ability.* This important fiber property is used to fabricate fibers that maintain their polarization, as described below.

***PMD characteristics*** Two orthogonal linear-polarized modes, $HE_{11}$, which we see as one $LP_{01}$ mode, travel along a singlemode fiber at different velocities because of the fiber's birefringence. This effect results in the form of pulse spread called polarization-mode dispersion, PMD. (See Figure 5.16.) PMD stems from the ovalness of the fiber's core. This ovalness is caused by any stress the fiber experiences. Total birefringence, $B$, of a fabricated fiber changes steadily because of the coiling, cabling, and installation processes. What makes this change random are thermal and mechanical stresses over time. In other words, changes of birefringence occur randomly, meaning PMD is also a random process. Many measurements of PMD were done on submarine, terrestrial, aerial, and buried fiber cables and all of them confirm this point.

As with any dispersion, PMD is described by a pulse spread that can be determined by the following formula:

$$\Delta t_{PMD} = D_{PMD}\sqrt{L} \tag{6.20}$$

The reason why PMD depends on $\sqrt{L}$ but not on $L$ is inherent in the random nature of the PMD mechanism. Fortunately, $D_{PMD}$ is relatively small (between 0.5 and $0.2\ \text{ps}/\sqrt{\text{km}}$, typically, with $0.1\ \text{ps}/\sqrt{\text{km}}$ being the best value). Unfortunately, however, PMD is unpredictable; therefore it cannot be directly compensated for.

Incidentally, theory predicts—and measurements confirm—that $\Delta t_{PMD}$ obeys the Maxwellian, not the Gaussian, distribution. Also, both theory and measurements show that for today's accuracy, we have to take into account second-order PMD. However, we restrict our consideration here to first-order PMD only.

Adding to the PMD problem is the fact that, typically, a fiber-optic communications link includes many strands of fiber obtained from different manufacturers. This further exacerbates the situation because each fiber has its own polarization characteristics, causing even more severe and abrupt changes in $B$.

How severely does system performance suffer from PMD? That depends on the bit rate of the system. If $T$ is the bit duration, then by having $\Delta t_{PMD} < 0.1T$, one can build a distance-bit-rate limitation imposed by PMD [13]. These limitations are plotted in Figure 6.16.

***Polarization-maintaining fiber*** The most straightforward means to cope with PMD today is to use special fibers and other link components that allow us to preserve and control the state of mode polarization.

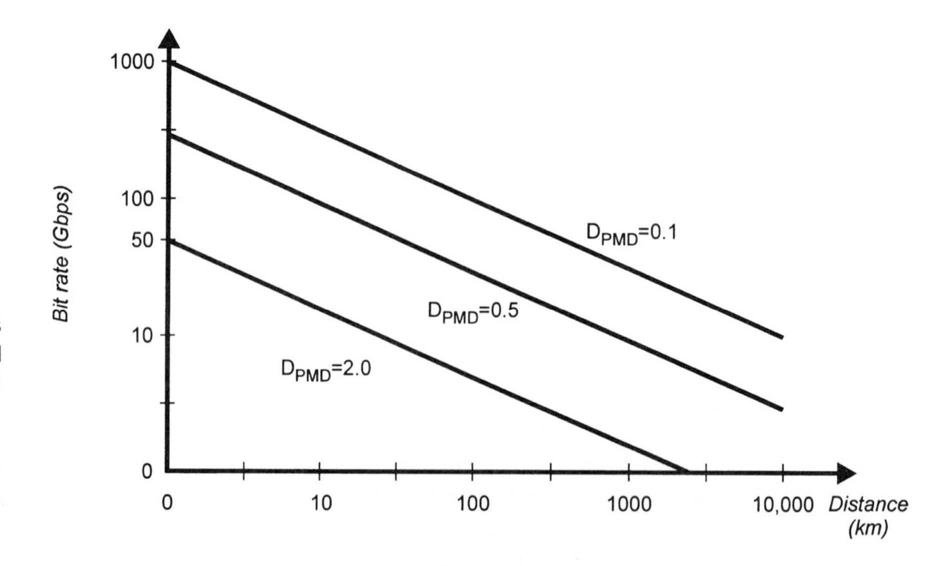


**Figure 6.16** Bit rate versus distance limitations imposed by PMD. *(Rajiv Ramaswami and Kumar Sivarajan,* Optical Networks: A Practical Perspective, *San Francisco: Morgan Kaufman Publishers, 1998. Reprinted with permission.)*

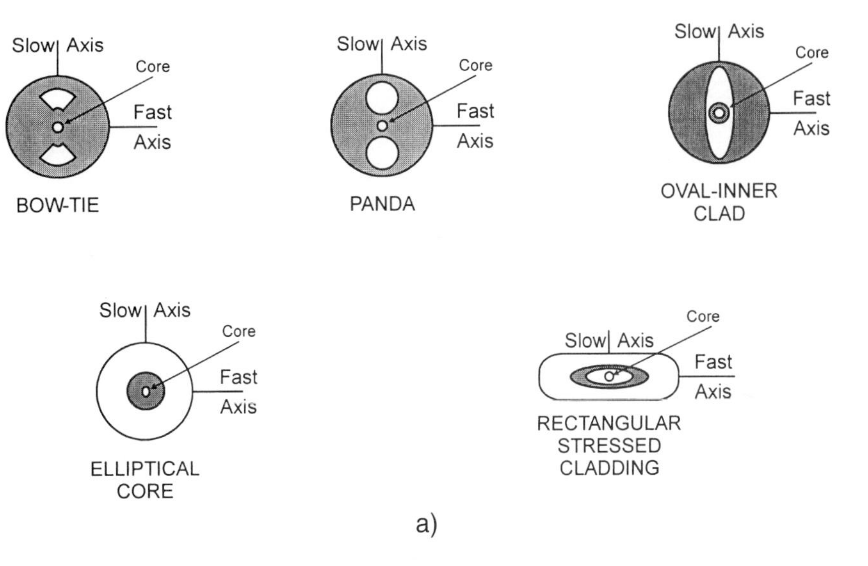


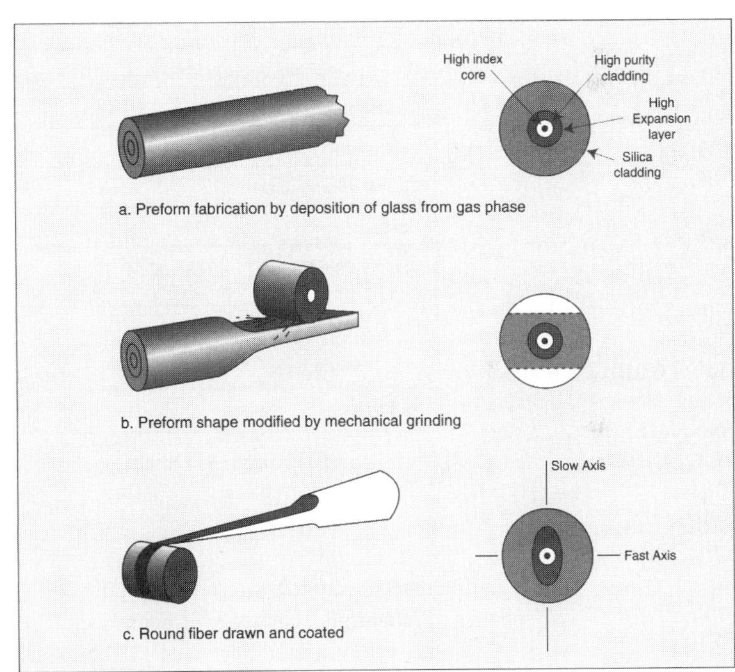


**Figure 6.17** Examples of structures of polarization-maintaining fibers: (a) Typical structures; (b) how a PM fiber is fabricated. *(Courtesy of 3M Specialty Optical Fibers.)*

Polarization-maintaining (PM) fibers, also called polarization-preserving or Hi-Bi fibers, have very high birefringence deliberately induced to enable them to function as the name implies. Figure 6.17 (a) gives some examples of the structures of PM fibers and Figure 6.17 (b) shows how a PM fiber is fabricated. (More about fiber fabrication in Section 7.1.)

As you can see, high birefringence is achieved by having very high asymmetry in the core and/or cladding. This asymmetry is made either in the core and cladding shape or in pre-induced mechanical stress. Both measures result in a fiber that has well-pronounced slow and fast—in terms of the velocity of the polarized light propagation—axes. Linear-polarized light launched along, say, the fast axis will keep its state of polarization during propagation along the fiber if

3M

# Polarization-Maintaining Fiber Tiger PM

| 3M™ Tiger PM Fiber Specifications — TI Series | | |
|---|---|---|
| Operating Wavelength | nm | 1550 |
| Mode Field Diameter (Petermann model) | μm | 10.5 ± 1.0 |
| Second Mode Cutoff Wavelength | nm | 1440 ± 80 |
| Attenuation[1] | dB/km | <1.5 @ 1550nm |
| Fiber Diameter | μm | 125 ± 2 |
| Cladding Ovality | | <1% |
| Core-to-Clad Concentricity Tolerance | μm | <0.8 |
| Coating Diameter | μm | 400 ± 20 |
| Beatlength | mm | 2.5 to 5.0 @ 1550nm |
| Typical *h* Parameter[1] | $m^{-1}$ | $<5 \times 10^{-5}$ |
| Coating Type | | Dual Acrylate |
| Temperature Range — Operating<br>— Storage | °C<br>°C | -40 to +85<br>-40 to +85 |
| Proof Test Level | kpsi | 200 |
| | | |
| Part Number | | FS-TI-7128 |

[1] As measured on standard 6-inch shipping spool.

**Figure 6.18** Data sheet of polarization-maintaining fiber Tiger PM. *(Courtesy of 3M Specialty Optical Fibers.)*

high inherent mechanical stress is pre-induced in a PM fiber. Thus, external stress, always being smaller in magnitude, cannot significantly change a fiber's birefringence.

Figure 6.18 is a data sheet of a PM fiber. You are now armed with all the information you need to analyze specifications sheets. But be careful. Be aware that the manufacturer uses the term *nominal* birefringence. This term covers the tolerance of this parameter. If you plug $B = 3 \times 10^{-4}$ from a typical data sheet into Formula 6.15, you'll obtain $l_p = \lambda/B = 1550\,\text{nm}/3 \times 10^{-4} = 5.17\,(\text{mm})$, whereas the specification given in Figure 6.18 shows $l_p$ from 2.5 to 5.0 mm. Keep in mind that manufacturers provide the measured, not the calculated, numbers for all parameters given in a data sheet.

It seems that PM fibers hold the solution to the PMD problem. Unfortunately, high attenuation and the high cost of a fiber—plus costly installation and maintenance (just image how difficult splicing alone would be)—preclude the widespread use of PM fibers in telecommunications networks.

***Other polarization-maintaining components*** To reduce pulse spread caused by PMD, we need to preserve the state of polarization throughout the entire communications link—from transmitter to receiver. This implies that, besides using PM fibers, we have to use all the other

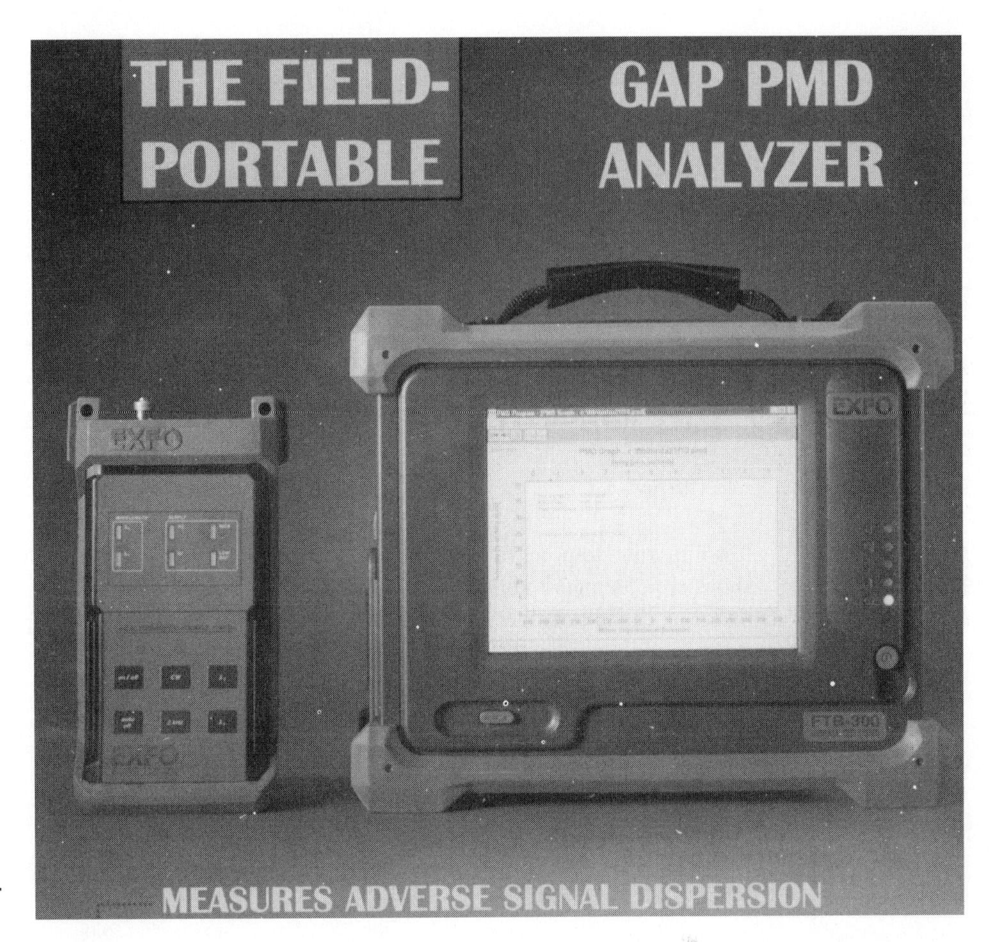


**Figure 6.19** Field-portable PMD analyzer. *(Courtesy of EXFO.)*

fiber-optic link components to maintain the state of polarization. Industry today offers a set of such components. This set includes PM connectors, fiber-optic polarizers (to couple two fibers), fiber-polarization controllers (to convert the state of polarization within a fiber), PM splitters (to split the signal between two PM fibers), and rotators/analyzers (to prepare light for coupling into the fiber and to measure ER).

One of the key devices in a polarization-maintaining fiber-optic system is a PMD analyzer, which measures PMD delay, PMD coefficient, and PMD distribution of the fiber-optic link. Such an analyzer is shown in Figure 6.19.

Measuring PMD will not increase the system's capacity but it can help with the PMD diagnosis. The point to remember is this: Using PM fibers and PM network components can help you cope with the PMD problem; however, a PM fiber cannot replace regular telecommunications fiber in terms of the functions that the latter performs. *PMD, which is inherent in conventional fiber, remains, unfortunately, one of the most severe limitations of the information-carrying capacity of fiber-optic communications systems.*

***PMD compensation*** Since PMD severely restricts a system's bit rate, many research and manufacturing efforts have been undertaken to cope with this dispersion through simple compensation similar to the technique developed for chromatic-dispersion compensation. A number of PMD compensators have been developed and marketed that show promising results [13].

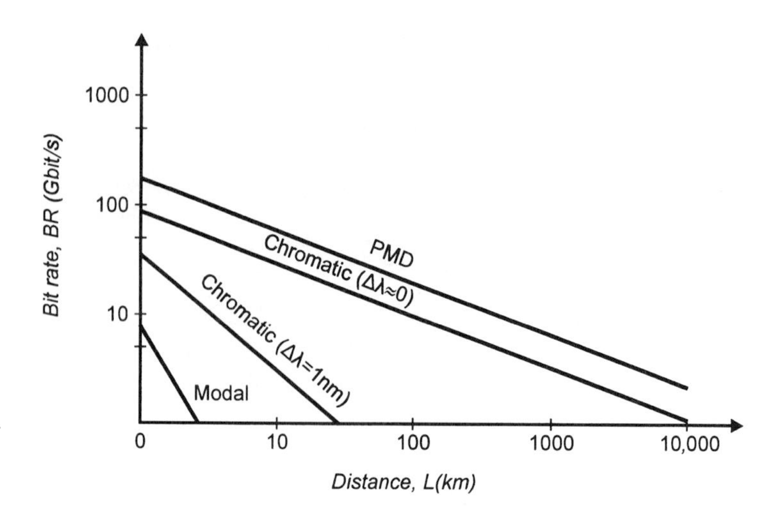


**Figure 6.20** Bit rate versus distance limitations imposed by different types of dispersion. *(Rajiv Ramaswami and Kumar Sivarajan, Optical Networks: A Practical Perspective, San Francisco: Morgan Kaufman Publishers, 1998.)*

## Polarization-Dependent Loss (PDL)

It is important to point out that the losses incurred by the many network components mentioned above are contingent upon the state of polarization of the light interacting with them. This type of loss is called polarization-dependent loss, or PDL. A network component attenuates light selectively, depending on its state of polarization, thus randomly changing the intensity of the propagating signal. This action results in what's called second-order pulse distortion, which puts yet another limit on the transmission capacity of an optical network.

**Table 6.1** Coping with dispersion

| Dispersion type | Fibers that are subject to modal dispersion . . . | Methods of coping with modal dispersion . . . | Comments |
|---|---|---|---|
| Intermodal (modal) | 1. Step-index MM<br>2. Graded-index MM | 1. Graded-index MM fiber<br>2. Singlemode fiber | A singlemode fiber is the ultimate solution to the modal-dispersion problem. |
| **Dispersion type** | **Fibers that are subject to chromatic dispersion . . .** | **Methods of coping with chromatic dispersion . . .** | **Comments** |
| Chromatic (intramodal) | Regular singlemode | 1. Dispersion-shifted fiber, DSF<br>2. Use of a laser diode with near-zero spectral width<br>3. Compensation with DCF and DCG<br>4. Slope compensation<br>5. Pre- and post-compensation techniques. | There are other methods currently in the research and development stage. |
| **Dispersion type** | **Fibers that are subject to PMD . . .** | **Methods of coping with PMD . . .** | **Comments** |
| PMD | All types of fiber except PM fiber. | 1. PM fiber and PM components<br>2. PMD compensation. | 1. PM fiber and PM components have to be used together to make a PM link. Final damage is done by PDL.<br>2. PMD compensation techniques have been developed and devices based on them are commercially available. |

PDL is an important characteristic of fiber-optic-network components, as we will learn in subsequent chapters.

**Brief Summary**

To sum up our consideration of dispersion problems, we refer to two graphics. First, Figure 6.20 shows the bit-rate-distance limitations imposed by all the dispersion types we've discussed.

Secondly, Table 6.1 summarizes our discussion of the different types of dispersion and the means we have to deal with them.

The figures displayed in this section and in Table 6.1 cover only linear dispersion. There is another type of dispersion, nonlinear, which we discuss next.

## 6.4 NONLINEAR EFFECTS IN A SINGLEMODE FIBER

In optics, the terms *linear* and *nonlinear* mean "power-independent" and "power-dependent" phenomena, respectively. From this standpoint, all the effects we have considered so far are linear because their mechanisms are not the function of power. Until recently, nonlinear effects in optical fibers were an area of academic research without—as it seemed then, at least—any practical importance. Several recent events have changed this situation dramatically:

- Using singlemode fibers with their small cross section of light-carrying area has led to increased power density inside a fiber.
- Using in-line optical amplifiers has resulted in a substantial increase in the absolute value of the power carried by a fiber.
- The deployment of multiwavelength systems (together with optical amplifiers) has created new nonlinear effects, such as four-wave mixing and cross-phase modulation.

The move to deploy high-bit-rate (>10 Gbit/s per channel) WDM systems cannot be done without considering nonlinear effects and reducing their impact on these systems. This is why nonlinear effects are today the most significant factor determining the performance of high-bit-rate long-haul fiber-optic communications systems and why design engineers must take them very carefully into account in their everyday work.

**Nonlinear Refractive Effects**

*An optical effect is called nonlinear if its parameters depend on light intensity (power).* Thus, higher-order dispersion, strictly speaking, doesn't fall into this category, but all the effects we are going to consider in this section do.

Light power is proportional—not equal—to light intensity but, for our purposes, we will use these terms interchangeably.

Two major classes of nonlinearities stem from the dependence of the refractive index and attenuation on the intensity of light. Thus, one may talk about dispersion and attenuation nonlinearities but, in the nonlinear world, attenuation affects dispersion and vice versa; therefore, we should be very careful when making strict classifications.

***Nonlinear refractive effects—general approach and basic terms*** The theory of nonlinear effects in optical fibers is well developed (see [1], [9], [14], and [15]), and we will refer to their implications in this section. First, however, we will consider the basics of nonlinearities in the development of fibers; as usual, our discussion will be from a practical standpoint.

The response of a dielectric medium to an applied electric field, **E,** is described by the polarization vector, **P.** (See Formula 4.56, repeated here.) As a result, we get

$$\mathbf{P}\,(\mathbf{r}, t) = \epsilon_0\, \chi_e\, \mathbf{E}\,(\mathbf{r}, t), \tag{6.21}$$

where $\chi_e$ is the electric susceptibility of the medium related to its refractive index as:

$$n = \sqrt{\epsilon_r} = \sqrt{(1 + \chi_e)} \tag{6.22}$$

Nonlinear refractive effects are caused by the dependence of electric susceptibility on the field strength, **E.** In the case of an optical fiber, nonlinear phenomena might be included in the consideration by the following approximate expression:

$$\mathbf{P}\,(\mathbf{r}, t) = \varepsilon_0 \chi_e\; \mathbf{E}\,(\mathbf{r}, t) + \varepsilon_0 \chi_e^{(3)} \mathbf{E}^3(\mathbf{r}, t) \tag{6.23}$$

Thus, a medium refractive index becomes

$$n(\omega, E) = n'(\omega) + n^* E^2, \tag{6.24}$$

where $\omega$ is the light angular frequency. The first member, $n'(\omega)$, is the linear refractive index and is responsible for material dispersion. The second member represents the nonlinear effect because it is proportional to the light intensity, $I = \frac{1}{2}\, \epsilon_0\, c\, n\, E^2$. The *nonlinear index coefficient, $n^*$*, is given by $n^* = 3/8n\, \chi_e^{(3)}$. Another way to represent nonlinear refraction is:

$$n = n'(\omega) + n^*(P/A_{\mathrm{eff}}), \tag{6.25}$$

where $P$ is light power and $A_{\mathrm{eff}}$ is the effective area of the fiber, both of which will be discussed shortly. Thus, nonlinear effects depend on the ratio of light power to the cross-sectional area of the fiber. For a standard silica fiber, typical numbers are $n^* \approx 3.2 \times 10^{-20}\, m^2/W$ and $A_{\mathrm{eff}} \approx 55\ \mu\mathrm{m}^2$. Assuming $P = 1$ mW, we compute $n^*(P/A_{\mathrm{eff}}) = 5.8 \times 10^{-9}$. This is the nonlinear contribution to a refractive index under regular conditions. Since the refractive index of silica is about 1.45, it is obvious why we could neglect the nonlinear effects in previous discussions.

The phase (propagation) constant, $\beta = \omega n/c$, must also depend on $E^2$. This dependence can be approximated by:

$$\beta = \omega n'/c + (3\omega/8cn)\chi_e^{(3)} E^2 \tag{6.26}$$

Another way to represent the nonlinear phase constant is:

$$\beta = \beta' + \gamma_n P, \tag{6.27}$$

where $\beta'$ is the linear portion of the phase (propagation) constant and $\gamma_n = (2\pi/\lambda)\, n^*/A_{\mathrm{eff}}$ *is the nonlinear propagation (phase) coefficient.* Assuming $n^* = 3.2 \times 10^{-20}\, m^2/W$, $A_{\mathrm{eff}} = 55\ \mu\mathrm{m}^2$, and $\lambda = 1550$ nm, we compute $\gamma_n = 2.35 \times 10^{-3}\ 1/m{\cdot}W$. Again, for $P = 1$ mW, the nonlinear contribution to the phase constant is about $10^{-6}$. Be aware in particular that the nonlinear member in Formula 6.27 depends on $P/A_{\mathrm{eff}}$; that is, $\gamma_n P = (2\pi/\lambda)\, n^*\, (P/A_{\mathrm{eff}})$, so that

$$\beta = \beta' + (2\pi/\lambda) n^* (P/A_{\mathrm{eff}}) \tag{6.27a}$$

It is informative to compare Formulas 6.27(a) and 6.25.

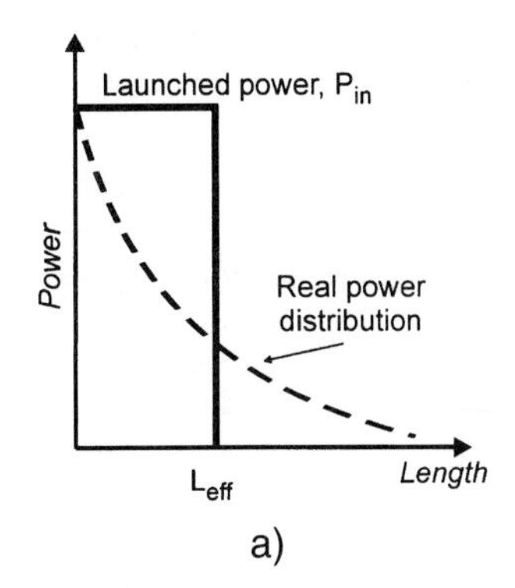


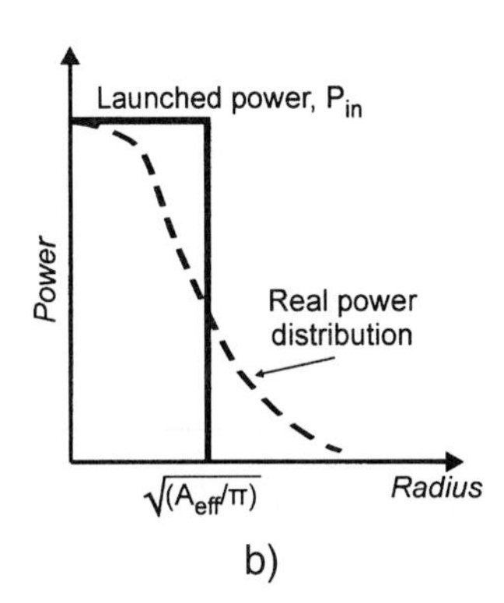


**Figure 6.21** Definitions of effective length and effective area: (a) Effective length; (b) effective area.

The nonlinear effects depend on the transmission length. The longer the fiber, the more the light interacts with the fiber material and the greater the nonlinear effects. On the other hand, if the power decreases while light travels along the fiber, the effects of nonlinearity diminish. To show this phenomenon, *effective length,* $L_{eff}$, is introduced, as Figure 6.21(a) shows.

To put the definition of $L_{eff}$ in formal terms, we can write:

$$P_{in}L_{eff} = \int_0^L P(z)\,dz \tag{6.28}$$

Since $P(z) = P_{in}\,e^{-\alpha z}$ (see Formula 4.68), where $\alpha$ is the fiber-attenuation constant in 1/km, we can obtain:

$$L_{eff} = 1/\alpha[1 - e^{-\alpha L}] \tag{6.29}$$

The real transmission length, $L$, in communications fibers is long enough so that $L >> 1/\alpha$, which results in $L_{eff} \approx 1/\alpha$. Typically, $\alpha = 0.2$ dB/km at $\lambda = 1550$ nm; that is, $\alpha = 0.046$ 1/km, so $L_{eff} =$ 21.7 km.

Similar to $L_{eff}$, *effective area,* $A_{eff}$ is defined as Figure 6.21(b) shows. Essentially, this is the cross section of the light path. Unfortunately, the formula for calculating $A_{eff}$ is very complicated. For an estimation of the order of value, you can use this simple formula:

$$A_{eff} \approx \pi w_0^2 \tag{6.30}$$

Or you can replace the MFD radius, $w_0$, with the core radius, $r$. Actually, the fiber manufacturer determines the effective area by measuring, not computing. Effective area is a new parameter that has been introduced to characterize a new generation of fibers. The effective area of a standard singlemode fiber is about 55 $\mu m^2$.

***Self-phase modulation (SPM)*** Light, as an EM wave, is described in Formula 4.21a as amplitude times $\cos(\omega t - \beta z)$. Since $\beta$ is now given by Formula 6.26, we have an additional phase shift caused by a nonlinear phase constant. This shift can be estimated by

$$\Phi = \int_0^L (\beta - \beta')\,dz = \int_0^L \gamma P(z)\,dz = \gamma\,P_{in}L_{eff} \tag{6.31}$$

It follows from Formula 6.27 that the nonlinear phase shift is proportional to the total amount of light power that a fiber carries over its length.

Another form of this dependence is given by

$$\Phi = (3w/8cn)\chi_e^{(3)}E^2L_{eff} \tag{6.32}$$

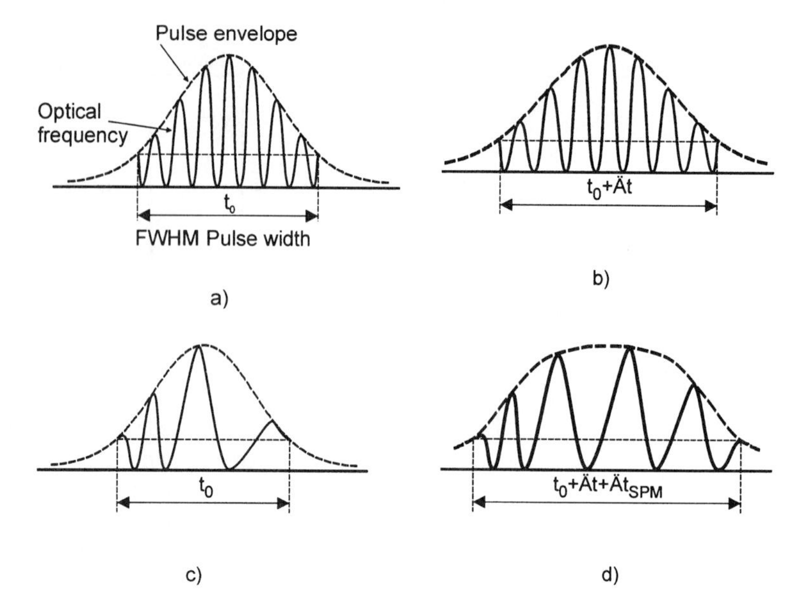


**Figure 6.22** Self-phase modulation effect: spreading of chirped pulse: (a) Regular unchirped pulse entering the link; (b) the same pulse distorted after traveling distance *L* along the fiber; (c) chirped pulse entering the link; (d) chirped pulse broadens after traveling distance *L*.

Both Formulas 6.31 and 6.32 underscore the dependence of the nonlinear phase shift on input light power or intensity. They also reveal the reason for changing $\Phi$: time changing $P_{in}$, or $E^2$.

To better understand the mechanism of self-phase modulation (SPM), consider a light pulse traveling over the fiber. This pulse is built like an envelope filled with modulated-carrying optical frequency, as shown in Figure 6.22.

The SPM mechanism works as follows: The nonlinear phase shift, $\Phi$, of the optical carrier signal changes with respect to time because pulse intensity (power) changes over time. The latter changes are caused by $P_{in}(t)$ varying and/or by pulse-amplitude time varying while the pulse propagates along the fiber. (See Figure 6.22 and Formulas 6.31 and 6.32.) Thus, $\Phi$ becomes a function of time, $\Phi(t)$. Since frequency, by definition, is the derivative of phase shift with respect to time, we have an optical frequency change caused by $d\Phi(t)/dt \neq 0$. This varying frequency is called *chirping.* (See Figures 6.22[c] and [d]. We met this term when we considered chirped-fiber Bragg gratings, you'll recall, in Section 6.3.) Now it should be clear why this phenomenon is called *self-phase modulation: Modulation is the frequency change caused by a phase shift induced by the pulse itself.*

You probably are wondering at this point how SPM can distort the performance of a communications system. Recall how, in chromatic dispersion, different wavelengths (frequencies) propagate at different velocities. This results, remember, in pulse spreading. Look again at Figure 6.22. Since a pulse carries different frequencies, it will spread, as Figure 6.22(d) shows. Thus, *SPM causes pulse spreading through chromatic dispersion.*

How serious this damage might be depends on the power transmitted, the length of the link, and the bit rate. Figure 6.23 shows that the pulse can be twice as wide at the end of a 200-km transmission as it was at the start.

The graphs in Figure 6.23 were calculated for the system operating at a bit rate of 10 Gbit/s with the initial pulse width ($t_0$) = 50 ps, $\lambda$ = 1550 nm, and a conventional singlemode fiber. There is a potential advantage to SPM, it should be noted: For high propagating power at the entering arm of a fiber, SPM can compress the pulse. The price, however, is significantly increased pulse spreading farther along the fiber. Therefore, if you want to increase the spacing between in-line amplifiers by increasing the power launched into the fiber, think twice and bear in mind the impact SPM can have.

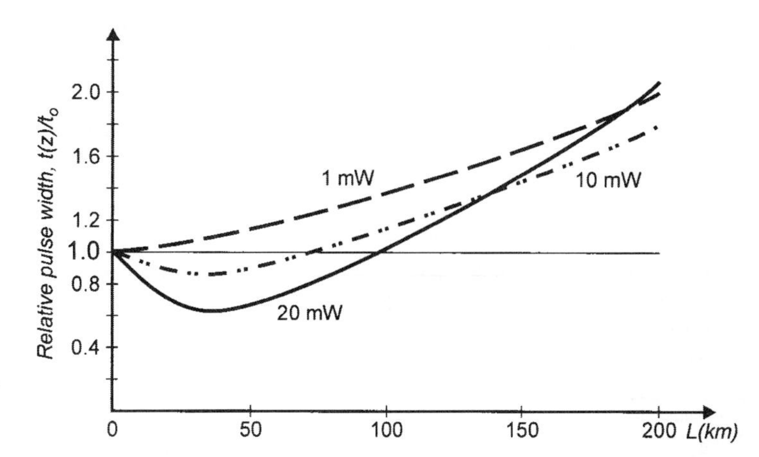


**Figure 6.23** Pulse spreading caused by SPM as a function of distance. *(Rajiv Ramaswami and Kumar Sivarajan,* Optical Networks: A Practical Perspective, *San Francisco: Morgan Kaufman Publishers, 1998. Reprinted with permission.)*

Nonlinear effects are measured in terms of the power limitations they impose on a communications system. To keep the SPM impact to a minimum, we need to keep the nonlinear phase shift small, that is, $\Phi << 1$. As Formula 6.31 shows, by having $L_{\text{eff}} \approx 1/\alpha$, we obtain:

$$P_{in} << \alpha/\gamma \tag{6.33}$$

For $\alpha = 0.2$ dB/km, that is, $\alpha = 0.046$ 1/km and $\gamma = \gamma_n = 2.35 \times 10^{-3}$ $1/m{\cdot}W$, the input power should be kept below 19.6 mW.

***Solitons*** Refer to Figure 6.31 again. There is a possibility of compressing the pulse width using SPM. On the other hand, all types of dispersion, as we know, cause pulse widening. *Fortunately, we can use SPM to compensate for dispersion-caused pulse spread to keep the pulse width constant over the entire transmission distance. Such a pulse, having a constant width, is called a soliton.* A soliton either keeps its width constant or changes it periodically, but its width never exceeds a given value. A soliton, then, is essentially a nonlinear effect because it is based on SPM.

It looks very attractive because it completely overcomes dispersion-induced pulse widening, totally eliminating the need to cope with any type of dispersion. Obviously, many problems arise when you try to implement a new approach. However brilliant it appears in theory, putting a new solution into practical use often poses its share of difficulties. This is currently an area of active research and results so far have been impressive. Recent experiments demonstrated 40-Gbit/s transmission over 70,000 km with a single-channel system [20] and an $N \times 10$ Gbit/s WDM system transmitting over a long distance [16]. Fiber specifically designed for soliton transmission is about to become ready for mass fabrication [17].

At this time we are not aware of any commercially deployed soliton system, but such systems are indeed on the brink of deployment.

***Cross-phase modulation (XPM)*** SPM is the major nonlinear limitation in a single-channel system. In a multichannel system, we meet another nonlinear phenomenon—cross-phase modulation, XPM. This effect occurs only in WDM systems. Refer to Formulas 6.21 through 6.32. When several optical pulses propagate within a fiber simultaneously, the nonlinear phase shift of, say, the first channel, $\Phi_1$, depends not only on the intensity (power) of this channel but also on the signal intensities of the other channels. Let's consider, for example, a three-channel transmission. Then $\Phi_1$ is given by [9]:

$$\Phi_1 = \gamma L_{\text{eff}}(P_1 + 2P_2 + 2P_3) \tag{6.34}$$

### Soliton Transmission

The soliton—a pulse able to keep its shape and width steady as a result of mutual compensation of dispersion-broadening and self-phase-modulated narrowing processes—has been briefly discussed above. There have been new theoretical and experimental studies of this phenomenon, and there are new field trials of soliton transmission currently under way. At SuperComm '98—a major conference on telecommunications—in Atlanta, Pirelli Cables and Systems (the Italian-based telecommunication company) unveiled its new TeraMux hyper-dense wavelength-division multiplexing system based on soliton transmission. This system offers 128 channels, with each channel capable of carrying 10 Gbit/s; its pulses are able to travel up to 6,000 km without regeneration.

Since solitons can be transmitted over a very long fiber link without amplification and dispersion compensation, they look like the most promising transmission technology. The urgent need for solitons stems from the necessity to cope with the most limiting effect in today's fibers—polarization-mode dispersion, PMD. Recall that PMD is random in nature and, hence, can't be compensated for directly. The major challenge in the development of commercial soliton-transmission systems is accomplishing soliton transmission over already installed multimillion-mile networks of a standard fiber. This is where the efforts of researchers and engineers are concentrated today.

In the long-term perspective, recent research has opened new horizons in the development of completely new optics technology based on solitons. For fiber-optic communications, these innovations have created opportunities that will completely change transmission and switching techniques, which, in turn, will one day lead to new network technology barely even imagined today.

XPM hinders system performance through the same mechanism as SPM: chirping frequency and chromatic dispersion. But XPM can damage the system even more than SPM because of coefficient 2 in Formula 6.34.

For a system with more than 10 channels and standard singlemode fiber, the power limitation might be as low as 1 mW per channel. However, by keeping chromatic dispersion low, one can reduce XPM impact to negligible values. Bear in mind that XPM impact essentially depends on transmission technology, that is, modulation and detection techniques. XPM becomes a very serious limitation in coherent systems. (See Chapter 14.) Another critical issue relating to XPM influence is the number of channels in the system. Theoretically, for a 100-channel system, XPM imposes a power limit of 0.1 mW per channel.

The effects of both SPM and XPM become truly practical limitations for high-bit-rate (>10 Gbit/s) systems. In fact, a dispersion-compensating technique aiming to compensate for linear and nonlinear dispersion has been developed [9].

## Four-Wave Mixing (FWM)

Whenever you work with WDM systems, you will have to eliminate the effect of four-wave mixing (FWM).This nonlinear phenomenon is bit-rate independent, which is why it imposes restrictions on a system's transmission capacity. It is a major hurdle that has to be surmounted for the high-bit-rate WDM systems.

The basic mechanism of FWM is as follows: When three EM waves co-propagate through one fiber, they generate a fourth EM wave because the fiber's electric susceptibility, $\chi_e$, includes the nonlinear part (thus, the term *four-wave mixing*). In a sense, three original waves can generate not one but many waves, as Figure 6.24 shows.

A formal description of the FWM phenomenon involves a polarization vector (see Formula 6.23): $\mathbf{P}(\mathbf{r},t) = \varepsilon_0 \chi_e \mathbf{E}(\mathbf{r},t) + \varepsilon_0 \chi_e^{(3)} \mathbf{E}^3(\mathbf{r},t)$. Now, taking the scalar form for simplicity, we have

$$E(\mathbf{r},t) = \sum_{i=1}^{n} E_i \cos(\omega_i t - \beta_i z) \tag{6.35}$$

so that the nonlinear part of the polarization vector can be written in this form [14]:

$$P_{nl} = \varepsilon_0 \chi_e^{(3)} E^3(\mathbf{r},t) = \varepsilon_0 \chi_e^{(3)} \sum_I \sum_j \sum_k E_i \cos(\omega_i t - \beta_i z) \\ E_j \cos(\omega_j t - \beta_j z) E_k \cos(\omega_k t - \beta_k z) \tag{6.36}$$

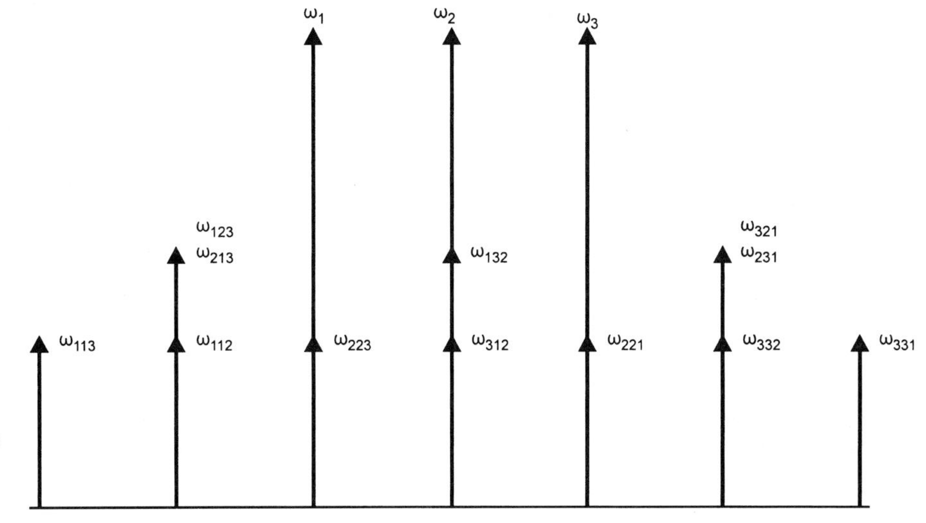


**Figure 6.24** Four-wave-mixing components caused by the beating of three evenly spaced channels.

We leave you to develop Formula 6.36 in explicit form. (See Problem 6.18.) When you do so, you'll meet a term that contains combinations like $\cos(3\omega_i\, t - 3\beta_i\, z)$ and $\cos((2\omega_i \pm \omega_j)t - (2\beta_i \pm \beta_j)z)$. Remember that, ideally, $\beta = \omega n/c$ and, consequently, $3\beta$ should be $3\omega n/c$. Since, in real fibers, $n = n(\omega)$ and $3\beta(\omega) \neq \beta(3\omega)$, any difference like $3\omega - 3\,\beta$ is called *phase mismatch.* Phase mismatch can also be thought of as the actual mismatch in phase between different signals traveling within the fiber at different group velocities. We can neglect all these EM waves because they have such a phase mismatch.

There are two terms in the final expression we should focus on. The first is:

$$\left[\left(3\varepsilon_0\chi_e^{(3)}\right)/4\right]\sum_{i=1}^{n}\left[E_i^2 + 2\sum_{1\neq j}E_iE_j\right]E_i\cos(\omega_i t - \beta_i z) \tag{6.36a}$$

This expression describes SPM and XPM effects.

The second term,

$$\begin{aligned}&\left[\left(6\varepsilon_0\chi_e^{(3)}\right)/4\right]\sum_i\sum_j\sum_k E_i\,E_jE_k\left[\cos\left((\omega_i+\omega_j-\omega_k)t-(\beta_i+\beta_j-\beta_k)z\right)\right.\\&\left.+\cos\left(\omega_i-\omega_j+\omega_k\right)t-\left(\beta_i-\beta_j+\beta_k\right)z)\right],\end{aligned} \tag{6.36b}$$

represents the phenomenon of four-wave mixing. This expression tells us that *three EM waves propagating in a fiber generate new waves with frequencies* $(w_i \pm w_j \pm w_k)$. This generation occurs because of the nonlinear part of the fiber's susceptibility, $\chi_e$, that is, the fiber's nonlinear refraction.

Four-wave mixing stems from the frequency combinations $\omega_i + \omega_j - \omega_k$ and the power of the fourth component reaches peak value when the channel's wavelengths (frequencies) are close to the zero-dispersion wavelength. It seems that using conventional standard fiber can help combat FWM, but this conflicts with the requirement to reduce dispersion. This is why fiber manufacturers designed and fabricated non-zero dispersion-shifted fiber (NZ-DSF), which introduces some residual dispersion at the appropriate wavelengths. (See Section 6.3.) Figure 6.25, which shows the power of the FWM component versus the dispersion parameter, underscores the point that the maximum damage done by FWM occurs near the zero-dispersion zone.

This figure also shows the way to cope with the problem: Use an NZ-DSF fiber with a large effective area.

FWM results in power tunneling from one channel to another, a phenomenon that has two major detrimental effects on system performance. First, power depletion of the channel might

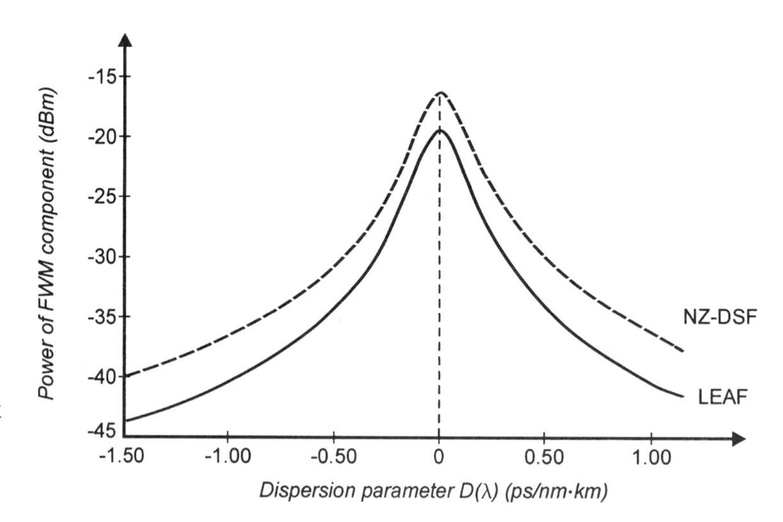


**Figure 6.25** Power of FWM component versus dispersion parameter. Key: NZ-DSF—non-zero dispersion-shifted fiber; LEAF—large-effective-area fiber.

increase bit error rate (BER) because many "ones" simply would not be provided with enough power. Since in-line amplifier spacings are designed for a link and not for a channel, power depletion of a specific channel might lead to the fading of this channel. Secondly, FWM itself is *interchannel cross talk,* which means that information from one channel interferes with information in another channel.

The impact of FWM on a communications system depends on the fiber's parameters ($A_{eff}$ and $L_{eff}$), the number of channels (remember, FWM occurs only in WDM systems), channel spacing, and, of course, the power transmitted. Remember, FWM is a bit-rate-independent effect. Maximum transmitted power versus distance imposed by FWM is shown in Figure 6.26.

The most straightforward method to reduce FWM is to space channels unevenly. Also, increasing channel spacings, reducing the power transmitted, and increasing operating wavelength can help. However, FWM today is still the most limiting nonlinear effect in WDM systems.

## Stimulated Scattering

We reviewed the role of scattering in our discussion of how Rayleigh scattering turns out to be one of the major sources of fiber attenuation. Even though the actual mechanism of Rayleigh scattering is rather sophisticated, we can visualize it simply by considering the scattering of light by some particles. This simplification is possible because with Rayleigh scattering only the direction of the light propagation changes; the frequency (wavelength) of the scattered light remains unchanged. When light power essentially increases, a new phenomenon—nonlinear, or stimulated, scattering—comes into play.

*Stimulated scattering is transferring energy from the incident wave to another, scattered wave at lower frequency (longer wavelength) with the small energy difference being released in the form of phonons.* (A phonon is an elementary particle analogous to a photon but differs from a photon in its quantum properties.) The incident wave can be seen as a kind of pump wave. Scattered waves are called *Stokes' waves.*

The result of stimulated scattering is a loss of energy by the incident wave, which is the signal carrier in fiber-optic communications systems. In other words, stimulated scattering is yet another attenuation mechanism in singlemode fibers.

Stimulated scattering is characterized by three major parameters: threshold power, $P_{th}$, gain, $g$, and the range of frequencies, $\Delta f$, within which scattering is effective. $P_{th}$ is the power of incident light at which the loss due to stimulated scattering is 3 dB—that is, half—over the fiber's length, $L$. The intensity of scattering light grows exponentially when the power of incident light exceeds $P_{th}$ [9]. Gain, $g$, refers to the peak gain of the stimulated scattering at the given wavelength.

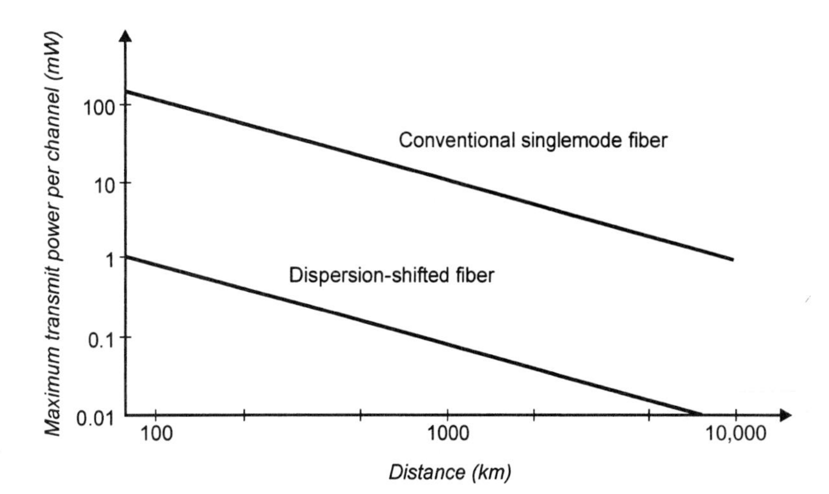


**Figure 6.26** Maximum transmitting power per channel versus distance imposed by FWM. (*Rajiv Ramaswami and Kumar Sivarajan,* Optical Networks: A Practical Perspective, *San Francisco: Morgan Kaufman Publishers, 1998. Reprinted with permission.)*

There are two major types of stimulated scattering: *stimulated Raman scattering (SRS)* and *stimulated Brillouin scattering (SBS).* SRS scattered light moves mostly *forward* and the phonons associated with this process are optical ones; SBS scattered light moves *backward* and the phonons associated with it are acoustic.

Let's consider the SRS effect. $P_{th}$ (SRS) can be estimated by the following formula [9]:

$$P_{th}(\text{SRS}) \approx 16A_{eff}/g_R\, L_{eff}, \tag{6.37}$$

where $g_R$ is SRS gain. If we approximate $L_{eff} \approx 1/\alpha$, as we did earlier (see Formula 6.29), we get

$$P_{th}(\text{SRS}) \approx (16\alpha\, A_{eff})/g_R \tag{6.37a}$$

The typical value of $g_R$ is about $1 \times 10^{-13}$ m/W at $\lambda = 1550$ nm. Thus, taking $\alpha = 0.046$ 1/km = 0.2 dB/km and $A_{eff} \approx 55\ \mu m^2$, one can compute $P_{th}$ (SRS) $\approx$ 405 mW for a single channel. *This number shows we can neglect SRS in a single-channel system.* The situation changes for WDM systems, where SRS causes power transfer from the shorter-wavelength channels to the longer-wavelength channels. This power depletion in the shorter-wavelength channels increases the BER and degrades other performance characteristics of the system. What makes the situation troublesome is the wide SRS bandwidth ($\Delta f_R \sim 10$ THz) so that many channels can be coupled with this phenomenon. The power limitation for a WDM system with channel spacing of 0.8 nm and amplifier spacing of 80 km imposed by SRS is shown in Figure 6.27. It is worth mentioning that dispersion doubles the reduction of the SRS effect.

Let's turn now to the SBS effect. Threshold power can be calculated as [3]

$$P_{th}(\text{SBS}) \approx 21A_{eff}/g_B\, L_{eff} \tag{6.38}$$

The crucial parameter here is SBS gain: $g_B \approx 5 \times 10^{-11}$ m/W, which is almost three orders of magnitude larger than $g_R$. Repeating calculations that we did with Formula 6.37, we can compute $P_{th}$ (SBS) $\approx$ 8 mW per channel. Fortunately, SBS occurs over a very narrow band of frequencies ($\Delta f_B \sim 20$ MHz). This reduces the damage done by SBS. $P_{th}$ (SBS) can be increased by the coefficient $1 + \Delta f_{source}/\Delta f_B$ so that, for $\Delta f_{source} = 200$ MHz, $P_{th}$ (SBS) can be about 14 mW ([9], [14]). Since $\Delta f_B$ is very narrow, SBS has almost no effect on WDM systems but puts the real limit—on the order of 100 mW of input power—on single-channel systems.

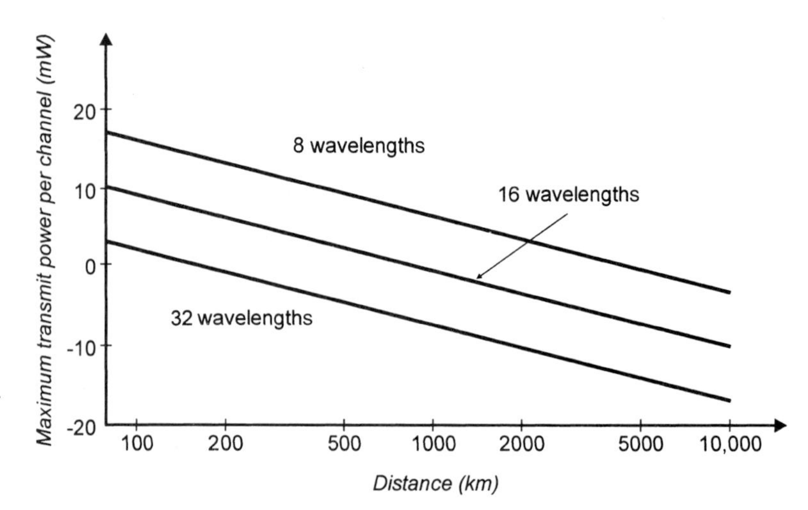


**Figure 6.27** Maximum transmitting power per channel versus distance imposed by SRS. *(Rajiv Ramaswami and Kumar Sivarajan,* Optical Networks: A Practical Perspective, *San Francisco: Morgan Kaufman Publishers, 1998. Reprinted with permission.)*

To summarize, stimulated scattering increases fiber attenuation at a high level of transmitted power. SRS has a negligible impact on single-channel systems but becomes an issue for WDM systems. By contrast, SBS is a problem for a single-channel system but has almost no effect on WDM systems.

It should be emphasized that nonlinear effects are a very special area today. It's where a whole gamut of activity is under way, activities ranging from research aimed at improving system performance characteristics in the presence of fiber nonlinearities to the development of mass manufacturing techniques to implement design developments that accomplish this goal.

## 6.5 TRENDS IN FIBER DESIGN

Optical fiber came on the telecommunications scene as a transmission medium promising unlimited information-carrying capacity. Almost immediately after the first successes in fabricating relatively low-loss fibers, it appeared that intermodal dispersion was restricting the bit rate by tens of Mbit/s—a very far from expected unlimited bandwidth. In response, graded-index fibers, with hundreds of Mbit/s of bit rate, and singlemode fibers were introduced as the ultimate solution to the intermodal-dispersion problem. Since the early 1980s, the mass installation of what we call conventional singlemode fiber has resulted in the deployment of millions of miles of this product.

The bandwidth requirement in long-haul systems since the early 1970s has been doubling almost every two years, and fiber, as a transmission medium, has easily met this demand [17]. Other components of fiber-optic communications systems, not the fiber, were the drawbacks limiting bit-rate increases. The situation changed in the 1990s. Deregulation and new types of services, plus the phenomenal growth of Internet communications, have boosted bandwidth demand into the Gbit/s scale. Straightforward attempts to respond to these demands by increasing transmission bit rate over installed fibers were not successful. Chromatic dispersion became the real limitation. In response, dispersion-shifted fiber (DSF) with zero dispersion at the 1550-nm operating wavelength was developed. At last, a workable technique for coping with dispersion was available.

Soon, two significant technological advances—optical amplifiers and wavelength-division multiplexing systems—were introduced, offering two major benefits: (1) a sharp increase in bandwidth by transmitting multiple channels (4, 8, 16, even 32) over the same fiber simultaneously and (2) in-line signal amplification without electrical-to-optical-to-electrical conversion regardless of the signal's format and modulation scheme.

But, as with most solutions to complex technological problems, the WDM systems came with a price. For one thing, there were severe restrictions on bit rate—obviously, at a much higher level. Also, the high optical power launched by the amplifiers in conjunction with the simultaneous transmission of many channels spawned new problems, such as the four-wave-mixing (FWM) phenomenon, where nonlinear refraction redistributes power between transmitting channels. The FWM effect reaches maximum at the zero-dispersion zone, making the use of DSF not simply worthless but even detrimental to the system. What's more, the trend to decrease spacing between channels, making dense wavelength-division multiplexing systems (DWDM), exacerbates the situation. To counter these adversities, non-zero dispersion-shifted fiber (NZ-DSF) was developed.

The move to increase the span between in-line optical amplifiers requires launching more power into each fiber. This, together with the DWDM technology, increases the nonlinear effects. Among them, self-phase modulation (SPM) and cross-phase modulation (XPM) result in pulse spreading, while stimulated Raman (SRS) and Brillouin (SBS) scattering bring on attenuation.

Fortunately, a solution to these problems was found in the form of a fiber with a large effective area, a development that vaulted fiber optics to the leading edge of today's technological innovations and made possible the astonishing strides we've made in transmission capability over just the past decade. One such example, Corning's NZ-DSF LEAF (large-effective-area fiber), has a typical $A_{eff}$ of 72 $\mu m^2$ in contrast to 55 $\mu m^2$ for regular NZ-DSF [19]. Physically enlarging $A_{eff}$ has been achieved by redesigning the core-index profile. The new profile includes a triangular central region and an outer ring, similar to that shown in Figure 5.12. The outer ring helps to distribute light outward from the central region, thus decreasing the peak power at the center of the core, in addition to simply increasing a cross section of the light path.

The major benefit of the new fiber is its ability to handle more power without being affected by nonlinearities, essentially increasing its information-carrying capacity. Specifically, the LEAF NZ-DSF suppresses the FWM component more effectively than does the NZ-DSF. (See Figure 6.25.) Moreover, it causes less SPM-induced pulse widening, allows an increase of 20% in in-line amplifier spacing, and is designed to work for dense WDM systems across the 1530- to 1565-nm band at a bit rate of 10 Gbit/s per channel. This fiber not only meets today's specifications but also has the potential to be used in future systems without modification.

Furthermore, enlargement of the effective area of a fiber is the major focus of activity in fiber design today. Commercial NZ-DSF fiber with an effective area of 90 $\mu m^2$ has been reported [20]. Experimental samples with an effective area of 140 $\mu m^2$ have been fabricated [21].

Activity is also moving at a fast pace in the design of dispersion-managed fiber (DMF), where alternating sections of positive and negative dispersions are built directly into a continuous fiber [22]. Such a fiber minimizes FWM and other nonlinear effects because of its large effective area and, at the same time, provides dispersion compensation together with a low dispersion slope.

Efforts to improve fiber parameters are continuing; in fact, special fibers to meet specific customer requirements, such as low PMD and extra-low bending sensitivity, are available from different manufacturers; this trend will no doubt continue as demand for proprietary fibers tailored to specialized systems increases.

To summarize, then, the key point to keep in mind is that the rapidly burgeoning demands for optical fibers with enhanced properties have spurred innovative approaches to fiber design and manufacture. The need to increase information-carrying capacity by using WDM and in-line optical amplifiers is largely responsible for the profusion of new activity, the result of which has been development of a new generation of optical fibers. The ramification is obvious: Fiber continues to surpass and leave in its wake other components of high-bit-rate optical-communications systems in meeting new, exponentially growing bandwidth requirements.

## SUMMARY

- It is impossible to overestimate the importance of singlemode (SM) fibers to modern telecommunications. About 90% of all installed fiber is of the singlemode type. The trend to deploy singlemode fiber in metropolitan and local networks enhances its role even more.
- To ensure that a singlemode fiber conducts only one mode, manufacturers make the core diameter smaller than 9 μm and the relative refractive index, Δ, less than 0.37%. In addition, they specify an operating wavelength longer than the cutoff wavelength ($\lambda_C$), which is typically more than 1260 nm. The *V*-number, which is less than 2.405 for a singlemode fiber, and $\lambda_C$ can be used interchangeably.
- The mode of an SM fiber is described by its field distribution over the fiber's cross section. The mode-field diameter (MFD), rather than the core diameter, is a representative characteristic of a singlemode fiber.
- A close look at attenuation in a singlemode fiber reveals that the bending sensitivity of a singlemode fiber is mainly the function of power confinement within the core. This confinement is greater for shorter operating wavelengths. Since the minimum operating wavelength is $\lambda_C$, a fiber designer doesn't have very much latitude here. A fiber's bending sensitivity can also be decreased by manipulating the refractive-index profile.
- Linear dispersion in an SM fiber is due mainly to chromatic dispersion, which is the major bandwidth-limiting factor for singlemode fibers. In response, dispersion-shifted fibers with zero dispersion at the 1550-nm operating wavelength were produced. However, using new technologies such as wavelength-division multiplexing and optical amplifiers increases the role of nonlinear phenomena, particularly four-wave mixing, which do not allow the use of an SM fiber with zero dispersion. Thus, some residual dispersion has to be incorporated into an SM fiber for modern transmission.
- Fortunately, chromatic dispersion is a time-stable phenomenon, enabling a network designer to compensate for the pulse spread caused by this dispersion. To cope with chromatic dispersion, many dispersion-compensating devices and techniques have been developed. They are quite effective, allowing system operators to improve network performance significantly.
- The specific implementation of these techniques depends, essentially, on the type of communications system used and its architecture; today, for instance, a dispersion-compensating fiber (DCF) compensates over a wide range of wavelengths, while a fiber Bragg grating (FBG) is more suitable for a very narrow band of wavelengths. To achieve even more capacity, not only dispersion compensation but also dispersion-slope compensation must be included.
- We need to use dispersion compensation and management only for conventional fiber-based networks installed years ago. The new generation of fibers already has dispersion-coping capabilities, so they not only satisfy today's bandwidth requirements but they also have upgrading potential to meet future needs.
- The high bit rate (more than 2.5 Gbit/s per channel), the high light power launched by in-line optical amplifiers (EDFA), and the ability now to transmit over many channels simultaneously (thanks to WDM systems) have brought about the *nonlinear* effects that were largely neglected before. (Nonlinear means that these effects depend on light power.)
- The most troublesome nonlinear phenomenon today in WDM systems is *four-wave mixing (FWM)*. Three waves copropagating over the same fiber generate the fourth wave, whose frequency is the combination of the three original wave frequencies. (In fact, not one but many new waves can be generated, but not all of them exist and only a few really impact on the transmission system.) As a result, power from one channel seeps into another, causing an increase in the bit error rate (BER) and inter-channel cross talk (the leakage of information from one channel to another).
- FWM is most detrimental near zero-dispersion wavelength; it is in direct conflict with attempts to eliminate chromatic dispersion by using dispersion-shifted fiber. This is why *non-zero dispersion-shifted fiber (NZ-DSF)* has been developed. This fiber's small residual dispersion in the range of wavelengths used for a WDM system essentially reduces damage done by FWM.
- To cope with FWM, *dispersion management* has been developed. This is a technique aimed at keeping dispersion low, but not at zero, at any point along a transmission link, with zero dispersion at the receiver end.
- The most straightforward method to reduce FWM is to space channels unevenly. Also, increasing channel spacings, reducing the power transmitted, and increasing operating wavelength can help. However, FWM today is still the most limiting nonlinear effect in WDM systems.
- Two nonlinear refractive effects—*self-phase modulation (SPM)* and *cross-phase modulation (XPM)*—in conjunction with chromatic dispersion result in additional pulse spread. These two effects can be called nonlinear dispersion. SPM stems from the fact that different parts of a propagating pulse have different levels of power, the re-

sult of which is a change in the phase of a carrying (or optical) signal. This self-modulation eventually leads to a change in the optical frequency. (Hence, the variation in pulse power produces a modulation in its frequency—thereby giving us the name of the effect.) Waves with different frequencies travel at different velocities, resulting in pulse widening.

- The mechanism of XPM is very similar to that of SPM except that modulation is induced by the power of the adjacent channel—from which is derived the name of the effect.
- Stimulated scattering is another nonlinear effect. This effect results in an increase in fiber attenuation. In stimulated scattering a high-power signal transfers its energy to another optical wave; that is, it attenuates. Stimulated scattering, in contrast with Rayleigh scattering, lowers the frequency of the scattered signal. There are two major forms of this phenomenon: stimulated Raman scattering (SRS) and stimulated Brillouin scattering (SBS). SRS is most troublesome in WDM systems, while SBS affects a single channel.
- All nonlinear effects impose transmission-distance limitations, depending on the transmitting bit rate, of course. To cope with nonlinear effects, new fibers—large-effective-area fibers—were developed.
- It should be emphasized that nonlinear effects are a very special area today. It's where a whole gamut of activity is under way, activities ranging from research aimed at improving system performance characteristics in the presence of fiber nonlinearities to the development of mass manufacturing techniques to implement design developments that accomplish this goal.
- Soliton—a pulse able to keep its shape and width steady as a result of mutual compensation of dispersion-broadening and self-phase-modulated narrowing processes—is an example of the positive use of fiber nonlinearities. Since solitons can be transmitted over a very long fiber link without amplification and dispersion compensation, they look like the most promising transmission technology. The urgent need for solitons arises from the necessity to cope with the most limiting effect in today's fibers—polarization-mode dispersion, PMD.
- Trends in fiber design include continuing work to improve the characteristics of fibers already being manufactured; development of new types of fiber with built-in dispersion-compensating capabilities; extensive efforts to increase a fiber's effective area, allowing more power to be launched without causing nonlinear effects; and development of special fibers, such as PM fibers, fibers with extra-low bending sensitivities, fibers with high numerical aperture, and fibers with desired cutoff wavelengths, to name just a few.
- A new generation of fibers is already in the works to meet not only today's high standards but also tomorrow's even higher-scale demands.

## PROBLEMS

**6.1.** Write the formula for a step-index refractive-index profile.

**6.2.** What is the $V$-number of the low-PMD SM from Plasma Optical Fibre BV, assuming that the core diameter is equal to 8.3 μm and the relative index equals 0.36%?

**6.3.** Manufacturers want to increase the relative refractive index, $\Delta = (n_1 - n_2)/n$, because experiments show this measure decreases a singlemode fiber's sensitivity to bending.
**a.** Explain why the sensitivity to bending decreases.
**b.** How can one increase $\Delta$?
**c.** Can manufacturers theoretically increase $\Delta$ up to 2%? Up to 10%?

**6.4.** For depressed-cladding fiber:
**a.** Calculate a fraction of $I(r)/I(0)$ for $a_1 = 5a$, where $a$ is the core radius.
**b.** Using Fresnel Formulas 4.31 and 4.32, calculate the portion of power lost owing to transmission (refraction) at the inner–outer cladding interface.

**6.5.** Calculate the distance limit imposed by a fiber's attenuation if $A = 0.25$ dB/km, $P_{out} = 0.5$ μW, and $P_{in} =$ 0.04 mW.

**6.6.** **a.** Briefly explain the mechanism of chromatic (intramodal) dispersion.
**b.** How do manufacturers make dispersion-shifted and dispersion-flattened fibers?

**6.7.** **a.** Since there are commercially available dispersion-shifted fibers (DSF), why do we need to cope with dispersion by any other means?
**b.** What methods of coping with dispersion do you know?

**6.8.** What length of DCF is needed to compensate for chromatic dispersion at a 120-km span of low-PMD fiber from Plasma Optical Fibre BV (see Figure 5.17) if $D_{DCF}(\lambda) = -127$ ps/nm·km and $\lambda = 1550$ nm?

**6.9.** Explain how chirped-fiber Bragg grating works as a dispersion-compensating device.

**6.10.** **a.** What means do we have to cope with PMD?
**b.** What bit rate can be achieved at a 120-km span with low-PMD fiber from Plasma Optical Fibre BV? (See Figure 5.17.)

**6.11.** Briefly comment on Figure 6.20.

**6.12.** **a.** What do we mean by higher-order dispersion?
**b.** How can we combat higher-order dispersion?

**6.13.** Why does nonlinear refraction depend on the ratio $P/A_{eff}$?

**6.14.** Derive Formula 6.34 from Formula 6.33.

**6.15.** Assuming that a rectangular pulse has a duration of 500 ps, calculate the number of light-wave cycles the pulse can accommodate at 1550 nm. Compare this number with the number of cycles shown in Figure 6.22.

**6.16.** **a.** Briefly explain the difference between the SPM and XPM phenomena.
**b.** What is a soliton?

**6.17.** **a.** Explain the four-wave mixing (FWM) phenomenon.
**b.** Why is FWM so important in today's optical networks?

**6.18.** Develop Formula 6.36 in explicit form.

**6.19.** **a.** Briefly explain the difference between the SRS and SBS phenomena.
**b.** Calculate $P_{th}$ (SRS) and $P_{th}$ (SBS) for a fiber with $A_{eff} = 72\ \mu m^2$ and $L_{eff} = 20$ km.

**6.20.** *Project:* Memory grid. (See Problem 1.20.) Sketch a grid summarizing what you have learned about single-mode fibers.

## REFERENCES[1]


1. John A. Buck, *Fundamentals of Optical Fibers,* New York: John Wiley & Sons, 1995.
2. Lars Gruner-Nielsen et al., "Large-volume manufacturing of dispersion-compensating fibers," Paper TuD5, *Proceedings of the Optical Fiber Communication Conference,* San Jose, Calif. (February 22–27, 1998).
3. Ashish M. Vengsarkar, "Tutorial on Dispersion Compensation," Paper TuL, *Proceedings of the Optical Fiber Communication Conference,* San Jose, Calif. (February 22–27, 1998).
4. S.V. Chernikov, F. Koch, J. R. Taylor, and L. Gruner-Nielsen, "Measurements of the effect of bending on dispersion in dispersion-compensating fibers," Paper TuD4, *Proceedings of the Optical Fiber Communication Conference,* San Jose, Calif. (February 22–27, 1998).
5. C. D. Poole et al., "Optical fiber-based dispersion compensation using higher order modes near cutoff," *Journal of Lightwave Technology,* vol. 12, 1994, p. 1746.
6. Massimo Artiglia, "Comparison of dispersion-compensation techniques for lightwave systems," Paper ThV2, *Proceedings of the Optical Fiber Communication Conference,* San Jose, Calif., (February 22–27, 1998).
7. *Chromatic Dispersion Compensator* (Specifications), Pirelli Cavi e Sistemi SpA, Milan, Italy, April 1998.
8. J-X Cai et al., "Dynamic dispersion compensation in a 10-Gbit/s optical system using a novel nonlinearly chirped fiber Bragg grating," Paper ThV3, *Proceedings of the Optical Fiber Communication Conference,* San Jose, Calif. (February 22–27, 1998).
9. Govind Agrawal, *Fiber-Optic Communication Systems,* 2d ed., New York: John Wiley & Sons, 1997.
10. Andrew Craplyvy et al., "Letters," *IEEE Photonics Technology,* vol. 7, 1995, p. 98.
11. Nidenori Taga et al., "Performance Evaluation of the Different Types of Fiber-Chromatic-Dispersion Equalization for IM-DD Ultralong-Distance Optical Communication Systems with Er-Doped Fiber Amplifiers," *Journal of Lightwave Technology,* vol. 12, 1994, p. 1616.
12. Nidenori Taga et al., "213 Gbit/s (20 × 10.66 Gbit/s) over 9000 km transmission experiment using dispersion slope compensator," Paper PD13, *Proceedings of the Optical Fiber Communication Conference,* San Jose, Calif. (February 22–27, 1998).
13. Henning Bülow, "PMD mitigation techniques and their effectiveness in installed fiber," Paper ThH1, *Proceedings of the Optical Fiber Communication Conference,* Baltimore, Md. (March 5–10, 2000).
14. Rajiv Ramaswami and Kumar Sivarajan, *Optical Networks: A Practical Perspective,* San Francisco: Morgan Kaufman Publishers, 1998.
15. Govind Agrawal, *Nonlinear Fiber Optics,* 2d ed., San Diego: Academic Press, 1995.
16. Linn Mollenauer, "Massive WDM in ultra-long-distance soliton transmission," Paper ThI5, *Proceedings of the Optical Fiber Communication Conference,* San Jose, Calif. (February 22–27, 1998).
17. A. F. Evans, "Novel fibers for soliton communications," Paper TuD3, *Proceedings of the Optical Fiber Communication Conference,* San Jose, Calif. (February 22–27, 1998).
18. K. Able, "Optical-fiber designs evolve," *Lightwave,* February 1998, p. 96.


[1] See Appendix C: A Selected Bibliography.

19. *Corning LEAF CPC6 Single-Mode Non-Zero Dispersion-Shifted Optical Fiber* (data sheet), Corning Incorporated, Corning, N.Y., 1998.

20. P. Nouchi, "Maximum effective area for nonzero dispersion-shifted fiber," Paper ThK3, *Proceedings of the Optical Fiber Communication Conference,* San Jose, Calif. (February 22–27, 1998).

21. Masao Kato et al., "A new design for dispersion-shifted fiber with an effective area larger than 100 $\mu m^2$ and good bending characteristics," Paper ThK1, *Proceedings of the Optical Fiber Communication Conference,* San Jose, Calif. (February 22–27, 1998).

22. V. A. Bhagavatula et al., "Novel fibers for dispersion-managed high-bit-rate systems," Paper TuD2, *Proceedings of the Optical Fiber Communication Conference,* San Jose, Calif. (February 22–27, 1998).

23. *Integrated Fiber Optic Solution: Optical Fiber, Cable, and Assemblies* (product catalog), SpecTran Specialty Optics Company, Avon, Conn., 1997.

# 7

# Fabrication, Cabling, and Installation

Now that we know how an optical fiber works, it's time to discuss how to work with it. A brief review of the fiber-fabrication process should help us to understand how much we can control a fiber's characteristics and how we should maintain the fiber in the field.

The fact is you'll never meet a bare fiber in your everyday work (except, of course, if you're attending an exposition or working in a laboratory). What you will meet on a daily basis is fiber cable. How manufacturers make these cables and how this process—cabling—influences a fiber's characteristics is a second topic covered in this chapter.

Ready fiber cable is installed within a building, outdoors, or under the sea. The place of installation may change a fiber's characteristics dramatically. Here, we'll discuss the basic installation technique.

It is not our intention to teach you here how to manufacture fiber-optic cables or how to install them in the field. Instead, we'll consider this material from the user's point of view. We'll assume you are involved in a project for installing or upgrading a fiber-optic network for your company. In such case, this chapter will help you work with your vendor and colleagues as an educated professional.

## 7.1 FABRICATION

The product of the fabrication process is an unbelievably transparent, flexible glass strand thinner than a human hair. As you already know from Chapters 3 through 6, manufacturers fabricate optical fibers with given properties, which are achieved by exercising strong control over the refractive-index profile and the fiber's geometric characteristics. The amazing thing about the fiber-fabrication process is the extremely small physical dimensions it can yield. All these operations have to be controlled within an 8-μm core for a singlemode fiber. An overview of the fabrication processes provided in this section helps you not only to familiarize yourself with this process but also to gain a deeper appreciation of the material discussed in the previous chapters.

**Two Major Stages**

***Vapor-phase oxidation technology*** Today's fiber-fabrication process includes two major stages: *The first stage produces a preform*—a cylinder of silica composition 10 to 20 cm in diameter and about 50 to 100 cm long. This preform consists of a core surrounded by a cladding with a desired refractive-index profile, a given attenuation, and other characteristics; in other words, this is a desired optical fiber, but on a much larger scale.

*The second stage is drawing the preform* into an optical fiber of the size desired.

The preform is made by vapor-phase oxidation, in which two gases, $SiCl_4$ and $O_2$, are mixed at a high temperature to produce silicon dioxide ($SiO_2$):

$$SiCl_4 + O_2 \rightarrow SiO_2 + 2Cl_2$$

Silicon dioxide, or pure *silica,* is usually obtained in the form of small particles (about 0.1 μm) called "soot." This soot is deposited on the target rod or tube. The depositing of the silica soot, layer upon layer, forms a homogeneous transparent *cladding* material. To change the value of a cladding's refractive index, some dopants are used. For example, fluorine (F) is used to decrease the cladding's refractive index in a depressed-cladding configuration.

The soot for the *core* material is made by mixing three gases—SiCl4, $GeCl_4$, and $O_2$—which results in a mixture of $SiO_2$ and $GeO_2$. The degree of doping is controlled by simply changing the amount of $GeCl_4$ gas added to the mixture. The same principle is used for doping other materials.

Since deposition is made by the application of silica layers atop one another, the manufacturer can control the exact amount of dopant added to each layer, thus controlling the refractive-index profile. Figure 7.1 explains the role of several widely used dopants.

The vapor-phase oxidation process produces extremely pure material whose characteristics are under the absolute control of the manufacturer.

***Drawing*** The preform, as mentioned above, is nothing more than an optical fiber but on a much larger scale. Drawing enables the manufacturer to obtain the fiber in the actual size desired. The major steps of a typical drawing process are shown in Figure 7.2.

The preform, after some ancillary steps we'll discuss shortly, is put into a draw furnace, where the bottom tip is heated to melting. This molten piece now starts to fall, forming a fiber with

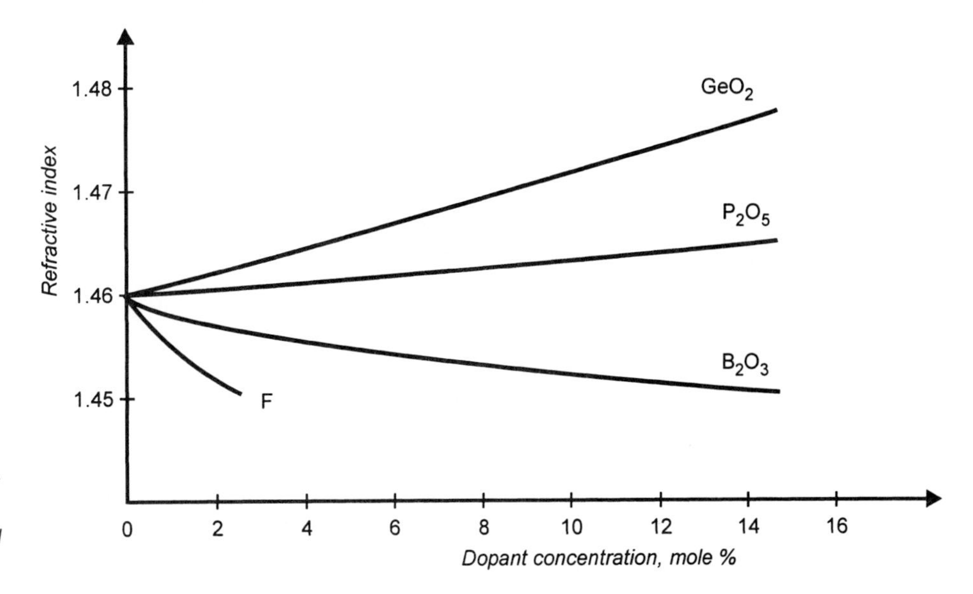


**Figure 7.1** Refractive index as a function of dopant materials and their concentration. *(Gerd Keiser,* Optical Fiber Communications, *2d ed., New York: McGraw-Hill, 1991. Reprinted with permission.)*

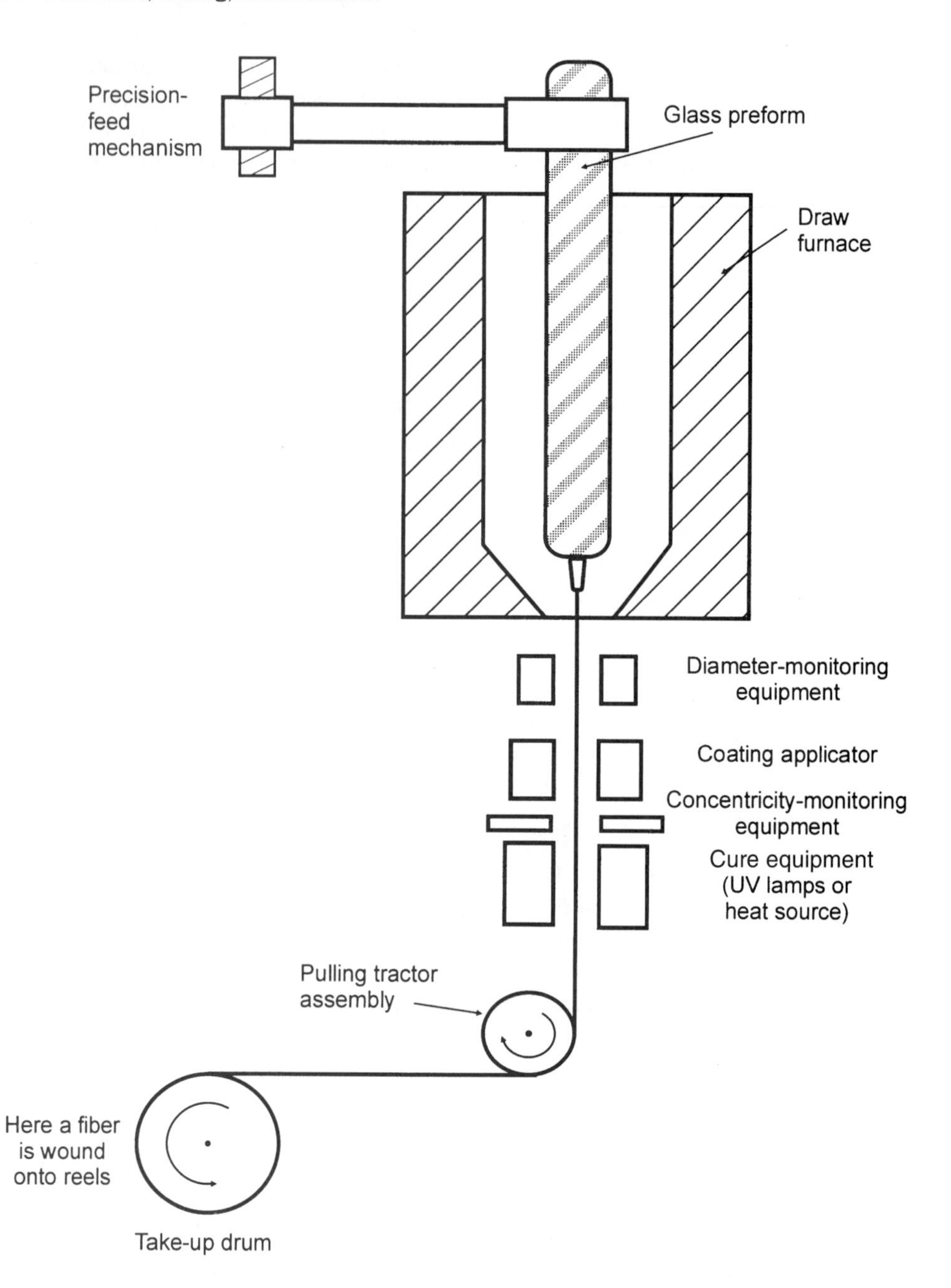


**Figure 7.2** Schematic of a typical drawing process.

a 125-μm outer diameter. Diameter-monitoring equipment controls the actual fiber diameter by changing, if necessary, the rate of drawing that is executed by a tractor assembly. A coating applicator applies a coating over the cladding. Concentricity-monitoring equipment controls this parameter. The coating is cured by ultraviolet lamps or some other heat source. The coated fiber is then wound onto ready-to-ship reels.

From this oversimplified description you might get the impression that drawing a fiber is very simple. It is not. It took many years of intensive, costly research and development efforts and the brainpower of thousands of scientists and engineers to make this process commercially possible. Critical to the process, of course, is the rate of drawing. And herein lies a dilemma. On the one hand, the slower the drawing, the better the manufacturer can control fiber quality. On the other hand, the faster the drawing, the more fiber one can produce in a given amount of time. It's clear, then, that there can be no "best" rate in an industrial operation, where both quality and productivity are demanded. So one has to find an acceptable compromise. Each of the processes described here has its own advantages and drawbacks associated with the drawing rate at which it runs. To give you some sense of how widely the numbers vary, the draw rate can run from 200 m/min to 2,000 m/min. To reach such a speed, all rotating parts of the drawing mechanism must be manufactured to extremely tight tolerances and the tension level to which the optical fiber is subjected must be controlled with a high degree of accuracy.

Even though Figure 7.2 shows only diameter-measuring equipment, the manufacturer actually measures and controls many of the fiber's characteristics during the fiber-draw stage, when some serious problems can crop up. Among them are internal problems, such as bubbles, contamination, and discontinuity, and external problems, like neckdowns, lumps, and flaws. Control of the external problems is very important because they can weaken a fiber. This is why manufacturers not only detect them constantly but also compare their size with threshold values [1].

If you analyze Figure 7.2 closely, you will find many problems that need to be resolved at each particular step and for the entire process.

Both stages of fiber manufacturing are fully automated and are performed in a clean, climate-controlled room. Obviously, the manufacturers use high-precision measuring equipment to automatically control each step of the fabrication process. For example, preform analyzers measure the critical characteristics of the optical-fiber preform. Also, specific measurement systems control fiber geometry, the refractive-index profile, and the coating geometry [2].

## Vapor-Phase Deposition Methods

There are four vapor-phase deposition methods used for fiber fabrication today. They were developed about 20 years ago but their major features are still the same. However, manufacturers are continuing to improve the techniques with the aim of achieving even better fiber characteristics than exist today. The techniques differ mainly in their methods of depositing fiber material on the target rod or tube. Let's review these techniques:

***Outside vapor-phase deposition (OVD)*** This was the first successful mass-fabrication process. It was developed by Corning in 1972. In fact, the first optical fiber with attenuation less than 20 dB/km was manufactured by Corning using this process. Today, Corning's fiber boasts attenuation as low as 0.25 dB/km along with other excellent characteristics, so you can readily imagine how much this process has been modified over the years.

*The process consists of four phases: laydown, consolidation, drawing,* and *measurement* [3]. During the *laydown* phase, the materials that make up the core and cladding are vapor-deposited around the rotating target rod. The result of this process is a soot preform. The refractive-index profile and fiber geometry are formed during this phase, as shown in Figure 7.3(a).

In the *consolidation* phase, the target rod is removed and the soot preform is placed inside a consolidation furnace. Here the soot preform is consolidated into a solid, clear glass preform and the center hole is closed. During consolidation, a drying gas flows through the preform to remove residual moisture. (Do you remember why this is important? If not, read the subsections entitled "Absorption" in Sections 3.2 and 4.5.) This phase is illustrated in Figure 7.3(b).

Phase *drawing* was described above. (See Figure 7.2.)

Finally, in the *measurement* phase, each reel is tested for compliance with the fiber characteristics given in the data sheets. (See Figures 3.18 and 3.20.) Specifically, a fiber is tested for strength, attenuation, and dimensional characteristics. MM fibers are also tested for bandwidth and *NA,* while SM fibers are tested for dispersion, MFD, and cutoff wavelength. The OVD process allows the production of a glass preform with a capacity of around 100 km of fiber.

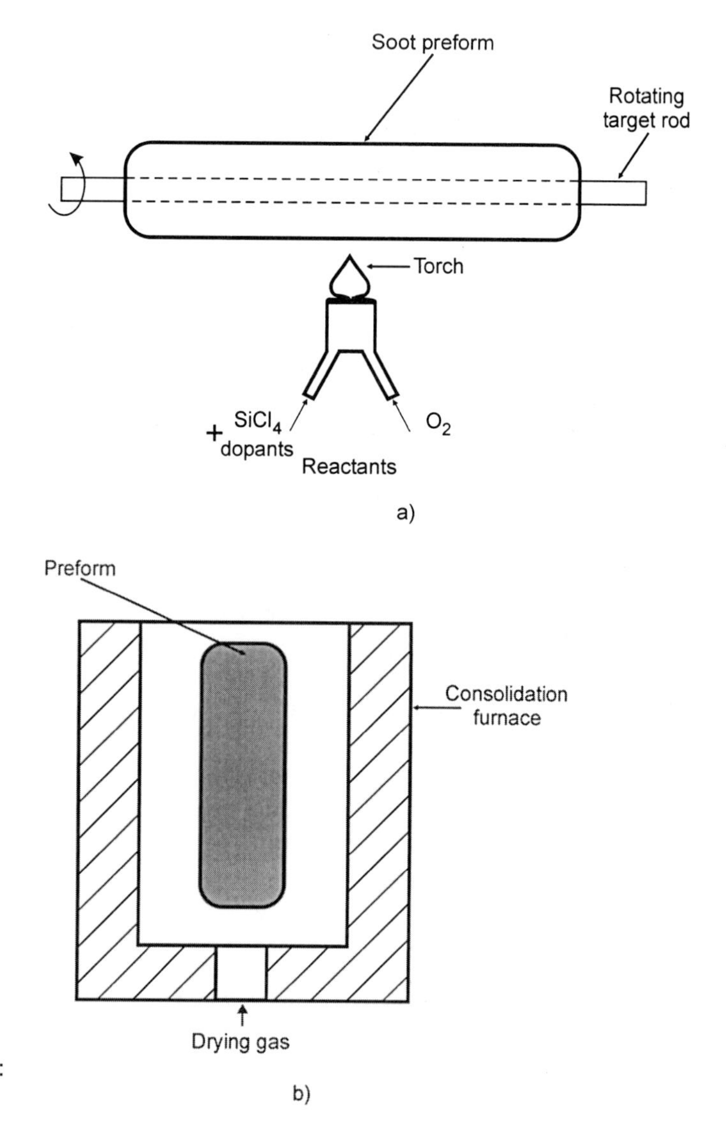


**Figure 7.3** Two phases of the OVD process: (a) Laydown; (b) consolidation.

***Modified chemical-vapor deposition (MCVD)*** This process was developed by Bell Laboratories in 1974 and has been widely accepted for the production of graded-index fiber. It has been modified over the years by AT&T and Alcatel, the French telecommunications company and the world's fourth largest producer of telecommunications equipment. In this process, the preform is also produced in two steps. First, reactant gases flow through a rotating glass tube made from fused silica while a burner heats its narrow zone by traveling back and forth along the tube. The $SiO_2$, $GeO_2$, and other doping combinations form soot that is deposited on the inner surface of the target tube. A burner heats the narrow zone of this deposit and sintering—heating without melting—occurs within this zone. The result is a layer of sintered glass. Operating temperature is about 1600°C. As in the OVD process, adding new dopants changes the refractive-index profile. The first step of the MCVD process is shown in Figure 7.4.

The second step involves heating the soot preform to 2000°C, thus collapsing the tube into a solid glass preform. Drawing and measurement complete the process. Preform capacity here is about 50 km of fiber.

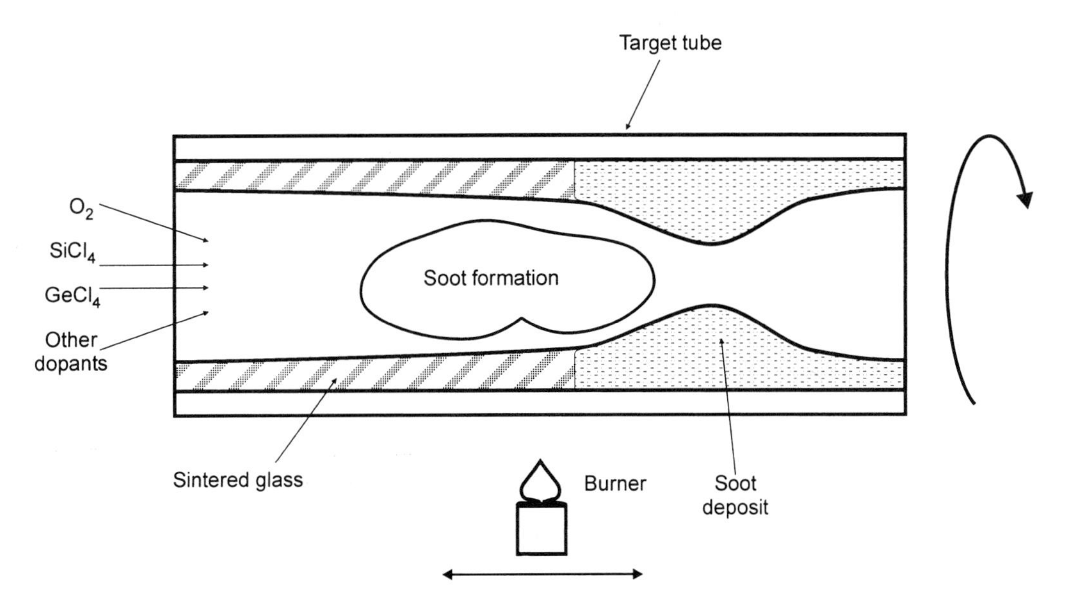


**Figure 7.4** Schematic of the MCVD process. *(Gerd Keiser,* Optical Fiber Communications, *2d ed., New York: McGraw-Hill, 1991.)*

***Plasma-activated chemical-vapor deposition (PCVD)*** This process was developed in 1975 by Phillips, a Dutch consumer-electronics and telecommunications company. The process differs from MCVD in its method of heating the reaction zone: Instead of delivering heat from the outside through a burner, PCVD uses microwaves to form ionized gas—plasma—inside the silica tube. A schematic of this process is shown in Figure 7.5.

Within the plasma, spot electrons acquire energy equivalent to 60,000°C while the actual temperature inside the furnace is 1200°C. Thus, the electrons move at a very high speed and, when they recombine with ions, considerable energy is released in the form of heat. This heat melts the soot particles obtained from the reaction between the gases $SiCl_4$, $GeCl_4$, $F_2Cl_6$, and $O_2$. As a result, deposition of the desired glass occurs directly on a target silica tube without the formation of soot. The consolidation stage is therefore not needed in this process. A resonator, within which the plasma zone forms, can move quickly along the target silica tube. Thus, very thin layers are formed because there is no restriction caused by the size of the soot particles (0.1 μm). This is a key advantage of the PCVD method, which enables manufacturers to approach, more closely than with other processes, the ideal refractive-index profile.

The final steps of the PCVD process are similar to those of the MCVD technique: The tube is collapsed, forming a glass preform, which is drawn and measured (the so-called fiber characterization stage).

The capacity of this preform is about 30 km of fiber.

***Vapor-axial deposition (VAD)*** This method was developed in 1977 by Japanese scientists. Depositing the glass particles—which were obtained from a reaction among gases in a heated zone—occurs at the bottom end of a target, or seed, rod that rotates and moves upward. This deposition forms a porous preform, the upper end of which is heated in a ring furnace to produce a glass preform. The drawing and measurement steps are similar to those of the other processes. (A schematic of the VAD process is shown in Figure 7.6.)

The profile of the refractive index is formed by using many burners, a technique that allows the manufacturer to change the direction of the flow of a specific gas mixture. A degree of latitude is inherent in this process in that the manufacturer can vary the distance between burners and pre-

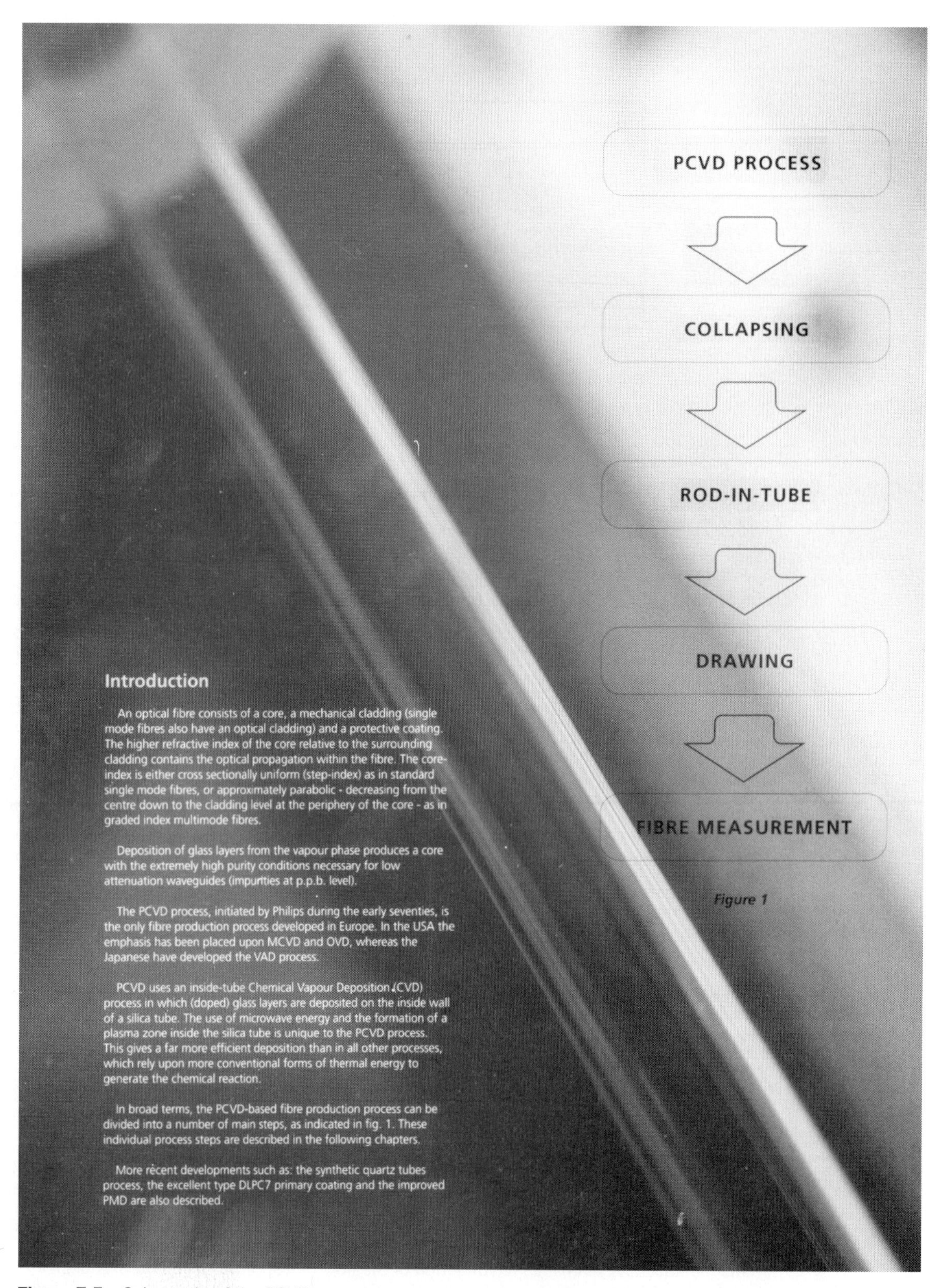

## Introduction

An optical fibre consists of a core, a mechanical cladding (single mode fibres also have an optical cladding) and a protective coating. The higher refractive index of the core relative to the surrounding cladding contains the optical propagation within the fibre. The core-index is either cross sectionally uniform (step-index) as in standard single mode fibres, or approximately parabolic - decreasing from the centre down to the cladding level at the periphery of the core - as in graded index multimode fibres.

Deposition of glass layers from the vapour phase produces a core with the extremely high purity conditions necessary for low attenuation waveguides (impurities at p.p.b. level).

The PCVD process, initiated by Philips during the early seventies, is the only fibre production process developed in Europe. In the USA the emphasis has been placed upon MCVD and OVD, whereas the Japanese have developed the VAD process.

PCVD uses an inside-tube Chemical Vapour Deposition (CVD) process in which (doped) glass layers are deposited on the inside wall of a silica tube. The use of microwave energy and the formation of a plasma zone inside the silica tube is unique to the PCVD process. This gives a far more efficient deposition than in all other processes, which rely upon more conventional forms of thermal energy to generate the chemical reaction.

In broad terms, the PCVD-based fibre production process can be divided into a number of main steps, as indicated in fig. 1. These individual process steps are described in the following chapters.

More recent developments such as: the synthetic quartz tubes process, the excellent type DLPC7 primary coating and the improved PMD are also described.



**Figure 7.5** Schematic of the PCVD process. *(Courtesy of Plasma Optical Fibre.)*

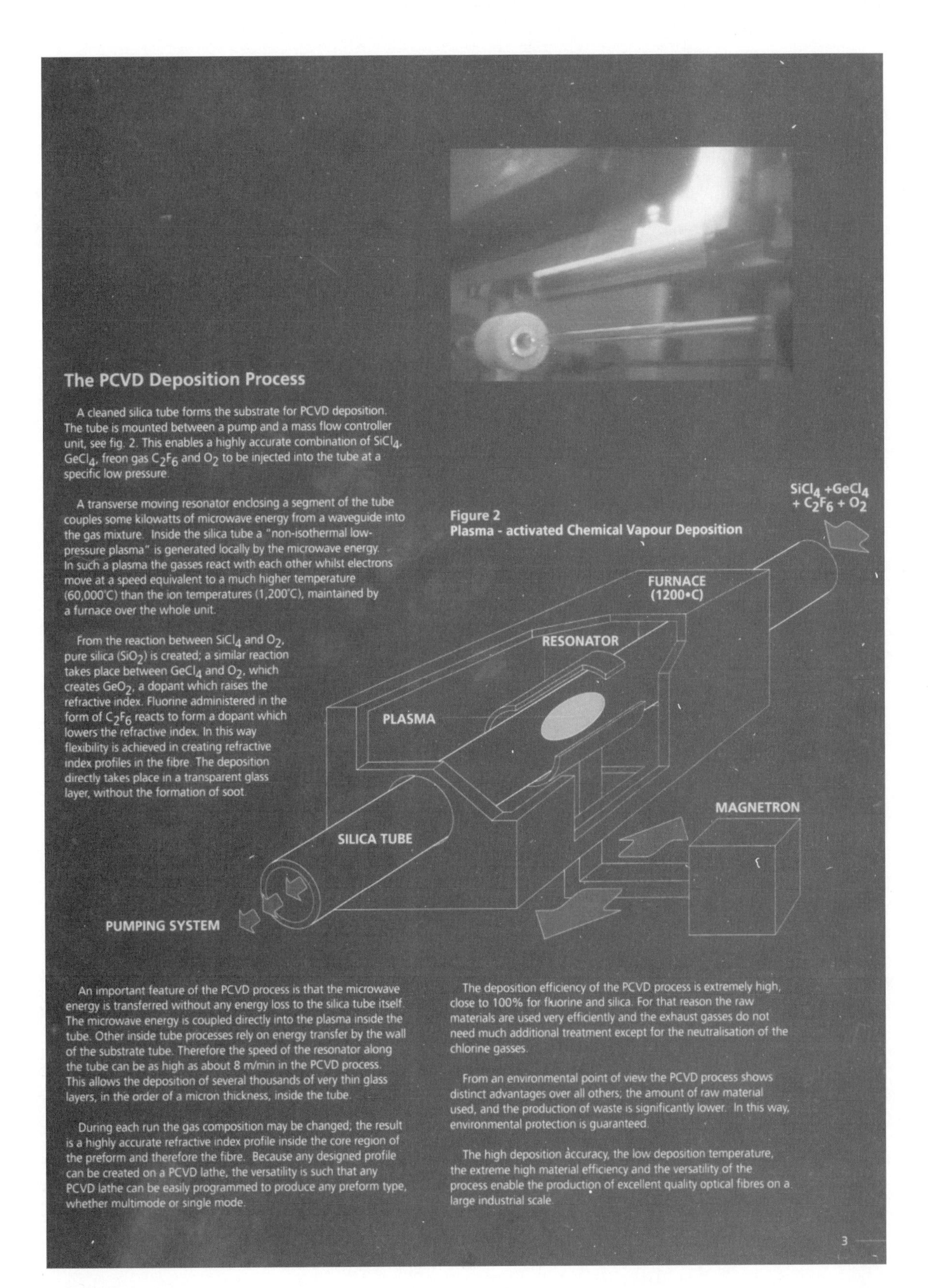

## The PCVD Deposition Process

A cleaned silica tube forms the substrate for PCVD deposition. The tube is mounted between a pump and a mass flow controller unit, see fig. 2. This enables a highly accurate combination of $SiCl_4$, $GeCl_4$, freon gas $C_2F_6$ and $O_2$ to be injected into the tube at a specific low pressure.

A transverse moving resonator enclosing a segment of the tube couples some kilowatts of microwave energy from a waveguide into the gas mixture. Inside the silica tube a "non-isothermal low-pressure plasma" is generated locally by the microwave energy. In such a plasma the gasses react with each other whilst electrons move at a speed equivalent to a much higher temperature (60,000°C) than the ion temperatures (1,200°C), maintained by a furnace over the whole unit.

From the reaction between $SiCl_4$ and $O_2$, pure silica ($SiO_2$) is created; a similar reaction takes place between $GeCl_4$ and $O_2$, which creates $GeO_2$, a dopant which raises the refractive index. Fluorine administered in the form of $C_2F_6$ reacts to form a dopant which lowers the refractive index. In this way flexibility is achieved in creating refractive index profiles in the fibre. The deposition directly takes place in a transparent glass layer, without the formation of soot.

**Figure 2**
**Plasma - activated Chemical Vapour Deposition**



An important feature of the PCVD process is that the microwave energy is transferred without any energy loss to the silica tube itself. The microwave energy is coupled directly into the plasma inside the tube. Other inside tube processes rely on energy transfer by the wall of the substrate tube. Therefore the speed of the resonator along the tube can be as high as about 8 m/min in the PCVD process. This allows the deposition of several thousands of very thin glass layers, in the order of a micron thickness, inside the tube.

During each run the gas composition may be changed; the result is a highly accurate refractive index profile inside the core region of the preform and therefore the fibre. Because any designed profile can be created on a PCVD lathe, the versatility is such that any PCVD lathe can be easily programmed to produce any preform type, whether multimode or single mode.

The deposition efficiency of the PCVD process is extremely high, close to 100% for fluorine and silica. For that reason the raw materials are used very efficiently and the exhaust gasses do not need much additional treatment except for the neutralisation of the chlorine gasses.

From an environmental point of view the PCVD process shows distinct advantages over all others; the amount of raw material used, and the production of waste is significantly lower. In this way, environmental protection is guaranteed.

The high deposition accuracy, the low deposition temperature, the extreme high material efficiency and the versatility of the process enable the production of excellent quality optical fibres on a large industrial scale.

3

**Figure 7.5** (continued)

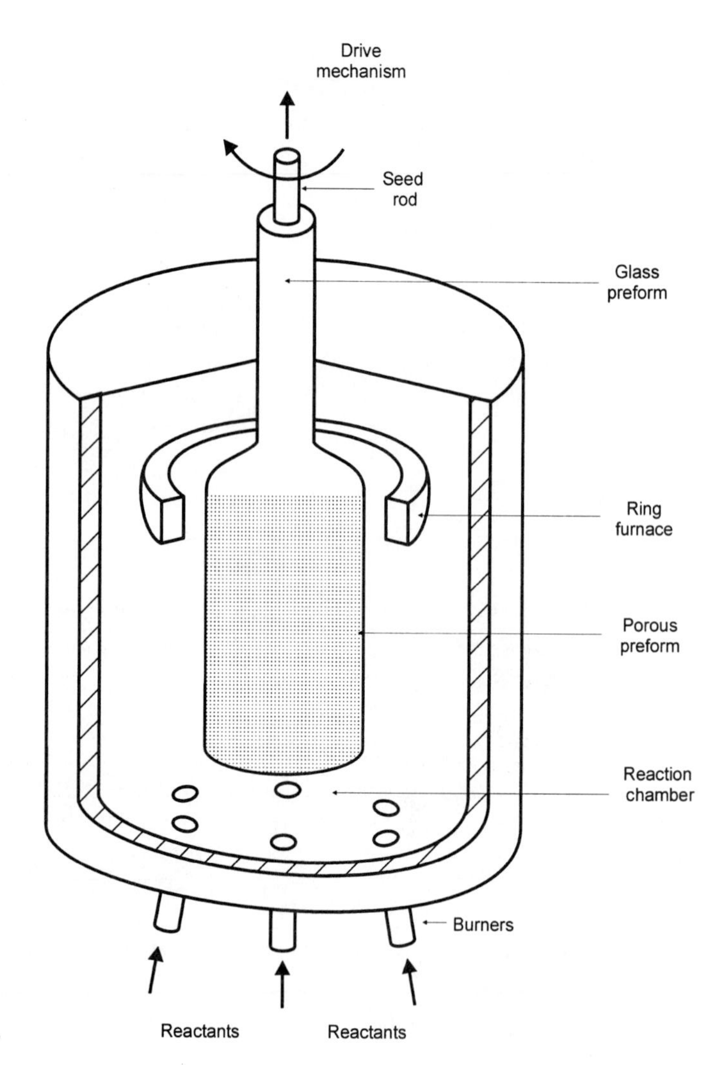


**Figure 7.6** Schematic of the VAD process.

form and vary the local preform temperature. The capacity of this preform is more than 100 km of fiber.

***Summing Up*** All these processes are fully automated and efficient in terms of the use of raw materials and other resources. Each achieves a high level of precision in obtaining the refractive-index profile and the required dimensional characteristics. Comprehensive descriptions and comparisons of all the fabrication methods are given in [4].

It should be noted that many methods were tried in the early days of fiber fabrication. Only these four survive and they continue to be improved.

## Coating

***Requirements*** As you know from the previous chapters, an optical fiber—a core surrounded by cladding—has to be coated. Figure 7.2 shows that manufacturers coat on-line. A liquid polymer is applied by the coating applicator as a fiber passes through the coating line. This liquid is solidified by heat or ultraviolet curing.

The main function of the coating is to protect the fiber from any external damage. But a closer look reveals that the coating has several other major functions:

- *Adhesion.* The coating, obviously, has to stick firmly to the glass surface of the fiber.
- *Ability to be stripped.* To connectorize a fiber, the coating has to be stripped. The stripping force has to be very small to facilitate handling of the fiber during installation. This force has to be stable in any environment—dry or wet. The range of the stripping force is between 1.4 N and 4.2 N. (See Figures 3.18 and 3.20.)

  If you think that the above two functions seem to conflict, you are right. It's a manufacturing problem optical fiber producers have to live with today. The conflict typifies the kinds of problems fiber makers have yet to resolve.
- *Toughness.* This quality is necessary to provide enhanced abrasion protection and to enable fiber handling and cabling without loss of strength. Toughness is gauged by elastic (Young) modulus testing, which determines whether a coating is soft or hard. For example, a soft coating may have a Young modulus of about 1.4 MPa; a Young modulus for a hard coating may be about 800 MPa.
- *Moisture resistance.* The coating is the sole line of defense protecting the fiber from moisture. This characteristic, in a sense, determines a fiber's aging and stability properties. Moisture resistance is determined by measuring the increase in attenuation during the fiber's exposure to water.
- *Glass transition temperature, Tg,* shows the low limit of the temperature at which induced attenuation is within a required margin. This characteristic may vary from 0°C to –55°C [5]. It shows how well the linear thermal-expansion coefficient of the coating complies with that of the fiber. Any significant incompatibility will result in stress put on the fiber by the coating. The consequences will be microbending and greater attenuation.

Many other parameters characterize coatings but these are the important ones to know. The point is simply this: *The coating is a critical component of an optical fiber. It determines bending sensitivity, abrasion resistance, static-fatigue protection, and many other important properties of the fiber.*

We've considered the role of the coating from the internal standpoint, in other words, how the coating protects the fiber. But there is another important consideration: how the coating works with a fiber cable. From this vantage, a coating should be smooth enough to be put inside a tight buffer and strong enough to protect the fiber from undergoing changes in its optical properties.

***Solutions*** To meet these complex requirements, manufacturers developed many coatings, which vary in design, materials used, curing processes, and so on. The overall point to bear in mind is this: The coating, along with its fiber, is designed for a specific application. For example, a tight buffer and a loose buffer apply different forces to the fiber, thus requiring different coating characteristics.

There are two main types of design: single-layer and double-layer coatings. In the single-layer design, the manufacturer tries to satisfy a wide range of requirements by choosing the proper material and coating thickness, among other characteristics. In the double-layer design, the inner layer is a soft coating. It provides good adhesion and cushions the fiber. The hard outer layer protects against an adverse environment, including abrasion.

The outer diameter of a coated optical fiber ranges from 245 μm to 900 μm.

Armed with this new knowledge, you will now have a better understanding of the manufacturers' data sheets.

## 7.2 FIBER-OPTIC CABLES

A bare fiber—a core with its cladding surrounded by a coating—is still very sensitive to the adverse environment regardless of how perfect the coating is. Thus, in practice, you will find a fiber only in some protective enclosure, which is called fiber-optic cable. Since optical fiber is the main transmission medium in modern telecommunications, the cabling business should be the most prolific sector of the fiber-optic communications industry. And it is, indeed. You can't begin to count the number of companies involved in this business.

This section introduces basic cable structure, types of cable in use, and other information necessary for reading manufacturers' data sheets. Since this book considers only fiber-optic technology, the simple term *cable* refers only to fiber-optic cable.

### Cables

***Basic structure*** Fiber-optic cable is the protective enclosure surrounding a bare fiber. Its function is to protect the fiber from any possible damage. The process of putting a fiber into such an enclosure is called *cabling.*

Imagine that you had to design cable to be laid on the ocean floor while a colleague had to design cable for installation inside several rooms your company occupies. Would there be a difference between these two cables? Indeed, there would be—a huge difference. Thus, the application determines the cable design. Fortunately, we can distinguish among four basic elements that almost every cable includes. (See Figure 7.7.)

First of all, a bare fiber is put into a *plastic buffer tube* (often called, simply, a *buffer*). This is the first shield protecting the fiber from environmental damage. Either many fibers or just one can be put into one buffer.

The second structural element of a cable is the *strength member.* Its function is to release the fiber from mechanical stress during installation and operation. A list of strength members includes, but isn't restricted to, flexible aramid yarns (such as Kevlar), flexible fiberglass roving, fiberglass rod, metal wire, and metal rope made from twisted steel wires.

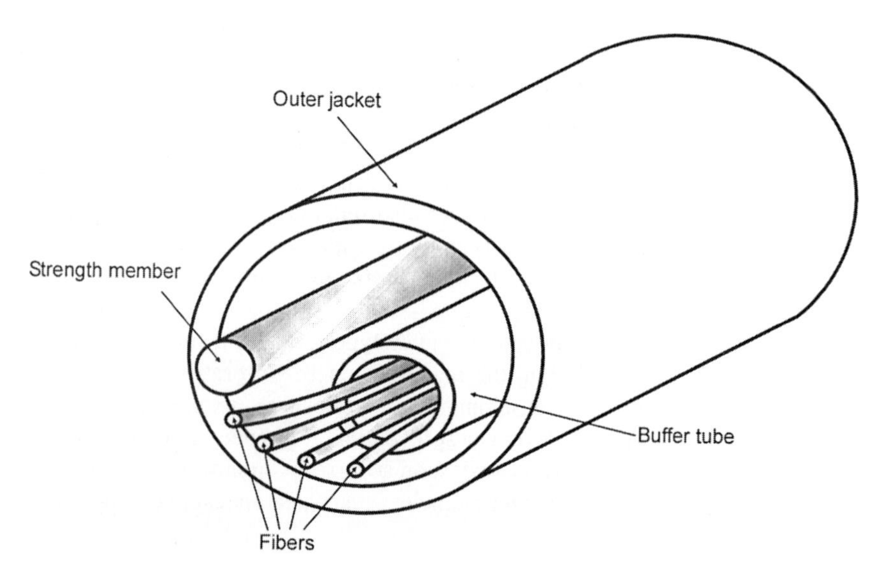


**Figure 7.7** Basic structure of a fiber-optic cable.

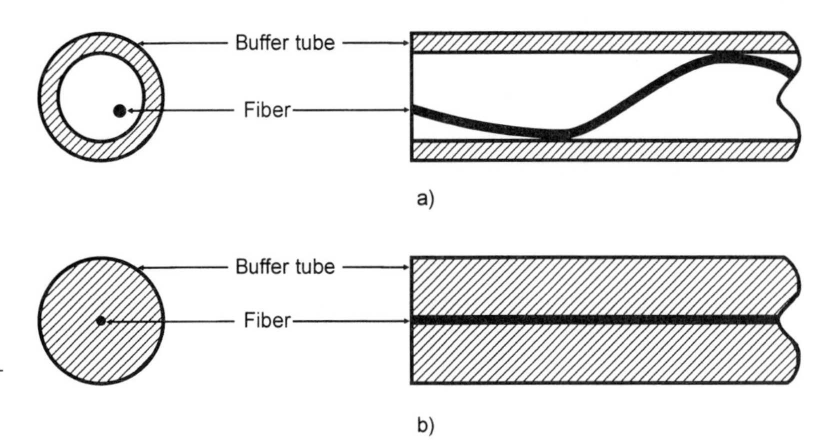


**Figure 7.8** Cross sections of loose- and tight-buffer designs: (a) Loose buffer; (b) tight buffer.

The third basic element is an outer jacket that assembles the entire construction and shields it from an adverse environment. Thus a fiber, a buffer, a strength element, and an outer jacket are the components needed to make up a fiber-optic cable.

All other cable elements are optional and are involved only so far as a specific cable application requires it.

***Loose buffer and tight buffer*** There are two major types of buffers: loose and tight. Cross sections of both types are shown in Figure 7.8.

A *loose* buffer's inner diameter is much larger than a fiber's outer (coating) diameter, as Figure 7.8(a) shows. Two major advantages stem from this design: perfect fiber isolation from mechanical forces (within some given range, of course) and protection from moisture. The first advantage comes from a so-called mechanical dead zone: A force imposed on a buffer does not affect the fiber until this force becomes large enough to straighten the fiber inside the buffer. A loose buffer can be easily filled with a water-blocking gel, which provides its second advantage. In addition, a loose buffer can accommodate several fibers, reducing the cost of the cable. On the other hand, this type of cable cannot be installed vertically (can you explain why?) and its end preparation for connectorization (splicing and termination) is labor-intensive. Consequently, the loose-buffer type of cable is used mostly in outdoor installations because it provides stable and reliable transmission over a wide range of temperatures, mechanical stress, and other environmental conditions.

A *tight* buffer's inner diameter is equal to the fiber's coating diameter, as Figure 7.8(b) shows. Its primary advantage is its ability to keep cable operational despite a break in the fiber. Since a buffer holds a fiber firmly, a small separation of the fiber ends won't interrupt the service completely, although it will definitely degrade signal quality. This is why the military was the first customer—and still is the largest—for this type of fiber cable [6]. A tight buffer is rugged, allowing a smaller bend radius. Since each buffer contains only one fiber and there is no gel to be removed, it's easy to prepare this cable for connectorization. Cables having a tight buffer can be installed vertically. In general, tight-buffer cables are more sensitive to temperature and mechanical and water impacts than loose-buffer cables; hence, they are recommended mostly for indoor applications. On the other hand, tight-buffer cables designed for special applications (such as military and undersea) are the strongest cables available.

***Examples of fiber-optic cable designs*** Let's take several examples of commercially available fiber-optic cables to see what components a cable might include and what cable designs you might meet in the field. (See the subsection "Reading Data Sheets.")

Figure 7.9(a) shows loose-tube Mini Bundle cable. A polyethylene (PE) outer sheath serves as the outer jacket—one of the four mandatory components discussed above. A corrugated steel armor tape protects the cable from rodents. It also provides good protection against any environmental hazard; its presence indicates, too, that the cable can be installed outdoors. The PE inner sheath isolates the steel tape from the cable's core. The dielectric-strength member here is made in the form of a tube, which absorbs all the outside mechanical stress and tension. A ripcord is used to break a strength member during the end-preparation work. All the space inside a dielectric-strength member is filled with water-blocking material. This cable includes six loose buffer tubes, each, in turn, filled with a water-blocking gel. Each tube includes six fibers; hence, this cable carries 36 fibers. A dielectric central member supports all the composites of the buffer tubes and provides extra strength to the entire cable.

Figure 7.9(b) shows a Figure-8 Mini Bundle cable. It has two separate units, which in cross-section, looks like the digit 8; thus, the name of the cable. The upper unit, the strength member of this cable, includes seven steel messengers, which release the lower unit from a heavy axial load. Its form indicates the cable is used in aerial installations. The lower unit is almost the same as the loose-tube cable discussed above. Corrugated steel armor tape is necessary to protect the fibers from squirrel and bird attacks.

MIC cables, shown in Figure 7.9(c), are examples of tight-buffer design. The uniqueness of the 4-fiber MIC cable is the design of the dielectric strength member: Four tight-buffered fibers are immersed into this member. This style is called *premise.* TBII stands for tight buffer design II, which is distinct from other designs by its low stripping force. Up to 48 fibers can be carried by this type of cable used for indoor applications.

FREEDM Mini Bundle ribbon cable (see Figure 7.9[d]) is a cable design using fiber ribbons rather than individual fibers as its basic unit. (Fiber ribbon cable is a flat tape on which several discrete fibers are arranged. This is very similar to a ribbon electric cable, which should be familiar to anyone who ever saw any type of electronic equipment—a computer, for example). Each ribbon here includes 12 fibers and each buffer tube carries one stack of six ribbons; the entire cable can carry up to six tubes, so that there can be a total of 432 fibers. Pay special attention to the outer sheath. It is flame retardant and has an ultraviolet-resistant jacket. Since ribbons don't allow the use of a water-blocking filling, a water-swellable tape is used. Dielectric rods serve to separate the buffer tubes and strengthen the cable. The beauty of this design is that it allows the manufacturer to make this cable suitable for both indoor and outdoor applications [7].

Figure 7.9(e) shows *fan-out,* also called *breakout, cable.* The key feature of this style is that each buffer tube is surrounded by its own strength element to increase the cable's ruggedness. Look at the column headed "Maximum Tensile Load" under "Mechanical Specifications" and compare that with the similar column for the MIC cable. Note that the allowed tensile load for this cable is much higher than it is for the MIC. The difference between long-term and short-term loads underscores the cable's ability to elongate without losing its mechanical and transmission characteristics. Now compare the data for "Minimum Bend Radius" for those two cables and observe the difference.

A *zipcord* cable is shown in Figure 7.9(f). Compare that with 2-fiber fan-out cable in Figure 7.9(e). A zipcord cable is used for short connections indoors.

In addition, there is *slotted* cable, where each buffer tube is placed in its own slot stranded around the central member.

***Cable classification*** No cable classification is accepted as standard by the entire industry. However, many terms are equally understood by different vendors and customers. Let's review these terms:

To begin with, all cable is divided into three types: *indoor, indoor/outdoor,* and *outdoor.* Indoor cable experiences less temperature and mechanical stress but has to be fire retardant, emit

LANscape™

# Loose Tube Mini Bundle® Cables

*Loose Tube Mini Bundle Cables*

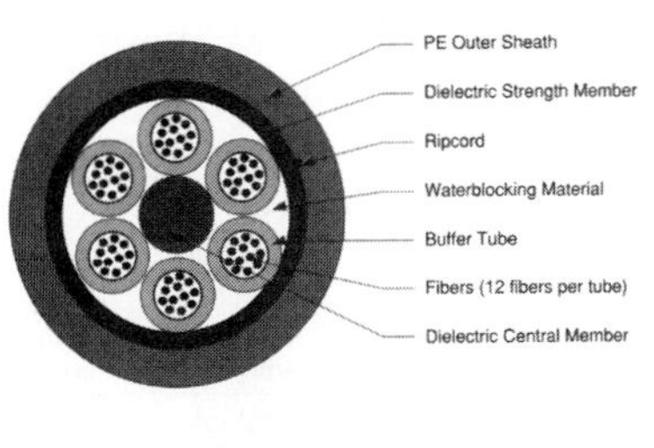


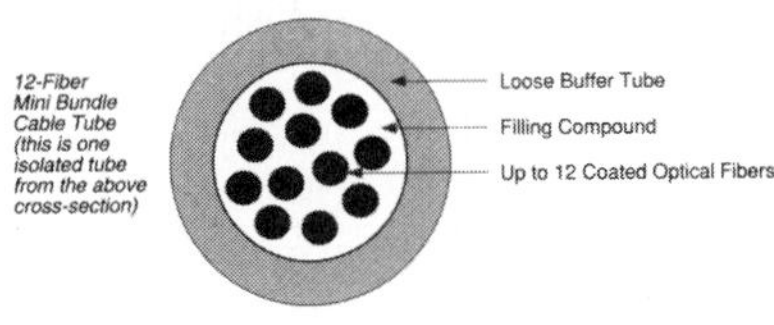


*72-Fiber Mini Bundle Cable Cross-Section*

## Description

Siecor's loose tube Mini Bundle® cables are for general purpose outdoor and indoor use. The loose tube cable construction, which Siecor pioneered, places varying numbers of fibers in each buffer tube. The loose tube design provides stable and highly reliable transmission parameters for a variety of voice, data, video, and imaging applications. The design also provides a high density of fibers within a given cable diameter while allowing flexibility to suit many system designs. These cables are suitable for outdoor duct, aerial, and direct buried installations, and for indoor use when installed in accordance with NEC Article 770.

## Features / Benefits

- Different fiber types available within a cable (hybrid construction)
- Lowest losses at long distances
- Loose tube cable with multimode fiber available in up to 8 km lengths
- Color-coded fibers and buffer tubes for quick and easy identification during installation
- Available with dispersion-shifted single-mode, single-mode, 50/125 µm, 62.5/125 µm, and 100/140 µm multimode fiber
- All-dielectric or steel central member
- Easily spliced using a variety of proven methods
- Ripcords for easy sheath removal

## Applications

- Interbuilding backbone
- For use in duct, aerial, and direct buried applications
- Wide range of fiber counts (up to 288)

*To order, please call a Siecor Authorized Distributor, or call Siecor at 1-800-743-2671.*

SIECOR Chapter 1: Fiber Optic Cables

Photo TOPSS1063
Drawing ZA-416

1.3

**Figure 7.9 (a)** Examples of fiber-optic cables. *(Courtesy of Siecor, Hickory, N.C.)*

Chapter 1: Fiber Optic Cables

### *Cable: Loose Tube*
### *Application: Duct/Aerial Lightning Resistant*
### *Construction: Dielectric Central Member/Non-Armored*

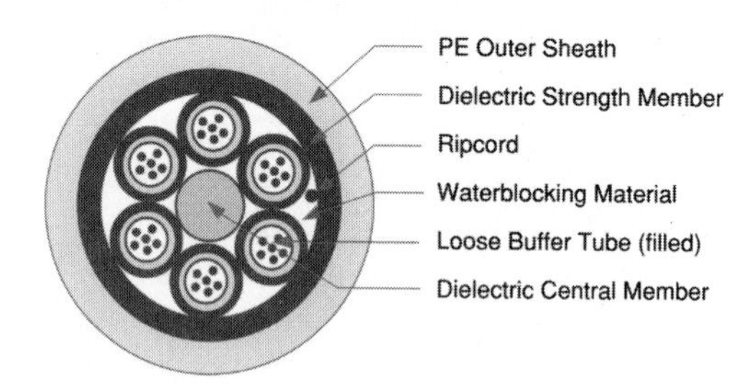


#### *Mechanical Specifications*

| | | |
|---|---|---|
| Maximum Tensile Loading | Installation:<br>Long Term Installed: | 2700 N[1] (600 lb)<br>890 N[2] (200 lb) |
| Operating Temperature | Storage:<br>Long Term: | -40° C to +70° C[1] (-40° F to +158° F)<br>-40° C to +70° C (-40° F to +158° F) |

| Fiber Count | Maximum Fibers Per Tube | Number of Tube Positions | Number of Active Tubes | Central Member | Nominal Weight kg/km (lb/1000 ft) | Nominal Outer Diameter mm (in) | Minimum Bend Radius Loaded cm (in) | Minimum Bend Radius Installed cm (in) |
|---|---|---|---|---|---|---|---|---|
| 2-4 | 2 | 5 | 1-2 | Dielectric | 84 (56) | 10.1 (0.40)[3] | 20.0 (7.9) | 15.0 (5.9) |
| 5-30 | 6 | 5 | 1-5 | Dielectric | 106 (71) | 11.3 (0.45)[3] | 20.0 (7.9) | 15.0 (5.9) |
| 31-36 | 6 | 6 | 6 | Dielectric | 114 (77) | 11.8 (0.46)[3] | 20.0 (7.9) | 15.0 (5.9) |
| 37-60 | 12 | 5 | 4-5 | Dielectric | 122 (82) | 12.2 (0.48)[3] | 20.0 (7.9) | 15.0 (5.9) |
| 61-72 | 12 | 6 | 6 | Dielectric | 138 (93) | 13.0 (0.51)[3] | 20.0 (7.9) | 15.0 (5.9) |
| 73-84 | 12 | 7 | 7 | Dielectric | 157 (106) | 13.9 (0.55)[4] | 20.0 (7.9) | 15.0 (5.9) |
| 85-96 | 12 | 8 | 8 | Dielectric | 178 (119) | 14.8 (0.58)[4] | 20.0 (7.9) | 15.0 (5.9) |
| 97-108 | 12 | 9 | 9 | Dielectric | 200 (134) | 15.7 (0.62)[4] | 25.0 (9.8) | 20.0 (7.9) |
| 109-120 | 12 | 10 | 10 | Dielectric | 225 (151) | 16.7 (0.66)[4] | 25.0 (9.8) | 20.0 (7.9) |
| 121-192 | 12 | 16 | 11-16 (High-Density) | Dielectric | 258 (173) | 18.0 (0.71)[4] | 25.0 (9.8) | 20.0 (7.9) |
| 193-216 | 12 | 18 | 17-18 (High-Density) | Dielectric | 280 (188) | 18.6 (0.73)[4] | 25.0 (9.8) | 20.0 (7.9) |
| 217-240 | 12 | 20 | 19-20 (High-Density) | Dielectric | 316 (212) | 19.6 (0.77)[4] | 25.0 (9.8) | 20.0 (7.9) |
| 241-264 | 12 | 22 | 21-22 (High-Density) | Dielectric | 347 (233) | 20.6 (0.81)[4] | 25.0 (9.8) | 20.0 (7.9) |
| 265-288 | 12 | 24 | 23-24 (High-Density) | Dielectric | 366 (246) | 21.6 (0.85)[4] | 25.0 (9.8) | 20.0 (7.9) |

*Notes:*
*[1] No irreversible change in attenuation.*
*[2] No measurable change in attenuation.*
*[3] Diameter represents an average figure and may vary by ± 10%.*
*[4] Diameter represents an average figure and may vary by ± 5%.*

*Drawing ZA-323*

1.4

**Figure 7.9 (a)** (continued)

## *Ordering Information (continued)*

| ❽ Select performance code for fiber type chosen. (Digits 10-11) | | | | | |
|---|---|---|---|---|---|
| **50/125 µm 850/1300 nm C in Digit 4** | **Attenuation[Note 1] (dB/km)** | **Bandwidth (MHz•km)[Note 2]** | | | |
| | | **400/400** | **600/600** | **600/800** | **400/1000** |
| | 2.7/1.0 | 01 | 02 | 20 | 14 |
| | 3.0/1.4 | 15 | 16 | 19 | 18 |
| **62.5/125 µm 850/1300 nm K in Digit 4** | **Attenuation[Note 1] (dB/km)** | **Bandwidth (MHz•km)[Note 2]** | | | |
| | | **160/500** | **160/600** | **200/500** | **200/600** |
| | 3.5/1.0 | 10 | 11 | AZ* | 13 |
| **100/140 µm 850/1300 nm N in Digit 4** | **Attenuation[Note 1] (dB/km)** | **Bandwidth (MHz•km)[Note 2]** | | | |
| | | **100/100** | | | |
| | 5.5/4.5 | 07 | | | |
| **Single-mode 1310/1550 nm R in Digit 4** | **Attenuation[Note 1] (dB/km)** | | | | |
| | .35/.25 | .4/.25 | .4/.3 | .4/.4 | .5/.4 |
| | 00 | 06 | 01 | 02 | 03 |

*Notes: [1]Attenuation specifications are maximum values for all fibers in a cable.*
*[2]Bandwidth specifications are minimum values for all fibers in a cable.*

**OptiNet performance.*

## *Product Ordering Example*

**Part Number Example:** 0 2 4 K 1 4 - 1 4 1 1 0 - 20 (❶ 024, ❷ K, ❸ 4, ❹ 10)

❶ 024 = 24 fibers

❷ K = 62.5/125 µm dual window fiber

❸ 4 = Dielectric central member

❹ 10 = Optical performance: Attenuation (dB/km): 3.5/1.0
Bandwidth (MHz•km): 160/500
@ Wavelength (nm): 850/1300

*To order, please call a Siecor Authorized Distributor, or call Siecor at 1-800-743-2671.*



**Figure 7.9 (a)** (continued)

SIECOR

Chapter 1: Fiber Optic Cables

# Figure-8 Mini Bundle® Cable

## Description

Siecor's Figure-8 Mini Bundle optical cable is a self-supporting aerial cable that is designed for easy and economical one-step installation. The Figure-8 optical cable design consists of Siecor's Mini Bundle fiber optic cable core integrated with a 1/4-inch EHS stranded steel messenger, and is jacketed with polyethylene extruded in a Figure-8 shape.

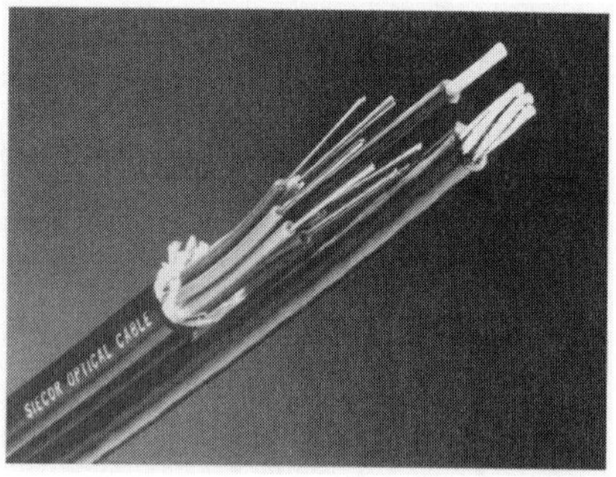

*Figure-8 Mini Bundle Cable*

## Features / Benefits

- One-step process provides ease-of-installation and savings on installation costs
- Uses standard Figure-8 cable hardware and installation methods
- Can be installed on pole spans exceeding 500 feet; cable installation and sag/tension guidelines are available
- Available in both single-mode and multimode fibers in a wide range of fiber counts (2 to 216 fibers)
- Available in armored and non-armored optical cable options
- Cable lengths available up to 6 km for multimode and single-mode cables
- Custom reel lengths also available
- Mini Bundle cable design isolates optical fibers from the rigors of installation
- Optical cable has ripcords to simplify cable stripping
- Optical cable contains yarn strength members for strain-relief in splice hardware

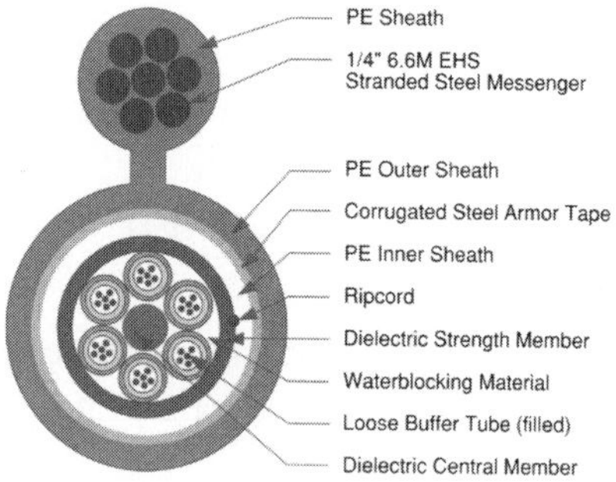

*36-Fiber Armored Figure-8 Mini Bundle Cable Cross-Section*

## Applications

- Interbuilding backbone
- For use in aerial applications

## Typical Sag and Tensions Information[1,2,3]

| Fiber Count | Cable Type | Span 150 ft Tension 800 lb | Span 250 ft Tension 1000 lb | Span 350 ft Tension 1200 lb |
|---|---|---|---|---|
| 2-60 | Armored | 1.6 | 2.3 | 3.7 |
| | Non-Armored | 0.8 | 1.8 | 2.9 |
| 133-144 | Armored | 1.6 | 3.4 | 5.6 |
| | Non-Armored | 1.2 | 2.7 | 4.4 |

[1] *Installation tensions at 59° F (15° C).*
[2] *Sag values are in feet and reflect installation sag.*
[3] *Sag and tension values are available for other pole spans, tensions, and fiber counts.*

*Photo TOPGS1061*
*Drawing ZA-407*

1.10

**Figure 7.9 (b)**

LANscape

# Figure-8 Mini Bundle® Cable

## Mechanical Specifications

| Operating Temperature | Storage:<br>Long Term: | -40° C to +70° C (-40° F to +158° F)<br>-40° C to +70° C (-40° F to +158° F) |
|---|---|---|

| Fiber Count | Maximum Fibers Per Tube | Number of Tube Positions | Number of Active Tubes | Cable Type | Nominal Weight kg/km (lb/1000 ft) | Nominal Diameter[1] mm (in) | Minimum Bend Radius cm (in) | Nominal Cable Height[1] mm (in) |
|---|---|---|---|---|---|---|---|---|
| 2-60 | 12 | 5 | 5 | Armored<br>Non-Armored | 438 (294)<br>344 (231) | 14.5 (.57)<br>11.9 (.47) | 22.5 (8.9)<br>22.5 (8.9) | 26.1 (1.03)<br>23.5 (0.93) |
| 61-72 | 12 | 6 | 6 | Armored<br>Non-Armored | 462 (311)<br>355 (239) | 15.2 (.60)<br>12.6 (.50) | 25.0 (9.9)<br>22.5 (8.9) | 26.9 (1.06)<br>24.2 (0.95) |
| 73-84 | 12 | 7 | 7 | Armored<br>Non-Armored | 490 (329)<br>376 (253) | 16.2 (.64)<br>13.6 (.55) | 25.0 (9.9)<br>22.5 (8.9) | 27.8 (1.1)<br>25.3 (1.0) |
| 85-96 | 12 | 8 | 8 | Armored<br>Non-Armored | 520 (349)<br>399 (268) | 17.2 (.68)<br>14.6 (.58) | 25.0 (9.9)<br>22.5 (8.9) | 28.8 (1.13)<br>26.2 (1.03) |
| 97-108 | 12 | 9 | 9 | Armored<br>Non-Armored | 551 (370)<br>422 (284) | 18.2 (.72)<br>15.6 (.61) | 25.0 (9.9)<br>25.0 (9.9) | 29.8 (1.17)<br>27.2 (1.07) |
| 109-120 | 12 | 10 | 10 | Armored<br>Non-Armored | 583 (392)<br>446 (300) | 19.2 (.76)<br>16.6 (.65) | 25.0 (9.9)<br>25.0 (9.9) | 30.8 (1.21)<br>28.2 (1.11) |
| 121-192 | 12 | 16 | 11-16<br>(High-Density) | Armored<br>Non-Armored | 623 (419)<br>475 (319) | 22.2 (.87)<br>18.5 (.73) | 25.0 (9.9)<br>25.0 (9.9) | 31.8 (1.25)<br>29.3 (1.15) |
| 133-144 | 12 | 12 | 12 | Armored<br>Non-Armored | 652 (438)<br>500 (336) | 20.3 (.80)<br>18.7 (.73) | 25.0 (9.9)<br>25.0 (9.9) | 31.9 (1.25)<br>30.3 (1.19) |
| 193-216 | 12 | 18 | 17-18<br>(High-Density) | Armored<br>Non-Armored | 657 (442)<br>507 (341) | 21.4 (.84)<br>19.8 (.76) | 25.0 (9.9)<br>25.0 (9.9) | 33.0 (1.30)<br>31.4 (1.23) |

[1] *Cable diameter and height represent average values and may vary by ± 5%.*

## Transmission Performance

| Performance Option Code* | 03<br>Single-mode<br>(1310/1550 nm) | 15<br>50/125 μm<br>(850/1300 nm) | 41<br>62.5/125 μm<br>(850/1300 nm) |
|---|---|---|---|
| Maximum Attenuation (dB/km) | 0.5/0.4 | 3.0/1.4 | 3.75/1.5 |
| Minimum Bandwidth (MHz•km) | — | 400/400 | 160/500 |

**Other fiber performances available upon request.*

## Ordering Information

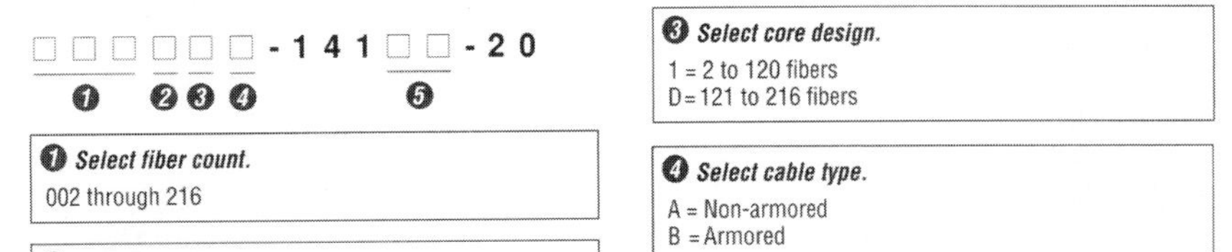


**❶ Select fiber count.**
002 through 216

**❷ Select fiber type.**
C = 50/125 μm
K = 62.5/125 μm
R = Single-mode

**❸ Select core design.**
1 = 2 to 120 fibers
D = 121 to 216 fibers

**❹ Select cable type.**
A = Non-armored
B = Armored

**❺ Select performance option code.**
03 = Single-mode
15 = 50/125 μm
41 = 62.5/125 μm

*To order, please call a Siecor Authorized Distributor, or call Siecor at 1-800-743-2671.*



**Figure 7.9 (b)** (continued)

SIECOR

Chapter 1: Fiber Optic Cables

# MIC® Cables

## Description

Siecor's MIC® cables are rugged, high-performance optical communications cables with a proven track record, designed for various indoor and outdoor requirements, including routing between buildings within ducts, inside buildings up riser shafts, in plenum spaces, and fiber-to-the-desk.

Available in a wide range of fiber counts, this cable family provides the necessary bandwidth capacity to transport all voice, data, video, and imaging signals normally required in today's evolving office and factory environment. Also, the cable's small diameter, light weight, and flexibility allow for easy installation, maintenance, and administration.

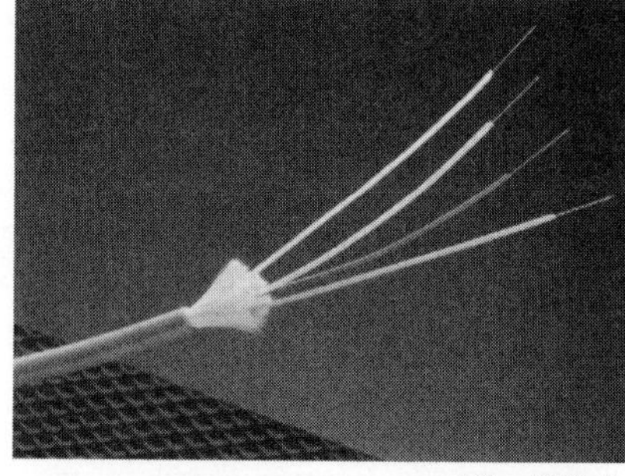

*4-Fiber MIC Cable*

## Features / Benefits

- Uses TBII buffered fiber
- Longer strip lengths
- Small diameter and bend radius allow easy installation in space-constrained areas
- Available with single-mode, 50/125 μm, 62.5/125 μm and 100/140 μm multimode fiber types; any combination of fiber types can be placed in a single cable
- Color-coded fibers for easy identification in accordance with EIA/TIA-598
- High-fiber count cables grouped in units for configuration control, ease-of-installation, and quick repair

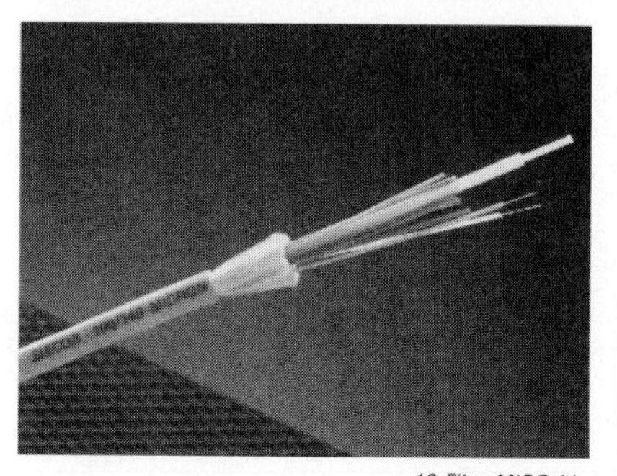

*12-Fiber MIC Cable*

## Features / Benefits

- Interbuilding backbone in conduits below the frost line
- Intrabuilding backbone
- Horizontal cabling
- UL and CSA listed; OFNR/FT4 for riser applications, and OFNP/FT6 for plenum applications

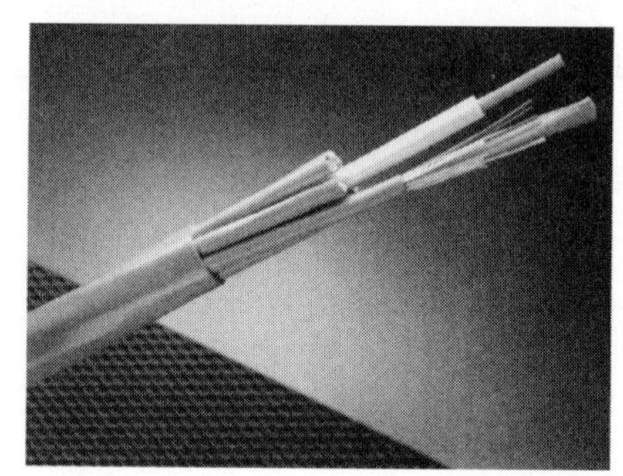

*High-Fiber Unitized MIC Cable*

Photo TOPSS411
TOPSS410
TOPSS409

1.16

**Figure 7.9 (c)**

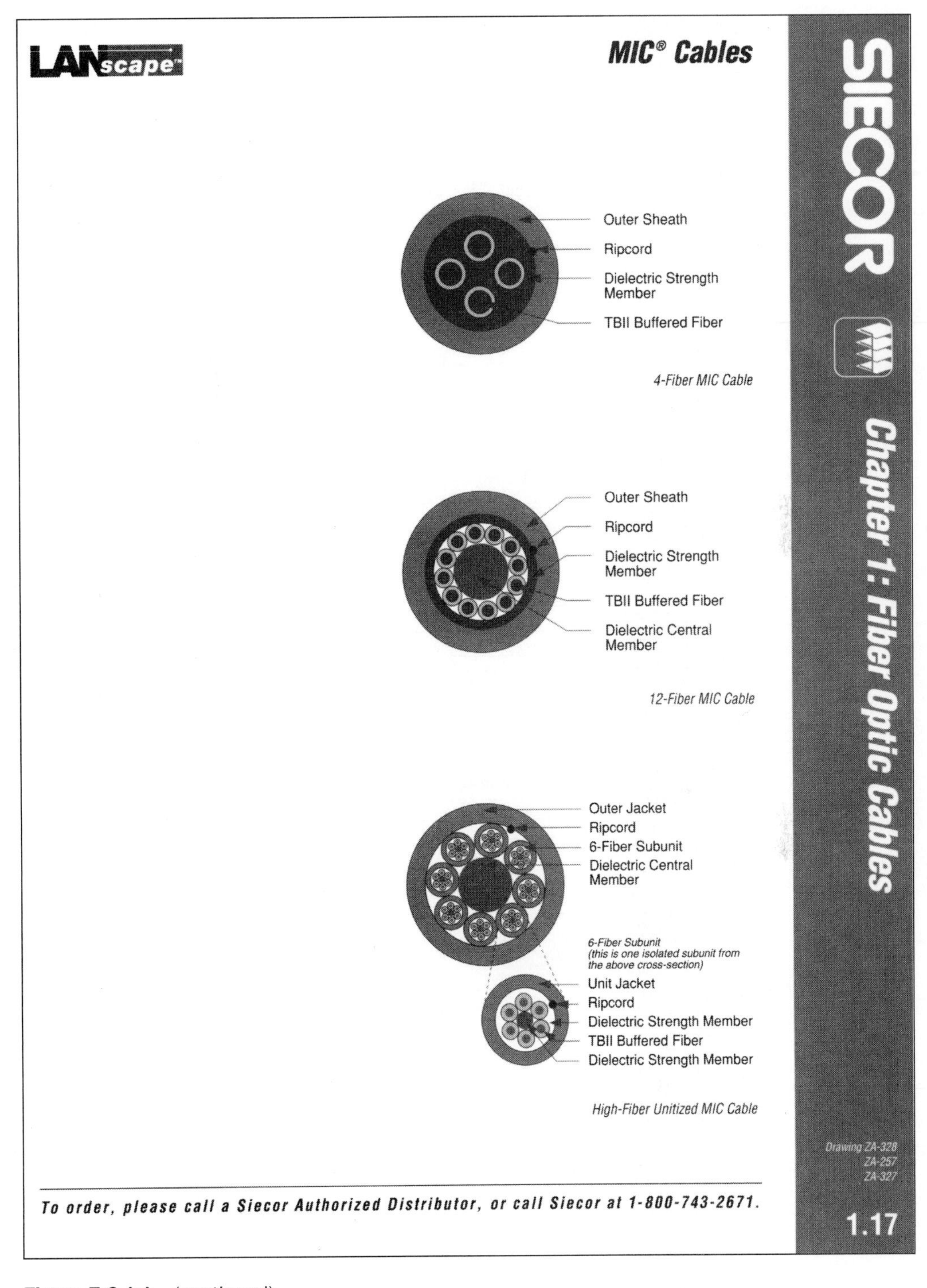


**Figure 7.9 (c)** (continued)

SIECOR

Chapter 1: Fiber Optic Cables

## MIC® Cables

### Mechanical Specifications

| | |
|---|---|
| Storage Temperature | -40° C to +70° C |
| Operating Temperature | -20° C to +70° C |
| NEC/CSA Listing | OFNR/FT4 |

### Up to 24-Fiber Single-Layer Riser

| Fiber Count | Nominal Outer Diameter mm | (in) | Nominal Weight kg/km | (lb/1000 ft) | Maximum Tensile Load Short Term N | (lb) | Long Term N | (lb) | Minimum Bend Radius Loaded cm | (in) | Installed cm | (in) |
|---|---|---|---|---|---|---|---|---|---|---|---|---|
| 2 | 4.7 | (0.19) | 19 | (13) | 800 | (180) | 200 | (45) | 7.0 | (2.8) | 4.7 | (1.9) |
| 4 | 5.2 | (0.20) | 24 | (16) | 1000 | (225) | 300 | (67) | 7.8 | (3.1) | 5.2 | (2.1) |
| 6 | 5.3 | (0.21) | 21 | (14) | 1000 | (225) | 300 | (67) | 8.0 | (3.2) | 5.3 | (2.1) |
| 8 | 6.2 | (0.24) | 37 | (25) | 1400 | (315) | 440 | (99) | 9.3 | (3.7) | 6.2 | (2.4) |
| 10 | 6.5 | (0.26) | 43 | (29) | 1400 | (315) | 440 | (99) | 9.8 | (3.9) | 6.5 | (2.6) |
| 12 | 7.0 | (0.28) | 50 | (34) | 1800 | (405) | 600 | (135) | 10.5 | (4.1) | 7.0 | (2.8) |
| 14 | 7.7 | (0.30) | 59 | (40) | 1800 | (405) | 600 | (135) | 11.5 | (4.5) | 7.7 | (3.0) |
| 16 | 8.3 | (0.33) | 69 | (47) | 2000 | (450) | 700 | (157) | 12.5 | (4.9) | 8.3 | (3.3) |
| 18 | 9.1 | (0.36) | 82 | (55) | 2000 | (450) | 700 | (157) | 13.6 | (5.4) | 9.1 | (3.6) |
| 20 | 9.7 | (0.38) | 95 | (64) | 2000 | (450) | 700 | (157) | 15.7 | (6.2) | 10.5 | (4.1) |
| 22 | 10.2 | (0.40) | 106 | (72) | 2700 | (607) | 1000 | (225) | 15.7 | (6.2) | 10.5 | (4.1) |
| 24 | 10.6 | (0.42) | 117 | (79) | 2700 | (607) | 1000 | (225) | 15.9 | (6.3) | 10.6 | (4.2) |

### 24 to 144 Fibers Unitized Riser

| Fiber Count | Nominal Outer Diameter mm | (in) | Nominal Weight kg/km | (lb/1000 ft) | Maximum Tensile Load Short Term N | (lb) | Long Term N | (lb) | Minimum Bend Radius Loaded cm | (in) | Installed cm | (in) |
|---|---|---|---|---|---|---|---|---|---|---|---|---|
| 24 | 12.2 | (0.5) | 120 | (81) | 2500 | (563) | 1000 | (225) | 18.3 | (7.2) | 12.2 | (4.8) |
| 30 | 13.6 | (0.5) | 151 | (101) | 3500 | (788) | 1700 | (383) | 20.4 | (8.0) | 13.6 | (5.4) |
| 36 | 15.2 | (0.6) | 188 | (126) | 4000 | (900) | 2000 | (450) | 22.8 | (9.0) | 15.2 | (6.0) |
| 42 | 16.5 | (0.6) | 225 | (151) | 4500 | (1013) | 2200 | (495) | 24.8 | (9.8) | 16.5 | (6.5) |
| 48 | 17.8 | (0.7) | 267 | (179) | 5000 | (1125) | 2500 | (563) | 26.9 | (10.6) | 17.9 | (7.1) |
| 60 | 21.1 | (0.8) | 388 | (261) | 5500 | (1238) | 3000 | (675) | 31.7 | (12.5) | 21.1 | (8.3) |
| 72 | 20.3 | (0.8) | 305 | (205) | 5500 | (1238) | 3000 | (675) | 30.5 | (12.0) | 20.3 | (8.0) |
| 84 | 21.7 | (0.9) | 347 | (233) | 7000 | (1575) | 3500 | (788) | 32.6 | (12.8) | 21.7 | (8.5) |
| 96 | 22.7 | (0.9) | 394 | (265) | 8800 | (1980) | 4000 | (900) | 34.1 | (13.4) | 22.7 | (8.9) |
| 108 | 24.1 | (0.9) | 449 | (302) | 8800 | (1980) | 4000 | (900) | 36.2 | (14.2) | 24.1 | (9.5) |
| 120 | 25.8 | (1.0) | 528 | (355) | 10000 | (2250) | 4000 | (900) | 38.7 | (15.2) | 25.8 | (10.2) |
| 144 | 28.5 | (1.1) | 655 | (440) | 10000 | (2250) | 4000 | (900) | 42.8 | (16.8) | 28.5 | (11.2) |

*Siecor MIC cables are designed and tested to ICEA S-83-596 Standard for Fiber Optic Premises Distribution Cable per requirements of ANSI/TIA/EIA-568A Commercial Building Telecommunications Cabling Standard.*



1.18

**Figure 7.9 (c)** (continued)

LANscape

## MIC® Cables

SIECOR



### Mechanical Specifications

| | |
|---|---|
| Storage Temperature | -40° C to +70° C |
| Operating Temperature | 0° C to +70° C |
| NEC/CSA Listing | OFNP/FT6 |

### Up to 24-Fiber Flexible Plenum

| Fiber Count | Nominal Outer Diameter mm | (in) | Nominal Weight kg/km | (lb/1000 ft) | Maximum Tensile Load Short Term N | (lb) | Long Term N | (lb) | Minimum Bend Radius Loaded cm | (in) | Installed cm | (in) |
|---|---|---|---|---|---|---|---|---|---|---|---|---|
| 2 | 4.7 | (0.19) | 20 | (13.4) | 440 | (99) | 110 | (25) | 7.1 | (2.8) | 4.7 | (1.9) |
| 4 | 5.0 | (0.20) | 25 | (16.8) | 660 | (148) | 165 | (37) | 7.5 | (3.0) | 5.0 | (2.0) |
| 6 | 5.1 | (0.20) | 28 | (18.8) | 660 | (148) | 165 | (37) | 7.7 | (3.0) | 5.1 | (2.0) |
| 8 | 5.8 | (0.23) | 35 | (23.5) | 660 | (148) | 165 | (37) | 8.7 | (3.4) | 5.8 | (2.3) |
| 10 | 6.1 | (0.24) | 41 | (27.5) | 660 | (148) | 165 | (37) | 9.2 | (3.6) | 6.1 | (2.4) |
| 12 | 6.4 | (0.25) | 39 | (26.2) | 1320 | (297) | 330 | (74) | 9.6 | (3.8) | 6.4 | (2.5) |
| 14 | 7.0 | (0.28) | 45 | (30.2) | 1320 | (297) | 330 | (74) | 10.5 | (4.2) | 7.0 | (2.8) |
| 16 | 7.0 | (0.28) | 48 | (32.3) | 1320 | (297) | 330 | (74) | 10.5 | (4.2) | 7.0 | (2.8) |
| 18 | 7.0 | (0.28) | 50 | (33.6) | 1320 | (297) | 330 | (74) | 10.5 | (4.2) | 7.0 | (2.8) |
| 20 | 7.9 | (0.31) | 57 | (38.2) | 1320 | (297) | 330 | (74) | 12.0 | (4.7) | 7.9 | (3.1) |
| 22 | 7.9 | (0.31) | 58 | (39.0) | 1320 | (297) | 330 | (74) | 12.0 | (4.7) | 7.9 | (3.1) |
| 24 | 8.1 | (0.32) | 62 | (41.6) | 1320 | (297) | 330 | (74) | 12.3 | (4.8) | 8.1 | (3.2) |

### 24 to 144 Fibers Unitized Plenum

| Fiber Count | Nominal Outer Diameter mm | (in) | Nominal Weight kg/km | (lb/1000 ft) | Maximum Tensile Load Short Term N | (lb) | Long Term N | (lb) | Minimum Bend Radius Loaded cm | (in) | Installed cm | (in) |
|---|---|---|---|---|---|---|---|---|---|---|---|---|
| **6-Fiber Subunits** | | | | | | | | | | | | |
| 24 | 13.7 | (0.54) | 157 | (106) | 4100 | (922) | 2050 | (461) | 20.6 | (8.1) | 13.7 | (5.4) |
| 30 | 15.2 | (0.60) | 198 | (133) | 4500 | (1012) | 2250 | (506) | 22.8 | (9.0) | 15.2 | (6.0) |
| 36 | 16.6 | (0.65) | 245 | (165) | 6300 | (1416) | 3150 | (708) | 24.9 | (9.8) | 16.6 | (6.5) |
| 42 | 18.2 | (0.72) | 302 | (203) | 7700 | (1731) | 3850 | (866) | 27.3 | (10.8) | 18.2 | (7.2) |
| 48 | 19.8 | (0.78) | 365 | (245) | 8500 | (1911) | 4250 | (956) | 29.7 | (11.7) | 19.8 | (7.8) |
| 54 | 21.5 | (0.85) | 440 | (296) | 9400 | (2113) | 4700 | (1057) | 32.3 | (12.7) | 21.5 | (8.5) |
| 60 | 23.4 | (0.92) | 541 | (364) | 10000 | (2248) | 5000 | (1124) | 35.1 | (13.5) | 23.4 | (9.2) |
| 66 | 25.0 | (0.98) | 629 | (423) | 11000 | (2472) | 5500 | (1236) | 37.5 | (14.8) | 25.0 | (9.8) |
| 72 | 21.7 | (0.85) | 390 | (262) | 11500 | (2585) | 5750 | (1293) | 32.6 | (12.8) | 21.7 | (8.5) |
| **12-Fiber Subunits** | | | | | | | | | | | | |
| 60 | 20.9 | (0.82) | 375 | (252) | 6900 | (1551) | 3450 | (776) | 31.4 | (12.4) | 20.9 | (8.2) |
| 72 | 22.3 | (0.88) | 455 | (306) | 8600 | (1933) | 4300 | (967) | 33.5 | (13.2) | 22.3 | (8.8) |
| 84 | 24.5 | (0.96) | 557 | (374) | 9700 | (2180) | 4850 | (1090) | 36.8 | (14.5) | 24.5 | (9.6) |
| 96 | 26.6 | (1.05) | 673 | (452) | 10500 | (2360) | 5250 | (1180) | 39.9 | (15.7) | 26.6 | (10.5) |
| 144 | 30.0 | (1.18) | 735 | (494) | 11600 | (2607) | 5800 | (1304) | 45.0 | (17.7) | 30.0 | (11.8) |

*Siecor MIC cables are designed and tested to ICEA S-83-596 Standard for Fiber Optic Premises Distribution Cable per requirements of ANSI/TIA/EIA-568A Commercial Building Telecommunications Cabling Standard.*

1.19

**Figure 7.9 (c)** (continued)

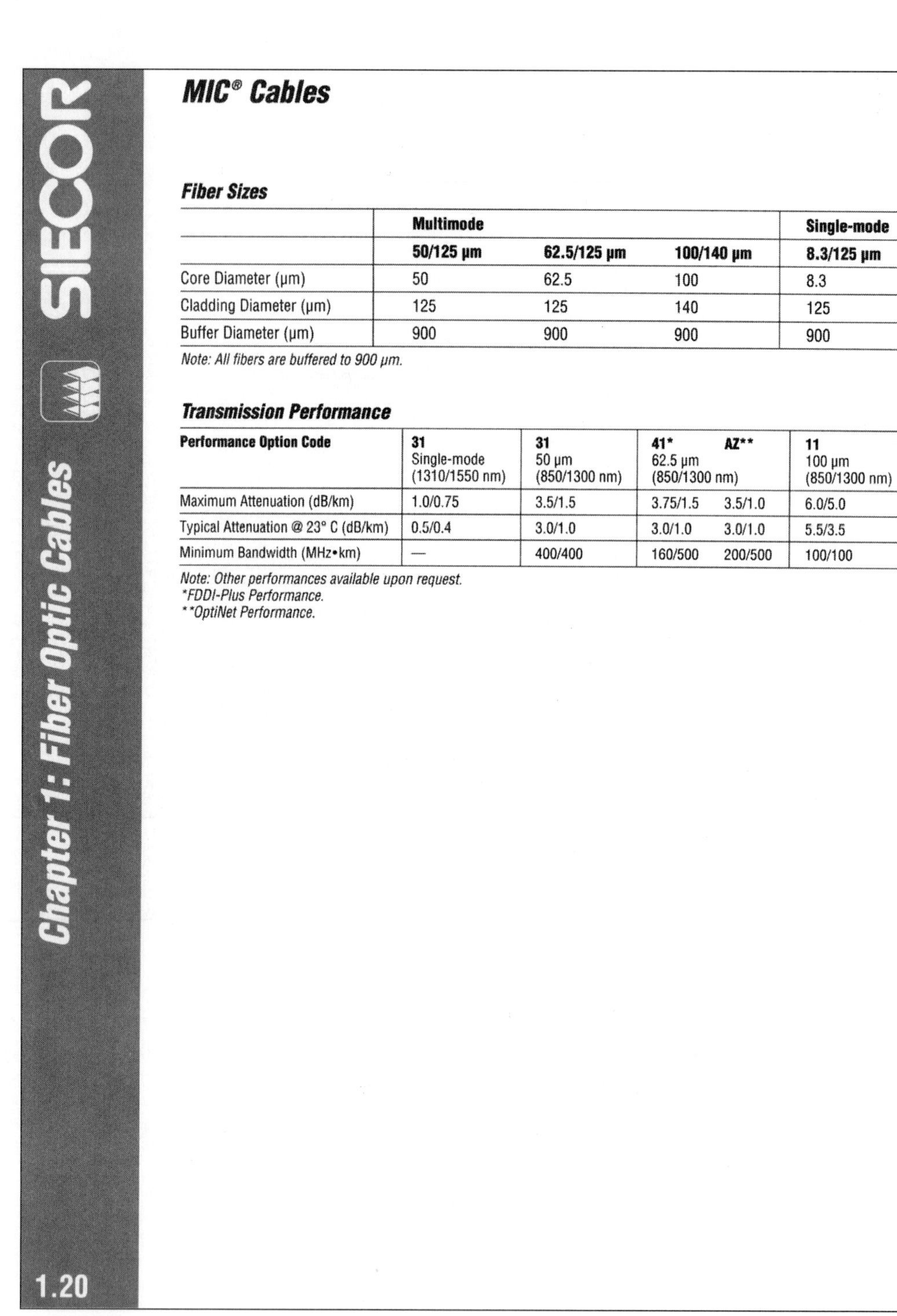


## MIC® Cables

### Fiber Sizes

| | Multimode | | | Single-mode |
|---|---|---|---|---|
| | 50/125 µm | 62.5/125 µm | 100/140 µm | 8.3/125 µm |
| Core Diameter (µm) | 50 | 62.5 | 100 | 8.3 |
| Cladding Diameter (µm) | 125 | 125 | 140 | 125 |
| Buffer Diameter (µm) | 900 | 900 | 900 | 900 |

*Note: All fibers are buffered to 900 µm.*

### Transmission Performance

| Performance Option Code | 31 Single-mode (1310/1550 nm) | 31 50 µm (850/1300 nm) | 41* 62.5 µm (850/1300 nm) | AZ** | 11 100 µm (850/1300 nm) |
|---|---|---|---|---|---|
| Maximum Attenuation (dB/km) | 1.0/0.75 | 3.5/1.5 | 3.75/1.5 | 3.5/1.0 | 6.0/5.0 |
| Typical Attenuation @ 23° C (dB/km) | 0.5/0.4 | 3.0/1.0 | 3.0/1.0 | 3.0/1.0 | 5.5/3.5 |
| Minimum Bandwidth (MHz•km) | — | 400/400 | 160/500 | 200/500 | 100/100 |

*Note: Other performances available upon request.*
**FDDI-Plus Performance.*
***OptiNet Performance.*

**Figure 7.9 (c)** (continued)

# FREEDM™ Mini Bundle® Ribbon Cables, 288-432 Fibers

SIECOR®

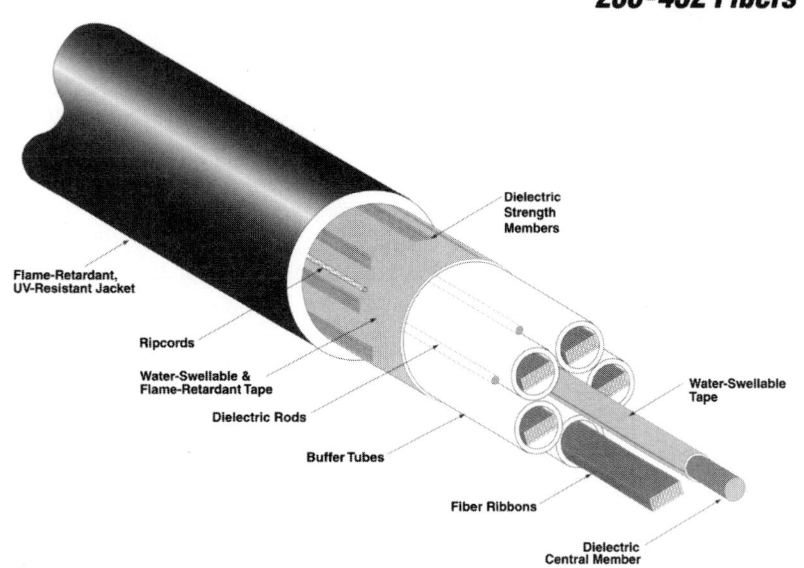


## Features / Benefits

- ***UL-listed OFNR and CSA-listed FT-4***
- ***Corning fiber for unsurpassed optical and mechanical performance***
- ***FREEDM™ Mini Bundle® Ribbon Cables provide up to 432 fibers in a rugged, compact design***
- ***Dry™ cable design improves cable handling and preparation***
- ***SZ-stranded, flexible buffer tubes isolate fiber ribbons from installation and environ-mental rigors***
- ***Fiber ribbons individually numbered for easy identi-fica-tion***
- ***Precise fiber and ribbon geometries result in excellent mass-splicing yields***
- ***Ideal for transition from indoor to outdoor environ-ments***

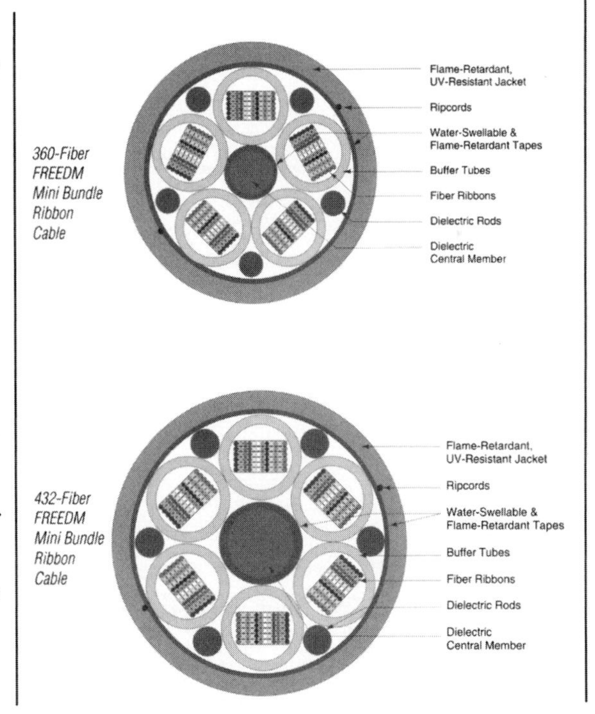


*Drawing ZA-1351 ZA-1352 ZA-1353*

Figure 7.9 (d)

# FREEDM™ Mini Bundle® Ribbon Cables, 288-432 Fibers

### Mechanical Specifications

| | | |
|---|---|---|
| Maximum Tensile Loading | Installation:<br>Long Term Installed: | 2700 N (600 lbf)<br>890 N (200 lbf) |
| Operating Temperature | Storage:<br>Long Term: | -40° to +70° C (-40° to +158° F)<br>-40° to +70° C (-40° to +158° F) |

| Fiber Count | Ribbons Per Tube | Nominal Weight kg/km (lb/1000 ft) | Nominal Diameter[1] mm (in) | Minimum Bend Radius Loaded cm (in) | Minimum Bend Radius Installed cm (in) | Representative Siecor Part Number[2] |
|---|---|---|---|---|---|---|
| 288 | 6 | 566 (380) | 25.4 (1.00) | 38.1 (15.0) | 25.4 (10.0) | 288RQF-14103-30 |
| 360 | 6 | 580 (390) | 25.4 (1.00) | 38.1 (15.0) | 25.4 (10.0) | 360RQF-14103-30 |
| 432 | 6 | 652 (439) | 27.0 (1.06) | 40.5 (16.0) | 27.0 (10.6) | 432RQF-14103-30 |

### Ribbon Positions

| Number of Fibers[3] | Number of Ribbons | Ribbon Sequence by Buffer Tube: Tube 1/ Blue | Tube 2/ Orange | Tube 3/ Green | Tube 4/ Brown | Tube 5/ Slate | Tube 6/ White |
|---|---|---|---|---|---|---|---|
| 288 | 24 | 1-6 | 7-12 | 13-18 | 19-24 | Filler | — |
| 360 | 30 | 1-6 | 7-12 | 13-18 | 19-24 | 25-30 | — |
| 432 | 36 | 1-6 | 7-12 | 13-18 | 19-24 | 25-30 | 31-36 |

**Notes:**

1 *Actual diameter may vary by ± 5%.*

2 *Part numbers are representative.*
*Please contact Siecor Customer Service to verify a specific design's part number when ordering.*

3 *Other fiber counts are available.*
*Please contact Siecor Customer Service to discuss your product requirements.*

*Siecor Corporation • PO Box 489 • Hickory, NC 28603-0489 USA • 1-800-SIECOR5 • 1-800-743-2675 • FAX: 704-327-5973 • International: 704-327-5000 • http://www.siecor.com* 

**Figure 7.9 (d)**
(continued)

# Fan-Out Cables

Heavy Duty 2-Fiber Fan-Out Cable with 2.9 mm Subunits

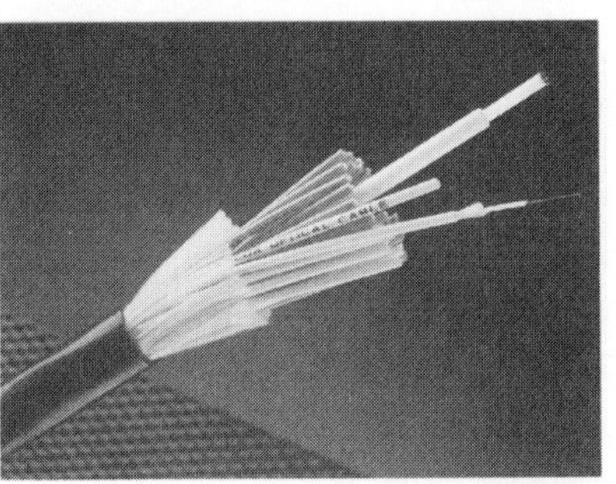

Heavy Duty 24-Fiber Fan-Out Cable with 2.4 mm Subunits

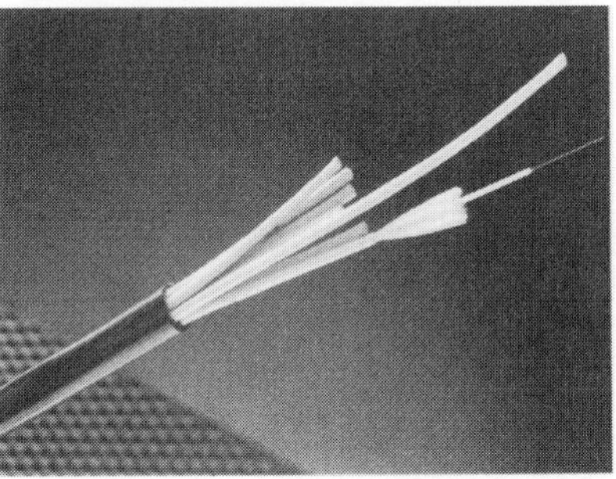

Heavy Duty 6-Fiber Fan-Out Cable with 2.0 mm Subunits

## Description

Siecor's Fan-Out cables are designed for voice, data, video, and imaging transmission in computer, process control, data entry, and wired office systems. The Fan-Out design enables the individual routing, or fanning, of individual fibers for termination and maintenance.

## Features / Benefits

- Uses TBII buffered fiber
- Easily strippable jacket and fiber buffer
- Available in three constructions:
  2.9 mm heavy duty subunits riser or
  2.7 mm heavy duty subunits plenum;
  2.4 mm heavy duty subunits;
  2.0 mm heavy duty subunits
- Can be routed between buildings within ducts below the frost line, inside buildings up riser shafts, under computer room floors, and fiber-to-the-desk
- Construction allows individual routing of fibers and provides for easy connectorization
- Subunits numbered for easy identification
- Fiber type and length markings appear on outer jacket
- Available with single-mode, 50/125 µm, 62.5/125 µm, and 100/140 µm multimode fiber types
- Available in a wide range of fiber counts
- All-dielectric construction insures EMI immunity
- Color-coded subunits available

## Applications

- Interbuilding backbone in conduits below the frost line
- Intrabuilding backbone
- Horizontal cabling
- UL and CSA listed; OFNR/FT4 for non-plenum applications, and OFNP/FT6 for plenum applications
- Available up to 24 fibers

Photo TOPSS337
TOPSS333
CPPSS1053





Figure 7.9 (e)

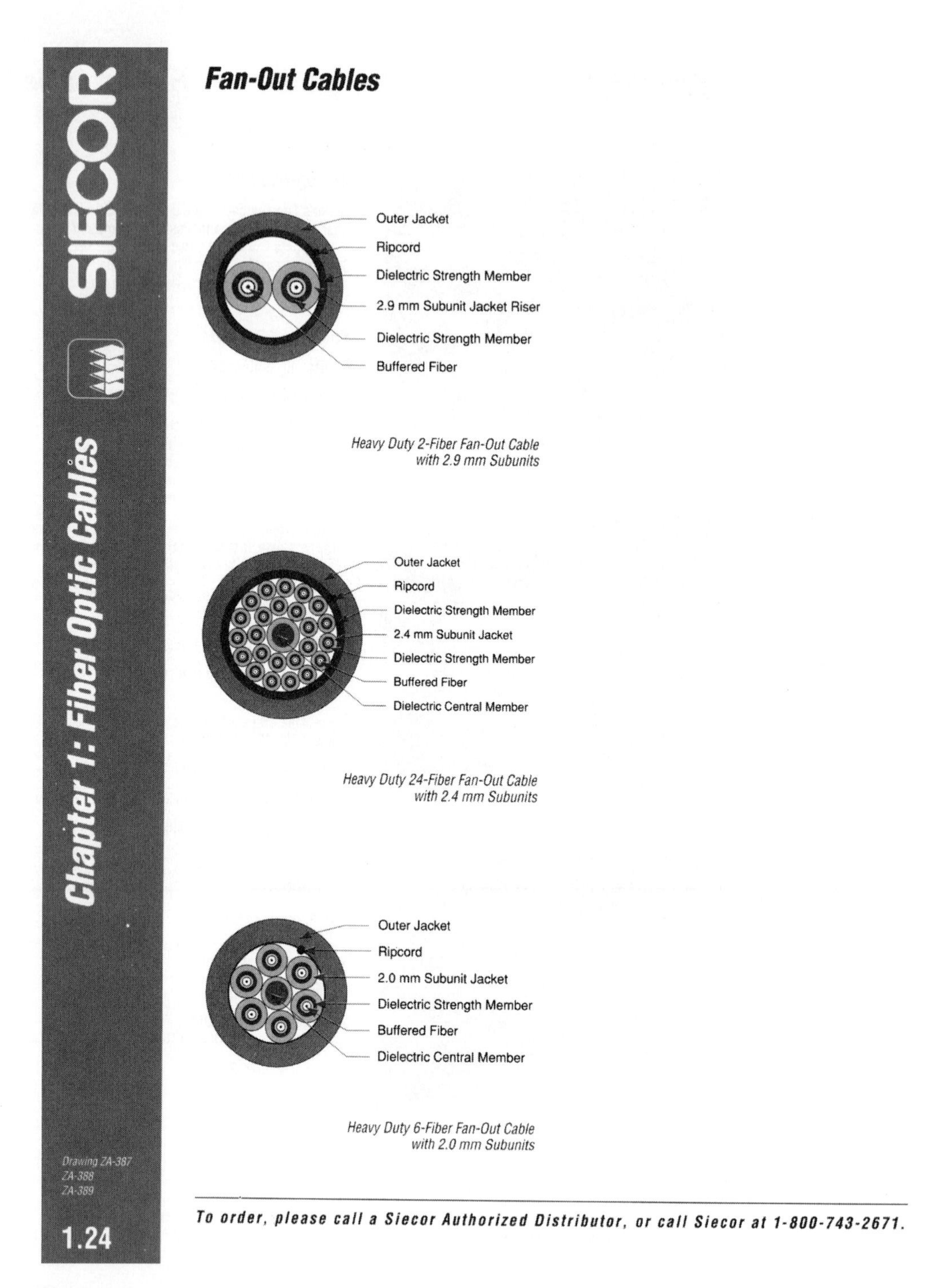


**Figure 7.9 (e)** (continued)

## Mechanical Specifications

### Heavy Duty Riser-Rated Fan-Out Cables with 2.0 mm Subunits

| Fiber Count | Nominal Outer Diameter mm | (in) | Nominal Weight kg/km | (lb/1000 ft) | Maximum Tensile Load Short Term N | (lb) | Long Term N | (lb) | Minimum Bend Radius Loaded cm | (in) | Installed cm | (in) |
|---|---|---|---|---|---|---|---|---|---|---|---|---|
| 2 | 6.9 | (0.3) | 35 | (24) | 900 | (203) | 300 | (68) | 10.4 | (4.1) | 6.9 | (2.7) |
| 4 | 7.7 | (0.3) | 45 | (30) | 1000 | (225) | 400 | (90) | 11.6 | (4.6) | 7.7 | (3.0) |
| 6 | 8.4 | (0.3) | 59 | (40) | 1200 | (270) | 350 | (79) | 12.6 | (5.0) | 8.4 | (3.3) |
| 8 | 9.6 | (0.4) | 80 | (54) | 1700 | (383) | 500 | (113) | 14.4 | (5.7) | 9.6 | (3.8) |
| 10 | 10.8 | (0.4) | 102 | (69) | 2000 | (450) | 700 | (158) | 16.2 | (6.4) | 10.8 | (4.3) |
| 12 | 12.1 | (0.5) | 130 | (87) | 2500 | (563) | 900 | (203) | 18.2 | (7.2) | 12.1 | (4.8) |
| 14 | 11.7 | (0.5) | 100 | (67) | 2000 | (450) | 750 | (169) | 17.6 | (6.9) | 11.7 | (4.6) |
| 16 | 12.1 | (0.5) | 109 | (73) | 2500 | (563) | 900 | (203) | 18.2 | (7.2) | 12.1 | (4.8) |
| 18 | 12.5 | (0.5) | 122 | (82) | 3000 | (675) | 1100 | (248) | 18.8 | (7.4) | 12.5 | (4.9) |
| 20 | 12.9 | (0.5) | 134 | (90) | 3500 | (788) | 1300 | (293) | 19.4 | (7.6) | 12.9 | (5.1) |
| 22 | 13.7 | (0.5) | 150 | (101) | 3800 | (855) | 1400 | (315) | 20.6 | (8.1) | 13.7 | (5.4) |
| 24 | 14.4 | (0.6) | 165 | (111) | 4200 | (945) | 1600 | (360) | 21.6 | (8.5) | 14.4 | (5.7) |
| Storage Temperature | | | -40° C to +70° C | | | | | | | | | |
| Operating Temperature | | | -20° C to +70° C | | | | | | | | | |
| NEC/CSA Listing | | | OFNR/FT4 | | | | | | | | | |

### Heavy Duty Plenum-Rated Fan-Out Cables with 2.0 mm Subunits

| Fiber Count | Nominal Outer Diameter mm | (in) | Nominal Weight kg/km | (lb/1000 ft) | Maximum Tensile Load Short Term N | (lb) | Long Term N | (lb) | Minimum Bend Radius Loaded cm | (in) | Installed cm | (in) |
|---|---|---|---|---|---|---|---|---|---|---|---|---|
| 2 | 5.6 | (0.2) | 28 | (19) | 900 | (203) | 300 | (68) | 8.4 | (3.3) | 8.4 | (3.3) |
| 4 | 6.9 | (0.3) | 40 | (27) | 1000 | (225) | 400 | (90) | 10.4 | (4.1) | 10.4 | (4.1) |
| 6 | 7.6 | (0.3) | 55 | (37) | 2700 | (608) | 350 | (79) | 11.4 | (4.5) | 11.4 | (4.5) |
| 8 | 9.6 | (0.4) | 80 | (54) | 2700 | (608) | 500 | (113) | 14.4 | (5.7) | 14.4 | (5.7) |
| 10 | 10.8 | (0.4) | 105 | (71) | 2000 | (450) | 700 | (158) | 16.2 | (6.4) | 16.2 | (6.4) |
| 12 | 11.4 | (0.4) | 135 | (91) | 2500 | (563) | 900 | (203) | 17.1 | (6.7) | 17.1 | (6.7) |
| 14 | 10.9 | (0.4) | 95 | (64) | 2000 | (450) | 750 | (169) | 16.4 | (6.4) | 16.4 | (6.4) |
| 16 | 11.2 | (0.4) | 106 | (71) | 2500 | (563) | 900 | (203) | 16.8 | (6.6) | 16.8 | (6.6) |
| 18 | 11.6 | (0.5) | 118 | (79) | 3000 | (675) | 1100 | (248) | 17.4 | (6.9) | 17.4 | (6.9) |
| 20 | 12.1 | (0.5) | 132 | (89) | 3500 | (788) | 1300 | (293) | 18.2 | (7.2) | 18.2 | (7.2) |
| 22 | 12.9 | (0.5) | 149 | (100) | 3800 | (855) | 1400 | (315) | 19.4 | (7.6) | 19.4 | (7.6) |
| 24 | 13.6 | (0.5) | 167 | (112) | 4200 | (945) | 1600 | (360) | 20.4 | (8.0) | 20.4 | (8.0) |
| Storage Temperature | | | -40° C to +70° C | | | | | | | | | |
| Operating Temperature | | | 0° C to +70° C | | | | | | | | | |
| NEC/CSA Listing | | | OFNP/FT6 | | | | | | | | | |

*Note: Higher fiber counts available upon request.*
*Siecor Fan-Out cables are designed and tested to ICEA S-83-596 Standard for Fiber Optic Premises Distribution Cable per requirements of ANSI/TIA/EIA-568A Commercial Building Telecommunications Cabling Standard.*



**Figure 7.9 (e)** (continued)

# Fan-Out Cables

## Mechanical Specifications

### Heavy Duty Riser-Rated Fan-Out Cables with 2.4 mm Subunits

| Fiber Count | Nominal Outer Diameter mm | (in) | Nominal Weight kg/km | (lb/1000 ft) | Maximum Tensile Load Short Term N | (lb) | Long Term N | (lb) | Minimum Bend Radius Loaded cm | (in) | Installed cm | (in) |
|---|---|---|---|---|---|---|---|---|---|---|---|---|
| 2 | 6.9 | (0.3) | 39 | (26) | 1200 | (270) | 500 | (113) | 10.4 | (4.1) | 6.9 | (2.7) |
| 4 | 8.2 | (0.3) | 56 | (38) | 2000 | (450) | 800 | (180) | 12.3 | (4.8) | 8.2 | (3.2) |
| 6 | 9.4 | (0.4) | 80 | (54) | 2000 | (450) | 800 | (180) | 14.1 | (5.6) | 9.4 | (3.7) |
| 8 | 11.0 | (0.4) | 109 | (73) | 2500 | (563) | 900 | (203) | 16.5 | (6.5) | 11.0 | (4.3) |
| 10 | 12.6 | (0.5) | 144 | (97) | 3000 | (675) | 1200 | (270) | 18.9 | (7.4) | 12.6 | (5.0) |
| 12 | 14.3 | (0.6) | 187 | (126) | 3500 | (788) | 1200 | (270) | 21.5 | (8.5) | 14.3 | (5.6) |
| 14 | 13.6 | (0.5) | 150 | (101) | 3500 | (788) | 1200 | (270) | 20.4 | (8.0) | 13.6 | (5.4) |
| 16 | 14.3 | (0.6) | 170 | (114) | 4000 | (900) | 1500 | (338) | 21.5 | (8.5) | 14.3 | (5.6) |
| 18 | 14.8 | (0.6) | 187 | (126) | 4500 | (1013) | 1800 | (405) | 22.2 | (8.7) | 14.8 | (5.8) |
| 20 | 15.5 | (0.6) | 207 | (139) | 4500 | (1013) | 2000 | (450) | 23.3 | (9.2) | 15.5 | (6.1) |
| 22 | 16.2 | (0.6) | 229 | (154) | 5000 | (1125) | 2000 | (450) | 24.3 | (9.6) | 16.2 | (6.4) |
| 24 | 17.1 | (0.7) | 254 | (171) | 5500 | (1238) | 2000 | (450) | 25.7 | (10.1) | 17.1 | (6.7) |
| Storage Temperature | | | -40° C to +70° C | | | | | | | | | |
| Operating Temperature | | | -20° C to +70° C | | | | | | | | | |
| NEC/CSA Listing | | | OFNR/FT4 | | | | | | | | | |

### Heavy Duty Plenum-Rated Fan-Out Cables with 2.4 mm Subunits

| Fiber Count | Nominal Outer Diameter mm | (in) | Nominal Weight kg/km | (lb/1000 ft) | Maximum Tensile Load Short Term N | (lb) | Long Term N | (lb) | Minimum Bend Radius Loaded cm | (in) | Installed cm | (in) |
|---|---|---|---|---|---|---|---|---|---|---|---|---|
| 2 | 6.7 | (0.3) | 36 | (24) | 1200 | (270) | 500 | (113) | 10.1 | (4.0) | 10.1 | (4.0) |
| 4 | 7.9 | (0.3) | 52 | (35) | 2000 | (450) | 800 | (180) | 11.9 | (4.7) | 11.9 | (4.7) |
| 6 | 9.1 | (0.4) | 77 | (52) | 2000 | (450) | 800 | (180) | 13.7 | (5.4) | 13.7 | (5.4) |
| 8 | 10.7 | (0.4) | 109 | (73) | 2500 | (563) | 900 | (203) | 16.1 | (6.3) | 16.1 | (6.3) |
| 10 | 12.3 | (0.5) | 148 | (99) | 3000 | (675) | 1200 | (270) | 18.5 | (7.3) | 18.5 | (7.3) |
| 12 | 13.7 | (0.5) | 192 | (129) | 3500 | (788) | 1200 | (270) | 20.6 | (8.1) | 20.6 | (8.1) |
| 14 | 12.5 | (0.5) | 131 | (88) | 3500 | (788) | 1200 | (270) | 18.8 | (7.4) | 18.8 | (7.4) |
| 16 | 13.2 | (0.5) | 151 | (101) | 4000 | (900) | 1500 | (338) | 19.8 | (7.8) | 19.8 | (7.8) |
| 18 | 13.7 | (0.5) | 169 | (114) | 4500 | (1013) | 1800 | (405) | 20.6 | (8.1) | 20.6 | (8.1) |
| 20 | 14.6 | (0.6) | 191 | (128) | 4500 | (1013) | 2000 | (450) | 21.9 | (8.6) | 21.9 | (8.6) |
| 22 | 15.7 | (0.6) | 232 | (156) | 5000 | (1125) | 2000 | (450) | 23.6 | (9.3) | 23.6 | (9.3) |
| 24 | 16.6 | (0.7) | 260 | (175) | 5500 | (1238) | 2000 | (450) | 24.9 | (9.8) | 24.9 | (9.8) |
| Storage Temperature | | | -40° C to +70° C | | | | | | | | | |
| Operating Temperature | | | 0° C to +70° C | | | | | | | | | |
| NEC/CSA Listing | | | OFNP/FT6 | | | | | | | | | |

*Note: Higher fiber counts available upon request.*
*Siecor Fan-Out cables are designed and tested to ICEA S-83-596 Standard for Fiber Optic Premises Distribution Cable per requirements of ANSI/TIA/EIA-568A Commercial Building Telecommunications Cabling Standard.*



**Figure 7.9 (e)** (continued)

SIECOR

Chapter 1: Fiber Optic Cables

# Fan-Out Cables

### Fiber Sizes

| | Multimode | | | Single-mode |
|---|---|---|---|---|
| | 50/125 µm | 62.5/125 µm | 100/140 µm | 8.3/125 µm |
| Core Diameter (µm) | 50 | 62.5 | 100 | 8.3 |
| Cladding Diameter (µm) | 125 | 125 | 140 | 125 |
| Buffer Diameter (µm) | 900 | 900 | 900 | 900 |

*Note: All fibers are buffered to 900 µm.*

### Transmission Performance

| Performance Option Code | 31 Single-mode (1310/1550 nm) | 31 50 µm (850/1300 nm) | 41* 62.5 µm (850/1300 nm) | AZ** 62.5 µm (850/1300 nm) | 11 100 µm (850/1300 nm) |
|---|---|---|---|---|---|
| Maximum Attenuation (dB/km) | 1.0/0.75 | 3.5/1.5 | 3.75/1.5 | 3.5/1.0 | 6.0/5.0 |
| Typical Attenuation @ 23° C (dB/km) | 0.5/0.4 | 3.0/1.0 | 3.0/1.0 | 3.0/1.0 | 5.5/3.5 |
| Minimum Bandwidth (MHz•km) | — | 400/400 | 160/500 | 200/500 | 100/100 |

*Note: Other performances available upon request.*
**FDDI-Plus Performance.*
***OptiNet Performance.*



**Figure 7.9 (e)** (continued)

SIECOR

# 2-Fiber Flexible Plenum Zipcord Cable

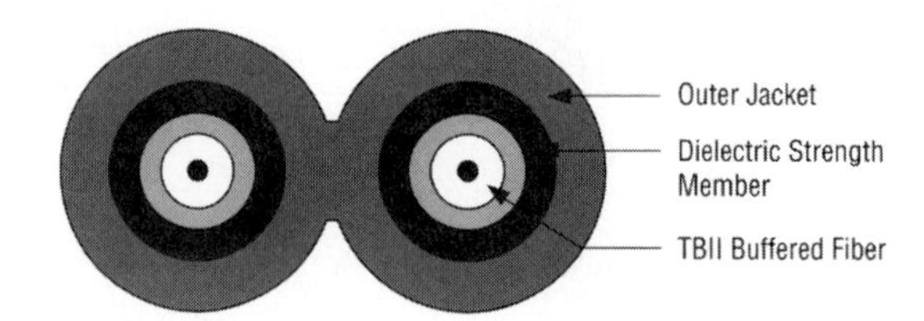


### Mechanical Specifications

| Fiber Count | Cable Outside Diameter mm (in) | Cable Weight kg/km (lb/1000 ft) | Maximum Tensile Load Loaded N (lbf) | Maximum Tensile Load Unloaded N (lbf) | Minimum Bend Radius Loaded cm (in) | Minimum Bend Radius Unloaded cm (in) | Crush Resistance N/cm (lb/in) | Impact Cycles | Flex Cycles |
|---|---|---|---|---|---|---|---|---|---|
| 2 | 2.9 x 5.8 (0.11 x 0.23) | 15 (10) | 600 (135) | 300 (67) | 5.0 (2.0) | 3.0 (1.2) | 750 (428) | 1,000 | 5,000 |

Drawing ZA-395

Siecor Corporation • PO Box 489 • Hickory, NC 28603-0489 USA • 1-800-SIECOR5 • 1-800-743-2675 • FAX: 704-327-5973 • International: 704-327-5000 • 

Figure 7.9 (f)

# 2-Fiber Flexible Plenum Zipcord Cable

SIECOR

## Cable Description

Siecor's zipcord cable utilizes two 900 µm TBII™ buffered fibers surrounded by aramid yarn strength members with a flexible flame-retardant jacket. These cables meet the application requirements of the National Electrical Code (NEC Article 770) and are UL-listed OFNP and CSA-listed FT-6. Zipcord cables are ideal for interconnect applications within plenum areas.

## Fiber Information

| | |
|---|---|
| Fiber Types (Core/Cladding Diameters) | 8.3/125, 50/125, 62.5/125, 100/140 µm |
| Buffering Diameter | 900 µm |

## Transmission Performance

| | Single-mode (1310/1550 nm) | 50 µm (850/1300 nm) | 62.5 µm (850/1300 nm) | 100 µm (850/1300 nm) |
|---|---|---|---|---|
| Performance Option Code | 31 | 31 | 41 | 11 |
| Maximum Attenuation (dB/km) | 1.0/0.75 | 3.5/1.5 | 3.75/1.5 | 6.0/5.0 |
| Typical Attenuation @ 23° C (dB/km) | 0.5/0.4 | 3.0/1.0 | 3.0/1.0 | 5.5/3.5 |
| Minimum Bandwidth (MHz•km) | — | 400/400 | 160/500 | 100/100 |

*Other performances available upon request.*

## Environmental Specifications

| | |
|---|---|
| Storage Temperature | -40° C to +70° C |
| Operating Temperature | 0° C to +70° C |

## Application Information

| | |
|---|---|
| NEC/CSA Listing | OFNP/FT-6 |
| Flame Resistance | UL-910 (For plenum, riser, and general building applications) |

## Shipping Information

| | |
|---|---|
| Maximum Reel Length | 4,000 m (13,123 ft) |

*Other lengths available upon request.*

## Ordering Information

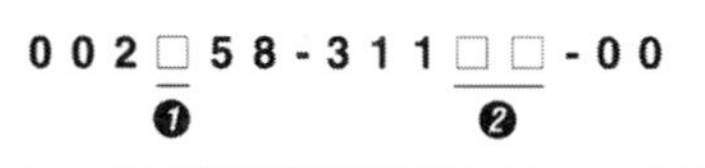


❶ ***Select fiber type.***
C = 50/125 µm
K = 62.5/125 µm
N = 100/140 µm
R = Single-mode

❷ ***Select performance option code.***
31 = Single-mode
31 = 50 µm
41 = 62.5 µm
11 = 100 µm

**Figure 7.9 (f)** (continued)

a low level of smoke, allow a smaller bend radius, be amenable to vertical installation, and handle easily.

Outdoor cable involves a very broad group, starting with cable running between two buildings in local area network (LAN) applications and ending with submarine cable lying on the ocean floor. Thus, the requirements for this type of cable are unique.

Indoor/outdoor cable is usually used in LAN applications to enable use of the same cable as the transmission link between and within buildings.

Most outdoor and indoor/outdoor cables use a loose-buffer design, while most indoor cable relies on a tight-buffer design.

Another classification is *light-duty* and *heavy-duty* cable. Fiber ribbon alone is an example of light-duty cable, but FREEDM Mini Bundle ribbon cable (see Figure 7.9[d]) is a heavy-duty cable. Still another example of heavy-duty cable is shown in Figure 7.9(e).

In addition, there is auxiliary cable used with installation hardware, such as patch cords and jumpers. (See Section 8.2.)

***Standards*** Two types of standards apply to cable. The first is *safety codes,* particularly as applied to the interior of buildings. There are local and national building codes. All fiber-optic cable has to satisfy the requirements introduced by Article 770 of the National Electric Code (NEC).

The NEC identifies three locations within a building where cable can be installed: plenums, risers, and general-purpose areas. A *plenum* is the compartment housing the air-distribution system. We usually associate a plenum with the air-handling spaces in ceilings or in raised floors. A *riser* is an opening through which a cable can pass vertically from floor to floor; a *general-purpose* area is any indoor space except plenums and risers.

Article 770 of the NEC lists fiber cable for use in these applications. This standard divides fiber-optic cable into two types: (1) nonconductive, that is, cable that contains no conductive elements such as metallic armor or metallic strength elements (often referred to as OFN), and (2) conductive, or OFC. Each of these types, in turn, is subdivided into three groups: OFNG/OFCG, OFNR/OFCR, OFNP/OFCP. Each type of fiber cable may be used in specific areas and has to pass a standard fire-resistance test. Thus, OFN/OFC is listed as the cable for *general-purpuse use* except in risers, plenums, and other spaces used for environmental air. Such cable must be resistant to the spread of fire and pass a vertical-flame test. OFNG/OFCG is listed for use in general-purpose areas and must meet its own set of fire-resistance requirements. OFNR/OFCR is suitable to use in risers. OFNP/OFCP is used in plenums without conduits. Both OFNP/OFCP and OFNR/OFCR can be used for general-purpose applications.

The rule for the project manager is this: *When you purchase any type of fiber-optic cable, make sure it has an NEC rating; when you install it, make sure to comply with local building codes and NEC requirements. Otherwise, your fiber-optic network will not pass inspection.*

Another type of standard created by manufacturers' groups, primarily the Telecommunications Industry Association (TIA) and the Electronics Industries Alliance (EIA), helps regulate, among other things, the test procedures for all fiber-optic cable. For example, the FOTP-88 (the fiber-optic test procedure) establishes the fiber-optic cable-bend test; FOTP-96 regulates the long-term storage-temperature test for fiber-optic cable operating in extreme environments.

For more about sources on standards, see Appendix D.

## Reading Data Sheets

The data sheet for the loose-tube Mini Bundle cable (Figure 7.9[a]) starts with mechanical specifications. The meaning of the *maximum tensile-loading* specification is that, during installation, tensile loading can reach a very high value and so fiber attenuation might also be very high at that moment. But nobody is going to use fiber as a transmission medium during installation, so the important thing is attenuation *after* installation. If the installer does not exceed the given maximum load (2700 N here), fiber attenuation will return to the given specification once the cable is in place.

The *operating-temperature* specification shows within what temperature range the manufacturer guarantees the cable's stated mechanical and transmission properties.

The table in Figure 7.9(a) shows the cable's characteristics. The term "Fiber Count" refers to the number of fibers the given cable carries. If you multiply "Maximum Fibers per Tube" by "Number of Active Tubes," you'll calculate the number of fibers in the cable. The "Number of Tube Positions" determines the maximum number of tubes the given cable can accommodate. Thus, any cable can include spare tubes for future upgrades. The rule for fiber count is as follows: For a low fiber count (< 24), multiples of six are used (that is, 6, 12, 18, and 24). For a larger fiber count, multiples of 12 are used (that is, 36, 48, 60, 72, and so on). Thus, if you need to use 138 fibers, purchase a cable with a fiber count of 144; otherwise, you will have to order a special cable.

Among other characteristics, the nominal weight (kg/km) and the nominal outer diameter allow the user to compare cables from different vendors with the same transmission capacity. "Minimum Bend Radius" is information for installers. "Loaded" minimum bend radius must be observed during the installation process and "Installed" radius must be followed after installation.

The data sheet for "Figure-8 loose-tube Mini Bundle cable" (Figure 7.9[b]) starts with typical sag and tension information because this cable is intended for aerial installation. So, for example, if the span between two poles is 250 ft, the tension for the installed cable is 1000 lb, and the armored cable carries 144 fibers, then the sag must be no more than 3.4 ft as measured immediately after installation. (To visualize sag and span, see Figure 7.10[b].)

Pay special attention to *transmission characteristics.* The maximum *attenuation* for installed cable with a singlemode fiber at 1550 nm is 0.4 dB/km, while the maximum attenuation for a singlemode fiber itself is typically not more than 0.3 dB/km. (See Figure 5.17.) The same figures for multimode graded-index fibers contrast even more sharply: 1.5 dB/km for a cable versus 0.7 dB/km for a fiber. (See Figure 3.18.) These examples are typical of how cabling and installation change the characteristics of bare fiber.

Data for *MIC cable* are quite understandable. The meanings of the terms *riser* and *plenum* are explained above in the subsection "Standards." To visualize general, riser, and plenum types of cable, see Figure 8.16.

The data sheet for the *FREEDM Mini Bundle ribbon cable* (Figure 7.9[d]) contains two tables. The first covers the entire cable and is self-explanatory. The second describes the ribbon positions. Let's study this table in more detail:

The first column, "Number of Fibers," shows the total number of fibers in this cable. The second column, "Number of Ribbons," displays the total number of ribbons in the cable. Recalling that each ribbon carries 12 fibers, it's easy to verify all these figures; for example, if the number of ribbons is 36 (see the last row), then $36 \times 12 = 432$, which is the total number of fibers for this cable. (See the last row of the first column.) Each ribbon has been assigned a number and each fiber also has its own number, so that any fiber can be easily identified. So, for instance, if an installer needs to find fiber #268 among the 432 fibers, he needs to divide 268 by 12—the number of fibers per ribbon. The answer, 22.3, means the fiber he or she is looking for is in the twenty-third ribbon. Looking at the data sheet, the installer finds ribbon #23 in tube 4, the last row of the column headed "Tube 4/Brown." The color—here, brown—makes it even easier to find the appropriate buffer tube. This is how each individual fiber is identified for connectorization.

These explanations clearly in mind, it should now be easy for you to read and understand the data sheets for the *fan-out cable* (Figure 7.9[e].)

Read and analyze every word and every figure in the given data sheets. Doing so will make you an educated customer and enable you to interact at the highest professional level with cable vendors.

## 7.3 INSTALLATION—PLACING THE CABLE

Installation means setting up a fiber-optic cable for transmission of information. In a broad sense, installation means building a complete network. The essential steps of this work involve placing the cable and preparing the cable for connection to the transmitter and receiver, a process called *termination.* To place the cable between given points, it is usually necessary to splice it—that is, to connect two pieces of cable. Testing and certification of the installed plant complete the installation. Prior to all this work, of course, the entire network must be designed. Thus, installation consists of many separate routines.

In this section, we'll discuss only placement of the cable; the other activities will be taken up in subsequent sections.

### Classification

Installation falls into two categories: *indoor* and *outdoor.* Indoor installation varies from working in a new building to placing a fiber-optic cable in the crammed rooms of very old edifices. Outdoor installation covers a myriad of conditions—from placing a cable between two adjacent buildings to submarine emplacement.

A fiber-optic cable can be installed *outdoors* in two general ways: *buried or aerial.* Buried cable, obviously, goes underground. It can be buried directly or housed in a buried conduit or duct. When buried directly, a cable should be protected from moisture, chemicals, and rodent attacks. Loose-tube Mini Bundle cable (Figure 7.9[a]) gives you an idea of how such cable should look. Burying cable inside a conduit protects it much more securely than burying it directly but it's more expensive. Sometimes several cables are placed in one conduit; innerducts are usually used in this situation. Figure 7.10(a) illustrates these structures. To make pulling a cable through a duct easier, the installer uses lubricants.

For aerial installation, a self-support cable like Figure-8 Mini Bundle cable can be used. (See Figure 7.9[b]) Aerial installation requires cable protection from lightning and atmospheric moisture. Figure 7.10(b) demonstrates the idea of aerial installation, illustrating the meaning of the terms *sag* and *span,* which are discussed in Section 7.2.

The above considerations are true for standard outdoor installations. Special work, like submarine installation, is a completely separate topic. Incidentally, submarine fiber-optic communications is a separate branch of the industry with its own research, development, and manufacturing activities.

Fiber-optic cable can be installed *indoors* either directly or using conduits or cable trays. For direct indoor installation, one has to use armored and fire-protected cable. When placing a cable, one has to use pull boxes to break (especially near the corners) long-run spans. These boxes facilitate cable pulling. Vertical installation requires special attention because there are obvious restrictions (cable load, rise conditions, and so forth) for this work. Only special cables—risers—rated as OFNR/OFCR can be used in a vertical run. Vertically installed cable must be clamped and this imposes another set of requirements. All connections should be placed in special housings, which are considered in Section 8.3.

When installing a cable using an innerduct, duct, or conduit, this simple rule should be observed: $d^2/D^2 < 50\%$, where $d$ is the cable diameter and $D$ is the innerduct/duct/conduit diameter.

### Installation Procedure

Installation, in the narrow sense we are discussing the term here, boils down to pulling and placing the cable in the appropriate places. To pull a cable, a pulling grip—also called a pulling eye—has to be connected to the strength member and a swivel has to be used to prevent the cable

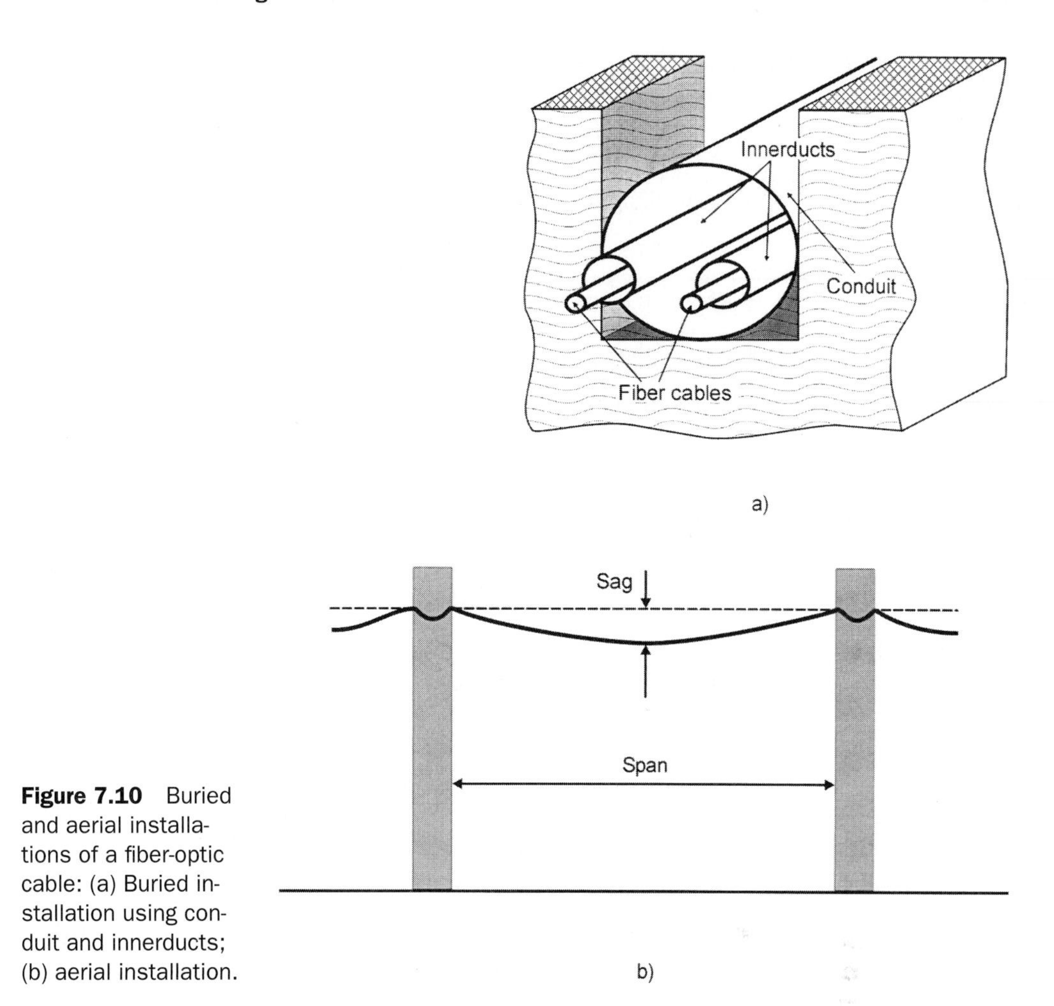


**Figure 7.10** Buried and aerial installations of a fiber-optic cable: (a) Buried installation using conduit and innerducts; (b) aerial installation.

from twisting. A dynamometer is used to control tension during the pull. A detailed description of the installation procedure can be found in references [8], [9], and [10].

A good installer should observe the following recommendations:

**Always:**

- Plan ahead. A detailed plan prior to installation saves time and money.
- Survey the route prior to installation. Observe any possible problems and find solutions before the work starts.
- Keep detailed documentation.
- Pull on the strength member only.
- Constantly monitor tension.
- Use sheaves and guides to maintain a minimum bend radius.
- Constantly maintain two-way communication between installers.
- Include spare fibers and innerducts.
- Test every cable before and after installation.

**Never:**

- Pull directly on the fiber.
- Push the cable.
- Allow tight loops, kinks, or knots to form.
- Exceed the maximum pulling tension (force).
- Exceed the minimum bend radius.
- Exceed the maximum vertical run.
- Twist the cable.

And keep in mind this general but most important rule: *Use common sense when installing fiber-optic cable.*

Following these basic guidelines will help you, as a project manager or engineer, to supervise the installation performed by the contractor.

Installation of fiber-optic cable calls for skills that everyone can acquire relatively quickly. You can even open your own business [11]. Books like those noted at the end of this chapter—references [8], [9], and [10]—combined with a brief training period will enable you to get started in this field and do the work successfully.

## SUMMARY

- The fabrication process for silica optical fibers is a prominent achievement of modern technology. It allows manufacturers to produce fibers with exceptional performance characteristics and to control fiber dimensions to within tenths of a micron. A variety of fabrication processes meet the need for a diversity of fiber properties so that almost any customer's requirements may be satisfied. But new demands for network bandwidth pose new requirements on optical fiber and manufacturers have to be ready to meet the new challenge.
- Optical fibers cannot be used without protective sheaths, that is, fiber cables. Making a fiber cable is a complex process. On the one hand, a cable must provide sufficient protection from environmental and operational hazards, and, on the other hand, this process must introduce minimum damage to the performance characteristics of the bare fiber. The cabling industry provides a wide spectrum of fiber-cable styles and designs to satisfy the myriad needs of today's fiber-optic networks ranging from LANs to transcontinental and transoceanic networks.
- A thorough look at cable installation shows that though this seems to be a very simple process, it requires a professional approach based on knowledge of certain rules and standards.

## PROBLEMS

**7.1.** What are two major stages of the fiber fabrication process? Briefly describe each.

**7.2.** A drying gas flows through a preform during the consolidation phase of the OVD process. Why is this important?

**7.3.** What do all four vapor-deposition processes have in common and how do they differ?

**7.4.** Assume you are the manager responsible for building a new plant for manufacturing an optical fiber. What vapor-deposition method will you choose and why? Do you need additional information? If so, what kind?

**7.5.** In double-layer coating, what properties does each layer have to possess and why?

**7.6.** List four basic components of a fiber cable and explain their functions.

**7.7.** What is the difference between a loose and a tight buffer in terms of operations and applications?

**7.8.** Classify all the cables in Figure 7.9 as either loose buffer or tight buffer.

**7.9.** What do the terms *riser* and *plenum* mean?

**7.10.** You need to install 125 fibers in one cable. What will be the fiber count of the cable you purchase?

**7.11.** Analyze the "rules" an installer must "always follow" and "never follow" (Section 7.3) and give your opinion as to why each rule exists. (For example, the rule to "use spare fibers and innerducts" can be justified as follows: In case of a break in, or damage to, a working fiber, a spare fiber allows us to easily keep all the cable in operational condition. In addition, spare fibers enable a system to be upgraded without installing a new cable. Thus, this rule is crucial from reliability and cost standpoints.) Such a problem allows you to use all your knowledge and helps develop your creativity.

**7.12.** *Project 1:* Sketch a memory map. (See Problem 1.20.)

**7.13.** *Project 2.* Choose one row from any table of any cable data sheet and analyze the meaning of each number. Make a comparison of the different types of cable and compare attenuation and bandwidth data with bare fibers (consult the fiber data sheets).

## REFERENCES[1]


1. H. Turunen, "Tension behavior of optical fiber in finishing processes," 67th Annual Convention of the Wire Association International, Atlanta, April 1997.
2. *Optical Fiber Process Controller* and *High Speed Optical Fiber Flaw Detector* (data sheets), Beta LaserMike, Dayton, Ohio, 1997.
3. *Corning Premises Optical Fiber Tutorial,* Corning Incorporated, Corning, N.Y., 1997.
4. Suzanne R. Nagel, "Fiber Materials and Fabrication Methods," in *Optical Fiber Telecommunications-II,* ed. by S.E. Miller and I.P. Kaminow, Boston: Academic Press, 1988, pp. 121–216.
5. *DLPC7 Coating, a New Coating for Optical Fibres with Superior Aging Behaviour and Improved Cabling Performance* (application notes), Plasma Optical Fibre B.V., Eindhoven, the Netherlands, 1996.
6. *Product Catalog,* Optical Cable Corporation, Roanoke, Va., 1999.
7. (1) *LANscape: Select fiber optics products, 1997;* (2) *LANscape: Design guide, Release 4, 1999;* (3) *LANscape: Premises fiber optics products catalog,* 7th ed., 1999, Siecor, Hickory, N.C.
8. Bob Chomycz, *Fiber Optic Installations (A Practical Guide),* New York: McGraw-Hill, 1996.
9. Eric Pearson, *The Complete Guide to Fiber Optic System Installation,* Albany, NY: Delmar Publishers, 1997.
10. *Lennie Lightwave's Guide to Fiber Optic Installation,* Medford, Mass: Fotec, 1997.
11. William Graham, "Expanding Your Fiber Optic Business in 10 Easy Steps," *Fiber Optic Installer News,* vol. 1, no. 3, 1997.


[1] See Appendix C: A Selected Bibliography

# 8

# Fiber Cable Connectorization and Testing

## 8.1 SPLICING

Splicing is the permanent connection of two pieces of optical fiber. Two types of splicing—mechanical and fusion—are in use today. Splices are of two types: midspan (the connecting of two cables) and pigtail. (A pigtail assembly consists of a fiber that has been factory-installed into a connector at one end, with the other end free for splicing to a cable.) The quality of splicing is measured by the insertion and reflection losses introduced by the splice.

At the beginning of the fiber-optic communications era, splicing was a procedure everybody talked about but nobody was willing to undertake commercially. Telecommunications providers invested countless hours and staggering sums of money to develop splicing techniques and to train their personnel in their use. But today splicing is a largely automated, routine operation that does not require highly skilled professionals. The theory developed in the early years of this technology helped lead to today's new splicing machines and to establish manufacturing tolerances for the fiber's geometry. With the job done, we don't need to delve into this theory very deeply.

Splicing is a very common mass procedure. To establish a fiber-optic network, an installer, depending on the scale of the network, has to make thousands—or even hundreds of thousands—of splices. Thus, any efforts to make splicing more cost-effective and improve the quality of the splices will significantly increase the efficiency of fiber-optic communications technology.

This section reviews the basics of splicing and focuses on the splicing techniques in use today, again from the user's point of view.

### Connection Losses

***Intrinsic losses*** For any type of splicing—midspan or pigtail—connection losses introduced at any splice stem from the fact that *not all light from one fiber is transmitted to another.* The basic reason for this is mismatch and misalignment between the two fibers. Mismatch results when the fiber's mechanical dimensions are out of tolerance and so the problem cannot be resolved by an improvement in splicing technique. We usually call these *intrinsic connection losses.* (See Figure 8.1 and review the basic formulas.)

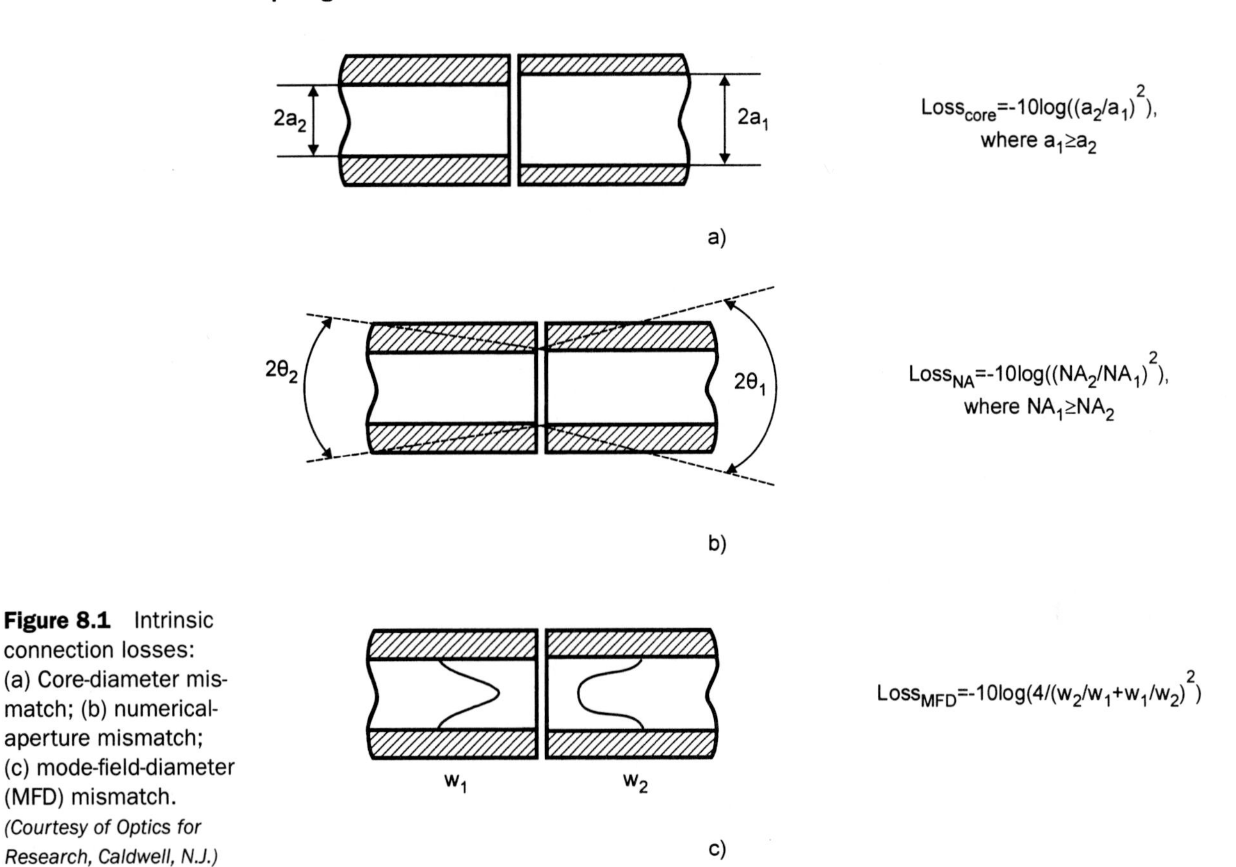


**Figure 8.1** Intrinsic connection losses: (a) Core-diameter mismatch; (b) numerical-aperture mismatch; (c) mode-field-diameter (MFD) mismatch. *(Courtesy of Optics for Research, Caldwell, N.J.)*

The formula for the core-diameter mismatch, shown in Figure 8.1(a) is applicable for graded-index multimode fiber. All these losses are caused by imprecise tolerances of the fiber diameter, *NA*, and MFD. These tolerances can be read from the data sheets.

## Example 8.1.1

***Problem:***

Calculate the intrinsic loss for two 62.5/125 graded-index multimode fibers manufactured by SpecTran (Figure 3.20) that is caused by (a) diameter mismatch and (b) *NA* mismatch. Also, calculate the intrinsic loss caused by the MFD mismatch of two SpecTran singlemode fibers (Figure 6.5).

***Solution:***

(a) A core diameter of the given fiber is specified as 62.5 ± 3 μm. Let's take two extreme cases: 62.5 + 3 μm and 62.5 – 3 μm. Connecting two fibers with diameters of 65.5 μm and 59.5 μm, we can expect $Loss_{\text{core}} = -10\log(a_2/a_1)^2 = 0.83$ dB.

(b) Reading the data sheets (Figure 3.20), we find the margin for *NA:* 0.275 ± 0.015. Again, taking the worst-case scenario, one should connect fibers with *NAs* of 0.290 and 0.260. Thus, the loss is: $Loss_{\text{NA}} = -10\log(NA_2/NA_1)^2 = 0.95$ dB.

Repeating this procedure for the two SpecTran singlemode fibers (Figure 6.5), one finds: $MFD_1 = 9.3 + 0.5$ μm and $MFD_2 = 9.3 - 0.5$ μm. Thus, the loss caused by the MFD mismatch of the two fibers is: $Loss_{MFD} = -10 \log(4/(w_2/w_1 + w_1/w_2)^2) = 0.05$ dB.

In fact, only the formula for MFD loss gives a number close to the real value. Formulas for $Loss_{core}$ and $Loss_{NA}$ estimate only the upper limit of the real loss.

---

Study Figures 8.1(a) and 8.1(b). These sources of insertion loss depend on the direction of light propagation. Indeed, the loss of light coming from a larger core into a smaller core will be higher than it would be were the situation reversed. The same is true for *NA*. If you need to know average insertion loss of a connected fiber, you have to measure loss in both directions.

There is yet another cause of intrinsic loss: Fresnel reflection. Every time light strikes the boundary of two surfaces with different refractive indexes (that's exactly what happens between the ends of two connectorized fibers) some reflection occurs. We have discussed this phenomenon several times already (Formulas 4.31 and 4.31a describe the Fresnel reflection). Loss from this cause can be reduced by using a special gel whose refractive index matches that of the fiber to fill the space between the two ends. Another measure is to place the two ends in physical contact.

***Extrinsic losses*** Extrinsic losses are caused by some imperfections in splicing that, theoretically at least, can be eliminated. Extrinsic losses caused by misalignment, along with formulas for their estimation, are shown in Figure 8.2. (See [1] and [2].)

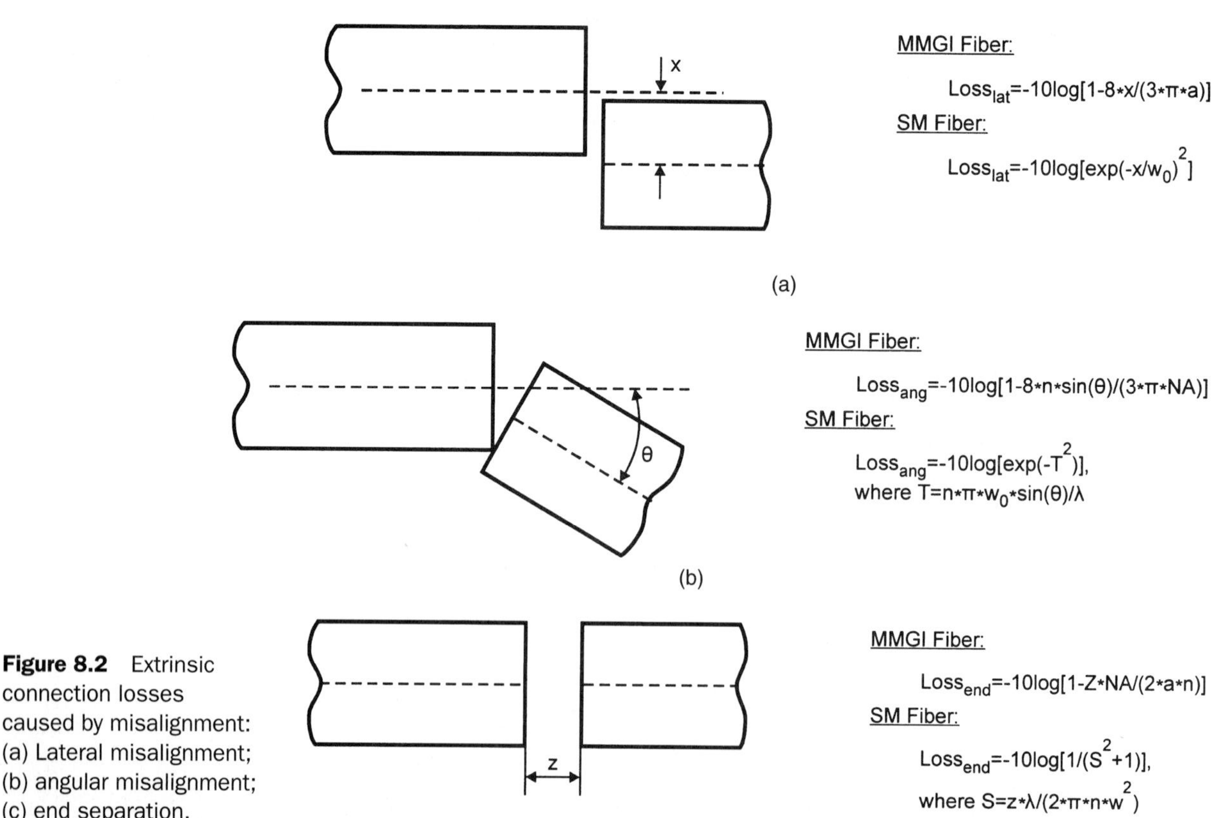


**Figure 8.2** Extrinsic connection losses caused by misalignment: (a) Lateral misalignment; (b) angular misalignment; (c) end separation. *(Courtesy of Optics for Research, Caldwell, N.J.)*

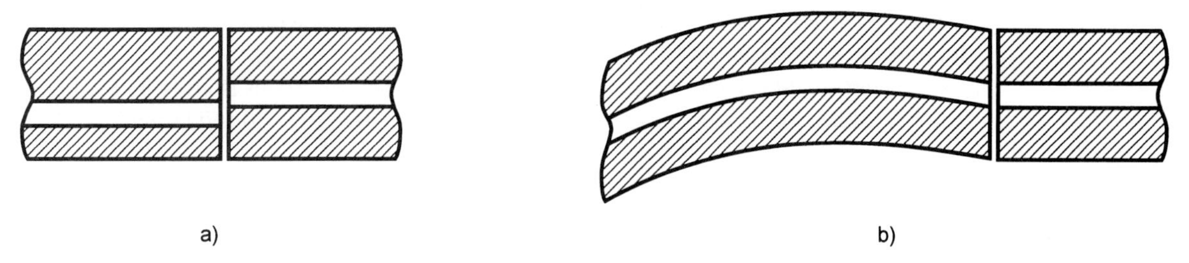


**Figure 8.3** Lateral and angular misalignments caused by tolerances of fiber geometry: (a) Nonconcentricity; (b) fiber curl.

Other reasons for extrinsic losses include improperly prepared fiber ends and foreign particles on the surface of fiber ends. The former can be reduced by a special process for cutting the fiber ends called *cleaving*. Foreign particles can be eliminated by thoroughly cleaning the fiber ends.

Looking at Figure 8.2, it is easy to understand the importance of the fiber-geometry characteristics discussed in Sections 3.5 and 5.4. Since the cladding's outer diameter is the reference point in the alignment of the two fiber ends with reference to each other, fiber curl and a lack of concentricity will cause lateral and angular misalignments, as illustrated in Figure 8.3.

The fiber at the left in Figure 8.3(a) exhibits the absence of core-clad concentricity. (See Figure 3.19.) This defect results in lateral misalignment despite the fact that the outer cladding diameters coincide perfectly. The fiber at the left in Figure 8.3(b) exhibits curl, which results in angular misalignment regardless of how precise the cladding's outer diameter is. (See Section 5.4.)

These examples demonstrate the importance of fiber geometry to splicing quality. We recommend you reread the sections (3.5 and 5.4) devoted to data sheets, where fiber geometric characteristics are described, and analyze the data with regard to their importance in terms of connection losses. Questions to ask yourself include: Why is the tolerance of a cladding diameter important? What roles do the core and the cladding noncircularities play? Why does the manufacturer specify these geometric characteristics but not others? Answers to these questions will definitely help you to better understand the problem.

In view of this discussion, it should be clear why fiber manufacturers pay special attention to fiber geometry. Splicing is one of the mass routines carried out during fiber installation, and reducing its cost means increasing your competitiveness. The better the fiber geometry, the lower the cost of splicing and the more competitive your fiber. It's as simple as that.

***Reflection loss*** Since connection implies the contact of two surfaces, some light will be reflected back. This is called the Fresnel reflection. (You'll recall that it was described in Section 4.2.) Let's take the simplest situation: connecting two ideal fiber ends with some end separation, as shown in Figure 8.4.

Light propagating from fiber 1 toward fiber 2 will be reflected twice: at the fiber-1 air interface and at the fiber-2 air interface. Reflected beams are shown as dotted lines in Figure 8.4.

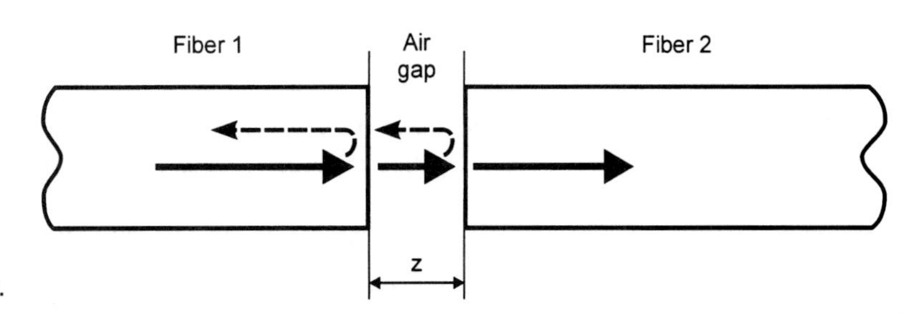


**Figure 8.4** Reflection loss.

Obviously, incident and both reflected beams travel along the same path; we show them separately for illustrative purposes only.

The ratio of the intensity (power) of reflected and incident beams is given by Formulas 4.31 and 4.32. When the incident beam is orthogonal to the interface, $\Theta_i = 0$ and Formula 4.32 yields:

$$R = P_{refl} / P_{in} = ((n_1 - n_2)/(n_1 + n_2))^2 \tag{8.1}$$

Reflection loss, *RL,* is:

$$RL(\text{dB}) = 10 \log R \tag{8.2}$$

*Reflection loss is also called return loss, or backreflection, or backreflection loss, or reflectance.* Assume the gap is filled with air, so that $n_2 = 1.0$. A germanium-doped fiber core has, typically, an $n_1$ of about 1.46; thus, one can calculate $R = 0.035$ and $RL = -14.56$ dB. (This result should be familiar to you from your study of the Fresnel reflection in a basic optics course. About 4% of incident light is reflected at the glass–air interface.)

*Be especially aware of how reflection loss is measured: It is negative and it is in dB.*

Figure 8.4 shows that two reflected beams are traveling along the same line. Consequently, they will interfere with each other, which might result in their canceling each other out, enhancing each other, or some situation in between. This circumstance is formally described as follows: If $R_1$ and $R_2$ are reflections at the first and the second interfaces, respectively, then total reflection is given by [3]:

$$R = R_1 + R_2 - 2\sqrt{R_1 R_2} \cos(4\pi z/\lambda) \tag{8.3}$$

For an air gap, $R_1 = R_2$, hence, $R = 2\,R_1 - 2\,R_1 \cos(4\pi\, z/\lambda)$. Since the cosine may vary from +1 to −1, $R$ might be 0 to 4 $R_1$.

Reflection loss is minimized by several methods. Fusion splicing doesn't leave a gap; thus, there is no reflection loss. When mechanical splicing is used, the gap is filled with some adhesive material having a refractive index close to $n_1$. It is called *index matching material* and it reduces reflection dramatically. Another means is to bring the fiber ends into physical contact. Doing so requires perfect end surfaces. This can be achieved by polishing. Typical reflection loss for contemporary mechanical splices is in the range of –45 dB to –55 dB.

Reflection obviously changes the amount of power transmitted toward a receiver. *This insertion loss, caused by reflection, is called Fresnel loss.* Since transmitted power equals $P_{trans} = P_{in} - P_{ref}$, we can easily calculate this loss from Formula 8.1:

$$Loss_{Fresnel} = -10\log\left(4n_1n_2/(n_1 + n_2)^2\right) \tag{8.4}$$

For 2 interface reflections, Fresnel loss is given by:

$$Loss_{Fresnel} = -10\log(1 - R), \tag{8.5}$$

where R is determined by Formula 8.3.

## Splicing Procedure

It is not our intention to train you to be a professional splicer, so the following discussion of splicing procedure will be short and serve only to introduce the basics to the potential supervisor.

***End preparation*** Regardless of splicing technique, the first step in splicing is end preparation. This involves *stripping* the cable jacket, the buffer tube, and the coating. Each of these stripping steps is done with specific tools. For example, stripping the cable jacket means stripping the jacket and aramid yarns. Since aramid (Kevlar) is extremely strong, a special tool for cutting it should be used. If a strength member is a tube, an especially strong tool is required for fast, safe stripping. For gel-filled loose buffer cable, a gel cleaning solution is needed.

Stripping the coating is the most delicate procedure because one doesn't want to damage the fiber. Fiber manufacturers make every effort to fabricate a coating strong enough for good fiber protection and, simultaneously, soft enough to be easily stripped. (See Section 7.1.)

All the tools necessary for end preparation are commercially available. The variety of tools is amazing, ranging from simple manually operated pocket strippers that enable you to make only one operation (stripping a coating, for example) to a fully automated machine that enables you to do all the operations in one step—inserting an entire cable and coming out with a bare fiber ready for splicing. Strippers are available to strip any type of cable and any number of fibers, including ribbon cable containing up to 12 fibers.

Stripping exposes a bare fiber. To prepare fiber ends for connection, you need to *cleave* them, that is, cut the fiber with a special tool to make the end surface flat, clean, and perpendicular to the fiber's centerline. The importance of good cleaving to the quality of splicing—reducing connection loss—cannot be overemphasized. Any surface not properly prepared will cause additional loss due to the separation of its ends. This point is visualized in Figure 8.5.

Keep in mind that a splicer works with fibers whose minimum size is measured in microns and whose tolerances are within half microns.

There is a wide variety of tools (cleavers) on the market—from a pen-size manual instrument to a fully automated machine that performs cleaving in a one-step operation. Most use diamond blades for cutting. The newest devices add ultrasonic vibration to the blade, thus making a cut without applying undue compressive stress to the fiber. Most cleavers guarantee an average cleave angle of less than 0.5°; in fact, 95% or more of cleave angles are less than 1°. This means a cleave angle measured at any point along 95 percent of the end surface is not more than 1°. (See Figure 8.5[a].) In other words, these numbers—95% and 1°—characterize how flat the cleaved surface is and are used to evaluate the quality of a cleaver.

*Cleaning* the fiber with a special solution and cloth is necessary to keep the end surface free of any foreign particles.

You can readily find on the market splicer tool kits equipped with all the necessary tools required for preparing the fiber for splicing.

The new generation of fibers (see Section 6.4) requires even greater accuracy in stripping, cleaving, and cleaning.

***Mechanical splicing*** As mentioned in the introduction to this section, the splicing techniques are fusion and mechanical splicing. The schematic of a mechanical splice (Figure 8.6) shows how two pieces of stripped, cleaved, and cleaned fibers are inserted into the splicing device, where their ends are directed toward each other. They come into optical contact and are held by an adhesive or clamp (Figure 8.6[a]). The heart of the device is an alignment guide: the better the fiber's alignment, the higher the quality of the splice, that is, the less the connection loss. A number of arrangements are used for alignment guides, with the V-groove (Figure 8.6[b]) being the most popular. Many different designs of mechanical splices are commercially available.

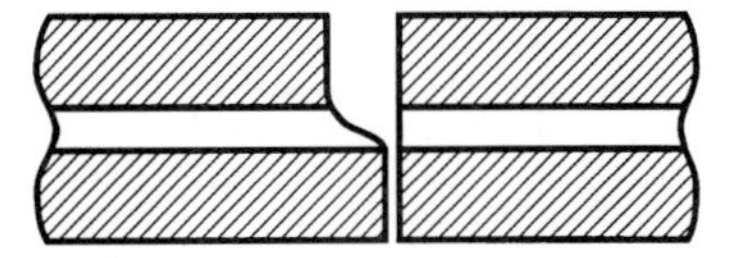

a)

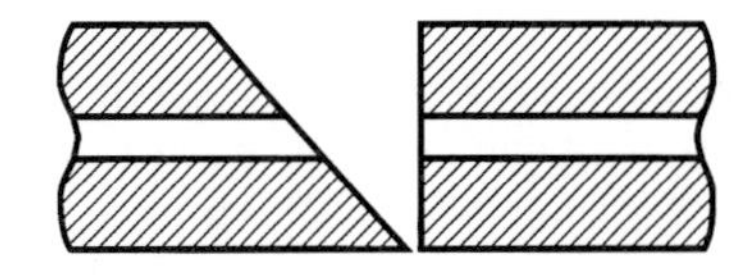

b)

**Figure 8.5** Connection loss caused by bad cleaving: (a) Nonflat ends; (b) large cleave angle.

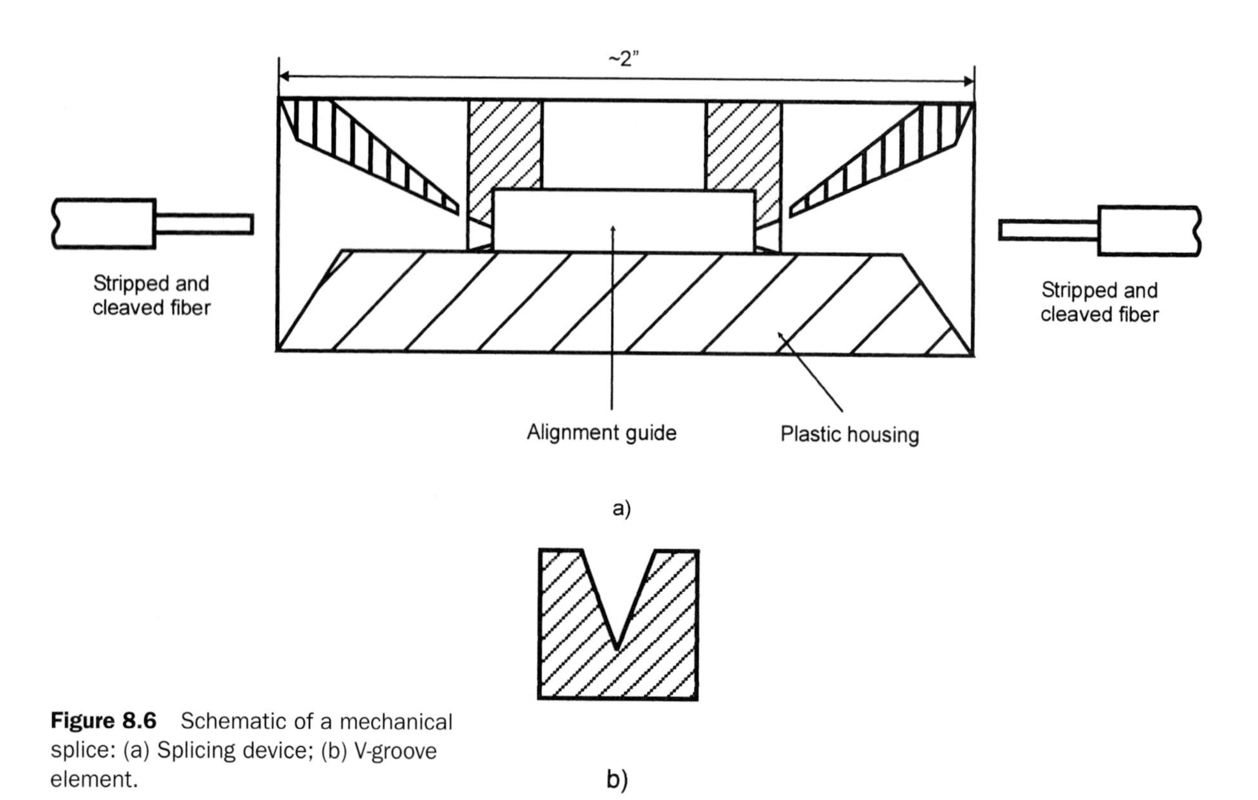


**Figure 8.6** Schematic of a mechanical splice: (a) Splicing device; (b) V-groove element.

High-precision optical-fiber geometry and improvements in mechanical-splice technology make mechanical splices very competitive products. Splices with insertion loss as low as 0.2 dB per splice are quite common and many manufacturers claim that their splices provide 0.1 dB. Their reflection loss varies from –45 dB to –55 dB. Some are preloaded with index-matching gel, whose function is to minimize reflection loss and fix fiber ends. Other mechanical splices use UV curing for adhesion. Some types of mechanical splicing do not use any adhesive but, instead, fix fiber ends mechanically.

The cost of an individual mechanical splice is relatively high. It varies from several dollars to several hundred dollars. Typically, an experienced professional can execute a splice in 30–40 seconds. Mechanical splicing is usually used for a quick repair and when only a small number of splices are required.

***Fusion splicing*** Fusion connects two fiber ends by melting them. This process is similar to welding metallic wires and is usually accomplished by use of an electric arc. A schematic of fusion splicing is shown in Figure 8.7(a).

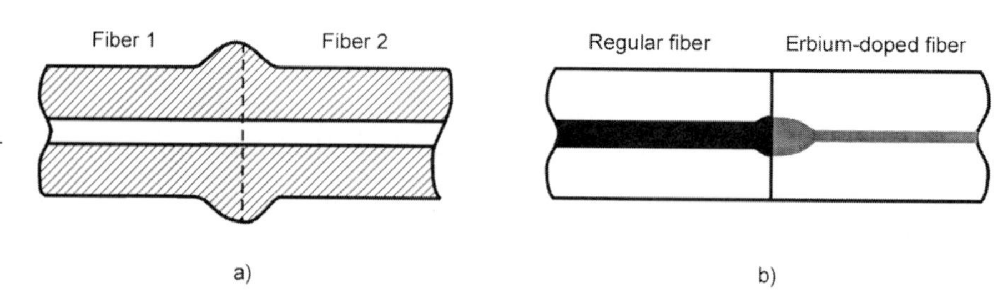


**Figure 8.7** Schematic of fusion splicing: (a) Regular fusion splicing; (b) fusion splicing of erbium fiber. *(Part b courtesy of Ericsson Cables AB, Network Technologies, Stockholm, Sweden.)*

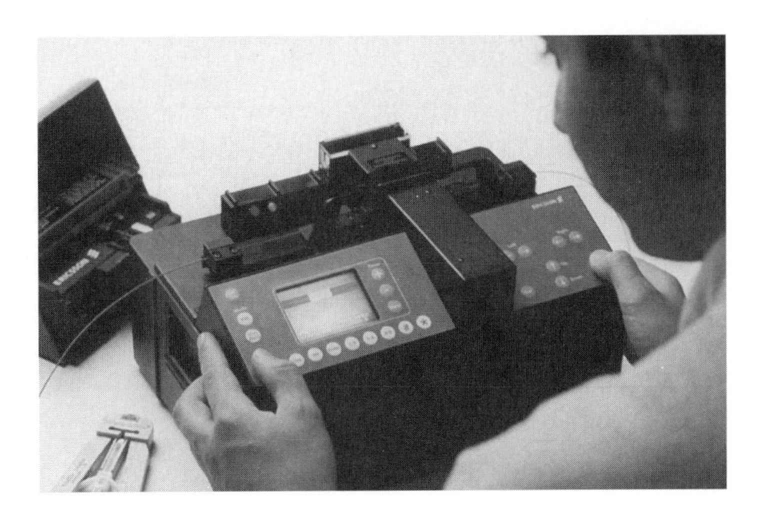

**Figure 8.8** A fusion splicer. *(Courtesy of Ericsson Cables AB, Network Technologies.)*

Fusion splicing connects fibers without a gap; therefore, no reflection loss is introduced. Insertion loss, also minimal, is in the range of 0.01 dB to 0.15 dB.

Fusion splicing is done by special machines called fusion splicers. A fusion splicer is a fully automated desk-size apparatus controlling the quality of splices during its operation. (See Figure 8.8.) The cost of fusion splicers varies from several thousand dollars to tens of thousands of dollars, the ones most widely used ranging from $25,000 to $40,000. This costly investment can be justified when you are doing a large number of splices or are looking for a high-quality splice. As you can now see, mechanical and fusion splicing have their separate, specific applications.

Commercially available fusion splicers can be found with *passive* or *active* fiber alignment. Passive alignment is done mechanically, usually with very precise V-grooves. This operation is very similar to the alignment for mechanical splicing. Tightening of standards on fiber geometry has increased the popularity of these splicers. They are less costly, easy to operate, and less bulky than *active*-alignment splicers, but you cannot expect exceptionally low splicing loss from them. Mass-fusion splicers and mini-splicers are based on passive alignment, with typical loss from 0.03 dB to 0.07 dB when they are automatic and up to 0.15 dB when they are operated manually.

Active-alignment splicers use two technologies to align the fibers for splicing: a *profile-alignment system (PAS)* and *local injection and detection (LID)*. Power-alignment technology (PAT) is the latest generation of the LID method. Schematics of these systems are shown in Figure 8.9.

In the PAS system, video cameras capture images of the two fiber ends, a microprocessor interprets these pictures, and positioners align the fibers under microprocessor control (Figure 8.9[a]). The goal of a PAS system is to align the cores of fibers. Since the dimension of a wavelength of visible light—between 0.4 μm and 0.7 μm—is comparable to the precision of alignment, the PAS system has a physical limit. This limit is governed by the following physical principle: The minimum size we can measure using an optical instrument is more than the operating wavelength of this device.

In LID or PAT systems, fibers are bent and light is locally injected and detected where the bending occurs (Figure 8.9[b]). A microprocessor commands the positioners to move the fibers until maximum power transmitted through a connection is attained. The precision of this system is limited by the technology, such as how light emerges from the fiber, how sensitive the receiver is, and so forth.

A detailed comparison of PAS and LID splicers shows that the LID outperforms the PAS

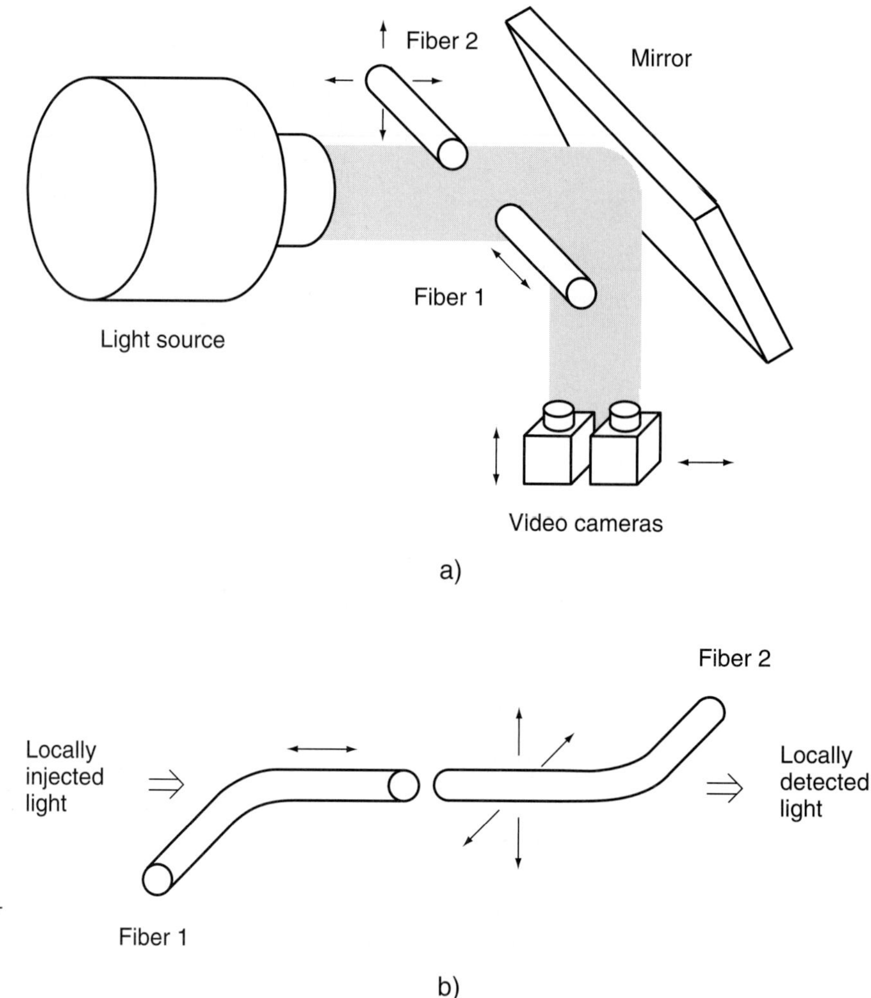


**Figure 8.9** Schematics of fusion-splicer systems: (a) Profile-alignment system (PAS); (b) power-alignment technology (PAT).

in precision of alignment and loss estimate but that only the PAS can perform end inspection and find and reject fibers with defects caused by bad end preparation (cleave angle and dirt particles) [4]. Both active-alignment splicers demonstrate excellent quality of splices with insertion loss between 0.01 dB and 0.02 dB.

Regardless of the alignment system, a modern fusion splicer can splice singlemode and multimode fibers, operate with single fiber and multifiber cables (including ribbon), provide on-screen viewing of the fusion process, estimate fusion loss, store data, and interface with a regular computer for analysis. Splicers are programmable, which allows you to choose the optimal regimen for splicing different types of fibers.

New fibers—that is, ones that have a large effective area and are non-zero dispersion shifted, polarization maintained, and erbium doped—pose a new challenge for splicing technology. For example, the erbium-doped fiber has a mode-field diameter of about one-half that of a regular telecommunications fiber. This could result in a huge connection loss during splicing. (See Figure 8.2[a].) A fusion splicer, under the control of special software, makes core dopants diffuse into the cladding, creating a tapered region, as shown in Figure 8.7(b). PM fibers, as well as being aligned in the *x, y,* and *z* directions, are also rotated around their centerlines to align their fast and slow axes.

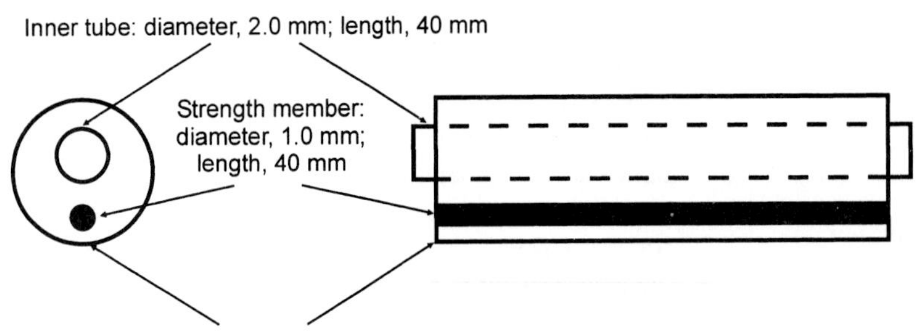


**Figure 8.10** Basic structure of a fusion-protection sleeve.

To conclude the list of features of modern fusion splicers, we should mention their capacity to provide both attenuation splicing, which introduces controlled connection loss to make a splice joint work as an attenuator, and tapering, which produces a tapered fiber with a semispherical end acting as a microlens [5].

***Fusion splice protection*** As noted previously, a fiber's coating is stripped before the fiber is spliced. The mechanical splice is itself a container protecting the spliced fiber, but fusion splicing leaves the spliced area unprotected. Thus, the equipment for fusion splicing includes fiber-recoater machines, which coat the spliced area. Alternatively—or in addition—fusion-splice *protection sleeves (bushings)* are used. These are small layered tubes whose basic structure and typical dimensions are shown in Figure 8.10.

After the protective sleeves are placed on the spliced area, they are heated. An inner tube is made from heat-shrinkable material so that the sleeve is fixed firmly to the protected area. A strength member prevents this area from bending.

### Conclusion

Splicing is inevitable and is the most frequent operation during the installation of a fiber-optic network. Where splicing takes place in the field during the installation of long-haul links, special trailers are used. For a LAN installation, portable equipment is obviously more suitable. The variety of splicing techniques matches the variety of fiber-optic applications. Mechanical and less-precise fusion techniques are used for multimode-fiber splicing, while precision fusion splicing is used for connecting singlemode fibers. Today, splicing is a routine operation supported by a well-developed range of equipment that provides excellent connection characteristics. All splices must be protected in splicing enclosures. This topic is discussed in Section 8.3.

When choosing a splicing method (mechanical or fusion) and technique, you must weigh both performance (end-to-end attenuation) and cost considerations to reach the right decision for the application at hand.

## 8.2 CONNECTORS

The installation process includes the connectorization of two fibers. This is accomplished by splicing them permanently and linking them temporarily through connectors. Since connectors constitute the end of a fiber link, attaching connectors is called cable termination. There are two means by which to attach a connector: field installation and pigtail. The former entails attaching a separate connector to a fiber cable directly in the field; the latter is a factory-installed connector with an attached piece of fiber that requires splicing the connector's fiber to your cable. Field installation is much more flexible and cheaper but introduces higher loss. Pigtail termination is more expensive but provides a higher-quality connection. This procedure is used primarily for single-mode fibers. We'll focus in this section on connectors themselves and their field installation—fiber-optic cable termination.

We need a connector to temporarily connect an optical fiber to another fiber or to a transmitter/receiver (transceiver). Thus, a connection system includes a connector and a receptacle whose function is to accept an optical signal with minimum loss. For this reason, a connector is also called a *plug*.

In fiber-optic communications systems, the most common point of failure is at the connector, yet about 30% of the installation cost is for labor associated with connectorization. Thus, connectors are vital components of the system both from performance and economic standpoints.

## Connectors —A Basic Structure

There are many types of connectors but, regardless of the specific design, each has the following basic components: a ferrule, a latching mechanism, a backshell, a crimp sleeve, and a boot. The basic structure of an assembled connector is sketched in Figure 8.11.

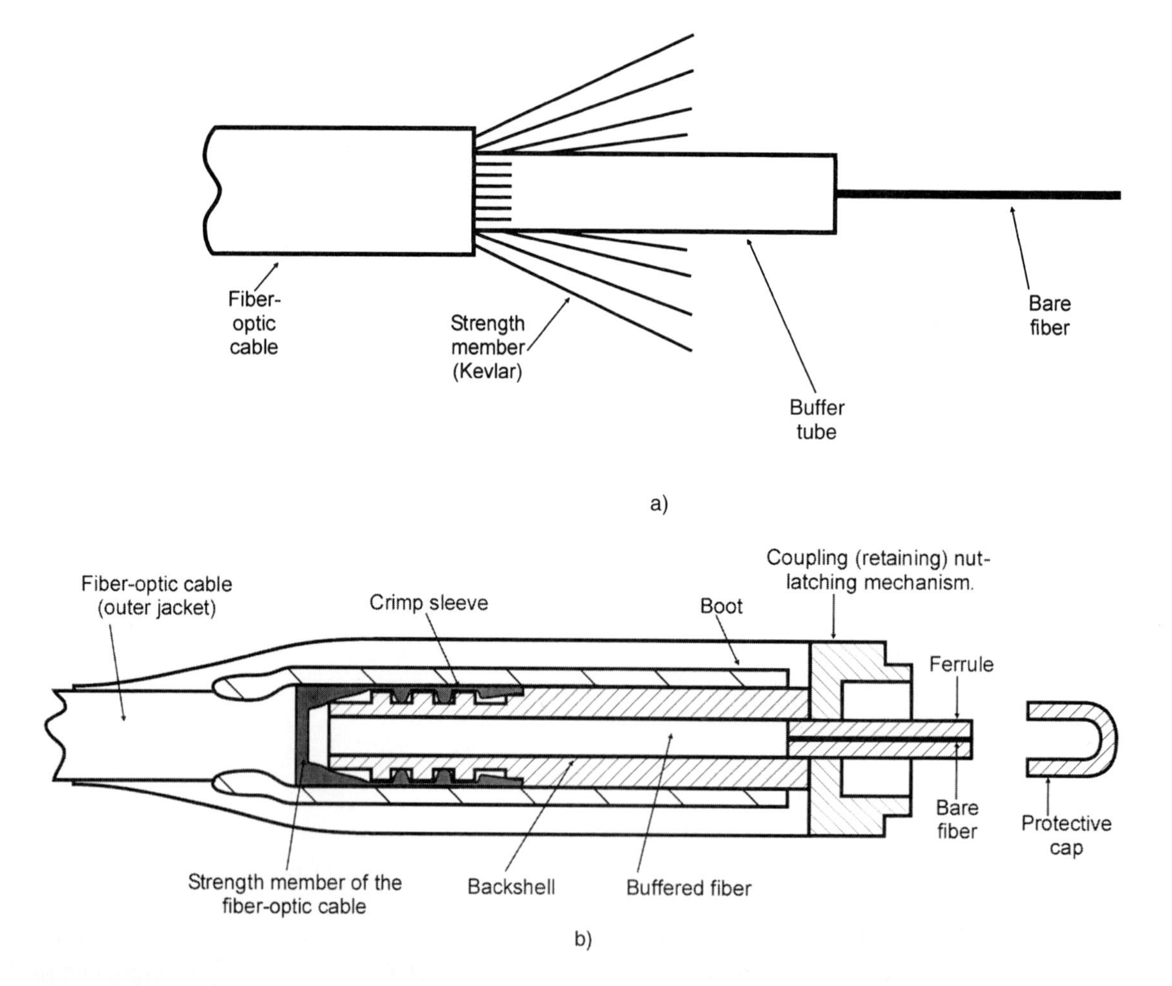


**Figure 8.11** Typical structure of a connector: (a) Fiber-optic cable prepared for termination; (b) connector.

Figure 8.11(a) shows how a fiber-optic cable is prepared for termination: The outer jacket is stripped, the strength member (here, aramid yarns) is revealed, and the buffer is stripped to disclose the bare fiber. Such a prepared cable end is inserted into the connector so that the ferrule houses the bare fiber. The buffer is placed within the backshell, the strength member is laid over the backshell, and the crimp sleeve affixes the strength member to the backshell (Figure 8.11[b]). The boot covers the entire assembly from the latching mechanism back to the cable, preventing the cable from bending while the technician is working with the connector. The specific implementation of this procedure can be quite different from installer to installer but the basic ideas are the same: We have to affix the cable with respect to the connector. This is done by positioning the strength member between the backshell and the crimp sleeve, and we have to guide the fiber itself, which is done by using the ferrule inside (to which the fiber is attached with epoxy).

A latching mechanism—a coupling, or retaining, nut shown in Figure 8.11(b)—fixes the connector with respect to the receptacle so that light is transmitted from the fiber cable to the receiver. Thus, the fiber core is aligned with respect to the ferrule through the ferrule hole, the ferrule is aligned with respect to the latching mechanism, and the latter is aligned with respect to the receptacle. If you want to use a connector-receptacle pair to connect two singlemode fibers, you have to align their cores with the precision of a core tolerance, that is, between 1 and 0.5 microns. This is why some manufacturers use active alignment to reduce the core offset to 0.25 microns.

Connectors come from their manufacturers with protective caps, which screen the endface of the ferrule for dust, dirt, scratches, and any other possible blemishes.

Analyzing the basic structure of a connector, one can easily understand how it performs its three major functions: retention, end protection, and alignment.

## Major Characteristics

***Insertion loss*** The first—and most important—characteristic of a connector is insertion loss. Indeed, you might have an ideal fiber, do perfect splicing, yet lose your entire signal at the connection point between the fiber and the transceiver. Minimization of insertion loss starts with the use of a boot to minimize the bend loss that occurs when one connects or disconnects a cable. A second measure to reduce loss is to attach the strength member of the cable—for example, the aramid yarn—to the connector, thus releasing the fiber itself from any tension. The third, and probably most critical, thing to do is use the ferrule to protect the bare fiber and direct it to the receiver.

Insertion loss is given by the manufacturer in two numbers: average (or typical) and maximum. The latter helps you reject a certain connector if your system cannot afford this maximum loss. For regular connectors, the average insertion loss today is about 0.25 dB. This figure can vary from 0.1 dB (ideal) to 1 dB. The maximum insertion loss is about 0.5 dB, varying approximately from 0.3 dB to 1.5 dB.

***Return loss (reflectance or backreflection)*** Return loss is a problem for a connector used for singlemode fibers, but not a factor in multimode systems. Refer to Section 8.1, where the physics of return loss and measures to deal with them are discussed. For a connector, the problem of return loss arises from a simple conflict: To minimize insertion loss, one needs to polish the fiber end as thoroughly as possible, thus increasing return loss.

Recalling that backreflection occurs at the core–air interface (see Figure 8.4), installers came up with this effective solution: Eliminate the air gap by bringing two connectors into *physical contact (PC).* Most of today's connectors are arranged this way. Since making a perfectly flat surface for ideal physical contact is impossible, manufacturers finish ferrule endfaces with different profiles, the most popular of which are shown in Figure 8.12(a).

As you can see from Figure 8.12(a), to improve physical contact, one must reduce the contact area because the quality of a smaller area can be controlled more effectively. The polishing process has been so improved that manufacturers have been able to reduce the return loss of PC connectors from –40 dB, which it was just a few years ago, to –55 dB today, simultaneously holding average insertion loss to an acceptable 0.2 dB level. Some manufacturers do additional polishing to achieve an even better performance. So today we have what are called ultra-polishing connectors (UPC) or enhanced ultra-polishing connectors (EUPC), each of whose backreflections are about –60 dB.

Return loss can be reduced even more by polishing the ferrule endface at an angle of 8° with respect to the centerline of the hole. Such connectors are designated *angled physical contact (APC),* or *angled polishing connectors.* The ferrule end of an APC connector is shown in Figures 8.12(b) and (c). It's clear that with this endface, reflected light will be directed into the fiber cladding and be lost. Average return loss at the –75 dB level is common for APC connectors but, as you are aware, achievement carries a price. Refer again to Figure 8.12(b). This endface tends to increase the insertion loss, but today manufacturers manage to keep the insertion loss of APC connectors between 0.2 dB and 0.5 dB, which is the same as for PC connectors.

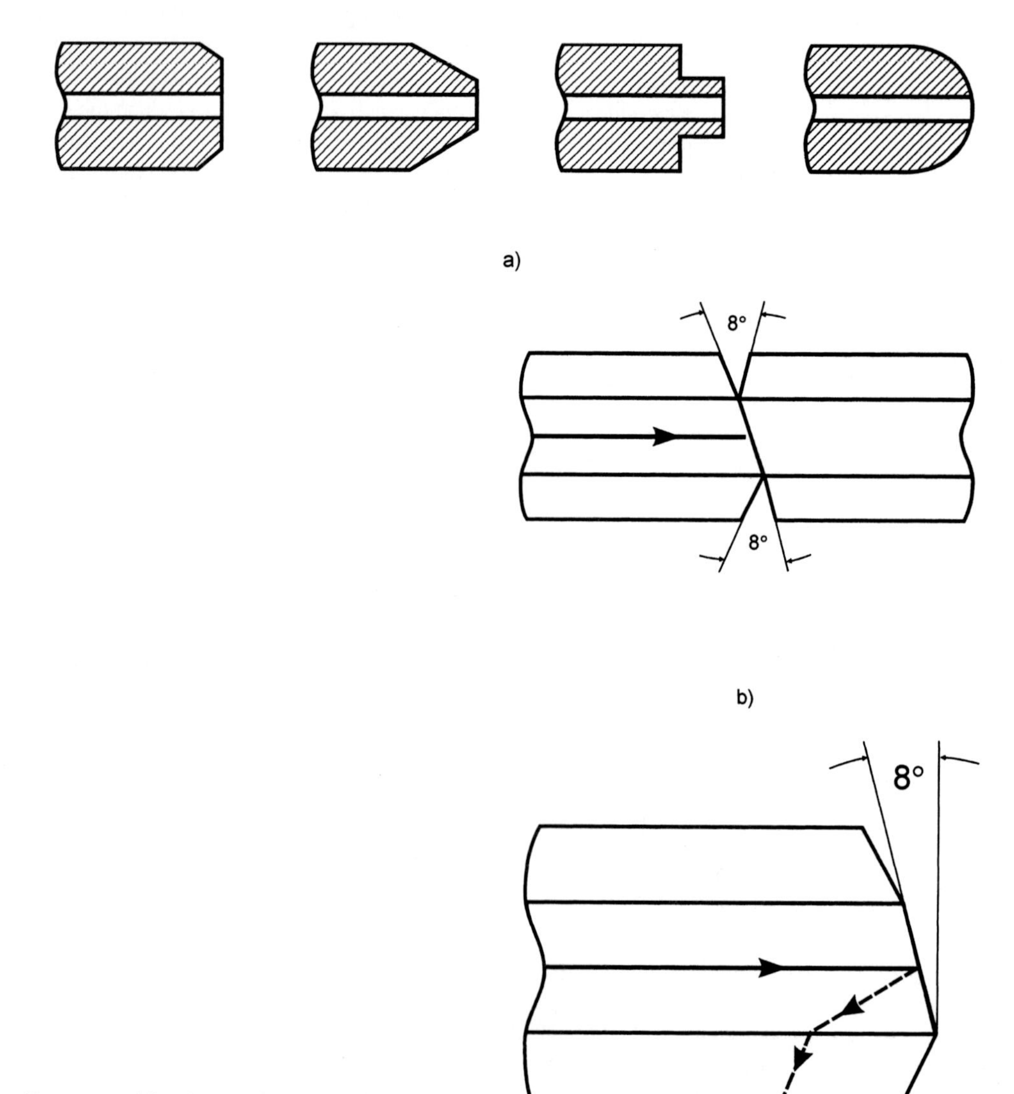


**Figure 8.12** Ferrule endfaces used for decreasing return loss: (a) Ferrule endfaces for improvement of physical contact; (b) angled physical contact (APC); (c) APC in large scale.

The shortcoming of APC connectors is that they must be oriented properly to mate with another APC. This is done by incorporating an antirotating key in the latching mechanism. The problem is, however, that this key makes connector standardization even more difficult. (See the discussion of connector standards on pages 265 and 266.)

### Example 8.2.1

***Problem:***

Why is the polishing angle for APC connectors 8°?

***Solution:***

Reflected light is simply light that attempts to propagate along a fiber core. The typical numerical aperture of a standard singlemode fiber is 0.13 (see Figure 5.18), which corresponds to an acceptance angle of 7.5° ($NA = \sin \Theta_a$). A polishing angle of 8° causes reflected light to propagate at an angle larger than the acceptance angle, thus making this light eventually disappear. (See Figure 8.12[b].)

---

***Repeatability (Durability)*** A connector is used as a temporary connection, so it should be able to hold its characteristics after many connect-disconnect operations, called *matings.* Thus, repeatability, or durability, is one important connector characteristic. It shows that insertion loss increases after a certain number of cycles, or matings. Typically, this increase is < 0.2 dB for 500 matings.

A clean ferrule endface, immune to scratches and other microdamage, can hold up during many matings. Just remember, it is the installer's responsibility to keep a connector clean from dirt and dust.

***PM connectors*** There are special connectors designed for specific applications. One example is a polarization-maintaining (PM) connector, whose characteristics include extension ratio, *ER* (see Section 6.3), and axis of orientation. Typically, *ER* is 25 dB, and either SLOW or FAST is shown as the orientation axis.

## Connector Styles—Yesterday, Today, and Tomorrow

If you shop around the country looking for connectors to transmit data through telephone lines, you'll find only one—the well-known RJ-45 jack. But if you examine fiber-optic connectors, you'll find a number of them, and this is the unfortunate element in our connector story. Today, we have many fiber-optic suppliers, each developing its own proprietary connector that is totally incompatible with anyone else's. And this poses a serious industry problem. Although your fibers may come from different manufacturers and have slightly different characteristics (see Figures 3.18 and 3.20), you can splice them, which means you *can* make them compatible. Unfortunately, this is not the case for connectors. Figure 8.13 shows sketches of the most popular connectors on the market today along with the newest connectors that will dominate the field tomorrow.

Let's briefly review the styles shown in Figure 8.13:

- The *biconic* style was developed by AT&T in the early 1980s. It has been actively used in telephony, its numbers totaling in the millions. But its time has passed because its characteristics cannot be improved to meet today's specifications.
- *SMA* (D-4) connectors, developed about 20 years ago, are still being manufactured to meet the requirements, obviously, of today's technology.

**Biconic Connector**

- Rugged Prevex hardware
- Conical ferrule design
- High precision, tapered ends
- Special tip length gauges provide low insertion loss
- Meets TIA 604.1 Fiber Optic Connector Intermateability Standard (FOCIS-1)

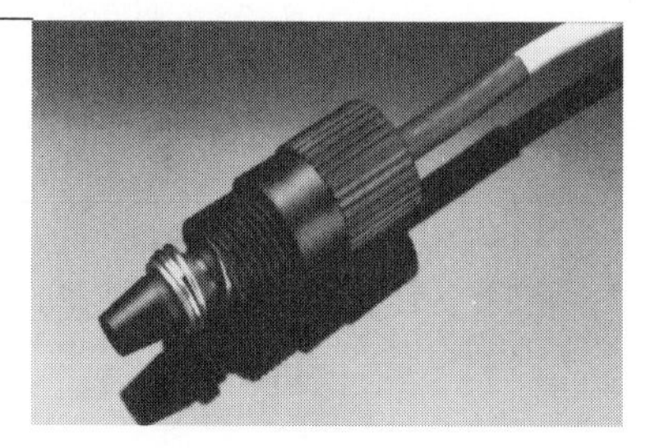

**D4 Connector**

- Cylindrical metal coupling nut with keyed sleeves
- 2.0 mm ceramic ferrule
- Pull-proof ferrule for durability
- For long-haul and local network applications
- Manufactured to Japan Industrial Standard (JIS) C 5971 specifications
- Ultra PC polish available

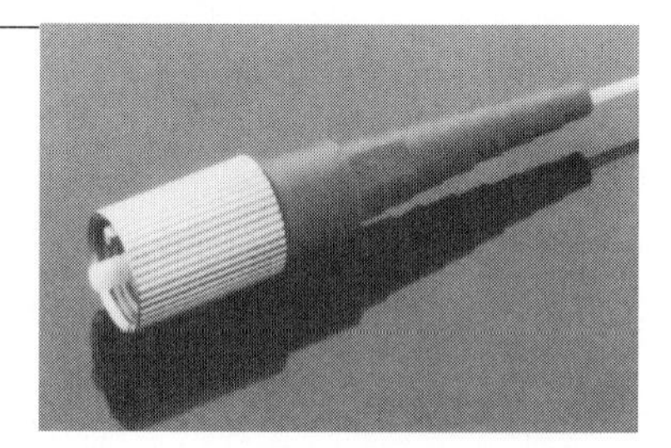

**ST® Compatible Connector**

- Twist-lock bayonet coupling
- 2.5 mm keyed ferrule assembly meets durability and repeatability requirements
- For long-haul and local network applications
- Ultra PC polish available

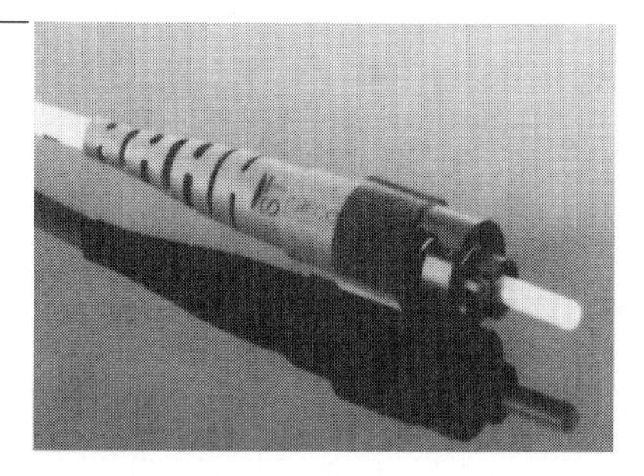

**Figure 8.13** Connectors for today and tomorrow. *(Courtesy of 3M Telecom Systems Division, Austin, TX)*

- *ST, SC,* and *FC* connectors, with their impressive characteristics, dominate the market today.
- *FDDI* connectors were developed for application in FDDI networks, which are essentially LANs. Their characteristics are not very impressive (typical insertion loss is about 0.5 dB) but they are convenient and reliable.
- The *ESCON* connector was developed by IBM for the its own network system carrying the same name.
- *MTP* connectors are used for ribbon-fiber cables.

The ferrules of conventional connectors described above have outer diameters of 2.5 mm, which results in 12.5 mm fiber-to-fiber spacing (pitch) of a duplex connector. The size of a connector becomes an issue because we need to connect hundreds and hundreds fibers at the distribution frames (see Section 8.3). To meet these new requirements, manufacturers reduce the connectors' size. The following list represents the new generation of connectors. They are usually called *small form factor* connectors.

### SC Connector

- Square, push-pull latching mechanism
- Keyed, molded housing provides optimum protection
- 2.5 mm ferrule, pull-proof design
- Available as a duplex connector (568SC) compliant to requirements of TIA/EIA-568A and Fibre Channel standard
- Manufactured to Japan Industrial Standard (JIS) C 5973 specifications
- Composite ferrule available for multimode
- Ultra PC and Angled PC polish available

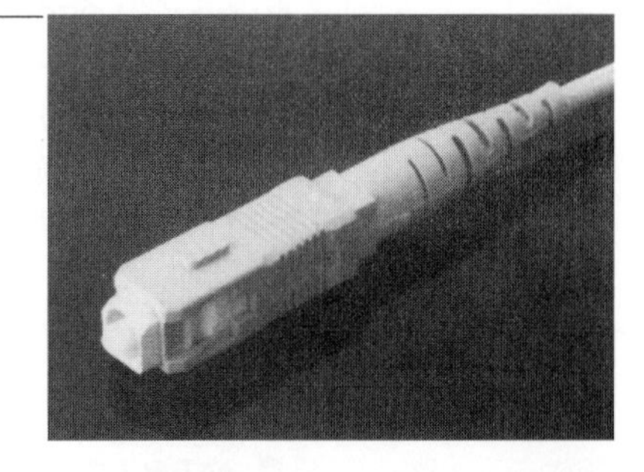

### FC Connector

- Cylindrical metal coupling nut with keyed sleeves
- 2.5 mm ceramic ferrule
- Pull-proof ferrule for durability
- For long-haul and local network applications
- Manufactured to Japan Industrial Standard (JIS) C 5970 specifications
- Ultra PC and Angled PC polish available

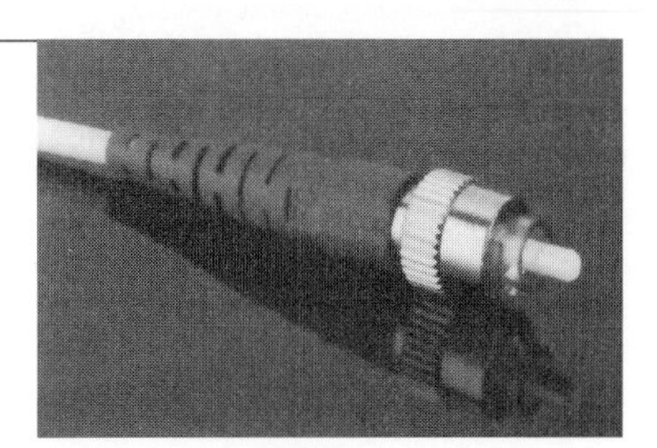

### FDDI (ODC® II) Connector

- Duplex FDDI connector
- Push-pull design for quick and easy connections
- Three keys for versatile assignments in the network
- 2.5 mm ferrule

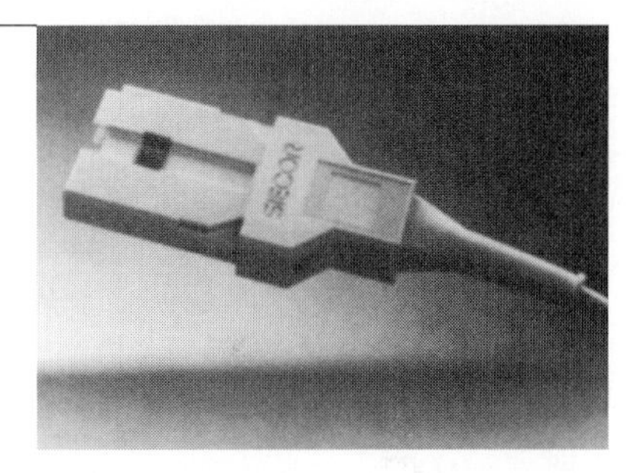

### ESCON® Compatible Connector

- Designed for IBM ESCON and IBM ESCON compatible systems
- Duplex connector
- Retractable shroud protects ferrules

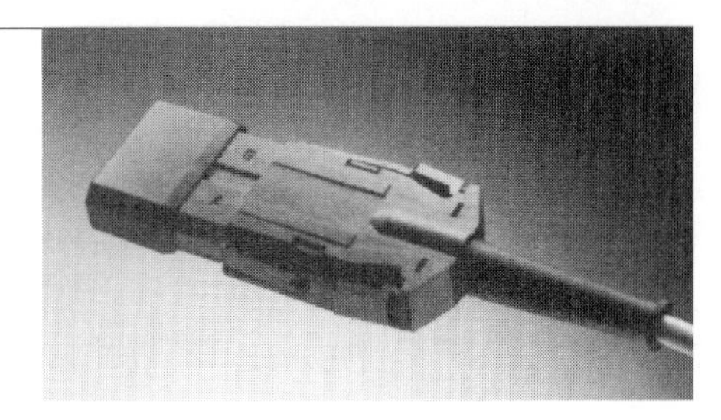

**Figure 8.13** (continued)

- *MT-RJ* and *FJ* (fiber jack) are inserted into the telephone RJ-45 jack. Interestingly, the MT-RJ connector has been introduced through the combined efforts of AMP, Siecor, Hewlett-Packard, USConec, and Fujikura, each a major player in the manufacture of fiber-optic equipment.
- *SC-DC* is the newest Siecor development and, in addition to Siecor, it is being manufactured by Siemens and IBM. This connector has a ferrule that can fit two (DC, or duplex) or four (QC, or quadruple) fibers within a traditional SC body.

### MTP™ Connector

- Multifiber connector (4, 6, 8, 12)
- High-density interconnect and OEM applications
- Angled PC polish available
- Single-mode and multimode designs

### MT-RJ Connector

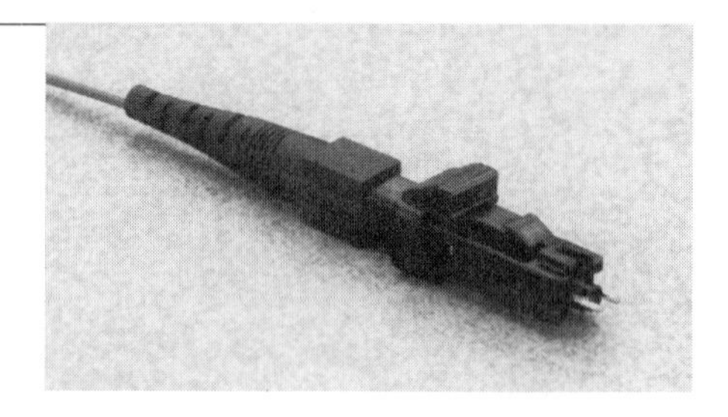

- 2-fiber, dual connect, single ferrule design
- RJ-style snag-free latch
- Multimode and single-mode performance to TIA specifications
- High density interconnect compatible with MT-RJ small form factor transceivers

### SC-DC™ Connector

- 2-fiber, dual connect, single ferrule design
- 2.5 mm composite ferrule
- Familiar SC housing with push-pull mating mechanism
- Multimode and single-mode performance to TIA specifications

### MU Connector

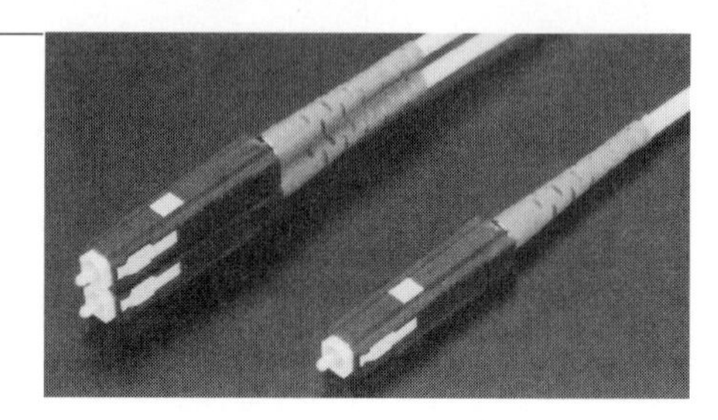

- Square, push-pull mating mechanism
- 1.25 mm ferrule, pull-proof design
- Reduced footprint
- High-density interconnect and OEM applications

**Figure 8.13** (continued)

■ MiniMAC Connector

- Form a high-density push-pull fiber interconnect system with a shorter housing.
- MiniMACs use precision etched silicon chips.
- Can accommodate up to 32 single mode of multimode fibers.
- Is available in 2, 4, 8, 12, 16, 18, and 32 fiber configurations.

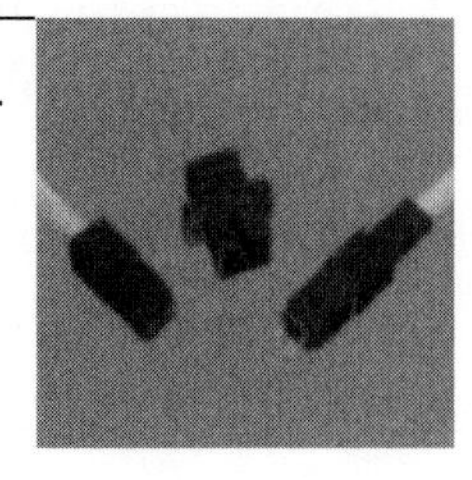

**Figure 8.13** (continued)

- The *MU* connector was developed by NTT (Japan). Its ferrule is only 1.25 mm in outer diameter and its package is about one-quarter the size of a regular connector.
- The *Mini-MAC,* from Berg Electronics, is designed for use with ribbon-fiber cable; it can handle up to 32 fibers. This is a miniature version of an MTP connector.
- The *LC* connector (not shown) was developed by Lucent Technologies. Again, the outer diameter of its ferrule (1.25 mm) is half the size of the currently used ST, SC, and FC connectors. This enables the fitting of two fibers into the same connector frame.

The newest trend is the development of *ferruleless* connectors. Such connectors come from the manufacturer with a short piece of the fiber pre-installed inside the connector. The termination process here simply entails inserting a connected fiber into the connector, where a V-groove alignment hole makes the optical contact. Thus, termination becomes similar to mechanical splicing. This development is based on substantial achievements in the tightening of the dimensional characteristics of fiber. The advantages of this type of connector are simplification of the field-installation process and a reduction in the number of connector parts. This connector, introduced by 3M Corporation, is designated VF-45 (Figure 8.13).

The new connectors meet most of today's requrements even though they are produced by relatively new manufacturing technology [6].

All these connectors represent major trends in connector design: miniaturization, the fitting of many fibers into one connector (increasing fiber port density), compliance with the RJ-45 requirements, and, eventually, decreasing installation cost with a concomitant increase in the quality and reliability of connection.

Another industry trend requires that connectors have singlemode capability because singlemode fibers move from backbone to desktop applications. They must be strong mechanically because, increasingly these days, relatively untrained people are using them, and so they should be capable of being installed easily and quickly. These requirements stem from the fact that new connectors are oriented toward the residential market, where they have to compete with their copper counterparts.

## Standards

There are no mandatory standards in the world of connectors. All existing connectors, as pointed out above, are completely different and attempts to standardize them—to make the market accept one universal connector—have proved to be exercises in futility. Today, the situation with regard to the development of standards can be summed up as follows [7]:

The Telecommunications Industry Association (TIA) has a working group developing Fiber Optic Connector Intermateability Standards (FOCIS). This group initially developed a standard covering biconic connectors (FOCIS 1), then standards for ST, SC, and FC connectors (FOCIS 2, 3, and 4, respectively). Now the group is trying to develop standards for the eight new connectors, several of which were described above. The trend in standardization has gone from one biconic connector originally to three connectors in the 1990s to eight connectors, which will represent the state of the art, at least for the next few years. So, as you can readily see, the trend in connectors is running against attempts at "standardization." In all likelihood, market forces, not committees, will decide which connector(s) will be "standard" in the future. This means that if you ever need to install a fiber-optic network for your company, you'll have to familiarize yourself with the current situation in connector technology, not rely on arbitrary "standards," to make the right decision. (More about standards can be found in Appendix D.)

## Reading Data Sheets

The data sheet for a connector is a small document (there are no standards, remember) that usually includes the following information:

- Style of the plug: ST, SC, FC, and so on.
- Manufacturer's brand name.
- The style(s) with which the connector is compatible.
- Insertion loss, reflectance (return loss), and repeatability (durability).
- The mechanical property, which is given as the maximum force that one can apply to a cable causing not more than a specified increase of insertion loss. This can be given as *cable retention* or as *tensile strength* (or perhaps even some other expression) in pounds with specified loss (for example, <0.3 dB change with 22 lb).
- Environmental characteristics, given as the *operating-temperature range* (for example, –40°C to +80°C).
- Temperature cycling. This refers to the allowed number of temperature changes within a specified range (for example, 40 cycles from –40°C to +70°C) or the increase in insertion loss (for example, < 0.5 dB).

All other information may vary from one manufacturer to another—and even from one style to another—but, armed with the knowledge you've acquired here, you should be able to decipher any connector data sheet you encounter throughout your career.

Although every component of a connector is important, the ferrule is the really crucial one. It should be strong enough to protect a bare fiber from mechanical damage, soft enough to be polished easily, and amenable to active alignment (that is, sufficiently elastic).

To attain good results in polishing, the grain size of the ferrule material should be very small (0.3 microns or less).

And don't forget that repeatability depends heavily on the ferrule. These rather evident requirements give you some insight into how difficult it is to make a good connector. The data sheet always indicates the *ferrule material.* A specific ceramic, called zirconia, is the most popular material in use today, but stainless steel and other materials are also widely used.

## Termination Process

The field installation of a connector is still a time-consuming process and therefore expensive. It includes preparing the fiber (Figure 8.11[a]), affixing the fiber to a connector by means of epoxy, cleaving the fiber at the ferrule endface, curing the epoxy, and polishing the ferrule endface. This is primarily a manual operation.

Excellent fiber termination kits are available that include everything you need to do this job.

There is a wide selection of portable curing ovens to choose from to fully automate the epoxy-curing operation.

Also, you can find excellent polishing machines, including ones with a microscope for visually inspecting the end product. These polishers range from simple hand-held units to automated mass-production machines.

Installed connectors must be kept clean, and products to do just that are plentiful.

Despite the trend to automation, the field installation of a connector is still labor-intensive and a relatively slow process, especially during mass installation. Some manufacturers claim that their epoxy-free connectors require only two minutes to install. True enough, but this does not take into account the time needed to polish, inspect, and test the connector. Remember, too, we always have to compare the installation of a fiber optic network with that of its competitors—such as copper wire—at residential sites, where most of the installations are going right now. You may wonder why we're placing so much emphasis here on field installation. Well, consider this: Imagine that you need to terminate a cable carrying 144 fibers. If each connector takes 3 minutes to put in place, you would have to spend 432 minutes to install a single cable. That's more than 7 hours—yes, an entire work shift for one cable! To repeat this crucial point: *Cable termination—both pigtail splicing and connector installation—is a highly labor-intensive and time-consuming process.*

### Receptacles, Adapters, and Special Connectors

To direct an optical signal from a fiber to a receiver, we have to use a plug-receptacle pair. A *receptacle* includes a guiding mechanism to provide direction for the connector's latching device. This is the key component responsible for directing the core of the incoming fiber to a receptacle hole from where light travels to its destination. Figure 8.14(a) shows a typical receptacle.

*Adapters* join two fiber cables having the same or different plug styles (such as SC and ST connectors). When fixed on a patch panel, they are called *bulkhead adapters* (Figure 8.14[b]). An adapter can include a lens to collimate light when fibers with different optical characteristics (such as regular and large-effective-area singlemode fibers) are joined. Universal adapters are available that accept several different connector styles. In addition, bare fiber adapters can be purchased that allow you to connect a bare fiber to a receptacle temporarily for testing or communicating purposes. Obviously, tool kits for installing the adapters are available in a variety of nomenclatures. Various adapters are shown in Figure 8.14(c).

If you need to terminate fiber cable that's not in use, you can do so with a so-called "fiber-stop" device, available in plug and receptacle styles. Its function is to terminate the optical line without backreflection, acceptance of external light, or enduring environmental stresses.

On the market are a plethora of special devices to satisfy even the most exotic of customer needs. For example, fiber-optic *rotary joints* allow you to connect two cables to the rotating bodies so that signal transmission continues while these bodies rotate. The challenge is to provide such a connection with minimum loss and errors. These joints find applications in robotics, avionics, ship facilities, and, of course, in the military for communications purposes [8].

Another example is a *reusable connector* [9]. Instead of relying on epoxy, the fiber is fitted into a tapered connector through a patented new design that includes nylon sleeves and this tapering feature. The typical insertion loss is at the acceptable 0.2 dB level and, also on the positive side, installation is fast and easy.

### Tests and Measurements

After a connector is installed, one needs to *visually inspect the device and to measure the insertion and return losses.* Most termination kits include a microscope for inspection of the ferrule's endface. Moreover, optical-fiber scopes are available that are especially designed for inspection of a connector. These are microscopes having a receptacle for fitting a connector to facilitate the

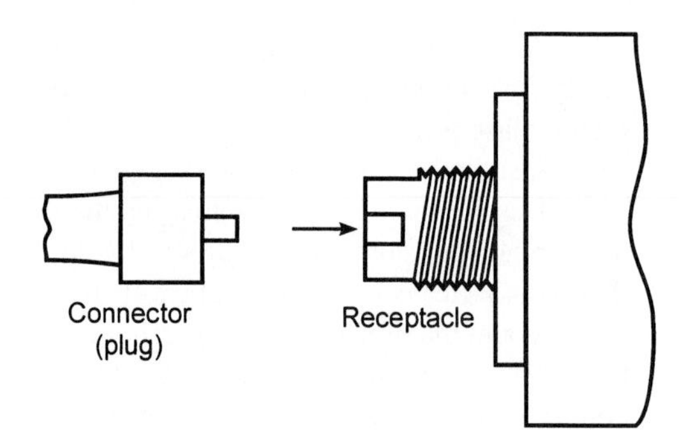


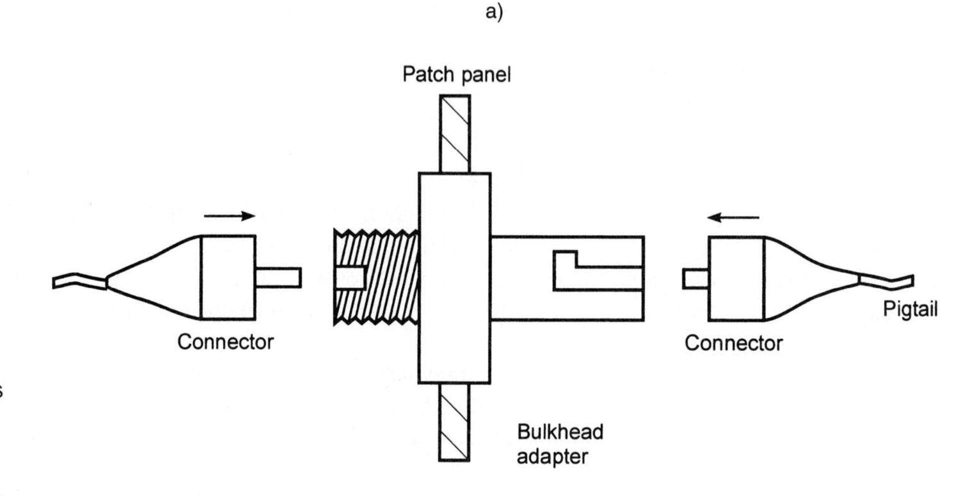


**Figure 8.14** Receptacles and adapters: (a) A connector and a receptacle; (b) connectors and an adapter; and (c) photos of various adapters. *(Part c courtesy of Wave Optics, Inc., Mountain View, CA.)*

inspection procedure. Recent developments include a tool that allows the inspection of connectors right at their installation sites—that is, at the patch panel or the transmission equipment—without the need to remove them [10]. One can understand how valuable this tool has become by just imagining tens of cables and connectors at the fiber-cabinet enclosure, which would otherwise have to be disturbed just to remove one, say, for inspection.

Measuring insertion loss is covered by TIA/EIA recommendation FOTP-34. The schematic of a typical measuring arrangement is shown in Figure 8.36(a). The procedure requires, first, measuring the power transmitted through the short piece of fiber, $P_{\text{fiber}}$ (dBm). This step counts the loss introduced by the fiber and its couplers to the source and power meter. Second, we repeat this measurement while a connector under test is included in-line. Thus, the second reading is $P_{\text{con}}$ (dBm). The difference between these two readings, $P_{\text{fiber}} - P_{\text{con}}$ (dB), is the *connector insertion loss.* The same procedure is used to measure the *splicing insertion loss.*

And remember: When measuring a multimode fiber, you have to excite all the modes. (See Section 4.4.)

The test arrangement for measuring return loss is shown in Figure 8.36(b). A coupler is the passive device that distributes an optical signal between input and output ports in a given

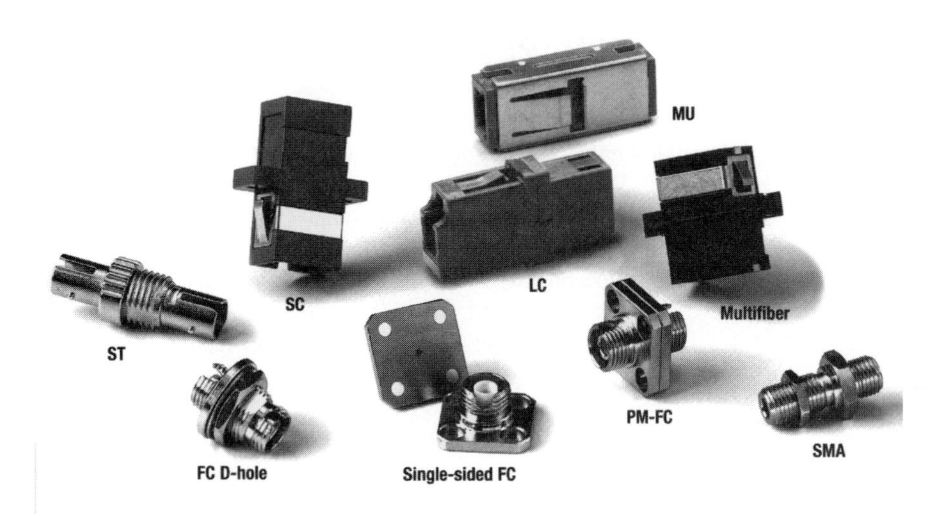


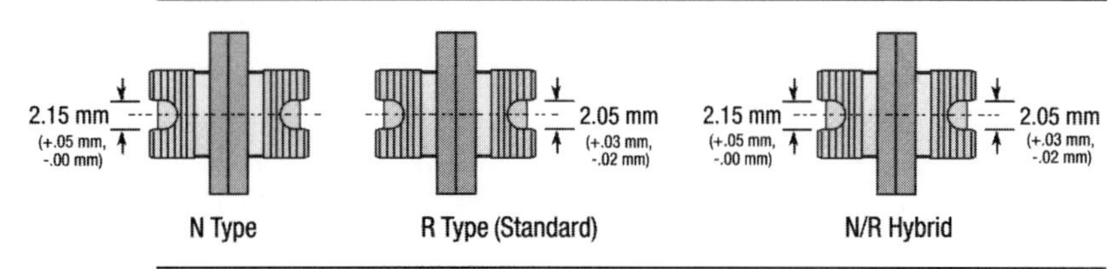


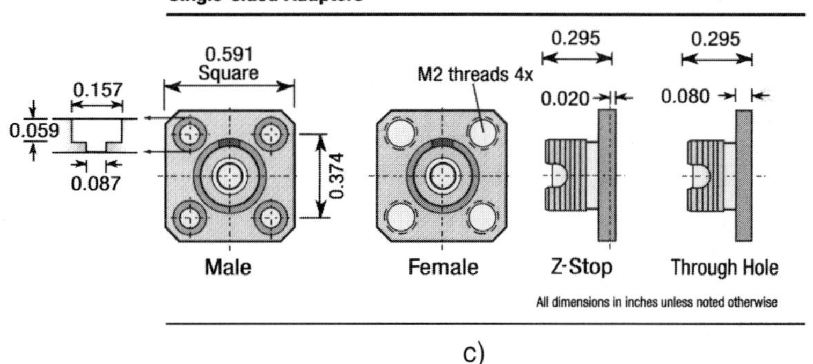


**Figure 8.14** (continued) c)

ratio, $R$. The procedure for measuring return loss includes two steps: First, we measure the transmitted power, $P_{in}$ (dBm), from the source to the cable end, where a connector will be installed. Second, we measure the power reflected from the connector pair back to a power meter, $P_{refl}$ (dBm). (See Figure 8.36[b].) Thus, the return loss is $RL$ (dB) $= P_{refl} - (R \times P_{in})$. (See Formula 8.2.) The formal return-loss test is covered by TIA/EIA FOTP-107.

Nobody performs a return-loss test in the field. There are many hand-held meters that are either dedicated to measuring return loss or that combine both insertion-loss and return-loss features. They may be used in conjunction with software that performs all the tests automatically. Another means to measure reflectance is with an optical time-domain reflectometer (OTDR). This device is discussed in Section 8.5.

## 8.3 INSTALLATION HARDWARE

Imagine that you are a fiber-optic cable installer and have to place cables both outside and within buildings, terminate them with connectors, and eventually bring an individual fiber to a single user. Do you want to connect two users with a dedicated fiber? This would seem the natural and simple thing to do, but what happens if tomorrow one user relocates to another room, another building, another city, or another country? What if your system doesn't work well and you need to troubleshoot it? For these and similar reasons, you would need to make hundreds, if not thousands, of connections. In order to place and protect these connections, you need frames, housings, organizers, and enclosures that, when considered as a single entity, are called fiber-installation hardware, the topic of this section.

### Why Installation Hardware

Let's consider connecting users in one building to users in a remote building a hundred feet or even a hundred miles apart. Each user eventually has to have his or her personal outlet where network equipment—a telephone, a computer, a fax machine, and so on—has to be connected. Even though copper or coaxial cable may be involved, we will consider only the fibers' paths. (Copper wire—in unshielded twisted pairs [UTP]—is still widely used for voice and data communications.) So, we have hundreds of fibers in one building that have to be distributed to many other locations. This situation, along with the main hardware components, is shown in Figures 8.15 and 8.16.

The idea is simple: You cannot connect two remotely located users with a dedicated fiber link. For practical purposes, you must place all the individual fibers gathered from one location inside a building into a cable, install this indoor cable, gather all the indoor cables, connect them to outdoor cable(s), and install these outdoor cable(s). Every time you switch from one cable to another, you must make an intermediate connection. The point is this: *Installation is impossible without intermediate connections.* To understand the reasons why, let's examine Figures 8.15 and 8.16.

First of all, you need to connect the nearest house to the aerially installed fiber cable (Figure 8.15). (This is fiber-to-the-curb, FTTC, architecture. Cable TV companies are making such

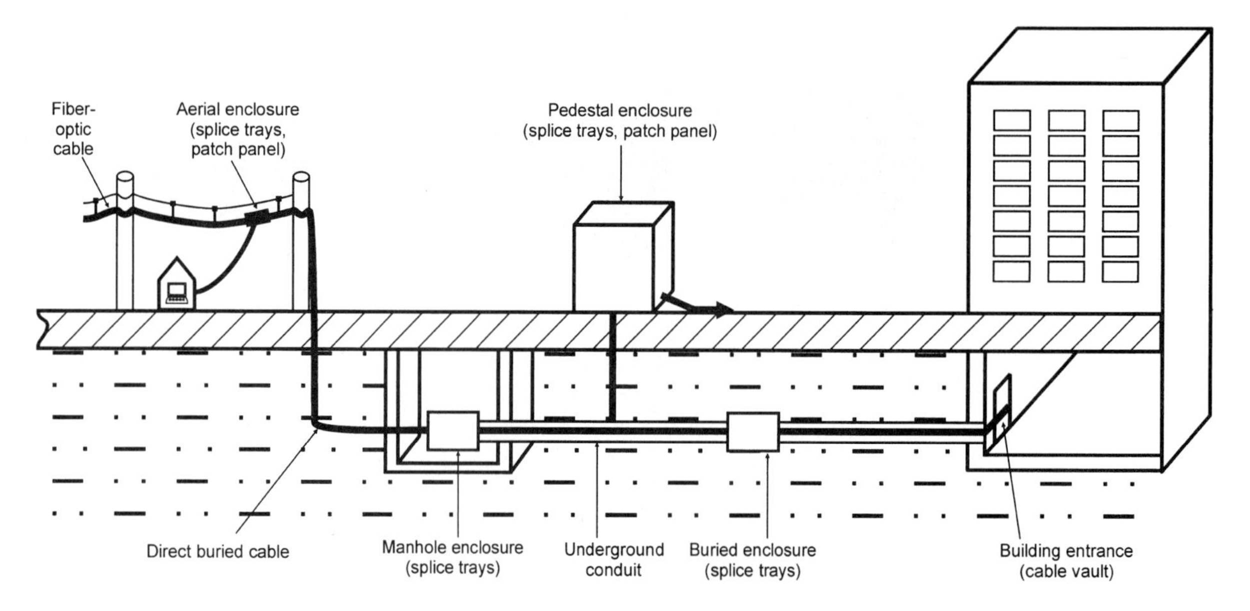


**Figure 8.15** Outdoor fiber-installation hardware. *(Adapted from* Fiber Optic Management Systems, *Catalog 269000, June 1997; Courtesy of AMP Incorporated, Harrisburg, PA.)*

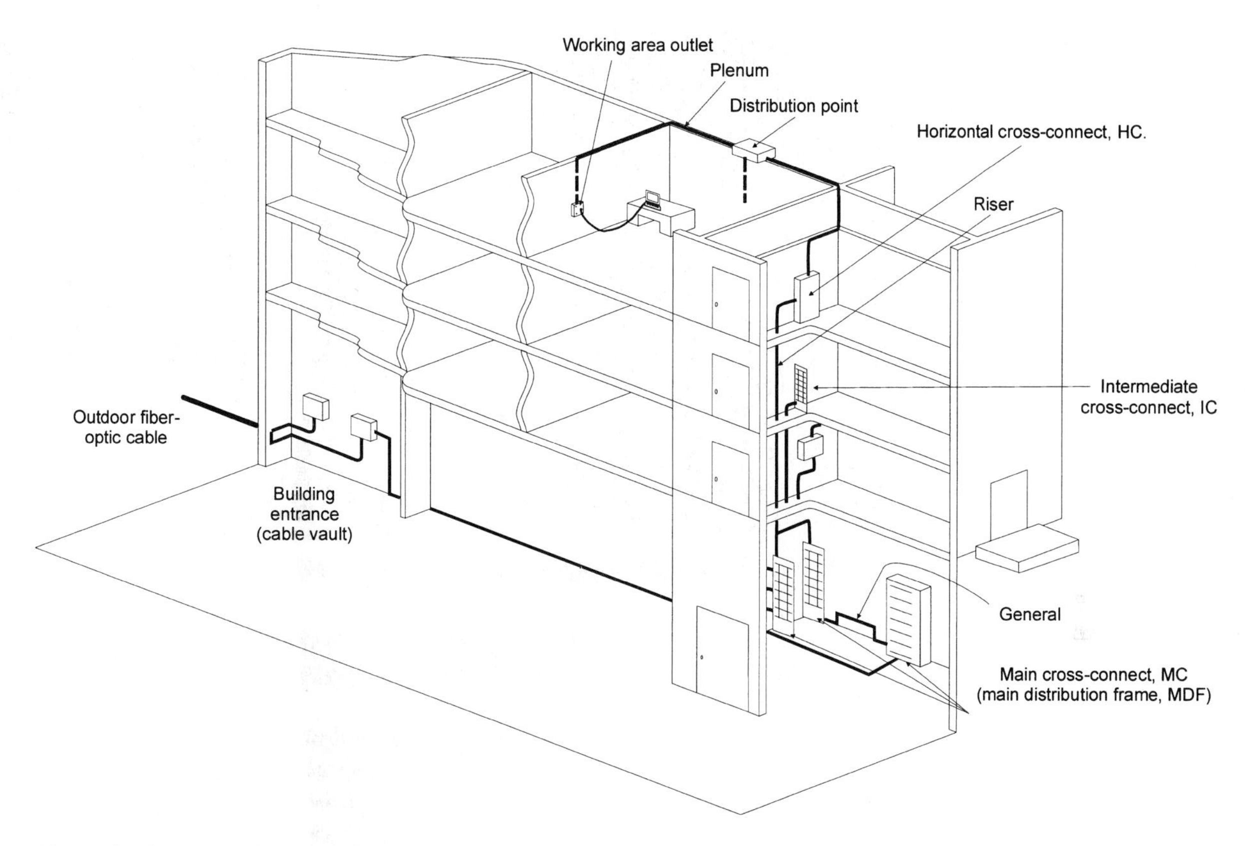


**Figure 8.16** Indoor fiber-installation hardware. *(Adapted from Fiber Optic Management Systems, Catalog 269000, June 1997; Courtesy of AMP Incorporated, Harrisburg, PA.)*

connections throughout the country. In reality, cable-TV fiber is a separate link not associated with data- and telecom-fiber cables.)

Secondly, the length of a cable is finite so you need to splice it.

Third, when entering a building (Figure 8.16), you need to distribute fibers from an outside cable—which, with all its assemblies, is called an outside cable plant (OSP)—to inside cables.

Fourth, over a period of time you will need to disconnect one customer and connect a new one. If anything goes wrong, you will need to check your connectivity. Obviously, it is reasonable to check this span by span (that's how all troubleshooting is normally performed), which is the fifth reason why intermediate connections must be made.

Finally, all wiring inside a building has to meet regulations imposed by the TIA/EIA 568A standard. This means that for fiber optics, your indoor backbone cable has to be a maximum of 2000 m for 62.5/125 fibers (3000 m for singlemode fibers) between the main cross-connect and the telecommunications closet with one intermediate cross-connect. (See Figure 8.16.) As you can see, then, experience confirms the need for intermediate connections.

For discussion purposes, imagine that you are faced with the task of distributing some 144 fibers from an incoming outdoor cable among all users inside a building. Should you try to splice or connect a fiber from a desktop directly to a fiber in the OSP? No, indeed. Just picture what a mess you will have, with your wiring room or closet becoming a rat's nest of tangled cables. (See Figure 8.17[a].) It's clear that your network never will work properly because you will never be able to maintain and troubleshoot this network. Only a well-organized fiber-distribution system—known as a *fiber-optic cable-management system*—makes a fiber-optic

a)

**Figure 8.17** You need installation hardware: (a) Do you want this total mess? (b) cable-management system. *(Courtesy of Siecor, Hickory, NC.)*

network superior to an electrical one. (See Figure 8.17[b].) But bear in mind that these management systems can be implemented only with appropriate connectivity equipment. Thus, the need for installation hardware becomes quite clear.

The result of these considerations is this: *To install a fiber-optic communications link, we have to make a great many intermediate connections. To place and protect these connections we need installation hardware.*

Incidentally, if you are familiar with electrical wiring, you will find much in common between electrical and optical installations. In fact, many fiber-optic hardware components borrow their functions and names from the electrical world.

We'll concentrate here on indoor hardware because, as we've emphasized several times, the installation of local area networks is one of the fastest growing areas today in the total picture of fiber-optic installations.

## Hardware Systems and Components

There are tons of installation hardware and reams of manufacturers' literature about it. Our goal is to arm you with the ability to understand your needs in order to be certain the appropriate hardware is in place at your job site. This is why we'll cover the main systems and components that can be found in most indoor-installed fiber-optic networks. Figure 8.18 shows what this equipment looks like.

***Systems*** The *main cross-connect, MC,* also called the main distribution frame (MDF), provides the interface (cross-connect and interconnect) between incoming outdoor cable(s) and out-

b)

**Figure 8.17** (continued)

# Fiber Main Distributing Frame

## ADC's Next Generation Frame Family of Products

Today's service providers must build a sound network infrastructure that will provide a solid base for delivering new and enhanced broadband services. Increasingly, that network infrastructure is based in optical fiber.

With unprecedented growth and widespread deployment of optical fiber, the challenges facing the industry include fiber frames and troughs overflowing with large masses of optical fiber patch cords, and increased termination density at the fiber distributing frame due to rapid deployment of SONET, fiber in the loop and video initiatives.

***Designed specifically to provide higher density while improving fiber cable management, ADC's Fiber Main Distributing Frame (FMDF) allows for virtually unlimited growth.***

**FEATURES**

- Provides 33% more terminations than existing fiber frames
- Saves space and money
- Allows for unlimited expansion
- Eliminates fiber pile-ups
- Innovative design that is available now
- Allows for long-term fiber growth
- Customer-driven solution
- Allows you to grow your network your way

a)

**Figure 8.18** Connectivity and distribution equipment: (a) Main cross-connect (MC); (b) interconnect and cross-connect; (c) intermediate cross-connect (IC); (d) horizontal cross-connect (HC). *(Part a courtesy of ADC Telecommunications, Inc., Meriden, CT. Parts c and d courtesy of Siecor, Hickory, NC.)*

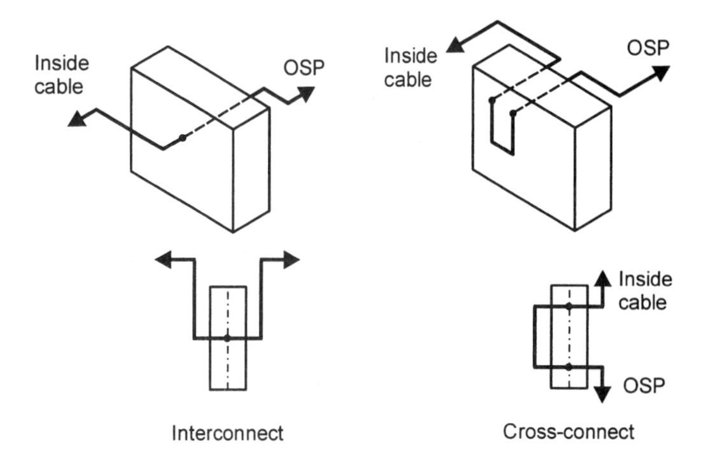


**Figure 8.18** (continued)

going indoor cable(s). An MC is shown in Figure 8.18(a). *Interconnect* means that the outside plant cable, OSP, is connected directly to the inside fiber cable. With a *cross-connect,* an intermediate cable—a jumper—is used; a cross-connect, which is how most connections on an MC are done, provides much more flexibility. (See Figure 8.18[b].) All the equipment is broken into modules that can be mounted in rack or wall cabinets. A specific room, usually in the basement, is set aside to house an MC.

An *intermediate cross-connect (IC)* provides the interface between an MC and a horizontal cross-connect (HC) when the latter is more than 2000 meters from the MC. (See Figure 8.18[c]). An IC performs the same cross-connect functions as an MC but on a smaller scale.

A *horizontal cross-connect* (Figure 8.18[d]) provides the interface between indoor backbone cable (the riser) and the horizontal cable plant (plenum and floor cables). ICs and HCs are usually placed in a specific room, called a *telecommunications closet (TC),* which is sometimes called a telephone room. TCs are usually found on each floor of the building.

*Work-area outlets* provide the interface between desktop equipment (mostly computers) and horizontal wiring. This wiring can also include *distribution points* between the HC and outlets for single users.

Look again at Figure 8.16(b), which depicts the functions of all these components.

***Interconnect cables*** Before progressing further, we need to digress briefly to clarify the types and usage of interconnect cables. These are short pieces of a fiber cable called *cable assemblies* when they terminate at both ends with factory-installed connectors or *pigtails* when they open at one end for splicing. Their outer jackets are either orange (for multimode fiber) or yellow (for singlemode fiber). (See Figure 8.19.) Depending on their usage, interconnect cables may be referred to by different names. For example, if these cables are used for cross-connect, they are called *jumpers* (Figure 8.19[a]). If they are used for connecting a patch panel to the terminal of transmitter/receiver equipment, they are called *patch cords* (Figure 8.19[b]). However, each manufacturer may use its own terminology so don't be surprised if in the field you come upon an expression like "patch-cord jumper."

***Hardware equipment*** MC, IC, and HC include mostly the same modules of hardware equipment. These modules are, essentially, connection enclosures that hold the connectivity points,

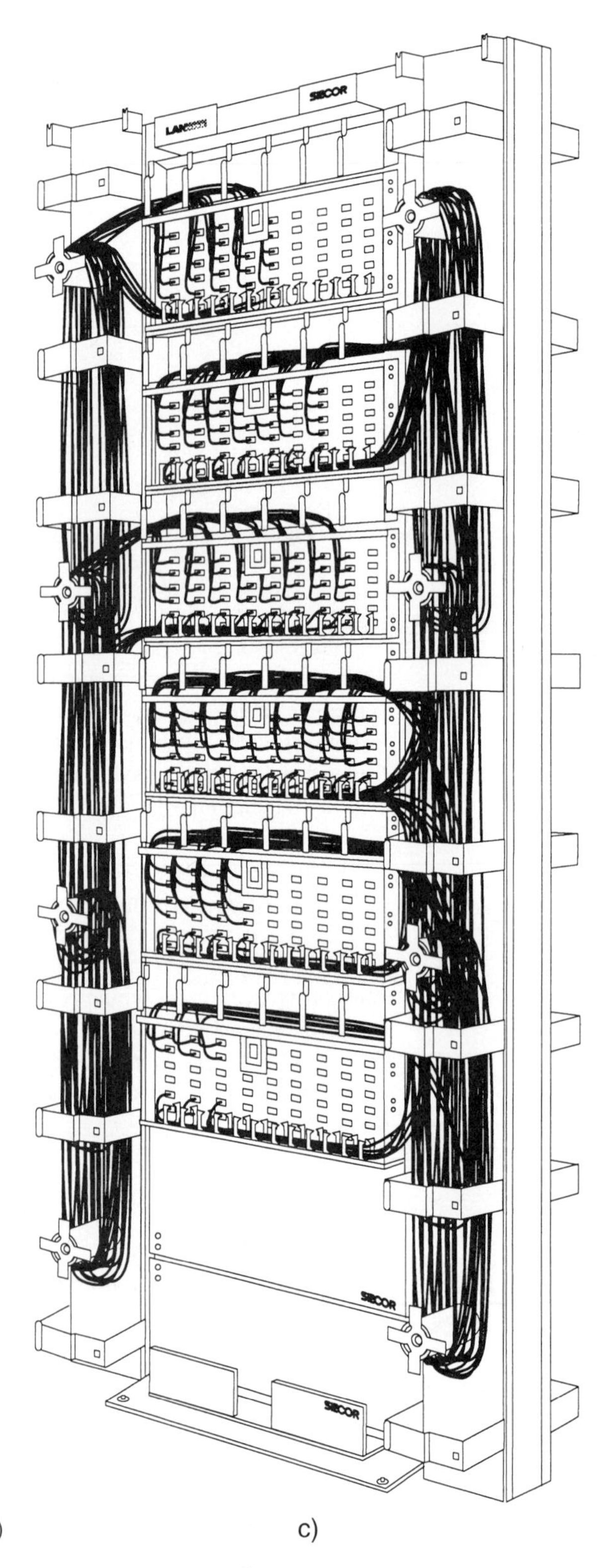


**Figure 8.18** (continued) c)

**Figure 8.18** (continued) d)

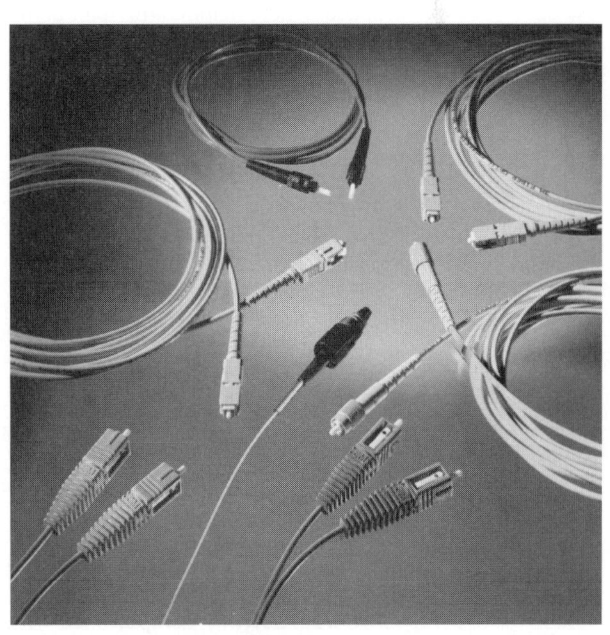

**Figure 8.19** Cable assembly, pigtails, jumpers, and patch cords: (a) Cable assemblies; (b) jumper; (c) patch cords. *(Part a courtesy of 3M Telecom Systems Division.)*

a)

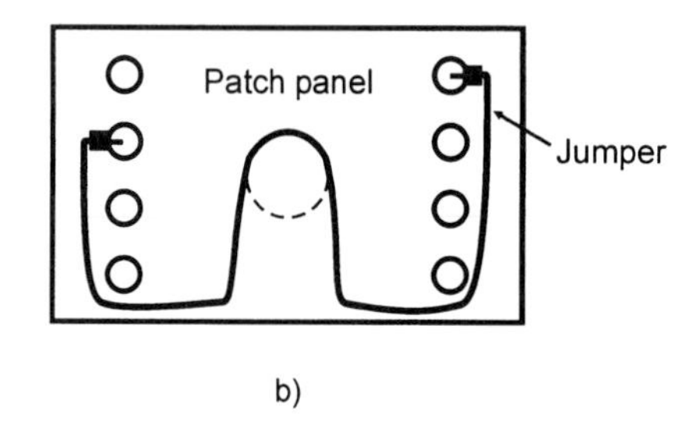


b)

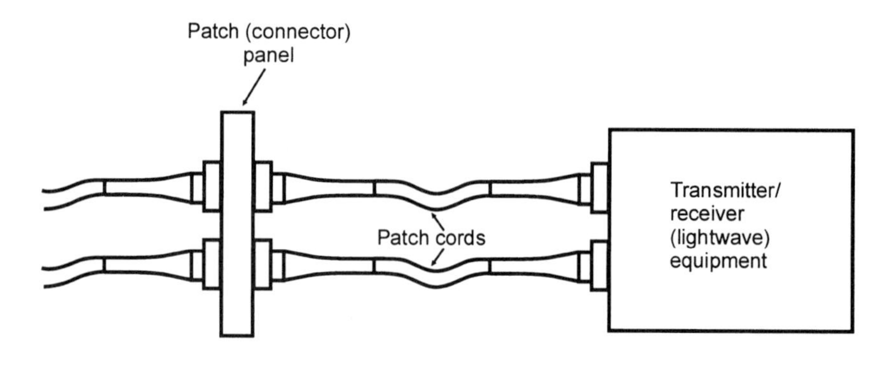


c)

**Figure 8.19** (continued)

protect them, and provide easy access to them. Physically, they are metallic boxes equipped with specific panels, trays, bars, guides, plates, troughs, and similar elements. The typical components of a fiber-management system are shown in Figure 8.20.

*A patch panel* (Figure 8.20[a]) is the key component of a fiber-cable management system. This is where most of the connections are physically made. This housing includes storage space, where a splice tray can be placed, and a patch panel itself. The latter, also called a connector panel, is a plane panel housing a set of adapters so that two plugs from opposite sides can be connected to one bulkhead adapter. These adapters can connect plugs of the same style or different styles. (See Section 8.2.) You will find a patch panel in almost every connection housing.

The functions of this interconnect center are to (1) provide the ability for flexible, changeable connections, (2) protect connectivity points, and (3) allow access to individual fibers for troubleshooting, testing, monitoring, and restoring communications channels.

A *splice panel* (Figure 8.20[b]) connects individual fibers of a cable from one side to pigtail connectors on the other side. This housing derives its name from the fact that pigtails have to be spliced. Its function is to provide fixed, permanent connections. Typically, it is used to change an outside cable to a riser or plenum cable. The essential component of the housing is a *splice tray,* which protects the splicing from environmental hazards.

MCs, ICs, and HCs consist primarily of patch and splice panels, even though these connection housings come in different designs, styles, and physical appearance.

A *wall outlet* (Figure 8.20[c]) is the termination point of the long distance that an optical signal travels. Its function is to connect the desktop-network equipment of an individual user to the entire network. This is done by running a single fiber to the wall outlet.

***Problem-solving components*** Looking at a couple of common problems installers run into will give you some insight into what to expect when you enter the field.

**The Fiber Connect Panel (FCP) interconnects, protects, labels, color-codes, and manages fiber optic cabling. It is available as a stationary panel or as a removable drawer with a smooth sliding action. Fiber managers are used to store slack fibers while maintaining minimum bend radius requirements. Optional splice trays are available which can accommodate up to 24 splices per tray and can be stacked to enable a total of 48 splices. There is a snap-on front shield which provides a labeling area as well as protection for fiber jumpers. The panel accepts up to eight colored bezels, each of which has two- or four-port SC, ST or ST/SC "universal"* adapters, allowing the panel to patch any quantity of fibers up to 32 (including the commonly used 24-port). A smoked polycarbonate cover protects contents, has a label for port designations, and is easily removed with a 1/4 turn of screwdriver for quick access. Colored bezels allow color-coding capability and integrate the panel into the Siemon Cabling System. A blank bezel is also available for unused ports.**

**The drawer version slides in and out of the rack mount, providing complete access to all areas of the panel. Defeatable latches allow the drawer to be completely removed from the rack. To replace the drawer, simply slide it back into the rack mounts.**

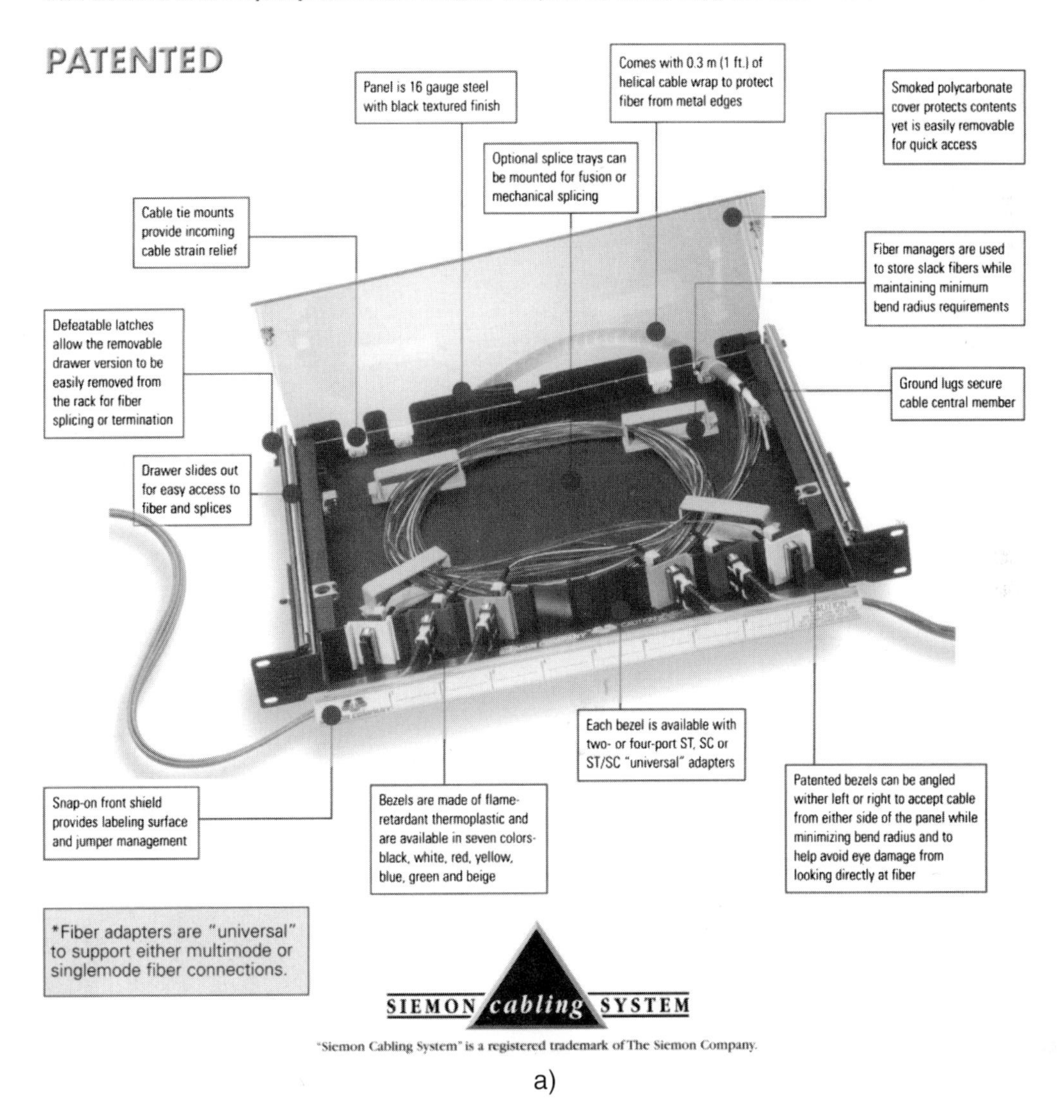


**Figure 8.20** Typical components of a fiber cable management system: (a) Patch panel; (b) splice panel; (c) wall outlet. *(Courtesy of The Siemon Company, Watertown, CT.)*

## RACK MOUNT INTERCONNECT CENTER (RIC)

The new Rack Mount Fiber Interconnect Center (RIC) provides a high-density fiber solution for terminating up to 144 fibers in a 4 RMS space on a 19" or 23" rack. Fiber entry is made available on both sides, at the front and the rear of the enclosure and conveniently located cable-tie anchor points are provided for securing incoming cables or innerduct. A sliding tray facilitates front and rear access to connections. Optional splice trays can be mounted in the rear of the unit where slack storage is available and minimum bend radius can be maintained.

Patchcord management is accomplished by recessing the adapter mounting plane to provide plenty of space for effective management in front of the unit to organize jumpers exiting from each snap-in 6- or 12-port adapter plate.

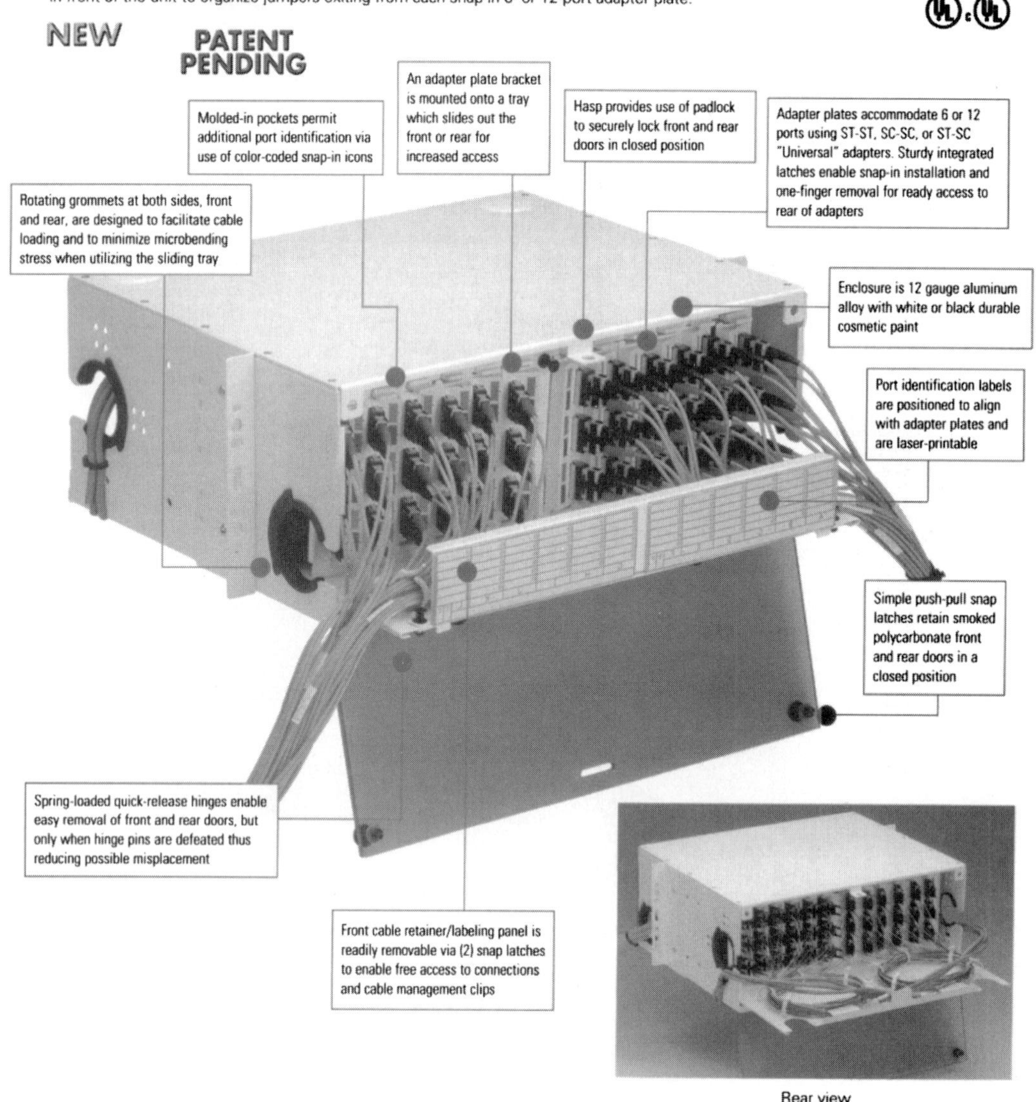


a)

**Figure 8.20** (continued)

**The Siemon Wall-Mount Interconnect Center (SWIC™) is a cost-effective fiber connection enclosure designed to manage and connect from 2 to 48 fibers. Fiber bezels are available in several colors for color-coding connectors and come with two- or four-port ST, SC, or ST/SC "universal"* adapters. A blank bezel is also available for unused ports.**

**The enclosure is fabricated from durable 18 gauge steel with textured black or ivory finish. The low profile compact design makes it ideal for telecommunications closets or other installation areas where wall space is a premium. Cable-tie anchor points are built into the base of the unit for securing fiber cable or innerduct. Cable can enter from the top, bottom or rear of the unit. Dust-proofing grommets are also included. Four unique dual-level fiber management clips are included to keep fibers organized and in-line with upper and lower connections while maintaining minimum bend radius requirements. Provisions have been made to mount a splice tray to the base of the unit for splicing pigtails or providing a fan-out point for field connectorization. A hinged jumper guard is available to provide jumper protection. Units can also be end-stacked for easy future expansion.**

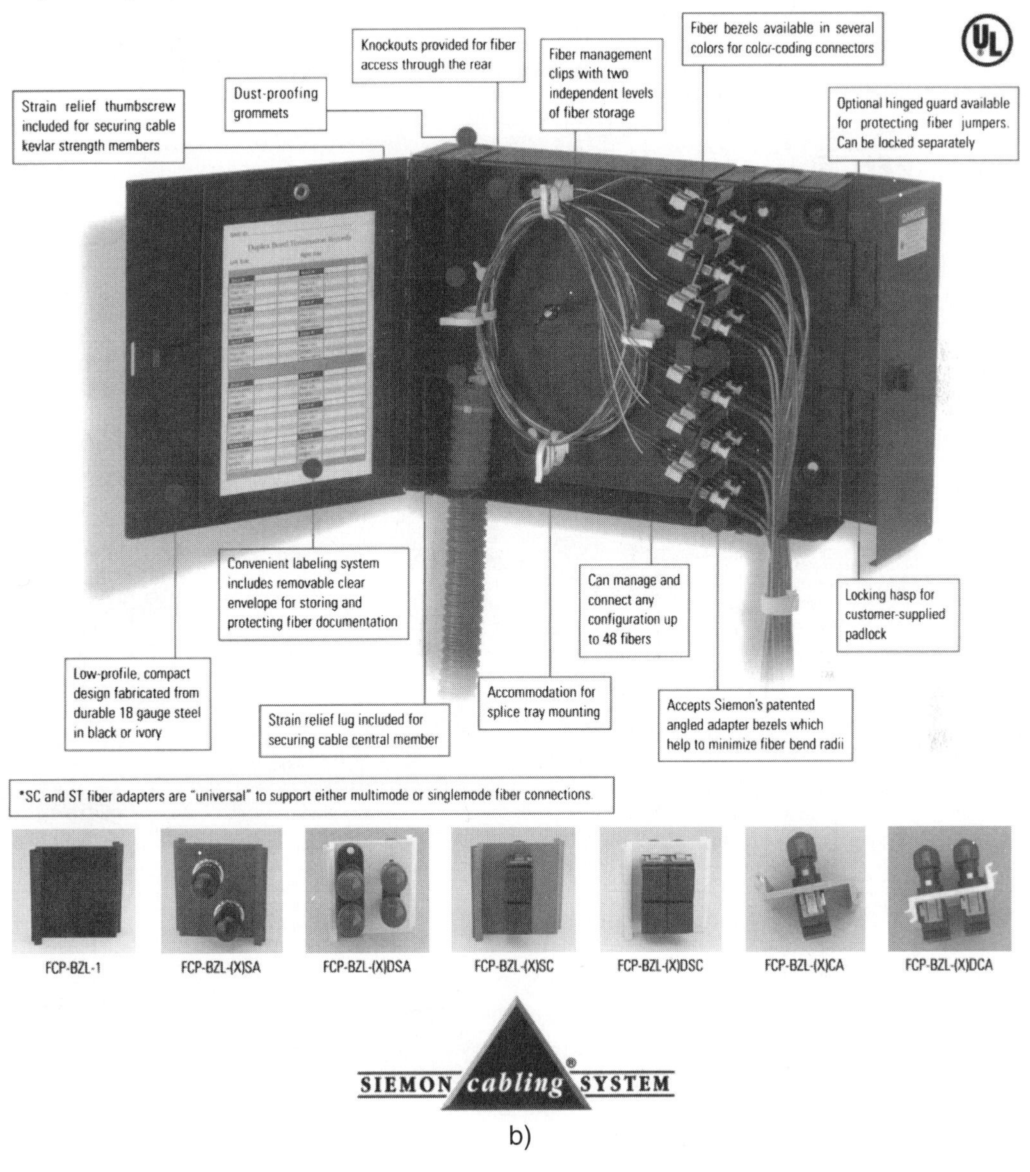


b)

**Figure 8.20** (continued)

**Siemon's new low-profile Fiber Outlet Box (FOB) offers a well-defined method for managing fiber cabling at the work area by providing a connection point for 2, 4 or 6 fiber connectors utilizing ST, SC or hybrid "universal" adapters mounted to slide-in bezels. There are also provisions for coaxial and twisted-pair connections with the integration of a single-gang faceplate which can be mounted to the FOB base, accommodating up to 6 MAX™ series or 4 CT® series outlets. In addition, BNC, F-type and RCA coaxial bezels are available to be mounted at the bottom of box in conjunction with, or in place of, the fiber bezels to provide additional multimedia capability.**

**The base of the FOB provides incoming fiber cable strain relief in addition to storage and management for up to 1m (3 ft.) of slack for as many as 6 buffered fibers. Due to the unique snap-on cover design, fiber connections can be accessed without disturbing the copper connections (or vice versa). For added security, an optional extended cover version may be used to conceal and protect the externally mated fiber connectors.**

**A concealed cover-to-base screw accommodation is included for added security if desired. Molded-in icon pockets and designation label holders permit fiber port identification with the use of color-coded, snap-in icons (3 red, 3 blue and 6 matching color icons are included).**

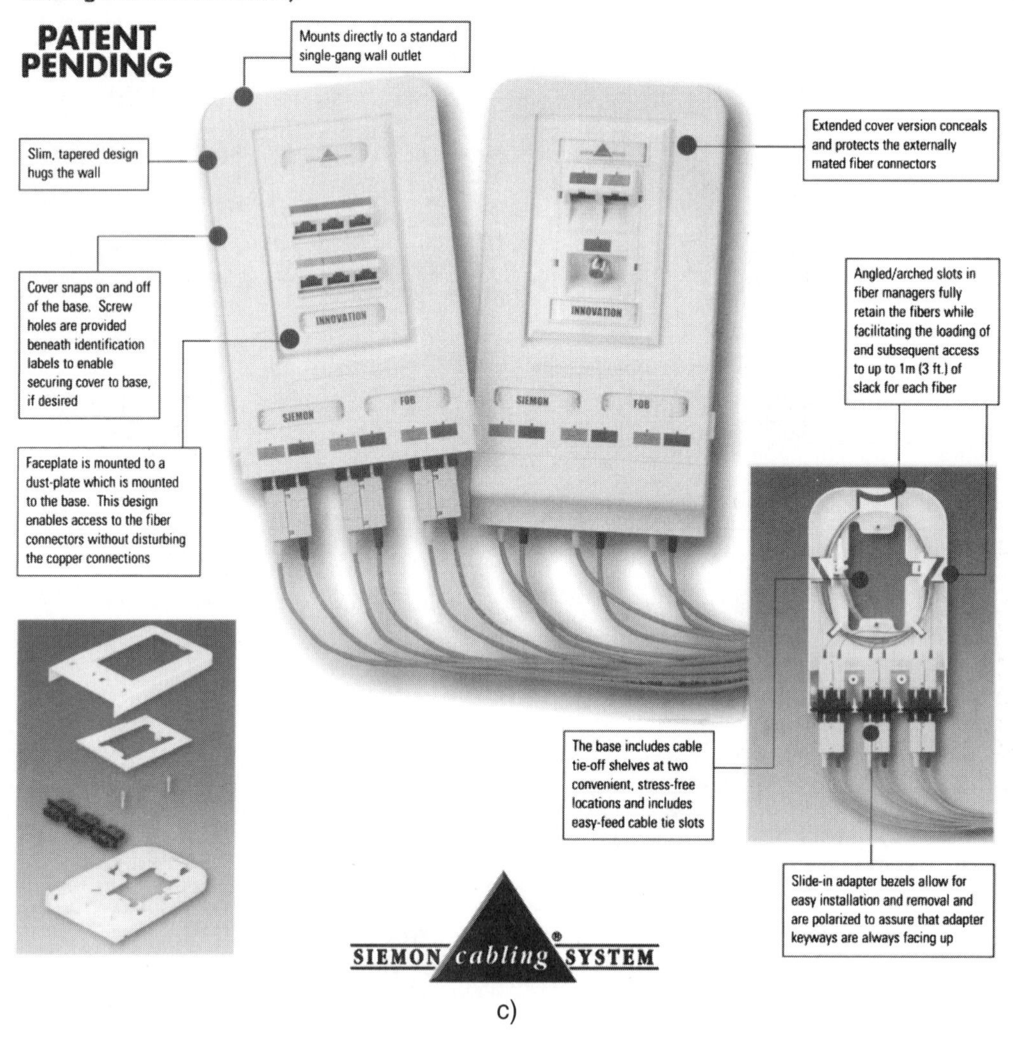


c)

**Figure 8.20** (continued)

*Problem:* The customer complains that his or her line doesn't work. The *wrong* solution would be to disconnect the customer, that is, interrupt all his or her service to troubleshoot the line. The *right* solution is to be sure, during installation, to add several switches and, of course, several spare fibers to your customer's distribution frame. (In fact, it is never too late to do this if your MC has spare space.) Then, when the need for troubleshooting arises, simply switch your customer to a spare line and do the troubleshooting without interrupting his or her service.

*Problem:* The customer complains that his or her line doesn't work properly; for example, he or she complains that the bit-error rate, BER, is too high. The *wrong* solution, again, would be to interrupt service and monitor the line in search of the source of the error. The *right* solution, again, is to be certain, during installation, to add a splitter to each transmitter and receiver connection at your distribution frame instead of using standard bulkhead adapters. Use the splitter to divert, say, 10% of the incoming and outgoing signals for the constant monitoring of transmitter/receiver activities.

***Software*** The main cross-connect has to accept hundreds of fibers and connect them to hundreds of others. Someone has to know which fiber is connected to which and what cables enclose the different fibers. These connections are dynamic, which makes keeping track of them even more difficult. Only a computer-based system can do this work effectively. You will find software packages on the market that help you keep these records clear and updated.

**Conclusion** If you review the mountain of manufacturers' literature, you'll find that the wide variety of designs and the many different functions of the hardware components we've been discussing stem from the diversity of customer needs. As mentioned above, it's almost impossible even to list, let alone discuss, every existing hardware component because they are countless both in nomenclature and function. Fortunately, there is no need for you to know about all of them at this point. Most of what you'll need to know to work with this equipment has been provided. Whatever else the job requires you'll pick up through good old common sense and further study once you enter the field.

## 8.4 DESIGN OF LOCAL-AREA-NETWORK INSTALLATION

A fiber-optic network is the most widespread type of network in the world and deployment continues at a rapid pace. The two hottest growth areas for fiber-optic-network installations are the new long-haul telecommunications networks, which provide terabit-per-second transmission of all forms of information, and local area networks (LANs), which provide on-site communications. Although LANs are traditionally considered computer (data) networks, today they support all other types of telecommunications services.

The technique of long-haul installation is well developed and well described. On the other hand, the technique of LAN installation is rapidly changing and today's description becomes obsolete tomorrow. That is why this section concentrates on LAN installation design, which represents only the physical layer of a local area network. The physical layout of fiber cables, the types of cables, the types of fibers, the types of connectorization, and the hardware—making the right decisions from among all these choices is the subject of this section. In other words, we're concerned here with fiber-optic-network installation in its broadest sense. In fact, this section integrates all the information we've covered in the previous sections of this chapter.

From a logical standpoint, design is the first step in network installation, but the designer has to be familiar as well with installation systems and components. It is not our intention to make you a professional network designer; rather, our purpose here is to give you, a future professional user of these systems, some insight into the area of design.

It should be noted at the outset that Regulation TIA/EIA 568A, the "Commercial Building Telecommunications Cabling Standard," is widely accepted as the industry standard for LAN installation. We'll refer often to this standard in this section. To start with, TIA/EIA 568A accepts only two types of fiber—62.5/125 μm graded-index multimode and a singlemode—as standard transmission media for local area networks.

## Link Consideration—Power Budget and Rise-Time Budget (Bandwidth)

***Power budget*** Physically, a network is a collection of nodes connected by links. (See Section 1.1.) Let's consider an individual fiber link that may include splices, connectors, and some other passive components. Because of the attenuation introduced by these components, a receiver gets much less light power than was launched by a transmitter. The question is this: Does the light signal arriving at the destination point of a link have enough power to be detected by the receiver? This is what power budget is about. Figure 8.21 demonstrates the concept: *Power-budget consideration allows us to calculate power at the receiving end and know the loss allocations along the link.*

An example shown in the graph in Figure 8.21 illustrates the power-budget concept: The power-vs.-distance curve shows the light power at each point along the fiber link. When a transmitter radiates a light signal with the power of −10 dBm (0.1 mW), this is called initial power. A connector coupling light from a transmitter to a fiber causes a 0.2 dB attenuation; hence, note the first decline in this curve. A patch-cord cable has an attenuation of 1.0 dB/km for 62.5 fiber at 1310 nm. This is the negative slope of the curve. Thus, after traveling 10 m, light attenuation is 1.0 dB/km × 0.01 km = 0.01 dB. At this point, a patch panel is used. The typical loss of a PC connector is 0.3 dB; hence, the curve drops at this loss level. From the patch panel, a regular fiber cable is used whose attenuation is 1.0 dB/km. Thus, the curve develops with a slope of −1.0 dB/km along the transmission distance. From the patch panel to the nearest fusion splicing point, the cable's loss is 1.0 dB/km × 0.49 km = 0.49 dB. The attenuation introduced by fusion splicing is 0.2 dB, thus, the appropriate curve drop is shown in Figure 8.21. As you continue to go along the link, you need to take into account all sources of attenuation with respect to their locations. This is shown in Figure 8.21, and an analysis of this figure will help you understand this idea.

(*You should recall the three basic terms used here:* light power *is measured in dBm and is negative when it is less than 1mW;* loss *is measured in dB as the difference between two dBm values of light power, and it is always positive by industry practice*; attenuation *is measured in dB/km as loss per distance.*)

Power-budget calculations can be made as follows:

- Cable loss 1.0 dB/km × 2.0 km = 2.0 dB
- Splicing loss

| | | |
|---|---|---|
| Fusion 0.2 dB × 2 | = | 0.4 dB |
| Mechanical | = | 0.3 dB |
| Total splicing loss | = | 0.7 dB |

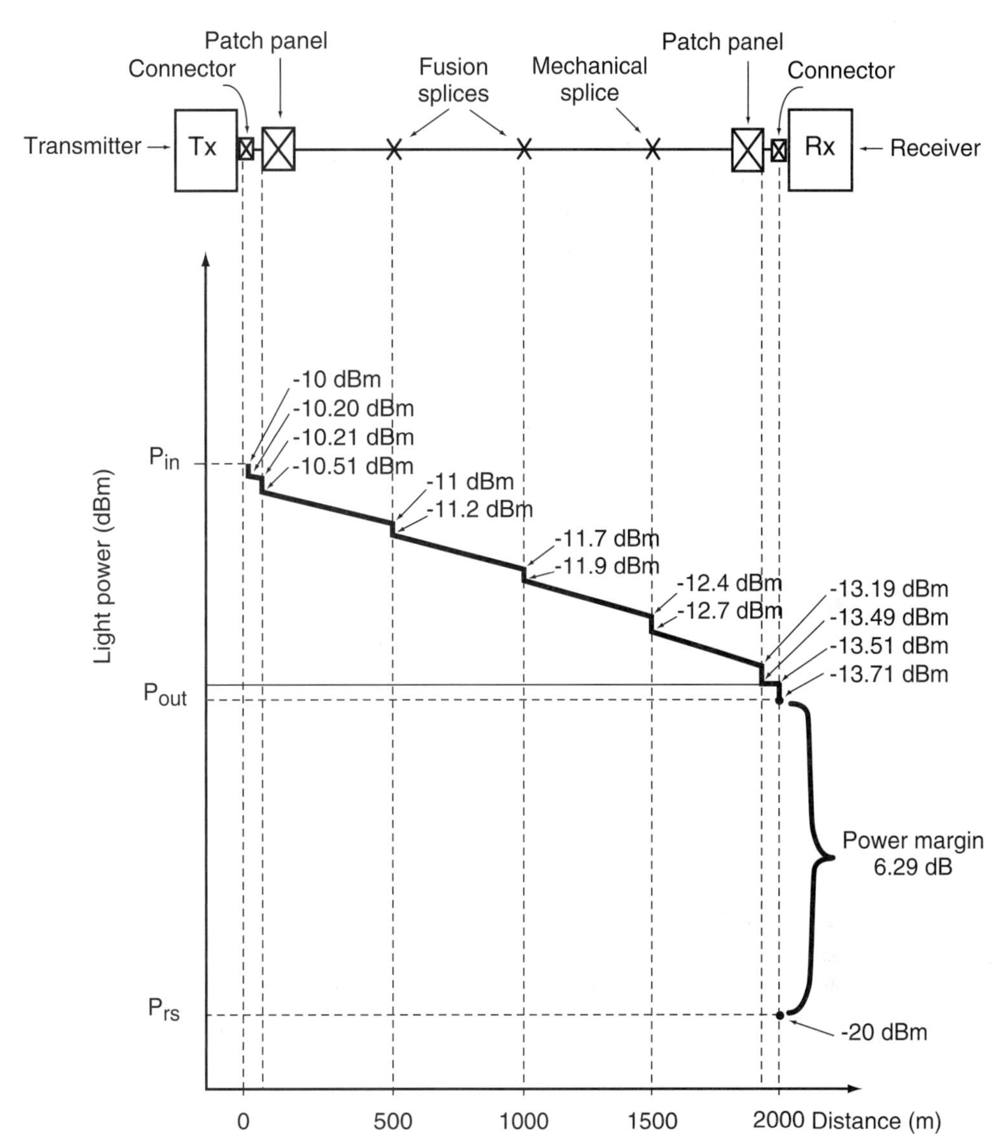


**Figure 8.21**
Power budget.

- Connector loss

| | | |
|---|---|---|
| 0.3 dB × 2 | = | 0.6 dB |
| 0.2 dB × 2 | = | 0.4 dB |
| Total connector loss | = | 1.0 dB |

- Total loss = 3.7 dB
- Transmitter power launched into a fiber, $P_{\text{in}}$ = −10 dBm
- Power at the receiving end, $P_{\text{out}} = P_{\text{in}} -$ total loss = −13.7 dBm

- Receiver sensitivity, $P_{rec}$ = −20 dBm
- Power margin = $P_{out} - P_{RS}$ = 6.3 dB

Again, these calculations introduce the idea of how power budget works for network design. It allows you to calculate the maximum distance you can achieve with a given fiber cable and the number of splices, connectors, and other components you can afford to run along your link. The loss allocation shown in Figure 8.21 helps you to find critical points and achieve a better design.

The first piece of advice for a designer: *Always keep reasonable power margins because your network will grow; therefore, you'll need to connect new users and include new components, steps that are accompanied by power loss.*

***Bandwidth*** The main advantage of an optical fiber over copper wire and other transmission media is its tremendous information-carrying capacity, or bandwidth. This general statement, however, has to be verified for each specific network. To do so, you can use two methods.

First, check the bandwidth of cable used in the installation. For example, refer again to the MIC cable from Siecor (Figure 7.9 [c]). Note in the "Transmission Performance" section of the data sheet that the minimum bandwidth is not specified for a singlemode fiber and it is equal to 160/500 MHz·km at the 850/1300-nm operating wavelengths. This assures you that *a singlemode fiber always meets your LAN bandwidth requirements, that is, that the fiber bandwidth always exceeds the electronic (lightwave-equipment) bandwidth.* For multimode fiber, the calculations are simple: For 2000 m—the maximum length dictated by the TIA/EIA 568A standard—this fiber provides 500 MHz·km/2 km = 250 MHz bandwidth at 1300 nm. This is modal (intermodal) bandwidth. At 1300 nm, the spectral-dispersion bandwidth is negligible. (See Sections 3.3 and 4.6.) To simplify this consideration, every manufacturer provides a table of recommendations, where a specific fiber type is assigned to a specific LAN type. (See, for example, [11].)

To make sure that the fiber bandwidth guarantees the required transmission speed, you can calculate link bandwidth. This is the second method by which we evaluate an optical fiber's bandwidth for LAN applications. Since you have optical and electronic components working together, your first step in the calculation is to convert the optical bandwidth into an electrical bandwidth (or vice versa). Then, using the well-known relationship $BW = 0.35/\tau_{rise}$, where $BW$ is electrical bandwidth and $\tau_{rise}$ is a component's rise time, convert the bandwidth of all the components into appropriate rise times. This allows you to compare the transmission ability of all the components on a unified basis. The total system's rise time, $\tau_{syst\text{-}rise}$, is equal to:

$$\tau_{syst-rise} = \sqrt{(\tau_{fib-rise}{}^2 + \tau_{ltw-rise}{}^2)}, \tag{8.6}$$

where $\tau_{fib\text{-}rise}$ is the fiber rise time and $\tau_{ltw\text{-}rise}$ is the rise time of the lightwave (electronic) equipment. This formula is called the *rise-time budget equation.* In the same way, you can include in Equation 8.6 the rise time of all the specific components constituting the fiber link.

### Example 8.4.1

***Problem:***

A local data link is to be installed having the following characteristics: maximum bit rate, 16 Mbit/s; installation length, 2000 m; operating wavelength, 850 nm; rise time of the lightwave equipment, 4 ns, and LED spectral width, 20 nm. Will MM 62.5/1125 μm fiber support the required bit rate?

***Solution:***

**Step 1.** Let's rearrange Formula 8.6 in this way [13]:

$$\tau_{fib-rise}^{2} = \tau_{syst-rise}^{2} - \tau_{ltw-rise}^{2} \qquad (8.6a)$$

Now the solution reduces to calculating and comparing both sides of Equation 8.6a.

**Step 2.** Calculate the right side of Equation 8.6a. System rise time can be found from the given bit rate, *BR,* as

$$\tau_{syst-rise} = 0.35/BW = 0.35/16 \text{ MHz} = 22 \text{ ns}.$$

(We used the $BW = BR$ relationship. See Section 3.4.) Thus, the right side of Equation 8.6a is equal to

$$\tau_{syst-rise}^{2} - \tau_{ltw-rise}^{2} = 21.6 \text{ ns}.$$

**Step 3.** Fiber rise time can be calculated as follows:

$$\tau_{fib-rise} = \sqrt{(\tau_{mod-rise}^{2} + \tau_{chrom-rise}^{2})}$$

where $\tau_{mod-rise} = \Delta t_{modal}$ and $\tau_{chrom-rise} = \Delta t_{chrom}$ are pulse spreads caused by modal and chromatic dispersion, respectively. (See Formula 4.73.)

**a.** The rise time caused by modal dispersion can be found as follows: The modal bandwidth-length product is given in the fiber's data sheet (Figure 3.18), as 160 MHz·km. With 2000 meters of installation length,

$$BW_{modal} = 160 \text{ MHz·km}/2 \text{ km} = 80 \text{ MHz}.$$

Convert this number into an electrical bandwidth (Section 4.6):

$$BW_{el\text{–}modal} = 0.707 \times BW_{modal} = 56.6 \text{ MHz}.$$

Fiber-modal rise time is:

$$\tau_{mod-rise} = 0.35/BW_{el\text{–}modal} = 6.2 \text{ ns}.$$

**b.** Rise time caused by chromatic dispersion is given by Formula 3.19: $\Delta t_{chrom} = \tau_{chrom\text{-}rise} = D(\lambda)L\Delta\lambda$. The chromatic-dispersion parameter $D(\lambda)$ for the typical characteristics used in Example 3.3.5, is equal to 0.1 ns/nm·km at 850 nm. Hence,

$$\tau_{chrom\text{-}rise} = 0.21 \text{ ns/nm·km} \times 2 \text{ km} \times 20 \text{ nm} = 8.4 \text{ ns}.$$

**c.** Go back to Step 3: $\tau_{fib\text{-}rise} = \sqrt{(\tau_{mod\text{-}rise}^{2} + \tau_{chrom\text{-}rise}^{2})} = \sqrt{(6.2 \text{ ns}^2 + 8.4 \text{ ns}^2)}$
$= 10.4 \text{ ns}$

**Step 4.** Compare the right and left sides of Equation 8.6a. Fiber rise time is equal to 10.4 ns, which is less than the required rise time of 21.6 ns; therefore, the chosen fiber will support this link.

In case the fiber won't have enough bandwidth, we can, theoretically, take two measures to reduce its rise time: shift to 1300 nm operating wavelength and switch to a laser diode. Both measures will lessen the effect of chromatic dispersion. (Can you explain why?) Can a laser diode work with MM fiber in practice? Recent developments allow a new type of laser diode, VCSEL (see Chapters 9 and 10), to be used with multimode fibers. The problems that once beset such combinations have been partly resolved with the development of new types of multimode fibers. To learn more about this topic, see Chapter 15.

It should be noted that the modal-bandwidth-length product given in the data sheet is measured by the manufacturer for its factory length. *If the installation length of the fiber, $L_{inst}$, is less than the factory length, $L_{fact}$,* then the installation modal bandwidth, $BW_{modal}$(inst), can be calculated by the following formula (see [3] and [12]):

$$BW_{modal}(\text{inst}) = BW_{modal}/L_{fact} \times (L_{fact}/L_{inst})^{\gamma}, \quad (8.7)$$

where $\gamma$ is called the cutback gamma and its value is between 0.5 and 1.0. Thus, if in this example $L_{fact} = 3.2$ km, $L_{inst} = 2.0$ km, and cutback gamma $\gamma = 0.9$, which is the typical value, then $BW_{modal}$ (inst) = [160(MHz·km)/3.2 (km)] × [3.2(km)/2.0(km)]$^{0.9}$ = 76.3 MHz. We should use this number for the calculations instead of what we did use, 80 MHz.

*If the installation length, $L_{inst}$, is less than the factory length, $L_{fact}$,* several fiber cables must be connected. In this case [3],

$$1/BW_{modal}(\text{inst}) = \sum_{I=1}^{N} (1/BW_{modal\text{-}I}^{(1/\gamma)}), \quad (8.8)$$

where $BW_{modal\text{-}I}$ is the bandwidth-length product of pieces of cable to be connected for installation and $\gamma$ is called concatenation gamma. The latter value is also between 0.5 and 1.0 with the typical range being from 0.65 to 0.75. If we make the link in this example from two cables— one where $BW_1$ = 100 MHz·km and $L_1$ = 125 km and the other where $BW_2$ = 60 MHz·km and $L_2$ = 0.75 km— then applying Formula 8.8 where $\gamma = 0.7$ gives us $1/BW_{modal}$ (inst) = $(1/BW_1^{(1/\gamma)} + 1/BW_2^{(1/\gamma)})$ = 1/ 164.5 MHz·km. We should use this number, 164.5 MHz·km, for our calculations instead of 160 MHz·km.

It is interesting to note that the bandwidth of connected cables is greater than that of a single cable. This effect is caused by mode coupling. (See Section 4.6.)

---

## Local Area Network—General Considerations

***Network topologies*** As you'll recall, a network is a collection of nodes connected by links. In the case of LANs, nodes are the main cross-connects (MC), the intermediate cross-connects (IC), the horizontal cross-connects (HC), and the outlets. The way in which these nodes are connected is called *topology.* The following network topologies are used to build LANs: *point-to-point, star, ring,* and *bus.* They are shown in Figure 8.22. (Also see Figure 1.2.)

Examples of how the same physical layout enables us to build different topologies by simply changing the cross-connects are shown in Figure 8.23. This example also underscores the advantage of cross-connect over interconnect systems.

An important point to understand is this: There are both *physical* and *logical topologies. A physical topology shows how network nodes are connected physically, while a logical topology shows how the flow of information is arranged.* For example, Figure 8.22(e) shows a network where nodes are physically connected in star topology but information flow makes a logical ring.

Star is the popular physical topology for LANs because it is a flexible configuration unable to support most logical topologies, it easily facilitates network expansion and upgrading, it has centralized cross-connect to make its administration and maintenance easy, and it meets the TIA/EIA 568A standard.

***LAN standards*** The building of computer (local area) networks is based on several standards. These standards define access, switching, signaling, and transmission techniques. A detailed

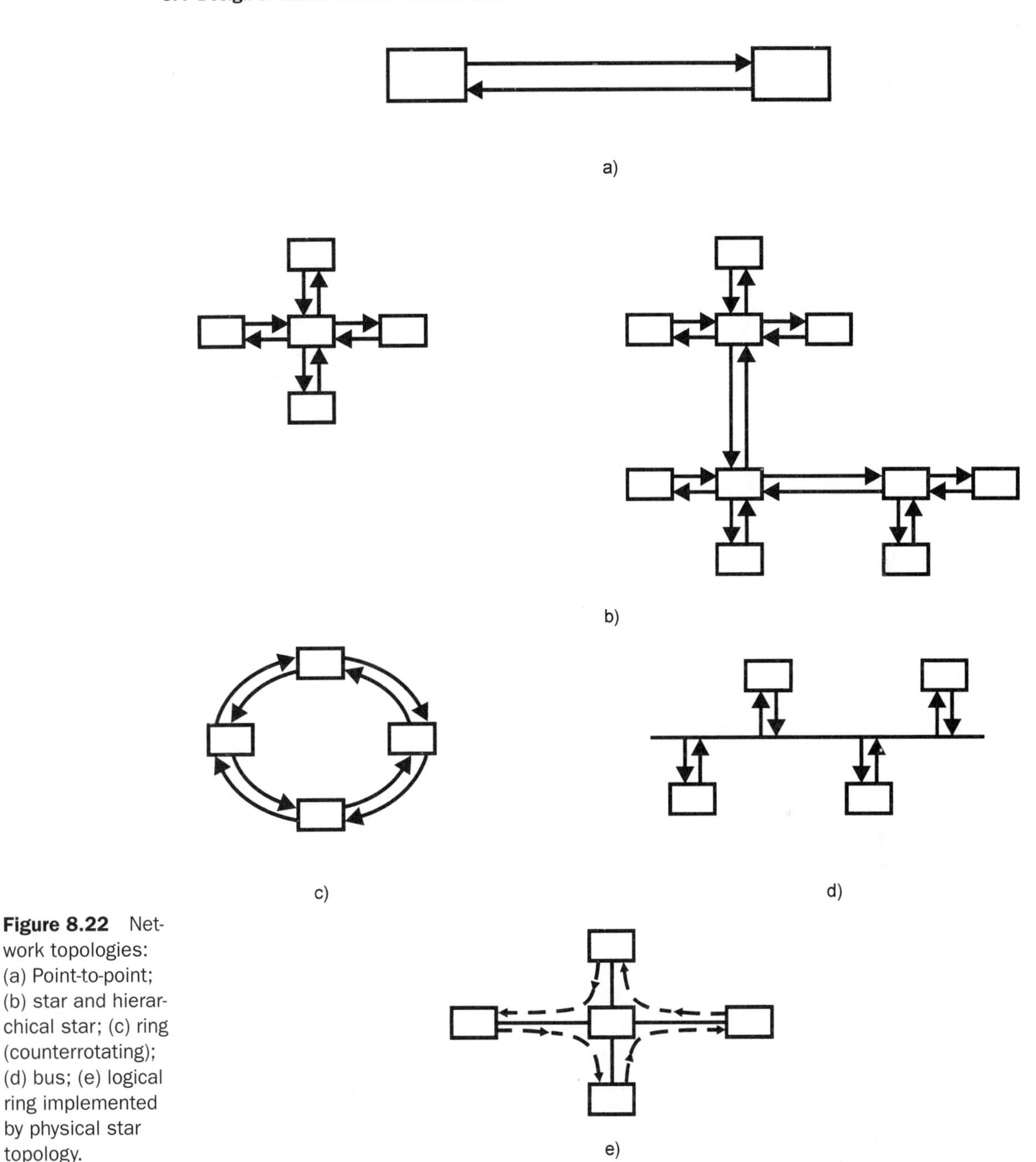


**Figure 8.22** Network topologies: (a) Point-to-point; (b) star and hierarchical star; (c) ring (counterrotating); (d) bus; (e) logical ring implemented by physical star topology.

description of these standards can be found in many references. (See, for example, [11], [13], [14], and [15].) We review them here to give you the background necessary for understanding fiber applications.

*Ethernet,* one of the first LANs, was developed by Xerox, Intel, and DEC in the mid-1970s. It is based on *bus topology.* This standard continues to develop and now there are three Ethernet fiber-optic standards based on star topology: 10Base-FL (Link), 10Base-FB (Backbone), and 10Base-FP (Passive). Fiber Ethernet operates at 10 Mbit/s and 100 Mbit/s. The newest standard within this group is Gigabit Ethernet.

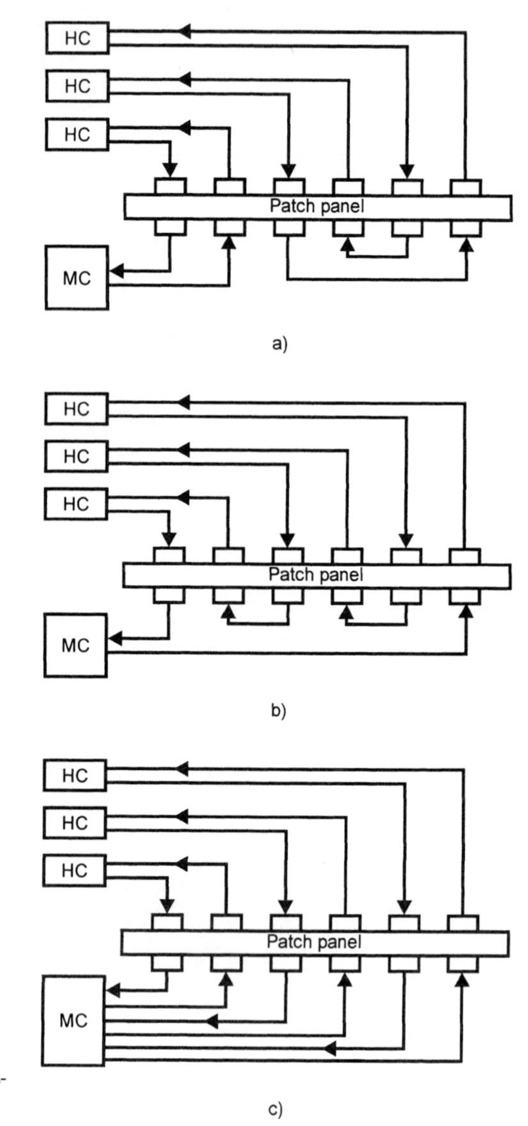


**Figure 8.23** Building different physical topologies by changing cross-connects: (a) Point-to-point; (b) ring; (c) star.

*Token Ring,* developed by IBM, is based on *logical ring topology.* Information flows from one station to another, carrying a token that indicates the destination station. It is a relatively slow LAN, operating at 16 Mbit/s.

*Fiber distributed data interface (FDDI)* was the first LAN developed especially for optical fiber as a transport medium. It is based on dual counterrotating ring topology. In case of a failure of any link or station, the network will short itself on both stations adjacent to the broken section, thus providing continuity of service. Each station regenerates a signal. The maximum distance between adjacent stations is 2 km; as a result, with its maximum number of stations, 500, FDDI can have a circumference of 100 km. It operates at 100 Mbit/s and may serve as a backbone for other LANs. This is one of the most popular fiber-optic LANs. There is a version of FDDI—FDDI-II—that provides even more services at a faster rate, but its future is uncertain.

*Enterprise systems connection (ESCON)* is a fiber-optic back-end channel connecting IBM mainframes with peripherals. Obviously, it is based on star topology. The network operates at 200 Mbit/s.

*Fibre Channel* is a high-speed fiber link providing transmission between mainframe and peripherals at 1.0625 Gbit/s. It supports both channel and network configurations. New versions with bit rates of 2.125 and 4.25 Gbit/s are under development.

***Other standards*** *Synchronous optical network (SONET)* is a high-speed fiber-optic telecommunications standard. The basic SONET rate, OC-1, is 51.840 Mbit/s. All other rates are simple multiples of this number. The highest rate now developed is OC-192 = 9.952 Gbit/s. This standard was developed by Bellcore (now called Telcordia Technologies) for the telephone industry to meet the demand for high-capacity long-distance networks. SONET is mentioned here because it supports the access of asynchronous LANs to synchronous public-telecom networks.

*Asynchronous transfer mode (ATM)* is a very efficient transporter of voice, data, and video signals. It packs all its signals in 53-byte cells, which are transmitted and switched at a bit rate dictated by the network. This flexibility makes ATM a very attractive technique, one that has been widely developed for today's telecommunications systems. It can support up to 622 Mbit/s and, of course, is based on optical fiber as its transmission medium.

SONET and ATM networks are discussed in greater detail in Chapter 14.

***Fiber versus copper*** Most of the above LANs can use either copper wire or fiber as the transmission medium. For years there was a heated debate over which link—copper or fiber—to choose for LANs. These transmission media have been compared in terms of bandwidth, cost, durability, ease of installation, size and weight, trained personnel required to work with them, compliance with FCC requirements, maintenance, and a host of other criteria. To learn more about this controversy, consult the references at the end of this chapter. In any event, the reality is simply this: If you look to the future and don't want to re-cable every three to five years, fiber is your choice.

## Cabling of Local Area Networks

***Physical layout*** A typical layout of a local area network is shown in Figure 8.24. We distinguish among three types of wiring: campus (interbuilding) backbone, building (intrabuilding) backbone, and horizontal cabling.

*Campus backbone* is a set of outdoor cables connecting buildings in a given complex. The topology shown in Figure 8.24(a) is a hierarchical star.

*Building backbone* is a set of indoor cables connecting building equipment with MC or IC. Figure 8.24(b) shows a two-level hierarchy of building-backbone wiring with one IC on the intermediate floor.

*Horizontal cabling* can be implemented in several layouts. Traditional cabling includes HC on each floor, with a fiber going to a single user's outlet (the upper floor in Figure 8.24[c]). In another layout, work-area (multi-user) outlet cabling delivers an outlet for many users from a floor HC. Individual users connect to this outlet by patch cords (the second floor in Figure 8.24[c]). Centralized cabling allows outlet connection directly to MC or IC without HC (the ground floor in Figure 8.24[c]).

Aware of this general picture of LAN cabling, we can now consider installation design more closely.

***Challenge in LAN design*** The object of LAN design is the proper physical placement of fiber cables for outdoor and indoor applications. After that, any changes in your network cabling will be a very expensive, time-consuming process. And this is the real challenge for a network designer: to design a network installation not only at minimum expense and maximum efficiency for current needs but also with built-in flexibility for future changes, upgrading, and expanding.

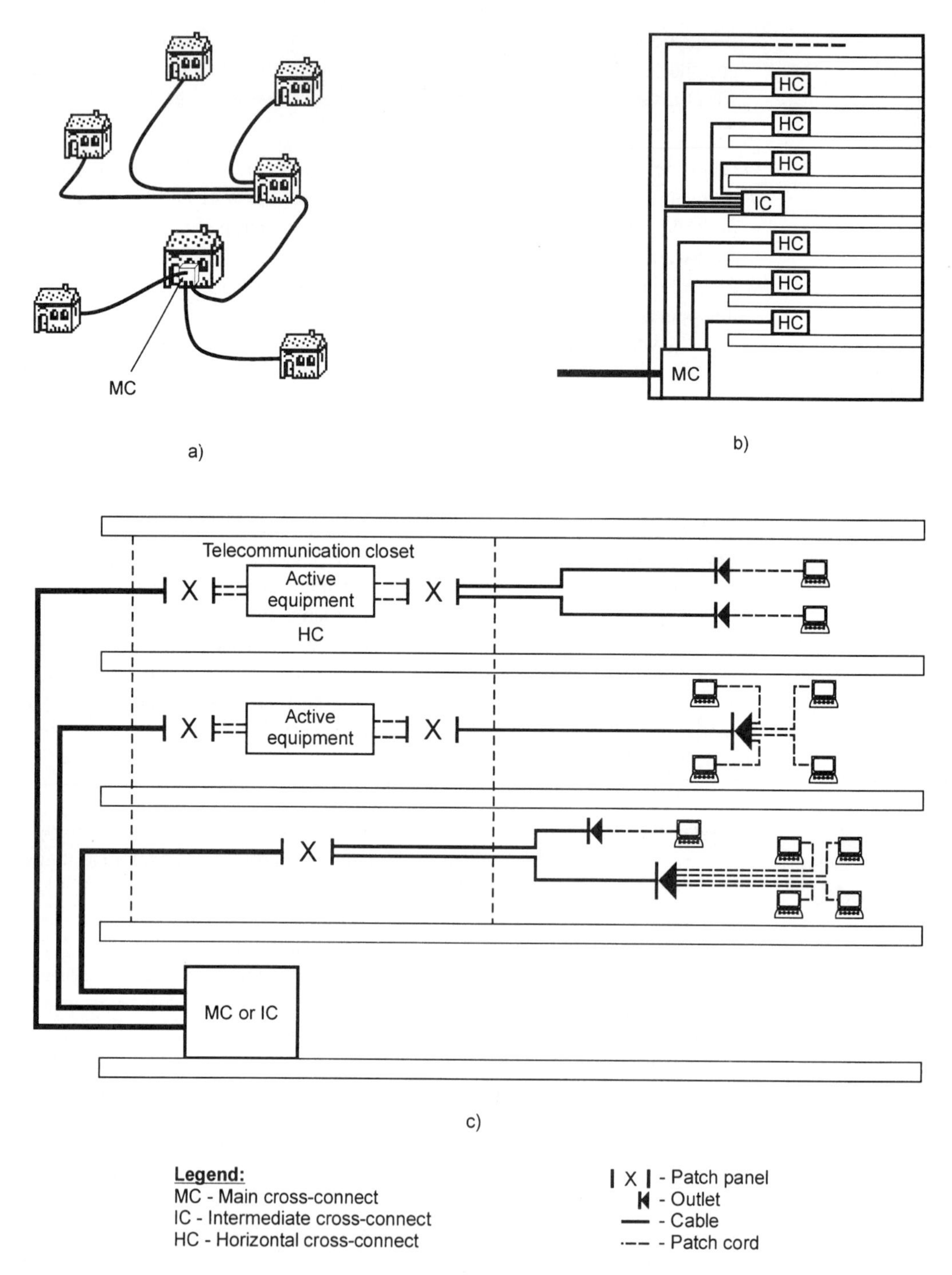


**Figure 8.24** Cabling for local area network: (a) Campus backbone; (b) building backbone; (c) horizontal cabling. *(Adapted from Siecor,* Design Guide, *Release 4, 1999, Hickory, NC.)*

Let's consider two examples:

**1.** When designing a campus backbone, the first step is choosing a physical layout, that is, the topology. All topologies have their pros and cons, but star topology offers maximum flexibility for supporting any logical topology and possibly future expansion. Figures 8.22(e) and 8.24(a) confirm this.

**2.** When designing an outlet installation, you have two options: Either connect a single user outlet directly to an HC (Figure 8.24[c], the upper floor) or connect an HC to a work-area outlet to which individual users can be connected (Figure 8.24[c], the second floor). Obviously, the latter solution gives the network much more flexibility for future moves whether they be additions or changes (often called MACs—moves, adds, and changes).

## Basic Recommendations

There are no rules or formulas for designing LAN installations. (The standards to be observed are considered on pages 265 and 266.) All we have is our professional common sense and the experiences of other installers. Thus, the following basic steps are only *general recommendations* but they might help you or your contractor one day avoid some of the more common and often obvious mistakes [11]. For more detailed guidelines, consult the references at the end of this chapter.

### *Campus backbone*

- Topology—Select the one- or two-level hierarchical star.
- Fibers—Use the multimode (MM) 62.5/125 μm and the singlemode (SM).
- Cables—Choose outdoor cables for aerial or buried installations; use hybrid cables with MM and SM fibers.
- Fiber count—This crucial factor determines the quality of your design. Most LANs require two fibers for duplex communications (simplex is one-way and duplex is two-way communications). Token Ring, FDDI, and SONET LANs need four fibers. Thus, the total number can be found by simply adding up the fibers for all types of networks involved in a campus LAN. For example, a campus backbone of a middle-sized LAN with one MC and three ICs has, typically, 32 MM and 6 SM fibers. But the golden rule here is this: *Always install spare fibers for future growth.* So, by adding 16 MM (50 %) and 6 SM (100 %), the total number of fibers will be 48 MM and 12 SM. Of course, you have to determine the fiber count for your own network, but always add extra fibers.
- Splicing—Choose fusion or mechanical splicing, depending on how many splices you have to make and the power budget of your links. Try to combine all splices at a single point to facilitate installation, splicing, and troubleshooting. But the best advice of all: *Avoid splicing if you can.*
- Connectors—SC and ST are the most popular LAN connectors. When installing SC adapters, observe the orientation (polarity) recommended in the manufacturer's data sheets. Consider small form factor connectors for high-density distribution frames.
- Cable termination method—Choose field connectorization or pigtail, depending on your network's power budget and the number of fibers. In prior years, the rule was this: field installation for MM fibers and pigtail for SM fibers. But today, with the high-quality, easy-to-install connectors on the market, most installers prefer field connectorization.

- Hardware—Campus backbone cable has to be terminated within 50 feet inside a building. (See the TIA/EIA 568A standard below.) Equip the entrance room with connecting hardware able to accommodate this and future cables.

***TIA/EIA 568A: Commercial-Building Telecommunications-Cabling Standard*** Before we consider building wiring, it is necessary to discuss this standard, which covers cabling practice not only as it pertains to fiber but as it relates to copper wire and coaxial cable as well. We'll restrict ourselves to mentioning several key points concerning only the fiber regulations.

- The standard distinguishes vertical runs between floors (building backbone) from horizontal runs from HC to outlets (horizontal wiring).
- The building backbone length between an MC and an HC must be no more than 2000 m for 62.5/125 μm and 3000 m for a singlemode fiber. A maximum of one IC can be included within a link.
- The distance between the HC and an outlet must be no more than 90 m. However, a direct IC-outlet connection is allowed within 300 m.
- A cross-connect on each floor (HC) is recommended.

This standard doesn't depend on LAN types; in fact, its regulations are based on performance. If your network can support 100 Mbit/s, it doesn't matter whether you use FDDI or Ethernet.

***Building backbone***

- Topology—A one-level hierarchical star is usually recommended.
- Fibers—Select the 62.5/125 μm fibers.
- Cables—If indoor nonconductive cables rated as risers are to be used, OFNR is generally recommended; if OFCR cables have to be used, make sure the metallic components are properly grounded.
- Fiber count—This decision depends on the type of wiring used. Generally, you need to install two fibers for each application (Ethernet, FDDI, etc.) from the IC to an outlet. Thus, with the centralized method (see Figure 8.24[c], the ground floor), your building backbone has to carry the number of fibers equal to the number of users times the number of applications times 2. With the distributed (traditional) method, when active electronics is placed on each floor (see Figure 8.24[c], the second and third floors), the building backbone can typically carry 24 MM fibers. Therefore, *the number of fibers versus active electronics is the difference between the centralized and the distributed methods.* Most users consider the centralized method the more effective one from cost and maintenance standpoints. However, the centralized layout is possible only with fiber because copper wire is restricted to a distance of 90 m between an outlet and an HC.
- Splicing—Avoid splicing if you can. If not, see the campus-backbone recommendations.
- Connectors—SC and ST connectors are recommended. Consider new-generation connectors.
- Cable-termination method—Field connectorization is recommended.
- Hardware—Installation hardware is always preferred. (See Section 8.3.)

***Horizontal cabling***

- Topology—Select a simple star topology.
- Fibers—Choose 62.5/125 μm fibers for fiber-to-the-desk (FTTD) application. Traditionally, a copper wire connects desk equipment, including computers, to the network. But today, for many reasons, including installation cost and maintenance, fiber is the medium of choice because it supports all existing and future high-speed applications. In addition, fiber doesn't have distance restrictions, while copper wire is restricted to a 3-m distance between an outlet and desk equipment.
- Cables—OFNP is the choice usually recommended; for outlet-electronic connections, single-fiber or two-fiber patch cords are recommended. (See, for example, Figure 7.9[f].)
- Fiber count—Two fibers per individual outlet are the minimum for data communications provided that voice communications are supported by copper cable.
- Splicing—Avoid if at all possible.
- Connectors—SC and ST connectors are recommended. Consider MT-RJ, LC, and other new connectors.
- Cable-termination method—If necessary, use field-installed connectors for HC connection.
- Hardware—Use HC outlet hardware. (See Section 8.3.)

These general rules should help you grasp the basic ideas regarding LAN-installation design and assist you in supervising the job being done by your contractor.

## Plastic (Polymer) Optical Fiber (POF)

A fiber-LAN designer has to keep abreast of ongoing developments in plastic optical fiber (POF). This transmission medium is now ready to compete with copper wire and MM silica fibers for the short-distance applications—horizontal cabling and fiber-to-the desk wiring. We'll summarize here the current state of plastic optical fiber:

***Technology*** POF is made completely of plastic. Developed by DuPont in 1968 with a step-index profile, this POF has worked for almost 30 years with a 650-nm LED in a very quiet corner of the fiber-optic communications world. POF exhibits several attractive properties: low cost, big sizes (1000 μm = 1 mm fiber diameter), a large *NA*, ease in its application. On the other hand, high attenuation (above 200 dB/km) and low bandwidth (about 5 MHz·km) have not allowed proponents of POF to regard it as a truly competitive transmission medium.

In 1992, researchers at Japan's Keio University developed a graded-index (GI) POF and in 1997 reported having developed a perfluorinated (PF) low-loss GI-POF. This fiber has an attenuation of about 50 dB/km and a bandwidth of 300–500 MHz·km, with the potential for up to 10 GHz·km [16]. Bear in mind that POF works, to date, at distances up to 100 m. The recent advances in this technology have awakened interest on the part of potential users.

***Market*** The primary area for POF applications is the horizontal and desk wiring of premises. The TIA/EIA 568A standard requires that horizontal cabling should support 100 Mbit/s at 100 meters. The newest POFs meet this specification. Another highly promising market is the ultra-short networks found in aircraft, automotive, and shipboard applications. Finally, the home-network area, with its exponentially growing demand for capacity driven by Internet connections, is eager for a new transmission link.

***Standards*** One serious drawback to POF is that it suffers from a lack of standards. In May 1997, the ATM Forum (the industry association for developing standards in asynchronous transfer mode, ATM, networks) approved POF for ATM application (155 Mbit/s) at a 50-meter run. This is the only industry standard that has been developed for POF so far.

***Conclusion*** POF promises great potential for short-distance networks but, at this stage, it hasn't captured a significant segment of the market.

## 8.5 TESTING, TROUBLESHOOTING, AND MEASUREMENT

*Testing* means verifying a fiber network's performance characteristics for its compliance with the specifications. *Troubleshooting* means finding faulty component(s) responsible for the incorrect operation of a system. *Measurement* means obtaining quantitative parameters by physically measuring an object or a phenomenon.

To test a system or a component, you need to take measurements. For example, you measure the value of attenuation of an installed link and, if this value is within given tolerances, the test is over. To troubleshoot, you also need to take some measurements. For example, again you would measure the attenuation of the link and, if this value is too low, you would need to find the source of the loss. Technically, the term *measurement* itself implies obtaining all the possible values of a specific parameter under certain conditions. For example, you measure attenuation as a function of a bend radius for a set of ambient temperatures.

In this section, we'll concentrate on testing and troubleshooting the installed LAN links and components, leaving measurements themselves for future discussion.

### Test Equipment

It is quite evident that we need some measuring equipment to perform testing and troubleshooting. Since many of the tools are used for various measurements, it is logical to briefly describe them first and, following that, discuss their usage.

***Microscope (optical fiberscope)*** We need to inspect the connector endfaces after installation to evaluate the quality of termination. The only test method is visual inspection through a microscope to see that the endface surface is well polished and crack- and dirt-free and that a fiber core is at the edge of a ferrule. There are special microscopes designed to do this job (Figure 8.25). They have a universal adapter in which to insert a connector for inspection as well as light and lens systems arranged to make the inspection in a quick and effective way. Typical magnification is 30× to 100×, though even larger magnification is available for inspecting singlemode-fiber connectors.

***Optical fiber identifier*** This device is a very sensitive photodetector. If you bend a fiber, some light radiates outside the core, as we learned in discussing macrobending loss. This light is detected by a fiber identifier, which allows the technician to identify a single fiber among many others in a multifiber cable or at a patch panel. A fiber identifier can detect the presence and direction of a signal (traffic) without interrupting its transmission. To make this job even easier, the test-light signal—typically modulated at 270, 1000, or 2000 Hz—is injected into a specific fiber at the transmitter end. (See Figure 8.26.)

Most fiber identifiers work with a singlemode fiber cable operating at 1310 or 1550 nm.

***Visual fault locator (tracer)*** This unit is a powerful source of visible (red) light based on a laser diode. When light is injected into a fiber, fiber breaks, faulty connectors, sharp bends, bad

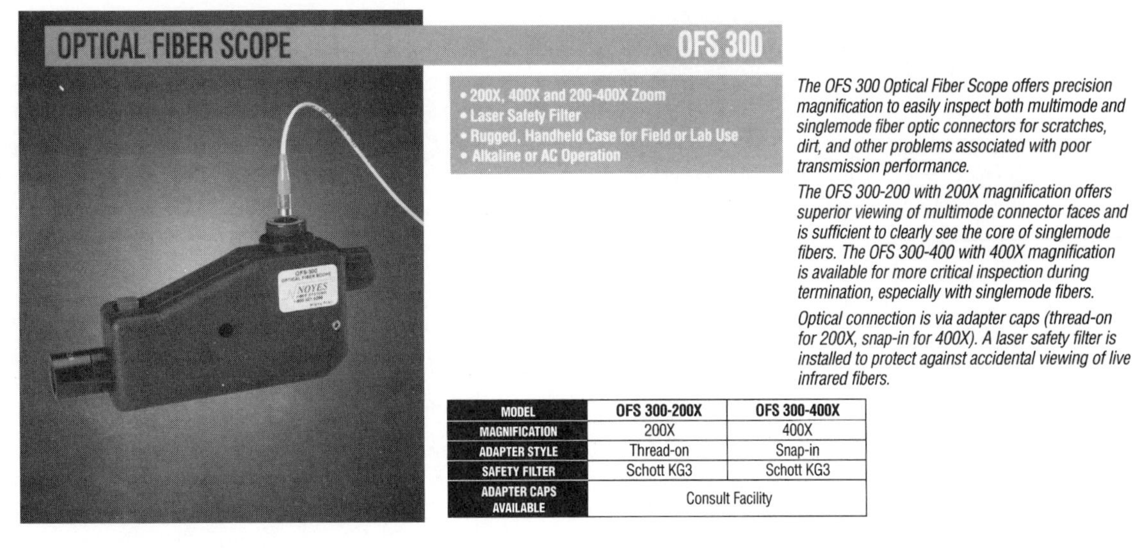


**Figure 8.25** Optical fiberscope. *(Courtesy of Noyes Fiber Systems, Laconia, NH.)*

splices, and similar faults can be visually located by light radiated outside the fiber. Yellow and orange cables transmit red light (so you can see the glow) but black and gray jackets do not; thus, the usage of a visual fault locator is restricted in its application. Figure 8.27 shows an example of a fault locator. It is also available as a pen-size device.

Visual fault locators radiate in continuous-wave (CW) or pulse modes. Frequencies at 1 or 2 Hz are very popular but these devices can also operate in the kHz range. Output power at 0 dBm (1 mW) or less is typical and the working distance is usually in the range of 2 to 5 km. A universal connector or any of the popular connector types can be obtained.

***Optical calibrated light source*** To measure attenuation in a fiber cable, we have to inject a calibrated, constant amount of light power, $P_{in}$. This is why we need a special light source, either an LED or a laser diode, the same types used in transmitters but with complex electronics to stabilize

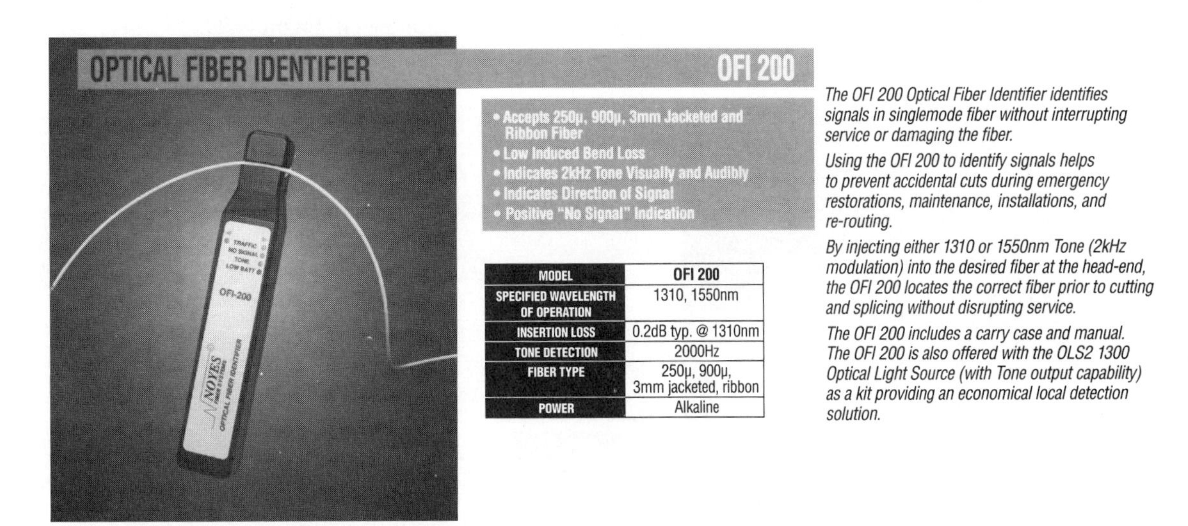


**Figure 8.26** Fiber identifier. *(Courtesy of Noyes Fiber Systems, Laconia, NH.)*

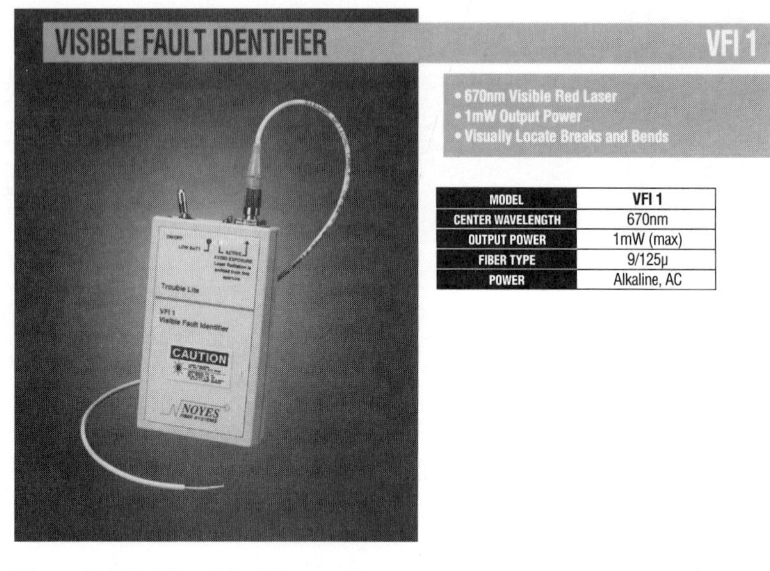


| MODEL | VFI 1 |
|---|---|
| CENTER WAVELENGTH | 670nm |
| OUTPUT POWER | 1mW (max) |
| FIBER TYPE | 9/125μ |
| POWER | Alkaline, AC |

*The VFI 1 Visible Fault Identifier generates a visible red laser pulsed at 2Hz to allow visual inspection for breaks or bends in optical fiber splices, jumpers and other devices used in optical fiber transmission systems.*

*When connected to optical fiber cable, the pulsing red laser can be visually detected. Discontinuity, or faults, in the fiber will cause scattering of the laser light at the fault location. The scattering of the red laser light is easily detected and pinpoints the fault.*

*The VFI 1 is a cost effective means of locating fiber cable faults within a Central Office (CO) or a Controlled Environmental Vault (CEV), inspection of jumper cable ends, and visually aligning mechanical splices.*

**Figure 8.27** Visual fault locator. (*Courtesy of Noyes Fiber Systems.*)

the radiated power. These sources also have calibrated wavelengths: 660 nm for plastic optical fiber, 850 nm and 1300 nm for multimode fiber, and 1310 nm and 1550 nm for singlemode fiber. (See Figure 8.28.) Obviously, all types of connectors are available to order.

You may wonder why we need two light sources, one for testing and the other to locate faults. Let's compare their typical specifications (See Table 8.1, page 299.)

It is advisable to analyze these data, paying attention to the comparison of absolute power values (remember, 1 mW is 0 dBm) and their precision. Can you measure the loss of an optical fiber if the input power varies over 30%?

***Power meter (PM)*** This is the most powerful device in fiber-optic technology. It is as necessary to the fiber-optic field as a digital multimeter is to the electronics industry. It measures the average light power emanating from a fiber or directly from a source. The device consists of a photodetector, electronics, and a display. Several samples are shown in Figure 8.29.

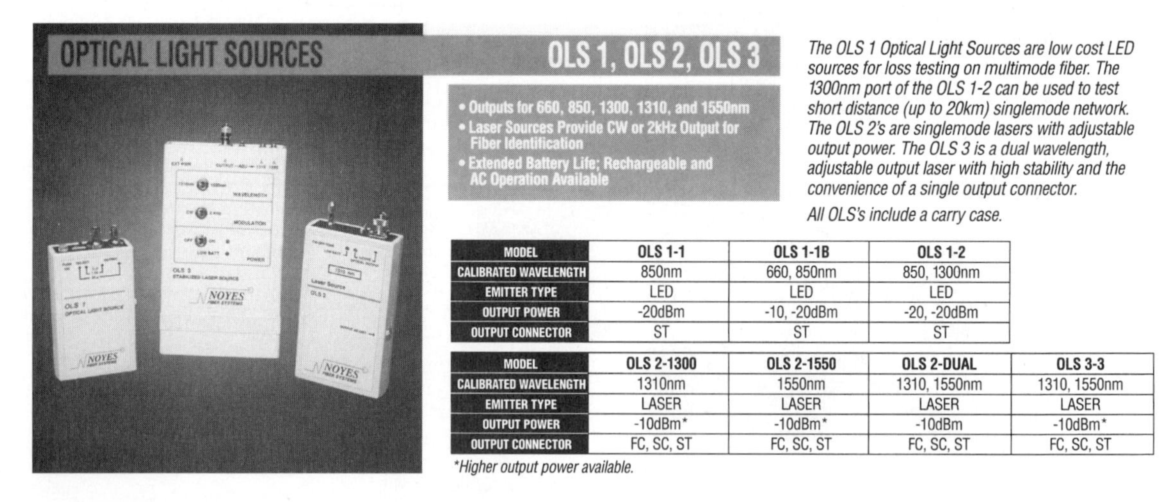


*The OLS 1 Optical Light Sources are low cost LED sources for loss testing on multimode fiber. The 1300nm port of the OLS 1-2 can be used to test short distance (up to 20km) singlemode network. The OLS 2's are singlemode lasers with adjustable output power. The OLS 3 is a dual wavelength, adjustable output laser with high stability and the convenience of a single output connector.*

*All OLS's include a carry case.*

| MODEL | OLS 1-1 | OLS 1-1B | OLS 1-2 |
|---|---|---|---|
| CALIBRATED WAVELENGTH | 850nm | 660, 850nm | 850, 1300nm |
| EMITTER TYPE | LED | LED | LED |
| OUTPUT POWER | -20dBm | -10, -20dBm | -20, -20dBm |
| OUTPUT CONNECTOR | ST | ST | ST |

| MODEL | OLS 2-1300 | OLS 2-1550 | OLS 2-DUAL | OLS 3-3 |
|---|---|---|---|---|
| CALIBRATED WAVELENGTH | 1310nm | 1550nm | 1310, 1550nm | 1310, 1550nm |
| EMITTER TYPE | LASER | LASER | LASER | LASER |
| OUTPUT POWER | -10dBm* | -10dBm* | -10dBm | -10dBm* |
| OUTPUT CONNECTOR | FC, SC, ST | FC, SC, ST | FC, SC, ST | FC, SC, ST |

**Higher output power available.*

**Figure 8.28** Optical calibrated light (test) source. (*Courtesy of Noyes Fiber Systems.*)

**Table 8.1** Specifications of a visual fault locator and an optical light source

| | Visual fault locator | Optical light source |
|---|---|---|
| Source type | Laser diode | LED or laser diode |
| Wavelength | 660 ± 10 nm | 850 nm, 1300 nm<br>1310 nm, 1550 nm |
| Spectral width | N/A | 50 nm (LED)<br>1 nm (LD) |
| Output power (LED) | 1 mW max, 800 μW typical | –15 to –25 dBm (LED)<br>–5 to –10 dBm (LD) |
| Calibration accuracy | N/A | ±0.5 dB |
| Short-term drift | N/A | 0.2 dB/hr |
| Distance range | Typically 3 km | N/A |
| Modulation modes | Continuous wave (CW),<br>2 Hz; square wave, 2 kHz,<br>other modulations | N/A |
| Connector | FC, ST, SC<br>(other styles available) | FC, ST, SC<br>(other styles available) |

Power meters use various semiconductors (silicon, indium gallium arsenide, and germanium are among the most popular) as photodetectors, and this enables them to work in a variety of wavelengths from 660 nm to 1550 nm. One unit can measure power over a wide *dynamic range* from + 3 dBm to –80 dBm. For cable TV (CATV) systems or optical amplifiers, PMs with a dynamic range from + 27 dBm to –60 dBm are available. (The term *dynamic range* defines the range of measured values within which a PM input–output characteristic is linear, with typical linearity being around ± 0.05 dB.)

Since power meters are used for measuring absolute (not relative) values, they must be calibrated. This means their measurements must be related to the national standard provided by the National Institute of Standards and Technology (NIST). Obviously, each of the thousands of PMs

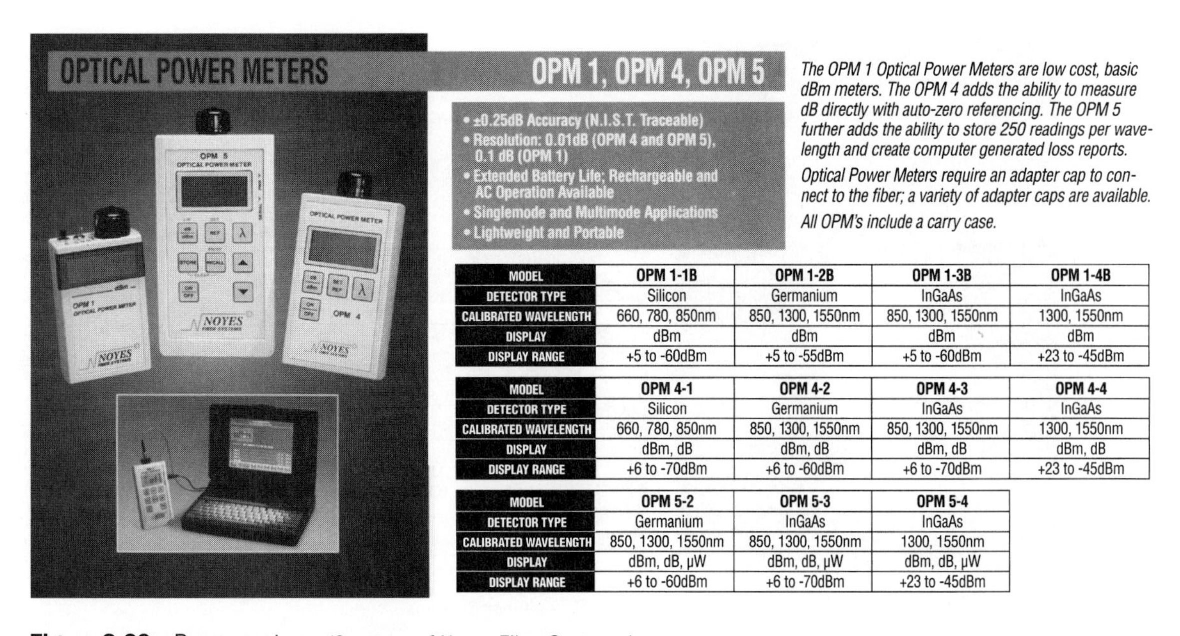


**Figure 8.29** Power meters. *(Courtesy of Noyes Fiber Systems.)*

manufactured every day cannot be directly compared vis-à-vis the national standard. Thus, some intermediate standards, tested against the original one, are used in the calibration of a specific PM. This characteristic of a PM is called *(calibration) traceability.* If you need to detect only the presence of light, traceability is not required, thereby lowering the cost of your PM significantly.

*Absolute accuracy* is the characteristic that tells you the preciseness of your measurements. For example, if the data sheet says ± 0.25 dB and your reading is – 21.9 dBm, this means the actual power measured is within – 21.65 dBm and – 22.15 dBm.

*Stability* describes how much the PM accuracy changes over the range of operating temperature. Numbers <± 0.05 dB are typical.

*Resolution* shows the minimum deviation in absolute values the PM can detect. This is given either in logarithmic scale (e.g., 0.01 dB) or in absolute scale (e.g., 0.01 μW).

The above discussion is based on a typical PM data sheet. The measurement of optical power is the most frequent and fundamental measurement taken in fiber-optic communications systems; therefore, a PM is one of the most important measuring devices in the field. Manufacturers' data sheets will assist you in choosing the right PM.

***Optical-loss test kits*** To measure loss in a fiber link, you need to inject calibrated light power and read the output power; thus, you need a light source and a power meter. These two devices constitute a loss-test kit. (See Figure 8.30.) You can also find these two devices combined in a single unit called an *optical-loss test set.* Which set—separate units or an integrated instrument—do you think is better? Imagine that you need to measure loss in your link. One person holds a light source at the transmitter end and another person operates a PM at the receiver end. Thus, loss measurement would be taken in one direction. Generally, you need to take it in both directions (directional connection loss, you'll recall); therefore, your technicians have to exchange instruments and repeat the measurement. But what if they are 10 floors or 10 km apart? It would be better if each of them had a light source and a PM optical-loss test kit and they were operating them simultaneously. Obviously, you can follow another line of reasoning, but the bottom line is this: To perform an optical-loss test, you need a calibrated optical source and a calibrated power meter.

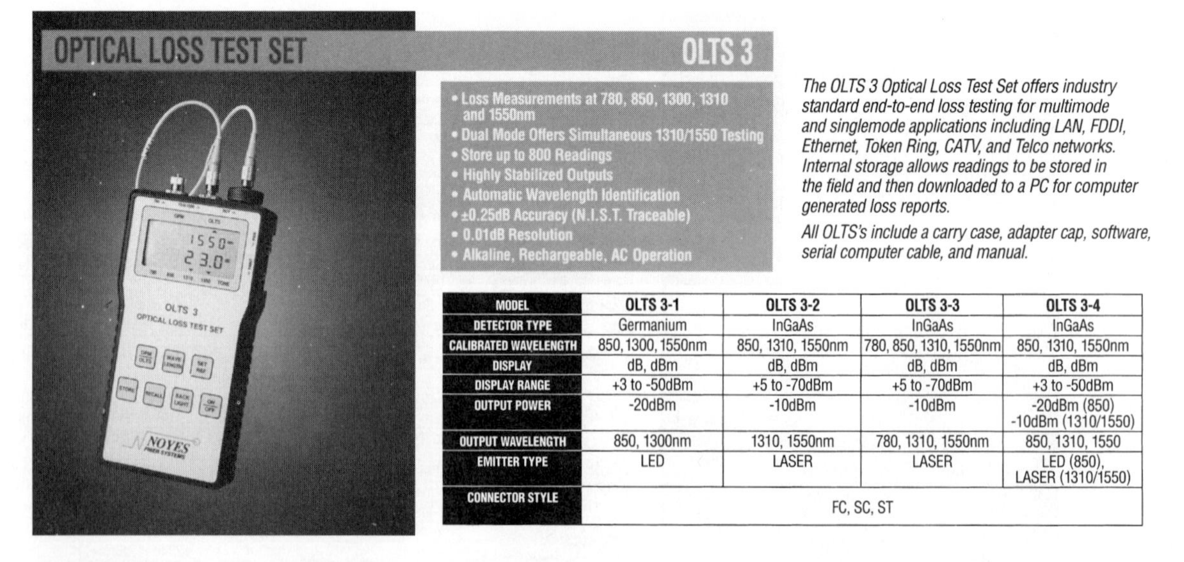


| MODEL | OLTS 3-1 | OLTS 3-2 | OLTS 3-3 | OLTS 3-4 |
|---|---|---|---|---|
| DETECTOR TYPE | Germanium | InGaAs | InGaAs | InGaAs |
| CALIBRATED WAVELENGTH | 850, 1300, 1550nm | 850, 1310, 1550nm | 780, 850, 1310, 1550nm | 850, 1310, 1550nm |
| DISPLAY | dB, dBm | dB, dBm | dB, dBm | dB, dBm |
| DISPLAY RANGE | +3 to -50dBm | +5 to -70dBm | +5 to -70dBm | +3 to -50dBm |
| OUTPUT POWER | -20dBm | -10dBm | -10dBm | -20dBm (850)<br>-10dBm (1310/1550) |
| OUTPUT WAVELENGTH | 850, 1300nm | 1310, 1550nm | 780, 1310, 1550nm | 850, 1310, 1550 |
| EMITTER TYPE | LED | LASER | LASER | LED (850),<br>LASER (1310/1550) |
| CONNECTOR STYLE | FC, SC, ST | | | |

**Figure 8.30** Optical-loss test set. *(Courtesy of Noyes Fiber Systems.)*

***Optical return-loss test kit*** A special procedure described on page 308 and in Figure 8.36 measures optical return (reflection) loss, ORL. But nobody wants to arrange a setup in the field; you just need a number. The optical-return-loss test kit gives you that number—the return loss in dB or μW. It can measure ORL in the range of 55 to 65 dB with a typical accuracy of about 0.5 dB. Its physical appearance is quite similar to that of an optical-loss test set; furthermore, it often can be a combination of an optical-loss and return-loss test kit.

***Fiber-optic talk set*** The above examples raise one more problem: Can technicians taking measurements (or doing splicing) in remote locations communicate? Radio-based communications might be unsuitable because of environmental conditions, including FCC restrictions. But why not employ ready-to-use transmission media such as an installed fiber-optic link? This is where a fiber-optic talk set is applied. This set (Figure 8.31) includes two transmitters and receivers (transceivers) that carry voice over the spare fiber in both directions. (This is known as full-duplex communication.)

Typically, this set can generate a 2-kHz tone for fiber identification and serve as a stabilized light source for loss measurement.

***Attenuator*** This device allows one to insert a calibrated attenuation into a fiber link to simulate distance or actual attenuation. (See Figure 8.32.) For example, using an attenuator, one can verify the power margin received by calculating a link's power budget. Using a variable optical attenuator (VOA), one can vary the light power injected from a light source into a fiber. Attenuator design is based on such phenomena as bending loss, end separation, and optical filtering. Attenuators must provide low reflection loss (–50 dB is typical) and low insertion loss (1.5 dB is acceptable). Other important characteristics are the attenuation range the device can provide (40 to 60 dB is typical) and resolution (< 0.1 dB is usual).

***Optical time-domain reflectometer (OTDR)*** An optical time-domain reflectometer (OTDR) is optical radar. This instrument sends a light pulse down the fiber and measures the time delay between launched and returned pulses. This time delay is a measure of the distance from the transmitter to a given reflection point. The longer the distance, the less light power returned

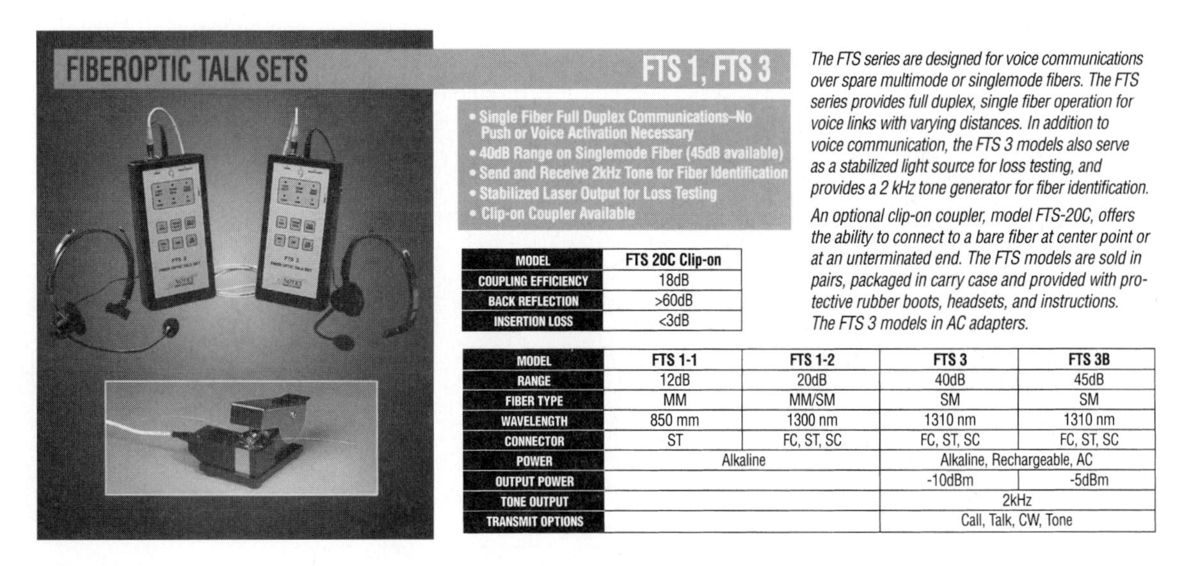


| MODEL | FTS 20C Clip-on |
|---|---|
| COUPLING EFFICIENCY | 18dB |
| BACK REFLECTION | >60dB |
| INSERTION LOSS | <3dB |

| MODEL | FTS 1-1 | FTS 1-2 | FTS 3 | FTS 3B |
|---|---|---|---|---|
| RANGE | 12dB | 20dB | 40dB | 45dB |
| FIBER TYPE | MM | MM/SM | SM | SM |
| WAVELENGTH | 850 mm | 1300 nm | 1310 nm | 1310 nm |
| CONNECTOR | ST | FC, ST, SC | FC, ST, SC | FC, ST, SC |
| POWER | Alkaline | | Alkaline, Rechargeable, AC | |
| OUTPUT POWER | | | -10dBm | -5dBm |
| TONE OUTPUT | | | 2kHz | |
| TRANSMIT OPTIONS | | | Call, Talk, CW, Tone | |

**Figure 8.31** Fiber-optic talk set. *(Courtesy of Noyes Fiber Systems.)*

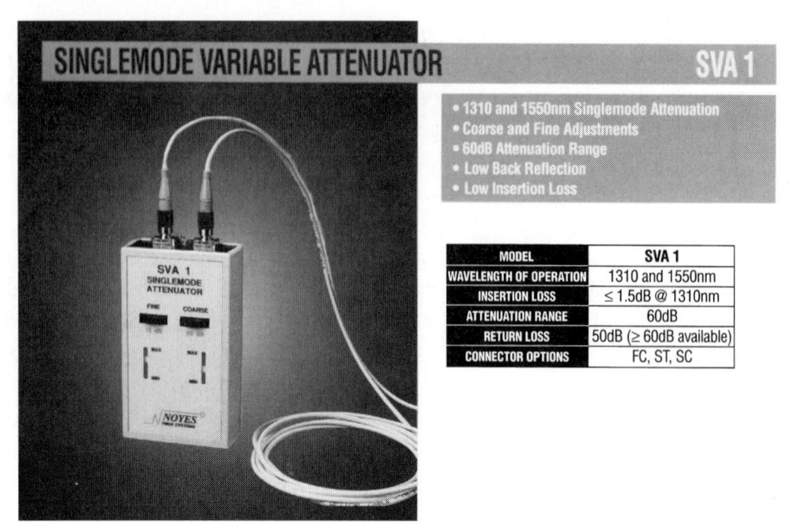


| MODEL | SVA 1 |
|---|---|
| WAVELENGTH OF OPERATION | 1310 and 1550nm |
| INSERTION LOSS | ≤ 1.5dB @ 1310nm |
| ATTENUATION RANGE | 60dB |
| RETURN LOSS | 50dB (≥ 60dB available) |
| CONNECTOR OPTIONS | FC, ST, SC |

*The SVA 1 Singlemode Variable Attenuator offers very low back-reflection with over 60dB of attenuation range at both 1310 and 1550nm in a convenient handheld package. The SVA 1 is available with a variety of connectors and reflectance options to better than 60dB (with APC connectors). With only two adjustments, COARSE and FINE, the SVA 1 is simple to understand and operate, allowing bi-directional signal transmission with no loss penalty.*

*The SVA 1 can be used for Bit Error Rate (BER) testing, systems margin analysis, and transmission receiver testing. Field applications include installation, new equipment turn-ups and routine maintenance.*

*The SVA 1 is suited for all singlemode applications including Telco, LAN, WAN, Video and CATV.*

**Figure 8.32** An optical attenuator. *(Courtesy of Noyes Fiber Systems.)*

because of attenuation; thus, the OTDR builds a trace of power versus distance. In other words, the OTDR traces the power allocation along the fiber link and displays this trace on its screen (See Figure 8.33.)

If you compare a power-budget graph (Figure 8.21) with an OTDR's trace, you'll see they are almost identical because they expound the same idea theoretically and practically.

From what points is light reflected when it emanates from an OTDR? To answer this question, first recall the Rayleigh reflection. Variations in a core's refractive index cause light reflection, including backreflection. This phenomenon allows the OTDR to build a trace of light power along an optical fiber. Second, any connections—mechanical splices and connectors—also cause backreflection. (Remember the Fresnel reflection? See Figure 8.4.) If you look at an OTDR's display closely, you'll see spikes at several points. These spikes don't show any amplification; they display a strong backreflection at the connection points. Incidentally, can you explain why there is a big spike at the fiber end?

OTDRs come in three versions: full size, mini models, and optical fault locators (finders). Full-size OTDRs are based on modern microprocessors and provide fully automated operation and processing. The results of their measurements are stored and printed immediately in any desired form, including a comparison with the required specifications. The basic characteristics of this type of OTDR are as follows:

**Figure 8.33** Optical time-domain reflectometer. *(Courtesy of Noyes Fiber Systems.)*

- They can work at any wavelength: 850, 1300, 1310, and 1550 nm with both multimode and singlemode fibers.
- Their dynamic range can reach 40 dB.
- Their inherent drawback is *dead zones,* the minimal distances where they start to work. They have event (initial-reflective) and attenuation (initial-nonreflective) dead zones. The first zone, where the first Fresnel reflection occurs, is between 3 and 5 m; the second, where Rayleigh scattering starts to show, is between 10 and 40 m. Their distance range runs from 5 km to 260 km and longer. But you have to fix the distance setting at the desired range; for example, you can work at a distance of 5 km, then switch to 10 km, then to 20 km, and so forth. Distance-measurement accuracy depends on the distance setting—for example, 0.01% ± 2 m for 5 km and 0.01% ± 4 m for 160 km.
- Loss resolution can be as low as 0.001 dB. Reading the basic data on the OTDR's specification sheet, you might have noticed that the accuracy of the measuring loss was not referred to there. This is because an OTDR's data sheet does not include such a characteristic.
- The accuracy, or reliability, of an OTDR depends on the backscatter index (coefficient), which is the cumulative characteristic of the scattering properties of a fiber or a connection. It changes, of course, from fiber to fiber and from connector to connector but —for a given fiber or connection—it can also change under environmental conditions. This is the *major shortcoming* of an OTDR.

The data given above are the best available today on OTDRs. Obviously, there are special designs for special applications. For example, there is a zero-dead-zone OTDR designed for measurements in aircraft fiber-optic networks, whose function is very short links (up to 1000 ft). There are special OTDRs working with submarine cables. A full-size OTDR is usually equipped with many devices, making it a universal multitester. It can perform alone the entire range of operations that it takes all the devices mentioned above to do together. Therefore, to test and troubleshoot a fiber-optic communications system, you need only a full-size OTDR. Unfortunately, the price—on the order of $25,000—may be more than the cost to do the entire test set.

Mini OTDRs offer the same functions but with fewer options. Optical fault locators, or fault finders, use the OTDR principle to locate breaks and/or points of high attenuation but their measuring abilities are sharply restricted.

***Environmental-test system*** Fiber LANs find applications in a variety of environments, such as avionics, aerospace, shipboard, and so on. But even for installation in a regular building, fiber-optic components and systems must be tested under certain environmental conditions, such as temperature. Equipment for such tests [17] includes complex computerized stations that allow you to artificially change temperature, humidity, and other environmental parameters within the test chamber. These stations, which are expensive, are used by manufacturers to test their products.

***Software*** There are many software packages that make testing and troubleshooting easy. They start with firmware (programmed digital circuitry) and range to sophisticated packages, enabling a designer to fully automate an optical-loss test kit. Such a kit (Figure 8.30) automatically certifies a network to fiber-industry standards like TIA/EIA 568A, FDDI, Ethernet, and others [18]. The kit includes a database system, which allows the engineer to manage numerous cable and fiber data and observe and print test records [19]. There are also packages that provide interface between a test kit and a desktop computer.

***Reading data sheets*** The above discussion, based on data sheets for test equipment, provides you with all the information necessary to read and understand the data sheets for any type of test equipment you will ever use for these applications.

***Safety rules***

NEVER LOOK AT THE FIBER END!

NEVER LOOK INTO THE OUTPUT HOLE OF THE LIGHT SOURCE!

REMEMBER, WE WORK MOSTLY WITH INVISIBLE LIGHT!

All light sources are made in compliance with safety requirements. This is true for fault locators, stabilized optical-light sources, optical-loss and return-loss test sets, and OTDRs. But, to ensure eye safety, you had better follow the safety tips above.

## What We Need to Test

After a local area network is physically installed and before it is put into operation, you need to verify its two major performance characteristics: *attenuation* and *bandwidth.* This procedure is called *certification.*

The testing of attenuation includes measuring insertion and return losses, inspecting connectors and splices, and measuring their losses. Bandwidth-associated measurements simply entail performing bit-error-rate tests. Bear in mind that these attenuation and bit-error-rate tests are major tests and they have to be done along the entire installed network. But even before doing these tests, you need to check network *connectivity (continuity)* because experience shows that most fiber-LAN troubles stem from improper connections.

***Connectivity (continuity) test*** This is the most fundamental but simplest test you may perform: Simply emit light from the transmitter end and check whether it appears at the fiber output. You can use any of the instruments described, starting with visual-fault locators and finishing with an OTDR. You can check the connectivity by segments (from one connector to another), but your eventual goal is testing the end-to-end connection. To do so, you need to identify the fibers under test.

***Fiber identification*** Imagine that you are standing in front of a main cross-connect with hundreds of fibers streaming in and out and you need to find a specific fiber. Obviously, you must first locate the appropriate cable—that's where a cable-management system and its database come into play—but your cable can carry up to 432 fibers. This is what a fiber identifier was designed for. It allows you to identify the fiber *without disconnecting* the cross-connect. If you are able to disconnect a fiber, then a fault locator, a talk set, a loss test set, or an OTDR can be used.

## Testing Network Attenuation

***Test arrangement and procedure*** Here are the steps necessary to perform end-to-end attenuation testing of an installed fiber link:

1. Connect test jumper 1 between a calibrated, stabilized optical-light source and a power meter and measure $P_{ref1}$ (dBm), as shown in Figure 8.34(a). Let's now assume that $P_{ref1} = -10.4$ dBm.

2. Connect test jumper 2 between the power meter and test jumper #1 and measure the power, $P_{ref2}$ (dBm). (See Figure 8.34 [b].) Let's assume here that $P_{ref2} = -10.8$ dBm.

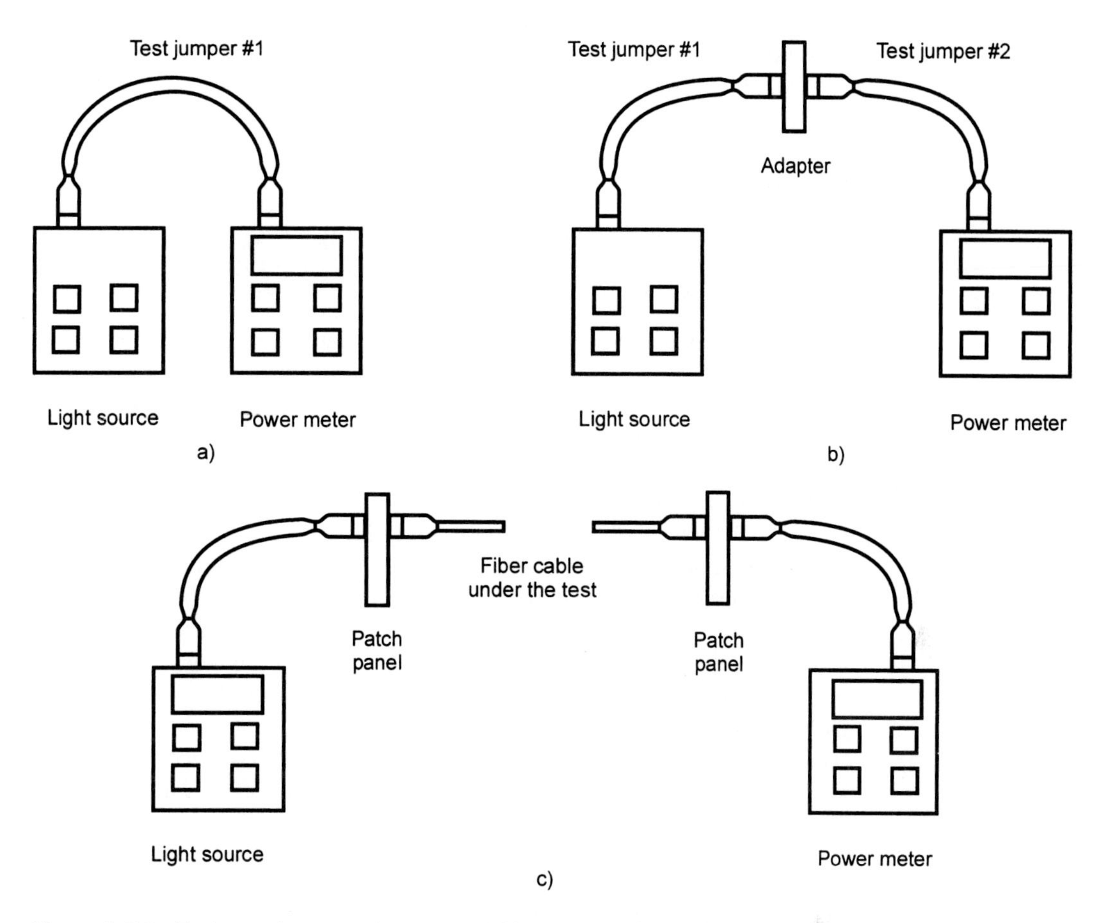


**Figure 8.34** End-to-end attenuation test: (a) Measuring reference power 1; (b) measuring reference power 2; (c) measuring cable attenuation. *(Courtesy of Fotec, Inc., Medford, MA.)*

3. Connect jumpers 1 and 2 to the cable under test and measure $P_{total}$ (dBm). (See Figure 8.34 [c].) Here we assume $P_{total} = -15.2$ dBm. Calculate the tested cable loss: $Loss_{cable}$ (dB) = $P_{ref2}$ (dBm) $-P_{total}$ (dBm); in our example, $Loss_{cable} = -10.8$ dBm – (–15.2) dBm = 4.4 dB.

4. Compare the measured loss with the expected cable loss calculated from the manufacturer's data sheet and given link length; for instance, the attenuation of a given cable found in the data sheet is equal to 3.75 dB/km at 850 nm, where the cable length is 2000 m; therefore, the expected loss is 7.5 dB.

#### *Key pointers in end-to-end attenuation testing*

- You can always rely on the numbers obtained when you use a calibrated source and meter.
- Be sure to measure the attenuation of a cable before installation (in a reel) to verify that the cable is in good condition and to have a reference value.
- Make sure that the test jumpers are made from the same fiber as the tested cable; otherwise, concatenations will cause additional loss.

- Keep the test jumpers connected to a light source and a power meter during the test; any disconnections or adjustments will change the reference power value.
- Calculate the loss of test jumper 2 to make sure it is in good condition: $Loss_{jump2}$ (dB) = $P_{ref1}$ (dBm) – $P_{ref2}$ (dBm) $< 0.5$ dB.
- Make sure to use an appropriate light source for testing multimode (LED) and single-mode (laser diode) fibers; a source's wavelength is a good indicator; for example, 850-nm and 1300-nm LEDs are used for MM fibers, while 1310-nm and 1550-nm laser diodes are used for SM fibers.
- Don't forget to measure cable loss at all operating wavelengths (the loss is obviously different at a different $\lambda$) and in both directions (if the fiber is used for full-duplex communications). For LAN cables, end-to-end attenuation testing is recommended for every connectorized fiber in backbone and horizontal segments.

***Analysis of the test procedure*** Step 1 is needed to obtain a reference power level in dBm. This is actual light power radiated by a light source minus the losses from the connectors terminating test jumper 1. In other words, this is power launched into the cable being tested. Since factory-terminated jumpers are used to conduct the test, a loss of less than 0.2 dB per connector is expected. The loss from a jumper's fiber is negligible because of its short length. Thus, if the light source radiated –10 dBm of power, the reference power should be no more than –10.4 dBm.

Step 2 ensures that the second test jumper is a good one. Its two connectors should introduce a loss of less than 0.5 dB; if its loss exceeds 0.5 dB, clean the connectors or replace the jumper. Thus, $P_{ref2} = -10.8$ dBm—obtained in the above example—is a good value because the loss from this jumper is: –10.4 dBm – (–10.8) dBm = 0.4 dB. The need to use two jumpers is called for by industry standards: TIA/EIA-526-14A for MM fiber, TIA/EIA-526-7 for SM fiber. Two jumpers are also referred to in the OFSTP-14 test procedure. At any rate, good old common sense says that you don't want to connect the light source and the power meter directly to patch panels. (See Figure 8.34[c].)

Step 3 is the actual measurement procedure. Some power meters store reference power (usually $P_{ref1}$ [dBm]) and display the result of the test directly in dB.

Step 4: When you compare the measurement obtained with the calculated loss, add the losses from the connectors and the splices that the tested cable includes.

A very important point to bear in mind when testing a multimode fiber cable is the *launch condition.* You will recall that the number of excited modes essentially changes the attenuation characteristics of an MM fiber. (See Section 4.5.) In testing installed cable, TIA-568A requires that the user select one of two launch conditions: overfill for links between 500 and 2000 m and 70/70 underfill for links under 500 m. Some manufacturers design light sources with the ability to control launch conditions. If no such control is available, one can place a 5-wrap mandrel between the light source and the cable undergoing the test to simulate the equilibrium-mode-distribution (EMD) launch condition. The mandrel's diameter is determined by Standard TIA/EIA FOTP-34A; for example, for 62.5/125 μm with a 0.9-mm buffer diameter, a mandrel's diameter is 20 mm. Differences in launch conditions can result in a measured loss difference of several dB. Unfortunately, there is no universal light source that can match any specific launch conditions. Again, you should try to simulate as closely as possible the actual launch condition of your transmitter.

***Using OTDR*** Since an OTDR measures power along the installed fiber cable, it can be used for testing cable attenuation. Its ability to test a fiber cable at one end makes an OTDR an even more attractive tool. The major advantage of an OTDR is that it builds a trace of attenuation—light power allocation along the fiber cable. Thus, one can easily calculate cable attenuation by subtracting output from input power in dBm, and an OTDR can do this for you automatically. Unfortunately, the accuracy of an OTDR power measurement cannot be calibrated; therefore, its reading cannot be used for certification of fiber cable. (More about OTDR drawbacks can be found in [20].)

Make no mistake about it, however. An OTDR is the most powerful testing tool in the industry. Its graphical display of loss over the entire fiber length provides the best analysis and documentation on the cable. An OTDR is used for on-the-reel cable inspection, testing, and documenting of all backbone cables, connector and splice loss measurement, fault finding, insertion- and return-loss evaluation, optimizing mechanical splices, and many other tasks.

To enhance an OTDR's measuring capability, some manufacturers integrate an OTDR, a power meter, and a visual fault locator into a unit called a *multifunction tester.*

An OTDR–power-meter combination allows you to perform end-to-end attenuation testing together with loss tracing.

Combining an OTDR with a visual-fault locator enables you to pinpoint a fault's position. If, for example, at the 160-km span, the OTDR locates a fault within 16 ± 4 m, then the visual fault locator helps determine the fault's precise position.

Any installed backbone cable should undergo OTDR testing, and the result (called the *cable trace signature)* should be stored as an important plant document concerning the cable.

## Testing Network Bandwidth

***Bandwidth measurement*** Why the need to perform end-to-end attenuation testing of an installed cable? Because an installation can adversely affect fiber attenuation. (Remember our discussion of macro- and microbending losses? If not, refer again to Sections 3.2 and 5.2.) However, there is no mechanism that can significantly change a fiber's bandwidth during the installation. Thus, measuring the bandwidth of an installed fiber cable is not necessary. Simply use the data provided by the manufacturer.

***Bit-error-rate test (BERT)*** For a data network, which a LAN by and large is, the bit-error-rate test may help to provide a deeper characterization of the network. The test arrangement is shown in Figure 8.35.

The test procedure is simple: Measure the reference light power, $P_{ref}$, at the receiver end without an attenuator or BER generator. Next, insert a variable optical attenuator, VOA, and turn on the BER generator. Increase attenuation until the bit-error rate exceeds the specified minimum (usually $10^{-9}$). Measure the attenuated power, $P_{att}$. The link power margin in dB is $P_{ref}$ (dBm) – $P_{att}$ (dBm). For example, $P_{ref} = -18.3$ dBm and $P_{att} = -21.6$ dBm; thus, the link power margin is 3.3 dB.

## Connector and Splice Testing

***Visual inspection*** After a connector has been installed, one needs to perform these testing procedures: *visual inspection, measurement of insertion,* and *return losses.* The visual inspection of a ferrule endface is done with a microscope or a visual fiberscope. Recent developments include using a tool that allows for the inspection of connectors directly at their workplaces, that is,

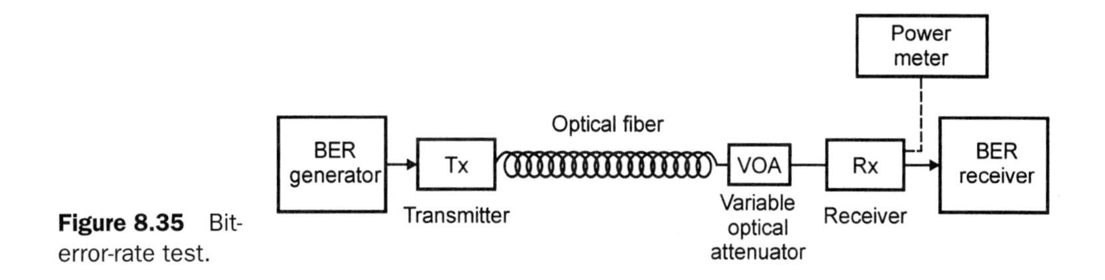


**Figure 8.35** Bit-error-rate test.

at the patch panel or transmission equipment, *without* removing them [10]. You can understand how important this tool is by just imagining yourself confronting hundreds of connectors at the MC, IC, or HC frames and the need to remove one connector after another for inspection.

Splices do not require visual inspection.

***Insertion loss*** The insertion loss of connectors and splices can be measured in the field by an OTDR during the testing of an installed cable. Since this is a pass/fail test, accuracy is not the issue. Typical LAN values are $\leq 0.75$ dB per mated connection, $\leq 0.2$ dB for multimode fiber, and $\leq 0.15$ dB for singlemode fiber per splice.

If you use a fusion splicer based on LID or PAS principles, you can obtain an estimate of the splicing loss directly from the splicer. These numbers can be used in the management of a cable plant.

If you want to make a precise measurement of a connector's insertion loss, follow EIA recommendation FOTP-34. The schematic of this measuring arrangement is shown in Figure 8.36 (a). The procedure requires, first, measuring the power transmitted through a test jumper, $P_{jumper}$ (dBm), which counts the loss introduced by the jumper's connectors to the source and power meter. The second step repeats this measurement while a connector or splice being tested is included in-line. Thus, the second reading is $P_{con}$ (dBm). The difference between these two readings, $P_{jumper}$ (dBm) $-$ $P_{con}$ (dBm), is the connector (splice) insertion loss in dB. For example, $P_{jumper} = -10.4$ dBm and $P_{con} = -10.8$ dBm; therefore, the insertion loss is 0.4 dB.

***Optical return loss (ORL)*** We use the return-loss test set to measure the optical return loss, as was pointed out on page 301. In addition, we can evaluate this loss in the field when inspecting the fiber cable with an OTDR. The amplitudes of spikes at any connection point are the measures of reflectance, or optical-return loss, of this point.

If you want to make a precise measurement, follow EIA recommendation FOTP-107. The test arrangement for measuring return loss is shown in Figure 8.36 (b).

A coupler is a passive device that distributes an optical signal between input and output ports in a given ratio, *R*, such as 50/50. The procedure for this arrangement involves, first, measuring

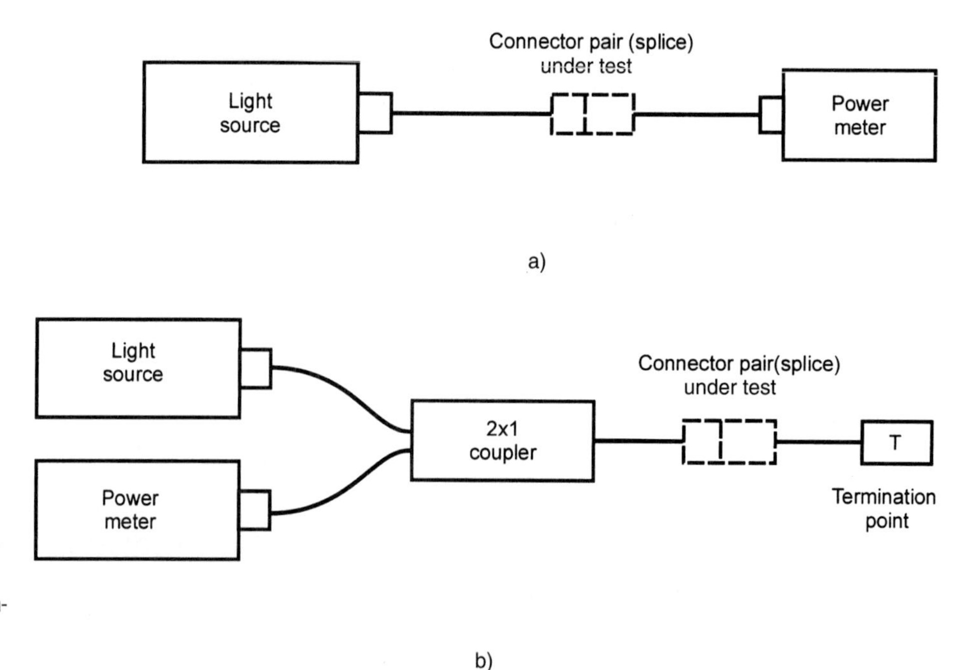


**Figure 8.36** Insertion and return-loss measurements: (a) Measurement of insertion loss; (b) measurement of return loss. (*Courtesy of Fotec, Inc.*)

power, $P_{in}$ (dBm), at the termination point without a tested connector (splice) and, second, measuring the reflected power with the mated connector (splice) in line, $P_{refl}$ (dBm). Thus, the optical-return loss is: $ORL$ (dB) = $P_{in}$ (dBm) – $P_{refl}$ (dBm) (See Formulas 8.1 and 8.2). For example, $P_{in} = -5.2$ dBm, and, $P_{refl} = -54$ dBm; hence, $ORL = 48.8$ dB. Note that ORL is given as a positive number.

***Documentation*** Without proper documentation, you cannot maintain your fiber-optic communications system. Any move—reconfiguration, upgrading, maintenance, and troubleshooting—relies on network documentation. The documentation has to include cable records (cable specifi-

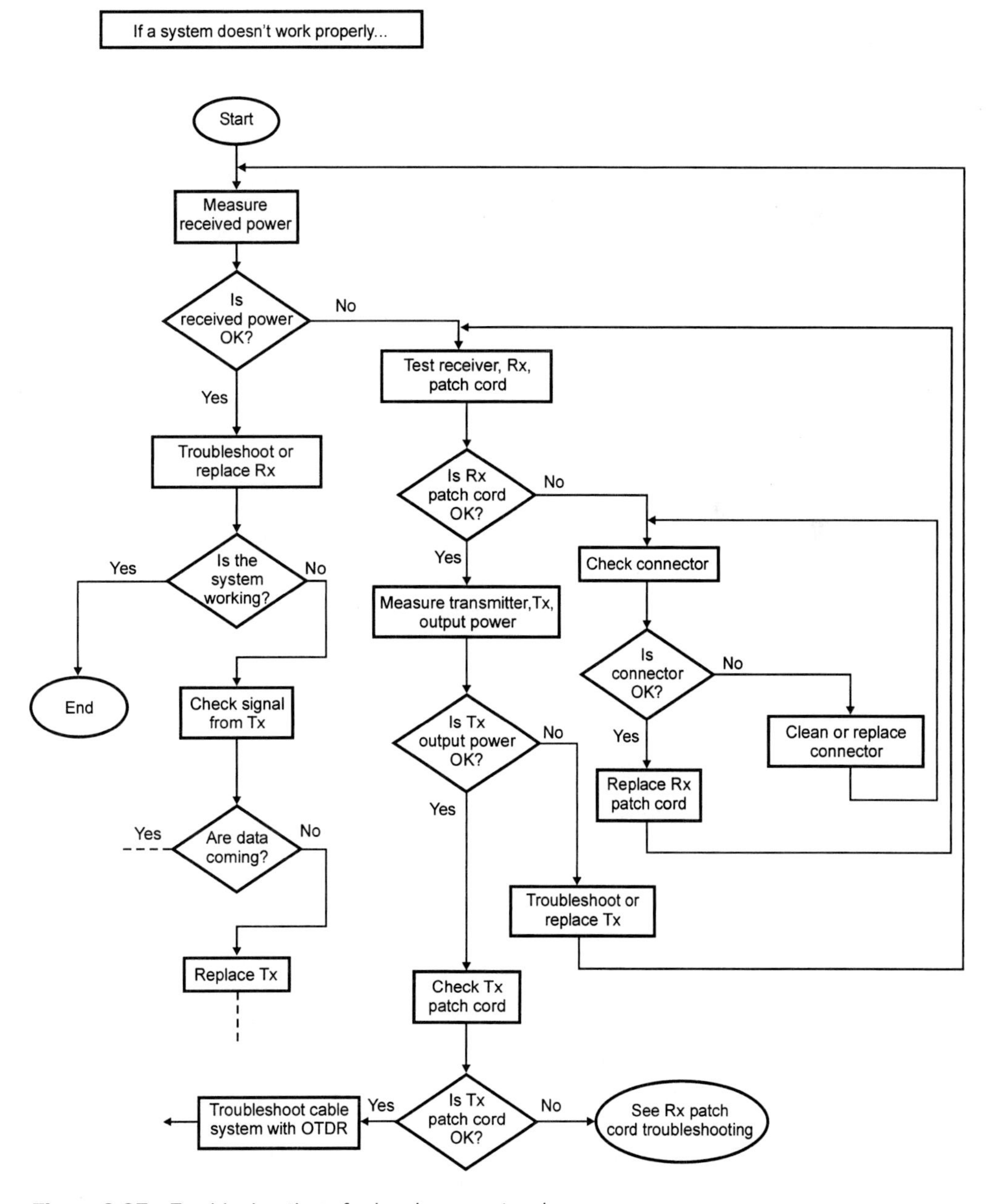


**Figure 8.37** Troubleshooting of a local area network.

cations, a cable-route diagram, and fiber numbering) and test results (an end-to-end attenuation test, an OTDR signature trace, a certificate for connectors, and splice losses). Make sure you have all the necessary documents because after the installer leaves, you are the one who will have to work with the system.

**Troubleshooting**

Troubleshooting is relatively easy when the system is completely down; usually, you just need to find a break. But it is much more difficult when the system is operating improperly; for instance, the BER may be too high or the signal may become unstable. There are no formulas or rules for troubleshooting but several general recommendations may help.

First, follow the flow of a signal through the system—but you cannot do this without comprehensive documentation.

Second, subdivide the entire network into terminated segments from one patch panel to another and check one segment at a time. Remember, though, that you cannot do so without properly installed hardware.

Third, use the appropriate test equipment described above.

Fourth, to fully appreciate how troubleshooting should be done, carefully study Figure 8.37.

It is evident that not every step in the troubleshooting procedure is shown in full detail, but this flow chart, along with the above general recommendations, will help you to find most sources of trouble.

## SUMMARY

- Splicing—a word that only 5 to 10 years ago symbolized a rapidly growing fiber-optic communications technology—is today the routine connection of two fibers, an operation performed in the field thousands of times a day with exceptional results. It is supported by excellent automated equipment that makes this process quick, easy to do, and efficient. Two types of splicing—fusion and mechanical—provide a variety of options for an installer. Still, splicing is a relatively time-consuming, labor-intensive process. What's more, it introduces insertion losses and, in the case of mechanical splicing, reflection losses as well.
- Cabled fibers are terminated by connectors for the temporary linking of two fibers and for connecting fibers to a transmitter and receiver. The problems with connectors are that ways must be found to minimize insertion and reflection losses along with high repeatability during hundreds of connect–disconnect (mating) operations. Installing connectors—whether it be in the field or splicing a factory-made connector assembly (pigtail) to a fiber—is called cable termination. This is one of the most frequent operations performed during the installation of a fiber-optic network; thus, manufacturers are striving to make this process as quick and easy to do as possible. Though there are many styles of connectors today, unfortunately, they are incompatible. New developments have improved connector characteristics and facilitated the termination process, but there is no indication that we will soon see a uniformity of connector styles.
- The installation of fiber-optic networks requires specific hardware: These are enclosures that protect connection points and make fiber inter- and cross-connection easy. Two basic elements of installation hardware are patch panels (for connecting two terminated fibers) and splice panels (for storing and protecting fiber splices). A main cross-connect (MC), an intermediate cross-connect (IC), and a horizontal cross-connect (HC) are the building blocks of hardware systems.
- Before installing a network, one must, of course, design the installation. This chapter concentrates on the design of an installation for a local area network (LAN), a rapidly changing application area relying on fiber optics. This design is based on evaluation of the network's power budget and bandwidth. The basic structure of fiber LAN cabling includes campus (outdoor) backbone, building (riser) backbone, and horizontal wiring. Cross-connects MC, IC, and HC are used for transition between this cabling. Star and hierarchical star are the leading physical topologies. A close look at this topic reveals many important details, which are discussed in Section 8.4.
- An installed network must be tested for its compliance with required performance characteristics. A large segment of the fiber-optic industry develops and manufactures test equipment, making network testing and troubleshooting

easy. The power meter (PM) and the optical time-domain reflectometer (OTDR) are the two basic instruments for this job, though there are a number of other test devices for performing specific measurements in the field. Connectivity testing, end-to-end attenuation testing, the inspection of installed fibers by use of an OTDR, testing connectors and splices—these are major ways to ensure the reliability of an installed fiber LAN.

- If the network does not work properly, troubleshooting must be done. The general recommendations for troubleshooting, along with a troubleshooting flow chart, are described in Section 8.5.

## PROBLEMS

**8.1.** a. Calculate the intrinsic connection losses for two 62.5/125 graded-index multimode fibers manufactured by Plasma Optical Fibre (Figure 3.18) caused by (a) diameter mismatch and (b) *NA* mismatch.
b. Calculate the intrinsic connection losses caused by the MFD mismatch of two singlemode fibers made by Plasma Optical Fibre (Figure 5.17).

**8.2.** Calculate the extrinsic connection losses for two 62.5/125 graded-index multimode fibers from Plasma Optical Fibre (Figure 3.18) caused by (a) lateral misalignment for $x = 3$ μm, (b) angular misalignment for $\phi = 1°$, and (c) end separation for $z = 3$ μm.

**8.3.** Calculate the extrinsic connection losses for two singlemode fibers manufactured by Plasma Optical Fibre caused by (a) lateral misalignment for $x = 0.5$ μm, (b) angular misalignment for $\phi = 30''$, and (c) end separation for $z = 0.5$ μm. (Refer to Figure 5.17.)

**8.4.** a. Why is cladding-diameter tolerance so important?
b. What is the role of the core and the cladding noncircularities?
c. Why do manufacturers specify these geometric characteristics but not some others?

**8.5.** What is the Fresnel loss for light traveling from fiber core to air gap?

**8.6.** A cleaver data sheet states that cleave angles are $<1/2°$ for 50% of all cleaves; $<1°$ for 85% of all cleaves; $<2°$ for 100% of all cleaves. Explain what these numbers mean.

**8.7.** For two mechanically spliced fibers, calculate the reflection loss with air in a gap if $z = 2$ μm and $\lambda = 1550$ nm.

**8.8.** List the steps necessary for end preparation.

**8.9.** List the advantages and drawbacks of mechanical splicing.

**8.10.** You are a project manager responsible for (a) installation of a long-distance fiber link and (b) installation of a local area network. What type of splicing should you choose in each case? Why?

**8.11.** Why does pigtail termination introduce less loss than field installation?

**8.12.** What connector component, or components, is responsible for (a) retention, (b) end protection, (c) alignment?

**8.13.** Return loss is not specified for connectors used for multimode fibers. Why?

**8.14.** What are the most popular connectors today?

**8.15.** What are the major trends in connector development?

**8.16.** Do we have a standard for today's and future connectors?

**8.17.** A guidance mechanism is the key component of a receptacle. Why?

**8.18.** What are bulkhead adapters?

**8.19.** Explain the meaning of MC, IC, and HC. What are the functions of these units?

**8.20.** What is the most widely used component in a fiber-cable management system?

**8.21.** A fiber link, including three mechanical splices and four connectors, has 0 power margin. What can you do to improve link performance?

**8.22.** A fiber link includes five splices at 0.02 dB/splice, four connectors at 0.2 dB/connector, transmitter power of −10 dBm, and receiver sensitivity of −25 dBm. What length of this link will be allowed if a singlemode fiber cable with attentuation of 0.3 dB/km is used and the required power margin is 3 dB?

**8.23.** A local data link to be installed has the following characteristics: maximum bit rate, 32 Mbit/s; line code—return-to-zero (RZ) format; fiber, 62.5/125 μm; installation length, 2000 m; operating wavelength, 1300 nm; rise time of lightwave equipment, 4 ns; and LD spectral width, 2 nm. Will this fiber support the required bit rate? Prove your answer.

**8.24.** You tried to use 62.5/125 μm fiber and found that it could not support the required transmission speed even at 1300 nm and even with a laser diode as the light source. What can be done to resolve the problem?

**8.25.** List the drawbacks of star topology.

**8.26.** What cable type do you want to use for campus backbone, building backbone, and horizontal wiring?

**8.27.** Discuss the general recommendations for LAN design.

**8.28.** Do you want to use plastic optical fiber (POF) for your LAN network?

**8.29.** A fiber identifier detects light radiating out of a fiber due to the effect of bending loss. But this work is done over a fiber *cable.* How does light penetrate all the protective layers of a cable?

**8.30.** Refer to Table 8.1. Compare the data there line by line for a visual fault locator and an optical light source and explain the differences.

**8.31.** List all the fiber test devices along with their functions.

**8.32.** You need to make an end-to-end test of an installed LAN. What test equipment—optical source/power meter or OTDR—will you use and why?

**8.33.** A measured connector's insertion loss is equal to 0.2 dB and the ORL is equal to 12 dB. Does the connector pass the test?

**8.34.** When you check an installed LAN at the receiver end, it appears that light emanates from it but data do not. What are the possible problems? What is your plan for troubleshooting the network?

**8.35.** *Project:* Sketch a memory map. (See Problem 1.20.)

## REFERENCES[1]


1. *Fiber-Optic Products Catalog,* Optics for Research, Inc., Caldwell, N.J., 1998.
2. Alan Mickelson, Nagesh Basvanhally, and Yung-Cheng Lee, eds., *Optoelectronic Packaging,* New York: John Wiley & Sons, 1997.
3. James Refi, *Fiber Optic Cable—A LightGuide,* Geneva, Ill.: abc TeleTraining, 1991.
4. L. Wesson, "PAS or LID for Fusion Splicing: Does It Matter?" *Cabling Business Magazine,* May 1998, pp. 38–46.
5. *Fusion Splicer FSU 925* (product catalog), Ericsson Cable AB, Network Products, Sundbyberg, Sweden, 1998.
6. Casimer DeCusatis et al., "Small Form Factor Optical Fiber Connectors: Performance Comparisons," *Optics and Photonics News,* December 1999, pp. 29–30.
7. Joyce Kilmer, "Connectors evolve for the premises market," *Lightwave,* May 1998, pp. 39–42.
8. *Fiber Optic Rotary Joints: Product Overview,* Focal Technologies Inc., Dartmouth, Nova Scotia, Can., April 1997.
9. "Pegasus, The Reusable Fiber-Optic Connector," *Cabling Business Magazine,* April 1998, pp. 70–73.
10. L. Upshaw, "Inspecting Fiber Optic Connectors," *Cabling Business Magazine,* March 1998, pp. 50–53.
11. *LANscape: Design guide, Release 4,* 1999, Siecor, Hickory, N.C.
12. Bob Chomycz, *Fiber Optic Installations (A Practical Guide),* New York: McGraw-Hill, 1996.
13. Eric Pearson, *The Complete Guide to Fiber Optic System Installation,* Albany, N.Y.: Delmar Publishers, 1997.
14. *Fiber Optic Design Guide,* Radiant Communications Corporation, South Plainfield, N.J., July 1996.
15. Donald Sterling, Jr., *Technician's Guide To Fiber Optics,* 2d ed., Albany, N.Y.: Delmar Publishers, 1993.
16. Noriyuki Yoshhara, "Low-loss, high-bandwidth fluorinated POF for visible to 1.3 um wavelengths," Paper ThM4, *Proceedings of the Optical Fiber Communication Conference,* San Jose, Calif. (February 22–27, 1998).
17. *Fiber Optics Environmental Test System* (product catalog), Rifocs Corp., Camarillo, Calif., June 1997.
18. *CertiFiber and Fiber Solution Kit—Product Catalog,* Microtest, Inc., Phoenix, Ariz., 1997.
19. *FORMS: Fiber Optic Record Management System—Product Catalog,* GN Nettest, Utica, N.Y., 1998.
20. James Hayes, *Fiber Optic Technician's Manual,* Albany, N.Y.: Delmar Publishers, 1996.


[1]See Appendix C: A Selected Bibliography

# 9

# Light Sources and Transmitters—Basics

Refer again to the basic block diagram of a fiber-optic link (Figure 1.4): A fiber-optic link starts at the point where an electrical signal is converted into an optical one. The device doing this conversion is a light source. The input for this source is an information signal in electrical form and the output is this signal in optical form. For digital communications, the input is a sequence of on–off electrical pulses and the output is on–off optical flashes. Thus, not only an optical source but also modulation and other supporting electronics—taken together, they're called a *transmitter*—are used to send an optical signal along the fiber-optic link.

A transmitter consists of a light source, coupling optics, and electronics. Only miniature semiconductor light sources—light-emitting diodes (LEDs) and laser diodes (LDs)—are used in fiber-optic communications technology. LEDs and LDs are the heart and soul of transmitters. This is why we'll concentrate on their principle of operation and key features while discussing electronics and coupling optics where necessary.

Progress in fiber-optic communications technology cannot be achieved without advances in optical-fiber technology, as pointed out in Chapters 3 through 6. But optical fiber itself cannot provide communications; a link must be completed with a transmitter and receiver. Fortunately, progress in both light sources and photodetectors has accompanied the progress made in optical fiber. At the same time, integrated electronics became part and parcel of transmitter/receiver specifications. However, since new demands have imposed new requirements on these components, technological developments continue unabated.

## 9.1 LIGHT-EMITTING DIODES (LEDs)

LEDs have been around for more than 30 years. They have found application in nearly every consumer-electronic device: TV sets, VCRs, telephones, car electronics, and many others. They are used in fiber-optic communications, mostly because of their small size and long life. However, their low intensity, poor beam focus, low-modulation bandwidth, and incoherent radiation—in comparison with laser diodes, that is—restrict their usage to a specific sector of communications technology: relatively short-distance and low-bandwidth networks. Local area networks are the largest application area for transmitters based on LEDs. Since fiber-optic LANs is a booming technology today, LEDs are in wide use. Thus, we need to take a thorough look at light-emitting diodes.

## Light Radiation by a Semiconductor

***Energy-band diagram*** You are probably familiar with semiconductor materials through your study of electronic devices such as diodes and transistors. Such background should help you to understand the workings of LEDs because an LED is, after all, a *semiconductor diode.* However, we'll discuss an LED's principle of operation on the assumption that you are unfamiliar with it or have forgotten much of what you learned some time ago.

First, you'll recall from Chapter 2 that all materials consist of atoms, which are nuclei surrounded by electrons rotating at stationary orbits. Each orbit corresponds to a certain energy value; thus, these atoms may possess only discrete energy values. We represent this idea through an energy-level diagram (Figure 2.8).

Semiconductors are solid-state materials consisting of tightly packed atoms. Atoms, in turn, are bonded by interatomic forces into a lattice structure. Each atom includes many electrons, but a material's properties are determined by its outermost electrons.

The important fact is that in semiconductors (and in solids in general) the possible energy levels are still discrete, but they are so close to one another that we depict them as an energy band rather than a set of separate levels. We think of an energy band as a wide, *continuous* region of energy, but if you had a magic magnifier to look at this band closely, you would see the discrete energy levels that make up the band. Figure 9.1(a) shows this. It should be noted that the vertical axis in Figure 9.1 represents an electron's energy, while the horizontal axis serves merely as a visual aid.

In semiconductors we distinguish two *energy bands: valence* (lower, meaning less energy) and *conduction* (upper, meaning higher energy). They are separated by an *energy gap,* $E_g$, where no energy levels (that is, no electrons) are allowed. In other words, electrons can be either at the valence band or at the conduction band but cannot be in between—at the energy gap.

An energy band consists of allowed, or possible, energy levels, which means the electrons may occupy them.

When the absolute temperature is zero and no external electric field is applied, all electrons are concentrated at the valence band and there are no electrons at the conduction band. This is because none of the electrons possess enough extra energy to jump over the energy gap. But when some external energy—either through temperature or by an external electric field—is provided to the electrons at the valence band, some of them acquire enough energy to leap over the energy gap and occupy energy levels at the conduction band. We say these electrons are "excited." These excited electrons leave *holes* (positive charge carriers) at the valence band, as Figure 9.1(b) shows.

***Light radiation—energy bands*** Recall again our discussion in Chapter 2 of how light is radiated: When an excited electron falls from an upper energy level to a lower one, it releases a quantum of energy called a photon. The relationship among $\Delta E_l$, $E_p$, and $\lambda$ is given by: $\Delta E = E_p = hf = hc/\lambda$, where $\Delta E$ is the difference between the two energy levels, $E_p$ is the photon's energy, and $\lambda$ is the wavelength.

The same idea holds for semiconductors. If an excited electron falls from a conduction band to a valence band, it releases a photon whose energy, $E_p$, is equal to or greater than the energy gap, $E_g$. Since not just one but many energy levels at the conduction and valence bands can participate in the radiation process, many close wavelengths, $\lambda_i$, can be radiated. This is why we said that $E_p \geq E_g$, which has another form: $\lambda_i \leq hc/E_g$. (If you measure $E_g$ in electron volts, eV, and

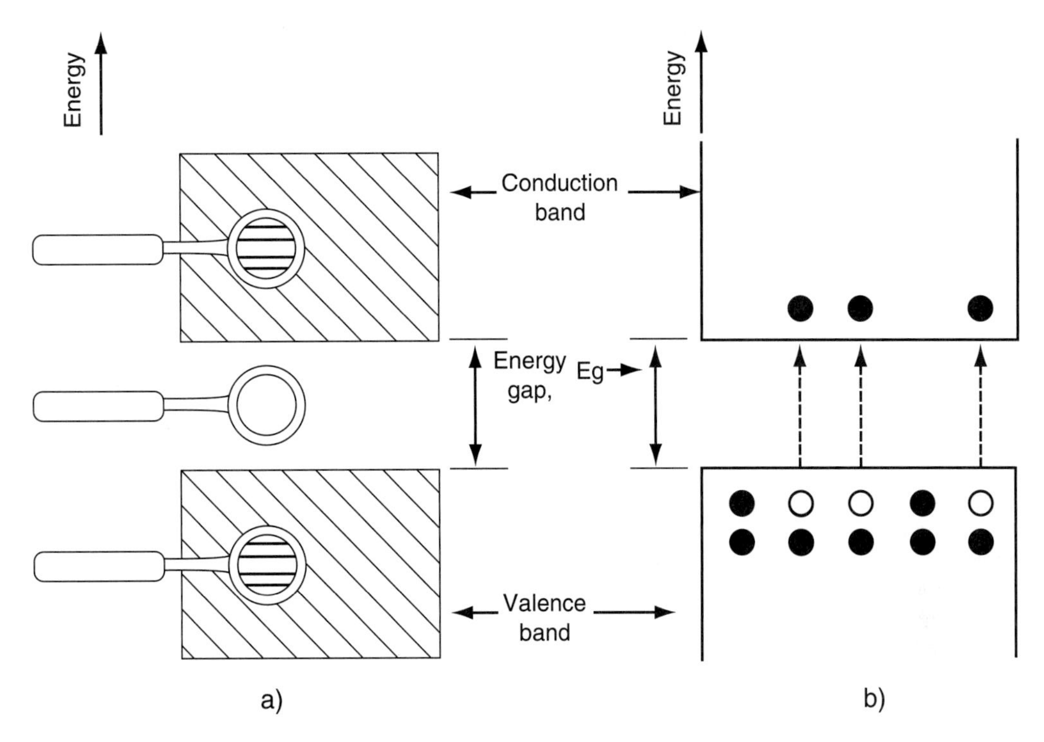


**Figure 9.1** Energy bands of an intrinsic semiconductor: (a) General representation; (b) for finite temperature.

$\lambda$ in nanometers, nm, then $\lambda_i \leq 1248/E_g$—see Formula 2.8.) The result of this multiwavelength radiation is a wide spectral width, $\Delta\lambda$, of light emitted by the semiconductor. This explanation is depicted in Figure 9.2.

Thus, to make a semiconductor radiate, it is necessary to excite a significant number of electrons at the conduction band. This can be done by providing external energy to the material. The most suitable form of this external energy is electric current flowing through a semiconductor.

***Light radiation—The* p-n *junction*** We can insert atoms of another material into a semiconductor so that either a majority of electrons (negative charge carriers) or a majority of holes (positive charge carriers) will be created. The former semiconductor is called the *n* type, where *n* stands for negative, and the latter is called the *p* type, where *p* stands for positive. We call these *n* type and *p* type *doped,* or *extrinsic,* semiconductors in contrast to a *pure,* or *intrinsic,* semiconductor, which consists of atoms of one material. The inserted foreign materials are called *dopants.* (Sound familiar? See Section 7.1, where the word *dopant* was used in the same sense but there applied to a fiber-fabrication process.)

When an *n*-type semiconductor is brought into physical contact with a *p* type, a *p-n* junction is created. At the boundary of the junction, electrons from the *n* side diffuse to the *p* side and recombine with holes and, at the same time, holes from the *p* side diffuse to the *n* side and recombine with electrons. Thus, a finite width zone, called the *depletion region,* forms. Here, there are no mobile electrons or holes. Since positive ions at the *n* side and negative ions at the *p* side within the depletion region are left without electrons or holes, these ions create an internal electric field called a *contact potential.* We characterize this field by *depletion voltage,* $V_D$. Figure 9.3(a) illustrates this explanation.

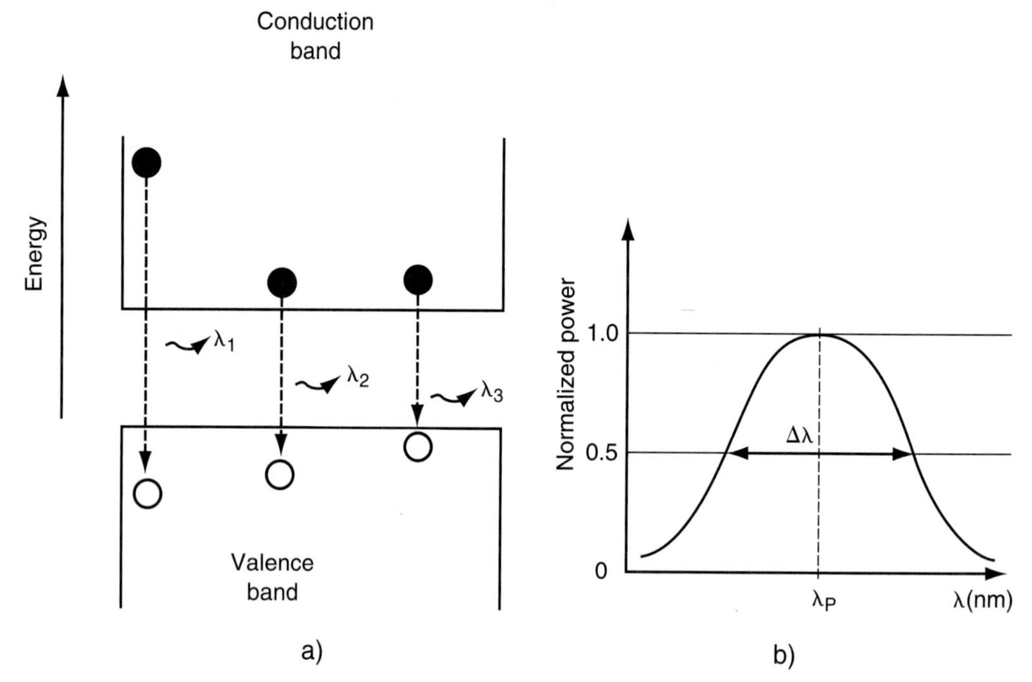


**Figure 9.2** Light radiation by the energy bands of a semiconductor: (a) Radiation process; (b) spectral width of radiated light.

The most important point to keep in mind is this: *An electron-hole recombination releases a quantum of energy—a photon.* In other words, to make a semiconductor radiate, it is necessary to sustain electron-hole recombinations. But the depletion voltage prevents electrons and holes from penetrating into a depletion region; therefore, external energy must be supplied to overcome this voltage barrier. This external voltage, called *forward biasing voltage, V,* is shown in Figure 9.3(b). Obviously, $V$ must be greater than $V_D$.

To achieve permanent light radiation, the following dynamic process must occur: Mobile electrons from the *n* side, attracted by the positive terminal of *V,* enter the depletion region. Simultaneously, mobile holes from the *p* side, attracted by the negative terminal of *V,* enter the same depletion region. Electron-hole recombinations within a depletion region produce light. Electric charges return through a biasing circuit.

(*Note:* In semiconductors, electrons are much more mobile than holes. This is why, when a dynamic process is described, it is customary to refer to electrons entering the active region and to ignore the movement of the holes. But holes are present even though they aren't mentioned explicitly and, again, only the electron-hole recombination produces light.)

***LED: Principle of action*** A light-emitting diode, LED, is a semiconductor *diode* made by creation of a junction of *n*-type and *p*-type materials. Thus, the principle of an LED's action works precisely the same way that we described the creation of permanent light radiation: The forward-biasing voltage, *V,* causes electrons and holes to enter the depletion region and recombine (Figure 9.3[b]). Alternatively, we can say that the external energy provided by *V* excites electrons at the conduction band. From there, they fall to the valence band and recombine with holes (Figure 9.2[a]). Whatever point of view you prefer, the net result is light radiation by a semiconductor diode.

This concept is displayed by the circuit of an LED (Figure 9.4[a]). If you are familiar with a semiconductor forward-biased diode, you will immediately recognize this circuit.

In fact, if you are at all familiar with electronics, you may even say, "Wait a minute. Electron-hole recombination is the process that occurs in regular diodes and transistors too. What's the

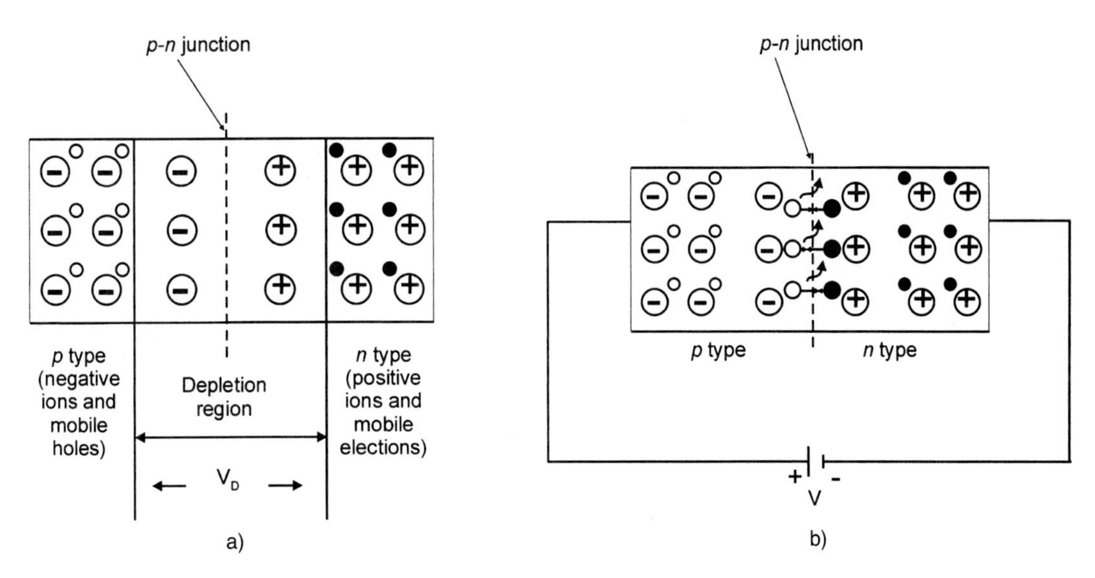


**Figure 9.3** Light radiation by the *p-n* junction of a semiconductor: (a) Depletion region and depletion voltage, $V_D$; (b) light radiation as the result of electron-hole recombinations.

difference between an LED and a regular diode?" The difference is that in a regular diode these recombinations release energy in the thermal—rather than the visible—portion of the spectrum. This is why these electronic devices are always warm when you turn them on. In an LED, however, these recombinations result in the release of radiation in the visible, or light, part of the spectrum. We call the first type of recombination *nonradiative,* while the second type is called *radiative* recombination. In reality, both types of recombination occur in a diode, when a majority of recombinations are radiative, we have an LED.

The forward current injects electrons into the depletion region, where they recombine with holes in radiative and nonradiative ways. Thus, nonradiative recombinations take excited electrons from useful, radiative recombinations and decrease the efficiency of the process. We characterize this by the *internal quantum efficiency,* $\eta_{int}$, which shows what fraction of the total number of excited (injected) electrons produces photons.

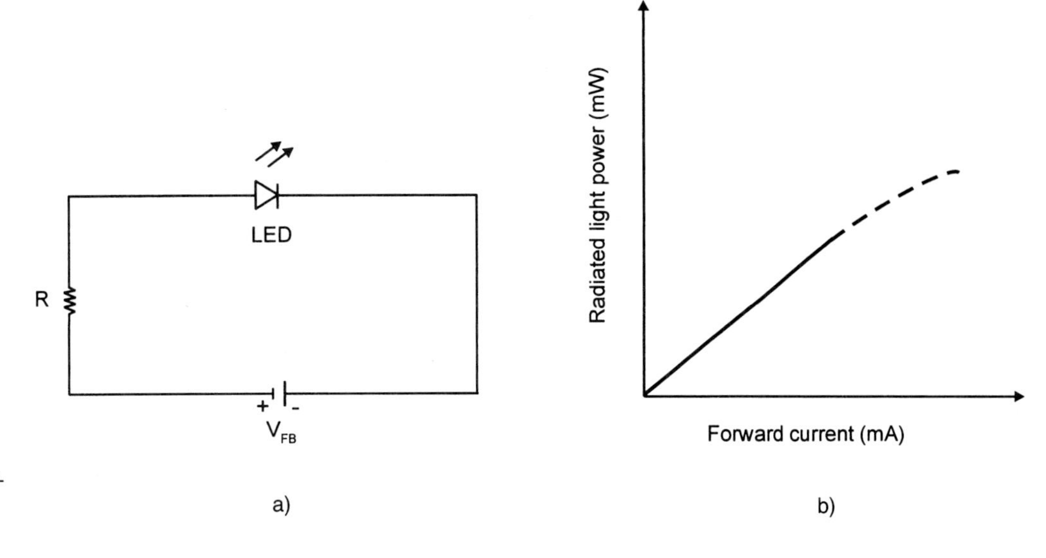


**Figure 9.4** An LED's principle of operation: (a) Electronic circuit; (b) an input–output characteristic.

If you understand the above explanation, you are able to sketch the input–output characteristic of an LED: power of radiated light as a function of forward current. It is evident that the greater the forward current, the greater the number of electrons that will be excited at the conduction band and the greater the number of photons (light) that will be emitted. An input–output characteristic is shown qualitatively in Figure 9.4(b).

The above reasoning can be quantified as follows: Light power, $P$, is energy per second, that is, the number of photons times the energy of an individual photon, $E_p$. The number of photons is equal to the number of excited (injected) electrons, $N$, times the internal quantum efficiency, $\eta_{int}$. Thus,

$$P = (N\eta_{int}E_p)/t \tag{9.1}$$

On the other hand, the number of electrons ($N$) times the electron charge ($e$) per second constitutes current ($I$):

$$I = Ne/t \tag{9.2}$$

and $N = It/e$. Hence, the radiated light power is:

$$P = (It/e)(\eta_{int}E_p)/t = [(\eta_{int}E_p)/e]I \tag{9.3}$$

Here, $E_p$ is measured in joules. If you measure $E_p$, in electron volts, eV, and $I$ in mA, then

$$P(\mathrm{mW}) = [\eta_{int}E_p(eV)]I(\mathrm{mA}) \tag{9.3a}$$

In sum, an LED's light power is proportional to the forward current, as Figure 9.4(b) shows.

### Example 9.1.1

***Problem:***

What power is radiated by an LED if its quantum efficiency is 1% and the peak wavelength is 850 nm?

***Solution:***

The key to solving this problem is given by Formulas 9.3 or 9.3a. Thus, we need to take two steps. First, we have to calculate the coefficient [$\eta_{int}$ $E_p$ (eV)], which is the slope of the graph showing power versus current. Second, we must calculate the amount of power at the given forward current.

If $\lambda$ = 850nm, then $E_p$ = he/$\lambda$ = 1248/$\lambda$ = 1.47 eV. (See Formula 2.8 and Example 2.2.) Hence, [$\eta_{int}$ $E_p$ (eV)] = 0.0147 mW/mA and from Formula 9.3a, $P$ = 0.0147 $I$.

To calculate the power value, we need to know the forward current. Typical values of $I$ for LEDs are in the range of 50 to 150 mA. Thus, for $I$ = 50 mA, the radiated power is $P$ = 0.735 mW.

---

One expects the saturation effect (see the dotted line in Figure 9.4), the point where all the available mobile electrons will be involved in radiation and further increasing the current value, will not produce additional photons.

## General Considerations

***Homostructure and heterostructure*** The $n$-type and $p$-type semiconductors discussed above are made from the same substrate. By adding various dopants, we can make either an $n$ type of semiconductor, with excessive electrons (that is, negative charge carriers) or a $p$ type of semiconductor, with excessive holes (that is, positive charge carriers). Both semiconductor types have the same energy gap. The *p-n* junction of such semiconductors becomes what's known as a homojunction. The possible structures of an LED made from such a semiconductor—homostructures—are shown in Figure 9.5(a) and 9.5 (b).

There are two basic arrangements of an LED: surface emitting (SLED) and edge emitting (ELED). The depletion region and surrounding area, where electron-hole recombinations take place, are known as an *active region.* Light produced by these recombinations radiates in all directions, but only a transparent window of the upper electrode (Figure 9.5[a]) or an open edge (Figure 9.5[b]) allows light to escape from the semiconductor structure. All other possible directions (in the case of SLED) and the opposite edge (in the case of ELED) are blocked from light by the LED's packages.

A homostructured LED has two major drawbacks. First, its active region is too diffuse, which makes the device's efficiency very low. This is because electron-hole recombinations take place in various locations, that is, over a large area, a situation that requires high current density to support the desired level of radiated power. (Remember, we are talking about the dimensions of a few microns, so the word *large* is relative here.) Second, this type of LED radiates a broad light beam. This makes the coupling of this light into an optical fiber extremely inefficient and is the reason why you cannot find an LED with a homojunction in practical applications.

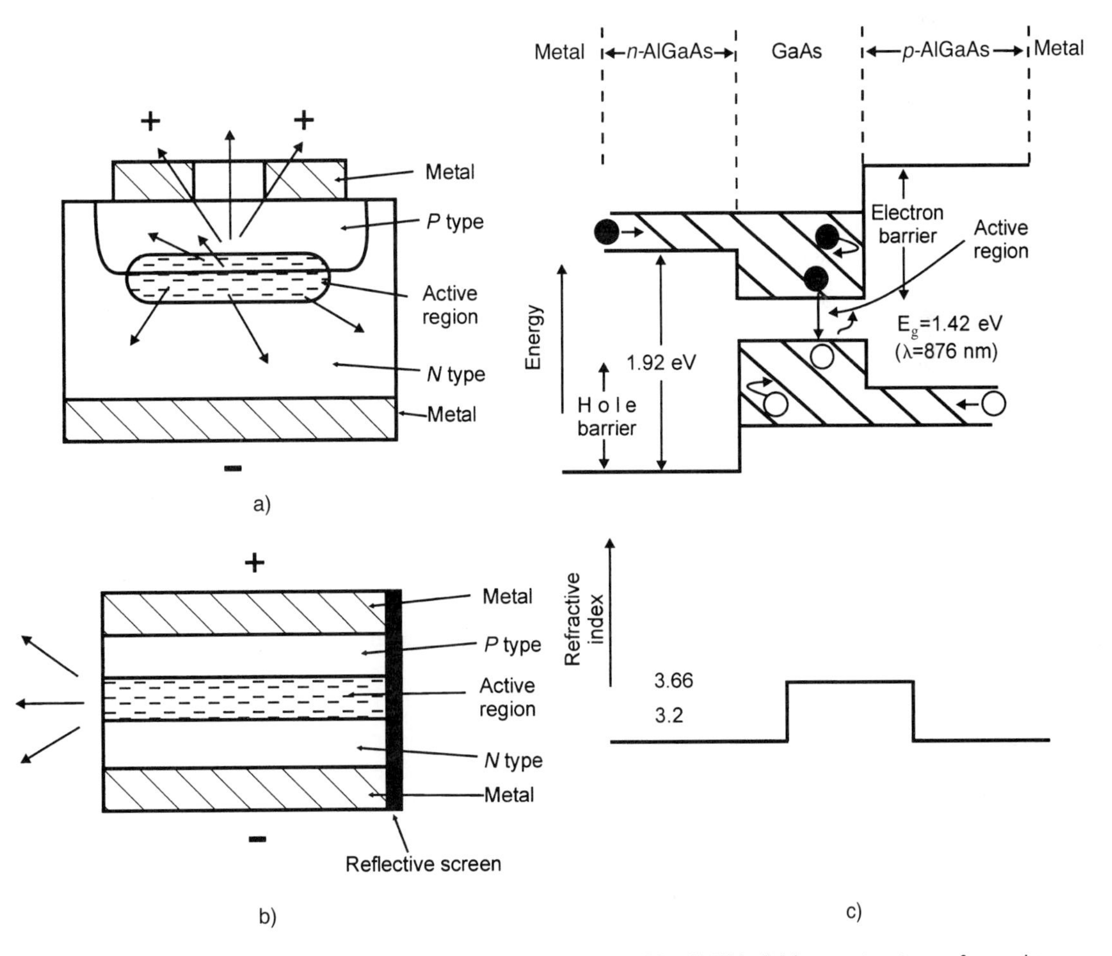


**Figure 9.5** LED structures: (a) Homostructure of a surface LED (SLED); (b) homostructure of an edge-emitting LED (ELED); (c) double heterostructure.

Commercially manufactured LEDs that radiate well-directed light with acceptable efficiency use heterojunctions. Heterostructured LEDs are made from different types of semiconductor materials, each type having a different energy gap. Figure 9.5(c) shows a heterostructure made from two different semiconductors.

*Two basic concepts are introduced with this heterostructure: the confinement of electron-hole recombinations within a highly restricted active region and the conduction of radiated light in one direction.*

The first is achieved by placing a semiconductor with a small energy gap between the two layers of the substrate semiconductor with the larger energy gap. Figure 9.5(c) shows that gallium arsenide (GaAs), whose $E_g = 1.42$ eV, is placed between the aluminum gallium arsenide (AlGaAs) layers, whose $E_g = 1.92$ eV. As one can see from Figure 9.5(c) electrons injected from $n$-type AlGaAs confront an energy barrier at the junction where GaAs and $p$-type A1GaAs meet and are reflected back into the active region. The same mechanism works for holes.

The conduction of light in one direction is achieved because the GaAs semiconductor has a higher refractive index (here, 3.66) than the substrate semiconductor (here, 3.2). Thus, the active region works as a waveguide similar to the way a fiber traps light within the core using the core-cladding interface. The same concept is implemented for another popular heterostructure, indium phosphide-indium gallium arsenide phosphide (InP-InGaAsP) [1].

Such a structure is also called a *double heterostructure (DH)*. Most commercial LEDs use not two but three different types of semiconductors to increase the light-radiation efficiency and to confine radiated light better.

***Radiant patterns—Spatial patterns of radiation*** Two basic types of light-emitting diodes—surface-emitting LED (SLED) and edge-emitting LED (ELED)—have different spatial-radiation patterns, as Figure 9.6 shows. SLED radiates light as a *Lambertian* source (named after Johann Lambert, an eighteenth-century German scientist). Its power distribution is described by the following formula:

$$P = P_0 \cos\theta, \tag{9.4}$$

where $\theta$ is the angle between the direction of observation and the line orthogonal to the radiating surface; thus, $P = P_0$ when $\theta = 0°$. Half of the power of the Lambertian source is concentrated in a 120° cone.

ELED radiates as a Lambertian source in the plane parallel to the edge and produces a much narrower beam in the plane perpendicular to the edge, as Figure 9.6(b) shows.

A Lambertian source is simply a reference model that describes in a general way a homostructured SLED. In reality, a heterostructured LED radiates a much better directed beam. Figure 9.6(c) depicts a sample of a real spatial pattern of radiation. Because of the form of its radiant pattern, a SLED is more suitable to use with a multimode fiber, while an ELED can be used with a singlemode fiber.

***Radiating wavelengths*** A radiating wavelength is determined by the energy gap of a semiconductor, as discussed above. We cannot change an energy gap just as we cannot change energy levels of a given material; therefore, to obtain another wavelength, we have to choose another material. In the case of semiconductors, a desired energy gap, $E_g$, is created by using compound semiconductors consisting of several components. For example, the energy gap for GaAs is equal to 1.42 eV, but if you use the composition AlGaAs, you obtain an energy gap from 1.42 eV to 1.92 eV. The value of the energy gap attainable depends on the ratio of the ingredients making up the composition. In our example, if the semiconductor is composed of 37% AlAs and 63%

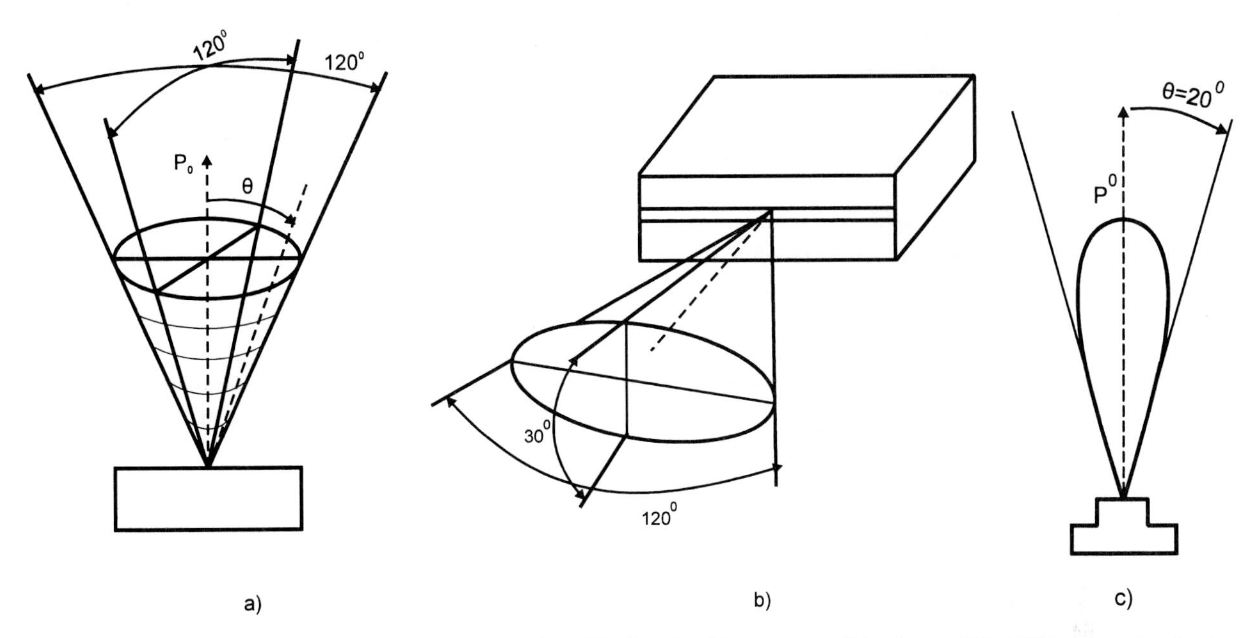


**Figure 9.6** LED radiant pattern: (a) Surface-emitting LED (Lambertian source); (b) edge-emitting LED; (c) real radiant pattern.

GaAs, $E_g$ equals 1.92 eV. If a smaller amount of AlAs is incorporated, the energy gap narrows. Table 9.1 displays the energy gaps and wavelengths of some popular semiconductors used for LED fabrication.

The first generation of fiber-optic communications systems used LEDs made from Al-GaAs, which radiate at around 850 nm at the first transparent window. The second and subse-

**Table 9.1** Energy gaps and wavelengths (T = 300k) of popular semiconductors used for LED fabrication

| Material | Energy gap, $E_g$ (eV) | Wavelengths (nm) |
|---|---|---|
| Si | 1.17 | 1067 |
| Ge | 0.775 | 1610 |
| GaAs | 1.424 | 876 |
| InP | 1.35 | 924 |
| InGaAs | 0.75–1.24 | 1664–1006 |
| AlGaAs | 1.42–1.92 | 879–650 |
| InGaAsP | 0.75–1.35 | 1664–924 |

Sources: Joseph Palais, *Fiber Optic Communication,* 4th ed., Englewood Cliffs, N.J.: Prentice Hall, 1998.

Rajiv Ramaswami and Kumar Sivarajan, *Optical Networks: A Practical Perspective,* San Francisco: Morgan Kaufman, 1998.

Tien Pei Lee, C.A. Burrus, Jr., and R.H. Saul, "Light-Emitting Diodes for Telecommunication," in *Optical Fiber Telecommunications-II,* ed. by S.E. Miller and I.P. Kaminow, Boston: Academic Press, 1988, pp. 467–508.

quent generations have used LEDs made from InGaAsP radiating at the second and third transparent windows (1300 nm and 1550 nm).

Surface- and edge-emitting LEDs radiate at different wavelengths: SLEDs at 850 and 1300 nm and ELEDs at 1300 and 1550 nm. There are LEDs radiating in the visible range of the spectrum that find use in ultrashort communications links with plastic optical fibers.

***Coupling light into a fiber*** It is quite evident that we are interested in having as powerful an input light signal as possible because, given fiber attenuation, a more powerful signal travels a greater distance. It would seem that to accomplish this, we would need a more powerful light source, but this is not the whole truth. *The key to the distance a signal travels is not just the power radiated by the source, but the power coupled into an optical fiber because this is the real input signal being transmitted.* With inefficient coupling, you may lose most of the light power radiated by your LED, thus making the quality of the LED absolutely unimportant from the transmission standpoint.

If you approximate the radiation pattern of a SLED by a Lambertian model, then light power ($P_{in}$) coupled into a step-index fiber with a numerical aperture ($NA$) can be calculated by the following formula:

$$P_{in} = P_0(NA)^2, \tag{9.5}$$

where $P_0$ is determined by Formula 9.4.

### Example 9.1.2

***Problem:***

What is the power coupled into a step-index multimode fiber whose $n_1 = 1.48$ and whose $n_2 = 1.46$ if the SLED radiates 100 μW?

***Solution:***

From Example 3.1.4, you know that for this fiber the $NA = 0.2425$. Therefore,

$$P_{in} = P_0\,(NA)^2 = 100\ \mu\text{W} \times 0.0588 = 5.88\ \mu\text{W}.$$

It is useful to calculate the power launched into a graded-index fiber. Even though, strictly speaking, Formula 9.5 is applied to a step-index fiber, we can extend its application to a graded-index fiber. We need bear in mind only that the result of our calculations gives us the order of magnitude, not the precise value.

Typical graded-index 62.5/125 μm fiber has an $NA$ of 0.275. Let's take this number for our calculations. The result:

$$P_{in} = P_0\,(NA)^2 = 100\ (\mu\text{W}) \times 0.0756 = 7.56\ \mu\text{W}$$

In other words, less than 10% of radiated power is coupled into a multimode fiber.

---

Formula 9.5 allows you to approximate the amount of power coupled, but by no means does it give you precise numbers. This is because of the inherent nature of the Lambertian model itself. Nevertheless, this formula underscores the basic idea: The amount of light power coupled into a fiber depends on the fiber's numerical aperture. Recalling that $NA = \sin \Theta_a$, where $\Theta_a$ is the fiber's acceptance angle (see Formula 3.4), you will appreciate the general coupling diagram in Figure 9.7(a).

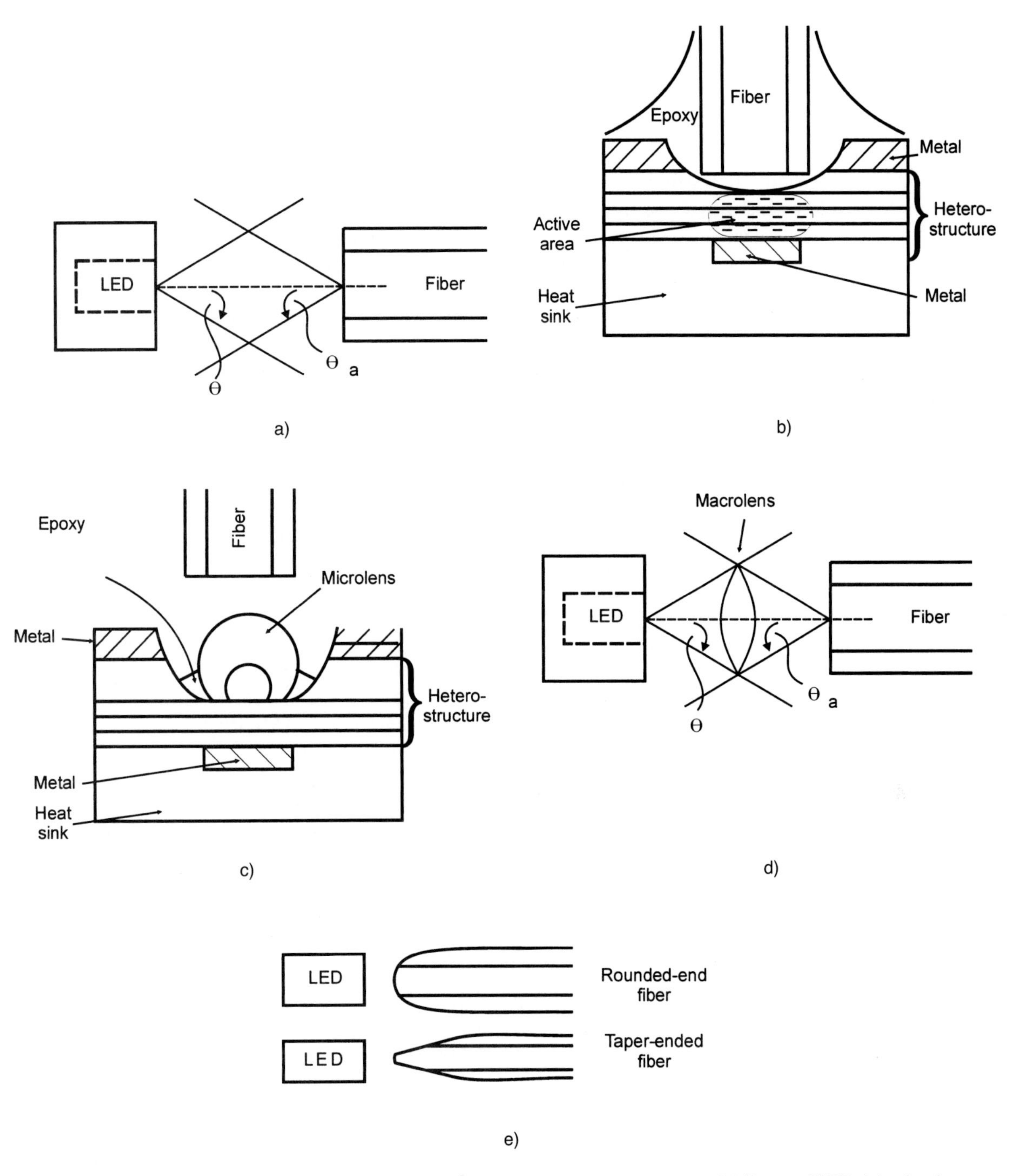


**Figure 9.7** Coupling light form an LED into an optical fiber: (a) General diagram; (b) Burrus SLED; (c) microlens coupling; (d) macrolens coupling; (e) rounded-end and taper-ended fibers. *([b] and [c] reprinted from C.A. Burrus and B.J. Miller, "Small-Area Double Heterostructure AlGaAs Electroluminescent Diode Source for Optical-Fiber Transmission Lines,"* Optics Communications, *vol. 4, 1971, pp. 307–309, with permission from Elsevier Science.)*

Any of several coupling techniques can be employed to improve coupling efficiency. The most popular one is direct coupling. A common example is the so-called Burrus SLED, developed by C. A. Burrus, Jr., at the Bell Telephone Laboratories in 1971. Here, a multimode fiber is inserted directly into a semiconductor structure in order to place the fiber end as close to an active area as possible. This arrangement is sketched in Figure 9.7(b). It is interesting to note that this SLED, commonly referred to as an example of good coupling design, was one of the first commercially successful GaAs/AlGaAs LED heterostructures. A novel feature of this design is its placement of an active layer very close to the surface, thus increasing the efficiency of the optical output by minimizing the absorption of radiated photons.

To improve coupling efficiency, various lensing techniques are also used. We can distinguish between the microlens and macrolens approaches. An example of the microlens technique is given in Figure 9.7(c), while the macrolens approach is illustrated in Figure 9.4(d). We have to keep in mind that a lens cannot improve the radiation property of an LED, but it can match the output angle of a light source to the acceptance angle of the optical fiber [2].

Many other lensing schemes—such as a double-lens optical system—are used to improve coupling efficiency. (See, for example, [3] and [4].) It is worth mentioning that rounded and taper-ended shapes of fiber ends (Figure 9.7[e]) are also effective means to achieving this goal.

Most of the techniques mentioned above are employed with SLEDs. Surface-emitting LEDs are used with multimode fibers but, without employing some coupling technique, the radiation from SLEDS would not fit into even the relatively large *NA*s of these fibers.

## Reading Data Sheets—Characteristics of LEDs

We will read the data sheets of LEDs in an unusual manner. First, we'll discuss the physics underlying each characteristic. Second, we'll consider not only the given specifications but also typical characteristics of other commercially available LEDs. We do this because modern LEDs come with a variety of characteristics that cannot be shown on one data sheet. Figure 9.8 displays the data sheet of a 1.3-μm SLED and ELED manufactured by AMP Inc.

***Packages*** Packages are shown in the photos in Figure 9.8. The basic package of an LED is the transistor-outline-style metallic *header* (case or can) shown in Figures 9.8(a) and 9.8(b). This case is usually hermetically sealed and may have a flat or lensed window cap. A SLED is packed with a variety of *connectors,* which is the typical packaging style for a surface-emitting LED. Packing LEDs with connectors guarantees a certain coupling efficiency because the user does not need to mount a fiber onto an LED; he or she need only connect the fiber through one of the standard connectors.

ELEDs are packed not only with a connector, as Figure 9.8(b) shows, but also in *pigtail* style. (See Figure 1.5.) This is because ELEDs are used not only with multimode fibers but also with singlemode fibers, which require much more accurate coupling. A factory-assembled pigtail package guarantees the maximum coupling efficiency and minimum insertion loss. (Connecting a pigtailed LED entails simply splicing a pigtail and a transmission fiber. You'll recall from Chapter 8 that the typical fusion-splicing loss is 0.01 dB, while the loss from a good connector is not less than 0.1 dB.)

Keep in mind, too, that an LED package includes a *heat-sink component.* As pointed out above, there are nonradiative recombinations that release a lot of heat in an active layer. This heat changes the junction temperature and thus the parameters of the light-conversion process. Therefore, a heat sink is a crucial component supporting an LED's operation.

***Output and coupled power*** The values of *coupled power* are given in the table of specifications and shown in the graph "Coupled Power vs. Drive Current" (Figure 9.8). Coupled power, obviously, depends on the type of fiber and on the LED's package. The typical power coupled

**1.3 µm LED**

Catalog 124002
6-97

**Features**

- **High coupled power, typically 75 µW into 62.5 µm fiber**
- **High reliability MTTW 2.3 x $10^8$ hours**
- **Wavelength centered at 1320 nm**
- **Hermetically sealed TO-18 style package installed in industry standard ADMs**
- **Functional over −40°C to 85°C operating temperature range**

The AMP InGaAsP SLED products offer high coupled powers for digital fiber optic transmission applications.

Compatible with industry standards, the AMP LED ADMs consist of hermetically sealed TO-18 style SLEDs which have been actively aligned for maximum coupled power. The devices are permanently fixed in place to assure stable performance over all operating conditions.

The SC and ST connectors are suitable for both panel/bulk-head and PC board mounting. The FC is panel mount.

Each unit is burned-in. Coupled power, capacitance, leakage current and spectral characteristics are measured on each unit. No data is supplied with the unit. A lot code is used for traceability.

For additional information on product qualification, reference Product Specification 108-55008.

For additional information on product performance, reference Application Note 82924.

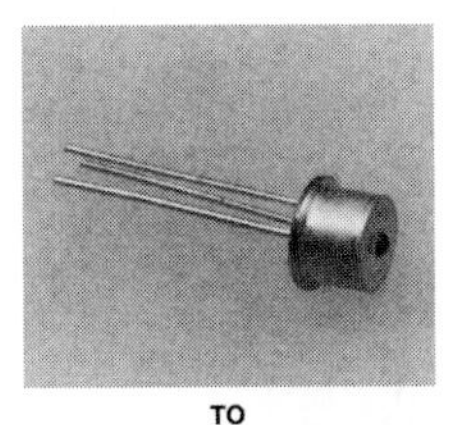

TO

ST Style

SC

FC

Low Profile ST

**Specifications: 100mA Forward Current, 25°C**

| Parameter | Part No. Suffix | Test Conditions | Units | Min. | Typ. | Max. |
|---|---|---|---|---|---|---|
| Coupled power<br>50 µm fiber | -1 | — | µW<br>dBm | 10<br>−20 | 20<br>−17 | — |
| 62.5 µm fiber | -1 | — | µW<br>dBm | 30<br>−15 | 45<br>−13 | — |
| 50 µm fiber | -2 | — | µW<br>dBm | 20<br>−17 | 30<br>−15 | — |
| 62.5 µm fiber | -2 | — | µW<br>dBm | 50<br>−13 | 75<br>−11 | — |
| Wavelength<br>Spectral FWHM | — | — | nm<br>nm | 1290<br>— | —<br>— | 1350<br>170 |
| Forward voltage<br>Capacitance<br>Leakage current | — | <br>f=1MHz, 0V<br>−2V | V<br>pF<br>µA | —<br>—<br>— | 1.4<br>15<br>— | 1.7<br>50<br>2 |
| Rise/fall time | — | 100mA 50% duty cycle 12.5 MHz | ns | — | 2.5 | 4 |
| Bandwidth | — | — | MHz | — | 115 | — |
| $\Delta\lambda/\Delta T$<br>$\Delta P_{out}/\Delta T$ | — | −40 to +85°C | nm/°C<br>dB/°C | — | .38<br>−.03 | — |
| Reliability MTTW | — | −1.5dB EOL | hrs | — | 2.3 x $10^8$ | — |

**Note:** dBm is rounded to nearest integer value.

**Absolute Maximum Rating**

| | Units | Min. | Max. |
|---|---|---|---|
| Operating temperature | C | −40 | 85 |
| Storage temperature | C | −40 | 125 |
| Reverse voltage | V | — | 2 |
| Forward current | mA | — | 150 |

53

For drawings, technical data or samples, contact your AMP sales engineer or call the AMP Product Information Center 1-800-522-6752.
Dimensions are in inches and millimeters unless otherwise specified. Values in brackets are metric equivalents.
Specifications subject to change. Consult AMP Incorporated for latest specifications.

a)

**Figure 9.8** LED data sheets: (a) SLED; (b) ELED. *(Courtesy of AMP, Inc., Harrisburg, Pa.)*

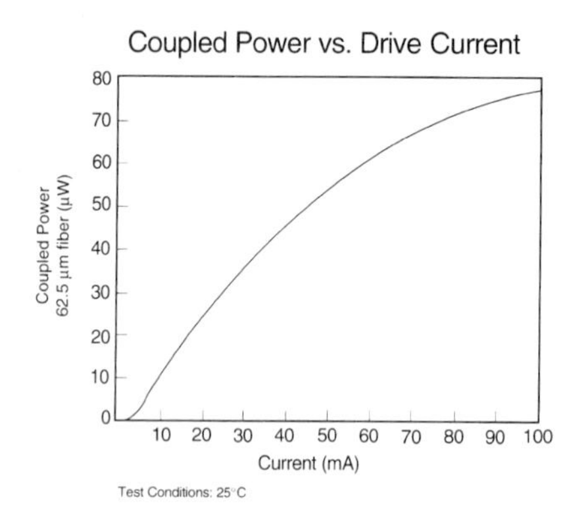

Coupled Power into 62.5 μm Fiber
vs. Temperature

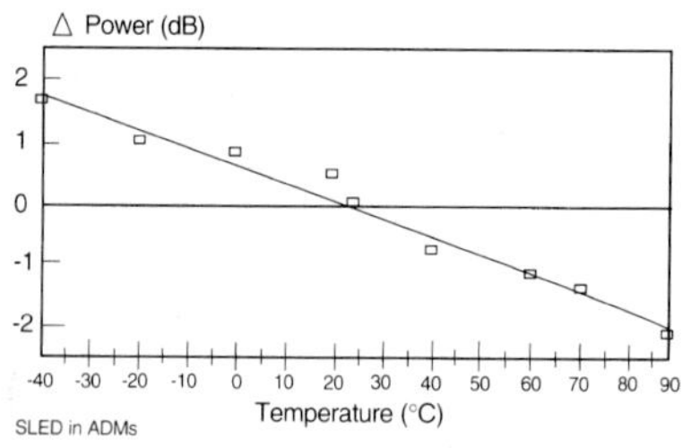

SLED FWHM in ADM
over Temperature

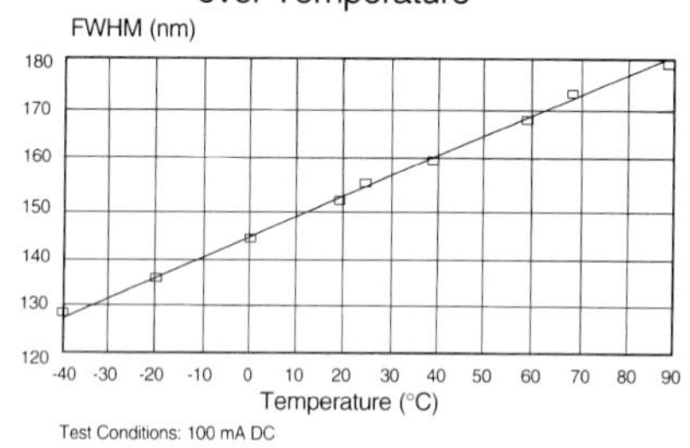

SLED FWHM vs. Drive Current
normalized to 100 mA DC

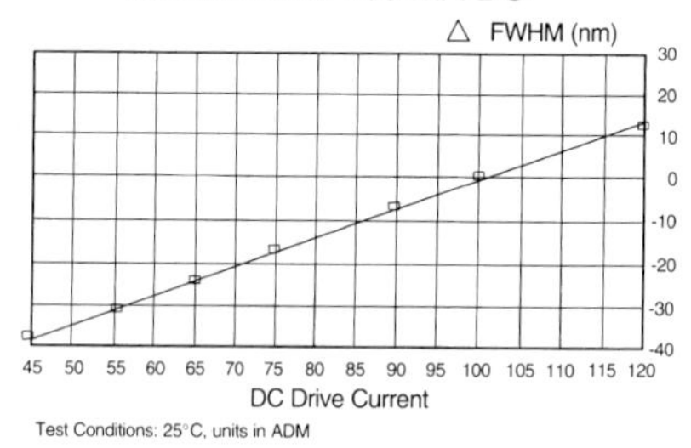

**Part Numbers**

| | Connector Interface | | | |
|---|---|---|---|---|
| | **TO** | **FC** | **SC** | **ST Style** |
| Standard | 259006-1 | 259014-1 | 269067-1 | 259012-1 |
| Premium | 259006-2 | 259014-2 | 269067-2 | 259012-2 |

**Note:** Coupled power in specifications.

**Mechanical Dimension Reference**

| | | | | |
|---|---|---|---|---|
| Figure # | 2 | 3 | 5 | 7 |
| Page No. | 85 | 86 | 86 | 87 |

For drawings, technical data or samples, contact your AMP sales engineer or call the AMP Product Information Center 1-800-522-6752.
Dimensions are in inches and millimeters unless otherwise specified. Values in brackets are metric equivalents.
Specifications subject to change. Consult AMP Incorporated for latest specifications.

**Figure 9.8** (continued)

## Features

- High coupled power, typically 15 μW into 9 μm fiber
- Wavelength centered at 1300 nm
- 0° to +65°C operating temperature
- Hermetically sealed package installed in industry standard ADMs
- High reliability

The AMP InGaAsP ELED product mounted in an ST style ADM offers high coupled power for digital fiber optic transmission applications.

Compatible with industry standards, this hermetically sealed AMP device is actively aligned for maximum coupled power and stable performance over all operating conditions.

The ST style assembly is suitable for both panel/bulkhead and PC board mounting.

No data is supplied with the unit. A lot code is used for traceability.

For additional information on product qualification, reference Product Specification 108-55008.

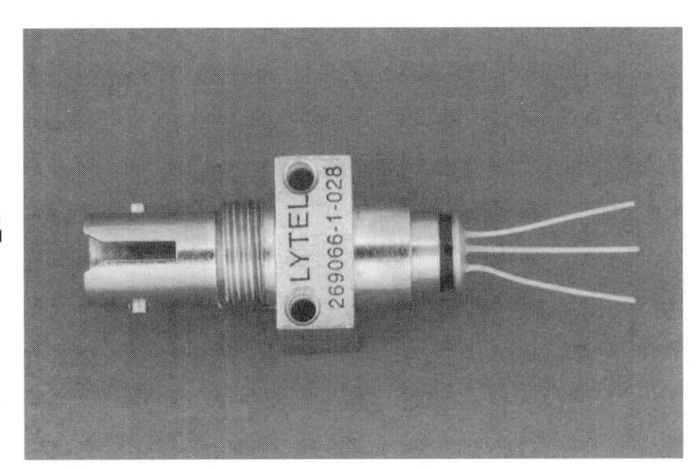

**Specifications: 100mA Forward Current, 25°C**

| Parameter | Part No. Suffix | Test Conditions | Units | Min. | Typ. | Max. |
|---|---|---|---|---|---|---|
| Coupled power<br>9 μm fiber | -1 | — | μW<br>dBm | 10<br>-20 | 15<br>-18 | — |
| Wavelength<br>Spectral FWHM | — | — | nm<br>nm | 1270<br>— | —<br>65 | 1330<br>100 |
| Forward voltage<br>Capacitance | — | f=1MHz, 0V | V<br>pF | — | 1.4<br>15 | — |
| Rise/fall | — | 100mA 50% duty cycle 12.5 MHz | ns | — | 2.5 | 4 |
| $\Delta\lambda/\Delta T$<br>$\Delta P_{out}/\Delta T$ | — | 0 to +65°C | nm/°C<br>dB/°C | — | .69<br>.114 | — |

**Note:** dBm is rounded to nearest integer value.

**Absolute Maximum Rating**

| | Symbol | Min. | Typ. | Max. | Units |
|---|---|---|---|---|---|
| Operating temperature | $T_O$ | 0 | — | 65 | C |
| Storage temperature | $T_S$ | −40 | — | 90 | C |
| Reverse voltage | $V_R$ | — | — | 1 | V |
| Forward current | $I_F$ | — | — | 150 | mA |

For drawings, technical data or samples, contact your AMP sales engineer or call the AMP Product Information Center 1-800-522-6752.
Dimensions are in inches and millimeters unless otherwise specified. Values in brackets are metric equivalents.
Specifications subject to change. Consult AMP Incorporated for latest specifications.

**Figure 9.8** (continued) b)

Light vs. Current

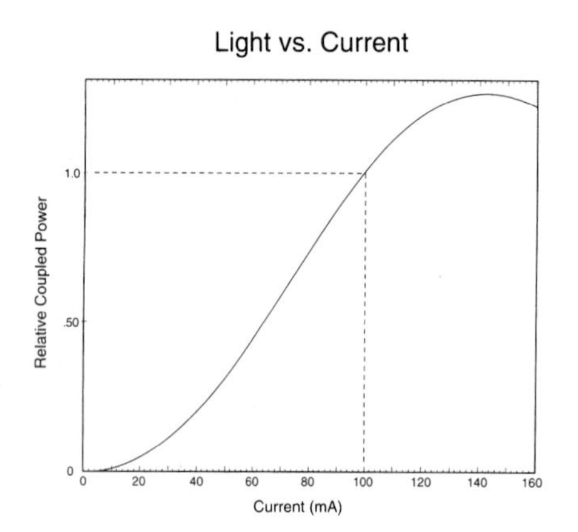

Coupled Power over Temperature Relative to 25°C

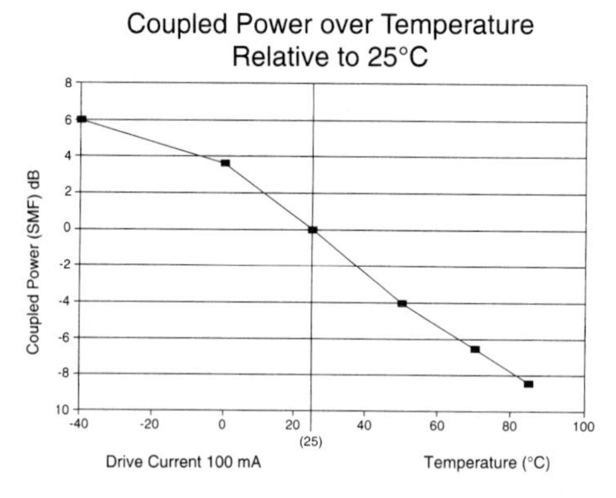

Rise/Fall Measurement

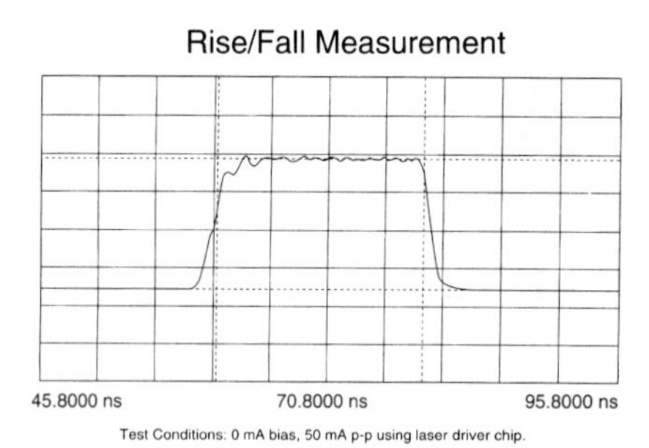

Test Conditions: 0 mA bias, 50 mA p-p using laser driver chip.

Spectral Width

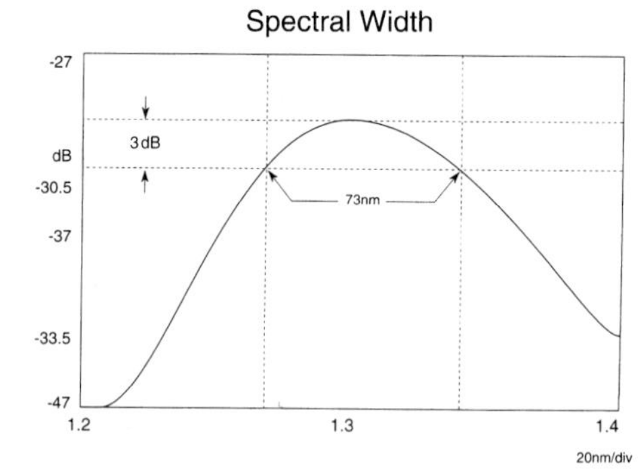

**Part Number**

| | ST Style |
|---|---|
| | 269066-1 |

**Mechanical Dimension Reference**

| | |
|---|---|
| Figure # | 9 |
| Page No. | 88 |

For drawings, technical data or samples, contact your AMP sales engineer or call the AMP Product Information Center 1-800-522-6752.
Dimensions are in inches and millimeters unless otherwise specified. Values in brackets are metric equivalents.
Specifications subject to change. Consult AMP Incorporated for latest specifications.

**Figure 9.8** (continued)

into 62.5/125-μm multimode fiber by an AMP SLED is 75 μW. The typical power coupled into a singlemode fiber by an AMP ELED is 15 μW.

Since a Lambertian source, by which we model a SLED, couples only a fraction of the output power, these data allow us to calculate output power. In the simplest approach, given by Formula 9.5, we have $P_0 = P_{in} /(NA)^2$. If the typical $NA$ for a multimode fiber is 0.275, then $P_0 = 13.2\ P_{in}$. Actually, this coefficient may be as much as two times less, which means a much larger portion of radiated power is coupled into a fiber.

Absolute numbers of output power range from units to tenths of milliwatts. To increase output power, one has to increase the current (more precisely, the current density) in the active area. This raises the number of nonradiative transitions, thus decreasing quantum efficiency and increasing the temperature of the junction. This, in turn, leads to a decrease in output power [5]. Thus, there is a limit to an LED's output power.

Most manufacturers prefer to specify not output power and radiant patterns but the net result: coupled power. This is what an end-user really wants to know: how much light is at the optical fiber's input. Values of coupled power range from units to hundreds of microwatts for SLEDs. ELEDs can couple into a singlemode fiber as little as 5 μW of light power and they need a cooled package to have more than 50 μW of coupled power.

The graph depicting light power versus driving current—*P-I* or *L-I*—shows one of the most important characteristics of LEDs. (See Figure 9.8.)

Pay particular attention to the nonlinearity of the curves in Figure 9.8.

The graph "Coupled Power vs. Temperature" shows a very important effect: Power decreases as temperature increases, with the slope approximately 2 dB per 65°C. Thus, if ambient temperature increases from 25°C to 90°C, coupled power drops to 79% of the original number; that is, $P_{in}$ (90°C) = 0.79 $P_{in}$ (25°C) because 2 dB = –20 log(0.79). This slope is given as a coefficient ($\Delta P_{out}/\Delta T$ = –0.03 dB°C) in the table of specifications. The coupled power of an ELED decreases with temperature even more steeply than a SLED's power does. (See Figure 9.8[b].)

***Wavelength and spectral width*** Radiated *wavelength,* often referred to as a peak wavelength, $\lambda_p$, is determined by an energy gap, $E_g$. Manufacturers usually specify minimum and maximum values of $\lambda_p$. For AMP's SLED, these numbers are 1290 nm and 1350 nm; for the ELED, they are 1270 nm and 1330 nm. Even though it doesn't show in Figure 9.8, $\lambda_p$ shifts to the longer wavelengths with increasing current and temperature but stays within a specified range.

A *spectral width,* $\Delta\lambda$, is measured as full width at half maximum, FWHM, as Figure 9.8(b) shows in the graph "Spectral Width." (Also see Figure 9.2[b].) For AMP's SLED, the spectral width is very wide: 170 nm. It is much narrower for the ELED: 65 nm. (In comparison, a laser diode's $\Delta\lambda$ is around 1 nm and less.) These values of $\Delta\lambda$ are typical for modern LEDs. They are much less for LEDs radiating at peak wavelength, around 850 nm, where the typical $\Delta\lambda$ is about 50 nm.

Spectral width depends on temperature, as the graph "SLED FWHM over Temperature" (Figure 9.8[a]) shows. In the range between 25°C and 90°C, spectral width increases from 155 nm to 180 nm; that is, the slope is 0.38 nm/°C. You can find this number in the table of specifications in Figure 9.8(a). The FWHM width also increases with the rise of forward (drive) current, with the slope equaling approximately 0.69 nm/mA.

You will recall that spectral width is the critical parameter that determines the chromatic dispersion—and, hence, bandwidth—of an optical fiber. Chromatic dispersion is proportional to both spectral width and distance (see Formula 3.19); therefore, these LEDs can be used for narrow-bandwidth, short-distance applications.

***Electrical characteristics*** The electrical characteristics—forward voltage, capacitance, and leakage current—are common to any electronic diode. Manufacturers sometimes specify the *forward voltage versus forward current* characteristic, which, typically, has a form shown in Figure 9.9(a). The value of the forward voltage usually does not exceed 2 volts.

*Capacitance, C,* specified in the data sheet, is inherent in an LED. There are two sources of *C:* (a) charge capacitance, associated with the *p-n* junction, and (b) diffusion capacitance, associated with carrier lifetime at the active region [5]. An LED's capacitance limits its practical modulation ability and, thus, restricts its bandwidth. For example, one manufacturer specifies a capacitance of 20 pF for a SLED whose bandwidth is 200 MHz (at a peak wavelength of 865 nm) and 200 pF for a SLED whose bandwidth is 125 MHz (at a peak wavelength of 1320 nm) [4]. This is the typical range of an LED's capacitance.

*Leakage current* is caused by the flow of minority charge carriers (electrons in the *p* region and holes in the *n* region). These charge carriers are created by thermal energy, which excites electrons even in the *p* region. This current is measured at some reverse-bias voltage (2 volts in Figure 9.8 [a]).

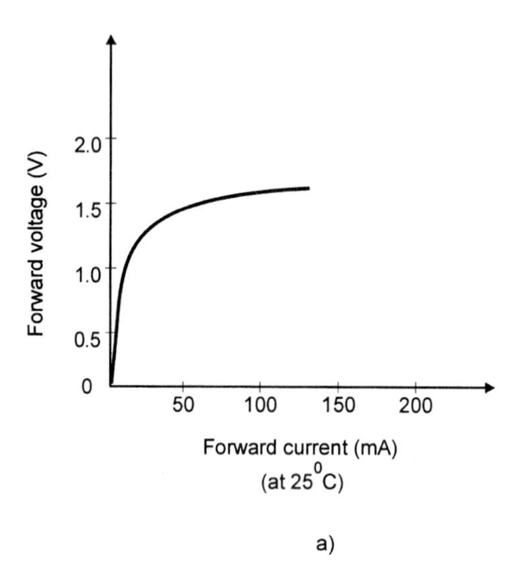


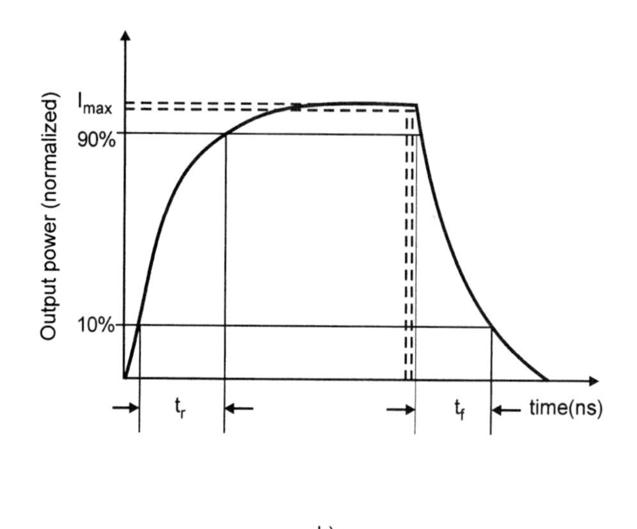


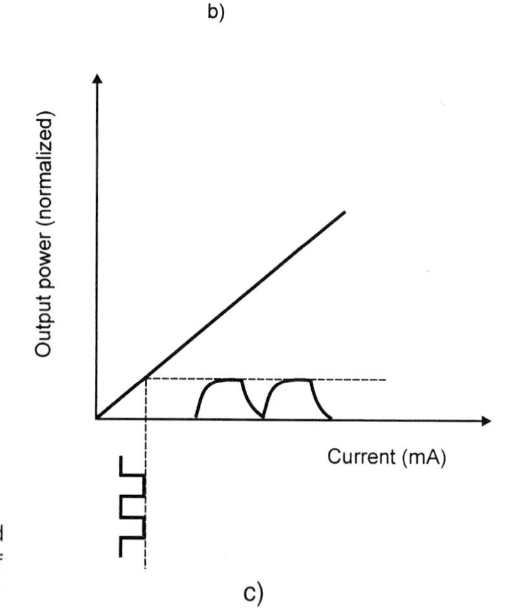


**Figure 9.9** Characteristics of an LED: (a) Typical graph of a forward voltage versus current; (b) rise, $t_r$, and fall, $t_f$, time; (c) modulation of an LED.

***Lifetime, rise/fall time, and bandwidth*** *Lifetime,* $\tau$, of the charge carriers is the time between the moment they are excited (injected into a depletion region) and the moment they are recombined. It is sometimes called *recombination lifetime* and it ranges from nanoseconds to milliseconds. We distinguish between radiative, $\tau_r$ and nonradiative, $\tau_{nr}$, recombination lifetimes so that the total carrier lifetime, $\tau$, is equal to [5]:

$$1/\tau = 1/\tau_r + 1/\tau_{nr} \tag{9.6}$$

Incidentally, internal quantum efficiency, $\eta_{int}$, which shows how many photons are radiated with respect to a specific number of injected electrons, can be quantified by the following formula:

$$\eta_{int} = \tau/\tau_r \tag{9.7}$$

*Rise/fall time,* $t_r$, is defined as 10 to 90% of the maximum value of the pulse, as Figure 9.9(b) shows. For an LED, this characteristic shows how an output light pulse follows the electrical-modulating input pulse. (See Figure 9.9[c].) An ideal step pulse is shown as two dotted lines in Figure 9.9(b). This enables you to visualize the pulse distortion caused by the rise/fall time.

Rise/fall time is determined by an LED's capacitance ($C$), input step current with amplitude($I_p$), and the total recombination lifetime ($\tau$) so that [3]:

$$t_r = 2.2\left[\tau + \left(1.7 \times 10^{-4} \times T°K \times C\right)/I_p\right], \tag{9.8}$$

where $T°K$ is absolute temperature in kelvin (0°C = 273°K).

This formula is important because it discloses the parameters on which rise time depends. With a high $I_p$, the second term on the right side of Formula 9.8 becomes negligible and *rise time is ultimately determined by the recombination lifetime.*

Manufacturers prefer to measure, not calculate, rise time, and typical values that can be found in data sheets range from 2 to 4 ns.

*Modulation bandwidth, BW,* is the range of modulating frequencies within which detected electric power declines at −3 dB. (See Figure 3.17.) In electronics, the general relationship between bandwidth and rise time is given by the well-known formula

$$BW = 0.35/t_r \tag{9.9}$$

This formula stems from the exponential response of an RC circuit to a step-input pulse. But if you plug into Formula 9.9 the value $t_r$ = 2.5 ns (from the data given in Figure 9.8[a]), you will not get $BW$ = 115 MHz, as given by this specification sheet. (You will, rather, obtain 140 MHz.)

This discrepancy occurs because if the forward current is modulated at angular frequency, $\omega$, an LED's output light intensity, $I(\omega)$, will vary as follows [5]:

$$I(\omega) = I(0) / \sqrt{\left[1 + (\omega\tau)^2\right]}, \tag{9.10}$$

where $I(0)$ is the LED's light intensity at constant current and $\tau$ is a carrier lifetime, as before. Detected electric power is proportional to $I^2$. (See our discussion of electrical and optical bandwidth in Section 4.6.) Taking $I^2(\omega)/I^2(0) = ½$, which is a −3 dB decline, one can find from Formula 9.10 that

$$BW = \Delta\omega = 1/\tau \tag{9.11}$$

This yields a very important principle: *An LED's modulation bandwidth is limited by the recombination lifetime of the charge carriers.* The physics governing this result is as follows: Suppose you excite an electron at the conduction band. It takes $\tau$ ns for this electron to fall to the valence band and recombine. During this interval you cannot change its status, so that if you turn off the

**Table 9.2** Typical characteristics of LEDs

| Active material | Type | Radiating wavelength λ (nm) | Spectral width Δλ (nm) | Output power into fiber (μW) | Forward current (mA) | Rise/fall time (ns) |
|---|---|---|---|---|---|---|
| AlGaAs | SLED | 660 | 20 | 190–1350 | 20(min) | 13/10 |
| | ELED | 850 | 35–65 | 10–80 | 60–100 | 2/2–6.5/6.5 |
| GaAs | SLED | 850 | 40 | 80–140 | 100 | — |
| | ELED | 850 | 35 | 10–32 | 100 | 6.5/6.5 |
| InGaAsP | SLED | 1300 | 110 | 10–50 | 100 | 3/3 |
| | ELED | 1300 | 25 | 10–150 | 30–100 | 1.5/2.5 |
| | ELED | 1550 | 40–70 | 1000–7500 | 200–500 | 0.4/0.4–12/12 |

Source: *Lightwave 1999 Worldwide Directory of Fiber-Optic Communications Products and Services,* March 31, 1999, pp. 58–61.

forward current, you must wait τ ns until radiation will actually cease. This τ ns interval is necessary to allow a charge carrier to reach its destination. In other words, you cannot stop an excited electron that is on its way from the conduction band to the valence band. Thus, *lifetime τ puts a fundamental limit on the modulation bandwidth of an LED.* (You can repeat this reasoning using a *p-n* junction model: While an electron is moving through an active region, you cannot stop it; that is, you cannot change its status until this electron recombines.)

This is why LEDs are restricted by bandwidth in the range of hundreds of MHz. Such restrictions determine their applications in local area and other low-bandwidth networks.

*Power-bandwidth product* is another important characteristic of an LED. It appears that the product of an LED's optical output power and its modulation bandwidth is constant:

$$BW \times P = \text{constant} \tag{9.12}$$

In other words, you can increase an LED's bandwidth but only at the expense of its output power. Alternatively, you can increase output power but then bandwidth decreases.

*Reliability* is one of the major advantages of an LED. The table of SLED specifications in Figure 9.8 shows that the mean time to failure is more than a hundred million hours. It's hard to imagine more impressive numbers describing the reliability of an opto-electronic device. (To characterize reliability, the industry determines the average time to failure of an LED, which it refers to as mean time to failure, or MTTF.)

As we have mentioned several times already, LEDs find their applications in LANs as Token-Rings, 100 Mbit/s Ethernets, Fibre Channels, FDDIs, and other datacom networks; they are also used in intraoffice telecom networks.

In conclusion, we have summarized in Table 9.2 the typical characteristics of LEDs. These numbers give you a general idea of what today's LEDs look like.

## 9.2 LASER DIODES (LDs)

Semiconductor laser diodes, developed in the 1970s, have found vast commercial applications in compact-disc (CD) players. With the advent of commercial optical fiber, such LD radiation properties as brightness, directivity, narrow spectral width, and coherence made them the best light sources for long-haul fiber-optic links. Over the years, the requirements for long-haul system capacity have continued to increase, as has the need to improve laser-diode quality. In response, quantum-well and distributed-feedback laser diodes with extremely narrow spectral width—on the order of tenths of a nanometer—have been developed. These laser diodes have become the most popular light sources for long-distance transmitters. Vertical-cavity surface-emitting lasers (VCSEL) are among the latest developments in the technology of light sources.

WDM, dense-WDM, and high-density WDM technologies pose a new challenge for laser-diode designers. The channel separation in these technologies—measured now in frequency, not in wavelength—is today so small (just imagine a 50-GHz separation) that a spectral width of 0.1 nm in a laser diode is now unacceptable. Thus, as is true of the entire fiber-optic communications industry, laser-diode technology continues to evolve dynamically. This section presents a review of the current situation.

## Principle of Action

***Laser*** The acronym *laser* means *l*ight *a*mplification by the *s*timulated *e*mission of *r*adiation.

The first working ruby laser was developed in 1960 by the American scientist Theodore Meiman. The theoretical and practical foundations for this development were made by the American Charles Townes and the Russians Alexander Prokhorov and Nikolay Basov, who shared the Nobel Prize for Physics in 1964 for their work.

Interestingly, the laser is not a light amplifier, as the term suggests, but, rather, a light generator. This was true for the first laser; it is true for today's devices. However, since the term exists and is well accepted, so be it. (It is also interesting to note that in the technical literature, derivative words like "lasing" and "to lase" have become common.)

The laser is a device that amplifies (or, as we now know, "generates") light by means of the stimulated emission of radiation. How a laser produces light amplification and what the words *stimulated emission of radiation* mean is our next consideration.

***Spontaneous and stimulated radiation*** We distinguish between two types of radiation: spontaneous and stimulated. *"Spontaneous" means that radiation occurs without external cause.* That's exactly what happens in an LED: *Excited electrons from the conduction band fall, without any external inducement, to the valence band, which results in spontaneous radiation.*

The *properties* of spontaneous radiation follow naturally from the way it occurs:

- First, the transition of electrons from many energy levels of conduction and valence bands contributes to the radiation produced, thus making the spectral width of such a source very wide. This is why a typical LED's $\Delta\lambda$ is about 60 nm at an operating wavelength of 850 nm and about 170 nm at an operating wavelength of 1300 nm.

- Second, since photons are radiated in arbitrary directions, very few of them create light in the desired direction, a factor that reduces the output power of an LED. This means that current-to-light conversion occurs with low efficiency and an LED has relatively low output power (intensity).

- Third, even those photons that contribute to output power do not move strictly in one direction; thus, they propagate within a wide cone, yielding widespread radiated light. For this reason, we model an LED with a Lambertian source.

- Fourth, this transition, and therefore photon radiation, occurs at any time, in other words, photons are created independently of one another. Hence, no phase correlation between different photons exists and the total light radiated is called *incoherent.*

These four main properties of spontaneous radiation—wide spectral width, low intensity, poor directiveness, and incoherence—make it impossible to use LEDs as light sources for long-distance communication links.

A different process occurs if you let an external photon hit an excited electron, as Figure 9.10(a) shows. Their interaction includes an electron transition and the radiation of a new photon. Now the induced emission is *stimulated* by an external photon. Thus, this radiation is called *stimulated.*

Stimulated radiation has four main properties.

- First, an external photon forces a photon with similar energy ($E_p$) to be emitted. In other words, the external photon stimulates radiation with the same frequency (wavelength) it has. (Remember, $E_p = hf = hc/\lambda$.) This property ensures that the spectral width of the light radiated will be narrow. In fact, it is quite common for a laser diode's $\Delta\lambda$ to be about 1 nm at both 1300 nm and 1550 nm.
- Second, since all photons propagate in the same direction, all of them contribute to output light. Thus, current-to-light conversion occurs with high efficiency and a laser diode has high output power. (In comparison, to make an LED radiate 1 mW of output power requires up to 150 mA of forward current; a laser diode, on the other hand, can radiate 1 mW at 10 mA.)
- Third, the stimulated photon propagates in the same direction as the photon that stimulated it; hence, the stimulated light will be well directed. If you compare the beam of a laser pointer—available in any stationery store—with any type of lamp, you'll appreciate the difference between spontaneous and stimulated radiation in terms of the way each directs light.
- Fourth, since a stimulated photon is radiated only when an external photon triggers this action, both photons are said to be synchronized, that is, time-aligned. This means that both photons are in phase and so the stimulated radiation is *coherent.*

### Spectral Width and Chromatism

You often hear that laser light is monochromatic, that is, consisting of one color. We know that color is scientifically determined by the range of wavelengths. (See Chapter 2.) Just to remind you, red is in the 650-nm wavelength range; violet is in the 450-nm range.

Do you know what wavelengths are radiated by the lamp in your classroom or office? If you see white light, it means your lamp radiates *entirely* in the visible-wavelength spectrum. And what about the spectral width of this light source? Do you know what that is? The visible-wavelength range is roughly from 400 nm to 700 nm, which means that the spectral width of white light is about 300 nm. If you look at an LED's radiation (you can find LEDs everywhere—in your home electronics, in your car, in any laboratory equipment), you see light of one color. It might be green, red, blue, or yellow, but it is one color. This is because the spectral width of an LED's visible radiation is about 30 nm. Compare 30 nm with 300 nm. The point is this: The narrower the spectral width of radiated light, the more monochromatic this light is.

Strictly speaking, monochromatic light should contain a single wavelength. But there is no light source in nature that, even theoretically, could radiate just one wavelength. Thus, we will always have light sources with finite spectral widths; therefore, a source's monochromaticity simply reflects the value of this width. For example, compare the red light radiated by a laser pointer and by an LED. You will see that the laser's light is much more color-saturated; that is, it is much redder than light radiated by an LED. This is because a laser's spectral width is in the range of 1 nm, while that of an LED is not less than 30 nm (typically, from 60 nm to 170 nm).

To sum up, then: The narrower the spectral width of the radiated light, the fewer the wavelengths involved in this radiation and the more "monochromatic" the light is.

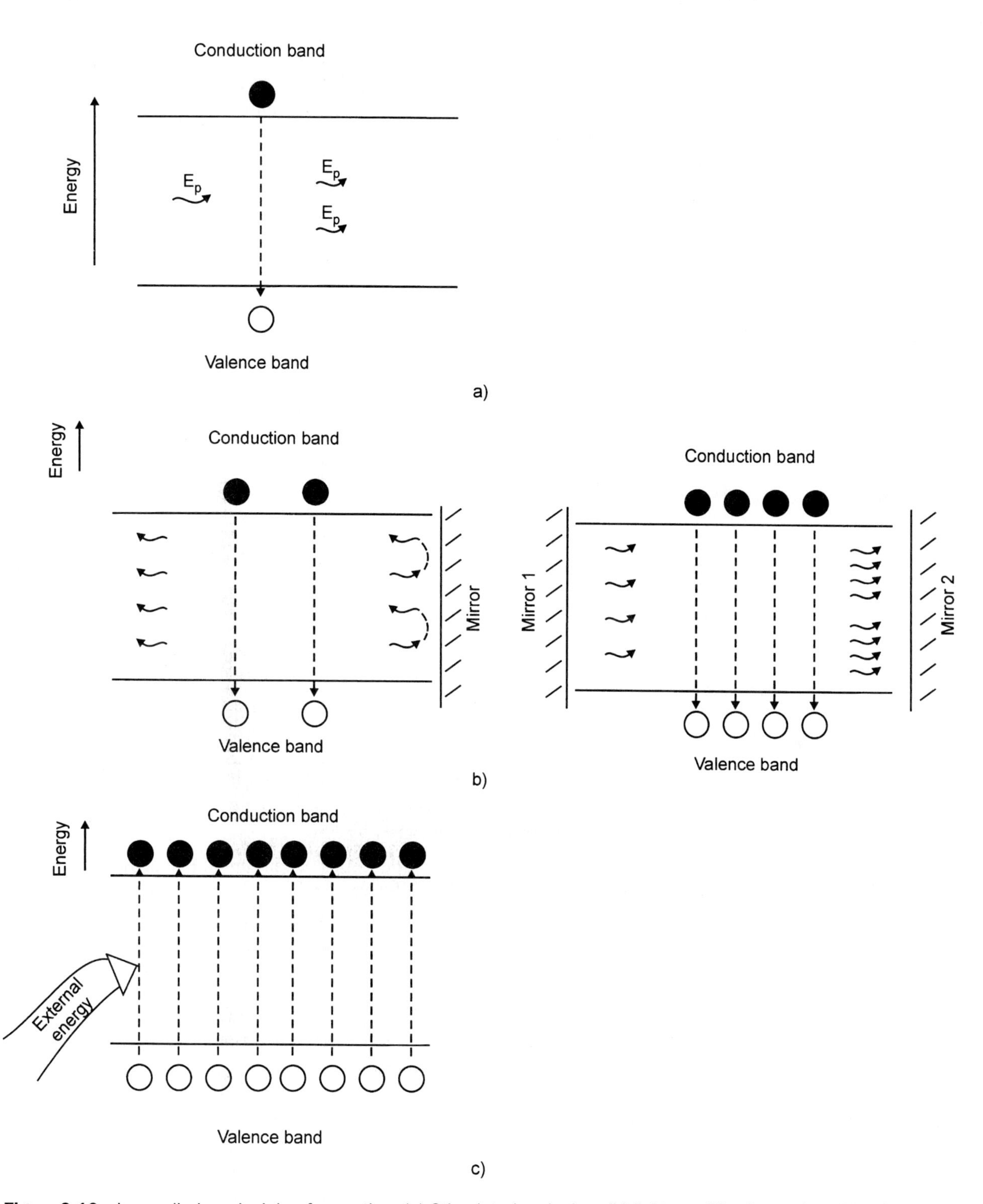


**Figure 9.10** Laser diode: principle of operation: (a) Stimulated emission; (b) light amplification and positive feedback; (c) pumping to create population inversion.

Thus, in contrast to spontaneous radiation, stimulated radiation has narrow spectral width, high intensity (power), a high degree of directivity, and coherence. This is why laser diodes, which radiate stimulated light, find use in long-distance communication links.

***Positive feedback*** To radiate stimulated light with essential power, we need not one photon but millions and millions. Here is how we can stimulate such radiation: We place a mirror at one end of an active layer, as Figure 9.10(b) shows. Two photons—one external and one stimulated—are then reflected back and directed to the active layer again. These two photons now work as external radiation and stimulate the emission of two other photons. The four photons are reflected by a second mirror, which is positioned at the other end of the active layer. When these photons pass the active layer, they stimulate emission of another four photons. These eight photons are reflected back into the active layer by the first mirror and this process continues ad infinitum. (Figure 9.10[b] illustrates these explanations.)

Thus, the two mirrors provide *positive optical feedback—positive* because the feedback adds the output (stimulated photons) to the input (external photons). (If the output is subtracted from the input, the feedback is called negative.) These two mirrors, then, constitute a *resonator.*

***More about the acronym "laser"*** Observe in Figure 9.10(b) that the number of stimulated photons increases. This means the power of the radiated light is intensifying; in other words, the active medium of this system amplifies light, thus the words *light amplification* in the acronym.

How can we obtain light amplification? By making the external photons pass through an active medium and stimulate emission of new photons—that is, by *stimulated emission.* But what does the active medium emit? A stream of photons—light, or electromagnetic radiation. As we have seen, then, all the words in the "laser" acronym prove out.

Let's now ask ourselves where the first external photon comes from. Remember, initially we have excited electrons that fall to the valence band and radiate spontaneously. Of these myriad spontaneous photons, at least one will head in the right direction (from mirror 1 to mirror 2). This external photon triggers the entire process.

Keep in mind, though, that our explanations are oversimplified. The important thing for you to remember is that we are discussing a dynamic and random process. A countless number of photons and electron-hole pairs are involved in the process; therefore, when we are describing the action of one photon or one electron, we are presenting only a bird's-eye view of the event, not a close-up account of this intricate, complex phenomenon in action. Excitation and radiation are governed by statistical laws. It was Albert Einstein[1] who, at the beginning of the twentieth century, was the one scientist who truly understood the difference between spontaneous and stimulated emission. He introduced parameters—Einstein's coefficients—to calculate the probabilities of both types of emission.

***Population inversion*** Refer to Figure 9.10(b) and note how fast the number of stimulated photons rises. To sustain this dynamic process, we need an incalculable number of excited electrons available at the conduction band. We know that using external energy—forward current for an LED—makes it possible to excite a number of electrons. But in lasers depletion of the conduction band occurs much faster than it does in LEDs; hence, we need to excite electrons at a much

[1]Albert Einstein (1879–1955), the prominent scientist who revolutionized our way of thinking about the universe, is well-known for his theory of relativity, but he also contributed to the creation of quantum physics. Interestingly, he developed the theory of photoeffect based on the concept of quantum mechanics and in 1921 was awarded the Nobel Prize for Physics for this work.

higher rate than we did in the LED process. In fact, for laser action (lasing) we need to have more electrons at the higher-energy conduction band than at the lower-energy valence band. This situation is called *population inversion* because, normally, the valence band is much more heavily populated than the conduction band. To create this population inversion, high-density forward current is passed through the small active area.

Population inversion is a necessary condition to create a lasing effect because the greater the number of excited electrons, the greater the number of stimulated photons that can be radiated. What's more, the emission intensity will be higher as well. In other words, the number of excited electrons determines the *gain* of a semiconductor diode. On the other hand, a laser diode introduces some *loss*. Two main loss mechanisms are at work: First, many photons are absorbed within the semiconductor material before they can escape to create radiation. Secondly, mirrors do not reflect 100% of the incident photons. In other words, the loss stems mainly from the absorption and transmission of the stimulated photons.

Look again at Figure 9.10(b). It seems that the number of stimulated photons continues to grow to infinity, as does the gain but, in fact, this is not true. This figure does not show the loss of these photons. As a matter of fact, at the beginning of the lasing process, the number of photons continues to grow at, literally, a nuclear-explosion rate; but, as the process continues, the more photons that are stimulated, the greater the number lost. Fortunately, loss is a constant for a given diode, but gain can be changed, as Figure 9.11(a) shows.

Increasing gain is done by increasing the forward current. Eventually gain becomes equal to loss, a situation called the *threshold condition.* (The corresponding forward current is called the *threshold current.*) At this threshold condition, a semiconductor diode starts to act like a laser. As we continue to increase the forward current (that is, the gain), the number of emitted stimulated photons continues to increase, which means the intensity of the output light also continues to increase. What we have, then, is a semiconductor diode that radiates monochromatic, well-directed, highly intense, coherent light. The point to remember? *To make a laser diode generate light, gain must exceed loss.*

***Lasing effect and input–output characteristic*** Taking all the above considerations into account, we conclude that a semiconductor diode functions like a laser (where gain exceeds loss) if the following conditions are met:

- Population inversion
- Stimulated emission
- Positive feedback

Fortunately, we know how to achieve these conditions.

Let's try to build an input–output characteristic of a laser diode (Figure 9.11[b]). Since the input here is forward current ($I$) and the output is light power ($P$), this graph demonstrates the *P-I* characteristic. (It is sometimes referred to as an "*L-I* graph," where "*L*" stands for "light.") When a small forward current is applied, a number of electrons are excited and the diode radiates like an LED. Hence, one can expect to see the same line that Figure 9.4 shows. But when the current density becomes sufficient enough to create population inversion and the threshold condition is reached (where gain equals loss), the diode starts to work like a laser. You will then see a much more intense, color-saturated, well-directed beam. This change is reflected by the graph in Figure 9.11(b), which shows that the laser diode emits much more power. After the threshold current, $I_{th}$, has been exceeded, increasing output power requires much less current to flow than before it was passed. In other words, the slope of the input–output characteristic, $\Delta P/\Delta I$, becomes

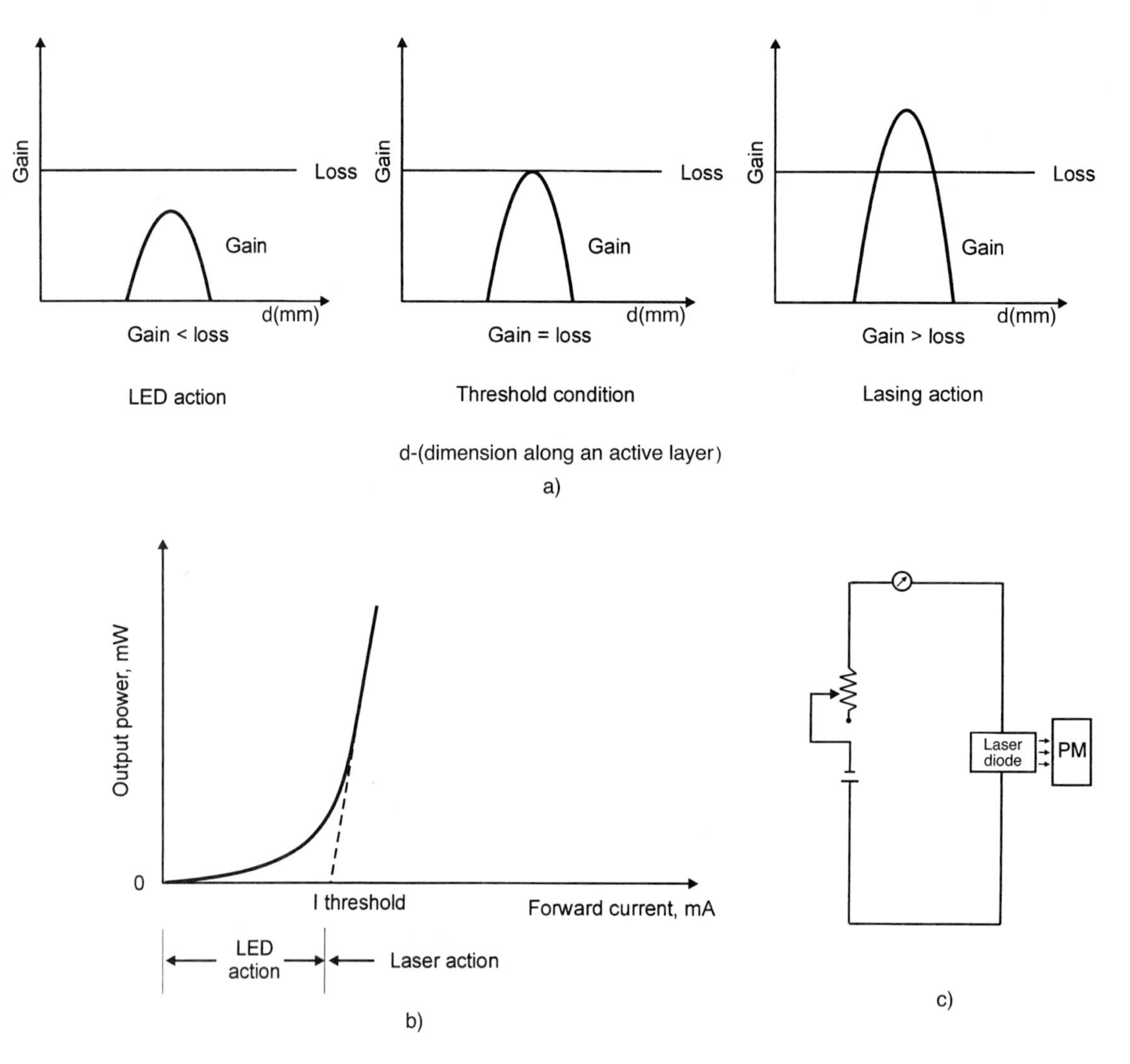


**Figure 9.11** Lasing effect: (a) Gain and loss; (b) input–output characteristic; (c) setup to measure input–output charateristic.

much steeper than that for an LED. A common laser diode with 1 mW output power has about a 30-mA threshold current and a 60-mA driving current.

One can easily build the input–output characteristic by measuring the laser diode's output power while varying the forward current. The arrangement for this measurement is shown in Figure 9.11(c).

***Laser-diode light: An analysis*** A laser diode radiates light that can be characterized as follows:

**1.** *Monochromatic.* The spectral width of the radiated light is very narrow. Indeed, the line width for a laser diode can be in tenths or even hundredths of a nanometer. (See Section 9.3.)

**2.** *Well directed.* A laser diode radiates a narrow, well-directed beam that can be easily launched into an optical fiber.

**3.** *Highly intense and power-efficient.* A laser diode can radiate hundreds of milliwatts of output power. A new type of laser diode, the VCSEL, radiates 1 mW at 10 mA of forward current, making current-to-light conversion 10 times more efficient than it is in the best LEDs.

**4.** *Coherent.* Light radiated by a laser diode is coherent; that is, all oscillations are in phase. This property is important for the transmission and detection of an information signal.

As you can see, these characteristics are very similar to those of stimulated emission. But remember: Only the combination of an active medium and a resonator, which together form a laser, produces light with these remarkable properties.

***p-n junction*** The above discussion of the principle of action of a laser is sufficiently general to apply to any solid-state laser. But in fiber-optic communications technology we use only laser *diodes,* that is, semiconductor devices. So let's now consider laser-diode action from the standpoint of a *p-n* junction.

We know by now that electrons and holes are injected by forward current into an active area, where they recombine, and that each recombination results in the radiation of a photon. (See Figure 9.3.) We know, too, that charges are carried out by the current, thus sustaining this dynamic process. In laser diodes, only heterojunctions are used. (See Figure 9.5[c].) A heterostructure confines the electron-hole's area of interaction (the active area) and serves as a waveguide for light. You'll no doubt recall this explanation from our discussion of the principle of action of LEDs in Section 9.1. What is new for a laser diode, however, is that its active area is much smaller. Its small size (thickness) results in a much higher current density and, thus, a much more intensive recombination process. A huge number of electrons injected into a small area leads to population inversion, to stimulated emission and, when gain exceeds loss, to laser action—the generation of monochromatic, coherent, powerful light.

The main point to remember is this: An active region of a laser diode is very small, ranging from a few micrometers to a few nanometers. This fact requires very high precision in the fabrication of a laser diode and accounts for its high price in comparison with that of LEDs.

***Basic structures and types of laser diodes*** The basic construction of a laser diode is shown in Figure 9.12(a). If it looks similar to the edge-emitting LED shown in Figure 9.5(b), it in fact is, except for two major differences: First, the thickness of an active region in a laser diode is very small, typically on the order of 0.1 μm. Second, a laser diode's two end surfaces are cleaved to make them work as mirrors. Since the refractive index of GaAs—the material making up the active region—is about 3.6, more than 30% of incident light will be reflected back into the active region at the GaAs–air interface. Thus, no special mirrors are required and these surfaces, called *laser facets,* provide positive feedback. This basic type of laser diode is called a *broad-area* LD.

To confine charge carriers—electrons and holes—even more securely within the laser diode's small active region, a strip contact is used. (See Figure 9.12[b].) This construction restricts the current flow within this narrow region. Since current flow produces gain in an active region, this type of laser diode is called *gain-guided.* A means to even further circumscribe the active region is to surround it with a material having a lower refractive index. Such an LD is called *index-guided.* Its structure is very similar to the core-cladding arrangement in an optical fiber. These surrounding layers are called cladding layers and the term *sandwich* is usually used to describe this structure. The most popular construction of an index-guided LD, one where the cladding layer's thickness varies, is known as a *ridge waveguide, RWG* (Figure 9.12[c]). In an index-guided laser diode, the small active region is buried between several layers having a lower refractive index. Such a structure is called a buried heterostructure (BH).

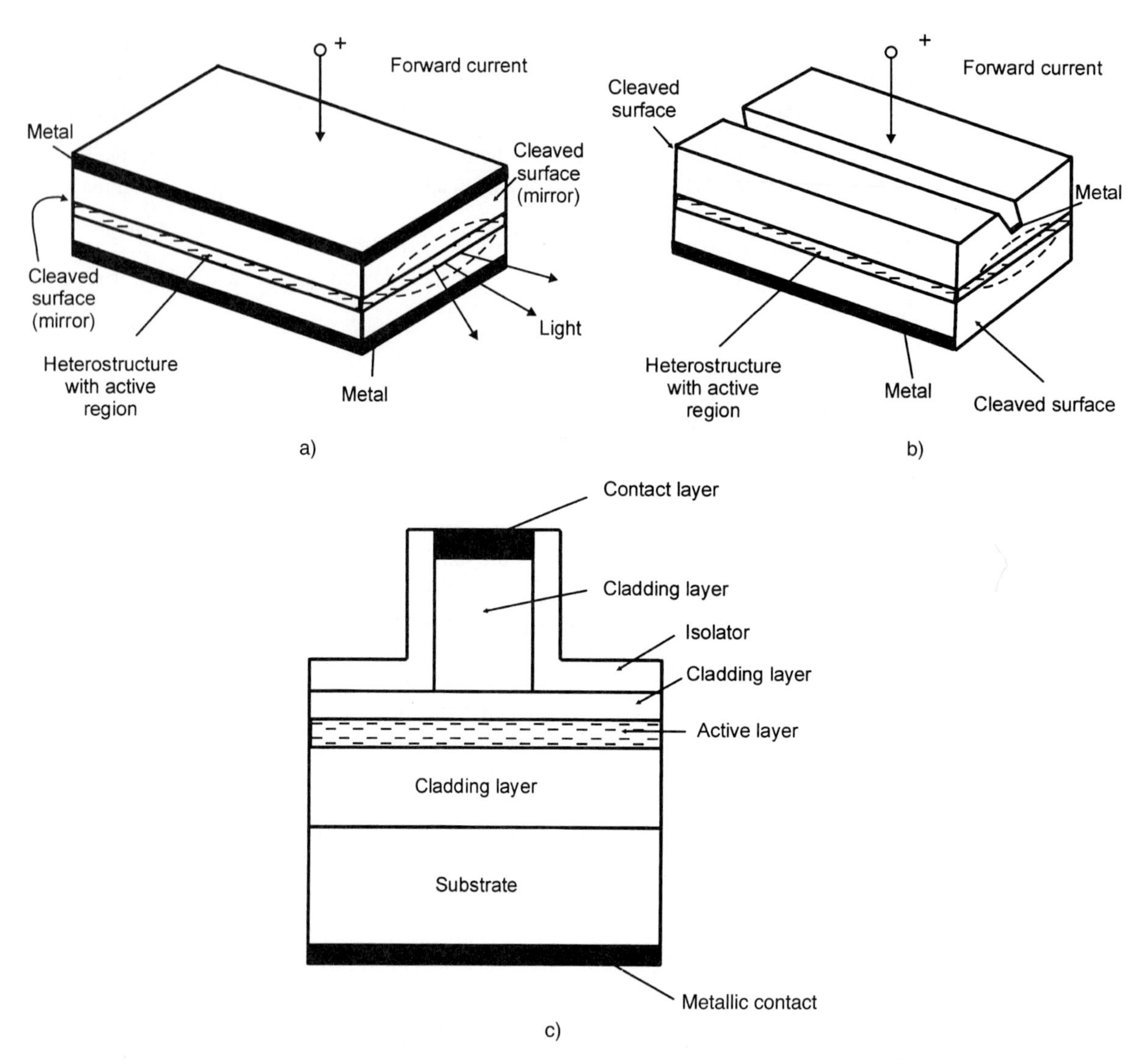


**Figure 9.12** Constructions of a laser diode (LD): (a) Basic structure of a broad-area LD; (b) gain-guided LD; (c) ridge-waveguide, RWG, laser diode.

***Quantum-well laser diodes*** To make lasing action more efficient, a special fabrication technique is used to form an especially thin active region, one on the order of 4 to 20 nm of thickness. Such devices are called *quantum-well (QW) laser diodes.* The quantum-well technique modifies the density of energy levels available for electrons and holes. The result is a much larger optical gain. From the *p-n* junction standpoint, a quantum-well diode is characterized by the lower potential energy of its electrons and holes, thus making their recombination easier. In other words, less forward current is required to reach and sustain lasing action in this type of laser diode. The main advantages of a quantum-well laser diode are more efficient current-to-light conversion, better confinement of the output beam, and the potential to radiate a variety of wavelengths.

From a practical standpoint, these advantages of the quantum-well structure dramatically reduce the threshold current and increase the possibility of changing radiating wavelengths by varying the thickness of an active layer. Figure 9.13 illustrates the concept of quantum-well structures.

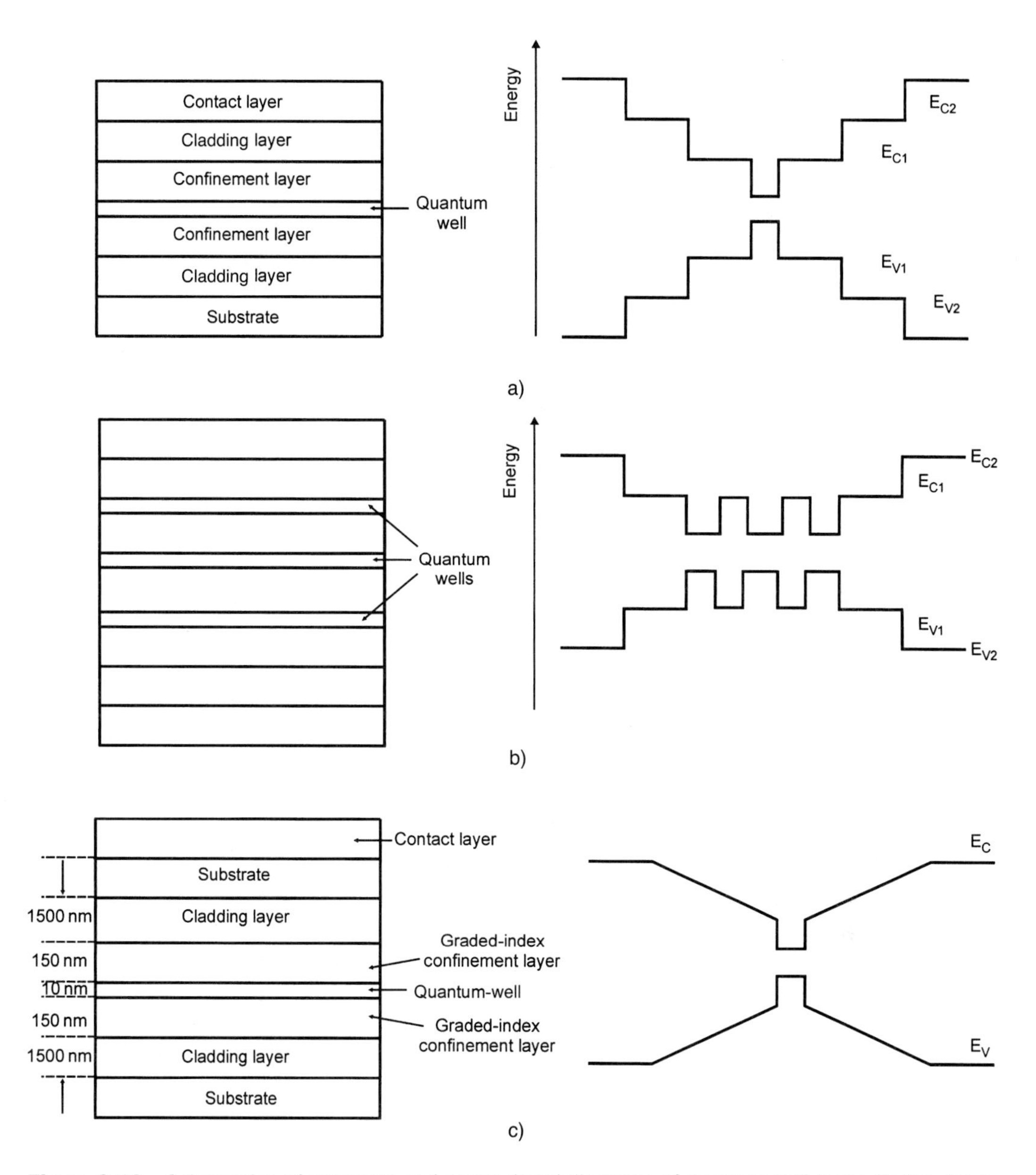


**Figure 9.13** Schematics of structure and energy-band diagrams of quantum-well laser diodes: (a) A single-quantum-well, SQW, laser diode; (b) a multiple-quantum-well, MQW, laser diode; (c) a graded-index separate-confinement heterostructure, GRINSCH. *(Shun Lien Chuang,* Physics of Optoelectronics Devices, *Copyright 1995. Reprinted with permission of John Wiley & Sons, Inc.)*

Quantum-well laser diodes are available with the following structures: single quantum well (SQW), multiple quantum well (MQW), and graded-index separate-confinement heterostructure (GRINSCH). An MQW laser provides powerful radiation (up to 100 mW).

Recent advances include *strained quantum-well* active media. By introducing a controlled strain of an active layer, a designer can control the quantum-well width and the potential barrier height. This results in the possibility of engineering the properties of a laser diode: controlling its wavelength, reducing its threshold current, and increasing laser efficiency. Two types of strain—compressive and tensile—are used to achieve this goal.

***Fabry-Perot laser diodes*** Look at Figure 9.10(b) one more time. The two mirrors and the active medium between them form a laser. This arrangement is shown again in Figure 9.14(a). Remember, we need mirrors to provide positive feedback, that is, the return of stimulated photons to an active medium to stimulate more photons. The two mirrors themselves form a resonator with length $L$.

Let an arbitrary wave travel from the left-hand mirror to the right-hand one, as Figure 9.14(b) shows. At the right-hand mirror, this wave is reflected; hence, the wave experiences a 180° phase shift. As you can see in Figure 9.14(b), the wave should have a break in its phase, which is impossible here. In other words, this resonator does not support this wave. Now let another wave, as shown in Figure 9.14(c), travel inside a resonator. At the right-hand mirror, the wave experiences a 180° phase shift and continues to propagate. At the left-hand mirror, this wave again has the same phase shift and continues to travel. Thus, the second wave shown in Figure 9.14(c) yields a stable pattern called a *standing wave.*

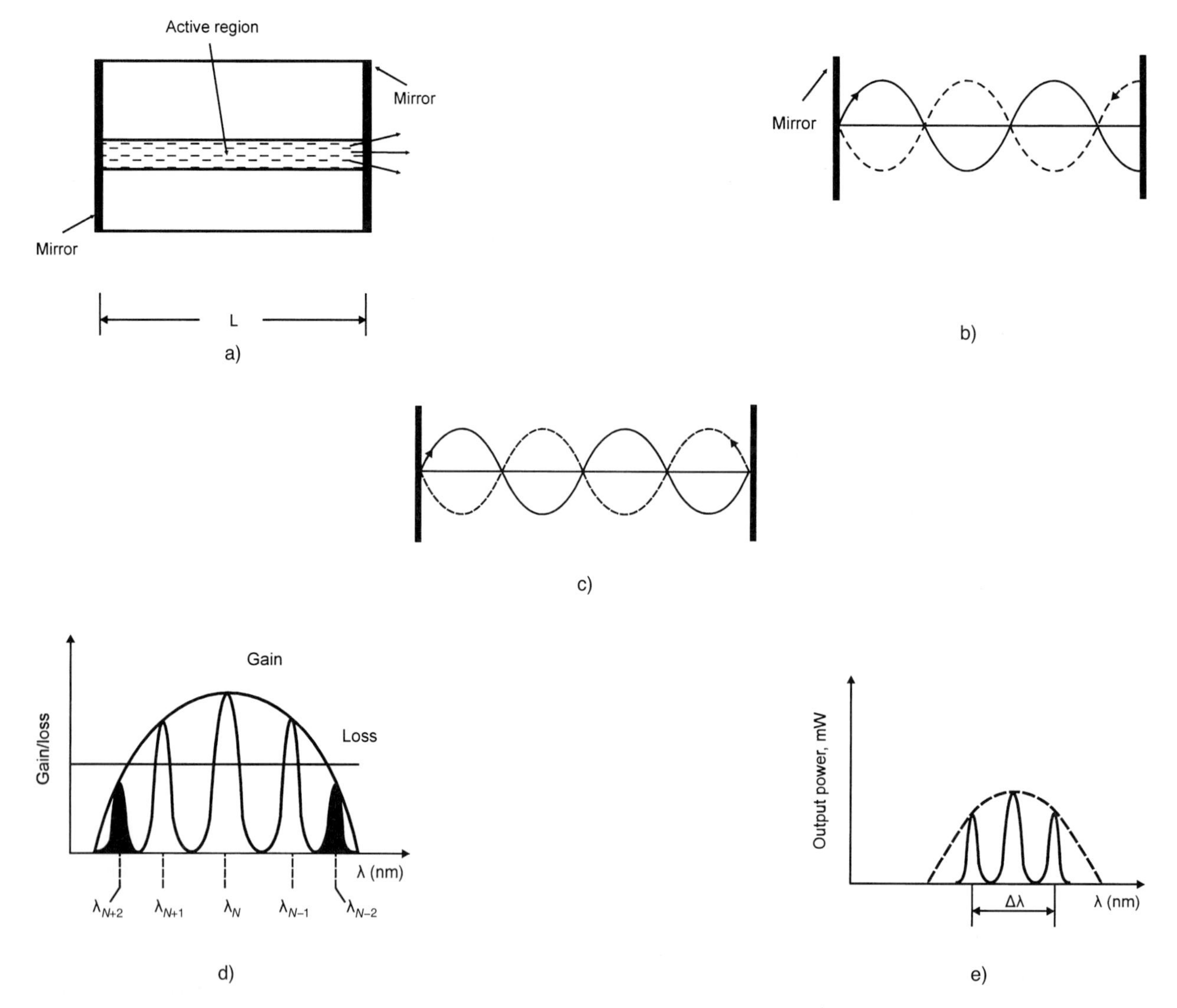


**Figure 9.14** Fabry-Perot laser diode: (a) A laser diode; (b) Fabry-Perot resonator with arbitrary wave; (c) Fabry-Perot resonator with standing wave; (d) gain-loss curve and possible longitudinal modes; (e) actual multimode radiation.

The only difference between the two waves shown in Figures 9.13(b) and 9.13(c) is their wavelengths. Thus, a resonator can support only a wave with a certain wavelength, the wave that forms a standing-wave pattern. This physical requirement can be written as

$$2L/\lambda = N, \tag{9.13}$$

where $L$ is the distance between mirrors and $N$ is an integer. For example, if $L = 0.4$ mm = 400 μm and $\lambda \approx 1300$ nm = 1.3 μm, then $N = 615$.

What is important to discern here is that this resonator supports a wavelength where $2L/N$ = 1300.8 nm. But this resonator also supports wavelengths equal to $2L/(N \pm 1)$, $2L/(N \pm 2)$, $2L/(N \pm 3)$, and so forth. In other words, many wavelengths that satisfy Equation 9.13 may exist. Wavelengths selected by a resonator are called *longitudinal modes.* When the length of a resonator increases or decreases, the laser switches from one longitudinal mode to another. This is called *mode hop.*

How many longitudinal modes can a laser diode radiate? A resonator can support an infinite number of waves whose wavelengths satisfy Equation 9.13. However, the active medium provides gain within only a small range of wavelengths. (Remember the energy-gap requirement: $\lambda < hc/E_g$.) Since a laser is formed by a resonator and an active medium and since radiation is the result of their interaction, *only several resonant wavelengths that fall within the gain curve might be radiated.* This is shown in Figure 9.14(d). Of course, light generation starts only when gain exceeds loss. Thus, eventually only those resonant wavelengths that are within the gain-over-loss curve will actually be radiated. Compare Figures 9.14(d) and 9.14(e). Waves with $\lambda_N$, $\lambda_{N\pm1}$, and $\lambda_{N\pm2}$ might be radiated, but only waves with $\lambda_N$ and $\lambda_{N\pm1}$ will be the actual laser output. Modes $\lambda_{N\pm2}$, depicted in black, are not generated.

To make this explanation more specific, let's introduce *spacing between two adjacent longitudinal modes,* $\lambda_N - \lambda_{N+1}$. Indeed, from Formula 9.13 we can obtain

$$\lambda_N - \lambda_{N+1} \approx 2L/N^2 = \lambda^2/2L \tag{9.14}$$

Thus, for a resonator whose $L = 0.4$ mm and that works at $\lambda$ around 1300 nm, we can compute $\lambda_N - \lambda_{N+1} \approx 2.1$ nm. Assuming the line width of a gain curve is equal to 7 nm, we find that this active medium can support three longitudinal modes.

The last matter to consider here is the actual spectral width of a laser diode. Is it the spectral width of the gain curve or the spectral width of an individual longitudinal mode? Neither. From the standpoint of a receiver, this is the width at half the maximum amount of light actually radiated. Thus, the spectral width is measured between two outermost longitudinal modes at half the maximum output power, as Figure 9.14(e) shows. The point is this: *The more longitudinal modes a laser generates, the wider its spectral width.* With all these modes, the typical spectral width of a laser diode is about 1 to 2 nm. (Remember, the spectral width of an LED is between 60 and 170 nm.)

The type of laser we are considering in our discussion is the *Fabry-Perot* laser, named after the French scientists Charles Fabry and Alfred Perot. This device provides lumped feedback, which results in many longitudinal modes and, eventually, in a relatively large spectral width.

***Distributed-feedback (DFB) laser diodes*** To reduce the spectral width, we need to make a laser diode merely radiate only one longitudinal mode. This has been done with *distributed-feedback (DFB)* laser diodes, whose principal arrangement is shown in Figure 9.15(a)

A DFB laser diode has the Bragg grating incorporated into its heterostructure in the vicinity of an active region. The Bragg grating (see Section 6.3) works like a mirror, selectively reflecting

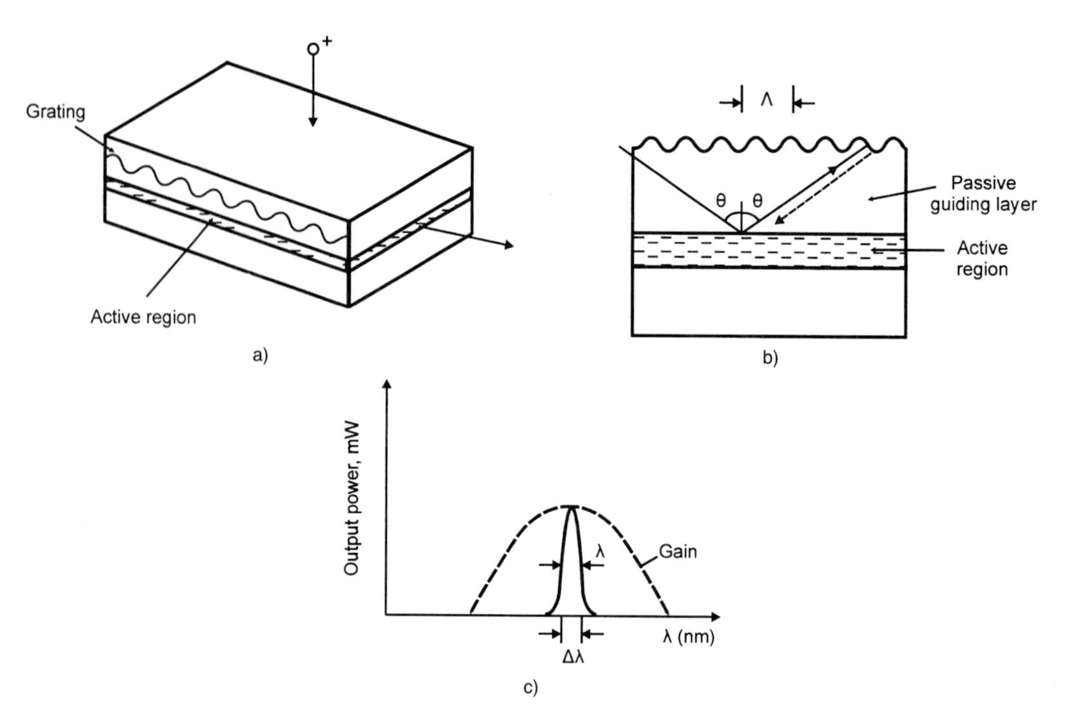


**Figure 9.15** Distributed feedback, DFB, laser diode: (a) A DFB laser diode; (b) how distributed feedback works; (c) actual singlemode radiation.

only one wavelength, $\lambda_B$. This wavelength can be found from the Bragg condition given in Formula 6.10 and repeated here:

$$2\,\Lambda\, n_{\text{eff}} = \lambda_B\,, \tag{9.15}$$

where $\Lambda$ is the period of grating, $n_{\text{eff}} = n \sin\theta$, and $n$ is the refractive index of the medium. All these quantities, including angle $\theta$, are illustrated in Figure 9.15(b).

As for the meaning of the term *distributed-feedback laser diode,* the word *feedback* emphasizes that we have the means to return stimulated photons to an active medium. This is done by reflecting a portion of the light at each slope of the grating, as Figure 9.15(b) shows for one beam. All the portions reflected at each slope of this corrugated structure then combine so that most of the light will be reflected back, provided of course that the condition shown in Formula 9.15 is maintained. The word *distributed* implies that reflection occurs not at a single point—a mirror, say, as in the Fabry-Perot laser—but at many points dispersed along the active region.

The net result of this arrangement is that the DFB laser radiates only one wave, with the wavelength equaling $\lambda_B$. Thus, its radiation contains only a single longitudinal mode and, as a result, the *spectral width* of this radiation is extremely narrow. (See Figure 9.15[c].) Actually, we distinguish between the spectral width of the entire output of light—Figure 9.14(e)—and the *linewidth* of each mode composing this light—Figure 9.15(c). In a singlemode operation, these two widths coincide.

DFB laser diodes were proposed in the early 1960s but were not developed commercially until the 1980s, when they were employed in long-distance fiber-optic communications systems. The newest developments allow manufacturers to fabricate DFB laser diodes radiating up to 30 mW at 1550 nm.

A modification of the idea of using a Bragg grating as a reflector is known as a *distributed-Bragg-reflector (DBR)* laser diode. In this laser, an active medium is placed between two Bragg gratings, which work as a reflector, thereby giving the laser its name.

***Vertical-cavity surface-emitting lasers—VCSEL*** All the laser diodes we have discussed so far are edge-emitting devices, as you can see from looking at Figures 9.14 and 9.15. They are characterized by the significant length of their active medium (on the order of hundreds of micrometers) and their asymmetrical radiation pattern, which is very similar to that of ELEDs. (See Figure 9.6[b].) Incidentally, the radiation patterns of these lasers usually produce an output beam cone on the order of $10° \times 30°$, which is much less than that of LEDs but is still asymmetrical. Special measures, such as the use of DFB technology, must be taken to ensure that such lasers generate only one longitudinal mode (i.e., singlemode operation).

Recent developments, however, have led to the fabrication of a new type of laser diode: a vertical-cavity surface-emitting laser (VCSEL) [6]. Since the space within a resonator is called the cavity, the words *vertical cavity* mean that the structure providing laser feedback is arranged in the vertical direction. The words *surface emitting* mean, in this context, that the laser's beam is emitted perpendicular to the wafer. (To recall the meanings of the terms *edge emitting* and *surface emitting,* see Figure 9.5.)

A basic arrangement of a VCSEL is shown in Figure 9.16(a). A semiconductor heterostructure (not shown) forms an active region. Several quantum wells are made within this active region to enhance light gain. This region is placed between Bragg reflectors—the stacks of layers with alternate high and low refractive-index material. (See Figure 9.16[b].) Each of these layers is $\lambda/4$ thick and is made from GaAs ($n = 3.6$) and AlAs ($n = 2.9$). These layers work like highly reflective mirrors, providing positive feedback.

Several significant advantages of VCSEL diodes make them among the hottest areas of activity in transmitter technology today:

- The size of the resonant cavity is very small, on the order of 2 μm. This results in huge *spacing between two adjacent longitudinal modes,* $\lambda_N - \lambda_{N+1}$. Indeed, using Formula 9.14 for $\lambda = 850$ nm $= 0.85$ μm and $L = 2$ μm, we compute $\lambda_N - \lambda_{N+1} = 0.07225$ μm $= 72.25$ nm. The spectral width of a gain curve is only a few nanometers; therefore, not more than one mode can be within the gain curve. Thus, *a VCSEL diode operates in a singlemode regime.* This is shown in Figure 9.16(c), where the numbers give you an idea of the order of magnitude of the spacing. The sketch is given without scale. The mode shown in black is not generated. (*Note:* Strictly speaking, Formula 9.14 is true only for Fabry-Perot lasers and should not be applied to VCSELs. However, it gives the order of magnitude correctly because, in reality, mode spacing in a VCSEL is about 100 nm.)
- VCSEL diodes have very small dimensions: A typical resonant cavity and diameter of the active region are about 1 to 5 μm, and the thickness of the active layer is about 25 nm = 0.025 μm. This allows manufacturers to fabricate many diodes on one substrate, thereby making *one-dimensional and two-dimensional (matrix) arrays* of diodes, precisely the constructions we need in multichannel systems.
- The small size of a VCSEL's resonant cavity leads to a concomitant key advantage: low power consumption and high switching speed. Thus, a VCSEL can radiate 3 mW output power at 10 mA forward current and it has an intrinsic modulation bandwidth up to 200 GHz. The first advantage stems from the fact that high current density is reached at low current value because of the small active area. (And don't forget the high quantum efficiency of this device.) The second advantage results from the short distance that electrons and holes have to travel within the active region before they recombine and the short

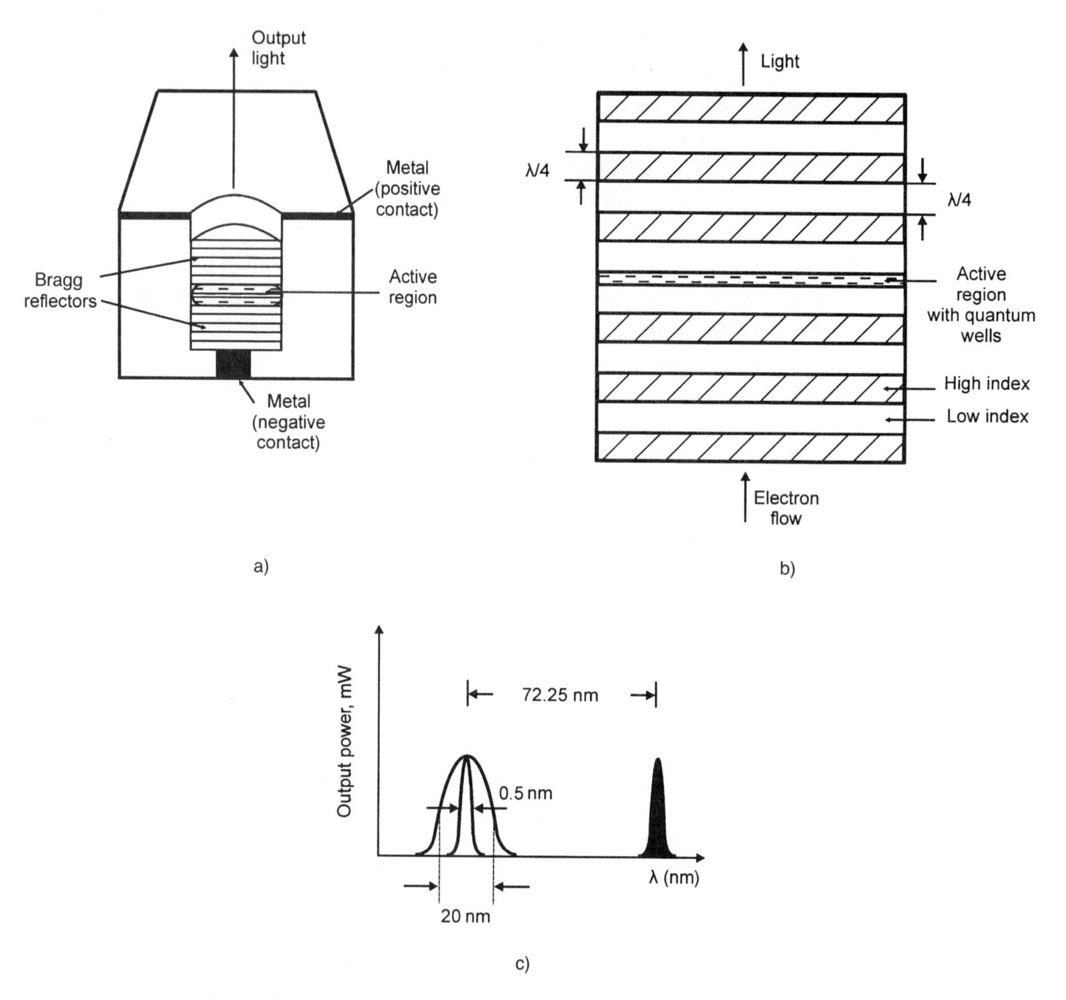


**Figure 9.16** Vertical-cavity surface-emitting laser, VCSEL: (a) Basic arrangement; (b) principle of operation; (c) gain curve and resonant modes (not to scale). *([b] from Christopher Davis,* Lasers and Electrooptics—Fundamentals and Engineering. *Copyright 1996. Adapted with the permission of Cambridge University Press.)*

distance a radiated photon has to travel before it escapes from the laser. This spawns a short lifetime, leading to the high modulation bandwidth of this device.

- A VCSEL diode radiates a circular output beam in contrast to that radiated by edge-emitting lasers.
- The fabrication technology for VCSELs is very similar to that for electronic chips, a fact that gives them the enormous range of advantages that chips have. VCSELs therefore manifest a useful new twist in the marriage of optics and electronics.

The one major drawback of VCSEL lasers, at least to date, is that they are commercially available in the wavelength range no longer than 850 nm; that is, they work only in the first transparent window of an optical fiber. Within this wavelength range, however, they have already found extensive applications in high-bit-rate LANs such as Gigabit Ethernet. The trend in VCSEL development is to extend their operating wavelengths to 1300 nm and 1550 nm. Research attests to this possibility and it is safe to predict that such devices will be commercially available. (Note:

As this book went to press, a VCSEL diode operating at 1300 nm was reported and is expected to be on the market soon [7].)

**Superluminescent Diodes (SLDs)**

A superluminescent diode is an optical source whose properties are intermediate between those of an LED and those of an LD. The device uses a double heterostructure to confine the active region under conditions of high current density. Hence, population inversion is created, so this source is able to amplify light. But an SLD does not have positive feedback; therefore, it radiates spontaneous emission. Thus, an SLD radiates a more powerful and more sharply confined beam than a regular LED, but an SLD's radiation is not as monochromatic, well directed, and coherent as a laser diode's radiation. So far, SLDs have found very few applications in fiber-optic communications.

## 9.3 READING DATA SHEETS—THE CHARACTERISTICS OF LASER DIODES

In this section we'll read the data sheets of different types of commercially available laser diodes and discuss their basic characteristics.

**Broad-Area Laser Diodes**

A typical data sheet for a broad-area laser diode is shown in Figure 9.17.

***Cooled and uncooled laser diodes*** Beneath the manufacturer's designation (QLM3S81 Series), the type of device is shown: *Uncooled DIL Laser.* "Uncooled" means that this laser diode doesn't require any cooling. The importance of this feature follows from the fact that a laser diode radiates a lot of heat. This heat comes from nonradiative transitions of excited electrons and, because of population inversion, these transitions are numerous. This is why a laser-diode package usually includes a thermoelectric cooler. But in this case the manufacturer fabricated an uncooled laser, which means it consumes less overall power than its cooled-laser counterparts.

Cooled lasers include a *thermoelectric cooler (TEC),* which is a heat pump that transfers heat from one place to another. Its function is to keep the laser diode at operating temperature. The principle of its action can be explained as follows: When electrons leap from a lower energy band to a higher energy band, they absorb external energy; in the reverse transition, they release energy. We discussed this mechanism previously as applied to light radiation, but this phenomenon is also true for heat radiation and absorption. Thus, when material with the appropriate energy gap—remember, $E_p = hc/\lambda$—is used, one can make a heat pump. When the ambient temperature becomes high, the cooler pumps heat from the LD to the heat sink. On the other hand, when the temperature is low, the cooler pumps heat to the LD. (See [8].)

A cooled laser diode can radiate increased light power at the expense of electric-power consumption. The reliability of a cooled diode decreases simply because extra components are involved in its operation. Obviously, the less reliable a laser diode is, the higher the level of maintenance required. An uncooled laser, on the other hand, is able to radiate less light power but it is more reliable than its cooled counterpart. Thus, each type of laser diode finds its specific areas of application. For example, cooled LDs are used in central-office (the local telephone switching center) equipment, while an uncooled LD is preferable for loop installation, that is, for lines connecting the central office to subscribers.

# QLM3S81 Series

## Uncooled DIL Lasers

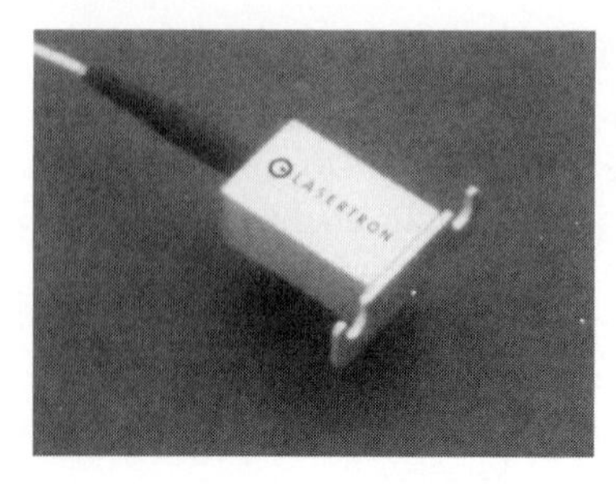


- *Telecom access, junction and trunk applications*
- *Rated output powers from 130 μW to 1 mW*
- *High-reliability ridge waveguide laser chip*
- *1300 nm wavelength*
- *Industry-standard package, single-mode fiber*
- *No cooler or temperature control required*

### Description

The QLM3S81-series laser modules are low-power, uncooled, 1300 nm GaInAsP devices packaged in industry-standard 14-pin dual-in-line (DIL) flanged packages. The single-mode fiber pigtail has a 0.9 mm Hytrel coating. Pigtail is available terminated with optional FC/PC, Biconic or ST connector. The QLM3S81-series is characterized over the temperature range 0 to 65°C. Four standard models address different performance requirements.

The QLM3S81-series laser modules are manufactured using proven production techniques and materials, including laser-welded fiber attachment, and a hermetically sealed, strain-relief fiber assembly packaging. The conventional and proven ridge waveguide (RWG) laser chip was chosen to deliver aggressive performance and reliability specifications in a coolerless design.

### Specifications *Contact Lasertron regarding special requirements.*

($T_{laser}$ = $T_{case}$ = 0 to 65°C, except as noted)

| ***Absolute Maximum Ratings*** | |
|---|---|
| **Laser Element (T=25°C)** | |
| Fiber-coupled optical power (mW) | 1 |
| Forward laser current (mA) | 150 |
| Reverse laser voltage (V) | 2 |
| Reverse laser current (mA) | 2 |
| **Monitor Detector (T=25°C)** | |
| Reverse photodiode photocurrent (mA) | 1 |
| Reverse photodiode bias voltage (V) | 15 |
| Forward photodiode current (mA) | 2 |
| **Package** | |
| Storage temperature (°C) | -40 to 70 |
| Operating temperature (°C) | 0 to 65 |
| Lead soldering temperature (°C) | 260 |
| Lead soldering duration (sec) | 10 |
| Fiber yield strength (N) | 10 |
| Fiber bend radius (mm, min.) | 30 |

| | QLM3S811 | QLM3S810 | QLM3S812 | QLM3S813 |
|---|---|---|---|---|
| **Rated Output Power, P (μW)** | 130 | 250 | 500 | 1000 |
| **Laser Spectral Characteristics** | | | | |
| Wavelength (nm, mean, 25°C) | 1283 to 1320 | 1283 to 1320 | 1283 to 1320 | 1280 to 1320 |
| Wavelength (nm, mean, 0 to 65°C) | 1270 to 1340 | 1270 to 1340 | 1270 to 1340 | 1267 to 1340 |
| Spectral width (nm, RMS) | ≤3 | ≤3 | ≤3 | ≤3 |
| Spectral width (nm, FWHM) | ≤6 | ≤6 | ≤6 | ≤6 |
| Spectral shift (nm/°C) | ≤0.5 | ≤0.5 | ≤0.5 | ≤0.5 |
| **Laser Drive Characteristics** | | | | |
| Threshold I (mA at 25°C) | ≤50 | ≤40 | ≤50 | ≤40 |
| Threshold I (mA at 0 to 65°C) | ≤100 | ≤90 | ≤100 | ≤90 |
| Threshold change with T (%/°C) | ≤2.5 | ≤2 | ≤2.5 | ≤2.5 |
| Output power at threshold (μW) | ≤8 | ≤8 | ≤25 | ≤50 |
| Modulation current (at P, mA at 25°C) | 10 to 40 | 10 to 40 | 10 to 40 | 10 to 40 |
| Slope efficiency (P/I, μW/mA at 25°C) | 3.25 to 13 | 6.25 to 25 | 12.5 to 50 | 25 to 100 |
| Slope efficiency change: (+dB, 25 to 0°C) | ≤1.5 | ≤1 | ≤1.5 | ≤1.5 |
| (-dB, 25 to 65°C) | ≤3 | ≤2.5 | ≤3 | ≤2.5 |
| Slope efficiency change (-%/°C, 25 to 65°C) | ≤1.7 | ≤1.4 | ≤1.7 | ≤1.4 |
| Forward voltage at P (V) | ≤1.5 | ≤1.5 | ≤1.5 | ≤1.5 |
| Series resistance (Ohms) | ≤8 | ≤8 | ≤8 | ≤8 |
| Optical rise/falltime (nsec, 10-90%) | ≤1 | ≤1 | ≤1 | ≤1 |
| Analog bandwidth (MHz, -3 dB response) | ≥800 | ≥800 | ≥800 | ≥800 |
| **Monitor Photodiode Characteristics** | | | | |
| Monitor detector responsivity (μA/μW) | 0.3 to 5 | 1 to 4 | 0.3 to 5 | 0.1 to 0.6 |
| Monitor detector dark current (μA at -5 V) | ≤0.25 | ≤0.25 | ≤0.25 | ≤0.15 |
| Tracking error (dB, relative to 25°C) | ≤1 | ≤0.6 | ≤1 | ≤1 |

### Pin Connections

1 NC
2 NC
3 NC
4 NC
5 Case Ground
6 NC
7 Monitor Cathode (+)
8 Monitor Anode (-)
9 Laser Cathode (-)
10 Laser Anode (+), Case Ground
11 NC
12 NC
13 NC
14 NC

(Bottom View)
1 7
1 meter fiber
14 8

a)

**Figure 9.17** Data sheet of a broad-area laser diode. *(Courtesy of Lasertron, Inc., Bedford, Mass.)*

QLM3S810-002: TYPICAL CHARACTERISTICS

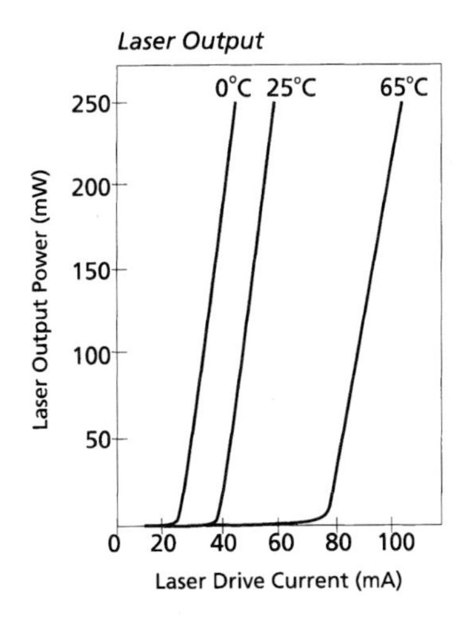


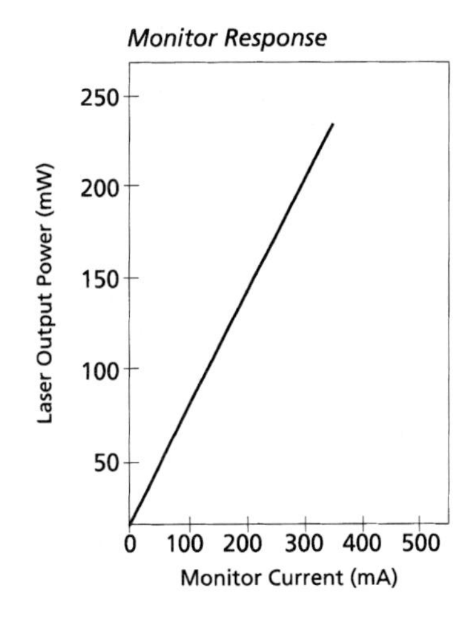


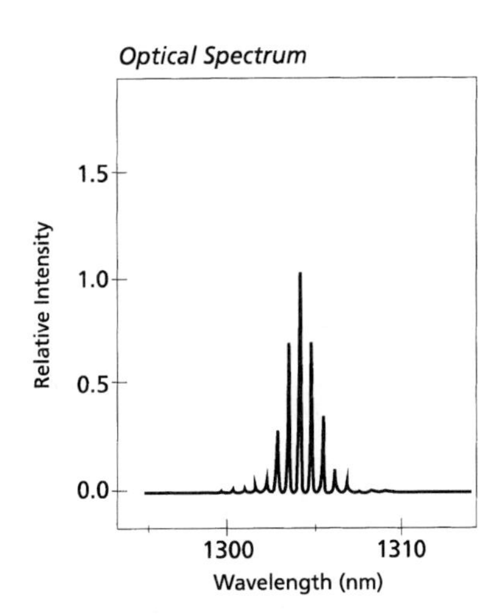


## Ordering Information

| Base Model | Suffix: No Connector | FC/PC | Biconic | ST |
|---|---|---|---|---|
| QLM3S811 (130 μW version) | -002 | -050 | -051 | -052 |
| QLM3S810 (250 μW version) | -002 | -050 | -051 | -052 |
| QLM3S812 (500 μW version) | -002 | -050 | -051 | -052 |
| QLM3S813 (1 mW version) | -002 | -050 | -051 | -052 |

b)

**Figure 9.17** (continued)

## 14-pin DIL "Longhorn" (Lasers, LEDs)

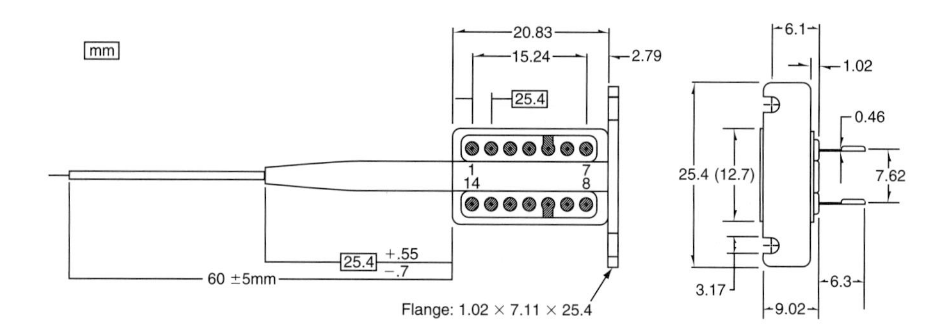


### Pin Connections

1 Cooler*
2 NC
3 NC
4 NC
5 Case Ground
6 NC
7 Monitor Cathode (Lasers only)
8 Monitor Anode (Lasers only)
9 Source Cathode
10 Source Anode, Case Ground
11 Thermistor*
12 Thermistor*
13 NC
14 Cooler*
* *NC for uncooled modules*

### Schematic Diagram

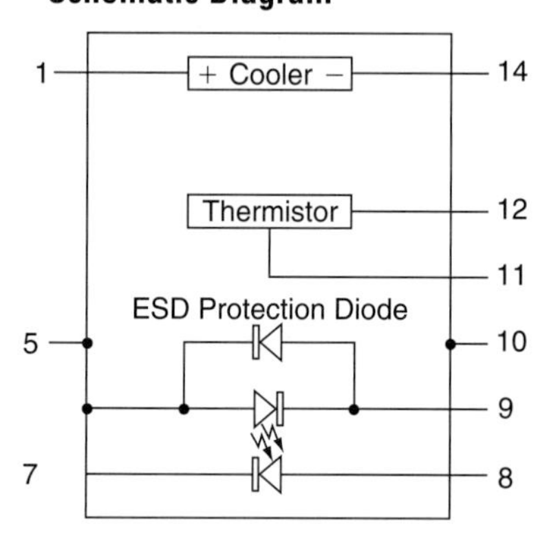


### Usage Notes:

1) The laser or LED is operated with a forward bias voltage (pin 9 at negative voltage relative to pins 5, 10 and Case Ground).
2) The monitor is operated with a reverse bias voltage (pin 8 at negative voltage relative to pin 7).
3) The cooler may be operated in a cooling mode or heating mode to maintain a fixed internal (submount) temperature. ***Cooling Mode:*** To maintain a submount temperature (Ts) that is lower than the case temperature (Tc), the cooler is operated with forward bias voltage (pin 1 at positive voltage relative to pin 14). ***Heating Mode:*** To maintain a submount temperature (Ts) that is higher than the module case temperature (Tc), the cooler is operated with reverse bias voltage (pin 1 at negative voltage relative to pin 14).
4) ***Lead Cutting:*** The improper cutting of module leads can result in sufficient physical shock resulting in fiber shift and misalignment of the power coupling. In addition, impact to the module with any metal object can deliver shock of sufficient force as to negatively effect the power coupling. Lasertron recommends that only commercially available anti-shock lead shearers be employed for the cutting of module leads and that due diligence and caution be used when handling and installing these devices. Lasertron recommends using an anti-shock lead shearer to cut leads. The Utica Swiss Model #630E or equivalent will produce flush cuts to leads with minimal shock to the module.

c)

**Figure 9.17** (continued)

***Applications and features*** Observe in Figure 9.17 that the list of applications and features is placed at the beginning of the data sheet. This is to draw the attention of potential customers to the basic properties of this laser diode. Let's review them:

- "Telecom access, junction and trunk applications" indicates the areas where this LD can be used. Applications include local loops, where this LD provides the access and junction of subscribers to the backbone network. This LD can also serve as a light source for trunks—that is, point-to-point links between, say, a remote terminal and a central office in telephone networks.
- "Rated output powers from 130 μW to 1 mW" has two meanings: First, *rated output power,* $P_R$, *is the maximum recommended power.* A laser diode performs within given specifications if it operates up to rated output power. The second meaning is obvious: A customer can choose an LD from this series with a wide range of output power.
- "High reliability ridge waveguide laser chip" underscores three points: First, this is a chip-like fabricated device. Second, it is an index-guided or, more specifically, a ridge waveguide (RWG) laser diode. (See Figure 9.12[c].) Third, the reliability of a laser diode is measured in mean time to failure (MTTF), the time when 50% of the original samples have failed. This manufacturer considers its diodes to have failed when the threshold current has increased by 30 mA, or 50% of its original value.
- By "1300 nm wavelength" the manufacturer means the peak wavelength radiated by the diodes at the normal condition.
- "Industry-standard package, single-mode fiber" assures customers that the 14-pin *dual-in-line package (DIP)* used for this LD adheres to industry-accepted specifications. This LD is factory-coupled to a singlemode fiber using pigtail technology.
- "No cooler or temperature control required" is an outstanding feature of this LD series, as discussed above.

***Absolute maximum ratings*** This table gives the limits that the customer can expect to reach during assembly or operation, but only for a very short period of time. Long-term device exposure under these conditions, or even exceeding these limits, might result in laser failure.

***Specifications*** Here we are given the electrical and optical characteristics of the entire series of these laser diodes. This table consists of three sections: "Laser Spectral Characteristics," "Laser Drive Characteristics," and "Monitor Photodiode Characteristics." Let's consider them in sequence:

"Laser Spectral Characteristics" describe the wavelength, spectral width, and spectral shift. Pay particular attention to how the range of mean wavelengths changes with temperature. Temperature dependence is a common problem with laser diodes.

Another term requiring explanation is "*mean wavelength.*" This is the weighted average of all the spectral lines, or modes, radiated by a laser. (See Figure 9.14[e].) All modes whose power is at least 2% of peak mode power are included. The mean wavelength is defined by:

$$\lambda_{\text{mean}} = \left[\sum (\lambda_n \cdot P_n)\right] / \sum P_n, \tag{9.16}$$

where $\lambda_n$ is the $n$th wavelength and $P_n$ is the power at the $n$th wavelength.

"Spectral width" is given in two numbers: RMS, the root mean square value, and FWHM, the full-width, half-maximum value. Compare the FWHM spectral width of a laser diode with the same characteristic of an LED: An LD has less than 6 nm and an LED has 170 nm at the same

1300-nm operating wavelength. Study the "Optical Spectrum" picture in Figure 9.17. You can clearly see that the output spectrum consists of several lines (modes). A measure of the spectral width includes all these modes, as Figure 9.14(e) shows.

"Spectral shift" indicates how much—here, 0.5 nm/°C—the peak wavelength, $\lambda_p$, moves with temperature. Thus, if $\lambda_p = 1300$ nm at 25°C and the temperature changes to 65°C, then $\lambda_p$ becomes 1320 nm.

"Laser-drive characteristics" are often designated as electrical characteristics. When looking at these characteristics, always consider carefully the *threshold* current. This number informs you immediately of the type and quality of the laser diode. (Compare this threshold current—from 40 to 50 mA—with the threshold current specified in the DFB laser data sheet, which is interpreted on page 358.

There are four methods by which to obtain the threshold-current value from the *P-I* curve [9]. (Note that the power-current, *P-I,* graph is also called a light-current, *L-I,* graph.) They are shown in Figure 9.18. The most precise method is to use the second derivative, $d^2P/dI^2$.

Threshold current depends heavily on temperature. To evaluate this dependence, one must calculate the *threshold-current change with temperature.* To do so, the following formula is used:

$$I_{\text{th}1} = I_{\text{th}2} \exp(T_1 - T_2), \tag{9.17}$$

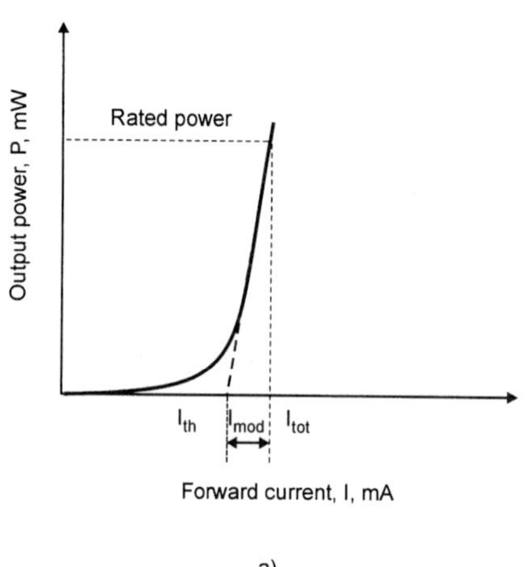


Rated power
Output power, P, mW
P_th
0
I_th
Forward current, I, mA
b)

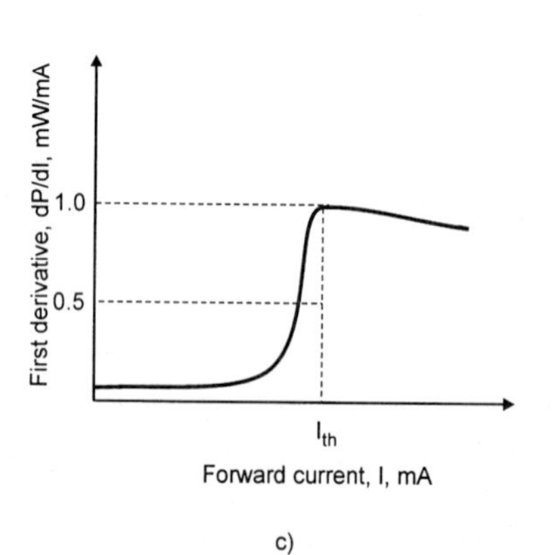


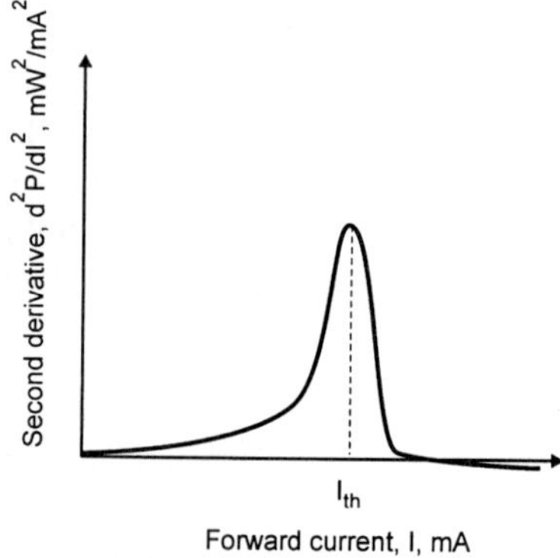


**Figure 9.18** Four methods to determine threshold current, $I_{th}$, from P-I curve. *(From Tyll Hurtsens' "Measuring Diode Laser Characteristics."* Lasers and Optronics, *February 1989. Reprinted with permission.)*

where $I_{th1}$ and $I_{th2}$ are threshold currents at temperatures $T_1$ and $T_2$, respectively. From Formula 9.17, one can easily derive the following relationship:

$$\textit{Threshold-current change with temperature } (\%/^{\circ}C) = [\ln(I_{th1}/I_{th2})/(T_1 - T_2)] \times 100 \quad (9.17a)$$

For the device under discussion, this number is not more than 2.5. You can also see this effect at different temperatures looking at the *P-I* graphs in Figure 9.17.

*Output power at threshold,* $P_{th}$, is the same as light power at threshold current and it is measured as shown in Figure 9.18(b).

"Slope efficiency," also called differential (quantum) efficiency, $S = \Delta P/\Delta I$ (μW/mA), is defined as:

$$S(\mu W/mA) = (P_R - P_{th})/(I_{tot} - I_{th}) \quad (9.18)$$

This characteristic shows how efficiently a laser diode converts electrical current into optical power. Compare these parameters for SLED and LD: From Figure 9.8(a), one can read $S_{SLED}$ = 75 μW/95 mA ≈ 0.79 μW/mA while $S_{LD}$ = 3.25 μW/mA minimum. In other words, this laser diode is at least four times more efficient than the SLED. If you take the best numbers for the LD, you find its efficiency is 100 times greater than that of the SLED. Keep in mind that, for the SLED, the data sheet gives coupled power but, for the LD, we get an output power; as a result, the efficiency advantage of a laser diode is, in reality, a little less than 100 times greater.

"Slope-efficiency change" with temperature can be measured in different ways. The basic temperature dependence is given by:

$$\textit{Slope-efficiency change with temperature } (\%/^{\circ}C) = [\ln(S_1/S_2)/(T_1 - T_2)] \times 100, \quad (9.19)$$

where $S_1$ and $S_2$ are slope efficiencies at temperatures $T_1$ and $T_2$, respectively. If you look at the *P-I* graphs in Figure 9.17, you'll see that $S = \Delta P/\Delta I$ is really different at 0°C, 25°C, and 65°C. The higher the temperature, the less efficient the laser diode. This is why the slope-efficiency change is a negative percentage.

Another means by which to calculate the slope-efficiency change with temperature is to use the classical dB definition:

$$\textit{Slope-efficiency change } (\text{dB}) = 10 \log(S_1/S_2) \quad (9.20)$$

Observe that the sheet gives these numbers from 25°C down to 0°C, where this change is positive, and from 25°C up to 65°C, where this change is negative.

Threshold current is much more sensitive to temperature change than is the slope efficiency. Look again at the graph "Laser Output" in Figure 9.17. If, for example, the driving current is 50 mA at 25°C, the laser diode radiates approximately 250 μW. If, however, temperature changes to 65°C, the laser diode will radiate nothing because it will require about 75 mA to start lasing. Meanwhile, for this range of temperature change, the slope efficiency changes very little.

Modulation current, $I_{mod}$, is given by:

$$I_{mod} = P_R/S, \quad (9.21)$$

where $P_R$ is the rated output power at 25°C. It is quite evident that Formula 9.21 defines the maximum $I_{mod}$. You can also find this parameter from the *P-I* graph in Figure 9.17.

"Forward voltage," *V,* is the voltage drop across the laser diode when the drive current supports the rated-output power. This voltage allows us to find the *series resistance* of a laser diode, $\Delta V/\Delta I$ (Ω).

"Optical rise/fall time," $t_r$ (ns), allows us to calculate the bandwidth for data transmission as we did for an LED in Formula 9.9. It is instructive to compare the laser diode's $t_r = 1$ ns and an LED's $t_r = 2.5$ ns. This is another advantage of a laser diode over a light-emitting diode. The manufacturer provides the customer with an analog bandwidth for possible application in cable television (community-antenna television, or CATV).

"Monitor photodiode characteristics" describe the following: A laser-diode package usually includes output-power stabilization circuitry. The heart of this circuitry is a photodiode, which detects light from a rear facet of the laser cavity and signals the feedback circuitry to increase or decrease the drive current to keep the output power stable. (The principle of operation and characteristics of a photodiode are discussed in Chapter 11. Here we need to remember only that a photodiode converts light into an electrical signal.)

"Monitor detector responsivity," $R$ (μA/μW), shows how efficiently a photodetector converts light into current. The graph in Figure 9.17 shows that photocurrent is proportional to light power; therefore, a single number—the slope of this line, $R$—is needed to describe this light–current relationship.

"Monitor-detector dark current" is the current flow through a photodiode when there is no light. This is a measure of a photodetector's accuracy or, more precisely, the measure of its offset.

"Tracking error" (dB) is defined as:

$$\textit{Tracking error}\ (\text{dB}) = 10 \log(\text{coupled power at } 0°\text{C or } 65°\text{C/coupled power at } 25\ °\text{C}) \quad (9.22)$$

The meaning of this parameter becomes clear from the fact that a monitor photodiode detects only the laser's output power; hence, the stabilization circuitry cannot control the laser power that is coupled into a fiber. In fact, in addition to changing the laser's output power with temperature, coupled power can also change because of variations in coupling conditions. Thus, tracking error is a measure of the stability of the light coupling from a laser module into a fiber under a range of operating temperatures. We refer to "tracking error" because coupled power measured from the fiber output differs from laser-diode power measured by the photodiode from a rear facet of the laser cavity.

*Note:* Some manufacturers introduce the ratio of the output (or coupled) power of a laser diode to the photocurrent of a monitoring photodiode. This measure is called the *tracking ratio* (mW/mA). This number enables you to run the device at the predetermined output power by setting the photocurrent to a specific value.

The package specifications are clear and easy to read and it's very instructive to do so. You will find here the mechanical dimensions, pin connections, and usage notes. Read and interpret the paragraph "Lead cutting" in the section entitled "Usage Notes." It gives you insight into the importance and sensitivity of laser-to-fiber coupling.

## Reading the Data Sheet of a DFB Laser Diode

Refer to Figure 9.19. In interpreting this data sheet, we will concentrate only on new terms and definitions not encountered during our discussion of a broad-area LD. In general, it would be very useful to compare this data sheet with one for a broad-area laser.

**Features** From the list of features, it is worth focusing on a new component: a built-in optical isolator. A laser diode, particularly a DFB laser diode, is very sensitive to backreflected light, that is, light reflected from a fiber back into the LD's active area. This light causes the deterioration of laser-diode performance. Among the most important of the adverse effects are *chirping* (a rapid change of the peak wavelength with time) and increasing *relative-intensity noise,* or *RIN* (the intensity fluctuations of a laser over time). These effects lead to deterioration in the quality of the fiber-link transmission, which is why a DFB LD usually includes an optical isolator. It prevents backreflected light from penetrating the laser's active region. The quality of an optical isolator is

## 2.Product Information
# 1.55μm DFB LD MODULE
### *SD3B905P*

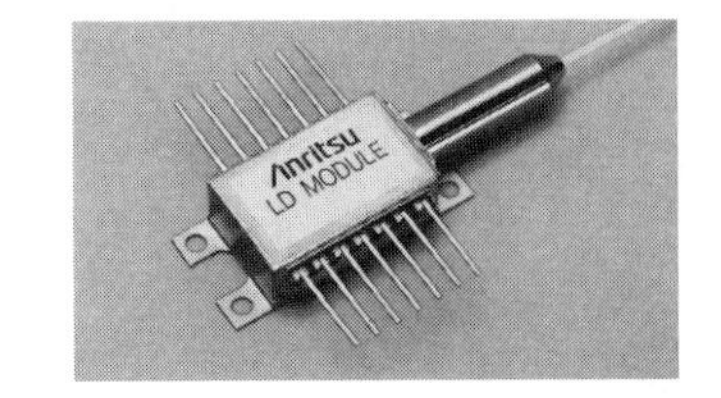


The SD3B905P is a 1.5μm-band InGaAsP/InP distributed feedback (DFB) laser diode developed as a light source for wavelength division multiplexing (WDM) optical communication. It is intended for use with an external modulator.

**FEATURES**

- Single longitudinal mode oscillation
- Accurate peak wavelength (±1nm)
- Built-in optical isolator (60dB)
- Built-in monitor PD
- Built-in thermo-electric cooler
- Polarization maintaining fiber

**APPLICATIONS**

- WDM light source
- Optical measurement
- Optical communication

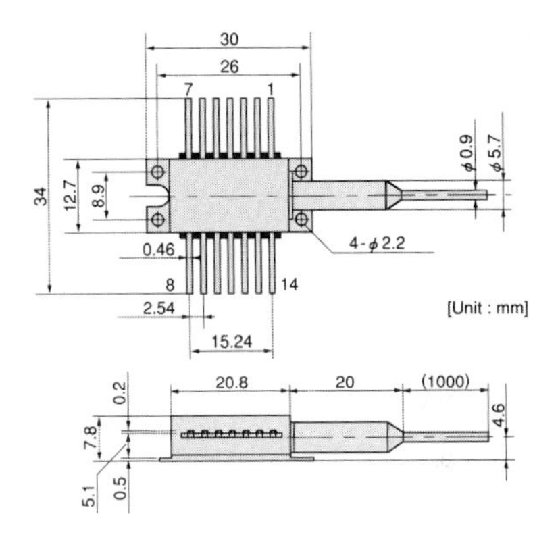


**Package outline (Unit:mm)**

**ABSOLUTE MAXIMUM RATINGS ($T_c$=25°C)**

| Item | Symbol | Rating | Unit |
|---|---|---|---|
| LD forward current | $I_F$ | 100 | mA |
| LD reverse voltage | $V_R$ | 2 | V |
| PD forward current | $I_{FD}$ | 10 | mA |
| PD reverse voltage | $V_{RD}$ | 15 | V |
| Operating case temperature | $T_c$ | -20~+65 | ℃ |
| Storage temperature | $T_{stg}$ | -20~+70 | ℃ |
| Cooler current | $I_c$ | 1.5 | A |
| Lead soldering condition | 250℃ x 10 sec. or less | | |

**OPTICAL and ELECTRICAL CHARACTERISTICS ($T_{LD}$=25°C, $T_c$=-20~65°C)**

| Item | Symbol | Test condition | Min. | Typ. | Max. | Unit |
|---|---|---|---|---|---|---|
| Forward voltage | $V_F$ | $P_f$=2.0mW | 0.9 | 1.2 | 1.4 | V |
| Threshold current | I | | | 25 | 40 | mA |
| Optical output power | Pf | $I_f=I_{th}$+40mA, CW | 2.0 | | | mW |
| Peak wavelength *3) | λp | $P_f$=2.0 mW | λ-1 | λ | λ+1 | nm |
| Side mode suppression ratio | SMSR | $P_f$=2.0 mW | 33 | 40 | | dB |
| Spectral linewidth | Δf | $P_f$=2.0 mW | | 10 | 40 | MHz |
| Tracking error | ΔP | $P_f$=2.0 mW, $T_c$=-20~65℃ | -0.5 | | +0.5 | dB |
| Monitor current | $I_m$ | $P_f$=2.0 mW, $V_{RD}$=5V | 50 | | | μA |
| PD dark current | $I_d$ | $V_{RD}$=5V | | | 1 | μA |
| Cooler voltage | $V_c$ | ΔT=40℃ *1) | | 1.2 | 1.5 | V |
| Cooler current | $I_c$ | ΔT=40℃ *1) | | 0.7 | 1.2 | A |
| Thermistor resistance | $R_{th}$ | $T_{LD}$=25℃ *2) | 9.5 | 10 | 10.5 | kΩ |

*1) $\Delta T = |T_c - T_{LD}|$ *2) B constant of thermistor B=3900±100K *3) peak wavelength λ=1545~1565nm (specified by 1 nm step)

a)

**Figure 9.19** Data sheet of a DFB laser. *(Courtesy of Anritsu Corporation, Richardson, Tex.)*

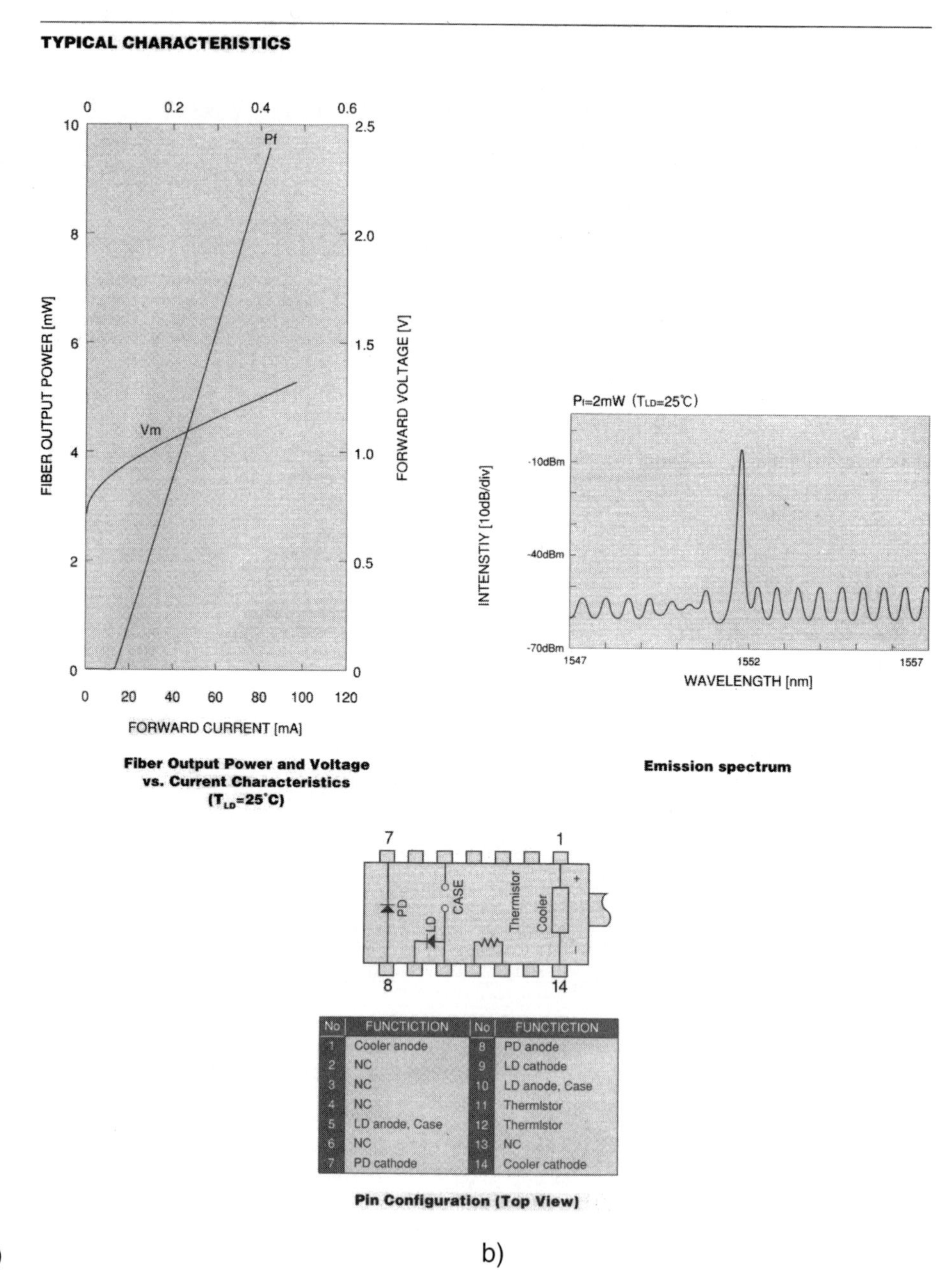


| No | FUNCTICTION | No | FUNCTICTION |
|---|---|---|---|
| 1 | Cooler anode | 8 | PD anode |
| 2 | NC | 9 | LD cathode |
| 3 | NC | 10 | LD anode, Case |
| 4 | NC | 11 | Thermistor |
| 5 | LD anode, Case | 12 | Thermistor |
| 6 | NC | 13 | NC |
| 7 | PD cathode | 14 | Cooler cathode |

**Figure 9.19** (continued)

measured as the ratio of light power falling at the isolator to light power penetrating the isolator (positive dB).

Another peculiarity of this laser is its ability to work with polarization-maintaining (PM) fiber. This implies that the laser radiates polarized light and that this polarization is preserved by the coupling mechanism.

***Applications*** A DFB laser's extra-narrow spectral width makes it the best light source for wavelength-division multiplexing (WDM) networks. As noted previously, WDM networks use several close wavelengths to transmit many signals over a single fiber. Thus, each wavelength has to have an extremely narrow spectral width to prevent interference (crosstalk) between them.

***Optical and Electrical Characteristics*** The typical *threshold current* is lower than that for a broad-area LD (25 mA versus 40 mA).

*Optical output power* is specified as the rated (maximum) output power. It is given as 2 mW at *current* = $I_{th}$ + 40 mA, that is, from 65 to 80 mA for this laser. It is tested at the *continuous-wave (CW)* operating condition.

*Peak wavelength* is specified within the 1545 to 1565 nm range; once established, however, its variations do not exceed 1 nm.

*Side-mode suppression ratio (SMSR)* is the measure of the intensity difference between the main longitudinal mode and the largest side mode. If you look at the "Emission Spectrum" graph in Figure 9.19, you will see the main mode and several satellite modes generated at the closest wavelengths. Thus,

$$SMSR(\text{dB}) = 10 \log(I_{\text{main}}/I_{\text{side}}) \tag{9.23}$$

The data sheet gives the *SMSR,* typically, as 40 dB, which means that the intensity of the main mode is ten thousand times larger than the intensity of the largest side mode.

*Spectral line width, Δf,* is measured here in MHz, not in nanometers, because this is an extremely small value. Again, for a broad-area laser, spectral width is an *FWHM* measure of an envelope comprising all the generated modes but, for the DFB laser, this is the width of only one mode (line) because the others are severely suppressed.

The relationship between linewidth measured in Hz and nm can be derived as follows: From basic Formula 2.1, $\lambda f = c$, we can obtain

$$\Delta f = -(c\Delta\lambda)/\lambda^2, \tag{9.24}$$

where $\Delta f$ and $\Delta\lambda$ are linewidths in Hz and m, respectively; $c$ is the speed of light in a vacuum, and $\lambda$ is the operating wavelength. The same idea can be represented in this way:

$$\Delta f/f = \Delta\lambda/\lambda \tag{9.24a}$$

Spectral linewidth has an inverse relationship to output power; for example, for a DFB laser radiating 30 mW, the linewidth can be as narrow as 1 MHz.

*Cooler voltage* and *current* determine the electrical parameters of a cooler circuit. Look at the picture of the "PIN Configuration" in Figure 9.19. A thermistor, which is a resistor whose resistance decreases when temperature increases, serves as a temperature sensor for a thermoelectric cooler. All other characteristics are similar to those of a broad-area laser diode so we don't need to comment on them again.

Note that no modulation bandwidth is specified in this data sheet. This is because this DFB LD works in the CW mode with an external modulator. (See Section 10.4.)

It is worth noting that there is a class of DFB laser diodes that radiate high output power (up to 30 mW). These diodes are used in long-distance and distribution networks. They are characterized by an extremely narrow linewidth (1 MHz), high drive current (up to 350 mA), and external modulation.

***VCSEL laser diodes*** Figure 9.20 displays a typical data sheet for a vertical-cavity surface-emitting laser.

| 840nm | 1A440 VCSEL Laser Diode | Datacom, General Purpose |
|---|---|---|

This Vertical Cavity Surface-Emitting Laser is designed for Fibre Channel, Gigabit Ethernet, ATM and general applications. It operates in multiple transverse and single longitudinal mode, ensuring stable coupling of power and low noise. And it matches the 1A354 PIN Photodiode.

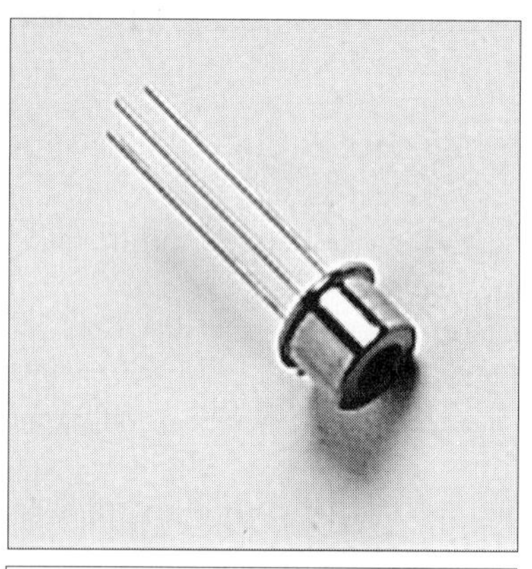

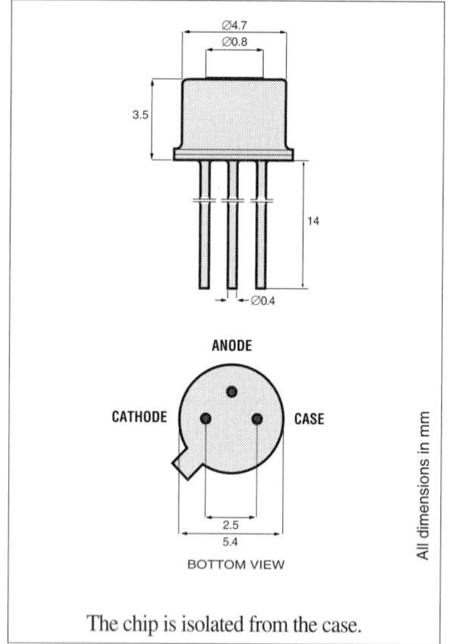


**TO-46 Package With Lens**

WARNING: Laser Radiation, avoid exposure to beam. Class 3B laser product, potential eye hazard. Warning labels in each box.

13430.11 1997-07-03

## Optical and Electrical Characteristics (25° C Case Temperature)

| PARAMETER | SYMBOL | MIN. | TYP. | MAX. | UNIT | TEST CONDITION |
|---|---|---|---|---|---|---|
| Fiber-Coupled Power | $P_{fiber}$ | | 1000 | | μW | $I_F$=10mA (Note 1) |
| Optical Power | $P_o$ | 500 | 1250 | 2200 | μW | $I_F$=10mA |
| Bandwidth (3dB$_{el}$) | $f_c$ | | 5 | | GHz | $I_F$=10mA |
| Peak Wavelength | $\lambda_p$ | | 840 | | nm | $I_F$=10mA |
| Spectral Width (FWHM) | $\Delta\lambda$ | | 0.5 | | nm | $I_F$=10mA |
| Forward Voltage | $V_F$ | | 1.8 | 2.1 | V | $I_F$=10mA |
| Threshold Current | $I_{th}$ | | 3.5 | 6 | mA | |

**Note 1:** Fiber: 50/125 Graded Index, NA=0.2 or 62.5/125 Graded Index, NA=0.275. An external glass ball lens with 2 mm diameter is required.

## Absolute Maximum Ratings

| PARAMETER | SYMBOL | LIMIT |
|---|---|---|
| Storage Temperature | $T_{stg}$ | −55 to + 125°C |
| Operating Temperature | $Top$ | 0 to +70°C |
| Electrical Power Dissipation | $P_{tot}$ | 35 mW |
| Continuous Forward Current (f≤10 kHz) | $I_F$ | 15 mA |
| Peak Forward Current (duty cycle≤50%, f≥1 MHz) | $I_{FRM}$ | 25 mA |
| Reverse Voltage | $V_R$ | 1.5 V |
| Soldering Temperature (2mm from the case for 10 sec) | $T_{sld}$ | 260°C |

## Thermal Characteristics

| PARAMETER | SYMBOL | MIN. | TYP. | MAX. | UNIT |
|---|---|---|---|---|---|
| Thermal Resistance - Infinite Heat Sink | $R_{thjc}$ | | 700 | | °C/W |
| Thermal Resistance - No Heat Sink | $R_{thja}$ | | 1000 | | °C/W |
| Temp. Coefficient - Wavelength | $d\lambda/dT_j$ | | 0.06 | | nm/°C |
| Optical Power - Variation 0 to 70°C | $\Delta P$ | | ±0.7 | | dB |
| Threshold Current - Variation 0 to 70°C | $\Delta I_{th}$ | | ±0.6 | | mA |

**Figure 9.20** Data sheet of a VCSEL laser diode. *(Reprinted with permission from Mitel Semiconductor's* Optoelectronic Solutions Product Catalog, *© 1997, Mitel Semiconductor, Kanata, Ont., Can.)*

You should become especially familiar with the following characteristics of this type of laser diode:

- The threshold current (6 mA) is much less than that of a DFB laser (25 mA) and that of a broad-area LD (40 mA).
- The driving current (10 mA at 1.25 mW output power) is much less than that of a DFB laser (65 mA at 2 mW output power) and that of a broad-area LD (65 mA at 1 mW output power).
- The spectral width (0.5 nm) is in between that of a DFB's linewidth (10 MHz) and that of a broad-area LD's width (6 nm).
- The modulation bandwidth (5 GHz) is higher than that of a broad-area LD (on the order of 1 GHz).
- The only operating wavelength available is 840 nm (or 850 nm) in contrast to that available for DFBs and broad-area lasers, which can work at 1300 nm, 1550 nm, and even longer wavelengths.

Bear in mind that the variety of laser diodes available is immense and that the discussion here has merely skimmed the surface of this continually developing technology.

## SUMMARY

- A transmitter is the component of a fiber-optic communications link that converts an electrical information signal into an optical signal. A transmitter consists of a light source, coupling optics, and electronics. Only miniature semiconductor devices—a light-emitting diode (LED) or a laser diode (LD)—are used as the light sources in fiber-optic transmitters.
- To better understand the physics behind LED and LD operations, review the energy-level concept introduced in Chapter 2. In doing so, it is important for you to recall that external electrical energy is used to pump electrons at the upper (higher) energy level. These electrons then cross to the lower energy level and radiate photons. This basic mechanism of light radiation holds true for semiconductor material also, but the energy levels are so close to one another that they constitute a bundle of energy called an energy band. The upper and lower bands are called conduction and valence bands, respectively. These bands are separated by a so-called "prohibited" region, that is, an area where no electrons are permitted, where no energy levels exist. This region—or state, if you prefer—is known as an energy gap, or bandgap.
- In LEDs, an information-containing electrical signal pumps electrons at the conduction band; they then fall to the valence band and radiate light. This is how an electrical signal is converted into an optical signal. Thus, on–off electrical pulses are converted into on–off optical flashes, which are transmitted down the optical fiber.
- Since an LED is a semiconductor diode, a radiating mechanism can be explained in terms of the *p-n* junction model. When an external electrical signal is applied, electrons and holes enter the depletion region and recombine, resulting in the release of many quanta of energy, that is, photons. In other words, electron-hole recombinations produce light. This light radiation occurs if, and only if, the LED is forward-biased, a phenomenon that forces electrons and holes to penetrate an active region and recombine.
- An LED radiates light at a wavelength not less than that dictated by the energy gap. The spectral width of this light is rather wide (on the order of tens of nanometers) because electron transitions from many levels of the conduction band to the valence band contribute to this light. The power of the radiated light is proportional to the forward current, as an LED's principle of operation suggests.
- An LED radiates rather dispersed light, which makes coupling this light into an optical fiber a problem. Special coupling techniques, including lens coupling, improve coupling efficiency. There are two types of LEDs: surface emitting (SLED) and edge emitting (ELED). The latter type radiates less divergent light, which, along with good coupling technique, allows the manufacturer to even

couple an ELED with a singlemode fiber. At 100 mA of forward current, a surface-emitting LED couples into an MM fiber at about 50 μW and an edge-emitting LED couples into an SM fiber at about 10 μW.

- A more efficient light source for fiber-optic communications is a laser diode. The term *laser* stands for *l*ight *a*mplification by *s*timulated *e*mission of *r*adiation. The key words here are "stimulated emission." When electrons are pumped at the conduction band, they can exist there for a while. If during that time, however, external photons enter the band, they cause the electrons already there to drop to the valence band and emit other photons. Thus, these new photons (light) have been stimulated by the external photons. This process takes place in the so-called "active" region, or area, of the semiconductor material. Such a material is called an "active medium."

- To make light amplification by the stimulated emission of radiation work, we must do two more things. First, we have to put two parallel mirrors at the ends of the active medium so that the stimulated photons return to the active region and stimulate new photons to be radiated. Thus, these two mirrors provide positive optical feedback. Second, since more and more stimulated transitions occur, the upper (conduction) band is depleted very quickly; hence, we need to pump more and more electrons at this band; in fact, we need to have more electrons at the conduction band than at the valence (lower) band. This situation is referred to as "population inversion" because, normally, the lower band contains more electrons than the upper one. To sum up, population inversion, stimulated emission, and positive optical feedback are the three conditions necessary to achieve lasing action.

- The input to the laser diode is forward (driving) current ($I_F$) and the output is light power ($P$). While the forward current is small, a laser diode works like a regular LED. But when the forward current reaches the threshold value, population inversion is then created and lasing action begins. Threshold value ($I_{th}$) is one of the critical characteristics of a laser diode. It ranges from approximately 3 mA to 40 mA. As with many other characteristics of an LD, threshold current depends on temperature. ($I_{th}$ increases with a rise in temperature.)

- This threshold effect can also be described in terms of a gain–loss relationship: When the optical gain of an active medium (determined by the value of population inversion) equals the loss of this medium (determined by the transmission of photons through the mirrors and absorption of these photons within the medium), the threshold condition is achieved and lasing action starts.

- The input–output characteristic of a laser diode is often depicted in a *P-I* (or *L-I*) graph, that is, a graph of power (light) versus current. The slope of this graph shows another critical characteristic of an LD: *slope efficiency* (or *differential (quantum) efficiency*), $S = \Delta P/\Delta I_F$. This slope shows how efficiently a laser diode converts input current into output light. For the best LDs, this efficiency can reach a value of 0.1 mW/mA, while the efficiency of the best LEDs is 100 times less. Unfortunately, the slope efficiency of a laser diode also depends on temperature (it decreases when temperature rises). To control a laser diode's characteristics during a temperature change, two methods are used: A thermoelectric cooler keeps the LD's temperature approximately constant and a monitor photodiode and control circuit stabilize the average output power.

- A laser diode radiates monochromatic, well-directed, highly intense, coherent light. These properties make a laser diode the light source of choice for long- and intermediate-distance fiber-optic networks.

- To improve the characteristics of a laser diode, laser scientists have made substantial innovations in two areas: First, they have made the active region as small as possible so that electron-hole recombinations take place in a small, well-controlled area. This allows the laser operator to decrease the driving current (or, more precisely, the current density), which results in an increase in diode efficiency. Diodes with double heterostructured and quantum-well active areas are good examples of such an achievement. Secondly, laser scientists have developed several types of optical feedback systems—optical resonators—that allow them to make a spectral width of radiated light extremely small—on the order of tenths of a nanometer. The distributed feedback (DFB) laser has such a linewidth and this is why a DFB laser is the most popular light source for long-haul optical links. The vertical-cavity surface-emitting laser (VCSEL), the latest development in laser-diode technology, combines both ideas—quantum-well and DFB—to introduce the best possible characteristics attainable today into a laser diode. The major drawback of a VCSEL diode is that it operates only in the first transparent window (around 850 nm). A minor drawback is that a VCSEL's spectral width is between that of a DFB and that of a regular laser.

## PROBLEMS

**9.1.** Draw a block diagram of a transmitter.

**9.2.** When describing semiconductors, why do we refer to an energy band, not to an energy level?

**9.3.** What is the difference between conduction and valence energy bands?

**9.4.** Can the conduction and valence bands overlap?

**9.5.** Under normal conditions (room temperature with no external energy provided), where do most electrons reside?

**9.6.** From the standpoint of an energy band, how does light radiation occur in semiconductor material?

**9.7.** From the standpoint of a *p-n* junction, how does light radiation occur in a semiconductor diode?

**9.8.** Draw the electric circuit of an LED and explain the function of each component.

**9.9.** Formula 9.3 predicts that the output power of an LED is proportional to the forward current but the graph "Coupled Power vs. Drive Current" in Figure 9.8(a) displays a nonlinear relationship for the real LED. What is the reason for this discrepancy?

**9.10.** Calculate the internal quantum efficiency of a 1.3-μm surface-emitting LED at a forward current of 100 mA (see Figure 9.8[a]), assuming that 7.5% of the LED's output power is coupled into an optical fiber.

**9.11.** What are the major drawbacks of a homostructured LED?

**9.12.** What are the major advantages of a heterostructured LED over a homostructured one?

**9.13.** What is the difference between a surface-emitting LED (SLED), and an edge-emitting LED (ELED)?

**9.14.** The LEDs in Figure 9.8 are made from InGaAsP, yet one of the most popular electro-optical semiconductor materials is GaAs. Why did the manufacturer use the former material, not the latter?

**9.15.** What light source is called Lambertian?

**9.16.** Evaluate power in mW and dBm (1) coupled into a 62.5/125 μm fiber and (2) radiated by a 1.3-μm SLED (Figure 9.8[a]) at 100 mA. Assume that the LED radiates as a Lambertian source.

**9.17.** Name and explain three basic techniques used to couple light from an LED into an optical fiber.

**9.18.** Formula 9.3 predicts that we can obtain any output power from an LED just by increasing the drive current. Is this true? Explain your answer.

**9.19.** The 1.3-μm SLED shown in Figure 9.8(a) has a spectral width of 155 nm at 25°C. What will be the spectral width if the temperature changes to 90°C?

**9.20.** What does "FWHM" stand for? How does it apply to an LED's characteristics?

**9.21.** Formula 9.8, strictly speaking, takes the form [3]

$$t_r = \ln\{\tau + [(2\,k_B T)/e]\,(C/I_p)\},$$

where $k_B = 1.38 \times 10^{-23}$ joule/kelvin is the Boltzmann's constant, $T = T°K$ is the absolute temperature in kelvin, $e = 1.6 \times 10^{-19}$ coulomb is an electron charge, $C$ is the charge capacitance of an LED in farads, $I_p$ is the amplitude of the step modulating current, and ln is a natural logarithm. (1) Derive Formula 9.8. (2) Evaluate the rise time for $\tau = 1$ ns. Use the other data from Figure 9.8.

**9.22.** An LED's modulation bandwidth calculated by the formula $BW = 0.35/t_r$ and as shown in the data sheet (see Figure 9.8[a]) turns out to be different. Why?

**9.23.** The typical modulation bandwidth of a 1.3-μm SLED shown in Figure 9.8(a) is equal to 115 MHz when the drive current is 100 mA. What will be the modulation bandwidth if the drive current becomes 35 mA?

**9.24.** What does the acronym *laser* stand for? Explain the meaning of each word.

**9.25.** How does spontaneous emission occur? How does stimulated emission occur?

**9.26.** List and compare the main properties of both stimulated emission and spontaneous emission.

**9.27.** What is the scientific meaning of the term *monochromatism?*

**9.28.** Name and explain three conditions necessary to attain lasing action.

**9.29.** The resonator of a laser diode is made by cleaving the opposite end surfaces of a diode fabricated from GaAs material. Calculate the Fresnel reflection at the GaAs–air interface if the refractive index of GaAs is 3.6.

**9.30.** What is the threshold condition in a laser diode in terms of the gain–loss relationship? How can you measure the threshold current of a laser diode?

**9.31.** Draw a graph depicting the typical input–output relationship in a laser diode and explain what mechanisms determine the course of this graph.

**9.32.** How many longitudinal modes can a Fabry-Perot laser diode generate if the length of its resonator is 0.3 mm and the operating wavelength is 1550 nm? The width of the gain curve is 9 nm.

**9.33.** Explain the principle of operation of a quantum-well laser diode.

**9.34.** How does a DFB laser diode operate? What does *DFB* stand for? What is the distinguishing characteristic of this type of laser diode?

**9.35.** What does *VCSEL* stand for? How does this type of laser diode operate? What is its distinguishing characteristic?

**9.36.** Why do we need to cool a laser diode?

**9.37.** Which laser diode—cooled or uncooled—does the industry prefer and why?

**9.38.** How can you determine the value of the threshold current?

**9.39.** The threshold current of a laser diode is determined to be 40 mA at 25°C. What is the threshold current at 65°C?

**9.40.** What is the threshold-current change with temperature for the laser diode considered in Problem 9.39?

**9.41.** Calculate the slope efficiency of a DFB LD whose specifications are given in Figure 9.19.

**9.42.** Calculate the slope-efficiency change with temperature for a broad-area laser diode whose specifications are given in Figure 9.17. Do your calculations in percent and decibels per degree centigrade. Compare the numbers obtained with those given in the data sheet.

**9.43.** Calculate the modulation-current value for the broad-area laser diode in Figure 9.17 and compare the numbers obtained with those given in the data sheet.

**9.44.** Do we want a small or big rise/fall time and why? What device—LED or laser diode—has the smaller rise/fall time and why?

**9.45.** Why does the manufacturer include the characteristics of a photodiode in a laser-diode data sheet?

**9.46.** What does the acronym "RIN" stand for?

**9.47.** Compare the spectral widths of the three laser diodes in Figures 9.17, 9.19, and 9.20 and explain why they are different.

# HFE4383-322

## Connectorized High Speed VCSEL

**FEATURES**

- Designed for drive currents between 5 and 15 mA
- Prealigned SC Connector sleeve
- Optimized for low dependence of electrical properties over temperature
- High speed ≥1 GHz
- Two different laser/photodiode polarities
- Packaged with a photodetector

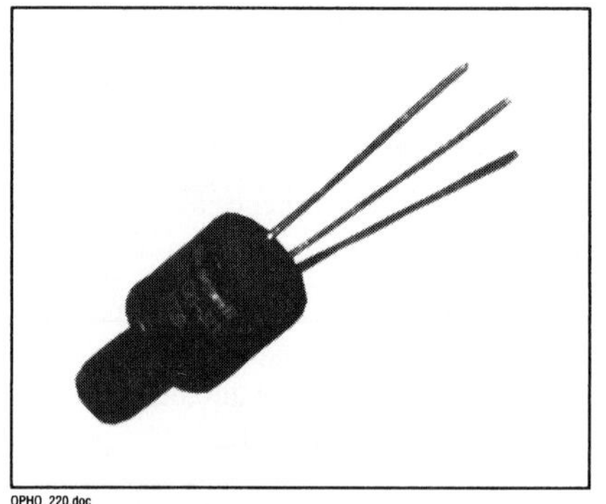

OPHO_220.doc

**DESCRIPTION**

The HFE4383-322 is a high-performance 850 nm VCSEL (Vertical Cavity Surface-Emitting Laser) packaged for high-speed data communications. This product combines all the performance advantages of the VCSEL with a custom designed power monitor diode.The power monitor diode can be used with appropriate feedback control circuitry to set a maximum power level for each VCSEL, simplifiying design for high data rate communication and eye safety.

Packaged in a fiber receptacle sleeve, this high radiance VCSEL is designed to convert electrical current into optical power that can be used in fiber optic communications and other applications. As the current varies above threshold, the light intensity increases proportionally. Data rates can vary from DC to above 2 Gb/s.

The HFE4383-322 is designed to be used with inexpensive silicon or gallium arsenide detectors, but excellent performance can also be achieved with some indium gallium arsenide detectors.

The low drive current requirement makes direct drive from PECL (Positive Emitter Coupled Logic) or ECL (Emitter Coupled Logic) gates possible and eases driver design.

The HFE4383-322 is a prealigned and focused fiber optic transmitter designed to interface with 50/125 and 62.5/125 multimode fiber.

**OUTLINE DIMENSIONS in inches (mm)**

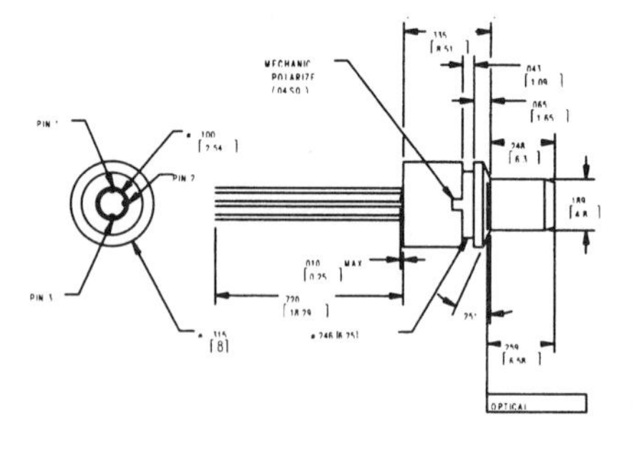

odim_272.doc

**Pinout**

1. $K_{LD}$ - VCSEL Cathode
2. $K_{PD}$ - Monitor Photodiode Cathode, $A_{LD}$ - VCSEL Anode
3. $A_{PD}$ - Monitor Photodiode Anode

**Figure 9.21** Data sheet of VCSEL laser diode from Honeywell. *(Courtesy of Honeywell Sensing and Control, Freeport, Ill.)*

# HFE4383-322

## Connectorized High Speed VCSEL

**ELECTRO-OPTICAL CHARACTERISTICS**

| PARAMETER | SYMBOL | MIN | TYP | MAX | UNITS | TEST CONDITIONS |
|---|---|---|---|---|---|---|
| Peak fiber coupled optical power | | | | | | $I_F$ = 12 mA peak |
| | $P_{OC}$ | 0.5 | 1.0 | 2.5 | mW | 50/125 μm fiber [1] |
| | | -3.0 | 0 | +4.0 | dBm | NA = 0.20 |
| | | | | | | $I_F$ = 12 mA peak |
| | $P_{OC}$ | 0.5 | 1.0 | 2.5 | mW | 62.5/125 μm fiber [1] |
| | | -3.0 | 0 | +4.0 | dBm | NA = 0.275 |
| Threshold Current | $I_{TH}$ | | 3.5 | 6 | mA | |
| Threshold current Temperature Variation | $\Delta I_{TH}$ | -1 | | 1 | mA | $T_A$ = 0°C to 70°C [2] |
| Slope Efficiency | $\eta$ | 0.06 | 0.15 | 0.3 | mW/mA | $P_{OC}$ = 1.0 mW [3] |
| Slope Efficiency Temperature Variation | $\Delta\eta/\Delta T$ | | -0.4 | | %/°C | $T_A$ = 0°C to 70°C |
| Peak Wavelength | $\lambda_P$ | 830 | 850 | 860 | nm | $I_F$ = 12 mA |
| λpTemp Coefficient | $\Delta\lambda_P/\Delta T$ | | 0.06 | | nm/°C | $I_F$ = 12 mA |
| Spectral Bandwidth, RMS | $\Delta\lambda$ | | | 0.85 | nm | $I_F$ = 12mA |
| Laser Forward Voltage | $V_F$ | 1.6 | 1.8 | 2.2 | V | $I_F$ = 12 mA |
| Laser Reverse Voltage | $BVR_{LD}$ | 5 | 10 | | V | $I_R$ = 10 μA |
| Rise and Fall Time | $t_R/t_F$ | | 100 | 400 | ps | Bias Above Threshold [4] (20%-80%) |
| Relative Intensity Noise | RIN | | -128 | -122 | dB/Hz | 1 GHz BW |
| Series resistance | $R_S$ | 15 | 25 | 50 | Ohms | $I_F$ = 12 mA |
| Coupled Power ratio | | | | | | $I_F$ = 12 mA peak |
| | CPR | | 9 | | dB | 50/125 μm fiber |
| | | | | | | NA = 0.20 |
| | | | | | | $I_F$ = 12 mA peak |
| | CPR | | 9 | | dB | 62.5/125 μm fiber |
| | | | | | | NA = 0.275 |
| Monitor current | $I_{PD}$ | 0.02 | | 0.10 | mA | $P_{OC}$ = 1.0 mW [5] |
| Monitor current temperature variation | $\Delta I_{PD}/\Delta T$ | | 0.2 | | %/°C | $P_{OC}$ = 1.0 mW |
| Dark Current | $I_D$ | | | 20 | nA | $P_O$ = 0 mW, $V_R$ = 3 V |
| PD reverse voltage | $BVR_{PD}$ | 30 | 115 | | V | Po = 0 mW, $I_R$ = 10 μA [6] |
| PD Capacitance | C | | 100 | | pF | $V_R$ = 0 V, Freq = 1 MHz |
| | | | 55 | | | $V_R$ = 3 V, Freq=1MHz |

Notes

1. Operating power is set by the peak operating current $I_{PEAK}=I_{BIAS}+I_{MODULATION}$.
2. Operation at temperatures outside the specified range may result in the threshold current exceeding the maximums defined in the electro-optical characteristics table.
3. Slope efficiency is defined as DPo/DIF at a total power output of 1.0 mW.
4. Rise and fall times are sensitive to drive electronics. 200 ps rise and fall times are achievable with Honeywell VCSELs.
5. Monitor current tested with a fiber in the receptacle. Reflections external to the 4383/4384 package can influence the monitor current.
6. To safeguard the VCSEL from current spike damage, short the VCSEL anode and cathode to each other during photodiode BVR verification testing. Additionally to safeguard the PIN photodiode, limit the photodiode reverse voltage in accordance with the absolute maximum rating.

# HFE4383-322

## Connectorized High Speed VCSEL

**ABSOLUTE MAXIMUM RATINGS**

| | |
|---|---|
| Storage temperature | -40 to 85°C |
| Operating temperature | 0 to +70°C |
| Lead solder temperature | 260°C, 10 src. |
| Continuous optical output power (indepedent of drive current) | 11 mW |
| Laser diode reverse voltage ($I_R$= 10 μA) | 5 V |
| Laser continuous forward current, heat sinked | 15 mA |
| PIN Photodiode reverse voltage | 30 V |

**Figure 9.21** (continued)

# HFE4383-322

**Connectorized High Speed VCSEL**

Fig. 1 Typical Performance Curve Emitted Power vs Current

OGRA_148.DOC

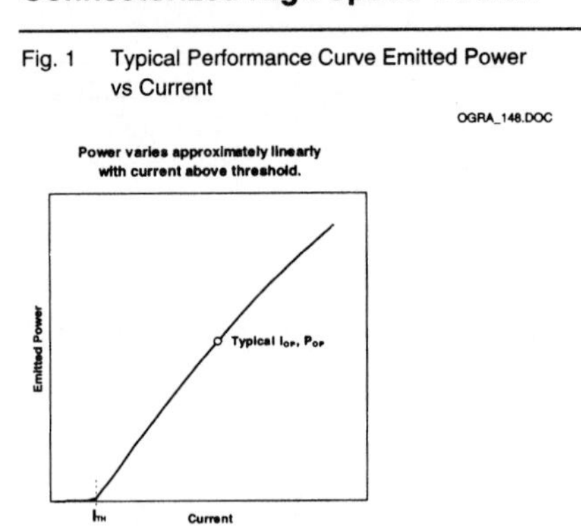


Fig. 2 Typical Performance Curve Threshold Current vs Temperature

OGRA_149.DO

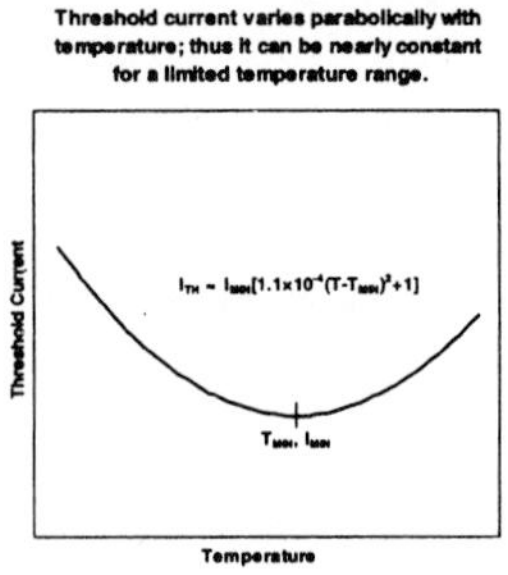

Threshold current varies parabolically with temperature; thus it can be nearly constant for a limited temperature range.

Threshold Current

Temperature

Fig. 3 Emission Intensity vs Wavelength

OGRA_150.DOC

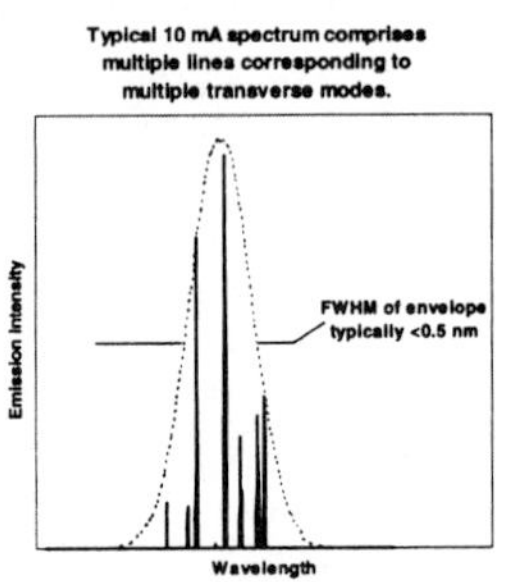


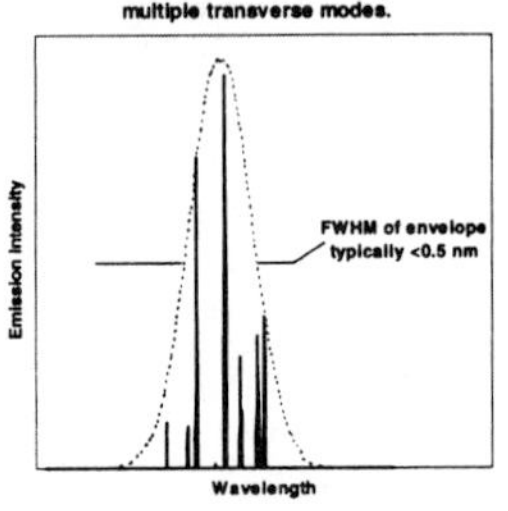

**Figure 9.21** (continued)

**9.48.** Explain the meaning of "SMSR."

**9.49.** Compare the threshold currents of the three laser diodes in Figures 9.17, 9.19, and 9.20 and explain why they are different.

**9.50.** *Project 1:* Figure 9.21 shows the data sheet of an HFE 4383-322 VCSEL from Honeywell. Analyze all the data in this specification sheet and compare these data with those shown in Figure 9.20.

**9.51.** *Project 2:* Build a memory map for this chapter. (See Problem 1.20.)

## REFERENCES[2]


1. Chin-Lin Chen, *Elements of Optoelectronics and Fiber Optics,* Chicago: Irwin, 1996.
2. Ronald Lasky, Ulf Österberg, and Daniel Stigliani, eds., *Optoelectronics for Data Communication,* San Diego: Academic Press, 1995.
3. Gerd Keiser, *Optical Fiber Communications,* 2d ed., New York: McGraw-Hill, 1991.
4. *Optoelectronic Solutions* (product catalog), Mitel Semiconductor AB, Jarfalla, Sweden, 1998.
5. Tien Pei Lee, C.A. Burrus, Jr., and R.H. Saul, "Light-Emitting Diodes for Telecommunication," in *Optical Fiber Telecommunications-11.,* ed. by S. E. Miller and I. P. Kaminow, San Diego: Academic Press, 1988, pp. 467–507.
6. P. Courley, K. Lear, and R. Schneider, "Surface-Emitting Lasers," *IEEE Spectrum,* August 1994, pp. 31–37.
7. Dave Welch, "Low-cost, singlemode transmission with long-wavelength VCSELs," *Lightwave,* February 1999, pp. 62–67.
8. "Thermoelectric Cooling Tutorial," *Diode Lasers & Instruments Guide,* Melles Griot Inc., Boulder, Colo., April 1998.
9. Tyll Hertsens, "Measuring Diode Laser Characteristics," *Lasers & Optronics,* February 1989, pp. 26–28.


[2]See Appendix C: A Selected Bibliography.

# 10

# Light Sources and Transmitters—A Deeper Look

## 10.1 MORE ABOUT SEMICONDUCTORS

The purpose of this section is to present a more formal description of the semiconductor materials used for the light sources discussed in previous sections. We are going to delve into semiconductor theory in order to quantify basic relationships. This approach will give you a deeper appreciation of the principle of operation of laser diodes and LEDs, which, in turn, will sharpen your understanding of the possible applications of these devices.

### Intrinsic Semiconductors: Fermi Energy Levels and Number of Charge Carriers

***Fermi energy levels and the Fermi-Dirac distribution*** Considering the valence band, the conduction band, and the energy gap in an intrinsic semiconductor, we have to introduce Fermi[1] energy, $E_F$. Let's consider the probability of finding an electron at energy level $E$ as a function of $E$, *assuming* that the energy levels are close enough to make a continuum that we represent as an energy band. When absolute temperature $T = 0$ K, all the electrons can be found at any energy level up to $E_F$. But when the absolute temperature is above zero, this thermal energy will excite some electrons so they will occupy energy levels higher than $E_F$. This consideration is formally described as:

When $T = 0$ K,

$$f(E) = 1 \qquad \text{for } E < E_F \tag{10.1}$$

$$f(E) = 0 \qquad \text{for } E > E_F$$

[1]Enrico Fermi (1901–1954), Italian-born physicist, Nobel Prize winner (1932), and one of the creators of quantum physics, played a prominent role in constructing America's first nuclear reactor in 1942. It led to development of the atom bomb.

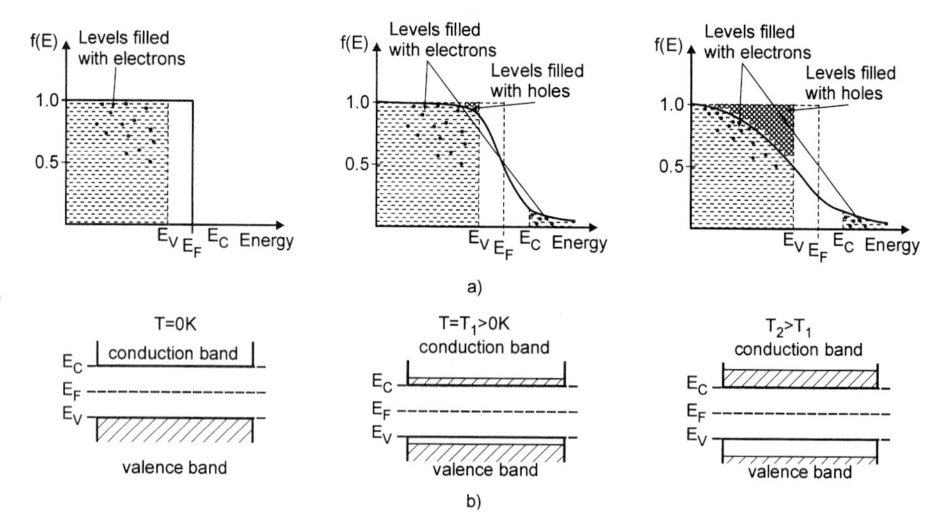


**Figure 10.1** Fermi energy level: (a) Fermi-Dirac distribution at different temperatures; (b) energy-band diagram. *([a] adapted from Christopher Davis'* Lasers and Electrooptics—Fundamentals and Engineering. *Copyright 1996. Reprinted with permission of Cambridge University Press.)*

When $T > 0$ K,

$$f(E) = (1 + \exp((E - E_F)/k_B T))^{-1} \tag{10.2}$$

where $k_B$ is the Boltzmann constant ($k_B = 1.38 \times 10^{-23}$ J/K). Formula 10.2 is called the *Fermi-Dirac*[2] *distribution.*

Formulas 10.1 and 10.2 can serve as a definition of Fermi energy and Figure 10.1 illustrates this definition. In Figure 10.1(b), the Fermi energy level, $E_F$, is midway between the edge levels of the valence band, $E_v$, and the conduction band, $E_C$. In other words, the Fermi energy level is in the forbidden region. (Remember, there are no energy states and, consequently, no electrons can be found within an energy gap, such as between $E_C$ and $E_V$). You can easily build the Fermi-Dirac distribution curve f($E$) shown in Figure 10.1(a) by using Formula 10.2.

At $T = 0$ K, all filled energy levels are below $E_F$, as Formula 10.1 states. Again, this means, physically, that no electrons can be found above the $E_V$ level. As the temperature increases, more and more electrons are thermally excited so that they acquire enough energy to leap over the energy gap and occupy higher energy levels at the conduction band.

## Example 10.1.1

***Problem:***

What is the probability of exciting electrons at a conduction band at room temperature in a gallium arsenide (GaAs) semiconductor? (GaAs is one of the most popular semiconductor materials used to fabricate LDs and LEDs.)

***Solution:***

We can apply Formula 10.2 but we need to know $E_C - E_F$. From Table 9.1, we can find that the energy gap for GaAs equals 1.424 eV. Since $E_F$ is in the middle of the energy gap, $E_C - E_F = 0.712$ eV.

Room temperature is commonly referred to as $T = 300$ *K*. Substituting the Boltzmann constant, we find that $k_B T = 0.025$ eV.

[2]Paul Dirac (1902–1984), the British physicist who contributed signficantly to the development of quantum physics, predicted the existence of the positron and other "anti-matter." He shared the Nobel Prize for Physics in 1933.

Now the exponential term becomes

$$\exp((E_C - E_F)/k_B T) = \exp(28.48) = 2.38 \times 10^{12}$$

which is much larger than 1. Hence, with a very close approximation, we can rewrite Formula 10.2 in this form:

$$f(E) \approx \exp-((E - E_F)/k_B T) \tag{10.3}$$

*Note:* We must be careful when using Formula 10.3: This approximation is acceptable for application in conventional electronics but it is not valid for the active region of a laser diode. This is because, as we will soon see, in laser-diode operation the Fermi level splits into two quasi-Fermi levels, and these levels are inside conduction and valence bands.

But to return to our example: The probability of exciting an electron at the edge of a conduction level of intrinsic GaAs by subjecting it to room temperature is $4.28 \times 10^{-13}$.

This example and Formulas 10.2 and 10.3 show that the greater the energy ($k_B T$) delivered to an electron residing in the valence band, the greater the probability of its becoming excited. To be sure of exciting an electron at the upper conduction band—that is, to obtain $f(E) \to 1$—one has to deliver infinite energy, that is, to make $k_B T \to \infty$.

Observe that $k_B T$, measured in joules, represents energy, as the units say, because the Boltzmann constant, $k_B$, is measured in joules per absolute temperature (*J/T*) and *T* is measured in absolute temperature (*T*). We usually measure $k_B T$ in eV, as is common in quantum physics ($1 \text{ eV} = 1.6022 \times 10^{-19}$ J).

---

***Number of charge carriers*** How many electrons will be excited at the upper band? In semiconductors, we are interested in knowing how many charge carriers—electrons and holes—are in the unit volume. We call this measure *charge-carrier density* and denote it as *n* (negative) for electrons and *p* (positive) for holes. It is quite obvious that the density of electrons excited at the conduction band is proportional to the probability, $f(E)$:

$$n = N_C \exp -((E_c - E_F)/k_B T) \tag{10.4}$$

The density of holes at the valence band can be found as $1 - f(E)$ so that

$$p = N_v \exp-((E_F - E_V)/k_B T), \tag{10.5}$$

where coefficients $N_C$ and $N_V$ are equal to [1]:

$$N_C = 2[(m_n * k_B T)/(2\pi \hbar^2)]^{3/2} \text{ and } N_V = 2[(m_p * k_B T)/(2\pi \hbar^2)]^{3/2} \tag{10.6}$$

Here, $\hbar = h/2\pi$ is the normalized Planck's constant and $m_n^*$ and $m_p^*$ are *effective masses* of electrons and holes, respectively. These effective masses describe the moving ability of electrons and holes in the lattice; therefore, these are characteristics used in semiconductors. Furthermore, an effective mass reflects two properties of electrons: the wave/particle duality of an electron and its interaction with a lattice. The effective mass is much less than its true mass, that is, the mass of a free electron. In addition, the effective mass of an electron is much less than the effective mass of a hole. For example, for GaAs, $m_n^* = 0.067m_0$ and $m_p^* = 0.55\ m_0$, where $m_0 = 9.0194 \times 10^{-31}$ kg is the true mass of a free electron. This relationship explains why electrons are much more mobile than holes. For instance, for GaAs at $T = 300$ K, electron mobility is 8600 $\text{cm}^2/\text{V·s}$, while hole mobility is 400 $cm^2/V{\cdot}s$. This is why when we refer to the motion of charge carriers, we usually think of the flow of electrons.

*Note:* Strictly speaking, there are two types of holes: light holes and heavy holes. They differ in effective masses. The number $m_p^* = 0.55\, m_0$ is true for the heavy, or normal, holes. We will take into account only these holes, leaving any deeper level of discussion to more specialized texts ([2], [4]).

Study closely Formulas 10.4 and 10.5, which show $E_F < E_c$ and $E_F > E_v$.

The net result of this discussion is that *the concentration of charge carriers at the desired band,* $E_C$ or $E_V$ *depends on* $E_C - E_F$ *and* $E_F - E_V$ *and on the delivered external energy,* $k_BT$, as described by Formulas 10.4 and 10.5.

Moreover, it is important to note that Formula 10.2 determines the probability that the energy state, $E$, is occupied by a single electron. What's more, the Pauli[3] exclusion principle asserts that not more than two electrons can be in the same quantum state simultaneously. Thus, Formula 10.2 contains the answer to the question of how many electrons are excited at the conduction band. It does so in a stochastic manner, as does the entire quantum theory.

We can also interpret Formulas 10.4 and 10.5 as follows: The density of charge carriers at the conduction and valence bands is determined by the densities of energy states times the probabilities of having these states occupied by electrons and holes. Coefficients $N_C$ and $N_V$ reflect the densities of the energy states.

## Doped Semiconductors

So far, our discussion has been restricted by reference to an *intrinsic semiconductor,* that is, a semiconductor where the concentration of electrons and holes is the same. This implies that the Fermi level of an intrinsic semiconductor lies in the middle of the energy gap. The perfect example of an intrinsic semiconductor is a pure material that consists of atoms of only one type, such as silicon or germanium. However, a compound semiconductor like GaAs can also be intrinsic: No matter how many different atoms constitute the lattice of a semiconductor, the only feature determining whether it is intrinsic is if there is an equal concentration of electrons and holes within it.

To facilitate leaping over the energy gap, one can insert atoms of other materials. These are called *dopants.* For example, atoms of arsenic (As) have five valence (outer) electrons and germanium (Ge) has four valence electrons. These materials have a very similar lattice structure. When arsenic atoms are added to germanium, each ion of As replaces an ion of Ge, thus leaving the crystal structure of the material unchanged but adding an electron that is ready to move when a small external voltage is applied. In such a material, therefore, the negative charge carriers become a majority and so this semiconductor is called an *n* type. On the other hand, when, say, boron atoms having three valence electrons are added to germanium, the holes—positive charge carriers—exceed the electrons and one obtains a *p* type of semiconductor. A doped semiconductor is called an *extrinsic* semiconductor.

Doping changes the energy configuration dramatically: *n*-doping moves the Fermi energy closer to a conduction band; this energy level is usually called a *donor's level,* $E_d$. However, *p*-doping moves the Fermi level closer to a valence band; this energy level is usually called an *acceptor's level,* $E_a$. (See Figure 10.2.)

As a result, electrons in an *n*-type extrinsic semiconductor have to leap over a much smaller energy barrier than they have to do in an intrinsic material. Thus, it is much easier to create negative charge carriers—free electrons at the conduction band—in *n*-type semiconductors than in an

[3]Wolfgang Pauli (1900–1958), Austrian-born theoretical physicist, worked both in Europe and the United States. He introduced the exclusion principle, named after him, and predicted the existence of the neutrino. He won the Nobel Prize for Physics in 1945. His name is also associated with those who devised the quantum theory.

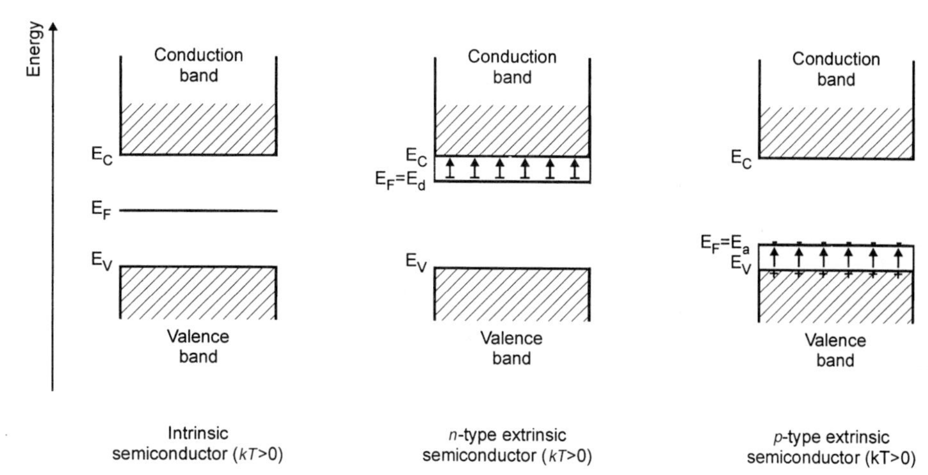


**Figure 10.2** Energy bands and Fermi energy levels of intrinsic and extrinsic semiconductors.

intrinsic material. A similar consideration holds for positive charge carriers—holes. The net result of doping is that *n*-type and *p*-type semiconductors have a high concentration of mobile electrons and holes, respectively. We call them *majority carriers* and denote them as $n_n$ and $p_p$. But there are a small number of mobile electrons on the *p* side and a small number of holes on the *n* side. We call them *minority carriers* and denote them as $n_p$ and $p_n$, respectively. The concentration of minority carriers is about $10^{11}$ times less than the concentration of majority carriers. In other words, $n_n/n_p \approx p_p/p_n \approx 10^{11}$.

Of course, there is a general charge equilibrium in doped semiconductors. This status is supported by ions, that is, atoms that have lost their carriers.

## *p-n* Junction

***Energy levels at equilibrium*** Now let us put in physical contact *n*-type and *p*-type semiconductors made from the same material. At first (for a couple of nanoseconds), excessive mobile electrons from the *n* side flow to the *p* side and excessive mobile holes from the *p* side move to the *n* side. But as soon as the electrons have moved, they leave immobile positive donor ions—with which they have made the material electrically neutral—on the *n* side. The holes leave immobile negative acceptor ions on the *p* side. As a result, an electric field, with potential $V_D$, is formed, as Figure 9.3(a) shows. This depletion voltage, $V_D$, prevents the charge carriers from further movement so that static equilibrium is reached.

The important point to underscore is this: *In equilibrium, the Fermi energy levels at both sides of a* p-n *junction are the same.* Indeed, if two materials, A and B, are brought into physical contact and they are in equilibrium, they must have the same Fermi-Dirac distributions. This means $f_A(E) = f_B(E)$. (For a formal derivation of this statement, see [1], pp. 1286–1287.) Since $k_BT$ is the same for both materials, it follows from Formula 10.2 that $E_{FA} = E_{FB}$. Thus, after a short transition period, energy bands at the *p-n* junction reach the positions shown in Figure 10.3(b). We can also express the same idea by saying that equilibrium is reached when the Fermi levels of the valence and conduction bands line up with each other. Note that for heavily doped material, the Fermi levels lie inside the conduction and valence bands, as Figure 10.3(c) shows.

***Densities of charge carriers*** Let's consider the densities of the charge carriers on both sides of a *p-n* junction. Formula 10.4 can be modified to show the concentration of majority, $n_n$, and minority, $n_p$, charge carriers as follows:

$$n_n = N_c \exp-((E_{cn} - E_F)/k_BT) \tag{10.7}$$

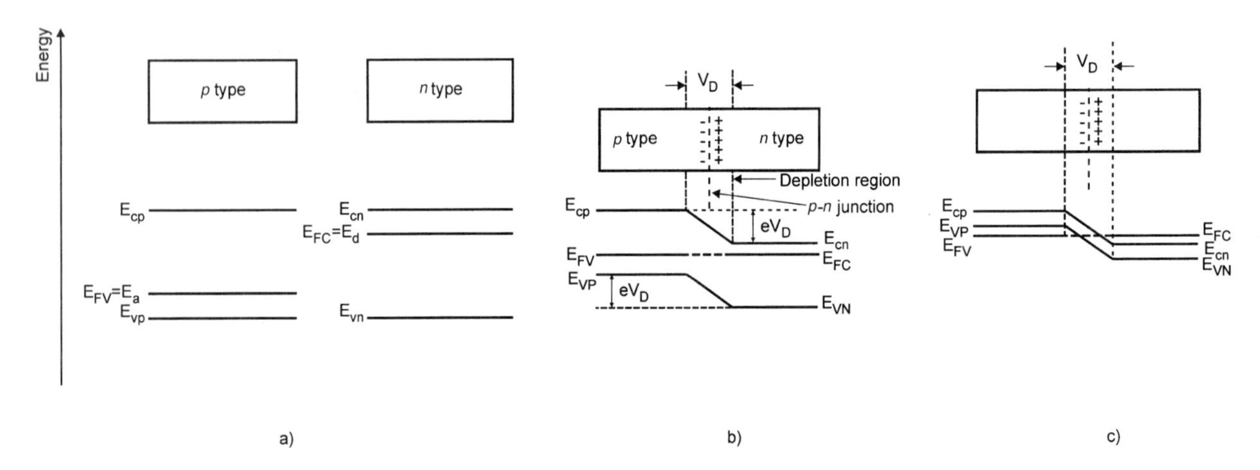


**Figure 10.3** Energy bands of a *p-n* junction: (a) Energy band of *n*-type and *p*-type semiconductors before connection; (b) energy bands after connection; (c) energy bands of a heavily doped semiconductor.

$$n_p = N_c \exp -((E_{cp} - E_F)/k_B T) \tag{10.8}$$

Working with the ratio $n_n/n_p$, we can obtain

$$n_n/n_p = \exp((eV_D)/k_B T), \tag{10.9}$$

where $eV_D = E_{cp} - E_{cn}$, as Figure 10.3 shows. (Remember, $e$ is an electron charge and equals $1.6 \times 10^{-19}$C.) We can repeat this consideration for the concentration of holes at the *p-n* junction and obtain

$$p_p/p_n = \exp((eV_D)/k_B T) \tag{10.10}$$

Formulas 10.9 and 10.10, in a sense, give us the ratio of concentration of majority to minority carriers at the *p-n* junction. The physical mechanism behind these formulas is this: There are very few electrons on the *p* side of the *p-n* junction, but they can move easily through the junction because they drift from a higher energy level, $E_{cp}$, to a lower energy level, $E_{cn}$. This flow is called *drift current.* There are many electrons on the *n* side of the *p-n* junction, but they have to climb a potential barrier—$eV_D = E_{cp} - E_{cn}$—to move to the *p* side. As a result, only a few of them have enough energy to diffuse from $E_{cn}$ to $E_{cp}$. This flow is called *diffusion current.* In equilibrium, diffusion current cancels drift current. This is another example of how a *p-n* junction works.

The reverse reasoning holds for holes. They have to overcome the potential barrier $-\,eV_D = E_{vn} - E_{vp}$, which has the same magnitude for holes as it does for electrons. As for positive charge carriers, they have to leap over this barrier to reach the *n* side. This potential barrier is another way of explaining why electrons and holes cannot flow through the depletion region.

### Example 10.1.2

***Problem:***

Calculate the ratio of majority to minority charge carriers in an *n* type and a *p* type of silicon (Si) semiconductor.

***Solution:***

Formulas 10.9 and 10.10 provide the solution. Recall that the depletion voltage for silicon is 0.7 V. Assume that $T = 300$ K and plug in the other numbers:

$$n_n/n_p = p_p/p_n = \exp((eV_D)/k_BT) = \exp((1.6\times10^{-19}\text{C}\times0.7\text{ V})/1.38\text{ J/K}\times10^{-23}\times300\text{ K})$$
$$= \exp(27) = 5.3\times10^{11}$$

Within the depletion region, the concentration of majority carriers is reduced by $5.3 \times 10^{11}$ compared with the bulk *n* region; this number is equal to the concentration of the minority carriers here.

---

When *n*-type and *p*-type semiconductors are brought into physical contact, the conduction and valence energy levels of the *n* type become lower than those of the *p* type, as Figure 10.3(b) shows. One may wonder why energy values on the *n* side are lower than those on the *p* side and not vice versa. The answer is this: Energy is determined by multiplying charge times voltage; that's where the electron volt comes from. But electrons have a negative charge; thus, the energy of electrons on the *n* side is reduced by $eV_D$.

## Biasing

When external voltage is applied to a *p-n* junction, the junction is biased. Forward biasing means applying positive voltage to the *p* side and negative voltage to the *n* side. (See Figure 9.3[b].) *The forward-biased* p-n *junction is not in equilibrium any longer.* This is because a significant number of electrons acquire sufficient energy from the biasing source to leap over the depletion barrier. We describe this in Section 9.1 as the flow of electrons through the depletion region when they are attracted by the positive voltage. The same consideration is true for holes. As a result, *Fermi levels will be different for the* n-*type and* p-*type sides of a forward-biased semiconductor,* as shown in Figure 10.4(a).

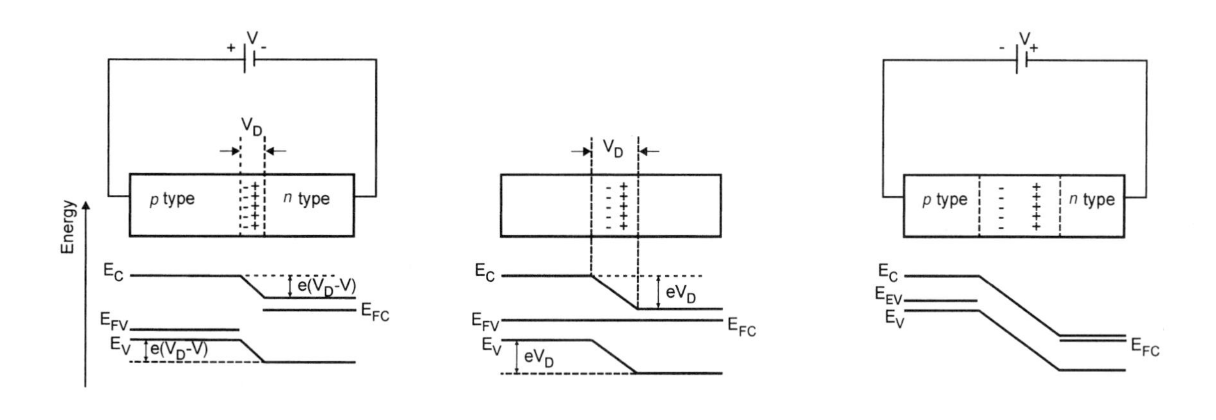


**Figure 10.4** Energy bands of a biased *p-n* junction: (a) Forward bias; (b) unbiased junction; (c) reverse bias.

Forward biasing reduces, if it doesn't entirely remove, the potential barrier, thereby making it easy for excessive electrons from the *n* side to flow through a depletion region to the *p* side. The same holds for the holes; however, when reverse bias is applied, the potential barrier increases dramatically, as Figure 10.4(c) shows. Thus, very few electrons can overcome this potential barrier. The result: almost no flow through the depletion region. Figure 10.4(b) repeats the energy-bands layout for an unbiased junction as the reference point.

The following interpretation can also be made regarding the results of biasing: The width of a depletion region is proportional to $\sqrt{(V_D - V)}$, where $V$ is a biasing voltage [2]. Forward biasing means that $V$ is positive; thus, forward biasing reduces the width of a depletion region. $V$ is negative for a reverse-biased diode; therefore, in this case the depletion region becomes wider. This is shown symbolically in Figure 10.4.

When we compare Figures 10.2, 10.3, and 10.4, it is necessary to direct our attention to two things: (1) the energy gap $E_g = E_c - E_v$ and (2) the location of Fermi levels $E_{Fc}$ and $E_{Fv}$ with respect to the edges of the conduction, $E_C$, and the valence, $E_v$, bands. These *two quantities determine the electrical and optical properties of a semiconductor material.* Keep this point in mind.

An important result of forward biasing is that, from the standpoint of the *p-n* junction, it not only forces a number of electrons to move through the depletion region (which, from the standpoint of the energy band, means exciting a number of electrons at the conduction band) but it also gives excited electrons directed velocities in the conduction band. In other words, forward biasing generates a directed flow of electrons—that is, current—through connected pieces of *p*- and *n*-type semiconductors. This result should be familiar to you from the basic discussion in Section 9.1.

## A Closer Look at the Bandgaps

***Conservation of energy and momentum*** The energy-band diagrams we have considered so far have helped us to understand the basic operation of a semiconductor laser. They were sketched in accordance with a simple rule: The vertical axis shows the energy value and the horizontal axis serves as a visual aid. (See, for example, Figure 10.4, a practical way to represent optical processes in a laser diode.)

When describing the absorption and emission of a photon, we operated with energy levels $E_2$ and $E_1$ within the conduction and valence bands, as shown in Figure 9.10. Thus, a photon's energy, $E_p$, is equal to $E_2 - E_1$. Since a photon's energy determines the frequency, $f$, or wavelength, $\lambda$, of emitted or absorbed light, it would seem that the only thing we need to know is the *energy conservation law:* $E_p = E_2 - E_1$. This is not enough, however, for optical transitions in semiconductors. Radiation and absorption processes in a laser diode have to satisfy conservation laws for both energy and momentum. But up to this point we have not considered the law of conservation of momentum. We'll do that now.

Let's introduce a wave vector of an electron, **k**, whose magnitude is equal to

$$|\mathbf{k}| \equiv k = 2\pi/\lambda \tag{10.11}$$

The direction of this vector is the direction of an electron's velocity, $v$. We'll also introduce an electron's momentum, **p**, as

$$\mathrm{p} = m^* v, \tag{10.12}$$

where $m^*$ is the electron's effective mass, as defined in Formula 10.6. As an elementary particle, an electron exhibits wave-corpuscular duality, which is formally described by de Broglie's famous formula:

$$p = h/\lambda, \tag{10.13}$$

where $h$ is Planck's constant ($h = 6.63 \times 10^{-34}$ J-s). Since both vectors—the electron's momentum, **p**, and the electron's wave vector, **k**—have the same direction, their relationship is given by:

$$\mathbf{p} = (\hbar)\mathbf{k}, \qquad (10.14)$$

where $\hbar = h/2\pi$. Therefore, the law of conservation of momentum for optical transition from the conduction band (the initial state) to the valence band (the final state) is reduced to the *k conservation rule:*

$$\Delta k = k_e - k_h + k_{phot} \approx 0, \qquad (10.15)$$

where $k_e$ is the wave vector of an electron (the initial state), $k_h$ is the wave vector of a hole (the final state), and $k_{phot}$ is the wave vector of a photon. The magnitude of a wave vector of a photon is two orders of value less than that of either the wave vector of a hole or of an electron; consequently, we can use the following approximation of Formula 10.15 [4]:

$$k_e - k_h \approx 0 \qquad (10.15a)$$

***The E-k diagram*** A relationship exists between the energy, $E$, and wave vectors, **k**, of all the elements involved in optical transition in a laser diode. What's more, the E-k relationship delivers important information about the availability of states in the conduction and valence bands.

The linkage between energy $E$ and wave vector **k** can now be derived as follows [4]:

$$E = \tfrac{1}{2} m^* v^2 = p^2/2m^* = (h)^2 (k^2/2m^*) \qquad (10.16)$$

If we now depict the energy bands—plotting energy $E$ along the vertical axis and wave vector **k** along the horizontal axis—we'll have parabolas representing both valence and conduction energy bands. But we must remember that we are in the realm of quantum physics, where both momentum **p** and wave vector **k** are discrete, or quantized, entities. Thus, Formula 10.16 gives us a set of separate energy levels for each value of **k.** You'll recall from Section 9.1 that these levels are so close to one another in a semiconductor material that we depict them as energy bands. Such a drawing is called an *E-k* diagram (Figure 10.5).

Remember, energy bands in *E-k* space change continuously, as is shown qualitatively in Figure 10.5 (c). The major properties of a semiconductor are determined by the electron transitions between the closest segments of the conduction and valence energy bands. This is why we can neglect other segments of the energy bands and concentrate only on their closest sections, as we have done in Figures 10.5 (a) and (b).

***Direct and indirect bandgaps*** In the mutual location of conduction and valence energy bands in the *E-k* space, only two possibilities exist. One is where the minimum portion of a conduction band is located directly against the maximum segment of a valence band. This situation is shown in Figure 10.6(a). Such a material is called a *direct-bandgap semiconductor.* The second possibility is when the minimum portion of a conduction band and the maximum segment of a valence band are kept apart. Such a material is called an *indirect-bandgap semiconductor.* It is shown in Figure 10.6(b).

The laws of both energy and momentum conservation are satisfied with a high degree of accuracy in direct-bandgap semiconductors. Indirect-bandgap semiconductors violate these laws to some degree. This is why direct-bandgap semiconductors exhibit strong and probable interaction—that is, stimulated emission—between external radiation and the material. Indirect-bandgap materials, on the contrary, undergo weak and less probable interaction between external radiation and the semiconductors. This is the reason only direct-bandgap semiconductors are used

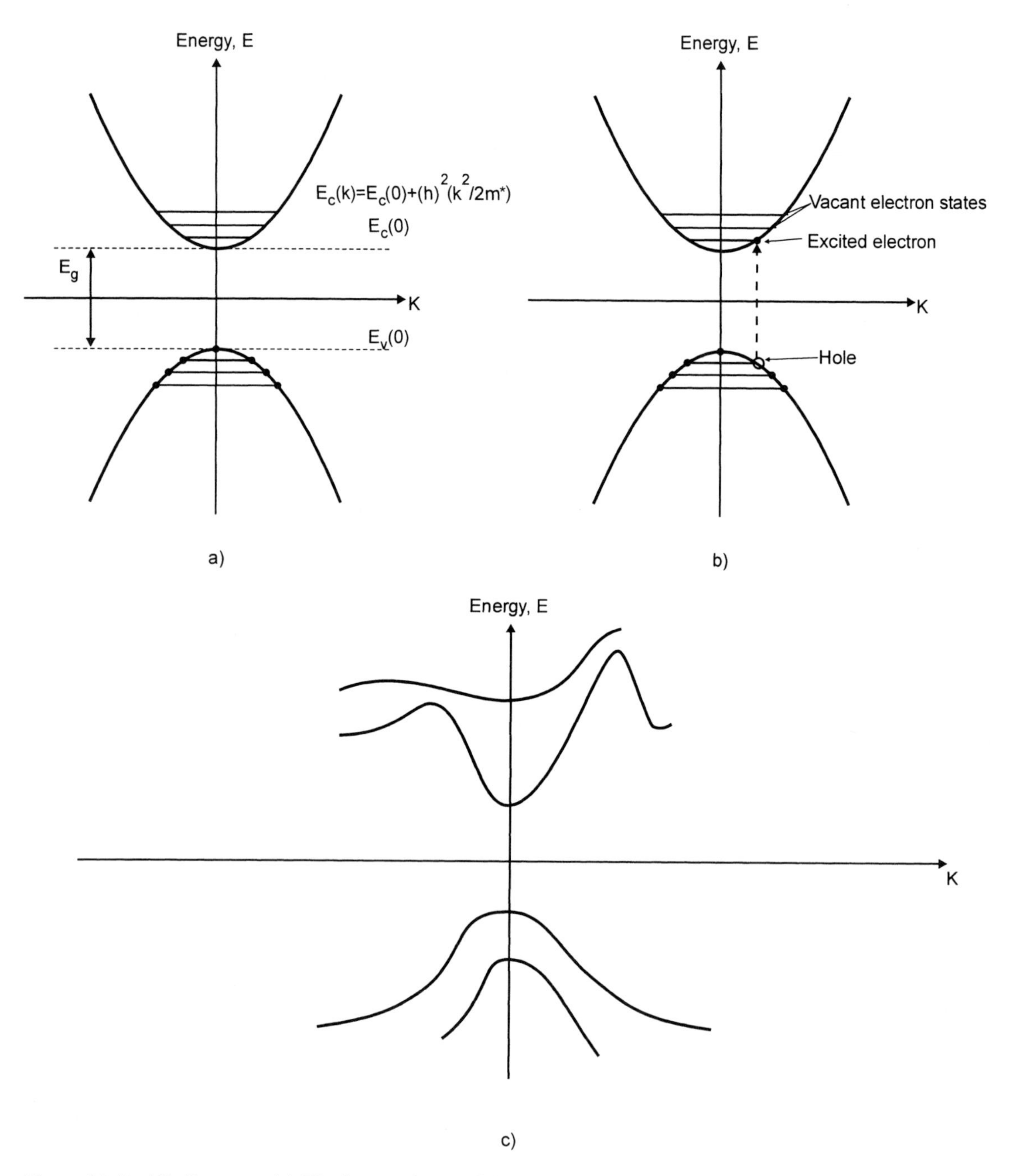


**Figure 10.5** E-k diagrams: (a) E-k diagram for $T = 0$ K; (b) E-k diagram for $T>0$ K; (c) energy-band diagram of GaAs. *([a] and [b] adapted from Christopher Davis'* Lasers and Electrooptics—Fundamentals and Engineering. *Copyright 1996. Reprinted with permission of Cambridge University Press.)*

to fabricate LEDs and laser diodes. You can now immediately understand that semiconductors like GaAs, AlGaAs, InP, and InGaAsP, mentioned previously, are direct-bandgap materials.

Don't forget: We want to use these semiconductors as a light source and we need to consider all processes from this standpoint. Hence, it is important to recognize that *the transitions of excited electrons from conduction band to valence band are purely radiative in direct-bandgap semiconduc-*

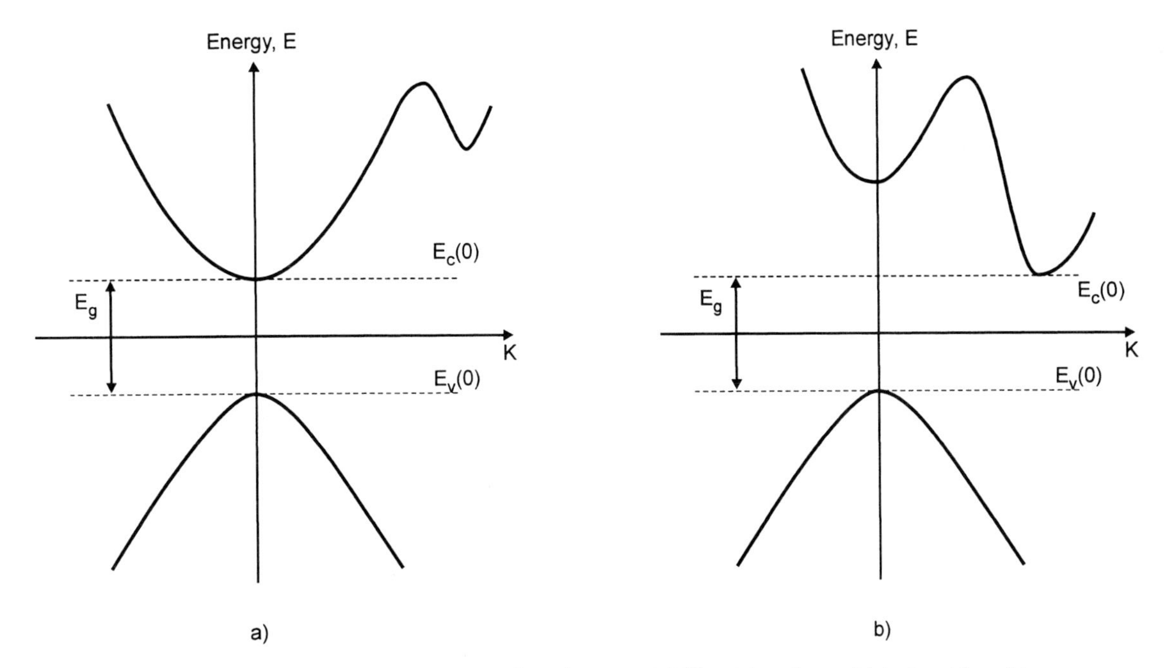


**Figure 10.6** Direct- and indirect-bandgap semiconductors: (a) Direct bandgap; (b) indirect bandgap.

*tors and are accompanied by nonradiative transitions in indirect-bandgap materials.* This is because to conserve the momentum of all the particles involved in the transition—an electron, a hole, and a photon—the radiation process in indirect-bandgap materials is accompanied by emission of a nonradiative phonon. The internal quantum efficiency of a semiconductor laser is essentially determined by the ratio of radiative and nonradiative recombinations. (See Formulas 9.6 and 9.7.) This is another reason why only direct-band semiconductors are used to fabricate semiconductor lasers.

Now that you have been provided with a solid foundation on semiconductor materials as they are used in laser diodes, let's turn our attention to the laser diodes themselves.

## 10.2 EFFICIENCY OF A LASER DIODE

What makes a laser diode an efficient converter of electric signals into optical ones is the subject of this section. We'll describe and quantify a laser's efficiency based on an in-depth consideration of the physical processes that occur inside these semiconductor diodes. Both energy-band and *p-n* junction approaches will be exploited to make the picture clearer. We concentrate on laser diodes because they have found a wide spectrum of applications and are now replacing LEDs in many fiber-optic communications systems. Apart from this, if you understand laser-diode theory and practice, you will have no problem understanding how an LED works.

### Input-Output Relationship

We are looking for the relationship between the output power of a laser diode, $P$, and the input forward current, $I_F$, above the laser threshold. Output power is equal to

$$P = (E_p \times N_{esc})/t, \tag{10.17}$$

where $E_p$ (J) = $hf = hc/\lambda$ is the photon's energy, $N_{esc}$ is the number of photons escaping from the laser's resonator (that is, the number outside the laser, which we count as the actual output), and $t$ (s) is the time of radiation.

The number of escaped photons, $N_{esc}$, is equal to the number of charge carriers injected into an active area times *external quantum efficiency,* $\eta_{ext}$. Thus,

$$\eta_{ext} = \textit{number of photons escaped/number of injected charge carriers} \qquad (10.18)$$

Since the radiation process in an active area is determined mostly by electrons, we can approximate the number of charge carriers by the number of electrons. The latter is equal to

$$\textit{number of injected electrons} = (I_F \times t)/e, \qquad (10.19)$$

where $I_F$ (A) is the forward current and $e$ (C) is the unit charge. From Formulas 10.19 and 10.18 we find

$$N_{esc} = ((I_F \times t)/e) \times \eta_{ext} \qquad (10.20)$$

Substituting Formula 10.20 into Formula 10.17, we can determine how the output power relates to the input forward-biasing current:

$$P = (E_p \times N_{esc})/t = (E_p/e) \times \eta_{ext} \times I_F \qquad (10.21)$$

For a given radiating wavelength, all coefficients in Formula 10.21 are constant and we can denote that as $S$. Since we are discussing the relationship between power and current above the laser threshold, we actually consider $P - P_{th} = \Delta P$ and $I - I_{th} = \Delta I$. (See Figure 9.11[b].) Therefore, we can write:

$$\Delta P = S\,\Delta I, \qquad (10.22)$$

where $S$ is the slope efficiency, which can be found in a laser diode's data sheet. Thus,

$$S(\mathrm{W/A}) = (E_p/e) \times \eta_{ext} \qquad (10.23)$$

Since $E_p \geq E_g$, where $E_g$ is the semiconductor bandgap, we can rewrite Formula 10.23 in the following form:

$$S(\mathrm{W/A}) = (E_g/e) \times \eta_{ext} \qquad (10.23a)$$

### Example 10.2.1

***Problem:***

Calculate the slope efficiency of a laser diode operating at $\lambda = 1300$ nm if its external quantum efficiency $\eta_{ext} = 0.1$.

***Solution:***

For $\lambda = 1300$ nm, $E_p = 0.0153 \times 10^{-17}$ J. Recalling that $e = 1.6 \times 10^{-19}$ *C,* we can compute $\Delta P = 0.956 \times \eta_{ext,} \times \Delta I$. (We'll discuss shortly the mechanisms determining external efficiency.) Now let's take the given value $\eta_{ext} = 0.1$. Then we have $S = 0.096$, where the dimension of the slope efficiency is *J/C* = *W/A*. Therefore, for the given laser diode, $\Delta P = 0.096\,\Delta I$.

In reality, the best slope efficiency that a VCSEL diode has displayed is 0.192 mW/mA, as the data sheet in Section 9.3 shows. This oversimplified approach allows us to estimate the order of magnitude of the slope efficiency of a laser diode. Understand, however, that this derivation is valid only for a steady-state lasing operation because the relationship obtained through Formula 10.21 does not reflect any phenomena below and at the threshold.

---

## Three Types of Efficiency

***Power efficiency*** As the above considerations show, the input–output relationship of a laser diode is determined by its external quantum efficiency. Before we investigate this parameter further, let's digress to introduce one more quantity—power efficiency—that also connects the input and output of a laser diode.

*Power efficiency,* $\eta_p$, is the ratio of optical output power radiated by a laser diode, $P$, to the dc electric power supplied, $P_{dc}$, so that

$$\eta_P = P/P_{dc} \tag{10.24}$$

$P_{dc}$ is the electric power dissipated across the laser diode. $P_{dc}$ consists of two parts. The first is power dissipated across the junction and equals $I_F \times V$. Voltage across the junction is on the order of $E_g/e$, as Figure 10.6 shows, and it is typically between 1 and 2 V. Hence, the first part can be presented as $I_F \times E_g/e$. The second part is power dissipated across the contacts and internal resistances of the bulk semiconductor. If we denote the equivalent resistance of all the above resistances as $R$, this part will be equal to $I_F{}^2 \times$ R. Equivalent resistance is on the order of 1 to 2 Ω. Thus, the power efficiency is given by [5]:

$$\eta_P = P/((I_F \times E_g/e) + I_F{}^2 \times R) \tag{10.25}$$

### Example 10.2.2

***Problem:***

**a.** Calculate the power efficiency of a VCSEL diode whose data sheet is given in Figure 9.20.

**b.** Calculate the power efficiency of a broad-area laser diode whose data sheet is given in Figure 9.17.

***Solution:***

**a.** From Figure 9.20, we see that the optical power radiated by this diode is 1.25 mW at $I = 10$ mA.

We need voltage across the junction, which is $V = E_g/e$. The value of $E_g$ can be found from $E_g$ (eV) = 1248/λ (nm). Since λ = 840 nm from Figure 9.20, $E_g = 1.48$ eV. Hence, $V = 1.48$ V.

*Note:* (1) $E_g/e$ gives the voltage in volts if $E_g$ is given in electron volts (eV). This follows from the definition of eV. One can verify this result by substituting into the formula the numbers measured in International System (SI) units. Thus, V (volts)= $E_g$(eV $\times 1.6 \times 10^{-19}$) J/$e$ ($1.6 \times 10^{-19}$ C).

(2) Typical forward voltage, measured at $I = 10$ mA, is given as 1.8 V. This is the voltage running across the laser diode while we've been calculating the voltage across the *p-n* junction. Note how close they are.

We also need the value of $R$. Let's assume $R = 1$ Ω. Plugging the numbers into Formula 10.24, we obtain:

$$\eta_P = P_{opt}/((I \times E_g/e) + I^2 \times R) = 1.25 \text{ mW}/\ (14.8 \text{ mW} + 0.1 \text{ mW}) = 0.084 = 8.4\%$$

This is the power efficiency for one of the best types of laser diodes available on the market.

**b.** Before repeating all the above considerations, we need to find what forward current is needed to obtain 1 mW of output power—the maximum power radiated by this LD. From Section 9.3, we can determine that the slope efficiency, $\Delta P/\Delta I$, for this LD is equal to 3.25 μW/mA. Hence, for 1 mW of optical power, we need about 300 mA of forward current.

The second number we need is the voltage across the junction, that is, $E_g$. For $\lambda = 1300$ nm, we find that 1248/1300 = 0.96 eV; hence, $V = 0.96$ V. (Again, note that the data sheet gives 1.5 V.)

Now the power efficiency is:

$$\eta_P = P_{opt}/((I \times E_g/e) + I^2 \times R) = 1 \text{ mW}/(300 \text{ mW} + 180 \text{ mW}) = 0.002 = 0.2\%.$$

Thus, you can see that a broad-area laser diode is much less efficient—in terms of power conversion—than a VCSEL diode.

---

***External quantum efficiency*** Let's return now to our point under discussion: laser-diode efficiency. We want to examine all the processes from the injecting of charge carriers to the escaping of photons. The various efficiencies of a laser diode help us do this.

*External quantum efficiency,* $\eta_{ext}$, is defined above (Formula 10.18) as the ratio of the number of photons escaping from a semiconductor to the number of charge carriers injected into an active region. To find the relationship between the external and power efficiencies, let's express $\eta_{ext}$ in terms of output power. From Formulas 10.22 and 10.23 we find:

$$\eta_{ext} = (e/E_g) \times \Delta P/\Delta I \quad (10.26)$$

Let's approximate the slope efficiency with $\Delta P = P - P_{th}$ and $\Delta I = I - I_{th}$, where $P$ and $I$ are rated (maximum) power and current, respectively. (Again, see Figure 9.11[b].) Note that threshold power is small compared with rated output power. For example, from Figure 9.17 we see that $P_{th} \leq 50$ μW and P = 1 mW. Thus, Formula 10.23 can be rewritten as follows:

$$\eta_{ext} \approx (e/E_g) \times (P/(I - I_{th})) \quad (10.27)$$

Formula 10.27 allows us to express power efficiency through external quantum efficiency:

$$\eta_P \sim \eta_{ext} \times ((I - I_{th})/I), \quad (10.28)$$

where further approximation, $I^2 \times R << I \times E_g/e$, is introduced. Be careful in using Formula 10.28 and check whether the data of your laser diode and the precision of your calculations allow you to apply approximation $I^2 \times R << I \times E_g/e$ or not. If not, derive and use the more accurate Formulas 10.25 and 10.27.

***Internal quantum efficiency*** External quantum efficiency covers two mechanisms: the radiation of photons from an active region and the transmission of these photons from the active region to the outside of the laser. The first mechanism is described by the laser's internal quantum efficiency, $\eta_{int}$, and the second one by the laser's cavity efficiency, $\eta_{cav}$. The term *cavity* here refers to the physical space from the point where a photon is radiated to the outside of the laser diode. Thus, the cavity includes the mirrors and the semiconductor material. (See Figure 9.12[a].) We can now write:

$$\eta_{ext} = \eta_{int}\eta_{cav} \quad (10.29)$$

*Internal quantum efficiency*, $\eta_{int}$, is the ratio of the number of photons radiated from an active region (radiated photons) to the number of injected charge carriers. Thus, we can write:

$$\eta_{int} = \textit{number of radiated photons/number of injected charge carriers} \quad (10.30)$$

A comparison of Formulas 10.18 and 10.30 shows that

$$\eta_{ext}/\eta_{int} = \textit{number of escaped photons/number of radiated photons} \quad (10.31)$$

In other words, not all the photons radiated in the active region will escape from the cavity of a laser diode and contribute to the optical-power output of that laser diode. In a good-quality LED, $\eta_{int}$ reaches 0.5 and, in a good laser diode, $\eta_{int}$ approaches 1.

*Cavity efficiency*, $\eta_{cav}$, is actually defined by Formula 10.31.

$$\eta_{cav} = \eta_{ext}/\eta_{int} = \textit{number of escaped photons/number of radiated photons} \quad (10.31a)$$

There are two main mechanisms triggering the loss of radiated photons: losses within the semiconductor material and partial transmission through the resonator mirrors. The term *material* covers all matter that a photon meets from its point of origin up to the mirrors. "Mirrors" here refers to any means to provide positive optical feedback in a laser diode, including Bragg gratings. Therefore, the second reason for the losses—the partial transmission through the resonator mirrors—simply takes into account the fact that not all the radiated photons will be reflected back into an active region by the resonator. We can write these considerations in the following form [2]:

$$\eta_{cav} = \eta_{ext}/\eta_{int} = (\textit{transmission loss/total loss}), \quad (10.32)$$

which is another way to say that external quantum efficiency is equal to internal quantum efficiency times the ratio of transmitted power to power generated by the active region [5].

In the case of lump reflectors, such as mirrors from which a Fabry-Perot resonator is formed, *transmission loss*, $\alpha_{tr}$, can be expressed in this form (see also the box "Gain and Loss of a Laser Diode" on page 382:

$$\alpha_{tr} = (1/2d)\ln(1/(R_1 \times R_2)), \quad (10.33)$$

where $d$ is the length of the active region and $R_1$ and $R_2$ are the power-reflection coefficients of the laser mirrors. *Total loss*, $\alpha$, is equal to

$$\alpha = \alpha_{mat} + \alpha_{tr}, \quad (10.34)$$

where *material loss*, $\alpha_{mat}$, is loss caused by absorption, scattering, and other possible mechanisms that prevent radiated photons from escaping from the laser cavity. Thus, the relationship between external quantum efficiency and internal quantum efficiency is given by [2]:

$$\eta_{ext} = \eta_{int}(\alpha_{tr}/(\alpha_{mat} + \alpha_{tr})) \quad (10.35)$$

*Note:* Some books give different meanings to some of the terms used here. For example, $\eta_{ext}$ is sometimes called *differential quantum efficiency* [7] because it shows how output power increases in response to an increase in input current. (See Formula 10.26.) So you must read carefully the definitions in relation to the context in which they're applied before you use a specific term.

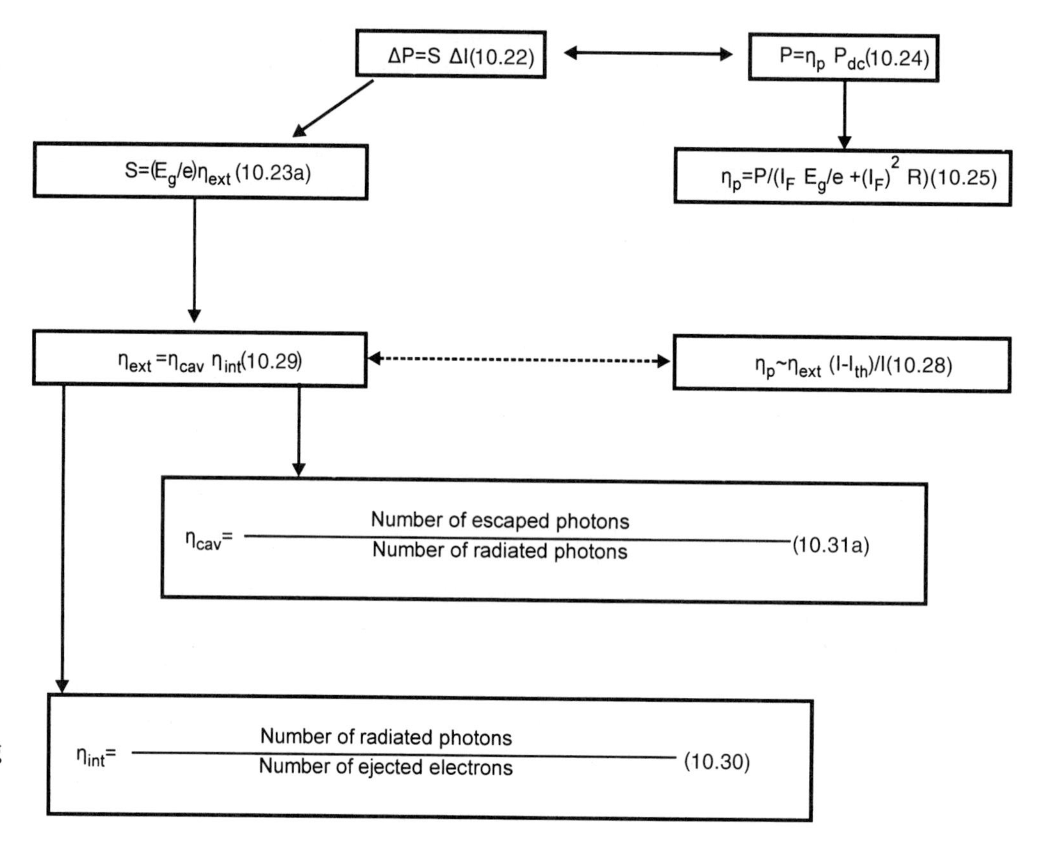


**Figure 10.7** Relationships among efficiencies of a laser diode.

It follows from Formula 10.35, then, that *cavity efficiency* is a useful term, one that has served us well in working out all these derivations.

The net result is this: Internal quantum efficiency recounts how efficiently the active region converts injected charge carriers into photons; cavity efficiency shows how efficiently a laser cavity allows these radiated photons to escape. Thus, these two efficiencies describe how well, how effectively, a laser diode converts injected charge carriers (input) into escaped photons (output). These two parameters are integrated into one laser-diode characteristic: external quantum efficiency. The diagram in Figure 10.7 shows the relationship among all the efficiencies of a laser diode.

### Gain and Loss of a Laser Diode

Consider a laser diode with a Fabry-Perot resonator. (See Figure 9.14.) Let $g(f)$ (1/m) be the *gain of an active region* and let $\alpha_{mat}$ (1/m) be the loss within a semiconductor material. *Both* g(f) *and* $\alpha_{mat}$ *are given per unit length.* In this case, an amplitude, $A$, of the electromagnetic wave propagating between two mirrors experiences a gain that equals $\exp(g \times d)$ and a loss that equals $\exp(-\alpha_{mat} \times d)$, where $d$ is the length (thickness) of an active region. After striking the mirrors with *reflectances* $r_1$ and $r_2$, the electromagnetic wave returns to its initial position. In a steady-state operation, initial amplitude $A$ has to be equal to the amplitude of the returned wave, that is [5]:

$$A = r_1 \; r_2 \; A \exp((g(f) - \alpha_{mat})\, d) \qquad (10.36)$$

To make a laser diode operate, the gain has to be equal to or more than the total loss:

$$g(f) \geq \alpha_{mat} + 1/2\,d \; \ln(1/(R_1 \times R_2)) \qquad (10.37)$$

where $R_1 = r_1^{\,2}$ and $R_2 = r_2^{\,2}$ are *power-reflection coefficients* of the resonator mirrors. This is where Formulas 10.33 for transmission loss and 10.34 for total loss come from.

One can easily derive the *threshold condition* from Formula 10.37. Threshold gain, $g_{th}$, is equal to total loss:

$$g_{th}(f) = \alpha_{mat} + 1/2d\ ln(1/(R_1 \times R_2))(10.37a)$$

The meaning of Formulas 10.37 and 10.37(a) is depicted in Figure 10.8.

The important point to underscore is that *gain is a function of frequency but loss is not.* Thus, the lasing effect can be reached only within a certain range of frequencies (from $f_1$ to $f_2$ in Figure 10.8), that is, where gain exceeds loss.

Don't confuse the meanings of Figures 9.11(a) and 10.8. The former depicts gain/loss as a function of the length (thickness) of an active region and the latter pictures gain/loss as a function of frequency.

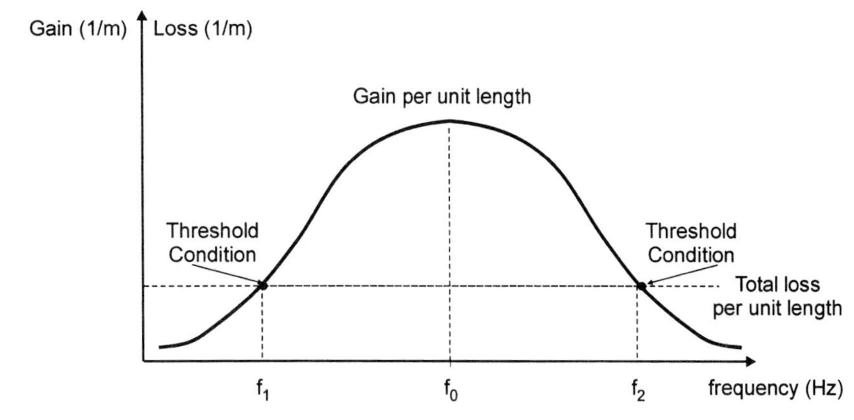


**Figure 10.8** Gain and loss of a laser diode as a function of frequency.

## More About the Efficiency of Laser-Diode Operation

***Internal quantum efficiency and recombination lifetimes*** The idea of internal quantum efficiency was introduced in Section 9.1. (See Formulas 9.1 and 9.7.) This parameter describes what fraction of electrons and holes entering an active region will actually produce photons. Since photons are released as a result of the recombination of electrons and holes, we can say that $\eta_{int}$ describes what fraction of the total recombination will result in radiation. This point is brought out in Formula 10.30.

The major factor that reduces internal quantum efficiency is nonradiative recombinations, which prevent electrons and holes that enter the active region from producing photons. Formula 9.7, which is repeated here, reflects this mechanism:

$$\eta_{int} = \tau_{nr}/\tau, \tag{10.38}$$

where $\tau_{nr}$ is the *nonradiative lifetime.* This is a measure of how long a charge carrier exists between the moment it enters the active region and the moment it recombines nonradiatively. Introducing *radiative lifetime,* $\tau_r$, we can define *total recombination lifetime,* $\tau$, as $1/\tau = 1/\tau_r + 1/\tau_{nr}$. (See Formula 9.6.)

Thus, internal quantum efficiency is, on the one hand, the ratio of the number of radiated photons to the number of injected electrons, as Formula 10.30 says, and, on the other hand, the ratio of the total recombination lifetime to the radiative lifetime, as Formula 10.38 states. The relationship between these two conditions can be found from the following consideration:

Let $n$ be the volume density of electrons entering the active region. Then the *recombination rate, d(n)/dt,* can be expressed as:

$$d(n)/dt = -n/\tau \tag{10.39}$$

Bear in mind that the same holds true for holes. This formula says that the rate at which recombinations occur is equal to the volume density of charge carriers, *n,* divided by their recombination lifetime, $\tau$. Indeed, this is the recombination lifetime that determines how long it takes for a charge carrier to recombine. The shorter the $\tau$, the faster the recombination and vice versa. The negative sign on the right side of Formula 10.39 tells us that $n$ decreases during the recombination process.

To count both radiative and nonradiative recombinations, we have to introduce appropriate lifetimes, $\tau_r$ and $\tau_{nr}$, respectively. Then, recombination rates of radiative and nonradiative processes are equal to $-n/\tau_r$ and $-n/\tau_{nr}$, respectively. The total recombination process is the sum of these two, so that

$$n/\tau = n/\tau_r + n/\tau_{nr} \tag{10.40}$$

Thus, the relationship given in Formula 9.6, $1/\tau = 1/\tau_r + 1/\tau_{nr}$, is obtained.

The internal quantum efficiency can be defined as the ratio of the recombination rates of the radiative and total processes, that is,

$$\eta_{int} = (n/\tau_r)/(n/\tau) = (1/\tau_r)/(1/\tau_r + 1/\tau_{nr}) = \tau_{nr}/\tau \tag{10.41}$$

This is where Formula 9.7 comes from. Incidentally, from Formula 10.41 we can derive another popular form of the internal quantum efficiency formula:

$$\eta_{int} = \tau/\tau_r \tag{10.41a}$$

The physics underlying this definition can be seen from the following representation of this formula:

$$\eta_{int} = \tau_{nr}/\tau = 1/(1 + \tau_r/\tau_{nr}) \tag{10.42}$$

In a good laser diode, a radiative recombination occurs almost immediately after charge carriers enter the active region. This means *the lifetime of the radiative process approaches zero.* In contrast, nonradiative recombination takes a relatively long time. This means the nonradiative lifetime is a large number compared to the radiative lifetime. Thus, the second term in the denominator, $\tau_r/\tau_{nr}$, approaches 0 and the internal quantum efficiency in a superior laser diode approaches 1.

***A well-designed double heterostructure is the key to the efficient operation of a superior laser diode*** The internal quantum efficiency tells us what fraction of the injected charge carriers is converted into photons. This is extremely important but not all that we want from a superior laser diode. We want a laser diode to radiate high-power, well-directed, and coherent monochromatic light. And these characteristics are not described by $\eta_{int}$ only. Thus, we need to investigate a laser diode's operation in more detail.

The core of a laser diode is its active region. The operating efficiency of an active region is determined mainly by two mechanisms: *the confinement of charge carriers* within the small active region and *the guiding of radiated photons* from the active region to the outside of the diode. We know from Sections 9.1 and 9.2 that good laser diodes are built using a *double heterostructure.* Now we will take a closer look at how such a heterostructure provides for the efficient operation of these two mechanisms.

The mechanisms of charge-carrier confinement and *population inversion* can be understood from an analysis of a detailed sketch of a double-heterostructure semiconductor and its energy-band diagrams, shown in Figure 10.9. Also, refer to Figures 9.5(c) and 9.12.

First of all, bear in mind that the dimensions of layers shown in Figure 10.9(a) are far out of scale. An active layer is typically about 0.1 μm thick—even around 0.01 μm for quantum-well lasers—while confining layers are about 2 μm in thickness. Contact layers and substrates (not

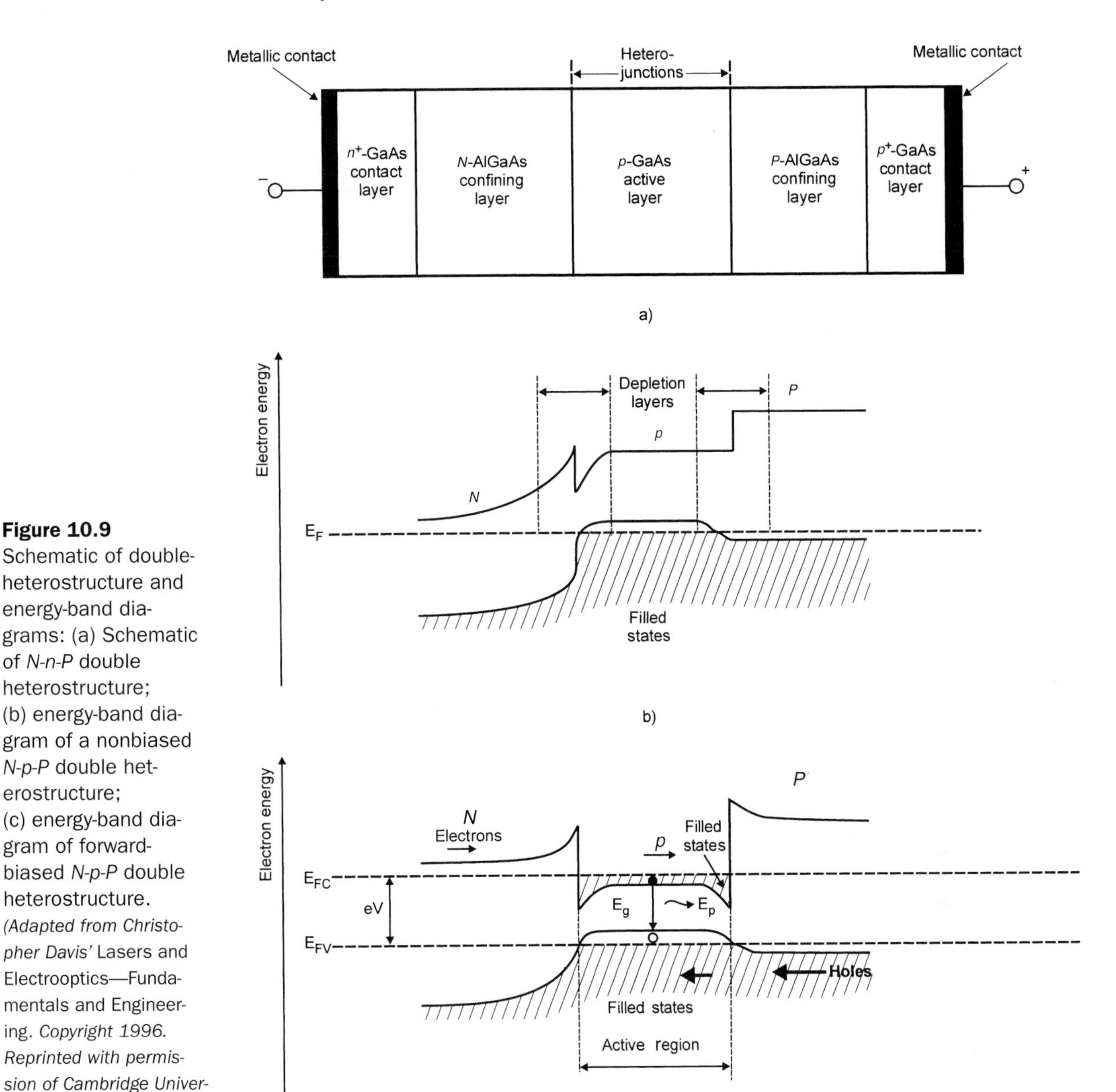


**Figure 10.9** Schematic of double-heterostructure and energy-band diagrams: (a) Schematic of *N-n-P* double heterostructure; (b) energy-band diagram of a nonbiased *N-p-P* double heterostructure; (c) energy-band diagram of forward-biased *N-p-P* double heterostructure. *(Adapted from Christopher Davis'* Lasers and Electrooptics—Fundamentals and Engineering. *Copyright 1996. Reprinted with permission of Cambridge University Press.)*

shown) are much thicker—they start at 3 μm. Thus, you can see how much the real dimensions are distorted. This has been done for illustrative purposes.

The notations used in Figure 10.9 mean the following: The lowercase letters *n* and *p* stand for regular-doped *n*-type and *p*-type semiconductors. Heavy-doped *n*-type and *p*-type semiconductors are denoted as *n*+ and *p*+, respectively. Such materials are used for contact layers because they exhibit relatively large conductivity. These properties facilitate metal-semiconductor contact, which is important for optical applications. Interestingly, in electronic or rectification diodes, this contact has to be abrupt.

In addition to their different conductivities, the materials shown in Figure 10.9 have different bandgaps. The *n* and *p* semiconductors, as well as the *n*+ and *p*+, have small bandgaps. The *N* and *P* semiconductors are the regular *n*-type and *p*-type semiconductors with large bandgaps.

Let's now consider the energy-band diagrams. When a laser diode is *nonbiased* (Figure 10.9 [b]), all contacting materials are in *thermal equilibrium.* This means they have the same Fermi energy level, $E_F$, as discussed in Section 10.1. All energy levels up to Fermi level $E_F$ in the valence band are filled. When a laser diode is *forward biased* (Figure 10.9 [c]), the materials used in heterojunctions are not in equilibrium any longer. Fermi levels become different for the conduction and valence bands. The most important effect is that the energy of the conduction band at the active layer becomes much less; as a result, this energy-level band makes a potential well. Electrons have filled all the states at the conduction band below the new position of Fermi level $E_{FC}$. They have also filled all the states below new Fermi level $E_{FV}$ at the valence band so that the top of this band is replete with vacant states for electrons. The new Fermi levels, $E_{FC}$ and $E_{FV}$, are called *quasi-Fermi levels.*

Look again at Figure 10.9(c). Electrons in the active region at the conduction band are trapped by the high energy walls—this is why this energy diagram is called a "potential well"—and they have only two "exits." The first enables them to drop to the valence band, where they recombine with holes and where the release of photons occurs. This is what we want from them. Since the energy gap between the conduction and valence bands in the active region is very small, the descent of electrons ("drops," if you like) is very likely to happen, so the probability of such a process taking place is very high. The second exit for electrons forces them to climb over a potential barrier, leaving this well to remain a *P*-confining layer. However, since this barrier is rather high, only a few electrons acquire enough energy to surmount it. Thus, most of the electrons will be confined to the conduction band of the active layer.

Now let's turn our attention to the valence band. The top of the valence band at the active layer contains numerous holes. These holes are trapped inside the active region by the energy walls. (Remember, holes are positive charge carriers, so additional negative energy is the barrier for them.) This is how charge carriers are efficiently confined within an active region. But a hole is the vacancy left by an electron, so we can say that the top of the valence band in the active layer is devoid of electrons. In other words, forward biasing in a double-heterostructured laser diode creates *population inversion* when the upper energy band contains more electrons than the lower one. Thus, confinement of charge carriers within an active region helps to create population inversion.

At this juncture, let's consider the second mechanism providing for the efficiency of a laser-diode operation—the guiding of radiated photons from the active region to the laser output. Refer to Figure 9.5, which shows the refractive-index profile of active and confining layers.

This structure works as a waveguide. (Have you remembered that total internal reflection governs the operation of an optical fiber?). Thus, most radiated photons will be directed along this waveguide to the output of the laser diode.

But how about the absorption of these photons as they travel along this waveguide? Look at Figure 10.9(c) again. Notice the difference in bandgaps between the active region and the confining layers. The bandgap of the active region, $E_g$, is much smaller than the bandgaps of both the confining layers, *N* and *P.* This means that most of the radiated photons with energy $E_p$, which is equal to or a little more than $E_g$, will pass through both confining layers without interaction because $E_p$ is much less than the values of both bandgaps. (Remember, a photon's absorption occurs if, and only if, $E_p \approx E_g$.) However, few photons will be radiated after the transition from the conduction-band energy levels above $E_{FC}$ to the valence-band energy levels below $E_{FV}$. These photons have energies that fit the bandgaps of the confining layers, *N* and *P.* In other words, these photons will be absorbed by the confining layers. This is the physical mechanism of the material loss, $\alpha_{mat}$, discussed above.

You should recall that this type of laser structure is referred to as "index guided." This is how most laser diodes are built today. There is also a gain-guided structure. Both types are discussed in Section 9.2.

***Efficiency in terms of an E-k diagram*** The efficiency of a laser diode can also be described in terms of an *E-k* diagram, as Figure 10.10 shows. In a normal situation, a semiconductor material absorbs photons and thus exhibits a loss of light power. This absorption is proportional to the difference in population between the valence and conduction bands. These populations are described by Fermi-Dirac distributions, as Formula 10.2, states. Thus, we can write:

$$\alpha \sim f_V(E) - f_C(E) \sim 1/(1 + \exp(E_V - E_{FV})k_BT - 1/(1 + \exp(E_C - E_{FC})/k_BT), \quad (10.43)$$

where $\alpha$ is total loss and $E_V$ and $E_C$ are the boundary energies of the valence and conduction bands, respectively, so that $E_C - E_V = E_g$, where $E_g$ is the bandgap. $E_{FV}$ and $E_{FC}$ are valence and conduction quasi-Fermi levels, respectively. All these notations are shown in Figure 10.10. Therefore, the loss is proportional to

$$\alpha \sim \exp((E_C - E_{FC})/k_BT) - \exp((E_V - E_{FV})/k_BT) \quad (10.44)$$

To make a laser diode radiate, we need to obtain gain, not loss. But gain is not more than negative loss. Hence, in order to obtain $\alpha < 0$, that is, gain, we need to have

$$\exp((E_C - E_{FC})/k_BT) < \exp((E_V - E_{FV})/k_BT) \quad (10.45)$$

Simple manipulations lead to the following condition of having *gain* in a laser diode:

$$E_{FC} - E_{FV} > E_C - E_V = E_g \quad (10.46)$$

This formula tells us that in order *to achieve gain in a semiconductor material, the quasi-Fermi levels must be inside the conduction and valence bands.* But this immediately implies that electrons fill the lower portion of a conduction band up to $E_{FC}$ and that they deplete the upper portion of a valence band down to $E_{FV}$ because they fill all valence-band levels up to $E_{FV}$. This is equivalent to saying that *the top of the valence band is available to accept excited electrons dropping from the conduction band.* This situation is shown in Figure 10.10.

Remember, gain can be achieved when population inversion is created. Thus, Formula 10.46 is another way to formulate the condition for population inversion and Figure 10.10 depicts this.

Here, indeed, is what makes a laser diode so efficient: *Electrons stimulated to drop to the valence band will find vacancies to complete their path with the radiation of photons. On the*

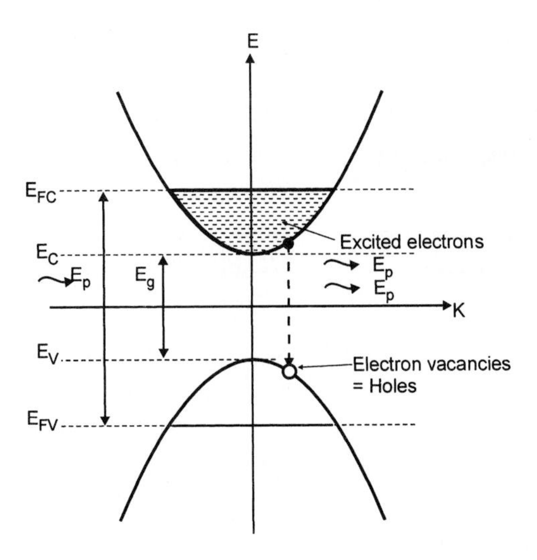


**Figure 10.10** E-k diagram of a laser diode.

*other hand, an external photon with energy* $E_p$ *most likely won't be absorbed by this diode with population inversion because the bottom of the conduction band is occupied; thus, there are no vacancies to allow for transition from the valence band to the conduction band. In such a system, the probability of stimulated emission is very high and the probability of absorption is very low.*

This is a dynamic process in which the time needed to complete all processes within a band is much less than the time needed for transition between the bands. In other words, the distribution of the occupation of energy levels within the bands develops much faster than the process of transition from conduction to valence bands occurs. Thus, we say that *interband transitions* take place much more slowly than do *intraband transitions.*

The role of the *E-k* diagram in this discussion can be explained by considering the *recombination rate.* The overall rate of recombination, *R,* is proportional to the number of electrons and holes available to recombine. (You may notice how we switch back and forth very quickly between an energy-band diagram and a *p-n* junction. This is done to make you more cognizant of both concepts and their relationship.) Thus, we can write:

$$R = r_{\text{rec}}\, np, \tag{10.47}$$

where $r_{\text{rec}}$ ($m^3/s$) is the recombination-rate constant and $n$ and $p$ are the volume density of the charge carriers ($1/m^3$). Formula 10.47 states that for two semiconductors with the same population, the radiation rate depends only on the recombination constant. The point to remember is this: *For direct-bandgap semiconductors, the radiation constant is larger than it is for indirect-bandgap semiconductors by three to five orders of magnitude.* This is why direct-bandgap semiconductors are a thousand to a hundred thousand times more efficient than indirect-band semiconductors. The *E-k* diagram helps us to identify direct- and indirect-bandgap semiconductors.

The only thing left to note is the numbers for $n$ and $p$. They are defined by Formulas 10.2 and 10.3. The Fermi-Dirac distribution now is given in Formula 10.43, so we need to compute coefficients $N_C$ and $N_V$ given by Formula 10.6. To give you an idea of the order of magnitude of $n$ for a homojunction laser with $E_{FC} - E_C = 0.05$ eV, $n$ was computed [6] as ~ $10^{17}$ ($1/cm^3$), while the radiative lifetime is about 0.5 ns.

## 10.3 CHARACTERISTICS OF LASER DIODES

The operation of a laser diode and its characteristics are our next concerns.

### Threshold and Operating Currents

Threshold current is the current at which a diode starts to operate as a laser. (See Figure 9.11[b].) This value is determined by the threshold gain given by Formula 10.37(a). In light of the waveguide effect discussed above, Formula 10.37a should be rewritten in this form [2]:

$$\Gamma\, g_{\text{th}} = \alpha_{\text{tr}} + \alpha_{\text{mat}}, \tag{10.48}$$

where $\Gamma$ is the confinement coefficient that accounts for the guiding of radiated photons by index- or gain-waveguide structures.

As the first approximation, we can consider the *threshold-current density,* $J_{th}$, to be proportional to the threshold gain (which is true for lasers operating in a singlemode regime). Thus,

$$J_{\text{th}} = (1/\beta)\,(\alpha_{\text{tr}} + \alpha_{\text{mat}}), \tag{10.49}$$

where coefficient $\beta$ includes the confinement factor. It is interesting to note that $\alpha_{mat}$ (1/cm) for GaAs changes from 100 for a homojunction to 15 for a double heterostructure. For the same material, coefficient $\beta$ (cm/A) changes from $3 \times 10^{-3}$ for a homojunction to $1.5 \times 10^{-2}$ for a double heterostructure [4].

Laser physicists and designers make every possible effort to reduce threshold-current density because $J_{th}$ eventually determines the power consumption of a laser diode. Progress in reducing threshold-current density can be described by the following figures [2]: The first homojunction semiconductor lasers operated at cryogenic temperature and required $J_{th}$ ~ 19 000 A/cm$^2$. Developing double-heterostructured diodes reduced $J_{th}$ to 1600 A/cm$^2$. Quantum-well laser diodes require a threshold-current density of only about 500 A/cm$^2$. Recent developments in new structures and progress in the fabrication of laser diodes have reduced $J_{th}$ to 65 A/cm$^2$. And there are signs of the possibility to further reduce threshold-current density. Review again the data sheets for the different types of laser diodes given in Figures 9.17, 9.19(a), and 9.19(b). The threshold current given in these data sheets is equal to the threshold-current density times the square of the active area of a laser diode.

The importance of threshold current, or its density, is clear from the prior discussion in this section of a laser diode's input–output characteristic. The lower the threshold current, the less the current required to reach the desired level of output power, as Formula 10.22 shows.

*Threshold current depends on temperature.* This dependence is approximated by the following empirical formula:

$$I_{th}(T) = I_0 \exp(T/T_0), \tag{10.50}$$

where $I_0$ and $T_0$ are characteristic constants whose value depends on the diode's material and structure. For a conventional GaAs laser diode, $I_0$ is on the order of tens of mA, and $T_0$ is on the order of 100°C. The laser-diode data sheets mentioned above display this temperature dependence in explicit forms. The physics underlying this temperature effect dictates that the rate of nonradiative recombination increases with rising temperature. This decreases the internal quantum efficiency, which means a larger number of charge carriers—that is, more current—are required to attain threshold gain. But a more accurate analysis discloses a complex picture, which is why we have to use empirical Formula 10.50 rather than deriving an exact formula based on a physical consideration.

### Example 10.3.1

**Problem:**

Find $I_0$ and $T_0$ for a GaInAsP laser diode whose data sheet is given in Figure 9.17.

**Solution:**

We have two ways of solving this problem: (1) using only the data from the specifications sheet or (2) using Formula 10.50.

Let's consider them in sequence.

(1) Looking at the specifications sheet, we find the following data (see the "Laser Output" box in Figure 9.17): (a) $I_{th}$ equals 25 mA, 40 mA, and 65 mA at 0°C, 25°C, and 65°C, respectively, and (b) threshold change with the temperature is less than 2.5%/°C. Let's denote threshold change with the temperature as $\delta$ and formalize the above relationship in this way:

$$I_{th} = I_0 + I_0\,\delta\,T, \tag{10.50a}$$

If we plug $T$ = 0°C into Formula 10.50a, we find $I_0$ = 25 mA. Thus, from Formula 10.50a we can compute $I_{th}$ = 40 mA at 25°C and $I_{th}$ = 66 mA at 65°C. The value $I_{th}$ = 40 mA coincides with that shown in the "Laser Output" graph, Figure 9.17, but the value $I_{th}$ = 66 mA doesn't. (The graph shows approximately 75 mA.) Hence, this approach works well for small changes in temperature.

(2) Let's now use Formula 10.50. Again, at $T$ = 0°C we find $I_0$ = 25 mA. Plugging in the numbers $I_{th}$ = 40 mA at $T$ = 25°C, we find $T_0$ = 53°C. If we substitute this value of $T_0$ into Formula 10.50 at $T$ = 65°C, we obtain $I_{th}$ = 88 mA, while the specifications sheet gives us $I_{th}$ equal to approximately 75 mA.

The conclusion we can draw from these exercises is that both Formulas 10.50 and 10.50a give us inaccurate results. We should simply rely on the data provided in the manufacturer's specifications sheet, such as the values of $I_{th}$ given in the "Laser Output" graph in Figure 9.17. Also, don't forget about the methods for determining the value of $I_{th}$ discussed in Section 9.3.

---

Above the threshold, gain and, consequently, output power are proportional to current up to a certain point. There are some mechanisms that lead to a saturation effect, but we will leave this topic to more specialized texts ([2], [4]).

## Radiating Wavelength and Spectral Width

***Radiating wavelength*** At what wavelength does a laser diode actually radiate? An active medium can support the range of possible wavelengths, which is given by Formula 10.46 and can be seen in Figure 10.10. To clarify this point, we can rewrite Formula 10.46 in the following form:

$$E_{FC} - E_{FV} > E_p > E_C - E_V = E_g, \tag{10.51}$$

where all notations are clear from Figure 10.10. Since $E_p = hc/\lambda$, the range of radiating wavelengths is determined by:

$$hc/(E_{FC} - E_{FV}) < \lambda < hc/E_g \tag{10.52}$$

It is possible to find all the numbers necessary to calculate the range of wavelengths from Formula 10.52 ([2], [4], and [5]). But this calculation, again, gives us the wavelength range that a given laser's active region can support. (See Figure 10.8.) The actual wavelength is determined by the interaction of a laser's active medium and its resonator. Referring to Figures 9.14(d) and (e), one can see how a resonator "chooses" actual radiating wavelengths from the range provided by an active medium.

Probably the best illustration of the above explanations is given in the data sheet for a broad-area laser diode. (See the "Optical Spectrum" graph in Figure 9.17.) It is clear from this figure how the actual radiating wavelength is inscribed into the gain curve, even though the curve is not shown explicitly.

All the above considerations are true in general for any laser diode, but they are most pronounced for the now classic double-heterostructured diode with a Fabry-Perot resonator. The new developments, especially quantum-well lasers and—based on them—VCSELs, have new, distinguishing features. The first concerns their active medium. It is possible to engineer the energy gap of the active medium. This results in a much narrower and well-controlled gain profile so that the radiating wavelength can be designed, to a greater extent, at the active-medium level [6]. The second important feature is that the use of a Bragg grating makes this "resonator" much more wavelength-selective than a Fabry-Perot resonator formed by lump mirrors. The outcome is singlemode operation for this type of laser diode. Simply compare the optical spectrum given in Figure 9.17 for a broad-area diode with the emission spectrum given in Figure 9.19 for a DFB laser diode and you'll immediately see this effect.

**Spectral width** We have to distinguish between two parameters: *spectral width* and *linewidth.* The spectral width of the radiation from a multimode laser is essentially determined by its gain curve because, again, this diode radiates many modes (lines) inscribed in the gain curve. (See Figure 9.17.) Measured as the full width at half maximum, FWHM, such a laser diode exhibits a spectral width of several nanometers. VCSEL diodes operate in the singlemode regime, which means they radiate only one mode, or line (Figure 9.20). Hence, the spectral width of light radiated by this diode is the spectral width of this line. The same holds true for a DFB laser diode, as Figure 9.19 shows. Thus, the spectral width of radiation of these singlemode laser diodes is the linewidth.

What parameters of a laser diode determine linewidth? Research shows that the shape of a radiated line is governed by the Lorentzian function and its FWHM width is determined by the following formula [2]:

$$\Delta f = R_{sp}\,(1+\alpha_e^2)/2N_{rad}\,, \tag{10.53}$$

where $R_{sp}$ is the rate of spontaneous emission, $\alpha_e$ is the linewidth enhancement factor, and $N_{rad}$ is the number of photons radiated inside the laser cavity. Repeating the same manipulations we did previously in this section to relate $N_{rad}$ to $P$, we can express linewidth in terms of output power, $P$, and other parameters of a laser diode [2]:

$$\Delta f = (R_{sp}(1+\alpha_e^2)/P)\,M, \tag{10.54}$$

where $M = (E_p v\alpha_{tr}/8\pi)$ and $E_p = h_\omega = hc/\lambda$ (a photon's energy), while $v\alpha_{tr}$ is the rate at which photons escape from a laser resonator. Here, $v = c/n_{ri}$ (m/s) is the group velocity of light within the active region, whose refractive index is $n_{ri}$. Other components of Formula 10.54 will be introduced shortly. Transmission loss, $\alpha_{tr}$, has been defined in Formula 10.33 as $(1/2d)\ln(1/R_1R_2)$, which allows us to show explicitly more laser parameters in Formula 10.54. The meaning of the terms combined in coefficient $M$ has already been discussed in this section. Let's concentrate, then, on other parameters.

First, linewidth is inversely proportional to output power, $P$. This general statement is true not only for other types of lasers (gas and solid state) but also for many generators, including microwave and electronic ones. For semiconductor laser diodes, this relationship holds for the low-level power devices (up to 10 mW) and tends to saturate when power increases significantly.

Secondly, linewidth is proportional to the *rate of spontaneous emission,* $R_{sp}$, which is quite understandable. Spontaneous transitions occur between different energy levels without control and this results in the emission of different wavelengths. This emission contributes to output-stimulated light and spreads the linewidth. (For a better understanding of this material, review the box "Spectral Width and Chromatism.")

The third important parameter is the *linewidth enhancement factor,* $\alpha_e$, which accounts for the coupling inside the laser cavity of the EM wave amplitude with its phase. Another way to understand the physics of this phenomenon is to examine the following formula [2]:

$$\alpha_e = (-4\pi/\lambda)(dn_{ri}/dn/dg/dn), \tag{10.55}$$

where $dn_{ri}/dn$ is the rate of change of refractive index $n_{ri}$ as a function of the density of the injected carriers, $n$, and $dg/dn$ is the differential gain. It follows from Formula 10.55 that the refractive index of an active medium, $n_{ri}$, is a function of the charge carriers' density; this derivative changes with differential gain. Formula 10.55 also shows that the linewidth enhancement factor depends on lasing wavelength. This dependence and the range of values of $\alpha_e$ for a couple of *multiple quantum-well (MQW)* lasers are shown in Figure 10.11.

As you can see from Figure 10.11, the linewidth enhancement factor changes from 1.5 to 12 within the working range of wavelengths. Thus, linewidth can increase by a factor of 3.25 to

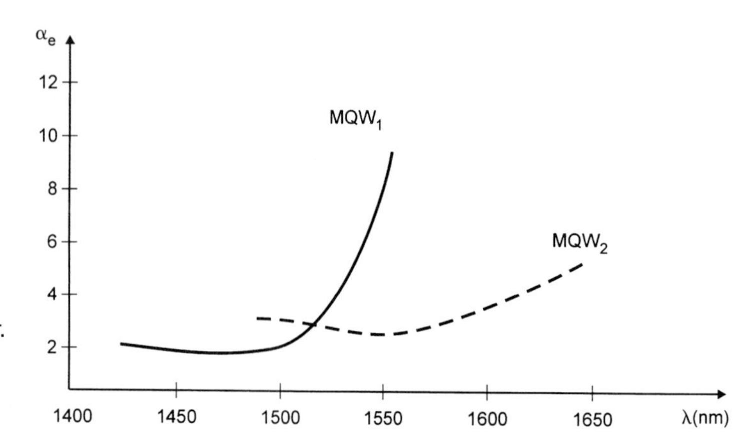


**Figure 10.11** Linewidth enhancement factor. *(Shun Lien Chuang,* Physics of Optoelectronics Devices, *Copyright © 1995 by John Wiley & Sons, Inc. Adapted with permission of John Wiley & Sons, Inc.)*

101. Designers make every effort to decrease this factor and a linewidth enhancement factor close to 1 has been reported [7].

Advances in the design and fabrication of laser diodes have resulted in amazingly narrow linewidths, ones so narrow, in fact, that it is customary to measure them in Hz rather than in nm. The relationship between $\Delta\lambda$ and $\Delta f$ follows from the fundamental formula $\lambda f = c$ (Formula 2.1). So we can easily obtain

$$\Delta f = -(c/\lambda^2)\Delta\lambda \tag{10.56}$$

The data sheet of a DFB laser given in Figure 9.19 shows the typical linewidth to be 10 MHz. A linewidth of 100 kHz has been achieved experimentally [7]. Minimizing the linewidth is crucial for decreasing spacing between channels in wavelength-division multiplexing systems.

## Radiation Patterns

***Beam radiated by a conventional laser diode*** A conventional laser diode radiates a well-directed beam if we compare it with an LED's beam. A typical radiation pattern of a gain-guided LD is shown in Figure 10.12(a).

We consider a stripe-contact—that is, a gain-guided—laser diode with an active region about 1 μm in thickness, 100 μm in length, and about 10 μm in width. These dimensions are typical but, obviously, they can vary in specific numbers while keeping the same order of magnitude. Such a diode radiates an elliptical beam which, just outside the diode, is inscribed into the rectangle made from the width times the thickness of the active region. This pattern is called "near field." Apart from the diode, the laser beam is still elliptical but the major axis is now vertical, as Figure 10.12(a) shows. This is because horizontal radiation keeps its dimension while vertical radiation diverges. Such a pattern is called "far field." The net result is that a conventional laser diode radiates an elliptical beam, so the designer of a fiber-optic system has a hard time coupling this beam to a circular core of an optical fiber. Typical divergence angles of this radiated elliptical cone are 30° × 10°.

What's more, the radiation pattern changes with an increase in laser gain. To understand why, we have to return to the idea of modes in a laser diode.

***Longitudinal and transverse modes*** Let us, for simplicity's sake, discuss a laser with a Fabry-Perot resonator. As was shown in Section 9.2, the resonator supports only those EM waves whose wavelengths satisfy the simple criterion $2L/\lambda = N$, where $L$ is the resonator length and $N$ is an integer. (See Formula 9. 13.) These EM waves are called *longitudinal modes* and they are shown,

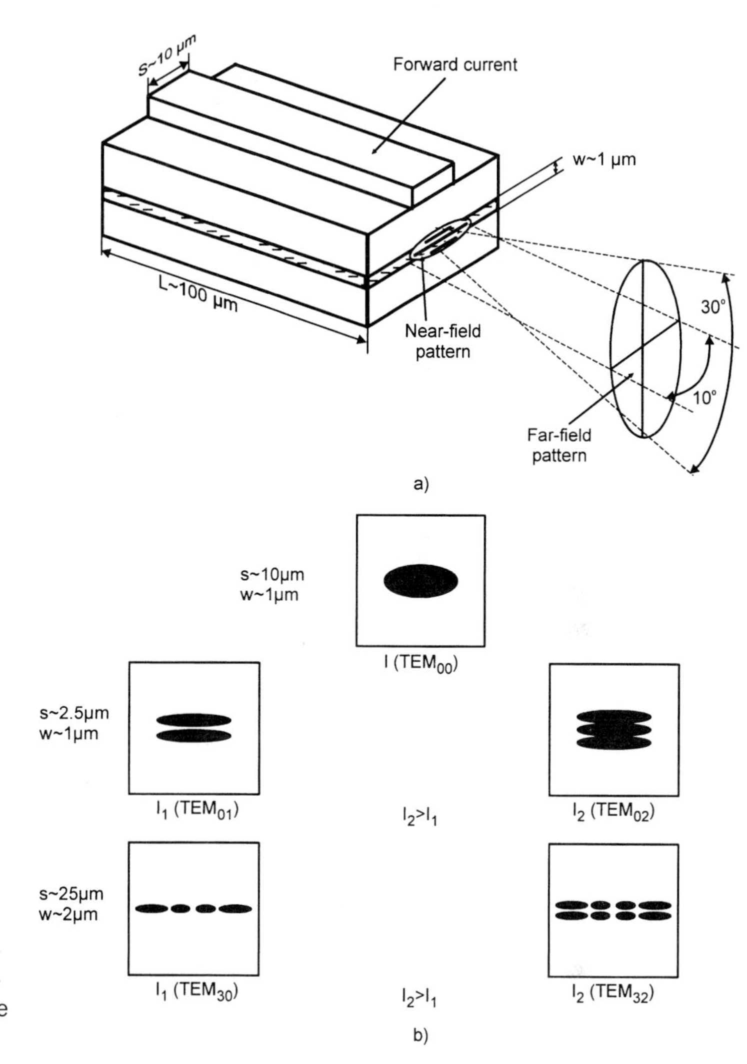


**Figure 10.12** Radiation patterns of a laser diode: (a) Near-field and far-field radiation patterns; (b) near-field radiation patterns with different forward currents and widths of an active region.

for example, in Figure 9.14(d). Longitudinal modes are composed of an EM field whose **E** and **H** vectors are perpendicular to the direction of propagation. (If you feel uncomfortable with these explanations, it would be a good idea to review Section 4.4.) All longitudinal modes propagate along the same path—the centerline of a resonator. They are distinguished only by their frequency or by the integer $N$ mentioned above.

However, longitudinal modes are not the only patterns of an EM field that a laser diode can support. EM waves whose **E** and **H** vectors are slightly tilted with respect to the resonator centerline can also exist. These patterns are called *transverse modes.* They propagate along different paths within the laser-diode cavity and form various radiation patterns. The number of transverse modes grows when gain increases. Gain, in turn, can be increased by increasing the forward current and/or by increasing the width times thickness dimension of an active layer. These points are illustrated in Figure 10.12(b), where the top picture displays the longitudinal-mode radiation pattern, $TEM_{00}$. The others show how near-field radiation is changed when more and more transverse modes are excited by increasing the size of an active region and the forward current. (It is very instructive to compare these radiation patterns with LP modes in a singlemode fiber—see Figure 4.11.)

In any event, the points you must keep in mind above all else are these: (1) *We need a laser diode to transmit a light signal along an optical fiber,* (2) *it takes a circular nondivergent beam to efficiently launch this light into the optical fiber, and* (3) *as we have seen, the conventional laser diode doesn't give us such a beam.*

***Radiation pattern of a VCSEL diode*** In Section 9.2 we discuss the solution to the problem of beam geometry. This solution is a vertical-cavity surface-emitting laser (VCSEL). A VCSEL diode radiates a circular well-directed beam that is easy to couple to an optical fiber. But, you'll recall, commercially available VCSEL diodes today radiate only in the 850-nm wavelength range. This is why they are used mostly with graded-index multimode (GI MM) fiber in local area networks.

The shift of a LAN's bit rate to the gigabit-per-second range raises a new problem: GI MM fiber installed in most of these networks cannot support such a communication bandwidth because of the restrictions caused by intermodal dispersion. (Read again the discussion on fiber bandwidth in LAN applications in Section 8.4.) Fortunately, there is an interesting solution to this problem. You learned that intermodal dispersion occurs when different beams (modes) propagate within the fiber along different paths, thus arriving at the destination point at different times. This causes the spread of the output pulse and, consequently, restricts the bit rate, or bandwidth, that an optical fiber can support. GI MM fiber reduces intermodal dispersion by forcing beams traveling the longer paths to propagate faster and those traveling the shorter paths to propagate more slowly. (See Section 3.3.) This situation is true when all possible modes are excited; in other words, the input light fills the entire cross-sectional area of the fiber. If we concentrate most of the input beam within the smaller area, a smaller number of modes will be excited, thus reducing the effect of intermodal dispersion. With the radiation patterns shown in Figure 10.12, this solution is, obviously, impossible. But with recent developments in VCSEL diodes, it becomes possible. Figure 10.13 demonstrates such a radiation pattern by a VCSEL diode, allowing us to use GI MM fiber in gigabit LAN applications. While it would seem quite logical to generate a beam concentrated along the fiber's centerline, as you can see in Figure 10.13, this VCSEL design makes the laser diode radiate light with a donut-like beam. This is because the actual refractive-index profile of a graded-index fiber deviates from the ideal shown in the data sheet. (See Figure 3.18.) This deviation occurs mostly at the center and/or at the edge of the fiber's core.

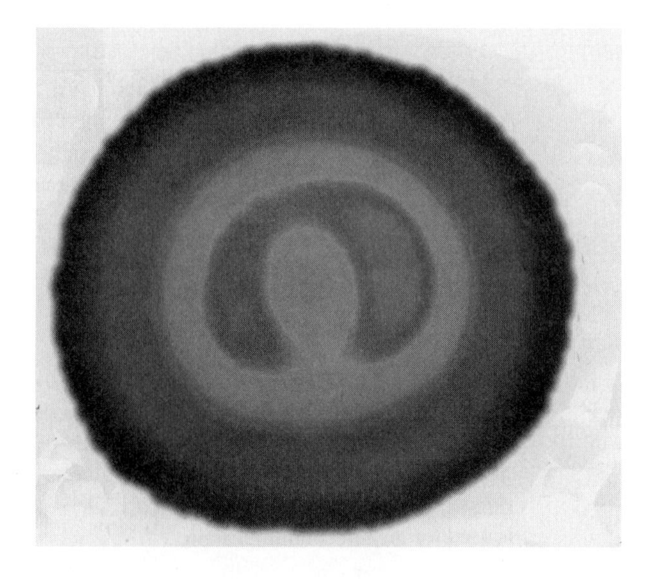

**Figure 10.13** Radiation pattern of a VCSEL diode designed to combat modal dispersion in a GI MM fiber. *(Jeff W. Scott, "Vertical-cavity lasers facilitate gigabit communications,"* Laser Focus World, *October 1998, pp. 75–78. Reprinted with permission.)*

Thus, the concentration of light into the donut radiation pattern shown in Figure 10.13 allows us to minimize modal dispersion because in the middle of the core's cross section the actual graded-index profile is closer to the ideal than it is at any other location. This arrangement allows the system designer to succeed almost completely in reducing modal dispersion in a GI MM fiber [9]. New types of GI MM optical fibers also help to attain the same result: the efficient reduction of modal dispersion. (See Section 6.5 and Chapter 15.)

The above discussion gives you some idea of how important radiation patterns are for the application of laser diodes in fiber-optic communications systems.

## Laser Modulation

***Physics*** There are two types of laser diode modulation: *digital* and *analog*. In the simplest digital-modulation scheme, logic 0 is represented by a dark period and logic 1 by a flash of light. To attain this state, an information signal changes the forward current of a laser diode from values below threshold to values above threshold. This is shown in Figure 10.14(a). In analog modulation, designers want to use the linear portion of a *P-I* curve to avoid nonlinear distortion of an output signal. This is attained by applying dc biasing current, $I_b$, along with the information signal, as Figure 10.14(b) shows.

Most communications signals today are transmitted in digital form; there's no question about this anymore. However, analog modulation is still used, mainly in cable-TV applications.

Let's consider what happens inside a laser diode when a step-like electric signal is applied. After the drive current exceeds the threshold value, population inversion is created, the transition of excited electrons from conduction to valence band occurs (or, equivalently, electron-hole recombinations take place), and photons are radiated. These radiated photons travel within the laser cavity and escape from there, creating an output information signal in the form of flashes of light. This summary of the activity taking place inside a laser diode when a step-like electrical signal is applied is necessary to know because it characterizes all the lifetimes involved in the modulation

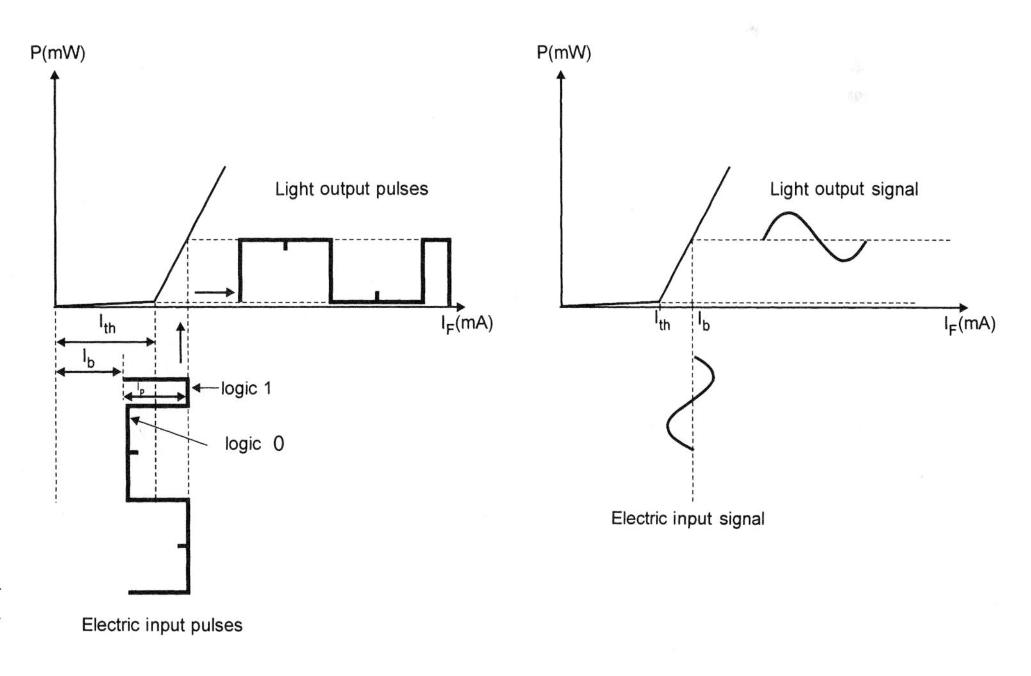


**Figure 10.14** Digital and analog modulation of a laser diode: (a) Digital modulation; (b) analog modulation.

process. To radiate a light pulse, for example, the following lifetimes are involved: time to create population inversion, $\tau_{pi}$; recombination lifetime, $\tau$; radiative lifetime, $\tau_r$ (which, in turn, is divided into spontaneous, $\tau_{sp}$, and stimulated lifetimes, $\tau_{st}$); and the photon's lifetime, $\tau_{ph}$. The latter is the time that a photon needs to escape from the laser diode as it travels from its place of origin.

A photon's lifetime puts an upper limit on the modulation capability of a laser diode because nothing can be done to change a photon's status once it is created. This lifetime can be related to the laser diode's characteristics through the following consideration [3]: Loss after traveling $t$ sec is equal to loss after traveling $x$ meters, that is:

$$\exp(-t/\tau_{ph}) = \exp(-\alpha\ x) \tag{10.57}$$

The time and the distance are related through the velocity of propagation as:

$$x = vt = (c/n_{ri})t \tag{10.58}$$

Formulas 10.57 and 10.58 yield:

$$\tau_{ph} = n_{ri}/(\alpha\ c) = n_{ri}/(g_{th}\ c) \tag{10.59}$$

For $\alpha = 10^3$ 1/m and $n_{ri} = 3.6$, a photon's lifetime is 0.12 ps. This implies that the theoretical upper limit on the modulation ability of a laser diode is about 8.3 THz. Unfortunately, reality is often far removed from theoretical predictions, as in this case. To understand what actually limits the modulation capability of a laser diode, we need to consider rate equations.

***Rate equations*** Rate equations describe how electron density, $n$ ($1/\text{m}^3$), and photon density, $s$ ($1/\text{m}^3$), change with respect to time, depending of course on the parameters of the laser diode in question (see, for example, [2] and [3]):

$$dn/dt = J/(ed) - n/\tau_{sp} - Dns \tag{10.60}$$

$$ds/dt = Dns + \zeta n/\tau_{sp} - s/\tau_{ph} \tag{10.61}$$

Equations 10.60 and 10.61 say that the change of $n$ and $s$ in the active region depends on the balance of positive and negative members on the right side of these equations:

- Electron volume density, $n$, increases as more electrons are injected into the active region. This trend is represented by the first expression on the right-hand side of Formula 10.60: current density, $J$ ($\text{A/m}^2$), divided by the electron charge, $e$ (C), and by the thickness of an active region, $d$ (m).
- Electron volume density, $n$, decreases as more electrons recombine. This is shown by two negative expressions: (1) rate of spontaneous emission, $n/\tau_{sp}$ ($1/\text{m}^3\cdot\text{s}$), and (2) rate of stimulated emission, $Dns$, where $Dn = vg$ (1/s). Thus, the constant, $D$, shows the strength of stimulated emission, which results in gain, $g$.

$$D = vg/n \tag{10.62}$$

  Strictly speaking, gain is a function of $n$ but we neglect this dependency in our analysis.

- Photon volume density, $s$, in a lasing mode increases as more photons are radiated through stimulated emission. This is shown by the positive term $Dns$ ($1/\text{m}^3\cdot\text{s}$). There is a small increment of $s$ through spontaneous emission into the lasing mode. This is represented by $(+\zeta n/\tau_{sp})$. The coefficient $\zeta$ is usually very small, reflecting the fact that very few spontaneously emitted photons move in the same direction as stimulated photons.

- Photon volume density decreases because of a loss of photons through absorption and emission outside the laser diode. This is represented by $(-s/\tau_{ph})$, where the photon's lifetime, $\tau_{ph}$, is given by Formula 10.59. Using Formulas 10.48 and 10.37, we can express $\tau_{ph}$ through the parameters of a laser diode as:

$$1/\tau_{ph} = (v/\Gamma)(\alpha_{mat} + \alpha_{tr}) = (v/\Gamma)(\alpha_{mat} + 1/2d\ ln(1/(R_1 \times R_2))) \tag{10.63}$$

Rate equations are a very informative source of a laser diode's behavior and many of the features of a laser diode discussed above can be derived from these equations. (See the box titled "Using Rate Equations," page 400.)

Steady-state solutions of these rate equations, $n_0$ and $s_0$, can be obtained by having $dn/dt = 0$ and also $ds/dt = 0$. *Small-signal modulation* can be investigated through Equations 10.60 and 10.61 if we substitute $n = n_0 + \delta n(t)$ and $s_0 + \delta s(t)$, where $\delta n(t)$ and $\delta s(t)$ are created by modulating the injection current. ("Small signal" means we want to restrict ourselves to the linear approach.) It can be shown that the rate equations become second-order differential equations describing the damping oscillations of $\delta n(t)$([3], [6], and [7]):

$$d^2(\delta n)/dt^2 + 2\chi\ d(\delta n)/dt + \omega_r{}^2\ \delta n = 0 \tag{10.64}$$

A similar equation can be derived for $\delta s(t)$. Here, $\chi$ is the *laser damping constant,* which equals

$$\chi = D\ n_0 + 1/\tau_{sp} \tag{10.65}$$

The laser *relaxation oscillation frequency,* $\omega_r$, is given by

$$\omega_r^2 = D^2\ n_{th}\ s_0, \tag{10.66}$$

where

$$n_{th} = 1/(D\tau_{ph}) \tag{10.67}$$

is the threshold carrier density.

Thus, Equation 10.64 shows that a laser diode behaves like any damping-oscillation system. The solution of Equation 10.64 is well known and can be approximated here as:

$$\delta n = (\omega_1/D)\exp(-\chi t)\sin\omega_1 t \approx \sqrt{(n_{th}s_0)}\exp(-\chi t)\sin\omega_1 t, \tag{10.68}$$

where $\omega_1 = \sqrt{(\omega_r^2 - \chi^2)}$ and approximation $\omega_1 \approx \omega_r$ is used.

Now we can investigate the response of a laser diode to a modulation signal. For *digital modulation,* we apply a *step-like input.* The resulting electron and photon volume densities will display damping-oscillation behavior, as Figure 10.15 shows. Bear in mind that output light is simply a stream of photons; therefore, the curve $s(t)$ is the output in another scale. The points to pay particular attention to here are that (1) *there is a delay time before a laser starts to generate light after the modulation pulse has been applied (1.5 ns in Figure 10.15)* and (2) *it takes about 8 ns for laser oscillation to settle down.* There is no way to eliminate oscillation time but, fortunately, we don't need to wait 8 ns for it to drop to logic 1. Thus, delay time is the critical issue.

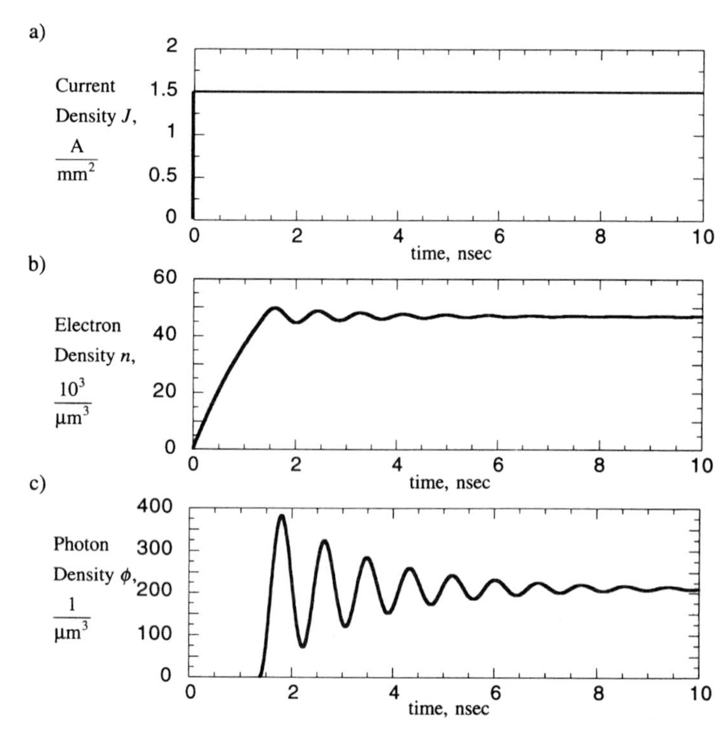


**Figure 10.15** Step-like modulation of a laser diode. *(Reprinted from Leonid Kazovsky, Sergio Benedetto, and Alan Willner's,* Optical Fiber Communication Systems, *Boston: Artech House, 1996. Used with permission.)*

Delay time, $t_d$, can be reduced by using dc bias current, $I_b$. If $I_p$ is the amplitude of modulation current, then $t_d$ can be calculated by the following formula [10]:

$$t_d = \tau \ln[I_p/(I_p + I_b - I_{th})], \tag{10.69}$$

where $\tau$ is the carrier recombination lifetime. (All currents are shown in Figure 10.14.) Thus, for digital modulation we have a limit in terms of the order of Gbit/s.

## Example 10.3.2

***Problem:***

Calculate the delay time for a broad-area laser diode whose data sheet is given in Figure 9.17. Assume the carrier recombination lifetime is equal to 2 ns.

***Solution:***

From Figure 9.17, we can see that the amplitude of modulation current at 1 mW power is from 10 mA to 40 mA and the threshold current is 90 mA. Thus, we need to choose $I_b$. (Note that this problem is not simply about calculations but also about design.) If we postulate that $I_p = 40$ mA and $I_b = 80$ mA, then $t_d = 0.56$ ns. If $I_b = 70$ mA, $t_d = 1.38$ ns, and if, $I_b = 90$ mA, $t_d = 0$ ns, which, of course, is obvious. Hence, theoretically, $t_d$ can be eliminated by the biasing, but how this elimination can be implemented will be discussed in the following section.

---

To investigate *analog modulation,* we represent modulation current and electron and photon volume densities in phasor form, such as:

$$\delta J(t) = Re[\delta J(\omega)\exp(-j\omega t)] \tag{10.70}$$

The ratio $\delta s(\omega)/\delta J(\omega)$ is known as the *transfer function* of a laser diode. It can be shown [2] that

$$\delta s(\omega)/\delta J(\omega) = [(\Gamma\tau_{ph}/ed)\omega_r^2]/[(\omega^2-\omega_r^2)^2+\omega^2(1/\tau+\tau_{ph}\omega_r^2)^2]^{1/2} \tag{10.71}$$

If you are familiar with the basics of communications theory, you'll immediately recognize Formula 10.71. This is a magnitude of the transfer function of a low-pass filter. We will not obtain the phase of the laser transfer function in an explicit form; it can be found in any textbook on communications theory. However, we present the magnitude and phase of the transfer function in graphical form in Figure 10.16(a). These graphs can be seen in textbooks on mechanics, electronics, and communications theory simply because they are visual representations of solutions to Equation 10.64 in the form given by Formula 10.70.

Plotting Formula 10.70 as a function of modulation frequency produces the *frequency-response* graph depicted in Figure 10.16(b). Studying Figure 10.16 reveals that the upper limit of the analog-modulation capability of a laser diode is established by the relaxation oscillation frequency, $\omega_r$. Figure 10.16(b) also reveals that this frequency depends on output power, as Formula 10.66 predicts. One can calculate a 3-dB bandwidth using Formula 10.71 but we leave this exercise to a communications course.

***Result*** Modulation of a laser diode can be accomplished by varying its driving current. This results in *intensity modulation* (IM), which is the most popular type of modulation used in deployed fiber-optic communications systems. Regardless of the type of modulation—digital or analog—the upper limit of modulation frequency (bit rate) for a laser diode is set by its relaxation oscillation frequency (also called resonant frequency). This frequency is determined by Formula 10.66 and, in conjunction with the definition of $n_{th}$ given in Formula 10.66a and the definition of $D$ given in Formula 10.62, it can be rewritten in this form:

$$\omega_r = D^2 n_{th} s_0 = (Ds_0)/\tau_{ph} = (1/\tau_{ph})(vgs_0/n) \tag{10.72}$$

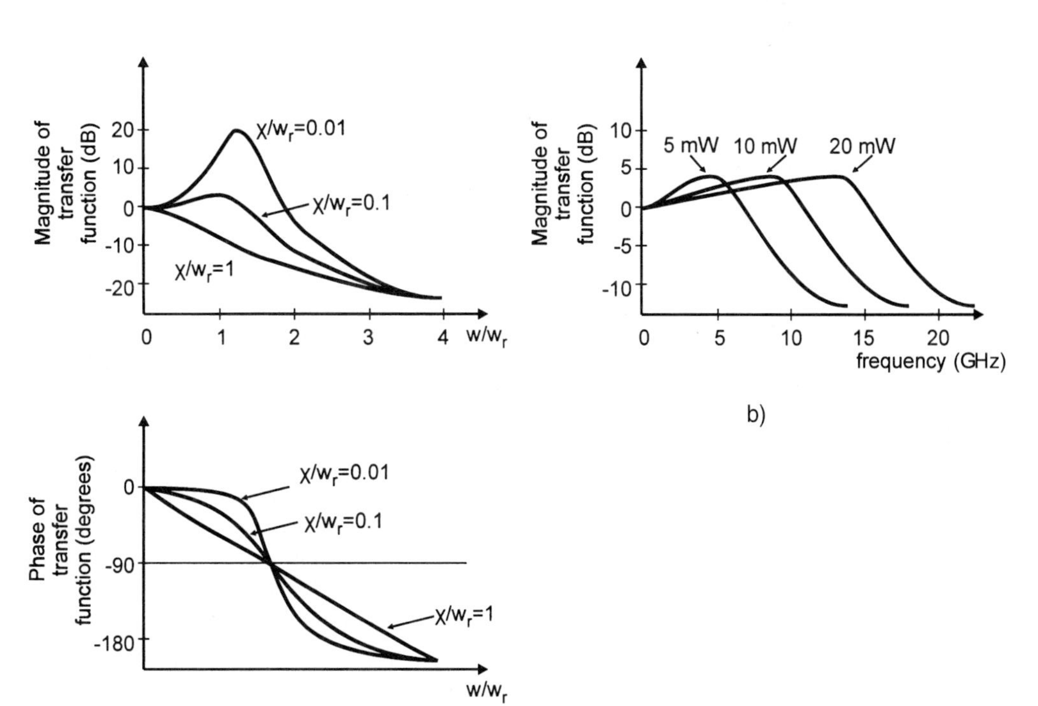


**Figure 10.16** Analog modulation of a laser diode: (a) Magnitude and phase of the transfer function; (b) frequency response of a DFB laser. *([a] reprinted from Leonid Kazovsky, Sergio Benedetto, and Alan Willner's* Optical Fiber Communication Systems, *Boston: Artech House, 1996. Used with permission. [b] Shun Lien Chuang's* Physics of Optoelectronics Devices, *Copyright © 1995 John Wiley & Sons, Inc. Reprinted by permission of John Wiley & Sons, Inc.)*

Since ratio $s_0/n$ is essentially the internal quantum efficiency, Formula 10.72 reveals that $\omega_r$ is determined by the photon's lifetime, $\tau_{ph}$, and depends on laser-diode gain, that is, on its power (again, see Figure 10.16[b]). Hence, Formula 10.72 takes the form [6]

$$\omega_r = (MP)/\tau_{ph}, \quad (10.73)$$

where $M$ is the constant and $P$ is the output power of a laser diode. Formula 10.73 shows that *the harder a laser diode is driven*—and, therefore, the higher its output power—*the faster its response.* It is also seen that the shorter a photon's lifetime, the higher its modulation frequency, as is stressed in the subsection "Physics."

### Using Rate Equations

The rate equations allow us to obtain most of the information already introduced without derivations in this section.

For example, let's now derive the threshold condition and the input–output power relationship in a laser diode using the rate equations [3].

The *threshold condition* can be obtained from Equations 10.60 and 10.61 as follows: At threshold, $dn/dt$ and $ds/dt$ are equal to zero, $J = J_{th}$, $n = n_{th}$, and $s_{th} = 0$. Hence, Equation 10.60 yields

$$J_{th}/(ed) = n_{th}/\tau_{sp} \quad (10.74)$$

Equation 10.61 yields

$$n_{th} = 1/(D\tau_{ph}) \quad (10.75)$$

From Formula 10.62, we can obtain

$$n_{th} = (vg_{th})/D, \quad (10.76)$$

which, along with Formula 10.75, yields

$$1/\tau_{ph} = vg_{th}, \quad (10.77)$$

as Formula 10.59 previously stated.

Substitute Equation 10.76 into Equation 10.74 and recall Formula 10.48. This results in

$$J_{th}/(ed) = (vg_{th})/(D\tau_{sp}) = (v\Gamma\alpha)/(D\tau_{sp}) \quad (10.78)$$

Comparing this with Formula 10.49, we see that

$$J_{th} = (1/\beta)\alpha, \text{ where } 1/\beta = (edv\Gamma)/(D\tau_{sp}) \quad (10.79)$$

We will leave the analysis of the physical meaning of these formulas for you to do as an exercise.

It is worth noting that above the threshold, the steady-state volume density of electrons at the conduction band, $n$, is always equal to its threshold value, $n_{th}$. This is because an increase in $n$, through an increase in the injection current, results in an increase in the number of stimulated transitions. In other words, the more electrons that are excited at the conduction band, the more electrons that will be stimulated and available to cross to the valence band and radiate more photons. This is why increasing the injection current increases the output power but does not increase the long-term population of the conduction band. This phenomenon is called *clamping.*

The *input–output power relationship* in a laser diode can be obtained from the rate equations by substituting Formula 10.74 into Formula 10.60 at the steady-state condition ($dn/dt = 0$) and using Formula 10.75. This results in

$$s = (1/Dn_{th})[(J - J_{th})/ed] = [\tau_{ph}/(ed)][J - J_{th}] \quad (10.80)$$

Formula 10.80 says that the number of photons radiated into a laser mode (inside the laser diode) is proportional to the difference between the actual and threshold current densities. It is advisable to compare Formulas 10.80 and 10.22.

These two examples show that the rate equations are key sources of information about the behavior of a laser diode.

## Chirp

So far we've considered only a linear model, where the frequency (phase) of radiated light does not depend on the amplitude of this light. In reality, this assumption is not true; hence, a change in light intensity during modulation causes a change in the frequency of the light. This phenomenon is known as *chirp.* (Sound familiar? You are right. We discussed "chirp" twice before: first in the subsection "Chirped fiber Bragg gratings," and again in the subsection "Self-phase modulation" both in Section 6.4.)

*Chirp is the deviation of laser frequency from its radiation-center frequency.* The order of chirp magnitude is about 100 MHz—1 GHz per 1 mA of drive-current change [3]. You must realize that we are talking about, say, a 1 GHz change of the $1 \times 10^{14}$ Hz central-frequency. In other words, the chirp of a laser diode is on the order of 0.001%, however, chirp is so important in terms of today's need for accuracy that we cannot leave it without a fuller discussion. The physical mechanism of chirp is that a change in carrier population—and, hence, gain—causes a change in the refractive index of an active region. For a formal description of chirp, refer to the more specialized textbooks listed at the end of this chapter ([2],[4], and [7]).

The value of chirp—instantaneous frequency deviation, $\delta f(t)$, from a radiation-center frequency, $f_0$—can be evaluated by the following formula [11]:

$$\delta f(t) = (\alpha_e/4\pi)[d/dt\ ln(\Delta P) + \chi\Delta P] \tag{10.81}$$

where $\alpha_e$ is the linewidth enhancement factor (see Formula 10.55), $\Delta P$ is the variation of optical power corresponding to switching from HIGH to LOW logic levels, and $\chi$ is the constant. The right-hand side of Formula 10.81 consists of two terms. The first one, called *transient* (or instantaneous) *chirp,* is caused by relaxation oscillation, which we discussed in the previous subsection. The second term, *adiabatic chirp,* is produced by a change in carrier densities because of power variation ($\Delta P$), a phenomenon discussed in Section 10.2.

*Chirp results in a broadening of a laser's linewidth.* You will recall that chromatic dispersion is proportional to a signal's spectral width; therefore, chirp degrades the performance of a fiber-optic communications system. This is obviously true for systems that employ intensity modulation—and all of today's deployed systems use this type of modulation.

There are several methods to combat this phenomenon, the most radical of which is to use external modulation. (See the following section.) Other methods are in use and they can be found in the list of references at the end of this chapter. (See in particular [7].)

## Noise

Light radiated by a laser diode fluctuates in its intensity and phase even when biasing current is ideally constant. These fluctuations, caused mostly by spontaneous emission, are random in nature. They are called *laser noise.* We have already accounted for spontaneous emission by the term "$\tau_{sp}$," but that is a deterministic approach. Now we will undertake a brief review of the world of stochastic noise.

*Phase fluctuations*—in short, noise—result in the broadening of a linewidth of radiated light. Linewidth has been discussed above and any further elaboration would add only some minor details but would not change the major features given in Formula 10.54.

*Intensity fluctuations* result in what's called intensity noise in a laser diode. This noise is usually measured as *relative intensity noise (RIN),* given by [12]:

$$\text{RIN}(1/Hz) = \langle P_N^2 \rangle / [\langle P \rangle^2\,\text{BW}], \tag{10.82}$$

where $\sqrt{\langle P_N^2 \rangle}$ is the average noise power, $\langle P \rangle$ is the average output power, and *BW* is the bandwidth. Since measurement of a laser diode's RIN requires a receiver and a link between them, BW here is the bandwidth of the receiver and the link. Note that RIN depends on bandwidth. It is normally measured in units of 1/Hz; however, RIN is often measured in dB/Hz, as Formula 10.83 shows:

$$\text{RIN(dB/Hz)} = 10\log\{\langle P_N^2 \rangle / [\langle P \rangle^2\,\text{BW}]\}, \tag{10.83}$$

Two major mechanisms produce RIN in laser diodes. The first, as already mentioned, is the spontaneous emission that is amplified within the laser cavity. The second is *backreflection,* which can occur at any component of a fiber link (connectors, splices, and even the fiber itself). Light reflected back into a laser diode is amplified by the active region and is added to the main stream, resulting in intensity fluctuation. This is why return loss, as discussed in previous chapters, is a very important parameter of any connection in a fiber link.

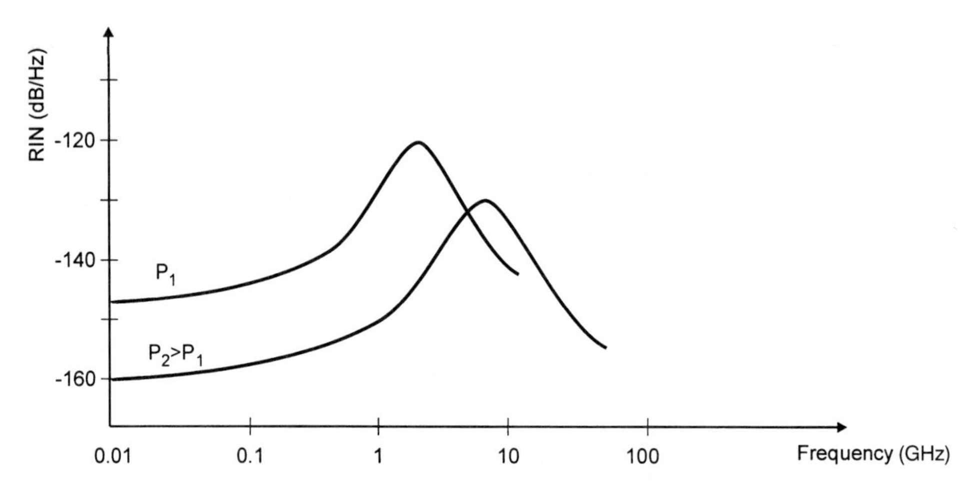


**Figure 10.17** RIN of a laser diode.

RIN describes the frequency content of the noise power and mainly depends on two characteristics of a laser diode: *relaxation oscillation frequency,* $\omega_r$ and *laser power, P.* See Figure 10.17.

As you can see from this figure, a laser diode works as a low-pass filter. RIN power dependence is a complete matter, and this dependence changes from ~ $P^{-3}$ at a low power level to ~ $P^{-1}$ at a higher power level [7].

**Example 10.3.3**

***Problem:***

Calculate the noise power detected by a receiver if (1) the relative intensity noise of a laser diode (RIN) is −160 dB/Hz, (2) the power received is 100 μW, (3) the receiver bandwidth, BW, is 100 MHz, and (4) the link bandwidth is 500 MHz.

***Solution:***

We can calculate noise power from Formula 10.82. But we are given RIN in dB/Hz; thus, we need to convert this into RIN in terms of 1/Hz. This conversion yields: RIN = $10^{-16}$ (1/Hz). You can see that the bottleneck here is the receiver's bandwidth; therefore, BW = 100 MHz is the figure we have to plug into Formula 10.82. Now, using Formula 10.82, we obtain

$$\langle P_N \rangle = \sqrt{\langle P_N^2 \rangle} = \sqrt{[\text{RIN}(1/\text{Hz})\langle P \rangle^2 \text{BW}]} = \sqrt{10^{-16}(1/\text{Hz}) \times 10^{-4}\,(\text{W})^2 \times 10^8(\text{Hz})} = 0.01\ \mu\text{W}$$

As you can see, average noise power is small compared with received power.

We will discuss other types of noise in the following chapters, where receivers and an entire fiber-communication link will be considered.

## 10.4 TRANSMITTER MODULES

The transmitter is the unit of the fiber-optic communications system responsible for converting an electrical information signal into an optical one. The major component of a transmitter—a light source in the form of an LED or a laser diode—is discussed in Chapter 9 and the previous sections of this chapter. Now it's time to turn to other components of this unit, whose main function is to maintain a high-quality optical information signal. As we will see in this section, consideration of optical, electronic, mechanical, and thermal issues is necessary to understand how a properly functioning transmitter is made.

It is common practice to mechanically combine a transmitter and a receiver into one module, a *transceiver.* This is done when a fiber-optic link provides duplex (two-way) communication, which is very common. However, we'll consider the transmitter and receiver separately because, from a functional standpoint, they are different units.

## Functional Block Diagram and Typical Circuits of a Transmitter

***Block diagram*** A functional block diagram of a transmitter is shown in Figure 10.18. As you can see from this figure, the transmitter includes a light source, coupling optics, a signaling circuit, and a power-control circuit. All these components are packed into one module, as shown in Figure 1.5. Photos of the transmitter boards, replete with their necessary components and coupled into an optical fiber, are displayed in Figure 10.19.

In our discussion, we will refer to the laser-diode transmitter, but most of the following considerations are generally true for an LED transmitter as well. The dotted line in Figure 10.18 shows the border of the transmitter module, which is the board shown in Figure 10.19. Data from outside electronic circuits enter this module along with a clock signal. A special unit converts the data into a format suitable to control a laser driver. The latter changes the forward current to modulate the output light radiated by a laser diode. And now you know how a transmitter performs its primary function. Obviously, however, in the case of analog transmission, the signal circuit will be much simpler.

Let's consider all the components of a transmitter in detail.

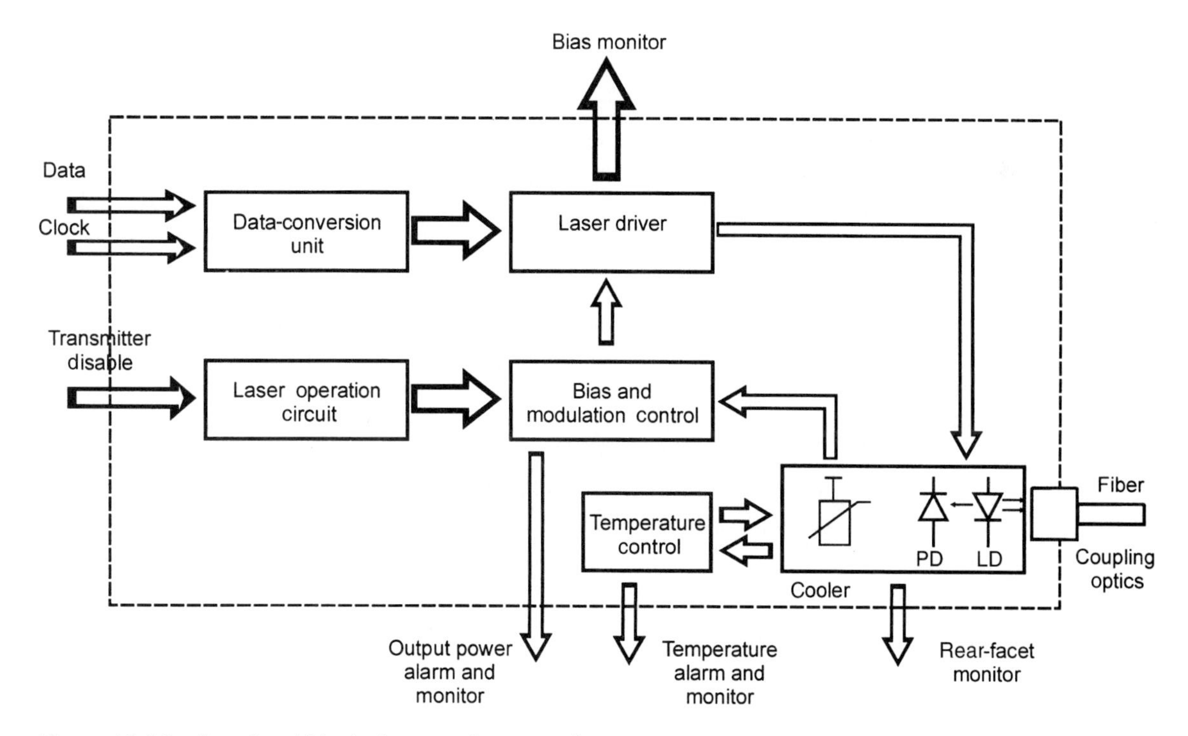


**Figure 10.18** Functional block diagram of a transmitter. *(Adapted from* Opto Electronic Products *with permission of Ericsson Microelectronics AB, Stockholm, Sweden.)*

**Figure 10.19** Photo of a transmitter assembly. *(Courtesy of Laser Diode, Inc., Edison, N.J.)*

***Data conversion unit*** The transmitter's *data conversion unit* performs three major functions: encoding, parallel-to-serial conversion, and reshaping the electric format of the data.

*Encoding* means representing data (binary numbers) in a physical format (pulses). This is necessary because data are transmitted in different line codes. The need for using different line codes can be clarified by the following example: All digital transmissions are governed by a clock signal, shown in Figure 10.20(a). The simplest way of representing logic 1 and logic 0 in electric form—this is what a line code is needed for—is the *non-return-to-zero (NRZ)* format, shown in Figure 10.20(b). Here, logic 1 is transmitted as an electric pulse with the amplitude ($A$) in volts and the duration ($\tau$) in nanoseconds, and so logic 0 is represented as 0 V during $\tau$ ns.

This code looks very natural and seems very convenient but it has poor transmission capability. Indeed, it carries a dc power component, which delivers no information but transmits a lot of heat, making both the transmitter and the receiver consume more electric power than they should. In addition, this signal doesn't carry any timing information, so any discrepancy between clock and data signals caused by noise or other line distortions will result in a false data reading. In short, this code has no self-synchronization ability.

To overcome these problems, the *Manchester code* (Figure 10.20[c]) can be used. Here, logic 1 is represented by the transition of an electric signal from the positive A/2-V level to the negative A/2-V level, and logic 0 is represented by the opposite transition. All the transitions occur in the middle of the clock pulse. This code has no dc component and its transitions can be used for synchronization purposes, so the signal itself carries synchronization information. Unhappily, there's a price to pay for obtaining these advantages: The Manchester code requires twice the bandwidth for transmission that the NRZ code needs.

Another line code, *return-to-zero (RZ),* represents logic 1 as a pulse with the amplitude ($A$) and the duration ($\tau/2$), while logic 0 is represented by a zero signal.

There are many line codes used for data transmission in different applications. The above example demonstrates the reasoning behind the need for such a variety. Data might enter a fiber-optic transmitter in any one of these numerous line codes, such as in the Manchester code. But the output of the transmitter—light pulses—represents logic 1 as a flash of light and logic 0 as a period of darkness. Fiber-optic communications systems mainly use the NRZ code; however, the trend is to increasing use of the RZ code. We'll return to this discussion in later chapters.

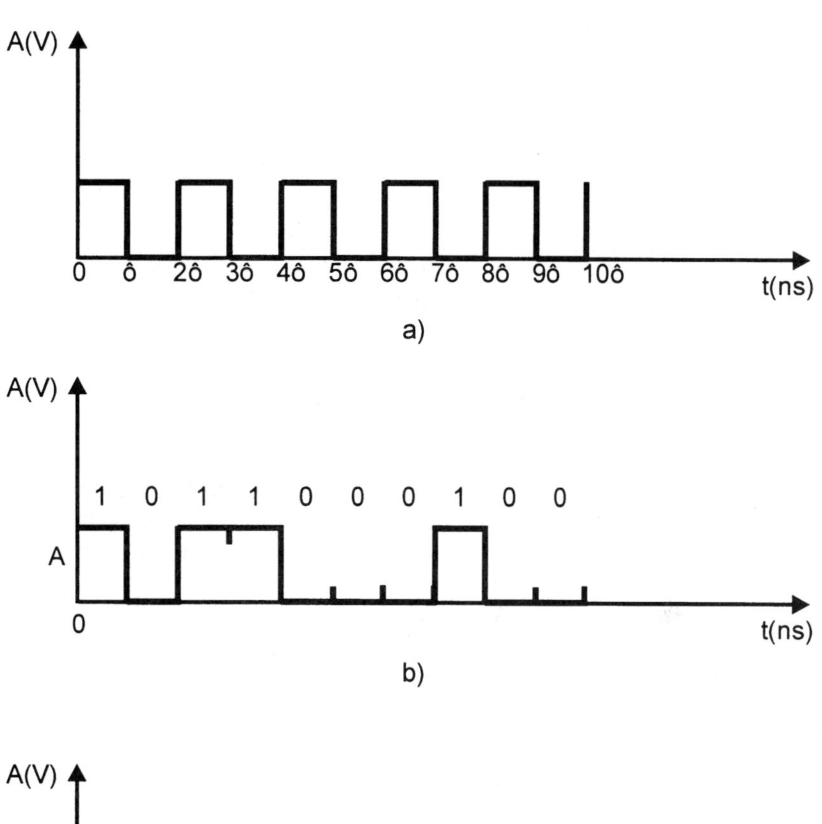


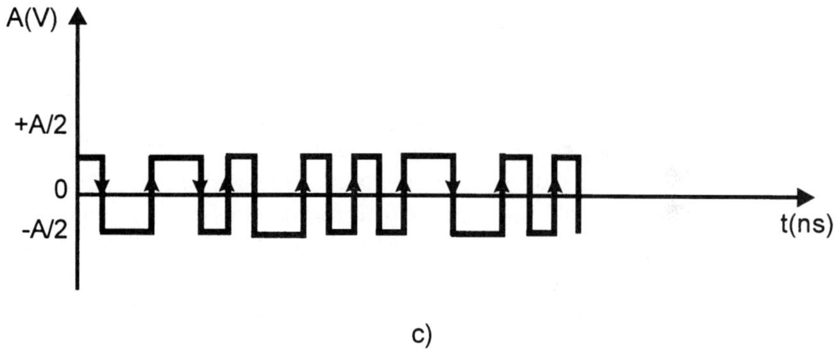


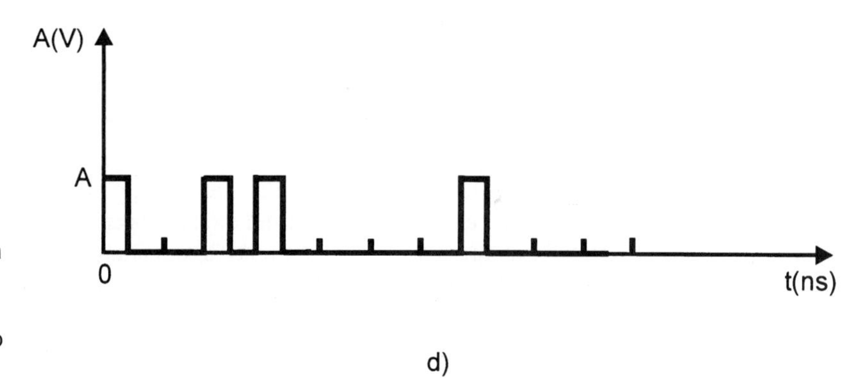


**Figure 10.20** Line codes for data transmission: (a) Clock signal; (b) non-return-to-zero (NRZ) code; (c) Manchester code; (d) return-to-zero (RZ) code.

Thus, the need for encoding the incoming data and converting it into a simple yes–no format becomes clear. This is the first function of the data-conversion unit and a specific *encoder* is used for this purpose. (See Figure 10.21[a].)

The second function of the data-conversion unit is parallel-to-serial conversion. Data enter in parallel format but a laser diode can be driven only by serial pulses of modulation current. Thus, a *parallel-in serial-out converter (PISO),* which is often called a *multiplexer,* is used to convert data into the serial format. (See Figure 10.21[a].)

*Note:* Often, these two blocks—the encoder and the multiplexer—are not part of a fiber-optic transmitter module and the data come into the transmitter in an appropriate format through differential or single-ended input.

The third function of the data-conversion unit is reshaping the electric format of data. Either a comparator or the buffer can be used for this purpose. Let's consider a *comparator* with differential input (Figure 10.21[a]). This is a circuit that compares two input signals: data and complementary. If the data signal is higher than the complementary signal, the output becomes almost equal to $V_{CC}$, that is, to the power-supply voltage. On the other hand, if the complementary signal is higher, the output becomes almost zero. In other words, the comparator produces high-voltage or low-voltage signals in response to input logic 1 or logic 0, regardless of the level of the electric voltage of the input signals. A comparator can also be driven by a single-ended input. This circuit is used to make the output compatible with digital logic in the units that follow. In addition, this circuit has high input impedance, which makes it a compatible load for the previous block, the multiplexer.

A *buffer* is a device that isolates the input from the output and amplifies the current while transferring the logic signal from the input to the output unchanged. Thus, a buffer can also serve to reshape the electric form of the input's logic signal. Which circuit needs to be employed—the comparator or the buffer—depends on the design of the specific transmitter.

In building electronic circuits, fiber-optic transmitters usually utilize *emitter-coupled logic (ECL).* This is a digital-logic family based on bipolar transistors operating in a nonsaturated regime. The features of ECL include high speed, low noise, and the ability to drive low-impedance circuits. ECL circuits are usually referenced to $V_{CC}$; this is why they normally are $V_{CC}$-grounded and the $V_{EE}$ is tied to –5.2 V. However, a designer can make ECL circuits operate from a +5 V supply, thus making their power-supply voltage comply with that of other circuits. In this case, these circuits are called *positive ECL (PECL)* logic. Other logic families used in fiber-optic transmitters are (1) transistor-transistor logic *(TTL),* which uses a bipolar transistor operating in the saturated regime, and (2) a complementary metal oxide semiconductor *(CMOS),* which uses metal oxide semiconductor field-effect transistors (MOSFET).

***Laser driver*** Data prepared for light transmission pass into a *laser driver.* We need this circuit because a laser diode is a current-driven rather than a voltage-driven device, while the power supply is always a voltage source. Thus, the first function of a laser driver is to convert outside voltage into the current needed to drive the laser.

As discussed in Section 10.2, driving current has to bias a laser diode to speed the modulation process. (See Figure 10.14.) So another function of a laser driver is to provide a *bias current.* The real problem with bias current is that it has to be very stable with respect to threshold current; otherwise, an error in data transmission can occur. The main factor causing a change in the relative value of threshold and bias currents is temperature. (The exact relationship between threshold current and temperature is known. Look at the LD data sheet in Figure 9.17.) The feedback signal from the temperature sensor that reaches the laser driver through the bias-control circuit closes the control loop.

An example of a circuit that performs the above-mentioned function is given in Figure 10.21(b). Control voltage, $V_{bias}$, is the input voltage for the *operational amplifier (op amp).* Current flowing through the resistor (*R*) depends only on the input voltage and does not depend

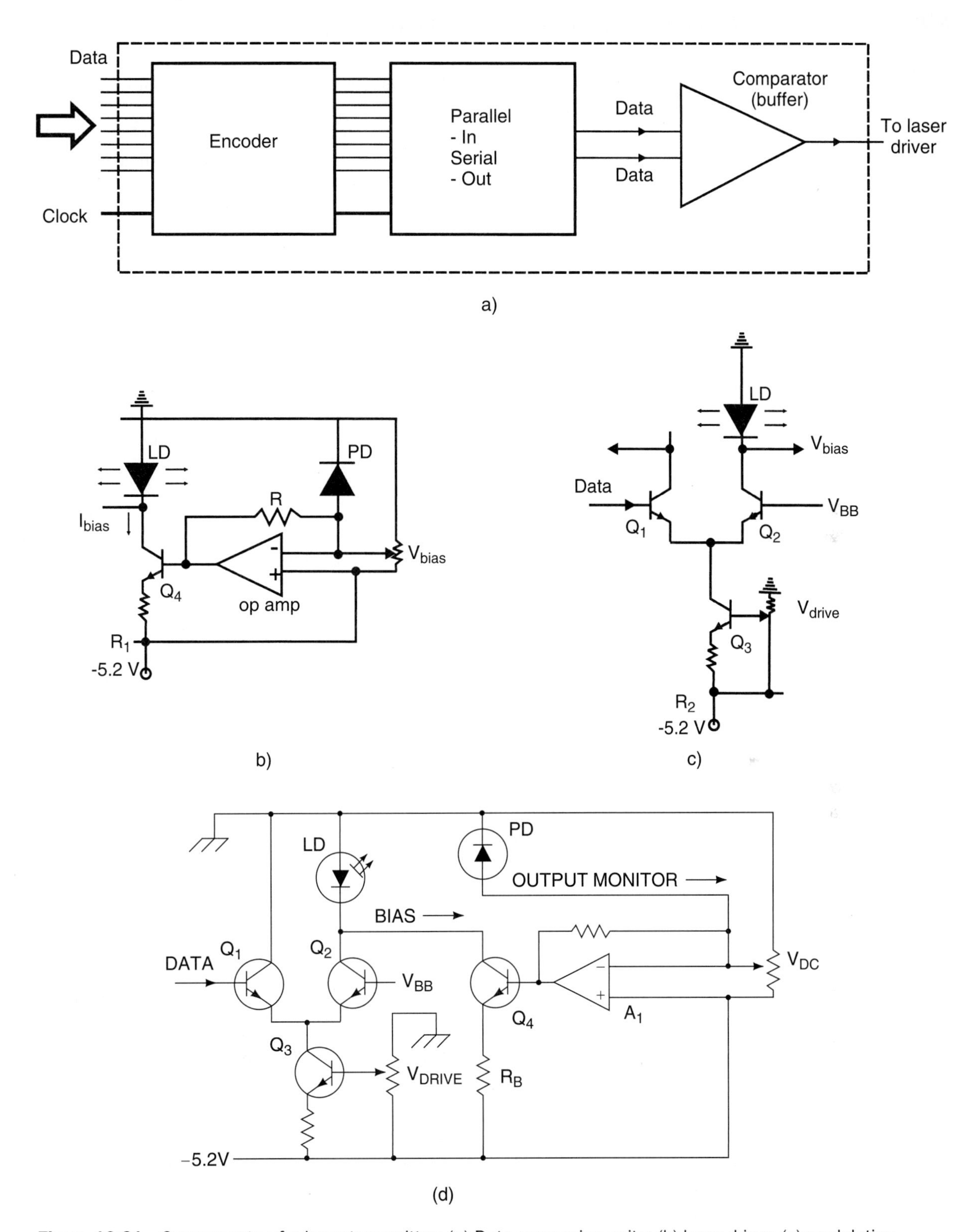


**Figure 10.21** Components of a laser transmitter: (a) Data-conversion units; (b) laser driver; (c) modulation circuit; (d) laser driver/modulation circuit. *([d] from Paul Shumate's "Lightwave Transmitters," in* Optical Fiber Telecommunications—*II, ed. by S. E. Miller and I. P. Kaminow, San Diego: Academic Press, 1988. Reprinted with permission.)*

on the load resistance—here, a laser diode. Thus, changing $V_{bias}$, one can control $I_{bias}$. Observe that $V_{CC}$ is grounded and that $V_{EE}$ is tied to –5.2 V, as explained above.

Since a laser diode is very sensitive to temperature changes, the output light's power has to be stabilized. It may seem that since the bias current is stabilized, this is sufficient to keep the output stable. Not so! *Recall that the threshold current and the slope efficiency of a laser diode also depend on temperature, so that even ideally stable driving current does not guarantee stability of the laser output.* Thus, when temperature varies, the feedback signal obtained from a *rear-facet photodiode (PD)* helps to stabilize the average output power by changing the bias current. Refer to the circuit in Figure 10.21(b). Note that a PD produces photocurrent proportional to the laser's output power. If the latter changes because of temperature variations, the photocurrent causes a change in the $I_{bias}$ that compensates for any deviation in the laser's light power.

***Modulation circuit*** Modulation is controlled by simply changing the driving current from the bias level to maximum, as Figure 10.14 shows. To do this, the circuit shown in Figure 10.21(c) can be used. When data are represented by a voltage greater than $V_{BB}$, transistor $Q_1$ conducts the current; hence, the LD is off. When data are represented by a voltage less than $V_{BB}$, transistor $Q_2$ conducts the current; hence, the LD is on. As you can see, then, this circuit produces *intensity modulation.*

Making the swing voltage small—for example, having the $V_{BB}$ at ± 400 mV and properly designing the transistor $Q_3$ circuit—one can keep transistors $Q_1$ and $Q_2$ from saturating. This results in two important advantages of this ECL gate: (1) A signal as small as 800 mV switches the gate from logic 1 to logic 0 and vice versa and (2) switching between transistors $Q_1$ and $Q_2$ occurs very fast because they are not saturated.

Transistor $Q_3$ (Figure 10.21[c]) works as a tail-current source, so that one can control the driving current by changing the $V_{drive}$. This is necessary because stabilizing the bias current (that is, the average light power) is not enough, as this characteristic actually represents output power only when the number of 1's and 0's over a given time period are equal. This situation is referred to as a 50% *duty cycle.* Bear in mind that since real transmission is far from being represented in this model, as Figure 10.20 shows, it is necessary to employ duty-cycle compensation of the output power. This is done by varying the $V_{drive}$ to change the driving current. One can well imagine the circuit that does this job automatically.

Another important concern is control of the extinction ratio of the output signal. The *extinction ratio (ER)* is the ratio of maximum to minimum light power representing logic 1 and logic 0. It is sometimes called the *on–off ratio* and is usually measured in dB. The extinction ratio is controlled by varying the driving current. If, for example, the ER drops because of a fall-off in power from an aging laser diode, one can increase the driving current by controlling the base current of transistor $Q_3$ (Figure 10.21[c]). The result is an increase in output light power and, therefore, in the extinction ratio. Such control can also be done automatically.

Figure 10.21(d) shows the circuit that performs both the above functions: It (1) stabilizes the average power by changing the bias current to compensate for aging and temperature-induced light-power variations and (2) compensates for fluctuations in output-light power caused by variations of the duty cycle when driving current changes. You can easily identify this circuit as a combination of the circuits shown in Figures 10.21(b) and (c). The simple circuit shown in Figure 10.21(d) gives you an idea of how such circuits are built. For a look at more sophisticated circuits, consult recent technical or trade journals, manufacturers' technical data sheets, and reference [10] at the end of this chapter.

***Controlling and monitoring circuits*** Refer again to the block diagram of the transmitter (Figure 10.18). We'll concentrate on the transmitter circuits that allow the user to control and monitor transmitter performance. The control signal *transmitter disable* allows the user to shut down the transmitter while keeping the module in the standby mode. This can be done, for example, by placing a high-voltage signal at the base of transistor $Q_1$ in Figure 10.21(c).

The voltages across resistor $R_1$ in Figure 10.21(b) and $R_2$ in Figure 10.21(c) allow the user to receive the bias and modulation monitoring signals shown at the top of Figure 10.18. The photocurrent produced by the rear-facet photodetector (PD) in Figure 10.21(b) lets the user know whether the laser diode is operating. The monitoring signal shown at the bottom of Figure 10.18 enables the user to *troubleshoot the transmitter.* The signal from the temperature sensor (the *thermistor*) permits the user to monitor the transmitter's temperature over the entire ambient-temperature range. A simple circuit (not shown) facilitates the monitoring of average light power. All these outputs are also provided with an alarm to signal if any of the monitored parameters deviate from their predetermined ranges.

***Coupling light from an LED transmitter*** Two approaches, discussed in Sections 9.1 and 9.2, are used for coupling from an *LED transmitter:* direct (butt) coupling and lens coupling. Look again at Figure 9.7.

We can introduce *coupling efficiency,* $\eta_C$, as the ratio of the source of maximum power, $P_0$, to the power coupled into a fiber, $P_{in}$, so that

$$\eta_C = P_{in}/P_0 \tag{10.84}$$

For a Lambertian source—the model we apply to an LED—the ratio $P_{in}/P_0$ has been determined to be equal to the square of the numerical aperture, $NA^2$. (See Formula 9.5.)

When a gap separates an LED's radiating surface from the fiber's endface, we have to include the loss caused by reflection at the fiber endface. Thus, we can rewrite Formula 10.84 in this form:

$$\eta_{CSI} = (1 - R)NA^2 = T\ NA^2, \tag{10.85}$$

where $R$ and $T$ are the reflectivity and the transmitivity at the fiber end, respectively. This formula is true for a step-index fiber, whose core is larger than the active radiating surface of an LED.

A graded-index fiber requires a more complex analysis because its *NA* changes over the cross section of the fiber, being at its maximum in the center of the fiber, with this result [13]:

$$\eta_{CGI} = T(NA)^2 = [1 - 1/(2D^2)], \tag{10.86}$$

where $D$ is the ratio of the diameter of the fiber core to the diameter of the source and $D$ must be $\geq 1$. When $D = 1$, which is a realistic assumption, the coupling efficiency of a GI fiber is less than twice that of an SI fiber.

## Example. 10.4.1

***Problem:***

Calculate the amount of light coupling in a 62.5/125 μm GI multimode fiber and in a singlemode fiber from an LED transmitter.

***Solution:***

To find the value of reflectivity, $R$, refer to Formula 8.1. For the air–silica interface, we calculated that $R = 0.035$.

The typical *NA* for 62.5/125 MM fiber is 0.275. (See Figure 3.20.) Thus, for $D = 1$, Formula 10.85 yields:

$$\eta_{CGI} = T(NA)^2\ [1 - 1/(2D^2)] = 0.036$$

The typical *NA* for SM fiber is 0.13. (See Figure 5.18.) Since SM fiber has a step-index profile, we have to use Formula 10.84, which yields:

$$\eta_{CSI} = T(NA)^2 = 0.016$$

In reality, the coupling efficiency is about 1% for SLED, whose radiation pattern is close to the Lambertian model, and about 10% for ELED, when coupling into MM fiber. Even an edge-emitting LED couples less than 1% into a singlemode fiber [7]. It is evident that we're discussing direct coupling. When using the lens technique, coupling efficiency will be much higher (up to 30% for ELED-MM fiber coupling).

---

***Coupling light from a laser transmitter*** Coupling light from a *laser transmitter* cannot be described as simply as we've described coupling light from an LED transmitter. The difficulties arise from the fact that a laser's radiating pattern is not governed by a Lambertian model. In addition, such features of laser radiation as coherence and high intensity require a markedly different approach to the problem.

When a laser transmitter is used with a *multimode fiber,* mode interference usually results in the formation of *speckle patterns*—in the form of bright and dark spots—within the fiber core because of the coherence of laser radiation. (See Figure 10.22.) The coherent modes that are in phase form bright spots; those that are out of phase form dark spots. Theoretically, speckling is not a problem itself because a photodetector collects the total amount of light power across the fiber's cross-sectional area. But, from a practical standpoint, it becomes an issue because all the connections and the fiber itself work as mode filters, thus changing the speckle patterns along the link. These changes result in variations in the speckle patterns over time and, eventually, in random variations of a link's loss. This phenomenon is known as *modal noise.* Modal noise, by the way, is not a concern when working with singlemode fibers or LED transmitters.

When the beam of a laser transmitter is launched into a *singlemode fiber,* the coupling efficiency eventually is determined by matching the field distributions of the incident beam and the fundamental mode of the fiber. This is shown in Figure 10.23(a).

The efficiency of *butt (direct) coupling* with a laser transmitter depends on the ratio of the laser-beam diameter to the fiber-core diameter. Typically, a laser transmitter is used with a singlemode fiber for which the mode-field diameter (MFD), not the core diameter, is the comprehensive characteristic. (See Sections 5.1 and 6.1.) Typical values of the SM fiber's MFD are 9–10 μm. (See Figures 5.17 and 5.18.) Since the intensity distribution of a laser diode is roughly Gaussian, its spatial distribution can also be described by its MFD. You will recall that the MFD is the beam diameter, where the beam's intensity drops to $1/e^2 = 0.0135$ of its peak value. This is why some technical documents contain an expression like "beam diameter, mm, $1/e^2$."

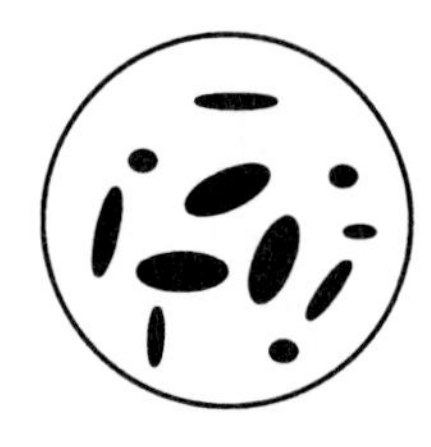

**Figure 10.22** Speckle pattern at the end of the fiber.

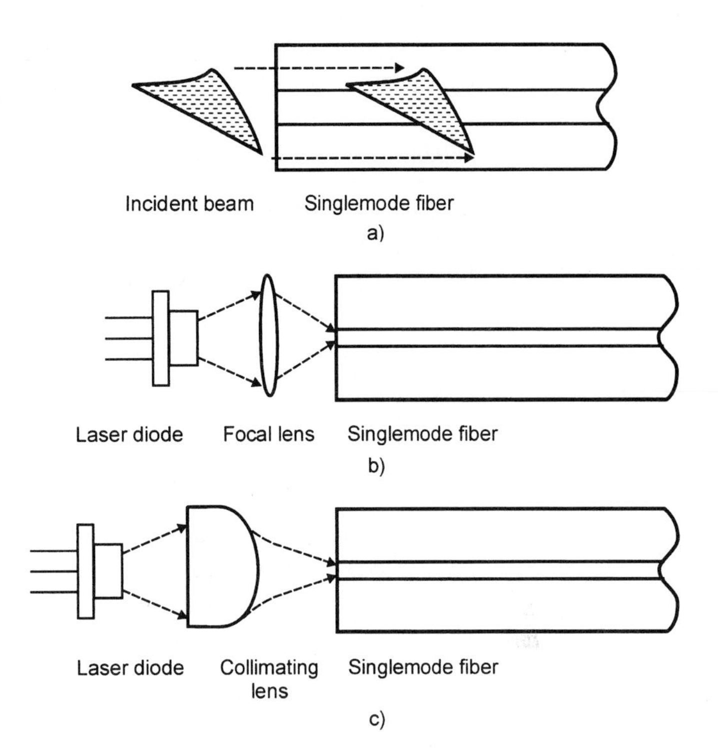


**Figure 10.23** Light coupling from a laser transmitter into a singlemode fiber: (a) Direct coupling; (b) using focal lens; (c) using collimating lens.

A beam diameter (that is, the MFD) of a laser diode can vary from tenths of a μm to tens of μm. Thus, in most cases there is a mismatch between the MFD of a singlemode fiber and that of a laser and the result is relatively low coupling efficiency. (Typical values are from 10% to 20%.) In comparison, the coupling efficiency of a laser transmitter in a 62.5/125 μm multimode fiber is always more than 70% and drops to 50% only at the 1550-nm operating wavelength.

To increase the coupling efficiency of a laser transmitter into a singlemode fiber, *lens coupling* is used. A focal lens can focus a laser beam at the fiber endface, that is, decrease or increase a beam's diameter, as required. This situation is shown in Figure 10.23(b). However, the objective of the design of a lens-coupling system is not to illuminate the fiber's endface but to match a laser beam to a fiber core for transmission. This is why a set of lenses providing the beam collimation is usually used in laser-fiber coupling, as Figure 10.23(c) shows. These lenses have a refractive index with a graded-index profile and are therefore called *GRIN lenses.* Lens coupling improves coupling efficiency up to 80%, but not without exacting a price: complexity of the transmitter package. The main concern here is the mechanical stability of the original alignment over the operating-temperature range, under the strain of vibrations, and in the midst of other environmental hazards.

It seems that the coupling efficiency from a laser transmitter into a singlemode fiber can be calculated using the numerical-aperture concept, as was done for an LED in Formulas 10.84, 10.85, and 10.86 above. Indeed, we know that a broad-area laser diode radiates an elliptical cone of light with a divergence angle of 30° × 10°. (See Figure 10.12.) Since $NA = \sin \theta_a$, we can calculate the largest *NA* of a laser diode as $NA = 0.5$ and compare this number with the *NA* of a singlemode fiber, which is typically 0.13. This approach was covered above with respect to LED coupling. (See Figure 9.7[d].) However, remember that the numerical-aperture concept stems from the beam theory of light, which does not provide sufficient explanation of the workings of a singlemode fiber. (See Sections 5.1 and 6.1.) Thus, the realistic results of laser-SM fiber coupling can be obtained from calculating the overlap integral describing mode-field distributions at both the incident (laser) and accepting (fiber) sides. But we leave these calculations to more specialized textbooks. (See, for example, [14].)

***Backreflection protection*** Another function of coupling optics is to protect the active area of a light source from *backreflected light.* Indeed, when laser light strikes a fiber's endface, it is reflected back to the source. In addition, connectors and mechanical splices also reflect light back to the laser diode. Even Rayleigh scattering results in the same effect. But light sources, particularly laser diodes, are very sensitive to this reflected light. (See the discussion of relative intensity noise, RIN, in Section 10.3.)

An *optical isolator,* placed between a laser transmitter and a fiber as part of a coupling's optics assembly, transmits light in one direction—from laser to fiber—and blocks light transmission in the opposite direction. (We'll discuss its principle of operation in a subsequent chapter.)

Another means to reduce backreflected light is to cover the surfaces of all optical components with an *antireflection (AR) coating.* This measure is well known in optics and, without even being aware of it, everyone encounters AR coatings every day in such items as camera film, eyeglasses, car windshields, and other common optical products. As the term suggests, this coating reduces the reflection of light back to its source. It is interesting to note, too, that *a highly reflective* coating is sometimes applied to the facets of a Fabry-Perot laser diode to increase the reflectivity of a resonator facet.

A schematic of a typical assembly of coupling optics is shown in Figure 10.24.

This assembly is placed within a transmitter package to protect it from environmental hazards. The crucial point in the design of this assembly is that it must ensure that the relative position of all the components is maintained over time and under operational stress. One of the major factors in maintaining this stability is to provide appropriate heat release. This is why the schematic in Figure 10.24 shows a *thermoelectric cooler* (TEC) and a *heat sink* (a heat-dissipation plate) as important parts of the entire assembly.

## Packaging and Reliability

***Packaging*** To put all the transmitter components together, we need to pack them in a single unit whose function is to provide solid, reliable connections among them. Packaging is important because it determines the main properties of a transmitter. In fact, it requires the consideration of three major components: the package itself, a transmitter board, and an assembly of a coupling optic. It is common practice to mount these three components separately. Since the coupling-optic assembly has already been discussed, we will concentrate here on the package and the transmitter board.

An example of just what a *transmitter package* looks like can be seen in Figures 1.5(c) and (d). The transmitter is hermetically sealed inside the package to protect it from environmental disturbances such as humidity, dust, and even normal air flow. There are several packaging

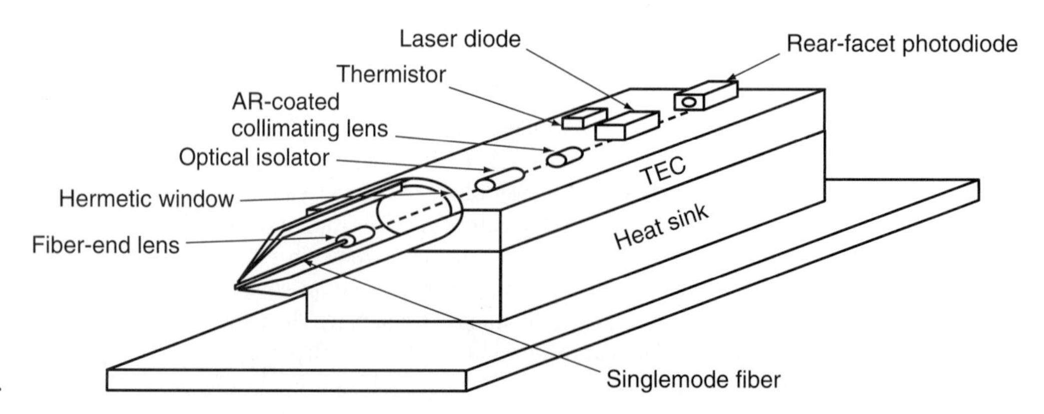


**Figure 10.24** Schematic of a typical assembly of coupling optics.

styles and they differ in pin layout. A dual-in-line package and a butterfly design are the most popular and are shown in Figures 1.5(c) and (d), respectively. Another difference in packaging style is in the way it's connected to an optical fiber. LED transmitters are usually connected to multimode fibers. Since the requirements for connection precision in this case are not very rigid, LED transmitters are usually terminated with a regular connector, such as ST, SC, or FC types, as shown in Figure 1.5(a). Laser transmitters usually work with singlemode fibers, where the precision of connectorization is critical. This is why laser transmitters usually terminate with a pigtail. (See Figures 1.5[c] and [d].)

***Designing a transmitter board*** An example of a *transmitter board* is seen in Figure 10.19. Many problems have to be resolved during the design and assembly stages of such a board. Most of the problems arise because of the high speed at which fiber-optic transmitters operate. They are problems common to any high-speed electronic board and must be confronted when doing a *board layout.*

The first problem cropping up with a transmitter board is designing the right electric connections. There are two common types of electric connections: microstrips and striplines. They are shown in Figure 10.25.

A microstrip circuit, the more popular of the two, is based on the transmission-line theory, which can be found in any textbook on electromagnetics. We will consider here only the practical applications of this theory. An electric signal, which is an electromagnetic wave, propagates along the microstrip line and eventually reaches the termination point. This wave can be reflected, which is undesirable, or totally absorbed by the load, which is what we want to happen. The ratio of reflected signal voltage, $V_r$, to incident signal voltage, $V_i$, is given by:

$$V_r/V_i = (Z_L - Z_0)/(Z_L + Z_0), \qquad (10.87)$$

where $Z_L$ is the load impedance and $Z_0$ is the characteristic impedance of a transmission line. Since $Z_L$ is given, the designer can work only with $Z_0$.

The characteristic impedance of a microstrip line is given by [15]:

$$Z_0 = [87/\sqrt{(\varepsilon_r + 1.41)}]\ln[5.98h/(0.8w + t)], \qquad (10.88)$$

where $\varepsilon_r$ is the relative *dielectric constant* of the dielectric (usually between 4.5 and 4.7); other dimensions are shown in Figure 10.25. By varying the dimensions of the microstrip and board, the designer can obtain the desired characteristic impedance to make $Z_0 = Z_L$, as Formula 10.87 requires. If a typical value of $Z_0$ is 50 Ω, the designer can build the termination circuit to make Formula 10.87 work. (For a more detailed discussion of the application of Formulas 10.87 and 10.88, see [13].) Simply keep in mind that we want to decrease the strip width, $w$, to increase the density of the board layout, but certain restrictions to making $w$ smaller are applied.

Another important matter when dealing with board layout is *propagation delay,* $t_{pd}$. It depends only on the dielectric constant, $\varepsilon_r$, and it is equal to [15]:

$$t_{pd}(\text{ns/ft}) = 1.017\sqrt{(0.475\varepsilon_r + 0.67)} \qquad (10.89)$$

For $\varepsilon_r = 4.5$, propagation delay is 1.70 ns/ft.

The designer has to take propagation delay into account when considering the *trace lengths.* It follows immediately from Formula 10.89 that the differential input signals that control the comparator (buffer) depicted in Figure 10.21(a) might have different phases if their traces were not equal to one another. Duty-cycle distortion, often called pulse-width distortion, results when the pulse duration varies with respect to the reference time of the signal. Another negative effect of the inequality of the trace lengths is an increase in power-supply noise.

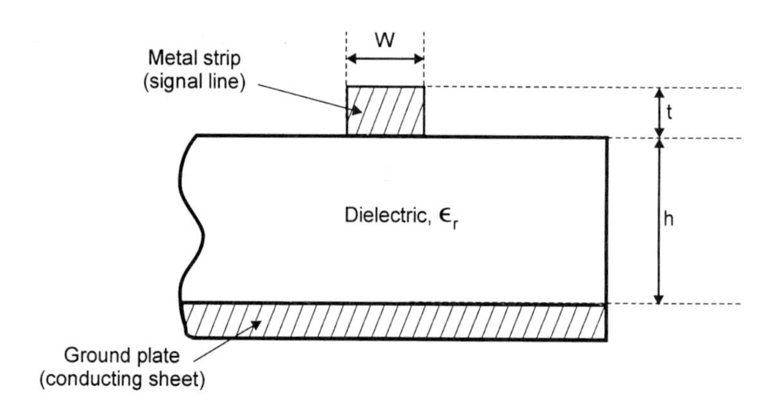


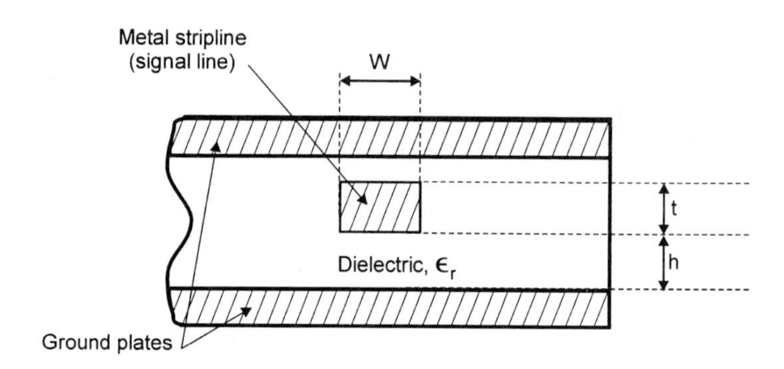


**Figure 10.25** Microstrip and stripline: (a) Microstrip circuit; (b) stripline circuit.

The last point to note when considering board layout is *ground plate,* also called *reference plate.* This plate provides a return path for the electric signals of all the circuits mounted on the board. Since many lines share this plate, it must be large enough to accommodate different electric signals with a minimal change in reference voltage.

***Reliability, testing, and troubleshooting*** The reliability of a transmitter is measured in terms of either *failures in time (FIT)* or *mean time to failure (MTTF).* FIT is the number of failures per billion hours of cumulative operation and MTTF is the average time until a given component fails. The order of values for fiber-optic transmitters is as follows [15]:

| Temperature | FIT | MTTF |
|---|---|---|
| 25°C | 294 | $2.4 \times 10^6$ hours (about 274 years) |
| 50°C | 1456 | $0.69 \times 10^6$ hours (about 78 years) |

Note how dramatically temperature influences the reliability of transmitters: Doubling the temperature within a given environment decreases the MTTF three-and-a-half times and increases the FIT nearly five times.

The least reliable part of a transmitter is its light source: an LED or a laser diode. Even though their reliability, if measured separately, is rather high (the MTTF for LDs is $10^6$ and even reaches $10^9$ for LEDs), when they are used in the transmitter package, the numbers drop to those shown above.

The most critical component of a laser diode is the active region itself. Another crucial component is the facet coatings, which tend to lose their reflectivity over time. As these components age, the driving current has to be increased to compensate for their lower gain and additional loss, which eventually result in device failure.

To evaluate a device's operational status, manufacturers rely on several criteria. Obviously, if a transmitter doesn't operate at all, no criterion need be considered. But, in practice, this seldom occurs; in most cases, *transmitter failure means the device is not operating properly.* Since the light sources are the critical components, they are the first ones to be evaluated. You will recall that manufacturers consider a laser diode to be out of order if its threshold current (or driving current) exceeds the initial value by a certain percent. The figures of 10% or 50% are commonly used. (See Section 9.3.)

So, then, just how do manufacturers *test* their transmitters? Nobody, of course, wants to wait for several years to see the threshold current increase by 10%. The common practice, therefore, is to perform *purge testing.* This procedure tests transmitters under extreme conditions: high temperature, high driving current, mechanical and other environmental stresses. Special chambers are used to perform such tests. Purge tests are performed not only on the individual components but also—and mainly—on the ready-to-ship transmitters.

Another widely used method for testing completed devices is called *burn-in.* In this test, transmitters operate at a 50% duty cycle at the normal level of electrical signals but at a high temperature (up to 100°C) over a 48-hour period. This test reveals early-life failures.

The meaning of both tests becomes clear if we notice that the rate of transmitter failure, as for any semiconductor device, changes over time. During the first stage, called *infant mortality,* the failure rate declines from a high to a low, stable level. The initial high failure rate occurs because of the inability of some components to withstand the stresses of real operation during this early period. Then, for a long period (as defined by the MTTF criterion) the failure rate remains low. After the MTTF limit is reached, however, the failure rate increases again because the devices are aging. This last stage is called *wearout.*

*Troubleshooting* a transmitter is relatively easy because manufacturers provide access to the electrical signals reflecting the state of crucial parameters. For example, the photocurrent of a rear-facet PD indicates the level of an LD's output power, which is sufficient to evaluate LD performance. All the other signals discussed in the subsection "Controlling and monitoring circuits" allow the user to do comprehensive troubleshooting.

When taking any measurements, accuracy and other performance characteristics, such as the reproducibility of the measuring instruments, are critical. Fortunately, industry provides a variety of measuring devices with excellent characteristics to meet any practical requirements. For the names of these specific devices, consult the manufacturers' technical documentation. (See, for example, [16], [17], [18], and [19].)

### Reading the Transmitter's Data Sheet

The detailed discussion of optical transmitters given above brings us to the point where understanding the manufacturers' specifications sheets should be easy Figure 10.26 is the data sheet for a specific transmitter. Read the list of *applications* first: SONET/SDH and ATM are transmitting standards; they are mentioned in Section 8.4 and will be considered in later chapters. All the other information listed has already been discussed.

Take a look at the list of "FEATURES." The only one we still have not covered is "multi-source pinout," which means that this chip has several pins through which source power can be

**DIGITAL FIBER OPTIC TRANSMITTER**
**TL-1160 SERIES**
**SPECIFICATIONS AND USER INFORMATION**

APPLICATIONS

- **SONET OC-1, OC-3 / SDH STM-1**
- **ATM/FDDI**
- **Data Rates to 300 Mb/s**
- **LAN, MAN's**
- **Subscriber Loop**
- **High Loss Budget Links**

FEATURES

- **Low Threshold FP Laser**
- **Automatic Power Control**
- **Transmitter Disable**
- **Optical Power Monitor**
- **Laser Bias Monitor**
- **20 Pin Multisource Pinout**

**DESCRIPTION**

The series TL-1160 transmitters are full function ECL/PECL-compatible modules designed for applications up to 300 Mb/s. The optical source is a low threshold Fabry-Perot laser coupled to single-mode fiber. The module is pin by pin compatible with the SONET multisource standard. A single 5 Volt supply allows for interfacing with ECL or PECL logic.The TL-1160 series transmitters are available with fibers terminated with FC, ST or SC connectors.

FUNCTIONS- Refer to fig. 1

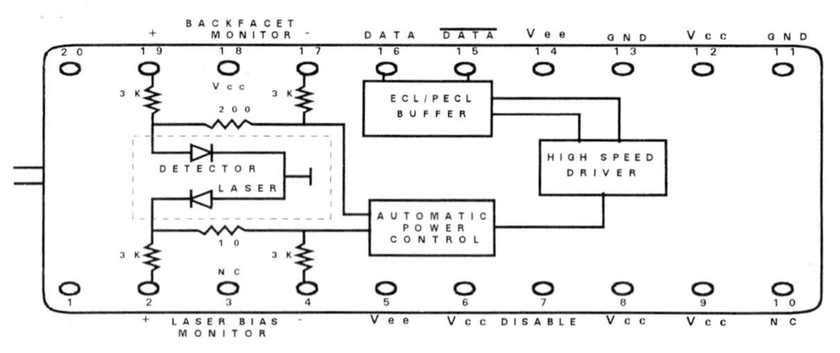


TRANSMITTER FUNCTIONAL DIAGRAM
FIG. 1

**Figure 10.26** Data sheet of a typical transmitter. *(Courtesy of Laser Diode, Inc.)*

supplied. (See the table labeled "PIN ASSIGNMENT.") By now, then, the "DESCRIPTION" of this transmitter in the data sheet should be perfectly clear to you. If such terms as *FC, ST,* or *SC* connectors or *Fabry-Perot* laser still puzzle you, reread the appropriate sections of Chapter 8 and this chapter.

Study the *functional diagram* of the transmitter depicted in "Fig. 1" of Figure 10.26. You will find it very informative to analyze this diagram by also reviewing Figures 10.18 and 10.21. Observe how all the transmitter functions discussed in these figures are implemented in the actual device.

Consider the *performance characteristics.* By now you have gained sufficient knowledge to understand all of these; therefore, let's review a few:

- *Input-data voltage:* A 1-V difference switches the transmitter from HIGH to LOW and back.
- *Duty-cycle distortion:* This is the distortion of pulse width caused by a variation in rise/fall time.

High Speed Driver

The high speed driver modulates the laser above threshold to maximize speed and minimize pulse width distortion. The high speed drive current is controlled by a temperature-sensitive network. This network compensates for the reduction in a laser's power slope at higher operating temperatures by increasing the drive current. The result is that a minimum extinction ratio is maintained over all operating conditions.

Laser Status Section

The laser's backfacet monitor network shown (pins 17 & 19) provides an external differential voltage proportional to the optical power emitted. This monitor is useful for determining whether the transmitter is functioning properly when locating faults along the fiber link.

The laser's bias current is also monitored with a differential voltage via pins 2 & 4. An increase in this voltage is normal when operating the module at elevated temperatures since the automatic control circuit will increase the bias to maintain constant optical power. Any connection to the monitor pins should be high impedance.

Enable/Disable

The disable function completely shuts down the transmitter. This feature may be used in applications where the output power must be off when the transmitter is in standby mode.

Automatic Power Control

The Automatic power control section maintains constant average power over the -40°C to +85°C operating range. The control also compensates for normal increases in the laser's operating current due to normal aging effects.

User Connections

Figs 2 and 3 demonstrate typical connections which the user can implement to interface with ECL and PECL logic. Good high frequency techniques should be used when laying out a PC board to insure good signal quality. A ground plane is recommended with the by-pass capacitors and terminating resistors located as close to the module as possible.

Application of Status Alarms

When implementing an external status alarm, a differential amplifier should be used to convert the differential monitor signal to a single-ended signal. This signal can then be fed to a simple comparator to detect a change in operating status. For example, the laser bias monitor can be used in conjunction with a differential amplifier and comparator to signal an increase in bias current of 50% or more at 25°C.

Evaluation Board

Test boards are available for simple interfacing to test equipment such as pattern generators. Contact the sales department for information.

**Figure 10.26** (continued)

- *Transmitter mask,* also called *eye-pattern* (or *eye-diagram)* performance. An eye diagram is commonly used when evaluating electrical and optical data transmission. This pattern is formed by superimposing 010 and 101 signals, as Figure 10.27(a) shows.

The word *eye* is used because of the pattern's similarity to the open human eye. An ideal eye, such as the one shown in Figure 10.27(a), can degrade because of common transmitter problems. This is shown in Figure 10.27(b). For example, "jitter"—a slight variation in the digital signal with respect to reference time—makes the signal diffuse; increasing rise (or fall) time closes the eye; pulses that overshoot and undershoot also close the eye and distort its shape. In other words, eye pattern is the integral evaluation of the quality of a transmitter. To quantify this evaluation, the manufacturer uses the template ("eye" mask) shown in Figure 10.27(c). *If the eye diagram doesn't touch or cross this template,* which is described as an open eye, *the quality of the transmitter's performance is acceptable.* The notation "500 wfm," which you will find at the top of the transmitter mask figure, means that 500 waveforms are used to perform this test. Figure 10.27(d)

PERFORMANCE SPECIFICATIONS

**Absolute Maximum Ratings**

| Parameter | Minimum | Maximum | Units |
|---|---|---|---|
| Supply Voltage | | 6 | V |
| Operating Case Temperature<br>• TL 1163<br>• TL 1165 | <br>-40<br>-40 | <br>85<br>50 | <br>°C<br>°C |
| Storage Temperature | -40 | 85 | °C |
| Lead Soldering Temperature/Time | | 250/10 | °C/sec. |

**Electrical Characteristics**

| Parameter | Minimum | Typical | Maximum | Units |
|---|---|---|---|---|
| Supply Voltage [Vcc- Vee] | 4.75 | 5 | 5.5 | V |
| Supply Current | | 70 | 130 | mA |
| Input Data Voltage (1)<br>• Low<br>• High | | <br>Vcc- 1.8<br>Vcc- 0.8 | | <br>V<br>V |
| Bias Monitor (@ 25 °C) | 0.01 | | 0.45 | V |
| Back Facet Monitor | 10 | | 200 | mV |

1. When Vee is -5 V, Vcc must be 0 V. With Vcc at +5 V, Vee must be 0 V.

**Optical Characteristics**

| Parameter | Minimum | Typical | Maximum | Units |
|---|---|---|---|---|
| Central Wavelength (2)<br>• TL 1163<br>• TL 1165 | <br>1270<br>1500 | | <br>1360<br>1576 | <br>nm<br>nm |
| Average Output Power<br>• 010 (TL 1163, TL 1165)<br>• 003 (TL 1163) | <br>-12<br>-5 | <br>-10<br>-3 | <br>-8<br>0 | <br>dBm<br>dBm |
| Spectral Width (RMS) | | | 3 | nm |
| Extinction Ratio | 10 | 12 | | dB |
| Tx Disable | Vcc- 3.2 | | Vcc | V |
| Duty Cycle Distortion | | 0.4 | 1 | ns |

2. Over operating temperature range. A narrower operating temperature range will result in a smaller wavelength spread.

**Figure 10.26** (continued)

shows a sample of a computer-simulated distorted eye pattern [20]. One can easily observe all types of eye distortions, which are shown schematically in Figure 10.27(b).

Now let's return to the data sheet (Figure 10.26). The mechanical dimensions, given in inches, enable you to realize how small this device is. Pin assignment is important to the designer and the user. When you design a transmitter board, you have to know all the requirements—such as traces, lengths, characteristic impedance—that you have to meet with a given pinout of the transmitter. If you are a user, you have to know where actually to draw the signals for testing and troubleshooting. Incidentally, now you know what a "20 pin multi-source pinout," indicated in the "FEATURES" section of this data sheet, means.

## External Modulators

***Internal and external modulation*** Let's pause here to summarize the topic of modulation, which we have already covered twice: in Section 10.3, under "Laser Modulation," and in this section where we discussed modulation circuitry. Again, let's concentrate on modulation as applied

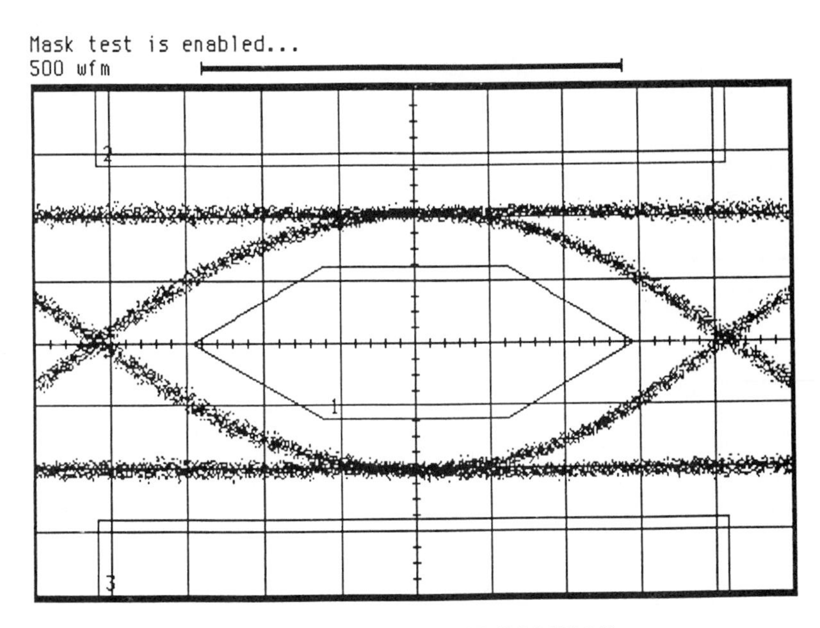


**Figure 10.26** (continued) TYPICAL TRANSMITTER MASK (OC-3,STM-1)

to a laser diode. First, we learned that the modulation of a laser diode can be accomplished by simply changing the driving current. Second, we found that we needed biasing to accelerate the switching of a laser diode from on to off and vice versa. Third, we considered the circuit that carried out these functions. (See Figure 10.21[c].)

Thus, the concept of modulating a laser transmitter becomes clear. Let's underscore once more that we have covered *internal* (or *direct*) *modulation* and, with this type of modulation, the intensity of the radiated power is changed from maximum to minimum; this is why it is called *intensity modulation (IM)* as described in Section 10.3.

You will recall that, despite the benefits that direct modulation offers, there are at least two serious drawbacks:

1. Bandwidth is restricted by the laser diode's relaxation frequency.

2. Chirp—the fast variations of the laser's peak radiating frequency in response to a change in driving (modulation) current—results to produce a broadening of the light pulse. Chirp is a problem for DFB lasers and it is a serious limiting factor in high-speed communications, where DFB lasers are primarily applied.

There is also a third drawback that we have not yet mentioned: Long-distance fiber-optic networks require that extra light power be launched into the fiber to increase the span between each optical amplifier. Launching high light power into a fiber is also a regular practice for the cable TV networks. Numbers of 30 mW and higher are common. Attaining such power requires that a laser diode operate at high driving current—on the order of 100 mA and higher. But this, in turn, means that switching to such a current at high speed becomes a major problem, as you will recall from our discussion of the physics of laser-diode operation.

The radical means to overcome all problems associated with internal modulation is to resort to *external modulation.* This approach leaves a laser diode to radiate a *continuous light wave (CW)* while a change in light power occurs outside the laser diode. A block diagram of external modulation is shown in Figure 10.28.

To understand the advantages of external modulation, we simply need to recall the shortcomings of internal modulation: restricted bandwidth and chirp. With external modulation, a

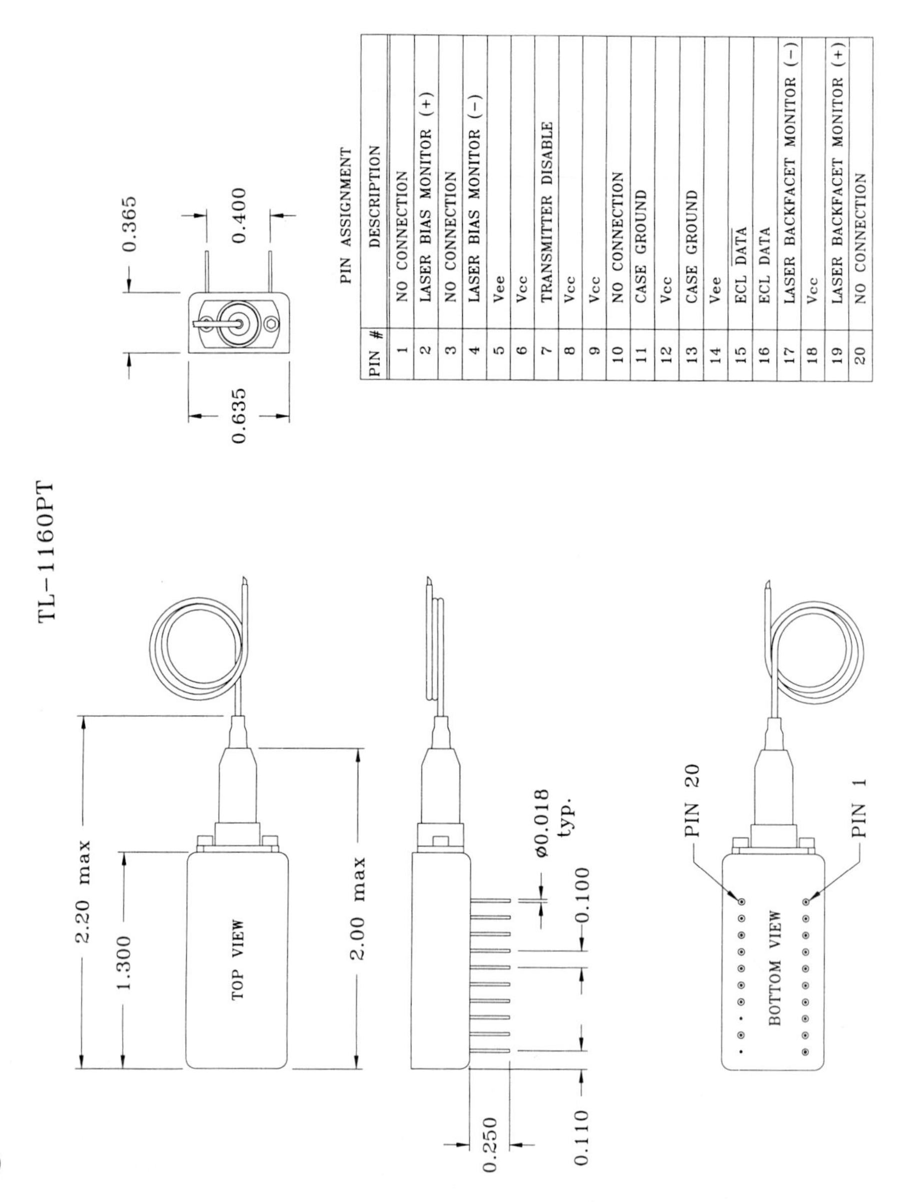


**Figure 10.26** (continued)

feedback loop with a photodiode provides a very stable level of power radiated by the laser diode because this circuit isn't loaded with extra tasks like biasing and modulation. In addition, extremely stable radiating frequency is achieved because the chirp problem is completely eliminated. This advantage is of great importance for *wavelength-division multiplexing (WDM) systems*, where the stability of the radiating wavelength is the chief requirement. There is no restriction on light power here and the bandwidth of this transmitter is determined exclusively by the external modulator.

What price must be paid for all these advantages? For one thing, we need to insert an external component into the optical loop. And you can well imagine that any extra connection or extra optical component in a fiber-optic communications link gives rise to a whole new set of problems.

***Mach-Zehnder (MDM) external modulators*** Two basic types of external modulators are mainly in use today. The first one is a stand-alone external modulator called a *Mach-Zehnder*

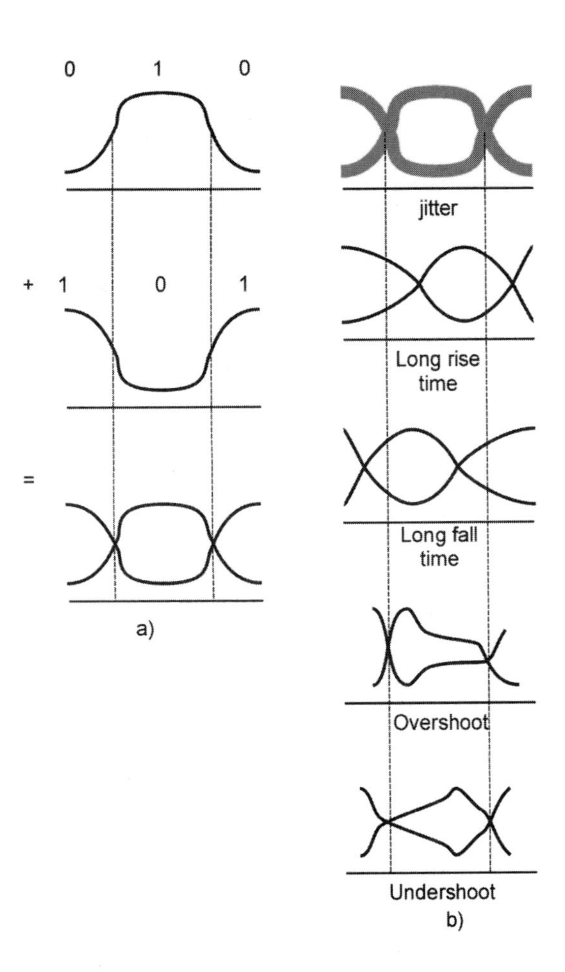


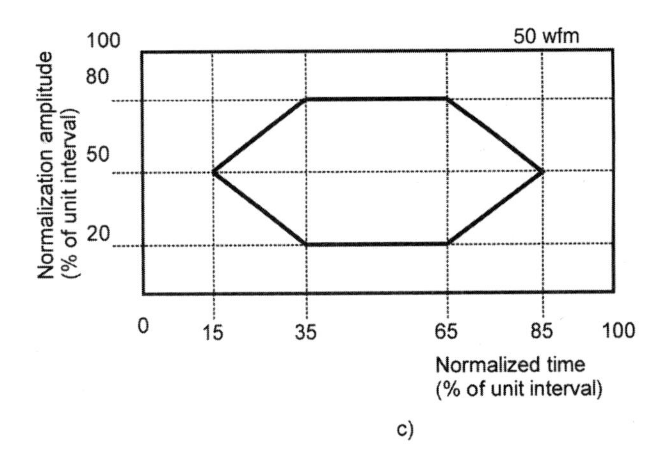


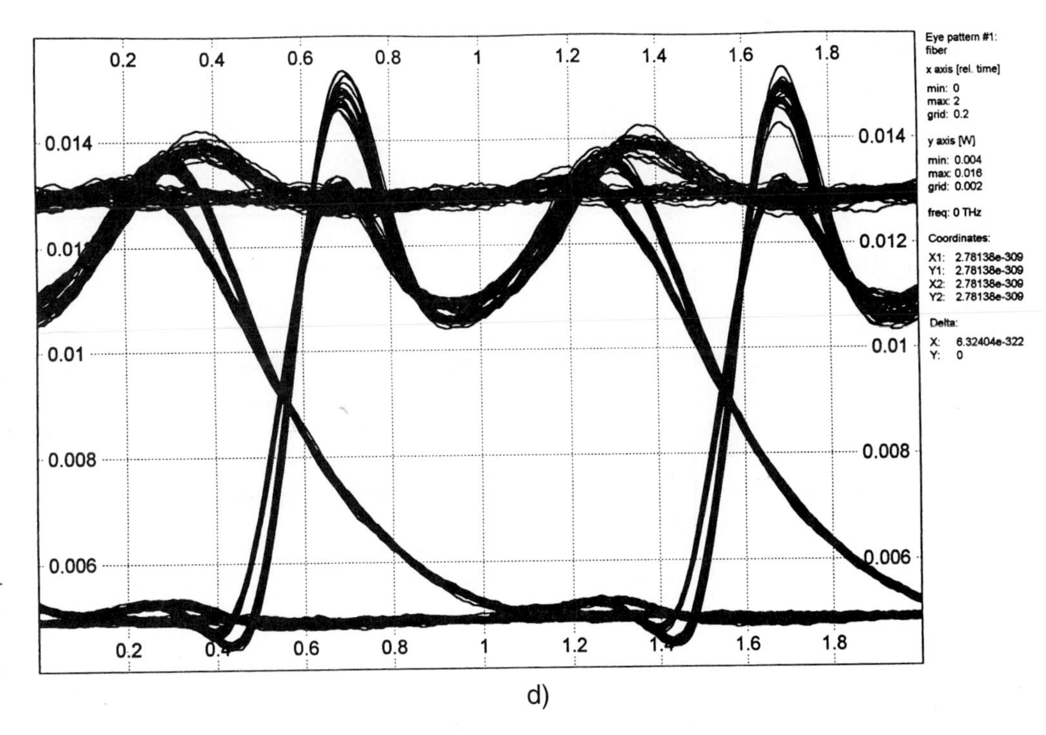


**Figure 10.27** Eye pattern: (a) Formation of eye diagram; (b) eye-diagram degradations; (c) transmitter "eye"-mask definition; and (d) computer simulation of a distorted eye pattern.

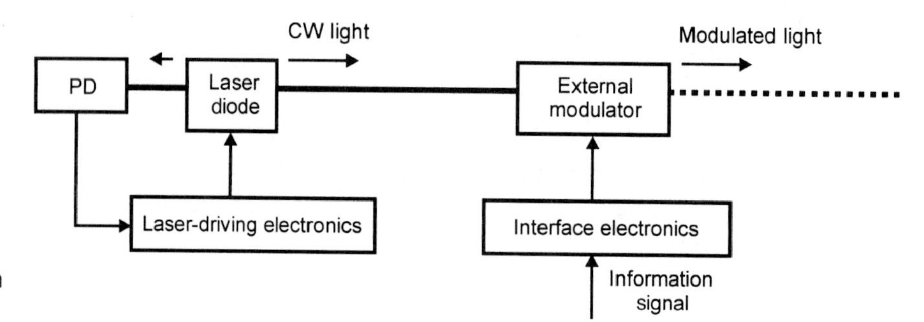


**Figure 10.28** Block diagram of external modulation.

*(MDM)*, or *lithium niobate ($LiNbO_3$), modulator.* It is based on the so-called Mach-Zehnder configuration. (See Figure 10.29.)

Its principle of operation is as follows: The refractive index of lithium niobate changes when an electrical voltage is applied. Light for this modulator emanates from the laser diode and splits equally when it enters the waveguide. When no voltage is applied, both halves of the incident wave have no phase shift and so they interfere constructively, forming the original wave, as Figure 10.29(a) shows. When voltage is applied, one-half of the incident wave experiences a phase shift of +90° because the refractive index of this portion of the waveguide decreases, increasing the velocity of the light propagation and lessening the delay. The other half of the waveguide receives a –90° shift because its refractive index increases, reducing the velocity of the light propagation and lengthening the propagation delay. When the

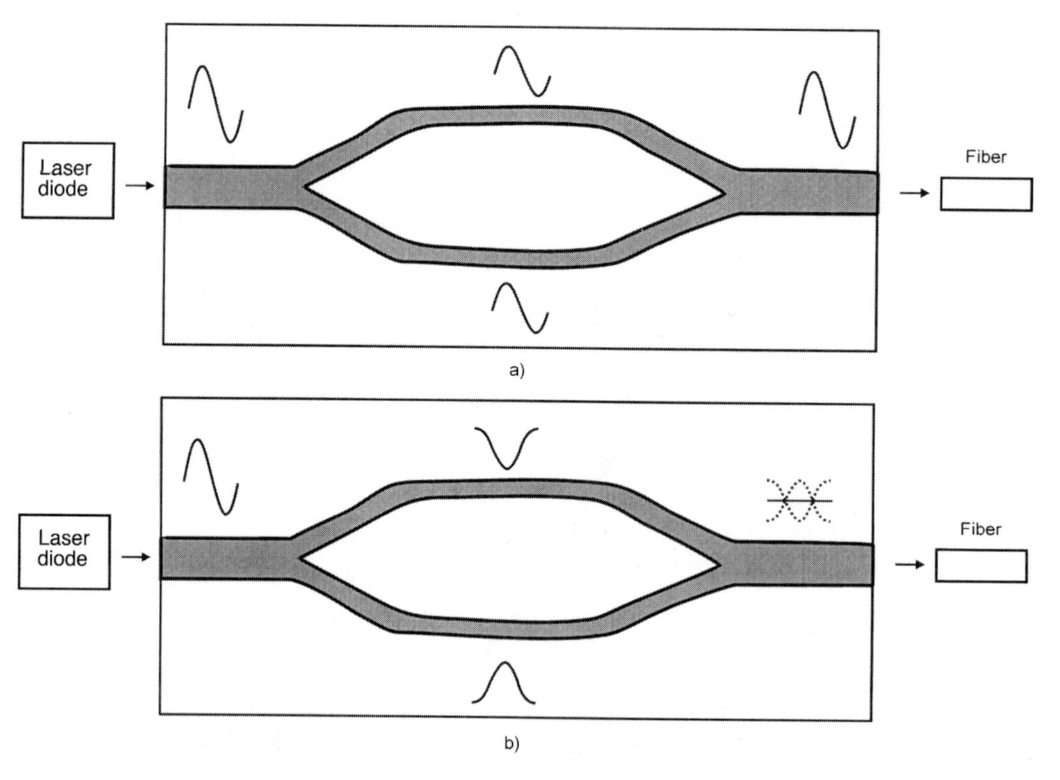


**Figure 10.29** Principle of operation of an external MDM modulator: (a) No signal—constructive interference; (b) signal applied—destructive interference; (c) typical modulator configuration and modulator transfer function. *(Courtesy of Uniphase Telecommunications Products, Inc.)*

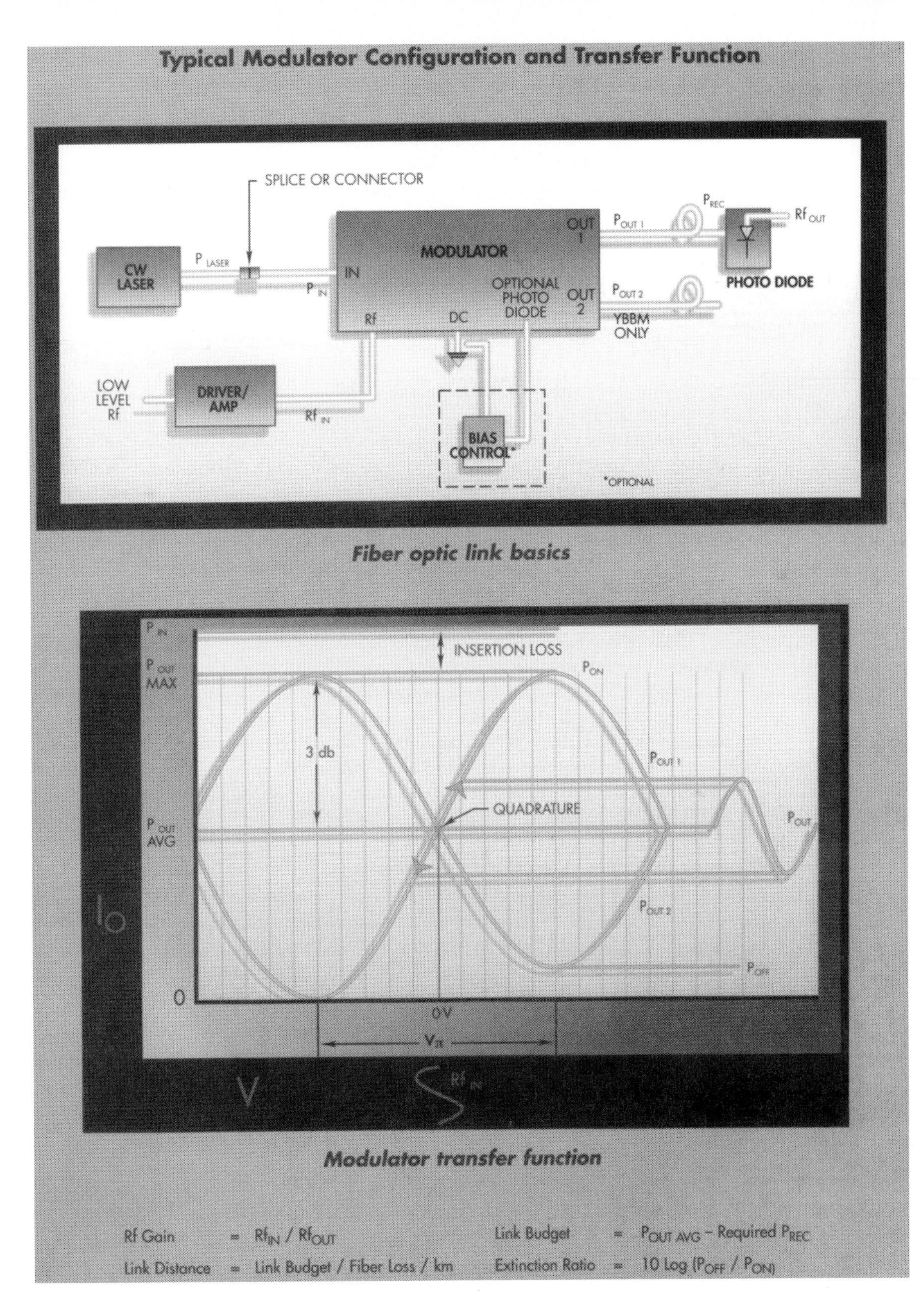


c)

**Figure 10.29** (continued)

halves recombine, they cancel one another, as Figure 10.29(b) shows. Such "destructive" interference demonstrates that we can control output intensity by applying external voltage. If you look again at Figure 10.28, the principle of external modulation now becomes absolutely clear. Note that we can introduce any phase shift we want between the halves of the propagating wave.

A typical modulator configuration and the modulator transfer function—the relationship between output light and input voltage—are shown in Figure 10.29(c). A radio-frequency (RF) signal produces an analog optical signal using the linear portion of the transfer function. A digital signal switches light between $P_{on}$ and $P_{off}$ while bias dc voltage controls the operating point shown as 0 V. In addition to the Mach-Zehnder modulator described above, there is its dual-output modification, known as a *Y-fed balanced bridge modulator (YBBM).*

Figure 10.30 shows the data sheet of an MDM external modulator. We leave the analysis of its specifications as an exercise for you. At this point, you should be able to understand the data given in Figure 10.30. (If, however, you have read this text out of sequence and have some trouble following the specifications, read Sections 8.1 and 8.2.) Specifically note the *loss* shown here. The optical insertion loss introduced by this external modulator is 4.0 dB, which is typical. This means that $P_{out} = 0.4\ P_{in}$. Thus, if the input power ($P_{in}$) = 50 mW, only 20 mW will actually be launched into the transmission fiber. This is why the data sheet specifies *minimum* in terms of input power. Pay attention, also, to the optical return-loss figure. If you understand the RIN problem discussed in Section 10.3, you will certainly appreciate the significance of this number.

You can find much more information on MDM modulators in the manufacturers' technical literature. (See, for example, references [21] and [22] at the end of this chapter.)

External MDM modulators have found their specific applications primarily in cable-TV optical networks; in fact, most of the fiber-optic communications systems already deployed use internal modulation. However, external modulators are playing increasingly important roles today in very fast fiber-optic networks using time-division and wavelength-division multiplexing. Indeed, the data sheet shown in Figure 10.30 affirms that this modulator "is ideal for SONET, SDH, and WDM applications."

***Electroabsorption (EA) external modulators*** The major drawbacks of an MDM modulator are high insertion loss (up to 5 dB) and relatively high modulation voltage (up to 10 V). (These numbers appear in the data sheet in Figure 10.30.) An already-mentioned disadvantage of an MDM modulator is that it is a stand-alone unit. This configuration gives the designer and fiber-optic-network managers migraines. What they want, ideally, is to have one transmitter unit with a built-in modulator integrated into the laser diode through use of a single chip—but without the chirping problem.

The closest we have come to this ideal is the *electroabsorption modulator (EA).*

Here's how it works: A DFB laser radiates a continuous wave of light, which runs through a waveguide made from a semiconductor material. Without applied voltage, this waveguide is transparent to the light emitted by the DFB laser because its cutoff wavelength, $\lambda_C$, is shorter than the wavelength of incident light. (See Sections 10.1 and 10.2.) *When modulation voltage is applied, a bandgap,* $E_g$, *of the waveguide material decreases.* This is called the *Franz-Keldysh effect* and it is the key to understanding the operation of an EA modulator. Since the bandgap decreases, the cutoff wavelength increases (remember, $\lambda_C = 1024/E_g$) and the waveguide material starts to absorb the incident light. Hence, by applying modulation voltage to a semiconductor waveguide, you are able to change the absorption property of this waveguide.

The beauty of this type of modulator is that a semiconductor waveguide can be fabricated onto one substrate with a DFB laser. The industry refers to this device as a "monolithically integrated" chip.

Digital

## 2.5 Gb/s and 10 Gb/s Digital Modulators

### Features

- *Ideal for SONET, SDH and WDM applications*
- *BIAS FREE™ operation (APE™) – <u>no</u> bias loop circuit required*
- *Chirp-free or fixed-chirp versions for maximum transmission distance*
- *Low drive voltage – compatible with commercially available drivers*
- *Low insertion loss – maximizes launch power*

10 Gb/s Package

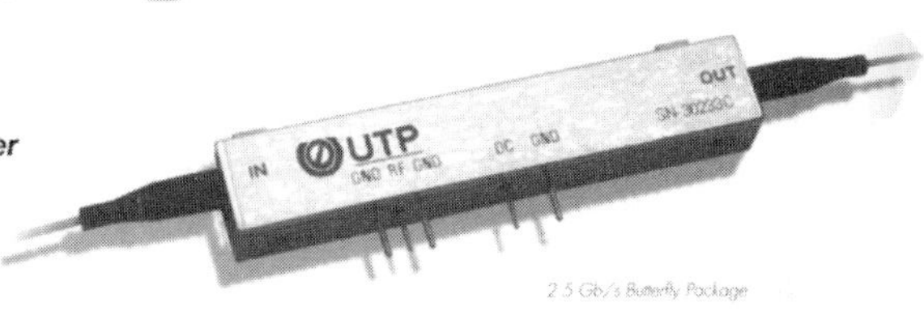


2.5 Gb/s Butterfly Package

#### General Specifications

Crystal orientation .................. x-cut, y-propagating

#### Package Dimensions

2.5 Gb/s APE ..................... See outline drawings AA, AB, AN
2.5 Gb/s TI ....................... See outline drawing AA
10 Gb/s ........................... See outline drawing AC

#### Absolute Maximum Ratings

Operating temperature
APE or TI w/ bias circuit .......... -5 °C minimum, 70 °C maximum
APE without bias circuit .......... 20 °C minimum, 70 °C maximum
Storage temperature ................ -45 °C minimum, 90 °C maximum
Bias port: applied DC voltage .......... 0 V recommended (APE), ± 15 V max.
Rf port: applied DC voltage ............ 0 V recommended, ± 2 V max.

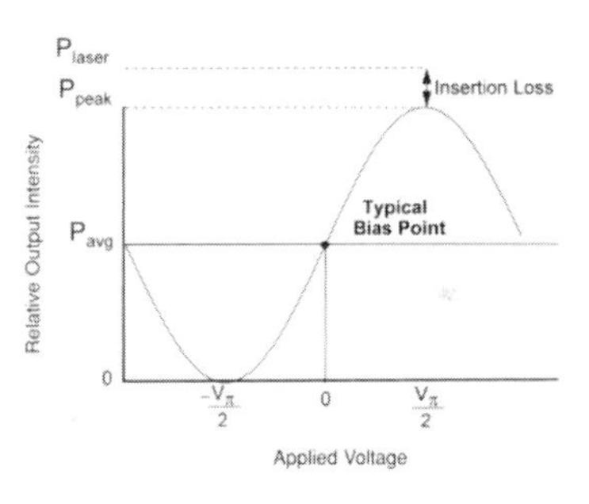


Modulator Transfer Function

#### Fibers

*Typical configuration is PM on input, SM on output*

1300 nm device
PM - Input .................. Fujikura SM-13-P-7/125-UV/UV-400
SM - Output ................. Corning SMF 28
1550 nm device
PM - Input .................. Fujikura SM-15-P-8/125-UV/UV-400
SM - Output ................. Corning SMF 28

### ORDERING INFORMATION

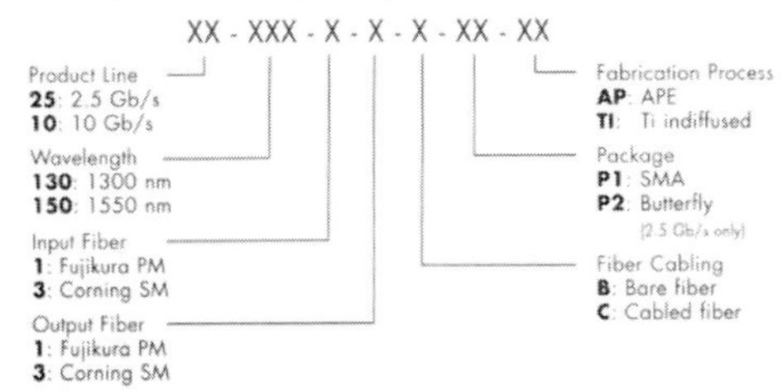


**Figure 10.30** Data sheet of an external modulator. *(Courtesy of Uniphase Telecommunications Products, Inc., Melbourne, Fla.)*

**2.5 Gb/s**

| | | Device Model # | 25-130- -AP | 25-150- -AP | 25-130- -TI | 25-150- -TI |
|---|---|---|---|---|---|---|
| | | Fabrication Process | APE | APE | TI | TI |
| OPTICAL | | Operating wavelength, minimum (nm) | 1300 | 1535 | 1300 | 1535 |
| | | Operating wavelength, maximum (nm) | 1330 | 1565 | 1330 | 1565 |
| | | Insertion loss, typical (dB) | 3.5 | 3.5 | 4.0 | 4.0 |
| | | Insertion loss, maximum (dB) | 4.5 | 4.5 | 5.0 | 5.0 |
| | | On/off extinction ratio, typical (dB) | 30 | 30 | 25 | 25 |
| | | On/off extinction ratio, minimum (dB) | 20 | 20 | 20 | 20 |
| | | Optical return loss, typical (dB) | >60 | >60 | >60 | >60 |
| | | Optical return loss, minimum (dB) | 50 | 50 | 50 | 50 |
| | | Input power, maximum (mW) | 200 | 200 | 50 | 50 |
| ELECTRICAL | | Bandwidth, typical (GHz) | **2.8** | **2.8** | **2.8** | **2.8** |
| | Rf PORT | Bandwidth, minimum (GHz) | 2.2 | 2.2 | 2.2 | 2.2 |
| | | $V_\pi$ at DC, typical (V) | 2.3 | 2.8 | 2.9 | 3.5 |
| | | $V_\pi$ at DC, maximum (V) | 2.6 | 3.2 | 3.3 | 3.8 |
| | | Rise time, typical (ps) | 110 | 110 | 110 | 110 |
| | | Rise time, maximum (ps) | 130 | 130 | 130 | 130 |
| | | Fall time, typical (ps) | 110 | 110 | 110 | 110 |
| | | Fall time, maximum (ps) | 130 | 130 | 130 | 130 |
| | | Return loss, typical (dB) | -8 | -8 | -8 | -8 |
| | BIAS PORT | $V_\pi$ at DC, typical (V) | 5.2 | 6.5 | 5.2 | 6.5 |
| | | $V_\pi$ at DC, maximum (V) | 5.8 | 7.2 | 5.8 | 7.2 |
| | | Impedance, minimum (Ω) | 1000 | 1000 | 1000 | 1000 |

**10 Gb/s**

| | | Device Model # | 10-130- -AP | 10-150- -AP | 10-150- -TI |
|---|---|---|---|---|---|
| | | Fabrication Process | APE | APE | TI |
| OPTICAL | | Operating wavelength, minimum (nm) | 1300 | 1535 | 1535 |
| | | Operating wavelength, maximum (nm) | 1330 | 1565 | 1565 |
| | | Insertion loss, typical (dB) | 4.0 | 4.0 | 4.0 |
| | | Insertion loss, maximum (dB) | 5.0 | 5.0 | 5.0 |
| | | On/off extinction ratio, minimum (dB) | 20 | 20 | 20 |
| | | Optical return loss, typical (dB) | >60 | >60 | >60 |
| | | Optical return loss, minimum (dB) | 50 | 50 | 50 |
| | | Input power, maximum (mW) | 200 | 200 | 50 |
| ELECTRICAL | Rf PORT | Bandwidth, typical (GHz) | **10** | **10** | **10** |
| | | Chirp, typical (α) ❷ | 0 | 0 | 0.6 |
| | | Bandwidth, minimum (GHz) | 8 | 8 | 8 |
| | | $V_\pi$ at DC, typical (V) | 4.5 | 5.0 | 5.5 |
| | | $V_\pi$ at DC, maximum (V) | 5.0 | 5.5 | 6.0 |
| | | Rise time, typical (ps) | 35 | 35 | 35 |
| | | Fall time, typical (ps) | 35 | 35 | 35 |
| | | Return loss, typical (dB) | -10 | -10 | -10 |
| | BIAS PORT | $V_\pi$ at DC, typical (V) | 7.0 | 7.0 | 8.0 |
| | | $V_\pi$ at DC, maximum (V) | 8.0 | 8.0 | 10.0 |
| | | Impedance, minimum (Ω) | 1000 | 1000 | 1000 |

❶ Wider operating temperature ranges are possible with a reduced extinction ratio.
❷ Other chirp values are available on a custom basis

**Figure 10.30** (continued)

A block diagram of a transmitter module with a built-in electroabsorption modulator is shown in Figure 10.31. Observe how much simpler the transmitter module with the built-in EA modulator is when compared with the direct-modulated transmitter shown in Figure 10.18.

The data sheet of a 10-Gbit/s transmitter module with an EA modulator is given in Figure 10.32

You should by this time be familiar with all the specifications in Figure 10.32. Pay particular attention to the following data:

1. The optical output power after the EA modulator is at 0 dBm (1 mW). In general, the output power from such a transmitter is less than it is from its direct-modulated counterpart. However, the power is not much less and can, in some situations, be even more. (See the data sheet in Figure 10.26, where the average output power is –3 dB.)

2. Modulator drive voltage, which is only 2 V, much less than the drive voltage for an MDM modulator, where it can be as high as 10 V, as Figure 10.30 shows.

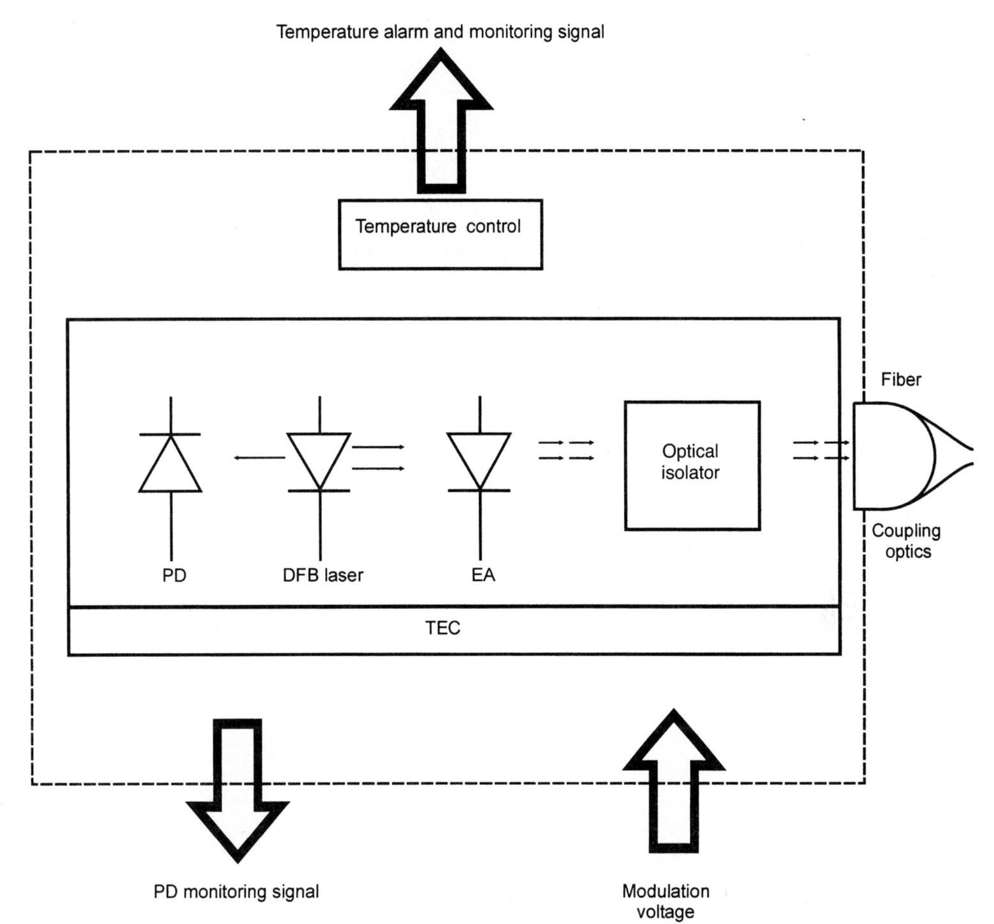


**Figure 10.31** Block diagram of a transmitter module with electroabsorption modulator.

**3.** *Dynamic extinction ratio (DER),* which is the ratio of maximum, $P_{max}$, to minimum, $P_{min}$, optical power for bit 1 and bit 0, respectively. Here, the minimum DER is 10 dB, which means that $P_{max}$ is at least 10 times greater than $P_{min}$.

It is worthwhile to compare all the other specifications in Figure 10.32 with the specifications of another transmitter, such as the one in Figure 10.26, and with the specifications of a DFB laser, such as that shown in Figure 9.19.

The electroabsorption modulator is, without a doubt, the most promising modulator for WDM applications. This device is the talk of the fiber-optic industry today—both in the R&D labs and manufacturing plants.

# Alcatel 1915 LMM

## 10 Gbit/s digital Laser Module with integrated electro-absorption Modulator

New Product

### Description

This Alcatel 1915 LMM contains an Alcatel DFB laser with monolithically integrated electro-absorption modulator (ILM).

This chip provides much lower dispersion penalties than a directly modulated DFB, without the complexity of LiNbO3 external modulators. The Alcatel 1915 LMM is optimized for 10 Gbit/s TDM transmission systems.

### Features

- 7-pin package with either GPO or K connector RF input
- Very low dispersion penalty over 80 Km for 10 Gbit/s operation
- InGaAsP monolithically integrated DFB laser and modulator chip
- High frequency RF connector package with 50 Ω RF impedance
- Low drive voltage (≤ 2 Vpp)
- Internal optical isolator
- High power available ($P_{AVE}$ ≥ 2 dBm)

### Applications

- STM-64 and OC-192 intermediate and long reach transmission systems
- Terminals for submarine transmission systems
- Short reach 10 Gbit/s applications

## Optical characteristics

| Parameter | Condition | Symb | Min | Typical | Max | Unit |
|---|---|---|---|---|---|---|
| Threshold current | CW, $V_{bias}$ = 0 V | Ith | 5 | 17 | 35 | mA |
| Operating current | CW, $V_{bias}$ = 0 V 0.5 | Iop | 50 | 80 | 100 | mA |
| Optical output power | $I_{OP}$, $V_{mod}$, [1], [4] | $P_{AVE}$ | - 3 | | | dBm |
| | $I_{OP}$, $V_{mod}$, [1], [2] | $P_{AVE}$ | 0 | | | dBm |
| | $I_{OP}$, $V_{mod}$, [1], [3] | $P_{AVE}$ | - 3 | | | dBm |
| Laser forward voltage | CW, $I_{OP}$, $V_{bias}$ = 0 V | Vf | | 1.3 | 2 | V |
| Modulator bias voltage | See [1] | $V_{bias}$ | - 2 | - 1.25 | 0 | V |
| Modulator drive voltage | See [1] | $V_{mod}$ | | 1.5 | 2 | V |
| Dynamic extinction ratio | $I_{OP}$, [1], [4] | DER | 9 | | | dB |
| | $I_{OP}$, [1], [2], [3] | DER | 10 | | | dB |
| Emission wavelength | | lm | 1530 | | 1570 | nm |
| Side mode suppression | @ $I_{OP}$ | SMSR | 35 | | | dB |
| Cut off frequency | - 3 dB, $V_{bias}$ = - 1 V | S21 | 10 | | | GHz |
| RF return loss | DC to 7 GHz | S11 | 10 | | | dB |
| Dispersion penalty | See [1], [2], [3] | Ds | | | 2 | dB |
| Tracking error | Tsubmount = 25 °C, Tcase = 70 °C<br>If = 100 mA, Q = 10 log [P(70 °C)/P(25 °C)] | TR | - 0.5 | | | dB |
| Rise time/ Fall time | See note [1], 10 %, 90 % | Tr/Tf | | 30 | 45 | ps |
| Optical return loss | Tc = - 5 to 70 °C | Ol | 25 | | | dB |
| Monitor diode current | $I_{OP}$, VM = - 5 V | Im | 0.2 | 0.8 | 1.5 | mA |
| Dark current | | Id | | | 0.1 | mA |
| TEC current | DT = 45 °C, $I_{OP}$ = 120 mA, $T_C$ = 70 °C, $V_{bias}$ = - 1 V | It | | 1 | 1.3 | A |
| TEC voltage | DT = 45 °C, $I_{OP}$ = 120 mA, $T_C$ = 70 °C, $V_{bias}$ = - 1 V | Vt | | 2 | 2.5 | V |
| Thermistor resistance | | $R_{TH}$ | 9.5 | | 10.5 | kΩ |

Notes : All limits start of life, TCase = 25 °C, TSubmount = 25 °C, monitor bias = - 5 V, unless otherwise stated.

[1] BER = $10^{-10}$; 2.488 Gbit/s modulation ; $2^{23}$ - 1 PRBS ; NRZ line code

[2] 800 ps/nm dispersion, assuming fiber with an average dispersion of 18 ps/nm/km

[3] 1600 ps/nm dispersion, assuming fiber with an average dispersion of 18 ps/nm/km

[4] 50 ps/nm dispersion, assuming fiber with an average dispersion of 18 ps/nm/km

Optical power in the fiber shall not exceed the linear transmission regime.

**Figure 10.32** Data sheet of 10-Gbit/s transmitter module with electroabsorption modulator. *(Courtesy of Alcatel Optronics, Noray, Cedex, France.)*

## Absolute maximum ratings

| Parameter | Min | Max | Unit |
|---|---|---|---|
| Operating case temperature | -5 | 70 | °C |
| Storage temperature | - 40 | 85 | °C |
| Laser forward current | | 150 | mA |
| Laser reverse voltage | | 2 | V |
| Modulator forward voltage | | 1 | V |
| Modulator reverse voltage | | 5 | V |
| Photodiode forward current | | 1 | mA |
| Photodiode reverse voltage | | 20 | V |
| TEC voltage | | 2.8 | V |
| TEC current | | 1.4 | A |
| ESD applied on modulator | | 500 | V |
| ESD applied on laser [1] | | 2000 | V |
| Lead soldering time (at 260 °C) | | 10 | s |
| Package mounting screw torque | | 0.2 | Nm |

[1] Human body model
Stresses in excess of the absolute maximum ratings can cause permanent damage to the device.
These are absolute stress ratings only.

September 99


Customized versions are available for large quantities.

Performance figures contained in this document must be specifically confirmed in writing by Alcatel Optronics before they become applicable to any particular order or contract. Alcatel Optronics reserves the right to make changes to the products or information contained herein without notice.

## Mechanical details

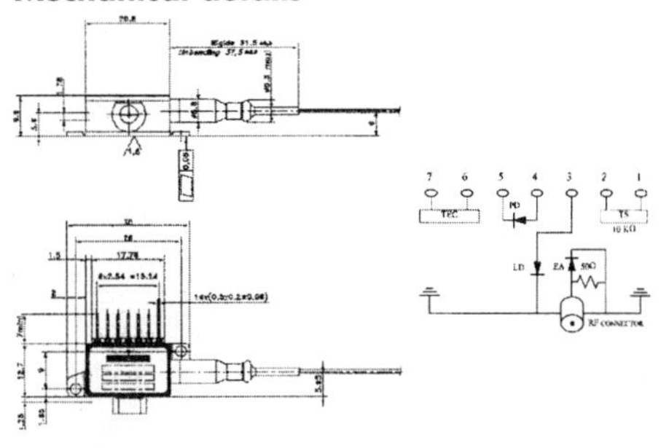

## Pin out

| N° | Description |
|---|---|
| 1 | Thermistor |
| 2 | Thermistor |
| 3 | Laser DC bias (+) |
| 4 | Photodetector Anode (-) |
| 5 | Photodetector Cathode (+) |
| 6 | TEC (+) |
| 7 | TEC (-) |

## Ordering information

Alcatel 1915 LMM

| Dispersion | Part number | RF connector Input Type | Pigtail connector |
|---|---|---|---|
| < 50 ps/nm | 3CN 00317 AA | K type | FC/PC |
| < 50 ps/nm | 3CN 00320 AA | GPO type | FC/PC |
| 800 ps/nm | 3CN 00315 AA | K type | FC/PC |
| 800 ps/nm | 3CN 00321 AA | GPO type | FC/PC |
| 1600 ps/nm | 3CN 00316 AA | K type | FC/PC |
| 1600 ps/nm | 3CN 00322 AA | GPO type | FC/PC |

## Standards

ITU-T G.691
IEC 68-2 and MIL STD 883 environment

**EUROPE**
Route de Villejust
F-91625 NOZAY CEDEX
Tel : (+33) 1 64 49 49 10
Fax : (+33) 1 64 49 49 61

**USA**
12030 Sunrise Valley Drive
RESTON - VA 22091
Tel : (+1) 703 715 3921
Fax : (+1) 703 860 1183

**JAPAN**
Yebisu Garden Place Tower
PO Box 5024
20-3, Ebisu 4 - Chome
Shibuya - ku TOKYO 150 - 6028
Tel : (+81) 3 5424 85 65
Fax : (+81) 3 5424 85 81

**Figure 10.32** (continued)

## SUMMARY


- An in-depth look at semiconductor materials allows us to describe these materials quantitatively. These quantitative characteristics are obtained from closer consideration of the physical mechanisms of pumping and radiation in today's semiconductors. This consideration also points the way to how we can engineer the properties of the semiconductors used to fabricate laser diodes.

- A laser diode is a very efficient converter of an electrical information signal into an optical one. We have quantified a laser diode's efficiency by the ratio of decrements of output light power to input forward current and we've named that *slope efficiency* ($S = \Delta P/\Delta I_F$). An in-depth look at this characteristic reveals that slope efficiency is eventually determined by the ratio of the number of radiated photons to the number of injected electrons. The physical mechanisms determining this ratio boil down, roughly, to the concentration of electron-hole recombinations in a very small active area and to directing the radiated photons along the desired path. This process is achieved physically by building a laser diode from thin layers of semiconductor materials with slightly different electro-optical properties.

- There are several characteristics of a laser diode that one has to keep in mind when evaluating the quality of the device: threshold and operating (forward or driving) currents, radiating wavelength and spectral width, radiation pattern, modulation bandwidth, and noise. We've discussed not only the meaning of these characteristics but also the physical mechanisms determining them.

- A transmitter consists of a light source, coupling optics, and electronics, as stated at the beginning of this chapter. Most of the chapter is devoted to light sources because the light source is really the heart of a transmitter. Ultimately, however, it's not just the laser diode but the entire unit—the transmitter—that conveys information along the fiber. The transmitter's electronics provide the following: (1) the conversion of an input signal into an acceptable format (such as converting the Manchester line code into a non-return-to-zero code and converting parallel transmissions into serial transmissions), (2) operation of an LED or LD with forward (operating or driving) current, (3) control of modulation and biasing current, and (4) temperature control. In addition, transmitter electronics give the user the ability to control and monitor the transmitter's operation. This latter feature is extremely important for troubleshooting a fiber-optic communications link. Coupling optics is the best technique to use to launch light from a light source into an optical fiber. An important factor when considering a transmitter is its packaging. What's preferred here is a package that is an assembly of all the parts together and that protects the assembled product from environmental hazards.

- The demand for high-quality fiber-optic communications systems—in short, the need for more bandwidth—puts a lot of pressure on each component of these systems. Fortunately, recent developments in transmitter technology have provided adequate response to this demand. However, the crossing of new frontiers in fiber-optic transmission—such as terabits per second—is driving physicists and engineers to work even harder to meet the fast-changing requirements.


## PROBLEMS

**10.1.** What are the Fermi levels in a semiconductor?

**10.2.** What does the Fermi-Dirac formula describe?

**10.3.** Prove that Formula 10.2 reduces to Formula 10.1 for $T = 0$ K.

**10.4.** What does an effective mass of electrons describe and what is its relationship to the true mass?

**10.5.** What parameters does the density of the charge carriers depend on?

**10.6.** Calculate the probability of exciting electrons at the conduction band at room temperature in an InP semiconductor.

**10.7.** Calculate the ratio of majority to minority charge carriers in $n$- and $p$-type germanium (Ge).

**10.8.** What is the position of the Fermi level with respect to energy bands for unbiased, forward- and reverse-biased semiconductors? Explain your answer.

**10.9.** What two quantities determine the electro-optical characteristics of a semiconductor material?

**10.10.** What is the meaning of an $E$-$k$ diagram? Draw a typical $E$-$k$ diagram.

**10.11.** Which semiconductor—direct or indirect bandgap—is used in the fabrication of light-emitting diodes and laser diodes? Why?

**10.12.** Give the definition and explain the meaning of both *external* and *internal* quantum efficiency. How do they relate to each other?

**10.13.** Which efficiency—external quantum, internal quantum, cavity, or power—determines the input–output relationship of a laser diode?

**10.14.** Calculate the external quantum efficiency of a laser diode if its slope efficiency is equal to 0.08 mW/mA and the operating wavelength is 1550 nm.

**10.15.** Calculate the power efficiency of the DFB laser diode whose data sheet is given in Figure 9.19.

**10.16.** Calculate the power efficiency of the SLED whose data sheet is given in Figure 9.8(a).

**10.17.** Calculate the external quantum efficiency of the broad-area laser diode in Figure 9.17, the DFB laser in Figure 9.19(a), and the VCSEL diode in Figure 9.20 and compare them. Evaluate, where appropriate, the error introduced by neglecting $P_{th}$ in Formula 10.26. (In other words, validate the usage of Formula 10.27.)

**10.18.** Derive the formula for power efficiency using Formulas 10.25 and 10.27.

**10.19.** Two interpretations have been given for internal quantum efficiency: (1) It is the ratio of radiated photons to injected electrons and (2) it is the ratio of recombination lifetime to radiative lifetime. How do those two interpretations relate to each other?

**10.20.** Prove that Formulas 10.41 and 10.41a are equivalent.

**10.21.** Based on Figure 10.9, explain how the double heterostructure of a laser diode concentrates electrons and holes in the active area and guides radiated photons along the desired path.

**10.22.** When an external photon whose energy is equal to or more than the energy gap enters the active medium, it can be absorbed or it can stimulate the radiation of another photon. Which event is more likely to happen and why?

**10.23.** An InGaAsP laser diode featuring MQW and RWG properties shows the following dependence of threshold current on temperature: $I_{th}$ = 10 mA at −10°C, $I_{th}$ = 19 mA at 25°C, and $I_{th}$ = 27 mA at 50°C at 1300-nm operating wavelength. Find $I_0$ and $T_0$. Also, explain what MQW and RWG stand for.

**10.24.** Figure 10.13 shows that the VCSEL diode radiates not Gaussian-distributed light but, rather, donut-like light. Why has the designer made this radiation pattern?

**10.25.** Derive Equation 10.64 from the rate equations.

**10.26.** Prove that Formula 10.68 is the solution to Equation 10.64.

**10.27.** Find the bias current to eliminate delay time for a DFB laser (Figure 9.20) if the laser is digitally modulated.

**10.28.** If a laser diode is intensity modulated, what parameter limits its modulation bandwidth? Explain.

**10.29.** Derive Formula 10.73 from Formula 10.72.

**10.30.** Analyze (1) the dimensions and (2) the physical meaning of Formula 10.78.

**10.31.** Derive Formula 10.80.

**10.32.** Explain the meaning of the term *chirp* as applied to a laser diode. How does this phenomenon affect the LD's transmission characteristics?

**10.33.** What does RIN stand for? Is it a good or bad phenomenon?

**10.34.** Calculate the laser noise power detected by a receiver for the following data: RIN = −150 dB/Hz, $P_{received}$ = 80 μW, and BW = 140 MHz.

**10.35.** Draw a detailed functional block diagram of a laser transmitter and write a brief explanation of the function of each of the transmitter's components.

**10.36.** Can we calculate the coupling efficiency of a laser transmitter using Formulas 10.85 and 10.86? If yes, how?

**10.37.** Why is modal noise not a problem for an LED transmitter or for a singlemode fiber? Can a laser diode be used with a multimode fiber?

**10.38.** Why is backreflection a problem for a laser transmitter but not for an LED transmitter?

**10.39.** What major problems does an engineer have to consider in designing the board layout of a laser transmitter?

**10.40.** Calculate the propagation delay for a microstrip laid on a dielectric whose relative dielectric constant equals 4.7.

**10.41.** What measures are usually taken in transmitter design to facilitate the troubleshooting of a fiber-optic communications link?

**10.42.** The data sheet of the laser transmitter depicted in Figure 10.26 starts with the section "FEATURES." Explain the meaning of each feature and indicate how these features are implemented in this transmitter. For example, consider the feature "Low-Threshold FP Laser." You might write: "A laser with a Fabry-Perot resonator is used for this transmitter. This laser has low threshold current. The specific values of the threshold current of this LD are . . . while the threshold current for a VCSEL LD is. . . ." You will find these values in the laser-diode data sheets.

**10.43.** What are the advantages and drawbacks of internal and external modulation?

**10.44.** Why is switching to a high-power light signal a problem for an internally modulated laser diode?

**10.45.** Analyze Figure 10.28 and write a brief explanation of the function of each component.

**10.46.** *Project 1:* Analyze the data sheet of a 2.5-Gbit/s DFB laser transmitter module from Broadband Communications Products (Figure 10.33).

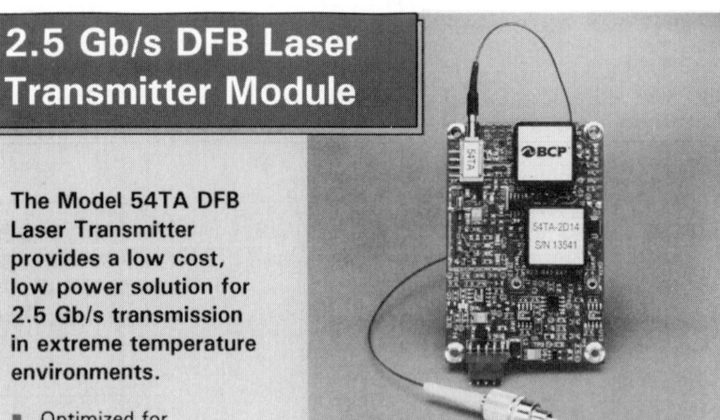


# 2.5 Gb/s DFB Laser Transmitter Module

**The Model 54TA DFB Laser Transmitter provides a low cost, low power solution for 2.5 Gb/s transmission in extreme temperature environments.**

- Optimized for 2.48832 Gbps (SONET OC-48 and SDH STM-16)
- Compliant with BellCore GR-253 and ITU G.957/958 over the full -40 to +85°C temperature range.
- Low Cost. Isolated, no thermoelectric cooler.
- Low Power. No cooler power required.
- Optional retimed input produces excellent optical eye pattern with imperfect input data waveform.
- Plug-and-play module. No PC card required. Fully connectorized.

BCP
305 East Drive
Melbourne, Florida 32904 • USA
TEL: +1-407-984-3671
FAX: +1-407-728-0487
Email: fiberlink@bcpinc.com
http://www.bcpinc.com

## General Description

The BCP Model 54TA module is a low-cost laser transmitter, designed to accelerate the use of OC-48/STM-16 2.5 Gbps technology into more hostile and cost sensitive environments, such as the outside plant local loop. When combined with the companion 54R receiver module, a complete, economical link solution is now available for the transport of serial digital video and ATM applications at 2.5 Gb/s, over fiber spans of up to 28 dB optical loss, minimum.

The 54TA design eliminates the need for high-cost thermoelectrically cooled lasers. The 54TA is designed to perform within specification over the full -40°C to +85°C temperature range, meeting all BellCore GR-253 and ITU G.957/958 eye mask and jitter generation requirements for LR-1/L-16.1 applications. An optical plant return loss of 25 dB or greater, per ITU specifications, should be maintained by the user for optimum performance.

The 54TA is a complete plug-and-play module solution. Fully connectorized and requiring only ±5VDC supplies, the module does not require a customized motherboard for operation. It may be installed in practically any convenient and physically protected location, compatible with the specified temperature range. Thermoelectric heating and cooling power is not required for the 54TA. This feature reduces system-level cost, by eliminating from 2 to 5 watts of external DC power normally required for transmitter cooler drive.

The 54TA transmitter is optimized for 2.5 Gb/s operation, but it is not bit rate specific and may also be used at lower data rates. For jitter-critical applications, an optional clock input is available, which enables retiming and resymmetry of the input data at the point of entry into the 54TA module.

Short-reach and intermediate-reach versions are also available. Contact BCP for details.

**Figure 10.33** Data sheet of a 2.5-Gbit/s DFB laser transmitter module. *(Courtesy of Uniphase Telecommunications Products, Inc.)*

**10.47.** Explain the principle of operation of an electroabsorption modulator.

**10.48.** Is an electroabsorption modulator an external or internal device? Explain.

**10.49.** *Project 2:* Analyze the data sheet of Alcatel 1915 LMM in Figure 10.32.

**10.50.** *Project 3:* Build a memory map. (See Problem 1.20.)

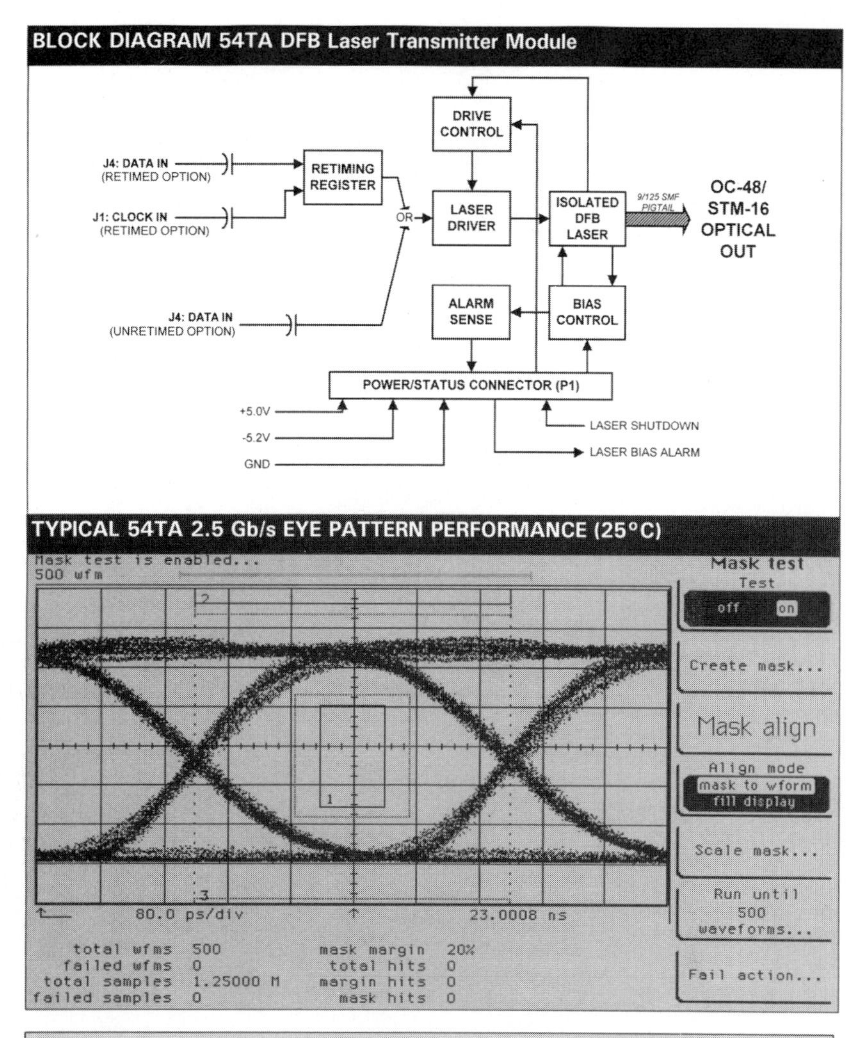


Note: Eye pattern shown is with an optical reference receiver using OC-48/STM-16 standard reference filter. BellCore/ITU standard eye mask limits are superimposed on the display as solid lines labeled 1, 2 and 3. Measured performance margins are indicated by the larger, dotted areas.

**Figure 10.33** (continued)

## Model 54TA DFB Laser Transmitter Module

### Specifications

| Absolute Maximum Ratings: Exceeding these limits may cause permanent damage. | | |
|---|---|---|
| Operating Case Temperature | -40°C to +85°C (Note 2). | |
| Storage Temperature | -40°C to +85°C | |
| Positive Supply Voltage | 0.00 to +6.00 VDC | |
| Negative Supply Voltage | -6.00 to 0.00 VDC | |
| Input Data/Clock Levels | 2.00 Vpp, maximum | |
| **Inputs: Electrical at 2.5 Gbps** | | |
| Data/Clock Amplitude | 0.70 to 1.50 Vpp | Both inputs AC-coupled |
| Data Rise/Fall Times | 150 ps (maximum) | 20-80% |
| Clock Rise/Fall Times | 100 ps (maximum) | 20-80% (optional retiming clock) |
| Impedance | 50 ohms | AC-coupled |
| Data Pattern | PRBS, SONET, SDH | 50% ones density |
| Positive Supply ($V_{CC}$) | +5.0 VDC, ±5%, 100 mA maximum | |
| Negative Supply ($V_{EE}$) | −5.2 VDC, ±5%, 550 mA maximum (with retiming option) | |
| Laser Shutdown Voltage | +3.5V minimum (laser off)<br>+0.2V max (laser on) | CMOS compatible |
| **Output: Optical (Full temperature range, 2.5 Gbps):** | | |
| Center wavelength | 1280 to 1335 nm | -40°C to +85°C |
| Spectral Width<br>Full width (3 dB)<br>Full width (20 dB) | <br>0.3 nm (maximum)<br>1.0 nm (maximum) | <br>Note 1<br>Note 1 |
| Sidemode Suppression Ratio | 30 dB (minimum) | Note 1 |
| Wavelength Temp. Coefficient | +0.1 nm/°C (maximum) | Laser not cooled |
| Average Power Output | −1.0 dBm to +2.0 dBm | |
| Extinction Ratio | 10 dB (minimum) | |
| Optical Eye Quality | Meets GR-253 and ITU G.957 | Note 1 |
| Jitter Generation<br>Peak-to-peak<br>RMS | <br>0.10 UI (maximum), typ ≤0.06 UI<br>0.01 UI (maximum), typ ≤0.007 UI | <br>Note 1<br>Note 1 |
| **General:** | | |
| Data/Clock Connectors | Huber-Suhner MMCX | Snap-in, surface mount type |
| Power/Status Connector | 5-pin header | |
| Optical Connector | FC, SC or ST, 40 dB R.L. (minimum) | Other connectors available |
| Laser Bias Alarm | +3.5V (minimum) - alarm state<br>+0.2V (maximum) - normal state | 10 K minimum load |
| Optical Pigtail | 0.85 meter minimum length, 9/125/900 μm singlemode fiber | |

**Notes:**

1. Performance guaranteed when connected to a 25 dB minimum optical return loss cable plant.
2. Case temperature is defined as the temperature measured at the laser case.
3. **Laser Safety:** This device is an eye-safe Class 1 laser product, and conforms to USA DHHS Regulation 1040.1 and to IEC 825-1 standards.

**Model 54TA Pin Functions**

| Pin No. | Function |
|---|---|
| 1 | Ground |
| 2 | +5.0 VDC |
| 3 | -5.2 VDC |
| 4 | Laser Bias Alarm |
| 5 | Laser Shutdown |

**Figure 10.33** (continued)

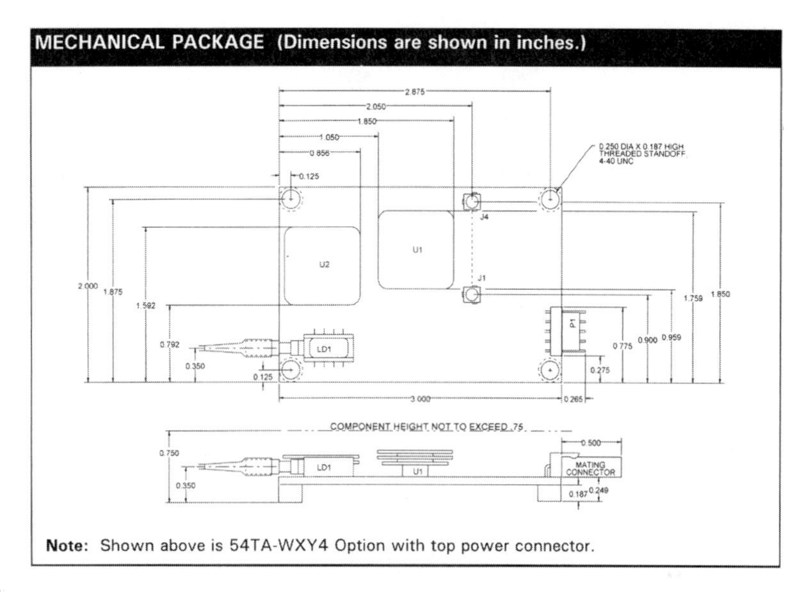


**Figure 10.33** (continued)

## REFERENCES[4]


1. Paul Fishbane, Stephen Gasiorowicz, and Stephen Thornton, *Physics for Scientists and Engineers,* Englewood Cliffs, N.J.: Prentice Hall, 1993.
2. Shun Lien Chuang, *Physics of Optoelectronics Devices,* New York: John Wiley & Sons, 1995.
3. Leonid Kazovsky, Sergio Benedetto, and Alan Willner, *Optical Fiber Communication Systems,* Boston: Artech House, 1996.
4. Christopher Davis, *Lasers and Electro-Optics—Fundamentals and Engineering,* New York: Cambridge University Press, 1996.
5. Chin-Lin Chen, *Elements of Optoelectronics and Fiber Optics,* Chicago: Irwin, 1996.
6. Joseph Verdeyen, *Laser Electronics,* 3d ed., Englewood Cliffs, N.J.: Prentice Hall, 1995.
7. Govind Agrawal, *Fiber-Optic Communication Systems,* 2d ed., New York: John Wiley & Sons, 1997.
8. Jeff W. Scott, "Vertical-cavity lasers facilitate gigabit communications," *Laser Focus World,* October 1998, pp. 75–78.
9. Jeff W. Scott, Ciello Communications Inc., Broomfield, Colo., private communication, October 22, 1998.
10. Paul Shumate, "Lightwave Transmitters," in *Optical Fiber Telecommunications*-II, ed. by S.E Miller and I.P. Kaminow, San Diego: Academic Press, 1988, pp. 723–758.
11. Tien Pei Lee, "Recent advances in long-wavelength semiconductor lasers for optical fiber communication," *IRE Proceedings,* March 1991, pp. 253–276.
12. Rajappa Papannareddy, *Introduction to Lightwave Communication Systems,* Boston: Artech House, 1997.
13. Tien Pei Lee, C.A. Burrus, Jr, and R.H. Saul, "Light-Emitting Diodes for Telecommunication," in *Optical Fiber Telecommunications-11.*, ed. by S.E. Miller and I.P. Kaminow, San Diego: Academic Press, 1988, pp. 467–507.
14. John A. Buck, *Fundamentals of Optical Fibers,* New York: John Wiley & Sons, 1995.
15. "Fiber Optic Transceivers," *Designer's Guide,* AMP Inc., Harrisburg, Pa., 1996.
16. *Fiber Optics Products* (product catalog), Optics for Research Inc., Caldwell, N.J., 1998.
17. *Photonics* (product catalog), Newport Corporation, Irvine, Calif., 1997/1998.
18. *Product Catalog,* vol. 8, New Focus Inc., Santa Clara, Calif., 1997/1998.
19. *Laser Diode Control: Fiber Optic Test & Measurement* (short-form catalog), ILX Lightwave Corporation, Bozeman, Mont., 1998.
20. *FOCUSS* (fiber-optic communication system simulator), Zentrum für Expertensysteme Dortmund e.V., Dortmund, Germany, 1998.
21. *Designer's Guide to External Modulation,* Uniphase Telecommunications Products, ElectroOptics Products Div., Bloomfield, Conn., 1997.
22. *Technical Data,* Integrated Optical Components Ltd., Eastways, Witham Essex, U.K., 1998.


[4]See Appendix C: A Selected Bibliography

# 11

# Receivers

If you look again at the basic block diagram of a fiber-optic link shown in Figure 1.4, you'll observe that an information signal arrives and leaves the fiber-optic link in electrical form. This is why we need a transmitter and a receiver. At the sending end, a transmitter converts an electrical signal into an optical one. At its own end, a receiver converts the optical signal back into electrical form, thus closing the optical path for information traveling along the fiber-optic communications link. The transmitter and optical fiber have already been discussed; now it's time to turn to the receiver segment of the link.

A receiver consists of a photodetector and electronics. Only miniature semiconductor photodiodes are used in fiber-optic communications technology to detect an optical signal. A photodiode is the heart of a receiver very much like an LED or an LD is the heart of a transmitter. This is why we concentrate our discussion on photodiodes before considering the entire receiver unit.

## 11.1 PHOTODIODES

If LEDs and laser diodes convert an electrical signal into light, the function of a photodiode is just the opposite: to convert light into an electrical signal. Thus, the principle of operation of a photodiode (PD) can be explained simply as operating in a manner exactly the opposite to the way an LED works. Indeed, the discussion of a PD's operation involves the same elements: energy bands and a *p-n* junction. This is why *rereading* the appropriate sections in Chapter 9 would be a good idea if you need to refresh your memory on how an LED works.

### *p-n* Photodiodes: How They Work . . .

***From the standpoint of energy bands*** You will recall that in semiconductors we deal with conduction and valence energy bands—two bundles of energy levels—separated by a forbidden region, an energy gap ($E_g$). The conduction band has higher energy than the valence band.

Electrons at the valence band are bonded and cannot move; thus, no current flows through the material. Electrons at the conduction band are free and when a small voltage is applied, they move, constituting current. In other words, *to induce material to conduct current, one needs to populate the conduction band with electrons.* But one obstacle stands in the way: the energy gap. The value of $E_g$ determines the conductive (resistive) properties of the material. Good conductors have no gap between the valence and conduction bands, good insulators have a big energy gap, and semiconductors have a gap somewhere in between. Indeed, diamond—a good insulator—has an energy gap around 6 eV, while silicon (Si) and germanium (Ge)—the most popular semiconductors—have gaps of 1.17 eV and 0.775 eV, respectively. (See Table 9.1.)

When a photon with energy $E_p = hf = hc/\lambda \geq E_g$ strikes the material, the photon is absorbed and its energy acquired by an electron. Thus, the electron is excited at the conduction band and is now able to move. This is how light power—a number of photons times a photon's energy per unit of time—is converted into electrical current. This explanation is visualized in Figure 11.1(a).

If we apply external voltage—bias—to this semiconductor, we make electrons flow in a much more pronounced manner, thus increasing the efficiency of the light-to-current conversion.

***From the standpoint of a* p-n *junction*** When a photon strikes a depletion region, its energy separates an electron from its hole, as Figure 11.1(b) shows. (Remember, electrons and holes have recombined at the *p-n* junction, thus creating a depletion region.) The separated electron and hole are attracted by the positive and negative potentials of the depletion voltage, respectively.

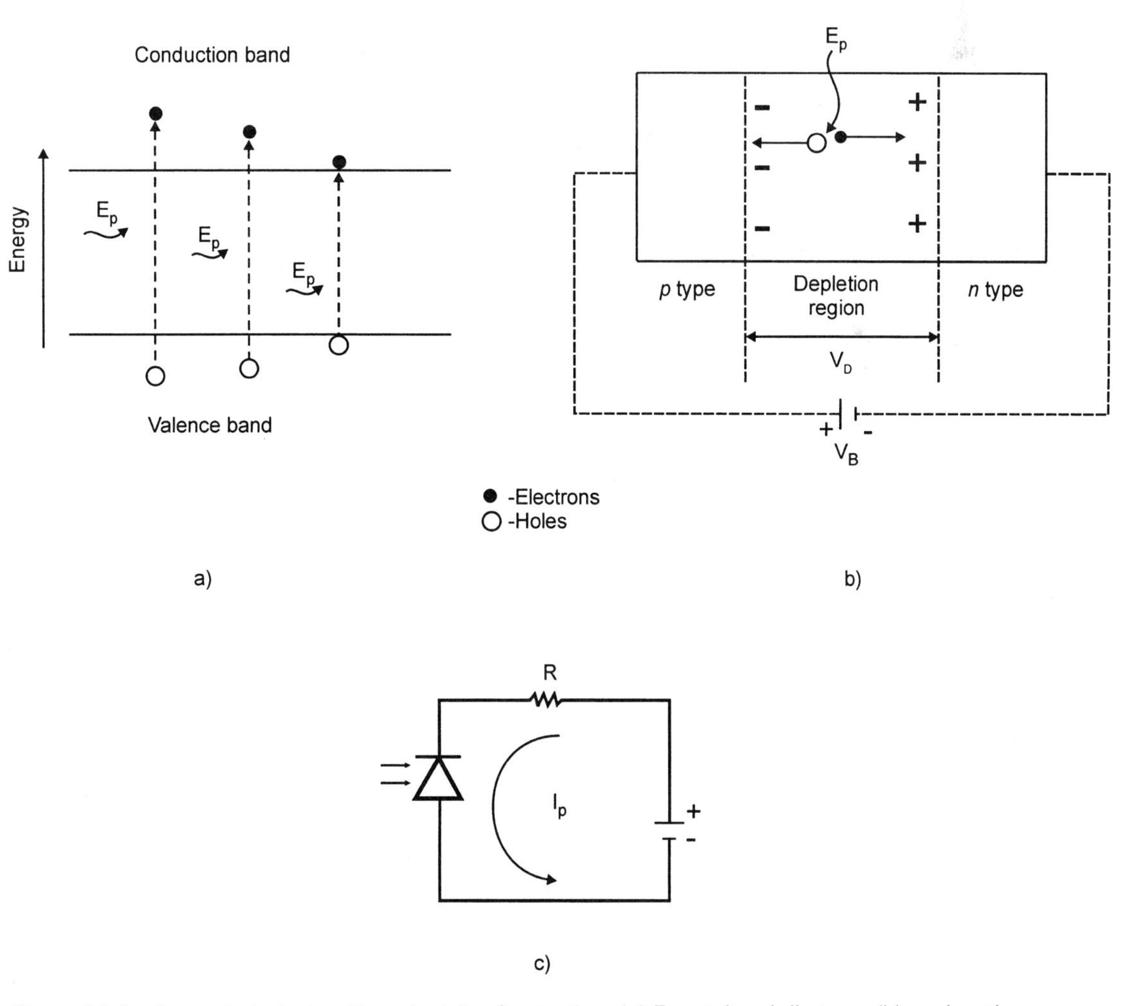


**Figure 11.1** A *p-n* photodiode—the principle of operation: (a) Energy-band diagram; (b) *p-n* junction; (c) electrical circuit.

Thus, a flow of charge carriers—current—is generated. Applying external voltage (reverse bias) enhances the flow of electrons and holes. Observe closely how the external battery is connected in Figure 11.1(b) to create *reverse bias.*

Let's summarize what we have discussed thus far: *External photons—that is, light—strikes the semiconductor and separates the electrons and holes. The flow of these free charge carriers produces current. External voltage (reverse bias) enhances this effect.* The electrical circuit of a photodiode is shown in Figure 11.1(c)

It is important to compare Figure 11.1 with Figures 9.2 and 9.3 to see the similarities and differences between the operations of a light source and a photodiode.

***Input–output characteristic*** The input for a photodiode is light power ($P$); the output is current, which is usually called *photocurrent* ($I_p$) because it is caused by light. It follows from the principle of operation that the more photons that strike the active area of a PD, the more charge carriers will be created; that is, the greater will be the photocurrent. Thus, $I_p$ is proportional to $P$:

$$I_p = R\,\mathrm{P}, \tag{11.1}$$

where $R$ is constant. This relationship is shown in Figure 11.2(a).

The slope of this graph is one of the major PD parameters and is called *responsivity, R (A/W).* It is defined by the following formula:

$$R(A/W) = I_p/P \tag{11.1a}$$

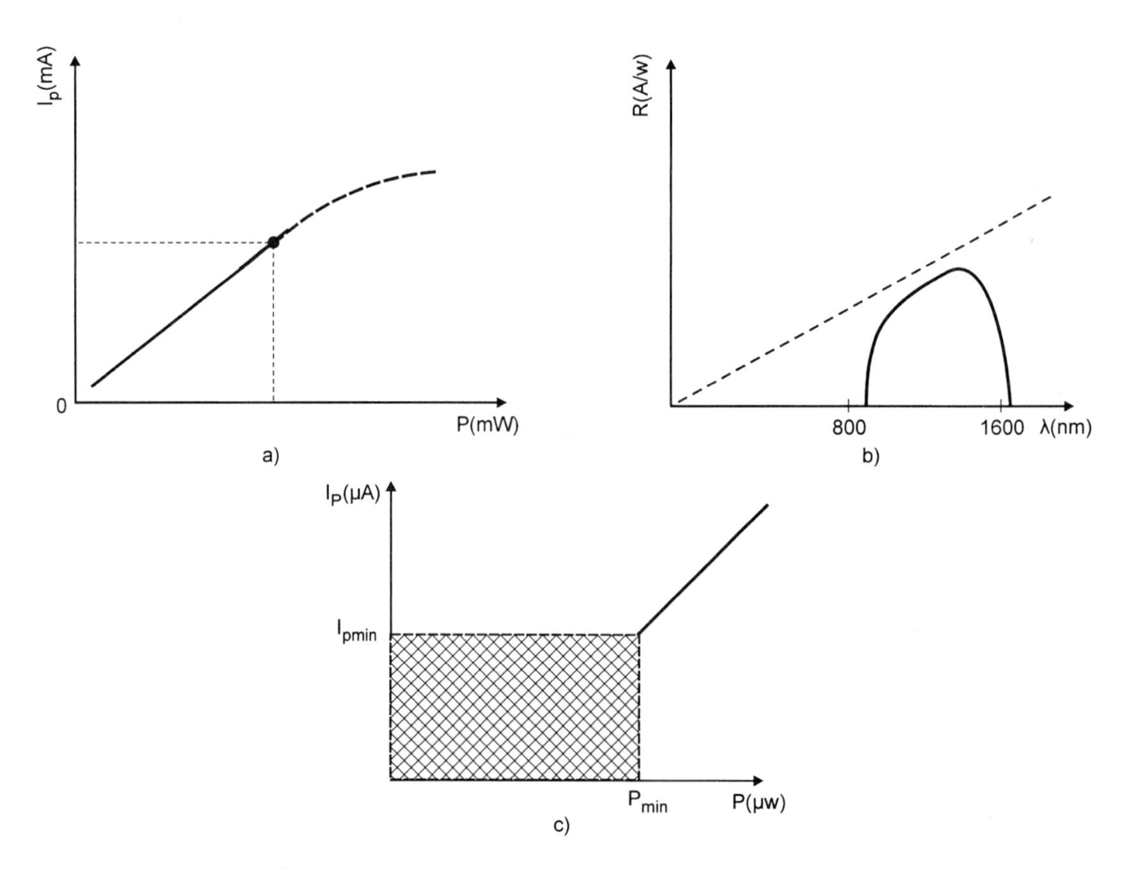


**Figure 11.2** Responsivity of a photodiode: (a) Input–output characteristic; (b) responsivity vs. wavelength; (c) dark-current sensitivity.

Typical values of $R$ range from 0.5 A/W to 1.0 A/W. This characteristic shows how efficiently a photodiode does its main job—converting light into an electrical signal. Since the value of $R$ is provided by photodiode manufacturers, you can calculate the output photocurrent with the given light-input power.

How far does this graph keep its linearity? It seems reasonable to expect that for a high level of light power—that is, when there are a tremendous number of photons per unit of time striking the PD—all available electron-hole pairs will be involved in producing photocurrent; therefore, one can expect to see a *saturation* effect. This is shown by the dotted line in Figure 11.2(a). Obviously, for a saturation region, Formula 11.1 cannot be applied.

### Example 11.1.1

***Problem:***

The responsivity of a photodiode is 0.85 A/W and the input-power saturation is 1.5 mW. What is the photocurrent if the incident light power is (1) 1 mW? (2) 2 mW?

***Solution:***

**1.** For input power 1 mW, we can apply Formula 11.1 and get

$$I_p = R\,P = 0.85 \text{ mA}$$

**2.** For input power 2 mW, Formula 11.1 is not valid, so we cannot find the photocurrent value.

## Power Relationship

We usually evaluate any communications device by looking for its response to the amplitude and the frequency of the input signal. The first parameter (the device's response to amplitude) describes the power input–output relationship, while the second (the response to the frequency of the input signal) tells us about the device's bandwidth. Thus, the photocurrent recounts all the power relationships in a photodiode.

***Responsivity versus wavelength*** Responsivity actually represents the power input–output characteristic of a photodiode, as Formula 11.1 states. You may wonder whether responsivity depends on the operating wavelength. The following simple derivation provides the answer. Responsivity, by definition, is equal to $I_p/P$. But photocurrent is the number of electrons, $N_e$, flowing per unit time, that is,

$$I_p = N_e/t \tag{11.2}$$

On the other hand, light power is light energy per unit of time, where light energy is equal to the energy of a photon ($E_p$) times the number of photons ($N_p$). Thus, we can write:

$$P = (N_p E_p)/t \tag{11.3}$$

Substitute $E_p = hc/\lambda$ and divide $I_p$ by $P$:

$$R = I_p/P = (N_e/N_p)(\lambda/hc) \tag{11.4}$$

The ratio of the number of produced electrons, $N_e$, to the number of falling photons, $N_p$, shows how efficiently the semiconductor material converts light into current. This ratio is called the *quantum efficiency of a photodiode,* $\eta$. (Sound familiar? It should, because we introduced and discussed this concept and a similar term in Chapters 9 and 10.) Thus,

$$\eta = N_e/N_p \tag{11.5}$$

The quantum efficiency of a regular communications photodiode ranges from 50% to almost 100%.

If we recall that the product $h \times c$ is the constant and is equal to 1248 (eV·nm), then Formula 11.4 takes the following simple form:

$$R(\text{A/W}) = (\eta/1248)\lambda(\text{nm}) \tag{11.6}$$

Thus, theoretically, the graph "Responsivity vs. Wavelength" should be a straight line, as the dotted line in Figure 11.2(b) shows. The slope of this line is equal to $\eta/1248$ (when $\lambda$ is expressed in nm).

The question that comes up at this point is, why is responsivity proportional to wavelength? If you closely examine the course of our derivation of Formula 11.6, you'll see that responsivity is inversely proportional to light power; the latter, in turn, is proportional to the number of photons and the energy of an individual photon, and this energy is inversely proportional to the wavelength. Thus, the longer the wavelength, the greater the number of photons needed to provide a certain amount of light power. Of course, increasing the number of photons also generates more electrons to produce more current. To sum all this up in a nutshell, then, *the longer the wavelength, the greater the amount of current produced from the same amount of light power.* If you now look at the definition of responsivity given in Formula 11.1, you will understand why $R \sim \lambda$.

### Example 11.1.2

***Problem:***

What is the responsivity of an InGaAs photodiode if its quantum efficiency is equal to 70%?

***Solution:***

We can find $R$ by using Formula 11.6 but we need to know the wavelength. Table 9.1 shows that the energy gap of InGaAs is equal to 0.75 eV, which corresponds to a wavelength of 1664 nm. Thus, from Formula 11.6 we find:

$$R = (\eta/1248)\lambda = 0.933 \text{ A/W}$$

This is a good number to obtain (each milliwatt of light power results in almost one milliampere of photocurrent) but, from a practical standpoint, it will turn out to be a little lower.

---

The real course of curve $R = f(\lambda)$ is shown in Figure 11.2(b) as a solid line. As you can see, in reality this graph is very far from the expected straight line. In fact, the graph shows *short* and *long cutoff wavelengths.* Let's discuss the reasons for this discrepancy.

As you can see from Figure 11.1(a) there is a cutoff wavelength, $\lambda_c$, determined by the energy gap, $E_g$, so that $E_g = hc/\lambda_c$. For wavelengths longer than $\lambda_c$, the energy of the photons is less than $E_g$; consequently, those photons will travel through this material without interaction. In other words, *for a given semiconductor material—that is, for a given energy gap—the photodiode can detect only wavelengths* $\lambda < \lambda_c = \text{hc}/E_g$. Looking at Table 9.1, you can appreciate what values of wavelength we are discussing. The most popular materials used in photodetectors are Si with

$\lambda_c \sim 1100$ nm and InGaAs with $\lambda_c \sim 1700$ nm. Si PDs are used in the first transparent window (around 850 nm), while InGaAs PDs are used in the second (around 1300 nm) and third (around 1550 nm) transparent windows and even higher. Thus, the cutoff wavelength, $\lambda_c$, determines the longest wavelength a PD can detect. This is why responsivity goes to zero at the longer wavelengths in Figure 11.2(b).

Why responsivity depends on wavelength can be further clarified by using the following approach [1]: Light falling on the active area of a photodiode is partially absorbed and partially transmitted. Assuming the common exponential dependence of absorption, we can write:

$$P_{abs} = P_{in}(1 - \exp[-\alpha_{abs} w]), \tag{11.7}$$

where $P_{in}$ and $P_{abs}$ are incident and absorbed power, respectively, $\alpha_{abs}$ is the *absorption coefficient,* and $w$ is the width (thickness) of the PD's active regions. When the absorption coefficient goes to zero, $P_{abs} \rightarrow 0$; when $\alpha_{abs}$ goes to infinity, $P_{abs} \rightarrow P_{in}$. In the case of a photodiode, each *absorbed* photon creates an electron; hence, the quantum efficiency is given by

$$\eta = P_{abs}/P_{in} = 1 - \exp(-\alpha_{abs} w) \tag{11.8}$$

Thus, quantum efficiency is not the constant, as was previously assumed, but the variable. As one can read from Formula 11.8, the bigger the product ($\alpha_{abs}w$), the closer $\eta$ is to its maximum. It follows from the physics of absorption (see Figure 2.10) that the absorption coefficient is the function of wavelenght; the graph exhibiting this dependency is shown in Figure 11.3. All these considerations elucidate the cause of the long-wavelength cutoff.

For wavelengths much *shorter* than the semiconductor's bandgap, photons will strike electrons at the valence band far from the energy-gap edge and the probability of exciting these electrons at the conduction band is very low. This is the main cause of the short-wavelength cutoff.

If you summarize all these considerations, you'll understand the reason for the discrepancy between the linear and the real graphs in Figure 11.2(b).

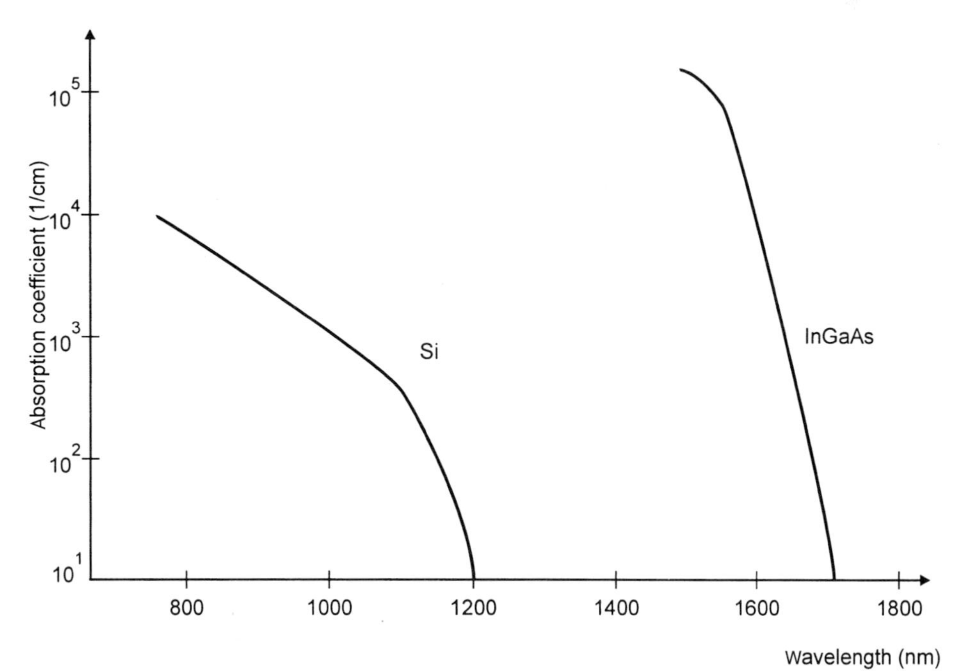


**Figure 11.3** Absorption coefficients as a function of wavelength.

### Example 11.1.3

***Problem:***

What width (thickness) of a depletion region of an InGaAs PD do we need to make its quantum efficiency 70%?

***Solution:***

The solution to this problem is, obviously, given by Formula 11.8. Solving Formula 11.8 with respect to *w,* we find:

$$w = [\ln(1 - \eta)]/(-\alpha_{abs}) \tag{11.8a}$$

To compute *w,* we need to know $\alpha_{abs.}$ We can use Figure 11.3, where, for InGaAs, $\alpha_{abs}$ is approximately $1 \times 10^5$ (1/cm) at the 1600-nm operating wavelength. Thus, we find *w* is equal to about 12 μm for InGaAs.

It is apparent that for a given absorption coefficient, *a wider (thicker) depletion region results in a higher quantum efficiency.* And this is an extremely important point. We need to make the depletion region wider to increase the quantum efficiency. This statement simply follows from the principle of operation of a PD. Indeed, the wider the depletion region, the greater the probability that most of the incident photons will fall here and thus the greater the likelihood they will be absorbed to create photocurrent.

(Note that InGaAs is much more efficient than Si as a material for a photodiode. One reason is that InGaAs is a direct-bandgap semiconductor, while Si is an indirect-bandgap material. Refer to the discussion of direct- and indirect-bandgap materials in Section 10.1.)

---

***Photovoltaic and photoconductive modes of operation*** Let's return to Figure 11.1. Observe that a photodiode can produce current without bias voltage because light conveys the external energy necessary to excite electrons at the conduction band (or to separate electrons from holes, if you prefer) and the depletion voltage ($V_D$) makes them flow. This mode of operation is called *photovoltaic.* It is how solar panels convert sunlight into electrical power. (Shaped like wings, these large panels, which contain a huge number of photocells, are wrapped around space satellites, with the photocells producing the photocurrent necessary to supply the satellite with electric power.) If external voltage is applied, the photodiode operates in the *photoconductive* mode. These two terms—*photovoltaic* and *photoconductive*—define the meaning of both operations: Without biasing, a photodiode works as the source of an electrical signal; with biasing, it's a good conductor of current originated by incident light. But remember: *A photodiode is actually a current source, with or without bias.*

***Advantages of reverse biasing*** For all practical purposes, we always use biasing because it dramatically improves the response of a photodiode. Without incident light, the depletion region of a photodiode does not contain free charge carriers (all electrons and holes are recombined, which is why we have a depletion region), whereas the *n* and *p* regions of a semiconductor have mobile charge carriers that are ready to flow. Hence, nearly all the bias voltage drops across the depletion region because this zone doesn't conduct. As soon as the incident photon creates electron-hole pairs, this voltage (or electric forces, to put it another way) helps to separate these free charge carriers and quickly removes them from the depletion region, thus generating photocurrent. This is the first—and the major—advantage of using reverse biasing.

What happens if the incident photon does not strike a depletion region but, rather, the *n* or *p* regions of a semiconductor? This can also create a free charge carrier, but the electric forces in these regions are weak so they will remove the electrons and holes there very slowly. Thus, a photogenerated electron-hole pair is separated by reverse voltage quickly and efficiently in the depletion region, but this separation occurs very slowly and inefficiently in the *p* or *n* regions because of the weakness there of electric forces. This is the second advantage of using reverse biasing. Incidentally, the photocurrent created in the depletion region is called *drift current.* Photocurrent created in *n* or *p* regions is called *diffusion* current.

A puzzling thought may cross your mind at this point: "If electrons and holes created by the incident photons and separated by reverse voltage have to drift through the depletion region before they reach the wire to flow to the battery, why don't they recombine again and radiate a photon?" Good question. Theoretically, electrons and holes can recombine again but, in reality, the loss of charge carriers due to secondary recombination is negligibly low. This is because the reverse voltage sweeps them from the depletion region faster than they can recombine again. In other words, the separation time of these carriers due to applied voltage is much less than their recombination lifetime. Thus, we have the third advantage of using reverse biasing.

The last, but not the least, advantage of reverse biasing is its ability to eliminate what's called *dark current.* Without incident light, some free charges in the depletion region can be created mostly by external thermal energy (temperature). The flow of these charges creates dark current, $I_d$. In other words, *dark current is current generated by a photodiode without light.* Clearly, dark current is a detrimental phenomenon because it eventually determines the minimum light power that can be detected, that is, a photodiode's *sensitivity.* How does reverse biasing help here? Since all voltage is applied across the depletion region, any free charge carriers that are occasionally created without light will be swept away by the reverse-bias voltage. This means that reverse biasing controls dark current.

So, from a practical standpoint, reverse biasing improves a photodiode's linearity, increases its speed and efficiency of operation, and reduces its dark current. All these advantages will be clarified in the course of this chapter.

**Dark-current sensitivity** *Sensitivity is the key parameter determining the quality of a photodiode.* As noted above, sensitivity refers to the minimum light power that a given photodiode can detect. It is measured in watts (in microwatts, actually) or in dBm, which is more common.

Here we're discussing only sensitivity determined by dark current, which is depicted in Figure 11.2(c). As you can see from this figure, there is an area of uncertainty around zero-input power in the $I_p$-$P$ graph. This is because some current flows through a PD's circuit, but we don't know whether it is dark current or photocurrent. Thus, until some minimal light power ($P_{min}$) truly generates photocurrent, we cannot rely on the output of a photodiode. As an example, it is easy to calculate that for $R = 1$ A/W, dark-current sensitivity is 5 nW for $I_d = 5$ nA.

The value of $I_d = 5$ nA at room temperature is typical for modern photodiodes. It is apparent from this discussion that dark current increases with temperature but it is still not more than 50 nA, typically, at $T \leq 70$°C. Compare this value with the value for photocurrent: If $R = 1$ A/W and input light power is 0.1 μW, then $I_p = 100$ nA. Thus, dark current is of concern at this typical level of photodetection.

Dark-current sensitivity is the major concern with photodetectors used in measuring devices such as power meters. For communication PDs, it is not a major issue. A more general means by which to evaluate a photodiode's sensitivity is noise description. We develop this approach later in this chapter.

**Power digest** A *p-n* photodiode converts light power into electric current. The efficiency of this conversion (1) diminishes at the air–semiconductor interface, where light is reflected, (2) decreases where photogenerated electrons and holes undergo a secondary recombination, and (3) increases within the active region, where light is better absorbed. Applying an antireflecting coating over the surface of the photodiode and using an angled fiber tip, we can resolve the reflection problem.

A widening depletion (active) region is the solution to two other problems. For instance, where power consideration is a major factor, we need a wide depletion region, a place where photons are absorbed, in order to achieve high quantum efficiency, which, in turn, provides high responsivity. But the width of a depletion region in a *p-n* junction photodiode is determined by the reverse voltage ($w \sim \sqrt{V}$) because the higher the reverse voltage, the more depleted the region around the *p-n* junction becomes. It might look, then, as though we need to apply high reverse-bias voltage to enhance the power response of a photodetector. But before jumping to any conclusion about the level of the reverse-bias voltage needed, we have to consider bandwidth.

## Bandwidth

Bandwidth, in terms of our current discussion, can be defined as *the maximum frequency, or bit rate, that a photodiode can detect without making essential errors.* (Again, strictly speaking, the term *bandwidth* [Hz] is applied only to analog signals; for digital transmission, we have the term *bit rate* [bit/s]. This said, it has become common practice today to use the term *bandwidth* to encompass both analog and digital technologies.)

There are two basic mechanisms restricting bandwidth in a photodiode. The first restriction stems from the fact that charge carriers created by a photon need some time to be collected. This time is often called *transit time,* $\tau_{tr}$. If we denote the maximum drift velocity of the charge carriers as $v_{sat}$, then for a depletion region with thickness *w,* transit time can be estimated as:

$$\tau_{tr} = w/v_{sat}\,, \tag{11.9}$$

where $v_{sat}$ is saturation velocity. With typical values of $w \sim 10\ \mu m$ and $v_{sat} \sim 10^5$ m/s, we can compute $\tau_{tr} \sim 100$ ps.

The second restriction on bandwidth derives from the inherent capacitance of a *p-n* photodiode ($C_{in}$). Indeed, a *p-n* junction can be considered as two charged plates isolated by a depletion region. This is the classical model of a capacitor. Hence, inherent capacitance is equal to

$$C_{in} = (\epsilon A)/w, \tag{11.10}$$

where $\epsilon$ is the permitivity of a semiconductor and $A$ is the active area (the photosensitive area) of the photodiode. This capacitance is parallel to the output of the photodiode.

To better understand this discussion, consider the equivalent circuit of a *p-n* photodiode. (See Figure 11.4 [2], [3].) Here the diode stands for an ideal diode operation; the current source ($I_p$) represents the flow of the photogenerated carriers; $R_j$ and $R_s$ correspond to the junction (shunt) and series resistance of a photodiode, respectively, and they form internal resistance $R_{in}$; $C_{in}$ is defined by Formula 11.10; and $R_L$ is the load resistance.

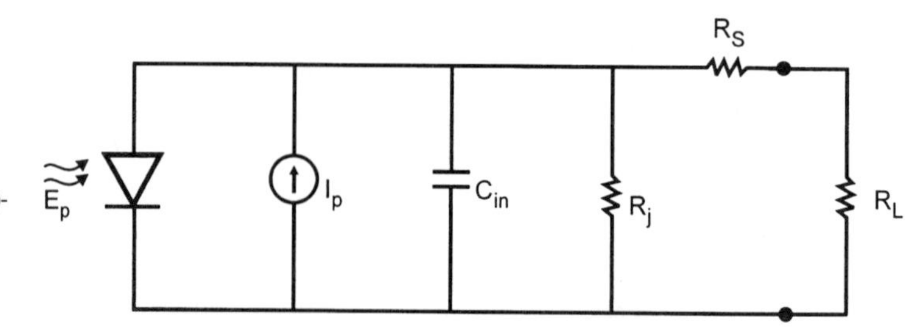


**Figure 11.4** Equivalent circuit of a *p-n* photodiode. *(Adapted from* Diode Lasers & Instruments Guide, *Melles Griot, Boulder, Colo., April 1998. Used with permission.)*

The junction (shunt) resistance ($R_j$) is the resistance of a photodiode's depletion region; by its very nature, this resistance is extremely high (from units to tens of MΩ). However, manufacturers can control the value of $R_j$. Resistance of the *p* and *n* layers and of the electrical contacts acts as a series resistance between the active region and the load circuit of a photodiode. We denote this resistance as $R_s$. The series resistance is very small, ranging from 5 to 10 Ω. So, from the point of view of the RC circuit, it is the load resistance ($R_L$) that determines the timing properties of a photodiode.

Let's return to the time constant ($\tau_{RC}$) induced by a capacitor. It follows from the above consideration that

$$\tau_{RC} = (R_s + R_L)C_{in} \approx R_L C_{in} \tag{11.11}$$

Typical values of $C_{in}$ are from 1 to 2 pF; hence, for $R_L = 50\ \Omega$, $\tau_{RC}$ is in the range of 50 to 100 ps.

Thus, the bandwidth of a photodiode is given by the following formula:

$$BW_{PD} = 1/[2\pi(\tau_{tr} + \tau_{RC})] \qquad 11.12$$

## Example 11.1.4

***Problem:***

A *p-n* photodiode has $\tau_{tr}$ = 100 ps and $\tau_{RC}$ = 100 ps.

**a.** What is its bandwidth?

**b.** Can we increase the bandwidth of a *p-n* photodiode by varying the thickness of the depletion region?

**c.** Can we increase the bandwidth by varying its load resistance?

**d.** What is the role of the active area of a photodiode?

***Solution:***

**a.** It comes directly from Formula 11.12 that BW = 0.796 Gbit/s.

**b.** We need to decrease both $\tau_{tr}$ and $\tau_{RC}$. To decrease $\tau_{tr}$, we have to increase the thickness of the depletion region, *w*. (We can do almost nothing about drift velocity.) To decrease $\tau_{RC}$, we need to decrease *w*, as you can see from Formulas 11.11 and 11.10. To clarify the point, let's rewrite Formula 11.12 in explicit form with respect to *w*:

$$BW_{PD} = 1/\{2\pi[(w/v_{sat}) + R_L(\epsilon A/w)]\}, \tag{11.13}$$

where $R_{in} = R_s + R_L \approx R_L$, as discussed above. If we rewrite Formula 11.13 in a different form,

$$BW_{PD} = (v_{sat}/2\pi)[w/(w^2 + v_{sat} R_L \epsilon A)], \tag{11.13a}$$

we see that the thickness of the depletion region (*w*) appears simultaneously in the numerator and the denominator of a bandwidth formula. Thus, one has to find a compromise for the value of *w* to achieve the optimal value of the bandwidth. Toward this end, let's take the derivative $\partial BW/\partial w$:

$$\partial BW_{PD}/\partial w = (v_{sat}/2\pi)[(-w^2 + v_{sat} R_L \epsilon A)/(w^2 + v_{sat} R_L \epsilon A)^2] \tag{11.13b}$$

Thus, the optimal thickness found from the condition $\partial BW_{PD}/\partial w = 0$ is given by

$$w_{opt} = \sqrt{(v_{sat} R_L \epsilon A)} \quad (11.13c)$$

All parameters under the square root, except $R_L$, are predetermined by the semiconductor material and the fabrication process of a *p-n* photodiode. Therefore, in a *p-n* photodiode we can control the thickness of the depletion region ($w$) only by reverse-bias voltage. But the value of reverse-bias voltage is determined mostly by the receiver package. (You don't want to have several different voltages in a small receiver chip, do you?) The fact is we don't have very much freedom when it comes to increasing the bandwidth of a *p-n* photodiode.

**c.** Load resistance ($R_L$) determines $\tau_{RC}$, as Formula 11.11 shows. This is why $R_L$ appears in Formula 11.13c. It seems, then, that we need to decrease the value of this resistance to increase the bandwidth. But we cannot do so because, first, $R_L$ must be at least ten times more than $R_S$ (see Figure 11.4) and, secondly, to minimize noise we need to increase the load resistance. (This topic is discussed in Sections 11.3 and 11.4.)

**d.** The active area ($A$) of a photodiode plays two roles. From the power standpoint, it's better to have this area large because it allows a PD to gather more light. (It is common practice to specify $A$ through the term *active-area diameter, $D_A$*.) Ultra-sensitive photodiodes for special (noncommunications) applications can have a huge active area. For example, large-area PDs can have a $D_A$ from 1 to 16 mm [4].

From the bandwidth standpoint, the active area must be small, as Formula 11.13 says. Indeed, the larger the active area, the higher the internal capacitance, which results in a large *R-C* time constant. (See Formulas 11.10, 11.11, and 11.12 above.) To fabricate very wide bandwidth photodiodes, manufacturers have to decrease the active-area diameters. To achieve a bandwidth up to 50 GHz, the manufacturer restricts the active-area diameter to 10 μm [5]. It's interesting to observe how a PD's capacitance changes with a change in the active-area diameter. For example, an InGaAs PIN photodiode exhibits the following relationships between active-area diameter and diode capacitance [4]:

| Active-area diameter, $D_A$ (μm) | 60 | 80 | 100 | 150 | 300 |
|---|---|---|---|---|---|
| Diode capacitance, $C$(pF) | 0.2 | 0.75 | 1.5 | 2 | 7 |

You can easily draw your own conclusion as to how the bandwidth of this photodiode changes in response to a rise in active-area diameter.

And so once again we confront a trade-off between power and bandwidth requirements. But don't forget: We need to couple light from an optical fiber into the active area of a photodiode. Therefore, we cannot make $D_A$ any arbitrary small value. In practice, you will find photodiodes used in fiber-optic communications technology with active-area diameters ranging from units of μm to hundreds of μm.

---

The above example shows that bandwidth is at its maximum when the thickness of a depletion region ($w$) is optimal. On the other hand, the above discussion shows that the power efficiency is proportional to $w$. In other words, *there is a trade-off between power and bandwidth efficiencies in a* p-n *photodiode.* We have to remember this fact because it shows that a *p-n* photodiode is not a very efficient device for communication purposes.

There is a third mechanism restricting the bandwidth of a *p-n* PD. Since the *p* and *n* regions are wider than the depletion region, many photons strike those regions, creating a diffusion photocurrent, as noted in the discussion of the advantages of reverse biasing. But diffusion flow is very slow compared with the drift flow of charge carriers. Therefore, the electric-output pulse of a *p-n* photodiode will be much wider than the optical-input pulse and its tail will be determined by the diffusion photocurrent. This is illustrated in Figure 11.5, where all the causes of the widening of the electric-output pulse are shown qualitatively.

***Bandwidth and p-n photodiode digest*** Increasing the bandwidth efficiency of a *p-n* photodiode requires a wide depletion region to reduce the diffusion current. Hence, it looks as though we need to increase the reverse bias because this voltage determines the width of a depletion region. Indeed, this is exactly what we need to do to increase the power efficiency. On the other hand, however, taking into account the transit and RC time constants, one has to find the appropriate reverse-biasing voltage to optimize the bandwidth of the photodiode. We cannot choose this voltage arbitrarily. For one thing, we need to remember that a photodiode is a part of a receiver, where low-voltage electronics is used.

The key point is simply this: *We need to increase the width of a depletion region without manipulating unnecessarily the value of the reverse-bias voltage.* The solution to this dilemma is a *p-i-n* photodiode.

## *p-i-n* Photodiodes

The basic structure of a *p-i-n* PD is shown in Figure 11.6. The major feature of this photodiode is that *it consists of a thick, lightly doped* intrinsic *layer sandwiched between thin* p *and* n *regions.* The word *intrinsic,* in semiconductor-industry parlance, means "natural," "undoped." Thus, we now have the full meaning of the letters *p-i-n*: positive-intrinsic-negative.

There are two major types of *p-i-n* photodiodes: front-illuminated (Figure 11.6[a]) and rear-illuminated (Figure 11.6[b]). In a front-illuminated PD, light enters the hole through the top contact. To reduce backreflection of the incident light, the active surface is covered by an antireflection coating. Then light passes through the thin *p* region and generates electron-hole pairs in the thick intrinsic layer.

In a rear-illuminated PD, light enters the active region through a heavily doped *n*+ layer. This layer is transparent to the incident light because its energy gap is larger than the energy of incident photons. All other processes are similar to those that take place in the front-illuminated PD.

To list the advantages of a *p-i-n* PD, we need simply to recall the drawbacks of a *p-n* photodiode. The first and the major feature of a *p-i-n* PD is that its intrinsic layer is its depletion layer, where the absorption of photons occurs. Since the intrinsic layer is naturally thick, most of the

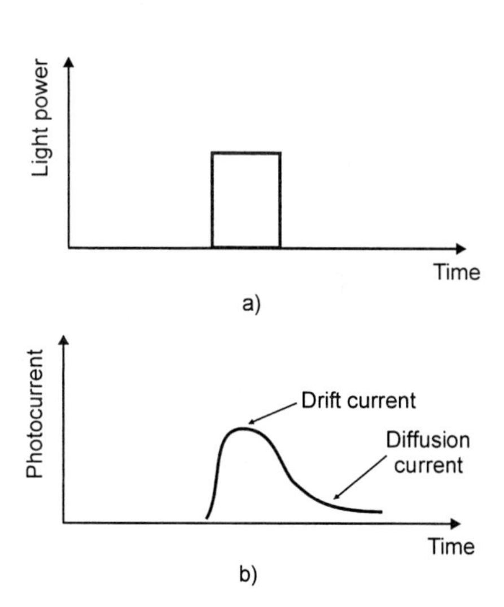


**Figure 11.5** Input and output pulses of a photodiode: (a) Optical input pulse; (b) electric-output pulse.

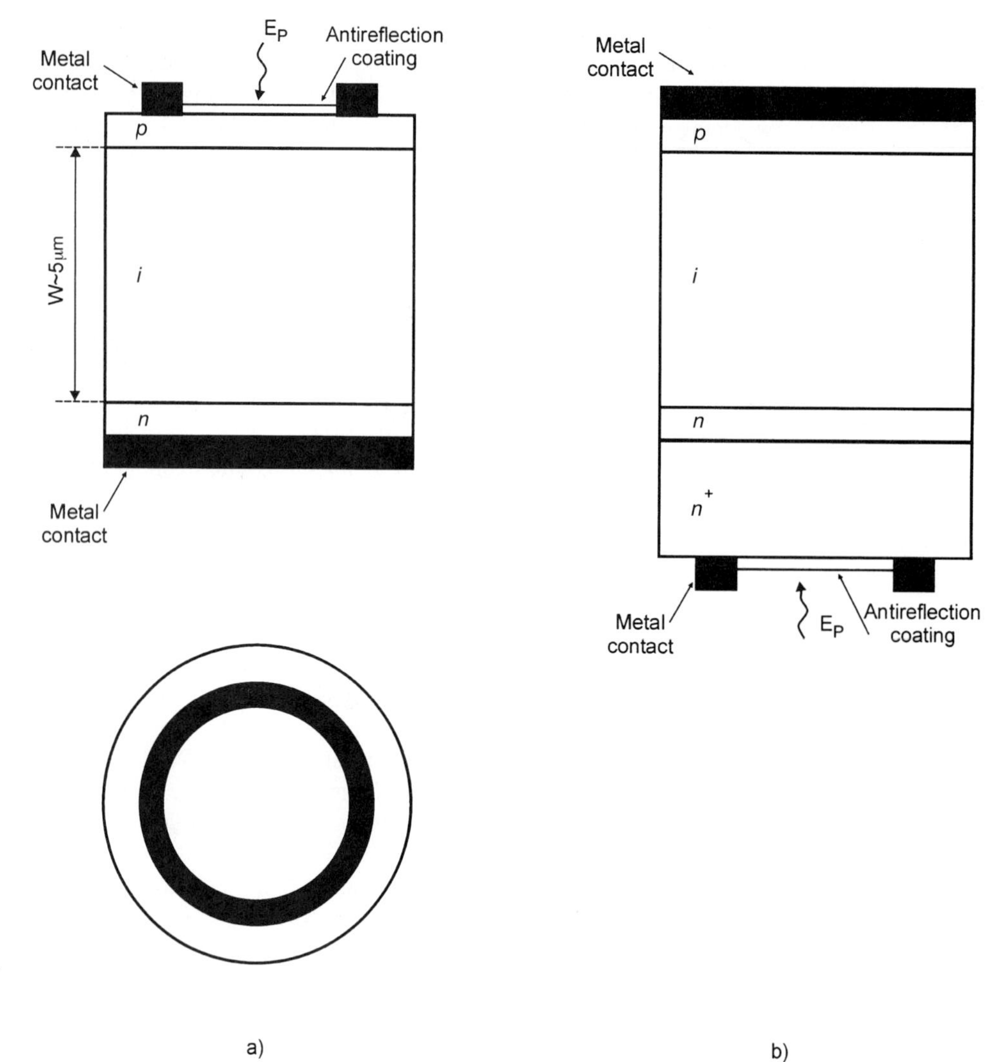


**Figure 11.6** *p-i-n* photodiode: (a) Front-illuminated PD; (b) rear-illuminated PD.

incident photons enter this layer and generate electron-hole pairs. This action results in the high quantum efficiency of this device. In addition, there is no need for tinkering with reverse voltage to increase the width of an absorbing layer. This is why both the power and bandwidth efficiencies are high, which is the major advantage of a *p-i-n* photodiode.

All other advantages are simply the consequences of the structure of a *p-i-n* PD. For example, since the intrinsic layer contains almost no free charge carriers, the electric field across this layer is large. The result is the efficient separation of electrons and holes generated by the incident photons. In addition, this field (reverse biasing, remember) decreases the dark current by sweeping away all thermally generated charge carriers. What's more, the diffusion current in a *p-i-n* PD is very small because the *p* and *n* layers are extremely thin compared with the intrinsic layer. Furthermore, incident photons are much more likely to enter the intrinsic (active) region than the *p* or *n* regions. The result is an increase in the bandwidth efficiency of the photodiode. One other advantage of a *p-i-n* PD is that the reverse-biasing voltage is small (usually, 5 V) because the thickness of the depletion region is controlled by the thickness of the intrinsic layer, not by reverse voltage.

There is one problem, however, that the *p-i-n* structure cannot resolve: the width of the intrinsic layer. Widening this layer will result in an increase in power efficiency but a decrease in bandwidth efficiency because of a rise in transit time, $\tau_{tr} = w/v_{sat}$. (See Formula 11.9.) Hence, some compromise must be found and the key to this compromise is given in Figure 11.3. A photodiode made from Si has to have a wider intrinsic layer because the absorption coefficient ($\alpha_{abs}$) at its operating wavelength (near 850 nm) is a little more than $10^3$ 1/cm, while InGaAs PDs have $\alpha_{abs} \sim 10^5$ 1/cm at $\lambda = 1600$ nm and, therefore, can have a smaller $w$. In practice, Si *p-i-n* photodiodes are fabricated with the width of an intrinsic layer on the order of 40 μm, while InGaAs PDs have a $w$ around 4 μm. This results in different bandwidths for Si and InGaAs photodiodes.

### Example 11.1.4

***Problem:***

What are the bandwidths of Si and InGaAs *p-i-n* photodiodes?

***Solution:***

The solution to this problem is obviously given by Formula 11.12 or Formula 11.13 but we need to know several specific parameters to compute the numbers. Instead of delving into reference sources in search of these parameters, we are better off observing that $\tau_{tr} \sim w$ and $\tau_{RC} \sim 1/w$. Hence, for a wide intrinsic layer, which is exactly the case here, we can assume $\tau_{tr} \gg \tau_{RC}$.

With this assumption, Formulas 11.12 and 11.13 take the following form:

$$BW_{PD} = 1/(2\pi\tau_{tr}) = 1/[2\pi(w/v_{sat})] \tag{11.14}$$

In Example 11.1.3 we used $v_{sat} \sim 10^5$ m/s. Thus, taking $w = 40$ μm for Si PD and $w = 4$ μm for InGaAs PD, we obtain $\tau_{tr}$ (Si) = 400 ps and $\tau_{tr}$ (InGaAs) = 40 ps, which results in:

$$BW_{Si} = 0.398 \text{ Gbit/s}$$

$$BW_{InGaAs} = 3.98 \text{ Gbit/s}$$

From Figure 11.3 and this example, one can see the areas of photodiode applications: A silicon PD is used in the first transparent window (near 850 nm), where relatively low-speed networks operate, while an InGaAs PD is suited for the second and third windows (near 1300 nm and 1550 nm, respectively), where high-speed networks operate.

---

Further improvement in the efficiency of a *p-i-n* operation can be accomplished by fabricating the photodiode in a double heterostructure, one similar to that used in LEDs and LDs. In fact, if you make the *n* and *p* regions of the diode shown in Figure 11.4 from InP and the intrinsic layer from InGaAs, your goal is achieved: All photons at operating wavelength will pass the *n* and *p* layers without interaction. This dramatically improves the quantum efficiency and eliminates the diffusion-current problem.

*A* p-i-n *photodiode is the most commonly employed light detector in today's fiber-optic communications systems because of its ease in fabrication, high reliability, low noise, low voltage, and relatively high bandwidth.* Many efforts have been made to improve its characteristics, in particular the bandwidth, where up to 110 GHz has been achieved [1].

## Avalanche Photodiodes (APDs)

What do we ultimately want from a photodiode? One of its major parameters is sensitivity— the minimum light power a PD can detect. This parameter determines the length of a fiber-optic link imposed by a power limitation. Indeed, refer to Section 8.4, where the power budget of a fiber communications link is considered. The more sensitive the photodiode, the longer the link a designer can afford to have with given losses. It seems that the remedy for this problem is quite apparent: Use an amplifier to magnify the photocurrent produced by the photodiode. In fact, a receiver circuit always includes an amplifier, as will be explained in the following sections. But an amplifier, as is true of any electronic circuit, introduces its own noise, thus reducing sensitivity. So, you are probably thinking, if only we were able to amplify photocurrent without an external amplifier and, therefore, without the noise associated with this circuitry! Well, we can; in fact, this is why the *avalanche photodiode (APD)* was invented.

**Power consideration** The basic mechanism of an avalanche PD is as follows: A special *p-i-n* structure of a photodiode is used. Incident photons generate primary electrons and holes, as they do in a regular *p-i-n* PD. Relatively high (around 20 V) reverse voltage is applied to the photodiode. This voltage accelerates photogenerated electrons and holes, which thereupon acquire high energy. These electrons and holes strike neutral atoms and separate other bonded electrons and holes. These secondary carriers gain enough energy to ionize other carriers, causing a so-called avalanche process of creating new carriers. Thus, one photon eventually generates many charge carriers, which means this photodiode internally amplifies photocurrent. This is equivalent to saying the APD's quantum efficiency is more than 1 (typically, it is from 10 to 100).

Referring to an energy-band diagram, one can say that primary photogenerated electrons strike bonded electrons at the valence band and cause them to rise to the conduction band; holes are left at the valence band to complete the picture of how secondary carriers are generated.

The process of creating many secondary carriers is called *impact ionization.* The basic diagram of an APD is shown in Figure 11.7.

Photons pass through the heavily doped $p^+$ region and enter the intrinsic layer, where they produce electron-hole pairs. Reverse voltage separates photogenerated electrons and holes and moves them toward the *pn+* junction, where a high electric field (on the order of $10^5$ V/cm) exists. This electric field accelerates the charge carriers, resulting in impact ionization.

The major advantage of an avalanche photodiode over a *p-i-n* PD is clear from the physics of its operation: The quantum efficiency of the APD is $M$ times larger than that of a *p-i-n* PD. ($M$ is called the *multiplication, or gain, factor.*) Indeed, an APD produces $M$ charge carriers in response to one photon. Thus, referring to Formula 11.6, one can write:

$$R_{APD} = MR_{p\text{-}i\text{-}n} = M(\eta/1248)\lambda \tag{11.15}$$

(Don't forget that the wavelength has to be measured in nm.)

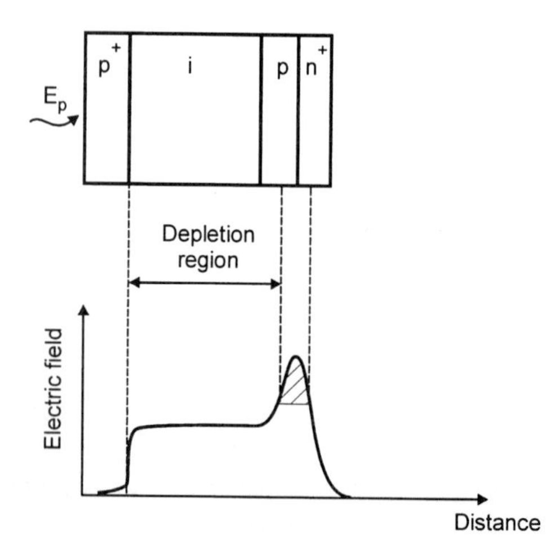


**Figure 11.7** Avalanche photodiode (APD).

Even though, theoretically, both electrons and holes can be involved in the ionization process, from a practical standpoint, an APD works better when only one type (usually electrons) is used. Thus, in the above explanation you can always substitute the word *electrons* where you see *charge carriers.*

The multiplication factor ($M$) depends on the accelerating voltage, the thickness of the gain region, and the ratio of electrons to holes participating in the ionization process. This implies that you can control the gain of an APD by varying the reverse voltage. $M$ values range from 10 to 500.

What we have to keep in mind is that since the ionization process is essentially random, so too is the multiplication factor. Thus, when concerned with the value of $M$, one works with an average number. It also follows from the physics of the avalanche process that it is noisy. However, this doesn't nullify the major advantage of an APD: internal amplification of photocurrent without the noise associated with external electronic circuitry.

***APD bandwidth*** Consideration of the bandwidth of an avalanche photodiode requires a different approach from that used when considering the bandwidth of a *p-i-n* PD. Since an APD introduces amplification, the most universal characteristic of such a device is the *gain-bandwidth product:* M × BW. For a typical APD, the gain-bandwidth product can be evaluated by [1]

$$M \times \mathrm{BW} = 1/(2\pi\ \tau_e), \tag{11.16}$$

where $M$ is zero-frequency gain and $\tau_e$ is effective transit time equal to

$$\tau_e = k_A \tau_{tr} \tag{11.17}$$

Here, $\tau_{tr}$ is transit time defined by Formula 11.9 and $k_A$ *is the ratio of holes to electrons involved in the ionization process.* (Strictly speaking, $k_A$ is the ratio of the impact-ionization coefficients of holes and electrons but, in a sense, it can be considered as simply the ratio of electrons to holes.) The ratio $k_A$ depends on the semiconductor material and it is in the range of 0.03 for Si, 0.8 for Ge, and 0.6 for InGaAs. The assumption $\tau_{tr} \gg \tau_{RC}$ is used in Formula 11.16.

To complicate the matter, the gain (the multiplication factor) of an APD depends on frequency [1]:

$$M(\omega) = M/\sqrt{(1 + (\omega \tau_e M)^2}, \tag{11.18}$$

where $\omega$ stands for radian frequency and $M$ is $M(0)$, as above.

The gain-bandwidth product is around 500 GHz for an Si APD and 120 GHz for an InGaAs APD. Since an Si APD has a gain as high as 500, its bandwidth is not more than 1 GHz, while an InGaAs APD has a typical gain of about 40, which yields a 3-GHz bandwidth.

As was the case for a *p-i-n* photodiode, we can conclude that Si APDs are useful for a moderate-speed (up to 1 GHz) fiber-optic network, which usually operates at 850 nm, while InGaAs APDs can be used in higher-speed fiber links (up to 3 GHz), which usually operate at 1300 nm and 1550 nm.

Overall, Si APDs demonstrate good performance characteristics from both power and bandwidth standpoints but they are inherently restricted to the use of the first transparent window, which is, for the most part, out of the commercial-application realm. InGaAs APDs—the devices for today's 1300-nm and 1550-nm wavelengths—have much worse power and bandwidth parameters. A major effort has gone into improving the characteristics of long-wavelength APDs. This has resulted in an increase in the gain-bandwidth product up to 150 GHz with a gain ($M$) of 10 [1].

Let's compare an APD's bandwidth of 15 GHz and gain of 10 with a *p-i-n* PD's bandwidth of 5 GHz and gain of 1. It is easy to conclude that an APD is at least 10 times more sensitive than a *p-i-n* PD with comparable bandwidth, which implies a 10-times-longer fiber-optic span between a transmitter and a receiver. But this advantage almost vanishes if you recall that an APD requires relatively high reverse voltage. From a practical standpoint, this means an increase in power consumption, implying less freedom for miniaturization of a receiver unit and, therefore, longer transmission lines with increasing noise and parasitic capacitance, not to mention the need for a separate power supply that is not compatible with other power units used in electronic circuits. So, when choosing a photodetector for your fiber-optic communications system, use a systems approach and take into account all the advantages and shortcomings of every type of device.

In conclusion, we have summarized in Table 11.1 the typical characteristics of *p-i-n* and avalanche photodiodes. These numbers give you a general idea of what today's PDs look like.

## MSM Photodetectors

An MSM (metal-semiconductor-metal) is another type of photodetector used in fiber-optic communications. This is not a *p-n* junction diode; however, its basic mechanism of light-current conversion is still the same: Photons generate electron-hole pairs whose flow makes current. The basic structure of an MSM photodetector is shown in Figure 11.8.

A set of flat metal contacts is deposited on the surface of a semiconductor. These contacts are called fingers and they are biased alternately so that a relatively high electric field exists between the fingers. Photons strike the semiconductor material between the fingers and create electron-hole pairs, which are separated by the electric field; thus, electric current is created.

Since both electrodes and a photosensitive region are fabricated on the same side of the semiconductor, this structure is called *planar*. The advantage of this photodetector, as compared with both types of photodiodes, is that a planar structure results in low capacitance and, consequently, in higher bandwidth. Indeed, the MSM photodetector promises to work at 300 GHz. Ease of fabrication is another advantage of a planar structure.

**Table 11.1** Typical characteristics of *p-i-n* and avalanche photodiodes

| Parameter | Symbol | Unit | Type | Material | | |
|---|---|---|---|---|---|---|
| | | | | Si | Ge | InGaAs |
| Wavelength | $\lambda$ | nm | | 0.4–1.1 | 0.8–1.8 | 1.0–1.7 |
| Responsivity | $R$ | A/W | *p-i-n* | 0.4–0.45 | 0.8–0.87 | 0.5–0.95 |
| Quantum efficiency | $\eta$ | % | *p-i-n* | 75–90 | 50–55 | 60–70 |
| APD gain | $M$ | — | APD | — | 50–200 | 10–40 |
| Dark current | $I_d$ | nA | *p-i-n* | 1–10 | 50–500 | 1–20 |
| | | | APD | 0.1–1 | 50–500 | 1–5 |
| Bandwidth | BW | GHz | *p-i-n* | 0.125–1.4 | 0–0.0015 | 0.0025–40 |
| | | | APD | — | 1.5 | 1.5–3.5 |
| Bit rate | BR | Gbit/s | *p-i-n* | 0.01 | — | 0.1555–53 |
| | | | APD | — | — | 2.5–4 |
| Reverse voltage* | $V$ | V | *p-i-n* | 50–100 | 6–10 | 5–6 |
| | | | APD | 200–250 | 20–40 | 20–30 |
| *k*-factor | $k_A$ | — | APD | 0.02–0.05 | 0.7–1.0 | 0.5–0.7 |

*Note: The reverse voltages listed here reflect only the orders of magnitude. Actual numbers found in the field vary widely. For example, you will find InGaAs *p-i-n* PDs with reverse bias up to 30 V, Si *p-i-n* PDs with reverse bias less than 30 V, and high-speed InGaAs APDs with reverse bias around 50 V.

Sources: Govind Agrawal, *Fiber-Optic Communication Systems,* 2d ed., New York: John Wiley & Sons, 1997, and *Lightwave 1999 Worldwide Directory of Fiber-Optic Communications Products and Services,* March 31, 1999, pp. 62–66.

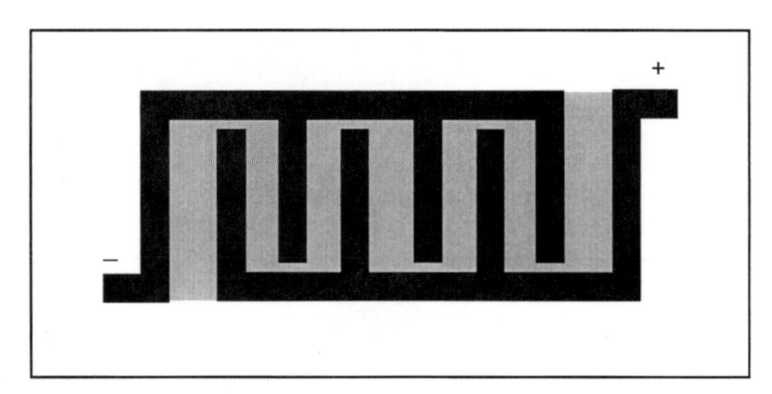


**Figure 11.8** MSM photodetector. Metal contact is black and active area is shaded.

There is, however, a drawback to an MSM photodetector: its relatively low responsivity, which ranges from 0.4 to 0.7 A/W. What's more, an essential area of the semiconductor material is taken up by metal contacts, thereby reducing the active area of the device.

As you can well imagine, then, MSM photodetectors are an area of intensive research and development today, so we can expect a number of these devices to be competing commercially in the near future.

## 11.2 READING THE DATA SHEETS OF PHOTODIODES

The data sheets of *p-i-n* and APD photodiodes are analyzed in this section, which will help you understand the meaning of each parameter.

### Data Sheet of a *p-i-n* Photodiode

The data sheet of a *p-i-n* PD by Mitel Semiconductor is shown in Figure 11.9. Focus in particular on how the manufacturer designates a *p-i-n* type of PD: PIN photodiode. There is no standard here but the research community usually employs the "*p-i-n*" notation, while the industry prefers the designation "PIN."

A photodiode comes in three assemblies for coupling an optical fiber to a PD. Two of them are receptacles for popular ST and SC connectors—for multimode fibers—and the third assembly is a pigtail—for singlemode fibers. (If you still are unfamiliar with these and the terms that follow, refer back to Sections 3.1 and 8.2.) Each of these assemblies has its own applications, as noted on the data sheet.

***Features and description*** If you come upon any unfamiliar terminology in the section headed "FEATURES," turn back to Sections 11.1 and 3.1. The only terms that haven't been discussed yet are the Bellcore specifications.[1]

[1] Bellcore (Bell Communications Research) was the company created to develop technical standards for the public telephone network after the AT&T divestiture in 1984. Until 1998 it served as the R&D arm of the Baby Bells (the seven Regional Bell Operating Companies established upon divestiture to provide local telephone service) in much the same way the old Bell Telephone Laboratories served AT&T until it became a unit of Lucent Technologies, a spinoff of AT&T, in 1997. Among the many renowned inventions at Bell Labs was the transistor (1948), for which William Shockley, John Bardeen, and Walter H. Brattain were awarded the Nobel Prize in 1956. Since its inception, Bellcore has developed a number of technical standards accepted by the telecommunications industry throughout the world. The Bellcore technical reference noted in the data sheet specifies the testing conditions and determines the performance criteria with which a device has to comply. In 1999 Bellcore was acquired by Science Applications International Corp. (SIAC) and its name was changed to Telcordia Technologies.

**MF432** Datacom, Telecom, General Purpose PIN Photodiodes

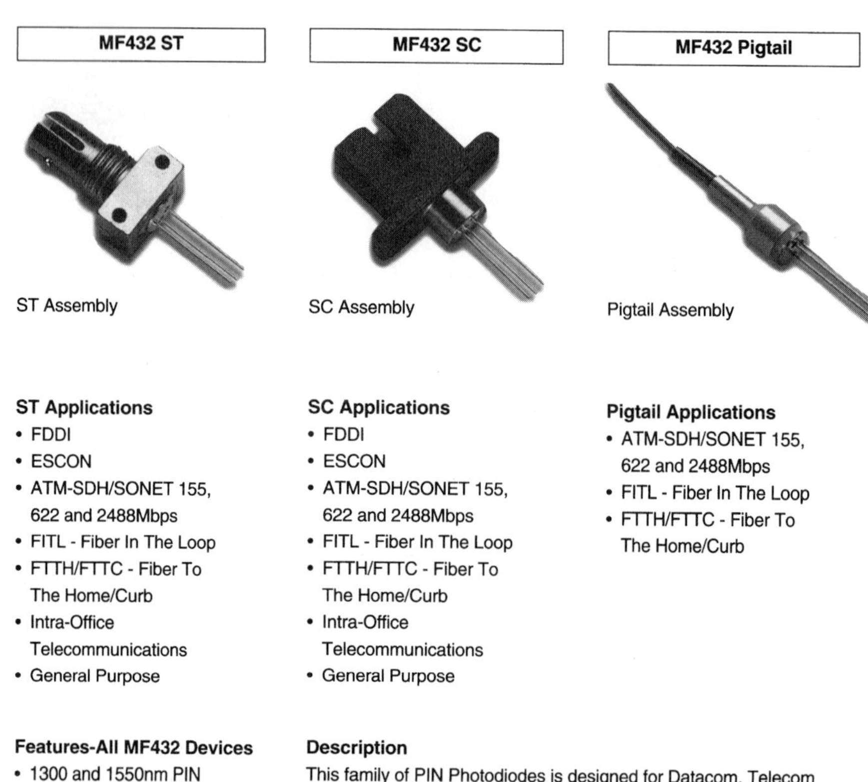


**ST Applications**
- FDDI
- ESCON
- ATM-SDH/SONET 155, 622 and 2488Mbps
- FITL - Fiber In The Loop
- FTTH/FTTC - Fiber To The Home/Curb
- Intra-Office Telecommunications
- General Purpose

**SC Applications**
- FDDI
- ESCON
- ATM-SDH/SONET 155, 622 and 2488Mbps
- FITL - Fiber In The Loop
- FTTH/FTTC - Fiber To The Home/Curb
- Intra-Office Telecommunications
- General Purpose

**Pigtail Applications**
- ATM-SDH/SONET 155, 622 and 2488Mbps
- FITL - Fiber In The Loop
- FTTH/FTTC - Fiber To The Home/Curb

**Features-All MF432 Devices**
- 1300 and 1550nm PIN Photodiode
- 2.5GHz Bandwidth
- Designed for Single-Mode and Multi-Mode Fiber
- Aligned in ST®, SC Receptacle or with a Single-Mode Fiber Pigtail
- Tested to Bellcore TA-NWT-000983
- High Return Loss in Pigtail Configuration

**Description**

This family of PIN Photodiodes is designed for Datacom, Telecom and General purpose applications. Their unique design combines high bandwidth with high responsivity for single-mode as well as multimode fibers up to 62.5μm core diameter. The MF432 PIN Photodiode is available in ST, SC, or Pigtail package.

Specially-designed connectors and clips for PC board assembly are included in deliveries of MF432 in SC and Pigtail configurations.

The MF431 LED is the recommended transmitter for these PIN photodiodes.

**Figure 11.9** Data sheet of a *p-i-n* photodiode. *(Reprinted with permission from Mitel Semiconductor's* Optoelectronic Solutions *Product Catalog © 1997, Mitel Semiconductor, Kanata, Ont., Can.)*

The "DESCRIPTION" section should be crystal clear to you. If it is not, reread Sections 11.1, 10.4, 8.2, and 3.1.

***Functional diagram*** This figure is self-explanatory. Its purpose is to show that a photodiode can be coupled to an optical fiber with different connector-receptacle pairs or with a pigtail assembly.

***Specifications*** The *absolute maximum ratings* are specified by the manufacturer for any optoelectronic device. You will recall our explanations of the data sheets of LEDs and LDs in Sections 9.1 and 9.3, respectively. There is no need, therefore, to repeat this discussion here.

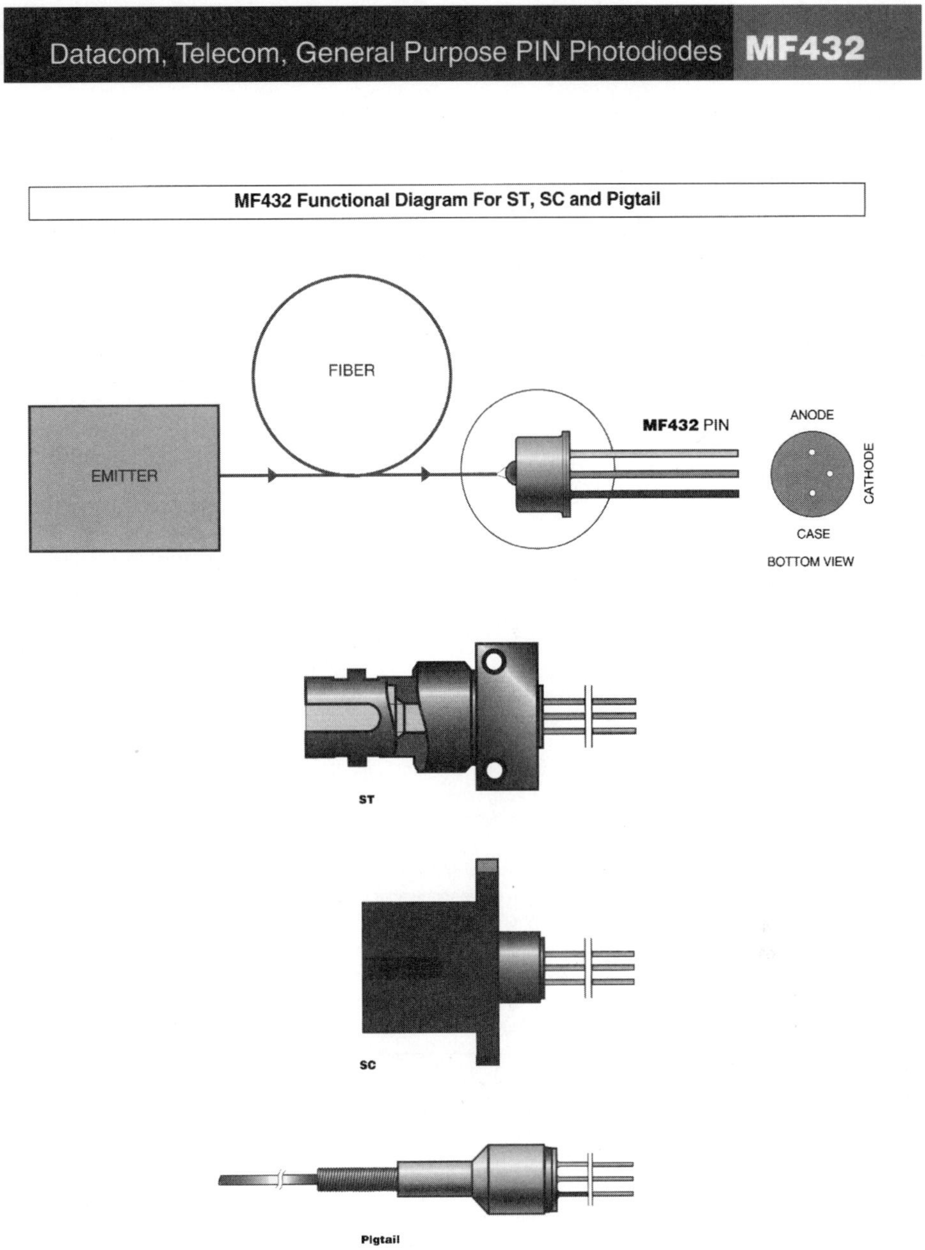

Datacom, Telecom, General Purpose PIN Photodiodes MF432

MF432 Functional Diagram For ST, SC and Pigtail

FIBER
EMITTER
MF432 PIN
ANODE
CATHODE
CASE
BOTTOM VIEW

ST

SC

Pigtail

**Figure 11.9**
(continued)

Optical and electrical characteristics are the data a designer looks at first when selecting a photodiode. The designer pays particular attention to the value of the reverse-bias voltage: V = 5 V.

The *responsivity* of this particular PD is rather high and reaches 1 A/W at 1550 nm. The given numbers (0.8 A/W and 1.0 A/W) are true when a photodiode is used with a specific fiber (see Note 1 on the data sheet), which means the responsivity has been measured with the assembly, not with a discrete photodiode.

Figures 1 and 2 of the data sheet show how responsivity depends on the axial (fiber-PD separation) and radial (lateral) displacements. Let's focus on Figure 1: It shows that responsiv-

**MF432** Datacom, Telecom, General Purpose PIN Photodiodes

**Absolute Maximum Ratings***

| Parameter | Symbol | Min. | Max. | Units |
|---|---|---|---|---|
| Storage Temperature | $T_{stg}$ | -40 | +85 | °C |
| Operating Temperature | $T_{op}$ | -40 | +85 | °C |
| Reverse Voltage | $V_R$ | | 20 | V |
| Soldering Temperature (Note 1) | $T_{sld}$ | | 260 | °C |

*Exceeding these values may cause permanent damage. Functional operation under these conditions is not implied.
Note 1: 2mm from the case for 10s.

**Optical & Electrical Characteristics** (Case Temperature -25 to +70˚C)

| Parameter | Symbol | Min. | Typ. | Max. | Units | Test Conditions |
|---|---|---|---|---|---|---|
| Responsivity (Fig 1, 2, 3) | R | 0.7 | 0.8 | | A/W | λ=1300nm (Note 1) |
| | | 0.8 | 1.0 | | | λ=1550nm $V_R$=5V |
| Bandwidth | $f_c$ | 2.5 | | | GHz | $V_R$=5V $R_L$=50Ω (Note 1) |
| Capacitance (Fig 4) | C | | 0.8 | 1.2 | pF | $V_R$=5V f=1MHz |
| Dark Current | $I_d$ | | | 3 | nA | $T_{Case}$=25˚C |
| | | | | 50 | | $T_{Case}$=70˚C $V_R$=5V |
| Return Loss | RL | 40 | 55 | | dB | (Note 2) |

Note 1: Data for 10/125μm single-mode fiber (NA=0.11) to 62.5/125μm graded index fiber (NA=0.275).
Note 2: With 10/125μm single-mode fiber pigtail (NA:0.11).

**Thermal Characteristics**

| Parameter | Symbol | Min. | Typ. | Max. | Units |
|---|---|---|---|---|---|
| Temperature Coefficient - Dark Current | $dI_d/dT_j$ | | 5 | | %/˚C |

**Figure 11.9** (continued)

ity largely depends on the separation between the fiber end and the photodiode assembly ($z$). The problem of axial displacement stems from a mismatch between the diameter of a fiber core (or beam, if you will) and the photosensitive diameter of the photodiode ($D_A$). Indeed, the MM-GI fiber core's diameter is 62.5 μm, while $D_A$ is typically 75 μm and higher. To comply with these dimensions, manufacturers use a dual-lens system, which allows them to obtain maximum light coverage of the photosensitive area of the photodiode [6]. This lens system has its focal distance, $F$, as Figure 11.9 shows. When the fiber end is placed at distance $F$, the light beam covers the photosensitive area of the PD to its maximum extent. You can easily see from Figure 11.9 what happens to responsivity when the fiber end is axially displaced from its optimal position. (It is instructive to draw such a figure.) The price for this arrangement is a sharp maximum relative response as a function of axial displacement. These explanations are shown in Figure 11.10.

Datacom, Telecom, General Purpose PIN Photodiodes **MF432**

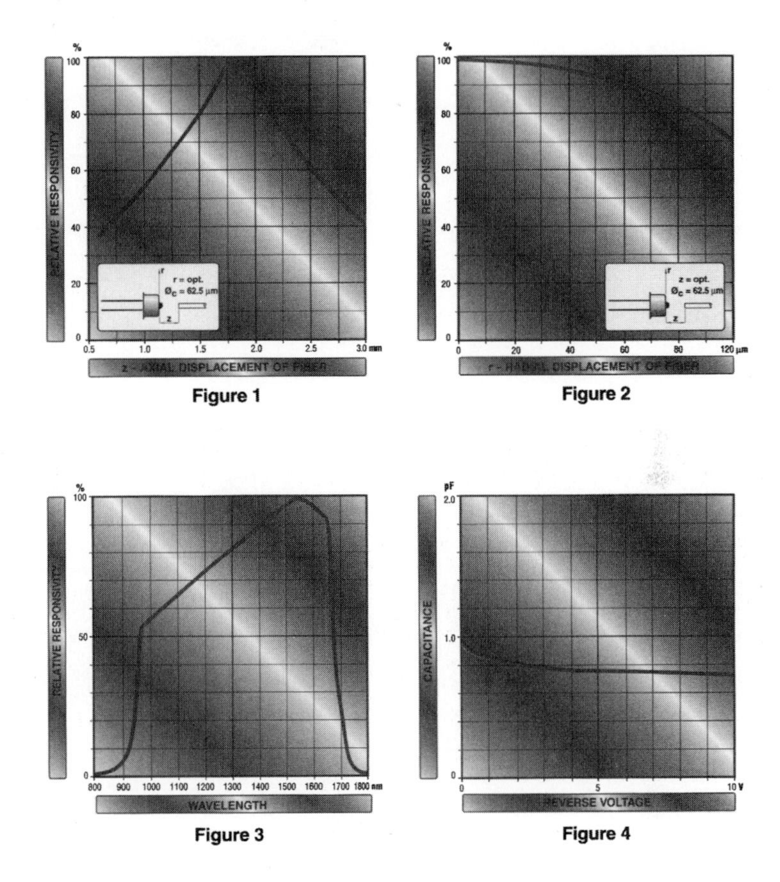


**Figure 11.9** continued)

In essence, Figure 11.10 shows how the problem of coupling light from a fiber into a photodiode can be resolved. Three basic techniques are used here: direct (butt) coupling, lens coupling, and fiber-tapered-end coupling. If you refer back to Figure 9.7, you will recall what's involved in these techniques and, at the same time, realize that the identical techniques are used for LED–fiber and PD–fiber couplings. We should emphasize that coupling is even more critical for a receiver than for a transmitter because of the very low power entering a receiver. The reflection problem is solved by using two means: an angle-taped fiber end and an antireflection coating.

But to return to our discussion of the data sheet: Angular displacement, obviously, decreases a photodiode's output. The direct angular response of a PD is described as

$$R = R_0 \cos \Theta, \tag{11.19}$$

where $\Theta$ is the angle of incidence and $R_0$ is the PD's response at the normal incidence of light. (By the way, you should immediately recognize Formula 11.19 and under what conditions it was previously discussed. Do you?)

With coupling optics, the angular response of a photodiode has changed. To evaluate the difference, you can simply plot the graph of Formula 11.19 and superimpose it onto Figure 2 of the data sheet in Figure 11.9.

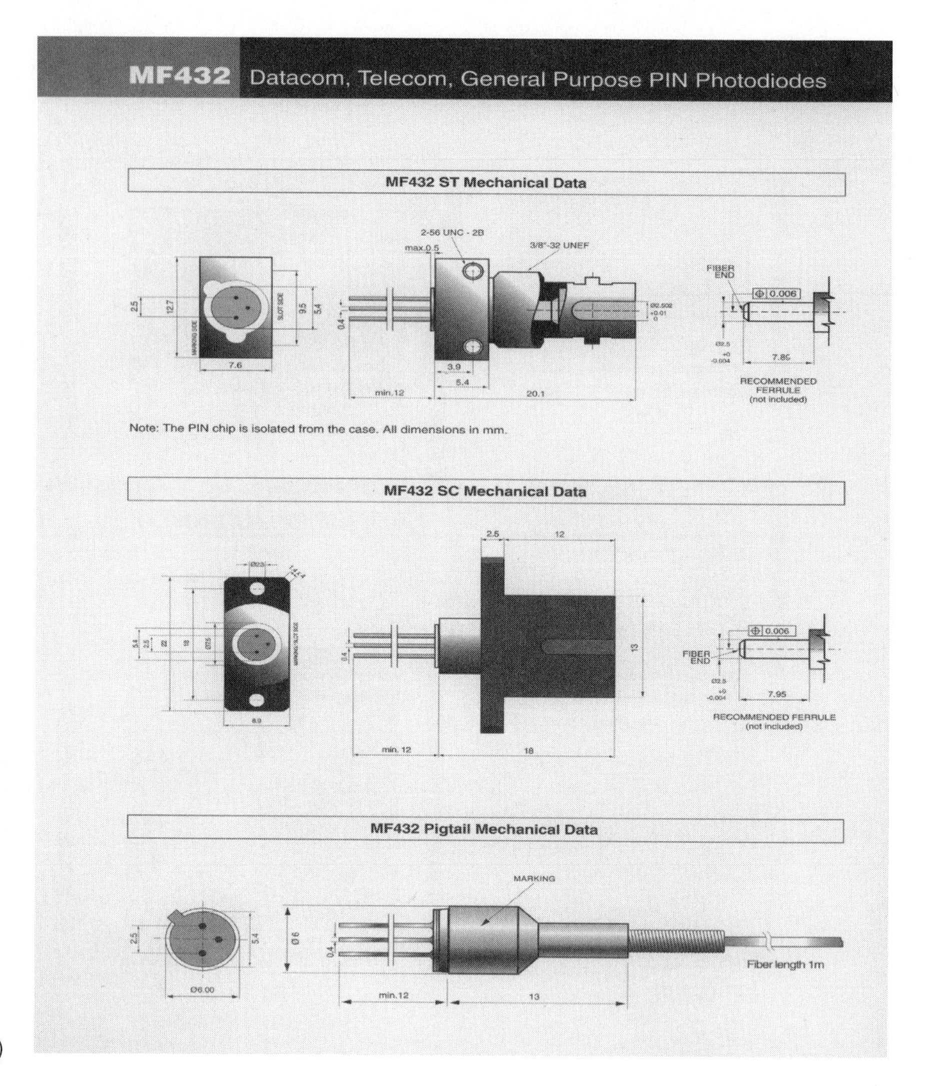


**Figure 11.9** (continued)

*Responsivity vs. wavelength* was discussed in Section 11.1. It is useful to compare Figure 3 given in Figure 11.9 with Formula 11.6 and Figure 11.2(b). Even though it isn't shown in the data sheet, responsivity depends on temperature, which usually decreases gradually at the rate of about 6% per 100°C.

The *bandwidth* of this photodiode is equal to 2.5 GHz, which is quite a good number. Note too that the manufacturer specifies the load resistance. Can you explain why?

Commercially available InGaAs photodiodes have bandwidths that range up to 10 GHz. Bear in mind that manufacturers do not always specify the bandwidth. Sometimes the data sheets contain only the rise/fall time values. If such is the case, you can derive the bandwidth using the well-known formula BW = 0.35/(*rise or fall time*).

The role of *capacitance* has been discussed in Section 11.1. (See Formulas 11.11 and 11.12.) Capacitance is measured at a specific frequency (1 MHz here). Can you explain why? Capacitance depends on reverse voltage (Formula 11.10); this is why the manufacturer specifies this value (5 V here) along with the frequency.

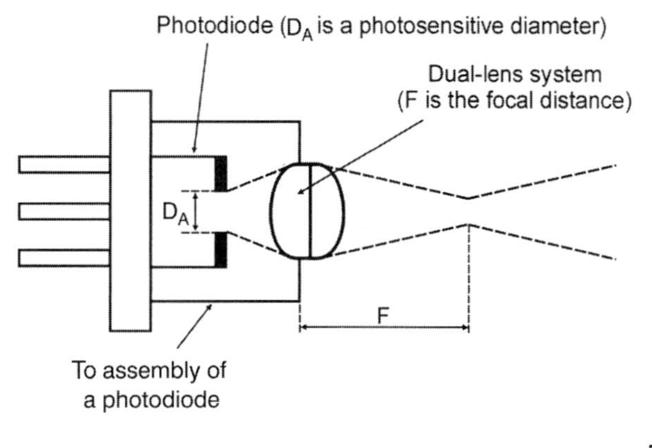


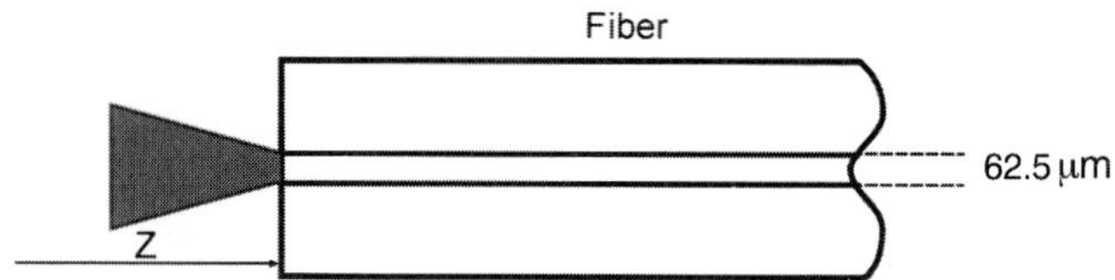


**Figure 11.10** Optimization of axial displacement.

*Dark current* is given at two temperature values and at a specific reverse voltage. This is done because dark current depends on both temperature and reverse voltage. By definition, dark current is current flowing in a photodiode devoid of input light. The only external-energy source in this case is temperature. Thermally excited electrons jump at the conduction band and flow when reverse voltage is applied. Thus, the higher the temperature, the greater the number of thermally excited electrons and the larger the flow of dark current. Look at the given numbers again: $I_d = 3$ nA at 25°C and $I_d = 50$ nA at 70°C. Dark current increases as temperature rises according to the following formula:

$$I_d(T_2) = I_d(T_1)\, 2^{(T_2-T_1)/C}, \tag{11.19a}$$

where $C$ is an empirical constant. For example, the typical value of $C$ is 8 for a silicon photodiode.

The dependence of dark current on reverse voltage is not so simple. On the one hand, thermally excited charge carriers are swept by the reverse voltage from the depletion region because most of the reverse voltage drops across this layer. (This is why a PD exhibits dark current.) We don't want those charge carriers to accumulate and flow randomly—this is the *positive* role reverse bias plays in controlling dark current. On the other hand, the higher the reverse voltage, the wider the depletion region and the larger the number of thermally excited charge carriers that will be created there. This, unfortunately, is the *negative* role of reverse voltage in terms of its effect on dark current. Overall, then, dark current increases as reverse voltage rises.

*Return loss* is specified only for pigtail assembly. This is because for a receptacle assembly return loss is determined not merely by reflection from a PD's surface but primarily by reflection from a connector. As for pigtail assembly, the manufacturer is responsible for the return loss of the entire unit. As pointed out in Section 11.1, the active surface of a photodiode is covered by an antireflection coating to minimize backreflection. But with a pigtail assembly, additional backreflection occurs from the pigtail-fiber end. However, the return-loss characteristic of this PD is very good. To appreciate these numbers (typically, 55 dB), turn back to Section 8.1, where we first broached the topic of return loss. It was mentioned there that a return-loss value of 55 dB is extremely good for both mechanical splicing and connectors.

The *mechanical dimensions* give you a feel for what a modern PD assembly looks like.

***Additional data*** In addition to the data given in Figure 11.9, you will some day come across these other characteristics of photodiodes:

- *Input power saturation.* This determines the maximum value of power to the point where the photocurrent is a linear function of the light power. Saturation power starts, typically, at 1 mW. (The saturation effect is discussed in Section 11.1.)
- *Photosensitive diameter.* This determines the active area of a photodiode. The role of this parameter in coupling light from a fiber into a PD was discussed above. Typically, the diameter of a regular *p-i-n* PD is 75 μm. Large-area PDs with diameters of 1 and 2 mm—and even ultra-large-area PDs with diameters up to 10 mm—are commercially available.
- *Reliability.* The number of hours of operation before failure is the measure of a device's reliability. (See Section 8.3.) The typical order of value for a *p-i-n* PD is $1 \times 10^8$ hr, which is more than for a laser diode ($10^6$ hr) but less than for an LED ($10^9$ hr).
- *Shunt resistance.* (See Figure 11.4.) This characteristic is specified as "minimum" and "typical" values. These numbers are about 2 MΩ and 20 MΩ, respectively. This resistance can be made even much higher.

Bear in mind that there are variants representing the same parameters in different data sheets. For example, a manufacturer can specify responsivity versus wavelength not by the graph, as in Figure 11.9, but by the minimum and maximum operating wavelengths. In this case this parameter is given as "Spectral Response" (e.g., min: 1100 nm and max: 1600 nm).

## Data Sheet of an Avalanche Photodiode

The data sheet of an avalanche photodiode differs from that of a *p-i-n* PD in only two characteristics: gain (the multiplication factor), $M$, and breakdown voltage ($V_{BR}$). Parameter $M$ is discussed in detail in Section 11.1. We'll concentrate here on $V_{BR}$.

The graph "Current vs. Voltage" is given in Figure 11.11. If you recognize this graph, that's not surprising. It is a typical *I-V* graph of an electronic diode. The only difference is that a photodiode operates at reverse bias (that is, at negative voltage), while the parameter for this graph is light-input power.

There is a breakdown voltage at which a depletion region breaks through and a diode starts to conduct like a wire. Thus, increasing the voltage beyond $V_{BR}$ causes no increase in current. It is important to know the breakdown-voltage value for two reasons. First, this is the absolute maximum rating. Exceeding the breakdown voltage will result in permanent damage to the diode. For the *p-i-n* PD, breakdown voltage $V_{BR} = 20$ V, as you can see from Figure 11.9. For an APD, the breakdown voltage ranges from 70 V to 200 V, depending on the material. In the case of InGaAs, a $V_{BR}$ of 70 V is typical.

Secondly, an APD operates at a high reverse voltage, whose value is measured in fractions of breakdown voltage. A typical bias-voltage value here is 0.8 to 0.9 of the $V_{BR}$.

This graph delivers another useful bit of information: The slope of the curve at $V = 0$ is the junction (shunt) resistance. That is, $dV/dI = R_j$.

An APD data sheet gives us all the characteristics in a specific gain value. For example, the graph "Responsivity vs. Wavelength" is given at a gain equal to a specific number (such as $M = 300$ [3]). Gain, in turn, is a function of reverse voltage and a graph depicting this can also be found in an APD data sheet. Noise (which is considered in Section 11.3) is a function of gain and this characteristic too is included in the data sheet of an avalanche photodiode.

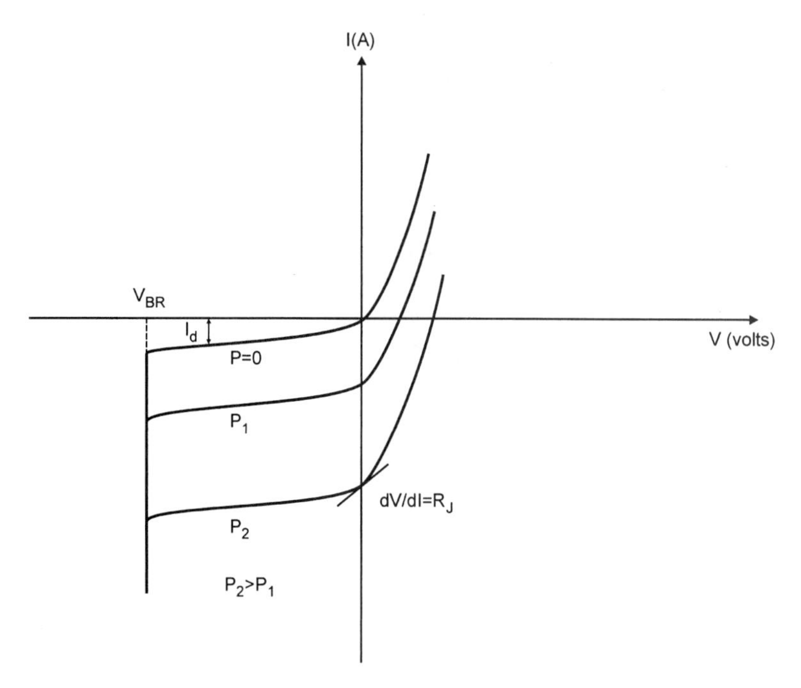


**Figure 11.11** Current versus voltage in a photodiode.

One of the popular applications of an APD is a photodiode for an optical time-domain reflectometer (OTDR). This is discussed in Section 8.5.

## Silicon Photodiodes

We have learned that silicon photodiodes have spectral responses ranging from 350 nm to 1100 nm (Section 11.1). This means that they can be used only in the first transparent window of an optical fiber, that is, around 850 nm. The characteristics of communications silicon photodiodes are optimized at this wavelength; in other words, they exhibit maximum sensitivity and responsivity at 850 nm.

Let's consider typical Si PD characteristics:

- Responsivity: 0.35 to 0.55 A/W
- Bandwidth: 100 MHz to 1 GHz (or rise time: 3.5 ns to 0.35 ns)
- Capacitance: 1 to 4 pF
- Dark current: 0.02 to 5 nA

In addition, it should be pointed out that Si photodiodes operate at a reverse voltage from 9 V to 20 V and a breakdown voltage that starts at 30 V [2].

One manufacturer indicates saturated light-power density as the limit of linearity. The typical value is 7.5 $mW/cm^2$ [7].

## Conclusion

It would be useful for you to compare the specific data discussed in this section with the typical photodiode characteristics given in Table 11.1.

Further information on photodiodes used in fiber-optic communications systems can be found in the technical documentation available from the diode manufacturers.

## 11.3 MORE ABOUT PHOTODETECTORS

What does a designer of a fiber-optic communications system want from a receiver in general and from a photodiode in particular? Sensitivity, sensitivity, sensitivity! This parameter determines the minimum optical power a receiver can detect while distinguishing between logic 1 and logic 0 with a given probability. The more sensitive a receiver, the longer the fiber-optic span a designer can allow; hence, the better—both technologically and economically—this communications system will be.

A receiver's sensitivity is determined by the parameters of a photodiode and the electronics discussed in this and the following sections. But even ideal electronics can amplify and shape only those electrical signals that a photodiode provides. Thus, a PD plays a key role in the operation of a receiver. At first glance, a photodiode's sensitivity is determined by dark current. However, a deeper look reveals that the noise generated by a photodiode puts a real limit on the sensitivity of a fiber-optic communications receiver. Thus, the real criterion of a PD's performance is its signal-to-noise ratio (SNR). And noise is the topic of this section.

### Noise Sources in a Photodiode

Suppose the light-power input is ideally constant. Does it mean the photocurrent will also be constant, as the basic photodetector formula, $I_p = R\,P$, states? No, because in reality photocurrent, as an output of the photodiode, contains noise components. There are several general sources of noise in a photodiode.

***Shot noise*** Again, suppose input power is ideally constant, which means the number of photons per unit of time, on average, is constant. But, in fact, the actual number of photon arrivals at a particular time is unknown and so is a completely random variable. Hence, the number of photogenerated electrons at any particular instance is a random variable. In addition, the number of electrons producing photocurrent will vary because of their random recombinations and absorptions. Therefore, even though the average number of electrons is constant, the actual number of electrons will vary. *Deviation of the actual number of electrons from the average number is known as shot noise.* Since electric current is a stream of electrons, we can repeat the same statement in terms of current.

Shot noise was discovered in 1918 and has been studied very thoroughly ever since. Poisson statistics describe its behavior as follows: If we denote the average number of generated electrons as $n^*$, then the probability of generating $n$ electrons per time interval $\Delta t$ is given by

$$P(n) = \exp(-n^*)(n^*)^n/n!, \tag{11.20}$$

which is the Poisson distribution.

The well-known features of Poisson statistics are that the *average values of* n *and* $n^2$ are given by:

$$\langle n \rangle \equiv n^* = \sum_{n=0}^{\infty} nP(n) \tag{11.21}$$

$$\langle n^2 \rangle = \sum_{n=0}^{\infty} n^2 P(n) = (n^*)^2 + n^*, \tag{11.22}$$

where the angular brackets indicate averaging.

It's easy to show that

$$\langle (n - n^*)^2 \rangle = n^* \tag{11.23}$$

If we recall that current is the number of electrons ($n$) per time interval ($\Delta t$), we can directly convert the above formulas in terms of current. For instance, the instantaneous current, $I(t)$, is equal to $ne/\Delta t$, where $e$ is the electron charge ($1.6 \times 10^{-19}$ C) and the average current ($I^*$) is $n^*e/\Delta t$. Shot-noise current, then, is determined in accordance with the definition as:

$$\langle (I(t) - I^*)^2 \rangle = \langle [(n - n^*)e/\Delta t]^2 \rangle = (n^*e^2)/(\Delta t)^2 = (I^*e)/(\Delta t) \tag{11.24}$$

The spectral density, $S_s(f)$, and *RMS* (root mean square) value of the shot-noise current, $i_s(t)$, are given by ([1], [8]):

$$S_s(f) = 2eI_p^* \tag{11.25}$$

$$i_s = \sqrt{[2e(I_p^*)\mathrm{BW_{PD}}]}, \tag{11.26}$$

where the average photocurrent is denoted as $I_p$* and $\mathrm{BW_{PD}}$ is a PD's bandwidth. (Dark current also contributes to the number of electrons; for this reason, it could have been included in Formulas 11.25 and 11.26 but we are treating this source of noise independently. Coefficient 2 in Formulas 11.25 and 11.26 appears because we are here considering one-sided spectral density.) Since spectral density is constant, shot noise is *white noise,* whose spectral density does not depend on frequency. (Reference [9] develops an even more comprehensive model of receiver shot noise, treating shot noise as an unstable, random process.)

Note that Formula 11.26 follows from Formula 11.25 because

$$\langle i_s^2(t) \rangle = \int_0^\infty S_s(f)df \tag{11.27}$$

It is quite evident that $i_s$ is the square root of the shot-noise variance, that is, the RMS value of shot-noise current, as defined above. *The RMS value of shot-noise current,* $i_s$ *(A), is the representative characteristic of a shot-noise phenomenon.*

In practice, noise is also represented by its current RMS value per unit of bandwidth—the *bandwidth-normalized RMS of noise current,* $i_N$. For shot-noise current, this is given by [2]:

$$i_{sN}(A/\sqrt{Hz}) = i_s/\sqrt{\mathrm{BW_{PD}}} = \sqrt{[2e(\mathrm{I}_p^*)]} \tag{11.28}$$

Be aware of the units of this measurement: amperes per square root of hertz.

***Thermal noise*** Electron motion due to temperature (that is, external thermal energy) occurs in a random way. Thus, the number of electrons flowing through a given circuit at any instance is a random variable. *The deviations of an instantaneous number of electrons from their average value because of temperature change is called thermal noise.* (Incidentally, we can repeat this same statement in terms of current, as we did with respect to shot noise.)

It follows from this definition that the only means to eliminate thermal noise is to have the temperature at absolute zero. Indeed, thermal noise deceases with a drop in temperature and vice versa.

Thermal noise is often called *Johnson noise* after the scientist who first investigated it experimentally or, less often, *Nyquist noise,* after the scientist who developed its theory. It is modeled after the random stationary Gaussian process with its one-sided spectral density, $S_t(f)$, which is equal to [1]:

$$S_t(f) = 2k_B T/R_L, \tag{11.29}$$

where $k_B$, the Boltzmann constant, equals $1.38 \times 10^{-23}$ J/K; $T$ (K) is the absolute temperature, and $R_L$ is the load resistance. (See Figure 11.4.) Strictly speaking, Formula 11.29 should include $R_j \| R_L$, as Figure 11.4 shows, but the junction resistance is much larger (it is on the order of MΩ) than the load (which is on the order of kΩ).

On the other hand, if you investigate just the thermal noise of a stand-alone photodiode, you should insert the junction resistance into Formula 11.29. This is how you can discern a lower limit of the thermal noise value of your photodiode.

Again, since spectral density is constant, thermal noise is an example of white noise but this is true only up to a frequency of approximately 1 THz.

Applying Formula 11.27, we can obtain the RMS value of thermal-noise current as follows:

$$i_t = \sqrt{[(4k_BT/R_L)BW_{PD}]} \tag{11.30}$$

Refer again to Figure 11.4, where an equivalent circuit of a PD is shown. If you want to take into account a series resistor ($R_s$) in spite of its small value (< 10 Ω), you have to calculate *not* current but thermal noise voltage by the following formula.

$$v_t = \sqrt{[(4k_BTR_s)BW_{PD}]} \tag{11.30a}$$

The bandwidth-normalized RMS value of thermal-noise current is, consequently, given by:

$$i_{tN}(A/\sqrt{Hz}) = i_t/\sqrt{(BW_{PD})} = \sqrt{[(4k_BT/R_L)} \tag{11.31}$$

***Dark-current noise*** We could include dark current in our shot-noise picture, but in case you need to treat it independently, you can repeat the steps given above. In other words, the *RMS* value of dark-current noise is given by

$$i_d = \sqrt{(2eI_d^*BW_{PD})} \tag{11.32}$$

and the RMS value per unit of bandwidth is equal to

$$i_{dN}(A/\sqrt{Hz}) = i_d/\sqrt{BW_{PD}} = \sqrt{(2eI_d^*)} \tag{11.33}$$

It was shown [10] that dark-current noise is essential at the low bit rate (around 100 Mbit/s), while at high speed (more than 1 Gbit/s) even 100 nA dark current does not produce an essential contribution to total noise.

***1/f noise*** A photodiode generates noise in complete darkness other than dark-current noise. The RMS value of this noise per unit of bandwidth is inversely proportional to frequency, which means this is not white noise. This normalized RMS value of $1/f$ noise current can be approximated by the following expression [8]:

$$i_{1/fN}(A/\sqrt{Hz}) = i_{1/f}(f)/\sqrt{BW_{PD}} = (K_{1/f}I^{\alpha})/f^{\beta} \tag{11.34}$$

Constants $K_{1/f}$, $\alpha$, and $\beta$ have to be found empirically. The approximate value for $\alpha$ is 2 and for $\beta$ between 1 and 1.5. The nature of this noise is not well understood. Fortunately, this noise is important only in the low-frequency range and we can neglect its influence at the output signal of a photodiode when the modulating frequency is above 100 Hz.

***Equivalent circuit of a photodiode*** It is a good idea to gather all the information about the sources of noise in photodiodes in the form of a generalized equivalent circuit of a photodiode.

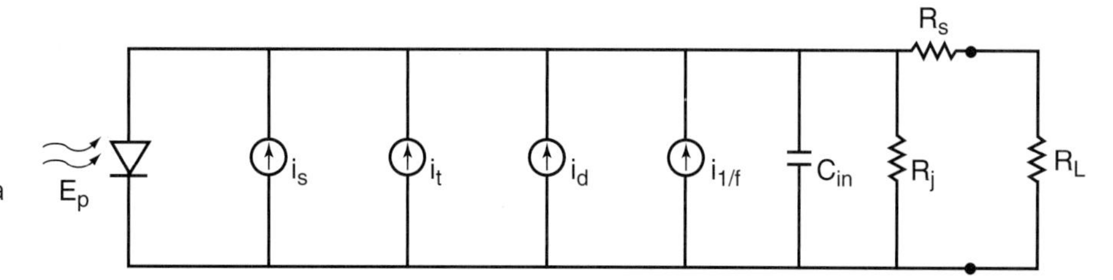


**Figure 11.12** Noise in a photodiode—generalized equivalent circuit.

We start with the circuit given in Figure 11.4. All noise sources are essentially the current sources and they work independently. Hence, they are connected in parallel, as Figure 11.12 shows.

Thus, we use an additive model of the noise sources; that is, we add each of the sources independently. The result is an RMS value of the total noise current in a photodiode:

$$i_{\text{noise}} = \sqrt{(i_s^2 + i_t^2 + i_d^2 + i_{1/f}^2)} \tag{11.35}$$

Here, individual noise sources are defined by Formulas 11.26, 11.30, 11.32, and 11.34. The assumptions made for Formula 11.35 are the following: Each noise component is an independent random process approximated by Gaussian statistics. If you see some discrepancy between the additive model used in the equivalent circuit in Figure 11.12 and Formula 11.35, resort to variances instead of the RMS of the noise current. In this case you can rewrite Formula 11.35 as

$$i_{\text{noise}}^2 = i_s^2 + i_t^2 + i_d^2 + i_{1/f}^2 \tag{11.35a}$$

### Example 11.3.1

***Problem:***

Calculate the RMS and bandwidth-normalized values of total noise current for an MF-432 PIN photodiode if the average input power is 0.1 μW, λ = 1550 nm, and the diode operates at room temperature.

***Solution:***

We need to calculate all the members of Formula 11.35 in sequence and then use Formulas 11.28, 11.31, 11.33, and 11.34.

(1) Shot-noise current

To apply Formula 11.26, $i_s = \sqrt{[2e(I_p^*)\text{BW}_{\text{PD}}]}$, we need to find two quantities: $I_p^*$ and $BW_{\text{PD}}$.

The average photocurrent is equal to (see Formula 11.1):

$$I_p^* = R\, P_{in}^* = 0.1\ \mu A$$

because the responsivity of MF-432 is 1.0 A/W at 1550 nm, as Figure 11.9 shows.

The bandwidth of MF-432 is specified as 2.5 GHz. Recalling that $e = 1.6 \times 10^{-19}$ C, we compute

$$i_s^2 = [2e(I_p^*)\text{BW}_{\text{PD}}] = 80 \times 10^{-18}\ \text{A}^2$$

and

$$i_s = 8.9\ \text{nA}$$

Using Formula 11.28, we obtain:

$$i_{sN} = i_s/\sqrt{(BW_{PD})} = \sqrt{[2e(I_p^*)]} = 1.78 \times 10^{-4}\ nA/\sqrt{Hz}$$

(2) Thermal-noise current

With a given temperature and bandwidth, this noise depends on the value of the load resistance ($R_L$). To illustrate the effect of this resistance, we will do calculations for two values of $R_L$: 50 Ω and 50 kΩ.

Applying Formula 11.30 for $T = 300°K$ we obtain:

$$(R_L = 50\ \Omega) \qquad i_t^2 = [(4k_BT)/R_L]BW_{PD} = 82.8 \times 10^{-14}\ A^2$$

$$(R_L = 50\ k\Omega) \qquad i_t^2 = 828 \times 10^{-18}\ A^2$$

and

$$(R_L = 50\ \Omega) \qquad i_t = 910\ nA$$

$$(R_L = 50\ k\Omega) \qquad i_t = 28.8\ nA$$

If the load resistance is 50 Ω, we can surely neglect all other sources of noise. But even with $R_L = 50\ k\Omega$, the thermal-noise current is much larger than the shot-noise current. However, if we increase the load resistance (or consider only the junction resistance of the photodiode itself), the number of $i_t(f)$ will decrease dramatically. For example, for $R_L = 2\ M\Omega$ (the minimum value of the junction resistance), $i_t$ becomes 4.5 nA. It is necessary for you to memorize this very important fact: *Increasing the load resistance will reduce thermal noise,* as Formula 11.30 tells us.

The bandwidth-normalized RMS value of thermal-noise current is given by Formula 11.31; hence,

$$i_{tN} = i_t/\sqrt{BW_{PD}} = \sqrt{(4k_BT/R_L)} = 5.76 \times 10^{-4}\ nA/\sqrt{Hz}$$

(3) Dark-noise current

To use Formula 11.32, let us refer to the specifications of an MF-432 photodiode: At room temperature the average dark-noise current is 3 nA. Thus, we compute:

$$i_d^2 = 2e\, I_d^* BW_{PD} = 2.4 \times 10^{-18}\ A^2$$

and

$$i_d = 1.5\ nA$$

From Formula 11.33 we obtain:

$$i_{dN} = i_d/\sqrt{(BW_{PD})} = \sqrt{(2eI_d^*)} = 0.31 \times 10^{-4}\ nA/\sqrt{Hz}$$

(4) 1/*f*-noise current

Since we use a photodiode in high-speed applications, this current will be negligible. There is thus no need to calculate this noise current.

(5) Total noise current

($R_L = 50\ \Omega$) The RMS value of the total noise current in this case is equal to that of the thermal-noise current, that is, 910 nA.

($R_L = 50$ kΩ) The RMS value of total noise current is computed by applying Formula 11.35:

$$i_{noise} = \sqrt{(i_s^2 + i_t^2 + i_d^2)} = \sqrt{[(80 + 828 + 2.4) \times 10^{-18} A^2]} = 30.2 \text{ nA}$$

and the variance of the noise current is given by Formula 11.35a:

$$i_{noise}^2 = 910.4 \times 10^{-18} A^2$$

You can see one more time how thermal-noise current is still the predominant factor in our case.

The RMS value of bandwidth-normalized total noise current is given by

$$i_{noiseN} = \sqrt{(i_{sN}^2 + i_{tN}^2 + i_{dN}^2)} = 6.04 \times 10^{-4} \text{ nA}/\sqrt{\text{Hz}}$$

This example gives you an idea of the current values induced by different noise sources.

---

One important note: All noise variances, except for the practically negligible $1/f$ noise, are proportional to the bandwidth of a photodiode ($BW_{PD}$). Therefore, we can—theoretically, at least—decrease the noise level by decreasing a photodiode's bandwidth. On the other hand, we need high-speed photodiodes, which means we need photodiodes with large bandwidths. What we have here, then, is a trade-off between the bandwidth and the noise of a photodiode. But bandwidth and noise both depend on load resistance, albeit in opposite ways: Noise increases (becomes worse) while bandwidth increases (becomes better) with a decrease in the $R_L$ value. (Compare Formulas 11.13 and 11.30.) A possible solution to this dilemma is to use a bandpass filter to restrict bandwidth within the desired range of frequencies. So there you have it: Along with load resistance, bandwidth is yet another issue of major concern to the designer of a *p-i-n* photodiode.

## Signal-to-Noise Ratio and Noise-Equivalent Power

***Signal-to-noise ratio (SNR) for a* p-i-n *photodiode*** The *signal-to-noise ratio (SNR)* is one of the most important criteria of a photodiode's performance. It can be found from the following reasoning. By definition, *SNR is the ratio of signal power to noise power.* In our case, signal power is proportional to the square of the average photocurrent ($I_p^*$), while noise power is proportional to the square of the *RMS* value of the noise-induced current ($i_{noise}$). Since both signal power and noise power are released at the same load resistance, we can write [1]:

$$\text{SNR} = \text{signal power}/\text{noise power} = I_p^{*2}/i_{noise}^2 \quad (11.36)$$

Since $I_p^* = R\,P^*$, Formula 11.36 takes the form

$$\text{SNR} = (R^2 P^{*2})/i_{noise}^2 \quad (11.37)$$

### Example 11.3.2

***Problem:***

What is the SNR of an MF-432 PIN photodiode if the average input power is 0.1 μW, λ = 1550 nm and the diode operates at room temperature?

***Solution:***

We will use the data from Figure 11.9 ($R$ = 1 A/W) and the numbers calculated in Example 11.3.1. Thus, we get

$$(R_L = 50\ \Omega) \qquad \text{SNR} = (R^2 P^{*2})/i_{noise}^{\ 2} = 0.01$$

$$(R_L = 50\ \text{k}\Omega) \qquad \text{SNR} = 10.98$$

The first result shows that where $R_L = 50\ \Omega$, we cannot unscramble the signal. Indeed, the average value of the photocurrent is 100 nA, while the RMS value of the noise current in this case is 910 nA. The signal is simply buried in noise.

The second result is much more favorable because the power of the signal is almost eleven times greater than the power of the noise. It was calculated [11] that where $SNR = 6$ bit-error rate, BER is not more than $10^{-9}$ (which means that one erroneous bit is received after the transmission of one billion bits). This is true with the assumption of Gaussian statistics of noise, which is an acceptable approximation of the Poisson statistics. Thus, for $R_L = 50\ \text{k}\Omega$, we have a signal-to-noise ratio that ensures that the photodiode will meet the common requirement that BER = $10^{-9}$.

---

Examples 11.3.1 and 11.3.2 give the numbers for a *p-i-n* photodiode. As you observed, the major contributor to noise current, $i_{noise}$, is thermal noise, $i_t$. In fact, *thermal noise is the dominant noise component in a* p-i-n *photodiode.*

The first measure we can take to reduce thermal noise and its influence on SNR is to increase the load resistance. Certainly, if we neglect all other terms in Formula 11.35, SNR takes the following form (this is the *thermal-noise limit*):

$$\text{SNR}_t = (R^2 P^2)/i_t^2 = (R^2 P^2 R_L)/[(4k_B T)\text{BW}_{PD}], \tag{11.38}$$

where Formula 11.30 was used. Thus, signal-to-noise ratio is simply proportional to load resistance. If we plug in the numbers from our example, we'll find that $\text{SNR}_t = 12$ for $R_L = 50\ \text{k}\Omega$; but if we can arrange $R_L = 500\ \text{k}\Omega$, we find that $\text{SNR}_t = 120$. Make note of this fact: *For a* p-i-n *photodiode, we need to increase the load resistance to improve the signal-to-noise ratio.* We'll soon see, however, that this measure conflicts with other design requirements of a fiber-optic receiver.

***Signal-to-noise ratio in an avalanche photodiode (APD)*** For an avalanche photodiode, the situation is as follows: The impact-ionization process amplifies the photocurrent, which means an increasing nominator in the SNR formula. On the other hand, the same process raises the shot noise. Formally, this follows from Formula 11.26, which shows that shot noise depends on photocurrent. The physics behind this enhancement is that secondary electron-hole pairs are generated at random times, which intensifies shot noise. Therefore, the multiplication factor ($M$) becomes random. To manipulate the ionization process with $M$ as an average multiplication factor, the new coefficient—excess noise factor $(F)_s$—is introduced. We can skip all the formal derivations and rely on the following formula for shot noise in an APD [1]:

$$i_s^2\ (\text{APD}) = M^2[2eF_s(RP)\text{BW}_{PD}], \tag{11.39}$$

where $F_s$ is determined by the following formula:

$$F_s = k_A M + (1 - k_A)(2 - 1/M) \tag{11.40}$$

Coefficient $k_A$ is the ratio of electrons to holes. (See the discussion of Formula 11.17.) For $M = 20$ and $k_A = 0.03$, as is the case for an Si APD, we get $F_s = 2.49$; for $M = 20$ and $k_A = 0.6$, as is the case for an InGaAs APD, we compute $F_s = 12.78$.

As a matter of fact, these same calculations should be repeated for dark-current noise but there we saw that this component was very small.

Thermal noise, on the other hand, does not depend on current and, therefore, is not changed by the ionization process. This is because thermal noise is created on resistors. As a result, then, the formula for the *RMS* value of an APD's thermal-noise current is still Formula 11.30.

Thus, the signal-to-noise ratio for an APD can be written as:

$$\begin{aligned} \mathrm{SNR(APD)} &= I_P^{*2}/I_{noise}^2 = (MRP)^2/(i_s^2 + i_t^2) \\ &= (M^2R^2P^2)/\{[2eM^2F_sRP + (4k_BT)/R_L]\mathrm{BW_{PD}}\} \end{aligned} \quad (11.41)$$

Let's now consider two limits: (1) Shot noise is much greater than thermal noise and (2) thermal noise is much greater than shot noise. In the first case, $i_s^2 \gg i_t^2$, we derive from Formula 11.41

$$\mathrm{SNR(APD)_s} = RP/(2eF_s\ \mathrm{BW_{PD}}) \quad (11.41a)$$

In the second case, $i_s^2 \ll i_t^2$, we obtain from Formula 11.41

$$\mathrm{SNR(APD)_t} = (M^2R^2P^2)/[(4k_BT/R_L)\mathrm{BW_{PD}}] \quad (11.41b)$$

As you can see, the advantage of using an APD photodetector in case of thermal-noise limit is clear: Its signal-to-noise ratio is raised by the square of the multiplication factor.

## Example 11.3.3

***Problem:***

Calculate $\mathrm{SNR_s}$, $\mathrm{SNR_t}$, and overall SNR for Si and InGaAs APDs if $M = 20$, $P = 0.1$ μW, $R = 0.9$ A/W, $R_L = 50$ kΩ, $\mathrm{BW_{PD}} = 2.5$ GHz, and $T = 300$°K.

***Solution:***

Using Formula 11.41a, we compute $\mathrm{SNR_s} = 45.2$ for an Si APD and $\mathrm{SNR_s} = 8.8$ for an InGaAs APD.

Using Formula 11.41b, we compute $\mathrm{SNR_t} = 3912.3$, which does not depend on the semiconductor material. You can see how much thermal-noise impact on the photocurrent is reduced in APDs.

Finally, using Formula 11.41, we compute the SNR of the APD as follows:

For an Si APD ($F_s = 2.49$), SNR = 44.67

For an InGaAs APD ($F_s = 12.78$), SNR = 8.78

It therefore follows from these computations that the SNR of an APD is determined primarily by its shot-noise signal-to-noise ratio. This is because Formula 11.41 treats both the shot-noise and the thermal-noise phenomena equally. Fortunately, thermal-noise influence is greater than Formula 11.41 recounts; hence, in reality, the SNR of an APD is somewhere between $\mathrm{SNR_s}$, and $\mathrm{SNR_t}$, which is larger than our calculations show.

If you look at Formula 11.41 again, you'll see that the SNR(APD) depends on APD gain ($M$). Since $M$ appears in both the numerator and the denominator of Formula 11.41, this dependence cannot be seen at first glance. A deeper analysis reveals there is an *optimum APD gain,* which can be approximated by the following expression [1]:

$$M_{opt} \approx [(4k_B T)/(k_A R_L eRP^*)]^{1/3} \tag{11.42}$$

If we plug in the numbers given in Example 11.3.3 with $k_A = 0.03$ for Si and $k_A = 0.6$ for InGaAs photodiodes, we compute $M_{opt} = 8.56$ for an Si APD and $M_{opt} = 3.15$ for an InGaAs APD. These are not very impressive numbers but we still have some amplification of the input signal. Again, the performance of an APD is better when $k_A$ is smaller.

***Noise-equivalent power (NEP)*** Another important characteristic of a photodiode's performance is *noise-equivalent power (NEP).* By definition, *NEP is the minimum signal power that produces SNR = 1.* In other words, this is the input power that produces the same output as the noise does.

The explicit formula for NEP can be obtained from Formula 11.37 when SNR = 1. That is:

$$\text{NEP(W)} = i_{noise}/R \tag{11.43}$$

Don't forget that $i_{noise}$ is the RMS value and, therefore, $i^2_{noise}$ is the variance of the total current induced by noise. This means that *NEP is the RMS of the optical power that produces SNR = 1.*

To calculate NEP, we use Formula 11.43. Thus, for an MF-432 photodiode, we compute

$$\text{NEP} = i_{noise}/R = 39.2 \text{ nW}$$

The industry likes to use the term *NEP per unit of bandwidth,* that is, bandwidth-normalized noise-equivalent power ($\text{NEP}_{norm}$) as a more representative characteristic of the phenomenon. It is defined as

$$\text{NEP}_{norm}(\text{W}/\sqrt{\text{Hz}}) = \text{NEP}/\sqrt{\text{BW}_{PD}} = I_{noise}/R \tag{11.44}$$

If we employ the numbers from Example 11.3.2, we'll find $\text{NEP}_{norm} = 0.756 \text{ pW}/\sqrt{\text{Hz}}$.

If you look at Formula 11.43 more closely, you'll see that NEP, in essence, is the minimum optical power that a noisy photodiode can detect. Without question, *NEP determines the weakest optical signal that can be detected in the presence of noise.*

In reality, NEP increases with a rise in frequency even though Formula 11.43 does not show that explicitly. (Incidentally, plotting NEP versus frequency is referred to as the *noise floor,* since it determines the bottom line for signal detectivity.) Therefore, NEP values are different for different ranges of bandwidth. When calculating NEP over the entire bandwidth of a photodiode, remember that we operate with RMS values of random variables. Hence, the rule of summation of independent variables should be observed:

$$\text{NEP} = \sqrt{\{[\text{NEP}_{BW1}\sqrt{(\text{BW1})}]^2 + [\text{NEP}_{BW2}\sqrt{(\text{BW2})}]^2 + \quad + [\text{NEP}_{BWn}\sqrt{(\text{BWn})}]^2\}}, \tag{11.45}$$

where $\text{NEP}_{BWk}$ are the values of $\text{NEP}_{norm}$ for the pertinent $k$ range of bandwidth.

Since the responsivity of a photodiode is involved in the definition of noise-equivalent power, NEP depends on the operating wavelength, so that [12]

$$\text{NEP}(\lambda) = \text{NEP}_{norm}[R_{max}/R(\lambda)]\sqrt{\text{BW}}, \tag{11.46}$$

where $R_{max}$ is the maximum value of responsivity and $R(\lambda)$ is the responsivity at the given wavelength.

### Example 11.3.4

***Problem:***

Calculate NEP and NEP(λ) for an InGaAs PIN PD with the following parameters [12]: $NEP_{norm} = 3.3\ pW/\sqrt{Hz}$ from dc to 10 MHz and $NEP_{norm} = 30\ pW/\sqrt{Hz}$ from 10 to 125 MHz; $R_{max}$ = 1.1 A/W at 1550 nm and $R(\lambda)$ = 0.9 A/W at 1300 nm. The photodiode operates at 1300 nm.

***Solution:***

To calculate NEP, we use Formula 11.45:

$$\begin{aligned} NEP &= \sqrt{\{[NEP_{BW1}\sqrt{(BW1)}]^2 + [NEP_{BW2}\sqrt{(BW2)}]^2\}} \\ &= \sqrt{\{[3.3(pW/\sqrt{Hz})\sqrt{(10\ MHz)}]^2 + [30(pW/\sqrt{Hz})\sqrt{(115\ MHz)}]^2\}} = 321.9\ nW \end{aligned}$$

To calculate *NEP*(λ), we use Formula 11.46, taking the upper value of noise-equivalent power, 30 pW/√Hz, and the entire bandwidth, 125 MHz:

$$NEP(\lambda) = NEP_{norm}[R_{max}/R(\lambda)]\sqrt{BW} = 409.9\ nW$$

Think about this number: We need a minimum of 400 nW, which is 0.4 μW, to provide SNR = 1. Since in reality we must have the *SNR* at least equal to 6, it's clear that we need to do something to improve the photodiode's performance. An analysis of Formula 11.45 reveals that the only means we can control is bandwidth. Obviously, since *noise in photodiodes is the white-noise type, it contributes at all frequencies of the entire bandwidth.* If we insert a bandpass filter, we will eliminate a substantial portion of the noise (along with the signal, of course). In our example, using a 10-kHz output filter, we obtain

$$NEP(\lambda) = NEP_{norm}[R_{max}/R(\lambda)]\sqrt{BW} = 3.67\ nW$$

Compare 3.67 nW with 409.9 nW minimum power and be sure to understand that *there is always a trade-off between bandwidth and NEP—the larger the bandwidth, the larger (the worse, that is) the NEP.* Recall our discussion of the effect of bandwidth on signal-to-noise ratio and compare that result with the conclusion given here.

---

Another parameter, *detectivity (D),* is sometimes used to evaluate a photodiode's performance. Unfortunately, there is no unique definition for this parameter, but what we can say with certainty is that detectivity is inversely proportional to NEP. You can define "detectivity" simply as 1/NEP [1] or you can use the following definition [13]:

$$D = \sqrt{[D_A(BW)]}/NEP, \quad (11.46a)$$

where $D_A$ is the photosensitive area of a PD and BW is its bandwidth. You may also come across other definitions of this term. Fortunately, detectivity is not a crucial parameter of a photodiode, so the industry's failure to establish a precise definition for this term will not affect your day-to-day work with photodiodes.

## Sensitivity and Quantum Limit

*Sensitivity,* the main criterion of photodiode performance, *is the minimum optical power that a photodiode can detect at a given bit-error rate.* So far, we've discussed dark-current sensitivity and noise-restricted sensitivity, NEP. Now we'll turn our attention to consideration of sensitivity induced by transmission technology, concentrating on digital signals.

***Bit-error rate (BER)*** We first need to introduce the formal definition of *bit-error rate (BER).* It is *the ratio of the number of erroneous bits to the total number of bits transmitted,* that is,

$$\text{BER} = \text{number of erroneous bits/number of total bits} \tag{11.47}$$

The error in this case is taking bit 1 for bit 0 and vice versa. Let's denote the probability of deciding 1 when bit 0 has actually been received as $P(1/0)$ and the probability of deciding 0 when bit 1 has actually arrived as $P(0/1)$. Then these conditional probabilities determine BER as follows [1].

$$\text{BER} = 0.5[P(1/0) + P(0/1)] \tag{11.48}$$

In this formula, we assume there is an equal probability that bit 1 or bit 0 will arrive. This is where coefficient 0.5 in Formula 11.48 comes from.

In digital transmission, the decision circuit determines which bit has arrived by comparing the level of photocurrent ($I_p$) with the threshold value of current ($I_{th}$). If $I_p > I_{th}$, the decision is in favor of bit 1; if $I_p < I_{th}$, the decision is in favor of bit 0. Nonetheless, the photocurrent fluctuates randomly from arriving bit to arriving bit, and the noise changes the value of the output current too. (See Figure 11.13.) Thus, the probability exists for making the wrong decision.

To include noise components, recall that all of them are well approximated with Gaussian statistics and, consequently, their variances can simply be added to represent the variance of noise, as we showed in Formula 11.35a. Since we are considering high-bandwidth applications, the $1/f$ noise can be neglected. Thus, the noise associated with bit 1 is represented as

$$i_1^2 = i_{s1}^2 + i_{t1}^2 + i_{d1}^2, \tag{11.49}$$

where $i_1^2$ is the variance of current representing bit 1, and subscription 1 in each term shows that $i_{s1}$, $i_{t1}$, and $i_{d1}$ are taken at value $I_1$ of the output current. The noise associated with bit 0 is different from the bit-1 noise because the *average value of bit-0 current* ($I_0$) *is different from the average value of bit-1 current* ($I_1$). Theoretically, $I_0$ should be 0 but, in reality, it is not. Thus,

$$i_0^2 = i_{s0}^2 + i_{t0}^2 + i_{d0}^2 \tag{11.49a}$$

Thermal noise, you'll recall, does not depend on current; therefore, $i_{t1}^2 = i_{t0}^2$. (For the explicit expressions of all members in the above formulas, see Formulas 11.26, 11.30, and 11.32.)

If we neglect dark-current noise and assume $i_{s0} = 0$, we can derive the simplified expressions for the variances of the photocurrent:

$$i_1 = \sqrt{(i_{s1}^2 + i_{t1}^2)} \tag{11.49b}$$

and

$$i_0 = i_{t0} \tag{11.49c}$$

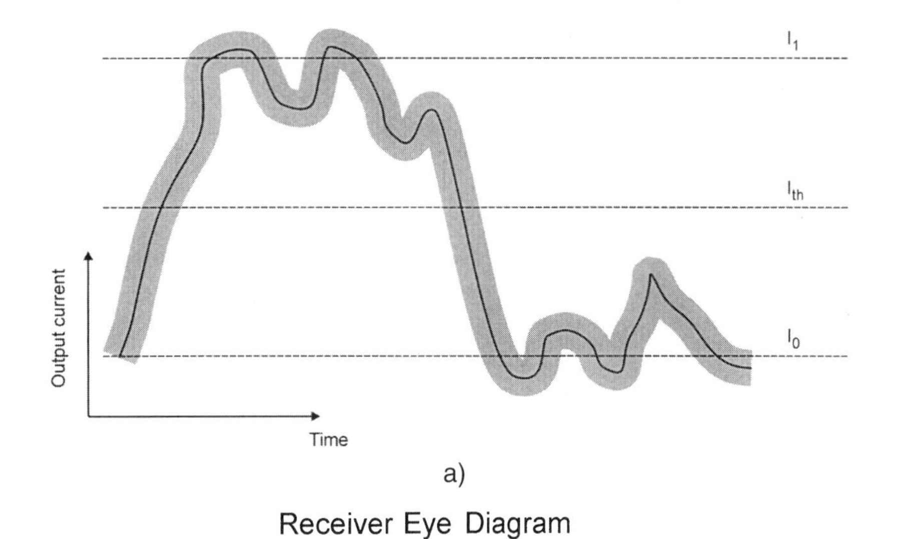


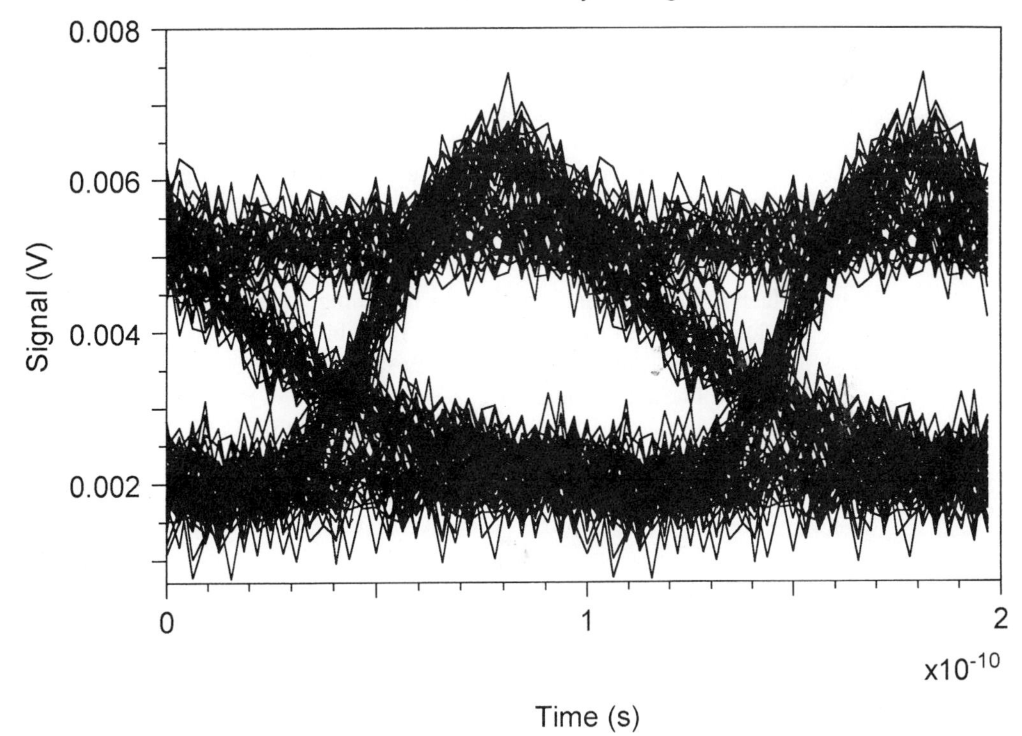


**Figure 11.13** A digital signal with noise: (a) The waveform of a digital signal with noise; (b) computer simulation of an eye diagram at the receiver end. *(Courtesy of RSoft Inc., Ossining, N.Y.)*

The conditional probabilities $P(1/0)$ and $P(0/1)$ are determined through *complementary error function (erfc)* as

$$P(1/0) = 0.5\ \text{erfc}[(I_1 - I_{\text{th}})/(i_1\sqrt{2})] \tag{11.50}$$

and

$$P(0/1) = 0.5\ \text{erfc}[(I_{\text{th}} - I_0)/(i_0\sqrt{2})].$$

where

$$\text{erfc}(x) = 2/\sqrt{\pi \int_{x}^{\infty} \exp(-y^2)dy}, \tag{11.50a}$$

so that BER is given by [1]:

$$\text{BER} = 1/4\{\text{erfc}[(I_1 - I_{th})/(i_1\sqrt{2})] + \text{erfc}[(I_{th} - I_0)/(i_0\sqrt{2})]\} \tag{11.51}$$

Can we decrease BER? The only independent variable in Formula 11.51 is the threshold current, $I_{th}$. Therefore, the only means to minimize BER is varying this current. BER is at minimum when $I_{th}$ satisfies the following condition [1]:

$$(I_1 - I_{th})/i_1 = (I_{th} - I_0)/i_0 \equiv Q, \tag{11.52}$$

where Q is the parameter (and is discussed below). To clarify the meaning of this condition, let's rearrange Formula 11.52 in this way:

$$I_{thopt} = (I_1 i_0 + I_0 i_1)/(i_1 + i_0), \tag{11.53}$$

where $I_{thopt}$ is the optimal value of the threshold current drawn from Equation 11.52. For a *p-i-n* photodiode, thermal noise is the dominant factor, so we can neglect all other components in Formulas 11.49 and 11.49a. Recalling that $i_{t1}^2 = i_{t0}^2$, we obtain

$$I_{thopt}(p\text{-}i\text{-}n) = (I_1 + I_0)/2, \tag{11.54}$$

which is almost evident because the first thing you want to do is set the threshold current in the middle of the average currents for bit 1 and bit 0.

For an avalanche photodiode, the noise situation is not so obvious and so we have to use Formula 11.53 to calculate $I_{thopt}$.

Returning to BER, if the condition in Formula 11.52 is satisfied, the expression for minimized BER takes this form [1]:

$$\text{BER}_{min} = \tfrac{1}{2}[\text{erfc}(Q/\sqrt{2})] \approx [\exp(-Q^2/2)]/(Q\sqrt{2\pi}) \tag{11.55}$$

The meaning of parameter $Q$, introduced in Formula 11.52, can be clarified if we derive the following expression from Formulas 11.52 and 11.53:

$$Q = (I_1 - I_0)/(i_1 + i_0) \tag{11.56}$$

The difference $(I_1 - I_0)$ is the excess of average current available for distinguishing bit 1 from bit 0 and the sum $(i_1 + i_0)$ is the RMS value of current induced by noise at both the 1 and 0 electric levels. If we accept as fact that the average current of bit 0 is zero (that is, $I_0 = 0$), parameter $Q$ takes on an even clearer form:

$$Q = (I_1)/(i_1 + i_0) \tag{11.56a}$$

Thus, $Q$ is simply the ratio of signal current to noise current, which is another way to represent the signal-to-noise relationship. Parameter Q is called *digital SNR* [9]. At this point we can better explain the meaning of Formula 11.55. To do so, substitute Formula 11.56 into Formula 11.55 and obtain

$$\text{BER}_{\min} = 1/2\,[\text{erfc}(Q/\sqrt{2})] = 1/2\{\text{erfc}[(I_1 - I_0)/(i_1 + i_0)], \tag{11.57}$$

where $(I_1 - I_0)$ is the average signal current and $i_1 + i_0$ is the sum of the RMS values of the noise currents of bit 1 and bit 0, respectively.

The dependence of the bit-error rate on the parameter $Q$ is shown in Figure 11.14. This graph is plotted by calculating Formula 11.55. Obviously, the larger $Q$ is, the less will be the bit-error rate. Look closely at the specific numbers: To achieve BER < $10^{-12}$, which is the common requirement in modern fiber-optic communications systems, we need to obtain $Q > 7$. For a modest BER ≈ $10^{-9}$, we need $Q \approx 6$. (*Note:* We previously referred to the need for SNR to be at least equal to 6 to obtain BER ≈ $10^{-9}$.

Figure 11.14 is a graphical device for finding the relationship between BER and the $Q$ parameter. For precise calculations, use Formula 11.55.

***Minimum optical power—photodiode sensitivity*** With all these tools in hand, we now can find *the minimum optical power required to provide the given bit-error rate—the real sensitivity of a photodiode used in a fiber-optic communications system.* But, first, let's review what we've learned so far: BER depends on the $Q$ parameter, as Formula 11.55 shows; the $Q$ parameter depends on average $I_1$ and noise currents, as Formulas 11.56 and 11.56a reveal; average and noise currents, in turn, depend on optical input power, as stated in Formulas 11.1, 11.26, 11.30, 11.32, and 11.49. Following this logical progression leads us to the explicit expression for the sensitivity of a communications photodiode.

At this point we must introduce the concept of *minimum received power* ($P_{min}$), which is the average optical power received in both bit 1 and bit 0, that is [1],

$$P_{\min} = (P_1 + P_0)/2 \tag{11.58}$$

For simplicity's sake, assume $P_0 = 0$ and neglect $i_d$; this implies that Formulas 11.49c and 11.49d hold true and

$$I_1 = R\,P_1 = 2R\,P_{\min} \tag{11.59}$$

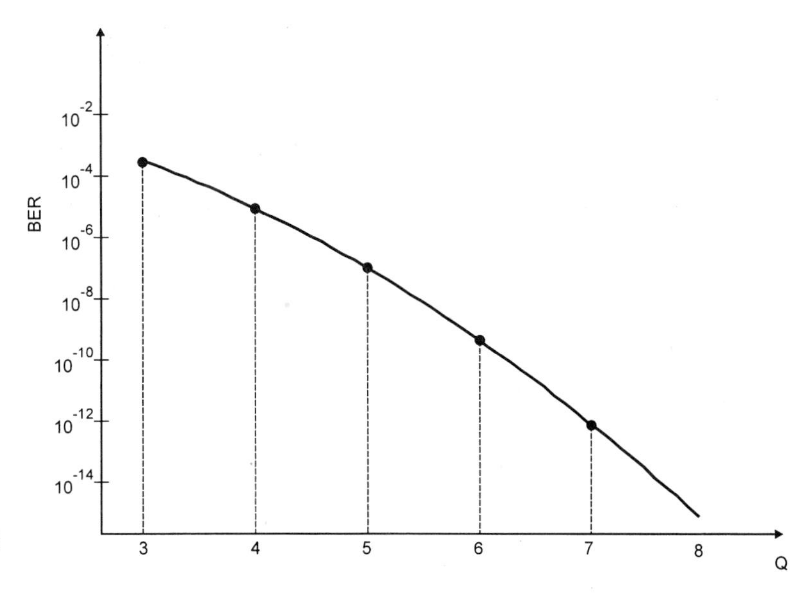


**Figure 11.14** BER as a function of Q parameter.

Using Formulas 11.26, 11.30, and 11.58, we obtain (don't forget: here $I_P{}^* = I_1$)

$$i_s^2 = 2eI_1\text{BW}_{\text{PD}} = 4eRP_{\min}\text{BW}_{\text{PD}} \qquad (11.60)$$

and

$$_t^2 = (4k_BT/R_L)\text{BW}_{\text{PD}} \qquad (11.61)$$

If we apply Formula 11.56a, we relate parameter $Q$ to the minimum received optical power as [1]

$$Q = (2RP_{\min})/\{\surd[(4eRP_{\min}\text{BW}_{\text{PD}}) + ((4k_BT/R_L)\text{BW}_{\text{PD}})] + \surd[(4k_BT/R_L)\text{BW}_{\text{PD}}]\} \qquad (11.62)$$

This equation can be solved with respect to $P_{\min}$ with this result:

$$\text{P}_{\min} = (Q/R)\{eQ(\text{BW}_{\text{PD}}) + \surd[(4k_BT/R_L)\text{BW}_{\text{PD}}]\} \qquad (11.63)$$

To make this formula easier to follow, we can rewrite it in this simpler form:

$$\text{P}_{\min} = Q/R[e(Q)\text{BW}_{\text{PD}} + i_t] \qquad (11.63a)$$

**Example 11.3.5**

***Problem:***

What is the sensitivity of an MF-432 photodiode at room temperature when BER = $10^{-9}$ and $R_L$ = 50 kΩ?

***Solution:***

The solution is given by Formula 11.63. From Figure 11.13 we get $Q = 6$ for BER = $10^{-9}$. Thus,

$$P_{\min} = Q/R\{e(Q)\text{BW}_{\text{PD}} + \surd[(4k_BT/R_L)\text{BW}_{\text{PD}}]\} = 0.00014 \text{ nW} + 5.45 \text{ nW} = 5.45014 \text{ nW}$$

We see again that for a *p-i-n* photodiode thermal noise is the dominant component. As a result, we can simplify Formula 11.63a even more [1]:

$$P_{\min}(p\text{-}i\text{-}n) = (Q/R)i_t = \{Q\surd[(4k_BT/R_L)\text{BW}_{\text{PD}}]/R\} \qquad (11.64)$$

For our photodiode, in thermal-noise limit, we get $P_{\min}$ (*p-i-n*) = 5.45 nW, or –32.6 dBm.

---

Both Formulas 11.63 and 11.64 (we can use either) give us the result we were searching for: the minimum received optical power, depending on such parameters of a photodiode as responsivity ($R$), load resistance ($R_L$), and bandwidth ($\text{BW}_{\text{PD}}$). But, in practice, *the designer of a fiber-optic communications system needs to know the relationship between* $\text{P}_{\min}$ *and BER.* What's more, the minimum received power (sensitivity) is specified only along with the bit-error rate. In measuring a receiver's sensitivity, engineers reduce the received optical power and measure the bit-error rate simultaneously to establish the relationship between these two characteristics of a receiver.

The formal relationship between BER and $P_{\min}$ can be obtained if we substitute parameter $Q$, expressed in terms of $P_{\min}$ (Formula 11.62), into Formula 11.55 for BER. We then get

$$\text{BER}_{\min} = \tfrac{1}{2}\,\text{erfc}(2RP_{\min})$$
$$/\{\sqrt{[(4eRP_{\min}\text{BW}_{\text{PD}}) + ((4k_BT/\text{R}_\text{L})\text{BW}_{\text{PD}})]} + \sqrt{[(4k_BT/\text{R}_\text{L})\text{BW}_{\text{PD}}]}\} \quad (11.65)$$

For practical calculations, you can use the exponential approximation of the erfc function given in Formula 11.55. A typical result of such a calculation is shown in Figure 11.15.

Formula 11.65 and Figure 11.15 give us the final word on the *sensitivity* of a photodiode: It is the *minimum received optical power that provides the given bit-error rate.* (Remember, of course, that $P_{\min}$ is the *average* power.) Because the sensitivity of a receiver depends not only on the properties of the photodiode but also on the entire design of the receiver, we'll discuss this topic in more detail in the next section.

***Minimum number of photons per bit*** Since the average optical power is related to the average number of photons, the photodiode's sensitivity can be expressed in terms of minimum number of photons per bit. Indeed, recalling Formula 11.3, we can write

$$P^* = N_\text{p}^*\, hf/T_{\text{BR}}, \quad (11.66)$$

where $P^*$ is the average received optical power per bit, $N_\text{p}^*$ is the average number of photons contained in a bit, $h$ is Planck's constant, $f$ is the photon's frequency, $T_{\text{BR}} = 1/\text{BR}$ is the duration of a bit, and BR is the bit rate. If we neglect, for simplicity's sake, the average received power for bit 0, then Formula 6.66 gives us the average number of photons for bit 1.

To find a photodiode's sensitivity in terms of number of photons, we need to calculate error probabilities, as noted in the definition of BER (Formula 11.48). This approach has already been well developed [9].

***Quantum limit*** Our concern here is how many photons we need to detect bit 1. The answer is at least one, provided that it produces an electron-hole pair. If we want to count single photons, we cannot approximate their distribution by Gauss's law, as we did for the high volume of photons. We have to use Poisson statistics, described in Formula 11.20. Therefore,

$$P(n) = \exp(-N_\text{p})(N_\text{p})^n/n! \quad (11.67)$$

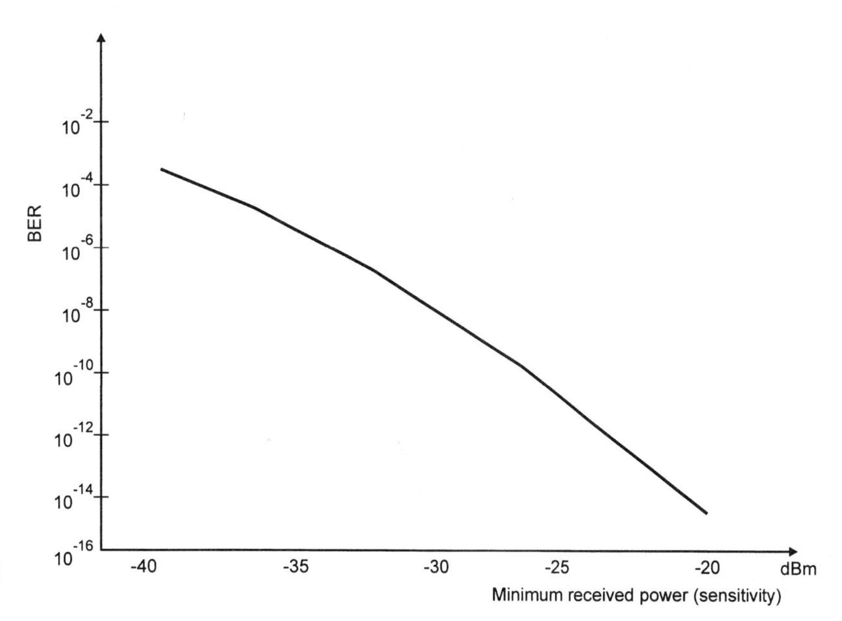


**Figure 11.15** BER versus received power (sensitivity).

Let's consider the probabilities $P(1/0)$ and $P(0/1)$, assuming that no photons have been received with bit 0 and one photon has been received with bit 1. The probability of identifying a bit as 1 when bit 0 arrives as $P(1/0)$ is zero because no photons have been received in this case. The probability of identifying a bit as 0 when bit 1 arrives is equal to $P(0) = \exp(-N_p)$ because $n = 0$. Refer again to the definition of BER given in Formula 11.48. In this case, BER is given by [1]

$$\text{BER} = 0.5[P(1/0) + P(0/1)] = 0.5\exp(-N_p) \tag{11.68}$$

So, a plain exponent represents the dependence of BER on the absolute minimal number of photons necessary to detect a bit. Since this relationship determines the absolute minimum of $N_p$, it is called the *quantum limit.* If one photon is received, BER is 0.18 (which means that 18 out of 100 bits received are interpreted incorrectly); if $N_p = 10$, BER is $2.27 \times 10^{-5}$. You can continue to calculate these numbers. For $N_p = 20$, BER $= 1.03 \times 10^{-9}$. Remember, BER $= 10^{-9}$ is considered the minimum acceptable bit-error rate in modern fiber-optic communications systems. The quantum limit for BER $= 10^{-12}$ is $N_p = 26$. (You may, in your work, come across quantum limit calculated in terms of average number of photons per bit ($N_p^*$), which is equal to $N_p/2$. So don't be confused if you see $N_p^* = 10$ for BER $= 10^{-9}$.)

Wrestling with quantum limit may seem to you to be an academic exercise, but it is not. In calculating the power budget of a fiber-optic communications system, the designer usually has to know how far (or how close) his or her system is from its quantum limit. This is another measure of system quality. For more information on this concept, see reference [9] at the end of this chapter.

## 11.4 RECEIVER UNITS

A receiver is a unit that converts an optical input signal into an appropriately formatted electric output signal. Since we are concentrating here on digital transmission, we can say that a receiver converts a stream of light pulses into a stream of electric pulses capable of driving the electronics that follow in the system. (See the general diagram of a fiber-optic communications system in Figure 1.4.)

As we have already seen with transmitters, the light source—either an LED or a laser diode—is the heart of these devices, but the properties of a transmitter depend also on the characteristics of its electronics and its packaging. The same holds true for receivers. Photodiodes determine the major features of the devices but even an ideal PD cannot make an ideal receiver. We need good electronics, packaging, and design in order to make a good device. In other words, a high-quality photodiode is the only condition necessary to attain a high-quality receiver. The output signal emanates from the entire receiver, not just from its photodiode. At this point, aware of how a photodiode works and fully cognizant of its characteristics, we will discuss the receiver unit in detail.

It is suggested that you reread Section 10.4, "Transmitters," because both units—transmitter and receiver—have much in common in terms of design approach, operation with a signal, and practical application. What's more, such matters as packaging, printed circuit board layout, reliability, testing, and troubleshooting are almost identical and so they will not be repeated in this section.

### Functional Block Diagram and Typical Circuits of a Receiver

***Block diagram*** Many types of receivers are used in modern fiber-optic communications systems. They differ in their architecture and in the components they use; however, most of them have very much in common functionally. A typical functional block diagram of a receiver is shown in Figure 11.16.

To refresh your memory as to what a receiver looks like, refer again to Figure 1.7, where two types of receiver units—DIP and butterfly—are shown. To see what a receiver board popu-

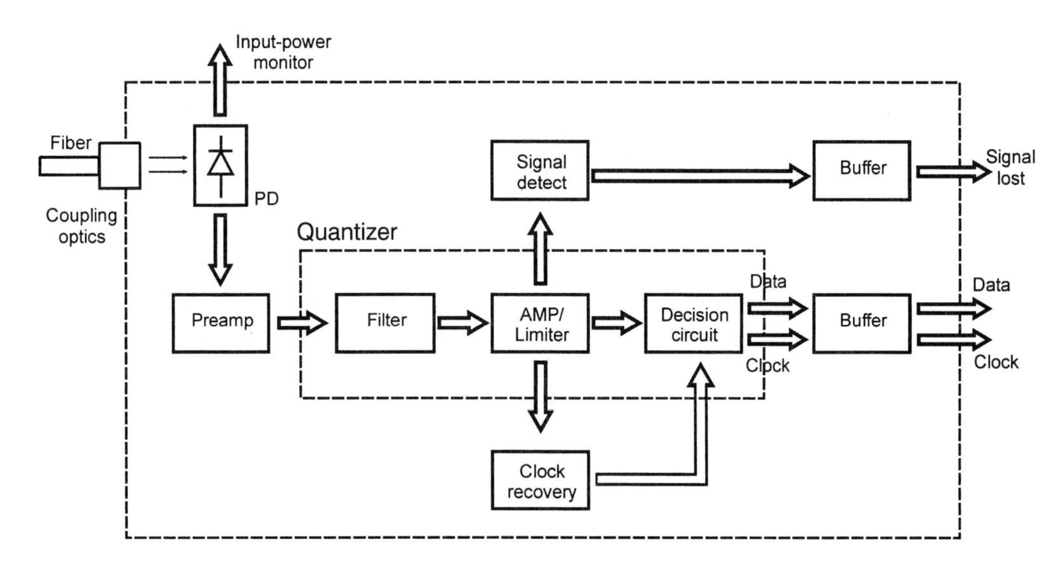


**Figure 11.16** Functional block diagram of a receiver. *(Adapted with permission from Opto Electronic Products by Ericsson Microelectronics AB, Stockholm, Sweden.)*

lated with IC chips and discrete components looks like, see Figure 10.40. (Even though Figure 10.40 exhibits a transmitter assembly, the physical appearance of both units is very similar.)

Light from an optical fiber passes through coupling optics and falls on the sensitive area of a photodiode. (A detailed discussion of coupling optics is presented in Chapter 9; additional discussion can be found in Section 11.2.) A photodiode (PD) converts light into photocurrent. This photocurrent is converted into voltage and amplified by a preamplifier, denoted as "Preamp" in Figure 11.16. The output voltage from the preamplifier enters a block (usually called a "quantizer") whose output is data and clock signals in an appropriate format. (This block will be discussed shortly.) The data and clock signals then enter a buffer, which converts them into ECL-compatible signals capable of driving the succeeding digital circuitry. Thus, the input of a receiver is an optical signal and the output is the electric voltage in a specific format with specific electrical characteristics. Now let's look at the components of a receiver in detail.

***Optical front end*** A photodiode along with the preamplifier linked to it is called the receiver's *optical front end.* The function of this section is to convert light into electric voltage of the required amplitude. This is done in two steps: First, the photodiode converts light into photocurrent; secondly, the preamplifier converts the photocurrent into voltage, amplifies the signal, and presents it to a quantizer. A receiver's photodiode can be chosen from a variety of commercially available devices. The critical concern in the design of an optical front end is choosing the right electronic amplifier.

You will recall that the load resistance ($R_L$) of a photodiode plays a very important role in both noise and bandwidth considerations. On the one hand, thermal noise, which is the dominant component of noise current in *p-i-n* PDs, is inversely dependent on the load resistance. This is why load resistance appears in all the major formulas determining SNR and BER as a function of the sensitivity of a photodiode. (See Formulas 11.30, 11.38, and 11.65.)

It is important to underscore one more effect of $R_L$: As with thermal noise, a photodiode's sensitivity is also inversely dependent on load resistance. (See Formula 11.64.) Hence, to decrease thermal noise (and increase the photodiode's sensitivity), we need to increase the load resistance. Nothing could work better in achieving this goal than connecting a photodiode directly to an amplifier because the input impedance of an electronic amplifier is very high (usually on the order of units of MΩ). This connection is known as *high-impedance design* and its equivalent circuit is shown in Figure 11.17(a).

On the other hand, the bandwidth of a photodiode is inversely proportional to load resistance, as Formula 11.13 shows. (In that formula we considered a PD's junction resistance [$R_{in}$] as the load resistance.) Thus, to increase the bandwidth of a receiver, we need to decrease the load resistance. The upshot: *There is a trade-off between bandwidth and noise (sensitivity) of a receiver* and the right design is determined by the load resistance of the photodiode.

Keep in mind that a preamp must not only amplify the signal but also convert current into voltage. The latter function is performed by an amplifier with negative feedback. Thus, another connection between a PD and a preamp—a so-called *transimpedance design*—is used in optical front ends. This design is shown in Figure 11.17(b).

Input impedance of an electronic amplifier with negative feedback is called *transimpedance* ($R_z$); this is the actual load resistance for a photodiode. Transimpedance is the function of the amplifier's gain ($A$) and feedback resistor ($R_F$). A designer can vary these two parameters to obtain the desired value of $R_L$, which is usually on the order of tens of kΩ. The output voltage of this preamp is given by

$$v_{out} = I_p \; R_z, \tag{11.69}$$

where $I_p$ is the photocurrent produced by the photodiode, and $R_z = R_L$, again, is the transimpedance of the preamp with negative feedback.

*The transimpedance design is the most common type you will meet in the field.* In nearly every commercial receiver's data sheet, you will find the term *transimpedance amplifier,* which means an amplifier with negative feedback.

A photodiode is usually integrated with a transimpedance amplifier in a component called a *PINAMP.* When a preamp is linked to field-effect-transistor (FET) circuitry, the front-end unit is called a *PINFET.* Typical characteristics of such an optical front end can be found in [14].

But even the transimpedance design itself cannot satisfy a variety of requirements that receivers must meet in today's applications. One important new characteristic of a receiver that is critical in fiber-optic-network applications is the *dynamic range* which *is the difference between the highest and the lowest input signals at which a preamp can operate.* The dynamic range of a receiver is measured in dB and its typical values vary from 35 to 45 dB.

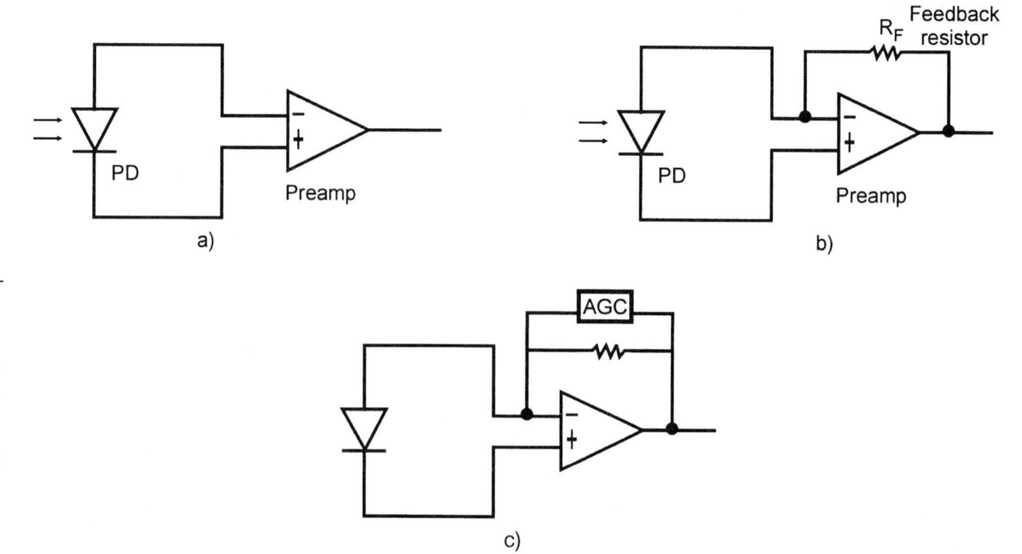


**Figure 11.17** Equivalent circuits of an optical front end: (a) High-impedance design; (b) transimpedance design; (c) transimpedance design with automatic gain control (AGC).

In fiber-optic networks, the input signal can fluctuate widely. A transimpedance amplifier has a much wider dynamic range than a high-impedance one. This is because a preamp's output voltage is proportional to its input impedance, as Formula 11.69 states, and this value for a transimpedance design is much smaller than for a high-impedance design. But this is still not enough. To further increase the dynamic range of a transimpedance preamp, manufacturers include *automatic-gain-control (AGC)* circuitry in the feedback loop, as Figure 11.17(c) shows. This circuitry, as the name implies, controls the amplifier gain to keep the output voltage stable; this means the transimpedance ($R_z$) is also varied, so that it is high at a low-input signal and it is low at a high-input signal. This is how a preamp can handle an input signal in a wide dynamic range.

To conclude, an optical front end is the first stage of a receiver. It accepts an optical signal and presents amplified voltage to the next block—the quantizer.

***Quantizer*** A quantizer typically includes three components: a noise filter, a power amplifier/limiter, and a decision circuit.

The *noise filter* improves the signal-to-noise ratio or, ultimately, the receiver's sensitivity, as was mentioned in Section 11.3. The design consideration for a noise filter centers on the bandwidth requirements. For example, in designing an optical front end, a manufacturer wants to achieve maximum bandwidth to, obviously, increase the sales potential of the device. But this has a negative side for the user. Specific applications probably don't need all the bandwidth available. We have already pointed out, remember, that noise is directly dependent on bandwidth so that excess bandwidth contributes to additional noise. This, in turn, reduces a receiver's sensitivity.

Consider this situation: Assume your network operates at BR = 155.520 Mbit/s, which is an OC-3 signal in the SONET standard. When converting the bandwidth (BW) and bit rate (BR) for a receiver, the rule of thumb is this: BW ~ 0.7BR. Hence, the required bandwidth is equal to approximately 109 MHz. You have on hand an MF-432 PIN photodiode, whose data sheet was discussed in Section 11.2. This PD has a bandwidth of 2.5 GHz. Thus, for our network, whose BR = 155.52 Mbit/s, we can use a low-pass noise filter to restrict the bandwidth of the receiver to 109 MHz. By restricting bandwidth, a noise filter also reduces *intersymbol interference (ISI).*

You can purchase a receiver for a specific application with a pre-installed filter or you might customize the filter. In the latter case, the manufacturer usually provides you with a circuit, a table of suggested values of capacitors, and an inductance to build an L-C noise filter according to your application. Needless to say, a noise filter is an optional component and some receivers, to keep their cost down, don't contain it.

An *amplifier/limiter* provides power amplification of a signal obtained from a preamp through the noise filter. Amplification is necessary to attain a signal with enough power to drive the decision circuit. (Remember, the preamp is the voltage amplifier.) If the amplified signal is high enough, this circuit clips the signal—thus the name "limiter." In other words, the gain of this amplifier is a function of the amplitude of the input signal: the larger the amplitude, the less the gain. This power amplifier can also include automatic gain control—the circuit we discussed above. This unit might also include an *equalizer* with a specific gain-versus-frequency characteristic to correct any bandwidth-caused signal distortion.

The *decision circuit* is the unit that determines the logical meaning of the received signal. Typically, this is a comparator driven by the input signal. The basic circuit is shown in Figure 11.18(a). When the received signal is above threshold, the comparator's output is high. This means the decision is made that this signal carries logic HIGH, or 1. When the signal is lower than threshold, the comparator's output is low. This means the decision is made that the received signal carries logic LOW, or 0. (See Figure 11.18[b].)

An electric signal representing a bit varies over even a single-bit time interval, as Figure 11.13 shows. The question is, when do you want to take a sample of data ($v_{ins}$) to compare with threshold voltage ($v_{th}$)? Obviously, it is better to take such samples in the middle of each bit, where the probability of having the best sample is the highest. (See Figure 11.18[c].) Thus, the

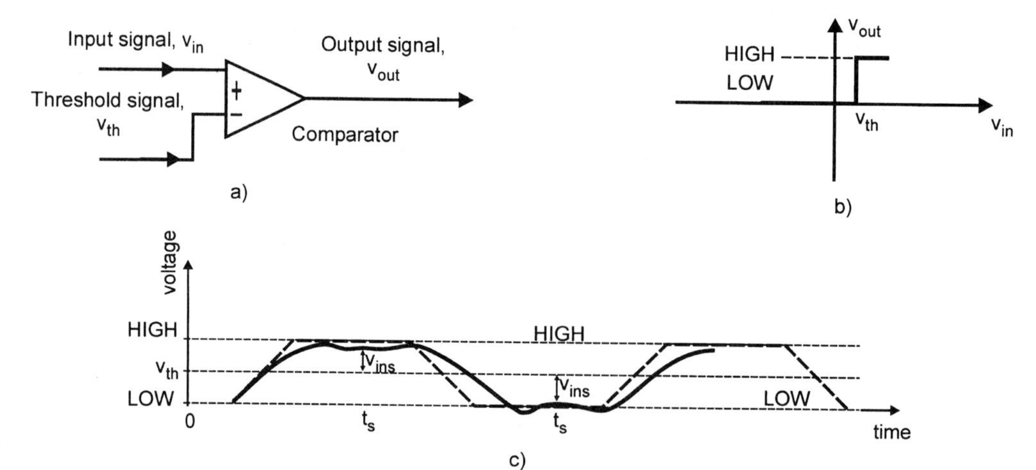


**Figure 11.18** Principle of operation of a decision circuit: (a) Basic circuit; (b) comparator output; (c) decision-making process.

need for precise timing of the signal arises. If your circuit makes a mistake in determining this midpoint of bit time ($t_s$), the probability of error increases. Incidentally, an eye diagram helps in finding the best position for $t_s$; this is where an eye diagram is opened to its maximum.

We need to emphasize that a comparator works with voltages, although, in our theoretical considerations, we did indeed describe decision-making techniques in terms of currents. (See Formulas 11.52, 11.53, and 11.65.) This is because in Section 11.3 we dealt with a photodiode itself but, in this section, we are considering the entire receiver unit. To convert photocurrent ($I_p$) into voltage, we need to multiply $I_p$ by some impedance ($Z_{cas}$) representing a cascade of transformations that photocurrent experiences before entering a comparator. In short, we need to consider the transformation of an optical input signal into voltage entering a comparator ($v_{inc}$) through the convolution integral. This would lead to our obtaining a transform function of the circuit performing this transform, including all noise sources. Such an approach is proposed in reference [9]. Instead of developing this theory, we will simply rely on a practical engineering approach: Measured voltage ($v_{inc}$) includes all the transformations from an optical signal to the input of a comparator.

The critical problem in designing a decision-making circuit is what signal should be used as the threshold (reference). Since this design is so crucial in the consideration of a receiver, we have set aside a separate subsection (pages 484–489) to discuss it.

***Buffers*** A *buffer* transfers a logic signal from the input to the output unchanged but reshapes the electrical form of this signal. Typically, this is an emitter-follower circuit. In this case, a buffer provides output in ECL-compatible format. (For more on buffers, see Section 10.4.) A receiver can contain several buffers, as Figure 11.16 illustrates.

***Clock recovery*** *Clock recovery* extracts timing information from the data stream and helps the decision circuit to generate clean and reshaped differential DATA and NON-DATA outputs.

You may ask why we need this circuit. As you know, synchronous digital circuits work under control of a clock signal. This timing signal must be the same at the transmitter and receiver ends to synchronize all operations. In our case, for example, we need to take a sample of data entering a comparator ($v_{ins}$) and compare it with the threshold voltage ($v_{ins}$) exactly in the middle of a bit, as Figure 11.18(c) shows. But bit timing is determined by the transmitter clock. If the receiver clock has a different time, we'll experience a data-sampling error. This error will lead to an error in determining the meaning of the received signal's logic; in other words, it will increase the bit-error rate.

The clock signal is produced by a frequency generator. Since the frequency stability of such a generator is finite (even though it is on the order of $10^{-6}$ and higher for practical miniature circuits), two generators will inevitably have a significant clock discrepancy over their operating time. Let's consider a popular crystal (quartz) generator [15]: Its typical stability is about 25 parts per million. This means that at 20 MHz the oscillator's frequency can vary up to $\pm$ 25 Hz. Therefore, two such generators, one in the transmitter and the other in the receiver, can have a frequency difference up to 50 Hz, producing an unacceptable BER.

The only way to ensure the same time frame along a transmission path is to have the data itself carry a timing signal. This is done by using specific line codes, a topic discussed in Chapter 8. For now, we only need to understand that the typical non-return-to-zero (NRZ) code shown in Figure 11.18 does not carry timing information in explicit form. Therefore, a special measure has to be taken to extract a clock signal from the data. This is accomplished by a clock-recovery circuit.

The principle of operation of this circuit can be seen in Figure 11.19. A *voltage-controlled oscillator (VCO)* generates approximately the same frequency as a transmitter generator. A phase of received data is compared with the phase of a signal generated by the VCO and their difference is converted by a low-pass filter into a dc signal. This signal makes the VCO change its frequency in order to eliminate any discrepancy between received and generated frequencies. (Frequency, as pointed out previously, is the derivative of a phase.) The corrected frequency signal—the clock signal in Figure 11.19—controls the operation of the decision circuit.

If you have some background in electronics, you may at this point say, "Wait a minute. In a sense, Figure 11.19 shows a typical *phase-locked-loop (PLL) circuit,* does it not?" Yes, indeed. This is a typical phase-locked loop and many receivers' data sheets refer to the clock-recovery circuit as simply a PLL circuit. The key feature of this digital circuit is that it compares the edge transitions of the data and the VCO pulses, which we referred to as the "phase comparison." Not all receivers contain clock-recovery circuits. These circuits are also available as stand-alone units [16].

***Signal detect*** *Signal detect* is essentially an alarm circuit. It monitors the level of the incoming signal and generates a logic LOW signal when the signal-to-noise ratio is not sufficient. The output of this circuit is a logical flag, which indicates when the level of the input signal drops below the acceptable (threshold) level. Suppose the input signal hovers around the threshold level. In this case, the flag signal will toggle between HIGH and LOW, making you nervous because you don't know whether you should take some kind of action or just wait until the alarm becomes steady. To prevent this situation, a signal-detect circuit generates an alarm signal only when the increasing input signal rises a certain level above the threshold. When the input signal decreases, the detect circuit waits until the signal drops below threshold to a certain predetermined level. In other words, the signal-detect circuit works with hysteresis, which is usually on the order of 1.5 dB.

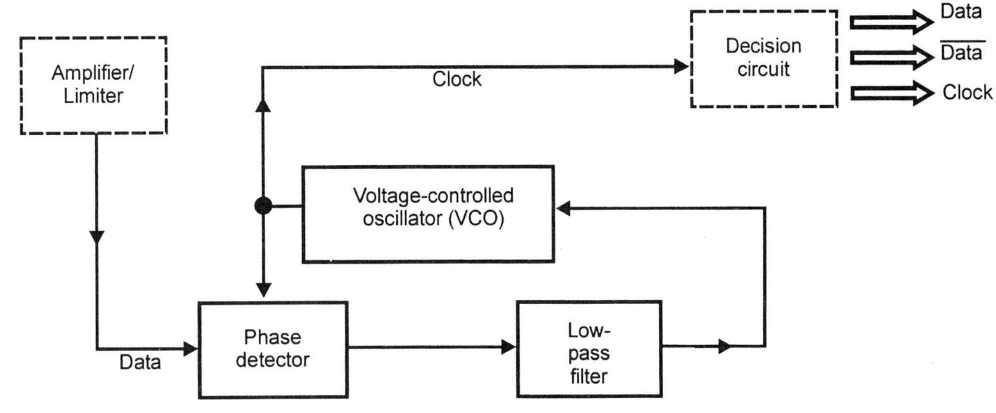


**Figure 11.19** Typical clock-recovery circuit.

***Monitoring circuits*** Look again at Figure 11.16. Notice the two monitoring circuits in that typical receiver. The first, monitoring the voltage drop produced by photocurrent flowing through a resistor, allows the engineer to keep tabs on input power. The second, a flag signal from a signal-detect circuit, watches for a possible signal-lost situation.

## Decision-Circuit Design

***Statement of the problem*** The main problem in designing a decision circuit, as noted above, is to determine what the level of its threshold signal should be. Theoretically, this is determined by Formula 11.53. For a *p-i-n* photodiode, this formula reduces to the simple requirement given by Formula 11.54: The threshold signal should be midway between the maximum and minimum levels of the received signal. But, in practice, these maximum and minimum levels vary, which presents a challenge to a designer of this circuit. Let's consider the situation in greater detail [17].

The first reason for setting the threshold signal at the halfway mark of the peak amplitude of an input signal, $^1/_2\ v_{in}$ (max), is so noise is the same for both bit 1 and bit 0, as we assumed when we derived Formula 11.54. But there is another reason: For an ideal signal, the receiver's output pulse width ($T_{RX}$) is the same as the transmitter's input pulse width ($T_{TX}$). This ideal-signal situation is depicted in Figure 11.20a.

In reality, the input signal is far from the ideal waveform, as Figure 11.20(b) shows. In addition, the threshold signal may not even be found in the middle-level position (Figure 11.20[c]). These two problems cause *pulse-width distortion (PWD),* which is determined as

$$\text{PWD} = [(T_{RX} - T_{TX})/T_{TX}] \times 100\% \tag{11.70}$$

For example, if $T_{TX}$ = 50 ns and $T_{RX}$ = 60 ns, then PWD = 20%.

PWD affects clock recovery because it causes a phase shift in the received signal. (Refer to Figure 11.19, where a phase detector compares the rising edges of the received and generated signals.) Another detrimental effect of PWD is that it changes the time frame during which a logic level exists, a factor leading to increasing bit-error rate.

Depending on how you establish a threshold signal, you can sharply curtail the PWD problem. There are four major methods of determining the proper threshold signal [17]:

**1.** *Fixed-threshold circuit*—In this design, the decision is reached by comparing the received signal with the fixed voltage. This circuit is shown in Figure 11.21(a). All the problems associated with PWD discussed above are inherent in this circuit. But this circuit has a big advantage: It works with any received signal. In other words, this circuit does not require prior knowledge of the duty cycle of an incoming data stream and any pulse duration and bit rate can be received. (A *duty cycle,* you'll recall, *is the average ratio of bit 1 to bit 0 of a data stream.*)

**2.** *Edge-detecting circuit*—This circuit is shown in Figure 11.21(b). Explanation of its operation is depicted in Figure 11.21(c). The circuit detects both positive and negative edges of incoming data by differentiating the input signal. The comparators generate SET and RESET pulses, which latch the last data transition in a flip-flop arrangement. (*SET/RESET flip-flop,* remember, is set by the input impulse and holds this logic state until the next impulse inverts [resets] it into the opposite logic state.) With an ideal input signal, the time duration between SET and RESET pulses is exactly the same as the duration of the transmitted signal. This is why the duration of the receiver signal ($T_{RX}$) is equal to the duration of the transmitted signal ($T_{TX}$), as the top portion of Figure 11.21(c) shows. With the real input signal, there is some pulse-width distortion, as the lower portion of Figure 11.21(c) shows, but this PWD is much smaller than in the case of a fixed-threshold circuit.

This improvement, however, requires setting up the initial state of the receiver output since the circuit detects data changes, not data levels. To do so necessitates an understanding of the logic

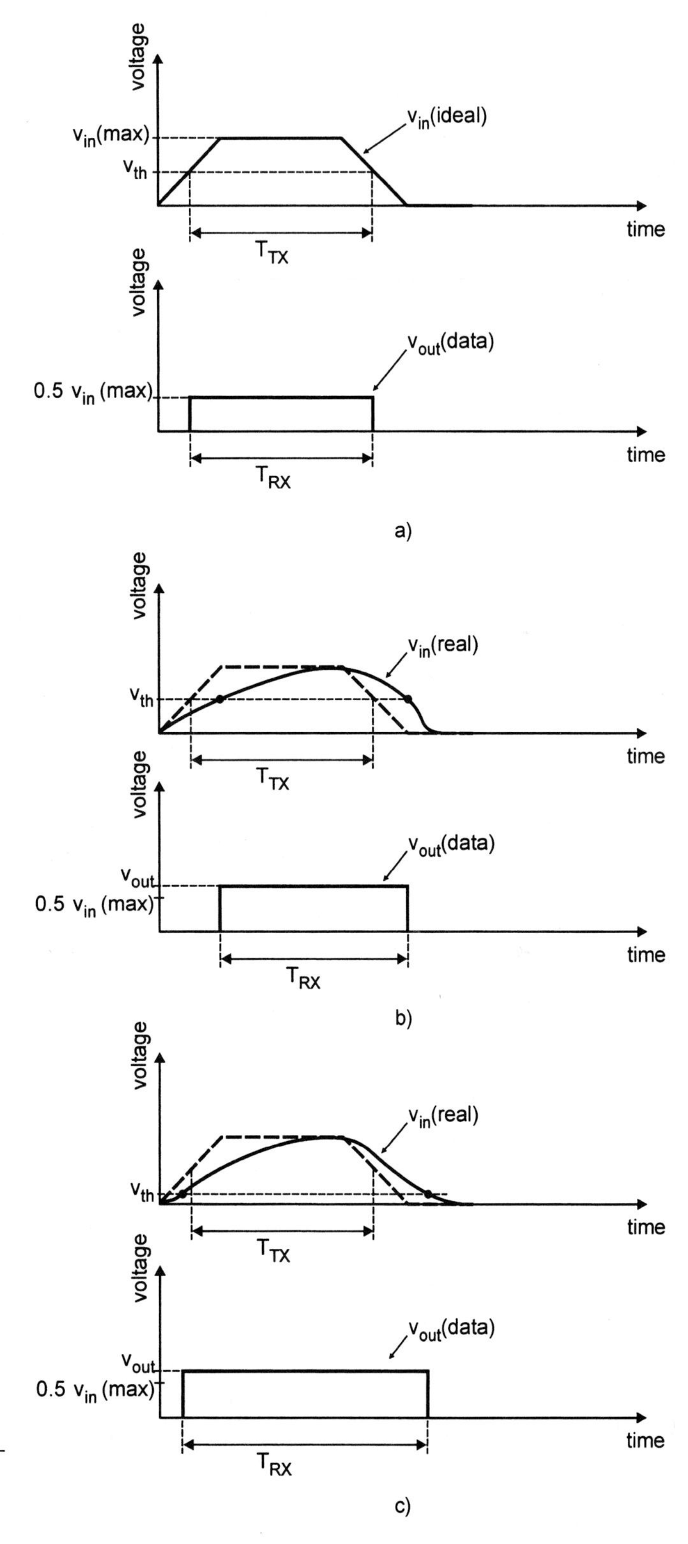


**Figure 11.20** Impact of threshold signal on pulse distortion: a) Ideal signals; b) real input with threshold equal to 0.5 $v_{in}$(max); c) real signal with threshold less than 0.5 $v_{in}$(max).

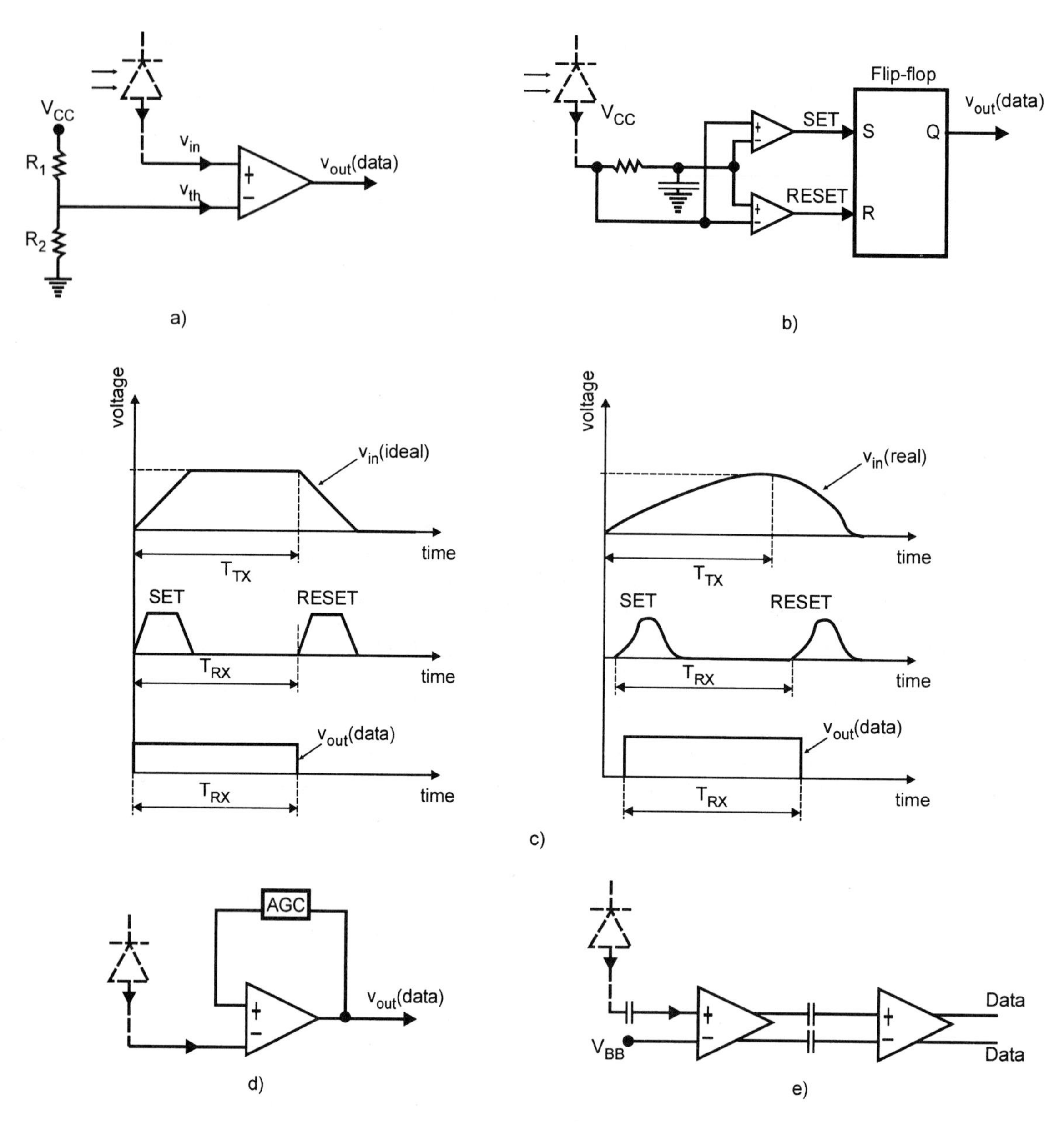


**Figure 11.21** Design of a decision circuit: a) Fixed-threshold circuit; b) edge-detecting circuit; c) operation of an edge-detecting circuit; d) automatic gain-control (AGC) circuit; e) capacitor-coupled circuit.

state of the first received bit. Fortunately, most communication protocols provide the receiver with such knowledge, so this is not a major problem.

**3.** *Automatic gain-control circuit*—A decision circuit with *automatic gain control (AGC)* optimizes the threshold-signal level and thus provides the best solution to a PWD problem. The circuit is shown in Figure 11.21(d). The AGC circuit first finds the average level of the received data stream and next sets up the threshold signal in the middle of the actual received signal. Hence, both PWD and BER are minimized because the threshold level is always at 50% of the actual input.

There are several drawbacks to this optimal circuit. First, this circuit works well only for a 50% duty cycle. Second, the circuit itself is relatively complex but, given today's level of integrated electronics, this is obviously a relatively minor problem. The third shortcoming is more serious.

To determine the average level of the incoming signal, a receiver has to initially accumulate a data stream. In other words, there is some *start-up delay* (also called *packet-response time*) inherent in this circuit. This problem is not an issue in point-to-point communications, where the same data stream is usually transmitted along the same path and where, therefore, we need to accumulate data only once. But it is a detriment to network applications, where data come from different transmitters in different electric forms and at different power levels. This start-up delay becomes a very serious problem when a data stream has essential interruptions in transmission. In such a case, it takes much longer to set up the average threshold level. A partial solution to this problem is to continue to transmit idle pulses (so-called *keep-alive pulses*) even when actual communication is not in progress. Keep-alive pulse transmission helps because it is much faster to change an average threshold level than to set it up from scratch.

**4.** *Capacitor-coupled circuit*—This circuit keeps the pulse-width distortion low while minimizing start-up delay. It cascades differential gain stages coupled with capacitors, as Figure 11.21(e) shows. When an input signal starts to increase, the gain stages begin to clip the signal. The last stage clips first because it receives the signal at the highest level. The amplifiers are differential; that is, they amplify the difference between the constant voltage ($V_{BB}$) and the input signal. In other words, these gain stages amplify only the transitions of the input signal, not its steady level. Coupled capacitors, in turn, average the input signal at 50% of its peak-to-peak levels if the duty cycle of the data stream is 50% over the capacitors' time constant.

From the point of view of a control system, an AGC receiver employs a closed-loop circuit, whereas a capacitor-coupled network uses an open-loop circuit. This is why the time constant of the capacitor-coupled circuit is much less than that of an AGC circuit. This constant is typically on the order of microseconds [17].

***Summing up*** The major concern of a decision-circuit designer is optimizing the threshold signal. To attain this optimum level, the designer faces a trade-off among the requirements of the duty cycle of a data stream, pulse-width distortion, and start-up delay. Depending on the specific applications (point-to-point link or network and type of data stream), some of these criteria become more important than others. The advantages and disadvantages of all circuits discussed above are summarized in Table 11.2.

**Table 11.2** Receiver circuits: pros and cons

| Receiver circuit | PWD | Required data stream duty cycle | Start-up delay |
|---|---|---|---|
| Fixed-threshold | High | No requirements | Small |
| Edge-detecting | Moderate | No requirements | Small |
| Automatic gain-control (AGC) | Low | 50% | Long |
| AGC with keep-alive pulses | Low | No requirements | Long but less than that of a simple AGC |
| Capacitor-coupled | Low | 50% | Moderate |

***Importance of duty cycle*** As Table 11.2 shows, AGC and capacitor-coupled receivers exhibit good characteristics but they require a 50% duty cycle. What happens if the actual duty cycle deviates from the ideal 50%? First, start-up delay for those two types of receivers increases. But, more importantly, this deviation results in threshold drift. Indeed, the threshold is determined by averaging the peak-to-peak values of a received data stream. If the number of, say, 0's will be more than the number of 1's in a received data stream, then the average will be less than half. This is how a duty cycle affects the threshold. Threshold drift, in turn, causes two major detrimental events: sensitivity degradation and an increase in pulse-width distortion.

Indeed, receiver sensitivity ($P_{min}$) depends on digital SNR ($Q$) and the latter depends on the threshold signal. (See Formulas 11.63 and 11.52, respectively.) Thus, the deviation of a duty cycle from 50% will cause a decrease in receiver sensitivity. The result will be degradation of BER at the same received power or, to compensate, more power will be required to maintain the given BER. To combat this effect, some manufacturers have developed so-called "compensated" receivers, which have a much higher tolerance for duty-cycle variations [18]. The dependence of receiver sensitivity on duty cycle for two types of receivers is shown in Figure 11.22(a).

To understand why pulse-width distortion depends on the duty cycle, review Figure 11.20. There you see that PWD depends on threshold and, thus, on the duty cycle. Figure 11.22(b) shows how PWD degrades with a deviation of the duty cycle from 50% for two types of receivers.

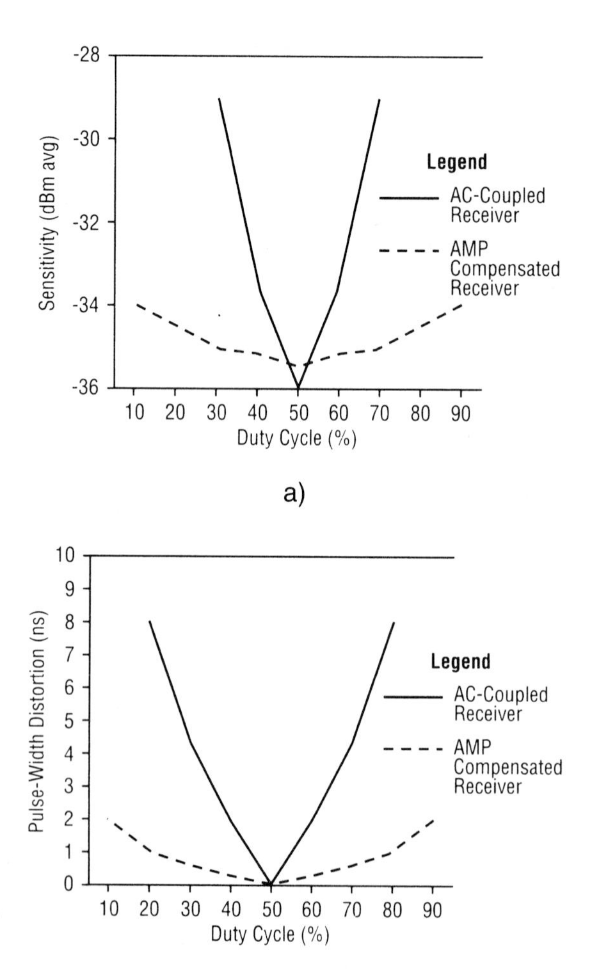


**Figure 11.22** Sensitivity and pulse-width distortion as a function of a duty cycle: (a) Sensitivity versus duty cycle; (b) pulse-width distortion versus duty cycle. *(Courtesy of AMP Incorporated, Harrisburg, Pa.)*

You should know, too, that PWD also depends on received power, bit rate, and case temperature [17].

It is hoped that, with the above information in mind, you will gain a thorough appreciation of the intricacies involved in decision-circuit design, certainly one of the more complex aspects of receiver technology.

## Reading a Receiver's Data Sheet

We will now focus on an analysis of the data sheet of a receiver that works with the transmitter discussed in Section 10.4. This approach helps us to close the communications link and avoid repetition in discussing their similar characteristics.

Figure 11.23 shows the data sheet of a specific receiver. The sections labeled "APPLICATIONS," "FEATURES," and "DESCRIPTION" are very similar to those of the transmitter discussed in Section 10.4.

First, let's compare Figure 1 in the data sheet—the receiver's *functional diagram*—with Figure 11.16. In doing so, we see how all the receiver functions discussed in general

**DIGITAL FIBER OPTIC RECEIVER**
**RT-1554 SERIES**
**SPECIFICATIONS AND USER INFORMATION**

**APPLICATIONS**

- ► **Data Rates up to 52 and 155/Mb s**
- ► **SONET OC-1,OC-3/SDH STM-1**
- ► **ATM/FDDI**
- ► **1300 nm and 1550 nm Operation**
- ► **High Loss Budget Links**

**FEATURES**

- ► **Single 5V Supply Operation**
- ► **Wide Dynamic Range**
- ► **PECL Operation**
- ► **20 Pin Multisource Pinout**
- ► **Differential Data and Flag Outputs**

**DESCRIPTION**

The RT-1554 series receivers are designed for use in SONET and SDH systems for short and long-haul applications. The wide dynamic range transimpedance design accepts a maximum optical input power of 0 dBm. A single +5 Volt supply is required without the need for additional negative detector bias. Data outputs are differential PECL compatible. The package configuration is the 20 pin multisource standard. The receivers are available with multimode fiber as well as FC and ST receptacle interfaces.

FUNCTIONS- Refer to fig. 1

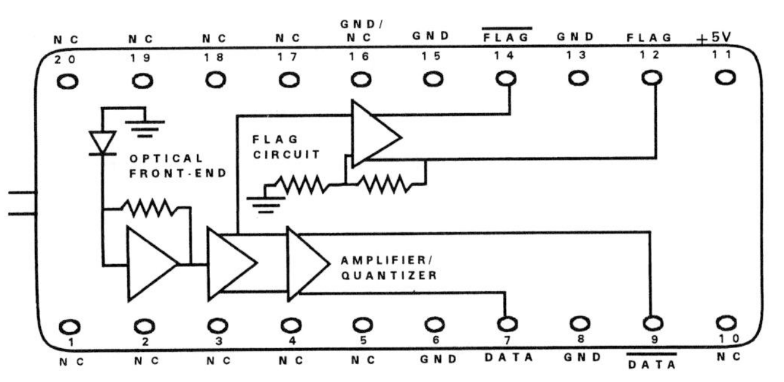


**Figure 11.23** Data sheet of a receiver. *(Courtesy of Laser Diode, Inc., Edison, N.J.)*

Optical Front End

The optical front-end consists of an InGaAs photodetector coupled to a wide dynamic range transimpedance amplifier for low noise and high bandwidth. An internal AGC circuit allows for input power up to 0 dBm.

Amplifier and Quantizer

The amplifier stage has high gain and limits large amplitude signals to maintain wide dynamic range. The quantizing section provides PECL output levels to the data and flag outputs.

Data Outputs

Data outputs are differential 10KH PECL (Positive or Pseudo-ECL). The recommended terminations are shown in figs. 2 and 3. It is recommended that differential outputs be used for connecting to other ECL families to eliminate possible noise margin problems, especially at elevated temperatures.

Flag Outputs

The flag circuit differentiates between the presence or absence of a minimum acceptable optical signal with a minimum hysteresis of 1 dB. When an acceptable signal (See Specification) is present the Flag output (pin 12) is logic high.

User Connections

Figs 2 and 3 demonstrate typical connections the user can implement to interface with ECL and PECL logic. Good high frequency techniques should be used when laying out a PC board to insure good signal quality. A ground plane is recommended with the bypass capacitors and terminating resistors located as close to the module as possible. Additional power supply filtering may be required if excessive noise is present due to nearby switching power supplies or other noise sources.

Power Supply

A +5 Volt DC source is the only supply required. Pin 10 may be grounded or left unconnected without any effect on performance. Pins 1,2,3,4 and 16 may be left unconnected or grounded.

Photodetector Monitor

The receiver may be ordered with the photodetector monitor option for applications requiring access to an analog signal proportional to the input optical power. This option is implemented by connecting pin 10 to -5 Volts through a series resistor. The signal is the voltage monitored at pin 10.
Note: Please call sales department for part number and pin configuration.

Evaluation Board

An evaluation board is available to provide connections to test equipment such as oscilloscopes or BER testers. Please contact the sales department for information.

**Figure 11.23** (continued)

are implemented in the actual device. Descriptions of all the components in this functional diagram are given in Figure 11.23. Equipped with the information already provided in this chapter, you should now be able to understand fully the diagram in the data sheet.

Now look at the data sheet's "PERFORMANCE SPECIFICATIONS." The sections "Absolute Maximum Ratings" and "Electrical Characteristics" do not require elaboration here.

The section of main interest to us is "Optical Characteristics." Observe how "Measured Average Sensitivity" depends on bit rate: The higher the bit rate, the greater the minimum received average power required. In this instance, we need, typically, –42 dBm at 55 Mbit/s and –38 dBm at 155 Mbit/s provided that the BER is maintained at $10^{-10}$ and the duty cycle is 50%.

PERFORMANCE SPECIFICATIONS

**Absolute Maximum Ratings**

| Parameter | Minimum | Maximum | Units |
|---|---|---|---|
| Supply Voltage | 4.75 | 5.5 | V |
| Operating Case Temperature | -40 | 85 | °C |
| Storage Temperature | -40 | 85 | °C |
| Lead Soldering Temperature/Time | | 250/10 | °C/sec. |

**Electrical Characteristics**

| Parameter | Minimum | Typical | Maximum | Units |
|---|---|---|---|---|
| Supply Voltage [Vcc] | 4.8 | 5 | 5.3 | V |
| Supply Current | | 150 | 200 | mA |
| Output Data Voltage (1)<br>• Low<br>• High | <br>-1.8<br>-1.1 | <br>- 1.75<br>-1.0 | <br>-1.5 | <br>V<br>V |
| Output Flag Voltage (1)<br>• Low<br>• High | <br>-1.8<br>-1.1 | <br>- 1.75<br>-1.0 | <br>-1.5 | <br>V<br>V |

1. Measured from Vcc with a 50 Ohm load to Vcc-2 Volts.

**Optical Characteristics**

| Parameter | Minimum | Typical | Maximum | Units |
|---|---|---|---|---|
| Measured Average Sensitivity (2)<br>• 52 Mb/s<br>• 155 Mb/s | <br>-38<br>-34 | <br>-42<br>-38 | | <br>dBm<br>dBm |
| Maximum Input Power | -3 | 0 | | dBm |
| Link Status Threshold<br>(Flag - Logic Low)<br>Decreasing Light Input<br>• 52 Mb/s<br>• 155 Mb/s | <br><br>-45<br>-45 | | <br><br>-40<br>-36 | <br><br>dBm<br>dBm |
| Link Status Threshold<br>(Flag - Logic High)<br>Increasing Light Input<br>• 52 Mb/s<br>• 155 Mb/s | <br><br>-44<br>-44 | | <br><br>-39<br>-35 | <br><br>dBm<br>dBm |

**Figure 11.23** (continued)

2. Measured with an optical input using a $2^{23}$-1 pseudo random pattern with a 50% duty cycle, for a BER of $10^{-10}$

The next optical characteristic of concern to us is "Maximum Input Power." This number tells us at what power level a receiver will reach saturation. For example, an AGC circuit at the optical front end allows a very high range of the received power (from –42 dBm to 0 dBm, that is, from 63 nW to 1 mW) to get through. The difference between the maximum and minimum received power is called the *dynamic range* of a receiver. Thus, if the maximum power is –5 dBm and the sensitivity (the minimum power) is –45 dBm, the dynamic range is 40 dB. This is an important characteristic of receivers used in fiber-optic networks, where a variety of power levels of transmitted signals is usually encountered.

The "Link Status Threshold" shows the minimum power level that is acceptable for this receiver: from –45 dBm to –44 dBm. Observe the difference of 1 dB between these numbers for increasing and decreasing light input; this is the hysteresis discussed in the subsection "Signal

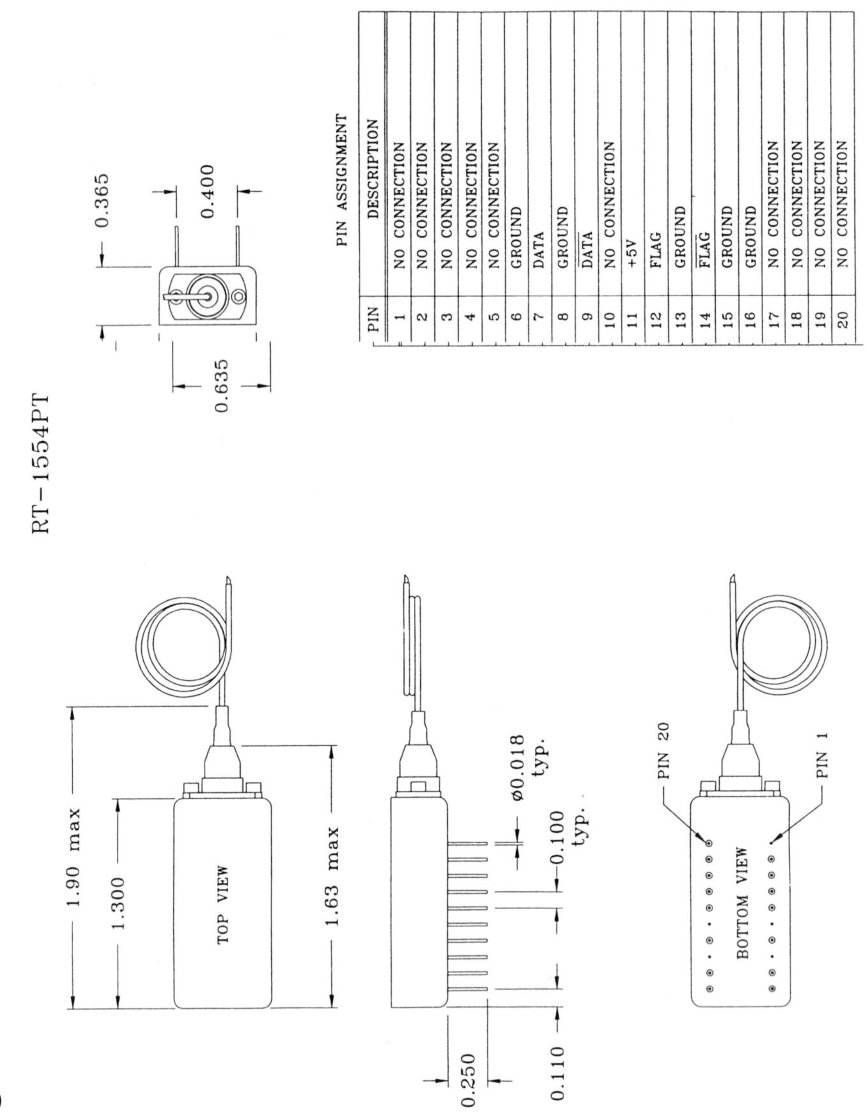


**Figure 11.23** (continued)

detect circuit" of this section. The mechanical dimensions segment of the data sheet gives you an idea of the actual size and arrangement of the receiver.

## Opto-Electronic IC (OEIC)

You may wonder why manufacturers do not integrate the entire receiver into one chip, as is the case with most other semiconductor devices. Doing so would minimize noise and eliminate all parazitics, thus making the design of a printed circuit board (PCB) unnecessary. The answer lies in the fact that, for one thing, a photodiode is made from InGaAs, whereas most semiconductor chips are made from Si and GaAs. Integrating different materials into a single chip is still impossible. On a positive note, however, there are several intermediate solutions to this problem and opto-electronic integrated circuits (OEIC) is one area of active research.

## SUMMARY

- A receiver is a unit of a fiber-optic communications link that converts an optical information signal into an electrical signal. A receiver consists of coupling optics, a photodetector, and electronics. Only miniature semiconductor devices—photodiodes (PDs)—are used as photodetectors in fiber-optic receivers.

- To better understand the physics behind a photodiode's operation, review Chapter 9. There you learned that the energy of an optical signal is used to pump electrons at the upper (higher) energy conduction band. These free electrons, set in motion by reverse-biasing voltage, produce photocurrent. This is the basic mechanism for converting a light signal into an electrical signal. On–off optical flashes converted into on–off electric pulses drive the electronics that follow in the system.

- Since PDs are semiconductor diodes, an understanding of the photodetecting mechanism can be ascertained through an analysis of a *p-n* junction. When an external photon (optical signal) strikes a depletion region, electrons and holes are separated. Reverse-biasing voltage removes them from the depletion region, thus releasing a flow of charge carriers. This is photocurrent, so called because light triggers its origin. (It is initiated by the absorption of photons.)

- Both models—energy bands and the *p-n* junction—describe the same two-step process: In the first step, external photons split electron-hole pairs in the depletion region; this is equivalent to saying that electrons from the valence band are pumped to the conduction band. In the second step, these electrons (and holes) are made to flow. Electrons and holes are swept from the depletion region either by depletion voltage (a photovoltaic mode of operation) or by reverse-bias external voltage (a photoconductive mode of operation). A photodiode can work without any bias but reverse bias makes optical-to-electrical conversion much more effective.

- The input to a photodiode is light power ($P$) and the output is photocurrent ($I_p$.). It follows from the principle of operation that the more photons that strike the active area of a PD, the more charge carriers will be created—that is, the higher will be the photocurrent. Thus,

$$I_p = R\,P, \qquad (11.1)$$

  where constant $R$ (A/W) is called *responsivity.* This is one of the main characteristics of a photodiode because it shows how efficiently a PD converts light into an electrical signal. Responsivity typically ranges from 0.4 A/W to 1.0 A/W.

- Responsivity is proportional to the wavelength of the received light and restricted from both the short and the long wavelength sides. The shorter the wavelength, the farther the photogenerated electrons are from the edge of the valence band and, therefore, the less probability there is of their becoming excited at the conduction band. Within the working region, responsivity is almost proportional to wavelength. This is so because the longer the wavelength, the greater the number of photons required to produce the same amount of light power; hence, a larger number of electrons are made to flow. The phenomenon known as *long-wavelength cutoff* occurs when the energy of the photons becomes less than the energy gap of a semiconductor; at that point, photons will not be absorbed.

- Responsivity describes the efficiency of the conversion of light power into photocurrent. Analysis reveals that the wider the depletion (active) region, the more efficient the conversion process. For a simple *p-n* photodiode, the only way to increase this width is to increase the reverse bias. But this measure conflicts with bandwidth requirements. Thus, there is a trade-off between power and bandwidth in a *p-n* photodiode.

- The bandwidth of a photodiode is inversely proportional to two time constants: transit time ($\tau_{tr}$) and R-C constant ($\tau_{RC}$) so that

$$BW_{PD} = 1/[2\pi(\tau_{tr} + \tau_{RC})] \qquad (11.12)$$

  where $\tau_{tr} = w/v_{sat}$ (Equation 11.9) and $\tau_{RC} \approx R_L C_{in} = (R_L \in A)/w$ (Equations 11.10 and 11.11). Here, $w$ is the width of the depletion region, $R_L$ is the load resistance, and all other members are constants for a given photodiode. Under these circumstances, we need to have an optimal value for the width of the depletion region. The third mechanism restricting bandwidth in a *p-n* photodiode is the diffusion current created by the charge carriers generated outside the depletion region. The point of all these considerations is that we need to increase the width of the depletion region but without manipulating the reverse-biasing voltage.

- The solution to this dilemma is a positive-intrinsic-negative, *p-i-n,* photodiode, where a thick intrinsic semiconductor layer is sandwiched between the *p* and *n* regions. This structure provides a wide depletion layer without manipulating reverse bias. This results in a number of advantages of *p-i-n* photodiodes over *p-n* PDs. The *p-i-n* PDs are the most commonly employed photodiodes in today's fiber-optic communications systems.

- Another popular type of PD is the avalanche photodiode (APD). Here, relatively high (from 20 to 100 V) reverse voltage accelerates photogenerated electrons so that they collide with neutral atoms and generate secondary electrons; these, in turn, generate other electrons by collision, thereby creating an avalanche process. In other words, an APD exhibits inherent gain without the extra noise associated with an external amplifier. This ability results in the very high sensitivity of an APD—the highest among all types of photodiodes. The bandwidth of an APD is a function of many factors and, in general, is comparable to the bandwidth of a *p-i-n* photodiode. Avalanche photodiodes are used when high sensitivity is the most desired feature of a photodetector.

- A metal-semiconductor-metal (MSM) photodetector is being developed today. This is not a *p-n* junction photodiode but the basic mechanism of light-current conversion is still the same: Photons generate electrons and holes and the electric field created makes them flow. MSM photodetectors demonstrate a high-bandwidth feature (up to 300 GHz), while their responsivities remain at a moderate level (from 0.4 to 0.7 A/W).
- Several characteristics of a photodiode should be kept in mind when evaluating the quality of the device. These are responsivity, bandwidth, capacitance, dark current (the current flowing through a photodiode circuit without incident light), and reverse voltage. The graph "Responsivity vs. Wavelength" presents the operating range of $\lambda$. All these characteristics are found in the data sheets of a photodiode. In addition, such information as optimal axial and angular displacements, current versus voltage, dark current versus temperature and voltage, capacitance versus voltage, response linearity, and rise–fall time can also be found in the data sheets.
- The range of operating wavelengths of photodiodes is determined by the energy gaps of their materials. Thus, Si PDs work only in the first transparent window of an optical fiber (around 850 nm) and InGaAs photodiodes cover the second (near 1300 nm) and the third (around 1550 nm) windows. In contrast to light sources, we want PDs to have wide spectral widths.
- Photodiodes exhibit noise current that originates from shot-noise, thermal-noise, and dark-current-noise mechanisms. For *p-i-n* photodiodes, thermal noise is the main source of trouble. To decrease thermal noise, we need to increase the load resistance ($R_L$). This requirement is exactly the opposite of the need to decrease the load resistance for a better bandwidth. To complicate the matter, noise is proportional to bandwidth. Fortunately, this dependence gives the designer an opportunity to increase the signal-to-noise ratio by filtering out excess bandwidth.
- Signal-to-noise ratio (SNR) is the ratio of the signal to the noise power. This is one of the main criteria for judging the quality of a photodiode. For a given *p-i-n* photodiode, the only means to improve this ratio is to increase the load resistance. An APD photodiode in the thermal-noise limit raises the signal-to-noise ratio by the square of the multiplication factor, that is, more than a hundred times. We typically need an SNR of more than 6 to maintain the bit-error rate (BER) at $10^{-9}$.
- Another important characteristic of a photodiode is noise-equivalent power (NEP), which is the minimum light power that results in SNR equaling 1. NEP is usually normalized by bandwidth so that you find NEP measured in watts and in watts/$\sqrt{\text{Hz}}$. Typical NEP values are expressed in units of nW and normalized NEP is less than 1 nW/$\sqrt{\text{Hz}}$.
- Bit-error rate is the number of erroneous bits per total number of bits transmitted. Obviously, this is a statistical characteristic and we are talking about average numbers. BER depends on signal-to-noise ratio, but it can easily be expressed as a function of digital SNR, which is the ratio of signal current to the RMS value of noise current.
- Digital SNR helps to introduce another important characteristic of a photodiode: sensitivity. A photodiode's (or a receiver's) sensitivity is the minimum received power required to maintain a given bit-error rate. Obviously, to achieve a better BER requires a device with greater sensitivity. Typical values of a receiver's sensitivity range from −30 to −45 dBm.
- Most of this chapter is devoted to photodiodes because this device is the heart of a receiver. But, ultimately, not only the photodiode but the entire unit—the receiver—acquires the information and transmits it to the customer. A receiver typically includes the following major components: an optical front end, a quantizer, a clock-recovery circuit, a signal-detect circuit, and buffers.
- The optical front end consists of a PD and preamplifier. The most popular type of preamplifier is the transimpedance amp, which employs an amplifier with negative feedback. The transimpedance design is the optimum solution to the power bandwidth trade-off. A quantizer usually includes a noise filter, a power amplifier/limiter, and a decision circuit. The design of the decision circuit is one of the major concerns in the design of a receiver. (See Section 11.4.) The clock-recovery circuit retrieves timing information from a data stream that it receives. The signal-detect circuit actually monitors the power level of the input signal and alarms when this level is below an acceptable threshold. Buffers reshape electrical signals in the required manner while the signals' logic remains unchanged.

Since many of the parameters of a receiver depend on the duty cycle of the data stream received, deviation of the duty cycle from 50% usually leads to degradation of the receiver's sensitivity and of other characteristics. (The duty cycle is the ratio of the average number of 1's and 0's in a data stream.)

- A typical receiver data sheet contains most of the characteristics discussed above. The only addition we can include is "maximum received power." Not all manufacturers' data sheets list this. The number given for maximum received power restricts the upper limit of received power because exceeding this limit will adversely affect the receiver's operation. The difference between maximum and minimum received power is called the receiver's dynamic range.

- There is ever-increasing demand for higher-quality fiber-optic communications systems, particularly in terms of the need for more bandwidth to take the pressure off the individual components of the receiver. Recent developments in receiver technology are providing adequate response to this demand, yet the need to meet the new goals—terabits per second, for instance—in fiber-optic transmission is pushing physicists and engineers to strive with all available resources to keep pace with the industry's unending demand for greater and greater transmission speed.

## PROBLEMS

**11.1.** What is the major function of a photodiode?

**11.2.** Draw a block diagram of a receiver.

**11.3.** How does a PD convert light into an electrical signal?

**11.4.** What is the input–output characteristic of a photodiode?

**11.5.** The responsivity of a PD is 0.9 A/W and its saturation power is 2 mW. What is the photocurrent if the received power is (1) 1 mW; (2) 2 mW; (3) 3 mW?

**11.6.** How does responsivity depend on the wavelength of an optical signal?

**11.7.** Why does responsivity depend on wavelength?

**11.8.** What is the responsivity of an InGaAs photodiode if its quantum efficiency is 95%?

**11.9.** What is the responsivity of an Si photodiode if its quantum efficiency is 90%?

**11.10.** What is the cutoff wavelength of a photodiode? What is the physical reason for the existence of a cutoff wavelength?

**11.11.** What is the reason for responsivity falling to zero at the shorter and longer wavelengths?

**11.12.** Why will short-wavelength photons, which penetrate deeper into a semiconductor material than long-wavelength photons, not be absorbed at the depletion region?

**11.13.** What must the thickness of the depletion region of an Si photodiode be to make its quantum efficiency 85%?

**11.14.** What is the difference between photovoltaic and photoconductive modes of operation of a photodiode?

**11.15.** List four advantages of using reverse bias in a photodiode.

**11.16.** a. Give your definition of the bandwidth of a photodiode.

b. What factors restrict the bandwidth of a *p-n* photodiode?

**11.17.** Why is the junction (shunt) resistance of a *p-n* photodiode very high?

**11.18.** Calculate $w_{opt}$ if $v_{sat} = 10^5$ m/s, $D_A = 75$ μm, $\epsilon_r = 11.7$, and $R_L = 50$ kΩ.

**11.19.** What does the notation *p-i-n* stand for?

**11.20.** List the advantages of a *p-i-n* photodiode over a *p-n* photodiode.

**11.21.** In the rear-illuminated *p-i-n* photodiode, the *n*+ layer is transparent for the incident light. (See Figure 11.6[b].) How is this transparency achieved?

**11.22.** What factor must you take into account to choose the correct width of an intrinsic layer in a *p-i-n* photodiode?

**11.23.** What is the bandwidth of a germanium photodiode with $w = 20$ μm and $v_{sat} \sim 10^5$ m/s?

**11.24.** What is the meaning of "avalanche" in the term *avalanche photodiode?*

**11.25.** What is the major advantage of an APD over a *p-i-n* photodiode?

**11.26.** What are the drawbacks of an APD?

**11.27.** The text says that bandwidth-gain product is an important characteristic of an APD. Why don't we use this characteristic for a *p-i-n* photodiode?

**11.28.** An APD and a *p-i-n* photodiode are radically different in their principle of operation. Why, then, is the same assumption—$\tau_{tr} \gg \tau_{RC}$—used for both photodiodes?

**11.29.** What semiconductor material is the photodiode shown in Figure 11.9 made from?

**11.30.** In Figure 11.9, why is the responsivity of the PD at 1300 nm and at 1550 nm different?

**11.31.** Explain why the graph "Responsivity vs. Radial Displacement" in Figure 11.9 does not exhibit any maximum or minimum points.

**11.32.** Refer to the data sheet of the MF-432 photodiode (Figure 11.9):
a. How does the optimal radial displacement ($z_{opt}$) relate to the focal distance ($F$)?
b. What is the value of $z_{opt}$?
c. What is the value of $r_{opt}$? Explain.

**11.33.** Why does the manufacturer specify $R_L$ when characterizing the bandwidth of a photodiode?

**11.34.** Show qualitatively how the capacitance of a PD depends on reverse voltage.

**11.35.** Explain the dependence of dark current on temperature using the *p-i-n* junction model.

**11.36.** Find the formula governing the graph $I_d$ vs. $T$ for a PIN InGaAs photodiode (Figure 11.24).

**11.37.** Discuss how you would approximate dependence $I_d$ versus $V$ given in Figure 11.24.

**11.38.** Calculate $\tau_{RC}$ of an MF-432 PIN photodiode. (See Figure 11.9.)

**11.39.** Reverse biasing improves a photodiode's linearity, increases its speed of operation, and reduces its capacitance. Explain the physical mechanisms underlying these effects of reverse biasing.

**11.40.** *Project:* Analyze the data sheet of the PIN photodiode given in Figure 11.24.

**11.41.** Explain how an MSM photodetector works.

**11.42.** List the sources of noise in a photodiode in the order of their importance in forming the output signal.

**11.43.** Derive Formulas 11.21 and 11.22.

**11.44.** Derive Formula 11.23.

**11.45.** Derive Formula 11.26 from Formula 11.25.

**11.46.** Briefly explain the physical mechanisms causing shot noise, thermal noise, dark-current noise, and $1/f$ noise in a *p-i-n* photodiode.

**11.47.** Why are all the sources of noise currents connected in parallel in Figure 11.12?

**11.48.** Calculate the RMS and bandwidth-normalized values of total noise current for a photodiode whose data are given in Figure 11.24 if the average input power is 0.1 μW, $\lambda$ = 1550 nm, $R_L$ = 50 kΩ, and the diode operates at room temperature.

**11.49.** Derive the formula for SNR involving the quantum efficiency of a photodiode.

**11.50.** Calculate SNR for the photodiode operating at room temperature described in Figure 11.24 if $P$ = 0.1 μW.

**11.51.** Calculate the thermal limit of the SNR PD described in Figure 11.24.

**11.52.** Calculate SNR for a photodiode whose data are given in Figure 11.24 if the average input power is 0.1 μW, $\lambda$ = 1550 nm, and the diode is operating at room temperature with the following parameters:
a. $R_L$ = 50 Ω, $R_L$ = 500 kΩ, and $R_L$ = 50 MΩ.
b. $BW_{PD}$ = 50 kHz, $BW_{PD}$ = 50 MHz, $BW_{PD}$ = 5 GHz.

Draw conclusions as to how load resistance and bandwidth affect SNR.

**11.53.** Briefly describe the physics behind shot noise and thermal noise in an avalanche photodiode.

**11.54.** Calculate $SNR_s$, $SNR_t$, and overall SNR for Si and InGaAs avalanche photodiodes if $M$ = 10, $P$ = 0.1 μW, $R$ = 0.9, $R_L$ = 50 kΩ, $BW_{PD}$ = 2.0 GHz, and $T$ = 300°K.

**11.55.** Example 11.3.3 shows that an Si APD has a much larger SNR than an InGaAs APD. Why, then, do we still need to use an InGaAs APD?

**11.56.** Give the definition and explain the meaning of the term *noise-equivalent power.*

**11.57.** Derive Formula 11.43.

**11.58.** Calculate NEP and NEP($\lambda$) for an InGaAs PIN PD with the following set of parameters: $NEP_{norm} = 5\ pW/\sqrt{Hz}$ from dc to 20 MHz and $NEP_{norm} = 20\ pW/\sqrt{Hz}$ from 20 to 155 MHz; $R_{max}$ = 1.0 A/W at 1550 nm and $R(\lambda)$ = 0.8 A/W at 1300 nm. The photodiode operates at 1300 nm.

**11.59.** a. What is BER?
b. How can you improve BER?

**11.60.** BER is defined through $P(1/0)$ and $P(0/1)$. What are the explicit expressions for $P(1/0)$ and $P(0/1)$?

**11.61.** Derive Formula 11.53.

**11.62.** Derive Formula 11.56.

**11.63.** Why does BER depend on the $Q$ parameter?

**11.64.** What is the sensitivity of a photodiode used in a fiber-optic communications system?

**11.65.** Does the RMS value of noise current depend on minimum received power?

**11.66.** Construct the graph BER versus $P_{min}$ using the following data: $BER_{min} = 1/2[\mathrm{erfc}(Q/\sqrt{2})] \approx [\exp(-Q^2/2)]/(Q\sqrt{2\pi})$, where $Q = (2RP_{min})/\{\sqrt{[(4eRP_{min}BW_{PD})((4k_BT/R_l)BW_{PD})]} + \sqrt{[(4k_BT/R_L\ BW_{PD}]}$. Values: $e = 1.6 \times 10^{-19}$ C; $R$ = 0.9 A/W; $BW_{PD} = 156 \times 10^6$ bit/s; $k_B = 1.38 \times 10^{-23}$ J/K; $T$ = 300°K; $P_{min}$ = {–40; –39; –38; –37; –36; –35; – 34} dBm.

**AMP** Singlemode InGaAs Photodetector

Catalog 124002
6-97

## Features

- High reliability passivated planar structure
- High responsivity
- Low dark current
- Low capacitance
- Hermetically sealed TO-18 style package installed in industry standard ADMs
- –40° to +85°C operating temperature

The AMP singlemode design uses an InGaAs PIN photodetector to offer high responsivity for nearly all digital and analog fiber optic applications. The AMP unique design balances high-speed performance with noise-free linear output. Spectral response has been optimized for the long wavelength region of 1150 nm to 1600 nm. Every component delivered has passed extensive high-temperature screening to ensure long-term reliability.

Compatible with industry standards, the AMP Active Device Mount components incorporate hermetically sealed TO-18 style PIN packages which have been actively aligned for optimal performance.

Products available include a TO-18 can, FC, SC or ST style active device mounts.

Each unit is burned-in. Responsitivity, dark current and capacitance are measured on each unit. No data is supplied with the unit. A lot code is used for traceability.

For additional information on product qualification, reference Product Specification 108-55009.

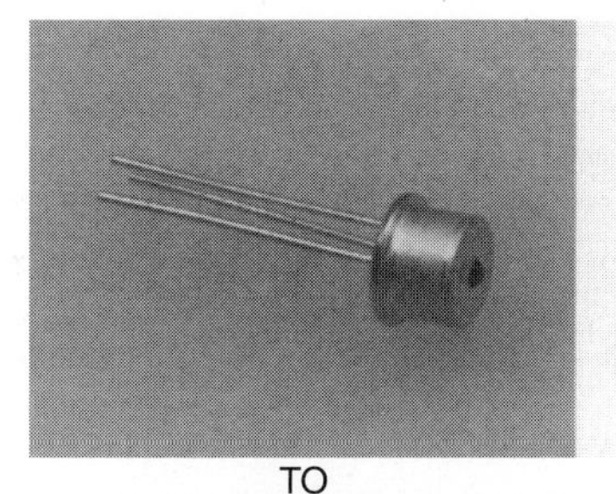
TO

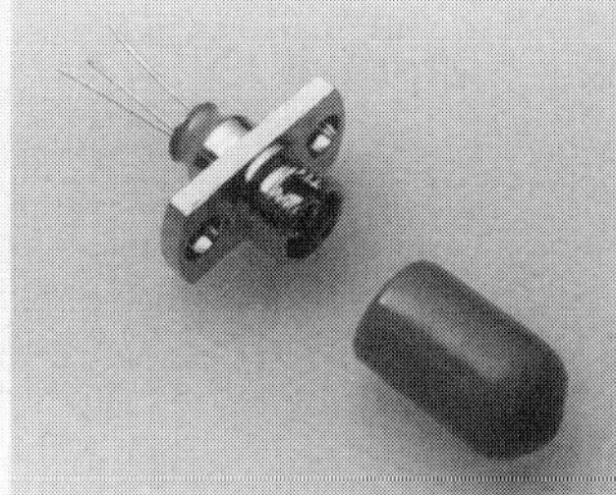
FC

SC

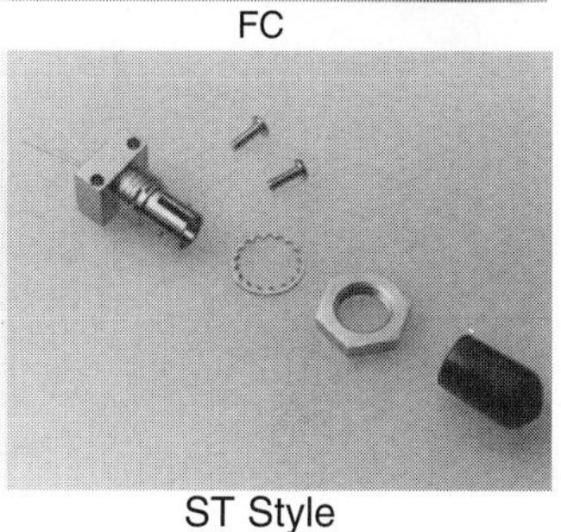
ST Style

**Specifications: 25°C, –5 Volts, 9 μm Fiber**

| Parameter | P/N Suffix | Test Conditions | Units | Min. | Typ. | Max. |
|---|---|---|---|---|---|---|
| Responsivity | -3 | Laser source of 10 μW | A/W | .75 | .86 | — |
| Spectral Response | -3 | — | nm | 1150 | — | 1600 |
| Capacitance | -3 | f=1 MHz | pF | — | .9 | 1.1 |
| Dark current | -3 | — | nA | — | 1.5 | 5.0 |
| Rise/fall | -3 | — | ns | — | — | 1 |
| Bandwidth | -3 | — | GHz | — | 2.0 | — |
| Reliability | -3 | $I_D$>5nA | hrs | — | $2.0 \times 10^8$ | — |

**Absolute Maximum Rating**

| | Units | Min. | Max. |
|---|---|---|---|
| Operating temperature | C | –40 | 85 |
| Storage temperature | C | –40 | 125 |
| Reverse current | mA | — | 1 |



For drawings, technical data or samples, contact your AMP sales engineer or call the AMP Product Information Center 1-800-522-6752.
Dimensions are in inches and millimeters unless otherwise specified. Values in brackets are metric equivalents.
Specifications subject to change. Consult AMP Incorporated for latest specifications.

**Figure 11.24** Data sheet of a singlemode InGaAs photodiode. *(Courtesy of AMP Inc.)*

**Singlemode InGaAs Photodetector**

Catalog 124002
6-97

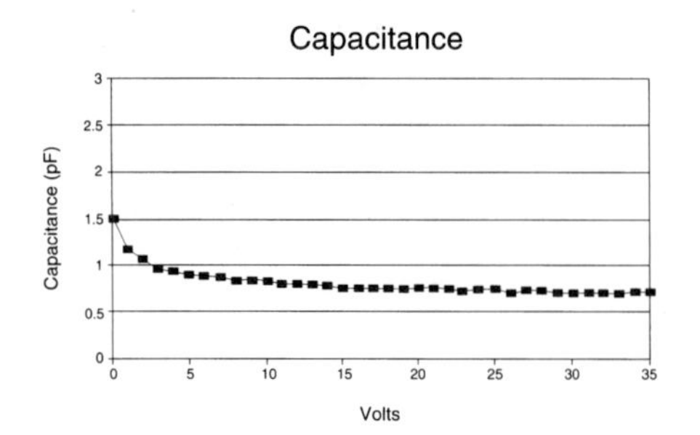

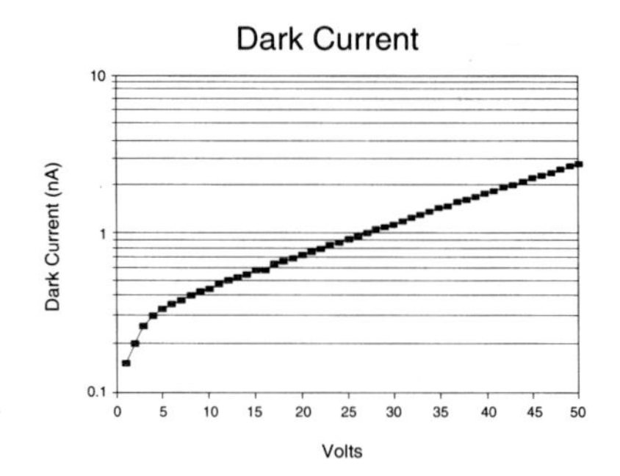

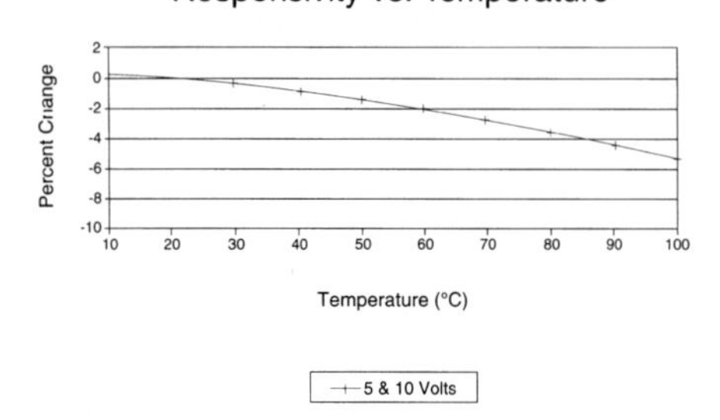

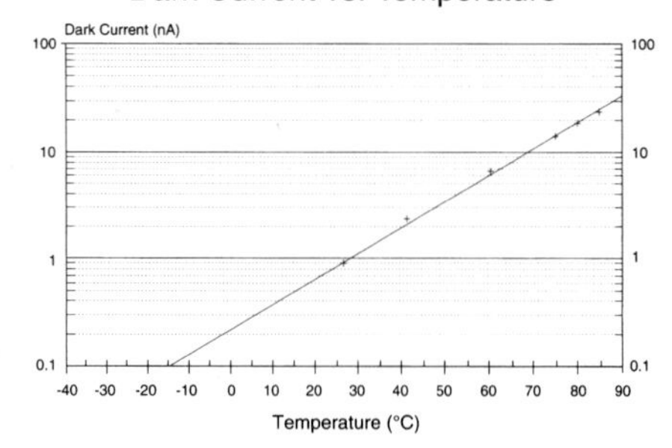

**Part Numbers**

| | Connector Interface | | | |
|---|---|---|---|---|
| | TO | FC | SC | ST Style |
| Elite | 259007-3 | 259015-3 | 269027-3 | 259013-3 |

**Mechanical Dimension Reference**

| | | | | |
|---|---|---|---|---|
| Figure # | 2 | 3 | 5 | 7 |
| Page No. | 85 | 86 | 86 | 87 |

For drawings, technical data or samples, contact your AMP sales engineer or call the AMP Product Information Center 1-800-522-6752.
Dimensions are in inches and millimeters unless otherwise specified. Values in brackets are metric equivalents.
Specifications subject to change. Consult AMP Incorporated for latest specifications.

**Figure 11.24**
(continued)

Plot the graph for: (1) $R_L = 50 \times 10^3\ \Omega$; (2) $R_L = 50\ \Omega$; (3) $R_L = 5 \times 10^6\ \Omega$. (Show the value of the parameter on each curve.)

**11.67.** Derive Formula 11.63.

**11.68.** What is the sensitivity of an AMP photodiode where $BER = 10^{-9}$, $R_L = 50\ k\Omega$, and operation is at room temperature? (See Figure 11.24.)

**11.69.** Why does receiver sensitivity depend on a photodiode's bandwidth?

**11.70.** Plot and analyze the graph $BER_{min}$ *(p-i-n)* versus $N_p$.

**11.71.** What is the quantum limit for a photodetector? What is its importance to the designer of a fiber-optic communications system?

**11.72.** What is a receiver? What is the difference between a receiver and a photodiode?

**11.73.** Draw a functional block diagram of a receiver and briefly explain the function of each component.

**11.74.** a. What is an optical front end of a receiver?

b. What is the major factor involved in the design of an optical front end?

**11.75.** What designs of an optical front end are you aware of?

**11.76.** What trade-off does the designer of an optical front end have to consider when choosing a type of preamp?

**11.77.** What do the acronyms PINAMP and PINFET stand for?

**11.78.** Draw a block diagram of a quantizer.

**11.79.** Describe the principle of operation of a decision circuit.

**11.80.** What is the main problem in designing a decision circuit? Explain.

**11.81.** Why is threshold such a critical factor in designing a decision circuit?

**11.82.** Briefly describe the possible designs of a decision circuit. List the advantages and drawbacks of each design.

**11.83.** What is the function of buffers?

**11.84.** Why do we need a clock-recovery circuit?

**11.85.** Explain the principle of operation of a clock-recovery circuit.

**11.86.** What is the function of a signal-detect circuit?

**11.87.** A signal-detect circuit employs hysteresis in its operation. Explain what hysteresis is and why we need to use it.

**11.88.** What control and monitoring circuit does a typical receiver usually include?

**11.89.** Why is a duty cycle important to the operation of a receiver?

**11.90.** What does OEIC stand for?

**11.91.** Build a memory map. (See Problem 1.20.)

## REFERENCES[1]


1. Govind Agrawal, *Fiber-Optic Communication Systems,* 2d ed., New York: John Wiley & Sons, 1997.
2. *Diode Lasers & Instruments Guide,* Melles Griot, Boulder, Colo., April 1998.
3. *Optoelectronics Data Book*, Advanced Photonix, Inc., Camarillo, Calif., 1999.
4. *InGaAs Photodiodes*, Fermionics Opto-Technology, Simi Valley, Calif., 1998.
5. *Lightwave 1999 Worldwide Directory of Fiber-Optic Communications Products and Services*, March 31, 1999, pp. 62–66.
6. *Optoelectronic Solutions*, Mitel Semiconductor, Ontario, Can., 1997.
7. *Silicon Photodiodes*, Centro Vision Inc., Newbury Park, Calif., March 1998.
8. Chin-Len Chen, *Elements of Optoelectronics and Fiber Optics,* Chicago: Irwin, 1996.
9. Leonid Kazovsky, Sergio Benedetto and Alan Willner, *Optical Fiber Communication Systems,* Boston: Artech House, 1996.
10. Bryon Kasper, "Receiver Design," in *Optical Fiber Telecommunications-II,* ed. by S. E. Miller and I. P. Kaminow, San Diego: Academic Press, 1988, pp. 589–722.
11. S. R. Forrest, "Optical Detectors for Lightwave Communication," in *Optical Fiber Telecommunications-II,* ed. by S. E. Miller and I. P. Kaminow, San Diego: Academic Press, 1988, pp. 569–599.
12. *Product Catalog*, vol. 9 (1998/99), New Focus, Inc., Santa Clara, Calif., 1998.


[1] See Appendix C: A Selected Bibliography

**13.** Alan Mickelson, Nagesh Basavanhally, and Yung-Cheng Lee, eds., *Optoelectronic Packaging,* New York: John Wiley & Sons, 1997.

**14.** *PINAMP receiver modules* and *PINFET optical receivers* (data sheets), Laser Diode, Inc., Edison, N.J., 1997.

**15.** *Frequency Control Products,* Vectron International, Norwalk, Conn., 1997.

**16.** *High speed clock recovery module* (data sheet), Broadband Communications Products, Inc., Melbourne, Fla., September 1996.

**17.** *Fiber Optic Products* (Catalog 27), Honeywell Inc., Micro Switch Div., Freeport, Ill., June 1998.

**18.** "Fiber Optic Transceivers," *Designer's Guide,* AMP Inc., Harrisburg, Pa., 1996.

# 12

# Components of Fiber-Optic Networks

Our discussions in the previous chapters concentrated on the point-to-point transmission aspects of fiber-optic communications systems. We considered how to convert an electrical information signal into an optical one, how to transmit an optical signal over a fiber, and how to convert an optical information signal back into electrical form. Detailed discussions of these aspects revealed serious problems in fiber-optic transmission, such as bandwidth and distance limitations imposed by the properties of transmitters, optical fibers, and receivers. We learned that all transmission devices are designed to minimize signal deterioration and provide the best technological performance of fiber-optic communications systems as economically as possible.

Even though point-to-point transmission technology continues to develop rapidly, the main focus in today's fiber-optic communications is on networks. This is why the following chapters are devoted to this critical technology.

## 12.1 FIBER-OPTIC NETWORKS: AN OVERVIEW

### Point-to-Point Links

When optical fiber became a commercially feasible transmission medium, long-distance and regional telephone companies started to deploy fiber on a scale of millions of miles. In the late 1980s and early 1990s, they installed many million miles of singlemode fiber to connect their switching facilities located throughout the country. (See Figure 1.8.) In this way, telcos and long-distance carriers have utilized the tremendous transmission capacity of an optical fiber, while the nodes—switches—remain electronic. Thus, they've developed point-to-point fiber-optic communications links.

Let's consider, for instance, long-distance telephone networks. The long-distance carriers use *mesh topology* for their networks. (See Figure 1.2.) The advantage of mesh topology can be easily understood from the following simple example:

Suppose you need to place a call from New York City to Los Angeles. Your request for a connection goes to your central *end* office; from there it is directed through a central *tandem* office to the long-distance carrier's nearest switching center. You may recall that the *central end office (CO)* is your nearest switching center. Equal access to your telephone line for any long-distance carrier is gained through what's called the *point of presence (POP)*. From a long-distance tandem switch, your request goes directly to another long-distance switch, the one that is nearest to Los Angeles; from there it is directed to the appropriate central office and then to your calling party. This calling process is shown schematically in Figure 12.1.

Thus, your call is transmitted from long-distance switch 1 directly over optical fiber to long-distance switch 2. If something goes wrong and the line from switch 1 to switch 2 is not

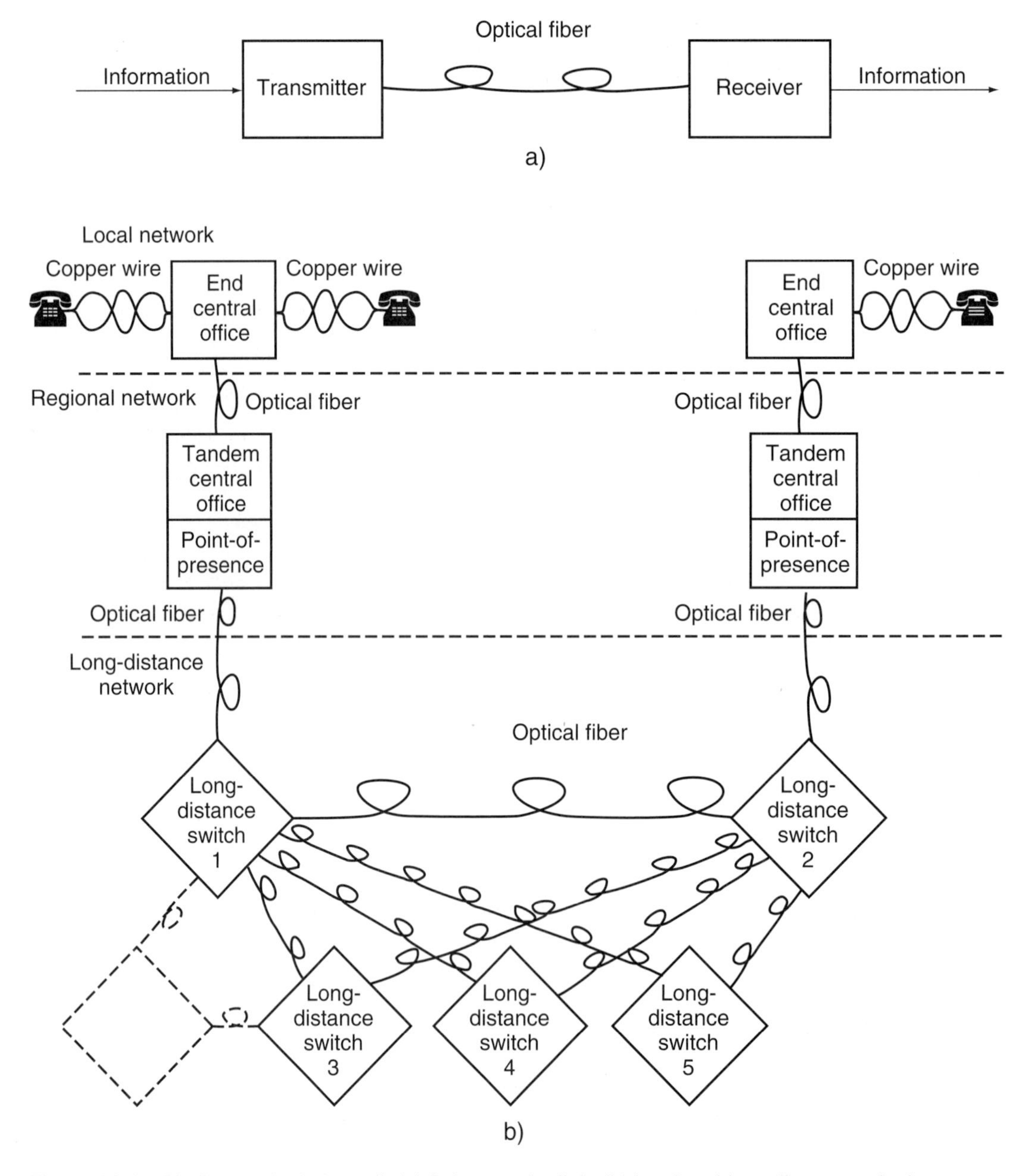


**Figure 12.1** Placing a telephone call: (a) Point-to-point link; (b) local and long-distance telephone networks.

available (it might be overloaded or in the throes of a catastrophic break), your call is directed from switch 1 through switch 3 to switch 2. Switch 3 in this case serves as a *tandem switch.* If the line through switch 3 doesn't work, switch 4 will be used. In any case, not more than one tandem switch is required to transmit your call to any location in the country. AT&T's long-distance network, for instance, contains more than 100 switches located throughout the United States and all of them are interconnected.

If you need to place a call within your region, your telco's central office directs your call to the appropriate central office nearest the area you are calling without involving the long-distance network. Such a call is termed an intra-LATA connection. (*LATA* stands for *local access and transport area.*) It is also known as a *regional call.* This regional call is also transmitted over optical fiber. This connection can be seen at the top of Figure 12.1(b).

It is important to emphasize that both long-distance and local fiber-optic networks are essentially sets of *point-to-point links.* (Look again at Figure 1.1, where a point-to-point link is shown. This is repeated in Figure 12.1[a].) A signal placed on such a link is not removed or redirected; it is going from its point of origin to a specific destination.

Such sets of point-to-point links were the first fiber-optic networks and are still dominant today. However, when in recent years the demand for bandwidth started growing at an explosive rate, the situation began to change. Optical fiber, the only transmission medium able to satisfy this demand, started to inch closer and closer to offices and homes in telecommunications networks. At the same time, data networks began to rely exclusively on optical fiber at the metropolitan and local area levels. All networks—local, regional, and global—began to move from point-to-point architecture to more sophisticated topology, such as rings. A typical example is Project Oxygen, which was discussed in Chapter 1. Aiming to provide a truly global fiber-optic connection, it is actually a set of giant rings. These networks are quite different from a set of point-to-point links, as we will now see.

## Networks

Recall that a *network is a set of nodes connected by a link.* The connections can be effected in different ways, called *topologies.* (Typical network topologies used in telecommunications networks are shown in Figure 1.2.) Let's consider how, for example, a star network, such as that shown in Figure 12.2(a), works. Suppose a message from node 1 has to be delivered to node k. This message enters the server, where its destination node is read and the message is redirected to node k.

Similar functions have to be performed for a ring network (Figure 12.2[b]). Suppose a message goes from node 1 to node k and the direction of the information flow is clockwise. Then the message passes through nodes 2 and 3, where it is read and directed farther along the network without processing until it reaches its destination—node k. Thus, *redirecting a message is the most distinguishing feature of a network.* This function is called *switching,* or *routing,* depending on the specific application.

It is important to distinguish between the *physical topology* and the *logical (virtual) topology* of a network. For example, your circuitry can be connected in a star layout but information can flow in a ring topology, as Figure 12.2(c) shows. Again, this is achieved by redirecting (switching) the signal flow.

There are two basic *switching technologies:* circuit switching and packet switching. They are shown schematically in Figure 12.3.

Let's consider again a star network. Suppose once more you need to send a message from node 1 to node k. In *circuit-switching* technology, the server has a switch that physically connects a line from node 1 to a line of node k and your message flows physically and logically to its destination. If a server has the ability to physically connect every circuit it serves, it's called a *non-blocking switch.*

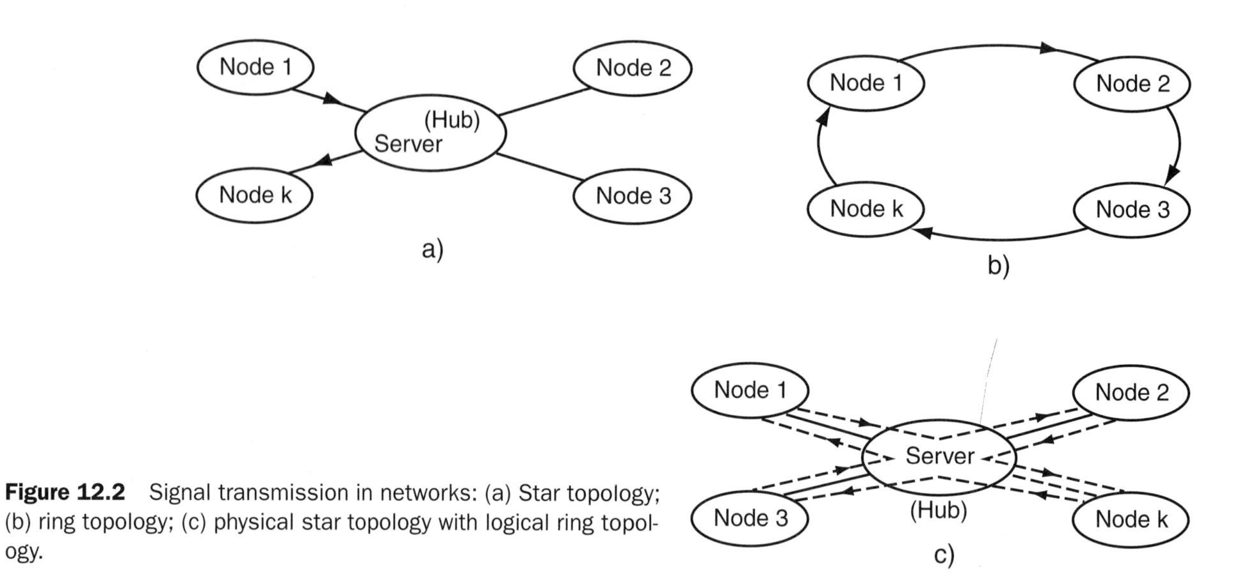


**Figure 12.2** Signal transmission in networks: (a) Star topology; (b) ring topology; (c) physical star topology with logical ring topology.

In *packet-switching* technology, a message is sent as a set of small packets (a certain number of bits). Thus, there is a continuous data stream from node 1 to a server; this data stream contains packets flowing from node 1 to node 2, denoted here as 1–2, along with messages from node 1 to node 3 (1–3) and from node 1 to node k (1-k). A server has control circuitry that enables it to recognize and direct all packets to their destination points. All lines from serving nodes are physically and permanently connected to one another within a server so, in this case, no physical switch is involved in directing messages. In our example, control circuitry separates the 1–2, 1–3, and 1-k packets and directs them to the 2, 3, and k nodes, respectively. Obviously, all nodes serve simultaneously as points of origin and destination.

In a circuit-switching network, the lines are connected by demand and stay connected regardless of whether transmission is actually going on or not. This is an ineffective way of using available bandwidth. In addition, in order to be connected, the nodes have to exchange "handshaking" messages, which may take even longer than the data transmission itself. For these reasons, circuit-

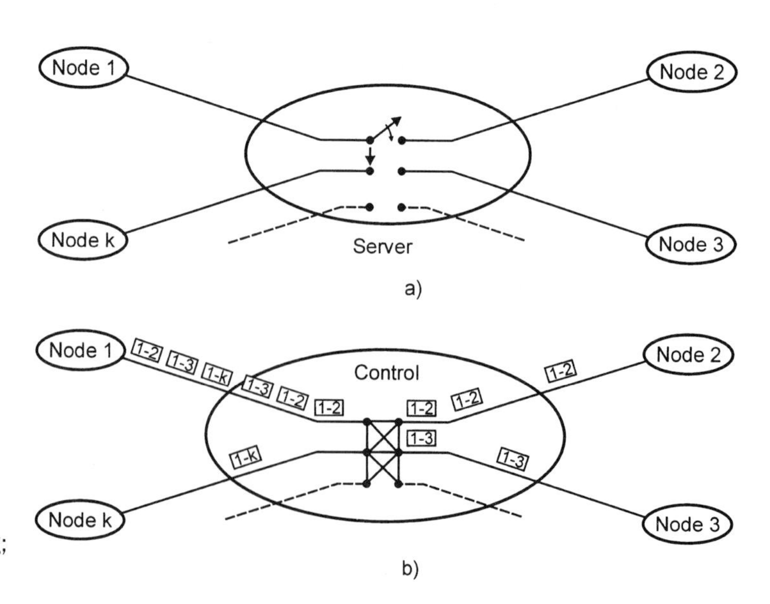


**Figure 12.3** Circuit-switching and packet-switching technologies: (a) Circuit switching; (b) packet switching.

switching networks are appropriate for transmission of rarely switched low-speed information such as telephone calls. In contrast, packet-switching networks transmit information only when necessary, thus using bandwidth very effectively. In addition, each packet carries service—address and synchronization—information with it. Hence, no "handshaking" delay is involved. What's more, each packet can be transmitted through an individual path; in other words, many packets, each containing a portion of a given message, may be transmitted in parallel, that is, simultaneously. This makes packet-switching transmission much faster than its circuit-switching counterpart. The price for these advantages is a more sophisticated and intelligent network controlling the processes of dividing a message into packets, transmitting the packets by different paths, and assembling the message from the arrived packets. But this intelligence is quite within the reach of modern digital technology and for these reasons packet-switching networks are becoming more and more popular.

One further note: Circuit-switching networks require a switching time in the range of a few ms, but packet-switching networks need a switching time around a few μs and even ns. This is because packet-switching time must be a small portion of the duration of an individual packet.

If you ponder our discussion, you will certainly conclude that, in general, a network exhibits a certain degree of "intelligence," such as the ability to recognize and redirect a message. One important conclusion that has to be drawn from this brief overview is that *a network is always a combination of physical and logical layers very much as a computer is a combination of hardware and software.*

***Time-division and wavelength-division multiplexing in fiber-optic communications*** The first phase in the deployment of fiber-optic networks occurred in the late 1980s, when the industry began to replace copper with optical fiber as the transmission medium. (Here, the term *copper* denotes all-electrical-based transmission technologies: copper wire, coaxial cable, and microwave.) It seemed at that time that optical fiber, with its tremendous information-carrying capacity (theoretically rated at 50 Tbit/s), would meet all possible bandwidth demands. However, today it is clear that assumption was wrong. Of course, a fiber's bandwidth is hundreds of times larger than copper's; however, demand for bandwidth is so high that today fiber-optic communications systems in their early forms could never satisfy these requirements.

A multimode fiber has severe bandwidth restrictions because of modal dispersion. But even singlemode fiber-optic systems cannot directly carry the traffic generated by the needs of modern society. This is because several obstacles impede efforts to increase the bit rate of a transmitted signal. First, you recall that transmission distance in most installed singlemode fiber is limited today by its dispersion and nonlinear effects (see Sections 6.3 and 6.4). But the main obstacle can be immediately understood if you look again at the point-to-point fiber-optic link shown in Figure 12.1(a). The theoretical bandwidth of a singlemode fiber is about 50 THz, while modern electronic transmitters and receivers can operate at no more than 10 GHz. Therefore, the bit rate of such a point-to-point transmission system is restricted by the system's electronic equipment. In other words, this transmission system utilizes only a small—1/5000—portion of the transmission capacity of a singlemode fiber.

The solution to this problem has existed in telecommunications even before the advent of fiber-optic technology: *multiplexing, which entails the transmission of several signals over the same channel.* For example, telephone companies have used time-division multiplexing (T1 and higher multiplexes) since the early 1960s to increase the capacity of installed copper lines.

Two multiplexing techniques are in use in fiber-optic communications: time-division multiplexing and wavelength-division multiplexing.

The concept of *time-division multiplexing (TDM)* can be understood by looking at Figure 12.4(a). Signals from Tx1 ("Tx" stands for transmitter), Tx2, and other transmitters enter the *multiplexer (MUX).* A multiplexer takes a sample of each signal, assigns a specific time slot to this sample, combines (multiplexes) these samples, and transmits them over the same line. This

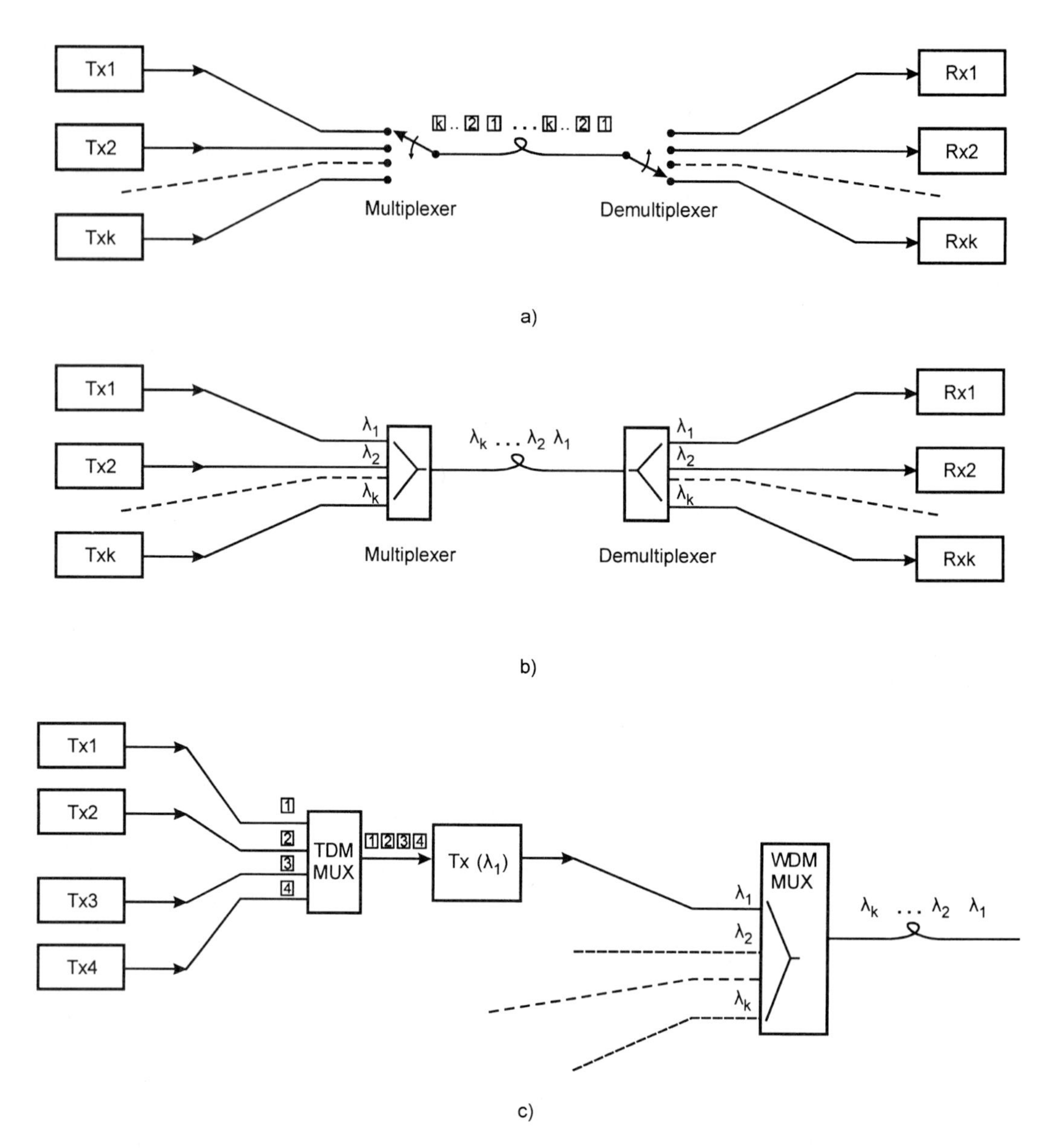


**Figure 12.4** Multiplexing: (a) Time-division multiplexing (TDM); (b) wavelength-division multiplexing (WDM); (c) TDM+WDM system.

procedure is shown schematically as the operation of a mechanical switch, which picks up one signal after another for a short time. Of course, no mechanical switches can be seen in the field nowadays, but this is a convenient way to illustrate the principle of TDM operation. Note that samples of signals from Tx1, Tx2, and the others are transmitted over the same line in a time sequence. At the receiver end, a *demultiplexer (DEMUX)* separates these samples and directs a signal from Tx1 to, for example, Rxk. ("Rx" stands for receiver.) Thus, a receiver obtains a signal from a transmitter as a sequence of time-slotted samples and then restores these to produce a coherent message.

How much bandwidth of an optical fiber does TDM use? Consider this: If you have four Tx's, each transmitting at 2.5 Gbit/s, your fiber has to carry a signal at a bit rate of 10 Gbit/s. Hence, *TDM simply increases the bit rate of a transmitted signal.* Assuming bandwidth (BW) is equal to bit rate (BR), you can readily see the leap that has been accomplished. The next question is, can we increase the number of multiplexed channels? Yes, but our multiplexer and demultiplexer would then have to operate at a higher bit rate. Both the TDM MUX and DEMUX are electronic circuits with their limits today of about 10 Gbit/s. Actually, 10-Gbit/s TDM systems are only now beginning to be deployed [1].

Since TDM is applied to fiber-optic communications networks, it is sometimes called *optical time-division multiplexing (OTDM).*

*Wavelength-division multiplexing (WDM)* is based on this fundamental physical principle: Several light beams at different wavelengths can simultaneously propagate over the same optical path without interference. "Two photons can occupy the same space" is the way one observer puts it [2]. The principle of operation of a WDM system is shown in Figure 12.4(b). Tx1 generates a data stream at wavelength $\lambda_1$, Tx2 uses $\lambda_2$, and so forth. All these signals are combined by a multiplexer and are transmitted simultaneously over the same optical fiber. A demultiplexer at the receiver end separates signals with different wavelengths and directs them to the appropriate receivers. From the transmission standpoint, *WDM divides the entire fiber-optic bandwidth into many segments and each signal (wavelength) uses its individual bandwidth segment.*

A common analogy exists between the methods to increase the capacity of fiber-optic and highway networks: TDM is analogous to increasing the speed of cars moving down a highway, while WDM is analogous to adding more lanes on the same road. This analogy underscores the principle by which each technology increases the speed of transmission: TDM does so in series, whereas WDM uses parallel transmission.

Are TDM and WDM, then, competitive technologies? No, they supplement and enhance each other. It is easy to see that each wavelength channel can carry several time-division multiplexed channels. This is schematically shown for one channel in Figure 12.4(c). Four individual channels are multiplexed by TDM MUX and presented to Tx ($\lambda_1$). The same holds for other WDM channels. As a result, each wavelength carries several time-multiplexed signals.

TDM and especially WDM technologies are the mainstream of today's developments in fiber-optic networks. Again, it is important to remember that the TDM bit rate is restricted by the speed of today's electronic components, which is about 10 Gbit/s. Unfortunately, we cannot increase bit rate beyond this until we have the technology to design new electronic components capable of operating at higher speeds. WDM alleviates the problem somewhat by taking the 10 Gbit/s bit rate and multiplexing many channels—for example, 16—to thereby increase the bit rate to 16 × 10 Gbit/s, or 160 Gbit/s. A WDM link with 128 wavelengths capable of transmitting at 1.28 Tbit/s is the commercial reality [3]. So, to a certain extent at least, it compensates for the limitations imposed by the electronic devices. But wavelength-division multiplexing does even more. Using WDM gives the network designer a tremendous flexibility because now he or she can associate an individual channel with a single wavelength and direct this wavelength channel from any point of origin to any destination point in the network. In short, WDM gives the designer an additional degree of freedom in designing and operating the fiber-optic network.

If you followed our discussion closely, you certainly noticed that, even though we mentioned networks several times, essentially our consideration of multiplexing related mostly to point-to-point transmission. This is because TDM and WDM are taking over as leading transmission technologies in point-to-point links. Fiber-optic networks with built-in TDM and WDM features are still in the early stages of deployment.

We will discuss these technologies in greater detail in Chapter 14. For now, we need only understand the principle of their operation.

*Important note:* Bear in mind that there are two types of WDM—*broadband* and *narrowband.* Broadband WDM uses the 1300-nm and 1550-nm wavelengths for full-duplex transmission. Therefore, if a signal is sent in one direction on the 1300-nm wavelength, it can be sent back on the 1550-nm wavelength over the same fiber. This type of WDM has been used for a number of years. *Narrowband,* also called *dense WDM,* is the multiplexing of 4, 8, 16, 32, or more wavelengths in the range of 1530 to 1610 nm with a very narrow separation between wavelengths. Today, "WDM" applies primarily to dense wavelength-division multiplexing (DWDM). For our purposes, we'll use "WDM" to refer to DWDM systems and sometimes we'll use both terms—WDM and DWDM—interchangeably. You should be aware of these nuances when you enter the profession.

***Add/drop problem*** Multiplexing, as we've seen, is a very effective solution to increasing traffic flow over optical fiber, but all benefits come with a price. If you look again at Figure 12.4, you'll see that this is still a point-to-point link. The real network starts with the ability to deliver a signal from an individual sender to an individual recipient. This ability implies that a network has to be capable of adding and dropping any signal. This is exemplified in Figure 12.5.

*If we employ TDM,* we need to have an add/drop node, and this node has to extract the desired signal from the main data stream. But we have made an effort to push the main data stream at a high bit rate while the individual channel operates at a much lower bit rate. Hence, the main data stream needs to be demultiplexed to the bit rate of the individual channel. Suppose each of the four transmitters in Figure 12.5(a) generates traffic flow at 622 Mbit/s. Thus, after time-division multiplexing, the total bit rate of the main data stream is equal to 4 × 622 Mbit/s, or 2.488 Gbit/s, the number we usually approximate as 2.5 Gbit/s. If we need to drop channel 2 and then add channel 5, as Figure 12.5(a) shows, each add/drop node has to reduce the bit rate of the main data stream to 622 Mbit/s, add or drop an individual channel, and then multiplex all of them again to attain 2.5 Gbit/s.

The TDM add/drop components must also include receivers and transmitters because they have to convert an optical signal into an electrical signal, multiplex and demultiplex this electrical signal, and then convert the processed signal back into optical form.

*If we use WDM,* two basic approaches to the add/drop problems can be taken. The first is called *broadcast-and-select,* or simply *broadcast.* In this case, each node transmits signals on its wavelength and all these signals are presented to a WDM MUX. *The MUX broadcasts all signals to all nodes. Each node selects the desired wavelength by filtering the entire signal.* This approach is shown in Figure 12.5(b). A WDM MUX is a passive device that does wavelength-division multiplexing and transmits a multiplexed signal farther along the fiber. Observe that the WDM MUX splits the signal power equally among all the nodes; bear in mind, of course, that since the signal power is always limited, it can be split only among a finite number of nodes. Note also that the same wavelength can be picked up by several receivers. This is called *multicast (point-to-multipoint)* transmission. On the other hand, this system can be independently accessed by many nodes; this is why such a network is called *multiaccess.*

Another approach to solving the add/drop problem in WDM transmission is *wavelength routing.* (See Figure 12.5[c].) A signal from an individual node on a specific wavelength is transmitted through a set of active switches. Any wavelength can be routed to any node—this is the major advantage of this technique. What we sacrifice in taking this approach is a technologically and architecturally sophisticated network.

The wavelength-routing network is based on what's known as the "lightpath concept." *A lightpath is the temporary optical communication channel between two nodes.* A lightpath may include several physical connections and nodes. For example, there is a lightpath between Tx6 and Rx3 in Figure 12.5(c). The information signal using this lightpath is carried by wavelength $\lambda_6$. It originates at Tx6 and goes through switches 7, 1, 6, and 3 to receiver Rx3. The lightpath between Tx1 and Rx2 includes switches 1 and 2. If you need to send a message from Tx1 to Rx4, you have to establish a new lightpath.

Another important feature of wavelength-routing networks is *wavelength reuse,* which allows a network manager to use the same wavelength for different lightpaths. An example given in

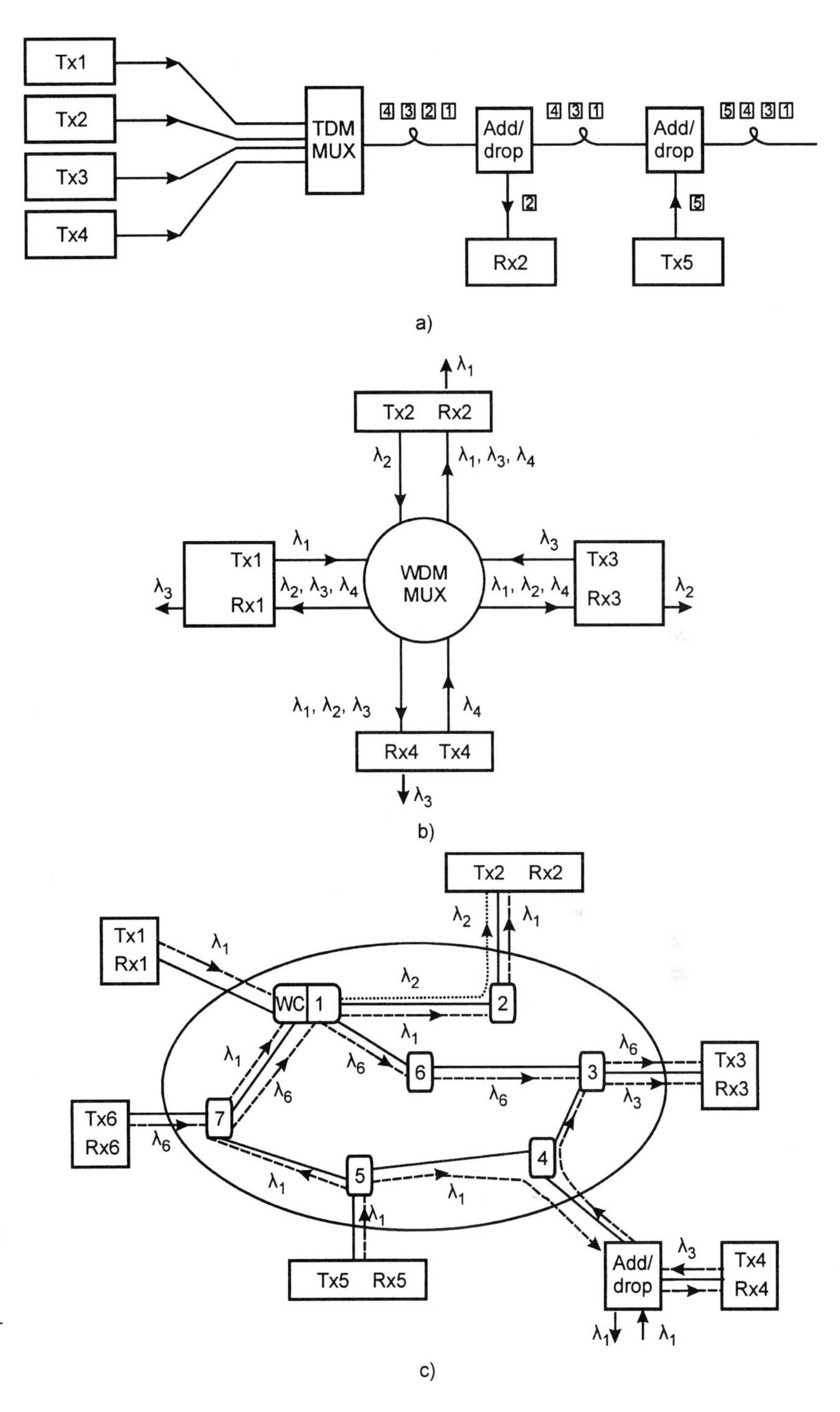


**Figure 12.5** TDM and WDM networks: (a) Add/drop operation of a channel in a TDM network; (b) broadcast-and-select in a WDM network; (c) wavelength-routing in a WDM network. *([c] adapted from Biswanath Mukherjee's* Optical Communication Networks, *New York: McGraw-Hill, 1997. Used with permission.)*

Figure 12.5(c) shows that wavelength $\lambda_1$ is used for two lightpaths: $Tx_1$-$Rx_2$ and $Tx_5$-$Rx_4$. This is possible because these lightpaths do not share any links and switches in the network.

Since there is no power splitting between links of a network, a wavelength-routing network allows many more connections. In addition, this network eliminates broadcasting a signal to unwanted receivers. These are other advantages of this approach.

There is, however, one fundamental restriction in the use of a wavelength-routing network: *If two or more lightpaths use the same fiber, they have to use different wavelengths to avoid interfering with one another.*

So, you correctly ask, what will happen if two transmitters send signals on the same wavelength to the same receiver? This situation is exemplified in Figure 12.5(c), where Tx1 and Tx5 are sending messages on $\lambda_1$ to Rx2. In this case, node 1 has to perform *wavelength conversion (WC)*. One possible scenario is that wavelength $\lambda_1$, coming from Tx5, is converted to wavelength $\lambda_2$ by node 1. Now from node 1 to Rx2 wavelength $\lambda_2$ carries the same information that wavelength $\lambda_1$ has carried from Tx5 to node 1. Thus, the information from Tx5 to Rx2 is carried by wavelength $\lambda_1$ and $\lambda_2$, while information from Tx1 to Rx2 is carried by wavelength $\lambda_1$ all the way down to the receiver.

***Repeaters and amplifiers*** You will recall that optical fiber has a finite attenuation and therefore an optical signal damps while traveling down the fiber. We also learned that the transmitting distance is limited by attenuation to less than 26 km in multimode fiber (see Example 3.2.2) and to less than 80 km in singlemode fiber (see Example 5.2.1). Hence, we need to boost an optical signal to transmit information over the longer distance. There are two means by which to strengthen an optical signal: repeaters (regenerators) and optical amplifiers.

A repeater (regenerator) accepts an optical signal, converts it into an electrical signal, makes a decision whether it is bit 1 or bit 0, generates a new electrical pulse, converts it back into an optical signal, and transmits the reshaped signal farther along the fiber. This description is illustrated in Figure 12.6(a). If this explanation sounds familiar to you, good. It means you've read Chapters 9, 10, and 11 very carefully. All these functions are discussed in detail in those chapters, where we consider transmitters and receivers in fiber-optic communications. For now, we can

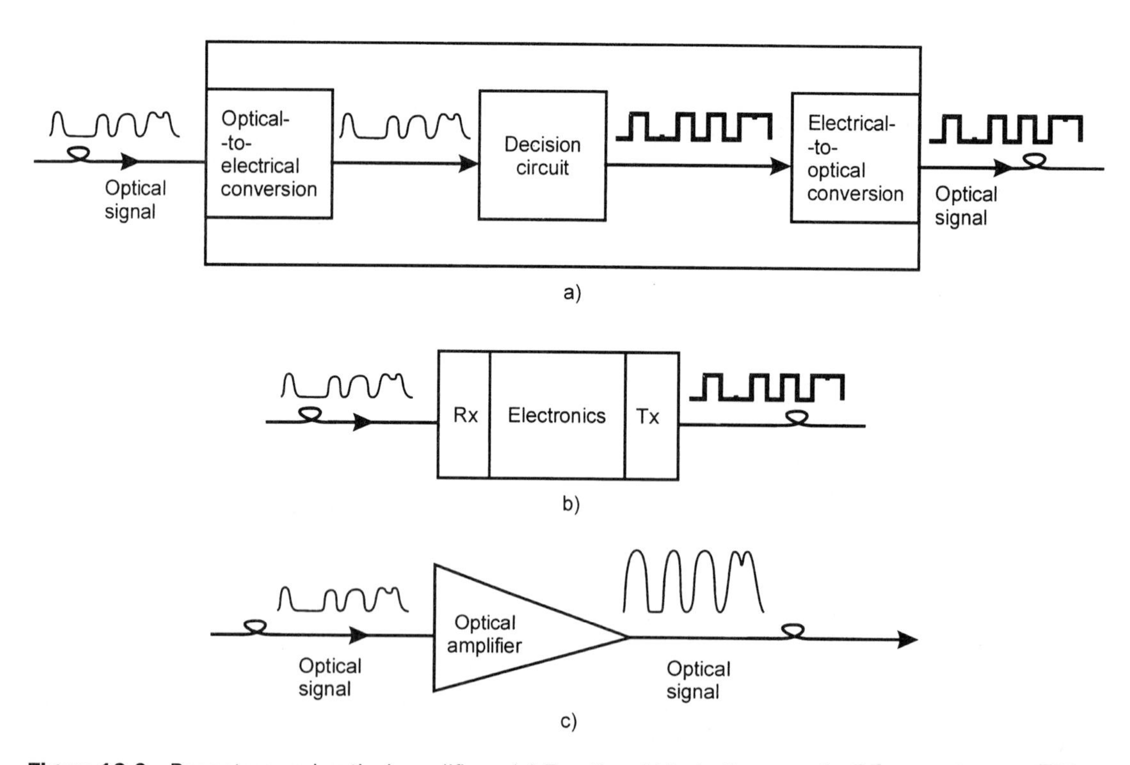


**Figure 12.6** Repeaters and optical amplifiers: (a) Functional block diagram of a repeater; (b) simplified diagram of a repeater; (c) optical amplifier.

simply say that a repeater consists of a receiver, a transmitter, and some processing electronics, as Figure 12.6(b) shows.

Repeaters, by the way, were the only signal boosters in the early days of fiber-optic communications. In fact, TAT-8, the first transatlantic link, which was installed in 1988, is still operational, employing 103 repeaters, one mounted every 35 to 50 km [4].

There are three basic methods to regenerate a signal [5]. The first, called *3R,* is *regeneration with retiming and reshaping.* This is the standard and most widely used method. As the name implies, this type of repeater extracts clock information from a signal. (See Section 11.4, where clock recovery is discussed.) Then the signal is completely reclocked and reshaped. In fact, this type of repeater generates a new electrical signal, which carries the same logic as the original. (That's why the term "regenerator" is often used in lieu of "repeater.")

In a sense, *a 3R repeater produces a "fresh copy" of a received signal.* Its concept sounds attractive, indeed, but poses one big problem: It can work only with a specific bit rate and signal format. Hence, the transmission characteristics of both transmitters and receivers at the ends of a fiber-optic link are predetermined.

The second method, called *2R,* involves *regeneration with reshaping* but without retiming. With this approach, the restrictions on the bit rate and frame format are not very stringent but this method does not work well at the higher bit rates (above a few hundred Mbit/s). This is because without retiming, jitter accumulates over all the regeneration steps.

The third method, *1R,* performs only *regeneration without retiming and reshaping.* This is the only method that can handle an analog transmission such as that used in cable TV. Nevertheless, its performance characteristics are worse than those of the two methods described above.

Repeaters are acceptable signal boosters for point-to-point links, such as TAT-8, where bit rate and signal format are determined by use of a single transmitter. However, *repeaters don't work for fiber-optic networks, where many transmitters send signals to many receivers at different bit rates and in different formats.* Thus, the need for an optical amplifier has arisen.

Optical amplifiers simply strengthen the optical signal, as shown schematically in Figure 12.6(c). Optical amplifiers work without having to convert an optical signal into electrical form and back. This feature leads to two great advantages of optical amplifiers over repeaters. First, optical amplifiers support any bit rate and signal format because, again, they simply amplify the received signal. This property is usually described by saying that optical amplifiers are *transparent* for any bit rate and signal formats. Secondly, they support not just a single wavelength, as repeaters do, but the entire region of wavelengths. For example, *an erbium-doped fiber amplifier (EDFA),* which is discussed in detail in Section 12.4, amplifies all wavelengths from approximately 1530 nm to approximately 1610 nm.

Again, only optical amplifiers can support TDM and WDM fiber-optic networks with a variety of bit rates, modulation formats, and wavelengths. Indeed, wavelength-division multiplexing became the real workhorse of the industry only with the advent of optical amplifiers, EDFA in particular. EDFA is by far the most popular optical amplifier. Its advent turned theories about WDM and fiber-optic networks into reality.

Looking at Figure 12.6, you may be puzzled if you have already read Section 11.4. After all, using a repeater will produce a perfect new signal, as Figures 12.6(a) and (b) show, while an optical amplifier strengthens both signal and noise equally. In other words, a repeater drastically improves the signal-to-noise ratio (SNR), while an amplifier at least keeps SNR at the same level it receives it. Moreover, a repeater narrows a pulse back to its original duration, while an amplifier does nothing about pulse spread. As a result, pulse widening increases over the transmitting distance, thus allowing dispersion to remain a constant threat. This is all true—but only to a point. We discuss this topic further in Sections 12.3 and 12.4. For now, it is important to stress that optical amplifiers facilitate the transmission of an information signal over a distance of 500 to 800 km, but beyond that electronic repeaters are still necessary. Such a configuration is typical in today's long-haul networks [6]; however, new systems claim to extend the distance between conventional electrical regenerators up to 6,000 km [3].

To sum up, then: *Optical amplifiers open two doors: (1) They increase the capacity of fiber-optic links by using WDM and (2) they provide the capability to develop real fiber-optic (all-optical) networks, not just powerful point-to-point links.*

***Fiber-optic networks: component needs*** If you analyze our previous discussion thoroughly, you'll realize that wavelength-division multiplexing not only increases the capacity of an individual link severalfold but it also makes it possible to build real fiber-optic networks. Do we need a network? Absolutely. Even the oversimplified diagram in Figure 12.1 highlights the main idea of modern communications: All customers need to be connected with one another at the highest possible bit rate. Such a comprehensive connection can be done *only* through fiber-optic networks.

Therefore, building fiber-optic networks is the next stage in fiber-optic communications. This stage, developing right now, is opening new horizons in fiber-optic communications and services that only communications of this type can provide. There is, however, one significant limitation to its growth: the need for new components. Fiber-optic networks are so much more sophisticated, from any point of view, than point-to-point links that the electronic devices available to them today do not meet the new requirements. The situation reminds one of the time when the electronics industry longed for a more powerful, more reliable, longer-lasting replacement for the vacuum tube. This, of course, turned out to be the transistor. It's this kind of revolutionary breakthrough that is needed now to serve the fiber-optic-network field.

The fact is that fiber-optic networks need a plethora of new designs for components, not just for transmitters, optical fibers, and receivers—although transmitters and receivers will have to be redesigned to handle the new types of transmission, such as WDM. What is important to realize, however, is that new fiber-optic networks are being developed using the millions of miles of already installed optical-fiber web. This is the key to the economic justification for developing this new technology. Of course, millions of miles of new optical fiber are being installed, but utilizing every single foot of already installed fiber is the real driving force behind the emergence of WDM and associated techniques. Indeed, it is much easier and cheaper to replace transmitters and receivers and to add several new components to the existing net of fiber than to build new networks from scratch.

This chapter is devoted to the new components we need to build tomorrow's fiber-optic networks. In fact, networks and their components are the fastest growing segment of this dynamic industry. It's growing so rapidly, in fact, that by the time this book reaches you, major technological advances will already have been made. Therefore, you very likely will encounter, on the job, components with better characteristics than those described here. However, the nomenclature of network components and their principles of operation will be similar to what we have recounted, at least until another technological revolution occurs.

As you study the content of this section, keep in mind that in order to build our ideal fiber-optic network, we need to have technological advances in the following components:

- Transmitters and receivers (capable of working with many wavelengths simultaneously)
- Optical amplifiers
- Passive components:

  Couplers/splitters

  WDM multiplexers and demultiplexers

  Filters

Isolators

Circulators

Attenuators

- Optical switches
- Wavelength converters
- Functional modules:

Optical add/drop modules

Optical cross-connects

Remember, network components are classified as *active* when they are required to be powered and *passive* when no external power is needed to make them functional. Thus, as we will see, transmitters, receivers, optical amplifiers, and optical switches are active components, while couplers, fixed filters, passive WDM MUXs/DEMUXs, isolators, and circulators are examples of passive components.

In fact, building a fiber-optic network requires even more components than we have listed, but the need for all the others will become clear as we proceed with a more detailed explanation of the technical aspects of networks.

A final thought on the importance of components: The quality of the components that make up a fiber-optic network determines the architecture of the entire network because a component's characteristics ultimately define how many wavelength channels can be used and how their use can be distributed throughout the network. One crucial overriding factor will ultimately determine the mass commercial deployment of WDM networks: *availability of components with the requisite characteristics at competitive cost and in large volume.*

## 12.2 TRANSCEIVERS FOR FIBER-OPTIC NETWORKS

You will recall that for full-duplex (two-way) communications, a transmitter and a receiver are combined in one unit: a transceiver.

The real challenge for designers arose when transceivers were applied to fiber-optic networks. For one thing, they had to handle the high-speed communications necessary for TDM networks. Secondly, WDM networks required transceivers able to radiate and accurately detect extremely close wavelength channels with unprecedented requirements in terms of wavelength stability. The properties of this new generation of transceivers are the focus of this section.

### Transmitters

***Laser diodes for WDM application*** Refer to Figure 12.4(b), where the idea of wavelength-division multiplexing (WDM) is presented. There you see that an individual transmitter generates a specific wavelength channel and that these wavelengths are combined for transmission over the same fiber by a WDM multiplexer. This is how today's WDM systems work.

This straightforward method uses DFB lasers as light sources. The question is, how do lasers generate at different wavelengths? Before we answer, let's consider some numbers.

The typical C-bandwidth of modern erbium-doped fiber amplifiers (EDFA) is 1530 nm to 1560 nm. (See Section 12.4 for a discussion of the C- and L-bandwidths of an EDFA.) Suppose we want to place 16 channels over this range of wavelengths. The spacing between the channels has to be 30 nm/16 channels, or 1.875 nm. It is standard today to measure channel spacing in terms of frequency rather than wavelength. Recall the fundamental relationship (Formula 2.1):

$$\lambda f = c \tag{12.1}$$

Therefore (see Formula 10.56),

$$\Delta f = (c\Delta\lambda)/\lambda^2 \tag{12.2}$$

We omitted the negative sign in Formula 12.2 because it is not important for crunching numbers. Hence, 1.875 nm is equivalent to a spacing between channels of approximately 250 GHz. (Bear in mind that the real channel spacings may be quite different and they are determined by ITU standards—see the box "ITU Grid.")

These numbers are indicative of the fine tuning that is necessary to make DFB lasers radiate at appropriate wavelengths. If, say, the first laser radiates at 1530 nm, the second one has to radiate at 1531.875 nm, and so forth.

### Example 12.2.1

***Problem:***

Calculate the channel spacing in nm if this spacing equals 100 GHz and the networks operate at 1550 nm.

***Solution:***

The solution is straightforward: Rearrange Formula 12.2. Indeed, from Formula 12.2 we find

$$\Delta\lambda = c\Delta f / f^2, \tag{12.2a}$$

where, again, we have omitted the negative sign. The frequency, ($f$), corresponding to $\lambda = 1550$ nm, is immediately found from Formula 12.1: $f = 193.548 \times 10^{12}$ Hz. Hence,

$$\Delta\lambda = 0.8 \text{ nm}$$

You'll encounter these numbers—100 GHz and 0.8 nm—shortly, when we discuss the ITU frequency grid. It's easy to see from Formula 12.2a that a channel spacing of 50 GHz is equivalent to a wavelength separation of 0.4 nm at a 1550-nm operating wavelength.

---

We have already noted that the general range of wavelengths over which a laser radiates is determined by the energy gap of its material (Figure 9.1), while the precise value of a radiating wavelength is determined by a laser resonator. (See Figures 9.14 and 9.15.) But now we really need to fine-tune the wavelength. The simplest way to do this is to fabricate Bragg gratings with slightly different periods for each individual DFB laser. (See Section 9.2.) Remember, too, that the refractive index of the active medium of a laser depends on drive current and temperature (Section 10.3); hence, changing these parameters can help. In practice, a slight change in laser temperature can be used to precisely adjust the radiating wavelength.

Today, WDM systems employ a set of individual DFB laser diodes to generate a set of required wavelengths. These diodes are mechanically assembled together to facilitate construction of the WDM transmitter. Each LD is driven by its own individually modulated current. Even though this arrangement is often called *laser-diode array,* this is essentially, again, an assembly of individual components.

Recent progress in the development of VCSEL lasers (see Sections 9.2 and 9.3) presents an exciting new opportunity: the fabrication of a real array of lasers. This would mean the fabrication of many independently modulated active areas onto one substrate very much as modern electronic chips with different functional regions are fabricated. A 32 × 32 laser array has already been achieved with independent addresses for every individual laser, implemented one at a time. Switching time is restricted only by the time it takes for switching electronically from one laser to another [7]. However, the problem of coupling the light from all the active lasers of the array into a single fiber has to be solved. Another major disadvantage of VCSEL lasers—that they can radiate only around the 850-nm window—still keeps this development out of the mainstream of fiber-optic network technology.

***Transmitter requirements in WDM networks*** The above discussion raises the question of what specific requirements apply to WDM transmitters compared with the general requirements for fiber-optic communications transmitters. Section 12.1 and this subsection provide you with enough information to understand the following specifications for WDM transmitters ([10], [11]):

**1.** *Quality of generated light.*

- **1.1.** The *linewidth* has to be as narrow as possible. Indeed, to attain 40-channel multiplexing, we need a channel spacing of 100 GHz, thus, the linewidth cannot be more than 1 GHz to avoid channel crosstalk. Fortunately, modern DFB lasers have linewidths on the order of tens of megahertz. Refer to the data sheet of the DFB laser in Figure 9.19(a), where the typical linewidth is 10 MHz and the maximum is 40 MHz.
- **1.2.** The *sidemode suppression ratio (SMSR)* has to be as high as possible. Refer again to Figure 9.19(a) and the accompanying discussion to recall that even a DFB laser can support several transverse modes. All these side modes have to be suppressed to avoid any crosstalk in WDM systems. Fortunately, again, modern DFB lasers have very good SMSR (40 dB is the typical value, as shown in Figure 9.19[a], and 35 dB is common in DFBs with an electroabsorption modulator, as shown in Figure 10.32).
- **1.3.** A laser has to operate in a *single longitudinal mode.* This is achieved in DFB lasers by using Bragg gratings to provide positive optical feedback. (See the "Emission spectrum" in Figure 9.19[a].)
- **1.4.** *Chirp* has to be eliminated. You'll remember that the central radiating frequency (wavelength) varies when a DFB laser is driven by variable current, as the direct-modulation technique requires. Remember, too, that chirp is in direct conflict with the above requirements; therefore, it is unacceptable in WDM systems.

All this leads to a simple conclusion: *Direct modulation is unsuitable in WDM systems.* We have to use external modulation. (See Section 10.4.) This is a critical requirement for WDM networks, and DFB lasers with integrated external modulators are the trend in transmitter technology today.

**2.** *Stability.*

- **2.1.** Variations in *output power* result in variations in linewidth—a detrimental phenomenon. In addition, power instability causes several hidden problems associated with such nonlinear effects in singlemode optical fibers as four-wave mixing (FWM) and stimulated Brillouin scattering (SBS).
- **2.2.** Variations in a *peak wavelength* are unacceptable because we need to keep the channel spacing stable; otherwise, we will mix channels and lose information. More precisely,

## ITU Grid

The ITU grid, also known as the ITU-T grid, is an international standard for wavelength usage in WDM systems. ITU stands for the International Telecommunication Union, which is based in Geneva, Switzerland. The "T" after the hyphen stands for the Telecommunication Standardization Sector of the union. (There is also a Radiocommunication Sector. Its recommendations are signified by the letters "ITU-R.") This international organization develops standards for the telecommunications industry in general.

Do we really need standards? Of course, and for many reasons. One of the most important reasons is to ensure the compatibility of equipment from different equipment manufacturers with the service from different service providers. Consider WDM systems, for instance. The range of usable wavelengths from 1530 to 1560 nm is determined by the physical parameters of an erbium-doped fiber amplifier; however, within this range we can choose any wavelength. Now imagine what would happen if no such standard existed. In a word: chaos. Your WDM system would not be compatible with any other WDM system and we would never be able to build even a metropolitan area network, let alone wide area and global networks. This is why ITU has developed and continues to develop standards on wavelengths usable in WDM systems. And this is why many vendors proudly specify on the data sheets of their WDM-system products: "Compatible with ITU grid."

Technically, ITU specifies the *frequency grid* with reference to a frequency standard [5]. The reference frequency is the krypton line at 193.1 THz (1552.52 nm in a vacuum). The grid is defined by intervals equal to integer multiples of 100 GHz (approximately 0.8 nm). The current grid is wider than the bandwidth of modern EDFAs, which means the standard is open to the possibility of even wider intervals.

The ITU grid is shown schematically in Figure 12.7. For your convenience, Table 12.1 displays the table of frequencies and wavelengths of the ITU grid [8]. Keep in mind that frequencies are accurate values while wavelengths depend on the fiber properties.

**Table 12.1** ITU Grid—Frequencies and Wavelengths

| Frequency (GHz) | Wavelength (nm) |
|---|---|
| 196,100 | 1528.77 |
| 196,000 | 1529.55 |
| 195,900 | 1530.33 |
| 195,800 | 1531.12 |
| 195,700 | 1531.90 |
| 195,600 | 1532.68 |
| 195,500 | 1533.47 |
| 195,400 | 1534.25 |
| 195,300 | 1535.04 |
| 195,200 | 1535.82 |
| 195,100 | 1536.61 |
| 195,000 | 1537.40 |
| 194,900 | 1538.19 |
| 194,800 | 1538.98 |
| 194,700 | 1539.77 |
| 194,600 | 1540.56 |
| 194,500 | 1541.35 |
| 194,400 | 1542.14 |
| 194,300 | 1542.94 |
| 194,200 | 1543.73 |
| 194,100 | 1544.53 |
| 194,000 | 1545.32 |

the instability of a peak wavelength causes variations in the *channel crosstalk* level. This, in turn, results in a rise in the bit-error rate (BER).

**2.3.** What we have said in 2.2 holds true for linewidth stability as well.

**2.4.** The *relative intensity noise (RIN)* has to be minimized because this is another form of output-power instability. Since one of the major reasons for RIN is *backreflection* (Section 10.3), a laser has to be suitably protected.

We have to distinguish between short-term and long-term stability. Short-term variations are caused by variations in temperature and other ambient influences, while long-term stability is associated with the aging of laser and transmitter components. (To refresh your memory on laser and transmitter components, see Section 10.4.)

**3.** *Reliability.* (Here again, refer to Section 10.4 to brush up on this topic.)

**4.** *Power consumption.* We always want to minimize power consumption to reduce the heat radiated by a laser diode. But this requirement becomes critical for WDM transmitters, where many laser diodes operate under very congested conditions.

**Table 12.1** (continued)

| Frequency (GHz) | Wavelength (nm) |
|---|---|
| 193,900 | 1546.12 |
| 193,800 | 1546.92 |
| 193,700 | 1547.72 |
| 193,600 | 1548.51 |
| 193,500 | 1549.32 |
| 193,400 | 1550.12 |
| 193,300 | 1550.92 |
| 193,200 | 1551.72 |
| 193,100 | 1552.52 |
| 193,000 | 1553.33 |
| 192,900 | 1554.13 |
| 192,800 | 1554.94 |
| 192,700 | 1555.75 |
| 192,600 | 1556.55 |
| 192,500 | 1557.36 |
| 192,400 | 1558.17 |
| 192,300 | 1558.98 |
| 192,200 | 1559.79 |
| 192,100 | 1560.61 |

Source: *Dense Wavelength Division Multiplexer (DWDM) 100GHz 16-Channel Module* (data sheet), E-TEK Dynamics, San Jose, Calif., 1999.

ITU's recommendation is that all wavelengths used in WDM networks should be chosen from this grid. Furthermore, ITU has proposed, in addition to a frequency grid, *specific channel frequencies* for 4-, 8-, and 16-channel systems. Note that channels can be spaced not only equally but also unequally because *four-wave mixing (FWM)* is severely detrimental to equally spaced channels. (See Section 6.4.) In any case, an $n \times$ 100-GHz interval is observed, where $n$ is an integer. An example of such channel frequencies for an eight-channel system with unequal spacings is given in Figure 12.7(a). Keep abreast of the trend: reducing channel spacing to 50 GHz by increasing the number of channels to well above 100. It is important to observe this movement because two consequences follow from it. (1) Fiber-optic communications systems with 256 channels have been demonstrated [15] and (2) such technology requires unprecedented accuracy in frequency along with superior stability. This is why ITU specifies the *deviation of a nominal central frequency.* For example, for channel spacing of 200 GHz the deviation is specified as ± 40 GHz. For each 100 GHz increase in channel spacing after this, the deviation increases by ± 20 GHz, so that for a spacing of 300 GHz, the deviation is ± 60 GHz and so on [5]. There is no other way to satisfy these requirements except by carefully monitoring each wavelength used in WDM networks. This monitoring implies using an *absolute frequency reference (AFR)* and an *absolute frequency-controlled multicarrier generator.* In fact, such a technique is under development [9].

**5.** *Tunability.* Having 128-channel systems commercially available—and 256 channels having been demonstrated—obviously makes it necessary to use a large number of individual light sources, which can be very costly from both manufacturing and maintenance standpoints. Thus, the ability to tune lasers to all WDM wavelengths will soon become the most critical issue confronting designers of WDM transmitters. The key features of tunable lasers are the tuning speed and the ability to emit several wavelengths simultaneously. We have to bear in mind that light sources are necessary for WDM networks not only as end transmitters but also as elements of add/drop components.

Keeping these requirements in mind will help you choose an appropriate laser diode when designing a transmitter for your WDM fiber-optic system.

***Tunable lasers*** It is desirable to have a single laser tunable over the range of wavelengths instead of having an array of lasers. Just imagine possessing a laser that meets all the above specifications and, in addition, that can be tuned over 100 nm around the 1550-nm center wavelength at picosecond tuning speed. Such a laser would resolve all transmitter problems for WDM fiber-optic networks. Unfortunately, this is still only a dream but a dream based on real achievements toward developing tunable laser diodes. Here is a review of their principle of operation and the state of the art of the characteristics available in today's tunable lasers:

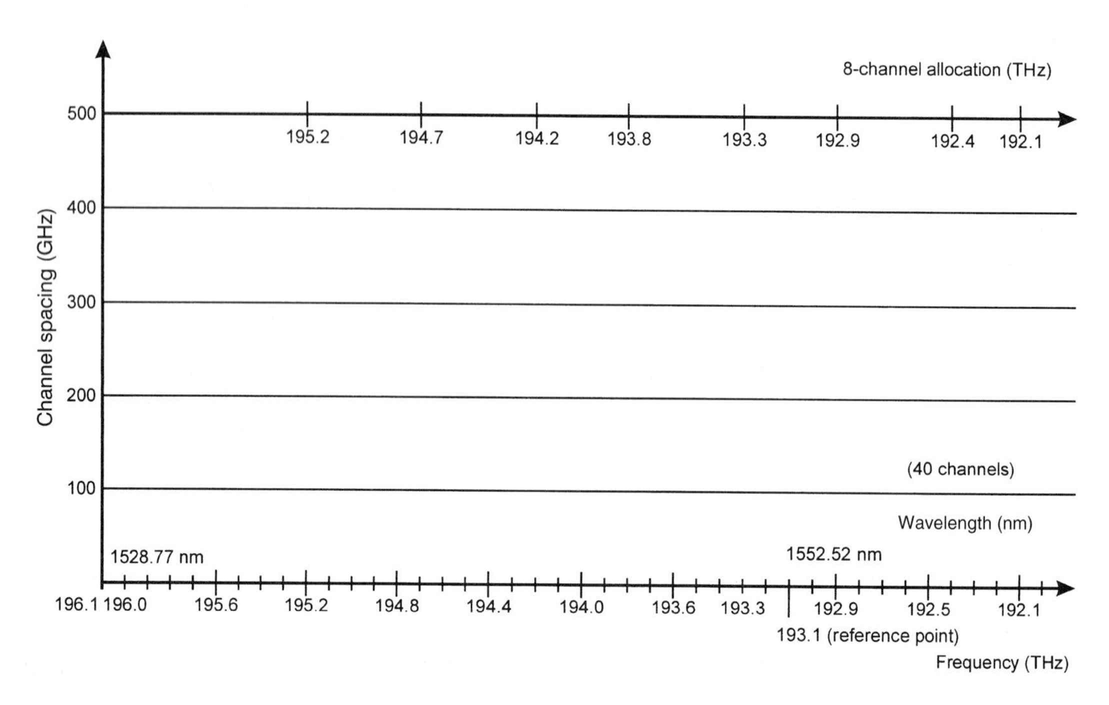


**Figure 12.7** ITU frequency grid for WDM networks: graph of frequencies, wavelengths, and channel spacings.

**1.** *Temperature-tuning lasers* exploit the dependence on temperature of the refractive index of an active area. This characteristics results in the drifting of the peak frequency radiated by a laser diode. (See the data sheets in Figures 9.17, 9.19, and 9.20.) Both the tuning range (only up to 10 nm) and the tuning speed (a few milliseconds) of these lasers are not very good. However, the temperature-tuning effect is used to adjust the radiating wavelengths in some types of commercially available tunable lasers.

**2.** *External-cavity tunable lasers* change their wavelengths by varying the resonance condition for the lasing wavelength. This practical approach is based on using diffraction gratings as one of a laser's reflectors. (See Figure 12.8[a].)

A *diffraction grating* works as a highly reflective mirror reflecting a wavelength ($\lambda_i$) that satisfies the condition:

$$d \sin \Theta = m\lambda_i, \tag{12.3}$$

where $d$ is the period of a diffraction grating, $\Theta$ is the angle of incidence (tilting angle), and $m$ = 0, ±1, ±2, ±3. (More about diffraction gratings is given in Chapter 13.) Only those reflected wavelengths that reproduce themselves after a round trip within the laser cavity will be radiated. We call this requirement the *resonant condition.* For the arrangement shown in Figure 12.8, the resonant condition depends on the tilting angle and the length between the grating and the active medium. Changing these two parameters, one can tune a wavelength generated by a laser.

Figure 12.8(b) shows a modification of external-cavity tuning known as the *Littman-Metchalf method.* Here, the grating is stable but the additional mirror being used is tilted, making tuning even more precise. Automatic gain control allows the user to obtain the desired light power from output 2. Since the driven current and cavity length are stable, this type of tunable laser demonstrates exceptional characteristics: a tunable range from 1500 nm to 1580 nm;

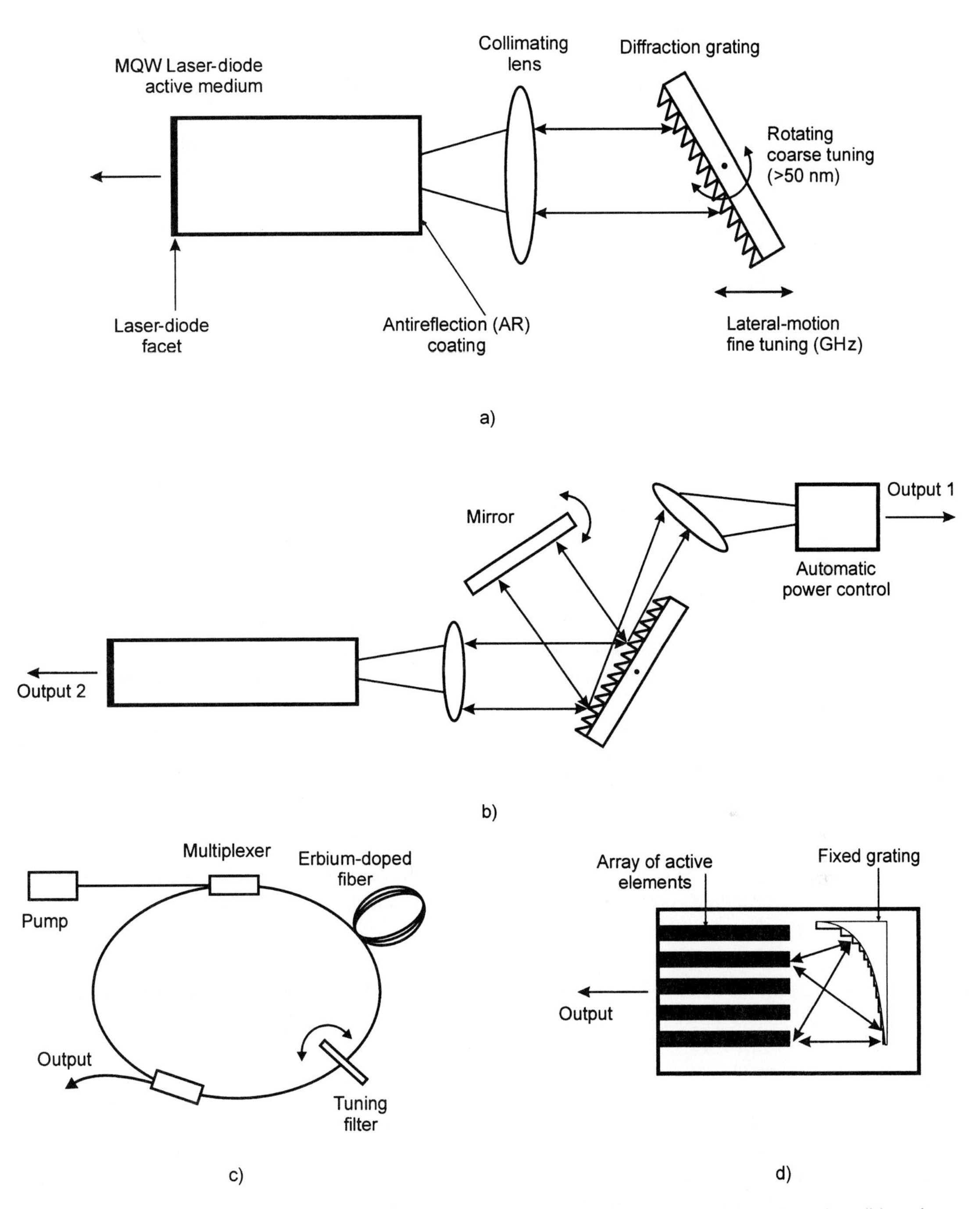


**Figure 12.8** External cavity tunable lasers: (a) Using diffraction grating for wavelength tuning; (b) tuning by Littman-Metchalf method; (c) erbium-doped fiber-tunable laser; (d) using fixed grating and array of active elements. *([a] and [d] adapted from Dennis J. G. Mestdagh's* Fundamentals of Multiaccess Optical Fiber Networks, *Boston: Artech House, 1995, with permission. [b] reprinted with permission of Anritsu Corporation, Kanagawa, Japan; [c] courtesy of EXFO, Vanier, Que., Can.)*

wavelength-setting resolution of 1 pm; absolute wavelength accuracy of 0.1 nm; linewidth of 700 kHz that is tightly controlled; and power stability from output 2 of ±0.01 dB/hr [12].

The major disadvantage of an external-cavity tunable laser is the tuning speed, which is on the order of milliseconds. Such a low speed is inherent in these lasers because of the mechanical motion involved in the tuning process. This is why external-cavity tunable lasers are used today in WDM systems not as transmitters, but for testing and measurements. Such a laser is shown schematically in Figure 12.8(c). Power from a pump laser excites the atoms in the erbium-doped amplifier section of the fiber ring. A tunable filter selects a wavelength within the EDFA band. Tuning is done by mechanically tilting the filter [9]. (If you require a deeper understanding of this light source, read Section 12.4 on erbium-doped fiber amplifiers before continuing.)

External-cavity tunable lasers have a number of additional drawbacks. Among them are a relatively high noise level caused by spontaneous emission, relatively low sidemode suppression, and far-from-ideal wavelength accuracy. To achieve the best performance possible when these lasers are used in test equipment, the manufacturers have to introduce special circuitry so that parameters won't deviate during testing. Such measures increase the cost of the equipment.

The Holy Grail of research in this field today is the quest for a tunable laser with high tuning speed. The invention shown in Figure 12.8(d) can be regarded as at least a step toward a tunable light source for WDM systems. Here, both the active medium and the grating are fixed, thereby preventing mechanical motion—the main factor slowing the tuning of laser wavelength. An active medium consists of an array of separate elements that can be individually addressed and activated. Tuning is done by electronically exciting one of the active elements at a time. Thus, tuning speed is restricted only by the electronic processes. With this progress, however, comes a setback: The laser loses its major advantage—pinpoint-tuning accuracy. The wavelength separation achieved experimentally is only 1.8 nm, not a very large accomplishment. The search continues.

**3.** *Sectional distributed-Bragg-reflection (DBR) tunable lasers* are tuned at speeds on the order of tens of nanoseconds, which are much higher than the tuning speed of external-cavity lasers. This is why these types of tunable lasers are used for transmission in WDM systems. The principle of their operation is as follows:

You will recall that the injection current changes the carrier densities in an active region of a laser diode and, in so doing, alters the refractive index of this medium. Variations in the refractive index are equivalent to altering the optical length (the product of the geometrical length and the refractive index) of a laser cavity. Thus, *a change in driving current leads to a variation in the radiating wavelength* because it changes the resonant condition of the laser diode. This is the basic mechanism leading to the tunability of a semiconductor laser. But—and here's the catch—if you simply change the driving current, you immediately change both the radiating wavelength and the output power, leaving the laser scientist to confront this challenge: how to use the driving current to change the radiating wavelength while, at the same time, maintaining steady output power.

For our needs, we can skip several intermediate solutions to the problem of finding a way to use driving current for wavelength tuning. (Those interested in pursuing this matter can consult reference [13], where these solutions are discussed in detail.) What is of more pressing concern to us is the practical—the commercial—solution: a three-section DBR tunable laser. Its basic structure is schematically shown in Figure 12.9(a). The first section is an active region, where light actually obtains gain. By changing current $I_1$, one can set up the desired power level of output light. The second section provides a phase shift of a reflected wave. To achieve a lasing effect, remember, a wave's phase shift must be equal to $2\pi$ after each round trip within a laser cavity. Thus, if we vary the shift phase of a traveling wave, we will change its wavelength. This is equivalent to saying that we will change the optical length of a cavity. Thus, by changing current $I_2$, one can change a radiating wavelength. The third section is the *distributed Bragg reflector (DBR)*, which is essentially a Bragg grating. As we have already seen, the Bragg grating reflects only wavelength $\lambda_B$, thereby meeting the Bragg condition. (See our discussion of Bragg gratings in

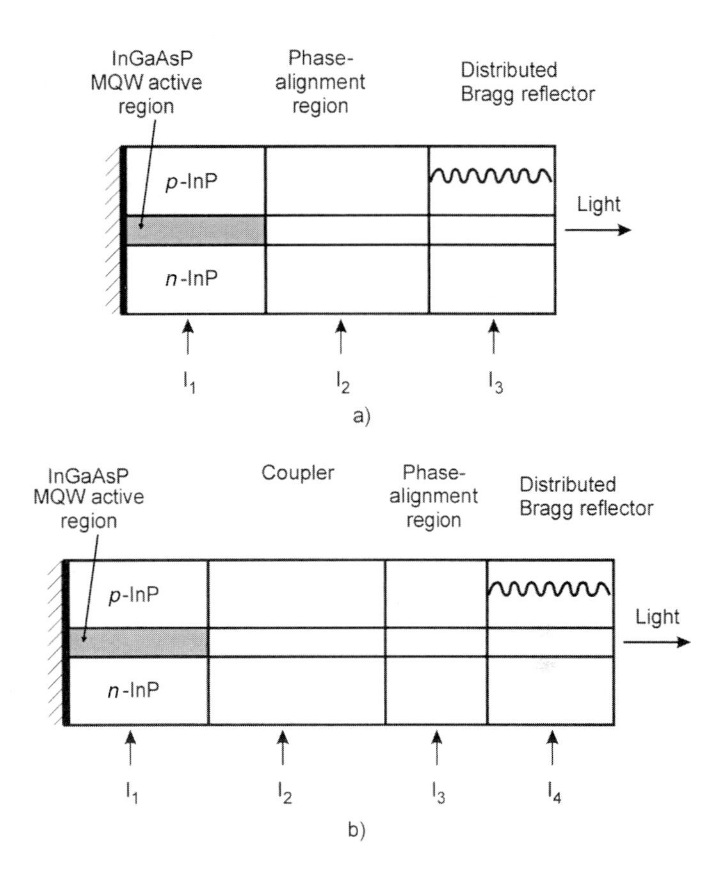


**Figure 12.9** Sectional distributed-Bragg-reflector lasers: (a) Three-section DBR laser; (b) four-section DBR laser.

Section 6.4.) By changing current $I_3$, one can change the effective refractive index and, hence, the Bragg's wavelength.

The result is this: Current $I_1$ controls the laser gain, currents $I_2$ and $I_3$ tune its operating wavelength, and all three currents can be changed independently. You can find three-section DBR laser diodes on the market as transmitters for WDM networks; however, they have one major drawback: Their tuning range is relatively small; it is limited to 10 nm.

Further development of this concept is found in a four-section DBR laser called the *grating coupler sampled reflector (GCSR)* laser [13]. Here, a vertical coupler section creates a set of wavelengths from among which an actual wavelength can be chosen by tuning the phase-shift and Bragg-reflection sections. Thus, the coupler section serves for coarse tuning, while the phase-shift section provides fine tuning of the operating wavelength. A GCSR tunable laser is shown schematically in Figure 12.9(b). The total tuning range of this type of laser exceeds 100 nm. Commercially available GCSR lasers exhibit these typical characteristics [14]: a wavelength range from 1528 to 1565 nm; a sidemode suppression ratio (SMSR) of 25 dB; laser-chip tuning time of about 20 ns (but the tuning time of the transmitter, which is determined by a control board, is not more than 1 ms); typical output power of 0 dBm; and RIN of –140 dB/Hz.

**4.** *Integrated-cavity lasers* hold promise of becoming the light sources of choice for WDM because they combine such advantages as high-speed tuning and the ability to emit several wavelengths simultaneously. These features are critical for high-speed packet-switching WDM networks. Their principle of operation can be seen in Figure 12.10.

A number of active media (optical amplifiers) are terminated by the common cleaved-mirror facet at one end. These amplifiers are optically connected to an optical multiplexer/filter. The multiplexer has a single output port optically connected to the second cleaved mirror facet. These facets (mirrors) constitute a laser cavity.

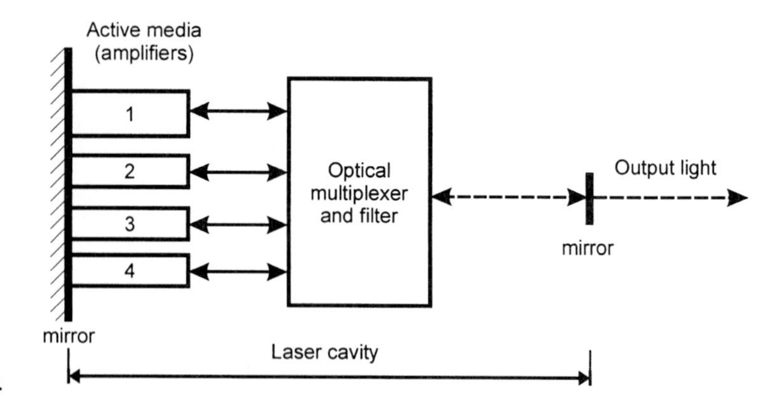


**Figure 12.10** Integrated-cavity tunable laser.

Let's summarize the operation of the device: Light amplified by the first active medium passes through the multiplexer, where the individual waveguide for this medium exists. Thus, the combination of the first amplifier and the unique path of its light within the cavity generates an individual wavelength. The same holds true, of course, for all amplifiers. In other words, there are many individual lasers within a single cavity, and a multiplexer combines all the beams into a single output. Note that there are no moving parts here and tuning is done by simply activating an individual active medium. No wonder such a laser demonstrates a tuning speed of less than 3 ns [11].

Besides high-speed tuning, this type of laser exhibits channel spacing down to 75 GHz. The major problem with intracavity lasers is the chip size: It is larger than a DFB laser array and it has to be increased with a rise in the number of radiating wavelengths.

To sum up, we can conclude that tunable lasers are among the most promising light sources for WDM networks. There are, of course, laser-array, temperature-tuning, external-cavity, and integrated-cavity light sources. They differ in many ways, but in terms of the most critical characteristics for commercial application—tuning speed and the ability to emit many wavelengths simultaneously—one of the most promising transmitters for WDM applications is the integrated-cavity tunable laser. However, the commercial reality is that today WDM networks rely mostly on transmitters radiating fixed wavelengths within the EDFA bandwidth. The search for the best WDM light sources continues as researchers explore the potential of various tunable and fixed-wavelength lasers [15].

## Receivers

***General considerations*** The basic requirement for WDM receivers is the ability to operate within the entire window of the wavelength range (1530 to 1560 nm and even to 1610 nm) used in today's WDM networks. The *p-i-n* and APD photodiodes are said to have very wide spectral characteristics, as we saw in Chapter 11. The electronics associated with them are also available with the appropriate bandwidth. And so, it would appear, the receiver modules discussed in Section 11.3 are able to work in a WDM environment as end units. But not so fast. This statement is not completely true, as we will soon see.

The problem starts with a network application in which we have to obtain multiple *access routes* to WDM networks. In such a case, we need a receiver capable of choosing an individual channel from the many transmitted over the network. In WDM networks, a channel is associated with a wavelength; thus, the problem of receiver-wavelength selectivity arises. This problem is generally solved by placing a filter in front of the receiver, but many other questions have to be answered for the practical implementation of a *wavelength-division multiple-access (WDMA)* network.

There are two basic approaches to selecting a desired channel (wavelength) from the network's main stream [16]:

The first uses a *tunable transmitter and fixed receiver (TTFR).* This scheme is shown in Figure 12.11(a). In this scenario, the receiver selects only a fixed wavelength. To address a specific receiver, the transmitter has to tune its emitting wavelength. For example, transmitter Tx1 addresses receiver Rx1 by tuning its radiating wavelength at $\lambda_1$.

The second—and entirely opposite—approach uses a *fixed transmitter and tunable receiver (FTTR).* In this scenario, the receiver is tuned to pick up the desired wavelength while each transmitter emits a fixed wavelength. For example, receiver Rx1 needs to pick up a signal from transmitter Tx1, which emits only one wavelength ($\lambda_1$). Thus, receiver Rx1 is tuned to wavelength $\lambda_1$ to accomplish the task. (See Figure 12.11[b].)

Since most of the transmission in today's WDM networks is done with fixed-wavelength transmitters, the second approach, FTTR, is the one primarily in use.

Two basic methods can be used to achieve the wavelength selectivity of a receiver: active and passive. In the *active* approach, a tunable filter actively seeks a desired channel. When the search is done, the designated wavelength is transmitted to a standard receiver. The search occurs in the *optical domain,* as shown in Figure 12.12(a). The *passive* approach uses an optical demultiplexer, which separates the received wavelengths and directs each of them to an individual photodiode; here, all the PDs are combined in one array. Selection of the desired signal is done by electronic components that switch from one PD to another. Here the search occurs in the *electronic domain,* as we see in Figure 12.12(b).

Optical demultiplexing can be done in two ways. For example, the first demultiplexer breaks 16 channels into 4 and then 4 demultiplexers separate each of the 4 channels into 16 individual wavelengths. A photodetector array is a chip inside which individual photodiodes are monolithically fabricated. The center-to-center space between individual diodes can be as small as 250 μm. The modulation bandwidth of each photodiode can reach 50 GHz [16].

Since switching from one wavelength to another in the second approach is performed electronically, the switching speed is rather high and the selection process can be done within a few nanoseconds. This advantage, along with the availability and reliability of passive optical demultiplexers, makes passive selection the method of choice today in WDM networks. (To understand why an actively tuned filter is a disadvantage, see Chapter 13.)

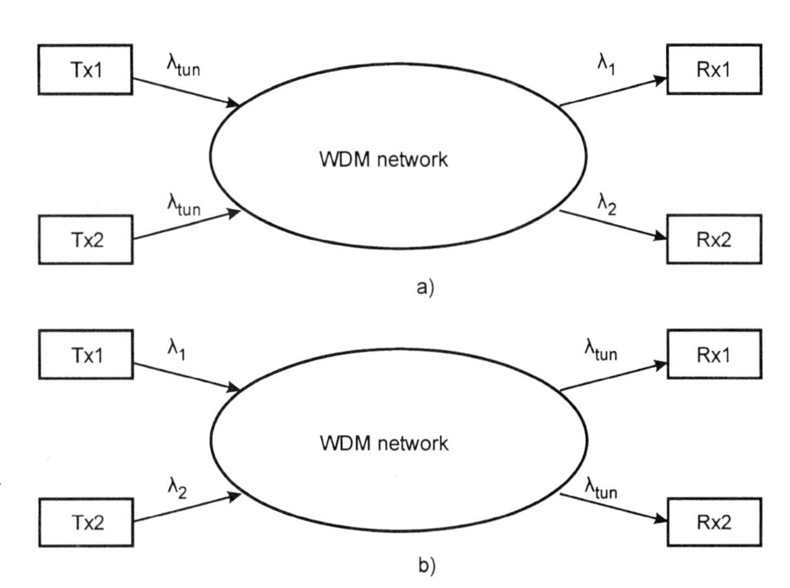


**Figure 12.11** Two schemes of multiple access in WDM networks: (a) Tunable transmitter and fixed receiver (TTFR); (b) fixed transmitter and tunable receiver (FTTR).

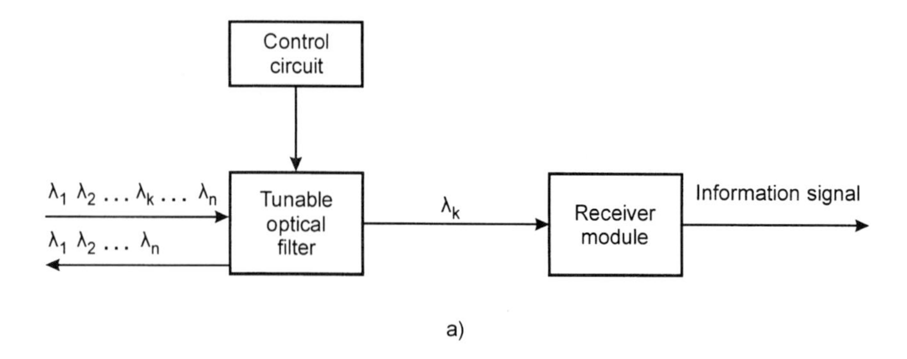


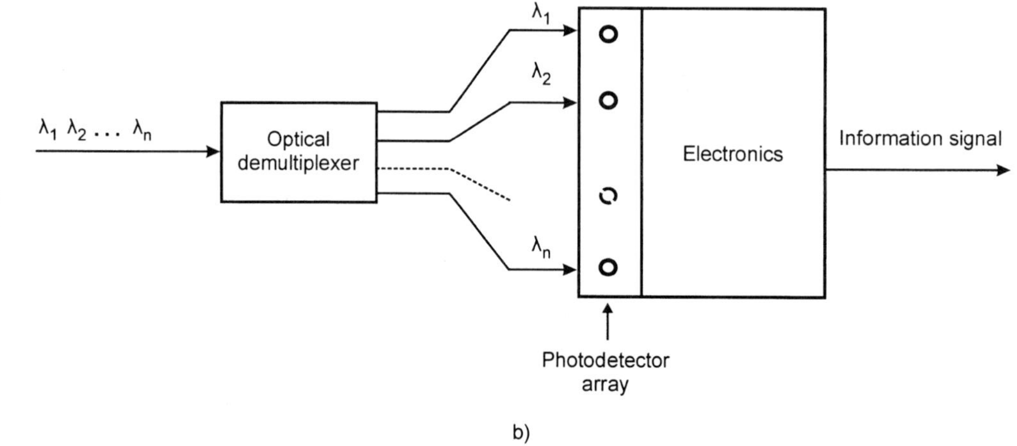


**Figure 12.12** Two schemes of selecting channels in WDM networks: (a) Active search with tunable filter; (b) passive optical demultiplexing.

***Receiver requirements in WDM networks*** Let's now consider the requirements for receivers in WDM networks [16]. Because the general requirements for receivers are discussed in detail in Chapter 11, we will concentrate here only on features specific to WDM applications.

**1.** *Spectral width.* The spectral width required today in WDM networks is determined by the range of frequencies—that is, the bandwidth—that a conventional EDFA can handle. As already mentioned, the usable bandwidth of an EDFA is considered to be approximately from 1530 to 1560 nm, although it recently has been extended to 1610 nm. If you read the photodiode data sheets (Figures 11.9 and 11.24) and the receiver data sheet (Figure 11.23), you certainly noted that they claim these devices have spectral response, or wavelength range, from 1150 to 1600 nm and even beyond. It may seem that we don't have any problem here. However, by studying carefully the "Relative Responsivity versus Wavelength" graph in Figure 11.9, you learn that a photodiode's response within this range varies from 60% to 100%. Even within the EDFA bandwidth, a photodiode's response changes up to 10%. Therefore, either the spectral response within the required range has to be flattened or unequal amplification has to be established to keep the output signal from different wavelength channels at approximately the same power level.

Moreover, the new bandwidth of EDFAs ranges up to 80 nm, allowing more channels for multiplexing. This puts even more stringent requirements on the spectral response of a receiver in WDM networks.

**2.** *Receiver sensitivity.* The presence of many channels changes the requirement for receiver sensitivity. Although a receiver's sensitivity may be good enough to handle a single channel with a given bit-error rate (BER), this doesn't guarantee that your receiver will work well with a WDM network. This is because *channel crosstalk* degrades the signal-to-noise ratio (SNR) in some wavelength channels, meaning more power is required to maintain the same BER. Furthermore, the narrower the *channel spacing*—and this is the trend in WDM networks—the more detrimental the crosstalk and the higher the receiver sensitivity that's required.

**3.** *Tuning time.* The tuning-time requirement for selecting a specific wavelength depends largely on the type of network. *Transport networks,* such as long-haul and backbone links, support connection-oriented traffic. In other words, the traffic of such a network goes from its point of origin to its destination without being diverted. Here, the tuning-time requirement is rather relaxed, staying within a few milliseconds. But in *packet-switching networks,* tuning time has to be a small fraction of the packet time. This is necessary for packet-switching efficiency, where a requirement in nanoseconds is common.

**4.** *Temperature sensitivity.* WDM receivers have to sustain operation over a wide temperature range. We know from our discussion of receivers in Chapter 11 (see the receiver data sheet in Figure 11.23) that modern receivers can operate over a temperature range from –40 to +85°C. Monitoring and control might be added to a receiver module to improve its performance but, remember, all additional circuits will boost a receiver's cost. And the components of today's WDM networks are very costly indeed, making it one of the key reasons why WDM technology has not penetrated the commercial market more rapidly.

**5.** *Polarization independence.* This is a necessary requirement because a received signal is usually polarized randomly; therefore, a receiver has to handle a signal with any type of polarization.

**6.** *Power consumption and packaging.* These are quite obvious requirements, referring to such characteristics as minimum power consumption, compactness and reliability of the receiver module, and immunity to electromagnetic interference and internal noise.

Building tunable receivers involves such components as *passive demultiplexers, tunable filters,* and *arrayed waveguide gratings (AWG).* These components are discussed in Chapter 13. The fabrication of multiwavelength receivers is also an important matter, but we leave this topic to more specialized sources. (See [16] and the references given there.)

In conclusion, we see that the trend in the development of WDM network transceivers is to build them as *opto-electronic integrated circuits (OEIC).* This indicates that transceiver arrays could one day be integrated in one chip containing all the necessary components very much the way big electronic chips are built today.

## 12.3 SEMICONDUCTOR OPTICAL AMPLIFIERS

The need for optical amplifiers is clear from the attenuation-induced weakening of an optical signal as it travels down the fiber. As discussed in Section 12.1, there are two means to boost a signal: repeaters and optical amplifiers. In spite of several attractive features, repeaters themselves cannot be used to build WDM links and networks. Thus, the shift from TDM to WDM links and from links to fiber-optic networks could not have been done without optical amplifiers. That's exactly the major problem that confronted the industry several years ago and, in response, optical amplifiers were developed. There are two major classes of optical amplifiers: semiconductor and fiber. The most popular is the erbium-doped fiber amplifier (EDFA), which has opened the door to the deployment of WDM fiber-optic communications systems. But semiconductor and other types of amplifiers are also in use and even more types are in the process of research and development. In this section we'll discuss the general concerns relating to optical amplifiers and concentrate on the semiconductor optical type.

## Optical Amplifiers: General Considerations

Before reading further, review Section 12.1 to be sure you clearly discern the difference between repeaters (regenerators) and optical amplifiers. In particular, look again at Figure 12.6.

***Optical-amplifier types: semiconductor and fiber*** Two major classes of optical amplifiers are in use today: *semiconductor optical amplifiers (SOA)* and *fiber optical amplifiers (FOA).* A semiconductor optical amplifier is, in essence, an active medium of a semiconductor laser. In other words, an SOA is a laser diode without, or with very low, optical feedback. Hence, its operating principle can easily be drawn from our discussions given in Sections 9.2, 9.3, 10.1, 10.2, and 10.3.

A *fiber optical amplifier* is quite different from a semiconductor optical amplifier. Essentially, it is a piece of specialty fiber spliced with a transmission fiber and connected to a pump laser. Like a semiconductor optical amplifier, a fiber amplifier works on the principle of *stimulated emission.* Energy delivered by a pump laser is used to excite atoms at the upper energy level, where they are stimulated by the photons of an information signal to fall to the lower level. Fiber amplifiers, specifically *erbium-doped fiber amplifiers (EDFA),* are the workhorses in today's WDM networks. For this reason, Section 12.4 is devoted exclusively to a detailed discussion of EDFAs.

Erbium-doped fiber amplifiers operate only in the 1550-nm window, while semiconductor optical amplifiers cover both the 1300-nm and 1550-nm transparent windows.

***Other types of optical amplifiers*** There are, of course, other types of optical amplifiers besides SOAs and EDFAs. These others use nonlinear effects for amplification rather than stimulated emission. Two types of optical fiber amplifiers that are close to reaching practical implementation use the *Raman and Brillouin effects.* (These are discussed in Section 6.4.) Using these effects would make it possible to build *distributed,* not lump, amplification of an optical signal. Just imagine the advantage: The same piece of fiber would serve as a transmission and amplification medium simultaneously. Even more remarkable, you would be able to place this piece at a remote location from your pump source, thus feeding and controlling amplification from a central station. These devices definitely promise to open new vistas in optical amplification.

A Raman amplifier recently appeared on the market in both distributed and discrete forms. Its most popular configuration today is a hybrid EDFA/Raman amplifier, a device in which the Raman amplifier compensates for the EDFA's lack of gain in the wavelength range from 1570 to 1630 nm [17].

Another important application of a distributed Raman amplifier is to prevent the deleterious effect caused by the large gain of an EDFA amplifier. Such a gain is needed to increase the span between two adjacent amplifiers, but it results in launching high-power light into the optical fiber immediately after the amplifier. This, in turn, causes severe nonlinear effects in an optical fiber (see Section 6.4). Placement of a Raman amplifier close to the end of each span between EDFAs allows the network designer to reduce the gain of the EDFA positioned before the Raman amplifier while keeping the total gain over the span at the desired level. (To learn more about the theory of a Raman amplifier, consult references [18] and [19].)

***Functional types of optical amplifiers*** Optical amplifiers are categorized in terms of the function they perform. The three basic types are boosters, in-line amplifiers, and preamplifiers. (See Figure 12.13.)

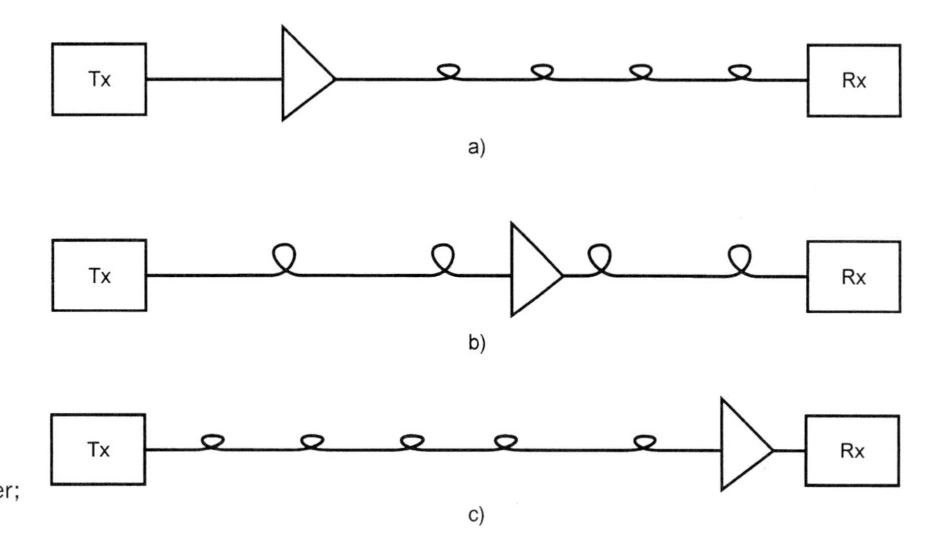


**Figure 12.13** Functional types of optical amplifiers: (a) Booster; (b) in-line amplifier; (c) preamplifier.

A *booster,* also called a *post-amplifier,* is a power amplifier that magnifies a transmitter signal before sending it down a fiber. (See Figure 12.13[a].) A booster raises the power of an optical signal to the highest level, which maximizes the transmission distance. The main requirement of this amplifier is to produce maximum output power, not maximum gain, because the input signal here is relatively large; it comes almost immediately from a transmitter. An additional benefit of using a booster is that it relieves the transmitter of the necessity of producing maximum optical power, enabling transmitter designers to concentrate on improving a broad range of transmitter characteristics rather than always worrying about power requirements.

An *in-line amplifier* operates with a signal in the middle of a fiber-optic link, as Figure 12.13(b) shows. Its primary function is to compensate for power losses caused by fiber attenuation, connections, and signal distribution in networks. Hence, the main requirement of this type of amplifier is stability over the entire WDM bandwidth. Since many in-line amplifiers may be cascaded, similarity in gain characteristics is also of concern when working with this amplifier. Keeping noise at the minimal level and performing good optical interaction with a transmission fiber are other requirements of this type of optical amplifier.

A *preamplifier* magnifies a signal immediately before it reaches the receiver. (See Figure 12.13[c].) This type of optical amplifier operates with a weak signal. Hence, good sensitivity, high gain, and low noise are major requirements here. Noise becomes an extremely important feature of a preamplifier because a receiver's performance is limited not by its own noise but by the noise of a preamplifier [20]. Using a preamplifier lessens the stringent demand that otherwise would be made on a receiver's sensitivity and eventually allows a network to operate at a higher bit rate.

The number of boosters and preamplifiers required for a specific network is determined by the number of transmitters and receivers it uses. But the number of in-line amplifiers needed depends both on the length of the fiber-optic link and the network's configuration. For long-haul links, such as transoceanic and transcontinental, in-line amplifiers are usually installed every 80 to 100 km. These amplifiers compensate for losses caused by fiber attenuation and splices. However, in-line amplifiers are also needed for short-distance networks to compensate for losses caused by signal distribution in a local area network, as shown schematically in Figure 12.14.

When we talk about boosters, in-line amplifiers, or preamplifiers, we usually imply the use of erbium-doped fiber amplifiers. However, semiconductor optical amplifiers can also be used in many such applications, as we'll soon see.

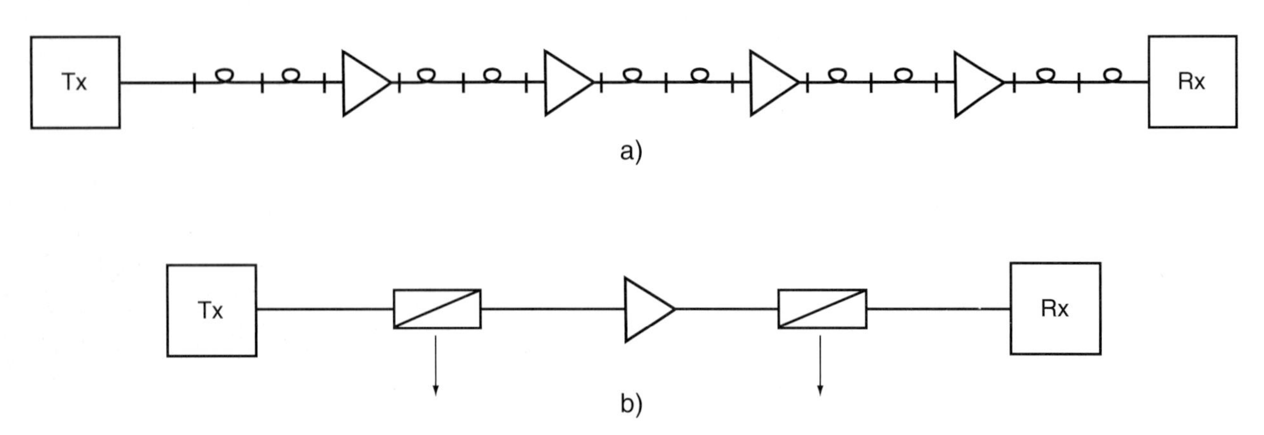


**Figure 12.14** Use of in-line amplifiers: (a) Compensation of fiber attenuation and splicing; (b) compensation of signal distribution.

## Principle of Operation of a Semiconductor Optical Amplifier (SOA)

An SOA uses the principle of stimulated emission to amplify an optical information signal. How an SOA is connected to a fiber link is shown schematically in Figure 12.15.

An optical input signal carrying original data enters the semiconductor's active region through coupling optics. Coupling is required because the MFD of a singlemode fiber beam is typically 9.3 µm, while the size of the active region is less and can even be on the order of tenths of micrometers. Injection current delivers the external energy necessary to pump electrons at the conduction band. The input signal stimulates the transition of electrons down to the valence band and the emission of photons with the same energy—that is, the same wavelength—that the input signal has. Thus, the output is an amplified optical signal. (If you are having trouble following this explanation, reread Section 9.2, which describes the properties of stimulated emission.)

## Gain of an SOA

***Fabry-Perot and traveling-wave amplifiers*** The two basic SOA types are the *Fabry-Perot (FPA) amplifier* and the *traveling-wave amplifier (TWA).*

The FPA has the same configuration as a Fabry-Perot laser. (See Figure 9.14[a].) Light entering the active region is reflected several times from cleaved facets and, having been amplified, leaves the cavity. This explanation is depicted in Figure 12.16(a), where different paths of a reflected beam are shown for illustrative purposes only, not to repeat our discussion of the structure

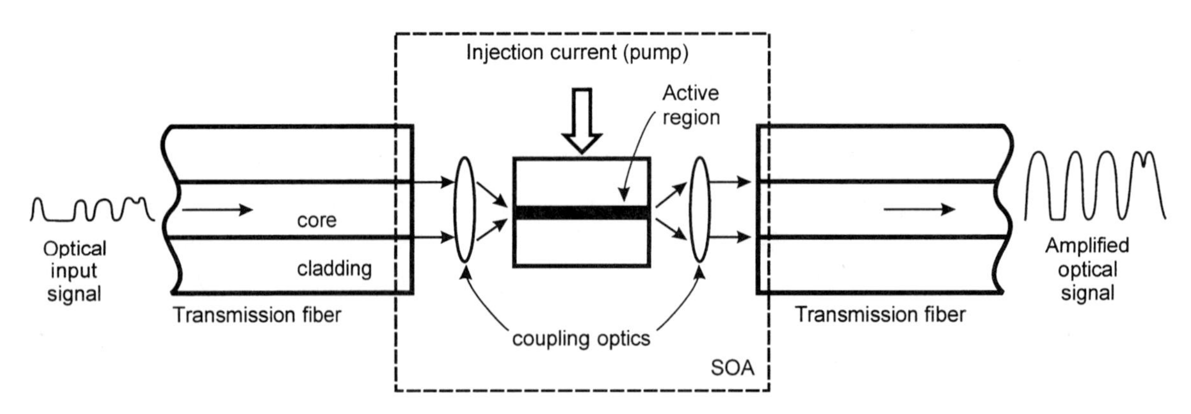


**Figure 12.15** Semiconductor optical amplifier (SOA).

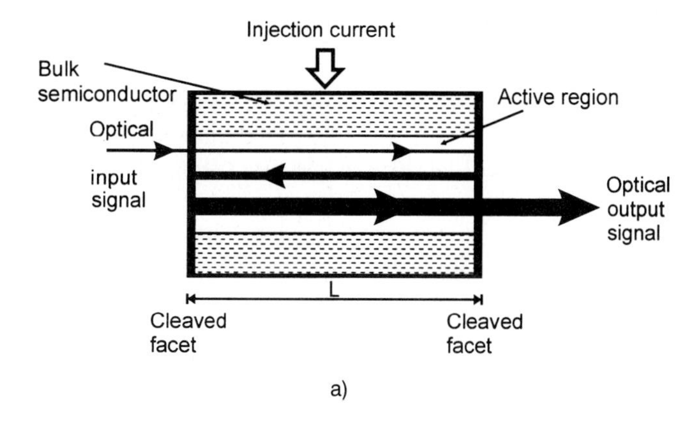


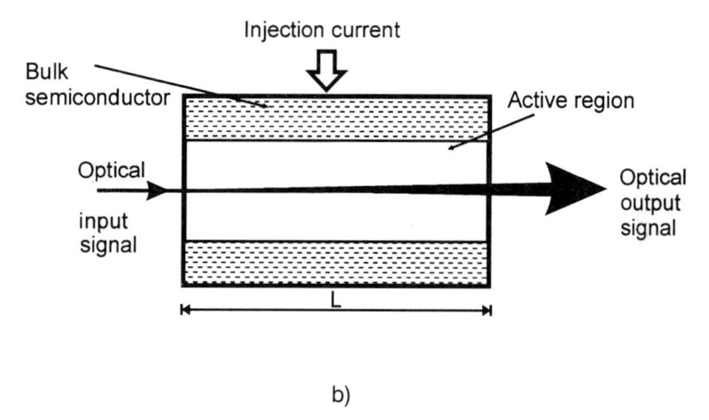


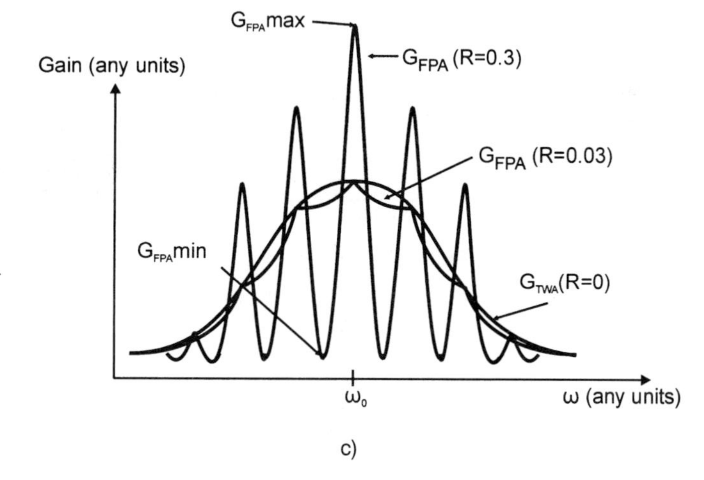


**Figure 12.16** Fabry-Perot and traveling-wave semiconductor optical amplifiers: (a) Fabry-Perot amplifier; (b) traveling-wave amplifier; (c) gain as a function of frequency. *([c] adapted from Dennis J. G. Mestdagh's* Fundamentals of Multiaccess Optical Fiber Networks, *Boston: Artech House, 1995. Used with permission.)*

and principle of operation of a Fabry-Perot semiconductor laser diode. (See Sections 9.2, 10.1, 10.2, and 10.3.)

***Gain of an FPA*** If we denote the power-reflection coefficients of cleaved facets as $R$, the length of an active region as $L$, the velocity of light within the active medium with refractive index $n$ as $v = c/n$, and the single-pass power amplification factor as $G_s$, then the gain ($G$) of an FPA can be expressed as [7],[18], [20]

$$G_{FPA}(\omega) = P_{out}/P_{in} = [G_s(1-R)^2]/\{(1-RG_s)^2 + 4RG_s \sin^2[(\omega-\omega_0)L/v]\}, \qquad (12.4)$$

where $\omega$ and $\omega_0$ are the current and center angular frequencies, respectively.

A single-passage amplification factor, $G_s(\omega)$, is assumed to have a Gaussian-shape dependence on frequency, as shown in Figure 12.16(c).

An FPA exhibits peaks of gain, called *gain ripple,* at resonant frequencies (wavelengths). These are frequencies that a resonator made from facets separated by distance $L$ can support. If we rearrange Formula 9.13, we obtain these resonant wavelengths ($\lambda_N$) as

$$\lambda_N = 2L/N, \qquad (12.5)$$

where $N$ is an integer and $L$ is the length of the active region that is equal to resonator length. Resonant frequencies from Formula 12.5 are given by

$$\omega_N = (2\pi vN)/(2L), \qquad (12.6)$$

where $v = c/n$ is the speed of light within a resonator cavity. You can easily calculate the spacing between adjacent resonant frequencies in the same way we did in Section 9.2.

With these explanations in mind, you can appreciate the dependence of FPA gain on frequency, which is shown in Figure 12.16(c). If we consider reflection occurring at the natural semiconductor–air interface, which is equal to about 0.32, we'll see peaks of FPA gain at the resonant frequencies. The smaller the reflectance, the less pronounced the gain peaks, as shown in Figure 12.16(c) for $R = 0.03$. Ultimately, we arrive at $R = 0$, which is a TWA semiconductor amplifier. This is simply a Gaussian curve, $G_s(\omega)$, as we have assumed.

From this discussion, we conclude that using a Fabry-Perot resonator, which provides optical feedback, can significantly increase the gain of an SOA. The higher the reflectance ($R$), the higher the gain at the resonant frequencies. But bear in mind that increasing the reflectance beyond a certain point can put us into an oscillation mode, that is, turn our amplifier into a laser. Indeed, the closer $RG_s$ is to 1, then, as follows from Formula 12.4, the higher will be the $G_{FPA}(\omega)$. In the extreme, when $RG = 1$, $G_{FPA}(\omega)$ goes to infinity, that is, to the point where an FP amplifier starts to generate light.

### Example 12.3.1

**Problem:**

Estimate the gain of a Fabry-Perot semiconductor optical amplifier if its cleaved facets have a reflectance of $R = 0.32$.

**Solution:**

The answer is given by Formula 12.4, but we need to know the value of the resonant frequencies and $G_s$. If you read the question carefully, you'll see that the solution can be obtained for the central resonant frequency, that is, $\omega = \omega_0$, where $G_{FPA}$ reaches its maximum. For this condition, Formula 12.4 yields

$$G_{FPA}{}^{max} \equiv G_{FPA}(\omega = \omega_0) = [G_s(1-R)^2]/(1-RG_s)^2 \qquad (12.7)$$

Now we need to determine the value of $G_s$. Theoretically, a Fabry-Perot amplifier works like an amplifier until $RG_s < 1$. Hence, $G_s$ (FPA) $< 1/R$. Since $R = 0.32$ and less, $G_s$ (FPA) must be less than 3. To reiterate, this is true only for a Fabry-Perot amplifier. What the value of $G_s$ is for a traveling-wave amplifier we'll see shortly.

Let's assume $G_s = 2$; then for $R = 0.32$, Formula 12.7 yields

$$G_{FPA}(\omega = \omega_0) = 7.1 \text{ or } G_{FPA}(\omega = \omega_0) = 8.5 \text{ dB}$$

If you consider $G_s = 3$, then $G_{FPA}$ $(\omega = \omega_0) = 867$, or 29.4 dB. Thus, you can appreciate how much you can increase $G_{FPA}(\omega = \omega_0)$ by varying $G_s$, which reflects the amplification characteristic of the active region.

Let's keep $G_s$ at 2 and vary the reflectance. We can physically increase or decrease the reflectances by applying a reflecting or antireflecting coating to the cleaved facets. Since $R = 0.32$ represents natural reflectances, the designer's main concern is changing the reflectances in a well-controlled manner. If we increase $R$ to 0.48, that is, obtain $RG_s = 0.96$, then $G_{FPA}$ $(\omega = \omega_0)$ becomes 867, or 29.4 dB; if we decrease $R$ to 0.03, then $G_{FPA}(\omega = \omega_0)$ reduces to 2.13, or 3.3 dB.

---

***Gain of a TWA*** A *traveling-wave amplifier* is essentially an active medium without reflective facets so that an input signal is amplified by a single passage through the active region, as shown in Figure 12.16(b). The *gain of a traveling-wave amplifier* is given by Formula 12.4, where the reflectance should be zero. Hence,

$$G_{TWA}(\omega) = P_{out}/P_{in} = G_s(\omega) \tag{12.8}$$

Thus, the gain of a TWA is the gain of an FPA with $R = 0$, which immediately follows from the definition of a TWA.

Let's turn now to an analysis of TWA gain. A single-pass gain ($G_s$) can be expressed through the parameters of an SOA as follows [7]:

$$G_s = \exp[(\Gamma g - \alpha)L], \tag{12.9}$$

where $\Gamma$ is the confinement factor (see Formula 10.48) that accounts for the guiding of radiated photons by the waveguide structure of an active region, $g$ (1/m) is the gain coefficient of an active region per unit of length, and $\alpha$ (1/m) is the loss coefficient of a cavity per unit of length. (See the box "Gain and Loss of a Laser Diode," Section 10.2.)

In a sense, Formula 12.9 defines the gain of a TWA amplifier. Now you can see that a designer has four means to increase the gain of an amplifier without reflective facets: increase $\Gamma$, $g$, and $L$, and decrease $\alpha$. (If you need to bone up in more detail on how to do this on a practical basis, return to Sections 10.1, 10.2, and 10.3.)

**Example 12.3.2**

***Problem:***

Calculate the gain of a traveling-wave semiconductor amplifier if maximum gain coefficient $g$ = 106 (1/cm), $\alpha = 14$ (1/cm), and $\Gamma = 0.8$.

***Solution:***

Our approach is straightforward. Let's plug the numbers into Formula 12.9 and obtain:

$$G_s = \exp [70.8 \text{ (1/cm)} \times L]$$

The gain depends exponentially on the length of the active region, which makes the length the most powerful parameter in controlling the gain of a traveling-wave amplifier. For a typical TWA length of 500 μm, we obtain $G_s = 34.5$, which means the power of the output signal becomes

34.5 times greater than the power of the input. If you manage to produce an active region 1000 μm long, you will attain a single-pass gain equal to 1187.9, that is, 30.7 dB.

If you notice a contradiction between the values of $G_s$ given in Examples 12.3.1 and 12.3.2, good. Indeed, in Example 12.3.1 we use $G_s = 3$ and in Example 12.3.2 we obtained $G_s = 34.5$. The answer to this quandary is that in the first case we dealt with an FPA, where the length, $L$, of the active region can be made very small and we can increase the overall gain of the FPA by increasing the reflectances, $R$, which is equivalent to reducing the loss of a resonator. In the second case we dealt with a TWA, where the length, $L$, is actually the only parameter for increasing gain. We can obtain $G_s = 3$ but $L$ should be equal to 155 μm. In addition, in the FPA case we are restricted by the value $G_s$ equals 3; otherwise, a Fabry-Perot amplifier becomes a generator (laser). In the case of a TWA, we don't have such restrictions because we will never obtain generation without an optical feedback.

---

The real question regarding a TWA is this: How can we make a practical active region without any reflective facets? Indeed, as soon as you produce a piece of semiconductor material, you have built-in facets, which are the physical boundaries of that piece. But the facets, even without any cleaving, reflect light back into the active region. Hence, special measures must be taken to reduce the reflectivity of these natural "mirrors." This technological problem is resolved by covering the facets with an antireflection (AR) coating, tilting the active region with respect to the facets, and using buffer material between the active region and the facets. All these means are shown in Figure 12.17.

How much must we reduce the reflectance? Look again at Figure 12.16(c). If $R = 0.03$, an amplifier functions almost as a traveling-wave device. But, in practice, we need $R$ as small as $10^{-4}$[20].

***Gain: The difference between FPA and TWA*** At this point, we have to realize that $R = 0$ can never be achieved. Hence, *in reality, the difference between Fabry-Perot and traveling-wave amplifiers is the value of their reflectances.* Indeed, as Figure 12.16(c) shows, if the ratio of maximum to minimum gain is large, we have an FPA; if this ratio is small, we have a TWA. Maximum FPA gain is when the sine in Formula 12.4 is zero, as given by Formula 12.8. Minimum FPA gain is when the sine in Formula 12.4 is 1, which yields

$$G_{FPA}{}^{min} \equiv G_{FPA}\,[(\omega - \omega_0)L/v = \pi/2] = [G_s(1-R)^2]/(1+RG_s)^2 \tag{12.10}$$

(Again, Figure 12.6[c] shows the maximum and minimum gain of an FPA.) Hence, the criterion for distinguishing FPA from TWA is the following [18]:

$$\Delta G = G_{FPA}{}^{max}/G_{FPA}{}^{min} = [(1+RG_s)/(1-RG_s)]^2 \tag{12.11}$$

If we take $\Delta G < 2$, we compute

$$G_s R < 0.17 \tag{12.12}$$

Formula 12.12 is considered a good criterion for determining whether a device that began operating as an FPA is now performing as a TWA.

To satisfy Formula 12.12, with a maximum of a single-pass gain ($G_s$) on the order of 30 dB, which is 1000, we need to attain $R < 0.00017$. You can imagine how difficult it is to achieve such a small reflectance, yet it can be done.

This discussion demonstrates the complexity of the technical problems challenging engineers these days. In the case of laser diodes, for example, we need to increase reflectances to

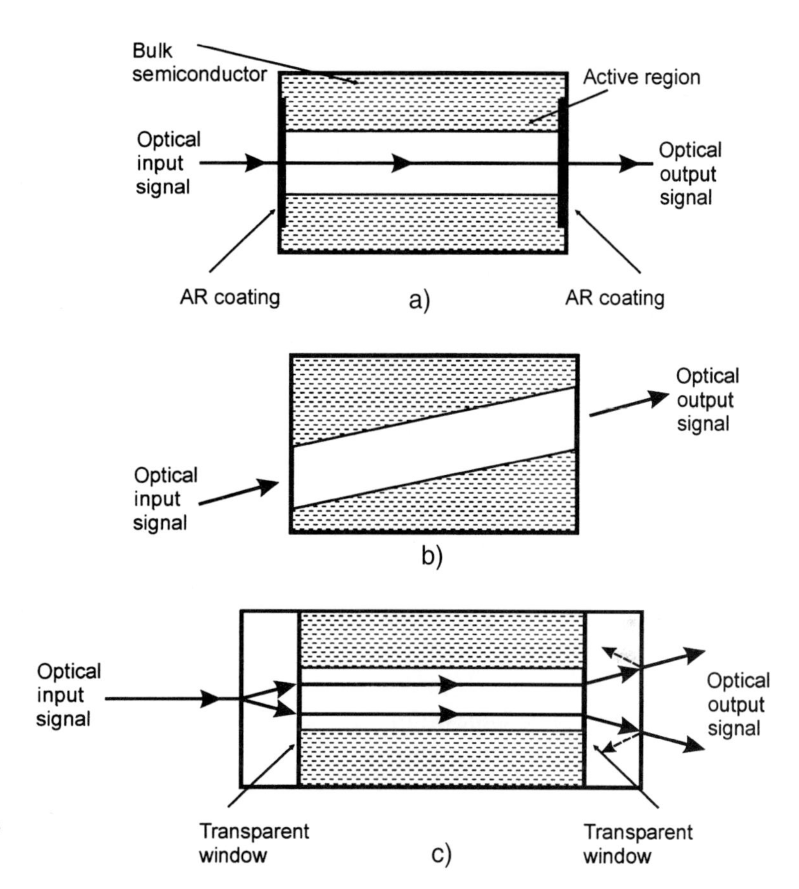


**Figure 12.17** Making a traveling-wave semiconductor optical amplifier: (a) Using antireflection (AR) coating; (b) tilting active region; (c) using transparent windows.

almost 1; yet in another case—traveling-wave amplifiers, for instance—we need to decrease the same reflectances to almost 0. Paradoxical? Yes, but if you'll permit us to paraphrase an old cliche, "Don't view them as technical challenges but as engineering opportunities—with huge financial awards awaiting the solutions."

***Gain saturation*** Looking at Formula 12.9, you may think that the solution is not so terribly complex after all, that this does not appear to bring on a major technological headache. To achieve high gain in a TWA, we need only make an active region long enough, right? Well, not exactly. Formula 12.9 hides the fact that gain coefficient ($g$) depends on the frequency ($\omega$) and power of the signal being amplified ($P$) so that $g = g(\omega, P)$. Let's concentrate first on power dependence, which is given by (see references [7] and [18]):

$$g(\omega, P) = g(\omega)/[1 + (P/P_{sat})], \tag{12.13}$$

where $P_{sat}$ is the saturation optical power. What Formulas 12.9 and 12.13 state is that by increasing length you can increase the power of a signal, but when this power becomes too high, the gain coefficient starts to decrease, thus reducing the power of the signal undergoing amplification. That is why this effect is called *gain saturation.* The physics behind this effect is that high optical power involves all the electrons from the conduction band so that a further increase in the number of external photons will not stimulate any further transition of electrons down to the valence band; that is, it will not produce additional stimulated photons.

Gain ($G_s$) is given by the following transcendental equation (see references [7] and [20]):

$$G_s = 1 + (P_{sat}/P_{in})\ln(G_s^{max}/G_s), \tag{12.14}$$

where $G_s^{max}$ is $G_s$ at $\omega = \omega_0$ and $P_{sat}$ is saturation power. The gain saturation is illustrated in Figure 12.18. The values of saturation output power for an SOA range from 10 to 15 mW.

Now we have to remember that Formula 12.9 is valid only for a small signal, one whose power is far from $P_{sat}$. This is why in the data sheets you will often see the specification "*small-signal gain,*" which means the gain has been measured for small input power.

To review this topic of gain, especially keep in mind that three terms carrying the word *gain* have been introduced here:

- *Gain (dB),* also called *total gain.* $G(\omega,P)$ is defined by Formula 12.4 for an FPA.
- *Single-pass gain,* $G_s(\omega, P)$, is also the *gain of a TWA.* This characteristic is measured in dB and is defined by Formula 12.9.
- *Gain coefficient.* $g(\omega, P)$—having the dimension 1/m—is defined by Formula 12.13, where $g(\omega)$ is the parameter of a specific active region.

## Bandwidth of an SOA

***FPA bandwidth*** The bandwidth of an amplifier, by definition, is the frequency range at which the gain drops twice (3 dB) its maximum value. To determine the 3-dB bandwidth of an FPA, we have to look at Formula 12.4 again. While detuning $(\omega - \omega_0)$ increases, gain ($G$) decreases. The value of $(\omega - \omega_0)$ at which $G$ is reduced by a factor of 2 is the FPA bandwidth. Mathematically, this is equivalent to requiring the denominator of $G$ in Formula 12.4 to double. Thus, we obtain:

$$4RG_s \sin^2[(\omega - \omega_0)L/v] = (1 - RG_s)^2, \tag{12.15}$$

which yields

$$BW_{FPA} = (\omega - \omega_0) = (v/L)\sin^{-1}\{(1 - RG_s)/[2\sqrt{(RG_s)}]\} \tag{12.15a}$$

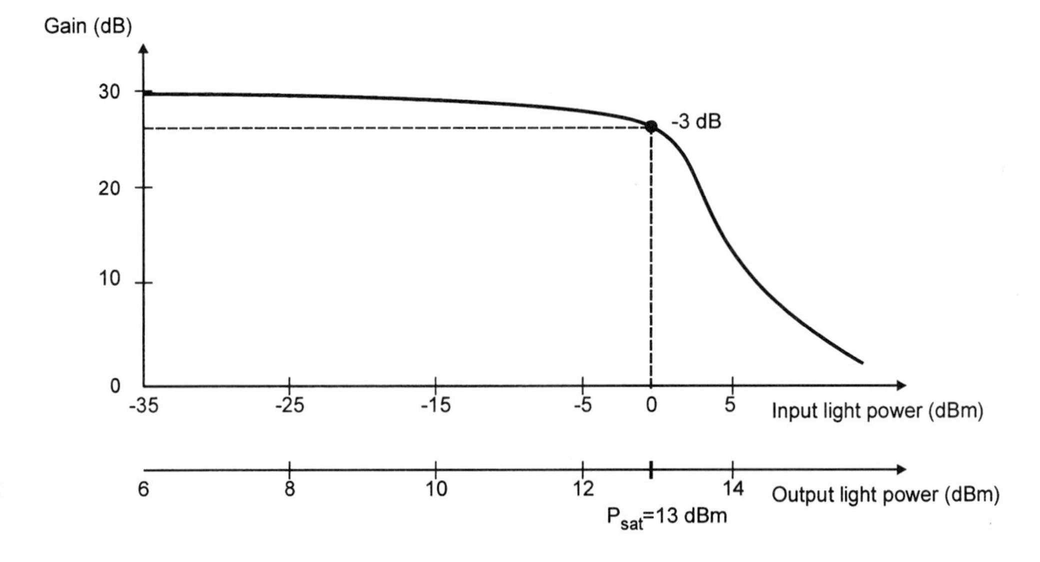


**Figure 12.18** Gain saturation.

and this formula is true for these conditions: $3 - 2\sqrt{2} < RG_s < 3 + 2\sqrt{2}$; that is, $0.17 < RG_s <$ 5.83. This formula determines the bandwidth of an FPA.

### Example 12.3.3

***Problem:***

Estimate the bandwidth of a Fabry-Perot semiconductor optical amplifier.

***Solution:***

Since no specific data are provided, we need to make some assumptions. Our prior discussion of FPA gain and Example 12.3.1 show explicitly that we need to keep $RG_s$ close to 1 to attain high gain. But this means the numerator of Formula 12.15 will become small and so too will the bandwidth of an FPA. Let's assume $RG_s = 0.96$ and $L = 500$ μm, as was the case in Example 12.3.2. Let's also take the refractive index of an SOA's active medium to be equal to 3.6, as we did in Section 9.2. Plugging these numbers into Formula 12.15, we compute

$$BW_{FPA} = 6.8 \times 10^{12} \text{ rad/s, or } 1.08 \text{ GHz,}$$

which is equivalent to 0.0086 nm at a 1550-nm operating wavelength.

In our discussions in Sections 12.1 and 12.2, we noted that the typical bandwidth of a WDM network is about 30 nm—that is, 3.746 THz—and is expanding to 80 nm. Thus, the bandwidth of an FPA is unacceptably small.

---

Can we increase an FPA's bandwidth? Yes, but at the expense of the gain because the only constituent to alter in Formula 12.12 is the $RG_s$ product.

We can demonstrate this gain–bandwidth trade-off in explicit form by studying the following derivation [7]. Since $1 - RG_s$ in an FPA is very small, we can approximate $\sin^{-1}\{(1 - RG_s)/[2\sqrt{(RG_s)}]\}$ by its argument, so that.

$$BW_{FPA} \approx (2c/L)\{(1 - RG_s)/[2\sqrt{(RG_s)}]\} \tag{12.16}$$

From Formula 12.8 we easily obtain

$$1 - RG_s = \sqrt{\{[G_s(1-R)^2]/G_{FPA}{}^{max}\}} \tag{12.17}$$

Substituting Formula 12.17 into Formula 12.16, we get

$$BW_{FPA}\sqrt{G_{FPA}{}^{max}} \approx (c/L)\sqrt{[(1-R)^2/R]} \tag{12.18}$$

From Formula 12.18, you can clearly see the trade-off between bandwidth and gain in an FPA.

***TWA bandwidth*** Let's consider the 3-dB bandwidth of a TWA. It would seem that we could estimate this bandwidth from Formula 12.15 by plugging in $R = 0$, but it doesn't work. Indeed, from Formula 12.15 we can find this very evident condition,

$$(1 - RG_s)/2\sqrt{(RG_s)} \le 1, \tag{12.19}$$

dictated by the property of a sine function. A simple manipulation yields

$$RG_s \ge 0.1715 \tag{12.20}$$

But the criterion given in Formula 12.12 states just the opposite: An amplifier is considered a TWA if $R\,G_s < 0.17$. Hence, Formula 12.15 cannot be applied to calculate TWA bandwidth.

However, qualitatively, a TWA's bandwidth property can be described based on Formula 12.18, modified to apply to a TWA's condition:

$$BW_{TWA}\sqrt{G_s} \sim (c/L)\sqrt{[(1-R)^2/R]} \tag{12.21}$$

From Formula 12.21, you can see the same bandwidth-gain trade-off as was the case for an FPA. Moreover, for a TWA we want to have the reflectances ($R$) much smaller than they are for an FPA; hence, the product of the gain and bandwidth for a TWA is larger than it is for an FPA. As a matter of fact, a TWA with a single-pass gain of about 30 dB and a bandwidth near 40 nm is commercially available. (See Figure 12.24.)

***Bandwidth: The difference between FPA and TWA*** Let's summarize our consideration of bandwidth by redrawing Figure 12.16(c) as Figure 12.19. Figure 12.19 shows explicitly the difference between the gain-bandwidth product of an FPA and the gain-bandwidth product of a TWA. An FPA has a large gain ($G_{FPA}$) but a small bandwidth ($BW_{FPA}$), while a TWA has a much larger bandwidth ($BW_{TWA}$) but a smaller gain ($G_s$). (To illustrate the point, only one central ripple of an FPA's possible gain is shown.) Each type of SOA has its own gain-bandwidth trade-off, as discussed above.

In Example 12.3.3, we've computed the bandwidth of an FPA as 0.0086 nm; the bandwidth of a TWA is about 40 nm. Clearly, then, is it any wonder that only TWA-type amplifiers are deployed in transmission systems.

## Crosstalk

We've seen that for amplification of a single channel (wavelength), SOAs are far and away the leading systems in use. But suppose we want to amplify several channels (wavelengths) simultaneously, as we do in WDM networks. The first thing we have to worry about in such a case is *crosstalk,* which is *any distortion of a channel caused by the presence of another channel.* (The term traces its origin to telephone circuitry. The penetration of one voice channel into another line was the type of crosstalk telecommunications engineers first encountered.)

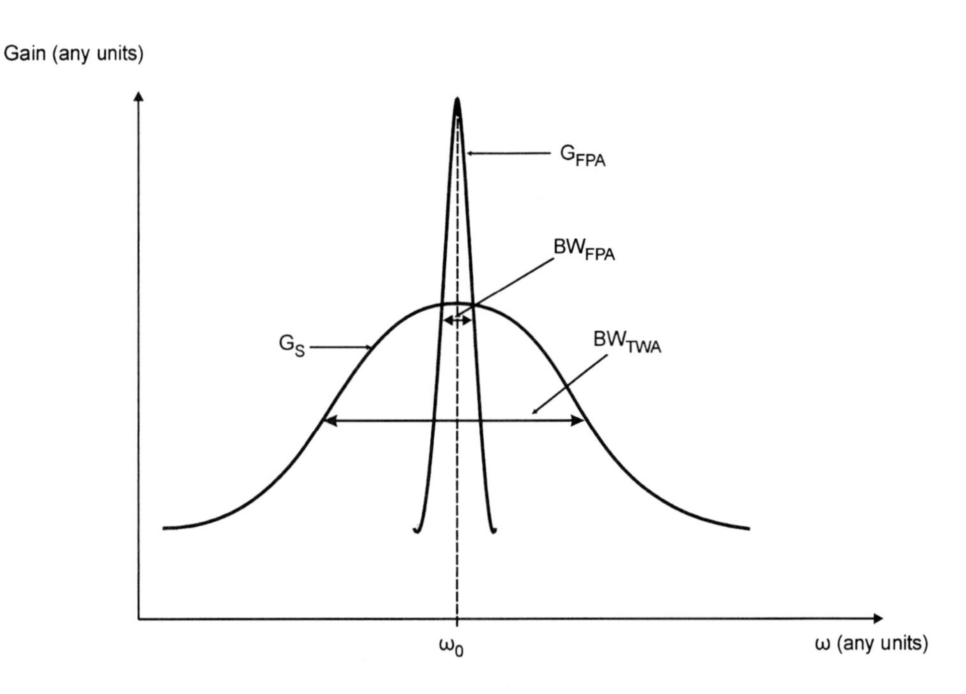


**Figure 12.19** Bandwidths of Fabry-Perot and traveling-wave amplifiers.

There are two types of crosstalk in SOAs: *interchannel crosstalk* and *cross saturation* ([18] and [20]). Interchannel crosstalk is essentially a four-wave mixing (FWM) effect, which is discussed in Section 6.4. When two wavelengths (channels) enter an SOA, their nonlinear interference produces new signals at the beat (combinations of sums and differences) frequencies. The physical reason for the generation of new signals is the modulation of the excited electrons at beat frequencies. But these new signals deplete the conduction band; that is, they "steal" some gain from the original signals. Thus, amplification of the original input signals becomes less, and new, undesired signals appear. This explanation is illustrated in Figures 12.20(a) and (b). (Everything is relative. We will see in Chapter 13 how this effect can be used in a constructive way.)

*Cross saturation* occurs when a semiconductor amplifier works in the saturated mode; that is, the power of the input signals is above the saturation value. When one channel changes from ON to OFF, the gain undergoes an opposite change. This gain change results in variations in the amplification of another signal because, again, all signals share the same gain produced by one active medium. Figure 12.20(c) illustrates this point.

How fast can the gain variation occur? In other words, can gain changes follow any input bit rate? The answer lies in the value of the *carrier lifetime,* or *spontaneous-emission time* ($\tau_{sp}$), which in semiconductors is around 1 ns. If the bit rate is less than $1/\tau_{sp}$, then the gain changes as the input signal dictates. Therefore, for an SOA, all bit rates less than 1 Gbit/s will cause severe crosstalk. In contrast, the lifetime of an erbium-doped fiber amplifier is about 10 ms, meaning

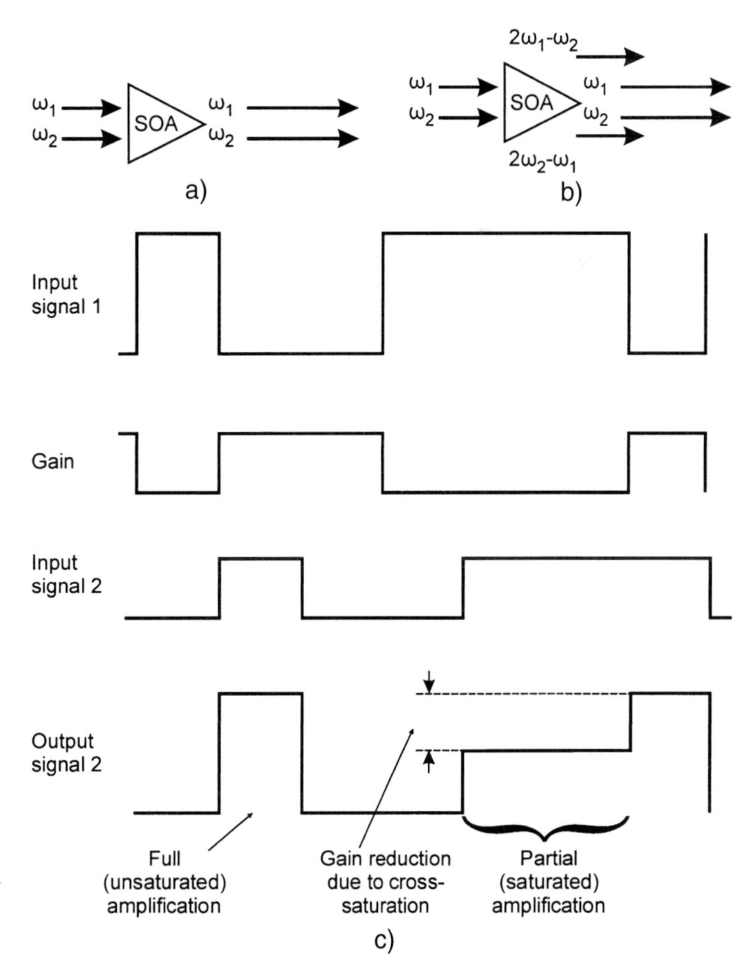


**Figure 12.20** Crosstalk in a semiconductor optical amplifier: (a) Signal amplification without crosstalk; (b) interchannel crosstalk; (c) cross saturation.

any bit rate above 100 kHz will not cause crosstalk. This is why crosstalk is not an important factor in EDFAs. The physics at work here tells us that an active medium does not react to changes in the magnitude of an input signal when the frequency of this signal is higher than $1/\tau_{sp}$.

To reduce crosstalk, we need to increase the saturation power and decrease the lifetime. Both of these goals were at least partially achieved by the development of TWAs with *multiple-quantum-well (MQW)* strained-laser structures. (See Sections 9.2 and 10.1 to refresh your recollection of these terms.) In fact, saturation power of about 16 dB (40 mW) and a lifetime of around 0.2 ns were achieved years ago [20].

## Polarization-Dependent Gain

The gain of semiconductor optical amplifiers depends on the state of polarization of the input signal. In other words, the amplification of transverse-electric (TE) and transverse-magnetic (TM) modes is different as shown in Figure 12.21(a). (See Figure 4.1 for an explanation of TE and TM modes.)

The physical reasons for the dependence of SOA gain on polarization include the rectangular shape and the crystal structure of the active region. These two characteristics make the gain ($g$) and confinement ($\Gamma$) coefficients dependent on polarization. The difference in gain between two orthogonal polarizations can reach 5 to 7 dB.

The problem with polarization-dependent gain stems from the fact that almost all of the deployed fiber is nonpolarization-maintaining fiber. This means that the state of polarization of an optical transmission signal cannot be maintained; its state is unpredictable. Now imagine, if you will, that at one instant a signal enters, say, in the TE mode and at the next instant the signal enters in the TM mode. The gain of this SOA would vary from 5 to 7 dB simply because of signal polarization, and you would see the signal amplified either 1000 (30 dB) times or 316 times (25 dB). Wouldn't we all like that?

There are several means to reduce polarization dependence in semiconductor optical amplifiers. One way is to attempt to make the active region as square as possible in cross section. Another way is to connect two SOAs in series or in parallel to compensate for the orthogonal polarization's unequal gain. (See Figures 12.21[b] and [c].) A double pass through the same active region also compensates for unequal gain because a *Faraday rotator* turns the polarization of the backreflected beam at a 90° angle with respect to the beam (Figure 12.21[d]).

The result of these measures is that polarization-dependent gain (or, more precisely, the variation in gain, depending on the polarization of the input light) is reduced to 0.5 dB in commercially available SOAs.

## Noise

The noise phenomenon in an SOA and in an EDFA, which is discussed in the next section, has very much in common. Thus, we will try to treat this topic here in general terms as much as possible, referring, as necessary, to the peculiarities of the noise in an SOA.

We know an optical amplifier magnifies the signal noise along with the signal itself, but actually it does much more than that. *An optical amplifier generates its own noise.* Therefore, an amplifier changes the signal-to-noise ratio of both the input and output signals. The noise performance of an optical amplifier is quantified through what's called a *noise figure* ($F_n$), which is defined as:

$$F_n = (\mathrm{SNR})_{in}/(\mathrm{SNR})_{out}, \tag{12.22}$$

where $(\mathrm{SNR})_{in}$ stands for the input and $(\mathrm{SNR})_{out}$ represents the output of this signal-to-noise ratio. Reflect for a moment on Formula 12.22. It's not the signal-to-noise ratio per se that we are interested in knowing but, rather, in changing this characteristic by means of an amplifier. Formula 12.22 accurately reflects the operation of an amplifier with respect to noise. What is not immediately clear from this formula is that an optical amplifier changes the signal-to-noise ratio

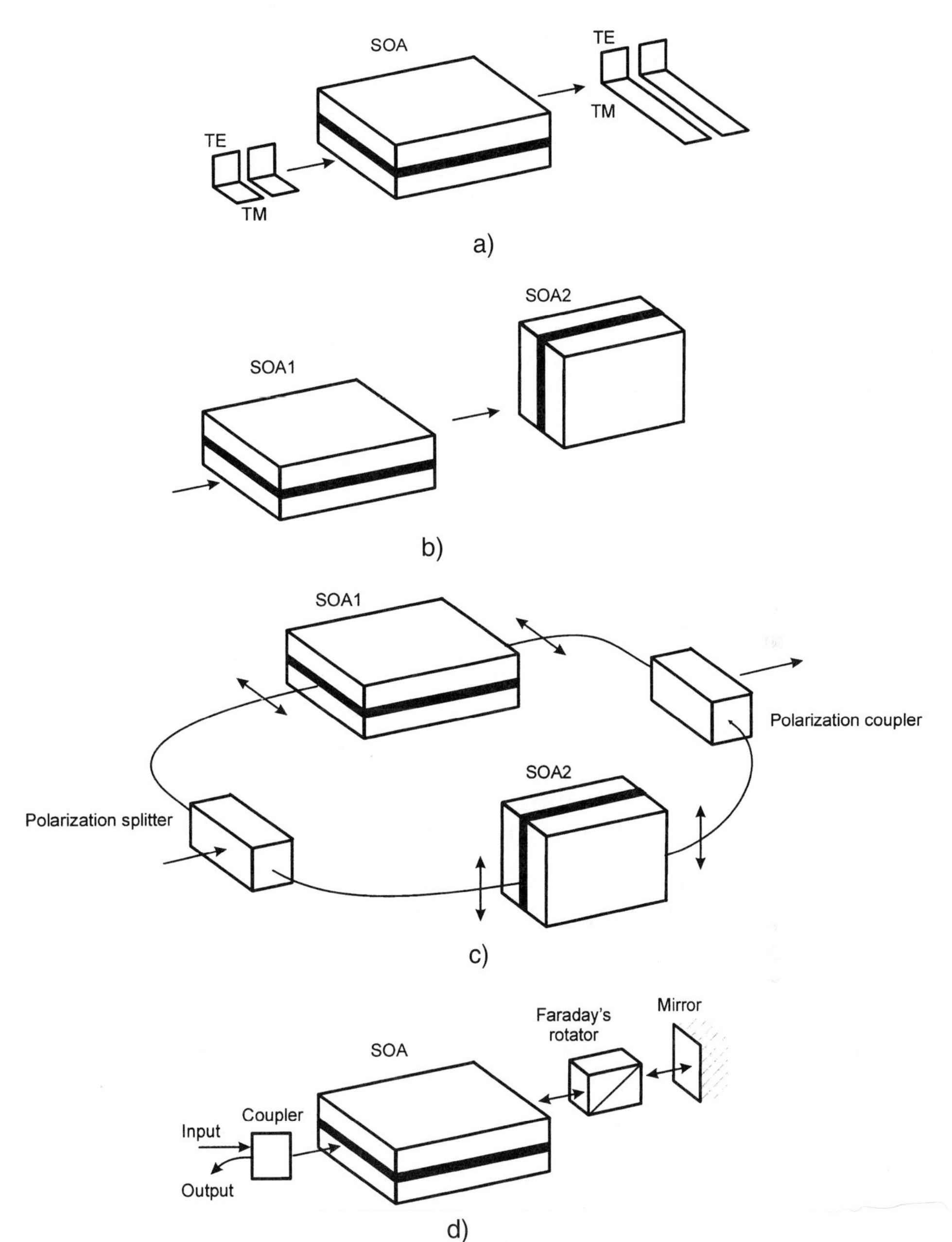


**Figure 12.21** Reducing polarization dependence in an SOA: (a) Polarization-dependent gain; (b) connection of two SOAs in series; (c) connection of two SOAs in parallel; (d) double pass through an SOA.

because it magnifies both the input signal *and* the noise (although we do not know in what proportion) *and* adds its own noise.

$F_n$ is often referred to as a *figure of merit* when one is evaluating the noise performance of an optical amplifier.

The noise-figure concept is further illustrated in Figure 12.22, where a measurement arrangement is shown [21]. The measuring characteristics of an optical source with a piece of regular fiber (Figure 12.22[a]) give the input signal-to-noise ratio, $(SNR)_{in}$, without an optical amplifier. The same measurement with an optical amplifier placed in line gives $(SNR)_{out}$, as Figure 12.22(b) shows. The use of the same connections in both measurements helps to eliminate additional sources of errors.

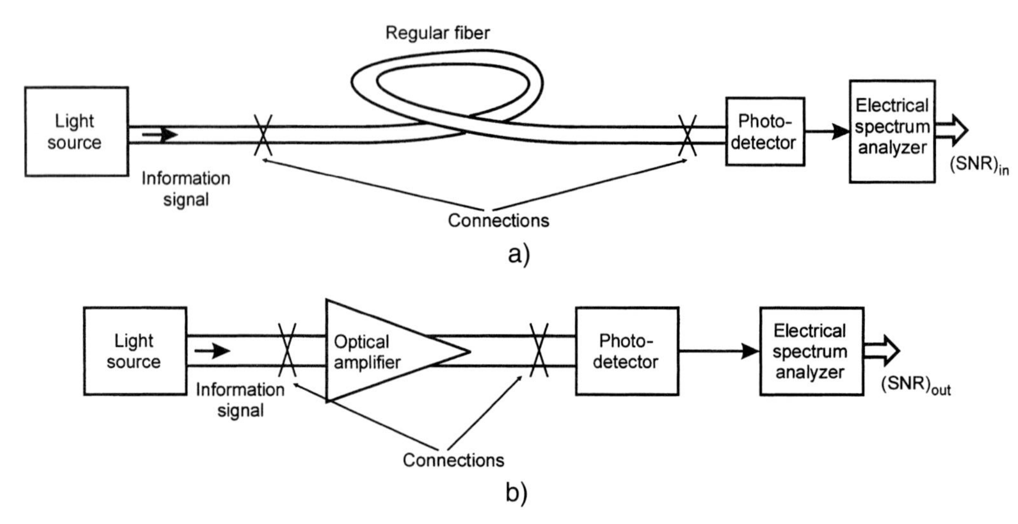


**Figure 12.22** Measurement of noise figure: (a) measurement of $(SNR)_{in}$; (b) measurement of $(SNR)_{out}$. *(Dennis Derickson, ed.,* Fiber Optic Text and Measurement, *Upper Saddle River, N.J.: Prentice Hall PTR. Copyright © 1998 Agilent Technologies. Used by permission.)*

As in any ratio, the noise figure is dimensionless. In the communications world, we usually measure a ratio in dB, and $F_n$ is not an exception. Typical noise figures for commercially available optical amplifiers range from 6 to 9 dB for an SOA and from 3.5 to 9 dB for an EDFA.

## Example 12.3.4

***Problem:***

Calculate the noise figure of an optical amplifier if the input-signal power is 300 μW, the input-noise power is 30 nW in a 1-nm bandwidth, the output-signal power is 60 mW, and the output-noise power is 20 μW in a 1-nm bandwidth.

***Solution:***

Let's calculate the input and output signal-to-noise ratios and then use Formula 12.22: $(SNR)_{in} = 10 \times 10^3$ and $(SNR)_{out} = 3 \times 10^3$; hence, $F_n = 3.33$, or 5.2 dB.

This example demonstrates a very important concept: *An optical amplifier does indeed decrease the signal-to-noise ratio, as the value of* $F_n$ *shows. However, an optical amplifier also raises the signal power to such a high level that we can tolerate this degradation of the SNR.*

If you have ever studied electronic-communication theory, you should be quite familiar with the term *noise figure.* In fact, when discussing electronic amplifiers, we define the term exactly the same way we did above.

***Amplified spontaneous emission (ASE)*** The noise generated by an active medium of an optical amplifier is caused primarily by *amplified spontaneous emission (ASE).* The physics underlying this phenomenon is as follows: The vast majority of excited carriers are forced by stimulated emission to fall to a lower level, although some of these carriers do so spontaneously. When they decay, these carriers radiate photons spontaneously. The spontaneously emitted photons are in the same frequency range as the information signal, but they are random in phases and directions. The spontaneously emitted photons that follow in the direction of the information signal are amplified by an active medium. These *spontaneously emitted and amplified photons constitute amplified spontaneous emission (ASE).* Since they are random in phase, they do not contribute to the information signal but generate noise within the signal's bandwidth.

You may recall that emission is termed *spontaneous* if it occurs without external stimulation. (If you need to refresh your memory on this topic, see Section 9.2.) It is easy to understand that spontaneous emission depends on the relative populations of the upper (excited) and lower energy levels. (In semiconductors, as we learned, the upper level is the conduction band and the lower level is the valence band.) A *spontaneous-emission factor, or population-inversion factor* ($n_{sp}$), can be defined as (see references [18] and [20])

$$n_{sp} = N_2/(N_2 - N_1), \tag{12.23}$$

where $N_2$ and $N_1$ are populations of the excited and lower levels, respectively. The spontaneous-emission factor reaches its minimal value, 1, when the population of the upper level is much higher than that of the lower level, which means that $(N_2/(N_2 - N_1) \rightarrow 1)$. In such a case, we would have an *ideal amplifier,* but a situation like this can never be attained and so the actual values of $n_{sp}$ range, typically, from 1.4 to 4.

The higher the spontaneous-emission factor, the greater the power of the amplified spontaneous emission generated by an optical amplifier.

The point to remember here is that spontaneous emission in an optical amplifier occurs within the same band of wavelengths (frequencies) in which signal amplification takes place. This is why spontaneous emission is the main mechanism adding noise to an amplified signal.

The average total power of amplified spontaneous emission ($P_{ASE}$) is equal to [21]:

$$P_{ASE} = 2n_{sp}hf\ G\ \mathrm{BW}, \tag{12.24}$$

where $hf$ is photon energy, $G$ is amplifier gain, and BW is the optical bandwidth of the amplifier. This formula clearly expresses the idea that *the greater the spontaneous emission, quantified by* $n_{sp}$, *the greater too will be the amplified spontaneous emission (ASE).*

## Example 12.3.5

***Problem:***

Calculate the ASE power generated by a TWA SOA operating at 1300 nm with a gain of 30 dB.

***Solution:***

The solution to this problem is given by Formula 12.24, but first we need to clarify several points. For one thing, what is the number for $n_{sp}$? We know that a spontaneous-emission factor in practice ranges from 1.4 to 4. Let's assume $n_{sp} = 3$. Secondly, what is the number for $hf$? For $\lambda =$ 1300 nm, we obtain $hf = 1.53 \times 10^{-19}$ J (remember $\lambda f = c$). Third, we need to convert the gain from dB into absolute numbers. This yields $G = 1000$. Fourth, we need to find the bandwidth and convert it from nanometers into hertz. Bandwidth, remember, is a range of frequencies; hence, $\mathrm{BW} \equiv \Delta f = (\Delta\lambda/\lambda^2)f$. We know that the bandwidth of a TWA is about 40 nm. For $\lambda = 1300$ nm, we find $\mathrm{BW} = 1.775 \times 10^{12}$ Hz.

Now we can insert all these numbers into Formula 12.24 and compute:

$$P_{ASE} = 1.6 \text{ mW}$$

This is a tremendous amount of noise power. Fortunately, this is the total amount of noise power allocated along the entire gain bandwidth. Therefore, within the spectrum of an individual information signal, $P_{ASE}$ will be much smaller. Indeed, if the linewidth of a DFB is 40 MHz (as we saw in Section 9.3), then the $P_{ASE}$ of this signal comes to 0.037 μW, which we compute using Formula 12.24.

---

ASE is a detrimental characteristic, so to attain a low noise figure, manufacturers include in the optical amplifier structure a special ASE-suppression device [22].

***Optical and electrical noise*** We must distinguish between the *optical noise* and the *electrical noise* of an optical amplifier. Optical noise is the noise measured in an optical field by optical-measurement devices such as an optical power meter and an optical spectrum analyzer. (Optical power meters are discussed in Section 8.5. An optical spectrum analyzer is an instrument that evaluates the spectrum of an optical signal in the optical domain. In other words, it is an optical analog to the electrical spectrum analyzer familiar to every electrical engineer.) A schematic of such a measurement is shown in Figure 12.23(a). Electrical noise is a characteristic of an optical amplifier measured at a point just beyond the photodetector, that is, after the conversion of an optical signal into an electrical signal, as shown schematically in Figure 12.23(b).

The importance in distinguishing between optical and electrical noise follows from the very nature of a photodetector: It converts light power (intensity) into electrical current. (See Section 11.1.) Here we have to emphasize a key point for consideration: *From the optical vantage point, interference between two light signals—the information signal and the amplified spontaneous emission—causes intensity fluctuations of detected light. From the electrical vantage point, these light fluctuations appear as photocurrent fluctuations, resulting in a certain signal-to-noise ratio.*

## Reading the Data Sheet of an SOA

Figure 12.24 shows the data sheet of a semiconductor optical amplifier. Let's start analyzing from the top. The manufacturer indicates that this device is fabricated from InGaAs semiconductor material using a multiple-quantum-well (MQW) structure. (If you have forgotten the meaning of these terms, refer back to Sections 9.2 and 10.1.)

The section headed "Typical Performance" gives, in a sense, the optical specifications of an SOA:

- *Operating wavelength* states the range within which an SOA provides a specified gain. In essence, this is the bandwidth of the SOA. You can see that this SOA spans the range from 1290 to 1330 nm; that is, its bandwidth is 40 nm.
- *Saturation output power* is specified as 13 dBm and it's measured at 0 dBm input power. (See Figure 12.18.)
- *Small signal gain* is 30 dB, which means that when the SOA is not saturated, its output power will be 1000 times larger than the input.
- *Polarization-dependent gain* shows that when gain is measured for two orthogonal polarizations, the difference will not be more than 0.5 dB.

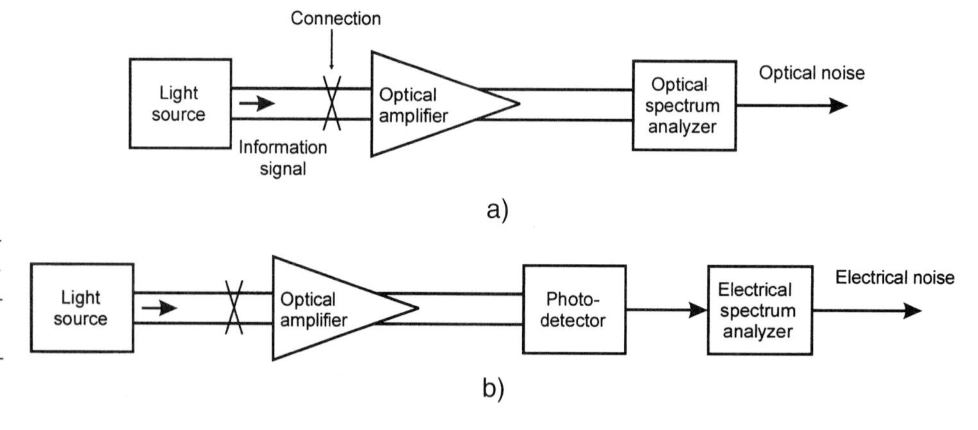


**Figure 12.23** Measurement of optical and electrical noise: (a) Measurement of optical noise; (b) measurement of electrical noise.

AFC's semiconductor amplifier at 1300 nm is a high gain high output power InGaAs MQW device. Packaged in a convenient 19" rack case, it provides a "plug & play" solution for upgrading the existing fiber networks at 1300 nm to a bit rates of 10 Gbit/s and beyond without the need for switching to 1550 nm.

**Typical Performance**

| | |
|---|---|
| Operating Wavelength (nm) | 1290 - 1330 |
| Saturation Power(1) (dBm) | 13 |
| Small Signal Gain(2) (dB) | 30 |
| Polarization Dependent Gain (dB) | 0.5 |
| Gain Ripple (dB) | 0.2 |
| Noise Figure (dB) | 8.0 |

(1) 0 dBm input, at 1310 nm. (2) - 30 dBm input, at 1310.

**Figure 12.24** Data sheet of a semiconductor optical amplifier. *(Courtesy of AFC Technologies Inc. Hull, Que., Can.)*

- *Gain ripple* refers to variations in gain caused by finite reflectances ($R$). (See also Figure 12.16 for a detailed discussion of gain ripple.) The small number for gain ripple is another indication that we are dealing with a TWA type of SOA. But this number also shows that we can never build an ideal TWA because we can never obtain $R = 0$.
- *Noise figure*—8 dB—is typical for an SOA, as already mentioned.
- An SOA's data sheet can also include *input/output isolation.* This is a measure of backward-propagated light. Light traveling from a transmitter to a receiver experiences backreflection at the splices, connectors, and facets of an SOA as well as along the fiber itself (see "Rayleigh scattering" in Section 3.2) and other elements of the fiber-optic transmission path. This backreflected light is amplified when entering the active region of an optical amplifier. Since the phase of backreflected light is random, it contributes to noise, which is why manufacturers take all measures to prevent backreflection. Typically, backreflected light is 1000 times weaker than the information signal traveling toward the receiver.

## SOA Applications

Semiconductor optical amplifiers find a number of applications in fiber-optic communications networks. They are used as transmission amplifiers in the 1300-nm window. Since all cable TV transmission is done at the 1300-nm wavelength, SOAs find applications, too, in CATV networks. In addition, SOAs are employed in switches, filters, modulators, wavelength converters, and tapping devices. (For a more in-depth discussion of these applications, see reference [20].)

## SOAs: Advantages and Drawbacks

In conclusion, let's reiterate the advantages and drawbacks of semiconductor optical amplifiers. The major advantage is their ability to operate at the 1300-nm and 1550-nm wavelengths—even simultaneously. Another advantage is a wide bandwidth (up to 100 nm has been achieved [5]). An SOA can be readily integrated, along with other semiconductor and photonic devices, into one monolithic chip called an *opto-electronic integrated circuit (OEIC).*

However, a number of serious drawbacks preclude SOAs from being the major player in the optical-amplifier field. A relatively high crosstalk level is one of its shortcomings. Polarization sensitivity is another significant drawback. Since the actual working type of SOA is the traveling-wave amplifier, the need exists to produce an active medium with reflectances as low as $10^{-4}$. This is the technological problem besetting amplifiers of this type and is what is keeping their price high. In addition, an SOA, like any semiconductor device, is temperature sensitive.

SOAs are used in some fiber-optic communications networks, mostly in CATV (as noted above), but the major type of optical amplifier in use is the erbium-doped fiber amplifier (EDFA).

## 12.4 ERBIUM-DOPED FIBER AMPLIFIERS (EDFAs)

We'll focus on erbium-doped fiber amplifiers because these fiber amplifiers are the ones normally deployed in WDM fiber-optic communications systems today.

### How Amplification Occurs

Figures 12.25(a) and (b) depict two special features of a fiber amplifier. A piece of fiber working as an active medium is heavily doped with ions of erbium (Er). External energy is delivered *optically,* not electrically, as it is for SOAs.

Pumping is done with a laser diode radiating powerful light at a wavelength other than an information signal's wavelength. Specifically, an information signal is transmitted in the vicinity of 1550 nm but pump lasers radiate either at 980 nm or at 1480 nm, or both. Both the optical information and optical pumping beams are put in the same fiber by a coupler. Those two beams propagate together along the doped section of the fiber, where the information signal is amplified while the pumping signal loses its power. In a sense, then, pumping light gives its power to an information signal and "dies."

A pumping signal can copropagate with an information signal, as Figure 12.25(a) shows, or it can counterpropagate, as it is doing in Figure 12.25(b). A *copropagating pump* features lower noise and lower output power, while a *counterpropagating pump* provides higher output power but produces greater noise, too. In the typical commercial amplifier, you often find a *bidirectional pump* with simultaneous copropagating and counterpropagating pumping.

A second *coupler* removes residual pumping light from the transmission fiber. An *isolator* prevents backreflected light from penetrating the amplifier fiber; otherwise, this light will also be amplified, a detrimental occurrence because such amplification can turn an amplifier into a laser, not to mention the unacceptable increase in noise level that occurs as well. A *filter* separates any remnant of light power from the information signal.

Amplification in an erbium-doped fiber amplifier occurs through the mechanism of stimulated emission. Energy from the pumping signal excites erbium ions at the upper energy band. The information signal stimulates transition of the excited ions to the lower energy band. These transitions result in the radiation of photons with the same energy—that is, the same wavelength—the input signal has. Since an EDFA has a relatively wide gain bandwidth, it can amplify many wavelengths (channels) simultaneously. Amplified signals (wavelength-separated channels) along with the noise associated with an EDFA are shown in Figure 12.25(c).

A fiber amplifier is usually a unidirectional device, as Figure 12.25(d) shows. However, an amplifier generates its own noise (in addition to amplifying the noise coming with the signal) and this noise propagates in both directions along the fiber. This EDFA characteristic justifies the use of an isolator in the input port of a fiber amplifier.

***Energy-level diagram*** Because an EDFA's operation is based on the stimulated-emission mechanism, we should first discuss the *energy-level diagram* of an active medium as it pertains to this amplifier. Free ions of erbium exhibit discrete energy levels. When erbium ions are incorporated into a silica fiber, each of their energy levels splits into a number of closely related levels so that we can consider them an *energy band.* (See Section 9.1.)

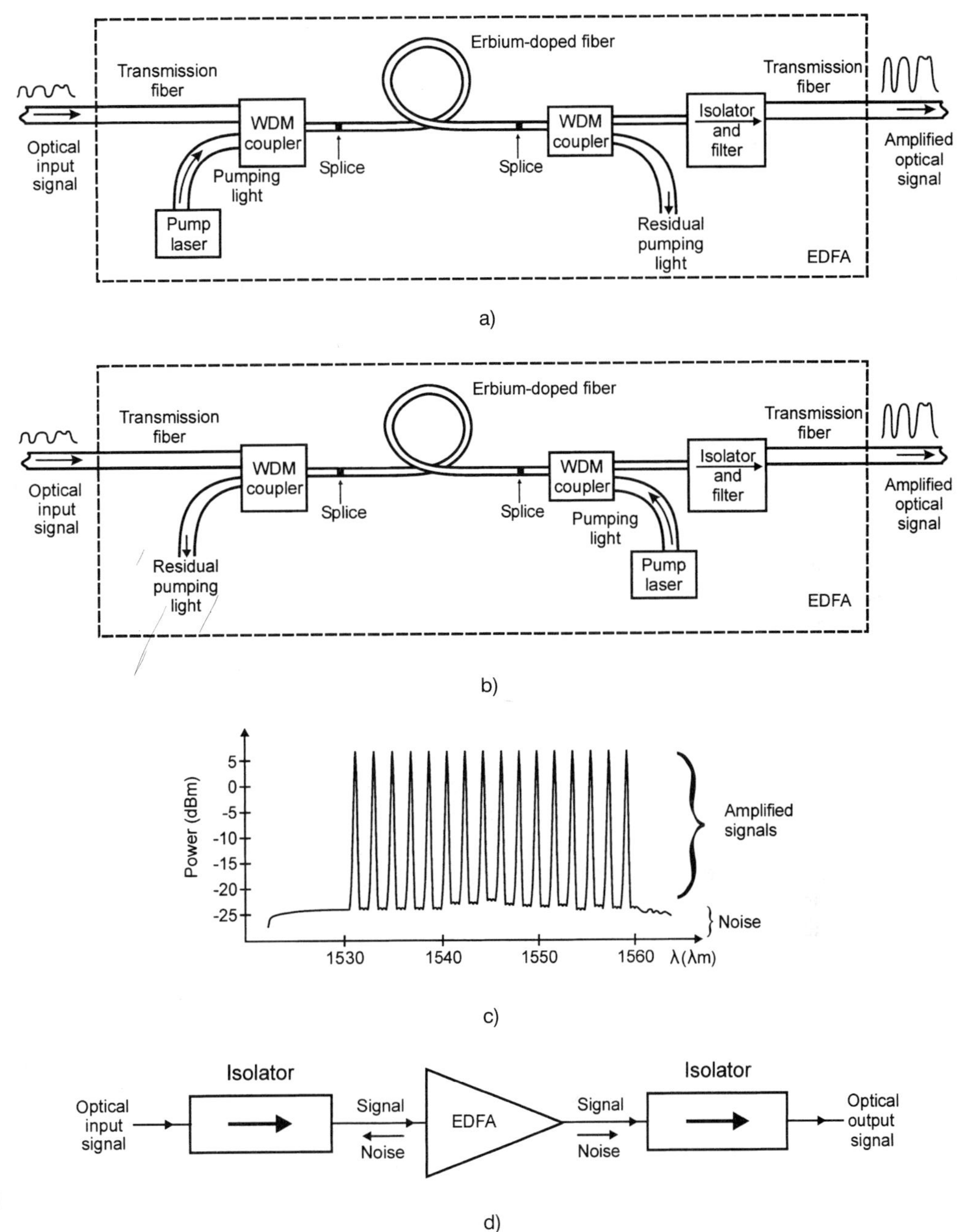


**Figure 12.25** Erbium-doped fiber optical amplifiers: (a) Copropagating pump; (b) counter-propagating pump; (c) amplified WDM channels; (d) signal and noise in EDFA.

In an EDFA, splitting energy levels into an energy band is beneficial. First and foremost, it gives the EDFA the ability to amplify not just a single wavelength but a set of wavelengths. Secondly, it eliminates the need to fine-tune a pumping wavelength.

The most important energy levels (bands) of erbium ions incorporated into a silica fiber are shown in Figure 12.26. It is a gift of nature that the transition between level 2 (the intermediate band) and level 1 (the lower band) occurs at a set of wavelengths around 1550 nm, where silica fiber exhibits minimum attenuation. This fortunate coincidence is why erbium-doped fiber amplifiers are so widely used. As you study Figure 12.26, note in particular the width of the energy

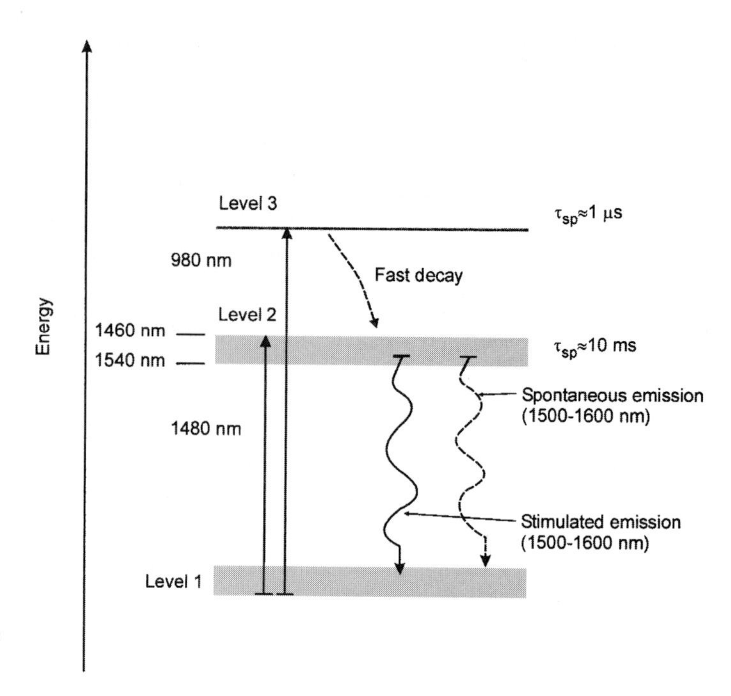


**Figure 12.26** Energy bands of erbium ions in a silica fiber.

bands, which determines the EDFA's ability to amplify the range of wavelengths from 1500 nm to more than 1600 nm.

***Pumping*** Our goal is to achieve population inversion, which, as you already know, means having more ions of erbium at the intermediate level (2) than at the lower level (1). To attain population inversion, we need to pump erbium ions at the intermediate level. There are two ways to do this: pumping them *directly* at the 1480-nm wavelength or *indirectly* at the 980-nm wavelength.

Let's consider the indirect method (980-nm pumping) first. In this case, erbium ions are continuously moved from the lower state to the upper level (3); there they nonradiatively decay to the intermediate energy level (2), from where they fall to the lower level (1), radiating the desired wavelength (1500 to 1600 nm). This is known as the three-level mechanism. The key to using this three-level mechanism is the lifetimes of the two upper levels. *Lifetime,* or *time of spontaneous emission* ($\tau_{sp}$), is the average duration atoms stay at a specific level before they move spontaneously down to the next energy level. The lifetime of erbium ions at the upper level (3) is only about 1 μs, while the lifetime of erbium ions at the intermediate level (2) is more than 10 ms. (With its long lifetime, such a level is called *metastable.*) Therefore, erbium ions pumped at the upper level will descend to the intermediate level very quickly and will stay at that level for a comparatively long time. In other words, erbium ions will accumulate at the intermediate level, creating *population inversion.*

When pumping is done directly (at 1480 nm), only two energy levels are involved. Erbium ions are continuously taken from the lower level by external optical energy at 1480 nm and placed at the intermediate level. Since the lifetime of erbium ions at this level is long, they accumulate here, again creating population inversion.

The result of both processes is that the intermediate level is populated with more erbium ions than the lower level. When an optical information signal operating at one of the WDM wavelengths passes through such an inversely populated erbium-doped fiber, it will stimulate the transition of erbium ions from level 2 to level 1. This stimulated transition will be accompanied by the *stimulated emission* of photons having the same wavelength, direction, and phase as the input photons have. Thus, *amplification of the input signal occurs.*

## C-Band and L-Band

It is worth noting that, potentially, an erbium-doped fiber can amplify light at wavelengths ranging from approximately 1500 nm to more than 1600 nm. However, practical gain windows have a much more restricted range of amplified wavelengths. These are called *gain bandwidths.* Two such bands are in use today. One, called *C-band,* ranges from 1530 nm to 1560 nm. Most EDFAs in the field work within this band. Recently, however, researchers extended EDFA's gain bandwidth even farther. The second gain band, called the *L-band,* occupies the spectrum from 1560 nm to 1610 nm. The C-band and L-band are also called the *red band* and the *blue band,* respectively; (The letter "C" stands for "conventional" and the letter "L" stands for "long-wavelength.") These bands are separated by a narrow low-gain region, as shown in Figure 12.27.

Using the L-band not only increases the gain bandwidth of an EDFA but also helps us cope with the four-wave mixing (FWM) problem in a transmission fiber. (See Section 6.4.) You will recall that the FWM phenomenon disappears when a singlemode fiber exhibits chromatic dispersion. Since dispersion-shifted fiber (DSF) has zero chromatic dispersion at 1550 nm, working a little bit away from this wavelength effectively reduces FWM to an insignificant level in those networks where DSF fiber is installed. However, FWM within an erbium-doped fiber becomes a concern in itself in an L-band. This is because the gain coefficient in an L-band is low; this, in turn, causes an increase in the length of an active fiber [23]. This contradiction is just one example of the kinds of technological headaches that designers and manufacturers have to cope with in their quest to attain significant extension of the bandwidth of optical amplifiers.

## Gain and Noise in an Erbium-Doped Fiber

What we've considered to this point are the amplification properties of a silica fiber doped with ions of erbium. This is the active medium of the entire optical amplifier. Even though, as a device, an EDFA includes several other elements, an active fiber is a key component of an erbium-doped fiber amplifier. This is why it's important that you become familiar with the basic properties of an erbium-doped fiber.

***Gain*** Gain: It's the first characteristic you look for when evaluating an amplifier. Let's consider its most important features with respect to an active fiber.

1. *Definition.* Gain is the ratio of output to input light power; that is,

$$\text{Gain} = P_{out} / P_{in} , \tag{12.25}$$

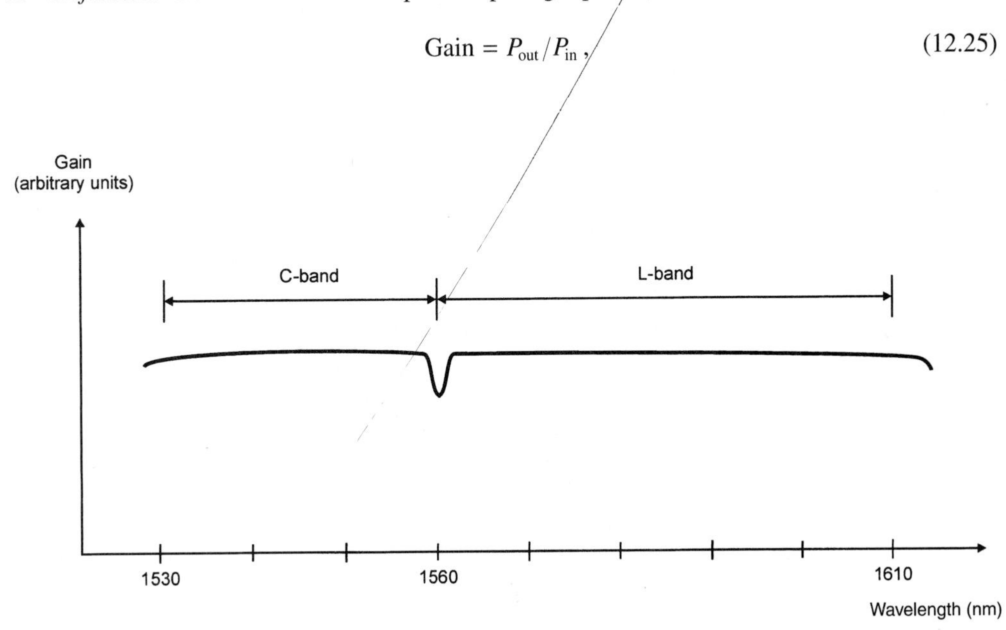


**Figure 12.27**
C-band and L-band of an EDFA bandwidth.

where power is measured in watts. Usually we measure gain in dB, which means

$$\text{Gain (dB)} = 10 \log_{10}(P_{out}/P_{in}) \quad (12.26)$$

(If these formulas look familiar to you, then you have remembered the definition of *loss* given in Formulas 3.9 and 3.10.)

If you look again at Figure 12.25(c), you'll realize that output power includes the power of both the signal and the noise. As a result, we need to take noise power into account when calculating gain; that is,

$$\text{Gain} = (P_{out} - P_{ASE})/P_{in} \quad (12.27)$$

or

$$\text{Gain (dB)} = 10 \log_{10}[(P_{out} - P_{ASE})/P_{in}] \quad (12.28)$$

### Example 12.4.1

***Problem:***

**a.** Calculate the gain of an erbium-doped fiber if the light input power is 300 μW and the output power is 60 mW.

**b.** Calculate the gain of the EDFA given above if $P_{ASE} = 30$ μW.

***Solution:***

**a.** The solution follows immediately from Formulas 12.25 and 12.26:

$$\text{Gain} = P_{out}/P_{in} = 200$$

and

$$\text{Gain (dB)} = 10 \log_{10}(P_{out}/P_{in}) = 23 \text{ dB}$$

This is moderate gain; however, notice how much the input signal has been enhanced in terms of absolute power.

**b.** Now we have to use Formula 12.28:

$$\text{Gain (dB)}=10 \log_{10}[(P_{out} - P_{ASE})/P_{in}] = 23 \text{ dB}$$

Actually, gain is equal to 23.010 dB without noise and to 23.008 dB with noise.

What we have to remember is that both the signal and noise powers ($P_{out}$ and $P_{ASE}$) in Formulas 12.27 and 12.28 refer to an individual channel (wavelength), not the entire bandwidth. Figure 12.25(c) should clarify this point.

---

The gain of modern EDFAs ranges from about 20 to about 40 dB, depending on their function, that is, whether they are designed as boosters, in-line amplifiers, or preamplifiers. (See Figure 12.13.)

**2.** *Gain as a function of wavelength (frequency) and gain flatness.* Gain depends on the wavelength (frequency) of the input signal. Indeed, the gain of an EDFA is restricted by the width of the radiating energy bands, as shown in Figure 12.26. Thus, there is no gain outside the specific range of wavelengths. But even within this range gain varies substantially. Erbium-doped silica fiber exhibits the gain shown in Figure 12.28. Actually, this figure displays the gain coefficient of an erbium-doped fiber measured in dB/m.

Looking at Figure 12.28, you might be surprised to see that the gain fluctuates twofold between 1530 and 1560 nm. Obviously, we cannot work with such an amplifier. We need to have a flat gain over its range of operating wavelengths. This characteristic of an EDFA is called *gain flatness.* We'll study the gain-versus-wavelength relationship of an industrial EDFA later in this section. At this point, it is necessary to understand only that Figure 12.28 displays the amplification property of an *erbium-doped fiber* itself. How the device—an *erbium-doped fiber amplifier*—behaves is another story, which we will discuss shortly. What you need to draw from analyzing Figure 12.28 is that fabricating an EDFA from an erbium-doped fiber is not a simple, straightforward process.

**3.** *Gain saturation.* Question: Does gain depend on the power of an input signal? Consider the following: A high-power input signal means a huge number of photons per unit of time will enter an erbium-doped fiber. These photons will then stimulate a vast number of transitions per unit of time from the intermediate level (2) to the lower level (1). This means that the intermediate level will be rapidly depleted of photons. (See Figure 12.26.) In other words, the greater the input of light power, the less populated the intermediate level (2) will become. But, as the principle of stimulated emission tells us, gain is proportional to the difference in the population of levels 2 and 1. (To review this concept, reread Section 9.2.) Hence, depleting level 2 means decreasing gain. This phenomenon, known as *gain saturation,* is discussed in Section 12.3 and is shown schematically in Figure 12.18. Keep in mind that Formulas 12.13 and 12.14 hold true for SOAs and cannot be applied directly to EDFAs; however, they describe the same phenomenon—gain saturation.

Gain saturation is an important characteristic of an EDFA, especially in a booster application, where an input signal comes almost immediately from a transmitter and is therefore strong. Gain saturation largely determines the maximum output power, often called the *saturated output power,* that an EDFA can handle. However, both graphs—"Gain versus Input Signal Power" and "Gain versus Output Signal Power"—can be found in a typical EDFA data sheet.

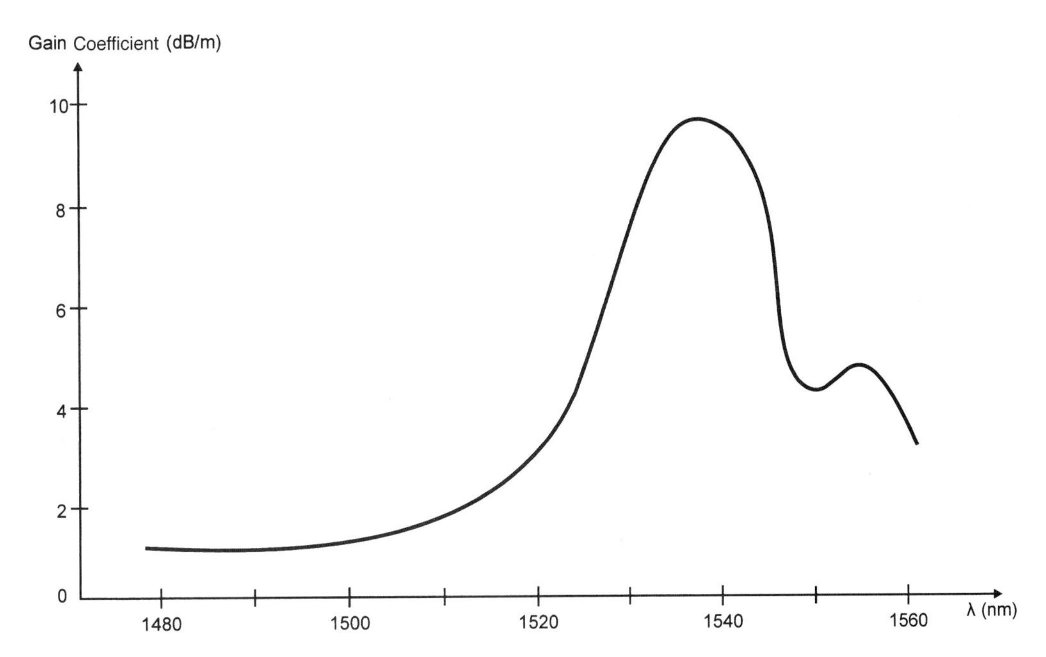


**Figure 12.28** Gain versus wavelength in erbium-doped silica fiber.

**4.** *Gain as a function of the length of an active fiber and optimal length.* You have to realize that *pumping in an EDFA is provided along the length of an active fiber.* (In all laser diodes, except VCSELs, we learned that pumping is done in the transverse direction with respect to light propagation—see Figures 9.12, 9.15, and 9.16.) Therefore, a pumping signal is strong at its input and becomes weaker as it propagates along an active fiber; an amplified signal, on the other hand, is weak upon entering but becomes stronger while propagating along the fiber. (See Figure 12.29, which illustrates the concept of amplification in EDFAs discussed above, to wit: Pumping light gives its power to an information signal and "dies.")

Since the power of the pumping signal damps along an active fiber, an information signal will experience less and less gain and eventually begin to undergo loss. The reason for this behavior is that population inversion along the fiber diminishes as pumping power decreases. Ultimately, population inversion reverts to normal population at the point where more erbium ions are found at the lower level than at the excited level. From this point on along the length of the fiber, more information-signal photons will be absorbed than added by stimulated emission. In other words, absorption prevails over amplification.

The message to be drawn from this consideration is now clear: An active fiber has an *optimal length.* This optimal length depends on the fiber's characteristics, such as doping concentration, gain bandwidth, and gain shape. But it also depends on the criteria we use to optimize this length; in fact, we've considered length to optimize the gain of an active fiber. But we can also change an active fiber's length to optimize noise performance or output saturation power. The upshot of all these considerations is that the typical length of an active fiber ranges from a few meters to 20 to 50 meters.

**5.** *Pumping power.* Optimal length depends, obviously, on pumping power. What level of pumping power do we need? The higher the pumping power, the larger the number of erbium ions that will be excited at the intermediate level (see Figure 12. 26) and the higher the gain of the EDFA. But don't think that high pumping power will sweep up all the erbium ions from the lower level and bring population inversion to its saturated state. Not at all. Remember, we are considering a dynamic process. The lower level is constantly populated with erbium ions stimulated by the photons of an information signal to drop there from a level of excitation.

We need to take into account that an EDFA amplifies many channels simultaneously and pumping power is shared by all amplified wavelengths. The more wavelengths that are multiplexed for transmission, the greater is the pumping power required for satisfactory EDFA operation.

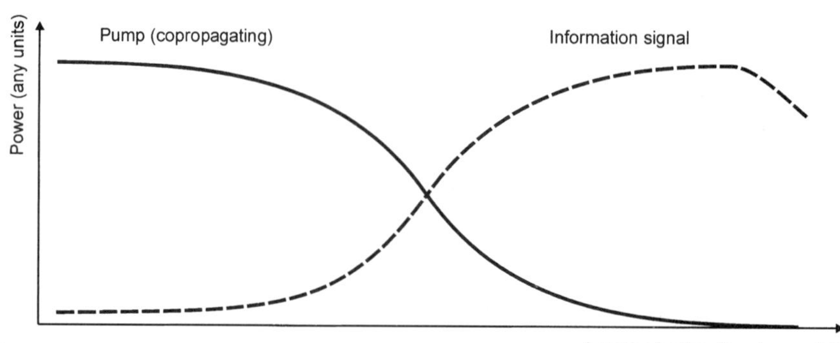


**Figure 12.29** Power of pumping and information signal versus length of an active fiber.

Pumping laser diodes with power up to 165 mW at 980 nm and up to 140 mW at 1480 nm are now widely available. These are sufficiently powerful for 8- and 16-channel WDMs. However, even 16-channel WDM systems start to experience a deficiency in gain because of insufficient pumping power; with the 32- and 40-channel systems that are already on the drawing boards, the problem becomes even more serious. As you can see, then, increasing the number of multiplexed channels—*channel counts*—requires more pumping power than today's single-laser-pump modules can produce. We'll address the solution to this problem later in this section.

***Noise*** Noise is the second most important characteristic of an optical amplifier (or any type of amplifier, for that matter). General definitions and explanations regarding noise in optical amplifiers are given in Section 12.3. The major peculiarities of noise in an EDFA are discussed here.

If you look again at Figure 12.25(c), you'll see that the gain of an EDFA is used to amplify a specific wavelength channel. Thus, it's important to consider gain and noise within the bandwidth of a single channel. Optical noise in an EDFA is termed *amplified spontaneous noise (ASE)*. The actual signal degradation comes from beating signals generated at *noise-noise* and *noise-signal interference*. This is shown in Figure 12.30. Each "slice" of noise can interfere with another "slice" to generate a beating signal at frequencies that are combinations of the sum and difference of the input frequencies. Fortunately, noise-noise beating can be easily removed by using a narrowband filter. The actual EDFA module usually includes an *ASE filter* to do this job. *The real damage is done by noise-signal interference.* Such noise cannot be filtered because it is within a signal's bandwidth, as Figure 12.30 shows. (See also Example 12.3.2.) This is the real contributor to the noise figure of an EDFA.

The noise figure based on signal-noise beating is given by [24]:

$$F_{\mathrm{n}} \approx P_{\mathrm{ASE}}(\lambda_{\mathrm{s}})/[hfGBW(\lambda_{\mathrm{s}})], \tag{12.29}$$

where $\lambda_s$, is the signal wavelength and all other notations are the same as in Formula 12.24. From Formula 12.24, we can readily see that the noise figure of an EDFA is given by

$$F_{\mathrm{n}} \approx 2n_{\mathrm{sp}} \tag{12.30}$$

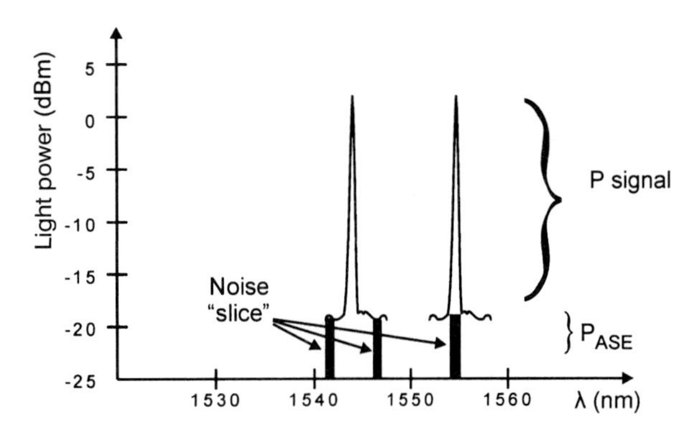


**Figure 12.30** Noise-noise and noise-signal beatings.

It follows immediately from Formula 12.30 that even for the ideal amplifier, when $n_{sp} = 1$, the SNR of an amplified signal decreases twofold (by 3 dB). Formula 12.30 shows that, unfortunately, an inherent property of an EDFA is to increase the noise level, but its use is justified because it amplifies the light signal to a degree that more than compensates for the rise in noise. Example 12.3.2 stresses this point. As mentioned previously, typical $n_{sp}$ values range from 1.4 to 4, which is why $F_n$ varies from 3.5 to 9 dB for EDFAs. What's more, don't forget that this noise figure is based on signal-noise interference only. Some data sheets explicitly underscore this fact.

***Gain and noise as a function of EDFA parameters*** To discuss together the two major characteristics—gain and noise—of an EDFA active fiber, let's consider their dependence on an EDFA's parameters.

Gain and noise as a function of the *length of an active fiber* are shown in Figure 12.31(a). Both gain and noise are higher for a counterpropagating pump; it is almost constant for a copropagating pump.

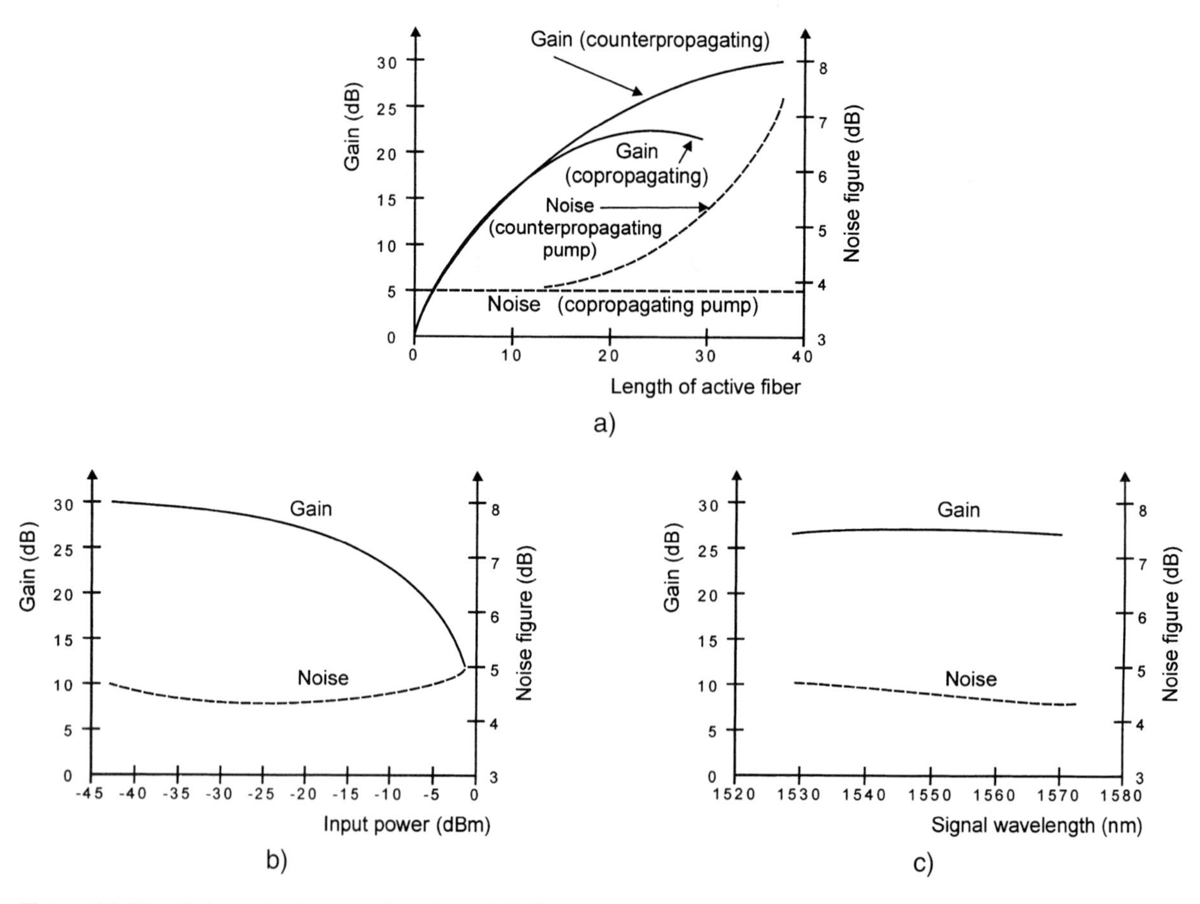


**Figure 12.31** Gain and noise as a function of EDFA parameters: (a) Gain and noise as a function of the length of an active fiber; (b) gain and noise as a function of input power; (c) gain and noise as a function of signal wavelength. *([a] adapted from Dennis J. G. Mestdagh's* Fundamentals of Multiaccess Optical Fiber Networks, *Boston: Artech House, 1995. Used with permission.)*

Gain and noise, as Figure 12.31(b) shows, depend on *the power of the input signal.* With an increase in input power, gain decreases (because of gain saturation, remember) while noise rises. Interestingly, there is a specific input power where noise is minimal.

How noise and gain are allocated along the signal bandwidth is dictated by the physics of their origins. Noise should be almost wavelength independent within the spectrum of an EDFA's stimulated emission. The same is true for gain. Figure 12.31(c) confirms this.

Look at the gain-bandwidth dependence more closely. If you study Figure 12.28 carefully, you'll detect a contradiction in what we have been saying. Let's deal with this. Figure 12.31(c) shows, for instance, that gain is very flat across the entire C-band of operating wavelengths—that is, from 1530 to 1560 nm—while Figure 12.28 depicts huge variations in gain over the same band. The key to this puzzle is the difference between an active fiber and an entire EDFA module, which we will now consider.

## Components of an EDFA Module

An erbium-doped (active) fiber is the heart of an erbium-doped fiber amplifier but it's not the only component. To make an EDFA work and to improve the gain and noise characteristics of an active fiber, other components are needed. The general configuration of an EDFA is given in Figures 12.25(a) and (b). As you can see, an EDFA includes an active fiber, a pump laser, a coupler, and an isolator with a filter. Let's analyze these components closely.

***Erbium-doped fiber*** An erbium-doped fiber is the active medium of an EDFA. It is fabricated in the same way as regular fibers (see Section 7.1), but its core is heavily doped with erbium ions. Since amplification is actually done by the erbium ions, it's important to have as high a concentration as possible of these ions in silica fiber. To increase the *density* of erbium ions—that is, the number of erbium ions in a volume unit of silica fiber—the manufacturer reduces the core diameter of the erbium-doped silica fiber, which reduces its *mode-field diameter (MFD).* The MFD of an erbium-doped fiber ranges from about 3 to about 6 μm; the MFD of a regular fiber, you may recall, typically ranges from 9 to 11 μm. Where the MFD of a doped fiber is small (meaning it has a small core diameter), the probability of a collision between the erbium ions and the photons of an information signal (that is, the signal being amplified) increases. In other words, a small core diameter increases the efficiency of the amplification process.

To achieve even more efficient amplification, the manufacturer not only decreases the core diameter but also concentrates most of the erbium ions in the center region of the small core. (See Figure 12.32.) The concentration of erbium ions in the center area varies from 100 to 2000 parts per million, as Figure 12.32 shows. An active fiber with concentration of erbium ions as high as 5000 ppm is on the market [25].

The size of both the cladding and the coating of an erbium-doped fiber is important. These sizes are standard as compared with multimode (Figures 3.18 and 3.20) and singlemode (Figure 5.17) transmission fibers. However, to repeat, the core sizes of a standard fiber and an erbium-doped fiber are different: 62.5 or 50 μm for MM fibers, 8.3 μm for an SM fiber, and 2.8 to 5.2 μm for an amplifier fiber. Other features of erbium-doped fibers are clearly described in the data sheet in Figure 12.33.

Let's specify the applications of each type of fiber in this data sheet. The EDF-PAX-01 type is designed for use in pre- and in-line amplifiers; it gives a flat and broad-gain bandwidth. The EDF-LAX-01 type is designed for use in in-line amplifiers; it gives high power-conversion efficiency and a low noise figure with medium pump power. The EDF-BAX-01 type is designed for use in boosters; it provides high output power.

Erbium-doped fibers with lower numerical apertures (*NA*) are used to build EDFAs with high gain and high output power. In Figure 12.33, the fiber designated EDF-BAX-01 has an *NA* of 0.22 and it is used as a booster. Fibers with *NA*s as low as 0.17 can be found on the market [26].

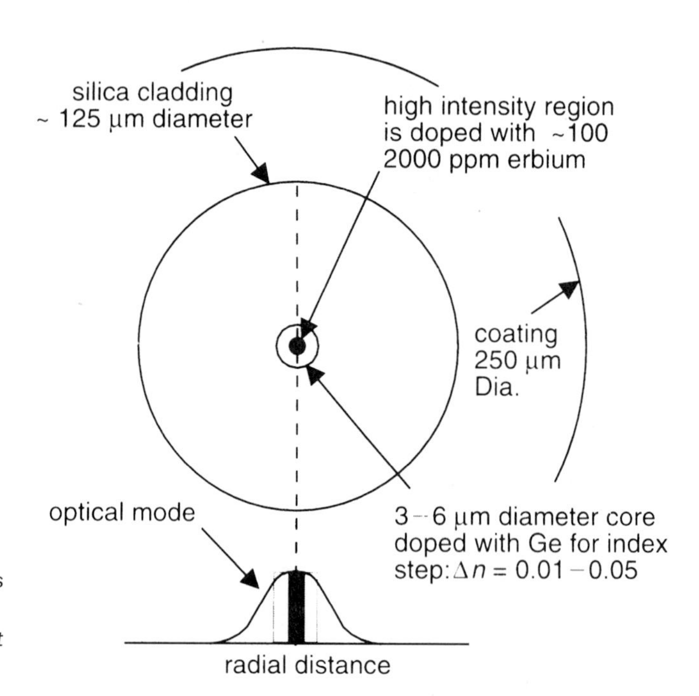


**Figure 12.32** Erbium-doped fiber geometry. *(Dennis Derickson, ed.,* Fiber Optic Test and Measurement, *Upper Saddle River, N.J.: Prentice Hall PTR. Copyright 1998 Agilent Technologies. Used with permission.)*

Review the specifications of these fibers. (If you need to refresh your memory on how to read the data sheet of an optical fiber, turn to Sections 3.5 and 5.4.) Depending on their applications, the fibers differ in *NA*. Fibers for pre-amp and in-line amplifiers have higher *NA*s (0.24) than booster fibers (0.22). Varying the *NA* enables the manufacturer to change the noise figure. The rule is this: *High NA results in lower noise.* To reduce noise, manufacturers produce erbium-doped fibers with *NA*s up to 0.29.

The above discussion covers two specifications in the data sheet: "Numerical Aperture" and "Typical application." Let's move on to some others:

- The specification "Cutoff wavelength" (nm) determines the minimum wavelength that the fiber can support in a singlemode configuration. Values less than 950 nm are typical even though cutoff wavelengths up to 1450 nm can be found [27].
- The "Mode Field Diameter" (μm) specification is familiar to us from our discussion of the features of a singlemode fiber in Chapter 5. The MFD values given in Figure 12.33 are typical even though you find MFDs up to 9.0 μm. Some manufacturers specify core diameter rather than MFD. Core diameters from 2.8 to 5.2 μm are common.
- The specification "Peak absorption wavelength" (nm) refers to the wavelength at which absorption is maximum, as the graph "Normalized cross section versus Wavelength" (Figure 12.33) shows. If you compare the emission line of this graph with that in Figure 12.28, you'll find they are very similar. In fact, you may encounter the very same curves in both the gain and cross-section axes.

To understand the graph in Figure 12.33, we need to introduce the concept of normalized cross section. A *cross section of emission* or *absorption* refers to the cross-sectional area of an atom in relation to the photon flux. Its dimension is $m^2$ and typical values for erbium ions incorporated in a silica fiber range from approximately 0.1 to $4.5 \times 10^{-25}$ $m^2$ [19]. *The importance of the cross section follows from the fact that the gain coefficient (1/m) is equal to the cross section ($m^2$) times the population-inversion density ($m^{-3}$).*

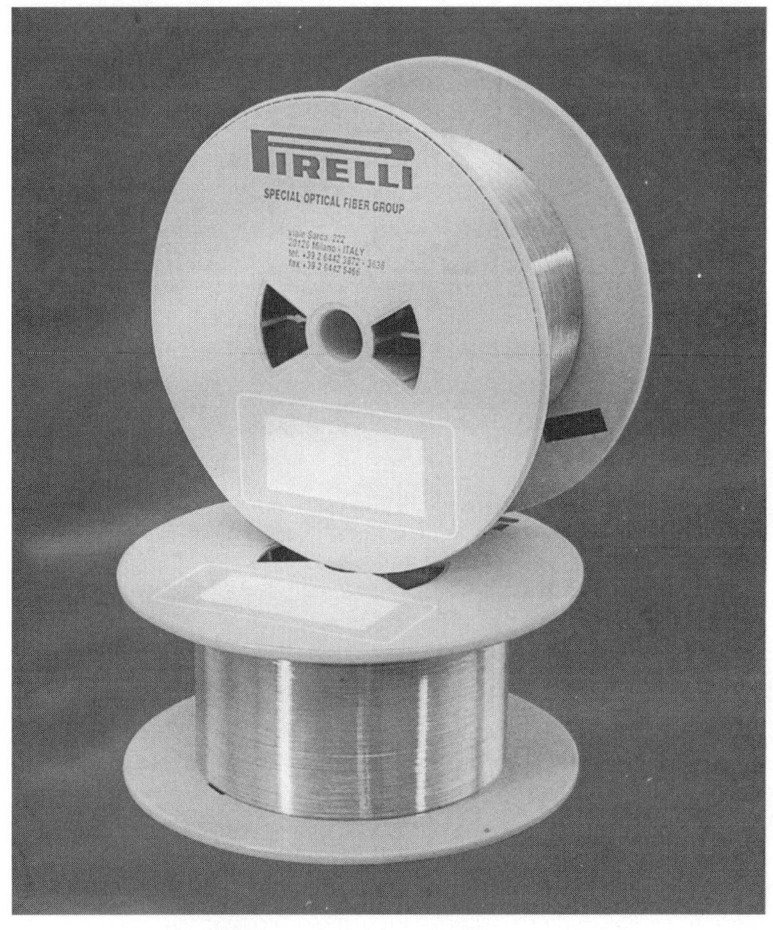


# Erbium Doped Fibers

**FEATURES**

- Support both 980 nm and 1480 nm pumping
- Broad gain bandwidth due to aluminium codoping
- High power conversion efficiency
- Low noise figure
- Low splicing loss (due to moderate NA)
- High Uniformity of Optical Parameters
- High Reliability
- Custom Design Availability

**APPLICATIONS**

- EDFA
- Sources

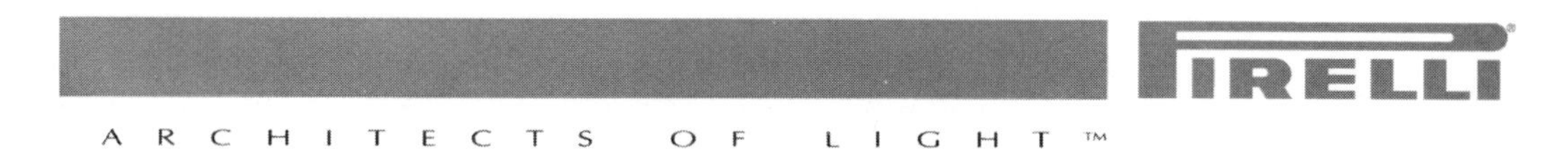


**Figure 12.33 (a)** Data sheet of erbium-doped fibers. *(Courtesy of Pirelli Cavi e Sistemi SpA, Milan, Italy.)*

Pirelli Cables & Systems has been a pioneer in the doped fiber fabrication and is known to be the manufacturer of the first commercial and in-field installed optical amplifier.

More than ten years experience in MCVD fiber technology combined with the R&D activity for EDFA development, have led to a fine optimization of a large selection of Erbium Doped Fibers in silica glass. The fibers have a multi-components codoped core optimized to meet different applications needs and match the stringent requirements of dense WDM and submarine optical transmission system.

These fibers have been manufactured in volumes for years and used within thousands of Pirelli EDFA deployed in the field.

Pirelli offers a range of standard erbium doped fibers to meet different applications needs.
Pirelli standard products are described below:

**EDF-PAX-01**, with high aluminium doping level, is specially designed for pre and in-line amplifiers used in Wavelength Division Multiplexing systems, giving broad and flat gain bandwidth.

**EDF-LAX-01** is generally used for standard applications where high power conversion efficiency and low noise figure are required with a medium pump power level.

**EDF-BAX-01** is the most suitable fiber for high output power applications as booster stages in optical amplifier systems.

**EDF-HCX-01**, with high erbium concentration level, is the right choice for laser and broadband sources.

Any fiber is supplied with optical parameters specified in a restricted range to ensure great uniformity and reproducibility of optical performances.

Special designed fibers can be developed according to customer requirements.

**Figure 12.33 (b)** (continued)

Another way to interpret cross section is to consider it to be a measure of the probability that an erbium ion will emit or absorb a photon. The amount of light power absorbed or emitted by the transition of an erbium ion is equal to the intensity of the incident light times the proper cross section.

- The specification "Peak attenuation" (dB/m) refers to the attenuation measured at the peak absorption wavelength, which is, for example, 1531.5 for EDF-BAX-01 fiber. Sometimes manufacturers combine the specifications "Peak absorption wavelength" and "Peak attenuation" and cite them as one. For example, one manufacturer [26] specifies peak absorption at 1530 nm. The value of this attenuation ranges, typically, from 2.4 to 9.0 dB/m.
- "Attenuation @ 980 nm" refers to attenuation at the pumping wavelength (here, 980 nm with $A = 3.5$ dB/m for EDF-BAX-01). Values from 2 to 7 dB/m are typical, but 23 dB/m at 980 nm can also be found [25].

## SPECIFICATIONS

| | Units | EDF-PAX-01 | EDF-LAX-01 | EDF-BAX-01 | EDF-HCX-01 |
|---|---|---|---|---|---|
| Numerical Aperture | | 0.24 ± 0.02 | 0.24 ± 0.02 | 0.22 ± 0.02 | 0.24 ± 0.02 |
| Cutoff wavelength (2) | nm | 935 ± 35 | 935 ± 35 | 920 ± 40 | 920 ± 40 |
| Mode field Diameter (1) | μm | 4.8 - 5.9 | 4.8 - 5.9 | 5.2 - 6.6 | 4.8 - 6 |
| Peak absorption wavelength | nm | < 1529.5 | 1530.5 ± 0.5 | 1531 ± 0.5 | 1531 ± 1 |
| Peak attenuation (2) | dB/m | 7 ± 2 | 7 ± 2 | 5 ± 2 | 12 ±2.5 |
| Attenuation @ 980 nm | dB/m | 5 ± 1.5 | 5 ± 1.5 | 3.5 ± 1.5 | 8.5 ± 2 |
| Background Loss (2) at 1200 nm | dB/km | < 35 | < 15 | < 15 | < 15 |
| Saturation Power @ 1530 nm typ. | mW | 0.17 | 0.15 | 0.18 | 0.20 |
| Typical application | | Pre-Amp. In-Line Amp. | In-line Amp. | Booster | Source |

(1) - Petermann II at 1550 nm (2) - Parameters uniformity measured every 500 m

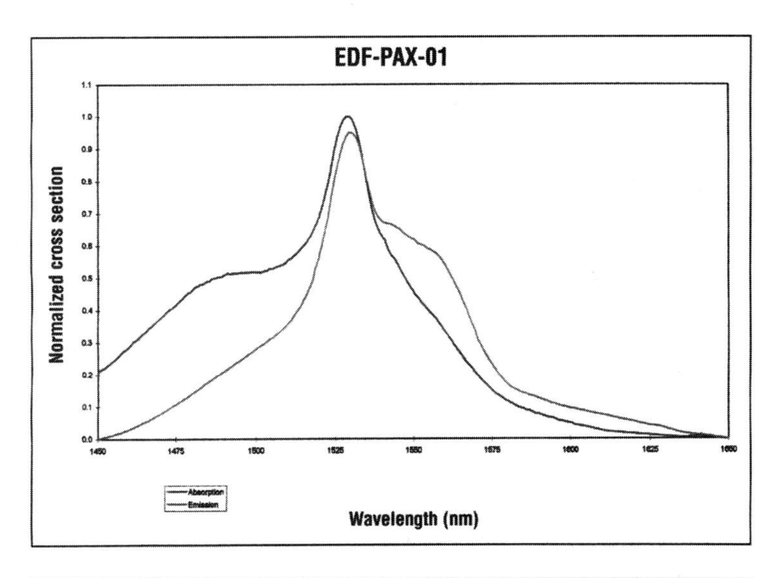


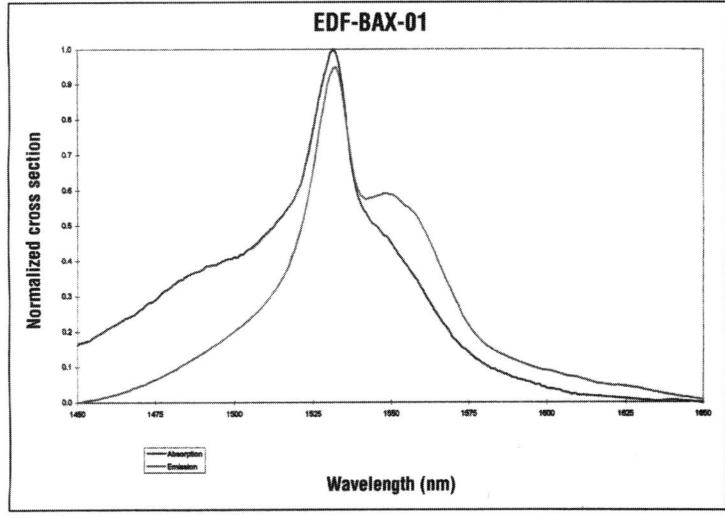


**Figure 12.33 (c)** (continued)

## PHYSICAL CHARACTERISTIC

| Parameters | Units | |
|---|---|---|
| Fiber outside diameter | μm | 125 ± 1 |
| Coating diameter | μm | 250 ± 10 |
| Coating type | | Double Layer Acrylate |
| Core/Cladding concentricity error | μm | < 1 |
| Screen test | % | 1 (100 kpsi) |

## ORDERING INFORMATION

| EDF-XXX-01-L | |
|---|---|
| XXX | Fibers series |
| L | Length in meters |

Please contact Pirelli for custom devices part number.

The previously mentioned products and/or the system including them may be covered by one or more of the following patents:

• US 4786135 • US 4790464 • US 4807958 • US 4846544 • US 4889399 • US 5129027 • US 5127076 • US 5570438
• US 5550947 • US 5640481 • US 5443536 • US 4395869 • US 4448484 • US 4497164 • US 4690496 • US 4676590
• US 4690498 • US 4690497 • US 4688889 • US 4673540 • US 4620412 • US 4722589 • US 4725121 • US 4741592
• US 4725123 • US 4703135 • US 4842438 • US 4805392 • US 4867527 • US 4902096 • US 4927294 • US 5140664
• US 5150444 • US 5185841 • US 5193134 • US 5455881 • US 4756600 • US 5229851 • US 5390273 • US 5509097
• US 4690627 • US 4859024 • US 5179619 • US 5533164 • US 5518516 • US 5444808 • US 5656090 • US 5658363
• US 4911742 • US 4923496 • US 4925269 • US 5054876 • US 5204923 • USRE 35697 • US 5119229 • US 5245467
• US 5138483 • US 5267073 • US 5638204 • US 5113459 • US 5218665 • US 5087108 • US 5161050 • US 5210808
• US 5278686 • US 5355250 • US 5233463 • US 5515200 • US 5646775 • US 5383051 • US 5491581 • US 5497265
• US 5381426 • US 5579153 • US 5497386 • US 5701194 • US 5712716 • US 5668909 • US 5677786 • US 5701378
• US 4881793 • US 4938561 • US 4768882 • US 5001338 • US 4986663 • US 5037179 • US 5150516 • US 5272433
• EP 448012 • EP 166138 • EP 193780 • EP 193779 • EP 200914 • EP 226188 • EP 242740 • EP 285917
• EP 297409 • EP 338482 • EP 466230 • EP 464918 • EP 488438 • EP 503469 • EP 632301 • EP 490803
• EP 210770 • EP 187060 • EP 487121 • EP 665191 • EP 518523 • EP 177206 • EP 350726 • EP 517169
• EP 308114 • EP 408905 • EP 409012 • EP 425014 • EP 426222 • EP 426221 • EP 431654 • EP 439867
• EP 441211 • EP 417441 • EP 442553 • EP 507367 • EP 509577 • EP 595395 • EP 595396 • EP 603925
• EP 567941 • EP 644633 • EP 294037 • EP 324541 • EP 346589 • EP 374614 • EP 458255

**Pirelli Cavi e Sistemi SpA** - Optical Systems Business Headquarter
Viale Sarca, 222 - 20126 Milano - Italy
Tel. +39 2 6442.3337/5956 - Fax +39 2 6442.9256
http: //www.pirelli.com



**Figure 12.33 (d)**
(continued)

You have to realize that the same amplifier fiber can be pumped at either 980 nm or 1480 nm, or both. This is why manufacturers give attenuation at a specific pump wavelength (980 nm or 1480 nm) or simply refer to attenuation at a pump wavelength.

- "Saturation Power" (mW) refers to input saturation power, as it follows from the given values. Output saturation power is typically in the range of 12 dBm (16 mW) to 16 dBm (40 mW).

You will find additional specifications in the data sheets of the various manufacturers, specifications dealing with mechanical dimensions and fiber geometry; nonetheless, all this aside, you are now armed with all the knowledge you need to decipher any data sheet of an erbium-doped fiber.

Study the data sheet provided here; we believe it is the shortest and most effective way for you to become familiar with the major features of an erbium-doped fiber.

***Splicing an erbium-doped fiber*** Splicing an erbium-doped fiber is a problem because of the great diameter mismatch between a regular singlemode fiber and an amplifier fiber. Indeed, as mentioned previously, the core diameter of a singlemode transmission fiber is between 8 and 10 μm, while the core diameter of an erbium-doped fiber varies from 2.8 to 5.2 μm. (The problem of splicing an erbium-doped fiber is discussed in Chapter 8 under fusion splicing. See Figure 8.7.)

## Example 12.4.2

***Problem:***

Calculate the connection losses when splicing a singlemode transmission fiber whose MFD equals 10.5 μm at 1550 nm and a Pirelli EDF-PAX-01 erbium-doped fiber.

***Solution:***

The solution can be found in Figure 8.1, where losses caused by mismatched core diameters and mode-field diameters (MFDs) are explained and the formulas for loss calculations are given. Since we are given only an MFD, not a core diameter, we need to use the following formula:

$$\text{Loss}_{\text{MFD}}(\text{dB}) = -10\log[4/(w_2/w_1 + w_1/w_2)^2] = -10\log[(2w_1w_2)/(w_1^2 + w_2^2)]^2$$

The MFD of a transmission fiber is given by $w_1 = 10.5$ μm at 1550 nm. The MFD of the Pirelli EDF-PAX-01 fiber is given in Figure 12.33: $w_2 = 4.8$ to 5.9 μm. Let's take the average MFD: $w_2 = 5.3$ μm. Inserting these numbers into the above formula, we compute:

$$\text{Loss}_{\text{MFD}} = 1.87 \text{ dB}$$

This loss means that the light power after splicing is only 65% of the power before splicing. (Don't forget to use the negative sign when calculating loss.) Remember, regular fusion splicing introduces a loss on the order of 0.01 to 0.07 dB, as we noted in Section 8.1. As you can readily see, this level of loss (1.87 dB) is too high and something has to be done to reduce it.

***Ways to reduce splicing loss*** To decrease splicing loss, several methods have been proposed ([19], [28]). All of them aim to reduce the core diameter (and therefore the MFD) mismatch. (See Figure 12.34.)

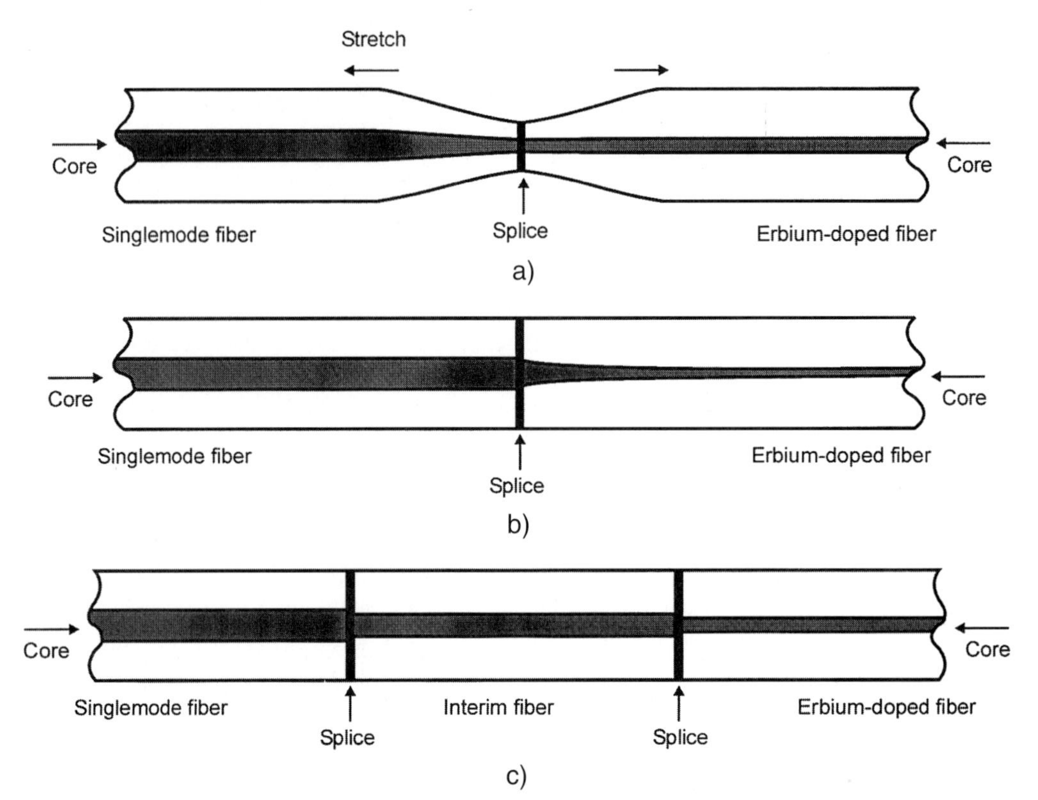


**Figure 12.34** Splicing an erbium-doped fiber: (a) Down-tapering; (b) up-tapering (TEC method); (c) using interim fiber.

The *down-tapering method* implies that both regular and amplifier fibers are stretched out when melted by fusion. This process reduces the core diameter of a regular fiber more than that of an erbium-doped fiber, thus lessening the core diameter's mismatch. (See Figure 12.34[a].)

The *up-tapering method,* also called the *thermally diffused expanded-core method (TEC),* on the other hand, increases the core diameter of an amplifier fiber by diffusing the dopants on the heated end of the fiber. Heating is done by the same fusion splicer used for splicing. The result: a reduction in the MFD mismatch, as shown in Figure 12.34(b).

When an *interim fiber* is used, as Figure 12.34(c) shows, its intermediate core diameter reduces the MFD mismatch at each splice but at the expense of additional splice loss.

You must understand that all these methods can be implemented *only* where both the regular and amplifier fibers have the same cladding and coating diameters. Both the down- and up-tapering methods reduce splicing losses caused by MFD mismatch tenfold so that, in Example 12.4.2, we would see losses of about 0.187 dB instead of 1.87 dB, which we've computed for direct splicing. Thus, using one of these methods solves the problem of how to splice an erbium-doped fiber.

***Pump laser diodes*** Pump lasers are semiconductor laser diodes radiating at 980 nm or at 1480 nm. Their basic feature is their high radiation power. As mentioned above, a single pump laser diode can radiate up to 165 mW of light power.

A pump laser module typically includes a laser diode (LD) radiating pump light, a back-facet photodiode (PD) for monitoring LD performance, and a thermoelectric cooler (TEC) with a thermistor for the control and stabilization of the laser's temperature. (All these components should be familiar to you from reading Section 10.4.) A functional block diagram of a pump laser module is given in Figure 12.35 (see references [29], [30], and [31]). Electrostatic-discharge (ESD) protection will be discussed shortly.

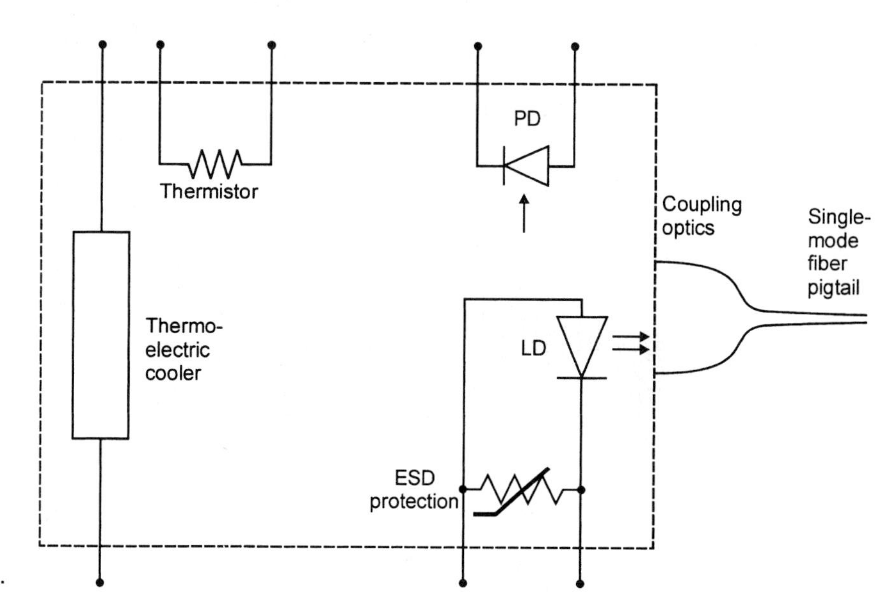


**Figure 12.35** Functional block diagram of a pump laser module.

A pump module may also include a built-in isolator to prevent any backreflection light from penetrating the laser diode and a wavelength stabilizer to "lock" in the pumping wavelength under variations of temperature, driving current, and optical feedback. A pump module is usually offered in a 14-pin butterfly package. (See Figure 12.37.)

The heart of a pump module is, obviously, its laser diode. The data sheet of a pump laser chip is presented in Figure 12.36.

This chip is an index-guided ridge-waveguide (RWG) laser diode based on a strained InGaAs quantum well. (See Section 9.3.) The layout and mechanical dimensions depicted in Figure 12.36(a) will help you understand how a laser chip is mounted on a substrate and connected to the outside circuits.

It is important to concentrate on *graphs* representing the essential characteristics of a pump laser diode. (See Figure 12.36[b].) The graph "P/I, V/I" shows the output power and voltage drop across a diode as a function of the driving current. We saw the same curves, remember, in Figures 9.17, 9.19, and 9.20. The only difference is that the values for the driving current and output power here are much higher than those for a regular transmission diode.

The "LINEARITY" graph shows the value for coupled (NA) power and the absolute value for a *kink* signal. "Kinks" are the nonlinearities in the P/I curve caused by the frequency fluctuations of a central single mode or the appearance and disappearance of higher modes with an increase in driving current. Kinks do not harm the pump laser itself but they can cause an increase in EDFA noise [32]. As you can see, at a driving current of less than about 220 mA, we can neglect the kink signal. This is what is meant by the expression "kink-free power," which is found on the data sheet of any pump laser.

The "FARFIELD" graph presents the radiation pattern of this laser diode. It shows the light intensity at some distance from the laser's radiation edge (called "farfield" in contrast to "nearfield") as a function of the spatial angle. Let's measure the full width of a radiated beam at one-half its maximum (FWHM). For this chip, we will find that beam diversion is typically 7° for the horizontal (parallel) measurement and typically 28° for the vertical (perpendicular) measurement. These data are taken from the chip specifications in Figure 12.36(c). To better understand this graph, see Figure 10.33.

The "SPECTRUM" graph shows the distribution of light output power along the wavelength axis, that is, the power spectrum of the laser. As you can see, this laser radiates a single mode with

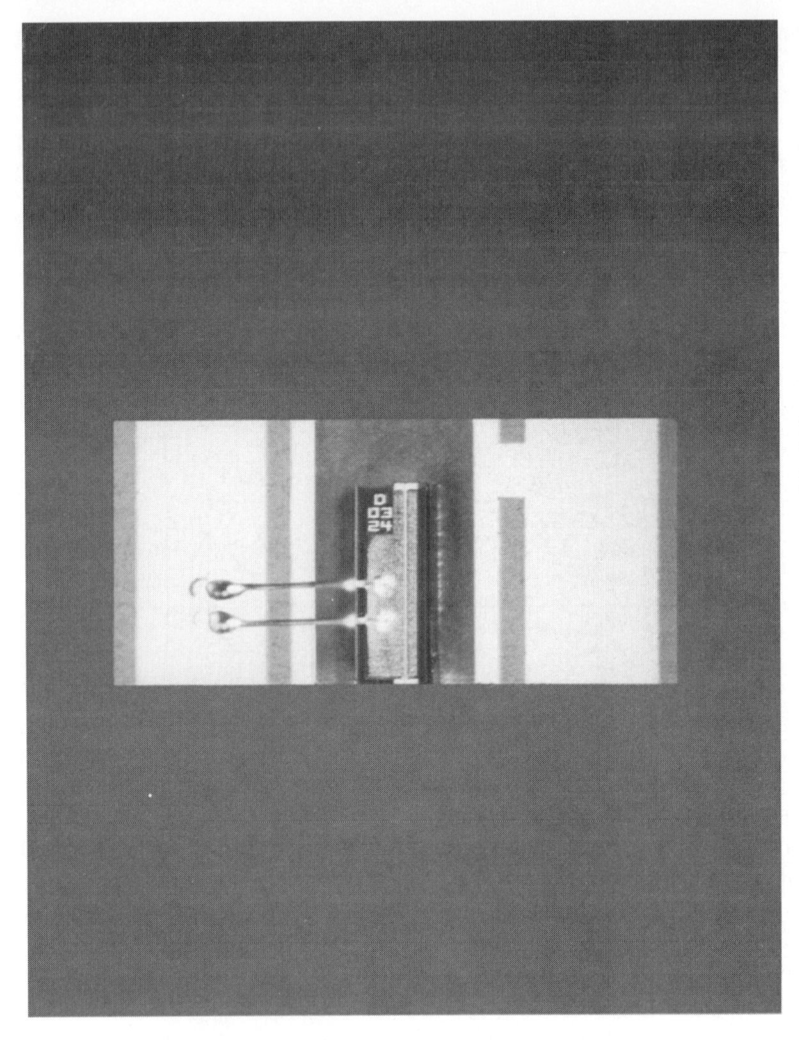

## 980 nm Laser Chip on Submount

### OVERVIEW

Pirelli is strongly committed in photonic research and has a world leading position in optical amplifiers.

The laser chip manufacturing facility has been specifically implemented for the E2-980 nm IBM technology transfer to Pirelli, choosing the most advanced and up to date equipment available on the market and is run by a skilled team of selected personnel with years of experience in epitaxial growth and laser chip manufacturing.

The Pirelli E2-980 nm pump laser chip represent the state of the art technology for single mode laser sources. It has been specifically developed to reliably pump the Erbium-doped fiber for low noise, long haul optical amplifiers.
Low power consumption, qualification data and E2 technology all concour to high device reliability.

### APPLICATIONS

- Laser modules for Erbium-Doped Fiber Optical Amplifiers.

### FEATURES

- E2 Laser chip technology
- Low power consumption
- Kink free power options up to 210 mW
- Reliable and proven technology

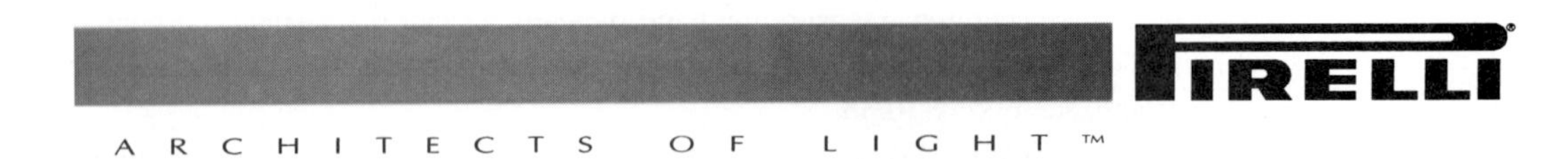


**Figure 12.36** Data sheet of a pump laser chip. *(Courtesy of Pirelli Cavi e Sistemi SpA.)*

The production is fully documented and a quality system guarantee the best process stability, reproducibility and full traceability of manufacturing information.

The Pirelli laser chip emits on single transversal mode with linear power ranging from 150 mW up to 210 mW.
Typical far field beam characteristics are reported below.

E2 technology allows operation free catastrophic optical mirror damage (COMD) and long therm reliability, as highlighted by the accelerated qualification data corresponding to more than 2 million devices/hours at standard operating conditions..

The low current degradation and the sudden failures elimination are crucial for the success of 980 nm pump lasers in optical amplifiers.

The laser chip is mounted, as a standard, on a silicon carbide substrate, ternary alloy plated on both sides to guarantee good electrical/thermal conductivity and adhesion.

Low drive voltage and low drive current allows these diodes to be soldered junction-side-up on the heat sink.

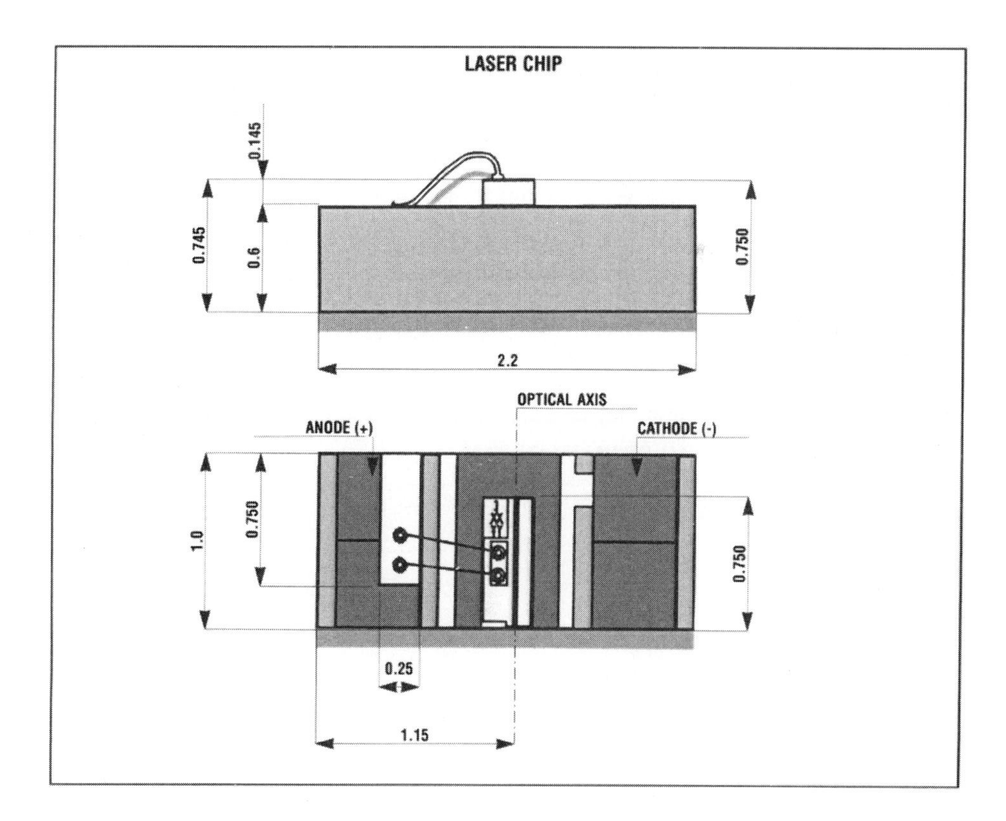


**Figure 12.36 (a)**
(continued)

**DESCRIPTION**

The Pirelli E2-980 nm is a index-guided Ridge Waveguide semiconductor laser based on strained InGaAs quantum well. The InGaAs/AlGaAs epitaxial structure is grown by Molecular Beam Epitaxy.

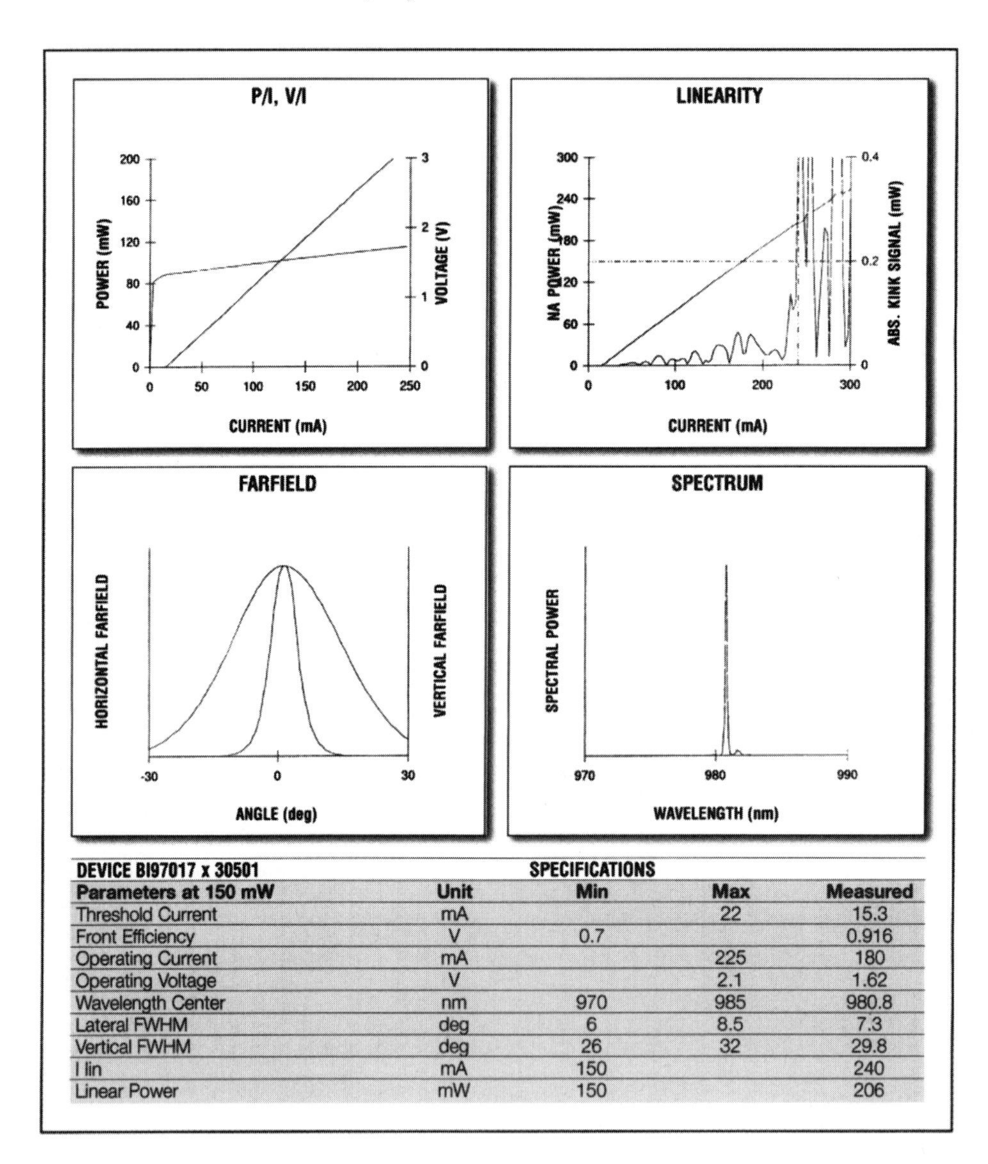


**DEVICE BI97017 x 30501** — **SPECIFICATIONS**

| Parameters at 150 mW | Unit | Min | Max | Measured |
|---|---|---|---|---|
| Threshold Current | mA | | 22 | 15.3 |
| Front Efficiency | V | 0.7 | | 0.916 |
| Operating Current | mA | | 225 | 180 |
| Operating Voltage | V | | 2.1 | 1.62 |
| Wavelength Center | nm | 970 | 985 | 980.8 |
| Lateral FWHM | deg | 6 | 8.5 | 7.3 |
| Vertical FWHM | deg | 26 | 32 | 29.8 |
| I lin | mA | 150 | | 240 |
| Linear Power | mW | 150 | | 206 |

**Figure 12.36 (b)** (continued)

a high sidemode suppression ratio (SMSR); this occurrence can be understood by looking at the small amplitude of the side mode shown in the graph.

The meanings of the "TECHNICAL SPECIFICATIONS" of a laser chip, given in Figure 12.36(c), should now be clear to you. Thus, with this background on the structure of a pump module and a solid foundation in the basic features of a laser diode, you should be able to read the data sheet of any manufacturer's *pump laser module.* Look at Figure 12.37. When reading this

**TECHNICAL SPECIFICATIONS**

| | Unit | Min | Typical | Max |
|---|---|---|---|---|
| **150 mW laser chip class** | | | | |
| Kink free power | mW | 150 | | |
| Threshold current | mA | | 20 | 22 |
| Operating current | mA | | 210 | 225 |
| Operating voltage | V | | 2.0 | 2.1 |
| Wavelength band | nm | 970 | | 985 |
| Power in wavelength band | nm | | | 10 |
| Front to back ratio | | 20 | | 80 |
| Beam parallel FWHM | degree | 5 | 7 | 10 |
| Beam perpendicular FWHM | degree | 26 | 28 | 32 |

180 mW and 210 mW minimum kink free output power devices are also available.

**ORDERING INFORMATION**

| Description | P/N |
|---|---|
| 120 mW min kink free Power | 28924 |
| 150 mW min kink free Power | 28923 |
| 180 mW min kink free Power | 28977 |
| 210 mW min kink free Power | 28978 |

The previously mentioned products and/or the system including them may be covered by one or more of the following patents:

• US 4786135 • US 4790464 • US 4807958 • US 4846544 • US 4889399 • US 5129027 • US 5127076 • US 5570438
• US 5550947 • US 5640481 • US 5443536 • US 4395869 • US 4448484 • US 4497164 • US 4690496 • US 4676590
• US 4690498 • US 4690497 • US 4688889 • US 4673540 • US 4620412 • US 4722589 • US 4725121 • US 4741592
• US 4725123 • US 4703135 • US 4842438 • US 4805392 • US 4867527 • US 4902096 • US 4927294 • US 5140664
• US 5150444 • US 5185841 • US 5193134 • US 5455881 • US 4756600 • US 5229851 • US 5390273 • US 5509097
• US 4690627 • US 4859024 • US 5179619 • US 5533164 • US 5518516 • US 5444808 • US 5656090 • US 5658363
• US 4911742 • US 4923496 • US 4925269 • US 5054876 • US 5204923 • USRE 35697 • US 5119229 • US 5245467
• US 5138483 • US 5267073 • US 5638204 • US 5113459 • US 5218665 • US 5087108 • US 5161050 • US 5210808
• US 5278686 • US 5355250 • US 5233463 • US 5515200 • US 5646775 • US 5383051 • US 5491581 • US 5497265
• US 5381426 • US 5579153 • US 5497386 • US 5701194 • US 5712716 • US 5668909 • US 5677786 • US 5701378
• US 4881793 • US 4938561 • US 4768882 • US 5001338 • US 4986663 • US 5037179 • US 5150516 • US 5272433
• EP 448012 • EP 166138 • EP 193780 • EP 193779 • EP 200914 • EP 226188 • EP 242740 • EP 285917
• EP 297409 • EP 338482 • EP 466230 • EP 464918 • EP 488438 • EP 503469 • EP 632301 • EP 490803
• EP 210770 • EP 187060 • EP 487121 • EP 665191 • EP 518523 • EP 177206 • EP 350726 • EP 517169
• EP 308114 • EP 408905 • EP 409012 • EP 425014 • EP 426222 • EP 426221 • EP 431654 • EP 439867
• EP 441211 • EP 417441 • EP 442553 • EP 507367 • EP 509577 • EP 595395 • EP 595396 • EP 603925
• EP 567941 • EP 644633 • EP 294037 • EP 324541 • EP 346589 • EP 374614 • EP 458255

**Pirelli Cavi e Sistemi SpA** - Optical Systems Business Headquarter
Viale Sarca, 222 - 20126 Milano - Italy
Tel. +39 2 6442.3337/5956 - Fax +39 2 6442.9256
http: //www.pirelli.com



**Figure 12.36 (c)** (continued)

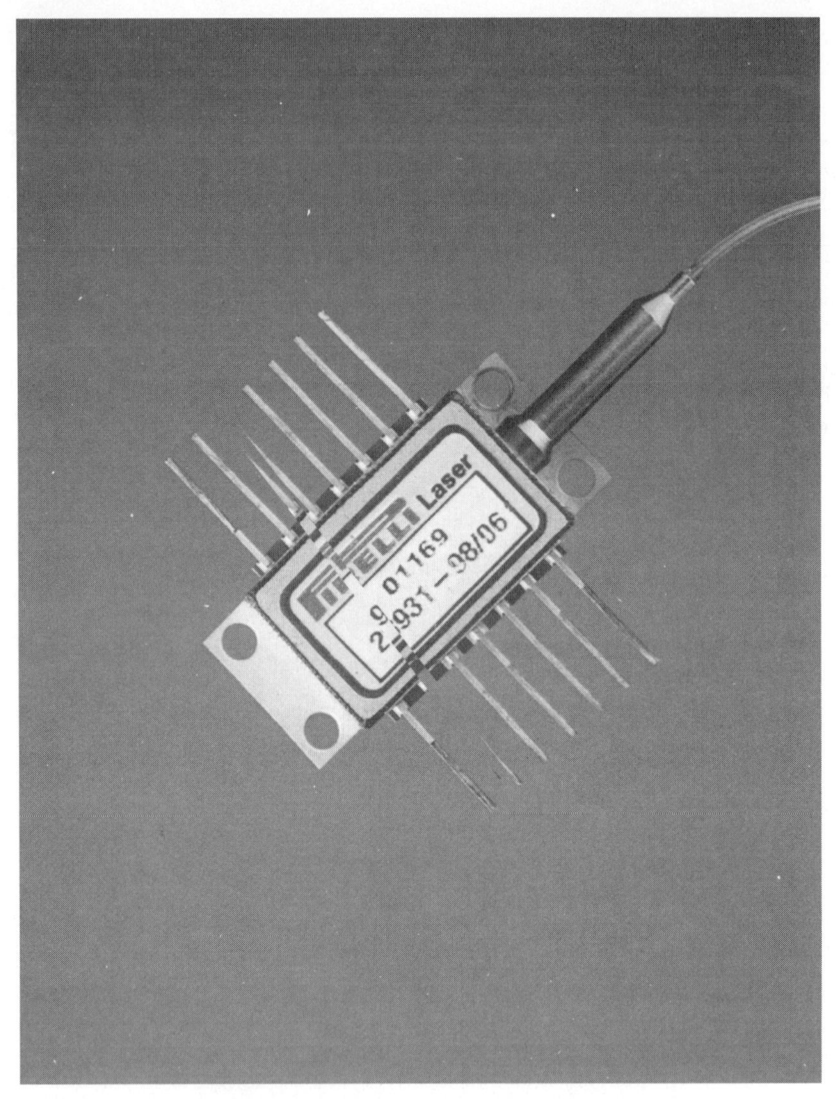


# 980 nm Laser Module

## OVERVIEW

The Pirelli E2-980 nm pump laser module is designed to provide the power to achieve very long, single span, single/multiwavelength transmission at any bit rate over single mode optical fiber cables. The low power consumption, the Pirelli proprietary optical fiber and the E2 technology, give further demonstration of this device reliability.

## APPLICATIONS

- Erbium-Doped Fiber Amplifier
- High-power, EDFA WDM

## FEATURES

- E2 Laser chip
- Epoxy free
- 14 pin butterfly hermetic package
- High coupling efficiency
- Low power consumption
- High ESD protection
- Eight power options
- Reliable technology

**Figure 12.37 (a)** Data sheet of a pump laser module. *(Courtesy of Pirelli Cavi e Sistemi SpA.)*

**DESCRIPTION**

The E2-980 nm hermetic Laser module have been specifically designed for optimum performance in the Erbium Doped Fiber Amplifier.

The Laser Module mounts a Pirelli E2 Laser chip, which offers high performance and high reliability by using the state of the art technology.

The standard option fiber pigtail is a single mode fiber which is Pirelli proprietary and its design makes the fiber a strategic component used in conjunction with the Pirelli Laser Modules to make a high coupling efficiency fiber alignment system.

The materials used in the laser module, have been chosen and designed to allow an high thermal stability on a low power consumption with very efficient Thermo Electric Cooler, whose qualification data shows the highest reliability recorded.

Great care has been taken to prevent Electro-Static Discharges, so the Pirelli Laser Module can withstand several discharges at 8.000 V, Human Body Model.

The E2 manufacturing technique, used in the Pirelli Laser Chips, is known to be the most reliable and field proven technology.

This reliability has been confirmed by the data obtained from the qualification program applied to the Laser Chips and Laser Modules. Both Chips and Modules have separately been qualified in compliance with the Bellcore and Military Standards requirements.

The reliability data are based on the qualification program carried out by Pirelli to obtain more than 1.000.000 devices/hours of ageing in operating conditions.

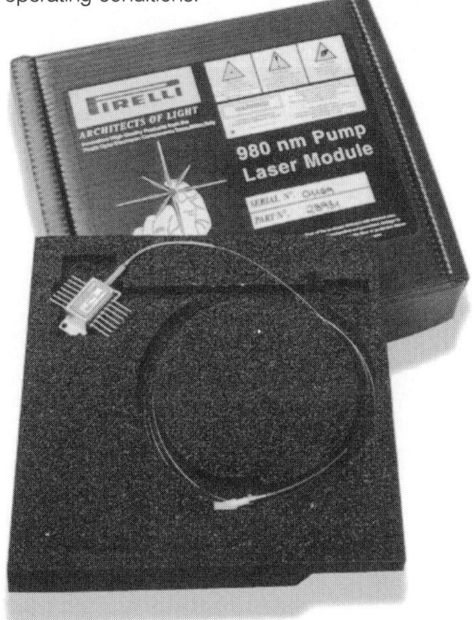


**Figure 12.37 (b)** (continued)

PIRELLI

## MAXIMUM ABSOLUTE RATINGS

| | | |
|---|---|---|
| **Laser** | | |
| Fiber output Power | mW | 160 |
| Foreward current | mA | 300 |
| Reverse voltage | V | 2 |
| Reverse current | mA | 2 |
| **Thermistor** | | |
| Current | mA | 2 |
| Voltage | V | 5 |
| **Thermoelectric cooler** | | |
| Current | A | 1,5 |
| Voltage | V | 2 |
| **Package** | | |
| Storage temperature | °C | -40 to 85 |
| Operating temperature | °C | 0 to 65 |
| **Fiber pigtail** | | |
| Fiber MDF | μm | 6.2 ± 0.5 |
| Cutoff wavelength | nm | 930 ± 40 |
| Tensile Strength | N | 5 |
| Bend radius | mm | 12.5 |

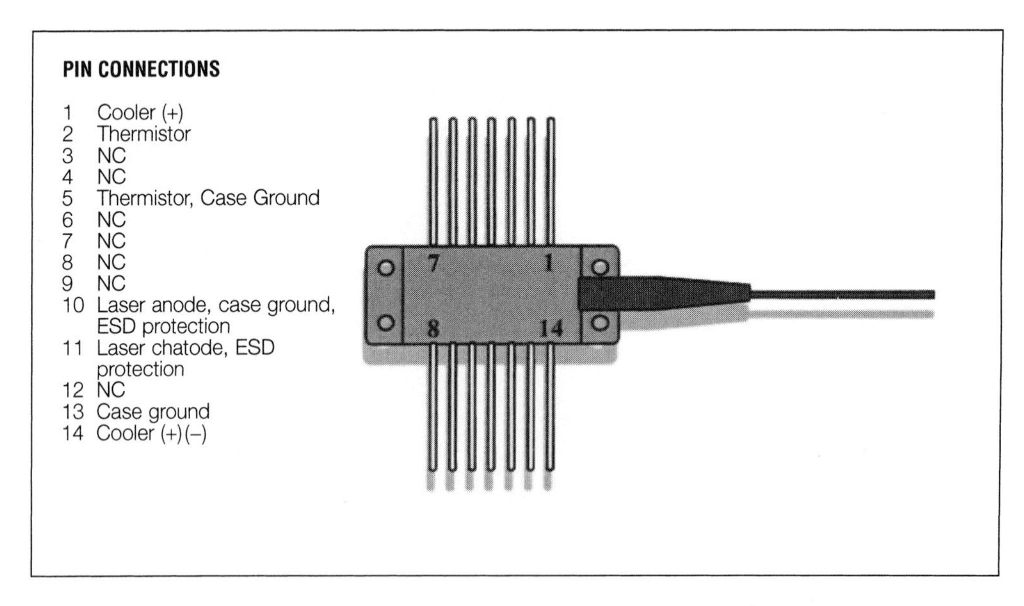

PIN CONNECTIONS
1 Cooler (+)
2 Thermistor
3 NC
4 NC
5 Thermistor, Case Ground
6 NC
7 NC
8 NC
9 NC
10 Laser anode, case ground, ESD protection
11 Laser chatode, ESD protection
12 NC
13 Case ground
14 Cooler (+)(–)
7
1
8
14


**Figure 12.37 (c)** (continued)

## TECHNICAL SPECIFICATIONS

| | Unit | Min | Max |
|---|---|---|---|
| **Operating drive current @ kink free power** | | | |
| @ 65 mW kink free Po | mA | | 200 |
| @ 75 mW kink free Po | mA | | 200 |
| @ 90 mW kink free Po | mA | | 225 |
| @ 100 mW kink free Po | mA | | 250 |
| @ 120 mW kink free Po | mA | | 250 |
| @ 130 mW kink free Po | mA | | 250 |
| @ 145 mW kink free Po | mA | | 280 |
| @ 155 mW kink free Po | mA | | 280 |
| Forward voltage | V | | 2.5 |
| Threshold current | mA | | 25 |
| Wavelength band Peak emission @ kink free Po | nm | 970 | 985 |
| Power in wavelength band | % | 90 | |
| Chip operating temperature | °C | 20 | 30 |
| Spectral shift over chip temp. | | | |
| Ts=20 to 30°c; Tc=25°C | nm/°C | | 0,7 |
| Thermoelectric Cooler current @ kink free Po, ΔT=40°C | A | | 1 |
| Thermoelectric Cooler voltage @ kink free Po, ΔT=40°C | V | | 1.6 |
| Thermistor resistance | kΩ | 9.5 | 10.5 |

Conditions: Ts 25°C; Tc 0,65°C

## ORDERING INFORMATION

| Description | P/N |
|---|---|
| 65 mW kink free Power Output | 28925 |
| 75 mW kink free Power Output | 28926 |
| 90 mW kink free Power Output | 28927 |
| 100 mW kink free Power Output | 28928 |
| 120 mW kink free Power Output | 28929 |
| 130 mW kink free Power Output | 28932 |
| 145 mW kink free Power Output | 28979 |
| 155 mW kink free Power Output | 28980 |

Option list:
1) (-BFM) Back Facet monitor
2) (-EXT) Extended operating temperature (-20, +80 °C)

The previously mentioned products and/or the system including them may be covered by one or more of the following patents:
• US 4786135 • US 4790464 • US 4807958 • US 4846544 • US 4889399 • US 5129027 • US 5127076 • US 5570438
• US 5550947 • US 5640481 • US 5443536 • US 4395869 • US 4448484 • US 4497164 • US 4690496 • US 4676590
• US 4690498 • US 4690497 • US 4688889 • US 4673540 • US 4620412 • US 4722589 • US 4725121 • US 4741592
• US 4725123 • US 4703135 • US 4842438 • US 4805392 • US 4867527 • US 4902096 • US 4927294 • US 5140664
• US 5150444 • US 5185841 • US 5193134 • US 5455881 • US 4756600 • US 5229851 • US 5390273 • US 5509097
• US 4690627 • US 4859024 • US 5179619 • US 5533164 • US 5518516 • US 5444808 • US 5656090 • US 5658363
• US 4911742 • US 4923496 • US 4925269 • US 5054876 • US 5204923 • USRE 35697 • US 5119229 • US 5245467
• US 5138483 • US 5267073 • US 5638204 • US 5113459 • US 5218665 • US 5087108 • US 5161050 • US 5210808
• US 5278686 • US 5355250 • US 5233463 • US 5515200 • US 5646775 • US 5383051 • US 5491581 • US 5497265
• US 5381426 • US 5579153 • US 5497386 • US 5701194 • US 5712716 • US 5668909 • US 5677786 • US 5701378
• US 4881793 • US 4938561 • US 4768882 • US 5001338 • US 4986663 • US 5037179 • US 5150516 • US 5272433
• EP 448012 • EP 166138 • EP 193780 • EP 193779 • EP 200914 • EP 226188 • EP 242740 • EP 285917
• EP 297409 • EP 338482 • EP 466230 • EP 464918 • EP 488438 • EP 503469 • EP 632301 • EP 490803
• EP 210770 • EP 187060 • EP 487121 • EP 665191 • EP 518523 • EP 177206 • EP 350726 • EP 517169
• EP 308114 • EP 408905 • EP 409012 • EP 425014 • EP 426222 • EP 426221 • EP 431654 • EP 439867
• EP 441211 • EP 417441 • EP 442553 • EP 507367 • EP 509577 • EP 595395 • EP 595396 • EP 603925
• EP 567941 • EP 644633 • EP 294037 • EP 324541 • EP 346589 • EP 374614 • EP 458255

**Pirelli Cavi e Sistemi SpA** - Optical Systems Business Headquarter
Viale Sarca, 222 - 20126 Milano - Italy
Tel. +39 2 6442.3337/5956 - Fax +39 2 6442.9256
http: //www.pirelli.com



**Figure 12.37 (d)** (continued)

data sheet, keep in mind the functional block diagram in Figure 12.35. At this point in your studies, you should be able to understand this technical document with ease. To test yourself, consider the following example.

### Example 12.4.3

***Problem:***

Draw the schematic of a Pirelli pump laser module.

***Solution:***

This schematic is a block diagram with pin assignment. The solution is based on the functional block diagram in Figure 12.35 and the pin connections shown in Figure 12.37. The schematic of a Pirelli pump laser module appears in Figure 12.38.

No unknown elements are found in this block diagram except for the ESD-protection circuitry. We need this component because pump lasers require high current (up to 280 mA, as you can see from Figure 12.37[d]) and they are extremely susceptible to damage from high-current electrical transients caused by the discharge of electrostatic voltage. Such transients are called *electrostatic-discharge (ESD) transients,* an unpleasant phenomenon familiar to us all through personally experienced static electricity in our everyday life. A surge-protection varistor connected directly to an LD provides exceptional ESD resistance.

The design of ESD-protection circuitry depends on the value of the voltage that a pump module has to withstand. This voltage, in turn, depends on the model used to evaluate the ESD process. Manufacturers usually consider a *human-body model (HBM),* meaning 1.5 kΩ resistance and 100 pF capacitance. Such a model can withstand about ± 500 V [29].

---

***Other components of an EDFA*** Refer back to Figure 12.25, a block diagram of an EDFA. The complete EDFA module includes a coupler combining pump and signal wavelengths into one

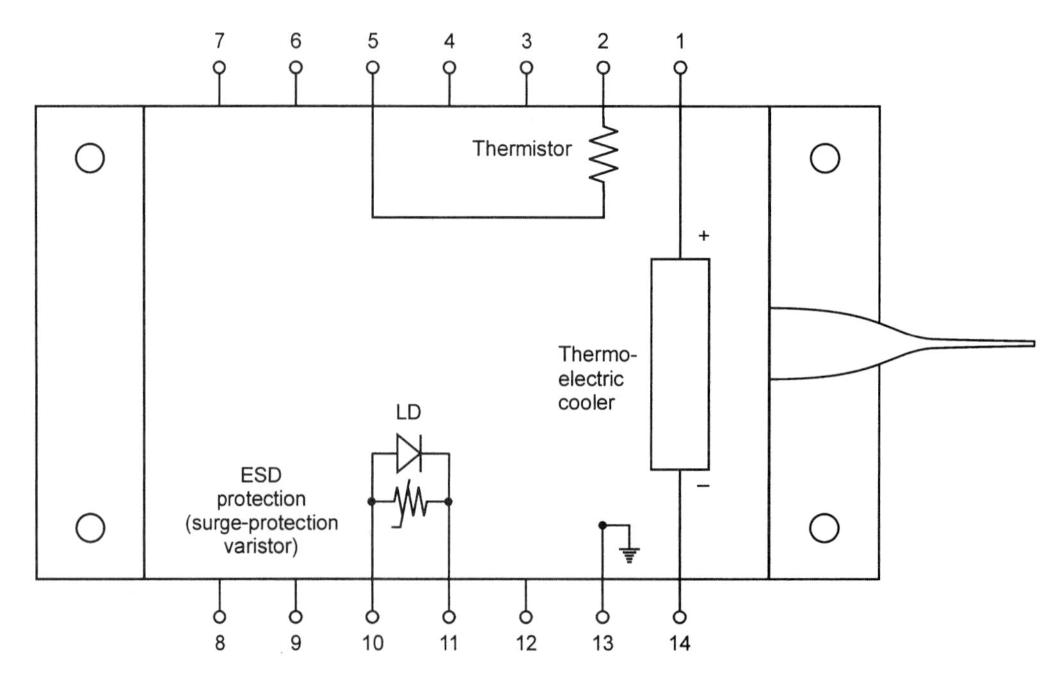


**Figure 12.38** Schematic of Pirelli pump laser module.

fiber, an isolator preventing the amplifier fiber from backreflected light, and optical filters flattening fiber gain and removing pump wavelength from the transmission fiber. A coupler, an isolator, and a filter are typical passive components. Bear in mind that these components determine to a great extent the performance of the entire EDFA unit. They are discussed in Chapter 13. Here, we need only highlight the filter's function as a gain-flattening device.

You have certainly noticed (and we have stressed this point several times) that the gain of an active fiber varies considerably along the gain bandwidth, as shown in Figures 12.28 and 12.33. On the other hand, using an EDFA in WDM communications systems requires equal amplification of all the wavelengths involved in signal transmission, that is, a flat gain over the gain bandwidth. The simplest way to achieve this is to use an optical filter whose transmission characteristic is tailored to compensate for the unequal gain of an active fiber. The principle of operation of an equalization filter and an optical scheme (that is, where the filter is placed in an EDFA module) are shown in Figure 12.39.

An *equalization filter,* also called a *gain-flattening*—or *notch—filter,* is usually a built-in component. It is available in a variety of styles, including those with variable spectral profiles. How optical filters are made and how they work will be discussed in Chapter 13.

## Reading an EDFA Data Sheet

An example of an EDFA data sheet is presented in Figure 12.40. It should be pointed out that data sheets of commercially available EDFAs vary significantly both in format and content, making it necessary for us to comment here on material in them that may be unfamiliar to you.

We'll start with the physical appearance of the device depicted in Figure 12.40(a). An EDFA is available in two packages: module and subrack. Let's concentrate on the module package. The fiber-gain module is a rectangular box whose dimensions are $5 \times 5.9 \times 1$ in., or $127 \times 149.9 \times 25.4$ mm. The unit requires only two optical ports—input and output—and Figure 12.40(a) shows these two fibers connected to the gain module. The unit has an electrical connector that provides direct access to all internal circuits. The amplifier needs only one standard power supply (usually $\pm$ 5 V, where the plus or minus sign indicates the polarity).

Let's now take a close look at what's inside the box. In addition to all the optical components shown in Figure 12.25 and discussed above—erbium-doped fiber, couplers, isolators, and pump lasers—manufacturers usually include the control and monitoring circuitry that was also mentioned above. A block diagram of an EDFA that includes such circuits is shown in Figure 12.40(b).

Using input and output taps, the manufacturer makes it possible to *monitor* input and output light power. Monitoring the performance of all EDFA components is done by a microcontroller connected to an optical fiber amplifier. The result is the ability of the unit to generate alarms indicating loss of input and output signals, output-power degradation, an overdriven pump laser, and out-of-range temperatures.

When such circuitry is built into the amplifier, the user can *control* some amplifier parameters, the most important of which is *automatic power control,* which keeps the output power constant when channels are added or dropped. (Gain stabilization can also be achieved optically by using special Bragg gratings [33].)

Manufacturers usually include two *shutdown circuits: automatic* (when temperature is out of range or output power is off) and *manual* (when operator intervention is necessary). Control circuitry is accessible through the RS-232 port. Figure 12.40(d) lists the alarm and control signals.

*Troubleshooting* an EDFA in case of improper performance is easy when all the circuits are activated. If, for example, you find there is no amplified (output) optical signal, you should take the following steps:

1. Check the input-power monitor PD. If a signal appears, then . . .
2. Check the output of the power-monitoring PD. If a signal does not appear, then . . .

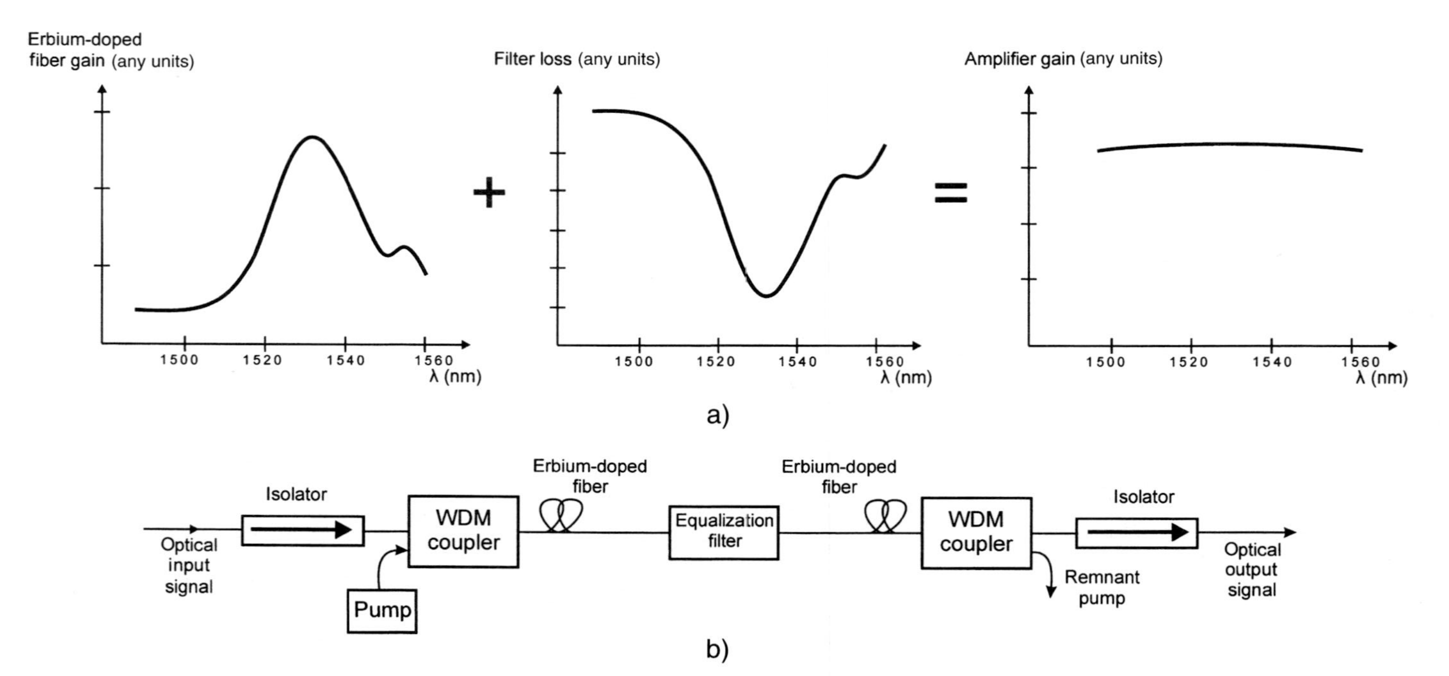


**Figure 12.39** Using equalization filter for gain flattening in EDFA: (a) Principle of operation; (b) optical scheme.

For Dense Wavelength Division Multiplexing Systems

## SR Series SystemReady™ Fiber Amplifiers

SR Series SystemReady™ Erbium Doped Fiber Amplifiers (EDFAs) and Extended Band Fiber Amplifiers (EBFAs) are specially designed for use in 1528 - 1563 nm and 1568 – 1603 nm DWDM optical transmission systems. High efficiency, compact size, as well as user-friendly interface are some of the features offered by this product line. An intelligent digital and analog circuitry provides precision bias and temperature control of the pump lasers and a full supervision of the device. Both module and subrack packages are offered for system integration or stand-alone in DWDM transmission systems.

**Figure 12.40 (a)** Data sheet of an erbium-doped fiber amplifier. *(Courtesy of AFC Technologies, Inc.)*

## Common Performance

(Preamplifier, Line amplifier or Booster amplifier)

### Absolute Maximum Ratings

| Parameter | Symbol | Minimum | Maximum | Units |
|---|---|---|---|---|
| Storage Temperature | Ts | −40 | +70 | °C |
| Operating Temperature | To | −5 | +55 | °C |
| Power Supply (Module) | Vm± | ± | 6.0 | V |
| Power Supply Current (Module) | Im± | ± 3.0 | | A |
| Power Supply (Subrack) | Vr | 36.0 | 72.0 | V |
| Power Supply Current (Subrack) | Ir | 0.6 | - | A |

### Operating Environment

| Parameter | Symbol | Minimum | Maximum | Units |
|---|---|---|---|---|
| Operating Temperature | To | 0 | + 50 | °C |
| Power Supply Voltage (Module) | Vm± | ±4.75 | ±5.25 | V |
| Power Supply Voltage (Subrack) | Vr | 46 | 50 | V |

### Optical Characteristics

| Parameter | Symbol | EDFA | EBFA | Units |
|---|---|---|---|---|
| Operating Wavelength | λ | 1525–1565 | 1565–1605 | nm |
| Return Loss (Input/Output) | RL | 30 | 30 | dB |
| Remnant Pump (Input/Output) | Rp | <−30 | <−30 | dBm |
| Polarization Sensitivity | PDL | <0.5 | <0.5 | dB |
| Polarization Mode Dispersion | PMD | <0.5 | <0.5 | ps |
| Temperature Dependent Gain | Gt | <0.5 | <0.5 | dB |



## Block Diagram

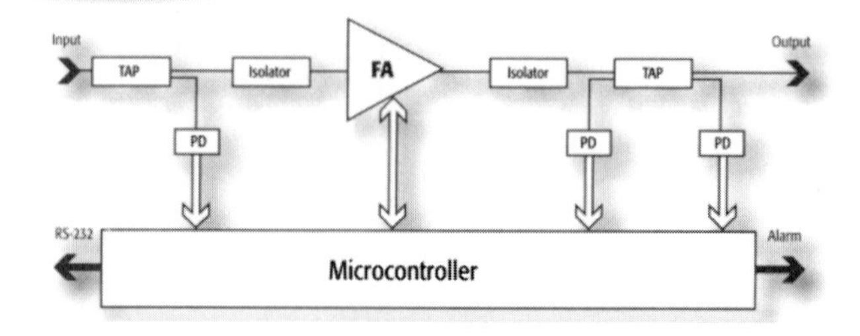


Figure 12.40 (b)
(continued)

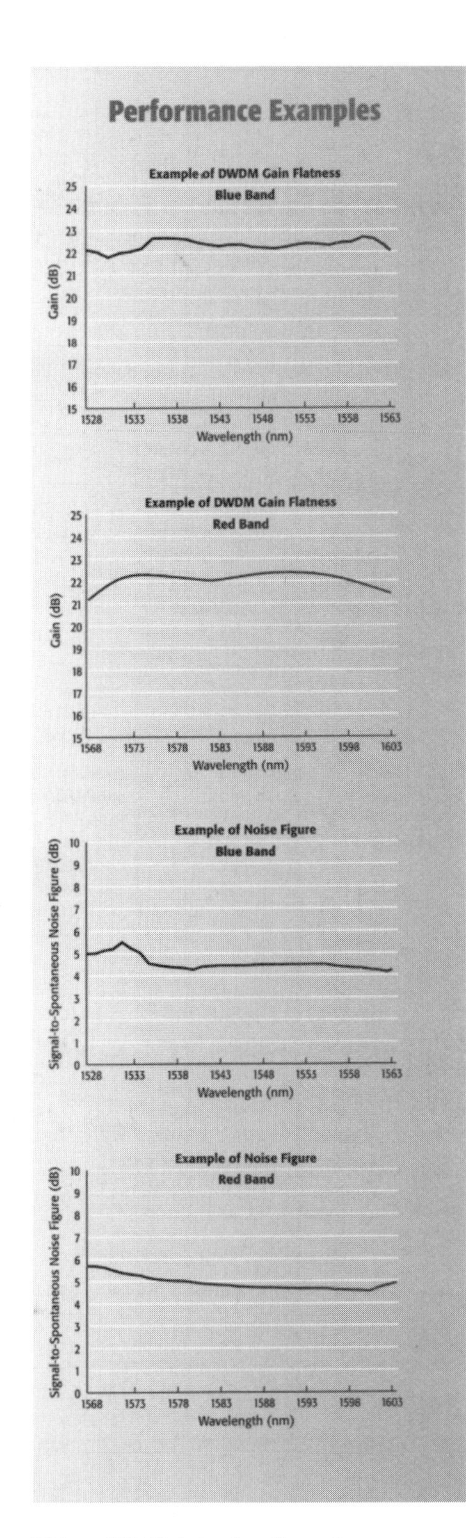


## Specifications

### Booster Amplifier

| Parameter | EDFA | EBFA | Units |
|---|---|---|---|
| Operating Wavelength | 1528–1563 | 1568–1603 | nm |
| Small Signal Gain | 30 typical | | dB |
| Typical Composite Input Level | −5 to +5 | | dBm |
| Input Per Channel ( x 32 channels) | −20 to −10 | | dBm |
| Input Per Channel ( x 16 channels) | −17 to −7 | | dBm |
| Saturation Power Available | 16 to 19 | 16 to 18 | dBm |
| Typical Per Channel Gain | 10 ~ 20 | | dB |
| Multi-channel Gain Flatness | < ± 1.0 dB | | |
| Noise Figure | 4.5 typical, 6.0 max | | dB |
| Custom Configuration | Yes, please contact AFC | | - |

### Inline Amplifier

| Parameter | EDFA | EBFA | Units |
|---|---|---|---|
| Operating Wavelength | 1528–1563 | 1568–1603 | nm |
| Small Signal Gain | 35 typical | | dB |
| Typical Composite Input Level | −20 to −5 | | dBm |
| Typical Input/Ch. ( x 32 channels) | −35 to −20 | | dBm |
| Typical Input/Ch. ( x 16 channels) | −32 to −17 | | dBm |
| Typical Per Channel Gain | 20 ~ 30 | | dB |
| Saturation Power Available | 14 to 18 | 14 to 18 | dBm |
| Multi-channel Gain Flatness | ± 0.5 typical, ± 0.8 Max. | | dB |
| Noise Figure | 4.0 typical, 5.5 max. | | dB |
| Custom Configuration | Yes, please contact AFC | | - |

### Pre-Amplifier

| Parameter | EDFA | EBFA | Units |
|---|---|---|---|
| Operating Wavelength | 1528–1563 | 1568–1603 | nm |
| Small Signal Gain | 35 typical | | dB |
| Typical Composite Input Level | −40 to −20 | | dBm |
| Typical Gain Constant | 25 to 35 | | dB |
| Typical Composite Output Level | −5 to 5 | | dBm |
| Multi-channel Gain Flatness | < ± 1.0 | | dB |
| Noise Figure | 4.0 typical, 5.5 max. | | dB |
| Custom Configuration | Yes, please contact AFC | | dB |

**Figure 12.40 (c)** (continued)

## Alarms and Controls

**Available Alarm Information**

| Item | Symbol | Description |
|---|---|---|
| 1 | LOS | Loss of Signal Alarm |
| 2 | ALTc | Case Temperature Alarm |
| 3 | LOO | Loss of Output Alarm |
| 4 | ALI | Pump Bias Alarm |
| 5 | ALTp | Pump Temperature Alarm |

**Subrack Front Panel Status Monitoring**

| Item | Symbol | Description | Threshold |
|---|---|---|---|
| 1 | PWO | Power On | - |
| 2 | LOS | Loss of Signal | User adjustable |
| 3 | FAIL | Unit failure | Current or Temperature Fault |

**Monitoring & Control Functions**

| Item | Symbol | Description |
|---|---|---|
| 1 | Po | Output Power Monitor |
| 2 | Pi | Input Power Monitor |
| 3 | AST | Automatic Shutdown (enabled or disabled) |
| 4 | AGC | Automatic Gain Control |
| 5 | APC | Automatic Output Power Control |
| 6 | AIC | Automatic Pump Current Control |
| 7 | Thr | User Adjustable Threshold for Alarms |
| 8 | Cst | User Accessible Constant Setting (Gain, Power or Current) |

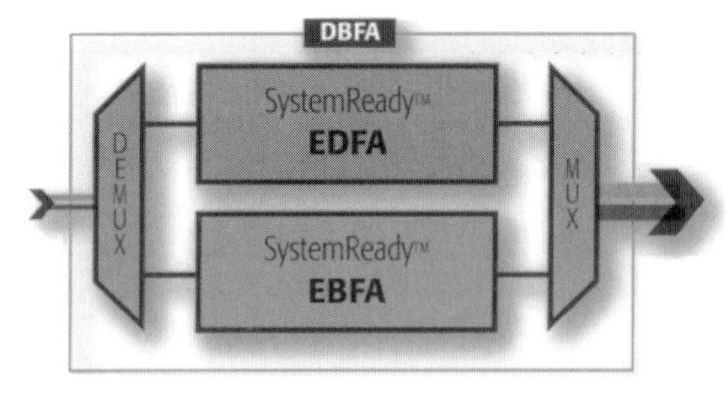


**Figure 12.40 (d)** (continued)

## Applicable Performance Standards

**GR - 1312 - CORE:** Generic Requirements for Optical Fiber Amplifiers

**GR - 2918 - CORE:** Dense Wavelength Division Multiplexing Systems with Digital Tributaries for use in Metropolitan Area Applications: Common Physical Layer Generic Criteria

**GR - 2979 - CORE:** Common Generic Requirements for Optical Add-Drop Multiplexers and Optical Terminal Multiplexers

**GR-63-CORE:** Network Equipment-Building System (NEBS) Requirement: Physical Protection

**GR-1089-CORE:** Electromagnetic Compatibility and Electrical Safety – Generic Critaria for Network Telecommunications Equipment

## Amplifier For Doubled Bandwidth

When a single band EDFA or EBFA is no longer enough to satisfy the bandwidth requirement, the two can be multiplexed together to offer doubled bandwidth. Multiplexer and demultiplexer for EDFA and EBFA bands can be integrated in each module or subrack. When incoming signal contains both blue and red band channels, they will be routed to corresponding amplifiers. After amplification, channels are multiplexed together so that everything will continue to transmit in a single fiber. Please contact your sales representative for more detail. Also see page 9.

3. Check the performance of the amplifier.
4. If the EDFA is working properly, check the output isolator. If this is also functioning properly, then . . .
5. Check all connections (splicing and connectors).

These are just a few of the procedures involved in troubleshooting an EDFA. You have to look at the specific configuration of the EDFA you will be working with to determine the specific troubleshooting algorithm you should follow. But remember: The right way to do this is to follow the signal flow through the device's components.

Since a number of configurations are available, some manufacturers include in their data sheets detailed optical schemes of the EDFA. (See [35], [36].)

It's noteworthy that the trend in EDFA design is integrating some amplifier elements into larger components. For example, on the input side of an EDFA, the isolator, the input tap coupler, and the WDM coupler are combined into an *integrated hybrid component* called an *ITW (isolator-tap-wdm);* on the output side of the EDFA, the output tap coupler and the isolator are integrated into a hybrid component called a *TAPI (tap-isolator).*

We will turn now to the sections headed "Common Performance" and "Specifications." The manufacturer offers an EDFA in three models: booster, in-line amplifier, and receiver amplifier (preamplifier). To recall the functions of these types of EDFAs, see Figure 12.13.

***Booster amplifier*** A booster amplifier, which is placed immediately after the transmitter, is designed to provide maximum output power. Indeed, the EDFA features 16 dBm (39.8 mW)—or even 19 dBm (79.4 mW)—of saturation output power. If you look at the saturation power of the other models given in the "Specifications" table, you'll see that the in-line amplifier has 14 to 18 dBm, whereas the data sheet does not show any saturation power for the preamplifier. Now you can see that the major function of a booster is to send maximum light power along a transmission fiber.

***In-line amplifier*** This type of amplifier features moderate gain, noise, and saturation output characteristics, as you can see from the "Specifications" table. Its main peculiarity is *gain flatness.* This means that this type of amplifier exhibits minimum variations in gain value across the gain bandwidth as compared to other types. Compare the "Multi-channel Gain Flatness" characteristics of all three types of EDFAs. You'll see that an in-line amplifier has the best value (±0.5 dB).

Gain flatness is the major requirement for in-line amplifiers because there are usually many of them (hundreds, in fact) cascaded along the long-distance link. You can easily imagine what a difference in the power of multiplexed channels (wavelengths) you'd realize if your in-line amplifiers would significantly vary in gain at different wavelengths.

It's worth mentioning yet another characteristic you might come across in the data sheets: *gain tilt.* Tilt is the gain change caused by adding or dropping a wavelength channel, that is, altering the input conditions. In the following formula, tilt is calculated as the gain change at any wavelength, $\lambda$, as compared with gain change at the reference wavelength, $\lambda_0$; that is [25],

$$Tilt(\mathrm{dB/dB}) = \Delta G(\lambda)/\Delta G(\lambda_0) \tag{12.29}$$

You should be aware that sometimes "gain tilt" is used interchangeably with "gain flatness."

***Preamplifier*** This type of EDFA is placed immediately in front of a receiver, its job being to amplify a weak optical signal coming in after a long transmission. Hence, the major features of a preamplifier should be low noise and high gain (not power). Look at the table headed "Specifications."

The noise figure of the preamplifier is 4.0 dB. You will recall that the theoretical limit of a noise figure is 3 dB, as Formula 12.30 states.

Sometimes a manufacturer specifies the cause of the noise: the beating between a signal and spontaneous emission. We already know why this is so: Signal-ASE beating is indeed the *major* source of amplifier noise.

Another important feature of a preamplifier is its *sensitivity,* that is, the minimum power of the optical input signal that the EDFA can handle. This preamplifier can operate with –40 dBm (0.1 μW) input power, certainly an impressive figure.

The specification under "Small Signal Gain" displays the gain an EDFA exhibits when the input signal does not saturate the amplifier. Some manufacturers even specify the value of the small signal (for example, $P_{in} = -6$ dBm) at which the gain has been measured. Note how gain changes depending on the amplifier's application: from 30 dB (1000) for a booster to 35 dB (3162) for a preamplifier.

Sometimes manufacturers provide data on large signal gain. In this case, the manufacturer specifies the level of the large signal (for example, 0 dBm at 1550 nm).

Many manufacturers include data on *polarization-dependent loss (PDL).* These numbers show you how much the loss changes for two orthogonal polarizations of the input signal. An EDFA compensates for any loss by its gain, so PDL is essentially the *gain-polarization sensitivity* of a fiber amplifier. Let's consider an example: If the input is –20 dBm, we will have two output powers—10.2 dBm and 10.7 dBm—depending on the state of polarization of the input signal. Then the polarization sensitivity, or PDL, is 10.7 dBm –10.2 dBm = 0.5 dB, which is given in Figure 12.40(b). EDFAs are not as sensitive to the state of polarization of an input signal as SOAs; this is clearly seen from their PDL number, which is typically 0.2 dB, whereas the PDL for SOAs is typically 0.5 dB. However, lest you forget, SOA designers take many measures to reduce the polarization sensitivity of this type of optical amplifier, as shown in Figure 12.21.

To describe the polarization-related properties of an EDFA, manufacturers sometimes even include data about *polarization-mode dispersion, PMD.* (See Section 5.3.) PMD values are generally on the order of 0.5 ps. (See Figure 12.40[b].)

You may have noticed that nothing about *crosstalk*—the phenomenon that draws so much attention in any discussion of semiconductor optical amplifiers—is found in the EDFA data sheets. This is because crosstalk is not a problem with EDFAs. Crosstalk becomes negligible at bit rates higher than the reciprocal of the lifetime of erbium ions at the level of excitation. In erbium-doped fiber, this lifetime is more than 10 ms; hence, at a bit rate higher than a few kbit/s, you don't need to worry about crosstalk in EDFAs.

The EDFA manufacturer presents the device's major characteristics graphically, similar to what is shown in Figure 12.31. These graphs are found in Figure 12.40(c) in the section headed "Performance Examples." Analyze and compare them with the graphs in Figure 12.31.

It is important to observe how the manufacturer increases the bandwidth of an EDFA. There are actually two EDFAs in one module. One of them operates in the C-band (called "Red Band" here) and the other works in the L-band ("Blue Band"). (See Figure 12.27.) An input signal is demultiplexed, amplified by the proper amplifier, multiplexed, and sent along the optical fiber. Using this technique, the manufacturer is able to extend the bandwidth of an EDFA module from 1528 nm to 1603 nm.

Other data we find in these specifications sheets include the "Absolute Maximum Ratings" and the "Operating Environment." In general, specifications sheets may also include the electrical characteristics of the amplifier's electronic components; the pump LD wavelength, which is 980 nm or 1480 nm; PIN connections similar to those shown in Figure 12.37; and, of course, purchasing information [34].

***Double pumping*** Some specifications sheets cite "double pumping" and "multiplexed pumping." Let's see what these terms mean. In addition to co- and counterpropagating pump configurations (see Figure 12.25), there is double (or dual) pumping. This occurs when an erbium-doped fiber is pumped with co- and counterpropagating light simultaneously. (See Figure 12.41[a].)

Double pumping distributes pump power along the erbium-doped fiber more evenly than single pumping. (Recall how pump power damps along the length of an EDFA, as seen in Figure 12.29.) Obviously, you can pump more power and therefore attain higher saturation power with double pumping. In fact, the double-pumping EDFA configuration is fast becoming the most widely used type.

For one thing, this setup makes it easier to exploit several pump lasers to increase the output power of an amplifier, an option discussed previously in this section. A configuration with four LDs (two *multiplexed pump lasers* at each side) is shown in Figure 12.41(b). With this pumping scheme, the manufacturer has attained saturation output power up to 24 dBm [36].

***EDFA design*** Suffice to say that software packages facilitating the design of an erbium-doped fiber are commercially available. An example of such software can be found in reference [19]. EDFA design is a highly specialized area lying outside the province of this book. (See [19] for a comprehensive review of EDFAs.)

## Other Types of Optical Fiber Amplifiers

Even though erbium-doped fiber amplifiers do their job well, new requirements are arising with the expansion of fiber-optic networks. More gain and more bandwidth—these are the two major forces driving efforts to improve EDFA characteristics.

There is one inherent problem with EDFAs: They work only in the 1550-nm transparent window. Therefore, they cannot work in the CATV industry, which operates exclusively at 1300 nm. Thus, a huge sector of the fiber-optic industry is left without fiber amplifiers. The only solution to this major problem lies in developing optical amplifiers working at 1300 nm.

How can we achieve this? Remember that gain in an EDFA is provided by the transition of erbium ions, and the range of amplified wavelengths is determined by erbium energy levels. Dopants that have energy levels resulting in amplification at around 1300 nm include the rare earth

### How to Get More Pump Power

As we already know, the more wavelength channels we want to multiplex, the more pump power we need. There are three basic ways to increase the pump power of an EDFA [37]:

1. Increase the power of a single pumping laser diode (up to 250 mW has been reached).
2. Pump an EDFA with lasers separated by the states of polarization of their radiation.
3. Use several pump lasers separated by different wavelengths, which results in the multiplexing of their power. Such pump modules are also called *multiplexed pump lasers.*

Multiple-wavelength (multiplexed) pumping modules use two or four laser diodes operating around 1480 nm with spacing of 10 or 15 nm. Thus, for a four-source module, the operating wavelengths are 1460 nm, 1470 nm, 1480 nm, and 1490 nm. This wide separation is possible because the width of an absorption line around the 1480-nm pumping wavelength is 80 nm. Refer again to Figure 12.16. Photons whose wavelength varies not more than 80 nm around 1480 nm will be absorbed and their energy used for pumping erbium ions at the excited level. A WDM coupler combines separate pump signals into one fiber. The result is a pump module using several laser diodes and producing high output power. In a sense, this is another application of wavelength-division multiplexing to increase the output of light power. Figure 12.41(b) shows a multiplexed pump module. Such multiplex pumping is available at 1480 nm with 20-nm, 10-nm, 7-nm, and 5-nm spacing. You can easily imagine four multiplex pump lasers at each side of an EDFA.

A multiplexed pump module can also be implemented around the 980-nm wavelength. However, since the absorption line here is about 15 nm in width, laser spacings must be much tighter. Such a pump module—using 970 nm, 975 nm, 980 nm, and 985 nm and producing up to 500 mW—is commercially available [38].

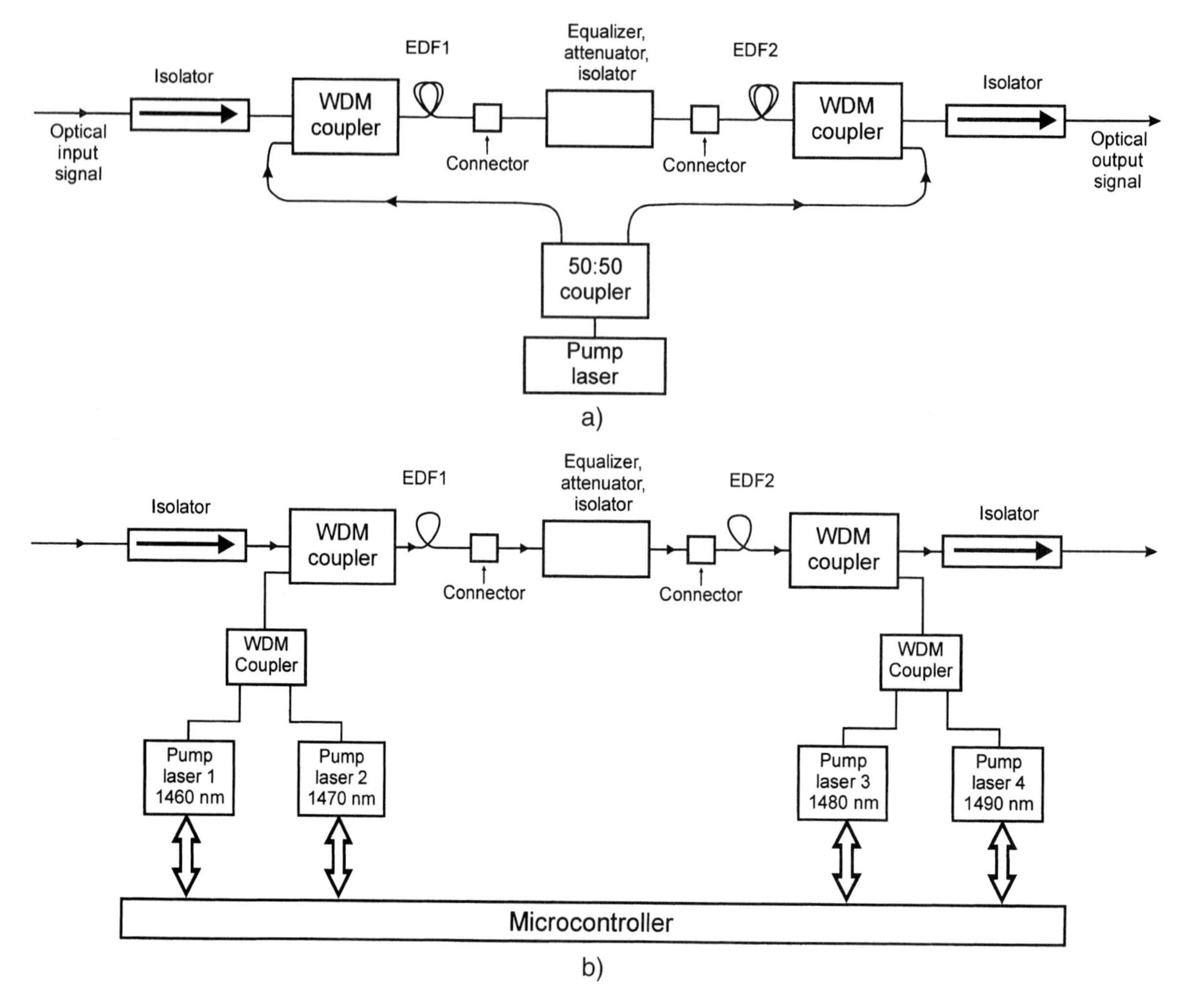


**Figure 12.41** Double-pumping configuration of an EDFA: (a) Single-pump scheme; (b) multiplexed-laser-pumping scheme.

metals praseodymium (Pr) and neodymium (Nd). That's the good news. The bad news is that neither Pr nor Nd works well with silica fiber. This leaves us with the need for another type of fiber.

Fortunately, a couple have been developed: fluoride (F) and telluride (Te). Fluoride-based fibers use a different combination of elements. The most popular type is called ZBLAN (pronounced "zee-blan") after the materials from which it is fabricated: $ZrF_4$-$BaF_2$-$LaF_3$-$AlF_3$-NaF [18].

Another composition uses ZrF-based and InF-based glasses [39].

Praseodymium-doped fluoride fiber amplifiers operate at 1300 nm and exhibit characteristics similar to those of EDFAs.

New fibers allow an amplifier designer to improve the characteristics of EDFAs. For example, erbium-doped fluoride fiber amplifiers demonstrate a flat spectral gain without a gain equalizer [40]. Erbium-doped telluride fiber amplifiers exhibit an ultra-broad—about 80 nm—gain bandwidth [39]. Another advantage is their higher efficiency in converting pump power into gain. For example, a flat gain of about 35 dB over the C-band, with pump power at only 30 dBm and fiber length of 7.5 m, is now available in commercial amplifiers [41].

However, difficulty in working with these new fibers and their high cost preclude their widespread acceptance at this time.

## SUMMARY

- We have to distinguish between a point-to-point transmission link and a network. A fiber-optic point-to-point link consists, basically, of a transmitter, an optical fiber, and a receiver. A fiber-optic network consists of nodes connected by optical fibers and includes many transmitters and receivers, that is, many senders and recipients. The main difference between these two means of telecommunications can be summed up as follows: In a point-to-point link, an information signal travels from a transmitter to a receiver without being diverted. In a network, the purpose, of course, is to have the signal from a sender reach a specific recipient among many. To attain this objective, a network has to have two main capabilities: to reroute (switch) the signal and to add and drop any signal at any node.
- Even though point-to-point links and networks are different entities, any connection between two nodes in a network is a point-to-point link. In other words, such a link is a basic unit of a network.
- For a point-to-point fiber-optic link, the major transmission problem today is the huge gap between information-carrying capacities of optical fiber and communication electronics. Indeed, the theoretical limit of the information-carrying capacity of an optical fiber is on the order of 50 Tbit/s. The bit rate that modern electronics can support is on the order of 10 Gbit/s. The only solution to bridging this huge gap is multiplexing, that is, sending several signals over the same optical fiber.
- Two multiplexing techniques are mainly exploited in fiber-optic communications: time-division multiplexing (TDM) and wavelength-division multiplexing (WDM). TDM operates on the principle of sending different signals over the same optical fiber in a given time sequence. This technique increases the transmission capacity of a link but is still limited by encountering the same restriction in bit rate dictated by today's electronic components. WDM assigns to each signal its own wavelength and sends all these wavelengths over the same optical fiber. A wavelength carries a transmission channel (signal), which occupies a certain portion of the bandwidth of an optical fiber.
- The two main advantages of WDM, also called dense wavelength-division multiplexing (DWDM), are the following: First, it allows much better utilization of the bandwidth of an optical fiber despite the limitations of existing electronics. Indeed, suppose we have 16 transmitters and receivers, each operating at 10 Gbit/s and at a particular wavelength. Then, after wavelength-division multiplexing, we utilize 160 Gbit/s of the fiber's capacity. It's still far from the theoretical limit but much better than TDM can accomplish. Second, each wavelength (channel) can be added and dropped independently. This feature gives the network manager an additional degree of latitude in operating the network.
- WDM is becoming the leading technology for point-to-point links. Transmission at 1.28 Tbit/s by combining 128 wavelengths is a market reality. Fiber-optic WDM networks are now being deployed.
- In order to build WDM transmission links and WDM networks, we need a number of new components with significantly advanced characteristics.
- Transmitters for WDM fiber-optic communications systems differ from regular transmitters mainly because they have to radiate as many wavelengths as a WDM system needs. This is done, by and large, by combining in one transmitter many individual laser diodes radiating at fixed wavelengths. The trend is to build tunable transmitters capable of radiating any desired wavelength by a simple tuning procedure. Unfortunately, tunable transmitters (and receivers, too) are much more expensive than their fixed counterparts. The transmitters for WDM communications have to meet a set of specific requirements in terms of the quality, stability, and reliability of the radiated light. A simple example explains the reasoning behind these requirements: 50-GHz spacing between individual channels requires a linewidth of at least 0.5 GHz to avoid interchannel crosstalk. Fortunately, modern transmitters meet most of these specifications; however, the ability to tune and integrate individual transmitters into a single unit capable of working with a WDM multiplexer is still among the scientific, technological, and economic problems yet to be completely resolved.
- Receivers for WDM systems have to cover the entire WDM bandwidth (up to 80 nm). In general, the spectral response of modern photodiodes (PDs) is even wider; however, their responsivity changes dramatically within the WDM bandwidth. Hence, additional measures for equalizing spectral response must be taken. Receiver sensitivity has to be designed taking into account the entrance of many channels (wavelengths) simultaneously and independently. The major concern here is interchannel crosstalk, which degrades the signal-to-noise ratio (SNR). Since transmitters radiate fixed wavelengths, WDM receivers have to be able to select the desired wavelength. This is done by selecting specific wavelengths in front of the PD, which raises the problem of tuning time. All other requirements for WDM receivers are similar to those discussed in Chapter 11.
- To compensate for the weakening of an information signal as a result of fiber attenuation, we can use repeaters (regenerators) and/or optical amplifiers. Repeaters convert

an optical signal into an electrical signal, determine whether it is bit 1 or bit 0, generate a new electrical pulse, convert it back to an optical signal, and transmit the reshaped signal farther along the fiber. In fact, repeaters generate a new copy of an incoming weak signal. Besides solving the problem of a deteriorating signal amplitude, a repeater eliminates pulse broadening caused by a fiber's dispersion. However, repeaters can work only with a specific wavelength, bit rate, and modulation format. These restrictions make it impossible to use repeaters in WDM transmission systems. Optical amplifiers boost an optical signal without optical–electrical conversion, a capability that allows them to work with a wide range of wavelengths, any bit rate, and any modulation format. This is why optical amplifiers are the only means to boost the signals in WDM systems.

- Semiconductor optical amplifiers (SOAs) and erbium-doped fiber amplifiers (EDFAs) are two basic types of optical amplifiers for WDM networks, with EDFAs being the major type. In addition, a Raman optical amplifier recently appeared on the market.

- All optical amplifiers have common basic characteristics, including gain, bandwidth, and noise performance. The value of gain, obviously, depends on the principle of the amplifier's operation but two general determining factors must be taken into consideration: First, gain depends on the frequency (wavelength) of the input signal. This dependence in general can be approximated by a bell-shaped curve. The gain-versus-frequency curve determines an amplifier's bandwidth. To get a flat gain within the working range of wavelengths, special measures for gain equalization must be taken. The second important phenomenon is that gain depends on the power of the input signal. When the input power is too high, gain decreases. This phenomenon is known as gain saturation.

- The noise performance of an optical amplifier is evaluated by the noise figure, which is the ratio of the signal-to-noise ratio (SNR) of an input signal to the SNR of an output signal. The noise figure accounts for the fact that an optical amplifier increases the input signal and input noise and generates its own noise.

- We distinguish among optical amplifiers depending on their applications. A booster, or a post-amplifier, is placed immediately after a transmitter. A booster magnifies a signal before sending it down a fiber. Its main function is to produce maximum optical power. An in-line amplifier is placed in the middle of a fiber-optic link to compensate for losses caused by fiber attenuation, connections, and signal distribution in networks. The main requirements for this type of optical amplifier are gain flatness and stability of all characteristics over the entire WDM bandwidth. A preamplifier magnifies an optical signal immediately before it reaches a receiver. Hence, good sensitivity, high gain, and low noise are major requirements here. These three functional types of optical amplifiers cover all applications that we need in WDM systems today.

- A semiconductor optical amplifier (SOA) uses the principle of stimulated emission to boost an information signal. Two main types of SOAs are suitable for use nowadays in WDM systems: the Fabry-Perot amplifier (FPA) and the traveling-wave amplifier (TWA). An FPA is simply a Fabry-Perot laser diode—that is, an active medium placed between two parallel mirrors—operating below the radiating threshold condition. In other words, an FPA SOA is an LD where gain is less than loss. A TWA is simply an active medium without any reflective facets (mirrors). This difference in construction implies some difference in characteristics. The gain of a TWA is equivalent to a single-pass gain of an FPA, while the actual gain of an FPA exhibits ripples whose amplitude depends on the value of the mirror's reflectances. The higher the reflectances, the larger the gain peak. If we make the reflectances equal to zero, we have a TWA. Gain for both FPAs and TWAs depends on wavelength (frequency). A TWA has much smaller gain but it covers a much wider spectral region. An FPA, in contrast, has very high gain within a narrow bandwidth. Therefore, there is a gain-bandwidth trade-off in SOAs. Since the gain of a TWA can be increased by extending the length of the active medium and since there is no means to extend the bandwidth of an FPA, the TWA is the more popular type of SOA today.

- The major advantage of SOAs is their ability to operate at 1300 nm and 1550 nm—even simultaneously. Keep in mind that EDFAs cannot work at 1300 nm, making SOAs the only optical amplifiers available for this—the cable TV—range. Other advantages include a relatively broad bandwidth (up to 100 nm can be covered by an SOA) and the possibility of integrating SOAs along with other semiconductor and photonic devices into opto-electronic integrated circuits. However, SOAs suffer from the following drawbacks: relatively high crosstalk, when the presence or absence of one channel changes the characteristics of another channel; sensitivity to the state of polarization of an incoming signal; technological difficulties in fabricating SOAs with low (up to $10^{-4}$) reflectances; and temperature sensitivity. These shortcomings preclude SOAs from being the major player in the optical-amplifier field.

- An erbium-doped fiber amplifier (EDFA) is the major means to boost an optical signal in WDM systems. The fact is that if it hadn't been for the advent of EDFAs, the actual deployment of WDM systems would have been impossible.

- An EDFA, as well as an SOA, works on the principle of stimulated emission. However, an EDFA has two distinguishing features: Its active medium is a piece of silica fiber heavily doped with ions of erbium and its external energy is delivered in optical—not electrical—form, as in an SOA. Thus, pumping is done by a powerful laser diode radiating at 980 nm or 1480 nm. The general structure of an EDFA includes a coupler that combines an information signal at 1550 nm with pumping light at 980 nm or 1480 nm; a piece of erbium-doped fiber, which is the active medium of the amplifier; another coupler, which separates an information signal from residual pumping light; a filter, which further separates remnant power light from the signal; and an isolator, which prevents any back-reflected light from entering the erbium-doped fiber. Still another isolator—at the input of an EDFA—keeps the noise generated by an EDFA from propagating in a backward direction with respect to the information signal.
- Amplification in an EDFA occurs as follows: With pumping taking place at 980 nm, an EDFA utilizes three energy levels of erbium ions incorporated into a silica fiber. External light pumps erbium ions at the highest level, at which point they decay and fall to the intermediate level. The lifetime of erbium ions at this level is very long (more than 10 ms). This is why these ions accumulate at the intermediate level. When pumping occurs at 1480 nm, only two energy levels (intermediate and lower) are involved but the net result is the same: Excited erbium ions accumulate at the intermediate level, creating *population inversion.* It is a fortunate coincidence that the energy gap between the intermediate and the lower levels in erbium corresponds to the 1550-nm wavelength range. Thus, when an optical information signal appears, it will stimulate the excited ions to drop to the lower level, and this transition will result in the emission of photons having the same frequency as the stimulating (i.e., the information) signal. These stimulated photons will add to the stream of incoming photons.
- EDFA gain depends on the wavelength (frequency) of the input signal. An erbium-doped fiber exhibits essential gain in the wavelengths between 1530 nm and 1560 nm, which is the EDFA bandwidth. However, this gain varies sharply within this range. Therefore, gain equalizers are included in the structure of an EDFA to attain the desired gain flatness within the EDFA bandwidth. Recent developments enable us to extend the EDFA bandwidth from 1530 to 1610 nm. An EDFA's gain decreases when the power of an input (information) signal increases, resulting in gain saturation.
- The main source of noise in an EDFA is the amplified spontaneous emission. Some of the excited erbium atoms drop to the lower level spontaneously, radiating photons within the EDFA bandwidth. These spontaneous photons stimulate other transitions, which means they are amplified. Since this amplified spontaneous light is out of phase and therefore not coherent with respect to the information signal, it constitutes noise. This noise is distributed evenly along the EDFA bandwidth and, potentially, could be filtered. The most detrimental portion of this noise is its so-called "slice," which lies within the spectral width of a specific channel (wavelength). This noise slice interferes with an information signal, and the resulting beat signal is the major contributor to EDFA noise.
- EDFAs meet most current requirements for use in WDM point-to-point links. The trend in this application is to use more channels (wavelengths) for multiplexing, which requires wider amplifier bandwidth and higher pump power. Fortunately, modern EDFAs meet these requirements and amplifiers with bandwidths from 1530 nm to 1610 nm are commercially available. To fulfill today's requirement of pumping more power, light from several pumping lasers radiating at close wavelengths is multiplexed. An important feature of an EDFA in its use in WDM networks is its ability to keep its characteristics constant when one or more input channels are added or dropped.
- Once again, it is EDFAs that made WDM transmission possible and it is wavelength-division multiplexing that has substantially increased the capacity of fiber-optic communications systems.

## PROBLEMS

**12.1** What is the difference between a point-to-point link and a network?

**12.2** Describe how a telephone call is connected within the United States.

**12.3** Explain the following terms: a *central office* (CO), b. *point of presence* (POP), and c. *LATA*.

**12.4** How does a signal travel in a star network? in a ring network?

**12.5** Does physical topology determine logical flow (the flow of information in a network)? Explain.

**12.6** Explain the difference between circuit-switching and packet-switching techniques.

**12.7** Which switching technique—circuit switching or packet switching—is faster and why?

**12.8** Explain how TDM systems work.

**12.9** Explain how WDM systems work.

**12.10** What is the relationship between TDM and WDM systems?

**12.11** a. What is the theoretical transmission capacity of a singlemode fiber? b. What is the maximum bit rate of modern transmitters and receivers? c. How are these two bit rates interrelated?

**12.12** What do we mean by the term *add/drop procedure?* Why is this a problem in TDM networks?

**12.13** What is the add/drop procedure in WDM systems?

**12.14** Describe the add/drop techniques you are familiar with in WDM networks.

**12.15** Explain how the broadcast-and-select technique works.

**12.16** Explain how the wavelength-routing technique works.

**12.17** What is a "lightpath"? Explain.

**12.18** What do we mean by "wavelength reuse"? What is a "wavelength conversion"? Why do we need these operations?

**12.19** What is the difference between a repeater and an optical amplifier?

**12.20** Explain how a repeater works.

**12.21** What do the terms 3R, 2R, and 1R mean?

**12.22** Explain the principle of operation of a repeater.

**12.23** Discuss the advantages and drawbacks of a repeater. For what applications is a repeater suitable and in what applications will it not work?

**12.24** What is the function of an optical amplifier?

**12.25** What are advantages and drawbacks of an optical amplifier?

**12.26** Why do we need new components for WDM systems?

**12.27** What new components are required to build a fiber-optic network?

**12.28** Classify all network components into two groups: active and passive.

**12.29** What is the channel spacing for a WDM with 40 channels?

**12.30** To allocate WDM channels, we need to keep the values of the operating wavelengths very accurate. Indeed, Figure 12.7(b), which shows the ITU grid, displays values like 1533.47 nm, 1534.25 nm, and so on. How can we fine-tune such radiated wavelengths?

**12.31** Figure 12.7 shows the ITU grid and an example of 8-channel allocation. It seems that for 8 channels the spacing should be equal to 30 nm/8 = 3.75 nm, that is, 468.75 GHz, where 30 nm is the standard bandwidth of an EDFA. However, the real allocation shown at the top of Figure 12.7(a) is quite different. Why?

**12.32** List five major requirements of transmitters used in WDM systems and explain why each requirement is needed. What laser diodes—fixed or tunable—do you prefer to use in WDM systems? Why?

**12.33** What techniques can be used to achieve tunability of laser diodes?

**12.34** Explain how a sectional DBR tunable laser works.

**12.35** Explain how an integrated-cavity tunable laser works.

**12.36** How can we get multiple access and how can we select the desired channel in WDM networks?

**12.37** List six major requirements of receivers used in WDM systems and explain why each requirement is needed.

**12.38** Explain why the concept of opto-electronic integrated circuits is so important in WDM systems.

**12.39** Name two major classes of optical amplifiers. To what class does a Raman amplifier belong?

**12.40** What are three functional types of optical amplifiers?

**12.41** How do a booster amplifier, an in-line amplifier, and a preamplifier differ in terms of functions and characteristics?

**12.42** Using a preamplifier lessens the stringent demand that otherwise would be made on a receiver's sensitivity and eventually allows a network to operate at a higher bit rate. Explain why using a preamplifier allows a network to operate at the higher bit rate.

**12.43** Explain the principle of operation of a semiconductor optical amplifier.

**12.44** How does the gain of an FPA type of SOA differ from the gain of a TWA?

**12.45** Calculate the spacing between adjacent resonant frequencies in an SOA if the refractive coefficient of the active medium is 3.6 and the resonator length is 50 μm.

**12.46** Calculate the maximum gain of an FPA if $G_s = 2$ and $R = 0.3$. How does this value change if $R = 0.03$? 0.003? What happens if $G_s$ becomes 3 and $R$ stays at 0.3?

**12.47** Calculate the gain of a TWA if $g = 52$ 1/cm, $\alpha = 14$ 1/cm, $\Gamma = 0.8$, and $L = 500$ μm. Compare your result with that of Example 12.3.2.

**12.48** Explain how the construction shown in Figure 12.17(b) reduces reflectances.

**12.49** What is gain saturation in optical amplifiers? Why is this phenomenon important?

**12.50** Define the bandwidth of an optical amplifier.

**12.51** Calculate the bandwidth of an FPA if $n = 3.6$, $RG_s = 0.96$, and $L = 50$ μm. Compare your result with the bandwidth value obtained in Example 12.3.3.

**12.52** There is a bandwidth-gain trade-off in an SOA, as Formulas 12.18 and 12.21 show. What SOA parameter plays the crucial role in this trade-off?

**12.53** Derive condition 12.20 from Formula 12.19.

**12.54** What is crosstalk in an SOA? What two major mechanisms cause crosstalk?

**12.55** Does the gain of an SOA depend on the state of polarization of an optical input signal? Explain.

**12.56** Do you want to reduce or increase the dependence of the SOA gain on polarization of the input signal? How can you achieve your goal?

**12.57** Define *noise figure* and explain its meaning with regard to amplifiers.

**12.58** Calculate the noise figure of an optical amplifier if the input-signal power is 250 μW, the input-noise power is 25 nW in a 1-nm bandwidth, the output-signal power is 50 mW, and the output-noise power is 15 μW in a 1-nm bandwidth.

**12.59** What is amplified spontaneous emission? What physical mechanism causes this phenomenon?

**12.60** A spontaneous-emission factor ($n_{sp}$), or population-inversion factor, is determined by Formula 12.23. Do we want the value of $n_{sp}$ to be large or small? Explain. What is the minimum value of $n_{sp}$?

**12.61** Calculate the power of ASE generated by an SOA operating at 1550 nm with a gain of 23 dB and a 40-nm bandwidth. What is the ASE power allocated within the bandwidth of an individual channel with a spectral width of 10 MHz?

**12.62** Is there any difference between optical and electrical noise?

**12.63** Look at the data sheet of the SOA presented in Figure 12.24. Is this an FPA or a TWA amplifier? Explain.

**12.64** The SOA data sheet shown in Figure 12.24 includes the value of saturation power. Is this input or output saturation power? Explain.

**12.65** Show the structure of an erbium-doped fiber amplifier and explain the function of each component.

**12.66** Explain how stimulation emission occurs in an EDFA.

**12.67** Why is erbium, not another material, used in an EDFA?

**12.68** What phenomenon determines the bandwidth of an EDFA?

**12.69** Which pumping wavelength—980 nm or 1480 nm—do you prefer and why?

**12.70** Calculate the gain of an EDFA if the input power is 250 μW, the output power is 55 mW, and the noise power is 20 μW.

**12.71** What is gain flatness? For which functional types of EDFA is gain flatness the critical parameter? Explain.

**12.72** In regard to an EDFA's gain saturation, the intermediate energy band is depleted rapidly. On the other hand, the lifetime of erbium ions at this level is about 10 ms, which is a very long time. How, therefore, can this level, where erbium ions survive so long, be depleted rapidly?

**12.73** For an SOA, the longer the length of an active medium, the higher the gain. Is this also true for an EDFA? Explain.

**12.74** Do we want a large or small amount of pumping power? Explain.

**12.75** How can we pump more power into an EDFA?

**12.76** Estimate the noise figure of an EDFA.

**12.77** Calculate the noise figure of an EDFA if the input-signal power is 300 μW, the input-noise power is 30 nW in a 1-nm bandwidth, the output-signal power is 60 mW, and the output-noise power is 20 μW in a 1-nm bandwidth.

**12.78** Calculate the ASE power generated by an EDFA working in the C-band with a gain of 30 dB.

**12.79** Compare the major peculiarities of an erbium-doped fiber with those of a regular singlemode transmission fiber.

**12.80** Calculate the connection losses caused by splicing (a) a low-PMD singlemode fiber manufactured by Plasma Optical Fibre and (b) Pirelli's EDF-BAX-01 erbium-doped fiber.

**12.81** How can we reduce the connection loss caused by splicing erbium-doped and regular singlemode fibers?

**12.82** Draw a functional block diagram of a pump laser module and explain the function of each component.

**12.83** Analyze the schematic of the Pirelli pump laser module given in Figure 12.37.

**12.84** How can you troubleshoot an EDFA?

**12.85** Based on the discussion developed in this section, draw a detailed block diagram of an EDFA.

**12.86** Are there optical fiber amplifiers other than EDFAs? If so, what types are they and why do we need them?

**12.87** *Project:* Build a memory map. (See Problem 1.20.)

## REFERENCES[1]


1. Gregory Borodaty, "The Arrival of 10 Gb/s," *Fiberoptic Product News,* July 1999, pp. 17–20.
2. Alan Willner, "Mining the optical bandwidth for a terabit per second," *IEEE Spectrum*, April 1997, pp. 32–41.
3. "Now armored to the teeth, the TeraMux still rips along at 1.28 Tb/s," advertisement by Pirelli Telecom Systems Div., Lexington, S.C., in *Fiberoptic Product News,* July 1999, p. 24.
4. Ahmed Khan, *The Telecommunications Fact Book and Illustrated Dictionary*, Albany, N.Y.: Delmar Publishers, 1992.
5. Rajiv Ramaswami and Kumar Sivarajan, *Optical Networks: A Practical Perspective,* San Francisco: Morgan Kaufman, 1998.
6. Rajiv Ramaswami, "Multiwavelength Optical Networking," *Tutorial 1,* IEEE Infocom '99, New York, March 21–25, 1999.
7. Dennis J. G. Mestdagh, *Fundamentals of Multiaccess Optical Fiber Networks,* Boston: Artech House, 1995.
8. *Dense Wavelength Division Multiplexer (DWDM) 100 GHz 16-Channel Module* (data sheet), E-TEK Dynamics, San Jose, Calif., 1999.
9. Hiroshi Yoshimura, Ken-ichi Sato, and Noboru Takachio, "Future Photonic Transport Networks Based on WDM Technologies," *IEEE Communications Magazine,* February 1999, pp. 78–81.
10. *Introduction to WDM Testing, EXFO*, Vanier, Que., Can., September 1997.
11. Martin Zirngibl, "Multifrequency Lasers and Applications in WDM Networks," *IEEE Communications Magazine*, December 1998, pp. 39–41.
12. *MG9637A/9638A Tunable laser source* (data sheet), Anritsu Corporation, Tokyo, Japan, 1997.
13. Robert Plastow, "Tunable lasers enable new optical networks to meet changing demands," *Lightwave,* December 1998, pp. 77–79.
14. *Anywave 40-ITU (Tunable Semiconductor Laser Module—ITU Grid),* (data sheet), Altitun AB, Kista, Sweden, February 1998.
15. Jens Buus, "WDM Sources," Tutorial ThC, *Tutorial Sessions of the Optical Fiber Communication Conference*, Baltimore, March 5–10, 2000.
16. Frank Tong, "Multiwavelength Receivers for WDM Systems," *IEEE Communications Magazine,* December 1998, pp. 42–49.
17. Hiroji Masuda, "Review of wideband hybrid amplifiers," Paper TuA1, *Technical Digest of the Optical Fiber Communication Conference,* Baltimore, March 5–10, 2000.
18. Govind Agrawal, *Fiber-Optic Communication Systems,* 2d ed., New York: John Wiley & Sons, 1997.
19. P. C. Becker, N. A. Olsson, and J. R. Simpson, *Erbium-Doped Fiber Amplifiers,* San Diego: Academic Press, 1999.
20. Leonid Kazovsky, Sergio Benedetto, and Alan Willner, *Optical Fiber Communication Systems,* Boston: Artech House, 1996.
21. Dennis Derickson, ed., *Fiber Optic Test and Measurement*, Upper Saddle River, N.J.: Prentice Hall PTR, 1998.
22. *EFA-R35 General-Purpose Preamplifier* (data sheet), MPB Technologies Inc., Pointe Claire, Que., Can., 1997.
23. Felton Flood and George Wildeman, "Moving Up in the Wavelength World," *Photonics Spectra,* February 2000, pp. 114–116.
24. André Girard, "Review of WDM Testing Issues," *Lightwave Test & Measurement Reference Guide,* EXFO Electro-Optical Engineering, Inc., Vanier, Que., Can., 1999.
25. *Rare-Earth-Doped Optical Fibers* (data sheet), INO, Sainte-Foy, Que., Can., 1998.
26. "Singlemode and PM Amplifier Fibers," *Fiber optic assemblies & components* (catalog), Wave Optics, Inc., Mountain View, Calif., 1998.
27. "DF1500F Amplifier Fibre," *Advanced fibre optic products* (catalog), Fiberscore Ltd., Southampton, Hampshire, U.K., 1998.
28. Norio Kashima, *Passive Optical Components for Optical Fiber Transmission*, Boston: Artech House, 1995.
29. *263-Type 0.98 μm Pump Laser Module* (data sheet), Lucent Technologies, Microelectronics Group, Allentown, Pa., April 1998.
30. *FOL1402PA Series 1480 nm High Power Pump Laser Diode Module* (data sheet), JDS Fitel, Nepean, Ont., Can., 1997.
31. *980 nm Laser Chip on Submount* and *980 nm Laser Module* (data sheets), Pirelli Cavi e Sistemi SpA, Milano, Italy, April 1998.


[1] See Appendix C: A Selected Bibliography

**32.** Curtis Eugene Pass, Lucent Technologies, Optoelectronics Products, Breinigsville, Pa., Private communication, April 1999.

**33.** *Flat and Stabilized Gain EDFA for DWDM Transmissions* (data sheet), HighWave Technologies, Lannion Cedex, France, 1999.

**34.** *Alcatel 1920 OFA* (data sheet), Alcatel Optronics, Nozay Cedex, France, 1998.

**35.** *KOFA 1550-L Linear Er-Doped Fiber Amplifier* (data sheet), Kumho Telecom, Inc., Kwangju, Korea, 1998.

**36.** *OA915 Series Optical Gain Module* and *ErFA-1200 Series Erbium-Doped Fiber Amplifiers* (data sheets), JDS Fitel, Nepean, Ont., Can., 1997.

**37.** Joseph Chon, "Multiple-wavelength pumping overcomes increased channel counts," *Lightwave,* February 1999, pp. 112–116.

**38.** *High power, high reliability multiplexed EDFA pump, SDLO-WM series* (data sheet), SDL, Inc., San Jose, Calif., November 1998.

**39.** "Fluoride & Telluride-Based Optical Amplifier Module and Fiber Module," *NEL Photonic Devices*, NTT Electronics, Ibaraki, Japan, July 1998, pp. 13–20.

**40.** *FluoroAmp 1310 Gain Flattened Optical Amplifier* (data sheet), Galileo Corp., Sturbridge, Mass., 1997.

**41.** *Er-Doped Fluoride Fiber Amplifier* (data sheet), KDD R&D Laboratories Inc., Saitama, Japan, 1998.

# 13

# Passive Components, Switches, and Functional Modules of Fiber-Optic Networks

Chapter 12, we trust, has convinced you that we need precisely attuned components to make a fiber-optic WDM network operational. Up to now we've only considered components that have to be powered to work: transmitters, receivers, modulators, and optical amplifiers. These are *active* components, a category that also includes switches.

Another class of elements used in WDM networks does not require external power. They use, rather, the light power of an input signal to function; this is why they are called *passive* components. This section is largely devoted to their construction and operation in WDM networks. In addition, we will consider other active components, including switches, optical add/drop modules, and optical cross-connects.

## 13.1 COUPLERS/SPLITTERS

Couplers are devices that combine light from several fibers. Splitters, as the name suggests, separate light into several fibers. These passive components themselves are absolutely necessary for fiber-optic networks. What's more, they serve as building blocks for functional modules. Understanding their principle of operation is important for appreciation of the working principles of many passive components.

### Fused Biconical Taper (FBT) Couplers—Their Principle of Operation

***Couplers: What they are*** *Couplers are devices that combine light from different fibers.* A simple $2 \times 2$ configuration of this device, with its internal structure, is shown in Figure 13.1.

Since this type of fiber allows light to pass through in one direction, it is called a *directional coupler.* The term *bidirectional coupler* indicates, obviously, that a component allows light to pass through in two opposite directions.

A coupler is manufactured by placing two or more fibers adjacent to one another, then fusing and stretching them, thus creating a coupling region ([1], [2]). The heated area is stretched until the desired coupling properties are achieved. This device is called a *fused biconical taper (FBT)* coupler.

When you elongate the heated fiber, you decrease its core diameter, which means the $V$ number becomes smaller. The $V$ number, you recall, is defined in Formula 4.50 as

$$V = [(2\pi a\ n)/\lambda]\ \sqrt{2\Delta}, \tag{13.1}$$

where $2a = d$ is the fiber core diameter, $\lambda$ is the operating wavelength, and $n$ and $\Delta$ are the average and relative refractive indexes, respectively. Thus, since all the constituents of Formula 13.1, except $2a$, are constant during heating and stretching, the $V$ number decreases as $2a$ becomes smaller. But the $V$ number determines how tightly light power is confined within a core. (See Figure 4.15.) In other words, the smaller the $V$ number, the more the mode-field diameter (MFD) exceeds the fiber-core diameter. (See Figure 5.3.) Hence, by decreasing the core diameter, we allow a greater portion of an optical mode to propagate in the cladding, where this mode will couple to the core of another fiber. This explanation is illustrated in Figure 13.1.

The coupling process occurs gradually as the field diameter of an input mode becomes larger in the down-taper region. Within the coupling region, the optical mode from one core couples to another core because both cores are very close to each other, as Figure 13.1 shows. In the up-taper region, where the fiber-core diameters increase, modes become more and more confined within the cores and eventually the two separated modes leave the outputs of the separated fibers.

Sometimes two fibers are twisted before heating and stretching [3]; another approach is to polish the fiber side, which allows the designer to control the coupling effect very precisely [4].

What portion of input light will be coupled to the second fiber depends on the distance between the two cores, the core diameters within the coupling region, and the operating wavelengths. Hence, by carefully sizing the coupling region, you will control the ratio of the output powers, called the *coupling ratio*. A coupling ratio of 50:50 is very popular; however, any ratio from 1:99 to 50:50 can be employed. Tap couplers with a typical ratio of 1:99 are used to monitor the input and output signals in EDFAs, as we already learned. (See Figure 12.40.)

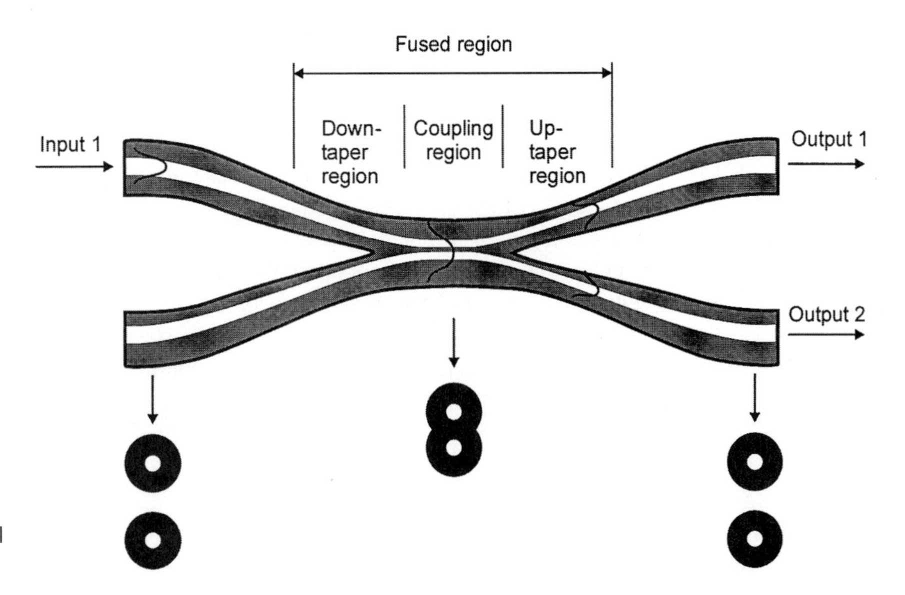


**Figure 13.1** Fused biconical taper (FBT) coupler.

How can we achieve, say, a 50:50 split? Simply make the coupling region so tiny that no cores can be distinguished. In such an arrangement, the optical mode will propagate through the combined cladding of the two fibers and will be separated in the up-tapered region. To attain, say, a 1:99 split, we do just the opposite; that is, we keep both cores well pronounced but move the receiving one close to the transmitting fiber. By controlling the distance between them, we can control the ratio.

***Advantages of FBT couplers*** FBT couplers boast three key advantages. First, the fiber coupling is a low-loss process. In fact, the conversion from core mode to coupling mode and back is itself theoretically lossless; hence, we have to account for losses caused by light propagation through a short cladding length. However, the insertion loss of an assembled coupler is rather high and depends on the coupling ratio.

Second, light never leaves the fiber structure during the coupling process, so it never encounters any interface. Thus, this type of coupler is intrinsically free from backreflection. In fact, data sheets for this type of coupler do not include this specification.

Third, since FBT couplers are made from regular fiber, connectorizing an FBT coupler with a transmission fiber is an easy, low-loss procedure.

***Port configurations of couplers*** You can easily imagine a number of fiber combinations that could be coupled by such a device. Some are shown in Figure 13.2.

A $2 \times 1$ coupler is used to combine two light inputs into a single fiber, as Figure 13.2(a) shows. When you change the direction of light propagation, this device will split one optical signal into two, as you can see in Figure 13.2(b). In this operation, the coupler is called a *splitter* in accordance with the function it performs. There are couplers that couple, or split, $1 \times N$ or $N \times 1$ ports, respectively, as Figure 13.2(c) shows. They are called *tree* couplers and may have an $N \times M$ configuration. An important coupler type for a WDM network is a *star* coupler, shown in Figure 13.2(d), where the same number of ports serve as inputs and outputs. In a sense, a star coupler is an $N \times N$ *bidirectional* coupler. However, a star coupler can be built as an $N \times M$ *unidirectional* coupler.

A coupler with a 50:50 output ratio—that is, a splitter—is called a *3-dB* coupler for obvious reasons. This simple device can be—and actually is—used as the basic building block of tree and star couplers. However, this is not the best approach to take because you need $M = (N/2) \log_2 N$ 3-dB couplers to make an $N \times N$ star coupler and only $1/N$ portion of the power launched into each port will appear at every output [5]. This is why modern tree and star couplers for broadcast WDM networks are fabricated directly using the FBT technique.

***A few words about coupler theory*** A coupler, like any waveguiding device, can be described based on Maxwell's equations [6]. One of the results of such an approach—power-transfer, or coupling, coefficient—is given on pages 599 through 605. Another theoretical approach, one actually being used in practice, is based on couple-mode theory and it enables us to determine a coupler's power-transfer functions, that is, the coupler's transmittances. These are usually presented in the form of a scattering matrix ([6], [7]). Using the principle of conservation of energy, we draw two important conclusions: First, *lossless coupling is impossible.* (We specify on page 595 the types of losses associated with the coupling process.) Secondly, *two outputs of a 50:50 (3-dB) coupler/splitter have the same amplitude but a relative phase shift of $\pi/2$* [7]. This is a very important point to bear in mind when analyzing a coupler's data sheet.

## Reading a Data Sheet

Let's now work our way step by step through a typical coupler data sheet. Figure 13.3 is a good example. If you can read it readily, you should have no problem with any other you will encounter in your job.

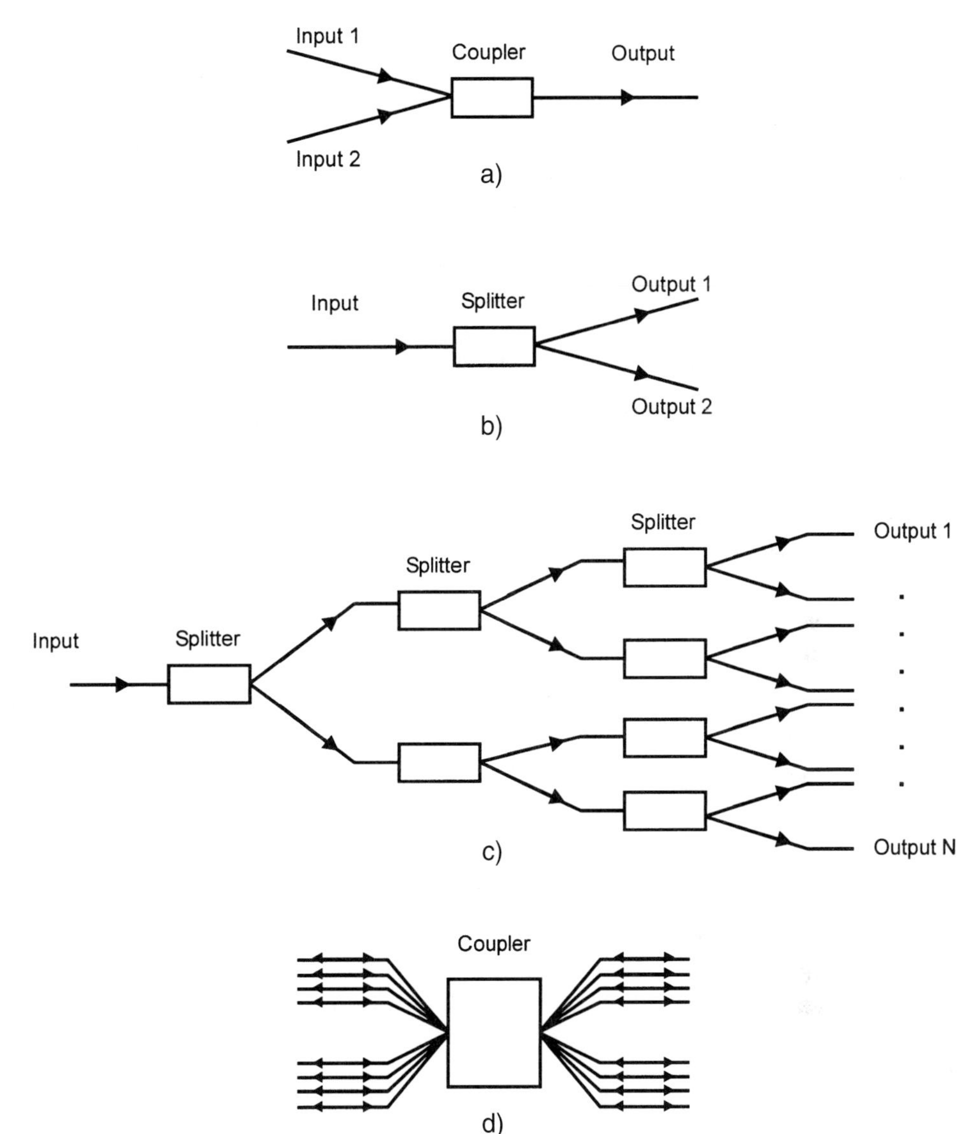


**Figure 13.2** Coupler port configurations: (a) 2 × 1 coupler; (b) 1 × 2 splitter; (c) tree coupler; (d) star coupler.

The "DESCRIPTION" section is self-explanatory, but pay attention to the operating wavelength: A coupler works at either 1310 nm *or* 1550 nm. The spectral performance of the device is described by the graph "Insertion Loss versus Wavelength." Sometimes a data sheet includes a specification like "a wide 80-nm bandwidth," which implies that the characteristics of this coupler apply not for a single wavelength but for the 80-nm range of wavelengths around either 1310 nm or 1550 nm. By all means possible, the manufacturer wants to inform you that all parameters of the device are within their specifications over a certain range of operating wavelengths. We'll discuss wavelength dependence shortly.

Observe the physical appearance of the device. Miniature versions of couplers are widely available.

***Characteristics of a WDM coupler*** To clarify the performance specifications shown in the data sheet, we'll turn now to Figure 13.14, where a four-port coupler operating at two wavelengths is considered. Using this example of a wavelength-division multiplexing (WDM) coupler, we'll introduce the following characteristics, which describe the properties of a WDM coupler:

## DESCRIPTION

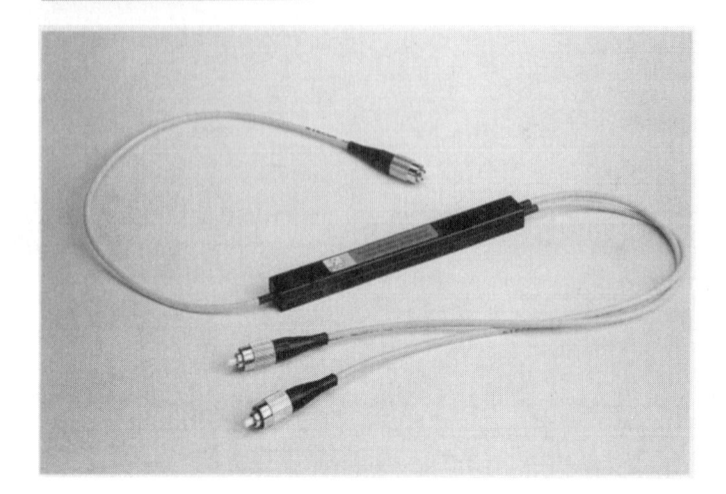

FOCI's Standard Singlemode Couplers are used to split light with minimal loss from one into two fibers, or to combine light from two fibers onto one fiber. Designed to operate at a single wavelength, typically 1310nm or 1550nm, these components are excellent for duplex transmission on a single fiber, CATV systems or within fiber optic testing sets and trunk/loop branching.

Available in both 1 x 2 and 2 x 2 configurations with splitting ratios customer specified between 1% : 99% and 50% : 50%. Various types of pigtailing, connector termination and packaging configurations are available to meet your requirements.

These ultra reliable devices feature low backreflection, low insertion loss and high port isolation over a wide range of temperatures and wavelengths.

## FEATURES

- Low insertion loss
- High port isolation
- High directivity
- Custom defined specifications
- Environmentally stable

## APPLICATIONS

- Telecommunications
- Local area network
- Fiber to the home
- Video transmission
- Fiber optic sensing
- Testing instruments
- CATV
- Point to point systems
- Wide area networks

**Figure 13.3** A coupler data sheet. *(Courtesy of FOCI Fiber Optic Communications, Inc., Hsinchu Taiwan, R.O.C.)*

STANDARD
SINGLEMODE COUPLERS

## PERFORMANCE SPECIFICATIONS

| ITEM | | Standard Singlemode Couplers | |
|---|---|---|---|
| Operating Wavelength, nm | | 1310 or 1550 | |
| Grade | | Super ( S ) | High ( H ) |
| Typical Excess Loss, dB | | 0.06 | 0.15 |
| Uniformity, dB (50:50) | | 0.5 | 0.9 |
| Thermal Stability, dB (peak-peak) | | <0.2 | <0.3 |
| Polarization Stability, dB | | <0.1 | <0.15 |
| Port Configuration | | 1 x 2 or 2 x 2 | |
| Coupling Ratio | | 1:99 to 50:50, (50:50 standard) | |
| Insertion Loss, dB | | Please refer to the coupling ratio vs. Insertion loss chart | |
| Directivity, dB | | >50 (1 x 2), >60 (2 x 2) | |
| Reflectance, dB | | <-55 | |
| Operating Temperature, °C | | -40 ~ +85(*) | |
| Storage Temperature, °C | | -55 ~ +85 | |
| Package Options (for different pigtailing) | 1. coated fiber (250 μm) | T2, MA, MB | |
| | 2. loose tube (900 μm) | TB, MA, MB | |
| | 3. PVC cable (3.0 mm) | A1, MA, MB | |

Note 1. The packaging option codes are explained in Packaging Dimensions below.
2. *-20°C ~ +70°C for PVC cable.

## PACKAGING DIMENSION

The dimensions of various packaging options are given below. For further detailed mechanical drawings, please refer to Appendix A:

- T2 .......Ø3.0 x 53 (mm)
- TB .......Ø3.8 x 70 (mm)
- A1 .......101 x 12 x 10 (mm)
- MA ......Inter-rack 4U
- MB ......154 x 110 x 16 (mm)

**Figure 13.3** (continued)

## COUPLING RATIO VS. INSERTION LOSS

| Coupling Ratio (%) | Insertion Loss (dB) | |
|---|---|---|
| | Super Grade ( S ) | High Grade ( H ) |
| 50 / 50 | 3.4 | 3.6 |
| 40 / 60 | 4.4 / 2.5 | 4.7 / 2.8 |
| 30 / 70 | 5.7 / 1.8 | 6.0 / 2.0 |
| 20 / 80 | 7.5 / 1.2 | 8.0 / 1.4 |
| 10 / 90 | 10.8 / 0.7 | 11.5 / 0.9 |
| 5 / 95 | 14.6 / 0.4 | 15.5 / 0.6 |
| 1 / 99 | 21.6 / 0.2 | 22.0 / 0.3 |

## ENVIRONMENTAL RELIABILITY TEST

The performance of the couplers has been carefully designed to meet the toughest Bellcore standards

***TA-NWT-001221***

in the tests of:

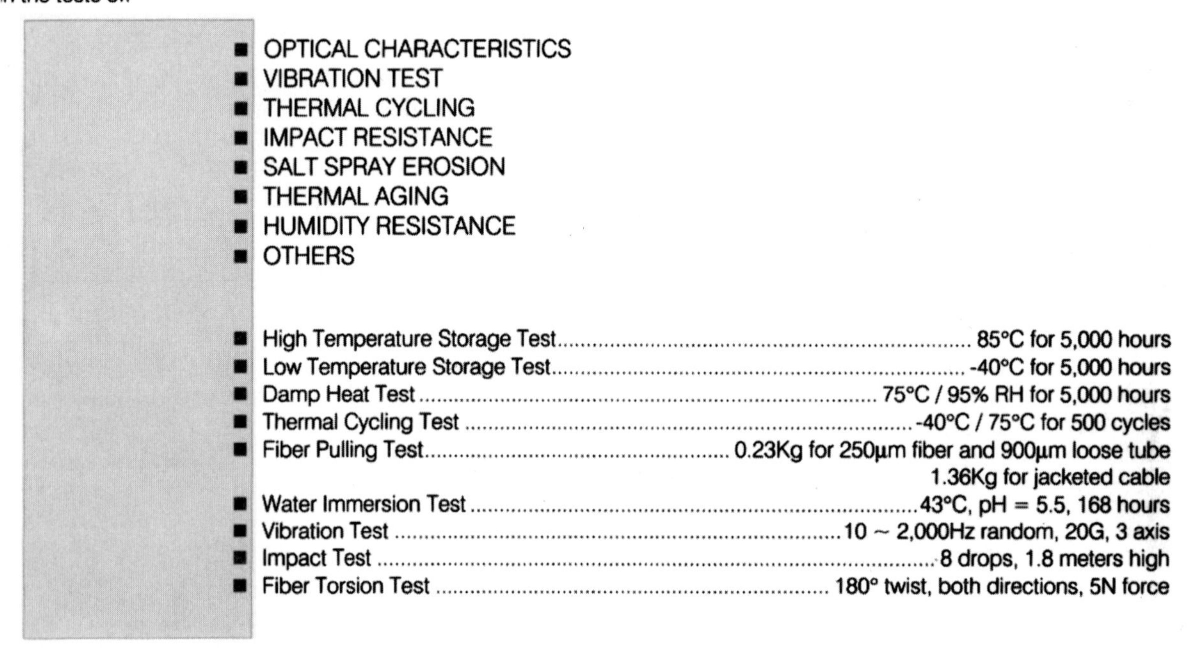

OPTICAL CHARACTERISTICS
VIBRATION TEST
THERMAL CYCLING
IMPACT RESISTANCE
SALT SPRAY EROSION
THERMAL AGING
HUMIDITY RESISTANCE
OTHERS
High Temperature Storage Test .......... 85°C for 5,000 hours
Low Temperature Storage Test .......... -40°C for 5,000 hours
Damp Heat Test .......... 75°C / 95% RH for 5,000 hours
Thermal Cycling Test .......... -40°C / 75°C for 500 cycles
Fiber Pulling Test .......... 0.23Kg for 250μm fiber and 900μm loose tube
1.36Kg for jacketed cable
Water Immersion Test .......... 43°C, pH = 5.5, 168 hours
Vibration Test .......... 10 ~ 2,000Hz random, 20G, 3 axis
Impact Test .......... 8 drops, 1.8 meters high
Fiber Torsion Test .......... 180° twist, both directions, 5N force


**Figure 13.3** (continued)

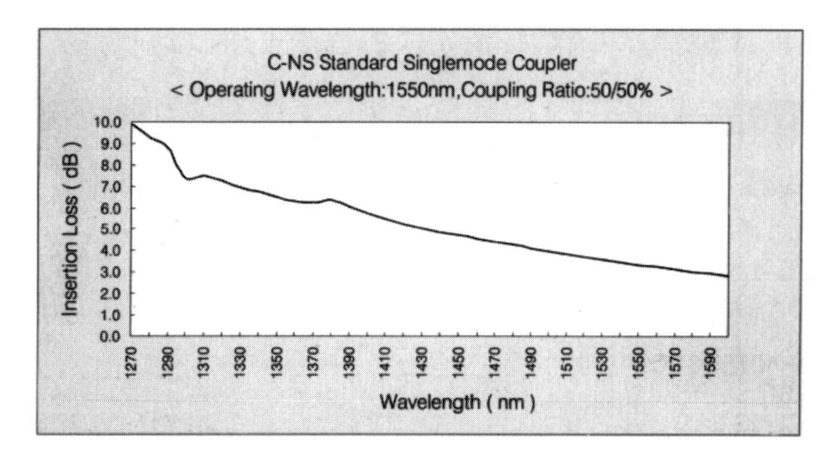


**Figure 13.3** (continued)

*Excess loss* ($P_{ex}$) is defined as

$$P_{ex}(\text{dB}) = -10\log[(\sum_{j} P_j)/P_i], \tag{13.2}$$

where $P_j$ is the output power at port j and $P_i$ is the input power. This definition is illustrated in Figure 13.4(a); here, excess loss is equal to

$$P_{ex}(\text{dB}) = -10\log[(P_2 + P_3)/P_1] \tag{13.2a}$$

Ideally, the sum of all output power should be equal to the input power. Excess loss quantifies the deviation from this ideal state; therefore, excess loss should be as small as possible. For the coupler under discussion, a typical excess loss varies from 0.06 dB to 0.15 dB, depending on its type.

The *insertion loss (IL)* of a coupler is the ratio of output to input light power at a particular wavelength:

$$IL_{12}(\text{dB}) = -10\log[P_2/P_1] \tag{13.3}$$

Insertion loss is the loss that a coupler inserts between the input and output ports. (See Figure 13.4[b].) Insertion loss may or may not include connector loss; this fact is usually specified in the data sheet. A coupler's insertion loss is rather high. Compare the typical 3.4-dB loss for a coupler, 0.01-dB for a fusion splicer, and less than 1 dB for a connector. Insertion loss depends on the coupling ratio, as Figure 13.3 shows.

The *coupling ratio* is formally defined as (Figure 13.4[a])

$$CR(\text{dB}) = -10\log[P_2/(P_2 + P_3)] \tag{13.4}$$

This characteristic is often used to describe the property of a coupler. It can be given as an absolute number or as a percentage, as in Figure 13.3. In the latter case,

$$CR\ (\%) = [P_2/(P_2 + P_3)] \times 100 \tag{13.4a}$$

Note that

$$IL = CR + P_{ex} \tag{13.5}$$

*Uniformity* is the coupler characteristic used for equal split ratios. For example, an ideal $1 \times 2$ coupler would split the input power equally into two output ports. In reality, however, the power at each output port will vary from the 50:50 ratio. The physical reason for this inequality

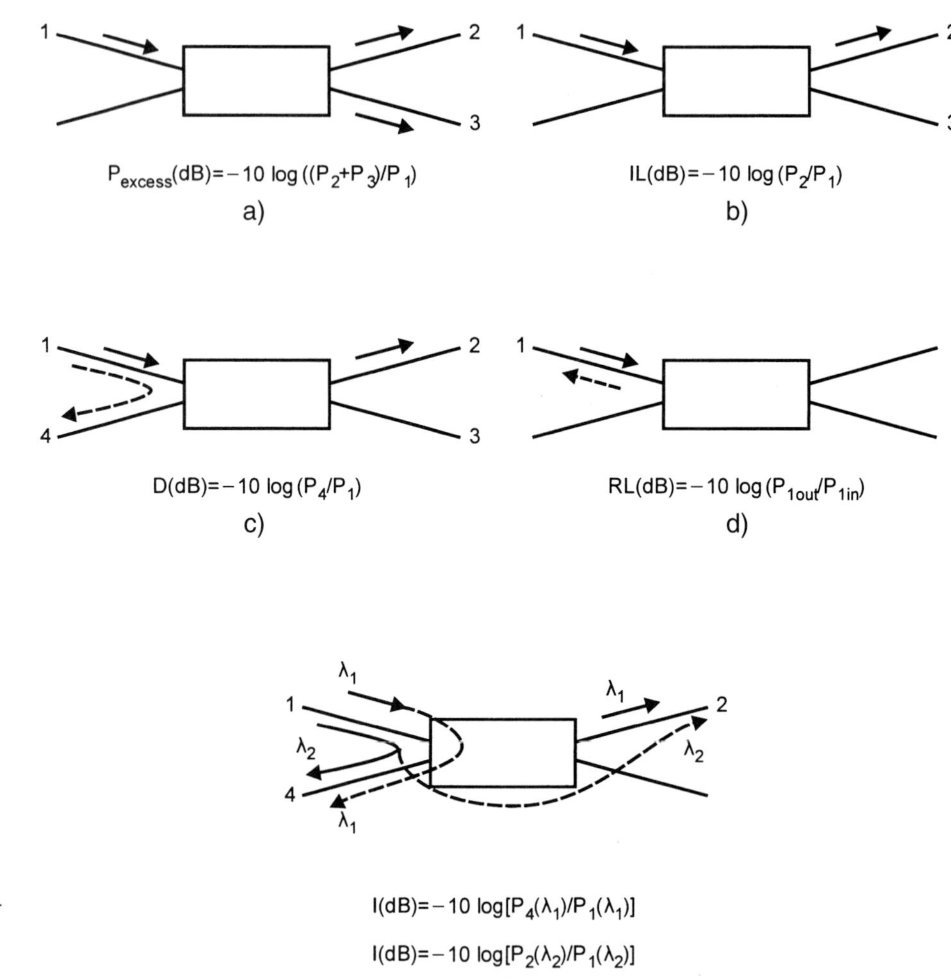


**Figure 13.4** Characteristics of a WDM coupler: (a) Excess loss; (b) insertion loss; (c) directivity; (d) return loss; (e) isolation.

is different insertion losses for the different couplings introduced during the fabrication process [8]. Thus, uniformity is defined as

$$Uniformity\ (\text{dB}) = IL_{\text{max}} - IL_{\text{min}} \tag{13.6}$$

From the definition of insertion loss, we get for a 50:50 coupler:

$$Uniformity(\text{dB}) = -10\ \log(P_2/P_1) - [-10\ \log(P_3/P_1)] = 10\ \log(P_3/P_2) \tag{13.6a}$$

This formula underscores the meaning of uniformity: It is a measure of the inequality of the split. Indeed, for the coupler discussed, the best uniformity is 0.5 dB, which means $P_3/P_2 = 1.12$, and the worst is 0.9 dB, which yields $P_3/P_2 = 1.23$. Quite obviously, we want the uniformity to be as small as possible; if uniformity is equal to 0, $P_3/P_2 = 1$, which is an ideal split.

It is important to note that we have discussed *power uniformity.* You will no doubt come across another characteristic of a coupler that is also called "uniformity," but that will refer to *spectral uniformity.* This characteristic describes how the coupling ratio changes over its operating wavelength range [9].

*Polarization stability,* or *PDL,* which stands for *polarization-dependent loss,* is a measure of the maximum variation of the insertion loss caused by changes in the polarization of the input light [10]. As you can see from Figure 13.3, variations in the insertion loss do not exceed 0.15 dB for any possible polarization state of the input signal.

*Directivity, D,* shows what portion of the launched power appears at the unwanted output. In Figure 13.4(c), light is directed from input 1 to output 2; however, some light penetrates another input port (4); this is an unwanted signal. Hence, directivity is given by

$$D\,(\text{dB}) = -10\log[P_4/P_1] \tag{13.7}$$

Backreflection causes this return signal.

Directivity can be given as a positive or negative number. It is a positive number in the data sheet shown in Figure 13.3; hence, the manufacturer uses a negative sign in Formula 13.7.

Be careful when faced with this characteristic because you may come upon a plethora of definitions of directivity in the course of your career. Sometimes directivity is defined as a measure of *near-end crosstalk* [10]. In this case, directivity in Figure 13.4(c) should be defined as

$$\textit{Near-end crosstalk}\,(\text{dB}) = -10\log\,[P_3(\lambda_1)/P_1(\lambda_1)], \tag{13.8}$$

Sometimes directivity is considered as another form of *isolation*—a characteristic that will be discussed shortly.

We want high directivity, which means we don't want to lose our input power to unwanted direction. Values from 50 dB to 60 dB, given in the data sheet in Figure 13.3, are typical for modern couplers.

*Reflectance,* or *return loss (RL),* of a coupler does not differ from the return loss of any connection. As discussed in Section 8.1 and shown in Figure 13.4(d), return loss is given by

$$RL\,(\text{dB}) = -10\log[P_{1out}/P_{1in}] \tag{13.9}$$

Again, we do not want input power to flow backward; hence, the return loss must be high. For the coupler shown in Figure 13.3, *RL* is not more than –55 dB.

When a coupler is designed to work with two or more wavelengths simultaneously (we call this a *WDM coupler*), you need to consult the specification in the data sheet called "isolation." *Isolation (I)* is a measure of how much light from one path is prevented from reaching another path. In the example shown in Figure 13.4(e), light at $\lambda_1$ is directed from port 1 to port 2, but, unfortunately, penetrates port 4. Similarly, propagation of light at $\lambda_2$ from port 1 to port 4 is desired but its appearance at port 2 is unwanted. "Isolation" is defined as ([10], [11], and [12]):

$$I_{41}\,(\text{dB}) = -10\log[P_4(\lambda_1)/P_1(\lambda_1)] \tag{13.10}$$

$$I_{21}\,(\text{dB}) = -10\log[P_2(\lambda_2)/P_1(\lambda_2)]$$

In other words, isolation shows how much of the signal at $\lambda_1$ reaches the output of signal $\lambda_2$ and vice versa. Since this is an unwanted effect, isolation should be a large dB number (typically, on the order of 30 to 40 dB).

Plain FBT couplers work, for the most part, at one fixed wavelength, that is, as *achromatic couplers.* This is the reason why the data sheet in Figure 13.3 does not give an isolation specification, even though the manufacturer has mentioned high port isolation in the "FEATURES" section.

It would seem that talking about isolation for a device whose main function is coupling light from one path to another is meaningless. That is true enough for an achromatic coupler. But, again, when you want to couple two or more different wavelengths simultaneously, you need to isolate them, as the wavelength-division procedure implies. Look at the EDFA WDM coupler in Figures 12.25 and 12.41. If its isolation is 30 dB, this means 0.1% of the light power at 980 nm penetrates the lightpath of the 1550-nm signal. But typical pump power is about 160 mW, which means 0.16 mW at 980 nm will propagate along the transmission fiber. Do we want that? Certainly not, because this foreign signal will eventually result in an increase in noise.

The wavelength and bandwidth specifications have already been discussed. All the other specifications in this data sheet are clear and do not require detailed discussion.

Let's turn, then, to the section headed "FEATURES" to clarify the meaning of each term. To begin, we will describe other coupler types in the field.

***Classification of fiber couplers*** First, there are *multimode and singlemode fiber couplers.* They are distinguished by their ability to work either with multimode or singlemode fibers. Both types are available in a variety of configurations, as Figure 13.2 illustrates.

Secondly, we find *polarization-maintaining (PM) couplers,* which work with PM fibers [13]. One such device is called a "polarization-maintaining fixed-ratio evanescent-wave coupler." (The term *evanescent wave* refers to the tail of an optical mode that propagates in a cladding and thus penetrates an adjacent fiber in the coupler—see Sections 4.3 and 6.2.)

As you can see from the title of the data sheet, couplers are either *fixed-ratio or variable-ratio devices.* Most couplers in the field are of the fixed-ratio type; that is, you may have 50:50 or 1:99 or any other predetermined ratio.

***All-fiber, micro-optic, and waveguide fabrication techniques*** The following discussion and terminology are applied to all passive components of fiber-optic networks. FBT couplers are usually called *all-fiber components* because, as we have seen, no other elements but fiber are involved in the fabrication of such couplers. However, couplers, as well as any other passive components, can be fabricated using techniques different from FBT. The most popular are *micro-optic* and *waveguide couplers.*

*Micro-optic* generally refers to a technique based on common (bulk) optics technology but implemented in a "micro" version. For example, micro-optic couplers are built using microlenses and a beam splitter ([14], [15], and [16]). See Figure 13.5(a).

You can easily place port 3 at the same side of a coupler by adjusting the mirror inside the unit. This is just an example of how you can tinker with the micro-optic technique to design a coupler with the required configuration and characteristics. This technique is well developed; in fact, it was the first technique used to fabricate passive components for fiber-optic communications networks. The main problem with it stems from the nature of micro-optic components: They are different from a fiber in geometrical and optical parameters. Thus, problems with coupling and losses abound.

*Waveguide* is usually referred to as a solid dielectric light conductor. Light waveguides are fabricated on a substrate very much the way a semiconductor chip is made. (See Figures 13.5[b] and [c].) To make a waveguide coupler, two planar waveguides are fabricated on a substrate. The geometry of the waveguides is very similar to an FBT coupler's configuration; the same is true for its principle of operation. Waveguide couplers and other waveguide components are easy to fabricate and their parameters are easy to control, but they have one inherent disadvantage: Coupling a cylindrical beam from a fiber into a planar (essentially rectangular, as Figure 13.5[c] shows) waveguide structure and back is an enormous challenge.

Micro-optic and waveguide couplers are very popular, too. To read more about them, see [6], [14], and [17].

The advantages of the three major coupler types are summarized in Table 13.1.

Of course, all couplers are not free of disadvantages. For example, it is hard to achieve a high port count in FBT and micro-optic couplers and the fabrication of a planar waveguide requires a large capital investment. In general, FBT couplers are most suitable for low-cost, low-port-count

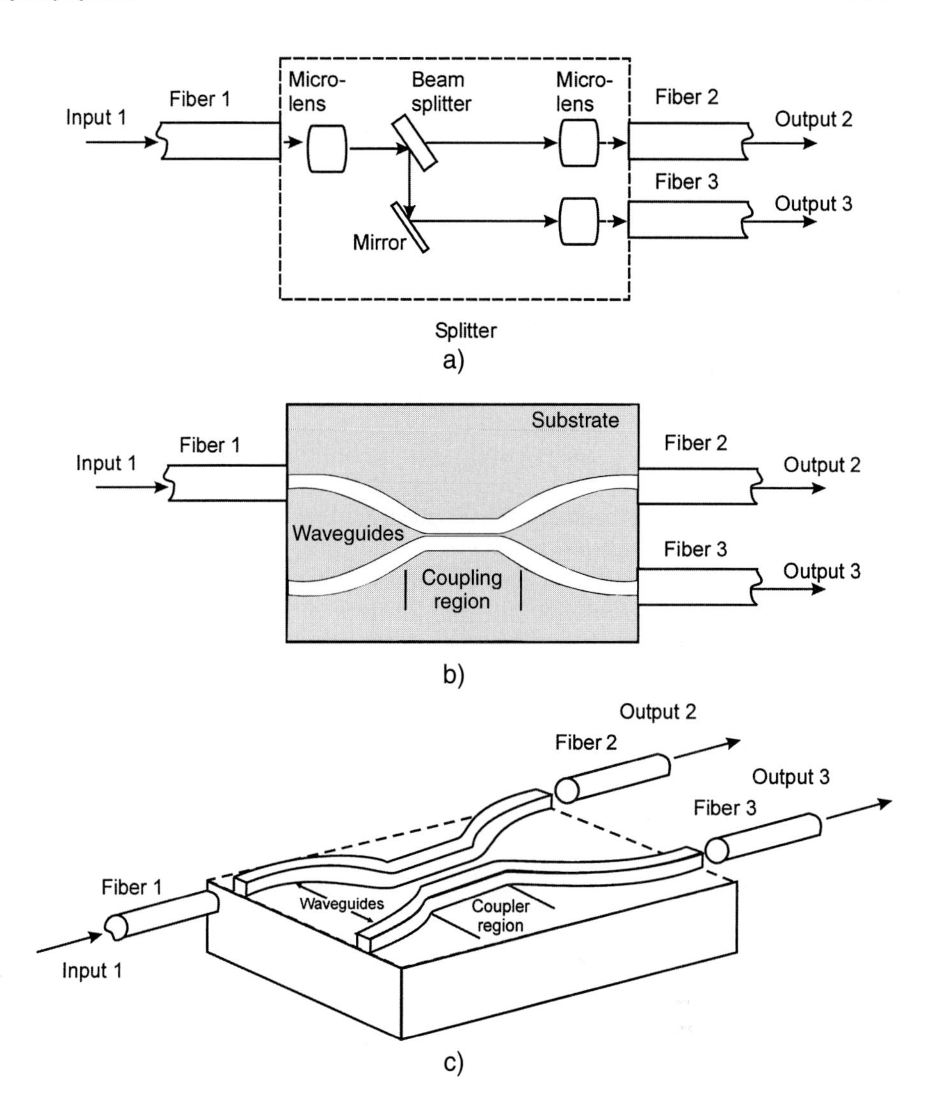


**Figure 13.5** Micro-optic and waveguide couplers: (a) Micro-optic couplers; (b) waveguide couplers; (c) construction of a waveguide coupler.

applications. Waveguide couplers work best in networks having 16 or more ports. Micro-optic couplers are good for high-performance communications systems. FBT couplers are best when it comes to the capability of being integrated into fiber-optic networks.

## FBT Couplers: How to Make a WDM Coupler

The devices we have discussed the most so far are the *achromatic couplers* because they couple signals having the same wavelength. Their advantages include a simple structure, low insertion and return losses, and ease in coupling to transmission fibers. The big question is whether we can use these couplers for wavelength-division multiplexing (WDM). The answer is yes. To understand how, we need to relate what happens to the input power as it propagates along the coupling region of an FBT coupler.

Consider the 1 × 2 FBT lossless coupler shown in Figure 13.1. Solving Maxwell's equations for a lossless coupler yields([6], [18]):

$$P_1 = P_{in} \cos^2[k(\lambda)z] \tag{13.11}$$

$$P_2 = P_{in} \sin^2[k(\lambda)z], \tag{13.12}$$

**Table 13.1** Comparison of FBT, micro-optic, and waveguide couplers

| | FBT | Waveguide | Micro-optic |
|---|---|---|---|
| Cost | Low | Medium | High |
| Insertion loss | Low | Low | High |
| PDL | Low | High | Medium |
| Isolation | Low | Medium | High |

where $P_1$ and $P_2$ are the powers of light within the fibers, $P_{in}$ is the launched power, and $z$ is the distance along the coupling region. For a lossless coupler,

$$P_1 + P_2 = P_{in} \tag{13.13}$$

Here, coupling of one optical mode to two identical fibers is assumed. The constant $k(\lambda)$ is called the *coupling coefficient.*

The main point here is this: *Along the coupling region, power from the first fiber may be completely transferred to the second fiber and back.* In other words, there is a periodic power exchange between two coupled fibers. Indeed, the raised sine and cosine functions can be expressed as

$$\cos^2\theta = \tfrac{1}{2}(1+\cos 2\theta) \tag{13.14}$$

$$\sin^2\theta = \tfrac{1}{2}(1-\cos 2\theta), \tag{13.15}$$

where $\theta = k(\lambda)z$. If you want to obtain $P_1 = 0$ and $P_2 = P_{in}$, make $k(\lambda)z = \pi/2$. Since $\cos\pi = -1$, you will get the desired result. If you want a 50:50 power distribution, make $k(\lambda)z = \pi/4$ and see from the above formulas that $P_1 = P_2 = \tfrac{1}{2} P_{in}$. These explanations are illustrated in Figure 13.6.

At this point, let us pause to clear up some possible confusion: When we explained the principle of operation of an FBT coupler, we said that coupling occurs because two cores are made small in diameter and are placed close to each other so that we can control the coupling ratio by changing the size of the core diameters and the distance between them. True enough, but now we are saying we can control the coupling ratio by changing the length of the coupling region and, in addition, take into account the wavelength dependence of coupling parameters. Are we contradicting ourselves? No.

Paradoxical as it may seem, both statements are true. When we discussed the principle of operation of an FBT coupler, we intentionally did not mention the role of the length of a coupling region because we did not wish to complicate the explanation. Indeed, the coupling ratio at the given *cross section of a coupling region* really depends on the size of the core diameters, the distance between the cores, and the wavelength. Apart from all this, the *coupling ratio changes along the length of the coupling region;* that is, the light power periodically transfers from one coupled fiber to another and back, as Formulas 13.11 and 13.12 show and as Figure 13.6 illustrates.

The coupling ratio at a given cross section of the coupling region is determined by the coupling coefficient, $k(\lambda)$. (Does this clarify that $k(\lambda)$ depends on the size of the core diameters, the distance between the cores, and $\lambda$?) But what you actually get at the outputs of the fibers depends on the length of the coupling region. Indeed, in the situation shown in Figure 13.6(a), we see that all the input power can be transferred to output 1 ($P_1$) and nothing to output 2 ($P_2$). The minimum distance at which power from one fiber is transferred completely to another fiber is when $k(\lambda) z = \pi/2$, as Figure 13.6(a) shows. Hence,

$$L_c = \pi/[2k(\lambda)], \tag{13.16}$$

which is called the *coupling length* [18].

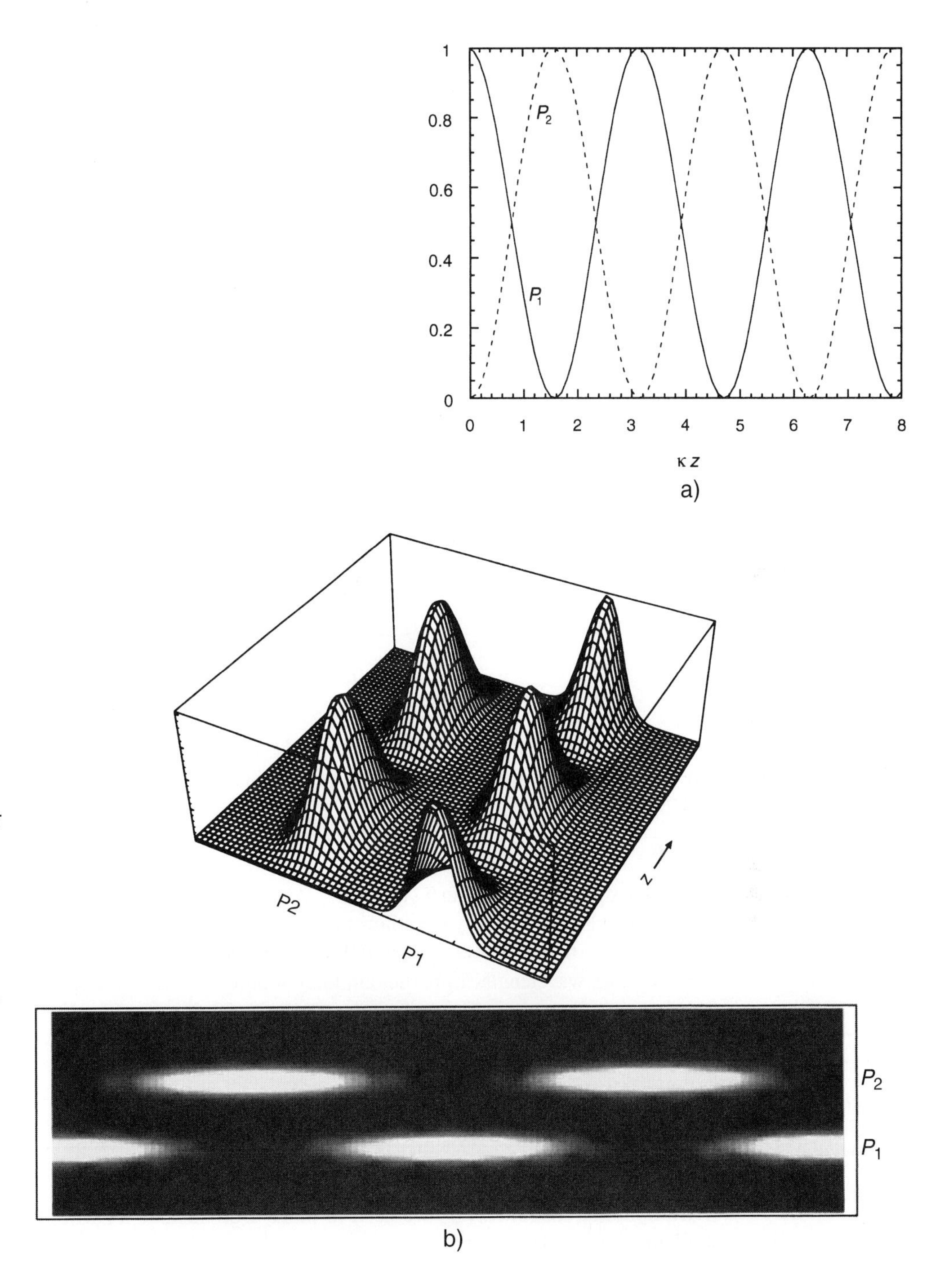


**Figure 13.6** Power variations in the coupling region of an FBT coupler: (a) $P_1$ and $P_2$ as a function of $k(\lambda)\, z$; (b) three-dimensional plot of power variation and density plot. *(Reprinted from Ajoy Ghatak and K. Thyagarajan's* Introduction to Fiber Optics. *Copyright 1998. Used with permission of Cambridge University Press.)*

Therefore, we have two parameters—that is, two options—in designing an FBT coupler: $k(\lambda)$ and $z$. Manipulating $k(\lambda)$ and $z$, we can effect power transfer from one fiber to another with a coupling ratio that is different for different wavelengths; in other words, we can obtain a WDM coupler. Look at Figure 13.6 again. At the same length of the coupling region ($z$), the power distribution for wavelengths $\lambda_1$ and $\lambda_2$ can be made different because $k$ depends on $\lambda$. Wavelength dependence appears here because MFD depends on the $V$ number and the $V$ number, in turn, depends on $\lambda$, as Formula 13.1 tells us. This dependence is crucial for producing WDM couplers and is how we make a WDM coupler using the FBT technique.

Indeed, having $k(\lambda_1)z = \pi/2 + 2\pi n$, where $n$ is an integer, we obtain from Formulas 13.11 and 13.12:

$$P_1(\lambda_1) = 0 \qquad \text{and} \qquad P_2(\lambda_1) = P_{\text{in}}$$

When $k(\lambda_2)z = \pi + 2\pi m$, where $m$ is another integer, we obtain:

$$P_1(\lambda_2) = P_{\text{in}} \qquad \text{and} \qquad P_2(\lambda_2) = 0$$

To make any calculations, we need to know the value of the coupling coefficient, $k(\lambda)$. For coupling two identical step-index fibers, a good practical formula for $k(\lambda)$ is [18]

$$k(\lambda) = (\pi/2)(\sqrt{\delta}/a)\exp[-(A + Bw + Cw^2)], \tag{13.17}$$

where $a$ is the fiber core radius, $w = u/a$ ($u$ is the distance between two fiber axes), and $\delta$ is the relative refractive index and equals $(n_1^2 - n_2^2)/n_1^2$ ($n_1$ and $n_2$ being the refractive indexes of a core and a cladding, respectively). Parameters $A$, $B$, and $C$ are defined as follows:

$$A = 5.279 - 3.663\,V + 0.384V^2$$

$$B = -0.777 + 1.225V - 0.015V^2$$

$$C = -0.018 - 0.006V - 0.009V^2$$

The parameter $V$ is well known to us (see Formula 3.14 or 13.1):

$$V\ \text{number} = (2\pi/\lambda_0)a\sqrt{(n_1^2 - n_2^2)}, \tag{13.18}$$

where $\lambda_0$ is the wavelength in a vacuum.

Formula 13.17 clearly describes the nature of the coupling coefficient $k(\lambda)$ we have mentioned several times: The coupling coefficient depends on the fiber cores, their separation, and the operating wavelength, In particular, the coupling coefficient drops rapidly as the separation between two fibers, $w$, increases.

Now we can quantitatively analyze wavelength dependence in a WDM coupler.

### Example 13.1.1

***Problem:***

What coupling length ($L_c$) do we need to make a broadband WDM 3-dB FBT coupler split at 1300 nm and 1550 nm?

***Solution:***

The solution can be found from an analysis of Formulas 13.11 and 13.12 and Figure 13.6. When the argument of a raised sine and cosine reaches $\pi/4 \approx 0.785$, the input power will be equally split between two fibers. Hence,

$$k(\lambda)z = 0.785$$

This is the condition necessary to build a 3-dB splitter at the given wavelength. If we compute $k(\lambda)$, we will find $L_c$ immediately.

In accordance with Formula 13.17, to compute $k(\lambda)$ we need to know a number of parameters of the fibers used to fabricate an FBT coupler. Since these parameters are not given, let's make some assumptions. First, we'll consider identical fibers because Formula 13.17 holds true only for this case. Secondly, let's take SpecTran's singlemode communications fiber as the material for an FBT coupler. (See Figure 6.5) This is a step-index fiber, as Formula 13.17 requires.

Let's first do the calculations for $\lambda = 1310$ nm. From Figure 6.5, you can find the core and cladding refractive indexes ($n_1 = 1.4513$ and $n_2 = 1.4468$), which allow you to compute $\delta = 0.0062$. Note that

$$\delta = (n_1^2 - n_2^2)/n_1^2 \approx 2(n_1 - n_2)/n_1 = 2\Delta \tag{13.19}$$

and $\Delta = 0.0031$ here, which is typical for an SM fiber.

Further, we know from Figure 6.5 that the MFD of this fiber is equal to 9.3 μm. A typical core diameter in such a case is 8.3 μm; let's assume

$$a = 4.0\ \mu\text{m}$$

The distance between the fiber axes should be approximately equal to the two core radii and some of the cladding between them. Let's now take $w = u/a = 3$; that is, $u = 12.0$ μm. Thus, we compute

$$V = 2.19$$

and

$$k(1310\ \text{nm}) = 411.06\ 1/\text{m}$$

Since $k(\lambda)\, L_c = 0.785$, the coupling length ($L_c$) is equal to

$$L_c = 1.91\ \text{mm}$$

Let's now do similar computations for $\lambda = 1550$ nm. We obtain:

$$V = 1.852$$

$$k(1550\ \text{nm}) = 852.47\ 1/\text{m}$$

and the coupling length is equal to

$$L_c = 0.92\ \text{mm}.$$

Thus, the output port positioned at 0.92 mm from the input port will gather light at 1550 nm and another port located at 1.91 mm from the input port will gather light at 1310 nm.

The conclusion to be drawn from this example is that power transferred from one fiber to another in an FBT coupler varies along the coupling region, as Figure 13.6 shows. These variations depend on the coupling parameters and the wavelength. Therefore, an FBT coupler can be used for wavelength-division multiplexing.

## Phase Mismatch

The above explanation assumes lossless transfer—an ideal state we never can reach. (Remember the principle of the conservation of energy?) A more realistic model introduces some losses. Let's see, then, how doing so changes the result.

Refer again to Sections 4.1 and 4.2, where Maxwell's equations and their applications are discussed. In the case of two waveguides where coupling occurs, Maxwell's equations yield the couple-mode equations describing the coupling effect. For two nonidentical fibers supporting the $LP_{01}$ modes, the coupling equations result in the following relationship among power $P_{in}$ launched at $z = 0$ and powers $P_1(z)$ and $P_2(z)$ transferred into two fibers at any given length $z$ of the coupling region [18]:

$$P_1(z)/P_{in} = 1 - k^2/\gamma^2 \sin^2 \gamma z \tag{13.20}$$

$$P_2(z)/P_{in} = k^2/\gamma^2 \sin^2 \gamma z, \tag{13.21}$$

where $k(\lambda)$ is the coupling coefficient as above, $\gamma^2 = k^2 + 1/4\,(\Delta\beta)^2$, $\Delta\beta = \beta_1 - \beta_2$, and $\beta$ is the propagation (phase) constant. You should recall from our discussion in Section 4.2 that $\beta = \omega/v = 2\pi n/\lambda$, where $n$ is the effective refractive index of the propagation medium. (See Formulas 4.19 through 4.21.)

The quantity $\Delta\beta$ is called the *phase mismatch.* For two identical fibers, $\Delta\beta = 0$; we can easily obtain Formulas 13.11 and 13.12 from the above equations.

For two nonidentical fibers, the phase mismatch is never zero; therefore, $\gamma \neq k$, which results in the partial transfer of power to a coupled fiber. For example, maximum power ($P_2$) transferred to fiber 2 in Figure 13.6 is given by

$$P_{2\max}/P_{in} = k^2/\gamma^2 = 1/[1 + (\Delta\beta/2k)^2], \tag{13.22}$$

which means one never fully transfers input power to a coupled fiber.

Using Formula 13.21, we can build power variations for the case of phase mismatch. These variations for different values of $\Delta\beta/2k$ are shown in Figure 13.7. Additional details can be found in [18].

By varying phase mismatch, we can control the portion of maximum transferred power (in addition to controlling the coupling ratio with $k[\lambda]z$). If we want to transfer 1% of the input power to fiber 2, we need to use $\Delta\beta = 19.9\ k(\lambda)$. Taking the coupling coefficients from Example 13.1.1, we obtain:

$$\Delta\beta = 8180.09\ 1/\text{m} \qquad \text{for } \lambda = 1310\ \text{nm}$$

$$\Delta\beta = 16964.15\ 1/\text{m} \qquad \text{for } \lambda = 1550\ \text{nm}$$

The question that comes up now is, how can we physically implement phase mismatch in an FBT coupler? Since

$$\beta = (2\pi n)/\lambda, \tag{13.23}$$

as we have seen, the appropriate phase mismatch can be achieved by coupling fibers with different effective refractive indexes; that is,

$$\Delta\beta = (2\pi/\lambda)\Delta n \tag{13.24}$$

For $\lambda = 1310$ nm and $\lambda = 1550$ nm, we computed $\Delta\beta$, which yields $\Delta n = 1.70 \times 10^{-3}$ and $4.18 \times 10^{-3}$, respectively. You will recall that the typical relative difference in the core and cladding refractive indexes for a singlemode fiber is about 0.4%. Consequently, we have another option when it comes to designing an FBT coupler: We can vary the difference in the core and cladding refractive indexes. For more information on this interesting topic, see [18].

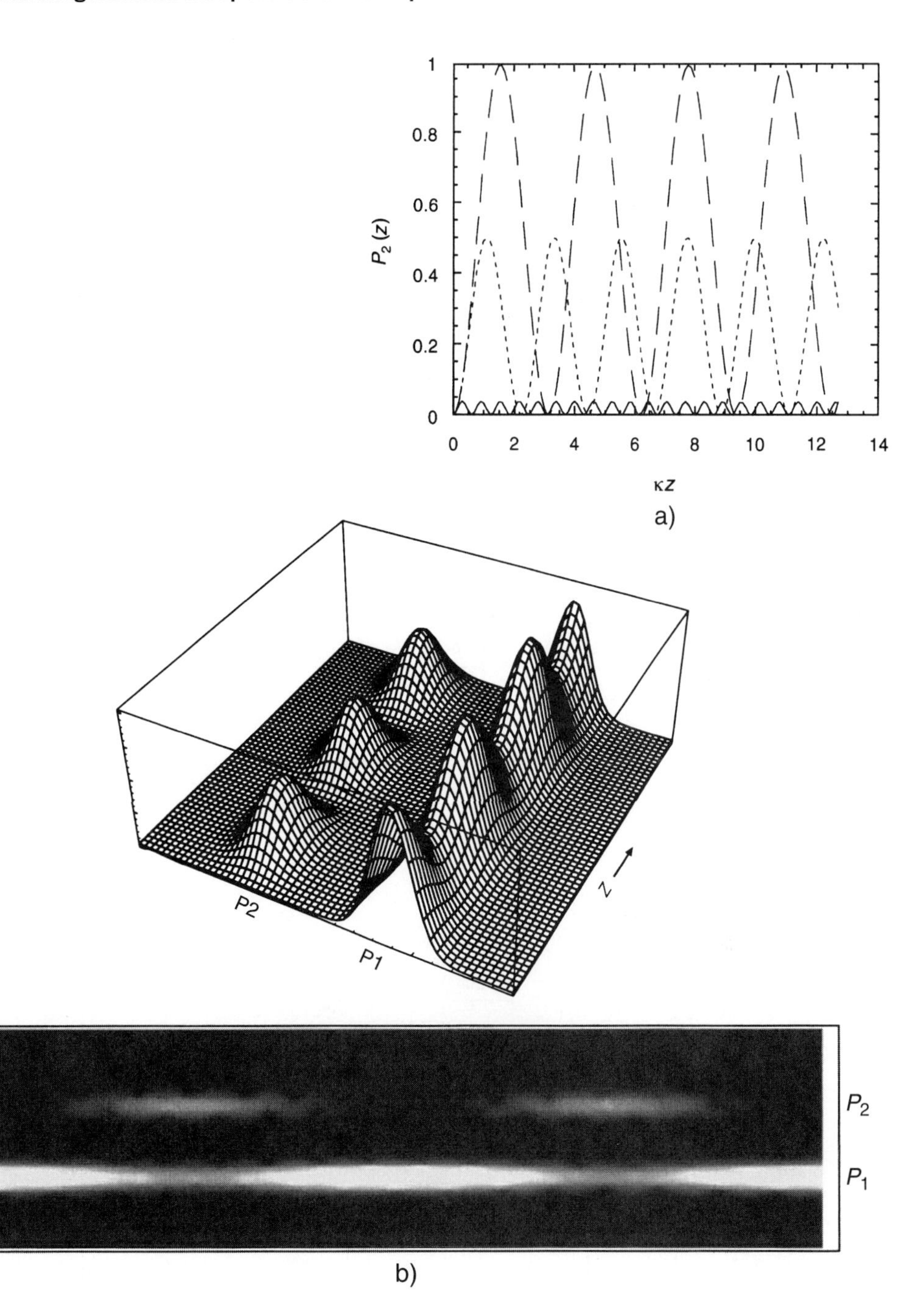


**Figure 13.7** Power variations in the coupling region of an FBT coupler (phase mismatch is non-zero): (a) $P_2$ as a function of $k(\lambda)z$ for various $\Delta\beta/2k$; (b) three-dimensional plot of power variation and density plot. *(Reprinted from Ajoy Ghatak and K. Thyagarajan's* Introduction to Fiber Optics. *Copyright 1998. Used with permission of Cambridge University Press.)*

## 13.2 WAVELENGTH-DIVISION MULTIPLEXERS AND DEMULTIPLEXERS

While couplers/splitters combine or separate light mostly at one wavelength, wavelength-division multiplexers/demultiplexers do this at different wavelengths. You can't imagine wavelength-division multiplexing systems without these devices. As is true with couplers, wavelength-division multiplexers/demultiplexers are crucial to WDM networks not only in themselves but also as building blocks of the network modules. This section discusses these devices.

## WDM MUX/ DEMUX and Couplers

Wavelength-division multiplexers and demultiplexers (also called WDM MUXs and WDM DEMUXs) are devices that combine (couple) and separate (split) signals at different wavelengths. A WDM MUX combines several wavelength channels into one fiber; a WDM DEMUX does just the opposite. WDM multiplexers and demultiplexers are crucial elements of WDM fiber-optic networks. Their function is shown in Figures 13.8(a) and (b). Application of a WDM MUX/DEMUX combination in a wavelength-division multiplexing network is shown in Figure 12.4(b).

Sometimes one device can perform both multiplexing and demultiplexing; that is, we have a bidirectional WDM MUX/DEMUX; at other times a device is designed specifically for multiplexing or demultiplexing.

There are three basic types of WDM MUXs and DEMUXs in terms of the wavelength spacing with which they are dealing: *Broadband WDMs (BWDMs)* combine and separate 1310- and 1550-nm channels—or even 850- and 1310-nm channels They have been in use for a number of years. *Narrowband WDMs (NWDM)* combine and separate wavelength channels with cen-

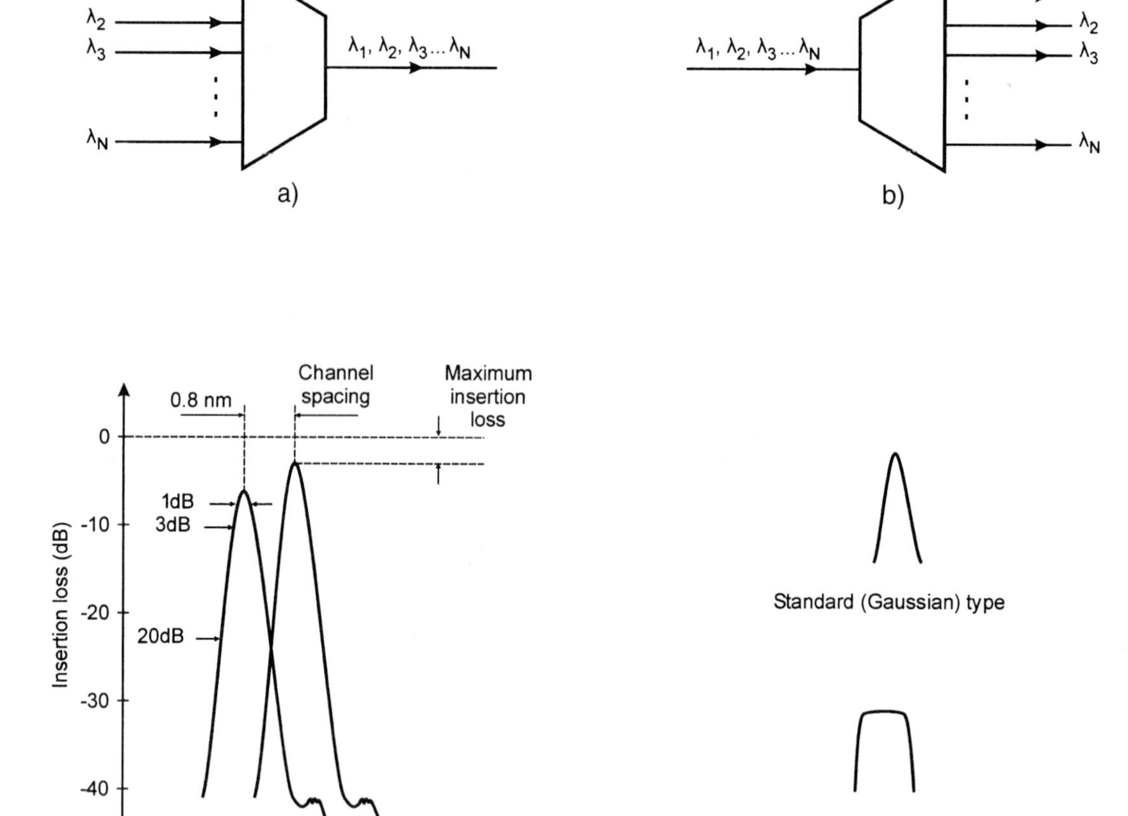


**Figure 13.8** WDM multiplexers/demultiplexers: (a) Function of a WDM multiplexer; (b) function of a WDM demultiplexer; (c) characterization of a WDM MUX/DEMUX; (d) channel-profile types.

ter-to-center spacing greater than 200 GHz. *Dense WDMs (DWDMs)* are combined MUX and DEMUX devices operating with wavelength-channel spacing not more than 200 GHz [19].

Bear in mind this classification is widely accepted but is not the standard because standardization has not yet been established by the industry for wavelength-division multiplexing and demultiplexing terminology. The result is that in the technical literature, particularly in the data sheets, you see the same abbreviation, WDM, used to refer to the wavelength-division multiplexing process and to the wavelength-division multiplexing devices themselves, such as the MUXs and DEMUXs. So always read the document closely to grasp the precise meaning of the term.

As you may have already gathered, a coupler/splitter and a MUX/DEMUX do the same job: They combine several signals in one fiber and vice versa. What differentiates them is that *a coupler combines the signal at the same wavelength* and *a WDM MUX couples different wavelength channels.* Therefore, the coupler used in an EDFA is a WDM MUX because it combines 980-nm (or 1480-nm) and 1550-nm signals. On the other hand, this device is different from a dense WDM MUX, where channel spacing could be as narrow as 0.4 nm (50 GHz). Thus, a term like *WDM coupler* appears to designate the component of an EDFA.

Data sheets of a typical WDM MUX/DEMUX include many characteristics you are familiar with from previous discussions, particularly from reading Figure 13.3. However, there are several new parameters used to characterize these devices:

- The *number of channels* usually describes the number of wavelength channels the device can multiplex/demultiplex. Typical numbers are 4, 8, 16, 32, 40, and 48. Further increasing the number of channels is the trend in the development of today's fiber-optic communications systems.

- The *center wavelengths (frequencies)* of these channels are given by the ITU-T grid shown in Table 12.1. Two wavelengths from this grid are shown in Figure 13.8(c) as an example.

- *Channel spacing* describes the minimum distance between channels a WDM MUX/DEMUX can handle. Standard numbers are 0.4 (50 GHz), 0.8 (100 GHz), and 1.6 nm (200 GHz). Channel spacing is shown in Figure 13.8(c).

- The *bandwidth* (also referred to as *channel width* or *passband width)* is actually the linewidth of a specific wavelength channel. A manufacturer usually specifies this linewidth at the 1-dB, 3-dB, and 20-dB level of an insertion loss, as Figure 13.8(c) shows. Typical values range from 0.7 nm for a 4-channel device to 0.3 nm for a 32-channel device at a loss of 1 dB, but it could be as low as 0.1 nm at 3 dB. Since bandwidth depends on the number of channels, the manufacturer sometimes specifies bandwidth in terms of channel spacing. For example, you may read in a data sheet that the bandwidth is equal to 30% of the channel spacing at 1 dB and 60% at 3 dB.

  It must be pointed out, too, that bandwidth also depends on the line profile. Two typical profiles are used in a WDM MUX/DEMUX: standard (Gaussian) and flat. They are shown in Figure 13.8 (d).

  A comparison of the linewidths at 1-dB, 3-dB, and 20-dB loss levels describes the steepness of the channel line. For example, the data sheet of one demultiplexer may state that the bandwidth is 0.3 nm at 1 dB, 0.4 nm at 3 dB, and 0.8 nm at 20 dB.

- *Maximum insertion loss* has the same meaning as given in Formula 13.3 but, since insertion loss varies from channel to channel, the manufacturer uses the maximum value to characterize the loss inserted by the device. (See Figure 13.8[c].) Typical values range from 1.5 dB for a 4-channel device to 6 dB for a 32-channel MUX/DEMUX.

- In addition to the maximum value of insertion loss, the manufacturer usually specifies the *loss uniformity over the channels,* that is, the difference between maximum and minimum loss. (Refer to Formula 13.6.) The typical value here is < 1.5 dB.
- *Isolation* (Formula 13.10) values depend on the device's function. For a demultiplexer, isolation is usually more than 30 dB, while for a multiplexer it could be as low as 18 dB.

## WDM MUXs and DEMUXs: How They Work

The same basic techniques—micro-optic, integrated optic, and all-fiber—that we have already described for the manufacture of couplers are used to fabricate WDM MUXs and DEMUXs. All MUXs and DEMUXs use dispersive elements, which make different wavelengths propagate at different directions.

***All-fiber WDMs*** Fused fibers, essentially FBT couplers, are examples of these devices. In their original forms, shown in Figure 13.1, they can be used for broadband WDM. Let's see how EDFA WDM couplers work.

Consider multiplexing of pumping light at 980 nm with a 1550-nm signal. You already know that a mode-field diameter (MFD) depends on wavelength: The longer the wavelength, the larger the MFD. (See Figure 5.4.) Thus, choosing the distance between two cores, we can make light at 1550 nm—where the MFD is larger—penetrate another core, while light at 980 nm—where the MFD is smaller—will stick to its core. This is illustrated in Figures 13.9(a) and (b).

The isolation of a pump coupler typically ranges from 14 to 24 dB. The data sheets of an EDFA WDM coupler may also include its *spectral characteristic (transmission performance),* where a loss-versus-wavelength graph is presented. The multipumping technique for increasing pump power discussed in Section 12.4 requires new pump couplers with wavelength (channel) spacing from 20 nm to 5 nm and even less. Narrowband (NWDM) couplers can be used to accomplish this.

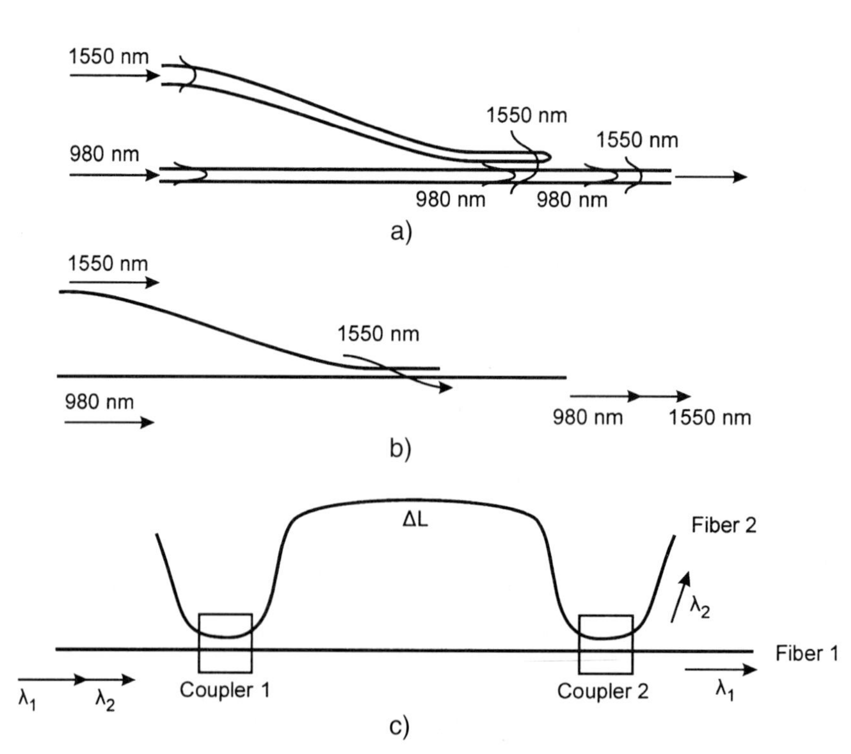


**Figure 13.9** All-fiber wavelength-division multiplexer/demultiplexer: (a) Coupling process in EDFA WDM coupler; (b) EDFA WDM coupler diagram; (c) WDM DEMUX using Mach-Zehnder interferometer.

We can also make NWDM and DWDM MUXs and DEMUXs using the FBT technique, but to do so two essential steps must be undertaken to fabricate good WDMs. First, manufacturers have to reduce the relatively high polarization-dependent loss by carefully controlling the fabrication process. Secondly, to achieve high isolation between very closely spaced channels, wavelength filters have to be added to the entire structure of the component [20]. These filters can be based on thin-film techniques or, as an alternative, fiber Bragg gratings can be used [21].

Another example of an all-fiber interferometric WDM coupler is a *WDM MUX/DEMUX based on the unbalanced Mach-Zehnder interferometer.* Such a structure is shown in Figure 13.9(c). (Also, see Figure 10.29.) In this case, not only the couplers themselves but the entire structure provides excellent wavelength-division multiplexing/demultiplexing. Take, for example, the demultiplexing of the two wavelengths shown in Figure 13.9(c). The first coupler splits the input signal equally and directs it along two paths having different lengths. The longer arm of the interferometer, having the additional length ($\Delta L$), introduces an additional phase shift for both wavelengths. This phase shift can be calculated using the following formula (see references [6], and [7]):

$$\Delta\theta_i = [2\pi n_{eff} \Delta L]/\lambda_i = \beta\Delta L, \tag{13.25}$$

where the propagation constant ($\beta$) is defined by Formula 13.23. The key point here is that a light wave acquires an additional phase shift and waves at different wavelengths traveling the same extra distance ($\Delta L$) experience different phase shifts. At coupler 2, two beams at $\lambda_1$ that have traveled different distances constructively interfere with each other. The result of this interference is that the maximum intensity of light at $\lambda_1$ is directed along fiber 1. At the same time, at coupler 2, the interference of two beams at wavelength $\lambda_2$ results in directing light at $\lambda_2$ along fiber 2. This is how wavelength-division multiplexing occurs.

Consider the power of the signals at both wavelengths. In conjunction with Formulas 13.11 and 13.12, we can write [7]:

$$P_1(\lambda_1)/P_{in} = \cos^2[\Delta\theta_1/2] \tag{13.26}$$

$$P_2(\lambda_2)/P_{in} = \sin^2[\Delta\theta_2/2] \tag{13.27}$$

Hence, making $\Delta\theta_1 = 2\pi\, n$, where $n$ is an integer, we can direct all the power at wavelength $\lambda_1$ to output 1; on the other hand, making $\Delta\theta_2 = \pi\, m$, where $m$ is another integer, we can direct all the power at wavelength $\lambda_2$ to output 2. Changing $\Delta\theta_i = [2\pi\, n_{eff}\, \Delta L]/\lambda_i$ is simple: We just control the length ($\Delta L$) of the additional arm.

Since an interferometer is very sensitive to wavelength, the finesse of the wavelength division that can be achieved with this device is much better than what can be achieved with a simple FBT coupler.

***Arrayed-waveguide-grating WDMs*** An arrayed-waveguide grating (AWG), also called a phased-arrayed waveguide (or phaser), is an interesting device. It is a kind of offshoot of a Mach-Zehnder interferometer but works like a diffraction grating. Its basic arrangement is shown in Figure 13.10.

An AWG is usually fabricated as a planar structure. (See Figure 13.5[c].) It consists of input and output waveguides, input and output WDM couplers, and arrayed waveguides, as Figure 13.10 shows. The length of any arrayed waveguide is distinguished from its adjacent waveguide by a constant $\Delta L$. Wavelength channels enter the AWG, where an input WDM coupler splits them equally among the arrayed waveguides. Each portion of the input light traveling through an arrayed waveguide includes all the wavelengths that have entered the device. Each wavelength, in turn, acquires an individual phase shift determined by Formula 13.25. In addition,

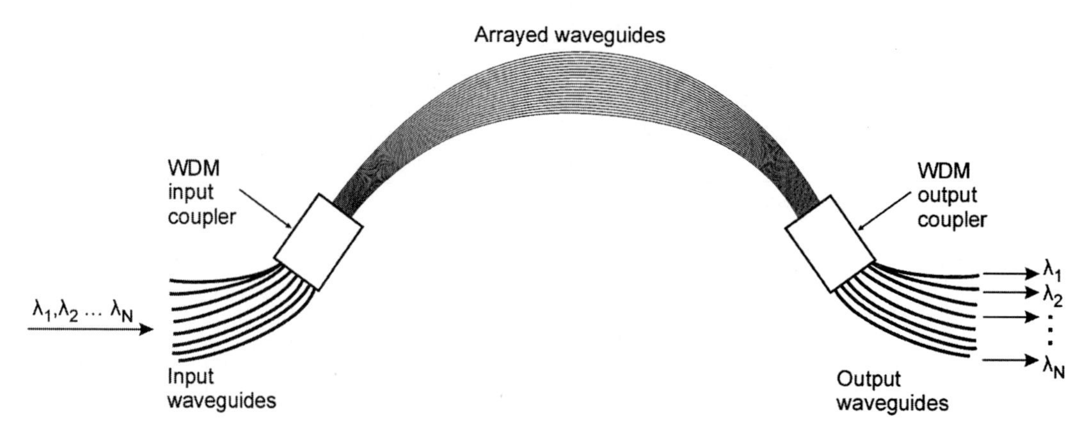


**Figure 13.10** Arrayed-waveguide grating (AWG).

each wavelength receives phase shifts at the input and output couplers. As a result, every portion of light at a given wavelength acquires different phase shifts and all these portions interfere at the output coupler. The net result is a series, or set, of maximum light intensities. The direction of each maximum intensity depends on the wavelength. (From this standpoint, an AWG works very much like a diffraction grating, which will be discussed shortly.) Thus, each wavelength is directed into an individual fiber at the output of the device.

AWG MUXs and DEMUXs can combine and separate 48 channels (and more) and they provide multiplexing/demultiplexing of wavelength channels with spacing as low as 0.4 nm (50 GHz). All the other characteristics you will encounter in an AWG data sheet should be familiar to you through your reading of Figure 13.3 and our discussion of Figure 13.8.

***Diffraction-grating WDMs*** A diffraction grating is a set of closely spaced slits. It can provide transmission or reflection of incident light. The distance between the slits (*d*) is called the grating *pitch* (period).

Let's see how a diffraction grating works [22]. Figure13.11(a) describes the principle of operation of a transmission diffraction grating.

Light from a light source (LED or LD) falls on the transmission diffraction grating as a plane wave. After passing through the individual grating slits, the light spreads in all directions. This is shown as a dotted semisphere in Figure 13.11. The interference of light with the same wavelength at the imaging plane (screen) results in a pattern of maximum and minimum intensity. The direction of the principal maximum intensity is given by

$$d \sin \Theta = m\lambda, \tag{13.28}$$

where $m = 0, \pm 1, \pm 2, \pm 3$, and so on. It is obvious that Formula 13.28 holds true for any wavelength.

Let's consider the first-order principal maxima, that is, $m = 1$. Thus, for wavelength $\lambda_i$ we obtain the following expression from Formula 13.28:

$$\sin \Theta_i = \lambda_i / d, \tag{13.29}$$

which means that an individual wavelength has its principal maximum at a certain angle. In other words, the principal maxima of the different wavelengths are separated from one another by some angle. This is how a diffraction grating directs different wavelengths in different directions.

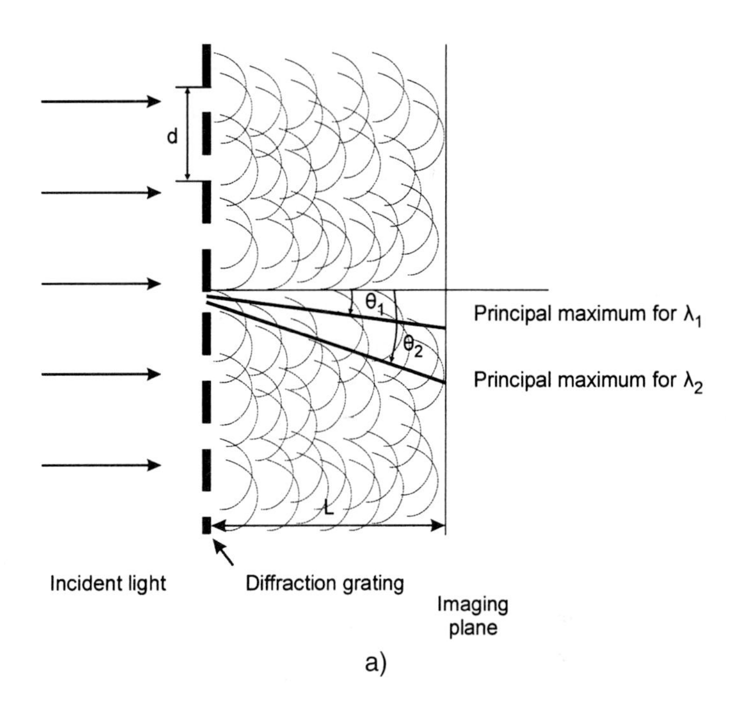


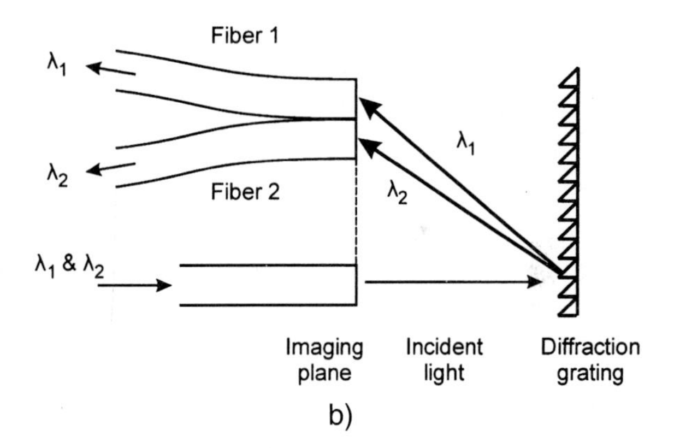


**Figure 13.11** Operation of diffraction grating: (a) Transmission grating; (b) reflection grating.

Angular separation, also called angular dispersion, can be easily obtained from Formula 13.28 by its differentiation:

$$\Delta\Theta/\Delta\lambda = m/d\ \cos\Theta, \tag{13.30}$$

where $\Delta\Theta$ is the angular separation of two principal maxima for two wavelengths that differ by $\Delta\lambda$.

To convert this angular separation into distance separation, we need to use the following formula:

$$y_i = L \tan\Theta_i, \tag{13.31}$$

where $L$ is the distance between the diffraction grating and the imaging plane (screen) and $y_i$ is the position of the first principal maximum of wavelength $\lambda_i$.

Figure 13.11(b) depicts the operation of a reflective diffraction grating with concentration on the net result: Light from an input fiber containing two wavelengths falls on the reflective diffraction grating and is separated at the output fiber ends. This separation is caused by the dispersive properties of the diffraction grating described above.

### Example 13.2.1

***Problem:***

**a.** What angular separation can we obtain when selecting two wavelengths—1540.56 nm and 1541.35 nm—if grating pitch $d = 5\ \mu m$?
**b.** What length ($L$) between a transmission diffraction grating and the fiber ends is required to separate these wavelengths with the same grating?

***Solution:***

**a.** This is a regular WDM channel spacing, according to the ITU-T grid (Table 12.1). We approximate it as a 0.8-nm separation. We can easily calculate the angular separation between these two wavelengths using Formula 13.29:

$$\Theta_1 = \sin^{-1} \lambda_1/d = 17.945° \qquad \text{for } \lambda_1 = 1540.56 \text{ nm}$$

$$\Theta_2 = \sin^{-1} \lambda_2/d = 17.955° \qquad \text{for } \lambda_2 = 1541.35 \text{ nm}$$

As you can see, the angular separation—0.01°—is very small. However, the critical point is the length separating the wavelengths because we need to direct these two wavelengths into two individual optical fibers.
**b.** To find the distance ($L$), we can derive from Formula 13.31 the following simple relationship:

$$L = (y_2 - y_1) / (\tan \Theta_2 - \tan \Theta_1) \qquad (13.32)$$

We need to know the minimum space separation, $y_1 - y_2$, between the fibers. Let's assume we are using regular singlemode fibers whose coating diameter equals 245 μm, as Figure 5.17 indicates. Therefore, the minimum distance between the centers of the two adjacent unstripped fibers is 245 μm.
Substituting $y_2 - y_1 = 245\ \mu m$ and $\Theta_2$ and $\Theta_1$ obtained above, we compute:

$$L = 1.323 \text{ m}$$

The typical packaging size of an 8-channel WDM MUX/DEMUX is about 250 × 200 × 20 mm and the size increases with an increase in the number of channels. Surely, such a miniature device cannot package a construction with the length measured in meters. The solution to this problem lies, therefore, in using focusing lenses or mirrors to shorten the distance between the centers of the two fibers. We consider such a construction in the accompanying text.

---

Two examples of the construction of WDM MUXs and DEMUXs based on diffraction grating are given in Figure 13.12. Figure 13.12(a) shows diffraction grating with a *graded-index (GRIN) lens,* or *rod,* used to direct light onto and from the grating. Figure 13.12(b) shows a similar device using a concave mirror to direct light. This device, known as a STIMAX [23], is a solid monoblock construction. Such a high-density WDM can combine up to 130 channels (fibers) with channel spacing of 0.4 nm.

Commercially available multiplex/demultiplex devices usually have 48 channels with 100 GHz (0.8 nm) channel spacing. However, by interleaving two units, 96 channels with a separation

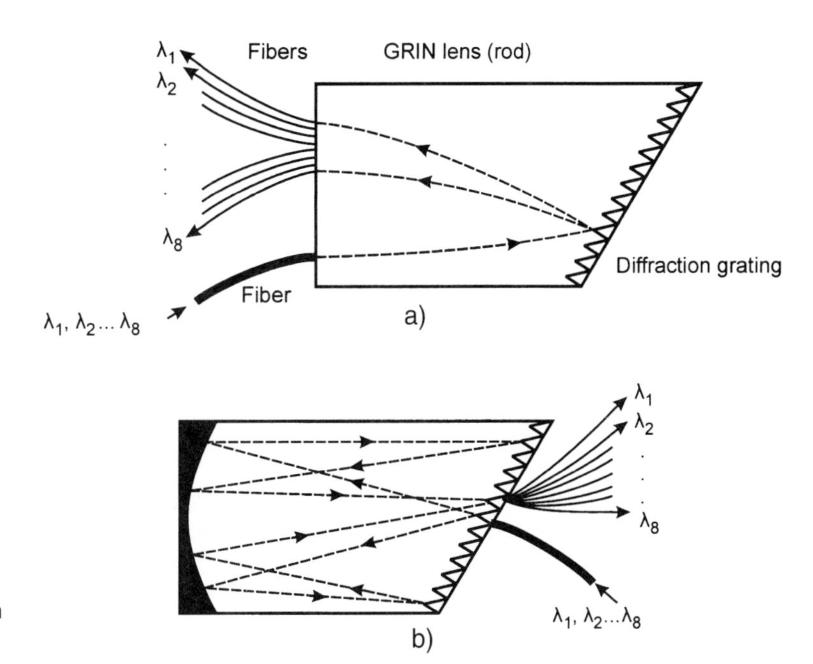


**Figure 13.12** Wavelength-division multiplexers/demultiplexers based on diffraction gratings: (a) Diffraction grating with a GRIN lens (rod); (b) diffraction grating with a concave mirror.

of 50 GHz (0.4 nm) can be multiplexed/demultiplexed. Such MUXs and DEMUXs introduce insertion loss from 5 to 8 dB. A key characteristic of such devices is *isolation.* Data sheets of WDM MUX/DEMUX units usually include several isolation characteristics. For example [23], the near-end crosstalk of adjacent and nonadjacent channel crosstalk is about –42 dB, while the far-end crosstalk of adjacent and nonadjacent channels is, typically, –25 dB and –32 dB, respectively.

***Fiber Bragg grating (FBG) WDMs*** Fiber Bragg grating (FBG) WDMs represent a further development of the diffraction-grating concept. In essence, the fiber Bragg grating is the diffraction grating implemented as a variation of the refractive index of the fiber core. It works as a highly reflective narrowband mirror, reflecting one wavelength and transmitting all the others. The principle of operation of an FBG is shown in Figure 13.13.

The wavelength being reflected is called the Bragg wavelength ($\lambda_B$) and it is determined by the following *Bragg condition (law):*

$$2\Lambda n_{\text{eff}} = \lambda_B, \tag{13.33}$$

where $\Lambda$ is the grating period (that is, the distance between two adjacent maximum points of the periodic core refractive index) and $n_{\text{eff}}$ is the effective core refractive index. Thus, an FBG can work as a WDM DEMUX, separating an individual wavelength from the wavelength-multiplexed light stream. (You might think of an FBG as a notch-type filter.) The cascade of FBGs having a different $\Lambda_B$ can separate one wavelength after another.

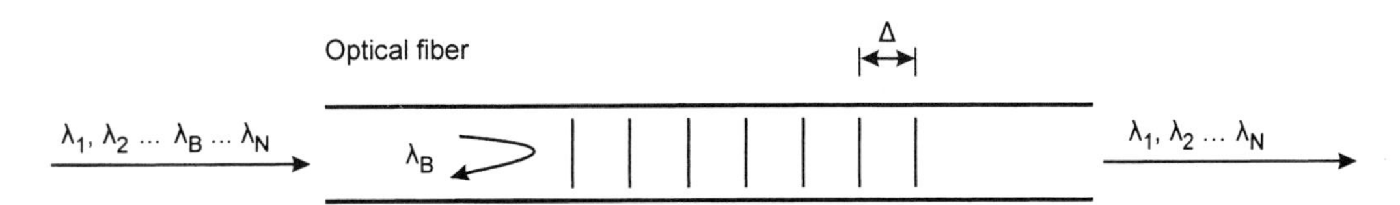


**Figure 13.13** Fiber Bragg grating.

***Dielectric multicavity thin-film filter (DMTF) WDMs*** This type of WDM MUX/DEMUX is based on dielectric multicavity thin-film filters, sometimes called dichroic MTFs. The term *dichroic* means that an optical filter transmits light selectively according to the wavelength [11]. Each of these filters allows only one wavelength to pass through while reflecting all the others. Let's consider Example 13.3.2 to understand how this type of WDM works.

### Example 13.3.2

***Problem:***

The "Features," "Applications," and "Schematic" sections of the data sheet of an "8200 Series WDM: 8 Channel DWDM" are shown in Figure 13.14(a). Analyze these sections and suggest a possible way to construct the device.

***Solution:***

We will start our analysis with the "Features" section. The words "Totally Passive Device" tell us no external power is required to operate this WDM. This means there is no need for active thermal-control units, which sometimes are used to provide the thermal stability required by such devices. All the other features are self-explanatory except for one: "Epoxy-Free Optical Path." Epoxy is a crosslinked polymer providing strong adhesion. It has been used widely for connecting optical components. Unfortunately, the optical properties of epoxy can change over time, so manufacturers avoid using epoxy along the optical paths. (See Section 8.2.)

The "Applications" section should give you no trouble at this point; in fact, it would be a good exercise for you to explain to your instructor or to a colleague the meaning of each sentence in this section. Remember that OC-48 and OC-192 are simply bit-rate standards equal to 2.5 Gbit/s and 10 Gbit/s, respectively.

Looking at the WDM schematic in Figure 13.14(a), let's now discuss the principle of operation of this device. The core of this WDM consists of thin-film filters designated MicroPlasma narrowband filters. The term *narrowband* means that each filter allows only one wavelength to pass through it, blocking all the others. Thus, when light, containing all the wavelengths, falls on the first filter, wavelength $\lambda_1$ will pass through the fiber but all the others will be reflected back and directed to the next filter. Wavelength $\lambda_2$ will pass through the next filter while the rest of the light will be directed to the following filter. This process continues until the last wavelength passes through the last filter. This is how wavelength-division demultiplexing occurs with this device. Lenses couple filtered light into a singlemode fiber (SMF), as illustrated in Figure 13.14(a).

Now what about the suggestion as to how we might construct such a device? Well, it might be constructed from a glass substrate with a number of thin-film filters attached to it. You could use epoxy for this assembly but be careful not to apply it along the optical paths. Lenses would facilitate the coupling of the light into the optical fibers. Such a construction would be solid and mechanically and thermally stable. This construction is shown in Figure 13.14(b).

---

***WDM MUXs and DEMUXs—some final observations*** Wavelength-division multiplexers/demultiplexers are crucial components in WDM fiber-optic communications systems. A number of their types and constructions are on the market, with many more in the works. Space limitations preclude our discussing every specific type of WDM MUX/DEMUX here; however, you should by now have gained all the knowledge you need to understand their principle of operation and to readily interpret their specifications sheets.

Table 13.2 summarizes the above discussion and compares the major characteristics of the different types of WDM MUXs and DEMUXs.

Technical Data Sheet

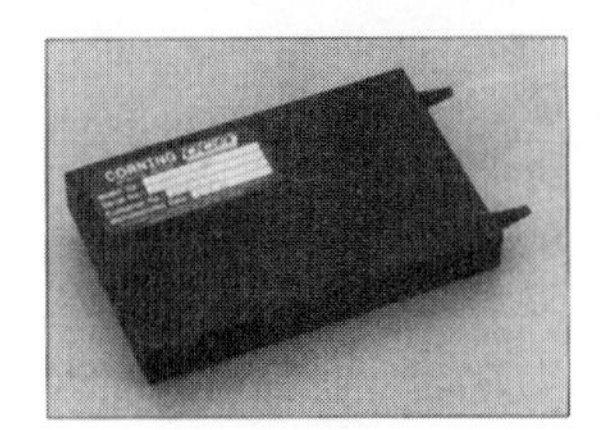

*Fiber Optic Products*

**8200 Series WDM: 8 Channel DWDM**

**Features:**

- Totally Passive Device
- 200 GHz Channel Spacing
- Low Insertion Loss and High Isolation
- Excellent Temperature/Humidity Stability
- Epoxy-Free Optical Path
- Custom Designs Available

**Applications:**

- Bi-directional and Unidirectional Networks
- High Speed Communications (OC-48/OC-192)
- Transport Protocol Independent
- CATV
- Optical Amplifiers

Corning OCA's 8200 Series WDM products are designed for use in advanced optical communications networks featuring 8 channels on the ITU grid with 200 GHz spacing. These products are typically customized for specific network configurations and sold to original equipment manufacturers (OEMs). Like all Corning OCA WDM products, 8200 Series WDMs can be combined with other components to create higher channel count networks.

The 8200 Series WDMs feature proven Corning OCA ESF™ filters, manufactured with our proprietary MicroPlasma® process. Corning OCA products incorporating ESF filters have demonstrated exceptional environmental stability and have passed rigorous qualification testing for use in telecommunications systems. These product attributes translate into improved performance for optical networks in real world operating conditions. The passive nature of Corning OCA WDM components simplifies network design by eliminating the need for thermal control mechanisms, required by other WDM technologies. The 8200 DWDMs have superior optical properties, passing broad optical channels and giving high isolation between channels. The inherently low loss ESF filter design maximizes the amount of power available for other system functions. In addition, Corning OCA 8200 DWDMs are assembled using manufacturing processes with a proven reliability record.

Corning OCA manufactures multiple product types to give system designers optimal configuration and cost flexibility. Two product types are available in the 8200 Series, the 8201 MUX and the 8202 DEMUX. The 8202 DEMUX WDM offers improved isolation as compared to the 8201 MUX WDM.

**8200 Series WDM Schematic**

SMF
Lens
Lens
SMF
Chan 1
Lens
Chan 3
Lens
Lens
Chan 2
Chan 4
MicroPlasma® Narrowband Filters

**Corning OCA Corporation**
170 Locke Drive • Marlborough, MA 01750
Phone: (508) 804-6200 • Fax: (508) 804-6540 • Web:http://www.oca.com

a)

**Figure 13.14** Operation of wavelength-division multiplexer based on dielectric multicavity thin-film filter: (a) Data sheet of an 8-channel DWDM; (b) suggested construction of a WDM. *([a] courtesy of Corning OCA Corporation, Marlborough, Mass.)*

## WDM MUX/DEMUX Applications—Add/Drop and Routers

WDM MUXs and DEMUXs find applications in two important components used in WDM fiber-optic communications networks: add/drop and passive wavelength routers. *Add/drop multiplexers* and *demultiplexers (ADMs)* are designed to add to the network signal stream and drop out one wavelength channel. The add/drop function is shown in Figure 13.15(a).

Popular physical implementations of ADMs are shown in Figures 13.15(b) and (c). The first ADM includes two fiber Bragg gratings connected in Mach-Zehnder interferometer configuration.

## Product Specifications:

All specifications apply for the given operating environment and include connectors.

### Channel plan

| | |
|---|---|
| Number of channels: | 8 |
| Channel center wavelength/frequency: | 1549.32 nm to 1560.61 nm<br>193.5 THz to 192.1 THz (ITU) |
| Channel spacing: | 200 GHz (approx. 1.6 nm) |
| Channel width: | ±30 GHz (approx. ±0.25 nm) |

### Optical parameters

(all values are specified over channel width and operating temperature)

| | |
|---|---|
| Channel isolation: | ≤ -25 dB - DEMUX<br>≤ -12 dB - MUX |
| Insertion loss: | ≤ 5.0 dB - DEMUX<br>≤ 4.5 dB - MUX |
| Ripple:<br>(within channel width) | ≤ 1 dB (peak-to-peak) |
| Polarization dependent loss: | ≤ 0.3 dB |
| Optical return loss: | ≥ 45 dB |
| Directivity: | ≤ -55 dB |

### Mechanical parameters

| | |
|---|---|
| Fiber pigtail length: | ≥ 0.5 m |
| Fiber pigtail type: | Corning® SMF-28™ optical fiber |
| Fiber pigtail jacket: | 0.9 mm or customer specified loose tube |
| Connector type: | FC/SPC, SC/SPC customer specified, or none |

### Environmental parameters

| | |
|---|---|
| Operating temperature range: | -5 to +70 °C |
| Storage temperature range: | -40 to +85 °C |

### Other Characteristics

| | |
|---|---|
| Average ESF filter center wavelength Stability vs. temperature: | ≤0.5 to pm/°C |

## Ordering Information:

When placing an order for the 8200 Series WDM, please have the following information available:

| | |
|---|---|
| Product type: | 8201 WDM–MUX<br>8202 WDM–DEMUX |
| Fiber pigtail jacket: | 0.9 mm loose tube |
| Connector type: | FC/SPC, SC/SPC customer specified, or none |

Contact Corning OCA Corporation for custom channel plans.

### Typical spectral response of 8200 Series WDM

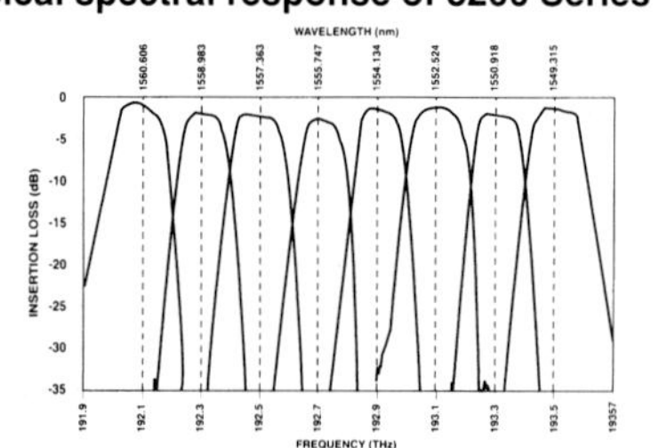


### Dimensions

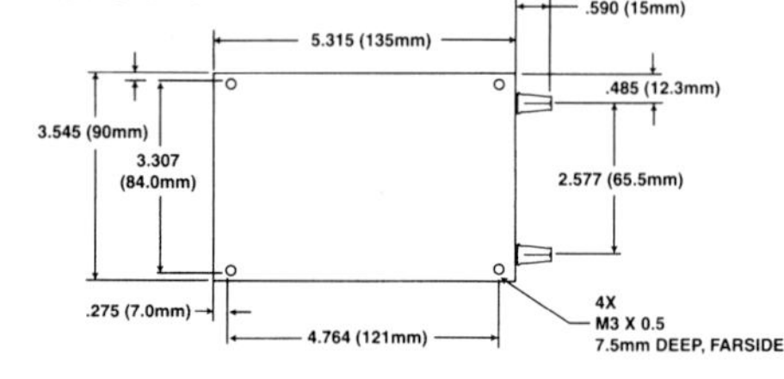


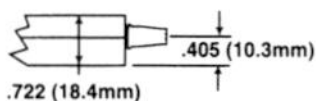


## For more information:

For additional information on Corning OCA Corporation fiber optic products, call (508) 804-6200 or fax (508) 804-6540.

**CORNING OCA**

**Technical Data Sheet**
**PI 750 Issued 9/98**

170 Locke Drive
Marlborough, MA 01752
Tel: 508 804-6200
Fax: 508 804-6540



**Figure 13.14** (continued) b)

The input stream is split by the first WDM coupler, and the Bragg (resonant) wavelength (here, it is $\lambda_5$) is reflected by the FBGs. This wavelength, designated $\lambda_5^D$, is taken out (dropped) through port 4. The wavelength being added, $\lambda_5^A$, enters the component through port 2, is reflected by the FBGs, and is added to the main stream by the output coupler.

The ADM shown in Figure 13.15(c) uses circulators to direct added and dropped wavelengths. Again, the FBGs reflect the resonant wavelength from both directions and circulators direct the $\lambda_5^D$ and $\lambda_5^A$ accordingly.

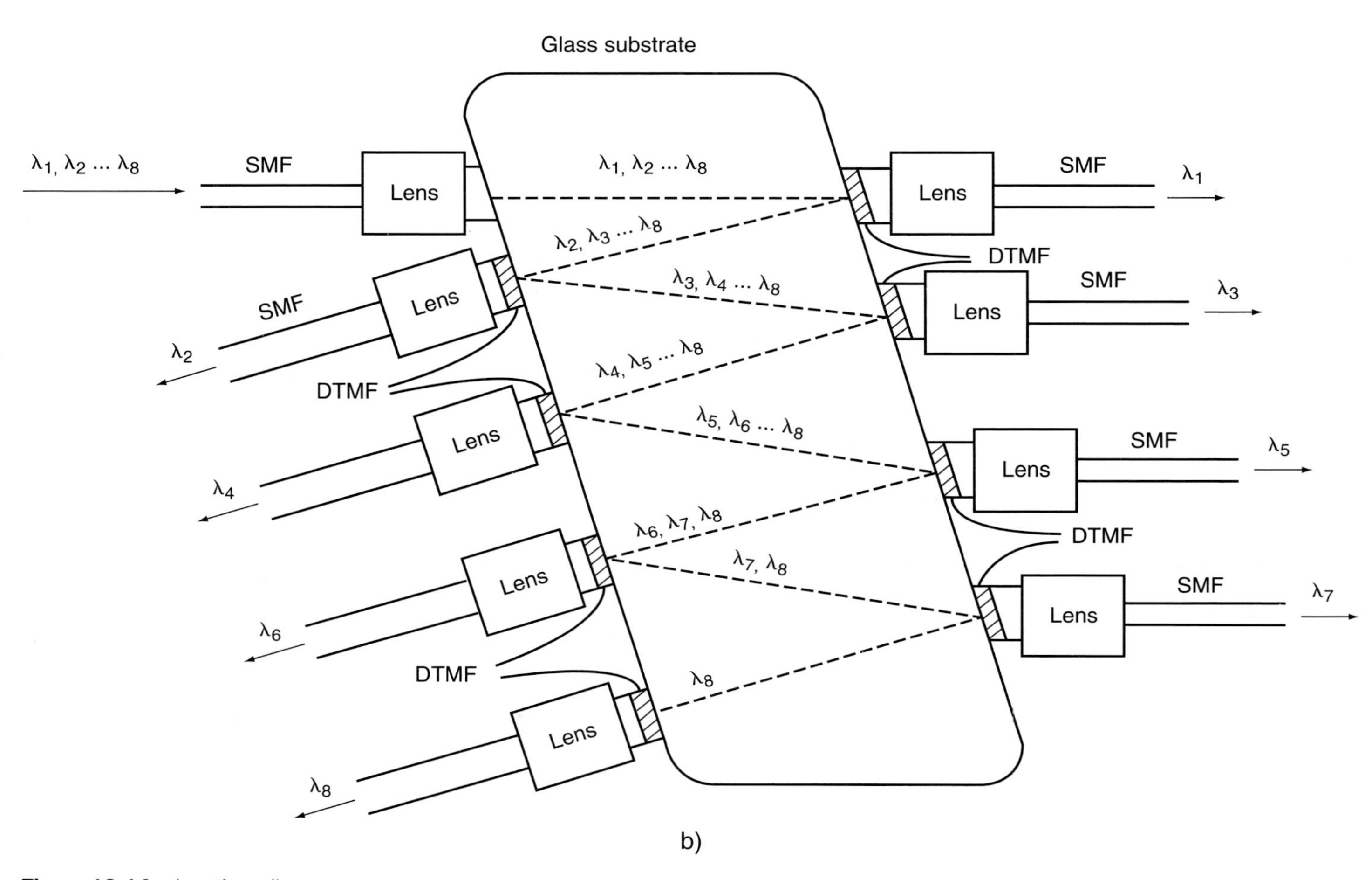


**Figure 13.14** (continued)

**Table 13.2** Comparison of specifications of MUXs and DEMUXs for 8-channel WDM with 1.6-nm spacing

| | Fused Biconic Taper (FBT) | Micro-optic (GRIN Rod and Filter) | Arrayed-Waveguide Gratings (AWG) | Diffraction Gratings (STIMAX) | Dielectric Thin-Film Multicavity Filters (DTMF) | Fiber Bragg Gratings (FBG) |
|---|---|---|---|---|---|---|
| Channel width (nm) at 1 dB | 0.5 at 0.5 dB | | 0.4 | 0.09 | 0.4 | 0.5 |
| Insertion loss (dB) | < 2.2 | < 4.3 | < 7 | <5 | 7 | < 0.5 |
| Uniformity (dB) | < 0.3 | < 1.5 | < 2.5 | | | |
| Isolation (dB) | > 30 | > 30 | > 22 | 35 | 25 | 35 |
| Return loss (dB) | > 55 | > 45 | > 40 | | | 15 |
| Directivity (dB) | > 60 | > 6 | > 40 | 55 | | |
| PDL (dB) | < 0.1 | < 0.1 | < 0.3 | | 0.2 | 0 |
| Temperature sensitivity (nm/°C) | < 0.001 | < 0.003 | <0.0015 | <0.014 | 0.0005 | 0.0007 |

Sources: Rajiv Ramaswami and Kumar Sivarajan, *Optical Networks: A Practical Perspective,* San Francisco: Morgan Kaufman Publishers, 1998. *Lightwave 1999 Worldwide Directory of Fiber-Optic Communications Products and Services,* March 31, 1999, pp. 282–290.

You can easily imagine constructions of add/drop multiplexers based on multicavity thin-film filters and all the other techniques discussed above.

*A wavelength router* directs input wavelengths to the desired paths (fibers). The function of a wavelength router is shown in Figure 13.16(a), where wavelengths entering the device are redistributed and directed to different output ports. Wavelength routers are the key components of the most effective class of WDM fiber-optic communications networks: wavelength-routing networks. (See Figure 12.5[c].)

An example of a practical application of such a router is shown in Figure 13.16(b). This arrayed-waveguide grating (AWG) is familiar to us from its previous discussion in this section. The input and output couplers are $N \times M$ and $M \times N$ star couplers, respectively. As already mentioned, each wavelength experiences a phase shift according to Formula 13.25. If $\Delta L = 0$, then, by symmetry, light entering the top input port will appear at the bottom, or conjugate, output port [24]. If $\Delta L \neq 0$, then, because of the phase shift, light at the given wavelength from any input port will appear at the conjugate (output) port.

Since such devices don't require any external power, they are called *passive,* or *static, wavelength routers.* Dynamic routers allow us to change the routing over time and they require the use of switches, that is, active components.

Figures 13.15 and 13.16 give examples of how WDM multiplexers and demultiplexers are used as building blocks to construct components of WDM fiber-optic networks.

## 13.3 FILTERS

In WDM fiber-optic communications networks, each channel is associated with an individual wavelength; thus, we have the term *wavelength channel.* The manipulation and selection of an individual channel require the optical selection of an individual wavelength—in other words, optical filtering. This need underscores the importance of optical filters in WDM networks.

### Optical Filters: What They Are

An *optical filter,* like the thin-film filter described above, allows only one wavelength to pass through it, blocking all the others. Its function is shown in Figure 13.17(a). There are two broad classes of filters: fixed and tunable. *Fixed filters* allow a fixed, predetermined wavelength to pass

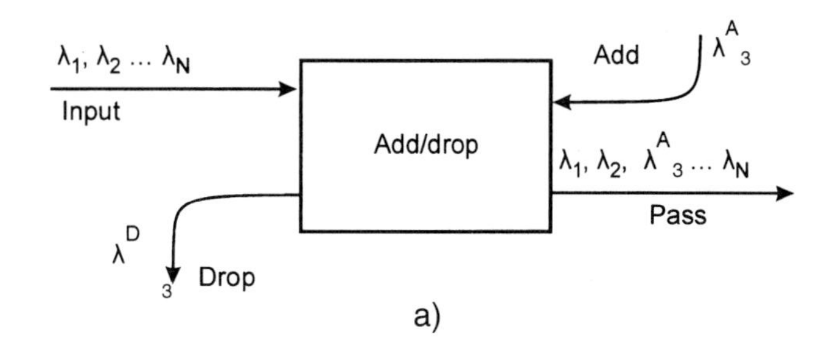


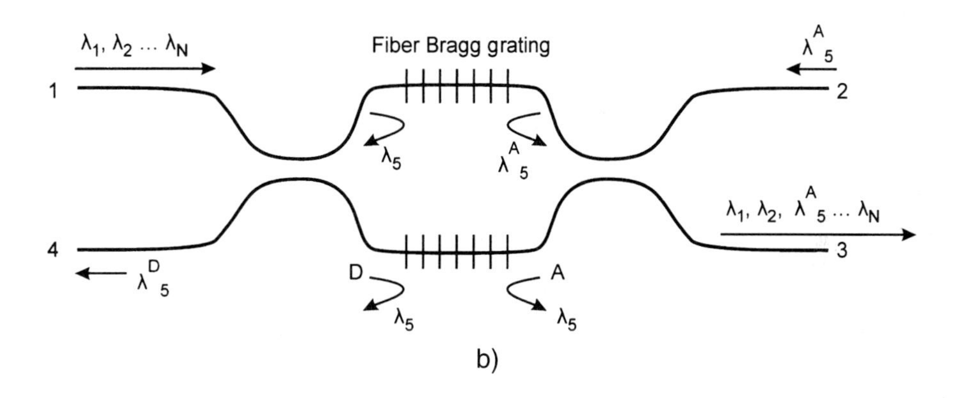


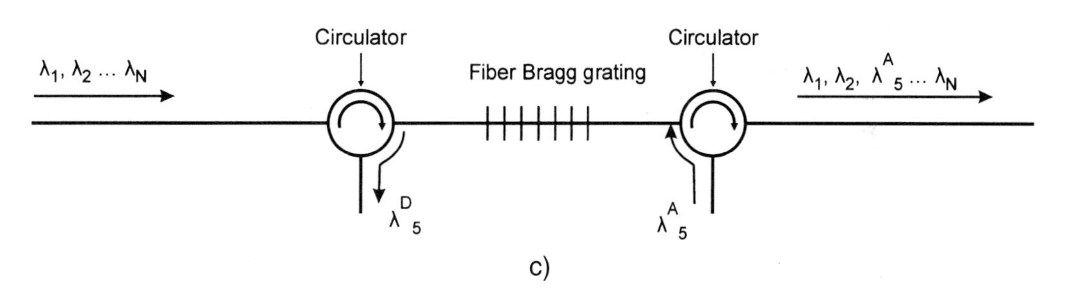


**Figure 13.15** Add/drop multiplexer/demultiplexer (ADM): (a) Function of add/drop MUX/DEMUX; (b) add/drop based on fiber Bragg grating and Mach-Zehnder interferometer; (c) add/drop based on fiber Bragg grating and circulators.

through, while *tunable filters* dynamically tune the selected wavelength. The filters can be further classified in terms of their applications. For instance, we've discussed the usage of gain-flattening filters for the improvement of the gain flatness of EDFAs (Section 12.4). In fact, however, all types of filters are characterized by the same parameters that we are considering here.

The characterization of a WDM filter is shown in Figure 13.17, and since it has much in common with a WDM MUX/DEMUX, depicted in Figure 13.8(c), it would be a good idea to review that figure again. In fact, many references use the terms *WDM MUX/DEMUX* and *WDM filter* interchangeably because both components do the same job and it's very hard to draw a clear distinction between them.

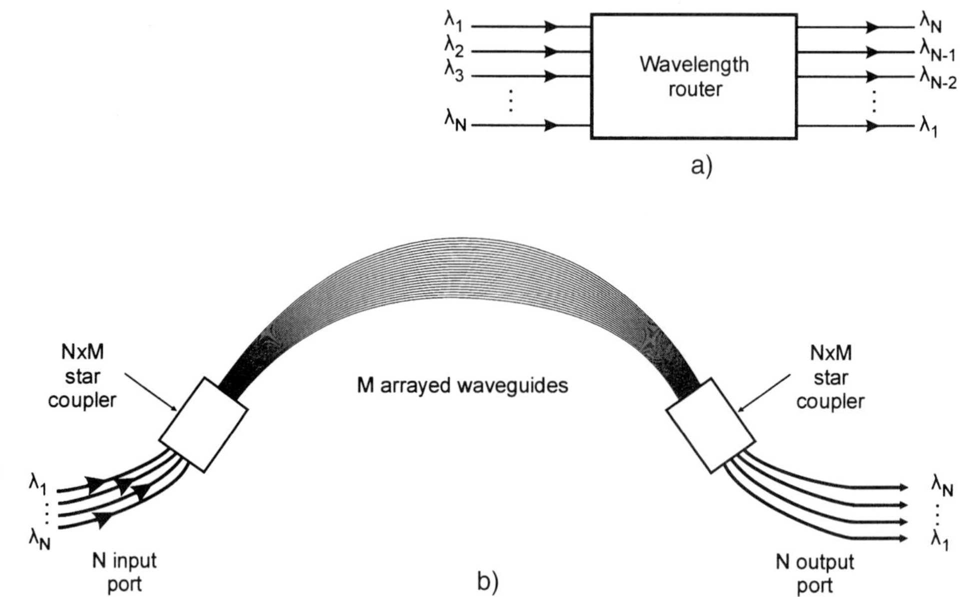


**Figure 13.16** Passive wavelength routers: (a) Function of a wavelength router; (b) AWG as a wavelength router.

A sketch of ideal and real characteristics of a WDM optical filter is given in Figure 13.17(b). The following major characteristics, shown in Figure 13.17, are used to describe a real WDM filter ([7], [10], [19]):

- *Center frequencies,* or *center wavelengths,* must comply with the ITU-T grid given in Table 12.1. The two center wavelengths (frequencies) shown in Figure 13.17 are presented, obviously, as examples only.
- *Pass bandwidth (BW)* is the width of a filter's transmission curve at the 0.5 dB level of the maximum insertion loss, as Figure 13.17(c) shows. This characteristic is also called *channel width,* or *passband width,* as mentioned in our discussion of Figure 13.8(c). Pass BW can also be measured *at the 1-dB or 3-dB levels of the insertion loss* (Figure 13.8[c]).
- *Stop BW,* or *stop bandwidth,* is the linewidth at the 20-dB level of the maximum insertion loss. A comparison of these two numbers shows the steepness of the *sidebands,* or *passband skirts.*
- *Isolation,* which we defined in Formula 13.10, shows how much power from an unwanted channel is present in a given output channel. This is illustrated in Figure 13.17(c), where the importance of the channel width and the steepness of the passband skirts are emphasized. Isolation can be specified for an adjacent channel and/or for all other channels in a device.
- *Ripple* is the peak-to-peak variation of the insertion loss within the channel width, as Figure 13.17(d) shows.
- The properly functioning filter has to transmit all light within its passband and reflect all other light. This ability is characterized by the *bandwidth-utilization* factor *(BUF),* which is the ratio of linewidths of transmitted light to reflected light at a certain level of insertion loss. Figure 13.17(e) clarifies this definition. An ideal filter is one whose BUF = 1; in reality, a value around 0.4 at –25 dB is typical.

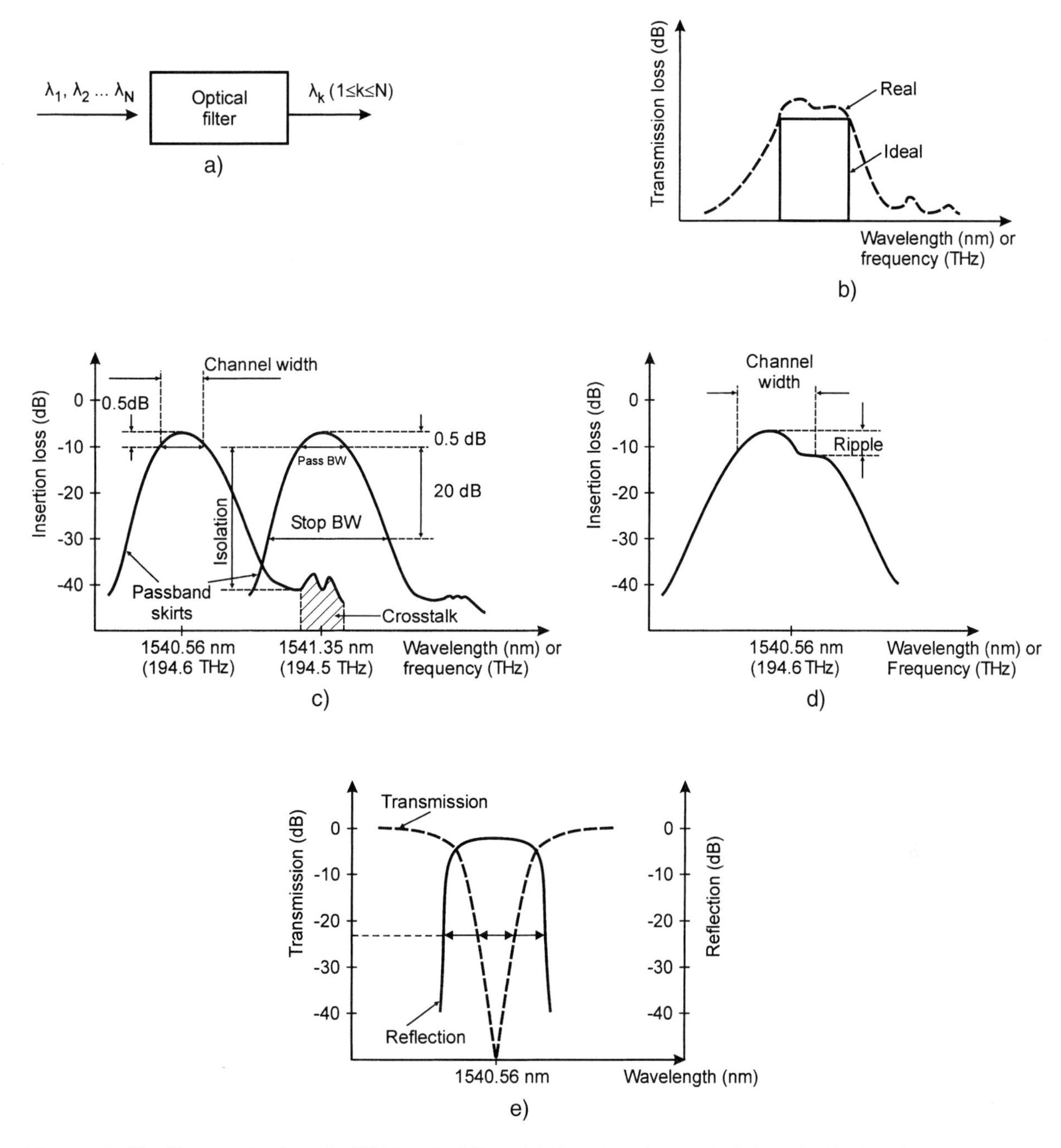


**Figure 13.17** Characterization of a WDM optical filter: (a) Function of an optical filter; (b) ideal and real transmission; (c) definition of basic terms; (d) definition of a ripple; (e) bandwidth utilization factor (BUF).

Bearing these major characteristics of a filter in mind, let's briefly consider the main techniques used to fabricate WDM filters.

## Fixed Filters

Most of the techniques used to fabricate WDM multiplexers and demultiplexers are also used to fabricate filters. The fused biconic taper (FBT), the Mach-Zehnder interferometer, diffraction grating, and the fiber Bragg grating are popular fixed-filter technologies. Let's review them:

***Thin-film interference filter*** This filter depends on the effect of interference among many light waves reflected from the sheaf of thin layers, as Figure 13.18(a) shows. If the thickness of each layer is equal to $\lambda/4$, then the light at wavelength $\lambda$ experiences a phase shift equal to $\pi$ after passing each layer when the angle of incidence ($\Theta$) is zero. (A general case of oblique incidence, shown in Figure 13.18(a), illustrates this idea.) Hence, the reflected wave will be out of phase with the incident wave and they will interfere destructively, that is, cancel each other. In other words, light at wavelength $\lambda$ will not be reflected, which means this light will pass through the filter. All other wavelengths will be reflected. This is how filtering occurs. The multilayer structure enhances the effect, making the filter's transmission characteristic close to ideal. This technique has been in use in optics for many years, one popular application being as an antireflective coating in cameras, eyeglasses, and similar optical instruments. In fiber-optic communications technology, such filters are produced by using *micro-optic* techniques.

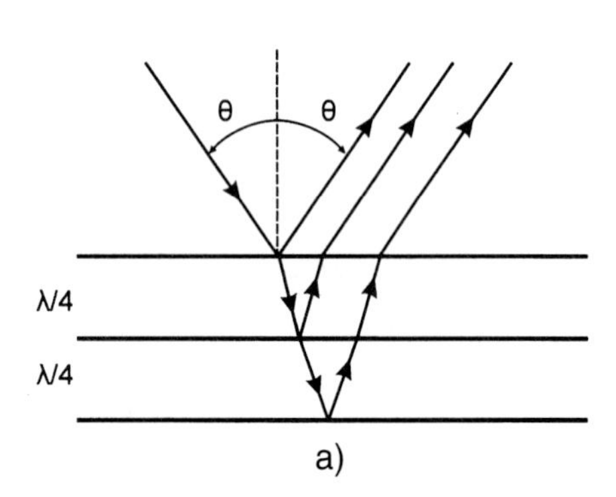


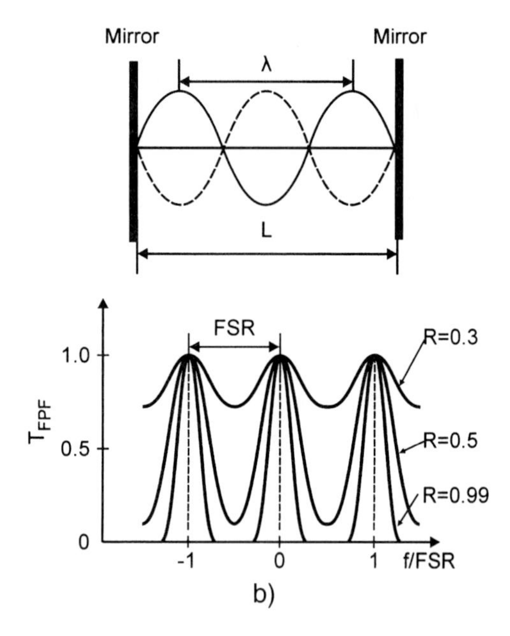


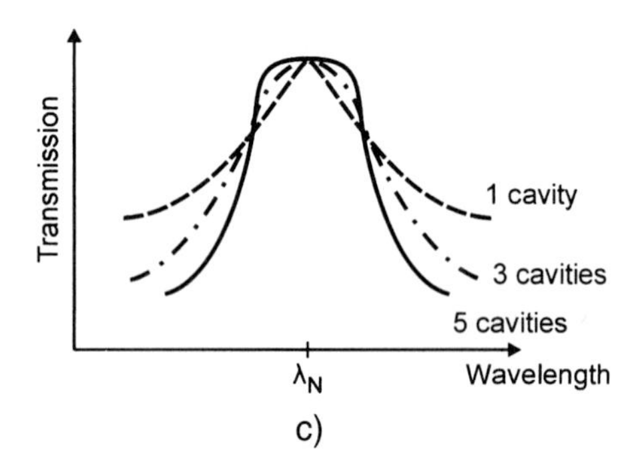


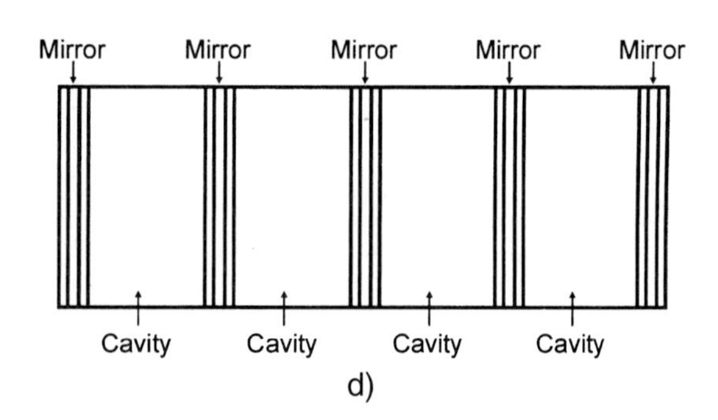


**Figure 13.18** Operation of optical filters: (a) Thin-film interference; (b) Fabry-Perot resonator and its transfer function; (c) transmission characteristics as a function of number of cavities; (d) multicavity thin-film filter. *([b] adapted from Dennis J. G. Mestdagh's* Fundamentals of Multiaccess Optical Fiber Networks, *Boston: Artech House, 1995. Used with permission.)*

If you make the thickness of each layer in Figure 13.18(a) equal to $\lambda/2$, then the reflected waves will interfere constructively; that is, they will be in phase and add to one another. The result will be a highly reflective mirror. Thin-film high-reflection mirrors are used in Fabry-Perot resonators to make cavity-type optical filters.

***Cavity filters (Fabry-Perot filters)*** These filters are based on the *Fabry-Perot (FP) resonator,* also called the *Fabry-Perot interferometer* or the *Fabry-Perot etalon.* (FP resonators are discussed in Section 9.2. See also Figure 9.14, which is reproduced in part in Figure 13.18[b].) An FP resonator supports only those wavelengths that complete their total path within the cavity with a zero phase shift, as Figure 13.18(b) shows. This is formally described by Formula 9.13:

$$2L/N = \lambda_N, \tag{13.34}$$

where $L$ is the cavity length and $N$ is the integer. Hence, an FP resonator allows through only those wavelengths that satisfy the resonant condition predicted by Formula 13.34. The spacing between the two adjacent resonant wavelengths is given by Formula 9.14, repeated here in slightly different form:

$$\lambda_N - \lambda_{N+1} = \lambda^2/2L$$

The transfer function of an FP filter (FPF) is obtained by the following formula [5]:

$$T_{FPF} = [\alpha_m(1-R)^2]/\{(1-R\alpha_m)^2 + 4R\alpha_m \sin^2[(\omega-\omega_0)L/v]\}, \tag{13.35}$$

where $\alpha_m$ is the internal loss, which includes losses caused by the medium and absorption by the mirrors; $R$ is the reflectivity of the mirrors (assuming it to be the same for both mirrors); $L$ is the cavity length; and $v$ is the light velocity within the cavity, which is equal to $c/n$, with $n$ being the refractive index of the medium. It is evident that $T_{FPF}$ is a periodic function. Its period is called the *free spectral range (FSR)* and it is equal to

$$FSR = v/2L = c/2Ln \tag{13.36}$$

The transmission function ($T_{FPF}$) is shown in Figure 13.18(b). Pay attention here to the role of reflectivity: The higher the reflectivity of the mirrors of an FP resonator, the sharper the transmission characteristics. The thin-film technique is used to increase the reflectivity of FP mirrors. FP-based filters are employed in fiber-optic communications systems through the use of micro-optic techniques.

(If some of what we are covering here seems familiar, including the formulas and figures, it should be. Much of it applies as well to the material in Section 12.3. For example, a graph akin to the one in Figure 13.18[b] is in Figure 12.16[c]. Formula 13.35 is identical to Formula 12.4, except for the fact that "gain" is replaced by "loss." These similarities stem from the simple fact that in all these cases we are essentially discussing the properties of a Fabry-Perot resonator.)

The 3-dB (FWHM) bandwidth of each peak of a $T_{FPF}$ is given by the following formula (compare with Formula 12.16 and observe the coefficient $\frac{1}{2}$ here):

$$BW_{FPF}(\text{FWHM}) = (c/2Ln)[(1-R)/\sqrt{R}], \tag{13.37}$$

where losses are neglected; that is, $\alpha_m = 1$. The figure of merit of a Fabry-Perot filter is called its *finesse* (F) and is defined as

$$F = FSR/BW_{FPF} = \pi\sqrt{(R)}/(1-R) \tag{13.38}$$

For highly reflective FPF, finesse ($F$) is approximately equal to the number of wavelength channels the filter can accommodate. Indeed, FSR is the frequency range that can be tuned without overlapping adjacent channels [5] and $BW_{FPF}$ is the width of the individual channel Typically, $F$ ranges from 20 to 120, but Fabry-Perot filters with higher values of $F$ are commercially available.

***Multicavity filters and dielectric thin-film multicavity filters (DTMF)*** These filters cascade several FP cavities, a feat that makes the transmission characteristic close to the ideal one shown in Figure 13.17(b). The effect of adding cavities to the cascade is shown in Figure 13.18(c).

A *dielectric thin-film multicavity filter (DTMF)* is a device constructed from several FP cavities, where reflection is provided by multilayer thin-film mirrors. A schematic of the DTMF structure is shown in Figure 13.18(d). Such a structure utilizes all the advantages of multilayer thin-film mirrors and the multicavity cascade to obtain a flat and steep transmission characteristic.

Let's consider, as an example, the characteristics of a DTMF filter designed for 1.6 nm (200 GHz) channel spacing [25]. The pass and stop bandwidths of this filter are approximately 0.88 nm and 1.76 nm, respectively. Channel isolation is not less than –15 dB to the nearest 200-GHz-spaced channel and not less than –30 dB to any other 200-GHz-spaced channel on the ITU grid. Insertion loss is not more than 2 dB and ripple is not more than 0.5 dB. As you can see, these characteristics are very impressive. Now refer again to Figure 13.14, which shows how the operation of a WDM MUX/DEMUX is based on DTMF filters. You see here how DTMF filters are used to make WDM devices.

## Tunable Filters

***What they are*** Tunable filters have the ability to change the wavelengths they select *dynamically* in contrast to fixed filters, which filter *statically.* Since tunable filters require some external power, they are, strictly speaking, not passive but *active* components. However, we are dealing with them in this section to keep all these similar-functioning devices under one heading.

The function of a tunable filter is shown in Figure 13.19(a). A set of wavelengths enters the tunable filter and a control mechanism guides the filter in its dynamic selection of the desired individual wavelength.

We need tunable filters for two purposes: for filtering wavelength channels in front of a receiver, as discussed in Section 12.2, and for building sophisticated fiber-optic switching (dynamic) networks.

What tuning speed we need depends on the type of network we're working in. Circuit-switching networks (see Figure 12.3) are currently deployed. These networks require relatively slow speeds, that is, switching times on the order of milliseconds. Packet-switching networks require tuning times on the order of μs and even ns [28]. In addition, tunable filters find a number of other applications, such as components in delay lines, tunable fiber lasers, and measurement instruments, where tuning-time requirements may vary.

The major parameters used to characterize tuning filters are ([7], [28], [30]):

- *Dynamic (tuning) range,* also called *wavelength tunability (nm).* This is the set of wavelengths a filter can handle. It is shown in Figure 13.19(a) as $\Delta\lambda$.
- *Bandwidth* (BW), or *pass* BW *(nm).* (See Figure 13.17[c].)
- *Number of resolvable channels.* This is the ratio of the dynamic range to the minimum channel spacing determined by the required level of crosstalk.
- *Tuning speed* (*s*), measured as the time needed to tune a filter at a specific wavelength.
- *Loss,* or *insertion loss (dB),* measured in accordance with Formula 13.3.

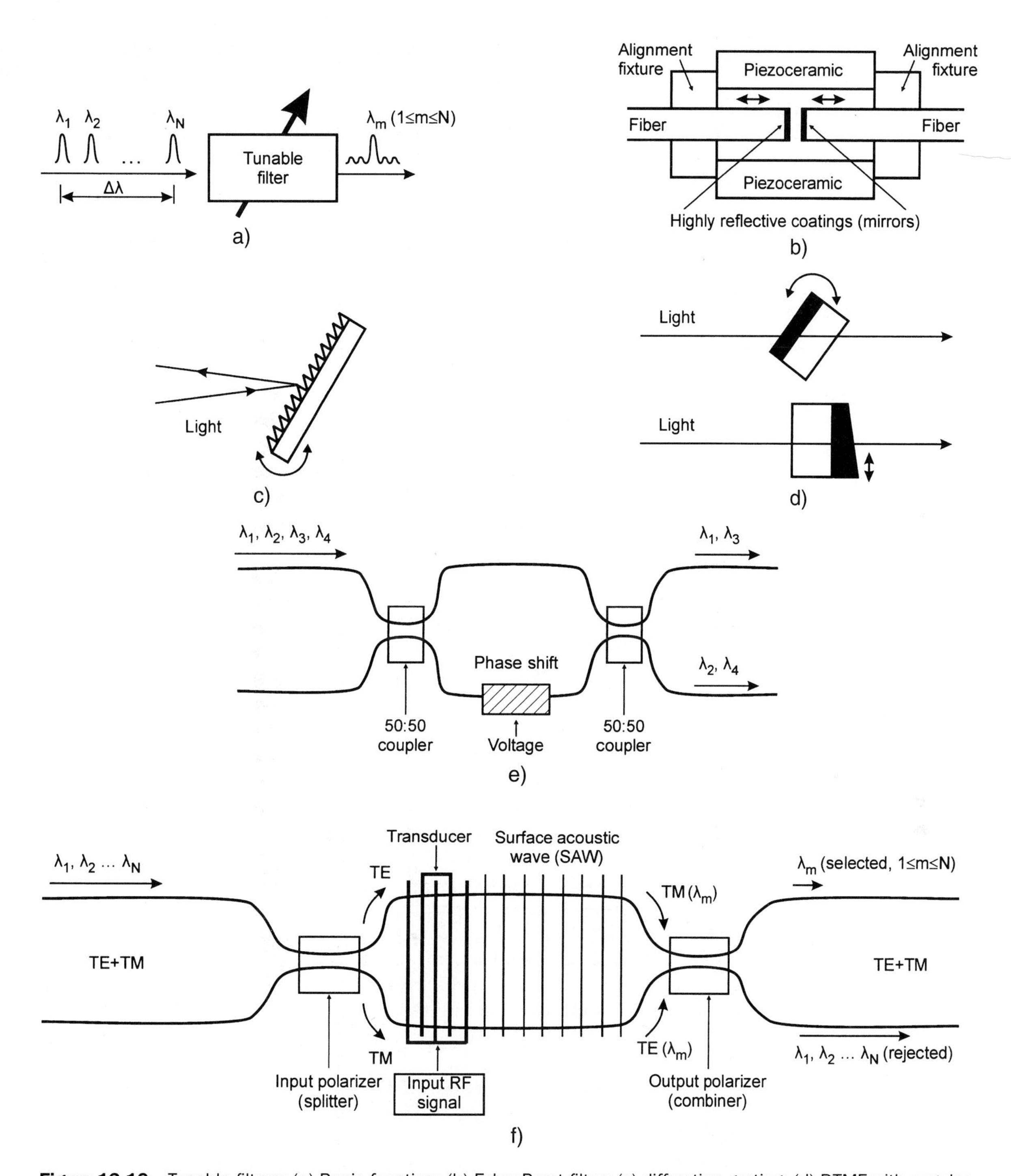


**Figure 13.19** Tunable filters: (a) Basic function; (b) Fabry-Perot filter; (c) diffraction grating; (d) DTMF with angular and lateral motions; (e) Mach-Zehnder filter; (f) acousto-optic tunable filter; (g) electro-optic tunable filter. *(Parts [c] and [d] courtesy of Santec Corporation, Hackensack, N.J.)*

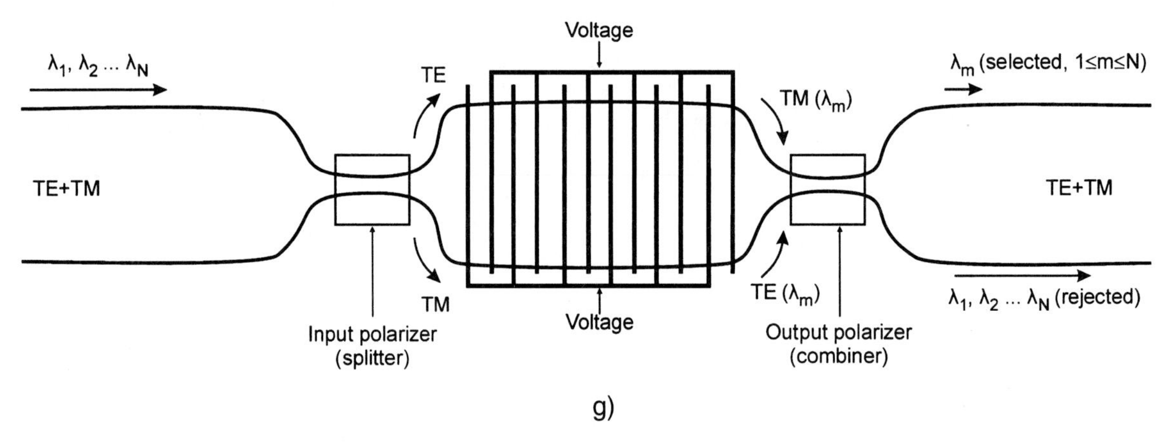


**Figure 13.19** (continued)

- *Polarization-dependent loss, PDL (dB),* as defined on page 596.
- *Sidelobe suppression ratio (SSR) in dB,* that is, the ratio of maximum power (intensity) of the first (largest) side peak to the power of the main peak (measured in dB).
- *Resolution (nm):* the minimum shift of a wavelength a filter can detect.

Such parameters as control mechanism, mechanical size, and price should also be taken into account when comparing the different types of tunable filters.

In considering the major types of tunable filters, we will restrict the discussion, as usual, to commercially available filters and those on the verge of commercial introduction.

***Fabry-Perot (FP) tunable filter*** We have noted that the center wavelength of a Fabry-Perot fixed filter is determined by the length of the FP cavity (resonator), so that $\lambda_N = 2L/N$, as Formula 13.34 states. Hence, you can tune the filtered wavelength simply by changing $L$. The most imaginative construction uses the endfaces of two fibers covered with highly reflective mirrors separated by an air gap. This gap is the cavity of an FP resonator. The length of the gap is controlled by piezoceramics, as Figure 13.19(b) shows. These piezoceramics work as electromechanical actuators: They change their lengths under applied voltage, thus tuning the filtering wavelength [26]. This type of tuning device is called a *fiber Fabry-Perot (FFP) filter.* A wide dynamic range, a narrow bandwidth, high tuning speed, and low PDL are its advantages, while relatively poor stability and a low sidelobe suppression ratio are its drawbacks. (For typical values, see Table 13.3.)

Recent developments in the design and fabrication technology of an FFP have resulted in significant improvement of the filter's characteristics. Finesse up to 2000 has been achieved with a one-stage fiber Fabry-Perot filter. This advance has enabled the FFP to find still another niche: use in WDM testing and measuring instruments [27]. The characteristics of an FFP, particularly its bandwidth and the number of channels, can be significantly increased by cascading two or more FP filters.

Formula 13.34, which determines the center wavelengths of an FP filter, $\lambda_N = 2L/N$, uses the *optical length of a cavity,* $L = L_0 n$, where $L_0$ is the geometrical cavity length and $n$ is the refractive index of the cavity medium. Therefore, if you can control the cavity's refractive index, you can tune the filter. A *ferro-electric liquid crystal,* whose refractive index can be changed under an applied electrical signal, is a good candidate for such a filter [28]. The filter demonstrates good characteristics (a tuning range of 50 nm, a bandwidth of 0.2 nm, tuning time in just several μs, a loss of 1 to 5 dB, and a PDL of 0.3 dB) and it now awaits mass production.

***Diffraction grating*** The tuning capability of a diffraction grating is determined by Formula 13.28: $d \sin \Theta = m\lambda$. The most practical way to change it is to vary angle $\Theta$, that is, to tilt the diffraction grating, as shown in Figure 13.19(c). A wide dynamic range is one of the main advantages of this type of tunable filter, but all its other characteristics are satisfactory.

***Fiber Bragg grating (FBG)*** Formula 13.33 shows how to tune an FBG: $2\Lambda\, n_{eff} = \lambda_B$. Observe that by changing the grating period ($\Lambda$), you will tune the filter at different wavelengths ($\lambda_B$). You can change $\Lambda$ by either applying a stretching force or heating the grating. Low loss, easy coupling, a narrow bandwidth, and high resolution are the advantages of an FBG tuning filter, while a narrow dynamic range is its major drawback, a disadvantage that can be overcome by cascading several FBGs.

***Dielectric thin-film multicavity filter (DTMF)*** You can tune a DTMF by tilting the filter with respect to propagating light, as depicted in Figure 13.19(d). Another possibility is to slide the triangular-shaped filter laterally. In either case, the angle of incident changes, which results in a change in the $\lambda/4$ condition in the filter. (See Figure 13.18.) In short, tilting or sliding the filter changes the center wavelength. Fair characteristics have been obtained with this type of filter. Its main advantage is low PDL.

***Mach-Zehnder tuning filter*** Refer to Figures 13.9(c) and 13.15(b), where the basic configurations of a Mach-Zehnder filter (MZF) are presented. To make a tunable MZF, a symmetric configuration of a Mach-Zehnder interferometer is used. Tuning is obtained by changing the refractive index of one of the arms. This can be done by heating the arm or placing electro-optic material, such as lithium niobite ($LiNbO_3$), in one arm and applying voltage to this phase shifter. (See Figure 13.19[e].) $LiNbO_3$ is discussed in Section 10.4, where an external signal modulator is considered. With an electro-optic phase shift, the tuning time can be on the order of tens of ns. The filter can be fabricated by a lithographic method, which is its major advantage. It is usually used in cascade, which makes its characteristics very good at the expense, however, of loss.

***Acousto-optic tunable filter (AOTF)*** This is a very interesting and very unusual filter. It is based largely on a Bragg grating but this grating is created by an acoustic wave, making the device versatile. Figure 13.19(f) illustrates this idea. As with a regular FBG, a selected wavelength will be filtered. In its operation, this filter converts a TE electromagnetic wave into a TM wave and vice versa. TE and TM are transverse electric and transverse magnetic waves, respectively. They are linear polarized and the planes of their polarization—two principal polarization planes—are orthogonal to each other. To review TE and TM waves, see Section 4.2 and Figure 4.1.

The principle of operation of an AOTF is as follows: Two titanium (Ti) waveguides making up the Mach-Zehnder configuration are engraved in an $LiNbO_3$ birefringent semiconductor. Light entering the filter is separated into TE and TM waves by an input polarizer. As an example, a TE wave travels along the upper arm in Figure 13.19(f) while a TM propagates through the lower leg. A transducer generates a *surface acoustic wave (SAW).* This SAW induces strain in the $LiNbO_3$, creating a periodic perturbation of the refractive index in the $LiNbO_3$. These perturbations work as a dynamic Bragg grating. Because of light-grating interaction, the light power from the TE mode, whose wavelength meets the resonant (phase-matching) condition, is transferred into the TM mode in the upper arm. Light power from the TM mode at this wavelength is transferred into the TE mode in the lower arm. An output polarizer combines the TE and TM modes. Those channels whose wavelengths do not satisfy the resonant condition will pass the structure without change.

The resonant condition in this case is [7]:

$$\lambda_m = \Lambda(\Delta n), \tag{13.39}$$

where $\lambda_m$ is the selected wavelength, $\Lambda$ is the period of grating created by the acoustic wave, and $\Delta n = n_{TE} - n_{TM}$ is the difference in the refractive indexes of $LiNbO_3$ for the TE and TM modes.

### Example 13.3.1

***Problem:***

Calculate (a) the frequency of a surface acoustic wave and (b) the tuning time of an acousto-optic tunable filter if the difference between the refractive indexes of the TE and TM modes $\Delta n = 0.07$, the velocity of sound in the $LiNbO_3$ is 3.75 km/s, the length ($L$) of the acousto-optic interaction is 22 mm, and the optical wavelength 1540.56 nm is to be selected.

***Solution:***

**a.** To calculate the frequency, we need to calculate the wavelength of a SAW. This wavelength is the period of a grating ($\Lambda$). We can relate the $f_{SAW}$ and $\Lambda$ through the well-known formula $\Lambda f_{SAW} = v$, where $v$ is the velocity of the wave. Thus, from Formula 13.39, we find

$$\Lambda = \lambda_m/(\Delta n) = 22.008\ \mu m$$

and

$$f_{SAW} = v/\Lambda = 170.4\ \text{MHz}$$

**b.** The tuning time ($t_{tun}$) is the time needed for a SAW to travel along the length ($L$) of an acousto-optic interaction. In other words, if $L$ is the length of an acoustically induced grating, then the minimum time required for changing the grating period is the time that it takes for a SAW to be set up along this length. Hence,

$$t_{tun} = L/v = 5.87\ \mu s$$

These numbers give you an idea of the values of an AOTF's parameters that are used in practice. An example of the values of the most significant parameters of an AOTF can be found in [29]. These values are: dynamic range, 50 nm; bandwidth at 3 dB, 2.2 nm; sidelobe level < 22 dB; SAW frequency, 172 MHz; loss < 6dB; PDL < 0.5 dB.

---

All tunable filters have the ability to change selected wavelengths but an AOTF does this readily by merely changing the frequency of a SAW generated by a transducer. Simple remote control of AOTF tuning is the great advantage of this filter for applications in WDM networks.

The unique feature of an AOTF, owing to its inherent nature, is its ability to select many wavelengths simultaneously. One can induce several gratings on the same interaction length by exciting several SAWs with different frequencies. (Several SAWs can coexist in the same space almost without mutual disturbance.) Because of this property and good selective characteristics, AOTFs find many applications, primarily as bandpass filters in front of receivers. The main advantage of an AOTF is its very wide dynamic range, covering practically the entire band of wavelengths—starting at 1300 nm—usable in fiber-optic communications.

**Table 13.3** Typical characteristics of tunable filters

| Type of tunable filter | FP | Diffraction Grating | FBG | DTMF | Mach-Zehnder | AOTF | EOTF |
|---|---|---|---|---|---|---|---|
| Dynamic range (nm) | 60 | 100 | 7 | 40 | 10 | 400 | 10 |
| Bandwidth (nm) at 3 dB | 0.5 | 1 | 1 | 1 | 0.01 | 1 | 2 |
| Number of channels | 10 | | | | 100 | 10 | 10 |
| Tuning speed (s) | ms | ms | μs | ms | μs | μs | ns |
| Loss (dB) | 2 | 4 | 0.1 | 1.5 | >5 | 6 | 5 |
| PDL (dB) | <0.1 | >1 | | 0.05 | | <0.5 | |
| SSR (dB) | | | | | | 20 | |
| Resolution (nm) | | | 0.02 | 0.05 | | | |

*Key:* FP—Fabry-Perot; FBR—fiber Bragg grating; AOTF—acousto-optic tunable filter; EOTF—electro-optic tunable filter; PDL—polarization-dependent loss; SSR—sidelobe suppression ratio.

Sources: Dennis Mestdagh, *Fundamentals of Multiaccess Optical Fiber Networks,* Boston: Artech House, 1995.
Dan Sadot and Efraim Boimovich, "Tunable Optical Filters for Dense WDM Networks," *IEEE Communications Magazine,* December 1998, pp. 50–55.
"Optical Filters OF Series," *1998 Santec Components Series,* Santec USA, Hackensack, N.J.
"Motorized Tunable Bandpass Filter," *Measurement Products* (catalog), DiCon Fiberoptics, Inc., Berkeley, Calif., 1999.
*Acousto-Optical Tunable Filter* (data sheet), Pirelli Cavi e Systemi SpA, Milan, Italy, April 1998.

***Electro-optic tunable filter (EOTF)*** The structure and the principle of operation of this filter are very similar to those of an AOTF. The periodic structure—that is, the grating—is created by an electro-optic effect in the $LiNbO_3$ material. Finger-like electrodes are used to induce this grating. Tuning is achieved by changing the voltage applied to the electrodes to vary the refractive-index difference ($\Delta n$), which eventually changes the period of grating seen by the propagating light.

Since tuning is done by an electrical signal, tuning speed is very high. On the downside, however, the dynamic range of an EOTF, usually on the order of 10 nm, is much less than that of an AOTF. Another drawback of an EOTF is its low sidelobe suppression ratio.

***Concluding remarks on tunable filters*** Tunable filters, important components of WDM fiber-optic communications systems today, will become even more crucial in the years just ahead, when WDM technology moves from point-to-point communications to the deployment of dynamic circuit-switching and then packet-switching networks. This is one area of this industry where you can expect to see rapid changes and significant technical achievements.

Table 13.3 presents the typical parameters of various types of tunable filters.

## 13.4 ISOLATORS, CIRCULATORS, AND ATTENUATORS

### Isolators

***Principle of operation*** In fiber-optic communication links, light is reflected from any components—connectors, mechanical splices, passive components, receivers, and so on—inserted into the optical path. We refer to this type of lump reflection as *Fresnel reflection.* In addition, light is reflected by scattering at an optical fiber itself and at fiber-made components. All this backreflected light is often referred to as *optical feedback.* Amplified spontaneous emission (ASE) also belongs to the optical-feedback category.

As you may recall, the performance of lasers and optical amplifiers severely degrades if backreflected light enters these devices. At the system level, optical feedback degrades the signal-to-noise ratio and consequently, the bit-error rate. So we need a device that prevents the propagation of backreflected light.

This is where isolators come into play. As the name suggests, they "isolate" lasers, EDFAs, SOAs, and other devices from backreflected light. The function of an isolator is shown in Figure 13.20(a). Of course, an isolator must allow the forward light to pass through with minimal (ideally, zero) loss and block backreflected light with maximum (ideally, infinity) loss. *Insertion loss* of a modern isolator could be as low as 0.15 dB, while *isolation* could be as high 70 dB.

In operation, an isolator uses the properties of polarized light. In the forward direction, an input polarizer converts unpolarized light into vertically polarized light. This polarized light passes through a Faraday rotator, which rotates a plane of light polarization 45°. An output polarizer (analyzer) allows 45° polarized light to pass and the light passes through the unit with minimum loss. In the reverse direction, an output polarizer converts unpolarized backreflected light into 45° polarized light. The Faraday rotator turns the plane of this polarization another 45° so that the backreflected light becomes horizontally polarized. But the input polarizer allows only vertically polarized light to pass through; therefore, horizontally polarized backreflected light will be rejected. This is how an isolator protects the optical path from backreflected light.

To what extent an input polarizer rejects horizontally polarized backreflected light depends on the *extinction ratio* of the device. This ratio, which we learned is the ratio of the in-

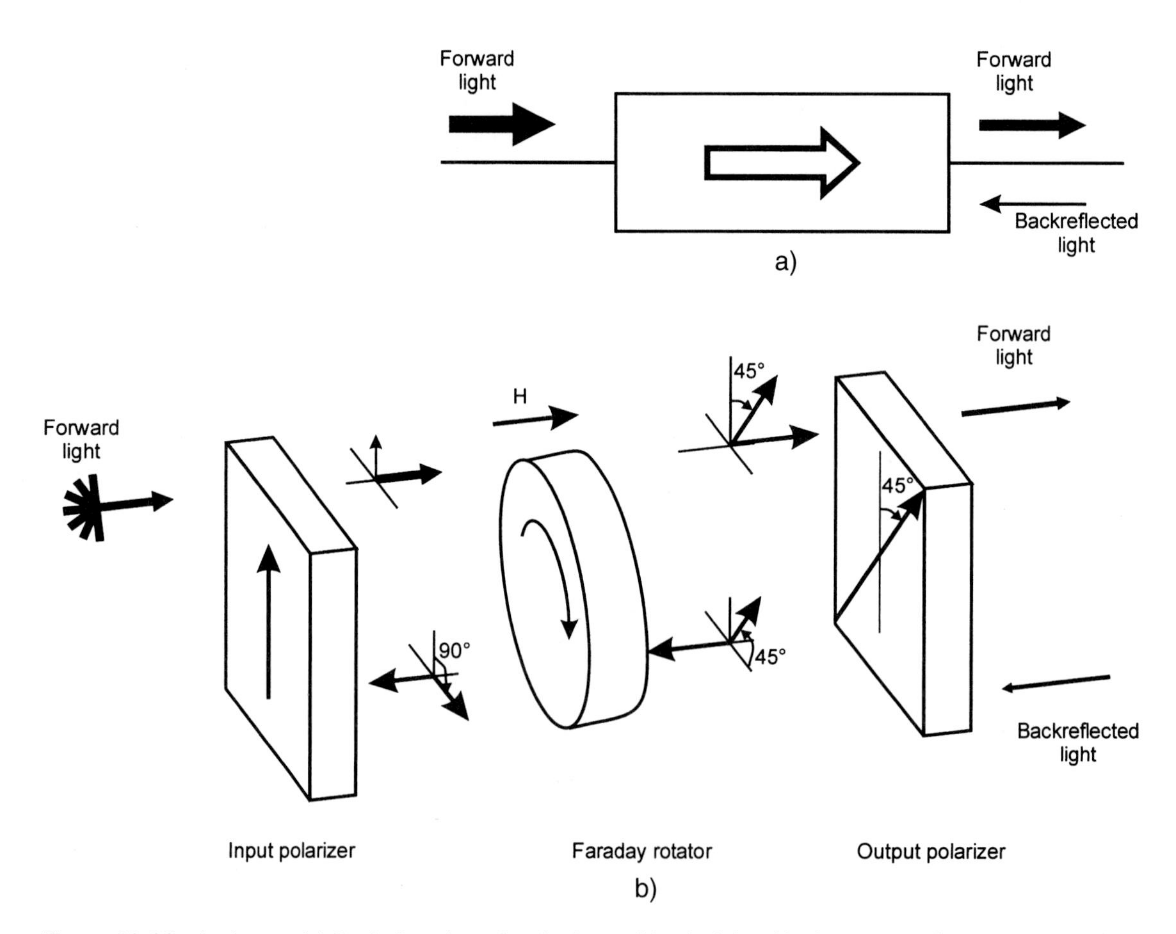


**Figure 13.20** Isolators: (a) Basic function of an isolator; (b) principle of isolator operation.

tensities of two orthogonally polarized beams, can be more than 100,000, resulting in an isolation >50 dB.

The core of this device is a *Faraday rotator,* which consists of a piece of optically transparent material and a magnet surrounding this material. The magnet applies a magnetic field parallel to the direction of the light propagation. Polarized light passing through a Faraday rotator undergoes the rotation of its plane of polarization. The angle of rotation ($\phi$) is given by

$$\phi = \rho HL, \tag{13.40}$$

where $H$ (*A/m*) is the strength of the magnetic field along the direction of the light, $L$ (*m*) is the length of the interaction between light and the magnetic field, and $\rho$ (angle degree/(*A/m*) – *m*) is the Verdet's constant of the material. We need to obtain $\phi = 45°$, which is accomplished by choosing the proper material and applying a magnetic field with the appropriate strength ($H$).

The main feature of a Faraday rotator is that it is *nonreciprocal,* which means that the direction of rotation of the plane of light polarization is determined by the direction of the magnetic field and *not* by the direction of the light propagation.

To make a miniature isolator, we need to choose the appropriate material, as Example 13.4.1 shows.

## Example 13.4.1

***Problem:***

You need to rotate the plane of light polarization to a 45° angle. Calculate the lengths of the Faraday rotators made from a silica fiber ($\rho = 0.0128$ angle min/Oe-cm) and from Bi-substituted iron garnet (BIG) crystal ($\rho = 9$ angle degree/Oe-cm) if the magnetic field at 1000 Oe is applied to both materials.

Note that *Bi* stands for *bismuth* and *Oe* stands for *oersted*—the units for the magnetic field strength ($H$). To convert Oe into amperes per meter (the SI units), use this formula:

$$1 \text{ (Oe)} = 10^3/4\pi \text{ (A/m)}$$

***Solution:***

Use Formula 13.40. For a silica fiber we compute

$$L_{\text{fiber}} = 210.9 \text{ cm}$$

(Don't forget to convert angle minutes to angle degrees or vice versa.) Thus, we need to have a straight span of fiber more than 2 m long—an absolutely unacceptable requirement for a miniature device such as an isolator, whose typical length must not exceed 50 mm. The reason we began with an optical fiber is that it would be very useful to have a fiber-based isolator. It would greatly facilitate our search for solutions to isolator-fiber coupling problems. By using specially doped fiber, this length can be shortened to acceptable values. Fiber-based isolators do exist.

For a Bi-substituted iron garnet crystal, we compute:

$$L_{\text{crystal}} = 0.05 \text{ mm}$$

Having a Faraday rotator of this size, we can now fabricate a truly miniature isolator. BIG crystal is a newly developed material for fiber-optic isolators and we can readily see the effect of its high Verdet constant in shortening the isolator's length.

Incidentally, if you are wondering why we need to apply a magnetic field with such high strength, the answer is that 1000 Oe *H,* for example, will saturate the crystal to the extent that the detrimental effect of external magnetic fields will be negligible. This is how manufacturers maintain EM immunity for their fiber-optic components. In practical terms, of course, the actual value of $H$ depends on the material the Faraday rotator is made of, and today's materials lend themselves to values much smaller than 1000 Oe.

---

***Problems in designing isolators*** Designers run up against three major problems in designing these devices. First of all, the unit is highly polarization sensitive, as you can gather from the above discussion. An optical isolator (Figure 13.20[b]) works perfectly when forward light is vertically polarized, but poorly with any other polarization or with unpolarized light. To make a *polarization-insensitive isolator,* the forward light is split into two orthogonally polarized beams. Both beams undergo Faraday rotation and are recombined before coupling into an optical fiber. In the backward direction, light is also split, the polarization planes are rotated, but the two beams are not recombined. The result is almost no transmission of backreflected light—in other words, high isolation.

Another problem is the dependence on wavelength of the characteristics of an isolator. As is true for any material, the transparency of a garnet crystal depends on the operating wavelength. By working with the appropriate material, manufacturers are able to optimize isolator performance at 1550 nm, 1300 nm, or other peak wavelengths. However, the range within which the characteristics of an isolator hold their values is always specified by the manufacturer. This range is often called the *bandwidth of the isolator.* Still a third problem is the temperature sensitivity of Faraday-rotator material. Manufacturers usually specify the temperature range within which the characteristics of their isolators hold true.

***Reading the data sheet*** Let's now review a typical specifications sheet of an isolator (Figure 13.21).

The words "polarization independent," as noted previously, mean "polarization insensitive." The "DESCRIPTION," "FEATURES," and "APPLICATIONS" sections require no explanations, so let's turn to the "PERFORMANCE SPECIFICATIONS." The manufacturer specifies three operating wavelengths around which the isolator holds its characteristics: 1310, 1480, and 1550 nm.

The specification "Stage" shows how many units, such as the one shown in Figure 13.20(b), make up the entire device. To increase isolation, manufacturers combine two units; in this case, the device is referred to as a *double-* (or *two-* or *dual-*) *stage isolator.* You can see from the specifications table that peak isolation increases from 40 dB for a single-stage isolator to 60 dB for a double-stage device. Even typical—*not* peak—isolation up to 70 dB can be found among commercially available double-stage isolators [30]. It should not be surprising to learn that the price for getting the higher isolation with a double-stage device is increased insertion loss. Indeed, typical values increase from 0.4 dB for a single-stage device to 0.6 dB for a double-stage unit. However, typical values of insertion loss of a single-stage isolator can be as low as 0.15 dB [30]. Note that both isolation and insertion loss—the main characteristics of an isolator—are specified for a given range of wavelength and temperature, as discussed above.

The specification "Grade" gives the customer three options—super (S), high (H), and average (A) grades—for the same device. Observe how the characteristics of this isolator differ from one grade to another.

Note especially the low values of "Polarization Dependent Loss" (PDL) the manufacturer has achieved. The isolator with a PDL of less than 0.1 dB is really polarization independent. But isolators with a PDL of less than 0.02 dB are available commercially [30].

As for "Polarization Mode Dispersion" (PMD), reread Section 5.3 and see Figure 5.16, where this problem is discussed in detail. With that background again fresh in mind, you'll readily understand why the manufacturer specifies PMD for an optical isolator.

POLARIZATION
INDEPENDENT ISOLATORS **M-II**

## DESCRIPTION

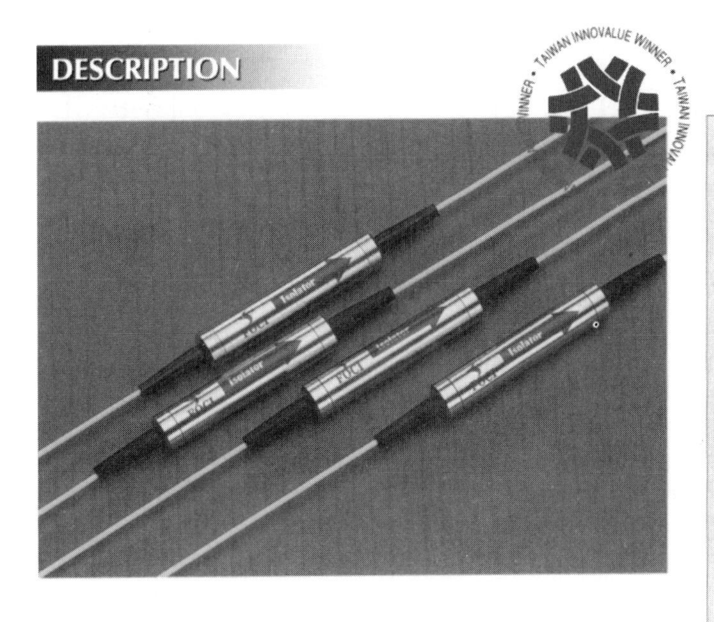


Unwanted backward travelling light can hinder laser and subsystem performance. Protect your Erbium-Doped Fiber Amplifiers, WDM systems and other instruments from back reflection with FOCI's Polarization Independent Isolators.

M-II's are passive non-reciprocal; high performance components designed to suppress optical feedback in laser-based fiber optic systems. FOCI's Polarization Independent Isolators are ultra reliable devices featuring high isolation and low insertion loss over wide ranges of temperature and wavelength.

With unbeatable PDL, PMD and other optical characteristics, the M-II series are the best quality isolators available. These highly reliable singlemode isolators are fabricated with reliable BIG and YVO4 crystals. Using laser welding technology, the M-II series are optical path epoxy free and come with your choice of single or dual stage packaging, plus your choice of pigtailings and connector terminations.

## FEATURES

- Ultra high isolation
- Minimum polarization dependent loss (PDL)
- Polarization mode dispersion (PMD) free
- Optical path epoxy free
- Low insertion loss
- Environmentally stable

## APPLICATIONS

- Optical amplification
- Optical transmission
- CATV
- Fiber laser
- High-bit rate optical communications
- High speed analog optical systems

**Figure 13.21** Data sheet of an isolator. *(Courtesy of FOCI Fiber Optic Communications, Inc.)*

## PERFORMANCE SPECIFICATIONS

| ITEM | Polarization Independent Isolators | | | | | |
|---|---|---|---|---|---|---|
| Operating wavelength, nm | 1310, 1480, 1550 | | | | | |
| Stage | Single | | | Dual | | |
| Grade | S | H | A | S | H | A |
| Peak Isolation, dB (typical) | >40 | >38 | >36 | >60 | >55 | >50 |
| Minimum Isolation, dB (over the center wavelength ± 20nm), (at 25°C, all SOP) | >32 | >28 | >26 | >48 | >46 | >44 |
| Minimum Isolation, dB (over the temperature range 0 ~ 60°C), (all SOP) | >31 | >27 | >26 | >47 | >45 | >44 |
| Typical Insertion Loss, dB (over the center wavelength ± 20nm), (at 25°C, all SOP) | 0.4 | 0.6 | 0.7 | 0.6 | 0.8 | 1.0 |
| Maximum Insertion Loss, dB (over the operating range ± 20nm), (-20 ~ 60°C, all SOP) | 0.5 | 0.7 | 1.0 | 0.7 | 0.9 | 1.3 |
| Polarization dependent Loss, dB | <0.1 | <0.15 | <0.2 | <0.1 | <0.15 | <0.2 |
| Return Loss, dB (input/output) | >65/60 | >60/55 | >60/55 | >65/60 | >60/55 | >60/55 |
| Polarization Mode Dispersion, ps | <0.5 | <0.5 | <0.5 | <0.05 | <0.07 | <0.15 |
| Operating Temperature, °C | -20 ~ +60 | | | | | |
| Storage Temperature, °C | -45 ~ +85 | | | | | |
| Maximum Output Power, mW | 300 | | | | | |

Note : SOP = States of Polarization

## SPECTRAL PERFORMANCE

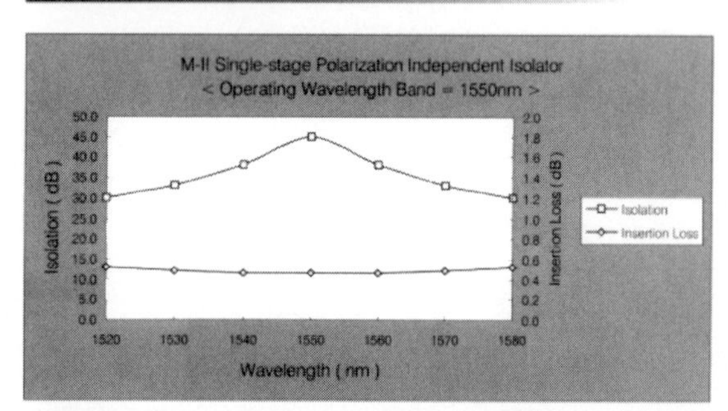


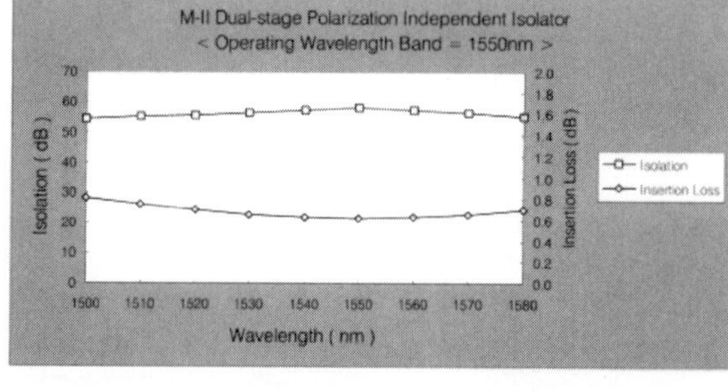


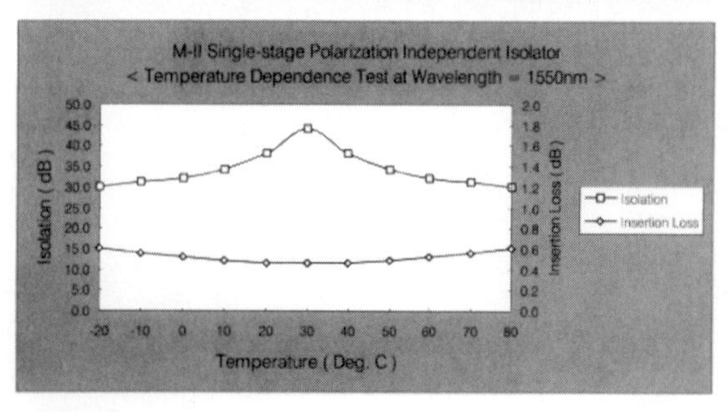


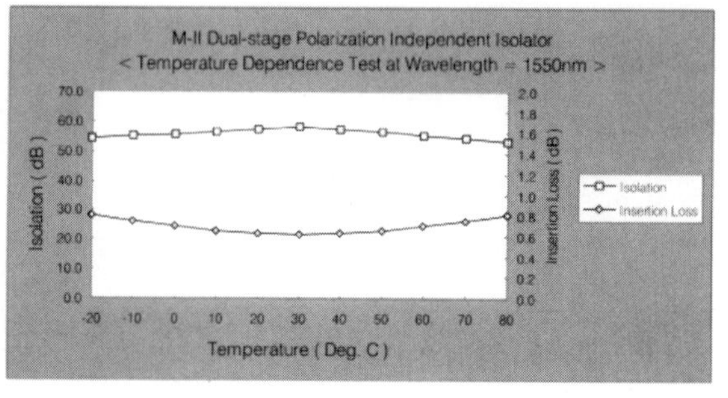


**Figure 13.21** (continued)

The manufacturer also specifies the "Maximum Output Power" the isolator can handle. This is because one of the applications for the device is in EDFAs, where high power is the main feature.

Typical mechanical dimensions of these devices are a diameter of 5 to 6 mm and a length of 30 to 40 mm (50 to 60 mm with pigtail sleeves), although lengths around 20 mm can also be found on the market.

The graphs showing isolation and insertion loss as functions of wavelength and temperature are self-explanatory and very informative. It would be a good idea for you to study these graphs.

Light entering an isolator meets the surface of the input polarizer and is reflected. Thus, an isolator itself is the source of backreflected light. To combat this detrimental feature, some manufacturers cover the surfaces of its entry point with antireflection (AR) coatings. Others tackle the problem by using angle-ended fibers, a measure similar to using angled physical-contact (APC) connectors. (See Section 8.2.) Both approaches result in high *return loss*—around 65 dB—that manufacturers usually specify in their data sheets.

Packaging isolators with lasers and coupling them into fibers are interesting topics, indeed, but fall outside the scope of this text [31].

## Circulators

Circulators are the nonreciprocal devices that direct a light signal from one port to another sequentially in only one direction, as Figure 13.22 shows. Three-, four-, and six-port circulators are commercially available for use in today's fiber-optic networks. We have already mentioned applications for these devices in dispersion-compensating units (Figure 6.15) and in add/drop pas-

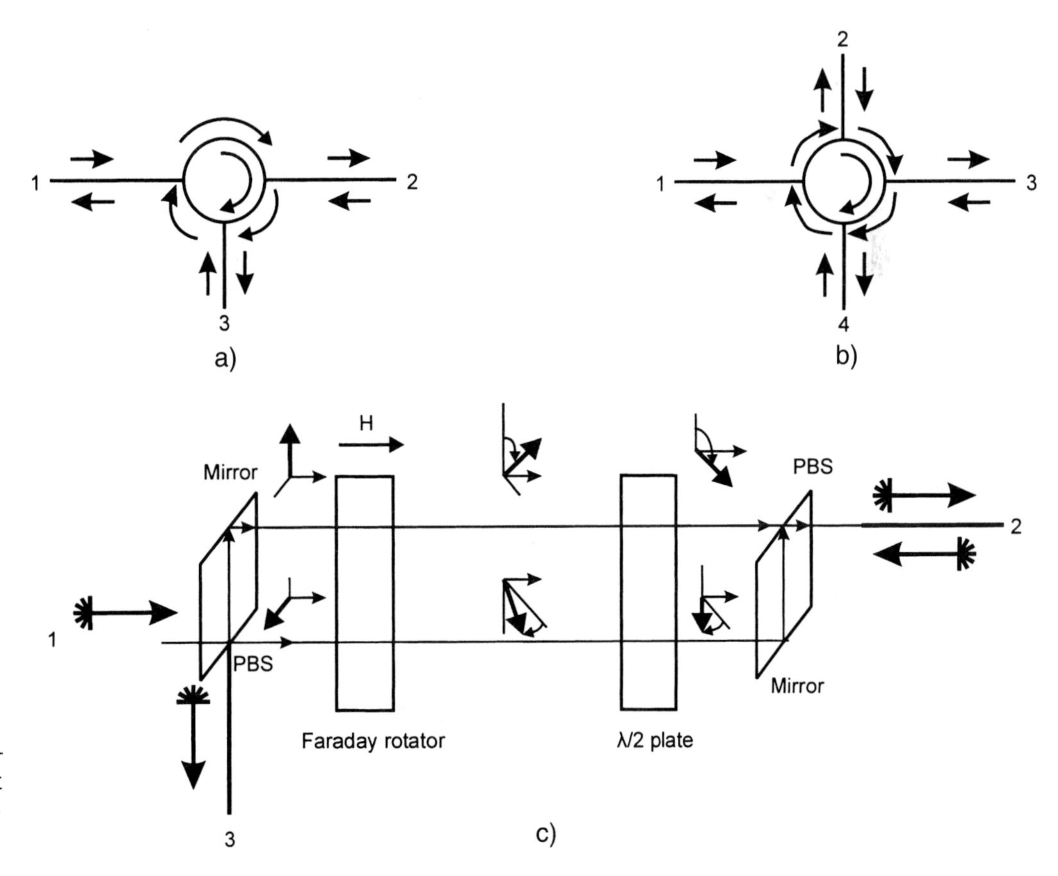


**Figure 13.22** Circulators: (a) Three-port device; (b) four-port device; (c) principle of operation.

sive multiplexers/demultiplexers (Figure 13.15). Circulators are necessary components in the construction of functional network modules.

Circulators can redirect a light signal from port 1 to port 2 and from port 2 to port 3, as shown in Figure 13.22(a). Figure 13.22(b) illustrates a four-port configuration. Light from any port can be redirected to any other port but it has to pass all the intermediate ports in sequential order.

The principle of operation of a circulator is described in Figure 13.22(c). Observe that unpolarized light entering port 1 is split by a *polarization beam splitter (PBS)* into vertically and horizontally polarized beams. Figure 13.22(c) shows that vertically polarized light travels along the top of the diagram and that horizontally polarized light travels along the bottom. A Faraday rotator turns the plane of polarization of both beams to a 45° angle. A quartz rotator, which is the $\lambda/2$ thick plate, turns the plane of polarization of both beams another 45° so that the vertically polarized light becomes horizontally polarized and vice versa. These two beams are recombined by another PBS and leave the device at port 2. Light entering port 2 experiences the same turn at the quartz polarizer—but an opposite turn at the Faraday rotator—so that both beams keep their original state of polarization. (The operation with this light is not shown in Figure 13.22[c].) The PBS recombines the two beams and directs them to port 3. The beams cannot go to port 1 because the PBS does not recombine them to head in that direction. As you can see, circulators have much in common with isolators.

To characterize an optical circulator, you need to consider the following main parameters: First, an optical circulator produces *isolation* in an undesired direction—from port 2 to port 1, for example (Figure 13.22[a]). The value of this parameter can be as high as 70 dB. Second, this device, like any device inserted in a light path, causes *insertion loss,* which can be as low as 0.6 dB among some commercially available circulators. Third, circulators with *polarization-dependent loss (PDL)* less than 0.05 dB are on the market. This is obviously very important because circulators must be polarization independent. *Return loss* is also an important factor because a circulator has reflecting surfaces, as Figure 13.22(c) shows. Typical return loss is more than 50 dB for commercial circulators. *Polarization-mode dispersion (PMD)* is important, too, because circulators operate with orthogonal states of polarization. Typical PMD values for commercial circulators are about 0.1 ps, but devices with numbers as low as 0.01 ps are also available. Circulators, as well as isolators, function over a *range of operating wavelengths (bandwidths).* Circulators are usually designed to work at around 1550 nm or 1310 nm with a typical bandwidth of ± 20 nm.

## Attenuators

*Attenuators are devices that reduce transmitted light power in a controlled manner.* At first glance, it seems a bit strange to decrease light power because our concern, by far, has been to *increase* transmitted power. However, a close look reveals a number of areas where we really need attenuators. The most important applications for attenuators in fiber-optic networks include the following:

- Preventing a receiver from reaching saturation (that is, keeping the input power within the dynamic range of the receiver).
- Wavelength balancing (channel-power equalization) before multiplexing in WDM systems and before amplification by an EDFA.
- Equalizing power among the various nodes in multifiber-distribution networks, such as CATV.
- Testing, which comprises EDFA, bit-error rate, power meters, system-loss simulation, receiver sensitivity, and general laboratory evaluations.

Consider, for example, the need for light-power equalization before amplification. Suppose two channels (wavelengths), $\lambda_1$ and $\lambda_2$, entering an EDFA have different light intensities. As discussed in Sections 12.3 and 12.4, any optical amplifier exhibits gain saturation (Figure 12.18). This means that the gain of an EDFA depends on the input power. Therefore, to amplify all channels equally, their input power needs to be equal. This can be achieved by using an attenuator, as Figure 13.23(a) shows. This straightforward consideration doesn't take into account more subtle effects in EDFAs, such as wavelength-gain dependence and gain tilt. Section 12.4 provides you with all the information necessary to evaluate how unequal power of the input channels would degrade EDFA performance.

There are two functional types of attenuators: plug-style (including bulkhead) and in-line. They are shown in Figures 13.23(b), (c), and (d). A plug-style attenuator is employed as a male–female connector where attenuation occurs inside the device, that is, on the light path from one ferrule to another. A bulkhead attenuator is designed to be placed on a patch panel. An in-line attenuator is connected to a transmission fiber by splicing its two pigtails. (To brush up on this terminology, review Sections 8.2, 8.3, and 12.2.)

***Fixed attenuators*** The principles of operation of attenuators are markedly different because they use various phenomena to decrease the power of the propagating light. The simplest means is to bend a fiber. Just coil a patchcord several times around a pencil while measuring attenuation

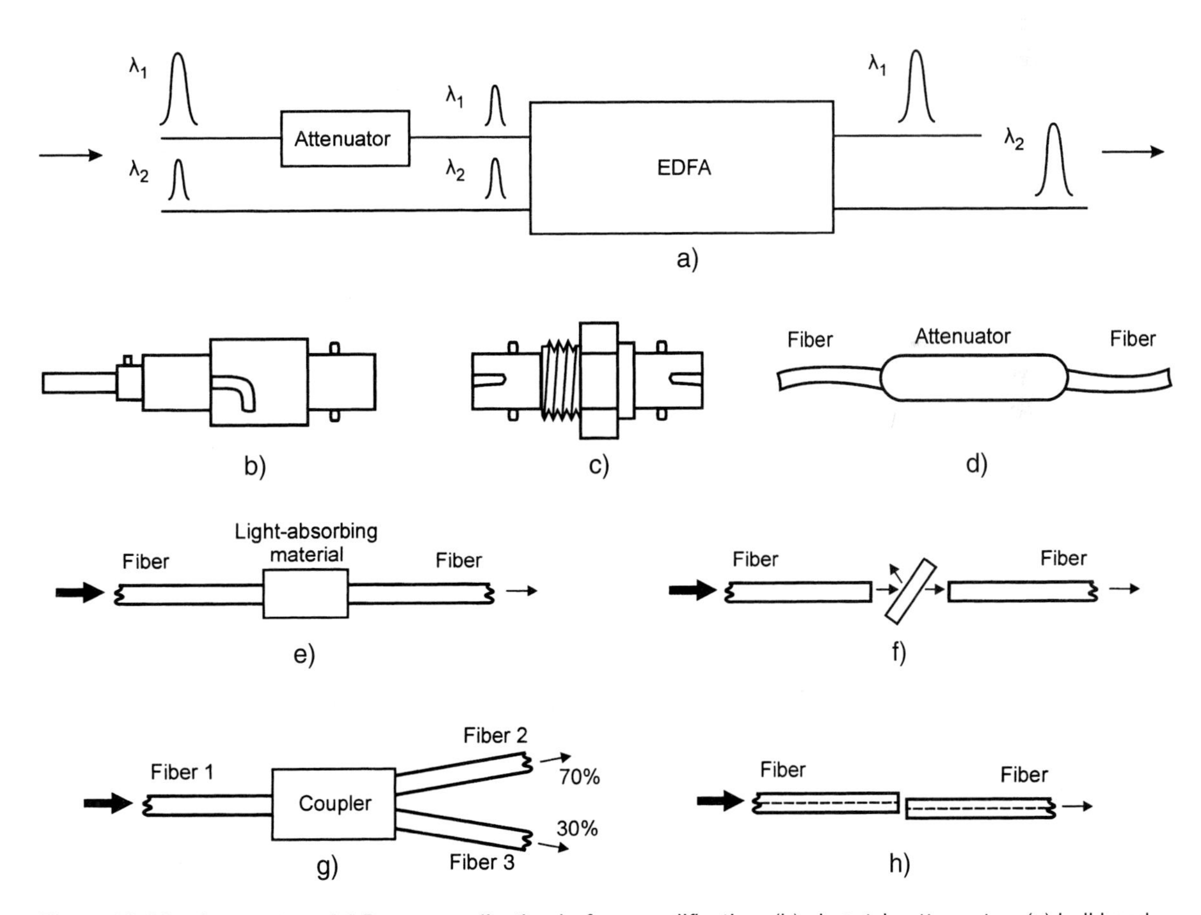


**Figure 13.23** Attenuators: (a) Power equalization before amplification; (b) plug-style attenuator; (c) bulkhead attenuator; (d) in-line attenuator; (e) light-absorbing attenuator; (f) light-reflecting attenuator; (g) coupler-type attenuator; (h) using offset cores and longitudinal gap for attenuation.

with your power meter; then tape this coil. Now you have a primitive but working attenuator. Quite obviously, this method is not good for the high-bit-rate WDM systems.

Manufacturers use various types of light-absorbing material to achieve well-controlled and stable attenuation. For example, a fiber doped with a transition metal that absorbs light in a predictable way and disperses absorbed energy as a heat is used in a commercial attenuator [3]. This principle is shown in Figure 13.23(e).

An attenuator can reflect some light, thus decreasing the power that is launched into a transmission fiber, as Figure 13.23(f) shows. A coupler can be used as an attenuator because the amount of light power that appears at each of the output fibers is always less than the input power. This is shown in Figure 13.23(g) where a 70:30 proportion is given as an example. End separation and lateral displacement, which are detrimental phenomena when connecting fibers, can be used to introduce attenuation, as Figure 13.23(h) shows.

Attenuators fall into two main categories: fixed and variable. *Fixed attenuators* introduce a predetermined amount of loss—for example, 5 dB, 10 dB, and so on. They are truly passive devices, their main advantages being small size and low cost, factors making them suitable for use in patch panels and splicing boxes. *Attenuation tolerance* is always given in the specifications sheets of fixed attenuators. Typical specifications like ± 0.5 dB at 5 dB attenuation, ± 1 dB at 10 dB attenuation, and <10% at 15 through 30 dB attenuation are listed there.

As noted above, an important characteristic of an attenuator is *return loss.* Typical values for commercial fixed attenuators are more than 55 dB. The range of operating wavelengths of attenuators is also included in the manufacturer's specifications sheet. This is shown, for example, as 1550 ± 25 nm.

Thermal stability and guaranteed number of matings (for plug-style and bulkhead attenuators) are key characteristics of fixed attenuators, too.

***Variable attenuators*** *Variable attenuators* allow a network installer and operator to change the attenuation, depending on the initially required amount, to compensate for the aging of a transmitter or amplifier or to respond to a new network's operating conditions. Therefore, variable attenuators are dynamic devices as compared to fixed attenuators, which are static devices. Variable attenuators are not passive components because they require external power to operate.

The principles of operation of these variable devices are quite different from those of their fixed counterparts, but these operating principles can be readily understood if you are familiar with the workings of fixed attenuators. First of all, variable attenuators can be broken into two main categories: those with moving parts and those without moving parts. The latter devices use thermo-optical, electro-optical, or magneto-optical effects to change the amount of light absorbed by the material. Thus, if you apply heat or cold or an electric or a magnetic field to the light-absorbing material shown in Figure 13.23(e), you can control the attenuation of the device. Obviously, you have to use the appropriate material to do this. Even though the concept of variable attenuators without moving parts appears very attractive, from a practical standpoint these types are handicapped by high power consumption, high insertion loss, polarization sensitivity, and low long-term reliability [32].

*Variable attenuators with moving parts* are actually opto-mechanical devices. Just imagine: You can change the angle of a partially reflecting mirror (Figure 13.23[f]) or the gap or core offset (Figure 13.23[h]). One of the uses for a commercially available variable attenuator is bending a coupler (Figure 13.23[g]) by means of a precision stepper motor [32]. This design results in a highly reliable electronically controlled variable attenuator with excellent characteristics. In addition, light from the second output serves as a monitoring signal. Overall, opto-mechanical variable attenuators today exhibit better characteristics than those without moving parts.

Some of the characteristics included in the data sheets of variable attenuators are the same as those for fixed attenuators, such as the range of operating wavelengths, PDL, PMD, return loss, and temperature range. These characteristics are in the same realm of values we found for fixed attenuators or slightly worse. In addition, variable attenuators are characterized by the following parameters:

- *Attenuation range,* also given as *maximum attenuation,* can be as high as 100 dB for special attenuators used in measurement equipment but, typically, it is about 60 dB.
- *Attenuation accuracy,* also known as *resolution,* refers to the precision to which an attenuation value can be fine-tuned. A typical resolution value is 0.5 dB but 0.1 dB or even 0.01 dB can be found among the commercial attenuators. Manufacturers of variable attenuators always specify insertion loss that is typically between 1.5 dB and 2.5 dB. In addition, variable attenuators have a limit on the optical power they can handle. Typical values of maximum light power range from 20 dBm to 25 dBm.

The above discussion provides you with the background you need on how attenuators work and the essentials necessary to understand their data sheets.

In concluding this section, it is pertinent to say a few words about *terminators.* We need them to close any open points in fiber-optic lines because, if we leave these lines open, unwanted backreflection will intrude from the fiber end. These open points are usually unused fibers. Terminators absorb all the stray light traveling along unused fibers and their main characteristic is backreflection, which is typically around –50 dB. Terminators are usually employed as single-side connectors (adapters).

## 13.5 OPTICAL SWITCHES AND FUNCTIONAL MODULES

### Optical Switches

No communications network can exist without switches. Indeed, the telephone network—the first telecommunications network—came into existence only when the set of dedicated wires among a number of customers was replaced by a switching center, a central office. (See Figure 12.1.) Switches, of course, are the vital components of a fiber-optic communications network. Since electronic switching is a well-developed technology, it would seem natural to use these switches in fiber-optic networks. However, the price for using this mature technology is optical-to-electrical (O/E) conversion with all the associated drawbacks of this method. You may recall that O/E and E/O conversions result in the loss of data-stream transparency; that is, such conversion can be done only for a specific bit rate and data format. (See the discussion of regenerators and optical amplifiers in Section 12.1.) In addition, the high power consumption and cost associated with the amount of electronic equipment required diminish the attractiveness of this switching technique. Nonetheless, electronic switching is the only switching technique in today's fiber-optic networks, such as telephone networks, where optical fiber serves as the transmission medium and all terminal equipment is still electronic.

We will concentrate in this subsection on *optical switches, which are the components of fiber-optic networks. These switches route an optical signal without electro-optical and opto-electrical conversions.* They are *active* components because they require external power for operation and they are activated, typically, by a 5-V dc electrical signal.

Optical switches have already found a number of applications in today's fiber-optic communications systems and they will be vital components in tomorrow's fiber-optic networks. (See Chapter 14.) We'll discuss these applications when we consider the specific types of switches.

Since optical switches stem from their electronic predecessors, it's quite natural that most of the terminology and optical-switch architectures come from electronic telecommunications switching technology. Thus, for example, the term *space switch* refers to the switch that physically connects one fiber to another; that is, it provides circuit switching.

We can break optical switches into two broad categories: single switches and multistage (large) switches. Since the latter are built as multistage assemblies of single switches, we will start our discussion with the former.

***Single switches*** Single switches fall into two main functional categories: on/off and passing. An on/off ($1 \times 1$) switch either allows or doesn't allow a light signal to pass, as Figure 13.24(a) shows. A $1 \times 2$ passing switch directs the light signal from fiber 1 to fiber 2 or to fiber 3, as Figure 13.24(b) shows. The $1 \times 2$ switch configuration in Figure 13.24(b) is of course no more than a simple example. Single $1 \times N$ passing switches are commercially available. A $2 \times 2$ passing switch can connect two fibers to two other fibers. Bear in mind that a $2 \times 2$ passing switch can be either in the *bypass (bar) state* or in the *cross (inserted) state,* as you see in Figure 13.24(c). The term *nonblocking* means that a switch can connect any of its inputs to any of its outputs. An example of a $2 \times 2$ blocking switch is given in Figure 13.24(d), where the switch connects only fiber 1 to fiber 4.

Several simple examples of switch *applications* underscore the importance of switches in fiber-optic communications systems [7]:

- On/off switches are used to *isolate sources and receivers* in test equipment. $1 \times 2$ switches permit channel selection and are used for *protection switching* (to redirect traffic when one fiber fails).

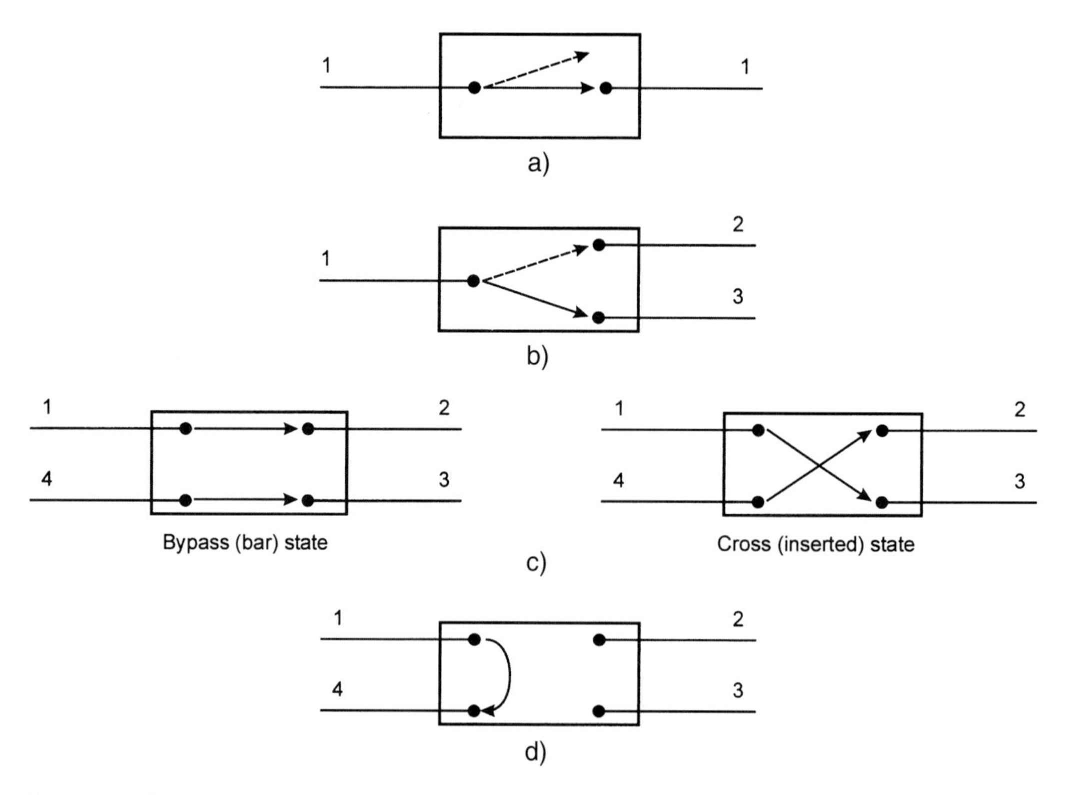


**Figure 13.24** Configurations of optical switches: (a) On/Off ($1 \times 1$) switch; (b) $1 \times 2$ passing switch; (c) $2 \times 2$ passing (nonblocking) switch; (d) $2 \times 2$ blocking switch.

- $1 \times N$ switches are used for the *testing and measurement of fiber-optic components,* the *remote testing* of fiber-optic communications systems, and fiber-optic network *restoration.*
- $2 \times 2$ switches are used for node bypassing in fiber-optic networks. A typical application of a $2 \times 2$ switch is in FDDI networks. When a station fails or loses power, the switch automatically changes to the blocking state, thus ensuring that the traffic stream avoids the failed node.

Single switches are manufactured in *latching* and *nonlatching* styles. The latching operation ensures that the switch remains in its selected position if power is lost.

If you want to become familiar with the principle of operation of commercially available switches, be aware that they come in *electro-optical, optomechanical, thermo-optical,* and *acousto-optical* forms. (The latter is simply a version of an acousto-optical tunable filter.) Some of the principles of switch operation are depicted in Figure 13.25.

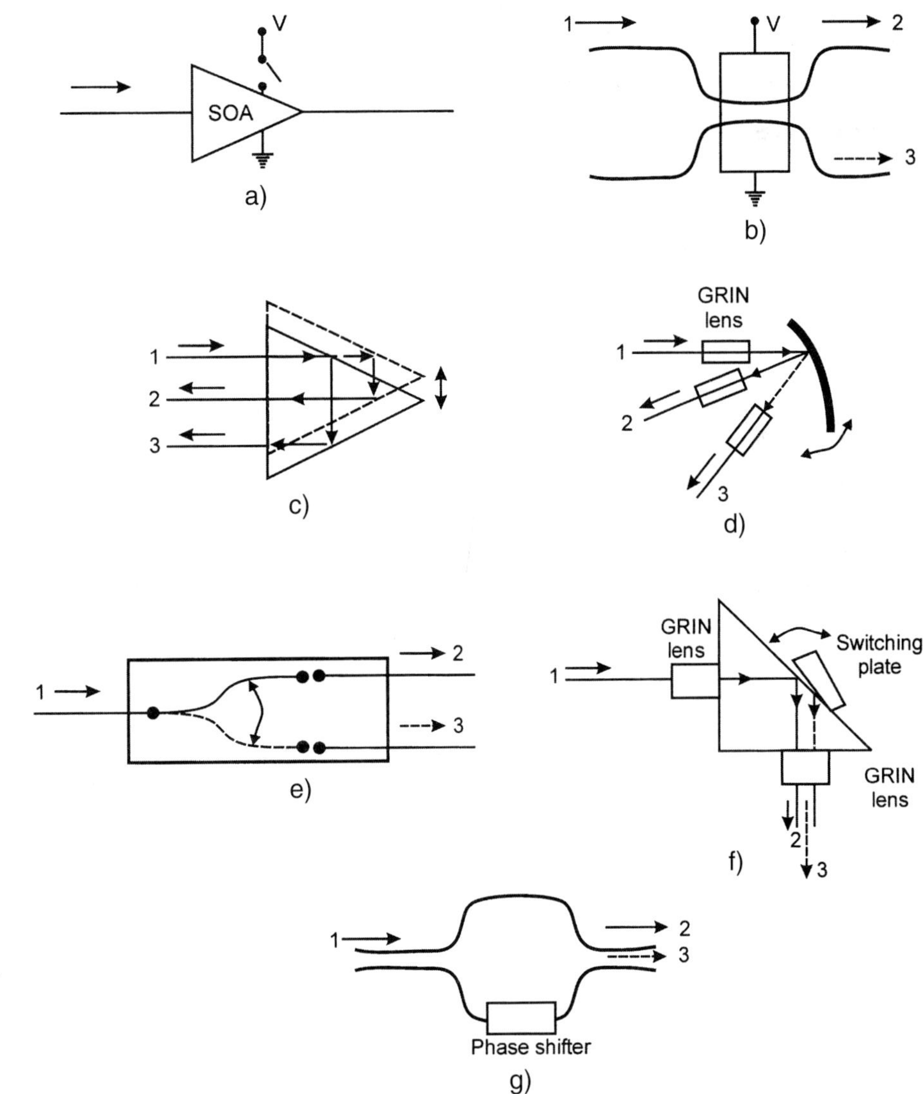


**Figure 13.25** Principles of operation of optical switches: (a) Semiconductor optical amplifier (SOA) as on/off switch; (b) waveguide coupler fabricated on $LiNbO_3$; (c) moving-prism switch; (d) spherical-mirror switch; (e) fiber-moving switch; (f) frustrated total internal reflection (FTIR) switch; (g) Mach-Zehnder interferometer thermo-optical switch.

If you switch the bias voltage of a semiconductor optical amplifier (SOA) on and off, you have created an electro-optical (EO) switch, as seen in Figure 13.25(a). This is so, remember, because an SOA amplifies light when it is biased and absorbs light when it is not biased. (See Section 12.3.) Another example of an EO switch is given in Figure 13.25(b), where the coupling ratio of a waveguide coupler is dependent on the applied voltage. Changing the coupling ratio is achieved by fabricating the coupler on top of $LiNbO_3$—the material whose refractive index depends on the value of the applied voltage. (See Section 10.4.)

*Opto-mechanical switches* are based on the mechanical motion of the optical components. For example, vertically moving the prism in Figure 13.25(c) enables the switching of an optical signal from fiber 2 to fiber 3. The same result can be achieved by slightly pivoting a spherical mirror, as Figure 13.25(d) shows. Graded-index (GRIN) lenses facilitate passage of the coupling light from and into optical fibers. You can no doubt imagine many optical schemes that would produce optical switches (which, by the way, is a good exercise to challenge your creativity). One such example is shown in Figure 13.25(e). A piece of input fiber is moving from one position to another, thus switching an optical signal. You can use a stepper motor or a solenoid to set a fiber into motion. If you use a solenoid, don't forget to apply a metallic sleeve to the fiber.

The *principle of frustration of total internal reflection (FTIR)* is used to manufacture the switch shown in Figure 13.25(f) [33]. You will recall that light partially penetrates another refractive medium even under the condition of total internal reflection. We call penetrated light an *evanescent wave,* a phenomenon described in Section 4.3. To utilize this effect, a switching plate is attached to the prism that provides total internal reflection. When the switching plate comes into close contact with the prism, total internal reflection is "frustrated," which results in the reflected beam's moving in a slightly different direction from the direction it would take were total internal reflection not being frustrated. Thus, by moving the switching plate toward the prism, you can direct a light signal into fiber 3. When the switching plate does not contact the prism, regular total internal reflection occurs and the optical signal enters fiber 2. You have to bear in mind that the mechanical motion of optical components is really microscopic and occurs quickly and reliably. In fact, optomechanical switches are among the most popular types of optical switches today.

An example of a *thermo-optical switch* is shown in Figure 13.25(g). A Mach-Zehnder interferometer incorporates a phase shifter in one of its legs. By applying heat, you can control the amount of phase shift, which means you can direct an optical signal either into fiber 2 or into fiber 3. Thermo-optical switches have faster switching speeds than their optomechanical counterparts and, most importantly, can be employed with planar solid-state technology as large switch matrixes [34].

When reading the data sheet of a single optical switch, you should know that these switches are characterized by the following main parameters:

- *Extinction ratio,* which is the characteristic of an on/off switch. This is the ratio of the light power of the on-state to the off-state position of the switch. It should be as high as possible; it's usually on the order of 45 to 50 dB.
- *Insertion loss,* which is, of course, the measure of power lost due to the presence of a switch; it is on the order of 0.5 dB
- *Crosstalk,* which is the ratio of the output power produced by the desired input to the output power produced by undesired input(s). It should be as high as possible; it is usually on the order of 80 dB.
- *Switching time,* an extremely important characteristic. As is true for tunable filters, the required switching time depends on the switch's applications. For today's circuit-switching networks, a switching time on the order of μs and even ms suffices, but for tomorrow's packet-switching fiber-optic networks we'll need switching times on the order of ns and

even ps. Optomechanical and thermo-optical single switches have switching times ranging from 2 to 20 ms, while electro-optical single switches offer speeds on the order of ns.

Other characteristics found in the switches' data sheets—such as the range of operating wavelengths, PDL, and ambient temperature—are already familiar to you, so there is no need to discuss them here.

***Multistage (large) switches*** Since we want to work with multichannel fiber-optic networks, we need switches capable of switching many individual channels simultaneously. These switches cannot be built as a physical assembly of single switches simply because space won't permit it, not to mention the power consumption and coupling problems involved. Thus, the need for *integrated switches,* or *switching fabrics,* arises. A switching fabric is an integrated switch enabling the simultaneous connection of $N$ inputs to $M$ outputs.

(For a full treatment of the topic of switching fabrics, consult the specialized sources [6], [7], and [35] listed at the end of this chapter. We restrict ourselves here, as usual, to commercially available devices.)

The architecture of a switching fabric is based on combinations of single optical switches. For example, a $4 \times 4$ switching fabric can be built using $2 \times 2$ optical switches. (See Figure 13.26[a].) Called a $2 \times 2$ four-array switch, it is implemented as a solid-state integrated unit [36]. The same architecture is used to manufacture a $2 \times 2$ eight-array switching matrix on a single chip [43].

Another example of an $8 \times 8$ optical-matrix switch is given in Figure 13.26(b), where only several characteristic connections are presented [43]. It is interesting to note that commercially

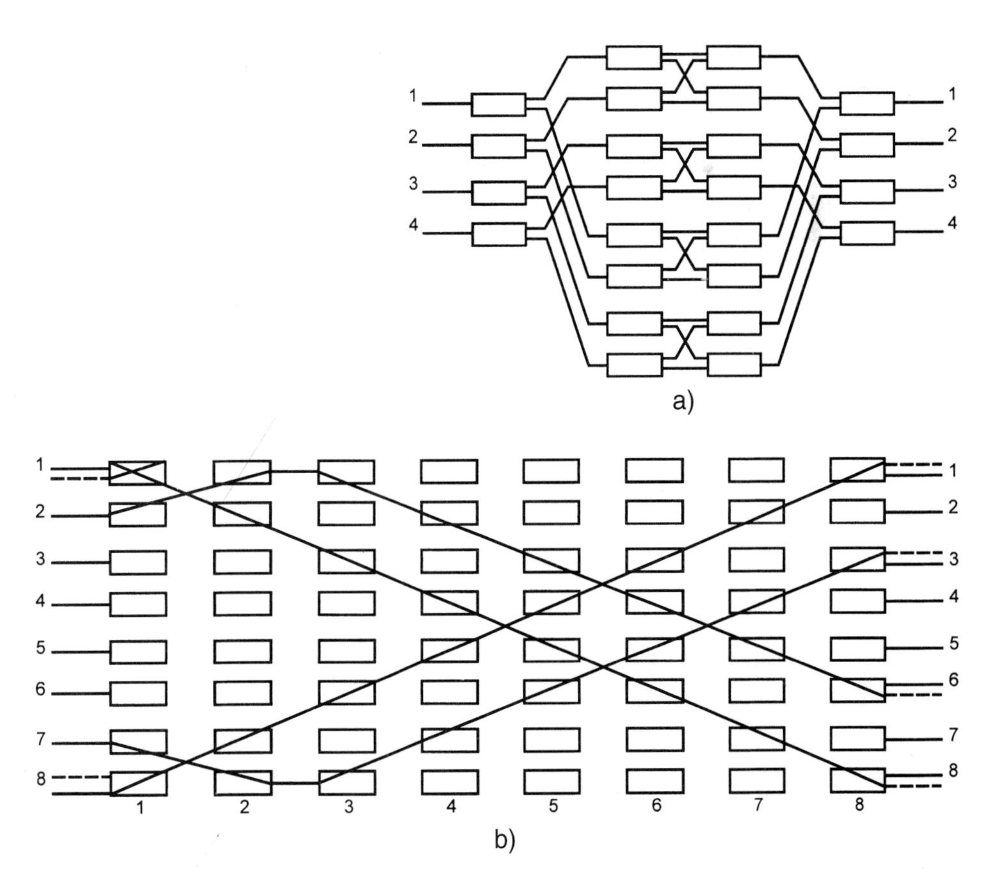


**Figure 13.26** Switching-matrix architectures: (a) $2 \times 2$ 4-array; (b) $8 \times 8$ optical matrix switch.

available switching fabrics are based on thermo-optical switches. For an in-depth look at such a switching matrix, see reference [37]. The matrix is formed by intersecting silica waveguides. Each intersection is filled with an index-matching fluid. Under normal conditions, this fluid allows light to pass through the intersections with minimum loss because its refractive index is closed to the refractive index of a waveguide. This is why it is called index-matching fluid. To redirect light to another output, a thermal element creates a bubble in the fluid. This bubble reflects the light by total internal reflection to the desired waveguide. The switch has the following characteristics: dimension, 32 × 32; insertion loss, 3.5 dB for directed light and 15 dB in a 512 × 512 matrix; switching speed, less than 10 ms; PDL, less than 0.25 dB; and crosstalk, less than –50 dB.

In general, $N \times M$ switches can be built for two functions: to switch each input fiber to each output fiber (a *directional switch*) and to switch each input fiber to one or more output fibers simultaneously in a broadcast configuration (a *distributional switch*).

We will return to optical switches several more times: when considering functional blocks later in this section and when discussing various fiber-optic-network issues in Chapter 14.

## Wavelength Converters

*Wavelength converters,* also called *wavelength translators* or *transponders, are devices that change transmission wavelengths while keeping data carried by these wavelengths unchanged.* The basic function of a wavelength converter is shown in Figure 13.27(a).

We need wavelength converters for several reasons. First, they connect incompatible equipment. Such an example is the conversion of 1300-nm-carrying wavelengths of first-generation fiber-optic networks to 1550 nm, which is the main operating wavelength of today's networks.

Another need for converters arises because we have different fiber-optic networks with different providers and different standards; therefore, we need wavelength conversion to traverse from one fiber-optic network to another. What's more, wavelength conversion helps us to reduce the number of wavelengths required. (See the discussion of Figure 12.5[c].)

There are two basic types of wavelength converters: opto-electronic and all-optical. An *opto-electronic wavelength converter* works as a regenerator. It converts an optical input signal

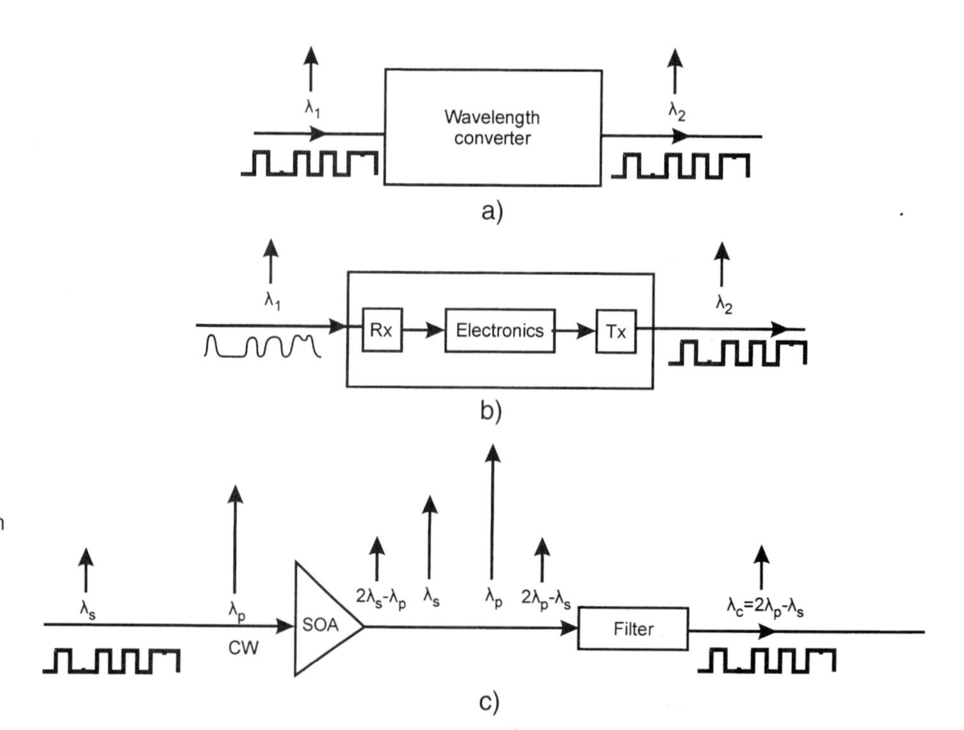


**Figure 13.27** Wavelength converters: (a) Basic function; (b) opto-electronic wavelength converter; (c) four-wave-mixing wavelength converter.

into electrical form, generates a logical copy of an input signal with a new amplitude and the shape of its electrical pulses, and uses this signal to drive a transmitter to generate an optical signal at the new wavelength. This specific type of wavelength converter is called a *transponder.* (To brush up on regenerators, turn to Section 12.1.) The principle of operation of an opto-electronic wavelength converter is illustrated in Figure 13.27(b).

All-optical wavelength converters use nonlinear effects such as cross-phase modulation (XPM) and four-wave mixing (FWM) to change operating wavelengths. (See Section 6.4.) Let's consider, for example, the principle of operation of an FWM converter [7]. An information signal at $\lambda_s$ along with a probe signal at $\lambda_p$ enters a semiconductor optical amplifier (SOA). We want to use an SOA to enhance the FWM effect because within the SOA's active medium light intensities are high. Because of the FWM effect, new wavelengths ($2\lambda_s - \lambda_p$ and $2\lambda_p - \lambda_s$) will be generated, so we can filter one of these wavelengths as a new signal carrier. Data initially carried by wavelength $\lambda_s$ will be completely preserved and transmitted farther along at the new wavelength, $\lambda_C = 2\lambda_p - \lambda_s$, as shown in Figure 13.27(c). An FWM wavelength converter is transparent for any bit rate and signal format but the efficiency of this conversion is very low.

All-optical wavelength converters are not commercially available yet but the demand for them will grow as all-optical fiber-optic networks move from the research laboratories into the field.

(Additional background on all-optical wavelength converters can be found in references [7], [38], and [39].)

## Functional Modules

The passive and active optical components described here allow system designers to build *functional modules,* also called *subsystems* or *subassemblies.* We'll consider optical add/drop and optical cross-connect modules because they are the fundamental functional blocks needed to build fiber-optic networks.

First of all, let's distinguish between two broad classes of fiber-optic networks: *static* and *dynamic.* An example of a static network is the WDM network shown in Figure 12.5(b). The main features of a static network are that it operates at fixed wavelengths and features rigid architecture. An example of a dynamic network is the wavelength-routing network in Figure 12.5(c). Through this network, the same information signal that is transmitted at different wavelengths and lightpaths from a sender to a recipient can be dynamically changed. In other words, a dynamic network operates at variable wavelengths and its architecture can be reconfigured, or rearranged, as required.

Let's consider *optical add/drop multiplexers (OADM)* and *optical cross-connects (OXC)* in both static and dynamic applications. (It might be advisable for you to refresh your memory on the add/drop problem by rereading Section 12.1.)

The basic function of an OADM is to take channel(s) out of, and add channel(s) to, a network's main data stream. A *static (fixed) OADM* is easy to construct: Simply use a filter to select a dropping wavelength and a multiplexer to add a new channel at the same wavelength because each wavelength in a WDM network is associated with an individual channel. This concept is illustrated in Figure 13.28(a). An example of a fixed (passive) OADM module is given in Figure 13.15.

The two main technologies used today for filtering dropping wavelengths are thin-film interference filters and fiber Bragg gratings [40]; however, other filters, some of which are discussed in Section 13.3, are also in use. Multiplexing can be done by any means discussed in this section.

A *dynamic OADM* is also called a *configurable OADM (COADM),* or a *rearrangable OADM.* It can be made by several means. One popular architecture is shown in Figure 13.28(b). The module includes a WDM demultiplexer (DEMUX), a set of 2 × 2 switches, a set of variable attenuators, and a multiplexer (MUX). The main data stream, in the form of a multiplexed set of wavelengths ($\lambda_1, \lambda_2 \ldots \lambda_N$), enters the module through an input fiber. A DEMUX demulti-

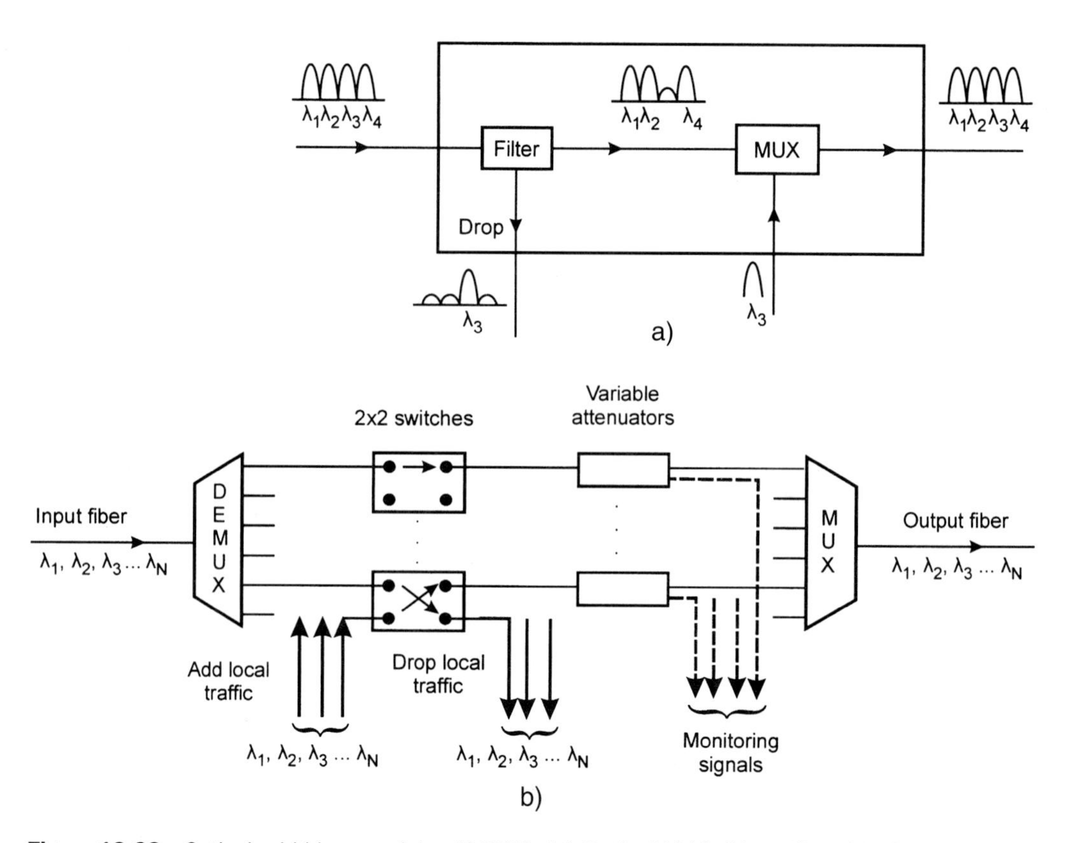


**Figure 13.28** Optical add/drop modules (OADM): (a) Static OADM; (b) configurable OADM.

plexes the incoming traffic into individual wavelengths. The $2 \times 2$ switch, in a bypass (bar) state, allows a channel to pass through while the switch, in a cross state, adds and drops a wavelength. This activity is shown in Figure 13.28(b). We need variable attenuators to equalize the light power of all the channels, a necessity to perform multiplexing efficiently and to transmit outgoing traffic to the following EDFA. (See Section 12.4.)

A COADM can be built using the components discussed above. For example, it can be based on an acousto-optical tunable filter and couplers [40]. (To learn more about OADMs, see references [5], [7], [39], and [40].)

Look again at Figure 12.1. The active nodes of telephone networks are central offices, where trunks and local links enter and where switching takes place. The same function for fiber-optic networks is performed by cross-connects. We'll concentrate on *optical cross-connects (OXCs)* even though electro-optical cross-connects exist and are in use [7]. You will also find the terms *wavelength cross-connect (WXC)* and *digital cross-connect (DXC)* used to denote these devices. (Bear in mind that *WXC* can also denote any node, including passive ones, of a fiber optic network.)

Originally, a static OXC was nothing more than a rack of patch panels where input and output fibers were mechanically connected. (See Figure 8.18[b].) Today, a static OXC is based on a passive wavelength-division multiplexer/demultiplexer or a wavelength router, which we have already discussed. (See Figures 13.8 through 13.16.) The architecture of a dynamic OXC is much more sophisticated than that of a static OXC and includes MUXs/DEMUXs and optical switches.

There are two main types of architecture for dynamic OXCs: one without wavelength conversion (Figure 13.29[a]), the other with wavelength conversion (Figure 13.29[b]). See also references [1], [18], and [41].

Consider Figure 13.29(a): Trunks 1 through M deliver long-haul and regional traffic to an OXC. Each trunk, carrying wavelengths (channels) from $\lambda_1$ to $\lambda_N$, enters a DEMUX, where incoming traffic is demultiplexed. Then each wavelength is directed to an appropriate $N \times N$ switch so that each switch operates with a single wavelength. For example, the first switch works with signals from all the trunks carried by wavelength $\lambda_1$; the second switch works with all the incoming signals at wavelength $\lambda_2$, and so on. After switching, all the wavelengths are directed to the appropriate MUXs, where they are multiplexed back to the individual data stream at wavelengths from $\lambda_1$ to $\lambda_N$. We need M DEMUXs and MUXs and M switches; the switches must have a minimum $N \times N$ configuration.

Local traffic can be added and dropped through the switches, an action that requires increasing the number of ports for the switches.

A dynamic OXC with wavelength conversion is shown in Figure 13.29(b). Information signals from all the trunks are demultiplexed and directed to an optical switch, where the actual switching occurs. Note that here we need a much larger switch with a minimum $MN \times MN$ configuration. The switched signals enter the wavelength converters, where they are converted into sets of appropriate wavelengths from $\lambda_1$ to $\lambda_N$. Each of these sets is multiplexed by MUXs and sent over the proper trunks.

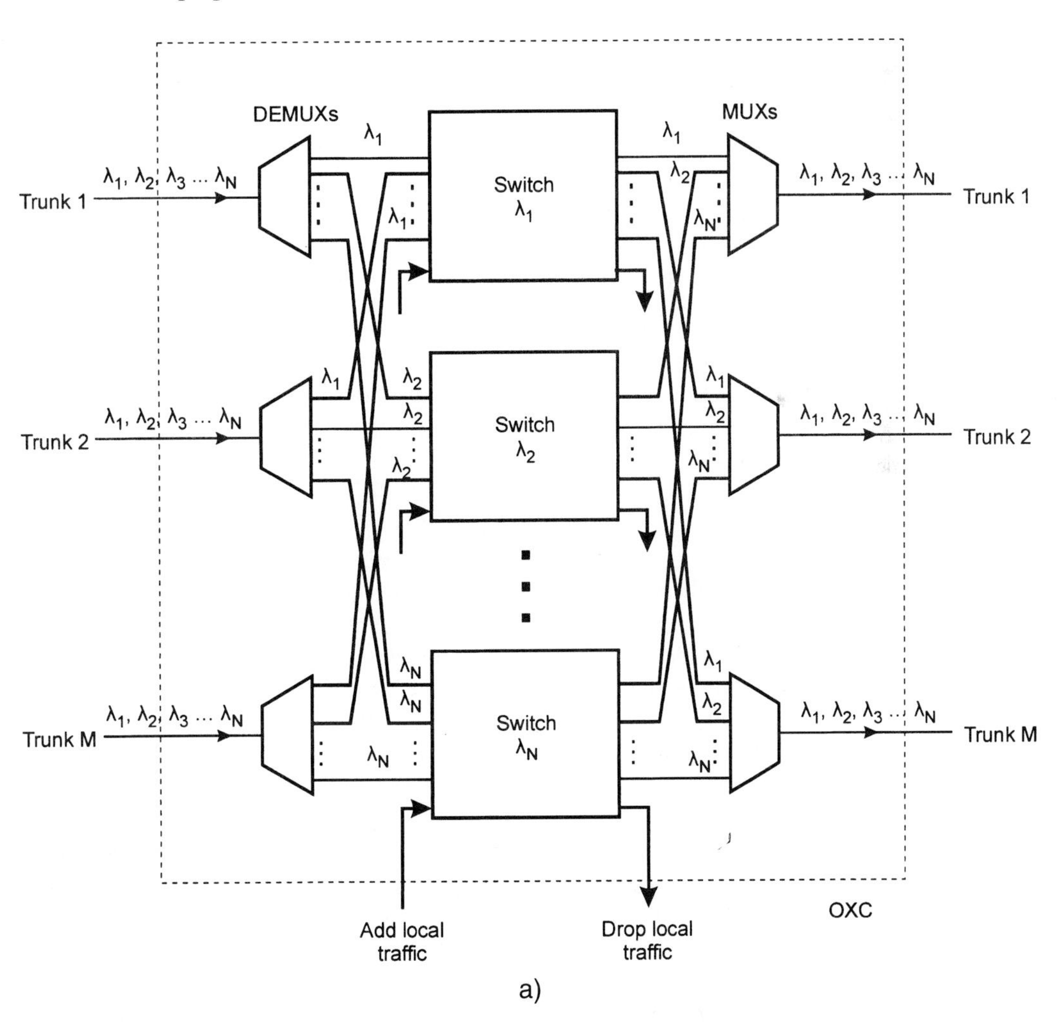


**Figure 13.29** Optical cross-connect (OXC): (a) OXC without wavelength conversion; (b) OXC with wavelength conversion.

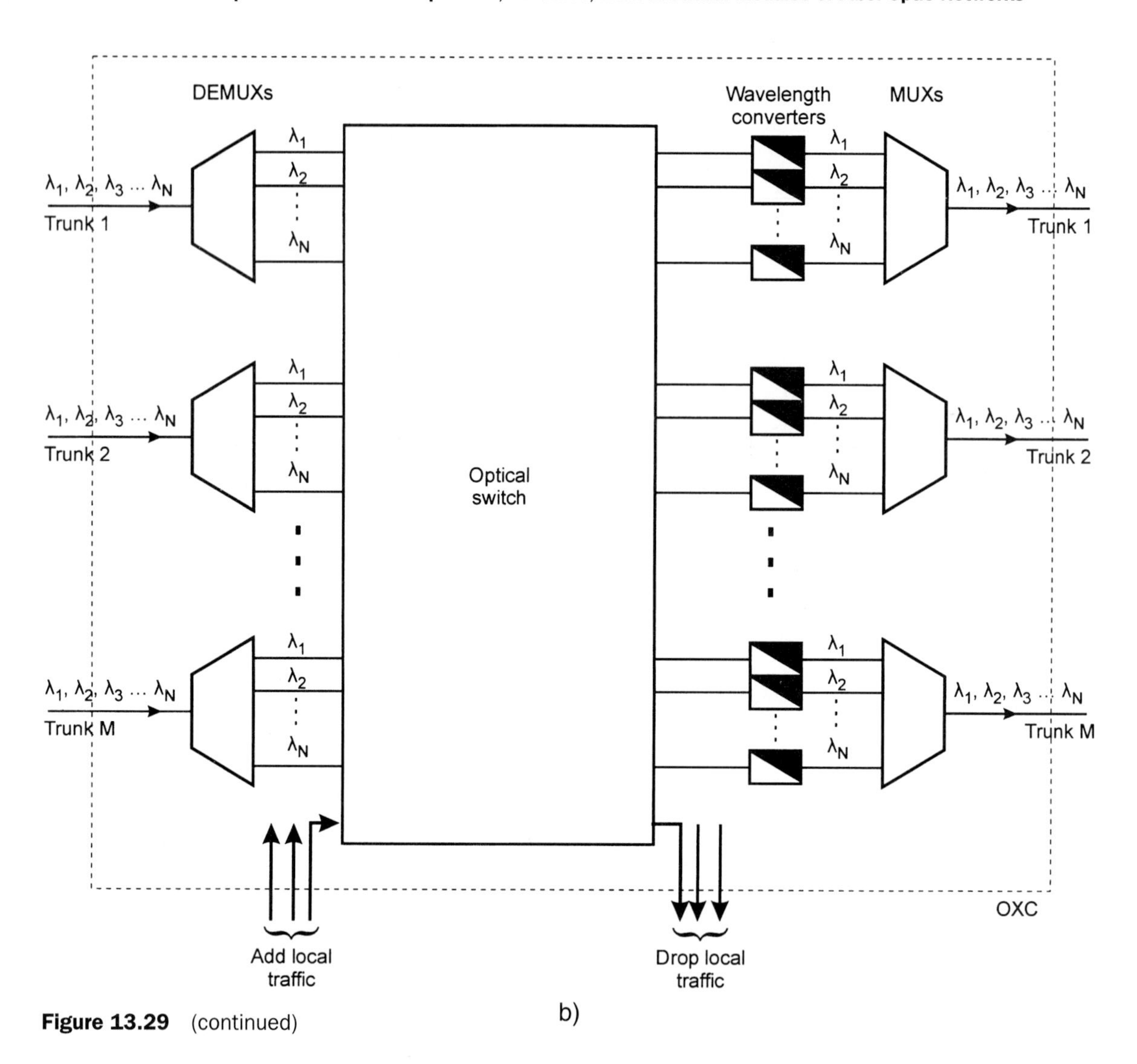


**Figure 13.29** (continued)

The advantage of an OXC with wavelength conversion is that any incoming wavelength (channel) can be switched to any output port at any wavelength. The price we pay for this type of OXC is considerable difficulty in its implementation.

Configurable optical add/drop modules and optical cross-connects are key subsystems in building all-optical networks [42]. COADMs and OXCs also serve here as examples of how functional modules can be built from the components described in this section.

## Conclusion

Today's fiber-optic-network components, which still work by and large by opto-electrical and electro-optical—and even all-optical—conversion, remind us of electronic tubes at the dawn of the electronics era. Not until we can mass produce, at low cost, all-optical integrated single-chip components will we make all-optical fiber-optic networks a worldwide reality. True, such components are indeed being developed in research laboratories and some are already commercially available and competing with their opto-electronic counterparts. But, remember, the real judge is the marketplace. We cannot emphasize enough that the deployment of fiber-optic networks will begin on a global scale *only* when high-performance, reliable, easy-to-manufacture all-optical low-cost integrated single chips replace today's dinosaurs. This is a development you should watch for to keep abreast of the cutting edge of fiber-optic communications technology.

It's important to understand that new WDM components are, on the one hand, the key to the future direction of fiber-optic networks and, on the other hand, the driving force behind the development of the fabrication, testing, and measurement techniques to meet the new demands of the industry as it strives for a higher technological level and tighter cost controls.

## SUMMARY

- All components for WDM systems can be divided into active and passive, depending on whether or not they need external power to operate. Transmitters, receivers, and optical amplifiers are active components. Other active components used in WDM transmission are optical switches, wavelength converters, and tunable filters and attenuators. The list of passive components includes couplers, isolators, circulators, and fixed filters and attenuators.
- Optical switches switch an optical information signal without electro-optical or opto-electrical conversions. They can perform on/off operations or reroute an information signal. Various physical principles are used to make optical switches; electro-optical, opto-mechanical, thermo-optical, and acousto-optical switches are commercially available. Among switch characteristics, switching time is one of the most important. Popular opto-mechanical and thermo-optical switches have switching times ranging from 2 to 20 ms, which is sufficient for today's circuit-switching networks. For tomorrow's packet-switching networks, however, switching on the order of a few ns and less will be required. Only electro-optical switches meet this target today. For WDM systems, switches with many input and output ports—a so-called $M \times N$ configuration—are required. Such a switching matrix, or array, is made by combining single switches in a multistage structure. The problems that still need to be solved for large switches are their integration and miniaturization.
- Wavelength converters (translators) change the carrying wavelength while keeping transmitted data unchanged. These devices fall into two categories: electro-optical and all-optical. Electro-optical converters work as repeaters (regenerators). They convert an optical input signal into electrical form, generate a logical copy of the input signal, and then use this signal to drive a transmitter to generate an optical signal at the new wavelength. All-optical converters use nonlinear effects to reproduce the information signal at the new wavelength. For example, one can enter two signals—an information signal and a probe—into an SOA's active medium. Because of the four-wave mixing (FWM) effect, a new wavelength carrying the input information will be generated. Wavelength converters are active components.
- Passive components work without external power. The following devices are needed to make WDM fiber-optic communications systems work:
  - Couplers, which couple light at the same wavelength from several fibers to several other fibers. They can also split light from one fiber into several fibers; in this case they are called splitters.
  - Passive multiplexers and demultiplexers (MUXs and DEMUXs), which couple and split signals at different wavelengths into, and from, several fibers. An example is a WDM coupler for an EDFA that couples pumping light at 980 nm or 1480 nm with an information signal at 1550 nm into one erbium-doped fiber. Passive MUXs and DEMUXs are key components of WDM systems because they combine and separate several wavelengths (channels) into, and from, one fiber.
  - Optical filters, which allow one wavelength to pass through and reject all others. They play an important role in WDM systems, where each wavelength is associated with an information channel.
  - Attenuators, which reduce transmitted light power in a controlled manner. They prevent receivers from reaching saturation, provide channel-power equalization, and are used for testing WDM systems.
  - Isolators, which allow light to pass in only one direction, thus isolating fiber-optic components from backreflected light. They find application in EDFAs.
  - Circulators, which are nonreciprocal devices that direct light from one port to another sequentially in one direction. They are used in dispersion-compensation units and in add/drop passive WDM MUXs/DEMUXs.
- Passive and active components are used to build functional modules, subsystems, or sub-assemblies. Examples of such functional blocks include optical add/drop multiplexers (OADMs) and optical cross-connects (OXCs). A configurable OADM (COADM) allows us to add and drop a variable number of wavelengths

(channels), depending on the WDM network's needs. A typical COADM includes a passive MUX and DEMUX, optical switches, and variable attenuators. An OXC is essentially a switching center for a WDM fiber-optic network. It includes several MUXs and DEMUXs, optical switches, and wavelength converters. COADMs and OXCs add functionality to WDM systems and allow their conversion into WDM networks.

## PROBLEMS

**13.1** What is the function of a coupler? A splitter?

**13.2** Explain the principle of operation of an FBT coupler.

**13.3** How can we change the output ratio of an FBT coupler?

**13.4** List, define, and comment on the major characteristics of a coupler.

**13.5** Prove that Formula 13.5 is correct.

**13.6** What techniques are used to fabricate couplers? Explain.

**13.7** Explain how to make a WDM coupler.

**13.8** Explain how and why light power varies along the coupling region in an FBT coupler.

**13.9** Calculate the length of the coupling region of a 3-dB coupler, enabling it to split 1550.12-nm and 1554.13-nm wavelengths.

**13.10** Formulas 13.11 and 13.12 follow from Formulas 13.20 and 13.21 under a certain condition. What is this condition? Prove your answer.

**13.11** Draw the power distribution along the coupling region for $\Delta\beta/2k = 10, 5, 1$, and 0.1.

**13.12** Explain why a WDM MUX/DEMUX is such an important component of WDM systems.

**13.13** List, define, and comment on the basic characteristics of passive WDM MUXs/DEMUXs.

**13.14** List the major types of WDM MUXs/DEMUXs.

**13.15** What will be the angle of the first principal maximum for the angular dispersion you obtained in Example 13.2.1?

**13.16** What angular separation can we obtain when selecting two wavelengths—1550.12 nm and 1550.92 nm—if the pitch of a diffraction grating is 5 μm? To separate these wavelengths, what length do we need between the transmission diffraction grating and the fiber ends?

**13.17** Analyze the characteristics of the 8-channel DWDM given in Figure 13.14(b).

**13.18** Analyze Table 13.21. State which type of WDM MUX/DEMUX you would use in your WDM systems and explain why.

**13.19** Look at Figures 13.15(b) and (c), where add/drop components based on fiber Bragg gratings are shown. Do added and dropped wavelengths have to be the same? Explain.

**13.20** Name and describe the two major classes of filters.

**13.21** What are the basic characteristics of a filter?

**13.22** Name and explain the principle of operation of each of the types of fixed filters you have studied.

**13.23** Prove that Formula 13.36 follows from Formula 13.35.

**13.24** Why do we need tunable filters? What major parameters are used to characterize tunable filters?

**13.25** List and explain the principle of operation of each of the types of tunable filters you have studied.

**13.26** Calculate the frequency of a surface acoustic wave and the tuning time of an acousto-optical tunable filter if the difference between the refractive indexes of the TE and TM modes is 0.07, the velocity of sound in $LiNbO_3$ is 3.75 km/s, the length of the acousto-optic interactive region is 20 mm, and the optical wavelength 1550.12 nm is to be selected.

**13.27** What is the function and the principle of operation of an isolator? Give examples of isolator applications.

**13.28** Calculate the length of a Faraday rotator made from a Bi-substituted iron garnet (BIG) crystal with a Verdet constant equal to 9°/Oe-cm if the strength of the magnetic field is 1000 A/m.

**13.29** What are the variations in isolation and insertion loss of an M-II isolator over the typical EDFA bandwidth (1530 to 1560 nm) and the typical ambient temperature range (0°C to 60°C)? (See Figure 13.21.)

**13.30** Give examples of reciprocal and nonreciprocal devices.

**13.31** Explain the function and the principle of operation of a circulator. Give examples of its applications.

**13.32** Explain the function, types, and principles of operation of attenuators. Give examples of their applications.

**13.33** Why do manufacturers not specify insertion loss for fixed attenuators?

**13.34** Why do manufacturers specify insertion loss for variable attenuators?

**13.35** Explain the function of optical switches. Describe the configurations of the optical switches you have studied.

13.36 List the types of single optical switches you are familiar with. Explain their principles of operation.

13.37 Give examples of large multistage optical switches.

13.38 What are the function and principles of operation of wavelength converters?

13.39 Why is the efficiency of FWM wavelength conversion so low?

13.40 Are wavelength converters passive or active components? Explain.

13.41 Explain the principle of operation of a configurable optical add/drop multiplexer. Why do we need this device?

13.42 Explain the principles of operation of optical cross-connects. Why do we need these modules?

13.43 *Project:* Build a memory map. (See Problem 1.20.)

## REFERENCES[1]


1. François Gonthier, "Fused couplers increase system design options," *Laser Focus World,* June 1998, pp. 83–88.
2. "Fused Biconical Taper Couplers," *Fiber Optic Component Catalog,* Gould Electronics Inc., Fiber Optics Div., Millersville, Md., 1998.
3. "Singlemode and PM Amplifier Fibers," *Fiber optic assemblies & components* (catalog), Wave Optics, Inc., Mountain View, Calif., 1998.
4. Kevin McCallion and Michael Shimazu, "Side-polished fiber provides functionality and transparency," *Fiberoptic Components,* supplement to *Laser Focus World,* September 1998, pp. S19–S24.
5. Dennis J. G. Mestdagh, *Fundamentals of Multiaccess Optical Fiber Networks,* Boston: Artech House, 1995.
6. Norio Kashima, *Passive Optical Components for Optical Fiber Transmission,* Boston: Artech House, 1995.
7. Rajiv Ramaswami and Kumar Sivarajan, *Optical Networks: A Practical Perspective,* San Francisco: Morgan Kaufman, 1998.
8. Alec MacGregor, AMP Inc., Harrisburg, Pa., private communication, May 1999.
9. Varis Hicks, EXFO, Vanier, Que., Can., private communication, May 1999.
10. "Applications of Filter Wavelength Division Multiplexer (FWDM)," *Application Note E-TEK-AN-095001* and *Application Note AN-096001,* E-TEK Dynamics, San Jose, Calif., 1998.
11. Martin Weik, *Fiber Optic Standard Dictionary,* 3d ed., New York: Chapman & Hall, 1997.
12. "MGC Optical Circuits Application Products," *Technical Brochure,* Mitsubishi Gas Chemical Co., Inc., Tokyo, Japan, 1999.
13. *Polarization Maintaining Fixed Ratio Evanescent Wave Couplers* (data sheet), Canadian Instrumentation and Research Ltd., Burlington, Ont., Can., April 1998.
14. *1.8%/98.2% Access Couplers for 1550 nm* (data sheet), JDS Fitel, Nepean, Ont., Can., December 1996.
15. Zee Hakimoglu, "Passive fiber-optic components are key to EDFA performance," *Lightwave,* April 1999, pp. 97–99.
16. Jeff Hecht, *Understanding Fiber Optics,* 3d ed., Upper Saddle River, N.J.: Prentice Hall, 1999.
17. Ronnie Chua, "Ultra-Wideband Fused Couplers Enable New System Applications," *Fiberoptic Product News,* November 1998, pp. 31–34.
18. Ajoy Ghatak and K. Thyagarajan, *Introduction to Fiber Optics,* New York: Cambridge University Press, 1998.
19. "WDM Component Terminology: Glossary of Terms," *Application Note PI 752,* Corning OCA, Marlborough, Mass., September 1997.
20. François Gonthier, "Fused-Coupler Technology for DWDM Applications," *Fiberoptic Product News,* September 1998, pp. 54–56.
21. *Dense Wavelength Division Mux/Demux* (data sheet), ADC AOFR, Symonston, Australian Capital Territory, Australia, August 1998.
22. Paul Fishbane, Stephen Gasiorowicz, and Stephen Thornton, *Physics for Scientists and Engineers,* Englewood Cliffs, N.J.: Prentice Hall, 1993.
23. *WDM STIMAX* (data sheet), Jobin Yvon - Spex, Instruments SA, Inc., Longjumeau Cedex, France, 1995, and *48 Channel Dense Wavelength Division Multiplexer 100 GHz - Fully Bidirectional Large FWHM: WDM-48/2/100ITU/S* (data sheet), Jobin Yvon - Spex, Horiba Group, Instruments SA, Inc., Edison, N.J., 1998.
24. Ivan Kaminow, "Waveguide Grating Router Components for WDM Networks," in *Guided-Wave Optoelectronics,* ed. by Theodor Tamir, Giora Griffel, and Henry Bertoni, New York: Plenum Press, 1995, pp. 297–298.
25. *1200 Series In-Line Filter* (technical data sheet), Corning OCA, Marlborough, Mass., September 1998, and *4 Channel*


[1]See Appendix C: A Selected Bibliography

*Dense Wavelength Division Multiplexer (DWDM)* (technical data sheet), Corning OCA, Marlborough, Mass, 1996.

**26.** J. Stone and L.W. Stultz, "Pigtailed high finesse tunable FP interferometer with large, medium, and small FSR," *Electronics Letters,* no. 23, 1987, pp. 781–783.

**27.** Calvin Miller and Lawrence Peiz, "Fabry-Perot tunable filters improve optical channel analyzer performance," *Lightwave,* March 1999, pp. 71–75.

**28.** Dan Sadot and Efraim Boimovich, "Tunable Optical Filters for Dense WDM Networks," *IEEE Communications Magazine,* December 1998, pp. 50–55.

**29.** *Acousto-Optical Tunable Filter* (data sheet), Pirelli Cavi e Sistemi SpA, Milan, Italy, April 1998.

**30.** *Lightwave 1999 Worldwide Directory of Fiber-Optic Communications Products and Services,* March 31, 1999.

**31.** "Laser interfaced fiber isolator (LIFI)," *Application Note E-TEK-AN-093006,* E-TEK Dynamics, San Jose, Calif., 1998.

**32.** Marcella R. Backer, "Electronic variable optical attenuators advance optical networking," *Lightwave,* February 1999, pp. 122–124.

**33.** Thomas Hazelton, "Get ready for the optical revolution," *Lightwave,* September 1998, pp. 43–52.

**34.** Stephen Montgomery, "Photonic Switches and the Fiber Optics Marketplace," *BeamBox Newsletter,* vol. 3, June 1998, Akzo Nobel Photonics, ABD bv, Arnhem, the Netherlands.

**35.** H. Scott Hinton, *An Introduction to Photonic Switching Fabrics,* New York: Plenum Press, 1993.

**36.** "BeamBox 2 × 2 4-Array Switch," *BeamBox Product Line* (catalog), Akzo Nobel Photonics ABD bv, Arnhem, the Netherlands, 1998.

**37.** *Agilent Photonic Switching Platform* (data sheet), Agilent Technologies, Inc., Santa Clara, Calif., March 2000.

**38.** Jaafar Elmirghani and Hussein Mouftah, "All-Optical Wavelength Conversion: Technologies and Applications in DWDM Networks," *IEEE Communications Magazine,* March 2000, pp. 86–92.

**39.** Biswanath Mukherjee, *Optical Communications Networks,* New York: McGraw-Hill, 1997.

**40.** Hector Escobar, "Acousto-optical tunable filter enables dynamic add/drop multiplexing," *Lightwave,* September 1998, pp. 97–98.

**41.** Rajiv Ramaswami, "Multiwavelength Optical Networking," *Tutorial 1,* IEEE Infocom '99, New York, March 21–25, 1999.

**42.** Satoru Okamoto et al, "Robust Photonic Transport Network Implementation with Optical Cross-Connect Systems," *IEEE Communications Magazine,* March 2000, pp. 94–103.

**43.** "Thermo-Optic Switches," *NEL Photonic Devices* (catalog), NTT Electronics, Ibaraki, Japan, July 1998, pp. 9–12.

# 14

# An Introduction to Fiber-Optic Networks

"Fiber-optic networks" is a nebulous term. One's understanding of its meaning depends on the perspective from which it is viewed. And this can range from today's telecommunications networks with electronic switches connected by optical fibers to tomorrow's all-optical networks, which will provide end-to-end user connection without electro-optical conversion. Telecommunications networks have existed for a number of years, of course, but the all-optical type is only now becoming more than just a gleam in the network designer's eye. If so, you may rightly wonder, why do we bother to discuss the topic since this book is supposedly limited to covering only commercially available fiber-optic products? The answer: The future all-optical networks, while not exactly around the proverbial corner, are indeed being developed today. These networks won't come overnight but, we anticipate, they will be added in stages—building block by building block—to existing networks. This, then, will be a step-by-step process, not a sudden, revolutionary changeover.

This chapter discusses today's networks and attempts to give you some idea of what we think networks will look like several years from now.

Shelves of periodical literature are readily available on this wide-ranging topic. The following references, listed at the end of this chapter, are textbooks devoted to networks: ([1], [2], [3], and [4]).

To fully understand fiber-optic networks, which are the major segment of today's general telecommunications networks, you should acquire at least a modicum of knowledge about networks in general. Only in this way can you come away from this chapter with a clear understanding of how fiber-optic networks function. We'll provide you with the basics necessary for understanding the material; in addition, some independent reading will surely facilitate and enhance your study. For your convenience, some sources of useful information at the introductory level include references [5], [6], [7], [8], and [9]. If you want to delve more deeply into the topic, we suggest you start with reference [10].

We also strongly recommend that you reread Section 12.1 to refresh your memory of some of the basic material covered there.

## 14.1 THE "WHAT" AND "HOW" OF DATA TRANSMISSION

The first thing we have to concern ourselves with is what kind of information we want to transmit over fiber-optic networks and how to do it. We will concentrate here on the physical aspect and some transmission characteristics of existing fiber-optic networks.

### What to Transmit: Voice, Video, and Data

All the information we ever need to exchange exists in three major forms: voice, video, and data. More precisely, from the standpoint of the telecommunications industry, these forms are what we need to deliver over our networks.

Voice is transmitted by *telecommunications networks.* (Here, the term *telecommunications* is used in its narrow sense—that is, in its "telephone" incarnation—while the broad meaning of the term is "communications.") Video is transmitted by broadcasting over airwaves and through cable TV networks. We are concerned only with the cable TV, or *community antenna television (CATV),* facet. Data are transmitted by computer networks, also called *data networks.*

This simple picture has been true for a number of years. It's important to know this because it is still true to some extent and it is the way modern networks have evolved. Today, however, your voice is transmitted as a stream of bits (that is, as data), your computer is connected to the Internet through telephone networks, and CATV gives you broad-bandwidth connections for both voice and data communications. At the same time, telephone companies can serve as your computer and—very soon—television provider, while CATV is ready to compete for both telephone and computer services. *The integration of technologies and services is the main feature of modern communications.* However, we will consider voice, video, and data transmissions separately to highlight their basic ideas and terms.

### Telephone Networks

To transmit *voice,* a telephone industry was established more than a hundred years ago. (Alexander Graham Bell was granted his patent on the first telephone apparatus in 1876, but it took a number of years before an actual telephone industry arose.) Today, the telephone network is the most prolific network in the world, enabling one to reach any point on the globe and serving as the linchpin for a wealth of electronic marvels, such as fax transmission, Internet connection, and home-lifeline security systems.

The structure of a typical telephone network is shown in Figure 12.1. To emphasize the basic architecture of the network, this graphic is redrawn as Figure 14.1.

All telephone lines from homes and offices are eventually connected to a *central office (CO)* either directly or through a *remote terminal (RT).* When RTs are in use, a local network is subdivided into a *feeder network* (from a CO to an RT) and a *distribution network* (from an RT to the customer's premises). All central offices, in turn, are connected to one another, forming a *regional network.* End offices are connected to a tandem central office. This type of central office opens the door to long-distance service, which is accomplished by linking a tandem central office to a *tandem switch.* Tandem switches with their links make up a long-distance network.

Before continuing with our discussion of network operations, let's first explain the meaning of some key terms applicable to this field: You are already familiar with the term *LATA* (local access and transport area), which denotes the specific area usually, but not necessarily, covered by one area code. (See Figure 12.1.) Other terms you may encounter while reading literature on telephony include the following

- *Local exchange carrier (LEC).* This refers to those companies providing your local and regional service.

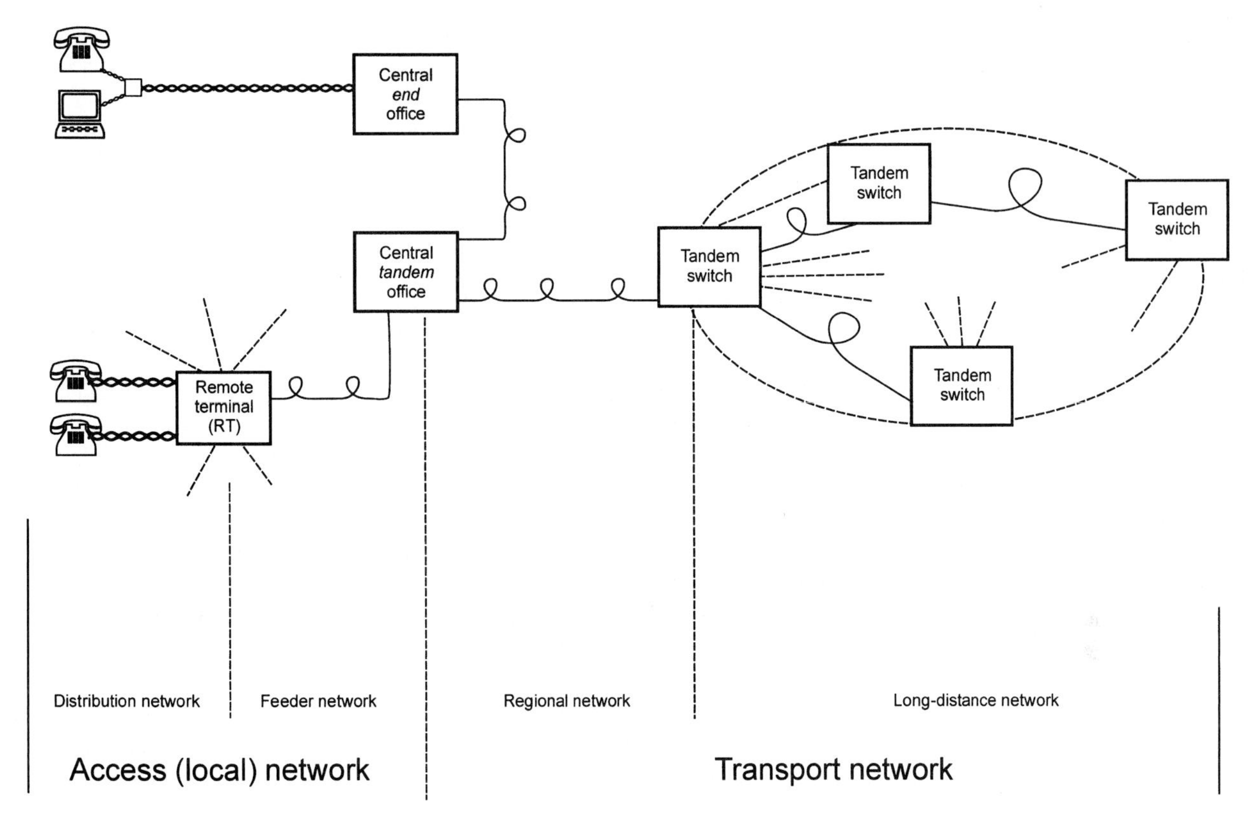


**Figure 14.1** Basic architecture of a telephone network.

- *Interexchange carriers (IECs,* also called *IXCs).* These are the long-distance companies. Note that the term *exchange* stands here for a switch, that is, a tandem central office. Note, too, that the Telecommunications Act of 1996 allows the LECs to provide long-distance service and the IXCs to provide regional service. That's why you now see, for instance, AT&T once again providing regional and local service and Bell Atlantic (and other Baby Bells, for that matter) offering long-distance service, both of which were prohibited before 1996.

In the United States, local, regional, and long-distance telephone networks are open to the public and they are called the *public switched telephone network (PSTN)* in contrast to private networks owned by specific companies.

Your home, business office, or other local places are called *customer premises.* All connections from customer premises along with local switches constitute an *access network.* All regional and long-distance connections along with appropriate switches make up a *transport network.* In other words, we include in transport networks all *interoffice* links. (But don't be confused: An end central office provides you with local and regional switching and a tandem central office connects you to a long-distance network. "Local" refers to the connection within a local access and transport area [LATA], that is, an intra-LATA connection; "regional" refers to the connection among LATAs, that is, inter-LATA connections; and "long distance" refers to the connection among the tandem switches.)

Transport networks themselves are sometimes subdivided into access, metro interoffice, and core (long-haul) sections. Such classifications are based on a line rate of each section. The access section of a transport network (don't confuse this with the access network that provides local connections) usually carries a line rate of 155 to 622 Mbit/s and connects PSTN access trunks to a transport network. The metro interoffice section carries up to 2.5 Gbit/s and the core section carries from 2.5 to 10 Gbit/s [11].

The concept of access and transport networks is very important in telecommunications because, on the one hand, they are parts of a single chain but, on the other hand, they are different in terms of their transmission techniques.

***Access network*** An access network is still based on the traditional twisted-pair copper-wire connections. It is estimated that there are approximately one billion telephone sets installed throughout the globe, almost all of them connected to the transport networks by these twisted pairs. The connection from a customer's premises to a remote terminal or a central office is called a *subscriber line,* which today has become a synonym for the twisted pair of copper wires. (Incidentally, in case you've ever wondered, the copper-wire pair is twisted to reduce crosstalk.)

Subscriber lines were originally designed to transmit low-frequency analog voice signals, which limits the length of such lines because of the role of attenuation. You should recall that the attenuation of copper wire depends on the signal frequency: the higher the frequency, the greater the attenuation. Telephone companies assign the 4-kHz bandwidth to an individual voice channel. Within this frequency range, the *attenuation-limited* average length of a subscriber line in the U.S. is 18 kilofeet. This length, obviously, varies widely, depending on the wire gauge, the use of loading coils, and other technical features.

More importantly, the subscriber line is severely limited in its *bandwidth.* Indeed, if you are an Internet user, you know how long it takes to download even a simple picture to your home computer from the Internet. This slow speed is caused by the narrow bandwidth of your connection to a central office. The *Hartley law* tells us that the amount of transmitted information is equal to the information-carrying capacity (that is, the bandwidth) of a transmission line multiplied by the time of transmission. It takes so long to transmit the information because the bandwidth of the subscriber line is small. If your computer were connected to the Internet by an optical fiber, it would take just a millisecond to download any kind of information, including images. But, in reality, your computer is connected to the Internet by a narrow-bandwidth copper wire. Since this wire cannot carry a digital signal directly, you need a modem to convert the digital signal from your computer into an analog signal for transmission over the twisted pair. But, since a regular modem is a slow-speed device, it takes what may seem to you like ages to communicate through the Internet. Example 14.1.1 should clarify the matter for you.

### Example 14.1.1

***Problem:***

How long does it take to download a static picture containing 4.16 Mbit from the Internet to a personal computer if a modem's bit rate is 56.0 kbit/s?

***Solution:***

Just apply Hartley's law:

$$H(\text{bit}) = C(\text{bit/s}) \times T(\text{s}) \qquad (14.1)$$

Simple calculations give this result:

$$T = 4.16\ (\text{Mbit})/56.0\ (\text{kbit/s}) = 232.9\text{s}$$

So you have to wait almost four minutes to get a simple picture on your screen. The actual downloading time will vary, of course, depending on the specific features of the picture, that is, the amount of information to be transmitted. The maximum bit rate currently available from a regular modem is 56.0 kbit/s.

---

Distance and bandwidth limitations inherent in copper wire have led to remote-control architecture, an arrangement where thousands of subscriber lines are terminated at a remote terminal, which itself is connected to an end central office by an optical fiber. This allows an access network operator to shorten the subscriber lines. Shortening a subscriber line helps to mitigate to some extent the severe attenuation and bandwidth limitations inherent in a twisted pair of copper wires. (The fundamental limitation in telecommunications is that bandwidth is inversely proportional to transmission length. Loss, as you know, is proportional to length.)

What we have to understand, then, is that a subscriber line is a bottleneck in today's telecommunications networks. Indeed, on the one hand, a transport network can transmit a tremendous amount of information at very high speed—today, on the order of Tbit/s. On the other hand, the demand on bandwidth from your desktop computer and other transmitting machines continues to grow at an exponential rate. The only network that cannot handle this desired high-speed transmission is an access network. The ultimate solution to this problem is to use an optical fiber for subscriber lines but, unhappily, this doesn't seem to be a realistic expectation for the foreseeable future.

Many efforts are being undertaken in leading R&D labs to "widen" this bottleneck without undertaking the radical step of replacing copper wire with optical fiber.

One approach uses advanced modulation and coding technology to transmit a high bandwidth signal over the twisted pairs. This includes the *Integrated-Services Digital Network (ISDN),* whose basic rate is 144 kbit/s, and its newest version, the *Broadband ISDN (B-ISDN),* with a transmission rate as high as 620 Mbit/s. The most promising set of new techniques in this category, however, is represented by the term *xDSL.* Here, *DSL* stands for *digital subscriber line* and *x* stands for various techniques, such as asymmetric (*A*), high speed (*H*), very high speed (*V*), universal (*U*), and so on. *ADSL,* which is commercially available, can transmit up to 6 Mbit/s downstream, that is, from the central office to the customer's premises. *VDSL* promises to raise this capacity to 20 Mbit/s and higher.

A second approach is based on the use of a cable TV network because of its relatively high bandwidth. This will be discussed shortly. A third method, using wireless technology developed for cellular telephones, connects the customer's premises to the PSTN. Wireless communications is another hot topic in the telecommunications field today but it is outside the scope of this book.

In short, these three approaches are being pursued with the goal of creating a *broadband access network.*

All these innovations are pushing the bandwidth boundaries of a twisted pair further and further, making it more difficult for optical fiber to compete with copper wire at the customer's premises, be it home or office. The market will determine in the near future which means of transmission—copper wire, optical fiber, cable TV, or wireless—will prevail at the subscriber line. We believe that optical fiber will eventually supplant these other technologies simply because the global network tends to be an optical network. (For a look at recent developments and trends in fiber-optic access networks, including a market analysis, see reference [41]. In the meantime, researchers continue to seek the best solution to the access-network problem. The scope of this search ranges from combining the best features of electronic and optical networks [11] to exploring the potential of all-optical access networks [12].)

***Transport networks*** Transport networks gather all the signals emanating from millions and millions of customer premises and then have to transmit this tremendous amount of information at high speed. In other words, the transmission load under which these networks operate puts the highest demand imaginable on network bandwidth. No wonder that transport networks have been a leading site for the deployment of the latest advances in fiber-optic communications technology.

Today, transport networks span the globe. We distinguish between *terrestrial* and *submarine (undersea)* networks by their location and the different technologies by which each operates.

As previously noted, transmission over subscriber lines is in analog form because of the narrow bandwidth of copper wire. But transmission in transport networks is exclusively in digital form because these networks have a wide bandwidth.

Refer again to Figure 14.1. A central office gathers signals from thousands of customers, multiplexes them and sends multiplexed (that is, high-bit-rate) signals farther along the transport lines. If you look back at Figure 12.1, you will see that in the AT&T long-distance network, a tandem switch connects, for example, the New York City metropolitan area with Los Angeles. Thus, the first tandem switch gathers multiplexed signals from all the central offices in its serving area, further multiplexes these signals, and then sends them to the Los Angeles switch. Imagine the high bit rate that has to be achieved after so many multiplexing steps. This is why transport networks now operate at the rate of terabits per second, and they will certainly operate at much higher bit rates tomorrow. This is also why designers of transport networks started to deploy optical fiber as their major transmission medium and then turned to singlemode fibers, WDM, and other advanced fiber-optic technologies before any other network designers considered doing so.

To understand the origin of the high bit rates used for transmission in transport networks, let's consider the multiplexing hierarchy.

#### *Multiplexing hierarchy in telecommunications*

**1.** *T-1 transmission system.* The first level of multiplexing a voice signal is known as the *T1 system.* This is a TDM system where multiplexing takes place as follows: First, a voice analog electrical signal is digitized using the *pulse-code modulation (PCM)* technique. In three consecutive steps—sampling, quantizing, and encoding—PCM converts the analog signal into a stream of bits. The PCM process is shown schematically in Figure 14.2(a). (See also Figure 1.3.) The human voice occupies the frequency range between 300 Hz and 3400 Hz. The telcos add some guard bandwidth; this is where the 4-kHz bandwidth for each voice channel comes from. This voice signal is sampled at the frequency of 8 kHz. (Remember the *Nyquist theorem?* The sampling frequency must be at least two times as great as the highest signal frequency.) Thus, the time frame in a T1 system is 125 μs.

To encode the decimal value of a single sample into a binary system, each sample is assigned 8 bits. The system multiplexes 24 voice channels, which means that 24 samples, or time slots, from the different channels are placed within a single time frame. Hence, one time frame contains 8 bits × 24 voice channels, or 192 bits of information. One framing bit is added to delimit a frame boundary, thus producing 193 bits/frame. The value of an 8-kHz sampling frequency means that the system produces 8,000 frames per second. Thus, the bit rate of a T1 system is as follows:

- 8 bits/sample × 24 samples/frame + 1 bit/frame = 193 bits/frame
- 193 bits/frame × 8,000 frames/s = 1.544 Mbit/s

Time-division multiplexing in a T1 system is shown schematically in Figure 14.2(b). You can read more about T1 systems in any telecommunications textbook; see, for example, references [13], [14], [15], [16], [17], and [18].

There are higher standard formats for bit rates. T2 combines 4 T1 signals and, by adding some overhead, transmits at 6.312 Mbit/s. T3 combines 7 T2 (or 28 T1) signals and transmits at 44.736 Mbit/s. The transmission-speed formats that have been developed for transmitting digital

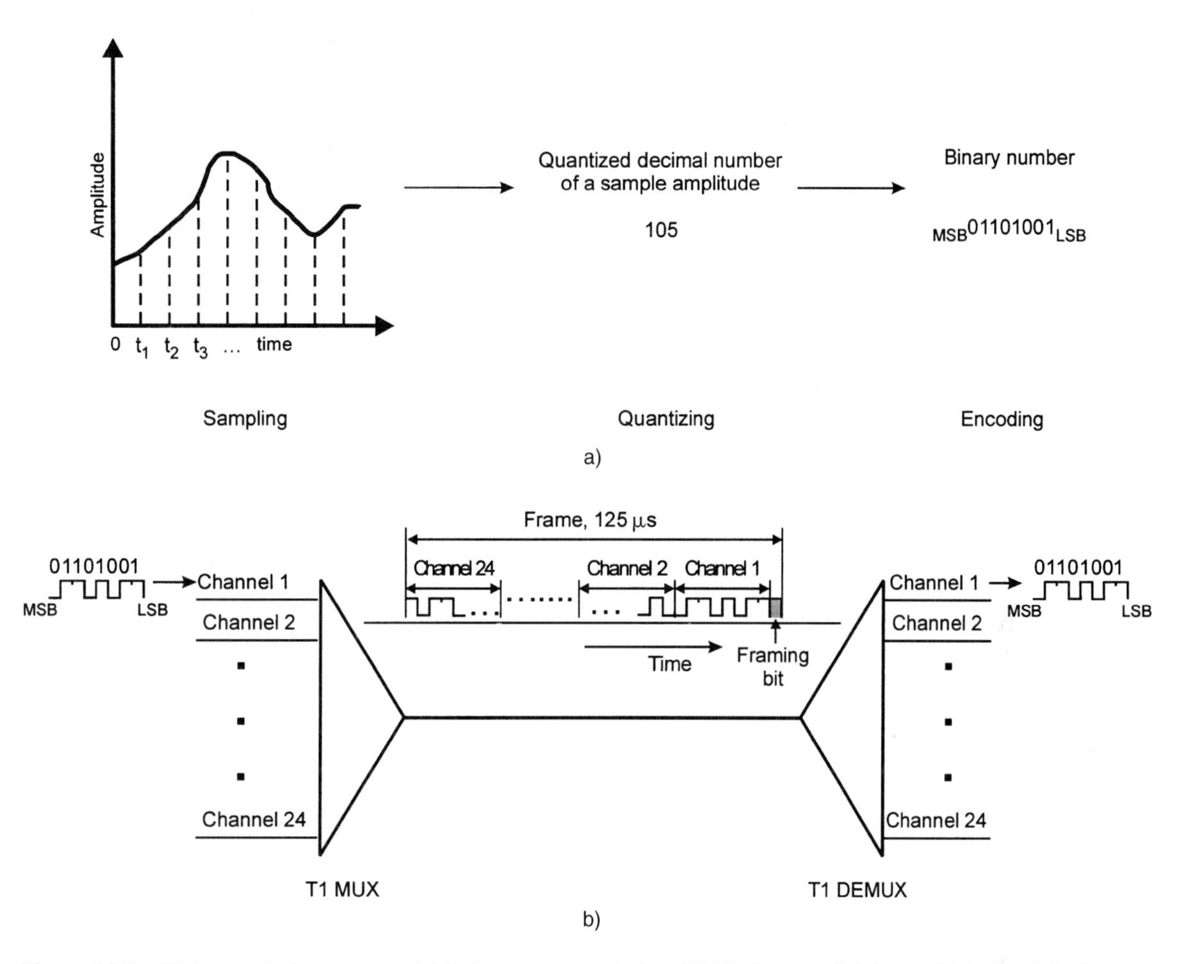


**Figure 14.2** T1 transmission system: (a) Pulse-code modulation (PCM); (b) time-division multiplexing (TDM).

signals are called *digital signal level* 1, 2, and so on, and are designated as DS-1, DS-2, etc. DS-1 transmits at 1.544 Mbit/s, that is, at T1 speed; DS-3 transmits at T3 speed, and so on.

The T carrier system is used in North America and Japan, while Europe and other parts of the world use the E TDM system. The E1 signal multiplexes 30 basic voice channels so each frame contains 30 samples, or time slots. Two time slots are added for framing and signaling and 8 bits are used for encoding. Hence 8 bits/sample $\times$ 32 samples/frame $\times$ 8,000 frames/s = 2.048 Mbit/s, which is the basic bit rate for the E1 signal. All other signals of this system are multiples of E1, as is true for the T system. The T and E multiplexing levels are presented in Table 14.1.

Observe that even DS-1 requires a higher bit rate than the simple calculation (24 voice channels $\times$ 64 kbit/s) shows. This is because the T system adds one bit to each frame. The next stage of multiplexing adds more service bits for packing, transport, and synchronization purposes. These service bits are called the *overhead,* while the user's data are called the *payload.* This is why Table 14.1 indicates, for example, a transmission rate of the T3 signal as "28 $\times$ T1 + overhead" and also why the transmission rate of the T3 signal is not a simple multiple of the bit rate of the T1 signal.

What is important to understand is that the T and E transmission systems have specific and, unfortunately, different transmission characteristics. For example, DS signals can use alternate-mark-inversion (AMI) line coding, which allows a receiver to synchronize its clock with a transmitter's clock. (See Section 11.4, where the clock-recovery technique is discussed.) If no data are transmitted, consecutive zeros pass through a transmission link, and a receiver loses synchro-

**Table 14.1** T and E transmission systems

| T System | | | | E System | | |
|---|---|---|---|---|---|---|
| T signal | DS signal | Bit rate | Multiples and numbers of voice channels (v.c.) | E signal | Bit rate | Multiples and numbers of voice channels (v.c.) |
| | DS-0 | 64 kbit/s | 1 | | | |
| T1 | DS-1 | 1.544 Mbit/s | 24 × DS-0 + overhead; 24 v.c. | E1 | 2.048 Mbit/s | 32 × 64 kbit/s; 30 v.c. |
| T1C | DS-1C | 3.152 Mbit/s | 2 × T1 + overhead; 48 v. c. | | | |
| T2 | DS-2 | 6.312 Mbit/s | 4 × T1 + overhead; 96 v.c. | E2 | 8.448 Mbit/s | 4 × E1 + overhead; 120 v. c. |
| T3 | DS-3 | 44.736 Mbit/s | 7 × T2 + overhead = 28 × T1 + overhead; 672 v.c. | E3 | 34.368 Mbit/s | 4 × E2 + overhead; 480 v.c. |
| T4 | DS-4 | 274.176 Mbit/s | 6 × T3 + overhead; = 144 × T1 + overhead; 4032 v.c. | E4 | 137.264 Mbit/s | 4 × E3 + overhead 1920 v.c. |
| | | | | E5 | 565.148 Mbit/s | 4 × E4 + overhead; 7680 v.c. |

*Sources:* John Belamy, *Digital Telephony,* 2d ed., New York: John Wiley & Sons, 1991.
Julie Petersen, *Data & Telecommunications Dictionary,* Boca Raton, Fla.: CRC Press, 1999.

nization. A specific technique called *B8ZS—binary 8 zero substitution*—substitutes four pulses that violate AMI rules when eight or more consecutive logic zeros occur in the data stream. These violations are used for synchronization purposes only; they are not the user's data. The B8ZS technique is a standard in the United States. The E transmission system also uses bipolar line coding; however, the E system has different voltages, pulse shapes, and line impedances. In addition, the E system uses the *high-density bipolar three-zero suppression (HDB3)* technique to provide synchronization. The T and E systems are also different in framing and many other transmission characteristics. This is important to bear in mind because all these differences combine to produce one major headache for developers of local fiber-optic networks.

**2.** *SONET.* SONET, or Synchronous Optical Network (Section 4.7), is actually a set of standards originally developed by Bellcore (now Telcordia Technologies) and then adopted by ITU-T (then known as CCITT) as an international standard designated *synchronous digital hierarchy (SDH).* SONET specifies the rates, formats, and parameters of all physical transmission media. Note that ITU-T stands for the Telecommunication Standardization Sector of the International Telecommunication Union and CCITT stands for the Consultative Committee on International Telegraphy and Telephony.

SONET is a TDM system providing synchronous transmission, meaning that each frame of a data stream starts at a specific moment with respect to the operating-time sequence of the entire network. Thus, clock synchronization for all transmitters, receivers, and other network components is the major concern of the network operator. (The term *asynchronous* means that data can be sent and can arrive at arbitrary moments of time. This implies that starting and ending bits are added to each cell of data transmitted.)

We will concentrate here on SONET's multiplexing format only. A wide range of literature is available on SONET, a truly vast technical subject. You may wish to begin an in-depth study of the subject by reading [1], [14], [15], [17], [19], and [20].

Even though SONET is called an optical network, it can support traditional electrical transmission. The electrical signals in SONET are called *synchronous transport signals (STS).* STS level 1, designated as STS-1, transmits at 51.84 Mbit/s; this is the basic building block of SONET multiplexing. The next level of multiplexing hierarchy is STS-3, which transmits at 155.52 Mbit/s (51.84 Mbits/s × 3). All other levels are multiples of STS-1, as Table 14.2 shows. Higher bit-rate signals, starting with STS-9, are not defined as electrical standards and they are never actually transmitted in electrical form. What is transmitted are their optical counterparts, called the level-N signals of the optical carrier, which are designated *OC-N.* OC-N signals are physically obtained from STS-N ones by scrambling and converting them to optical form.

A functional block diagram of SONET multiplexing is shown in Figure 14.3. For a better understanding of this figure, refer to Tables 14.1 and 14.2 to see the bit rates of all the signals involved in multiplexing.

Figure 14.3 shows that all signals are eventually converted to a base format of an STS-1 signal. The inputs with lower bit rates—such as DS-1, E-1, DS-1C, and DS-2—are first multiplexed into *virtual tributaries,* which will be discussed shortly. The inputs with higher bit rate, such as 155 Mbit/s ATM, are broken into several STS-1 signals. The N STS-1 signals are then multiplexed to form an STS-N signal. This is done by a byte-interleave synchronous multiplexer. Direct electrical-to-optical conversion forms an OC-N signal, which is transmitted farther over an optical fiber.

In time-division multiplexing, there are two basic methods of mixing different signals: *bit interleaving* and *byte (word) interleaving.* Look again at Figure 14.2. Each channel (that is, a byte containing 8 bits) is assigned an individual time slot within one frame. These bytes, or channels, intermingle within one frame. This is *byte interleaving.* If a time slot is assigned to a single bit in each channel, we have *bit interleaving.*

*SDH (synchronous digital hierarchy),* the international counterpart of SONET, is widely accepted in Europe. Signals in SDH, called *synchronous transport modules,* are designated *STM-N.* The basic STM signal, STM-1, transmits at 155.52 Mbit/s, the same rate at which OC-3 transmits.

**Table 14.2** SONET/SDH multiplexing

| SONET | | | |
|---|---|---|---|
| Electrical signal | Optical signal | SDH | Bit Rate (Mbit/s) |
| STS-1 | OC-1 | | 51.84 |
| STS-3 | OC-3 | STM-1 | 155.52 |
| STS-9 | OC-9 | STM-3 | 466.56 |
| STS-12 | OC-12 | STM-4 | 633.08 |
| STS-18 | OC-18 | STM-6 | 933.12 |
| STS-24 | OC-24 | STM-8 | 1244.16 |
| STS-36 | OC-36 | STM-12 | 1866.24 |
| STS-48 | OC-48 | STM-16 | 2488.32 |
| STS-192 | OC-192 | STM-64 | 9953.28 |

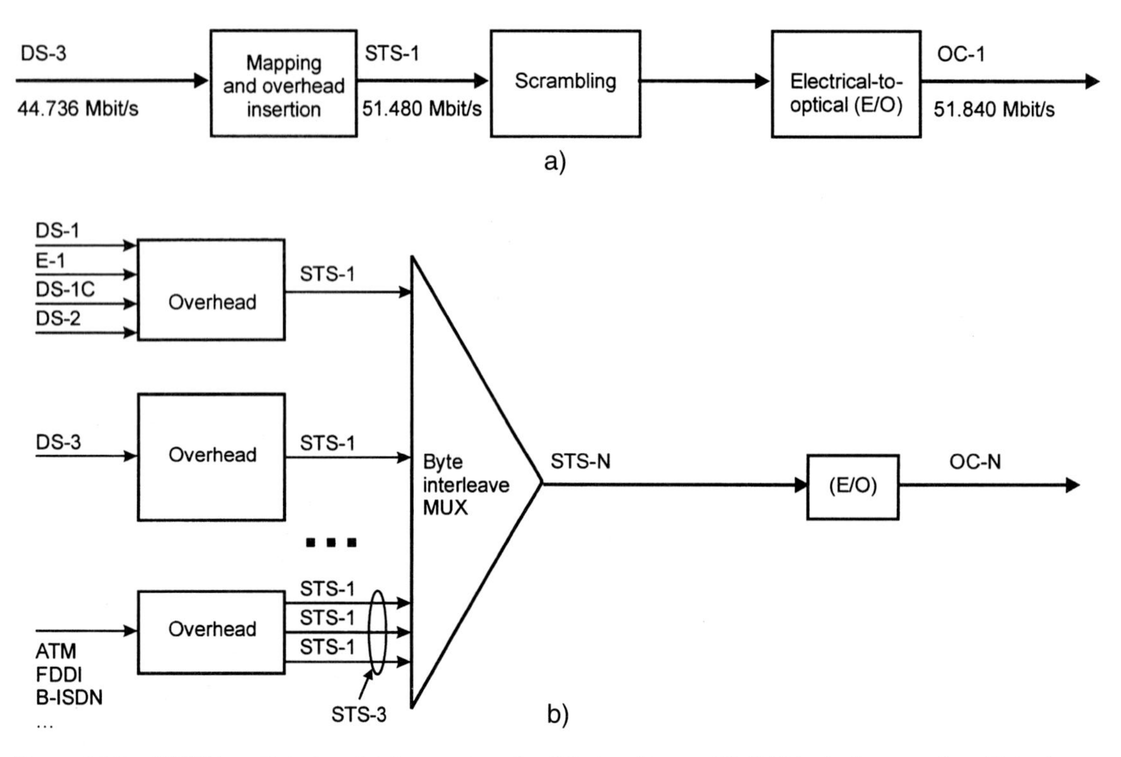


**Figure 14.3** SONET multiplexing: (a) Conversion of a DS signal to an OC; (b) block diagram of multiplexing.

SDH differs from SONET in small details and terminology. For example, SONET's synchronous payload envelope (SPE), which will be discussed shortly, is called a *synchronous container (SC)* in SDH. SONET's virtual tributary is called a *virtual container (VC)* in SDH. The SONET and SDH multiplexing hierarchies are shown in Table 14.2.

An STS frame consists of 9 rows and 90 columns of 8-bit bytes for a total of 810 bytes (6480 bits). The duration of each frame is 125 μs (remember the 8-kHz sampling frequency), which yields:

$$6480 \text{ bits}/125 \ \mu\text{s} = 51.84 \text{ Mbit/s}$$

This is where the basic bit rate of the STS-1 signal comes from.

You will recall that the process of organizing bits and bytes within a transmitting format is called *mapping.*

The STS-1 format includes a payload with a path overhead and a transport overhead. This is shown schematically in Figure 14.4(a). A payload with a path overhead is called a *synchronous payload envelope (SPE).* SPE is an information signal—it's also called a *tributary*—within STS-1. Once a *payload* is multiplexed into the SPE, it is transmitted through a SONET network without being examined or demultiplexed by intermediate modes. A *path overhead* is inserted into the SPE by the end transmitter; it remains attached there until the tributary is unpacked by the end receiver. Thus, a path overhead provides end-to-end communications. A *transport overhead* provides communication between STS-N multiplexers and adjacent nodes, such as regenerators.

To work with signals whose bit rate is lower than STS-1, SONET subdivides an SPE into smaller envelopes, called *virtual tributaries (VT).* The smallest VT is called VT1.5 and its bit rate is 1.728 Mbit/s. This envelope is designed to carry the DS-1 (1.544 Mbit/s) signal. There are VT2, VT3, and VT4 virtual tributaries, which are shown in Table 14.3. These VTs are mapped into a bigger structure, called the VT group. One SPE can accommodate seven VT groups.

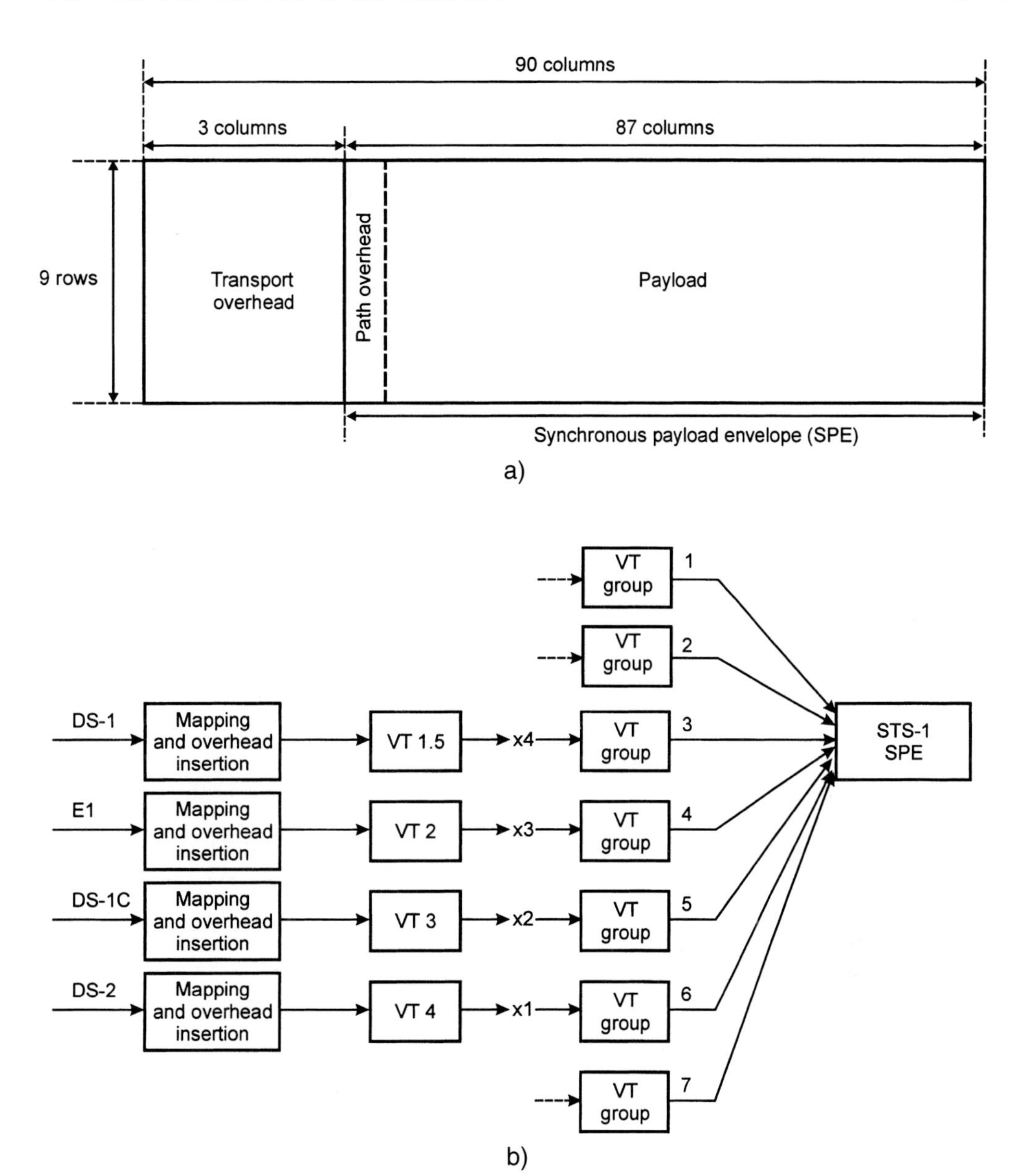


**Figure 14.4** The STS-1 signal: (a) The structure of an STS-1 format; (b) virtual tributaries (VT) and an STS-1 signal.

A possible way to map the DS and E signals into the STS through the VT structure is shown in Figure 14.4(b). This figure also displays how the STS-1 signal is made up of signals having lower bit rates as they pass through the VT structure.

These explanations should help you grasp the idea of how SONET transports signals of different bit rates, starting with DS-1 (1.544 MBit/s). The major advantage of SONET is its flexibility. It can carry not only synchronous data but also asynchronous-arranged data, such as data in the *Asynchronous Transfer Mode (ATM).* SONET's multiplexing structure readily accommodates add and drop signals, which is a crucial capability for networking. The standard is easily upgraded as technology progresses. All these features make SONET one of the major standards used in today's transport telecommunications networks.

**Table 14.3** SONET virtual tributaries

| VT designation | Carrying Number in a VT group | Maximum capacity (Mbit/s) | number in SPE |
|---|---|---|---|
| VT1.5 | 4 | 1.728 | 28 |
| VT2 | 3 | 2.304 | 21 |
| VT3 | 2 | 3.456 | 14 |
| VT4 | 1 | 6.912 | 7 |

*Sources:* John Bellamy, *Digital Telephony,* 2d ed., New York: John Wiley & Sons, 1991.
Daniel Minoli, *Telecommunications Technology Handbook,* Boston: Artech House, 1991.

As you can see from Tables 14.2 and 14.3 and Figures 14.3 and 14.4, both the SONET and SDH multiplexing hierarchies are based on the multiplexing of a voice channel. Moreover, the SONET 125-μs frame format comes from the 8-kHz sampling frequency, but this number is determined by a voice bandwidth. The point to bear in mind is this: SONET transmits only digital signals but its usage for data transmission is based on the voice-channel format. Until recently, voice traffic predominated in transport networks, but now the volume of data traffic exceeds that of voice traffic, a trend that will continue to increase very rapidly in the years ahead. SONET, a complex, sophisticated transmission technology, has had limited appeal because of its economic and technological shortcomings. Not only is it an expensive system to maintain, but its reliability leaves something to be desired and it is a difficult format to manage. For these reasons, there's been a lot of discussion within the industry as to whether the SONET format is still the most suitable for transmitting data. Its long-term future, therefore, remains uncertain.

The main conclusion we should draw from these considerations is that transport networks gather information from thousands of individual customers for transmission. Consequently, they must—and do—transmit a tremendous amount of information at high speed. In fact, the highest working bit rates in use today in SONET networks are 2488.32 Mbit/s (OC-48) and 9953.28 Mbit/s (OC-192), which we usually refer to as 2.5 Gbit/s and 10 Gbit/s, respectively. Since OC-1 carries 672 voice channels, it's easy to compute that OC-48 carries 32,256 voice channels and OC-192 carries 129,024 voice channels. But don't forget: SONET provides time-division multiplexing only. Thus, using WDM, we can increase the transmission capacity of fiber-optic transport networks by as many times as the number of wavelengths we multiplex. In fact, major long-distance carriers have realized that using OC-192 (10 Gbit/s) is much more difficult than moving to 32- and even to 64-channel WDM systems based on the OC-48 (2.5 Gbit/s) data format [21].

It should be mentioned that the multiplexing structure of network transmission is usually referred to as *granularity.*

## Computer Networks

***Types of networks*** Computer (or data) networks are designed and built to transmit digital signals, which implies that they have enough bandwidth. (Why a digital signal requires much more bandwidth for transmission than an analog one is discussed in Section 1.1.) We distinguish among data networks by their scale of operation. Thus, we have *local area networks (LANs), metropolitan area networks (MANs),* and *wide (world) area networks (WANs).*

**1.** *LAN.* A LAN is a network that connects, within one given area, computers, printers, faxes, modems, and other similar machines known as *computer resources.* The area could be a room, a building, or a campus. LANs arrange the circulation of the data through the networks and, in ad-

## Modulation and Multiplexing in TDM Fiber-Optic Communications Systems

Let's take a closer look at the modulation and multiplexing techniques used in high-speed fiber-optic TDM communications systems. Digital modulation, as we have already seen (Sections 9.2, 10.3, and 10.4), is simply switching light from on to off, with the "on" mode carrying logic 1 and the "off" mode carrying logic 0. If digital modulation is done by changing the driving current of a laser diode, we call it *direct modulation (DM).* However, you will recall that at a bit rate above 10 Gbit/s, direct modulation of a laser diode does not work because of the *chirp* phenomenon. Hence, for high-speed fiber-optic transmission, we have to use external modulation, where light is modulated after it exits the laser diode. With external modulation, a bit rate of 100 Gbit/s has been achieved.

Even though many line codes, or modulation formats, have been developed for electronic transmission (such as the bipolar and Manchester codes already mentioned), in today's fiber-optic communications systems we use mostly one: *non-return-to-zero (NRZ).* Consideration of the other modulation formats has been stimulated by new developments in WDM networks [22]. The most practical among the alternative formats seems to be the *return-to-zero (RZ)* line code. NRZ and RZ are shown in Figure 14.5(a).

With the NRZ format, light is kept on or off over the entire duration of a bit. The RZ format necessitates turning the light on at a portion—typically one-half—of the bit time when transmitting logic 1 and keeping it off all the time when transmitting logic 0. (See Figure 10.20.) The RZ code requires less power but more bandwidth for transmission. An actual waveform of a single pulse carrying logic 1 is shown in Figure 14.5(b). The pulse has finite rise and fall times. It is essentially an envelope of a light wave whose frequency is vastly enlarged in Figure 14.5(b) for illustrative purposes.

NRZ is the basic modulation format in already deployed fiber-optic communications systems. The code is simple and easy to implement. However, the problem with this format starts when we try to introduce high transmission speed (over 40 Gbit/s) because tails of the adjacent pulses start to overlap. Here, the RZ code is more suitable because it provides a time-guard zone between pulses. A comparison of these codes for use in various commercial applications continues in many research laboratories [23].

Above 100 Gbit/s, no transmitters exist that can directly produce pulses with appropriate widths. However, by manipulating the modulation and multiplexing techniques in conjunction with the use of an EDFA, we can even further increase the transmission bit rate. One such example is given in Figure 14.6

If you read Chapter 7, you are certainly aware that 10 Gbit/s is the bit-rate limit that modern commercially available electronics can support. Why, then, do we mention a bit rate of 100 Gbit/s here? This is the theoretically achievable level of fiber-optic systems using TDM. A lot of research has been carried out to increase the bit rate in TDM fiber-optic systems. Figure 14.6 provides an example of such efforts. But, again, why do we need to increase TDM speed when the bit rate can be increased much more easily by using the WDM technique? The answer (already given in Section 12.1): TDM and WDM are not competitive but complementary technologies. Thus, if you want to transmit 1 Tbit/s over a single fiber, you need to squeeze 100 wavelength channels, each carrying 10 Gbit/s. Alternatively, you may multiplex 10 wavelength channels, each carrying 100 Gbit/s, if this bit rate is to be achieved by TDM. Today, the efforts to increase bit rate in deployed networks are clearly focused on the WDM technique. The development of high-bit-rate TDM systems is still in the research stage.

dition, they allow all network components to share resources. For example, several computers can use the same printer or fax; on the other hand, one of the computers can manage the network database while another can manage a printer.

We discuss some types of LANs and LAN installations using optical fiber in Section 8.4 so you should read (or reread) that section before continuing here. The first LANs were developed long before fiber-optic links became commercially feasible but today, as noted in Section 8.4, the installation of fiber-based LANs is one of the fastest growing areas of the fiber-optic communications industry. All segments of a LAN—from a campus backbone through a building backbone to the desk connections—are now being wired with optical fiber.

Current LANs are standardized by type; the standards differ in terms of topology, transmission format, and protocol. Thus, the Ethernet family is based on bus topology, uses the Manchester line code, and has a protocol called "Carrier Sense Multiple Access with Collision Detection (CSMA/CD)," while the Token Ring family employs ring topology, uses differential Manchester encoding, and has a protocol that is based on carrying a token.

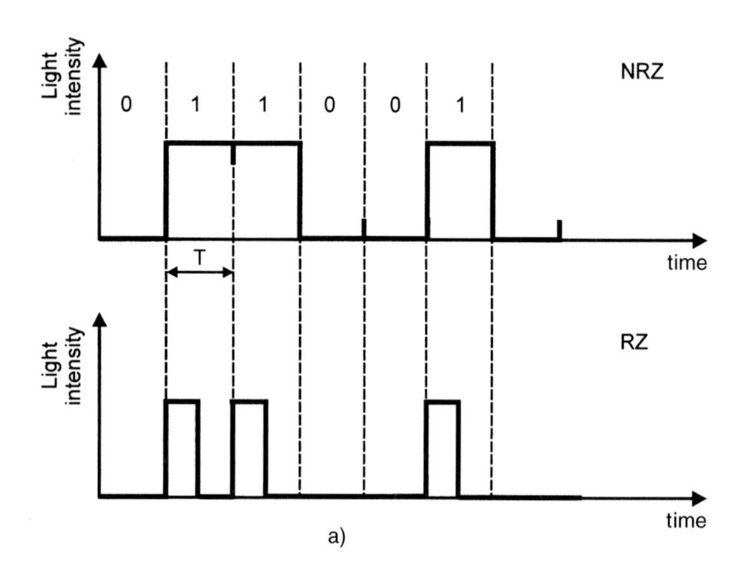


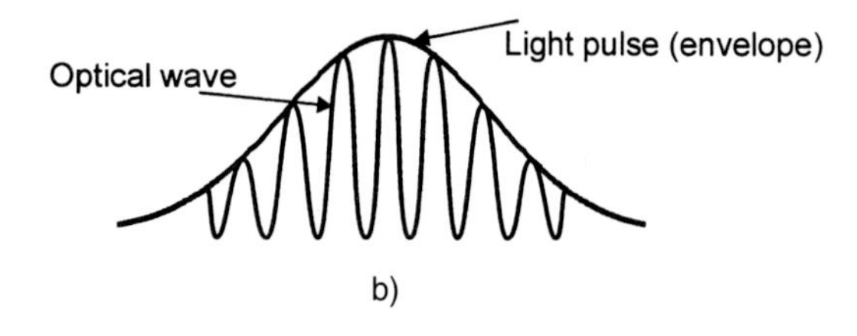


**Figure 14.5** Modulation in TDM fiber-optic systems: (a) Non-return-to-zero (NRZ) and return-to-zero (RZ) modulation formats; (b) actual waveform of a light pulse (not to scale).

Local area networks are characterized by relatively short transmission distances (on the order of tens of kilometers) and moderate bit rates (on the order of hundreds of Mbit/s). Few new LANs, such as Gigabit Ethernet and Fibre Channel, operate at the Gbit-per-second bit rate. (If you wish to explore this subject more deeply, there is a vast collection of technical literature on modern LANS. See, for example, reference [24].) Greatly increased computer power and the exponentially increasing volume of information being circulated are exerting new demands on LAN bandwidth. This is the motivating factor behind the drive to replace copper wire with optical fiber in LANs.

To understand how data are transmitted in computer networks, consider the frame format for data of a Fiber Data Distribution Interface (FDDI). FDDI is a good example to study because this standard, originally designed for LANs, has become the leading transmission technology for metropolitan area networks, which are discussed below. An FDDI frame format for data is shown in Figure 14.7. (See [5], [16], and [25].)

For starters, let's review the functions of each field in an FDDI frame format for data [25]:

- The preamble (PA) serves as an interframe gap. The PA's primary function is to synchronize the incoming signal with the receiver.
- The starting delimiter (SD) shows the starting boundary for the data frame.
- The frame control (FC) defines the type of frame (synchronous or asynchronous) and provides some other service information, such as the length of the address and the priority of the frame.
- The destination address (DA) contains the address and its type (that is, individual or group address).

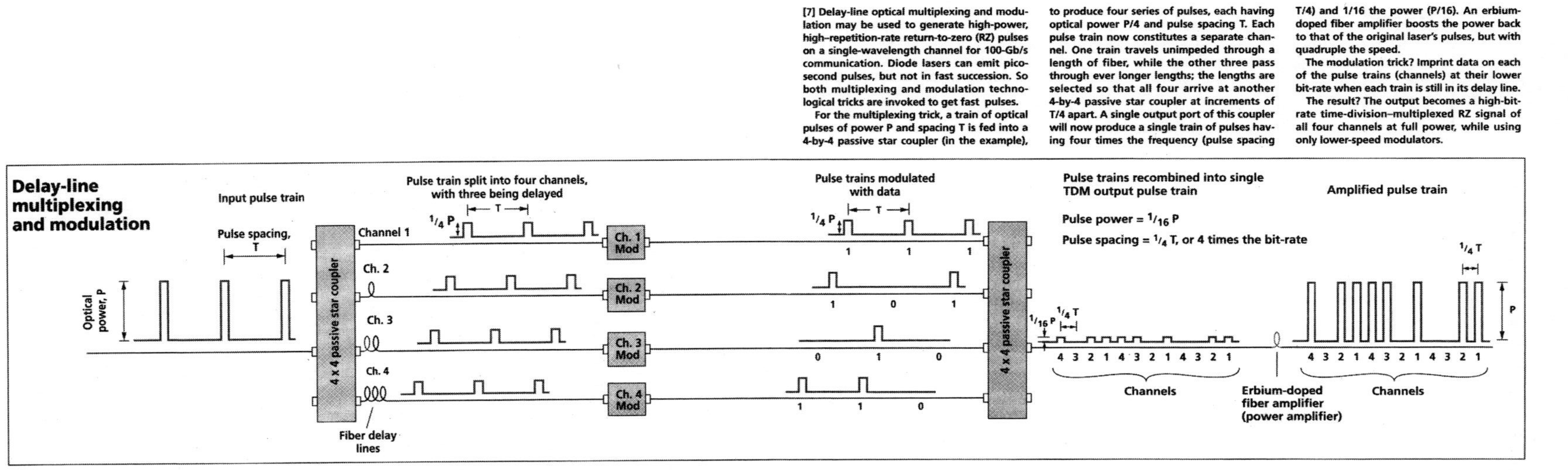


**Figure 14.6** Modulation and multiplexing technique to achieve 100 Gbit/s bit rate. *(Courtesy of Allan Willner, "Mining the optical bandwidth for a terabil per second,* IEEE Spectrum, *April 1997, pp. 32–41. © 1997 IEEE.)*

| Fields | PA | SD | FC | DA | SA | Info | FCS | ED | FS |
|---|---|---|---|---|---|---|---|---|---|
| Number of bits | 64 | 8 | 8 | 16 or 48 | 16 or 48 | 0 to 4478 bytes | 32 | 4 or 8 | 12 or more |

**Key**: PA -- preamble; SD -- starting delimiter; FC -- frame control;
DA -- destination address; SA -- source address;
Info -- information field; FCS -- frame-check sequence;
ED -- ending delimiter; FS -- frame status.

**Figure 14.7** FDDI frame format for data.

- The source address (SA) serves as a token, allowing the local node to strip the frame from the ring when the SA matches the local source address.
- The information field (Info) contains the data being transmitted.
- The frame-check sequence (FCS) is used to detect errors within the frame.
- The ending delimeter (ED) establishes the closing boundary of the frame.
- The frame status (FS) contains three control indicators to show whether an error has been detected, whether routing has been done correctly, and whether copying into a receiver buffer has been accomplished. FDDI protocol requires transmission of a data frame and a token.

As mentioned several times already, computer networks, and LANs in particular, are treated in the same way regardless of whether or not they use optical fiber. Obviously, you have to use different hardware and software but, apart from that, operation, administration, maintenance (OAM), and other functions are similar for all types of LANs. This is true as well for troubleshooting LANs. (You will find information on LAN troubleshooting in [5] and [26].)

**2.** *MAN.* A metropolitan area network (MAN), as the name suggests, is a network serving a certain geographic region. This network links different locations within a metropolitan area, which implies that a MAN connects many LANs. This puts a stringent requirement on the MAN bandwidth: a bit rate starting at 200 Mbit/s. To ensure their reliability, most MANs use double-ring topology. FDDI and SONET networks are good examples of MAN architecture.

The structure of a typical MAN is shown schematically in Figure 14.8. Each node of the MAN can be connected simultaneously to many LANs and to other MANs.

Today, MANs are employed in more and more broad-boundary networks, thus making it difficult to distinguish between metropolitan and wide area networks.

**3.** *WAN.* A world (wide) area network (WAN) is a national and global network that can link all the computers in the world. This network obviously includes LANs and MANs; however, the important thing to understand is that this is not a hierarchical structure. In other words, if you try to reach the computer of a friend living in another country, you don't have to go through all three networks—LAN first, then MAN, and finally WAN. Here's why:

A local area network is a dedicated network designed and built specifically for your room, building, or campus. This means that you have to install dedicated transmitters, receivers,

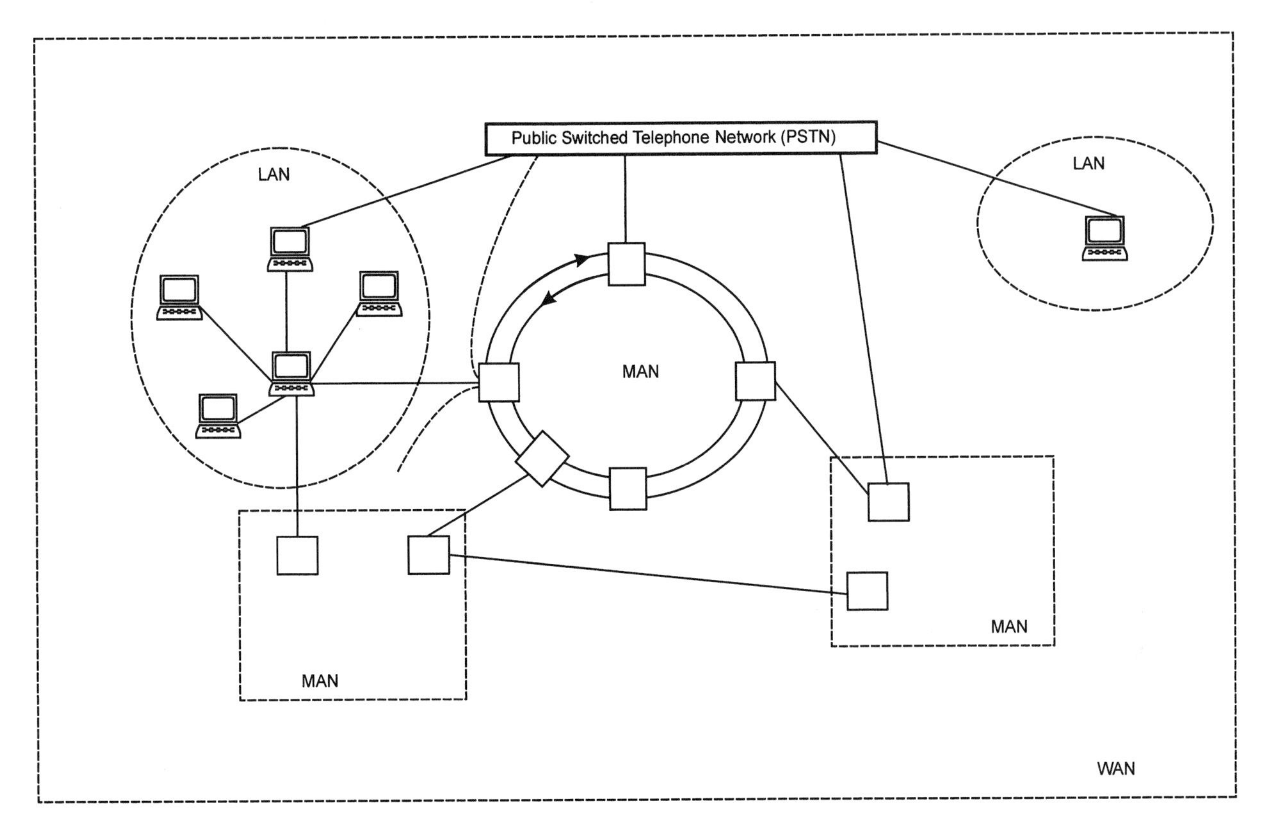


**Figure 14.8** LAN, MAN, and WAN.

optical fibers, and all the other necessary components to build your LAN. But metropolitan area and world area networks can be built in either of two ways: as dedicated networks or as adjuncts to the public networks—more specifically, the public switched telephone network (PSTN). Most of us communicate around the globe through e-mail by using PSTN. (If you check to see how your computer gains access to the Internet, you'll find that it is physically connected to the telephone jack.) However, the U.S. Department of Defense has built, and actively uses, its own dedicated WAN, which links U.S. military units throughout the world. The same is true for MANs. You can communicate with a colleague in another part of your city through PSTN or through a dedicated MAN if your company has built one. Figure 14.8 shows both options.

***Transmission systems (protocols)*** A network is a combination of hardware and software, that is, physical and logical entities. We've concentrated here mostly on the physical links, even though we inevitably have had to touch on the logic of transmission in discussing SONET and FDDI. What you have to understand clearly is that the transmission of information occurs through physical circuitry (these days, primarily through fiber-optic links) but this would be impossible without the establishment of rules of transmission called *protocols.* Thus, the FDDI standard specifies the hardware to use (LED transmitters, multimode fibers, operating wavelengths, and so on), but this standard also specifies the rules on how a node strips the message from the ring, how to arrange for self-repair of the ring in case of fiber or node failure, and so forth. The same holds true for SONET and any other transmission system. This is why we call these standards either protocols or transmission systems. Since data communication is particularly sensitive to the logic of transmission, the term *protocol* is commonly used in this area.

Let's consider two transmission systems (protocols) that play prominent roles in data communications.

**1.** *Asynchronous Transfer Mode (ATM).* Voice originates as an analog signal and, after multiplexing, is transmitted as a set of 64-bit frames. Data are transmitted as bursts of bits whose format depends on transmission protocols. The FDDI data frame discussed above illustrates this point very well. Voice and data transmissions were traditionally separate businesses having separate networks. However, as we've noted, voice is transmitted nowadays as a digital signal and voice and data use the same physical networks, such as PSTN. Thus, the need for a universal transmission system arises, and this is why the Asynchronous Transfer Mode (ATM) system was devised.

ATM, which is essentially a connection-oriented transmission system, conveys packets, or cells, through the network. By "connection oriented," we mean that a path for the cell must be established before the actual transmission starts. The term *asynchronous mode* means that the time when a given cell transmission may start is not predetermined.

The packet in an ATM is a fixed-length cell containing 53 bytes. This basic unit of transmitted information carries a 48-byte payload and a 5-byte header, or overhead. There are a number of advantages that make ATM one of the most important transmission systems in use today:

- It can carry any type of information. (Indeed, a 48-byte payload can carry voice, data, and even video traffic.) You cannot overestimate the importance of this feature because the need for such integrated transmission is growing rapidly.
- It supports both circuit switching and packet switching. In packet switching, each ATM switch is activated only when the cell is being transmitted. This enables effective use of the available network bandwidth but strictly regulates the speed of the switches. In circuit switching, the ATM transmits cells through an established line, just as any other circuit-switching system does. (To refresh your memory on packet- and circuit-switching technologies, reread Section 12.1 and see Figure 12.3.)
- It supports transmission on any scale from LANS to MANs to WANs without any additional transformation of the signal. This saves time, holds down the equipment cost, and makes OAM much easier to carry out.
- It provides quality-of-service guarantees, such as throughput, bandwidth, delay, and error control. This is certainly one of ATM's key advantages.

Despite all these benefits, ATM is still a complex, sophisticated transmission technology because its invention stems from the fact that bandwidth is scarce. It's quite possible that in the long run all-optical networks will replace ATM [20].

(A wealth of literature is devoted to ATM, enabling you to learn as much as you need to know about the system. See, for example, references [1], [5], [16], and [27].)

**2.** *Internet Protocol (IP).* Internet Protocol (IP) is the most widely used networking technology [1]. Every time you use e-mail or browse the Web, you employ IP, which provides all the necessary transmission functions in conjunction with a number of associated protocols. IP is used mostly with the Transmission Control Protocol (TCP), the reason why this type of transmission system is referred to as *TCP/IP.*

The basic unit of information with which IP operates is called a *datagram.* The entire message is divided into packets, which are transmitted through the network to their destination as independent units. The lengths of the packets vary, they are sent without having specific routes, they may be transmitted through different paths, and they may even arrive out of order. As you can see, then, IP is a "connectionless" transmission system, which means that the packets of information are sent without prior establishment of their transmission paths. (Compare this setup with a connection-oriented ATM system.)

IP's main advantage is its ability to work with almost all existing transmission systems (protocols). For example, when IP works with ATM, the IP packets are segmented into fixed-length cells of ATM, transmitted through the ATM network, and then reassembled into IP packets at the receiving end. In a similar manner, IP can work with FDDI frames. Another example is rapidly developing applications called *voice over IP (VoIP).* This technique allows us to transmit a regular telephone conversation over data networks. Of course, a voice signal must be digitized and then sent as regular data over IP transmission systems. The variable size of the IP data packets is advantageous in voice transmission. The motivating factor for developing this transmission system was the price of long-distance telephone calls: Sending voice over IP is much cheaper than using a regular telephone network.

In short, IP provides simple datagram service and this simplicity is its main advantage. But, as is often the case, a beneficial feature in one situation turns out to be a drawback in another. For example, IP does not provide the quality-of-service guaranties that ATM does. To overcome this shortcoming, new, more sophisticated protocols, such as the *Resource Reservation Protocol (RSVP),* have been developed, but, unfortunately, these innovations negate IP's advantage: its inherent simplicity.

We discuss ATM and IP systems further in Section 14.2.

(For a solid introduction to IP and TCP/IP systems, consult references [5] and [15]. Their specific applications in fiber-optic networks are considered in [1] and [21]. General computer networks based on fiber-optic systems are covered in [1], [2], and [3].)

## Cable TV

We are interested in dealing only with that aspect of cable TV that is concerned with fiber-optic networks. What follows here, then, is a brief review of a standard cable TV system. For an in-depth review of the field, see references [28] and [29].

Cable TV, or *community antenna television (CATV),* is a television system in which a signal is transmitted over a cable to an individual subscriber. This is in contrast to broadcast TV, a system in which a television signal is transmitted through the air so that any person possessing a TV set can receive it. The basic architecture of a cable TV system based on fiber optics is shown in Figure 14.9.

A television signal to be distributed by a CATV system is received through a satellite, a microwave antenna, or a fiber-optic trunk. It's processed and presented to a fiber-optic transmitter (Tx), from where it's transmitted in optical form. A postamplifier, or booster, raises the optical power of the transmitted signal and a splitter distributes it among many transmission fibers. A TV signal-processing unit, along with a fiber-optic transmitter (Tx) and a splitter, constitutes the *head end,* or control center, of a CATV system. The head end plays the same role in a CATV system that the central office plays in a telephone system.

The *transport segment* of a CATV system is within the realm of fiber optics. A single-mode optical fiber carries a television signal from the head end through a *secondary hub* to the optical-receiver node. The link from the head end to a secondary hub is sometimes called a trunk. To compensate for losses due to fiber attenuation and signal distribution, in-line optical amplifiers can be used. Keep in mind that the transmission links of CATV systems are usually on the order of tens of kilometers, so using in-line amplifiers is optional. CATV systems operate at 1310 nm.

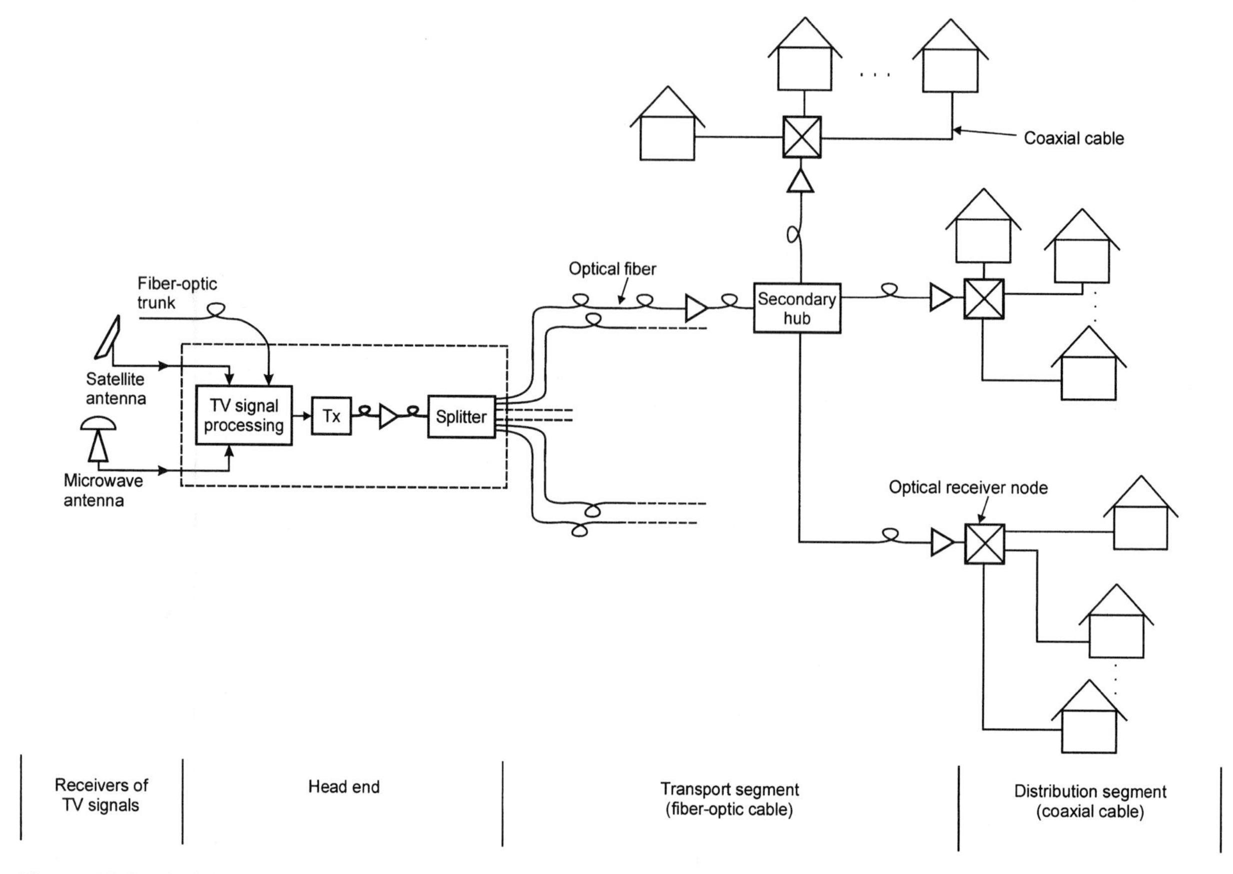


**Figure 14.9** Architecture of a hybrid fiber/coax (HFC) CATV system.

## Example 14.1.2

***Problem:***

A CATV transport line is 36 km long and uses a singlemode fiber operating at 1310 nm. The link power budget is 10 dB. Should we use an in-line (trunk) amplifier?

***Solution:***

We simply need to compare the actual link loss and available link budget. The typical loss of an SM fiber at 1310 nm is 0.6 dB/km. (See Figure 3.18.) Thus, the link loss is 0.6 dB/km × 36 km = 21.6 dB. Therefore, we need to use an in-line amplifier.

When designing a transport link, don't forget to take into account the splitter's losses.

Note that for a given link budget, there is a trade-off between the length of the transmission link and the number of distributed lines. Fortunately, optical amplifiers (OAs) help to resolve this dilemma without requiring any radical steps on the designer's part. A system using OAs can support both a sufficient length of transport link and the required number of connected customers.

It is worth noting that the typical loss of a coaxial cable is about 40 dB/km. What's more, this loss depends on signal frequency, which is significant for a TV signal. It's no wonder that transport segments of a CATV system are based on optical fiber.

The layout of a transport segment can be in either star or ring topologies.

The *distribution segment* starts with an *optical receiver node* and terminates at the customer's TV set. The optical receiver node converts an optical signal into an electrical one and sends it farther along a coaxial cable. A short coaxial cable, called a *drop,* connects your home wiring to a distribution cable.

We have to realize that modern CATV systems transmit an analog signal. For that reason, the system requires that the signal-to-noise ratio (SNR) be at a relatively high level. This justifies the use of a preamplifier in front of the optical-receiver node, as Figure 14.9 shows.

Turning to the bandwidth issues in a CATV, we know that a major bottleneck in a CATV network is the drop coaxial cable connecting your TV set to a distribution network. Its capacity is about 160 Mbit/s. As soon as the cable reaches an optical fiber, the bandwidth problem is resolved. This is why cable-network operators are interested in new architecture, such as fiber-to-the-curb (FTTC) and fiber-to-the-home (FTTH). FTTC implies that an optical fiber comes to your immediate neighborhood, leaving only the last connection for a coaxial cable. FTTH means an optical fiber runs from your home to a head end. FTTC architecture is used in cable networks. FTTH continues to be a hot topic for applied research. (It's important to note that the FTTC and FTTH architectures are used not only in CATV but in all other types of fiber-optic networks. What's more, LAN designers are now even considering fiber-to-the-desk architecture. Refer to our discussion of this topic in Section 8.4.)

The newest development in CATV systems is the integration of video, voice, and data transmission in cable networks. Cable modems enabling you to connect your personal computer to the Internet at a bit rate around 160 Mbit/s are commercially available, but this is not the only problem that has to be solved. CATV networks were designed for one-way traffic because TV transmission extends from a head end to your home. To make cable networks suitable for voice and data transmission, operators need to upgrade their infrastructure for full-duplex (two-way) transmission.

Integrating voice and data transmission with a video signal and preparing to provide new services such as video-on-demand and digital TV require more and more bandwidth. Thus, the capacity of already installed fiber-optic cable networks operating at 1310 nm is not enough to support a high-speed signal. Here, WDM technology is the only solution and cable-network operators have turned their attention to these systems. An example of how WDM can be used in a cable TV system is shown in Figure 14.10.

Using WDM in a CATV system allows a network operator to transmit, simultaneously, many video channels, OAM channels, and SONET signals with voice and data. Such systems are in the process of deployment by some cable TV operators [28].

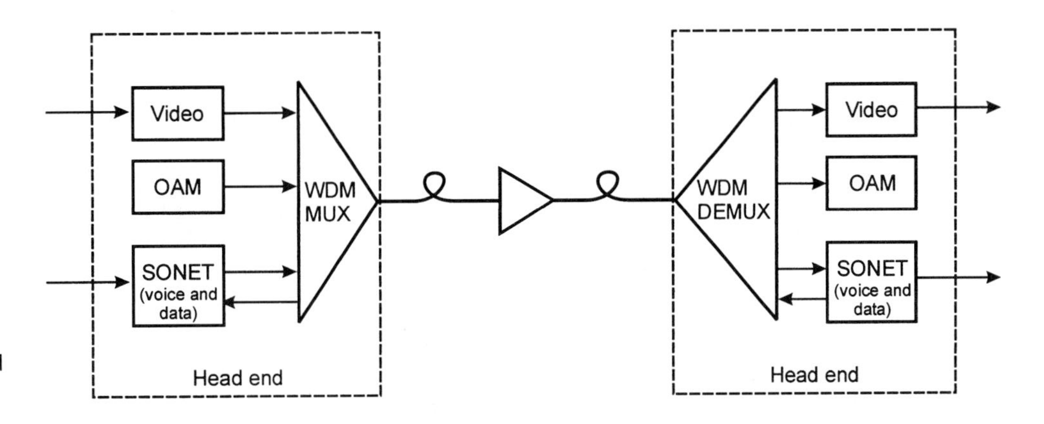


**Figure 14.10** WDM in a CATV system.

## Intensity Modulation/Direct Detection Versus Coherent Systems

The method of superimposing an information signal onto an optical carrier, such as the type we have been discussing throughout this book, is called direct modulation/demodulation. You may recall the basics of this technique:

An information signal changes the driving current and the latter modulates the amplitude of the output light. This is why the method is called *intensity modulation (IM).* At the receiver end, modulated light is directly converted into electrical current by a photodiode; that is, the receiver provides *direct detection (DD).* In short, the method is called *intensity modulation/direct detection (IM/DD).* Make no mistake: IM/DD is the *only* method employed today in fiber-optic communications systems. However, it is *not* the only possible method that can be used. One potential transmission technique is a coherent fiber-optic system.

*Coherent fiber-optic communications systems* use a principle similar to heterodyne detection in radio broadcasting. At the transmitter end, an information-carrying electrical signal modulates a microwave signal radiated by a local oscillator. This microwave signal, in turn, modulates a laser that presents modulated light for transmission. At the receiver end, the optical input signal mixes with light emitted by a local oscillator (laser) so that the information signal is detected at the *intermediate frequency (IF).* This concept is illustrated in Figure 14.11. (See references [1], [30].)

Let's consider this scheme in greater detail: The modulating information signal, which can be a stream of bits or an analog signal, is actually superimposed onto the microwave-frequency signal. The latter, in turn, superimposes the information onto an optical signal by modulating a laser diode. Thus, the microwave signal works as an intermediate carrier, called a *subcarrier,* and light serves as the main carrier. This is why this scheme is called *subcarrier modulation.* You can easily multiplex several information signals with microwave signals of different frequencies, which can range from 10 MHz to 10 GHz [1]. This kind of multiplexing, called *subcarrier multiplexing,* is used in CATV systems. Subcarrier modulation can be—and actually is—used without coherent detection.

Optical modulation can, of course, be internal or external. In the latter case, a local microwave oscillator is connected to an external modulator. Note, too, that not only a laser diode but an entire transmitter unit must be involved in preparing an optical signal for transmission. (See Section 10.4.) We can also use a *booster* (a *preamplifier*) to increase the power of the transmitted signal.

A coherent system works effectively if—and only if—the polarization of an incoming optical signal coincides at the receiver end with the polarization of a local optical signal. This is why we need *polarization-maintenance (PM) adapters* at the transmitter and receiver ends. We also need a PM fiber for transmission or we have to use some other means to control the polarization state of the transmitting signal. This is one of the major drawbacks of a coherent system.

When a received signal mixes with a local optical signal, the result is an optical signal that carries information at an *intermediate frequency* ($f_{IF}$). A photodiode converts an optical signal into an electrical signal and information is extracted from the carrier at $f_{IF}$. If $f_{IF} = 0$—that is, the detected signal is at the original (baseband, or modulating) frequency of the information signal—the detection is called *homodyne.* If $f_{IF} \neq 0$—that is, the detected signal is at the intermediate frequency—the detection is called *heterodyne.* "Homodyne detection" and "heterodyne detection" are generic terms familiar to any radio or telecommunications engineer.

The main motivation for developing a coherent system is to increase receiver sensitivity by 10 dB to 25 dB, that is, from 10 to 316 times. Theoretically, the use of a coherent system can reduce the quantum limit to 0.02 photons per bit at BER $10^{-9}$ [30]. You may recall that the quantum limit of an IM/DD system is 20 photons per bit. (See Section 11.3.) However, to achieve this increase in receiver sensitivity, a number of conditions must be met. The most important, as already noted, is the similarity of the polarization states of the received and local optical signals. You may recall that a commercial optical fiber does not maintain the state of polarization of a transmitted signal, so this condition is almost impossible to satisfy in practice.

Another major requirement is that the phases of both mixed optical signals must match, which is also almost impossible to maintain perfectly. Hence, the actual increase in receiver sensitivity is much less than is theoretically predicted.

Many technical stratagems have been developed to elude all these obstacles. The net result: To achieve a significant benefit, a coherent system becomes very complex, sensitive to environmental conditions, and eventually not justifiable economically for deployment. At the same time, the use of optical amplifiers all but eliminates the problem of receiver sensitivity. So there you have it—the reasons why coherent systems are not commercially deployed today and are no longer considered a promising approach to superimposing an information signal onto an optical carrier. (Consult references [1], and [30] for more on this topic.)

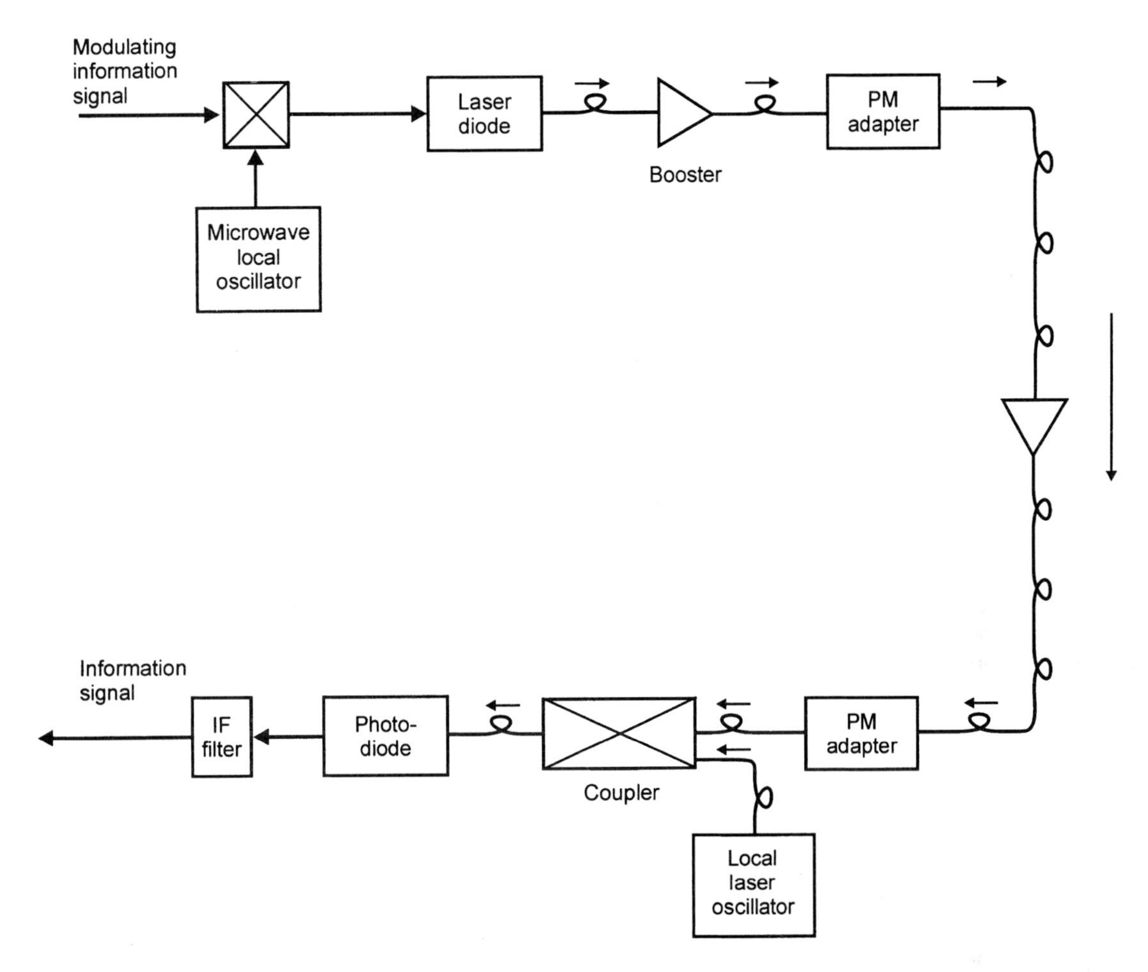


**Figure 14.11** A coherent fiber-optic system.

## 14.2 ELEMENTS OF THE ARCHITECTURE OF FIBER-OPTIC NETWORKS

Following our discussion of the kind of information we need to transmit over fiber-optic networks, we'll now explain how networks are organized, including the physical and logical levels of work that constitute what's termed *network architecture.* Full coverage of this broad topic is outside the scope of this book, but many excellent textbooks and monographs on the subject are listed at the end of this chapter. Here we'll concentrate on the key elements of the architecture of fiber-optic networks.

### Networks, Protocols, and Services

By now you should be convinced that a network has to include physical media *and* logical entities in order to function. (The relationship is analogous to that between computer hardware and software.) Up to this point we have been concerned largely with the physical media of fiber-optic networks. We'll turn here to a discussion of the logic of transmission.

As an analogy, consider the similarities existing between road and transmission networks that we mentioned in Section 12.1. To drive a car from our starting point to our destination point, for example, we might very well need to travel on local streets, regional freeways, and superhighways. And to reach our destination, we must obviously obey specific rules that regulate our actions on the roads.

In the same vein, to deliver a message from a source to a destination, we need, in general, to use local, regional, and global physical networks. What's more, the message also has to obey specific rules, ones that control its transmission over the network. For example, a "handshaking" message must be sent to establish a connection between the source and the destination before the actual transmission starts in a circuit-switching network. (See Section 12.1.) A set of rules that provides communication between the source and the destination point is called a *protocol,* a term used in its broadest sense to stand for an entity that represents the logic of a communications network. Among its many functions, a protocol prevents a collision between two messages entering the same node or link simultaneously, regulates the form by which the source is notified whether the transmission was successful, and does much more. Therefore, communications networks in general—and fiber-optic networks in particular—must have physical media and protocols for transmission. (See the discussion of transmission systems [protocols] in Section 14.1.)

Protocols—the sets of rules governing transmission—are not the only entities constituting the logic of communications networks. Another important aspect is *services,* a set of operations—or steps—that has to be put into effect if communication is to be successful. This term is specific to the logic level of a communications network, so don't mistake it for services provided by communications companies that deliver voice, data, and video communications.

In order to provide their services, source and destination systems require protocols. Because the services are not tied directly to their protocols in any physical way, we can update the protocols as technology evolves without affecting the services at all. At the same time, a set of actions needed to provide communication—services, remember—is very much independent of protocols and can be updated as technology changes.

With these points in mind, let's now consider the logical architecture of a communications network. We will start by studying an Open Systems Interconnection (OSI) reference model. We do so for two reasons: First, OSI is an international standard accepted by the International Standards Organization (ISO) and, secondly, it provides a perfect example of how all the ideas discussed above are implemented.

## Open Systems Interconnection (OSI) Reference Model

***Why an OSI reference model*** You will recall that, physically, a network is a collection of nodes connected by links. Links serve as the transmission media, while the nodes do all the intelligent work necessary to transmit the message through a network. Since this is a huge, complicated job, we'd better subdivide the work into several routines (functions), assigning a specific entity to perform a given function.

Your computer provides a perfect analogy to what takes place in a network. A computer consists, basically, of four blocks: input and output devices, a memory, and a central processing unit (CPU). Each block performs its function(s) and passes its output to the next block. For example, when you strike a key on your keyboard, you set into motion a series of electrical pulses representing the character you just pressed on your keyboard. This set of pulses enters the memory unit, where it's stored before being presented to the CPU. The CPU processes the information and commits the result to the memory unit for storage or commands the memory unit to pass the information to the output devices. But, of course, you won't see a set of pulses on your screen or printer. The output devices convert the internally processed information into a form appropriate for the user.

There are two main points we can draw from this admittedly oversimplified consideration: First, the entire task of processing information is subdivided among different units, each one performing its specific subtask, or function. Secondly, the recipient unit doesn't know—and doesn't need to know—how the sending unit performs its job; the recipient needs only to receive the result.

This said, we must now introduce a qualification: In our computer analogy, a specific function is associated with a physical unit. In actuality, however, this is not entirely true. The CPU, for instance, performs several functions, such as arithmetic and logic operations, and exercises control over the entire system. At the same time, the CPU, like the memory, can store information. So it is critical for you to understand that a function is not necessarily associated with a specific physical device. A given function can be performed by several units and one unit can perform several functions.

Applying the principles described above to networks, we can say that, very much like a computer, a node performs a transmission task. To facilitate the performance of this task, the job is subdivided into several subtasks (or functions, if you prefer). A wide variety of tasks have to be performed by each node in the same network and by various nodes in different networks. However, some general functions, or subtasks, can be distinguished and formulated. This is the basic idea behind an OSI reference model. *All functions that might be necessary for a node to perform in communications networks are presented in an OSI reference model as seven hierarchical layers.* The main idea underlying an OSI reference model is simply this: *Each layer performs specific functions and presents the result of its work to the layer above it* (never to the layer below).

The seven layers of an OSI are shown in Figure 14.12.

Before considering these layers more closely, it must be stressed that we will be dealing with the logical—not the physical—structure of a communications network.

***OSI layers*** To transmit a message over a network, the OSI model implies that seven functions have to be performed in general, but not all of them are needed for transmitting a particular message in a particular network. Each of these functions (in the form of layers, remember) can be performed at a sending node, a receiving node, and/or intermediate nodes; in other words, a layer is *not* associated with a specific physical device. Let's look more closely at these layers ([5], [6], [16], and [26]).

The *Physical layer,* or Layer 1, when transmitting, accepts data from the Data Link layer above it and sends the data through the physical media below it. (This is why the physical media are sometimes considered as Level 0.) When receiving data, Layer 1 accepts data from the physical media and sends this bit stream to the Data Link layer. Layer 1 specifies such items as the bit rate, modulation technique, pin-out of interfaces, and so on. The services it provides are modulation and demodulation.

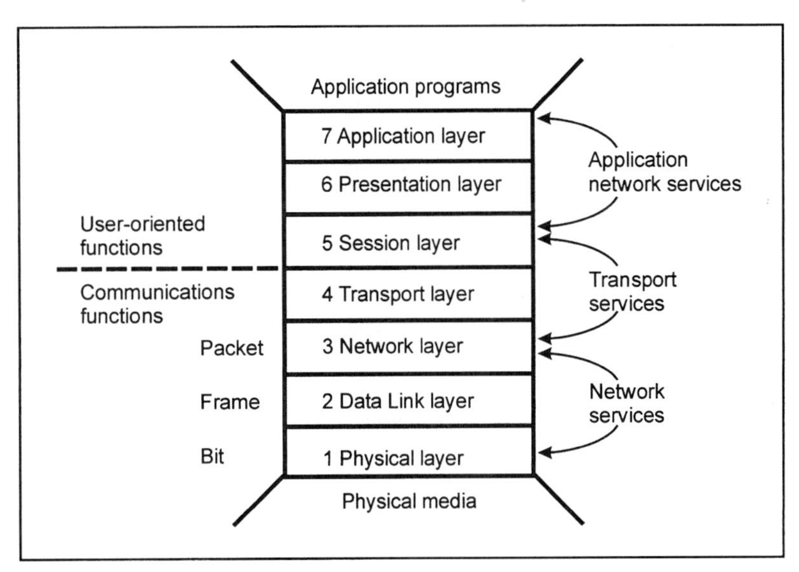


**Figure 14.12** Open Systems Interconnection (OSI) reference model.

It is customary to include the physical media in Level 1, as some protocols do ([5], [16]). Layer 1 is responsible for the transmission of bits, as Figure 14.12 shows.

The *Data Link layer* (DLL), or Layer 2, when transmitting, accepts data from Layer 3 (the Network), adds a header containing some housekeeping (service) information and a tailer containing error-control information. This added information is called the *Protocol Data Unit (PDU).* The information unit that the Data Link layer has made is called the *frame.* Hence, a header and a tailer are several bits added, respectively, at the beginning and at the end of a frame. (See Figures 14.12 and 14.14) When receiving data from Layer 1, the DLL examines the data for flow control and errors, removes this information, and sends what's left (a packet, or information unit) to Layer 3.

So, the Data Link layer's functions are basic transmission processing and error detection and control.

In local area networks, the DLL is split into two sublayers: Logical Link Control (LLC) and Media Access Control (MAC). LAN protocols specify only two layers: Physical and Data Link.

The *Network layer,* or Layer 3, when transmitting, accepts data from Layer 4 (the Transport layer), adds routing and error-control information, and sends the packets to Layer 2. The Network layer is the first layer actually routing messages through the entire network from a source to a destination, which is its major function. In addition, Layer 3 is also involved in error detection and correction (doing so in lieu of Layer 2 when a protocol requires it). It also does some transmission maintenance, such as counting packets, to avoid overloading a section in the network.

The Network layer deals with the packet. (See Figures 14.12 and 14.14.) However, it is able to work with both circuit-switching and packet-switching systems.

To clarify the function of the Netwok layer, we need to introduce the concept of *virtual communication,* which is the interchange between the same layers at the transmitting and receiving nodes. The peer layers, that is, the layers of the same level—Network and Network, for example—maintain virtual communication at different nodes by means of data packets (the Protocol Data Units mentioned above), while actual communication occurs through bits transmitted by the Physical layer.

Both the Physical and Data Link layers establish virtual communication between adjacent stations, while the Network layer maintains virtual communication between the source and the ultimate destination. Thus, the Network layer is the end-to-end layer.

The idea of virtual communication and the role of the Network layer are illustrated in Figure 14.13, where the links of virtual communications are shown as a dotted line while the path of physical transmission is shown as a solid line.

When transmitting, the *Transport layer,* or Layer 4, accepts information from the Session layer (Layer 5) and divides the data into packets. When in the receiving mode, Layer 4 reassembles the packets into data. Layer 4 is truly an end-to-end layer. When a break in virtual communication occurs between a source and the final destination at the Network layer, it is the Transport layer that bears the responsibility for effecting end-to-end communication. This is also true for error detection and correction functions. (Some protocols make the Transport layer directly responsible for end-to-end error control and recovery.) The role of the Transport layer is pictured in Figure 14.13(b).

The Transport layer isolates the lower layers (which have communications functions) from the upper layers (which have user-oriented functions).

The *Session layer* (Layer 5) is responsible for log-in and log-out procedures and for tracking the time of connection, which can be used for billing purposes. (Now you know what to blame for your monthly charges.) This layer is also responsible for terminating a connection.

The *Presentation layer* (Layer 6) presents the data in the appropriate format and code. Layer 6 receives data from Layer 5 as a bit stream and converts it into ASCII or EBCDIC codes (plain text codes) or into video, audio, or data formats. This conversion allows two different stations at the network ends to communicate regardless of their technical setup, or configuration.

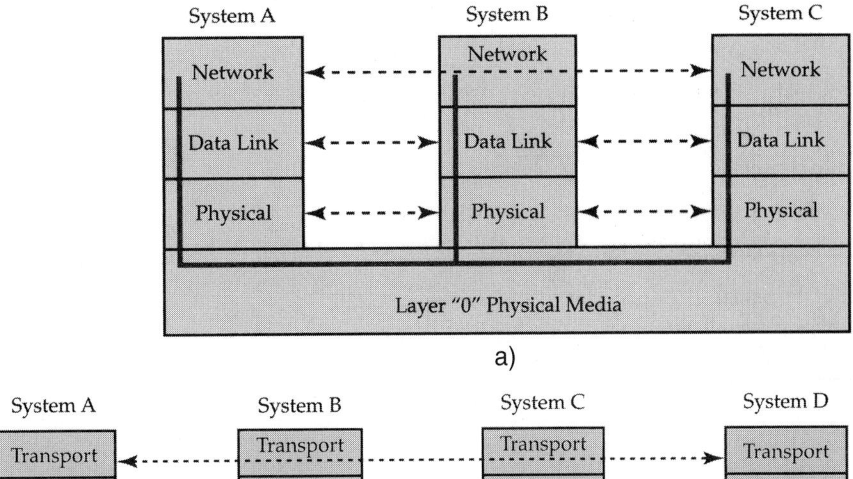


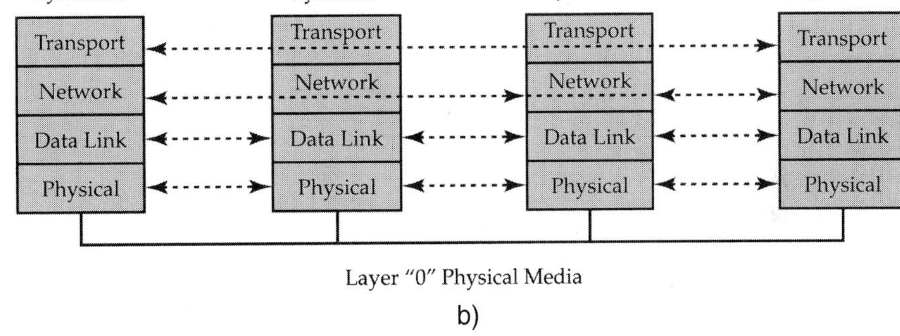


**Figure 14.13** Virtual communication in the OSI reference model: (a) Physical, data link, and network layers; (b) transport layer. *(Reprinted from Charles Thurwachter, Jr.* Data and Telecommunications, *Upper Saddle River, N.J.: Prentice-Hall, 2000. Used with permission.)*

When transmitting, the *Application Layer* (Layer 7) accepts data from a user for transmission through the network and, when receiving, it sends the data back to the user after the transmission is completed. This layer is responsible for how the network is used, a responsibility that is realized through translation of the user's request into specific network functions to be carried out. In other words, Layer 7 provides the user with access to the network, whose software determines the user's capabilities on the network. Figure 14.12 illustrates the ways in which the Application layer provides services for the user's application programs.

***PDUs in the OSI reference model*** When transmitting, each layer adds a Protocol Data Unit (PDU)—a number of bits constituting a packet of service information, remember—to the data received from the layer above and passes the new data composition to the layer below. When receiving information, the layer removes the PDU attached by the transmitting layer and sends the remaining data composition to the layer above. PDUs are actually the means by which layers perform their functions.

PDUs are usually added as headers (H) but the Data Link layer also adds a PDU as a tailer (T) for error detection and correction purposes.

Figure 14.14 shows how original data are transformed by adding headers and a tailer to all seven layers. The notations should be clear from the current discussion. For example, a header added by the Application layer is designated AH and its appropriate PDU is called APDU.

Thus, once again, PDUs are the means for virtual (service) communications among peer layers. We need them to manage the actual communication, which occurs through bits transmitted by the Physical layer.

***General structure of the OSI reference model*** Let's review the functions, services, and protocols of the OSI reference model:

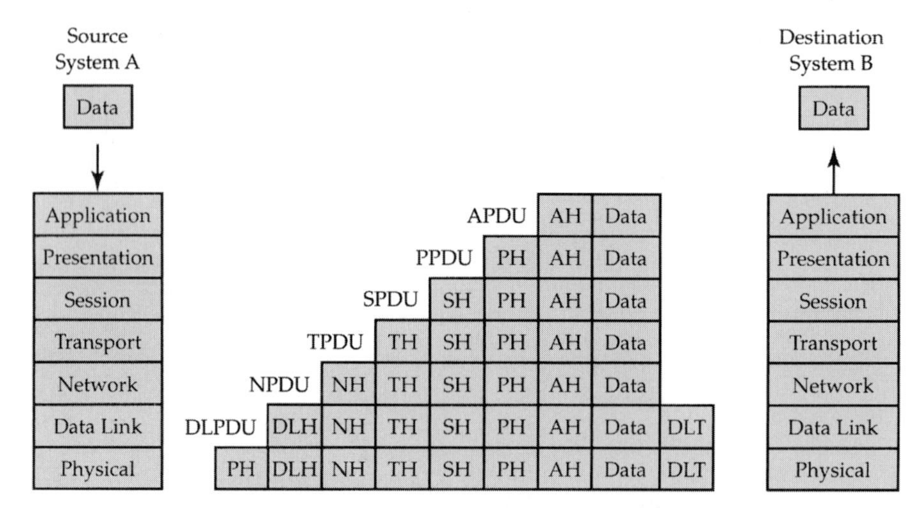


**Figure 14.14** Protocol Data Units (PDUs) in the OSI reference model. (*Reprinted from Charles Thurwachter, Jr.,* Data and Telecommunications, *Upper Saddle River, N.J.: Prentice-Hall, 2000. Used with permission.*)

**1.** *Functions.* Refer again to Figure 14.12. The entire OSI reference model can be broken down into two groups. The four lower layers carry out communications functions, while the upper three levels provide user-oriented functions. Indeed, the functions of the Physical, Data Link, Network, and Transport layers are to provide actual end-to-end communication. They do so by managing the transmission of data through the entire network. Specifically, they perform regular transmission functions, such as message routing and error control and recovery. The upper three layers—Session, Presentation, and Application—are primarily concerned with how the data are to be presented for transmission through the network. This involves such functions as log-in and log-out, data conversion, and network routing requests, among others. We should point out, too, that voice transmission over a telephone system can be executed entirely by the four lower layers.

**2.** *Services.* Services are the set of operations that each layer has to perform to fulfill its *raison d'être.* Each layer performs services for the layer directly above, so the layer above is called the *service user,* while the layer below is called the *service provider.* Physically, the service is done by passing *Interface Data Units (IDU)*—a number of other bits—through the boundary between the layers at what are called *Service Access Points (SAP).* It is worth noting the similarities and differences between PDUs and IDUs:

- A PDU carries information about the *functions* of a specific layer, whereas an IDU carries information about the *services* each layer provides.
- A PDU is an information unit that each layer adds before passing data *down* to the next layer. An IDU is an information unit that each layer adds before passing data to the layer *above.*
- A PDU is the means for virtual communication, that is, the means by which layers at the same level communicate at the transmitting and receiving nodes. An IDU is the means by which top and bottom layers at the same node communicate.

As Figure 14.12 shows, the OSI reference model provides three types of service: network, transport, and application network. As you can see from Figure 14.12, the four lower layers provide network services. For example, the Physical layer provides modulation and demodulation and the Logic Link Control sublayer of the Data Link layer insulates the higher layers from the differences among the protocols used in a LAN. Note that the same layer can provide two types of service. The Transport layer provides network and transport services and the Session layer provides transport and application network services.

Each layer insulates the layer above from the layer below. This insulation means that the above layer doesn't need to know what the layer below does or how it does it. Each layer need only obtain the result from its service provider. This is important because any changes in technology, which occur with lightning rapidity in our field, require an upgrading *only* of the layers affected; the others continue to do their job without change. You can readily imagine how efficient this structure is from both technological and economic standpoints.

**3.** *Protocols.* Protocols are sets of rules implemented as software programs. Several OSI protocols have been developed and standardized [26]. For example, the federal government has developed the Government OSI Protocol (GOSIP). General Motors and Boeing have developed a joint protocol called the Manufacturing Automation Protocol/Technical Office Protocol (MAP/TOP). This protocol has been designed for two purposes: to control the manufacturing process (MAP) and to facilitate the design process (TOP). (Other examples of OSI protocols can be found in references [5], [6], [9], [10], and [26].)

Even though the Transmission Control/Internet Protocol (TCP/IP) is a set of different protocols, we should point out that TCP operates at Layer 4 and IP operates at Layer 3 of the OSI reference model [6].

***Other reference models*** The OSI reference model shows the logical architecture of a communications network. The model subdivides all the transmission tasks into seven layers. But this is not the only way to represent the entire transmission operation. You can easily suggest another model, using more or fewer layers with some other functions, to achieve the same goal. This is why other reference models are used in data and telecommunications. These include System Network Architecture (SNA), Integrated Services Digital Network (ISDN), and Transmission Control Protocol/Internet Protocol (TCP/IP). SNA, the first reference model, was developed by IBM in 1972. Its advent actually stimulated the development of the OSI reference model. Don't be surprised to see ISDN and TCP/IP become standard reference models even though we have introduced them in Section 14.1 as transmission systems. Transmission systems, in fact, include reference models and protocols.

***Reference models: Why the need*** As we have seen, the OSI model does not describe specific devices and networks but, rather, operates with some generic entities, such as functions and services. This is exactly what makes this model so universal and has enabled it to survive since the early 1980s in spite of the revolutionary changes in technology since then. The reason for developing an OSI reference model was to allow products manufactured by different vendors to communicate with one another. And the OSI reference model works regardless of its specific implementation, that is, regardless of the specific devices used and their suppliers. This unique feature—adaptability to diverse hardware—allows the OSI model to establish communications among end-users despite the difference in their equipment and to evolve with the changing technology.

The OSI reference model is very general in structure and so its developers can include as many functions and services as the designer might need. In a specific network, you don't need all seven layers to operate. As we've pointed out, the voice-transmitting telephone network needs only the four lower layers.

It is important for you to become familiar with the OSI reference model for two reasons: First, it obviously plays a key role in networking technology and, second, it introduces the idea of layering in networking technology. This concept—layering—is critical and we will be referring to it often in this section in connection with fiber-optic networks.

(We should note that the dynamic change in networking methodology and technique has generated the need for modifying network models. The OSI reference model today seems to be too complex and the trend is to accept a simpler TCP/IP-centered model. The trend unquestionably is to replace layered architecture with object-oriented network models. [55].)

## SONET Networks and Layers

Even though we are restricting our discussion here to SONET, bear in mind that everything we are saying holds true for both SONET and SDH. In fact, most of today's technical literature refers to these transmission systems as SONET/SDH.

***Elements of a SONET network*** SONET was developed as a high-speed telephone network. (See Section 14.1.) Even though SONET relies almost exclusively on optical fibers as its transmission media, which implies using light sources and photodiodes, most of the other SONET equipment is still electronic. This is why SONET represents the *first generation of fiber-optic networks.*

In terms of elements, SONET uses *terminal multiplexers (TMs), add/drop multiplexers (ADMs), digital cross-connects (DCSs), and regenerators.* In terms of network topology, SONET uses point-to-point links and rings. Look again at Figure 14.1, which shows the basic architecture of a telephone network. SONET provides long-haul, interoffice, and access connections. We'll concentrate on the parts of this network providing access and interoffice connections called the *access* and *backbone* networks, respectively. This segment of a SONET network is shown in Figure 14.15 (See [1], [19], and [20].)

Low-speed links are connected to a SONET network through a terminal multiplexer (TM). If a TM is connected directly to a digital cross-connect (DCS), we have a *point-to-point link,* as Figure 14.15 shows.

SONET's ability to add and drop low-speed streams is realized through sets of add/drop multiplexers (ADMs) connected in sequence. In such a case, we have a *linear configuration,* as shown in Figure 14.15. ADMs are usually connected in a *ring* configuration, also shown in Figure 14.15. In this case, ADMs serve as ring nodes and, in addition to multiplexing/demultiplexing, also provide network management and protection functions.

SONET is able to transmit different data streams through its use of digital cross-connects (DCSs). Before the appearance of SONET, this function was performed by manually operated passive copper-wire patch panels similar to those discussed in Section 8.3. Digital cross-connects, which are actually digital switches, have not only automated this process but they also perform many other functions. In addition to cross-connecting (switching) different streams under software control, digital cross-connects monitor network performance and do multiplexing.

The SONET standard specifies the use of regenerators to boost a signal for long-distance transmission. Two such regenerators are shown in Figure 14.15 as an example; note in particular that the network can include as many regenerators as needed. The segment between two regenerators is called a *section.* (Optical amplifiers [OAs] are beginning to be used in SONET networks, so the term *section* can be applied to the segment between two adjacent OAs.)

Ring topology is used in the SONET access and backbone segments because it provides a high level of reliability in case of failure of any of the network segments.

Access rings operate at OC-3 and OC-12 transmission rates (see Table 14.2.), while backbone (interoffice) rings transmit at OC-12/OC-48/OC-192 bit rates.

SONET, now more than 15 years in operation, and its European counterpart, SDH, are still powerful transmission systems. For example, the new Pan-European I-21 Network is based on the SDH transmission system. This network, built as a set of interconnected rings, uses Corning's LEAF fiber and Alcatel's WDM technology [31].

It's worth noting that the general architecture of an optical-network model looks very similar to that of a SONET network [32]. However, new terminology is typically used for an optical

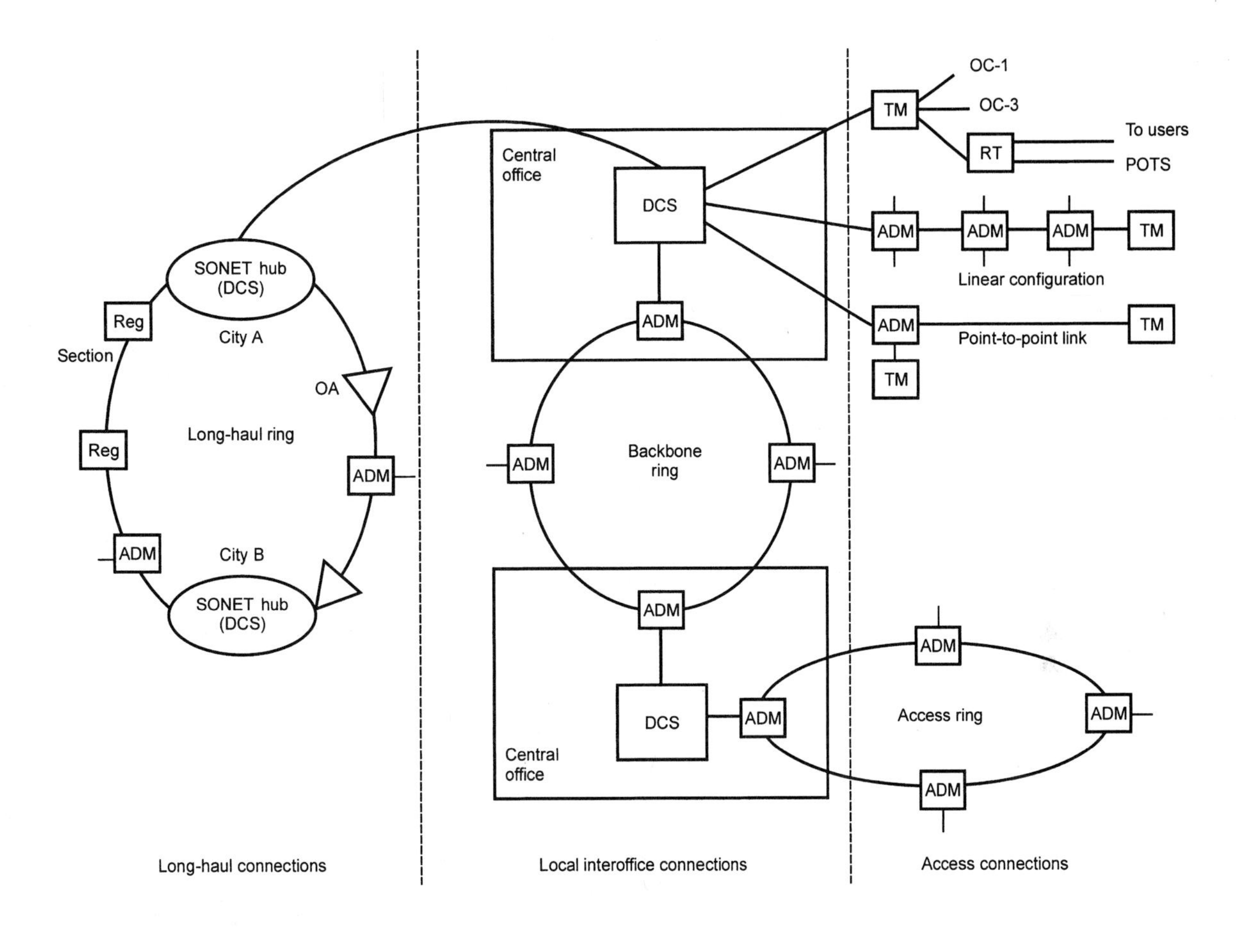


**Key**: ADM - add/drop multiplexer; DCS - digital cross-connect; OA - optical amplifier; OC - optical carrier; POTS - plain old telephone service; Reg - regenerator; RT - remote terminal; TM - terminal multiplexer.

**Figure 14.15** Elements of SONET network.

network. For example, the long-haul segment is called the "core," the interoffice section is called the "metropolitan" section (or simply "metro"), while, surprisingly, the access section keeps its original name. Obviously fiber-optic networks use optical add/drop multiplexers (OADMs) and optical cross-connects (OXCs).

***SONET layers*** Figure 14.15, the layout of a typical SONET network, helps us to understand how the network operates physically. As for the network's logic operation, that is, its protocol in the broad sense, we need to consider the SONET layers. A possible way to divide this SONET into layers is shown in Figure 14.16 [1].

As Figure 14.16(a) shows, an entire SONET network can be subdivided into sections bordered by nodes. Each node has its own layered structure; for example, a regenerator needs only Physical and Section layers, while the ADM node includes Physical, Section, and Line layers. The most complete layer structure is associated with a TM node; it includes Physical, Section, Line, and Path layers. (See Figure 14.16[b].)

As already pointed out, a SONET transmission frame includes a payload and an overhead. (See Figure 14.4.) Each layer adds its bytes to the overhead; this is how the layer performs its function. (For an overview of SONET layers with reference to the overhead bytes responsible for the functions of a specific layer, see [19].)

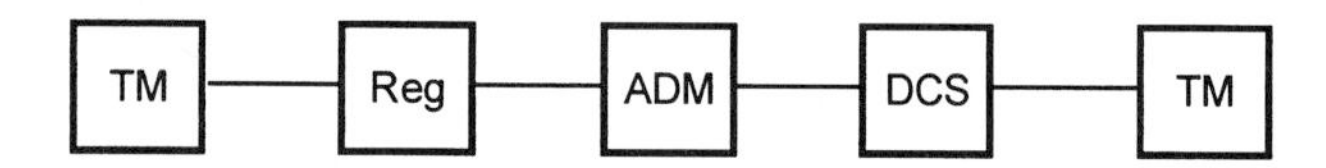


Key: TM - terminal multiplexer; Reg - regenerator; ADM - add/drop multiplexer; DCS - digital cross-connect.

a)

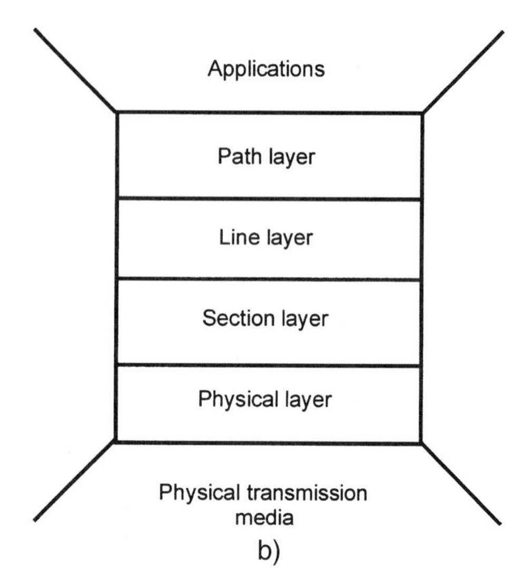


b)

**Figure 14.16** SONET layers: (a) Sections of a SONET network; (b) layered structure of a SONET network.

The *Physical layer* in SONET architecture plays the same role as the Physical layer in the OSI model: It is responsible for the transmission of bits over physical media. The SONET standard specifies such physical media as optical fibers and light sources with operating wavelengths that depend on the bit rates and distances for which the SONET network is designed.

SONET systems are subdivided, with regard to their loss, into short-reach (SR), intermediate-reach (IR), and long-reach (LR) networks. Thus, for example, for OC-3 and SR, SONET recommends the use of the 1310-nm operating wavelength, a multimode fiber, and an LED, while for OC-48 LR, SONET specifies 1550 nm, a singlemode fiber, and a single-longitudinal-mode DFB laser diode. (For more details on the physical interfaces of SONET, consult [1], [19], [20], and the standard itself.)

You will recall that a layer structure is associated with a node of a network; it follows, therefore, that a Physical layer is needed at each node throughout a network. The Physical layer of a SONET network corresponds to the Physical layer in the OSI reference model.

The *Section layer* is responsible for sending data to the Physical layer; it interfaces with the Physical layer by adding appropriate bytes to the frame overhead. It also provides error monitoring and control. In addition, the Section layer regenerates the signal and protects the multiplexing and switching operations [15]. A Section layer is needed at each regenerator.

The *Line layer* is responsible for multiplexing many path-layer connections onto a single link between adjacent nodes [1]. This layer also provides most of the operating, administrative, management, and provisioning (OAM&P) functions, including network protection. A Line layer is needed at ADMs and TMs.

The Line and Section layers together correspond to the Data Link layer in the OSI reference model.

The *Path layer* is the end-to-end layer. It is responsible for connections between a source and a destination. This layer, which is similar to the Network layer in the OSI reference model, also monitors and tracks the status of connections. A Path layer is needed at each SONET terminal multiplexer.

The advantage of the layering concept is that it can be generalized in the sense that the SONET network itself can be treated as a single layer of an entire general telecommunications network. We'll take this matter up later in this chapter.

## ATM Networks and Layers

***ATM networks*** An ATM network, whose basic features are discussed in Section 14.1, does not rely on any specific physical media; therefore, ATM transmission can be accomplished with any existing physical network, such as SONET, for example. What ATM requires in terms of hardware are ATM switches. ATMs, as you may recall, transmit cells, which are 53-byte packets consisting of a 48-byte payload and a 5-byte overhead. Each cell must be routed by ATM switches through the transmitting network.

The basic concept of an ATM network is shown in Figure 14.17. An ATM allows transmission through existing LANs, MANs, and WANs without the need to convert a data format, which is an ATM's great advantage. This transmission relies on the use of ATM switches.

ATMs are connection-oriented systems. This means, you'll recall, that the cell route is established before the actual transmission starts; however, in contrast to a circuit-switching technique, ATM connections remain in place only as long as the actual transmission lasts. This results in the effective use of a network's bandwidth. What's more, this approach increases the transmission speed in an ATM because there is no need to set up a connection for each individual cell. However, if needed, an ATM can provide connectionless service.

ATM transmission speed varies widely, which is another advantage of the system because it allows the ATM to work with any application and leaves room for improvement as technology advances. We are, of course, concerned with the upper limit of this speed.

Four basic techniques used in ATM systems allow the network to achieve high-speed transmission:

- ATM switches operate with the fixed length of transmitting cells. This relegates the switching function more to the system's hardware than to its software. Since hardware-switching is faster than software-switching, ATM networks can transmit at very high speeds.
- Since a route is established at the beginning of each conversation, the time for setting up the route is minimized. This is done by establishing a virtual channel and virtual path using the *Virtual Channel Identifier (VCI)* and the *Virtual Path Identifier (VPI),* which are the parts of a cell header. Reading these identifiers, each ATM switch simply directs the cell to the appropriate predetermined port rather than spending time to find out to which transmission port the cell should next be directed.
- ATM cells travel along only one path and arrive in the same sequence in which they were transmitted. This saves time for routing and for reassembling data from the arriving cell stream.
- ATM performs only some very basic error checking, which also saves transmission time. Thorough error monitoring is unnecessary because ATM transmission is done through presumably reliable networks, such as SONET, that have their own error-control capabilities.

Even though the system is called an *asynchronous* transfer mode, it has been intended for transmission over a SONET network, which is a *synchronous* network. Contradictory? Not really. Behind this apparent incongruity is the fact that the user's cells arrive at unpredictable times—that is, asynchronously—while the data stream is transmitted synchronously. This is done by filling

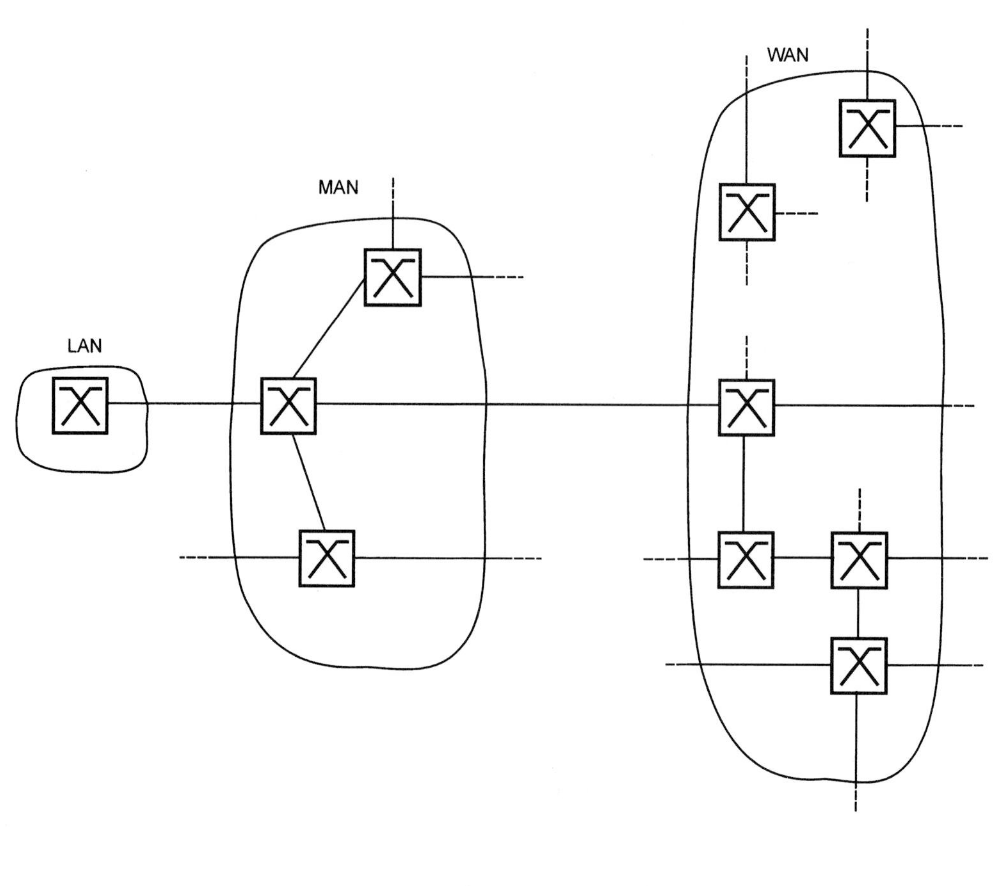


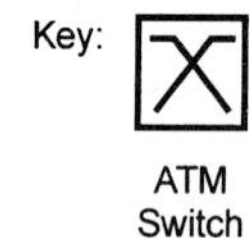


**Figure 14.17** Concept of an ATM network.

the idle intervals between the data with either timing bits or timing cells. Timing cells, of course, carry the headers that allow the network to distinguish them from the user's cells, a feature that adds another degree of sophistication to an ATM network.

***ATM layers*** Now that we've considered the physical operation of an ATM, we'll turn to its logical side. The layered structure of an ATM is shown in Figure 14.18.

The ATM's Physical layer is responsible for actual data transmission. From this standpoint, the ATM's Physical layer is similar to the OSI's Physical layer; however, the former transmits cells, whereas the latter transmits bits. SONET and T3 are usually the Physical layers of an ATM even though LANs can also be used for ATM transmission to the desk.

The *ATM layer* is responsible for handling actual transmission. This layer reads and interprets routing information, operating with VCI and VPI in the cells' headers. Headers, not payloads, are the concerns of the ATM layer. This part of the ATM layer's responsibility corresponds to the functions of the Physical layer of the OSI model, so that the ATM's Physical layer and the transmission part of the ATM layer together equate to the OSI's Physical layer.

The *ATM Adaptation layer (AAL)* both inserts and extracts the user's data. This layer converts the data into a payload and is responsible for end-to-end communication. The AAL is divided into

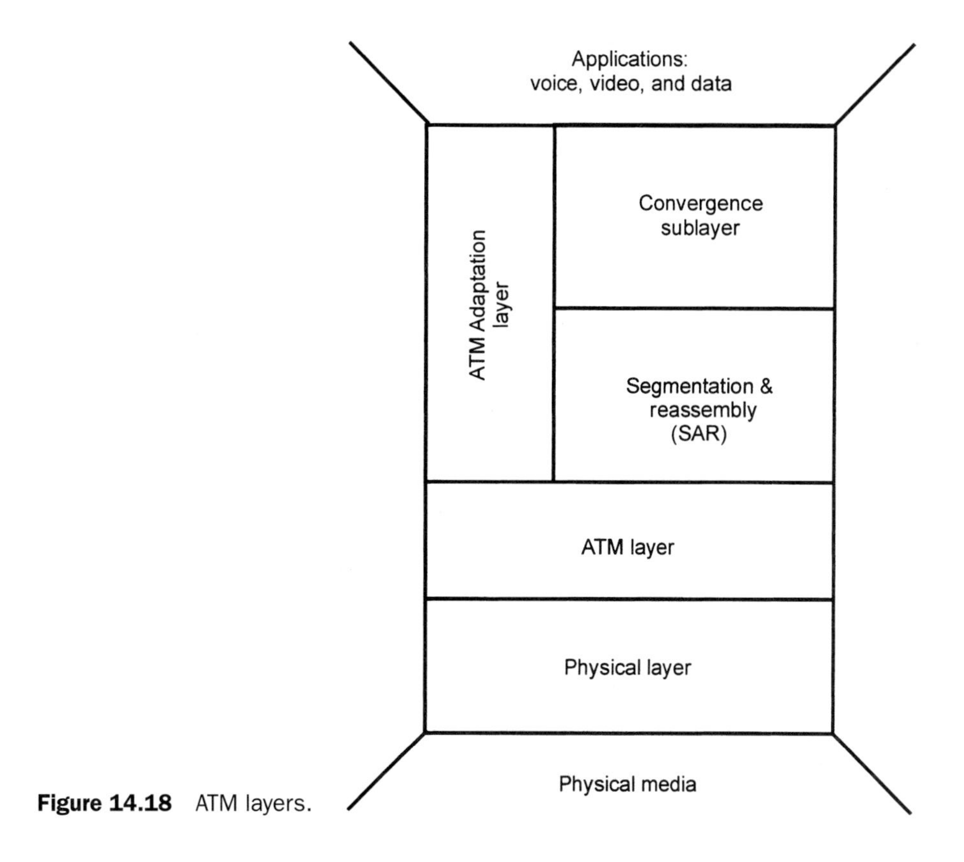


**Figure 14.18** ATM layers.

two sublayers: *convergence* and *segmentation and reassembly (SAR).* The segmentation portion of the SAR sublayer divides the entering data into 48-byte packets and sends them to the ATM layer. Obviously, this portion of the SAR sublayer works at the transmitting end. At the receiving end, the reassembly portion of the SAR reassembles the information from the stream of arriving cells and feeds this information to the convergence sublayer, which prepares the user's data for transmission over the ATM layer.

Such preparation is needed because different applications require different transmission formats. For example, voice transmission means that the cells will arrive in the same order in which they were sent, at a constant bit rate, and with a constant time interval between cells. TCP/IP, on the other hand, is a form of connectionless transmission that may use a variable bit rate and that does not require any timing (synchronization) between the end-users. All such requirements are grouped into four ATM classes of service. The convergence sublayer assigns the class of service to the user's information before transmission. (See reference [5] for more information on the ATM Adaptation layer.)

The ATM Adaptation layer and the upper portion of the ATM layer correspond to the OSI Data Link layer.

## Layered Architecture of Fiber-Optic Networks

***IP, ATM, and SONET networks*** The concept of layers can be used to help us develop a new point of view of a network. For instance, an entire network can be seen as a set of tiered layers, each layer representing a transmission system. Classic examples are SONET and ATM networks. ATM was developed for transmission over the SONET network [1]. Thus, from a logic standpoint, the entire network consists solely of an upper layer (an ATM) and a lower layer (a SONET network), as shown in Figure 14.19(a).

Physically, this concept can be put into effect by connecting ATM switches to SONET adapter cards, through which ATM gains access to the SONET network. A source presents data to an ATM network, where the data stream is converted into ATM cells. The ATM switches direct these cells to the SONET cards. SONET converts the ATM cells into STS frames and transmits these frames over the network. At the receiver end, SONET converts the frames back into ATM cells and presents these cells to an ATM switch, which directs the cells to their destination. This is shown schematically in Figure 14.19(b).

Still another example is IP transmission over an ATM network. In this case, IP resides over the ATM layer, which lies over the SONET layer (Figure 14.19[c].) Physical implementation is accomplished by connecting IP routers to the ATM network. IP packets emanating from a source are delivered to the IP routers. These packets are converted into ATM cells. The cells are transmitted through an ATM network and converted back into IP packets at the periphery of the ATM network. IP routers transport packets to their destination. (See Figure 14.19[d].)

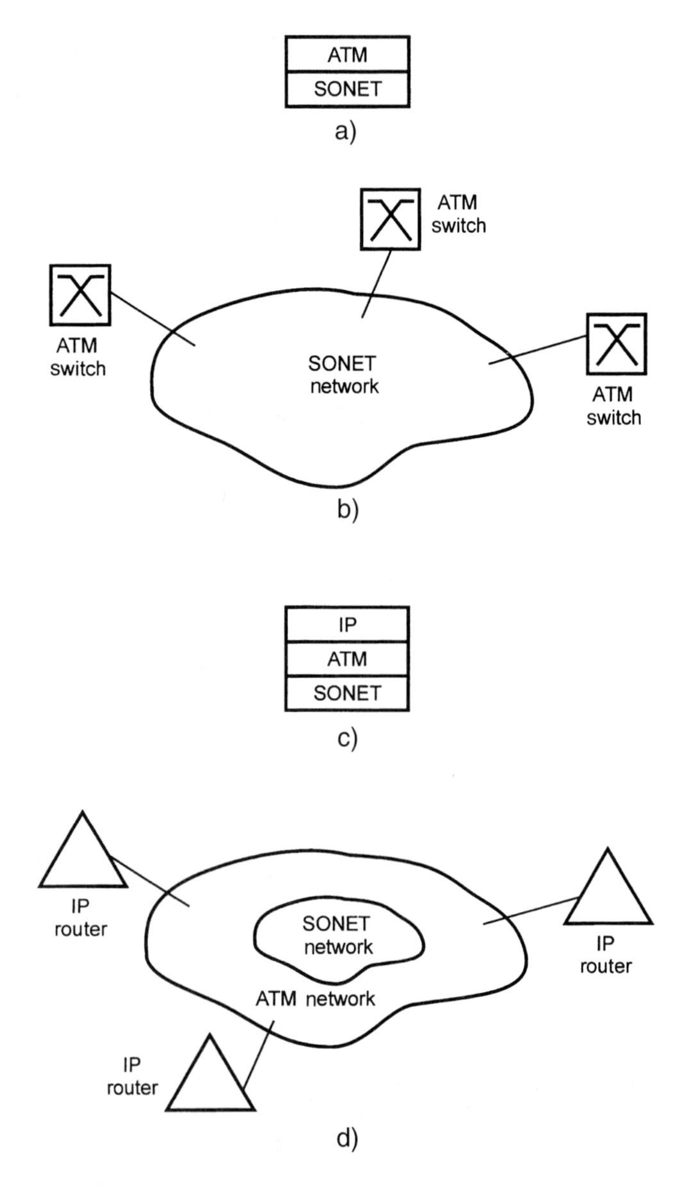


**Figure 14.19** IP, ATM, and SONET networks: (a) ATM over SONET; (b) physical connections of an ATM to the SONET network; (c) IP over ATM; (d) physical connections of the IP network to the ATM network.

Thus, in general, IP packets are encapsulated into ATM cells that, in turn, are wrapped into SONET/SDH frames; these STS frames are run across the optical layer using wavelength-division multiplexing.

We already know that both SONET and ATM networks have their own layered structure, as shown in Figures 14.18 and 14.16. An IP network also has its own layered structure. Thus, we can reveal the details of the layered architecture of the entire network protocol by including layers of the SONET, ATM, and IP networks in this picture [1]. This view is shown in Figure 14.20.

As you can see from Figure 14.19, SONET provides point-to-point connections for the ATM switches. Hence, SONET serves as the Physical and Data Link layers in terms of the OSI reference model (Figure 14.12), whereas ATM itself treats SONET as a Physical layer and a part of the ATM layer in terms of the overall ATM-layered structure (Figure 14.18). ATM, in turn, provides point-to-point connections among IP routers; therefore, the IP network treats ATM as its Physical layer and a portion of its Data Link layer in terms of the OSI reference model. To complete the end-to-end transmission of the user's data, IP includes—again, from the standpoint of the OSI reference model—its own Data Link and Network layers.

As traffic in fiber-optic networks grows, it becomes increasingly desirable to run IP directly over SONET. To make this architecture work, a set of problems has to be overcome, but we leave you to pursue this interesting topic in reference [33], for example, where you will find further details.

## Optical Layer

***Definition*** The Telecommunication Standardization Sector of the International Telecommunication Union (ITU-T) has introduced a new layer for a telecommunications network: the *Optical layer.* The layered structure of a modern network, including an Optical layer, is shown in Figure 14.21. Figures 14.21(a) and (b) give a general view [1], while Figure 14.21(c) combines both the physical and logical views to show how a SONET layer is placed over an Optical layer with connections among their nodes.

Physically, WDM links are point-to-point systems connecting SONET elements [21]. Thus, WDM is a Physical layer that transports SONET frames without affecting SONET overheads.

It is understandable if at this point you are wondering why we need another Optical layer, especially since all we've been saying up to now has been aimed at convincing you that almost all transmission today is done through optical fibers. Let's clarify this point: The physical transmission in SONET, ATM, and IP networks is performed by the fiber-optic communications systems that we have been discussing. The logical structure of these fiber-optic networks, however, has not included an Optical layer so far, as Figure 14.19(c) shows.

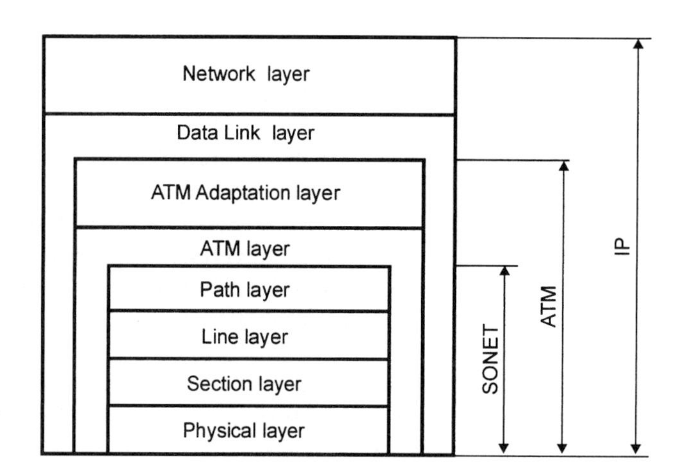


**Figure 14.20** Layered structure of IP over ATM over SONET network.

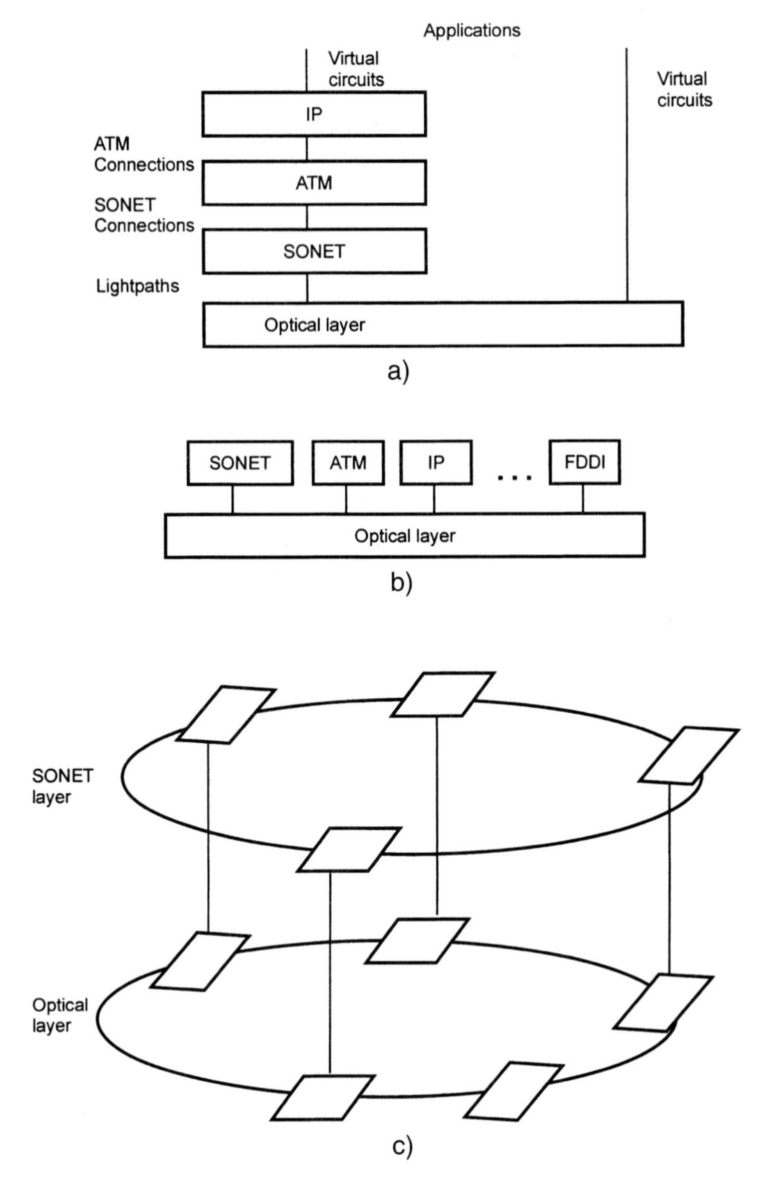


**Figure 14.21** Layered structure of a fiber-optic network including an Optical layer: (a) IP over ATM over SONET over Optical layer; (b) Network layers immediately over Optical layer; (c) SONET layer over an Optical layer.

Even though the transmission medium for modern networks is an optical fiber, most of the installed switches and add/drop multiplexers are still electronic. This is why we refer to these networks as first-generation fiber-optic networks. The new Optical layer is associated with the second-generation optical networks. This Optical layer serves not only as a transmission medium; it also carries out all the functions and services required by the layer concept.

The second-generation fiber-optic networks provide not only point-to-point optical connections but also routing and add/drop functions in the optical domain. Since these are WDM networks, all their functions are associated with wavelength channels.

The reason we need another layer with functions and services similar to those of existing layers like SONET or ATM is that each network layer—Optical, SONET, ATM, IP, and others—operates most effectively within its bit range. The transmission of terabits-per-second involves many wavelength channels, each composed from many smaller data streams by the TDM technique. It is much more efficient to delegate to an Optical layer all the operations at the Tbit/s level without getting involved in megabits-per-second transmission, where the Optical layer doesn't operate effectively. On the other hand, SONET is designed to operate most efficiently at the megabits-per-second level of the data stream but does not work effectively at the Tbit/s bit rate. This desire to optimize the performance of each layer is the fundamental driving force behind the development of the layered-structure concept of fiber-optic networks [1].

The overlayered structure shown in Figure 14.21(a) works well today but it has one inherent drawback: The greater the number of layer interfaces that are involved in the transmission, the more sophisticated the network becomes. This means less reliability and higher cost. However, the progress that has been made in fiber-optic networks now allows a network operator to run SONET, ATM, IP, FDDI, and other traffic directly over a common optical network. Logically, this means that all the layers mentioned above are placed immediately on top of an Optical layer, as Figure 14.21(b) depicts.

***Lightpaths*** An Optical layer provides *lightpaths* to a higher layer [1]. The concept of lightpaths is covered in Section 12.1. A lightpath, as we've seen, is a temporary optical-communications channel established throughout a network. In other words, it is an actual means to provide end-to-end transmission across an optical network. Using a lightpath, you can transmit your data using a single wavelength or even reuse the wavelength. (See Figure 12.5.) The flexibility of a lightpath enables an Optical layer to provide full utilization of a network's bandwidth.

Look at Figure 14.21 again. Bear in mind that an ATM or LAN—or any other network layer, for that matter—can reside on top of an Optical layer without the need for intervening layers.

***Services*** *[1]* Setting up a lightpath is actually the first type of service that an Optical layer offers to the layer above. Establishing a lightpath means setting up a connection among all the end-users in the entire network. We can draw an analogy here to a *public switching telephone network (PSTN)* in that setting up a lightpath in fiber-optic networks is similar to establishing a physical connection within the circuit-switching PSTN. Establishing a lightpath is done at the user's request and the connection is terminated when the conversation ends, an action very similar to the PSTN operation, where the connection between caller and party called is established at the caller's request and terminated by mutual consent.

A lightpath is usually associated with a single wavelength in a WDM fiber-optic network, as discussed in Section 12.1. An individual wavelength channel typically carries a high bit rate (2.5 Gbit/s). Therefore, a lightpath presents the entire bandwidth to the layer above.

The second type of service an Optical layer provides is the establishment of virtual circuits. A *virtual circuit* is a packet-switched connection between two nodes with a bandwidth smaller than the entire bandwidth of the link established between these nodes. The need for virtual circuits can be understood by considering this example: Assume that the bandwidth of a given link is 2.5 Gbit/s, which is typical for a single-wavelength channel today; however, you need only 155 Mbit/s. Without virtual circuits you would have to occupy the entire link, a very inefficient way to use a network's capacity.

To subdivide the entire bandwidth into smaller portions, an Optical layer uses the optical time-domain multiplexing (OTDM) technique. To return to our above example, your data are transmitted between two nodes as a set of packets at the bit rate of 155 Mbit/s. Your stream is optically time-multiplexed with other data so that the bit rate of the entire channel is 2.5 Gbit/s. This approach allows the network designer to fill the entire bandwidth of a link with smaller portions. But, for you, the network now operates as if you have a personal circuit at the transmission speed you need. Thus, this technique is called a "virtual circuit." Does it remind you of a concept you read about previously? If you answered "virtual tributaries (VT) in connection with SONET," you're right. (See Section 14.1 and Figure 14.4.)

There are two forms of optical time-division multiplexing: *fixed OTDM* and *statistical OTDM.* With fixed OTDM, each virtual circuit is granted a certain portion of the entire bandwidth so that the sum of the bandwidths of all virtual circuits is equal to the bandwidth of the link. With a guaranteed bandwidth, a fixed OTDM works better in networks having a steady flow of traffic, such as telephone networks. The downside of this technique is obvious: If you don't employ your virtual circuit, you are using the link inefficiently.

When the data stream moves in sporadic bursts, at any given time some virtual circuits may be idle while others are busy. The idea of statistical multiplexing is to use these idle circuits for transmission. This is done by subdividing the data from a virtual circuit into small packets and directing these packets along those virtual circuits that are idle at any particular moment. In this way, the bandwidth of a link is utilized most efficiently. The disadvantage? No guaranteed bandwidth for any virtual circuit. However, this approach works well in computer networks; the Internet is a good example of its practical application.

The third type of service an Optical layer provides to the layer above is *datagram* service, which entails the transport of small packets of data without establishing a connection prior to the actual transmission.

***Structure of an Optical layer*** All major network layers—SONET, ATM, IP, and others—consist of several sublayers. An Optical layer is not an exception and its structure is shown in Figure 14.22.

As we have done with the SONET architecture (see Figure 14.16), we will first consider the "horizontal" segments of a WDM network. From this viewpoint, the entire network is subdivided into segments between two adjacent WDM nodes and/or a WDM node and an optical amplifier node. Each node has its own layered structure, as shown in Figure 14.22. Let's consider this structure [1]:

An *Optical Transmission Section layer,* depicted as an *Amplifier layer* in Figure 14.22, exists at each node and provides an amplification function to the layer above. The Optical *Multiplex Section layer* represents the point-to-point connection between WDM nodes along the entire route of a lightpath. (It would be a good idea at this point to turn back to Figure 12.5, which shows the physical layout of lightpaths in a fiber-optic network.) A *Channel Section layer*—also called an *Optical Channel Section,* or *Lightpath layer*—is responsible for end-to-end lightpath routing across the entire network.

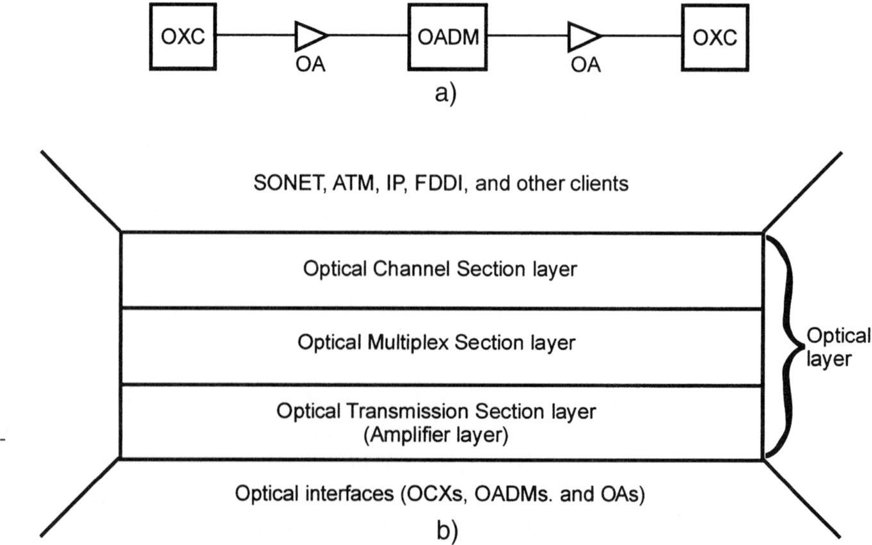


**Figure 14.22** Structure of an Optical layer: (a) Sections of an optical layer; (b) the sublayers of an Optical layer.

As you can see, it is difficult to make a direct correlation between the layers of the OSI reference model and the layers of an Optical-layer model. (By the way, this would be a good exercise for you to undertake.)

Refer again to Figure 14.19 and 14.20. The IP network is running over the ATM network, which, in turn, is running over the SONET, as Figure 14.19(c) shows. This layered architecture of an entire network is detailed in Figure 14.20, which presents all the sublayers of the network layers. Figure 14.21 continues to develop the representation in Figure 14.19 by showing all the major layers, including an Optical one. You can easily draw the detailed structure of a fiber-optic network similar to that shown in Figure 14.20 by studying both Figures 14.20 and 14.22. Bear in mind that an Optical layer serves as the Physical layer in a SONET network.

The model of an Optical layer introduced by ITU-T and described above does not reflect all the functions associated with second-generation fiber-optic networks. In particular, in broadcast and select fiber-optic networks (see Section 12.1), the packet-switching function is performed by a *Media-Access Control* layer; however, this layer is not shown in Figure 14.22. In wavelength-routing networks (Section 12.1), the packet-switching function is performed by a Network layer, which is not shown in Figure 14.22. As you can see, then, the model of an Optical layer needs to be developed further as fiber-optic networks evolve. Keep your eyes on the trade publications for reports on the latest developments in optical layers, a promising emerging technology in the fiber-optic communications scene [32].

## 14.3 NETWORK MANAGEMENT AND THE FUTURE OF FIBER-OPTIC NETWORKS

This section includes a review of network management issues and envisions the future of fiber-optic networking.

### The Functions of Network Management

Network management is a special topic in the area of communications networks and you can find many excellent reviews on this subject. (See, for instance, [34].) Here we'll present the basic ideas of fiber-optic network management, starting with a brief consideration of the major functions [1]. While most network-management functions are the same for all communications networks, we'll concentrate on the peculiarities associated with managing fiber-optic networks.

***Configuration management*** Configuration management is responsible for handling equipment and connections. Tracking installed equipment, adding new equipment and removing devices no longer needed, rerouting traffic caused by changes in equipment, establishing and terminating connections (lightpaths in a WDM network), and keeping track of connection setups—these are the major functions of configuration management. This management philosophy is usually broken down into two categories: *equipment management* and *connection management.*

***Performance management*** The general functions of this management undertaking are the monitoring and control of the performance characteristics of a fiber-optic network. Performance characteristics describe *attenuation, bandwidth,* and *bit-error rate (BER).* The most significant characteristic of attenuation is the power level of a signal: When this level is below threshold, an alarm has to be triggered. To monitor and control network bandwidth, we must maintain the signal dispersion on the one hand and the bandwidth of the equipment (such as transmitters, receivers, and optical amplifiers) on the other hand. It is also important to monitor the threshold level, as pointed out in Sections 10.1 and 11.4.

But almost all performance characteristics—such as the power and dispersion levels, the signal-to-noise ratio, and the influence of environmental parameters—can be integrated into the bit-error rate, which is the aggregate characteristic of a network's performance that reflects the signal deterioration caused by attenuation, dispersion, or any other problem. (See Sections 10.4 and 11.4.) Monitoring and controlling bit-error rate are the priority functions of performance management.

***Fault management*** Network protection and restoration are the two major functions of fault management. Obviously, monitoring and controlling a number of network performance parameters necessitate close cooperation of this management responsibility with all other management systems. Fiber-optic networks carry a tremendous amount of traffic and every failure is truly catastrophic. Since this topic is so important, fault management is dealt with more fully in a separate subsection beginning on page 694.

***Security management*** You, as a network user, want your data well protected from unauthorized access or corruption. This is the major function of security management. For some networks, such as those serving banks and the military, security is the overriding concern. The most popular means to provide the first level of security is encrypting data before transmission. This involves taking additional measures at both the physical and logical levels of the network.

***Accounting management*** To obtain your revenue through the fiber-optic network, you need to have a good billing system. This is the major function of accounting management. As mentioned above, all the functions of a fiber-optic network are similar to those of a communications network; this is especially true for accounting.

The five management functions mentioned above are sometimes abbreviated as FCAPS: Fault, Configuration, Accounting, Performance, and Security.

These management functions are not an industry standard and you may very well encounter other descriptions of them. For example, some emphasize that the major functions of network management are monitoring, control, troubleshooting, and statistical reporting [5]. Our list, however, gives you a good overview of what these functions are designed to accomplish.

***Safety management*** Safety management in electronic communications networks entails simple compliance with the National Electric Code (NEC). (See Section 8.4.) This is not the case with fiber-optic networks. In addition to electrical safety requirements, there are problems associated with light radiation in fiber-optic systems. One problem is that invisible (infrared) light is used and a second problem is that high light power is transmitted through the use of optical amplifiers. These reasons heighten safety concerns. Safety management's major function is to monitor and control the power level of the radiation exiting the equipment, radiation that could potentially injure personnel working with these networks.

To learn more about the major functions of network management, consult references [1], [3], [5], [6], [9], [35], [36], and [37].

## How Network Management Is Implemented

***Elements of network management*** The network-management functions discussed above can be performed through dedicated components melded into the unit and connected into a network both physically and logically. In other words, we need a network dedicated to the execution of specific functions, that is, a management network. Here's how such a network works [1]:

All the network components to be managed—both passive and active—are called *network elements.* Thus, optical fibers, transmitters, receivers, optical amplifiers, add/drop multiplexers, optical cross-connects—all are examples of network elements. To provide information about its status, each network element has a built-in *management agent.* One example of such an agent is shown in the block diagram of an EDFA (Figure 12.40). A microcontroller gathers information about the power of the input and output signals and the EDFA's status through tap couplers and other built-in components. The microcontroller presents the processed information to network management through alarm, control, and monitoring circuits. Thus, the hardware components along with the software embedded in the microcontroller constitute, in this case, what we call a management agent.

The management agent reports the status of each network element to a *network-element manager.* There are two basic ways to physically connect an agent and its manager: through a separate link or by using an existing optical fiber. In the latter case, an *optical supervisory channel (OSC)* is established within a communications network. A specific wavelength (usually 1510 nm or 1610 nm) is used for OSC. Since both of these wavelengths are outside the common band of working wavelengths, this type of OSC is called "out-of-band."

A network-element manager includes a *management-information base (MIB),* which is a database of a network element. MIB contains a set of variables to be monitored. In our EDFA example, the MIB contains an input–output characteristic and a gain-versus-wavelength graph of an EDFA. (See Section 12.4.) This allows a network EDFA manager to compare actual EDFA performance against its predetermined characteristics and issue a report of any discrepancy it detects.

Network-element managers supervising the same type of network elements are connected to make up an *element-network system,* which is the next level of the network-management hierarchy. This setup is supervised by a network-management system. The general architecture of a network-management system is shown in Figure 14.23.

***Management network*** Figure 14.23 presents the basic idea of the hierarchy of a network-management system. For the sake of clarity, not all the components of the system are shown. But, as you can see, there are essentially four levels of network-management hierarchy: network elements, network-element managers, a network-management system, and a network-management center. Each specific management network can be operated by centralized or decentralized methods.

When all the information is gathered in one network-management center, we have a complete picture of a network's status. However, it takes some time to collect all the information at the center and then send the command down to a network element. In other words, a centralized method of management is relatively slow, which is a serious drawback from the standpoint of network restoration. (Keep in mind that we are talking about a time interval in tens of milliseconds.) The decentralized method—where decisions are made at the lowest possible level of the hierarchy—is much faster than the centralized mode of operation, but it does not provide a general picture of the status of a communications network. Descriptions of fiber-optic *network-management systems (NMS)* can be found in the technical literature. (See [38] and [39].)

It is important to note that at each level of the network-management hierarchy—starting with the network-element managers—information databases store all the information necessary to control the performance of a communications network. Thus, the database of a network-management system contains the set of network-element parameters to be monitored and controlled.

As with any other network, a management network includes physical and logical levels. We have already discussed some of the elements of the physical connections and network "intelligence." Now it's time turn our attention to network protocols. In telecommunications, a variety of management systems and protocols exist. One of the most popular is the *common management information protocol (CMIP),* which runs over the OSI protocol stack. Another protocol, which is used for management in the Internet, is called a *simple network management protocol (SNMP);* it runs over a TCP/IP protocol suite. Management agents communicate with their

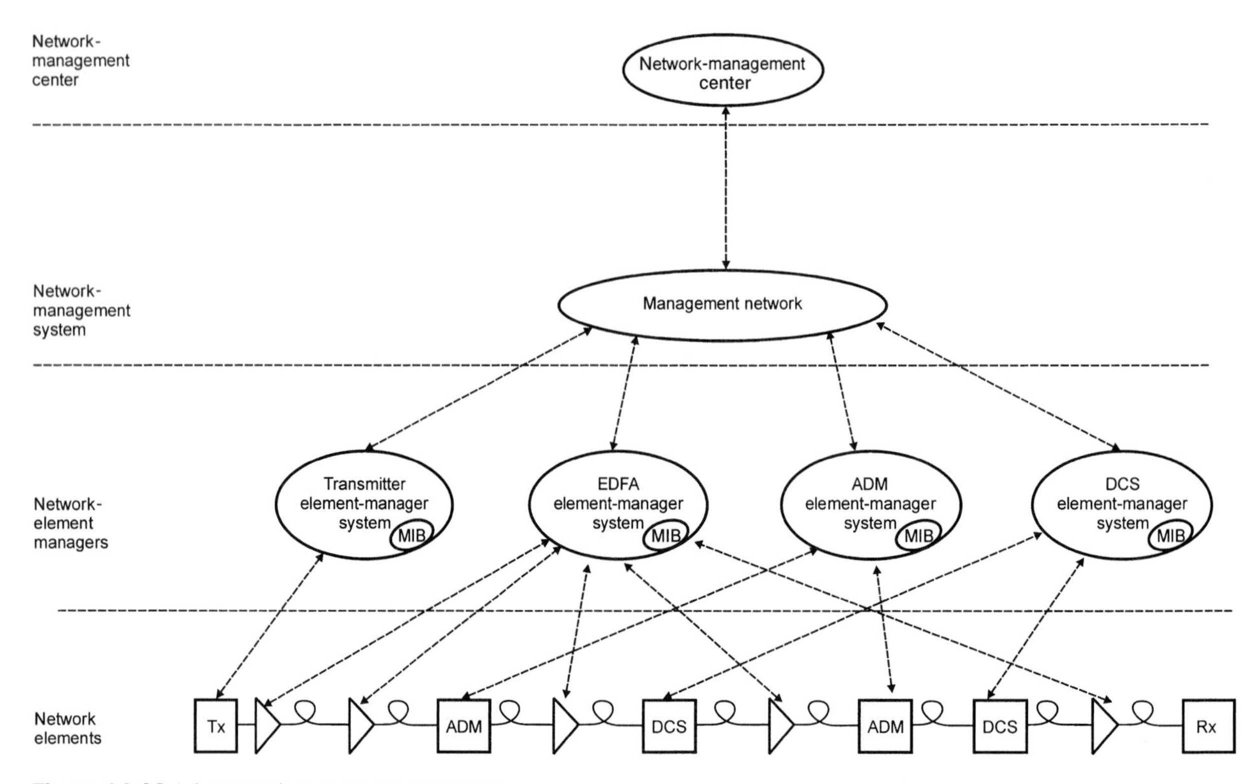


**Figure 14.23** A network-management system.

managers using SNMP. To learn more about network-management systems using SNMP, see reference [36]). Protocols for optical networks are developing rapidly and you should review the technical and trade periodicals to stay up to date on this topic. (See, for example, [40] and [41].)

Optical networks must have their specific management systems. Since all existing networks (SONET, ATM, IP, and others) run over an Optical layer, elements of the management system of an Optical layer *must* be accepted by all other networks [42].

Network management is a part of the more general *operating, administrative, management, and provisioning (OAM&P)* functions needed for every network.

With this brief review of network management fresh in mind, let's turn to one of the most important aspects of this management function: network survivability, that is, network protection and restoration.

## Fiber-Optic-Network Survivability (Protection and Restoration)

***Avoiding a catastrophe*** As already noted, the tremendous capacity of fiber-optic communications networks creates an unusual problem: Any failure becomes catastrophic. We can show this with a simple calculation. Each wavelength carrying an OC-192 channel carries 10 Gbit/s. We can multiplex 16 channels into a single fiber, a process that yields 160 Gbit/s per fiber. A modern trunk cable can include up to 100 individual fibers, which yields 16 Tbit/s. If this trunk were to fail, 16 trillion bits would be lost every second and many million telephone conversations would be interrupted. Fiber-optic communications networks span the globe, connecting Wall Street with other world financial centers, delivering national security and military data, linking national and regional centers that run transportation systems and utilities (airports, railroads, power plants, etc.), and transmitting all kinds of mundane business and personal information. You can see, then, how failure of such a network would cause chaos, not just in terms of customer frustration but also in possible catastrophic economic consequences and even in the potential threat to human life worldwide. It's no wonder, then, that network protection and high-speed restoration are top-priority considerations in network management.

Can we provide a fail-safe fiber-optic network? Unfortunately, no. With hundreds of thousands of network elements and links (the number, of course, depends on the scale of the network) and tremendous pressure from the environment (consider both human activity and natural disasters), failures are inevitable. The real question is not whether we can guarantee a failure-proof network but whether we can design a network with a high degree of survivability, that is, the ability to surmount any failure by quickly recovering its capacity to transmit.

A network's survivability depends largely on the reliability of its equipment, including its redundancy equipment, and well-designed network architecture. We'll concentrate here on the reliability aspect because network components are discussed throughout this book. (In fact, the reliability of a network begins with its power sources. For more on this, see [43].)

SONET has a well-developed protection and restoration technique ([3], [15], [19]). It is extensively applied for the protection and restoration of an Optical layer ([1], [3], [44]).

Bear in mind that either a transmission link or a node can fail; hence, different protection techniques should be involved in such cases.

***Point-to-point link protection*** Let's start with a point-to-point link. Two basic approaches are used to protect against a fiber cut: 1 + 1 protection and 1:1 protection. In the 1 + 1 scheme, the transmission goes simultaneously through two separate fibers usually placed on different routes. One fiber is called the *working (W)* fiber; the other, the *protection (P)* fiber. If the working fiber is cut, the receiver switches to the protection fiber. This switching is initiated and executed by the receiver without its knowing the actual status of the transmitter. Thus, no communication between receiver and transmitter takes place. The idea of 1 + 1 protection is shown in Figure 14.24(a).

In the 1:1 scheme, two fibers are also installed between the same two nodes, but one of them (the working fiber) actually transmits while the second (the protection fiber) does not. Thus, the protection fiber is simply the spare. In case the working fiber fails, traffic is automatically switched to the protection fiber. (See Figure 14.24[b].) This switching is initiated by the receiver but has to be done by the transmitter, so communication between receiver and transmitter is required. This signaling is effected through an *automatic protection switching (APS)* signaling channel and an APS protocol. Physically, an APS signaling channel can be established by using a wavelength outside the EDFA bandwidth—for example, 1510 nm or 1300 nm.

A variation in the 1:1 scheme is the so-called 1:*N* technique, where many (*N*) working fibers are backed up by one protection fiber. This is shown in Figure 14.24(c). The common ratio is 1:12, that is, 1 protection fiber for every 12 working fibers, but other ratios are found in the field.

The advantage of the 1 + 1 scheme is its simplicity and highest possible protection speed. Indeed, since a receiver can switch the traffic by itself, no communication with a transmitter is required, and so no protocol is involved. The drawback? Even during normal operation, two fibers are used for transmitting one traffic flow.

In contrast, the 1:1 scheme uses only one fiber during normal operation, meaning the protection fiber can be used for transmitting low-priority traffic. A second advantage of the 1:1 technique is realized in the 1:*N* scheme: Only one fiber is needed to protect many and this fiber can also transmit low-priority traffic. Thus, the protection safeguard is used very efficiently here, but we do pay a price for this more sophisticated network in slower protection response.

If, after repair, traffic remains on the protection fiber, we have what is called *nonreverting protection*. The 1 + 1 scheme allows this mode of operation. If, however, after repair, traffic is directed back through the working fiber, we are in the so-called *reverting-protection* mode. The 1:*N* scheme operates in this mode.

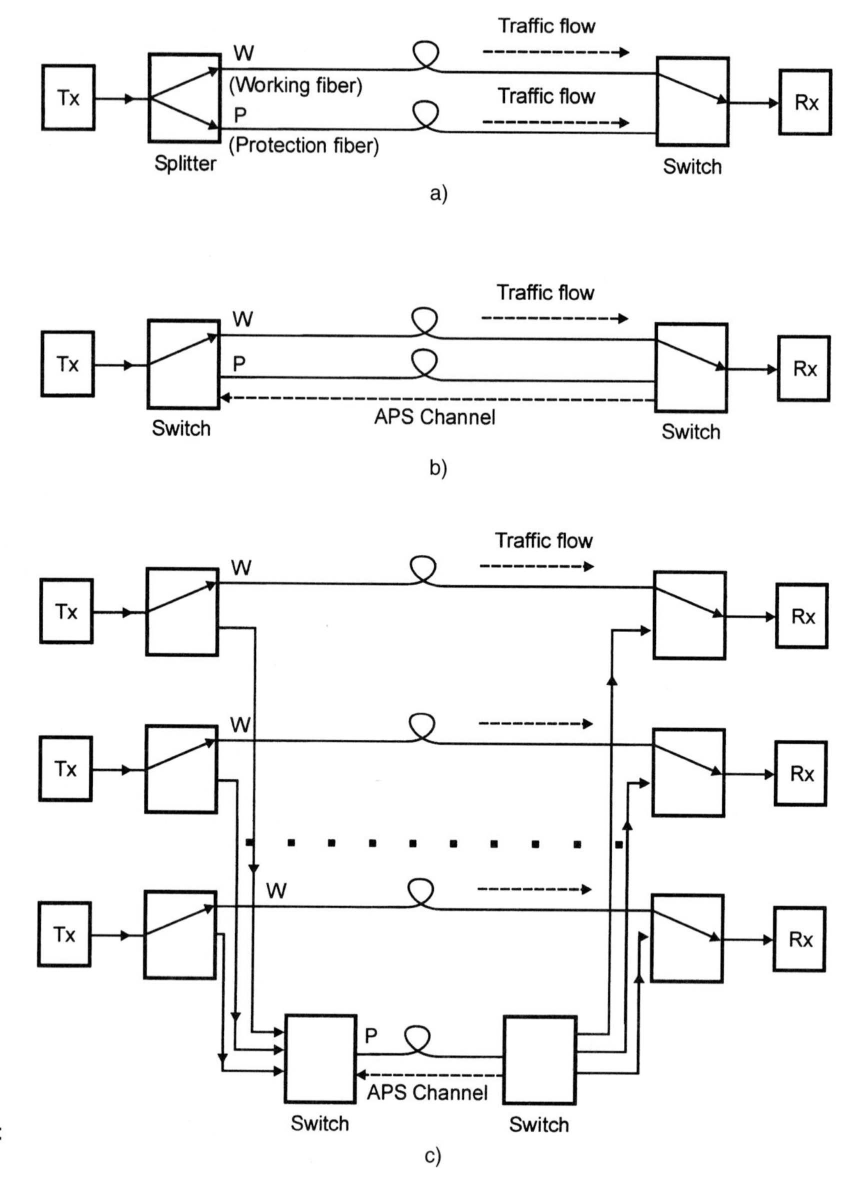


**Figure 14.24** Protection techniques: (a) 1 + 1; (b) 1:1; and (c) 1:N.

Note that protection is done in an *optical domain.*

Both the 1 + 1 and 1:1 techniques protect against only a fiber cut. In case of node failure, the traffic is lost. If a point-to-point link includes an optical add/drop multiplexer (OADM), this protection scheme will save only end-to-end traffic; the add/drop traffic will be lost.

***Path protection and line protection*** In networks, as compared with point-to-point links, there are two basic protection methods: *path protection* and *line protection.* In path protection, traffic is restored between the ultimate source and the destination; in line protection, traffic is restored between two nodes at the ends of a failed link. This concept is delineated in Figure 14.25.

Note that there is a slight distinction between the terms *protection* and *restoration. Protection* usually refers to the simple use of redundant equipment, as in the case of the 1 + 1 scheme. *Restoration* usually refers to the reconfiguration of a network, resulting in the eventual delivery of traffic flow.

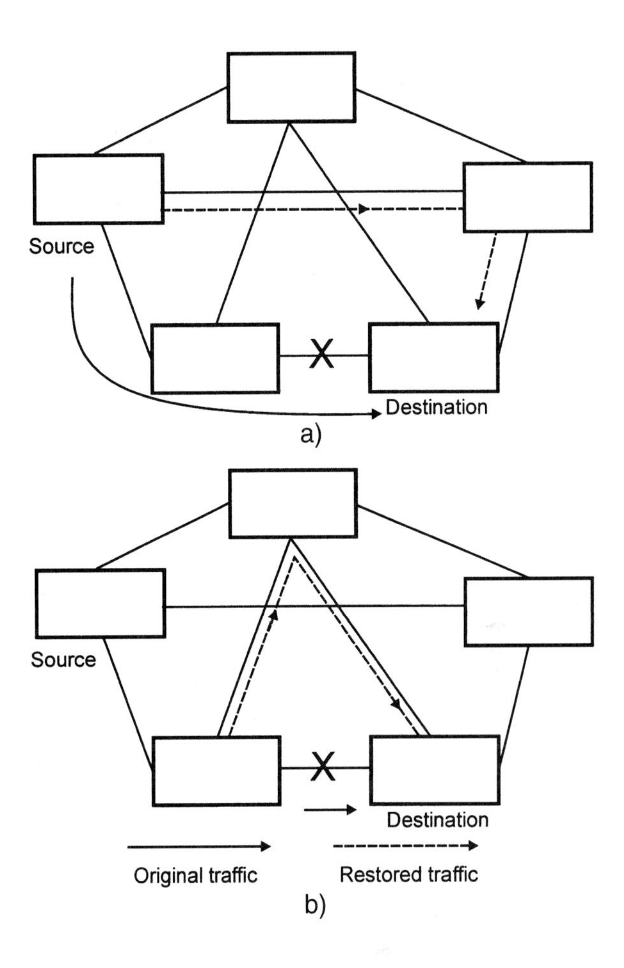


**Figure 14.25** Path protection and line protection: a) Path-protection method; (b) line-protection method.

***Self-healing rings (SHR)*** In terms of its ability to restore traffic, ring topology has proved to be one of the most reliable of all network configurations. A ring with the ability to restore traffic automatically is called a *self-healing ring (SHR).* This architecture is widely deployed in a SONET network; in fact, SONET SHRs can restore their traffic within 60 ms, including detection, initiation, and protection responses [45].

There are a variety of self-healing ring architectures; we'll concentrate on their two basic forms. Figure 14.26(a) shows a *unidirectional path-switched ring (UPSR).* In normal operation, traffic flows through a working (*W*) fiber in a clockwise direction. In case of a link failure, the receiver at the destination node switches to a protection fiber (*P*), where traffic flows counterclockwise.

Figure 14.26(b) shows a *bidirectional line-switched ring (BLSR).* This architecture uses four fibers—two working and two protection. In normal operation, traffic flows along the working fibers in opposite directions: clockwise along one and counterclockwise along the other. Incidentally, this architecture allows the fastest means for delivering traffic from one node to another, much faster than the two-fiber ring configuration. In fact, in normal operation, traffic from node 2 to node 1 in the two-fiber ring has to flow all the way around the ring, as Figure 14.26(a) shows, whereas in the four-fiber ring structure traffic from node 2 flows directly to node 1, as shown in Figure 14.26(b).

In case of link failure between nodes 1 and 2, switches at nodes 1 and 2 connect the working fibers to the protection ones. Thus, traffic between nodes 1 and 2 is restored by rerouting the flow completely around the ring, as Figure 14.26(b) shows.

Suppose the *add/drop multiplexer (ADM)* in node 1 fails. In such a situation, the switches connect the working and protection fibers at each end of node 1, thus rerouting traffic. In this case,

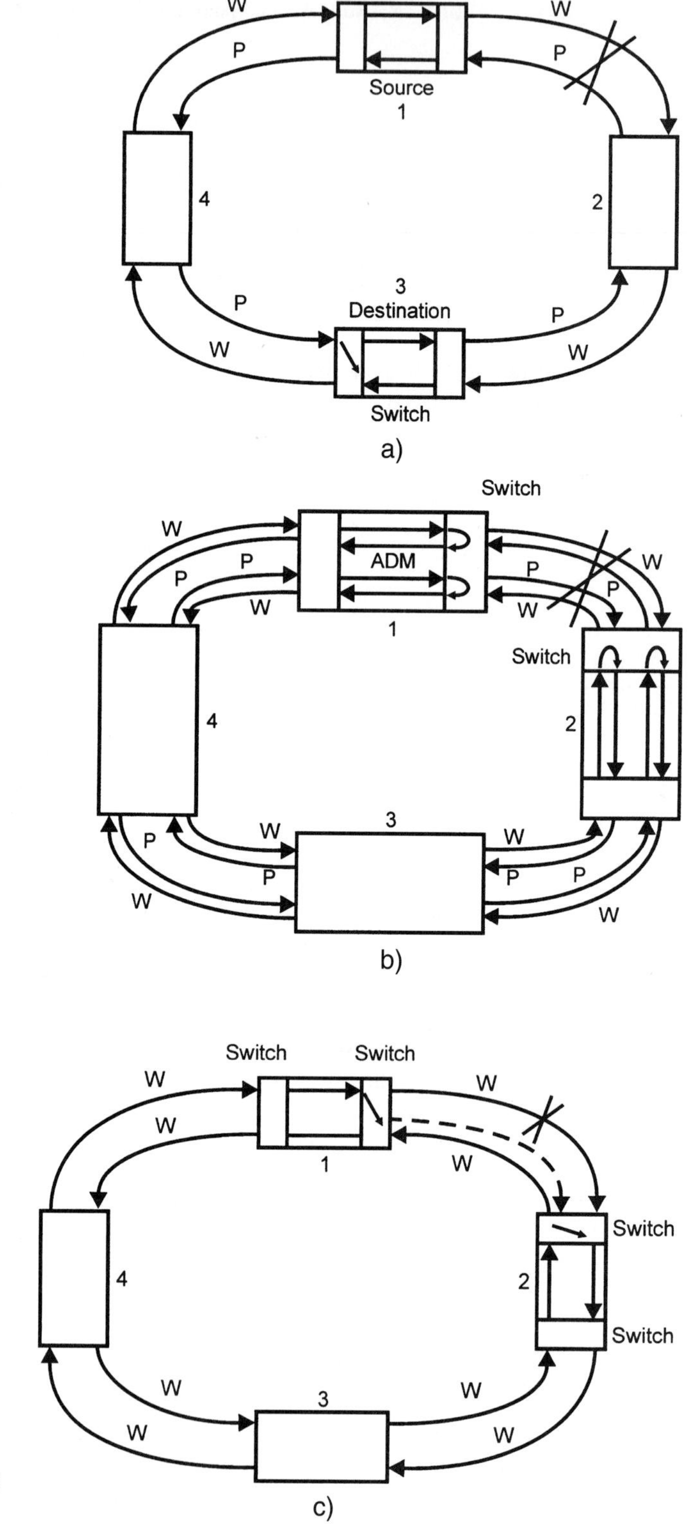


**Figure 14.26** Self-healing rings: (a) Unidirectional path-switched ring; (b) bidirectional line-switched ring; 698 (c) SPRing architecture.

the new traffic route avoids the failed ADM in node 1 and network operation is restored. (This operation and the ones discussed below are not presented in Figure 14.26 [b].)

Now suppose node 1 fails completely. Switches at nodes 2 and 4 respond by rerouting traffic around the ring. Hence, the BLSR ring can survive both link and node failures. However, if the failed node is a source, traffic will be lost.

If only one working fiber between two adjacent nodes—nodes 1 and 2, for example,—is cut, a protection fiber from another pair can be used to restore traffic. This technique is called *span protection.* Note that in span protection no rerouting traffic is needed; switches at nodes 1 and 2 will redirect traffic accordingly.

You can imagine, of course, some other types of architecture for ring restoration. You can, for instance, use two fibers to form a bidirectional line-protected ring, such as the setup pictured in Figure 14.26(c). In this case, both fibers are working ones. The trick is that both fibers carry working traffic at only half their capacities. Such architecture is called a *shared protection ring (SPRing).* In case one fiber fails, the other carries the total traffic. This protection technique requires that we separate the fibers to minimize the risk of their both being cut simultaneously.

Bidirectional rings are more reliable and efficient than unidirectional ones. Certainly, two-fiber rings are cheaper than the four-fiber networks. This is why two-fiber BLSR rings are widely deployed in transport networks; however Sprint—one of the major long-distance carriers—has deployed mostly four-fiber BLSR architecture in its networks [45].

As you can see, self-healing ring architecture does not protect a network from transmitter or receiver failure. These failures can be restored only by switching traffic at the protection (redundant) equipment. In case a significant hub—such as a *central office (CO)*—goes down, blocking the flow of a tremendous amount of traffic, simple redundancy of components will not resolve the situation. Special architecture, called *dual homing,* has been developed to guard against such an occurrence. In dual-homing architecture, two central offices are interconnected so that each of them can operate effectively with the traffic of the other. So in case one CO fails, its traffic is automatically rerouted to the other, which performs all necessary switching functions for its own *and* the new traffic. Quite obviously, each of the central offices has to have the capacity to handle duplicative traffic.

***Survivability: Architecture and other key issues*** We've considered protection and restoration in point-to-point links and WDM ring networks that are already deployed or in the process of deployment. Other, more sophisticated topologies, such as mesh, are already being installed. The protection and restoration of a mesh network poses a real challenge to a network designer. To learn more about this specialized technology, see references [1], [3] and [33] at the end of this chapter.

Many criteria, such as the speed of restoration and the effectiveness of the equipment being used, should be taken into consideration when choosing network architecture and a restoration method. These are important issues but, again, specialized ones beyond our scope, so if you wish to pursue these subjects, we refer you to textbooks [1] and [3] and articles published in the various periodicals. (See, for example, [11], [32], [44], [45], [46], and [47].)

The architectures we've just considered were developed for SONET networks. Fiber-optic networks are evolving from SONET networks (their starting points) and so today's optical-network architectures are still very similar to SONET architecture. Indeed, the first OC-192 WDM system, put into operation in the United States in June 1998 by Qwest Communications International, was the four-fiber BLSR ring [47].

However, developing optical networks cannot be done by merely copying SONET networks. In terms of survivability, the following simple example illustrates the point: In SONET APS, signaling is effected by the specific bytes in the overhead; however, no single overhead has yet been defined for all-optical networks [11]. In general, fiber-optic networks use an *optical supervisory channel (OSC)* to deliver management signaling across the network. The OSC can also be used for protection-switching signaling.

Another peculiarity in the protection of optical networks arises from the use of many wavelength channels in WDM systems. Even though a ring itself may be protected by some of the above architectures, inserted and extracted wavelengths require separate protection by either the 1 + 1 or 1:1 method [46].

You will recall that a fiber-optic network serves as an optical layer for the entire communications network. (See Figure 14.21.) This fact requires that consideration be given to the survivability of each sublayer of an optical layer. As a result, many new schemes for protection and restoration of optical layers have been proposed [48]. This matter has been the subject of some heated discussions in the standardization bodies of the telecommunications industry [32].

As already mentioned, a network's survivability depends not only on its architecture but also on many other factors. For example, replacing many electronic regenerators needed in traditional networks with just a few optical amplifiers in WDM networks increases the reliability of the optical networks in addition to providing cost savings, still another benefit in the long list of advantages of fiber-optic networks.

An important aspect of WDM network survivability is the reliability and performance of its elements. The fact is, you can lose a wavelength channel either because its transmitter is down or because the center wavelength radiated by the transmitter drifts outside the passband for the channel. Thus, for *monitoring* WDM networks we need to develop new approaches and new devices. Including a portable *optical spectrum analyzer (OSA)* in network-management equipment, for instance, helps to resolve a number of these problems ([48] and [49]).

*Troubleshooting* a faulty network element is another key facet of fiber-optic-network survivability. The general methods of network troubleshooting ([50] and [51]) can be applied to fiber-optic networks. Some specific techniques for troubleshooting fiber-optic networks are discussed in Section 8.5. Troubleshooting is important but not critical to network restoration; traffic must be recovered in a matter of milliseconds, of course, but locating and replacing a faulty element is a task that can be performed during regular working hours.

Although we have not covered all aspects of network reliability here, this overview should give you a good idea of how network operation has to be managed to prevent catastrophic breakdown.

## Conclusion: The Future of Fiber-Optic Networks

### *The current scene*

**1.** *Explosive demand for bandwidth.* Today we produce information at a rate we never would have dreamed about, say, just 25 years ago. As a result, the need to transmit more and more bits at higher and higher speeds grows every day. (See Section 1.1.) Specifically, the exponential growth rate of the Internet is the major driving force. Indeed, the Internet's growth rate exceeds 100% a year [52]. But the Internet is certainly not the only cause of this information explosion. In fact, the more bandwidth that becomes available, the greater the number of new services that are offered and, of course, the more bandwidth customers will then demand. (To cite just one example, fiber-optic networks make it possible to organize remote medical consultation with the leading experts in a specific specialty in a given area of the country and, eventually, even throughout the world. Such a consultation is based on high-resolution video of the body part being evaluated and, if needed, the specific area of surgery. To make this possibility become reality, a medical conference of this kind has to be held in real time. This means that to achieve this end, an enormous number of bits per second must be exchanged between the examining physician and the consultant.)

Another driving force is the new approaches—ISDN, ADSL, CATV, and others—that provide high-speed access to the customer's premises, thus applying even more pressure on all chains of the global communications network [45].

The point is simply this: The demand for bandwidth is soaring and the only means by which this demand can be met is the deployment of fiber-optic networks—more specifically, WDM networks.

**2.** *Requirements for developing WDM networks.* Several major criteria will play significant roles in the development of fiber-optic networks [53]. First and foremost are standards, of which the industry is in dire need. The telecommunications community is working diligently to develop new standards to facilitate deployment of WDM networks so as to ultimately replace SONET and possibly other layers with one Optical layer. (See the discussion of multilayer structures beginning on page 608. For a review of the current state of ongoing work on standards, see [54].) Unfortunately, the actual process of standardization is a long, drawn-out affair, so the telecommunications community is searching for various approaches, such as less formal de facto standardization provided by various industry associations, to accelerate this process [55].

The second criterion critical to the development of fiber-optic WDM networks is technology, the focus of this book. And, as we've already pointed out, the key to realization of these advanced networks is reliable, high-performance, cost-effective components. More specifically, *optical add/drop multiplexers (OADMs)* and *optical (wavelength) cross-connects (OXCs)*—that is, switches—are the crucial modules that determine the entire architecture and performance of WDM networks ([56]).

The third factor is the architecture and protocols of WDM networks, issues closely linked to developments in technology and that require special emphasis and considerably greater effort. (See [40] and [41].)

Finally, we have to take into account the relative pace of deployment of WDM networks in long-haul and local service areas. In the United States in the late 1980s the infrastructure of local networks vastly exceeded the capacity of long-haul pipes. Since then, the pace of deployment of WDM networks in long-haul service has far surpassed the deployment of these networks in local area service ([21], [45], [57]), *and* so today the situation is reversed: a plethora of long-haul transmission systems but a bottleneck in local (metropolitan) area interconnections [58]. The reason is understandable: Deploying WDM networks in the service and subscriber loop of a local exchange carrier is not an easy task [59].

The key to building the infrastructure of a truly global optical network lies in the deployment of WDM systems in access networks and eventually placing the fiber as close as possible to the desk and to the home. (See [60] and [61] for a look into the current situation. Other good sources of information on the current status of WDM networks are the short courses and workshops presented annually at the Optical Fiber Communication Conference [OFC], sponsored by the Optical Society of America [OSA].)

**3.** *Needs of WDM networks.* Nobody would have dared to predict the insatiable appetite for bandwidth when the first fiber-optic links were installed in the early 1980s [21]. The upshot was that in 1995 the carriers started to encounter a dearth of available fibers. Today, this shortage is the major force propelling the widespread deployment of WDM point-to-point links.

**4.** *WDM networks today.* Wavelength-division multiplexing is fast becoming the major technology in point-to-point links connecting SONET terminals in long-haul and metropolitan networks. It doesn't take a clairvoyant to predict that WDM will be the major technology for this application within the next couple of years.

The key modules for deployment of WDM networks—OADMs and OXCs—are now commercially available, but the industry continues to grapple with the high cost and mediocre performance of these modules and the other components discussed in the previous chapters. At the same time, new developments are on the horizon, presaging even more possibilities for WDM networks once these problems are dealt with and resolved [56].

The architecture and management of WDM networks are other concerns that need to be addressed and, indeed, are at leading research facilities.

***Future trends*** The chief requirement for telecommunications networks in the future can be summed up in three words: capacity, capacity, capacity. End-users continue to request more and more bandwidth as applications grow. To meet this demand and remain competitive, the carriers are seeking cost-effective, technologically advanced solutions, and are basing their future on WDM fiber-optic networks. They see WDM as the only solution enabling them to offer not only cheaper bandwidth—measured in transmission cost per bit—but also new money-making services. The catch? This solution creates its own headache—even greater demand for bandwidth. As one observer aptly put it, "The result is a self-reinforcing upward spiral in [demand for] bandwidth with no end in sight" [62].

So if the future of fiber-optic networks lies in WDM networks, the question then is, how soon will they reach the marketplace? WDM technology is only now gradually moving into the metropolitan area. Remember, WDM networks are still in their weaning period. To put a time frame on their maturing to the point of dominating the market through optical networking would take the insight of a Tiresias, the fabled blind seer of ancient Greek mythology. But hindsight provides us with some guidance: WDM technology rose in less than three years from an almost nonexistent entity to a billion-dollar business [46]. Considering this kind of growth rate and an exponentially increasing demand for bandwidth, it doesn't take a Tiresias to foresee WDM networking in the commercial marketplace on a widescale basis very soon. How soon? We'll venture an educated guess and say in the first decade of this new century it will become the workhorse of telecommunications.

The architecture and management of WDM networks will change, and one of these changes will no doubt be in the traffic pattern, which will become more data-centric (data-related), that is, moving in bursts in contrast to the constant bit-rate connections traditionally associated with voice, the flow pattern in use up to now. In fact, in the United States data traffic surpassed voice traffic several years ago and this will soon be the case worldwide [63]. The numbers tell the story: Voice traffic is growing at 8% annually, whereas data traffic is increasing at a 35% annual rate and Internet traffic is rising by more than 100% a year. One estimate has data traffic making up 70% of total network traffic early in this century [46]. The bursts of traffic associated with the Internet certainly increase the complexity of capacity planning, the result being a more sophisticated network in terms of its architecture and management.

The multilayer structure of modern networks, where one layer runs over another (see Figure 14.21), has, as already noted, one essential flaw: The many interfaces between layers slow the network's operation, compromising its reliability. The solution—the development of an *all-optical network (AON),* where one optical layer performs all the network's functions—is the current hotbed of optical-networking activity. Examples of this process can be found in the following references at the end of this chapter: [11], [45], [62], [58], [63], [21], [59], [64], [65], [56], and [66].

To sum up, we can say that the major trend in the development of telecommunication networks is the move to all-optical networks. (See [55] and [65].)

We began this chapter with the statement that the term *fiber-optic networks* is a nebulous one. In fact, the telecommunications community has been laboring for some time to come up with a formal definition, certainly a precursor to the development of appropriate standards [45]. Until this comes about, though, we'll have to accept what's generally understood by those in the business—that an optical network is a communications network transmitting a signal from one end to the other without electro-optical and opto-electrical conversion. Yes, this statement is kind of blurry, but at least it gives you a sense of what optical networks do.

***Networking: Coda*** Our discussion of fiber-optic networks in general and WDM networks in particular has, we trust, given you some idea of just how these systems are made up and how they work. This is, of course, an immense subject that should be studied in greater detail, especially the design of a fiber-optic WDM network. But these concerns lie outside our domain. We recommend the books and magazine articles listed starting on page 709. For those interested in design considerations, references [1] and [64] will be especially helpful.

## SUMMARY

- A network is a set of nodes connected by links. A communications network has to have two major features: (1) the ability to route a message through many links and nodes from a source to a destination and (2) the capacity to add/drop a signal at any intermediate node.
- The first generation of fiber-optic communications networks—telecommunications networks with electronic nodes and optical fibers as transmission links—has been employed for a number of years. The second generation of fiber-optic networks, now in its infancy, includes optical add/drop modules and switches. As of this writing, we are in the process of deploying these networks. The ultimate goal in the development of fiber-optic communications is all-optical networks (AON), where a signal will be transmitted from an ultimate source to its destination only in optical form—that is, without conversion into electrical form. All-optical networks are now in the early stage of deployment while, concurrently, research and development work continues at a frenetic pace.
- The next step in developing fiber-optic communications networks is the deployment of wavelength-division multiplexing (WDM) technology, which has become the leading technique in point-to-point communications links.
- "Telecommunications" is the exchange of information over communications networks. The information to be transmitted is presented in three basic forms: voice, video, and data. Originally, voice transmission was the realm of telephony, video signals were transmitted by television networks, and data were always associated with computers. Today, all three sectors of the telecommunications industry meld. This process was greatly accelerated by the Telecommunications Act of 1996, which essentially removed the legal barriers mandating the different services the communications companies could provide. However, the division of the industry into telephone, cable TV, and computer networks is still valid because these networks belong to different carriers and are, historically, based on different architectures and technologies.
- The telephone network is the most pervasive network in the world, enabling one to reach any point on the globe. This system consists of local and regional networks on the one hand and long-distance networks on the other hand. All telephone sets are eventually connected to a central office—the local switching center. These are the local connections, while connections among local access and transport areas (LATAs) make up a regional network. Long-distance networks are built separately and they also consist of switching centers, which are linked by fiber-optic cables. All the major long-distance carriers (AT&T, MCIWorldcom, Sprint and others) have their own physical networks. Interconnections among long-distance and local networks occur at tandem central offices. Classifications (arbitrary, not standard, remember) use the terms *access* and *transport* networks, "access" referring to local networks, "transport" including both regional and long-distance webs. (Bear in mind, however, that the classifications in use in the industry today, although widely accepted, do not yet constitute a set of standards.)
- The only access you have to the global network is through a twisted pair of copper wires running from your telephone set to the nearest switching facility. This connection is called a subscriber line (loop). A subscriber line is a bottleneck in today's telecommunications networks because of the physical transmission limits inherent in a twisted pair. While transport networks start to convey terabits per second, a subscriber line in its original form is capable of transmitting only kilobits per second. Several techniques are available to "widen" this bottleneck in the interim, but we believe the ultimate solution will be installation of an optical fiber running right up to your personal desk. Unfortunately, this does not seem to be feasible for the foreseeable future on a global scale but the process has already begun at the local level.
- Transport networks gather all the signals emanating from hundreds of millions of customer premises and then have to transmit this tremendous amount of information at high speed. Transport networks, therefore, place the highest possible demand on bandwidth. This is why they have been leading sites for the deployment of the latest ad-

vances in fiber-optic communications technology, advances such as wavelength-division multiplexing.

- Transmission over subscriber lines is done in analog form; all other transmissions over telephone networks today are digital.

- To transmit many individual signals over the same line, carriers use the multiplexing technique. Traditionally, telephone companies have used time-division multiplexing (TDM), where each individual channel is assigned a time slot within a transmission frame. The basic building block of a telephone TDM is a T1 transmission system. It multiplexes 24 voice channels and transmits at 1.544 Mbit/s. There are, of course, higher standard formats for bit-rate transmissions, such as T2, T3, and so on. T systems are used for transmission in local and regional networks.

- Much higher bit-rate transmission using TDM is specified by the SONET (Synchronous Optical Network) standard. The basic building block of SONET is OC-1, transmitting at a bit rate of 51.84 Mbit/s. The higher standard formats are multiples of OC-1. The highest data format deployed today is OC-48, transmitting at a bit rate of 2.5 Gbit/s. OC-192 (transmitting at a bit rate of 10 Gbit/s) is in the difficult start-up stage of commercial deployment.

- A further increase in line rate by using TDM doesn't seem practical. Only wavelength-division multiplexing, where each channel is carried by an individual wavelength over the same fiber, is seen today as the most promising approach. If each channel operates at 2.5 Gbit/s and the WDM system multiplexes 16 wavelengths, the total line rate transmitted over a single fiber is 40 Gbit/s. If a fiber cable contains 100 optical fibers, this cable transmits 4 Tbit/s. This is how modern transport networks operate.

- Computer networks, also called data networks, are designed and built to transmit digital signals only. We distinguish among data networks by their scale of operation. Local area networks (LANs), for instance, connect computer setups within a room, a building, or a campus. There are many types of LANs supported by commercially available hardware and software. These types differ in topologies, transmission formats, and protocols. Typical LAN data rates range up to hundreds of Mbit/s; however, new LANs, such as Gigabit Ethernet and Fibre Channel, operate at the gigabit-per-second bit rate. A metropolitan area network (MAN) serves a specific geographic region and connects many LANs. The bit rate used in these networks starts at 200 Mbit/s. FDDI and SONET are examples of MANs. A world (wide) area network links computers around the globe by connecting LANs and MANs. They operate at bit rates of 2.5 Gbit/s and higher.

- Several different transmission systems (protocols) are used in data networks. For example, the asynchronous transfer mode (ATM) transmits data in packets called cells. Each cell includes 53 bytes, with 5 bytes being an overhead and 48 bytes being the user's data (payload). ATM is a connection-oriented system, which means that a transmission path (connection) is established before the actual transmission starts. The fixed length of a cell and pretransmission connections make ATM a very fast transmission system.

- Another popular transmission system for data networks is the Internet Protocol (IP), which is used mostly with the transmission control protocol (TCP); in fact, the entire suite of protocols is referred to as TCP/IP. In this transmission system, the data to be sent is subdivided into packets with variable lengths. These packets are sent through different paths of a network so, as a result, they may not arrive in the order in which they were sent. IP is referred to as a "connectionless" system, which means that connections are not established when the actual transmission begins. TCP/IP is the most popular data-transmission protocol today even though it lacks a quality-of-service guarantee.

- Cable TV, or community antenna television (CATV), is another type of fiber-optic network. Specifically, the modern CATV network is a hybrid fiber/coax (HFC) network in which the fiber optics take up most of the network and the coaxial cable serves as the last drop from your home to an optical-receiver node. This is a very prolific network, serving millions of households, so the demand of cable TV is an essential factor in the fiber-optic communications industry. CATV fiber-optic networks operate at 1310 nm and transmit an analog signal. These features make this network unique from the standpoints of both its components and its architecture. Advances in fiber-optic communications technology such as optical amplifiers and WDM not only enhance the ability of CATV networks to transmit a video signal but also enable the use of these networks for both voice and data transmission.

- A network in general—and a fiber-optic network in particular—combines both physical and logical levels in order to operate. The physical level includes an optical fiber and components. The logical level includes protocols and services. A protocol is a set of rules needed for communications to flow between a source and a destination. Services are the sets of steps (operations) needed to provide communications between a source and a destination.

- The Open Systems Interconnection (OSI) reference model presents all the logical functions that are necessary for a network node to provide communications from a source to a destination. These functions are grouped and presented as seven hierarchical layers. Each layer performs specific functions and presents the result of its action to the next higher layer. These layers are (from bottom to top) designated Physical, Data Link, Network, Transport, Session, Presentation, and Application. The four lower layers carry communications functions, while the upper three levels have user-oriented functions.

- The Physical, Data Link, Network, and Transport layers provide actual end-to-end communication. They do so by handling data flow through the entire network. This they accomplish by performing such regular transmission functions as message routing, error control, and signal recovery.
- The upper three layers—Session, Presentation, and Application—are largely concerned with readying the user's data for transmission through the network. Such presentation is realized through log-in and log-out procedures, data conversion, a network-routing request, and other functions.
- The OSI reference model is the structure showing the logical architecture of a communications network. This is an internationally accepted universal model that works for any network regardless of the specific technology used for building the network. In addition, the OSI reference model introduces the concept of layered-structure communications networks.
- The Synchronous Optical Network (SONET) is a good example of first-generation fiber-optic networks. In terms of its physical makeup, SONET uses optical fiber and electronic equipment such as terminal multiplexers, add/drop multiplexers, digital cross-connects, and regenerators. Two topologies are used in SONET networks: ring and point-to-point. In terms of its layered-structure, the SONET model includes Physical, Section, Line, and Path layers. SONET and its European counterpart, SDH, are powerful transmission systems.
- The Asynchronous Transfer Mode (ATM) is another powerful transmission system, one that doesn't rely on any specific physical media. Transmission can be effected through any existing network (SONET or any LAN, MAN, or WAN) and without the need for conversion of a data format. In terms of hardware, this system needs ATM switches that route 53-byte cells through a network. ATM's layer structure comprises the ATM Physical, the ATM, and the ATM Adaptation layers. The basic features of ATM—the ability to transmit any type of information and high-speed transmission—make it a very popular transmission system today.
- The concept of layers can be generalized and each of the existing networks—SONET, ATM, and IP—can be considered as a single layer of the entire logical structure of communications networks. Each layer in a network runs over another layer. Developments in fiber-optic networks call for a change in this logical structure with the introduction of the Optical layer, which is the bottom layer in the network architecture.
- One relatively new structure consists of a SONET running over an Optical layer with an ATM superimposed over the SONET, and an IP riding over the ATM. Such a structure opens the door to new developments in access and transport networks. On the other hand, it also adds to network complexity as each additional interface reduces the network's reliability and increases its cost. The intermediate solution to this problem is placing each of the network's layers—SONET, ATM, and IP—immediately on top of the Optical layer. The ultimate solution to this problem, of course, would be the development of all-optical networks so that all functions would be performed by a single Optical layer.
- An Optical layer in its present form provides a lightpath to the layer above it. A *lightpath,* a temporary optical-communications channel established throughout a network, is an actual means to provide end-to-end transmission across an optical network. An Optical layer, in turn, is subdivided into several sublayers. Its current model does not reflect all the functions of which second-generation fiber-optic networks are capable and, as a result, needs further development.
- To work properly, fiber-optic networks need to be managed. Network management is a vital part of network operations and must be taken into account when one designs a network or estimates its cost. The major functions of fiber-optic-network management are configuration, performance, fault management, network security, accounting, and safety. Special components integrated into a management network perform these functions.
- A management network has its own hierarchical structure and uses its own protocols. A simple network-management protocol (SNMP) runs over a TCP/IP protocol suite, while a common management information protocol (CMIP) runs over an OSI protocol stack. Management networks help to put into effect the operational, administrative, management, and provisioning (OAM&P) functions needed for every network.
- Network survivability is the ability of a network to survive any failure by quickly recovering and restoring traffic transmission. To achieve this, network protection and restoration should be performed. The importance of the survivability—that is, the protection and restoration—of fiber-optic networks cannot be stressed enough because these networks carry tremendous amounts of information without which modern society could not function. With millions of miles of optical fibers and hundreds of thousands of components installed throughout the world, fiber-optic networks cannot be completely fail-safe. How-

ever, appropriate design and management can prevent a failure from being catastrophic.

- SONET has a well-designed and developed protection and restoration system which enables a SONET network to restore any type of failure within 60 ms. Many of SONET's protection and restoration features are used in other fiber-optic networks.
- Two basic techniques used for network protection in case of a fiber cut are 1+1 and 1:1 schemes. In 1+1 protection, the transmission signal travels simultaneously on two separate fibers—working and protection. In case of failure of the working fiber, the receiver switches to the protection fiber. In the 1:1 scheme, two fibers are also installed between the same nodes but the working fiber actually transmits while the protection one does not. In case the working fiber fails, traffic is switched to the protection fiber. A variation of the 1:1 scheme is the 1:*N* technique, whereby many (up to 12) working fibers are backed up by one protection fiber.
- In fiber-optic networks, as compared with point-to-point links, there are two basic protection methods: path protection and line protection. In path protection, traffic is restored between the ultimate source and the destination; in line protection, traffic is restored between two nodes at the ends of a failed link. These methods are used in the most popular network-protection architecture, the self-healing ring (SHR). Three basic schemes have been considered in this chapter: the unidirectional path-switched ring (UPSR), the bidirectional line-switched ring (BLSR), and the shared-protection ring (SPRing).
- Bidirectional rings are more reliable and efficient than unidirectional ones, but two-fiber rings are cheaper than the four-fiber networks. This is why two-fiber BLSR rings are widely deployed in transport networks; however, Sprint—a major long-distance carrier—has deployed mostly four-fiber BLSR architecture in its networks.
- Self-healing ring architecture does not protect a network from transmitter or receiver failure. Such failure can be recovered only by switching traffic to the protection (redundant) equipment. To compensate for a failure at a significant hub, such as a *central office (CO),* special architecture, called *dual homing,* has been developed.
- In WDM systems, a ring itself my be protected by some of the above types of architecture; however, inserted and extracted wavelengths require separate protection by either the 1 + 1 or the 1:1 method. Monitoring WDM networks requires inclusion of a portable optical spectrum analyzer (OSA) in network-management equipment.
- Nobody could have predicted the current demand for bandwidth when the first fiber-optic networks were installed. The result? The carriers started to encounter a lack of available fiber back in 1995. So what the industry refers to as "fiber exhaust"—a serious shortage of bandwidth of installed fiber—is the most important factor driving the deployment of WDM point-to-point links and the development of tomorrow's WDM networks.
- Several major issues confront the industry in its drive to develop WDM fiber-optic networks: standards, technology, the architecture and protocols of WDM networks, and the relatively slow pace of deployment of WDM networks in long-haul and local area service. The key modules enabling the installation of WDM networks—optical add/drop multiplexers (OADMs) and optical cross-connects (OXCs)—are now commercially available but the industry continues to struggle with the high cost and performance deficiencies of these modules. The other components discussed in this and the two preceding chapters are also crucial to the success of WDM networks and they too need to be further developed to achieve better performance at less cost. The architecture and management of WDM networks are in the process of development.
- The need for new architecture and ways to manage WDM networks arises from the changes in the traffic pattern. It is becoming more data-centered, that is, moving in bursts, in contrast to constant bit-rate connections traditionally associated with voice transmission. This increases the complexity of capacity planning. The result is the need for more sophisticated network architecture and management. The multilayer structure of today's networks, where one layer runs over another, has an essential drawback: The numerous interfaces between layers slow the network's operation and compromises its reliability. The solution: development of an *all-optical network (AON),* whereby one Optical layer performs all the network functions discussed in this chapter. This development is the current focus of optical-networking research.
- We can truly say that the major trend in the development of telecommunications networks is the move to all-optical networks. Future fiber-optic networks will certainly be WDM networks. True, WDM networks are still in their infancy, but with exponentially increasing demand for bandwidth and the unprecedented pace at which WDM technology is maturing, we can expect to see optical networks fully commercial and predominant in the field within the decade.
- The major requirement for tomorrow's telecommunications networks, as we've stressed so often in this book, will be *capacity,* as end-users continue to demand more and more bandwidth. To meet this demand and be competitive, the carriers are seeking cost-effective and technologically advanced solutions based on WDM fiber-optic networks. These networks allow the carriers to offer not only cheaper bandwidth—measured in trans-

mission cost per bit—but also new services, which, in turn, trigger even greater demand for bandwidth. The cycle seems endless.

## PROBLEMS

**14.1** What basic forms of information do you know?

**14.2** Draw the general architecture of a telephone network and explain how it works.

**14.3** Explain the meaning of LEC, IXC, and PSTN.

**14.4** What are *access* and *transport* networks?

**14.5** What parameters limit the length of a twisted pair that can be laid from your home to a central office?

**14.6** How long does it take to download a graphical message containing 236 Mbit from the Internet to your personal computer if a modem's speed is 56.0 kbit/s? 6.0 Mbit/s?

**14.7** Why do transport networks have to transmit at the highest possible bit rate?

**14.8** Why is the sampling frequency in a T1 system 8 kHz?

**14.9** Where does the figure 125 μs come from?

**14.10** Calculate the bit rate of a T1 signal.

**14.11** What is the bit rate of a single voice channel?

**14.12** What is the time slot of an individual bit in a T1 system? In an individual channel?

**14.13** T1 is the result of multiplexing 24 voice channels; however, if you multiply the bit rate of an individual voice channel (64 kbit/s) by 24, you won't receive a 1.544 Mbit/s bit rate of T1. Why not?

**14.14** What is the difference between T1 and DS-1?

**14.15** A T2 signal contains 96 voice channels, that is, 4 T1 signals. However, the bit rate of T2 is not equal to 1.544 Mbit/s $\times$ 4. Why not?

**14.16** What is the E system? What is the basic bit rate of the E system?

**14.17** What is meant by saying an overhead and a payload are in frame format for transmitted data?

**14.18** What procedure lies hidden behind the designation B8ZS?

**14.19** What does the acronym SONET stand for? What is the SONET?

**14.20** What procedures are performed by the component titled "Mapping and overhead insertion" in Figure 14.3(a)?

**14.21** Draw a diagram showing bit and byte interleaving for 4-channel TDM multiplexing.

**14.22** What DS and E signals can be carried by each type of VT?

**14.23** Compute the number of voice channels carried by the OC signals shown in Table 14.2.

**14.24** How many voice channels can a WDM link carry if it multiplexes 16 OC-192 SONET signals?

**14.25** After digitizing, or PCM conversion, an individual voice channel needs a bit rate of 64 kbit/s for transmission. SONET is a TDM system that multiplexes voice channels as basic units. However, if you divide the OC-1 transmission rate of 51.840 Mbit/s by 64 kbit/s, you obtain 810 voice channels, not 672 (as stated on page 662). What causes this discrepancy?

**14.26** What is the next OC signal after OC-192? What is its bit rate?

**14.27** There are two types of non-return-to-zero (NRZ) modulation formats: unipolar and bipolar, or polar. Unipolar NRZ is shown in Figure 14.5. Polar NRZ means that half of a high-pulse amplitude is positive and the other half is negative. Polar NRZ has an obvious advantage: It doesn't carry a dc component along with an information signal. However, in fiber-optic communications, we don't use polar NRZ. Why not?

**14.28** What do the acronyms LAN, MAN, and WAN stand for? What is the meaning of each of these terms?

**14.29** Explain the meaning of each field of the FDDI frame format for data. (See Figure 14.7.)

**14.30** In the FDDI frame format, some fields have fixed lengths; others, variable lengths. Why is that?

**14.31** Are computer networks dedicated webs or public networks?

**14.32** What does ATM stand for? How does an ATM work?

**14.33** List the advantages and drawbacks of an ATM.

**14.34** Explain the meanings of IP and TCP.

**14.35** List the advantages and drawbacks of an IP transmission system.

**14.36** Explain the major functions of each node of the cable TV network shown in Figure 14.9.

**14.37** Name the components of the fiber-optic transmitter (Tx) shown in Figure 14.9.

**14.38** Why do we need a booster (postamplifier) in the head end of a CATV system? (See Figure 14.9.)

**14.39** What is the main function and the principle of operation of a head-end splitter? (See Figure 14.9.)

**14.40** What type of optical amplifier—EDFA or SOA—is used in a CATV system?

**14.41** Draw a diagram of a transport section of a CATV network showing it in a ring topology.

**14.42** Why is a CATV system called HFC?

**14.43** In a CATV, a coaxial cable is used only for the short connection between an optical receiver node and your TV set. Why is this still necessary? In other words, why don't we replace this last electrical drop with an optical fiber?

**14.44** Project 1: Build a memory map. Show the connections among the telephone, computer, and cable TV networks.

**14.45** Network parlance uses the terms *protocols* and *services.* What do these terms mean?

**14.46** Why do we need the OSI reference model?

**14.47** The OSI reference model divides the job of each network node into several subtasks. Why do we mention only a "node" here and not a link?

**14.48** Give examples of the physical media mentioned in the OSI reference model (Figure 14.12.)

**14.49** Briefly describe the functions of each layer in the OSI reference model.

**14.50** What is meant by "virtual communication"? What does this form of communication physically provide?

**14.51** The OSI reference model operates with Protocol Data Units (PDU) and Interface Data Units (ITU). What is the difference between these units and what do these units have in common?

**14.52** The OSI reference model operates with bits, frames, and packets, as shown in Figure 14.12. Referring to Figure 14.14, show what these information units specifically consist of.

**14.53** How do peer layers physically provide virtual communication?

**14.54** Briefly describe the functions, services, and protocols of the OSI reference model.

**14.55** What are the advantages and drawbacks of the OSI reference model?

**14.56** What topologies and components are used in SONET networks?

**14.57** Do SONET networks rely on specific transmission media?

**14.58** At what bit rates, in terms of bits per second, do the access and backbone rings of a SONET network transmit?

**14.59** Why is a ring topology considered reliable?

**14.60** Name the SONET layers and briefly describe their major functions.

**14.61** Does an ATM network rely on specific transmission media? What physical components are needed for ATM networks?

**14.62** Name the basic features that make ATM transmission fast.

**14.63** Name the ATM layers and briefly describe their functions.

**14.64** Explain how IP runs over ATM and ATM runs over SONET networks.

**14.65** Why do we need an Optical layer?

**14.66** What services does an Optical layer provide to the layer above?

**14.67** Describe the Optical-layer service termed "establishing lightpath."

**14.68** Describe the Optical-layer service termed "establishing virtual circuits."

**14.69** Refer to Figure 14.22. Why does an Optical Transmission Section (Amplifier) layer exist at each WDM node and not just at the amplifier node?

**14.70** Show the layered structure of a fiber-optic network, including the Optical layer.

**14.71** Which fiber-optic networks belong to the first generation and which one belongs to the second generation? Give examples.

**14.72** List six functions of fiber-optic-network management.

**14.73** Describe the functions of equipment management and connection management.

**14.74** Name some equipment and connections that configuration management has to monitor and control.

**14.75** Why are the monitoring and control of bit-error rate (BER) the priority functions of performance management?

**14.76** Briefly describe how a management network is built and operated.

**14.77** Briefly describe the hierarchical structure of a management network.

**14.78** How does physical signaling occur in a fiber-optic network?

**14.79** What does the designation MIB stand for? Describe what a MIB contains.

**14.80** What protocols are used in network-management systems?

14.81 What is meant by the "survivability" of a fiber-optic network?

14.82 The protection technique is intended to protect WDM fiber-optic networks against the failure of links and nodes. Give examples of links and nodes.

14.83 What is the difference between the 1+1 and 1:1 protection techniques?

14.84 What is the difference between line protection and path protection?

14.85 Why is the network architecture shown in Figure 14.26 called a self-healing ring?

14.86 Describe how a ULSR network restores traffic.

14.87 Describe how a BPSR network restores traffic.

14.88 Refer to Figure 14.26. What method of protection—1 + 1 or 1:1—is used in a UPSR network? In a BLSR network?

14.89 Refer to the BLSR ring shown in Figure 14.26(b). Draw the traffic route if (a) the link between nodes 3 and 4 fails, (b) node 4 fails.

14.90 Why is the architecture shown in Figure 14.26(a) called path-protected while that shown in Figure 14.26(b) is called line-protected?

14.91 Refer to Figure 14.26(b). Suppose the ADM in node 2 fails. Show how the ring survives this failure.

14.92 Explain how a SPRing network survives a link failure.

14.93 What is dual-homing architecture? What is this architecture designed to protect?

14.94 What are the peculiarities of the protection and restoration methods of fiber-optic networks compared with those of SONET networks?

14.95 What is the role of troubleshooting in the survivability of fiber-optic networks?

14.96 Why do we need to develop WDM fiber-optic networks?

14.97 How fast is demand for Internet bandwidth growing?

14.98 What are the major issues that have to be resolved to spur faster development of WDM fiber-optic networks?

14.99 Why is wavelength-division multiplexing the major trend in fiber-optic networks?

14.100 What is the current status of WDM networks?

14.101 What is the trend in communications networking?

14.102 Project 2: Build a memory map. (See Problem 1.20.)

*Note:* The following suggested projects encompass material discussed throughout this text and, if undertaken, will help reinforce your newly acquired knowledge.

Project 3: Point-to-point WDM systems, used in conjunction with EDFAs, offer the following benefits:

1. Reduction in the amount of fiber needed.
2. Cost savings in electronic regenerators.
3. Savings in replacement cost for existing TDM equipment, which would no longer be needed.
4. Bandwidth flexibility, that is, expanding bandwidth as needed.

   Prove the validity of each of the above points by referring to material from the appropriate sections of this chapter.

Project 4: Consider the following statements:

1. For a given network, the physical architecture and logical topology can be the same but can also be quite different.
2. A data stream starts to be transmitted as light and it continues with or without conversion into electrical form until it reaches its destination.
3. There are two categories of fiber-optic networks: broadcast-and-select and wavelength-routed.
4. Broadcast-and-select networks require some form of tuning at either the source or the destination or at both.
5. Wavelength-routed networks require switching and add/drop abilities.
6. Fiber-optic networks need a set of specific components and modules.

   Explain each of these statements by referring to material from the appropriate sections of this book.

## REFERENCES[1]


1. Rajiv Ramaswami and Kumar Sivarajan, *Optical Networks, A Practical Perspective,* San Francisco: Morgan Kaufman, 1998.
2. Biswanath Mukherjee, *Optical Communications Networks,* New York: McGraw-Hill, 1997.
3. Thomas Stern and Krishna Bala, *Multiwavelength Optical Networks,* Reading, Mass.: Addison Wesley Longman, 1999.
4. Denis Mestdagh, *Fundamentals of Multiaccess Optical Fiber Networks,* Boston: Artech House, 1995.


[1]See Appendix C: A Selected Bibliography

5. Myron Sveum, *Data Communications, An Overview,* Upper Saddle River, N.J.: Prentice Hall, 2000.

6. William Beyda, *Data Communications,* 3d ed., Upper Saddle River, N.J.: Prentice Hall, 1999.

7. Tymothy Ramteke, *Networks,* Upper Saddle River, N.J.: Prentice Hall, 1994.

8. Marion Cole, *Telecommunications,* Upper Saddle River, N.J.: Prentice Hall, 1999.

9. Darren L. Spohn, *Data Network Design,* New York: McGraw-Hill, 1993.

10. Dimitri Bertsekas and Robert Gallager, *Data Networks,* 2d ed., Upper Saddle River, N.J.: Prentice Hall, 1992.

11. James Manchester, Paul Bonenfant, and Curt Newton, "The Evolution of Transport Network Survivability," *IEEE Communications Magazine,* August 1999, pp. 44–51.

12. I. Van der Voorde et al., "The SuperPON Demonstrator: An Exploration of Possible Evolution Paths for Optical Access Networks," *IEEE Communications Magazine,* February 2000, pp. 74–82.

13. Paul Young, *Electronic Communication Techniques,* 4th ed., Upper Saddle River, N.J.: Prentice Hall, 1999.

14. John Bellamy, *Digital Telephony,* 2d ed., New York: John Wiley & Sons, 1991.

15. Mike Sexton and Andy Reid, *Transmission Networking: SONET and the Synchronous Digital Hierarchy,* Boston: Artech House, 1992.

16. Julie Petersen, *Data & Telecommunications Dictionary,* Boca Raton, Fla.: CRC Press, 1999.

17. Daniel Minoli, *Telecommunications Technology Handbook,* Boston: Artech House, 1991.

18. William Flanagan, *The Guide to T-1 Networking (How to Buy, Install and Use T-1, from Desktop to DS-3),* 4th ed., New York: Telecom Library, 1990.

19. *SonetLYNX General Description,* 2d ed., Intelect Network Technologies, Richardson, Tex., September 1997.

20. *SONET 101—An Introduction to Basic Synchronous Optical Networks,* December 1995, Northern Telecom Canada, St. Laurent, Que., Can.

21. Robert Butler and David Polson, "Wave-Division Multiplexing in the Sprint Long Distance Network," *IEEE Communications Magazine,* February 1998, pp. 52–55.

22. Robert Jopson, "Alternative modulation formats," Paper W204, Optical Fiber Communication Conference, Baltimore, March 5–10, 2000.

23. Cornelius Fuerst et al., "RZ versus NRZ Coding for 10 Gbit/s Amplifier-Free Transmission," Paper MJ1, advance program of IEEE Lasers and Electro-Optics Society 1999 Annual Meeting, San-Francisco, Nov. 8–11, 1999.

24. *Fibre Channel, Technical Overview,* 2d ed., The Fibre Channel Assn., Dayton, Ohio, 1998.

25. Jay Kadambi, "Medium Access Control (MAC)," in *FDDI Technology and Applications,* ed. by Sonu Mirchandani and Raman Khanna, New York: John Wiley & Sons, 1993.

26. Charles Thurwachter, Jr., *Data and Telecommunications,* Upper Saddle River, N.J.: Prentice Hall, 2000.

27. William Flanagan, *ATM User's Guide,* New York: Flatiron Publishing, 1994.

28. Winston Way, "Hybrid fiber-coax (HFC) systems access system technologies," Paper SC145, Optical Fiber Communication Conference,, Baltimore, March 5–10, 2000.

29. Ernest Tunmann, *Hybrid Fiber-Optic Coaxial Networks,* New York: Flatiron Publishing, 1995.

30. Milorad Cvijetic, *Coherent and Nonlinear Lightwave Communications,* Boston: Artech House, 1996.

31. Kathleen Richards, "New pan-European network to support petabit speeds," *Lightwave,* August 1999, pp. 25, 38.

32. Paul Bonenfant, "Optical Networking Standards," Paper TU1, Optical Fiber Communication Conference, Baltimore, March 5–10, 2000.

33. James Manchester et al., "IP over SONET," *IEEE Communications Magazine,* May 1998, pp. 136–142.

34. Sabah Aidarous and Thomas Plevyak, eds., *Telecommunications Network Management: Technologies and Implementations,* New York: IEEE Press, 1998.

35. *Wave Watcher—Integrated Network Management* (data sheet), Ciena Corp., Linthicum, Md., 1997.

36. *Network Management Modules, NMS SNMP* (data sheet), Pan Dacom Telekommunitaion, GmbH, Hamburg, Ger., 1999.

37. *RPS-370 Intelligent DS3 Protection Switch* (data sheet), Bosch Telecom, Inc., Gaithersburg, Md., 1996.

38. *HP AccessFiber- Fiber Network Management System, Release 3.0* (data sheet), Hewlett-Packard, Test and Measurement Call Center, Englewood, Colo., 1998.

39. *FiberBase—Fiber Network Management Software* (data sheet), ADC Telecommunications, Minneapolis, April 1998.

**40.** Marco Marsan et al., "All-Optical WDM Multi-Rings with Differentiated QoS," *IEEE Communications Magazine,* February 1999, pp. 58–66.

**41.** John Senior, Michael Handley, and Mark Leeson, "Developments in Wavelength Division Multiple Access Networking," *IEEE Communications Magazine,* December 1998, pp. 28–36.

**42.** Andrew Schmitt, "New components enable management functions in the optical network," *Lightwave,* September 1998, pp. 115–118.

**43.** John Lewis, "When Disaster Strikes," *Cabling Business Magazine,* March 1998, pp. 22–30.

**44.** Piet Demeester, Tsong-Ho Wu, and Noriaki Yoshikai, "Survivable Communication Network," *IEEE Communications Magazine,* August 1999, pp. 40–42.

**45.** Mark Loyd Jones, Robert Butler, and William Szeto, "Sprint Long Distance Network Survivability: Today and Tomorrow," *IEEE Communications Magazine,* August 1999, pp. 58–62.

**46.** Rob Batchellor, "Putting the lens to optical-layer protection," *Lightwave,* September 1998, pp. 127–128.

**47.** John Nikolopoulos and Lois-René Parê, "Maximum reliability and performance for data-optimized backbone networks," *Lightwave,* September 1998, pp. 87–94.

**48.** John Marsh, "Reliability and maintenance considerations in the optical layer," *Lightwave,* March 1999, pp. 58–62.

**49.** Sami Hendow, "High-bandwidth networks bring new monitoring requirements," *Laser Focus World,* August 1998, pp. 171–178.

**50.** *Internetworking Troubleshooting Handbook,* Cisco Systems, San Jose, Calif., 1998.

**51.** *LAN Troubleshooting and Baselining,* Internetworking Wandel & Goltermann, Research Triangle Park, N.C., 1998.

**52.** John Ryan, "WDM: North America Deployment Trends," *IEEE Communications Magazine,* February 1998, pp. 40–44.

**53.** Paul Green, "Optical Networking Has Arrived," *IEEE Communications Magazine,* February 1998, p. 38.

**54.** Alan McGuire and Paul Bonenfat, "Standards: The Blueprints for Optical Networking," *IEEE Communications Magazine,* February 1998, pp. 68–78.

**55.** Andrzej Jajszczyk, "What Is the Future of Telecommunications Networking?" *IEEE Communications Magazine,* June 1999, pp. 12–20.

**56.** Yi Pan, Chunming Qiao, and Yuanyuan Yang, "Optical Multistage Interconnection Networks: New Challenges and Approaches," *IEEE Communications Magazine,* February 1999, pp. 50–56.

**57.** Patrick Trischitta and William Marra, "Applying WDM Technology to Undersea Cable Networks," *IEEE Communications Magazine,* February 1998, pp. 62–66.

**58.** Stephen Hardy, "Metro networks mesh from all angles," *Lightwave,* February 1999, p. 19.

**59.** Patricia Hatton and Frank Cheston, III, "WDM Deployment in the Local Exchange Network," *IEEE Communications Magazine,* February 1998, pp. 56–61.

**60.** Brian McCann, "Optical-networking platforms for metropolitan enterprise services," *Lightwave,* August 1999, pp. 58–62.

**61.** Joseph Linde, "Low Cost Fiber to the Desktop with 100BASE-SX," *Cabling Business Magazine,* January 1999, pp. 22–28.

**62.** Steven Hersey and Mark Wilson, "Realizing the terabit central office," *Lightwave,* February 1998, pp. 132–134.

**63.** John Adler and Stevan Plote, "Laying the foundation of the optical Internet," *Lightwave,* February 1999, pp. 126–131.

**64.** Rajiv Ramaswami, "Multiwavelength Optical Networking," *Tutorial 1,* IEEE Infocom '99, New York, March 1999.

**65.** Edward Traupman et al., "The Evolution of the Existing Carrier Infrastructure." *IEEE Communications Magazine,* June 1999, pp. 134–139.

**66.** Hiroshi Yoshimura. Ken-ichi Sato, and Noboru Takachio, "Future Photonic Transport Networks Based on WDM Technologies," *IEEE Communications Magazine,* February 1999, pp. 74–81.

# 15

# Conclusion

At this point in your study, you have no doubt come to the realization that both technical and commercial developments in this field are coming along at a mind-boggling pace—all driven by one dire need: a hungering for more and more bandwidth. What we'll do in this chapter is look ahead at some of the important newer developments in the works that attempt to cope with customers' endless appetites. Many of these developments have been touched upon elsewhere in the text. Here, we'll try to show how they interrelate and we'll put them in perspective in terms of how they're likely to affect the fiber-optics communications industry in the near future.

## 15.1 BANDWIDTH: THE INDUSTRY'S 'HOLY GRAIL'

First of all, consider what's been behind the prodigious growth of fiber-optic communications technology: the exponentially increasing demand for bandwidth, of course. Secondly, bear in mind what's driving this apparently unquenchable demand: the Internet, quite obviously. True, it's not the only glutton out there, but it is the most ravenous and it rose up the fastest. To appreciate the pace of the Internet's growth and its impact on the American economy alone, just look at the numbers [1]: The Internet surpassed the telephone as a business communications tool in 1998. Total revenue from Internet businesses exceeded $300 billion in 1998 and, industry observers believe, revenue could easily double every year over the next several years before leveling off. Competition among Internet service providers will surely result in a price reduction in communications services [2]. The current stability of the American economy can be partially attributed to the Internet. While no one can predict how the Internet will shape the American economy in the long haul, one thing we can say for certain: The Internet will grow faster and faster in the years ahead and, as it does, it will generate a steady, unrelenting demand for bandwidth.

The Internet and other bandwidth-famished services will determine the further development of fiber-optic communications technology. The most prominent features of the current stage of this development are wavelength-division multiplexing and all-optical networks. What the fiber-optic communications business itself will look like in the future as long-standing companies merge and start-ups emerge is almost impossible to foresee, but one course is already apparent: There will be—in fact, there must be—progress in developing fiber-optic components.

## 15.2 DEPLOYMENT OF NEW FIBER-OPTIC LINES

There are two basic means by which to meet the ever-increasing demand for bandwidth: increasing the capacity of the existing lines (WDM is the best example of such a solution) and deployment—that is, construction and installation—of new fiber-optic lines. Let's consider the deployment of the new lines:

On a *global scale,* the most prominent new installations are associated with the activity of two competitive companies: Global Crossing Ltd. and Project Oxygen.

Global Crossing is deploying giant rings in fiber-optic Transatlantic, Pacific, and South American networks as well as in terrestrial networks in Europe and other countries.

Project Oxygen's program will be carried out in two phases. The first, a 169,000-km backbone network, will be largely an undersea operation, with only 14,000 km of cable being terrestrial. Implemented in segments over a three-year period, it will comprise the following systems:

- A greater Atlantic ring.
- Major trans-Pacific and Southeast Asia rings.
- A terrestrial link acorss North America, a Mediterranean ring, and another trans-Atlantic link.
- Northern Europe, the Middle East, and South American rings, and an India-Thailand link.
- A Central American link, an Oceania ring, and additional Atlantic and Pacific links.

Phase two will increase regional connectivity to the core network [3].

As a result of all this activity—and other undersea fiber-optic networks too numerous to mention here—the South American continent will be enveloped by a number of fiber-optic rings. The African continent, moreover, will be encircled by a network called Africa One. A number of fiber-optic networks will link the Arabian Peninsula, the Indian subcontinent, East and Southeast Asia and the Far East. Australia and New Zealand are already well connected to the rest of the world and have several links under construction that will enhance these connections [4].

The fact is that undersea fiber-optic networks are becoming so prolific that this "spaghetti under the sea," as one wit puts it [5], already carries 10 times more global traffic than satellite communications.

On a *regional scale,* many competitive local exchange carriers (CLECs) are building their own fiber-optic networks. Such networks provide full service (local and long-distance telephone, data, Internet, etc.) to a given region. One example of such a network serves a five-state area in the Southwestern United States: Texas, New Mexico, Arizona, Nevada, and California [6]. The challenge in designing and building such a network is that it has to carry both extremely high-speed traffic to be a part of a long-haul national network and low-speed traffic to deliver service to the local area and, eventually, to an individual customer. For an excellent review of installed and planned fiber-optic networks, see reference [7].

One of the key factors facilitating the deployment of new networks is the decreasing cost of installing them. For example, in 1985 the cost of installing terrestrial long-distance networks was $12.40 per fiber mile; by 1998 this cost had dropped to just $0.07 per fiber mile [8]. Lower cost stimulates new demand, which, in turn, keeps prices low. Fortunately, fiber-optic network builders have been thriving under an exceptionally propitious economy, including low raw-material prices (that is, for optical fiber, fiber-optic cable, hardware) and steadily declining installation cost. How long will this continue? Your guess is as good as ours, but the trend has been this way for nearly a decade now if that's any precursor of what lies ahead.

## 15.3 OPTICAL FIBER: PROBLEMS GALORE, SOLUTIONS SOUGHT

As we have learned, singlemode fiber is used for long-haul transmission and multimode fiber for local area transmission. Now we're interested in knowing what improvements are on the horizon in terms of fiber development.

The basic requirements for long-haul *singlemode optical fiber*—minimum attenuation and maximum bandwidth, that is, minimum deterioration of the signal—have been the same for a number of years now; however, there's been a gradual shift in emphasis here. For example, with the advent of optical amplifiers, attenuation is no longer the main issue. The distance between adjacent regenerators—the major yardstick by which profit and loss is gauged—is largely determined, we know, by the amount of chromatic dispersion in a singlemode fiber. This has posed a real problem in a profit-squeezed industry. However, thanks to the development of dispersion-compensating techniques and their management, this problem has been moved off center stage.

Today *polarization-mode dispersion (PMD)* is the real limiting factor in fiber transmission. In fact, more than 20% of all installed fiber has a PMD above 1 ps/$\sqrt{\text{km}}$, which limits the line rate to 2.5 Gbit/s. Because of its random nature, PMD cannot be compensated for directly without great difficulty, and the most practical way to combat this phenomenon is to put ever more severe restrictions on the properties of new fibers. Both the ITU and TIA have developed new recommendations and test methods for defining PMD. PMD of not more than 0.5 ps/$\sqrt{\text{km}}$ has become the industry standard [9]. Meanwhile, research is continuing to find new and more effective techniques to reduce this type of dispersion. (See, for example, [10] and [11].)

To increase the span between adjacent EDFAs and to support more wavelength channels in WDM systems, network designers are now using optical amplifiers to launch considerably more light power than ever into the same optical fibers they've always used. This presents network designers with still another problem: It increases the level of nonlinear effects, sharply limiting fiber bandwidth.

The most severe limitation of these nonlinear effects today comes from *four-wave mixing (FWM)*. This phenomenon increases dramatically near the wavelengths where optical fiber has zero dispersion. To combat FWM, manufacturers have developed *non-zero dispersion-shifted fiber (NZ-DSF)*—fiber with a small positive dispersion around the 1550-nm operating wavelength. Approximately 80 percent of newly installed fiber is of the NZ-DSF type [12].

To realize the full advantage from using optical amplifiers, we need to incorporate more power into a fiber. To accomplish this, manufacturers have developed optical fiber with a *large effective area*—that is, fiber with the same geometric characteristics as the old standbys but capable of carrying a more powerful beam of light. What does this do? It increases the effective area of light propagation, thereby reducing the concentration of power around the fiber's centerline. The result is to launch more total power into the fiber. This technique has been successfully developed through a special design of the profile of a fiber's refractive index. Today new types of fiber, such as Corning's LEAF and Lucent Technologies' TrueWave, are used exclusively for new installations.

Using distributed Raman amplifiers helps allocate gain along the span of an optical fiber more evenly, thus to some extent alleviating the problem of launching high-power light at one point of the fiber. This new technique reduces nonlinear effects.

Designers of submarine networks, too, confront dilemmas. For instance, such networks that operate over long distances and that feature high bit rates and WDM suffer from another nonlinear phenomenon: *modulation instability*. This effect stems from the signal-to-ASE beating that occurs in EDFAs. To mitigate the phenomenon, the fiber has to have a slight negative dispersion shift. To the rescue, fortunately, come the Submarine LEAF and TrueWave XL optical fibers, which were designed exclusively for the extensive undersea installations discussed above ([13], [14]). Also confronting designers of submarine networks is the need to cope with dispersion and nonlinear effects in a very long link of an optical fiber. A new transmission format, *chirped return-to-zero* (CRZ), along with dispersion management (Section 6.3), helps to resolve this predicament [15].

Let's turn now to local area networks, where *multimode optical fibers* find their greatest use. Two major characteristics of optical fiber have to be taken into account right from the start: First, it enables employment here of centralized LAN architecture and, second, it allows high-speed transmission. With centralized LAN architecture, a single server is used for the entire campus. This, in turn, allows a longer transmission distance to exist among network elements. However, here's the rub: Such architecture becomes possible *only* with the advent of optical fiber. However, increasing transmission distances requires the use of laser diodes rather than LEDs, until recently the staple of light sources because of cost. But what we're facing is a doubling of Internet usage every three-and-a-half months, thus putting tremendous pressure on LAN bandwidth. Increasing bandwidth is possible only when using a laser as the light source. Hence, from any standpoint, designers of LANs, like it or not, must now turn to laser diodes—VCSELs in particular—instead of LEDs.

Using a laser diode with multimode fiber, however, poses a challenge. In this case—a multimode fiber—light is highly concentrated along the fiber's centerline, where the refractive-index profile is most likely to deviate from its ideal. An interim solution to this problem has been the use of a mode-conditioning patch cord that launches light off-center. A more advanced solution is to make a VCSEL radiates a beam with a donut-like profile, as discussed in Section 10.3. (See Figure 10.13.) Fortunately, a better solution has materialized: the development of new multimode fibers designed specifically for use with VCSEL laser diodes. (See [6] and [16].) The refractive-index profile around the center of the core of such a fiber does not exhibit any so-called "kicks," or abrupt changes—the key factor in ensuring high-quality transmission of a VCSEL's light.

## 15.4 FIBER-OPTIC COMPONENTS

***Major changes lie ahead*** Let's consider for the moment fiber-optic cable from the market standpoint [17]. Submarine construction has consumed 38% of this cable to date and is expected to account for as much as 50% of total global fiber-cable sales in the next few years.

In terms of terrestrial usage, North America consumed 28%, Europe 18%, and Japan and the Pacific Rim 12% of worldwide usage in 1997 alone. However, most observers see these numbers falling sharply over the next several years because of the growth of submarine networks, a $10.1 billion business in 1997 that is expected to more than triple within the next few years. To fully appreciate the significance of these numbers, keep in mind that the price of fiber-optic cable is constantly declining.

The major change we expect to see in the manufacturing of fiber-optic cables is an increase in the fiber count (that is, the number of optical fibers placed within a single cable). This trend is true for all types of networks. A fiber count as high as 800 is possible today and will soon become commonplace. As a result, the pressure on designers of connectors and hardware installers will increasingly be to accommodate more fibers per unit, which boils down to the miniaturization of both connectors and hardware.

As anyone associated with this business clearly knows, the main criterion for survival in this fiercely competitive marketplace hinges on the ability to meet the tough new application requirements with higher-performance characteristics in cable, connectors, splicing, and hardware.

***Transmitters and receivers*** One of the key issues confronting the industry is whether *transmitter* designers will be able to integrate all components in a single chip and, if so, how soon. The most promising laser diode today from this standpoint is the *vertical-cavity surface-emitting laser (VCSEL)*. Along with its low cost, a VCSEL is already the light source of choice for LANs. But a VCSEL diode suffers from one major drawback: It is available now only at the 850-nm operating wavelength. Fortunately, substantial research efforts are being expended to develop long-wavelength VCSELs and significant progress has been made in producing both 1300-nm and 1550-nm VCSELs [18].

VCSELs operating at 1550 nm that are capable of radiating 1 mW of CW power and that have a modulation speed of 2.5 G-bit/s have been reported [19]. What makes these devices even more attractive is that 1550-nm VCSELs are tunable, a factor that adds to the appeal of this laser diode for WDM applications [20].

Fixed distributed-feedback (DFB) lasers are today's primary light sources for WDM systems. It would be highly desirable to make them tunable because doing so would reduce the price of an entire system dramatically. However, tunable DFB lasers are commercially available now only for use in test equipment, which means they display very good performance characteristics but are too slow to be used for transmission purposes [21]. Nevertheless, this is a development to watch.

The emergence of WDM systems has generated the need for still another special device, a tunable transmitter capable of producing high-powered light. Such a transmitter, able to work at a bit rate of 10 Gbit/s, has already appeared on the market. It can cover up to 20 channels with 50 GHz spacing and generates 10 mW of CW power for all channels [22].

On the *receiver* side, the requirement now is not for more sensitivity but—and this will come as no surprise—for more bandwidth. Receiver sensitivity is not a major impediment today because EDFAs allow us to deliver enough power to the photodiode's active surface. It's the availability of high-speed photodiodes with good performance characteristics that builders of fiber-optic communications systems have been calling for. Manufacturers of photodiodes have responded well with a wide range of PDs having high bandwidth and good responsivity around the C- and L-bands [22]. For example, avalanche photodiodes (APDs), with a sensitivity of –26 dBm and capable of working at 10 Gbit/s are commercially available [24].

To sum up, from both performance and economic perspectives, the key to improving performance characteristics and reducing overall system cost lies in the integration of transceivers into one-chip modules [25]. This is a development that the entire fiber-optic communications industry awaits.

***Components: Smaller means better*** If you've read this text carefully, you know that we have been harping throughout on the limitations of the components in today's fiber-optic communications systems. Indeed, this is a critical matter. In the first generation of fiber-optic networks, opto-electronic components held sway much to everyone's chagrin. The replacement of opto-electronic regenerators with optical amplifiers (mostly EDFAs) marked a much desired leap toward the future generation of fiber-optic networks. Surely, this is the trend: the replacement of opto-electronic components with optical components. This major step heralds the move to *all-optical networks (AONs),* the industry's nirvana. But we must understand that AON will become a commercial reality when—and only when—all the necessary optical components become not simply commercially available but available at *competitive* prices.

For our second-generation fiber-optic networks, optical switches (optical cross-connects) and optical add/drop multiplexers are the crucial modules. While these components are commercially available now, their performance still leaves something to be desired, as does their high cost.

One of the more promising techniques for building optical switches today is *micro-electro-mechanical systems (MEMS).* This variant of micro-optics uses a movable micro-mirror controlled by an applied low voltage. Positioning the mirror allows an operator to direct a light beam at a specific fiber, that is, to switch light into the precise fiber specified by the customer [26].

Obviously, MEMs is not the only emerging technology used to develop optical components for fiber-optic networks. Another approach entails the use of an *acousto-optic tunable filter (AOTF)* to build an optical switch [27]. Other innovations in this area are discussed in Section 13.5.

Equally significant is that new companies are coming on the scene with terabit switches, certainly an important breakthrough. Why? Such a switch with its nonblocking switching capacity of 184 Tbit/s, will support OC 48 and even OC-192 interfaces [28]. The race continues.

## 15.5 WAVELENGTH-DIVISION MULTIPLEXING: A DIRE NEED MET

*Wavelength-division multiplexing (WDM)* came to the forefront in the middle of the 1990s as a means of coping with the fiber-exhaust problem in long-haul networks. It wasn't long before the industry realized that WDM did more than that. It proved to be a powerful tool that dramatically increased the capacity of fiber-optic networks in general. WDM technology has become the dominant technology in point-to-point connections since it's the best way to increase the capacity of the links at a fraction of the cost of the new installations. Having started with modest 4-channel multiplexing, WDM technology today reaches the 1021-channel level in the research laboratory [29].

Two developments served to thrust WDM technology very quickly into the forefront in its application to fiber-optic networks [30]. The first, the erbium-doped fiber amplifier (EDFA), made it feasible to use WDM by amplifying many wavelengths simultaneously. The second, the advent of the optical add/drop multiplexer (OADM), made adding and dropping an individual wavelength possible.

Today, there is only one answer to the question of how to meet the exponentially increasing demand for bandwidth by a network: WDM [31]. This technology gives rise to a new commercial option for providing telecommunications service: leasing an individual wavelength instead of leasing an entire fiber-optic link [32]. This feature will certainly propel the deployment of WDM systems to an even larger scale both in numbers and scope.

The future of WDM systems seems assured. The reason is simple enough: Only WDM technology enables the realization of the industry's holy grail: the all-optical network (AON), which, without WDM, would be extremely difficult, if not impossible, to build.

All-optical networks based on WDM systems offer many advantages compared with the traditional networks. You can, for instance, build an optical ring to any desired scale because the distance is no longer limited thanks to the use of EDFA, nor are the number of nodes restricted by a SONET/SDH overhead. These examples only scratch the surface of the wealth of possibilities that WDM technology offers for developing fiber-optic networks.

As we've noted several times already, however, the deployment of WDM systems is still limited to a large extent by a lack of appropriate components. The more (and the sooner) optical components supporting WDM come on the market at reasonable cost, the sooner all-optical networks become a reality.

## 15.6 NETWORKS

Second-generation fiber-optic networks are currently being developed and, to some extent at least, are even being deployed. Making this possible has been the incorporation into these networks of WDM MUXs/DEMUXs, EDFAs, OADMs, OXCs, and passive optical components—relatively new devices already on the market or in the process of being commercially produced. These second-generation fiber-optic networks, employing some optical elements as replacements for electronic equipment, bring us a giant step closer to the industry mecca: the AON. In fact, some companies even say they can build an all-optical network today [33]. Perhaps, but at what cost?

The following is a brief rundown of the current situation in applications and trends in the development of fiber-optic networks:

Networks in the United States, by some estimates, will have to carry up to 20 times more traffic within the next few years than they did in 1998. In fact, statistics show a stunning 80% annual growth rate in fiber-optic networks. Since there is only going to be less than 10% growth rate in voice traffic as far ahead as we can see, traffic will be very much data-centered ([34], [35]). This heavy volume and the nature of future traffic will demand new approaches to developing communications networks.

Along these lines, designers are taking a new look at network architecture. They're now viewing a network as a *set of concentric layers with the core network in the center surrounded by an edge network that, in turn, is itself surrounded by an access network with the final ring/layer being the customer-premises network* [34].

The *core* of future networks will be a WDM-based reconfigurable protocol-independent optical network that provides both transmission paths and protection switching. The bandwidth per fiber in this network will be greater than a terabit per second. Terabit-per-second ATM switches and IP routers will transmit and manage the traffic through this optical network. The interfaces of the core-edge networks will provide all necessary multiplexing/demultiplexing action as well as all the intelligence that will be necessary to support terabit-per-second traffic from multiple sources that flows into and out of the core. The core network will provide all the functions that the long-haul backbone networks now perform—and even then some.

The *edge* network will work, basically, as an interface between a very high-speed core network and a low-speed access network. This network will do everything that current metropolitan area networks now do—even better—while performing more functions. For example, we have mentioned that one of the future developments will be simplifying the multilayered structure of today's networks. Thus, running IP directly over SONET, without involving ATM, will be a step in this direction. This technique, called *packet-over-SONET (POS)* [34], is already being developed. The interface between the edge and the access networks will provide multiplexing/demultiplexing and the necessary intelligence by using existing protocols (transmission systems) such as IP, ATM, VoIP, T3/E3, T1/E1, and others.

To give you some insight into what kind of problem you may meet in developing such a network, consider the following: Today many service providers want to offer *voice over the Internet (VoIP)* because the Internet carries no access charge, while a fee for access to the public-switched telephone network (PSTN) still exists. Hence, VoIP is much cheaper than regular telephone service. However, VoIP, which is actually voice transmitted as data in real time, faces one serious obstacle: The Internet network was not designed to operate this way, which, unfortunately, makes this transmission highly unreliable. You pay less today for VoIP transmission at the expense of quality of service. Unquestionably, a lot of hard work lies ahead if this problem is to be solved.

The *access* network links customers to transport networks pretty much the way local networks do but, again, at a higher level of performance and with the capability of carrying out more functions. Using protocols (transmission systems) such as ADSL, ISDN, FTTC, FTTH, and others, the access network provides broadband access from the customer's premises to the edge network and, eventually, to the core network.

Concentric-layered architecture looks more attractive than the existing multilayered network structure. But the ultimate goal, of course, is the move to *all-optical networks (AON),* which drastically reduce the number of intermediate layers. AONs will also eliminate many of the existing equipment setups and interconnections that are major contributors to the high cost and performance shortcomings of the current networks.

On the logical side, the AONs, being transparent to any protocols at the transition stage, will stimulate the development of much simpler new protocols that will meet all possible demands.

From the service provider's standpoint, the AONs will require not only new equipment and logic but also new test, measurement, and management concepts and systems to support the operation of these new networks. From the end-user's standpoint, the AONs will integrate voice, video, and data and provide high-quality transmission. But development engineers and designers have some major obstacles to overcome before they can turn these ambitious plans into reality any time soon.

## 15.7 WIRELESS COMMUNICATIONS AND FIBER-OPTIC NETWORKS

Wireless communications—transmitting information without the use of cables—is a burgeoning area of telecommunications today. The cellular phone is a perfect example. Recent advances in cellular-phone technology include much-improved transmission quality and connections over an ever-larger geographic area. But the most noteworthy achievement of wireless communications is its ability to increase bandwidth, which enables users of this technology to transmit not simply voice but also data. Now, therefore, you can connect your laptop to a LAN, a MAN, or even to the Internet while you're traveling in your car (but, we hope, not actually driving it) or on a plane. Such transmission, of course, has necessitated devising new approaches to the technology, architecture, and management of wireless communications. To review the latest achievements in this area, see [36].

If you are wondering how wireless technology and fiber-optic communications are related, the answer is simply this: A wireless system provides you with access to the nearest receiver (antenna), from which a fiber-optic network carries your signal to a LAN, a MAN, or the Internet. In fact, a sector of the fiber-optic communications industry is specifically committed to serving the wireless business. So, as you can see, wireless communications is yet another means by which to gain access to fiber-optic communications networks.

### Summing Up

If you are going to work in some capacity in the field of telecommunications—and we assume you are or you probably wouldn't be reading this now—the technology you will be working with will be based on fiber optics. Why are we so certain? In 1998, the overall telecommunications market generated $467.2 billion and was growing at a rate of 11% a year [37]. As one wag quipped, "There are now three guarantees in life: death, taxes, and the *demand for more bandwidth*" [1].

Mark this well: There is *no* other means today to meet the demand for ever-more bandwidth except through fiber-optic communications systems. So our final word of advice: Work hard to master this technology. The potential payoff in financial and professional rewards is unlimited.

Fiber-optic communications technology is growing so rapidly and changing so swiftly and radically that innovations devised in research laboratories yesterday become commercial products today but will be obsolete by tomorrow. So keep abreast of the cutting edge of technology by reading the trade magazines, professional journals, and the manufacturers' technical literature. But to do so with understanding requires a solid grounding in the technology. We trust this text has done that for you.

## *REFERENCES*[1]


1. John Dix, "Net-business," *The Network World 200,* April 24, 2000, pp. 58–72.
2. Kathleen Richards, "Getting on the Internet wholesale," in *Fiber Exchange,* a supplement to *Lightwave,* June 1999, pp. 8–12.
3. "Battle at the bottom of the sea," in *Fiber Exchange,* a supplement to *Lightwave,* June 1999, pp. 2–6.
4. Robert Pease, "Undersea fiber business thrives on today's demand for global connectivity," *Lightwave,* September 1999, pp. 1 and 32–33.
5. Charles Petit, "Spaghetti under the sea," *U.S. News & World Report,* August 30, 1999, pp. 56–58.
6. Windsor Thomas, "InfiniCor CL Multimode Fiber: The Laser-Optimized Fiber Advantage," *GuideLines,* Corning Incorporated, Telecommunications Products Div., Corning, N.Y., Spring 1999, pp. 12–13.
7. *KMI's fiberoptic long-haul route maps,* KMI Corp., Newport, R.I., 1999.
8. "Global consumption of fiber cable to peak at 21.5% annual growth rate in 2002," *Lightwave,* September 1998, p. 150.
9. Étienne Cagnon, "The Growing Importance of PMD Characterization," *Lightwave Test & Measurement Reference Guide,* EXFO Electro-Optical Engineering, Vanier, Que., Can., June 1999, pp. 8–9.


[1] See Appendix C: A Selected Bibliography

10. Michael Chbat, "Mitigation of Polarization-Mode Dispersion," Paper TuB3, Advance Program of the IEEE Lasers and Electro-Optic Society's 1999 Annual Meeting, San Francisco, Calif., November 8–11, 1999.

11. Dipak Chowdhury, "PMD Induced System Impairments in Long-Haul Optical Communication Systems," Paper TuB1, Advance Program of the IEEE Lasers and Electro-Optic Society's 1999 Annual Meeting, San Francisco, Calif., November 8–11, 1999.

12. Gaynell Terrell, "Bring on the Bandwidth," *Photonics Spectra,* December 1999, pp. 72–74.

13. Kevin Able, "Corning Submarine LEAF Optical Fiber," *GuideLines,* Corning Incorporated, Telecommunications Products Div., Corning, N.Y., Winter 1999, pp. 10–11.

14. Jim Brennan et al, "Lucent develops undersea fiber," *Fiber Optics News,* October 26, 1998, pp. 1–3.

15. Neal Bergano, "Undersea Fiberoptic Cable Systems: High-Tech Telecommunications Tempered by a Century of Ocean Cable Experience," *Optics & Photonics News,* March 2000, pp. 21–25.

16. Preston Buck, "The Evolution of LANs," *GuideLines,* Corning Incorporated, Telecommunications Products Div., Corning, N.Y., Winter 1999, pp. 2–5.

17. "Global consumption of fiber cable to peak at 21.5% annual growth rate in 2002," *Lightwave,* September 1998, p. 150.

18. Vijay Jayaraman, "VCSEL makers pursue low-cost 1300-nm diodes," *Laser Focus World,* August 1999, pp. 159–168.

19. Christophe Starck, "Long-Wavelength VCSEL with Tunnel Junction and Metamorphic A1As/GaAs Conductive DBR," Paper TuA1, Advance Program of the IEEE Lasers and Electro-Optic Society's 1999 Annual Meeting, San Francisco, Calif., November 8–11, 1999.

20. Robert Pease, "New 1550-nm tunable lasers to cut DWDM system costs," *Lightwave,* June 1999, p. 25.

21. Olivier Plomteux, "Tunable DFB Laser Sources," *Lasers & Optronics,* September 1999, pp. 13–15.

22. "JDS Uniphase demonstrates high power tunable 10 Gb/s transmitter module," *www.jdsunph.com.,* JDS Uniphase, Nepean, Ont., Can., and San Jose, Calif., September 27, 1999.

23. Ock-ky Kim, "A Look at High-Speed Detectors," *Lasers & Optronics,* August 1999, pp. 21–22.

24. "High-Sensitivity 10Gb/s Avalanche Photodiode Receivers, –26 dBm!" (advertisement), Epitaxx Co., West Trenton, N.J., in *Fiberoptic Product News,* September 1999, p. 61.

25. Robert Schuelke, "Meeting the demands of the new optical module market," *Lightwave,* September 1999, pp. 117–120.

26. Robert Pease, "All-optical equipment emerging from the lab to the market place," *Lightwave,* June 1999, pp. 1 and 28–30.

27. N. A. Riza, "High-Speed Multi-Wavelength Photonic Switch," *Optics & Photonic News,* December 1998, pp. 18–19.

28. Stephen Hardy, "Terabit networking resources near market," *Lightwave,* June 1999, pp. 1 and 34.

29. Brandon Collins et al., "A 1021 Channel WDM System," *Optics & Photonics News,* March 2000, pp. 31–35.

30. André Girard, "Wavelength Division Multiplexing: Past, Present, & Future Directions for Telecommunications Network," *Fiberoptic Product News,* September 1999, pp. 21–28.

31. Patrick Doyle and Jane Li, "Connectiv Communications, Inc.," *GuideLines,* Corning Incorporated, Telecommunications Products Div., Corning, N.Y., Summer 1999, pp. 8–11.

32. Kathleen Richards, "Leasing capacity by color," in *Fiber Exchange,* a supplement to *Lightwave,* September 1999, pp. 2–6.

33. "All-Optical Network" (advertisement), Corvis Corp., Columbia, Md., in *Fiberoptic Product News,* September 1999, p. 31.

34. "Performance issues in optical internetworking," *Telecommunications News,* Hewlett-Packard, Santa Clara, Calif., September 1999, pp. 5–7.

35. "Maturing DWDM market expected to soar with increasing capacity demand," *Lightwave,* September 1999, p. 114.

36. Willie Lu and Qi Bi, "Wireless Mobile ATM Technologies for Third-Generation Wireless Communications," *IEEE Communications Magazine,* November 1999, p. 36.

37. "TIA report predicts double-digit growth for telecommunications industry," *Lightwave,* June 1999, p. 114.

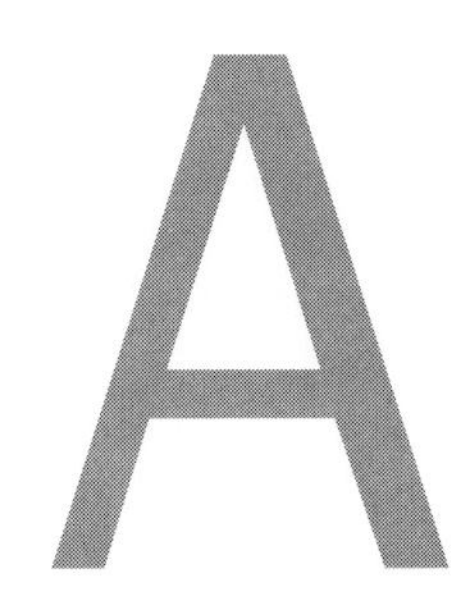

# List of Constants, Powers of Ten, International System of Units, Decibel Units, and the Greek Alphabet

CONSTANTS

| Constant | Symbol | Value |
| --- | --- | --- |
| Speed of light in a vacuum | c | $2.9979 \times 10^{8}$ m/s ~ $3 \times 10^{8}$ m/s |
| Electron charge | e or q | $1.6022 \times 10^{-19}$ C |
| Rest mass of the electron | $m_0$ | $9.109 \times 10^{-31}$ kg |
| Permitivity of free space | $\varepsilon_0$ | $8.8542 \times 10^{-12}$ F/m |
| Permeability of free space | $\mu_0$ | $4\pi \times 10^{-7}$ H/m |
| Planck's constant | h | $6.6261 \times 10^{-34}$ J-s |
| Electron volt | eV | $1.6022 \times 10^{-19}$ J |
| Boltzmann's constant | $k_B$ | $1.38 \times 10^{-23}$ J/K |
| Impedance of free space | $Z_0 = \sqrt{\mu_0/\varepsilon_0}$ | 376.7 Ω |

POWERS OF TEN

| Power | Prefix | Symbol |
|---|---|---|
| $10^{18}$ | Exa | E |
| $10^{15}$ | Peta | P |
| $10^{12}$ | Tera | T |
| $10^{9}$ | Giga | G |
| $10^{6}$ | Mega | M |
| $10^{3}$ | kilo | k |
| $10^{2}$ | hecto | h |
| $10^{1}$ | deca | da |
| $10^{-1}$ | deci | d |
| $10^{-2}$ | centi | c |
| $10^{-3}$ | milli | m |
| $10^{-6}$ | micro | μ |
| $10^{-9}$ | nano | n |
| $10^{-12}$ | pico | p |
| $10^{-15}$ | femto | f |
| $10^{-18}$ | atto | a |

INTERNATIONAL SYSTEM OF UNITS (SI)

| Quantity | Unit | Symbol | Dimension |
|---|---|---|---|
| **Basic units** | | | |
| Length | meter | m | |
| Mass | kilogram | kg | |
| Time | second | s | |
| Electric current | ampere | A | |
| Temperature | kelvin | K | |
| Amount of substance | mole | mol | |
| Luminous intensity | candella | cd | |
| **Supplementary Units** | | | |
| Plane angle | radian | rad | |
| Solid angle | steradian | sr | |
| **Other Units** | | | |
| Capacitance | farad | F | C/V |
| Conductance | siemens | S | A/V |
| Electric charge | coulomb | C | A•s |
| Energy | joule | J | N•m |
| Force | newton | N | $(kg\cdot m)/s^2$ |
| Frequency | hertz | Hz | $s^{-1}$ |
| Inductance | henry | H | Wb/A |
| Magnetic flux | weber | Wb | V•s |
| Magnetic induction | tesla | T | $Wb/m^2$ |
| Potential | volt | V | J/C |
| Power | watt | W | J/s |
| Pressure | pascal | Pa | $N/m^2$ |
| Resistance | Ohm | Ω | V/A |

## DECIBEL UNITS

Any ratio taken in natural scale can be presented in the logarithmic scale using the decibel (dB). For example, loss ($L$) is equal to

$$L = P_{out}\ (\text{W})/P_{in}\ (\text{W})$$

By definition, loss in dB is given by

$$L\ (\text{dB}) = 10 \log_{10} [P_{out}\ (\text{W})/P_{in}\ (\text{W})]$$

For instance, when $P_{out}\ (\text{W})/P_{in}\ (\text{W}) = 0.5$, $L = -3$ dB
Using the logarithmic scale, power can be measured in dBm as follows:

$$P\ (\text{dBm}) = 10 \log_{10} [P\ (\text{W})/1\ (\text{mW})]$$

Therefore, 0 dBm power is equivalent to 1 mW of power.

## THE GREEK ALPHABET

| Upper-case | Lower-case | Name |
|---|---|---|
| Α | α | Alpha |
| Β | β | Beta |
| Γ | γ | Gamma |
| Δ | δ | Delta |
| Ε | ε | Epsilon |
| Ζ | ζ | Zeta |
| Η | η | Eta |
| Θ | θ | Theta |
| Ι | ι | Iota |
| Κ | κ | Kappa |
| Λ | λ | Lambda |
| Μ | μ | Mu |
| Ν | ν | Nu |
| Ξ | ξ | Xi |
| Ο | ο | Omicron |
| Π | π | Pi |
| Ρ | ρ | Rho |
| Σ | σ, ζ | Sigma |
| Τ | τ | Tau |
| ϒ | υ | Upsilon |
| Φ | φ | Phi |
| Χ | χ | Chi |
| Ψ | ψ | Psi |
| Ω | ω | Omega |

# B

# Acronyms, Abbreviations, Symbols, and Units Used in This Book

| | |
|---|---|
| AAL | ATM Adaptation layer |
| ADM | Add/drop multiplexer |
| ADSL | Asymmetric digital subscriber line |
| ADSU | ATM data service unit |
| AGC | Automatic gain control |
| AL | Adaptation layer |
| AlGaAs | Aluminum gallium arsenide |
| AM | Amplitude modulation |
| ANSI | American National Standards Institute |
| AON | All-optical network |
| AOTF | Acousto-optic tunable filter |
| APC | Angled physical contact, or angled polishing connectors |
| APD | Avalanche photodiode |
| APDU | Application Protocol Data Unit |
| APS | Automatic protection switching |
| AR | Antireflection (coating) |
| ASCII | American Standard Code for Information Interchange (pronounced *as-kee*) |
| ASE | Amplified spontaneous emission |
| ATM | Asynchronous Transfer Mode |
| AWG | Arrayed-waveguide grating |
| B8ZS | Bipolar eight zero substitution |
| Bellcore | Bell Communications Research (now Telcordia Technologies) |
| BER | Bit-error rate |
| BERT | Bit-error-rate test |
| BH | Buried heterostructure |
| BIG | Bi-substituted iron garnet ("Bi" stands for *bismuth*) |
| B-ISDN | Broadband Integrated Services Digital Network |
| BLPR or BLSR | Bidirectional line-protected (or line-switched) ring |
| BPF | Bandpass filter |
| bps (bit/s, b/s) | Bits per second |

| | |
|---|---|
| BS | Beam splitter |
| BUF | Bandwidth-utilization factor |
| CATV | Community antenna television |
| CCITT | Comité Consultatif Internationale de Télégraphique et Téléphonique (Consultative Committee on International Telegraphy and Telephony. Changed its name to ITU-T in 1990.) |
| CDMA | Code-Division Multiple Access |
| CLEC | Competitive local exchange carriers |
| CMIP | Common Management-Information Protocol |
| CMOS | Complementary metal oxide semiconductor |
| CO | Central office |
| COADM | Configurable optical add/drop multiplexer |
| CPU | Central processing unit |
| CVD | Chemical vapor deposition |
| CW | Continuous wave |
| DARPA | Defense Advanced Research Projects Agency |
| DBR | Distributed Bragg reflector |
| dc | Direct current |
| DCF | Dispersion-compensating fiber |
| DCS | Digital cross-connect |
| DEMUX | Demultiplexer |
| DFB | Distributed feedback (laser) |
| DH | Double heterostructure |
| DIP | Dual-in-line package |
| DLL | Data link layer |
| DOD | Department of Defense |
| DSF | Dispersion-shifted fiber |
| DTMF | Dielectric thin-film multicavity filter |
| DWDM | Dense wavelength-division multiplexing |
| EA | Electroabsorption |
| EBCDIC | Extended binary-coded decimal-interchange code (pronounced *eb-see-dik*) |
| ECL | Emitter-coupled logic |
| EDF | Erbium-doped fiber |
| EDFA | Erbium-doped fiber amplifier |
| EIA | Electronic Industries Alliance |
| ELED | Edge-emitting LED |
| EM | Electromagnetic |
| EMI | Electromagnetic interference |
| EOTF | Electro-optic tunable filter |
| ER | Extinction ratio |
| ESD | Electrostatic discharge |
| EUPC | Enhanced ultra-polishing connectors |
| F | Finesse |
| FBG | Fiber Bragg grating |
| FBT | Fused biconical taper |
| FCAPS | Fault, configuration, accounting, performance, and security |
| FCC | Federal Communications Commission |
| FDDI | Fiber-Distributed Data Interface |
| FDM | Frequency-division multiplexing |
| FEC | Forward error correction |
| FET | Field-effect transistor |
| FEXT | Far-end crosstalk |
| FFP | Fiber Fabry-Perot (filter) |

| | |
|---|---|
| FIT | Failures in time |
| FM | Frequency modulation |
| FOCIS | Fiber Optic Connector Intermateability Standards |
| FOM | Figure of merit |
| FOTP | Fiber-optic test procedure |
| FP | Fabry-Perot |
| FRTT | Fixed receiver—tunable transmitter |
| FSR | Free spectral range |
| FTIR | Frustrated total internal reflection |
| FTTC | Fiber-to-the-curb |
| FTTD | Fiber-to-the-desk |
| FTTH | Fiber-to-the-home |
| FTTR | Fixed transmitter—tunable receiver |
| FWHM | Full width at half maximum |
| FWM | Four-wave mixing |
| GaAs | Gallium arsenide |
| GBIC | Gigabit interface converter (in Gigabit Ethernet and Fibre Channel LANs) |
| Gbps, Gbit/s | Gigabits per second ($10^9$ bits per second) |
| GCSR | Grating coupler sampled reflector |
| GI | Graded index |
| $GN/m^2$ | Giganewtons per square meter |
| GOSIP | Government OSI Protocol |
| GRIN | Graded refractive index |
| GRINSCH | Graded-index separate-confinement heterostructure |
| GVD | Group-velocity dispersion |
| HBM | Human-body model |
| HC | Horizontal cross-connect |
| HDSL | High bit-rate digital subscriber line |
| HFC | Hybrid fiber coaxial (cable TV) |
| Hz | hertz |
| IC | Integrated circuit |
| IC | Intermediate cross-connect |
| ILD | Injection laser diode |
| IM/DD, IM-DD | Intensity modulation-direct detection |
| InGaAsP | Indium gallium arsenide phosphide |
| InP | Indium phosphide |
| IP | Internet Protocol |
| IR | Infrared |
| IR | Intermediate reach (network) |
| ISDN | Integrated Services Digital Network |
| ISI | Intersymbol interference |
| ISO | International Organization for Standardization (formerly International Standards Organization) |
| ISP | Internet service provider |
| ITU | International Telecommunication Union |
| ITU-T | Telecommunication Standardization Sector of ITU |
| IXC | Interexchange carrier |
| kbps, kbit/s | Kilobits per second ($10^3$ bits per second) |
| kpsi | Kilopounds per square inch |
| LAN | Local area network |
| LATA | Local access and transport area |
| LD | Laser diode |

| | |
|---|---|
| LEC | Local exchange carrier |
| LED | Light-emitting diode |
| L-I | Light versus current characteristic |
| LID | Local injection and detection |
| $LiNbO_3$ | Lithium niobate |
| LLC | Logical Link Control |
| LPF | Low-pass filter |
| LR | Long reach (network) |
| MAC | Media-Access Control |
| MAN | Metropolitan area network |
| MAP | Manufacturing Automation Protocol |
| Mbps, Mbit/s | Megabits per second ($10^6$ bits per second) |
| MC | Main cross-connect |
| MCVD | Modified chemical vapor deposition |
| MDF | Main distribution frame |
| MDM | Mach-Zehnder modulator |
| MEMS | Micro-electromechanical system |
| MFD | Mode-field diameter |
| MIB | Management-information base |
| MIPS | Million instructions per second |
| MM fiber, MMF | Multimode fiber |
| MM GI | Multimode graded-index (fiber) |
| MONET | Multiwavelength optical network |
| MOSFET | Metal oxide semiconductor field-effect transistors |
| MPEG | Motion-Picture Experts Group |
| MQW | Multiple quantum well |
| MSM | Metal-semiconductor-metal |
| MSR | Mode-suppression ratio |
| MTTF | Mean time to failure |
| MUX | Multiplexer |
| MZI | Mach-Zehnder interferometer |
| NA | Numerical aperture |
| NE | Network element |
| NEC | National Electric Code |
| NEP | Noise-equivalent power |
| NIST | National Institute of Standards and Technology |
| NIU | Network interface unit |
| NMP | Network-Management Protocol |
| NMS | Network-management systems |
| NRZ | Non-return-to-zero |
| NSAP | Network-service access point |
| OA | Optical amplifier |
| OADM | Optical add/drop multiplexer |
| OAM | Operation, administration, and maintenance |
| OAM&P | Operation, administration, maintenance, and provisioning |
| OC | Optical carrier |
| OEIC | Opto-electronic integrated circuit |
| OFC | Optical Fiber Communication Conference and Exhibition |
| OFC | Optical fiber (cable)—conductive |
| OFCG | Optical fiber (cable)—conductive, general |
| OFCP | Optical fiber (cable)—conductive, plenum |
| OFCR | Optical fiber (cable)—conductive, riser |
| OFN | Optical fiber (cable)—nonconductive |

| | |
|---|---|
| OFNG | Optical fiber (cable)—nonconductive, general |
| OFNP | Optical fiber (cable)—nonconductive, plenum |
| OFNR | Optical fiber (cable)—nonconductive, riser |
| OSA | Optical Society of America |
| OSC | Optical supervisory channel |
| OSI | Open System Interconnection |
| OSP | Outside cable plant |
| OTDM | Optical time-division multiplexing |
| OTDR | Optical time-domain reflectometer |
| OVD | Outside vapor deposition |
| OXC | Optical cross-connect |
| PAS | Profile-alignment system |
| PBS | Polarization beam splitter |
| PC | Physical contact |
| PCB | Printed circuit board |
| PCM | Pulse-code modulation |
| PCS | Personal communications service |
| PCVD, or PACVD | Plasma-activated chemical vapor deposition |
| PD | Photodiode |
| PDL | Polarization-dependent loss |
| PDU | Protocol Data Unit |
| PECL | Positive emitter-coupled logic |
| PF | Perfluorinated |
| P-I | Power of light versus current characteristic |
| p-i-n, PIN | *p* type (positive), intrinsic, *n* type (negative) |
| PINAMP | PIN photodiode integrated with a transimpedance amplifier |
| PINFET | PIN photodiode linked with field-effect-transistor (FET) circuitry in a front-end unit |
| PISO | Parallel-in serial-out |
| PM | Power meter |
| PMD | Polarization-mode dispersion |
| PMD layer | Physical-Media-Dependent layer in the OSI reference model |
| POF | Plastic (polymer) optical fiber |
| PON | Passive optical networks |
| POP | Point of presence |
| POS | Packet over SONET |
| POTS | Plane old telephone service |
| PSTN | Public switched telephone network |
| QAM | Quadrature amplitude modulation |
| QOS | Quality of service |
| RBOCs | Regional Bell Operating Companies |
| RIN | Relative intensity noise |
| RMS | Root mean square |
| RRR, or 3R | Regeneration with retiming and reshaping |
| RSVP | Internet Reservation Protocol |
| RT | Remote terminal |
| RWG | Ridge waveguide |
| Rx (RX) | Receiver |
| RZ | Return-to-zero |
| SAP | Service access point |
| SAR | Segmentation and reassembly |
| SAW | Surface acoustic wave |
| SBS | Stimulated Brillouin scattering |

| | |
|---|---|
| SC | Synchronous container |
| SCM | Subcarrier multiplexing |
| SDH | Synchronous digital hierarchy |
| SHR | Self-healing ring |
| SI | International System of Units (Système Internationale d'unités) |
| SLED | Surface-emitting LED |
| SLM | Single longitudinal mode |
| SMF | Singlemode fiber |
| SMSR | Sidemode suppression ratio |
| SNA | Systems Network Architecture |
| SNMP | Simple Network Management Protocol |
| SNR (S/R) | Signal-to-noise ratio |
| SOA | Semiconductor optical amplifier |
| SONET | Synchronous Optical Network |
| SPE | Synchronous payload envelope |
| SPIE | The International Society for Optical Engineering (formerly the Society of Photo-Optical Instrumentation Engineers) |
| SPM | Self-phase modulation |
| SPRING | Shared protection ring (architecture) |
| SQW | Single quantum well |
| SR | Short reach (network) |
| SRS | Stimulated Raman scattering |
| SSP | Service switching point |
| SSR | Sidelobe suppression ratio |
| STS | Synchronous transport signal |
| Tbps, Tbit/s | Terabits per second ($10^{12}$ bits per second) |
| TCP | Transmission Control Protocol |
| TDM | Time-division multiplexing |
| TE | Transverse electric (wave) |
| TEC | Thermoelectric cooler |
| Telco | Regional and local telephone company |
| TEM | Transverse electromagnetic (waves) |
| TIA | Telecommunications Industry Association |
| TIA | Transimpedance amplifier |
| TM | Terminal multiplexer |
| TM | Transverse magnetic (wave) |
| TMN | Telecommunications management network |
| TOP | Technical Office Protocol |
| TTFR | Tunable transmitter—fixed receiver |
| TTL | Transistor-transistor logic |
| TTTR | Tunable transmitter—tunable receiver |
| TWA | Traveling-wave amplifier |
| Tx (TX) | Transmitter |
| UHF | Ultra-high frequency |
| UPC | Ultra-polishing connectors |
| UPSR (UPPR) | Unidirectional path-switched (or path-protected) ring |
| UTP | Unshielded twisted pairs |
| UV | Ultraviolet |
| VAD | Vapor axial deposition |
| VC | Virtual container |
| VC | Virtual channel |
| VCI | Virtual-channel identifier |
| VCSEL | Vertical-cavity surface-emitting laser |

| | |
|---|---|
| VOA | Variable optical attenuator |
| VoIP | Voice over the Internet Protocol |
| VP | Virtual path |
| VPI | Virtual-path identifier |
| VT | Virtual tributary |
| WADM | Wavelength add/drop multiplexer |
| WAN | Wide area network |
| WC | Wavelength conversion (converter) |
| WDM | Wavelength-division multiplexing |
| WGR | Waveguide-grating router |
| WXC | Wavelength cross-connect |
| WWW | World Wide Web |
| XCVR | Transceiver |
| XPM | Cross-phase modulation |
| YAG | Yttrium-aluminum-garnet |
| YBBM | Y-fed balanced-bridge modulator |
| YIG | Yttrium-iron-garnet |
| ZBLAN | $ZrF_4$-$BaF_2$-$LaF_3$-$AlF_3$-NaF (pronounced *zee-blan*) |

# A Selected Bibliography

## INTRODUCTORY-LEVEL BOOKS

### General


Crisp, John. *Introduction to Fiber Optics.* Woburn, Mass.: Newnes, 1997.
A well-written elementary text with good accompanying illustrations to enhance readability. Intended for readers who are interested in communications applications of fiber optics.

Goff, David. *Fiber Optic Reference Guide: A Practical Guide to Technology.* Boston: Focal Press, 1996.
A good, practical overview of the technology, although it requires some knowledge of fiber optics. Not recommended as a college textbook because it lacks mathematical descriptions and problems.

Hayes, Jim. *Fiber Optic Technician's Manual.* Albany, N.Y.: Delmar, 1996.
An overview of the technology and introduction to the fiber-optic business world. Advises on how to install, maintain, restore, and test fiber-optic cables.

Hecht, Jeff. *Understanding Fiber Optics.* 3d ed. Upper Saddle River, N.J.: Prentice Hall, 1999.
This classic introductory book provides a clear explanation of all topics related to fiber optics. However, this is not a college-course textbook because it is almost totally devoid of mathematical descriptions.

Hoss, Robert, and Edward Lacy. *Fiber Optics.* Upper Saddle River, N.J.: Prentice Hall, 1995.
Intended to help the technician or engineer who has to install and maintain a practical fiber system in today's environment. The book is practical and useful but contains no discussion of future technology.

Nelist, John. *Understanding Telecommunications and Lightwave Systems: An Entry-Level Guide.* 2d ed. Piscataway, N.J.: IEEE Press, 1996.
A brief but informative overview of fiber-optic communications systems with many practical examples, especially in Chapter 17.

Shotwell, Allen. *An Introduction to Fiber Optics.* Upper Saddle River, N.J.: Prentice Hall, 1997.
Very elementary explanations of fiber-optic principles and applications in communications. Briefly covers all basic materials, giving the reader an idea of how fiber-optic communications systems work. Making this book particularly useful is Chapter 10, "Optical Fiber Measurement and Testing," where the reader can find information about both laboratory and field measurement and testing.

Sterling, Donald. *Technician's Guide to Fiber Optics.* 3d ed. Albany, N.Y.: Delmar, 2000.
Covers almost all aspects of the topic in very simple language at the most elementary level. Almost no technical background is required.


### Specialized: Optical Fibers and Cables


Chomycz, Bob. *Fiber Optic Installations: A Practical Guide.* New York: McGraw-Hill, 1996.
A practical book for technicians with little or no background in fiber optics. Provides not only the basic concepts of fiber-optic technology but also gives many practical examples of how to work with this technology. Contains useful, practical information such as actual parameter values and characteristics of fiber-optic components and systems. Data sheets complement the text.

Pearson, Eric. *The Complete Guide to Fiber Optic Cable Installation.* Albany, N.Y.: Delmar, 1997.
A helpful, practical book that includes the basics of fiber-optic systems. Discusses physical layout of fiber-optic networks and all practical aspects of cable installation, including the preparation of ends, connector installations, loss measurements, certification, and troubleshooting fiber systems. Well written with plenty of instructive illustrations.

Reffi, James. *Fiber Optic Cable—A LightGuide.* Geneva, Ill.: abc TeleTraining, Inc., 1991.
An unusual but very good combination of manual and student guide. Concentrates on optical fibers and fiber-optic cables but also gives an overview of fiber-optic communications systems.

## UNDERGRADUATE-LEVEL BOOKS

### General

Green, Lynne. *Fiber Optic Communications.* Boca Raton, Fla: CRC Press, 1993.
A practical, useful design-oriented book devoted to point-to-point systems. A brief network overview is given in Chapter 2.

Keiser, Gerd. *Optical Fiber Communications.* 2d ed. New York: McGraw-Hill, 1992.
This book has been considered a classic text for a number of years. Provides many practical examples and presents an overview of technological achievements and trends. Does not deal with network problems.

Killen, Harold. *Fiber Optic Communications.* Englewood Cliffs, N.J.: Prentice Hall, 1991.
Briefly covers all the material necessary for an introductory undergraduate course: the physics of optical fiber, sources, detectors, etc. Hence, you get an idea of what a fiber optic point-to-point link is. What makes the book special is its concentration on the communications aspects of fiber-optic systems. Considers analog modulation, digital-system design, line codes, digital-video transmission, optical receivers, coherent optical communications, measurement in fiber telecommunications, and much more.

Meardon, Wymer. *The Elements of Fiber Optics.* Englewood Cliffs, N.J.: Regents/Prentice Hall, 1993.
Specifically oriented for students in a technology program. Covers the basic material for an introductory course. Well written with many examples, data sheets, questions, and problems. Could be used for an introduction to the subject but needs to be updated.

Palais, Joseph. *Fiber Optic Communications,* 4th ed. Upper Saddle River, N.J.: Prentice Hall, 1998.
A classic undergraduate textbook covering the physical media of intensity modulation-direct detection systems. In addition, gives an overview of some components of a fiber-optic network and describes an approach to the design of fiber-optic communications systems. Contains an extensive bibliography.

Zanger, Henry, and Cynthia Zanger. *Fiber Optics: Communications and Other Applications.* New York: Merrill, 1991.
This book is between the introductory and undergraduate levels. Clear explanations, simple examples, questions and problems—all these features make this book acceptable for technology programs. However, the book needs to be updated.

### Specialized

Kuhn, Kellin. *Laser Engineering.* Upper Saddle River, N.J.: Prentice Hall, 1998.
Covers the physics of lasers as well as such topics as Maxwell's equations, reflection and refraction, lenses, the elements of nonlinear optics. Clearly, a valuable complementary read.

Uiga, Endel. *Optoelectronics.* Englewood Cliffs, N.J.: Prentice Hall, 1995.
Covers the subject in general with just one chapter devoted to fiber optics. Useful for those who are concentrating on a specific topic, like radiation sources or detectors. Material is well organized and presented in a clear, concise manner.

## ADVANCED-LEVEL BOOKS

### General

Agrawal, Govind. *Fiber-Optic Communication Systems.* 2d ed. New York: John Wiley & Sons, 1997.
A high-level professional textbook on modern fiber-optic communications. Explains the physical meaning of a phenomenon first and then describes the phenomenon mathematically. Practical data

presented as examples make the book even more appealing, but understanding the material requires a solid background in physics and mathematics.

Gagliardi, Robert, and Sherman Karp. *Optical Communications.* 2d ed. New York: John Wiley & Sons, 1995.
Devoted to optical communications in general. Providing broad theoretical coverage of the topic, the authors consider fiber-optic communications as a subset of the subject. Of particular interest are Chapter 7 (covering fiber-optic communications) and Chapter 8 (dealing with fiber networks).

*Handbook of Photonics.* Edited by Mool C. Gupta. Boca Raton, Fla.: CRC Press, 1997.
Photonics employs optics and electronics in tandem to produce new devices and systems. This 800-page volume consists of three parts: photonic materials, photonic devices and optics, and photonic systems. Especially pertinent to those in the fiber-optic communications field are the sections on semiconductors and the chapter "Optical Communication," by Alan Willner, which is a brief but informative review of fiber-optic communications systems and a look at future trends.

Kashima, Norio. *Optical Transmission for the Subscriber Loop.* Boston: Artech House, 1993.
A two-part monograph: (1) basic technology and system examples and (2) optical devices. The first part describes fiber-optic communications technology in general, including such topics as modulation, the principle of coherent transmission, multiplexing, fiber nonlinearity, receivers, and optical amplifiers. The second part discusses fiber-optic transmission systems using time-compression multiplexing (TCM), wavelength-division multiplexing (WDM), subcarrier multiplexing (SCM), and coherent technologies. All the material is considered for subscriber loop applications.

Kazovsky, Leonid, Sergio Benedetto, and Alan Willner. *Optical Fiber Communication Systems.* Boston: Artech House, 1996.
Though designed primarily for graduate students, this book can be used in undergraduate programs with the careful guidance of the instructor. Text covers the most advanced, leading-edge developments in the field. Emphasizes the systems approach to fiber-optic communications, focusing on the newer technologies like optical amplifiers, solitons, and multichannel systems and networks. Also includes a general communications toolbox.

Lachs, Gerard. *Fiber Optic Communications: Systems, Analysis, and Enhancements.* New York: McGraw-Hill, 1998.
In addition to general coverage of fiber-optic communications systems, the book includes an introduction to electromagnetics, cylindrical waveguides, and laser and optical amplifiers. Coherent systems are discussed in great detail.

Maclean, D.J.H. *Optical Line Systems: Transmission Aspects.* Chichester, U.K.: John Wiley & Sons, 1996.
This is not a textbook but covers many aspects of fiber-optic communications systems. Includes general telecommunications notes (history, PSTN, the current situation in telecommunications, and standards) and thorough coverage and design consideration of optical fibers and cables, splices, connectors, transmitters, and receivers. Provides general system considerations, descriptions of installed long-haul systems (such as TAT-8), and aspects of system measurements. Presents the state of the art of British and European technology.

Miller, Stewart, and Ivan Kaminow, eds. *Optical Fiber Telecommunications II.* Boston: Academic Press, 1988.
Still informative and useful in spite of its publication date. Each chapter written by a leading expert in that field. Describes the state of the art of the technology of the time as well as presenting a glimpse into the future.

Ming-Kang Liu, Max. *Principles and Applications of Optical Communications.* Chicago: Irwin, 1996.
Covers almost all aspects of modern fiber-optic communications. Reflects the state of technology and networks at the time of publication. Thoroughly designed examples and problems enhance the book's value.

Papannareddy, Rajappa. *Introduction to Lightwave Communications Systems.* Boston: Artech House, 1997.
Brief graduate-level coverage of almost all topics in fiber-optic communications. A good review of wave propagation in optical fiber and other aspects of fiber optics, including soliton transmission.

van Etten, Wim, and Jan van der Plaats. *Fundamentals of Optical Fiber Communications.* Englewood Cliffs, N.J.: Prentice Hall, 1995.
Devoted to the technology aspects of fiber-optic communications with clear, thorough explanations of devices and circuits. A strong knowledge of physics and mathematics is required. Barely touches on transmission and networks.

## Specialized

(1) Measurement and software

Derickson, Dennis, ed. *Fiber Optic Test and Measurement.* Upper Saddle River, N.J.: Prentice Hall PTR, 1998.
Covers all aspects of measurement in fiber-optic communications, including the characterization of an optical fiber and system components. An instructive and practical guide, though not a textbook.

Morikuni, James, and Sung-Mo Kang. *Computer-Aided Design of Optoelectronic Integrated Circuits and Systems.* Upper Saddle River, N.J.: Prentice Hall PTR, 1997.
A review of simulation methods, with LEDs, laser diodes, and photodiodes serving as the authors' examples.

(2) Optical fibers and cables

Buck, John. *Fundamentals of Optical Fibers.* New York: John Wiley & Sons, 1996.
Devoted to optical fibers only. Requires a strong background in physics and mathematics. The author presents intermediate-level electromagnetic field (EM) theory on optical waveguides. One learns how light propagates over optical fiber from the point of view of EM theory. Instructive and interesting.

Ghatak, Ajoy, and K. Thyagarajan. *Introduction to Fiber Optics.* New York: Cambridge University Press, 1998.
A textbook for undergraduate seniors or beginning graduate students. All phenomena in an optical fiber are well described from both physical and mathematical standpoints. Discusses not only optical fiber but also all topics related to fiber-optic communications: light sources, detectors, optical amplifiers, system design, and such components as couplers, filters, and fiber Bragg gratings. Two chapters are devoted to measuring the parameters of optical fiber. Excellent illustrations throughout enhance comprehensibility.

Murata, Hiroshi. *Handbook of Optical Fibers and Cables.* 2d ed. New York: Marcel Dekker, 1996.
Not a textbook, but students should be able to read and understand most of the material. Provides comprehensive information about the properties of fiber-optic material, including the mechanical properties of fiber. Special emphasis is given to singlemode fibers. Other topics: fiber-optic cable (including some design theory), splicing, connectors, measurement, and cable installation.

(3) Components and devices

Becker, P.C., N.A. Olsson, and J.R. Simpson. *Erbium-Doped Fiber Amplifiers: Fundamentals and Technology.* San Diego: Academic Press, 1999.
A comprehensive monograph on EDFA theory and applications. Detailed discussion of every aspect of EDFA performance, including design issues. Contains computer-simulation software.

Chen, Chin-Lin. *Elements of Optoelectronics and Fiber Optics.* Chicago: Irwin, 1996.
Light sources (LED and LD), optical fiber, photodiodes, and passive components are the subjects of this textbook. Also included are modulation, integrated optics, and polarization effects in singlemode optical fibers and components.

Chuang, Shun Lien. *Physics of Optoelectronic Devices.* New York: John Wiley & Sons, 1995.
This high-level textbook covers electromagnetics and semiconductor theory. Discusses light propagation, light generation, and the modulation and detection of light. Informative but requires a strong background in physics and mathematics.

Davis, Christopher. *Lasers and Electro-Optics: Fundamentals and Engineering.* Cambridge, U.K.: Cambridge University Press, 1996.
A graduate-level textbook covering the fundamentals of optical systems, laser radiation, and light propagation within waveguides and nonlinear media. Includes semiconductor lasers, optical fibers, electro-optic and acousto-optic effects. Direct and coherent detection is discussed.

Ghafouri-Shiraz, H., and B.S.K. Lo. *Distributed Feedback Laser Diodes: Principles and Physical Modeling.* Chichester, U.K.: John Wiley & Sons, 1996.
A good introduction to coherent fiber-optic communications systems with an explanation as to why we need distributed-feedback laser diodes (DFB LDs). The authors present a clear, informative primer on DFB LD physics.

Kashima, Norio. *Passive Optical Components for Optical Fiber Transmission.* Boston: Artech House, 1995.
Aimed at a broad readership—engineers, researchers, and students—this book deals with optical fiber, connections, couplers, filters, optical and mechanical switches, and measurements for passive optical components. Not a textbook.

Lasky, R.C., U.L. Osterberg, and D.P. Stigliani, eds. *Optoelectronics for Data Communication.* San Diego: Academic Press, 1995.

A collection of practical articles devoted to different ways of using opto-electronics in data communications. Furnishes information on fibers, cables, coupling, light sources, detectors, basic optics, transceiver modules and packaging, connector-module interface, and the future of information technology. A practical book, but not a textbook.

Mickelson, Alan, Nagesh Basavanhally, and Yang-Cheng Lee, eds. *Optoelectronic Packaging.* New York: John Wiley & Sons, 1997.

A review of coupling from a packaging engineer's standpoint. Covers the packaging aspects of all components of fiber-optic communications systems and waveguide technologies. A useful, practical book.

Verdeyen, Joseph. *Laser Electronics.* 3d ed. Englewood Cliffs, N.J.: Prentice Hall, 1995.

A classic textbook devoted to general electromagnetic theory and laser fundamentals. Semiconductor lasers are covered well. Light propagation and signal detection are also included.

(4) Networks

Green, Paul, Jr. *Fiber Optic Networks.* Englewood Cliffs, N.J.: Prentice Hall, 1993.

The first book that refers to fiber-optic networks. Reflects the state of fiber-optic communications systems at the time of publication. Requires some background in physics, mathematics, and telecommunications.

Mestagh, Dennis. *Fundamentals of Multiaccess Optical Fiber Networks.* Boston: Artech House, 1995.

While the major elements of fiber-optic communications systems—optical fiber, sources, detectors, and passive components—are discussed, the emphasis is on multiaccess networks: basic topologies and all types of multiplexing (WDM, TDM, CDM, and SCM). Concentrates on the technology of networks, describing the physical media of networks rather than the protocols.

Mukherjee, Biswanath. *Optical Communication Networks.* New York: McGraw-Hill, 1997.

Dedicated to networks but includes a brief review of the overall technology of fiber-optic communications. Broadcast and wavelength-routed optical networks are explored in detail. The focus here is on WDM networks, but other types of multiplexing (TDM and CDM) are reviewed. This is a well-written computer-network-oriented textbook.

Ramaswami, Rajiv, and Kumar Sivarajan. *Optical Networks: A Practical Perspective.* San Francisco: Morgan Kaufman, 1998.

A comprehensive textbook. Reviews both technology (components and system performance) and network operations. Practical calculations and examples make the book especially useful.

Stern, Thomas, and Krishma Bala. *Multiwavelength Optical Networks: A Layered Approach.* Reading, Mass.: Addison Wesley Longman, 1999.

All major types of fiber-optic networks (broadcast and wavelength-routed) and their modifications are discussed. The peculiarities of WDM networks are the main focus, but survivability and protection are also discussed along with a look at industry trends.

(5) Nonlinear and coherent systems

Agrawal, Govind. *Nonlinear Fiber Optics.* 2d ed. San Diego: Academic Press, 1995.

A high-level graduate textbook that considers in depth all the nonlinear effects in optical fibers. Also discusses fiber amplifiers and fiber lasers.

Betti, Silvello, Giancarlo De Marchis, and Eugenio Iannone. *Coherent Optical Communications Systems.* New York: John Wiley & Sons, 1995.

A valuable source of information on regular fiber-optic communications systems through a discussion of intensity modulation-direct detection (IM-DD) systems. The book's main topic—coherent systems—is treated in detail, with emphasis on signal-detection problems.

Cvijetic, Milorad. *Coherent and Nonlinear Lightwave Communications.* Boston: Artech House, 1996.

Coherent fiber-optic communications systems are examined in the first part of this book, while nonlinear optical transmission is the focus of the second part. Internal and external modulations are included along with coherent detections.

Iannone, Eugenio, Francesco Matera, Antonio Mecozzi, and Marina Settembre. *Nonlinear Optical Communication Networks.* New York: John Wiley & Sons, 1998.

In addition to dealing with all the regular aspects of fiber-optic networks, the authors discuss the nonlinear effects at both the component and system levels. Soliton transmission and compensation of nonlinear dispersion are among the topics discussed. Computer simulation developed by the authors is widely used for evaluation of system performance.

(6) Dictionaries

Held, Gilbert. *Dictionary of Communications Technology.* 2d ed. Chichester, U.K.: John Wiley & Sons, 1995. Comprehensive explanations of terms used in communications technology. The word *technology* reflects the author's approach to choosing the topics. Includes, for example, the list of ITU-T recommendations and ISO standards.

Khan, Ahmed. *The Telecommunications Fact Book and Illustrated Dictionary.* Albany, N.Y.: Delmar, 1992. Contains short explanations with illustrations of basic telecommunications terms. Half the volume is actually a handbook presented as a set of appendixes. These appendixes include a listing of telecommunications agencies, organizations, magazines, CCITT recommendations and EIA standards, and many useful facts and data. But be aware that the book was published in 1992.

Petersen, Julie. *Data and Telecommunications Dictionary.* Boca Raton, Fla: CRC Press, 1999. Covers data and telecommunications in general, with no emphasis given to fiber optics. Nonetheless, a useful source of information, since fiber-optic systems are the heart of modern communications.

Weik, Martin. *Fiber Optics Standard Dictionary.* 3d ed. New York: Chapman & Hall, 1997. A comprehensive work on fiber optics, with concentration on the communications area. Contains about 16,000 entries.

# MAGAZINES

It is absolutely impossible to refer here to even the most important articles published in recent years. For current papers, it is recommended that you look through the following professional journals and magazines:

***Applied Optics***
Optical Society of America (OSA)
2010 Massachusetts Avenue NW
Washington, D.C. 20036
Phone: (800) 762-6960
Web: *www.osa.org*

***Cabling Business Magazine***
Cabling Publication, Inc.
12035 Shiloh Road, Suite 350
Dallas, Tex. 75228
Phone: (214) 328-1717
Web: *www.cablingbusiness.com*

***Fiber and Integrated Optics***
Taylor and Francis
11 New Fetter Lane
London EC4P 4EE
England
Web: *www.taylorandfrancis.com*

19 Union Square
New York, N.Y. 10003-3382
U.S.A.
Phone: (212) 414-0650

***Fiber Optics and Communications***
Information Gatekeeper
IGI Group, Inc.
214 Harvard Avenue, Suite 200
Boston, Mass. 02134
Phone: (800) 323-1088
Web: *www.igigroup.com*

**Fiber Optics News**
Phillips Business Information
1201 Seven Locks Road, P.O. Box 60043
Potomac, Md. 20859-0043
Phone: (888) 707-5808
Web: *www.TelecomWeb.com*

**Fiberoptic Product News**
Cahners
301 Gibraltar Drive, Box 650
Morris Plains, N.J. 07950-0650
Phone: (973) 292-5100
Web: *www.fpnmag.com*

**IEEE Communications Magazine**
The Institute of Electrical and Electronics Engineers
3 Park Avenue
New York, N.Y. 10016-5997
Phone: (212) 705-8900
Web: *www.comsoc.org/~ci*

**IEEE Lasers and Electro-Optics Society (LEOS) Newsletter**
The Institute of Electrical and Electronics Engineers
445 Hoes Lane, P.O. Box 1331
Piscataway, N.J. 08855-1331
Phone: (732) 562-3892
Web: *www.ieee.org/leos*

**IEEE/OSA Journal of Lightwave Technology**
The Institute of Electrical and Electronics Engineers
3 Park Avenue
New York, N.Y. 10016-5997
Fax: (212) 705-8900
Web: *www.opera.ieee.org.*

**IEEE Spectrum**
The Institute of Electrical and Electronics Engineers
3 Park Avenue
New York, N.Y. 10016-5997
Fax: (212) 419-7570
Web: *www.spectrum.ieee.org*

**IEEE Transactions on Communications**
The Institute of Electrical and Electronics Engineers
3 Park Avenue
New York, N.Y. 10016-5997
Phone: (212) 705-8900
Web: *www.opera.ieee.org.*

**Journal of Optical Communications**
Fachverlag Schiele and Schon Gmbh
D-10969 Berlin
Germany
Phone: (49) 30-25-37-520
E-mail: 0302537520-0001@t-online.de

**_Journal of the Optical Society of America B_**
Optical Society of America (OSA)
2010 Massachusetts Avenue NW
Washington, D.C. 20036
Phone: (800) 762-6960
Web: *www.osa.org*

**_Lasers and Optronics_**
Cahners
301 Gibraltar Drive, Box 650
Morris Plains, N.J. 07950-0650
Phone: (973) 292-5100
Web: *www.lasersoptrmag.com*

**_Laser Focus World_**
PennWell Publishing Company
98 Spit Brook Road
Nashua, N.H. 03062-5737
Phone: (603) 891-0123
Web: *www.optoelectronics-world.com*

**_Lightwave_**
PennWell Publishing Company
1421 South Sheridan Road
Tulsa, Okla. 74112
Phone: (918) 835-3161
Web: *www.light-wave.com*

**_Optical Engineering_**
The International Society for Optical Engineering (SPIE)
1000 20th Street
Bellingham, Wash. 98225
Phone: (360) 676-3290
Web: *www.spie.org*

**_Optical Fiber Technology_**
Academic Press
525 B Street, Suite 1900
San Diego, Calif. 92101-4459
Phone: (619) 231-0926
Web: *www.academicpress.com*

**_Optics and Photonics News_**
Optical Society of America (OSA)
2010 Massachusetts Avenue NW
Washington, D.C. 20036
Phone: (202) 223-8130
Web: *www.osa.org*

**_Optics Communications_**
Elsevier Science
P.O. Box 211
1000 AE Amsterdam
The Netherlands
Phone: (31) 20-48-53-757
Web: *www.elsevier.com*

655 Avenue of the Americas
New York, N.Y. 10010-5107
U.S.A.
Phone: (212) 633-3730

***Photonics Spectra***
Laurin Publishing Co.
Berkshire Common, P.O. Box 4949
Pittsfield, Mass. 01202-4949
Phone: (413) 499-0514
Web: *www.Photonics.com*

***R&D (Research & Development)***
Cahners
8773 South Ridgeline Boulevard
Highland Ranch, Colo. 80126-2329
Phone: (303) 470-4445
Web: *www.rdmag.com*

# D

# Products, Services, and Standards

## PRODUCTS AND SERVICES

Comprehensive information about products, systems, and services can be found in the following sources:

1. *Technology Reference*—an annual supplement to *Fiberoptic Product News,* Cahners, 301 Gibraltar Drive, Box 650, Morris Plains, N.J. 07950-0650. Phone: (973) 292-5100.
   This easy-to-read reference provides information about products and services, including a list of vendors, and about manufacturers and suppliers, including contact information on manufacturers, distributors, sales representatives, researchers, publishers, and societies.
2. *Lightwave 1999 Worldwide Directory of Fiber-Optic Communications Products and Services,* issued by *Lightwave,* March 1999. The publisher is Pennwell, 1421 South Sheridan Road, Tulsa, Okla. 74112. Phone: (918) 835-3161.
   This directory includes not only contact information on companies but also the basic characteristics of the products manufactured by the companies.
3. Sources of products and supplies and company contact information can be found in Eric Pearson's *The Complete Guide to Fiber Optic Cable System Installation* (Delmar, 1997), pp. 227–229.
4. A list of societies, conference sponsors, and trade magazines can be found in David Goff's *Fiber Optic Reference Guide.* (Focal Press, 1996), pp. 191–194.

## STANDARDS

There are two basic types of standards: regulatory and voluntary. Regulatory standards are introduced by government-affiliated agencies, while voluntary standards—they can be de facto or de jure—are offered by industry associations and individual providers. Thus, standards are developed by international and national standards bodies and industry associations.

The following organizations work to develop standards for the fiber-optic communications industry:

A. International

1. The International Telecommunication Union (ITU), a United Nations charter organization, is the leading worldwide body set up to devise standards. The Telecommunication Standardization Sector of this union (ITU-T) develops standards in information technology, including telecommunications and data communications. ITU-T had been called CCITT—the Consultative Committee on International Telephony and Telegraphy. Another ITU sector, ITU-R, is responsible for standards in radio.
2. The International Organization for Standardization (ISO), formerly the International Standards Organization, is a nongovernmental body that works on technical standards for both industry and the trades. ISO's activity in information technology is limited mostly to the standardization of data communications. (The OSI reference model—see Section 14.2 in text—was developed and is supported by the ISO.)

B. National

1. The National Institute of Standards and Technology (NIST), formerly the National Bureau of Standards, is the standards-formulating body of the United States government. By conducting its own research, NIST develops technical standards for industry, the sciences, and technology. (It developed the standards for power meters. Another example of NIST standard is the absolute-frequency value used in the ITU-T grid for WDM systems.)
2. The American National Standards Institute (ANSI) does not itself develop standards but, rather, administers and coordinates the efforts of the private-sector organizations in the United States to develop voluntary American national standards. More specifically, ANSI accredits organizations that develop such standards. ANSI also represents the United States in the International Organization for Standardization (ISO) and the International Electrotechnical Commission (IEC).
3. The Institute of Electrical and Electronics Engineers, a professional organization, develops standards for LAN and for measurements of parameters of communications systems. It presents its recommendations through ANSI to other national and international bodies. The IEEE's *Standards Products Catalog* is available online at *www.standards.ieee.org/catalog.contents.html.*
4. The Electronic Industries Alliance (EIA) represents manufacturers of communications products. The EIA is primarily concerned with electrical and interface standards, of which the RS-232 interface standard is the most widely known. The EIA is the most active organization in developing standards, mainly in the testing and measurement area, for the fiber-optic communications industry. (We refer to these standards in Chapters 7 and 8.) You can find a list of these standards in David Goff's *Fiber Optic Reference Guide* (pp. 181–184). See also Eric Pearson's *The Complete Guide to Fiber Optic Cable System Installation* (pp. 205–209).
5. The Telecommunications Industry Association (TIA) is the national trade organization providing a forum for manufacturers, suppliers, and service providers. This organization works closely with the EIA to develop standards for fiber-optic communications. Both bodies present their standards to ANSI for international consideration.
6. Telcordia Technologies (formerly Bellcore) develops standards for telecommunications in general and for fiber-optic communications in particular. This company develops technical advisories and technical requirements. Manufacturers of fiber-optic communications equipment usually state that their products are in compliance with Telcordia's standards as evidence of the quality of those products. Even though Telcordia is an American company, many foreign firms seek to be in compliance with its standards. A list of Telcordia's standards appears in David Goff's book (pp. 187–190) under the heading "Bellcore Standards." Also, see pages 219 and 220 in the Eric Pearson book noted above.

7. The United States Department of Defense (DOD) develops its own military standards. The Goff book presents a list of those standards (pp. 184–187). The Pearson book also discusses them (pp. 209–219).

The best way to learn more about standards, of course, is to contact the developers or providers of these documents directly:

* EIA Standards Sales Office
  2500 Wilson Boulevard
  Arlington, Va. 22201-3834
  Phone: (800) 854-7179
* Information Exchange Management
  Telcordia Technologies (formerly Bellcore)
  445 South Street
  P.O. Box 1910
  Morristown, N.J. 07962-1910
  Phone: (800) 521-2673
* Global Engineering Documents
  15 Inverness Way East
  Englewood, Colo.
  Phone: (800) 854-7179

Bear in mind that you have to purchase this documentation and a single package may cost several thousand dollars.

The Internet is also a good source of references on standards. Use any search engine to find the information you need.

General telecommunications standards, including standards on the seven layers of the OSI reference model, are referred to in R.J. Horrocks and R.W.A. Scarr's *Future Trends in Telecommunications* (John Wiley & Sons, 1995), pp. 343–367.

# FUTURE STANDARDS

Global standards for optical transport networks are being developed by ITU-T. You can read about this process in Alan McGuire and Paul Bonenfant's "Standards: The Blueprints for Optical Networking," *IEEE Communications Magazine,* February 1998, pp. 68–78, and in Paul Bonenfant's "Optical Networking Standards," Paper TUI, Optical Fiber Communication Conference, Baltimore, March 5–10, 2000.

One of the trends in the development of new standards is, obviously, tightening the required values of the specific parameters. Typical examples of the more stringent technical specifications include (1) a new definition of MFD, which has led to a more accurate measurement of a mode-field diameter; (2) a lower maximum PMD value; (3) a decreased standard value of the core-cladding offset; and (4) new network standards, a dire need of the young fiber-optic communications industry. You can certainly expect further developments in this area.

# Index

Bold-faced pages show the principal appearance of the entry.